INTEGRATED PRINCIPLES OF ZOOLOGY

FIFTH EDITION

INTEGRATED PRINCIPLES OF ZOOLOGY

WITH 922 ILLUSTRATIONS

CLEVELAND P. HICKMAN, Sr.
Department of Zoology, DePauw University, Greencastle, Indiana

CLEVELAND P. HICKMAN, Jr.
Department of Biology, Washington and Lee University, Lexington, Virginia

FRANCES M. HICKMAN
Department of Zoology, DePauw University, Greencastle, Indiana

THE C. V. MOSBY COMPANY   Saint Louis   1974

Fifth edition

Copyright © 1974 by The C. V. Mosby Company

All rights reserved. No part of this book may be reproduced in any manner without written permission of the publisher.

Previous editions copyrighted 1955, 1961, 1966, 1970

Printed in the United States of America

Distributed in Great Britain by Henry Kimpton, London

Library of Congress Cataloging in Publication Data

Hickman, Cleveland Pendleton, 1895-
  Integrated principles of zoology.

  1. Zoology. I. Hickman, Cleveland P., joint
author. II. Hickman, Frances Miller, joint author.
III. Title. [DNLM: 1. Zoology. QL47 H628i 1974]
QL47.2.H54 1974   591   73-14546
ISBN 0-8016-2184-4

TS/K/B   9 8 7 6 5 4 3 2 1

PREFACE

With the broadening of the authorship in this edition, it seems appropriate to reassure past users that the fundamental aim of the book remains unchanged. Our aim is to provide students of zoology with an acquaintance and appreciation of animals and the nature of animal life as it is presently understood. Zoology is a discipline that deals with an animal in all its aspects. Considered purely as a living organism, an animal is a self-perpetuating, self-regulating physicochemical system that is in continuous adjustment with its environment. This admittedly colorless description of an animal nevertheless serves to emphasize that understanding what an animal is, what it does, and where it comes from requires the fusion of numerous disciplines of study ranging from the burgeoning field of molecular biology to evolution and population biology. The zoologist must be concerned with an animal's morphology, its physiology, its behavior, its environmental relationships, its development, and its evolutionary history.

The amount of information that has been accumulated in any one of these areas is enormous; combined, it is beyond the comprehension of any one person to grasp. Furthermore, many aspects of life itself remain unsolved. Fortunately, despite their complexity and diversity, animals share a basic organization of form and function, and all have arrived where they are today through the same evolutionary process. Consequently, it is possible to order and integrate our present knowledge about animals into a logically developed presentation, as we have attempted to do in this undergraduate textbook of zoology.

A text such as this must be selective in its treatment. We have tried to balance our description of long-established principles of zoology with the more recent adventures in scientific exploration that have exploited the techniques of biochemistry, developmental biology, molecular biology, and submicroscopic histology. Nevertheless this text will undoubtedly reflect the interests and prejudices of not only the senior author, who has devoted a lifetime to studying and teaching zoology, but the junior authors, one an active participant in current physiologic research and the other with a primary interest in the invertebrates.

While retaining the aims and basic organization of previous editions, we have made significant changes in many sections of the book. The discussion of atomic structure and bonding has been rewritten. Chapter 2 also contains an expanded consideration of protein and nucleic acid structure and new illustrations have been added. The descriptions of membrane structure and function in Chapters 3 and 4 have been greatly expanded and other important changes have been made in discussions of cell division, membrane transport processes, and enzyme function. The section on cellular metabolism has been completely rewritten and reillustrated.

The invertebrate chapters have been revised to include a broader consideration of the diversity of life within each phylum and less anatomy of representative types, which are adequately covered in the laboratory manual. Many illustrations have been replaced and many fine photographs of living invertebrates in their native habitats have been added.

The six chapters on the phylum Chordata have been completely rewritten and reillustrated to incorporate new material, improve clarity, and reorder emphasis. Chapter 21 retains the morphologic descriptions of the ammocoete larva and amphioxus that many teachers have found useful, but the section on chordate ancestry and evolution has been thoroughly recast to emphasize theories of early vertebrate ancestry and the problems surrounding attempts of biologists to reconstruct lines of vertebrate descent. Considerations of the evolution of each vertebrate class have been shifted to the specific vertebrate chapters, where we believe they will be more helpful to the student. The chapter on fishes has been considerably expanded and completely reillustrated, and we have adopted a more modern classification for the fishes. As with the invertebrate chapters, emphasis in all the vertebrate chapters has been directed to adaptations, natural history, evolution, and diversity. Morphologic description has been reduced in keeping with our desire to avoid unnecessary overlap with the laboratory manual. Numerous new photographs have been carefully selected to illustrate the vertebrates and their adaptations.

Part Three, on the physiology of animals, was rewritten to include the rich harvest of recent discoveries in this area. Many old illustrations were replaced with new drawings and several scanning electron micrographs were generously contributed by biologists who are acknowledged in the legends. We have also rewritten the chapter on development to focus on gene expression and other recent discoveries that have so vastly enriched our current understanding of the developmental process.

Chapter 39 has been expanded in keeping with the current growth of interest in problems of the ecosystem. As in former editions, the final chapter is devoted to meaningful developments in the field of zoology. The historic background of fundamental discoveries is too often neglected in presenting zoologic information to students, yet most teachers realize that every important discovery rests upon foundations assembled by others.

The lists of references at the ends of the chapters have been reviewed and updated to make the selections as authoritative and pertinent as possible. We have for the first time separately listed a selection of readings from the popular and highly accessible *Scientific American* magazine.

Beginning with the first edition, many professional zoologists have read parts of the manuscript and offered useful suggestions and criticisms. In addition to those gratefully acknowledged for their assistance in previous editions we wish to thank Dr. J. Ralph Nursall and Dr. Joseph Nelson, of the University of Alberta, and Dr. Thomas G. Nye and Dr. Edgar W. Spencer, of Washington and Lee University, for reading portions of this manuscript and contributing valuable advice.

William C. Ober of Charlottesville, Virginia, prepared most of the many new drawings that appear in this edition. He has contributed not only unusual artistic skill but also many valuable suggestions for improving the instructive value of the illustrations. Other new drawings were executed by Sheila Ford, Provo, Utah; Barbara Hyams, Montreal, Quebec; and Katherine Payne, Alexandria, Virginia. We are indebted to many individuals who provided photographs for this edition. Their names are indicated in the pertinent legends. We wish especially to thank Bo Tallmark, of Uppsala University, Sweden; Tomas Lundälv, of the Kristinebergs Zoological Station, Sweden; and Leonard L. Rue, III, of Blairstown, New Jersey, all of whom spent many hours assisting in photograph selections with the second author. Many scanning electron micrographs were generously contributed by Dr. P. P. C. Graziadei, of Florida State University; and Professor C. G. Hampson, of the University of Alberta, provided several fine bird and mammal photographs that appear in Chapters 25 and 26. We are grateful to them all.

Cleveland P. Hickman, Sr.
Cleveland P. Hickman, Jr.
Frances M. Hickman

CONTENTS

PART ONE   INTRODUCTION TO THE LIVING ANIMAL

1   Life: General considerations and biologic principles, 3

2   Matter and life, 25

3   The cell as the unit of life, 46

4   Physiology of the cell, 65

5   Architectural pattern of an animal, 82

PART TWO   THE DIVERSITY OF ANIMAL LIFE

6   The classification of animals, 105

7   The acellular animals, 117

8   The lowest metazoans, 151

9   The radiate animals, 164

10   The acoelomate animals, 200

11   The pseudocoelomate animals, 225

12   The mollusks, 246

13   The segmented worms, 275

14   The arthropods, 299

15 The aquatic mandibulates, 318

16 Terrestrial mandibulates—the myriapods and insects, 340

17 The lesser protostomes, 379

18 The lophophorate animals, 390

19 The echinoderms, 398

20 The lesser deuterostomes, 418

21 The chordates: Ancestry and evolution, general characteristics, protochordates, 430

22 The fishes, 452

23 The amphibians, 484

24 The reptiles, 506

25 The birds, 524

26 The mammals, 564

PART THREE ACTIVITY AND CONTINUITY OF LIFE

27 Internal fluids: Circulation, respiration, and excretion, 605

28 Digestion and nutrition, 641

29 Support, protection, and movement, 658

30 Coordination: Nervous system, sense organs, and endocrine system, 678

31 The reproductive process, 719

32 Principles of development, 740

33 Principles of inheritance, 761

PART FOUR THE EVOLUTION OF ANIMAL LIFE

34 Origin of life (biopoiesis), 805

35 Organic evolution, 815

36 The evolution and nature of man, 847

PART FIVE THE ANIMAL AND ITS ENVIRONMENT

37 The biosphere and animal distribution, 857

38 Ecology of populations and communities, 865

39 Problems of man's ecosystem, 894

40 Animal behavior patterns, 909

Appendix: Development of zoology, 929

Glossary, 974

INTEGRATED PRINCIPLES OF ZOOLOGY

PART ONE

INTRODUCTION TO THE LIVING ANIMAL

■ In these five chapters, we discuss the basic nature of life as we understand it today. So far as can be determined, the same materials and physical and chemical laws of the nonliving world also apply to the living. The essential difference appears to lie in the organization of the elementary materials of the nonliving into the highly specific architecture of the living. In its simplest form, life is associated with a heterogeneous substance constructed of organic macromolecules. From this substance has evolved an organizational hierarchy of cells, tissues, organs, organisms, and, finally, the vastly complex community of life, the ecosystem. The living animal's activity, growth, reproduction, and use of its environment requires a still further organization of energy processes beyond that encountered in the strictly nonliving world.

CHAPTER 1
LIFE: GENERAL CONSIDERATIONS AND BIOLOGIC PRINCIPLES

CHAPTER 2
MATTER AND LIFE

CHAPTER 3
THE CELL AS THE UNIT OF LIFE

CHAPTER 4
PHYSIOLOGY OF THE CELL

CHAPTER 5
ARCHITECTURAL PATTERN OF AN ANIMAL

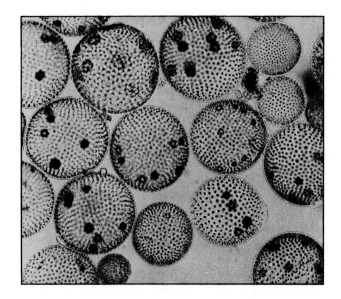

The colonial Volvox—*an organization level between the unicellular and the multicellular.*

CHAPTER 1

LIFE: GENERAL CONSIDERATIONS AND BIOLOGIC PRINCIPLES

■ WHAT IS LIFE?*

The concept that living things on this planet came from the nonliving is rather firmly established in the minds of biologists, even though it is highly speculative regarding how this phenomenon occurred. In recent years many scientists have advanced theories to explain the origin of life. Some of these theories have experimental evidence to back them up to some degree. Although it is difficult to define life, the difference between the living and the nonliving is becoming clearer as scientists understand more about the basic substance of life. Both the living and the nonliving share the same kind of chemical elements and both follow the law of the conservation of energy. The essential difference between the two appears to be organization. The living organism has combined atoms and molecules into patterns that have no counterparts in the nonliving. Examples of

*Refer to p. 17, Principle 2.

these patterns are the complex macromolecules of proteins and nucleic acids that give rise to the properties of life. Such combinations have dynamic systems of coordinated chemical or physical functions or activities that, taken as a whole, distinguish the living from the nonliving. The living differ from the nonliving not so much in qualities as in the way its substance is organized into unique patterns of position and shape. The organism is a system of interwoven and overlapping hierarchies of organization.

Almost any single criterion one may take of the living has its counterpart in the nonliving. Most characteristics of life are found to some extent in the inanimate world, which may indicate the close relationship of the two. Only animate things, however, have combined these properties into unique functional and structural patterns. One of the most fundamental properties of living matter is perhaps its uniqueness in reproducing itself, but there are other important aspects that deserve attention. Life illustrates a unique integrity under the impact of the environment so that it can adjust itself by its own energy.

Another important difference that the living possess is the intimate relationship between life and its environment. The evolutionary history of the organism has placed it specifically in certain environments and has determined its physical properties and different capabilities. Life and the environment are part and parcel of each other, and its innate behavior patterns and physiologic capabilities determine how well it gets along and adjusts itself to changing environmental conditions. This relationship between organism and environment is well shown by the behavior adjustments of correlative rhythms it must make to fluctuating surroundings. In other words, the interaction between the living and the nonliving has determined the character of life and to a certain extent the character of the earth planet on which we live. At all times in its daily living the organism is inseparable from its environment.

■ NATURE OF MATTER

Matter is the substance that comprises entities perceptible to the senses. The chief properties of matter are gravitation and inertia. Mass, which all material bodies have, is a measure of inertia. Matter and energy make up the universe, and these two components are interconvertible according to Einstein's theory of relativity, which is expressed in the equation Energy = mc^2, where m equals the mass and c equals the speed of light. This equation represents the theoretic basis for converting matter to energy. As we commonly understand matter and energy, they are separate. Matter in whatever form it may be (solid, liquid, or gas) occupies space and has weight. Energy has the ability to produce change or motion or the ability to perform work.

Both the living and the nonliving are made up of atoms. As will be pointed out later, living matter is more selective and has fewer kinds of atoms (although the same kinds are also found in the nonliving). An analysis of living matter after death shows no change in its atoms or elements. Living matter varies according to the general type of animal. Terrestrial animals have more nitrogen and sodium but less potassium, silicon, and aluminum than do aquatic forms. Marine organisms have more chloride and more water than do terrestrial organisms.

But the presence or absence of specific elements alone cannot explain the fundamental difference between the living and the nonliving. The chief difference is the arrangement of atoms into huge organic macromolecules that are able to display the basic properties of life.

■ LEVELS OF COMPLEXITY

Life is made up of a hierarchy of structure and function, ranging all the way from the atom and molecule to the highly developed and complex community, or even to levels of the ecosystem. This concept involves the nonliving as well as the living and is commonly divided into the following levels of organization: atom, molecule, organelle, cell, tissue, organ, organism, community, and ecosystem. Starting with the atom, it is seen that each level furnishes the building stones for the units of the next higher level. Of course, one can start with a lower level than the atom, the elementary particle, of which about 70 have been found. At each level are properties that are more or less unique for that level. Molecules of water have different properties from separate hydrogen and oxygen atoms. The large macromolecules may be considered as the lowest biologic level.

It will be noted that any given level contains all the lower levels as components. All the structural levels up to and including the molecular components are common to both the animate and the inanimate.

Cells are made up of organelles, and since cells are the lowest levels that can be considered alive, it follows that an organism must have at least one cell (unicellular). Organisms other then the unicellular ones are the multicellular organisms (metazoa). A level between the unicellular and the multicellular is the protistan colony, in which the cells are more or less alike. It is thus possible for the organism as a whole to have one or the other of five levels of organization complexity—the unicellular form, the colonial form, the cell-tissue level organism, the tissue-organ level, and the organ-system level.

Levels other than those already mentioned may be distinguished in the grouping of organisms, such as species, genera, families, populations, and communities. This structural hierarchy may represent the evolutional history of matter, both the living and the nonliving. This would indicate that, level by level, matter has become organized progressively to form the hierarchy as we now have it.

It will be noted that each level has fewer units than lower levels. For instance, there are fewer tissues than cells and fewer cells than organelles. Each level also has, in addition to its own complexity, the complexities of all the lower levels. In the shift from one organization level to another, energy is expended, and when a higher level has been attained, energy is necessary to maintain that level. A price in energy is required for the new properties that are acquired by having a higher level and the environmental advantages such a level confers.

This hierarchically organized matter reveals one of the basic aspects of life in that it provides the great concept of the evolution of matter from the nonliving to the living and gives a meaningful interpretation to the whole problem of the origin of life.

◼ COMPARISON OF THE LIVING WITH THE NONLIVING*

One of the characteristics of life is that it is an action-performing system of definite boundary, undergoing a continual interchange of materials with the environment, requiring energy to run the system and matter to repair it whenever needed. This may be called a metabolic criterion, which is actually made up of three vital processes; nutrition, respiration, and synthesis. It is true that some inanimate objects could very well simulate this metabolic criterion. A steady flame, such as that of an oil lamp, has a definite boundary and continually takes up oxygen, and gives off carbon dioxide and water, just as an organism would do. The similarity, however, is superficial, as the student will discover when he studies cellular metabolism and the complex metabolic pathways of enzymatic action. The metabolic criterion is almost nonexistent in suspended animation or in seeds and spores.

The metabolic mechanism by which chemical changes are brought about is made up of teams of enzymes. Although enzymes are biologic catalysts and catalysts exist in the chemical reactions of the nonbiologic world, the enzymes have an entirely different structure, being essentially proteins. Some enzymes have in addition a prosthetic group of a metal or a coenzyme of an organic compound. Indeed, all living matter may be considered an organized system of enzymes dispersed in an aqueous solution.

*Refer to p. 20, Principle 19.

Metabolism has two types—anabolism and catabolism. Anabolism is a chemical process in which simpler substances are built up into more complex ones for storage and growth and also to produce new cellular substances. On the other hand, catabolism is the breaking down of complex substances for releasing energy and the natural wear and tear on bodily structures by living processes. These two types are so woven together and so interdependent that it is difficult to separate them in the living process. As the student will see later, the body parts are in a state of flux by which interconversion of the basic substances is continuously occurring. Catabolism breaks down complex compounds for the release of energy that is needed in every aspect of life. Living matter is never the same at any two instances.

Another fundamental difference between the living and the nonliving is the specific organization of the former. In general, the organism has an upper size limit and a characteristic shape, whereas such factors are variable among nonliving substances. More basic still is the specialization of parts in the organism as expressed in the hierarchy described above. Organization, as we have seen, is one of the key criteria, and this involves different parts, each with its special function. Cells, tissues, and organs, so characteristic of life, have no counterparts in the nonliving world. Furthermore, a living system is a highly permanent form of matter, and its basic pattern was laid down perhaps in the first cells.

All living matter has certain basic properties of irritability, conductivity, movement, excretion, secretion, absorption, respiration, reproduction, and growth. As will be explained later, these functions form the basis for the development of the specialized tissues that are characteristic of all organisms. Each of these tissues is made up of the basic structures of cells, intercellular substances, and body fluids of various kinds that bathe the cells. Of all these properties, that of reproduction must be considered the most fundamental property distinguishing a living thing. This property of reproduction and its mutant types, or variables under the influence of natural selection, have made possible the course of evolution by which present living forms are found as they are.

◼ WHERE IS LIFE FOUND?

At present there is no convincing evidence that life exists elsewhere than on our planet Earth. Nor is there evidence that the physical conditions, which made life possible on this planet, are also duplicated elsewhere in the universe. But the immensity of the universe with its countless billions of stars and other bodies is beyond man's comprehension, and that similar conditions might be found elsewhere is within the realm of possibility. Matter is found within the universe as gas clouds, dust, rocky bodies, and stars. Stars are gathered together to form galaxies. According to astronomic calculations, stars are vast reservoirs where there are conversions of matter and energy. Within stars there has occurred the evolution of the elements with which we are acquainted here on our planet. Stars have life histories. Some are young and others old. Their properties change with time in their life history. Some stars die and return their matter and energy to interstellar space; others are born, probably from the same substances released by the disintegration of the older stars. Cosmic evolution is as real as is organic evolution on a smaller scale. But it is impossible to think of life within a star because of temperature and other conditions. If life is present elsewhere in the universe, it must be on satellites of stars. Only there would one find temperature and density conditions that might support life.

The old theory of the origin of life on earth from spores or similar bodies carried to this planet is not worthy of consideration. Such bodies could not withstand cosmic radiation, and it is simply "passing the buck" as far as explaining the origin of living matter is concerned.

It is possible that life may exist under different conditions from those under which it does on earth. All such matters are highly speculative and are not backed up by a shred of evidence. In recent years some study has been made of certain meteorites that contain a variety of hydrocarbons. Carbonaceous meteorites are not common, and the view that they may carry evidence of life from extraterrestrial sources has aroused some interest. Investigation has shown that organic material (hydrocarbons, fatty and aromatic acids, water, etc.) is found as microstructures embedded in the meteorite matrix. It is not known with certainty just where these meteorites originated—whether they are of lunar, comet, or solar nebula origin. At present it is impossible to determine whether such substances are abiologic or of extraterrestrial biologic origin.

■ BIOCHEMICAL BASIS OF LIFE*

Enough has been discussed in previous topics to indicate that living matter is made up of molecules that have biochemical roles. Although this aspect of life is discussed more fully later, it will suffice here to point out the significance of the biochemical background in a résumé of the general features of life.

Animals have only a fractional part of the chemical atoms found in nature. Hydrogen, oxygen, nitrogen, and carbon make up 98% to 99% of living substance (protoplasm). Some 25 to 30 other atoms are also found, many of which have been assigned restricted roles in the life process. But others, such as calcium, iron, and sodium, have wider and more general roles, yet are far less common than the "big four." The predominant elements of animals—C, N, O, H, P, and S—make up most of the molecules of life. The importance of life's constituents will be evaluated in a later section. The smaller life molecules are amino acids, sugars, fatty acids, the purine and pyrimidine bases, and the nucleotides. These smaller molecules are also constituents of the larger macromolecules of proteins, nucleic acids, glycogen, starch, and fats. The basic building blocks of all the macromolecules contain hydrogen, carbon, and oxygen. Some also have phosphorus, nitrogen, and sulfur.

Diversity in biochemical process is found especially in the metabolic functions and pathways of biosynthesis and other functions. This diversity is particularly striking in the different macromolecules used for the transportation of oxygen. Some organisms use copper-containing proteins and others iron-containing proteins, yet the different molecules perform similar tasks. Anaerobic breakdown of glucose may yield alcohol (yeast) or lactic acid (muscular contraction). Diversity is also shown in the ability to synthesize the various amino acids. Man cannot synthesize certain amino acids (essential amino acids). Rats require even more essential amino acids. Animals also vary in the ability to synthesize other molecules necessary for life.

Even though there is diversity, there is also unity or repetitious performance in metabolic patterns. The genetic code for DNA and RNA appears to be universal in all organisms. Proteins, nucleic acids, and amino acids are practically the same throughout the animal kingdom and the great energy package, ATP, is common to all.

*Refer to p. 17, Principle 2.

■ ENERGY RELATIONS

The reactions of living systems to make energy available represent a large part of their total activities. We have already mentioned the two fundamental components of the universe, energy and matter, and have indicated that under certain conditions they are interconvertible. There may be diversity in the different directions of their metabolisms, but all living systems require energy. Plants (autotrophic) use the energy from sunlight to construct organic substances such as carbohydrates, proteins, fats, and nucleic acids from carbon dioxide, water, nitrogen, and other compounds. Certain bacteria (chemosynthetic autotrophs) use reducible inorganic substances such as iron or sulfur as energy sources. Multicellular animals are heterotrophic and must depend on preformed organic food compounds for their energy. Some lower forms, such as the protozoa, some bacteria, and algae, must also depend on preformed compounds for energy purposes.

Even though photosynthesis is lacking in animals, their biochemical reactions are similar to those in plants. As will be seen under **cellular metabolism,** each biochemical reaction involves energy transfer and conversion, although not all reactions produce or utilize energy-rich substances. As far as animals are concerned, there are many energy transformations in cells, such as chemical to electrical energy in nervous processes, light to electrical in the retina of the eye, chemical to osmotic in the kidney and cell membranes, chemical to radiant in luminescence, sound to electrical in the ear, and chemical to mechanical in muscles and cilia.

Two kinds of energy are recognized by physicists: potential energy, as stored energy or energy of position, and kinetic, or the energy of motion. A familiar example of the two kinds is illustrated by a stone on an elevation where it has potential energy while at rest and kinetic energy when it rolls down to a lower level. Energy is stored as potential energy in the bonds that hold the atoms together in molecules of food. This potential energy can become kinetic energy when the animal transforms this food in its biochemical activities.

Thermodynamics, which is concerned with energy and its transformation, has two basic laws: the law of the conservation of energy, according to which energy can be transformed but not destroyed, and the law of entropy, which states that during any re-action or process there is a decrease in free energy in that some energy is dissipated to the environment as heat and unavailable energy.

The energy relations of the organism give a basic understanding of the true nature of an animal or organism. The numerous interacting systems of an animal can be maintained only by the continuous expenditure of energy. The biochemical interactions found in the transformation of the energy of oxidation necessary for physiologic work are complex, and the animal must be viewed as a highly adaptable dynamic system that is in continuous exchange of energy relations with its environment.

■ PROTOPLASM AS THE LIFE SUBSTANCE*

The substance making up the organism is often termed living matter, or protoplasm. The term "protoplasm" is less used by biologists than it was formerly. Instead of the concept that protoplasm is chiefly a colloid, the intrinsic meaning of the life substance is placed on a physicochemical appraisal of the types of forces and the chemical bonds that give living matter its rigidity and its fluidity. Consequently protoplasm is often now regarded as a vague and nebulous term.

Protoplasm is not homogeneous matter, and no drop of it can be called truly representative of the whole. It is a complex colloidal system, represented by phases of particles of widely varying sizes, physical natures, and chemical constitution. There is little to be learned by superficial examination of protoplasm from any source, and before the electron microscope and the era of molecular biology such descriptions are to a great extent worthless. Its organization, according to our present knowledge is not static but dynamic. It maintains itself by a continuous building up and breaking down of the substances that compose it. Much further investigation will have to be done before a rational evaluation of living matter can be made.

■ UNDERSTANDING THE NATURE OF LIFE

Much has been learned about the nature of life within the past few decades. How has this understanding come about? To understand living matter and the processes of life is difficult, for the life

*Refer to p. 17, Principle 1.

sciences do not lend themselves as easily to analytical methods as do the physical sciences. Any attempt to understand the nature of life involves a careful study of the processes by which the animal acquires the characteristics of living matter. If one selected an animal at random and attempted to learn all that is known about the basic principles of life from it and others like it, he would find that there would be many gaps in his final understanding. During the past few years the frontier of life sciences has been pushed more and more toward the molecular level. Biochemistry has come into greater prominence. Biophysical methods loom up at every step in analytical interpretation. Biologic investigation requires the methods and techniques from all scientific disciplines. This signifies that no one person has all the skills necessary to master all the analytical procedures required in most biologic investigations, even those of a modest scope.

One of the most outstanding advances made in the life sciences in recent years has been the development of our knowledge of the genetic code, or the manner in which the chromosomal genes carry the information for protein synthesis. By precise analytical methods it has been possible to determine that genetically controlled variants of a protein can be produced by changes in single amino acids. As the student will see in the discussion of the genetic code, an alteration in a tiny subunit of a gene is responsible for a different kind of amino acid. It is thus possible for information to be carried in the nucleotide sequence of the deoxyribonucleic acid (DNA) molecule and to be transformed into the amino acid sequence of a definite protein. The amount of work done by numerous investigators to prove how this process operates has been tremendous and includes many years of work.

■ THE "NEW BIOLOGY" AND ITS LIMITATIONS

The distinction between the life sciences and physical sciences is rapidly being broken down. No longer is the world of the zoologist set apart from the world of the physical scientist, nor are the phenomena of animal life considered to exist outside the realm of the inanimate.

This change in animal study has been so spectacular and so radical as compared with the classic study of zoology that the new approach has been called the "new biology." This innovation in zoologic study has had a revolutionary impact on methods of study and on the training of zoologists. Many persons who are not zoologists are now active in this field because the disciplines of biochemistry, physical chemistry, and physics are now considered by many to be necessary prerequisites for zoologic study. This has resulted in a tendency to exclude from the training of zoologists anything not directly related to the physical sciences.

Although this trend in zoologic study is pronounced, many zoologists regard it as extreme and altogether inadequate for an effective understanding of the life disciplines. Molecular biology is an important level of study, but there are also many others. The major aim of biology is to understand all aspects of life organization, not merely the molecular structure. The various disciplines of the life sciences cannot be arranged in a hierarchy of importance. One discipline may arouse a greater interest at one time, and much progress may be made in its development, for example, cytology in the last quarter of the last century and genetics and evolution in the first half of the present century; but for true biologic understanding, other disciplines such as taxonomy, ecology, and paleontology must make their contributions.

The true zoologist is interested in the whole organism, both functionally and structurally at all levels of organization. The functions of biologic molecules with which the biochemist is now so concerned can be understood and appreciated only in the light of their relations to the whole organism, its environment, its population structure, and its past history.

The major aim of most competent biologists is to discover facts and to establish biologic generalizations that can be applied to an understanding of the phenomena of life. Most scientific discoveries are not far removed from their practical applications. The rapid advance of biologic investigation and its exciting discoveries have already aroused many enthusiastic biologists to believe that man will soon be able to control his destiny in ways never dreamed of before. The possibility of directing mutations in certain microorganisms by means of specific agents, the genetic alteration or chemical hybridization of bacteria by the processes of transformation and transduction, the wide use of animal tissue cultures (including human) for analysis of genetic constitutions at the cellular level, and the development of techniques for extracting the major substance (DNA) of the gene have all been responsible for the belief that man will be able in the near future to remake or alter his own hereditary constitution. Similar enthusiastic beliefs have been advanced before when a major breakthrough in scientific achievement occurred, but in every case a cooling-off period followed under the impact of a more sober and realistic realization of scientific limitations.

However, certain aspects about the scope and future direction of biology can be predicted with reasonable assurance. Molecular biology will, of course, unfold more information of the inner working of the cell, such as the synthesizing processes of the many complex systems of the cell, under the impact of the decoding mechanisms of genetics. A breakthrough of great significance may be expected in the fields of development, differentiation, and growth of organisms. At the higher levels of community and population ecology, the application of the preceding studies as well as those of other biologic disciplines will no doubt contribute materially to the adaptation of animals (including man) to their physical and bi-

otic environments. Much of man's efforts will have to be expended in correcting or redressing his mistakes in upsetting natural communities. Out of these and other advancements, there are gradually emerging a keener recognition and analysis of man's cultural evolution, a more basic understanding of man as a biologic unit, and an effective motivating sense of his social responsibility.

■ WHAT IS AN ANIMAL?

The simple dictionary definition of an animal is not satisfactory to a biologist who appraises the living organism from its organization, properties, and historic character. The more we know about an animal the greater the difficulty in defining it. A definition of an animal that would exclude all plants cannot be made within the limits of a short, logical statement. Perhaps we should confine ourselves to defining any living organism and thus avoid debatable grounds that arise when different organic types are considered. For instance, certain basic differences between higher animals and higher plants are apparent and distinctive (Table 1-1), but among the lower forms of both the plant and animal kingdoms the

---

TABLE 1-1
Chief differences between plants and animals

Plants	Animals
Usually autotrophic or holophytic nutrition; simple minerals from soil and CO_2 from air; energy from sunlight to synthesize complex materials (photosynthesis)	Usually heterotrophic or holozoic nutrition; require complex, synthesized foods from plants or other animals
Usually rigid cell walls containing cellulose	Usually without cell walls; no cellulose
Oxygen as a waste product	Carbon dioxide, ammonia, urea, uric acid, and other simple substances as waste
Usually no movement or restricted movement	Locomotion usually characteristic; body parts with movement
Usually with variable body shape and size	Invariable body form; definite number of body parts
Organs added externally and growth not sharply restricted	Internal organs and restricted growth
Usually restricted response to stimuli	Usually pronounced response to stimuli
Carbohydrates stored as plant starch	Carbohydrates stored as glycogen

members grade imperceptibly into each other. The acellular or single-celled forms of both are now often lumped together under the Protista (Gr. *prōtistos,* first of all), a term proposed long ago by Haeckel. In general, plants are characterized by cellulose cell walls, synthesis of complex organic foodstuffs by photosynthesis (holophytic nutrition), inconstant body form, limited movement, and external organs; and animals are characterized by absence of cellulose cell walls, fairly constant body form, holozoic nutrition (ingestion and digestion of organic matter), mostly internal organs, pronounced movement, and definite irritability.

On the basis of what was stated about the living and the nonliving, we may tentatively define an organism as a **physicochemical system of specific and varying levels of organization patterns, self-regulative, self-perpetuating, and in continuous adjustment with its environment.**

■ DEFINITION OF ZOOLOGY

Zoology (Gr. *zōon,* animal, + *logos,* discourse) is the branch of the life sciences that deals with the animal organism as contrasted to botany, the science of the plant organism. Both zoology and botany make up the science of biology (Gr. *bios,* life, + *logos,* discourse), or the study of living things. The distinction between animals and plants is mainly one of convention rather than of basic differences. The biologic sciences are empirical; that is, knowledge about them is acquired by observation and experimentation. Theories and hypotheses must be testable and must be verified in a life science, as in any other science. At present a life science may be considered a descriptive science, in contrast to an exact science such as physics, but progress in molecular biology and the effective application of mathematics, biochemistry, biophysics, and other disciplines to biologic problems are providing the life sciences with a more exact status.

At the present time emphasis is placed on the following subdivisions of the life sciences, although much work is also done in related fields:

Genetics and molecular biology
Metabolic transformations
Cellular and subcellular structures and functions
Developmental biology
Function of tissues and organs
Behavior biology
Ecologic and pollution problems
Evolution and population biology

There is much overlapping and interrelation among the various fields of zoologic investigation. For example, cytogenetics represents the close dependence of two branches of study, cytology and genetics, which were formerly considered more or less separately. As specialization increases, branches of study become more restricted in their scope. We thus have protozoology, the study of protozoans; entomology, the study of insects; parasitology, the study of parasites; and many others.

The following are definitions of some of the important areas of zoologic study.

anatomy (Gr. *ana,* up, + *tomē,* cutting) The study of animal structures as revealed by gross dissection.

anatomy, comparative The study of various animal types from the lowest to the highest, with the aim of establishing homologies and the origin and modifications of body structures.

biochemistry (Gr. *bios,* life, + *chēmeia,* alchemy) The study of the chemical makeup of animal tissues.

cytology (Gr. *kytos,* hollow vessel) The study of the minute parts and functions of cells

ecology (Gr. *oikos,* house) The study of animals in relation to their surroundings.

embryology (Gr. *embryon,* embryo) The study of the formation and early development of the organism.

endocrinology (Gr. *endon,* within, + *krinein,* to separate) The science of hormone action in organisms.

entomology (Gr. *entomon,* insect) The study of insects.

genetics (Gr. *genesis,* origin) The study of the laws of inheritance.

helminthology (Gr. *helmins,* worm) The study of worms, with special reference to the parasitic forms.

herpetology (Gr. *herpein,* to creep) The study of reptiles, although the term usually embraces both reptiles and amphibians.

histology (Gr. *histos,* tissue) The study of structure as revealed by the microscope.

ichthyology (Gr. *ichthys,* fish) The study of fishes.

morphology (Gr. *morphē,* form) The study of organic form, with special reference to ideal types and their expression in animals.

ornithology (Gr. *ornis,* bird) The study of birds.

paleontology (Gr. *palaios,* ancient, + *onta,* existing things) The study of past life as revealed by fossils.

parasitology (Gr. *para,* beside, + *sitos,* food) The study of parasitic organisms.

physiology (Gr. *physis,* nature) The study of animal functions.

taxonomy (Gr. *taxis,* organization, + *nomos,* law) The study of the classification of animals.

zoogeography (Gr. *zōon,* animal, + *gē,* earth, + *graphein,* to write) The study of the principles of animal distribution.

■ METHODS AND TOOLS OF THE BIOLOGIST

The higher level of material organization represented by life is made up of properties that are the most difficult to understand in the universe. Protoplasm consists of very large molecules, but the organization of these molecules into particular patterns has given life many of its unique qualities. There are so many variables that reactions cannot always be predicted with certainty. Another difficulty of biologic investigation is that the whole is more than the sum of its parts. The mere study of one part of the living organism affords a restricted idea of the working of the whole integrated organism.

Biologic investigation, once purely descriptive, is now mostly experimental. The controlled experiment using a single variable factor is commonplace in investigations in the life sciences. However, in several fields such as evolution and taxonomy there has been less progress beyond the descriptive phase.

Biologists employ whatever available methods they deem necessary in doing their experimental problems. Some problems of great interest revolve around cellular biochemistry, energy relationships of metabolic pathways and processes, integrative relationships of physiologic processes, controlling factors of development, the basis of hereditary transmission, neuromuscular phenomena, and behavior patterns in animals. Many of these problems require complex techniques and apparatus, some of which have been drawn from physical and chemical laboratories. Ingenious biologists often modify such borrowed instrumentation to fit their particular purposes.

The microscope, with all its types and modifications, has contributed more to biologic investigation than any other instrument developed by man. Its major objectives are magnification, resolution, and definition. The following represents the chief advances in the improvement of the microscope.

1. First compound microscope (Janssen, 1590; Galileo, 1610)
2. Microscope with condenser (1635)
3. Huygenian ocular (Huygens, 1660)
4. Substage mirror (Hertzel, 1712)
5. Achromatic lens (Dolland, 1757; Amici, 1812)
6. Polarizing microscope (Talbot, 1834)
7. Binocular microscope (single objective with double oculars) (Riddell, 1853)
8. Water-immersion objective (Amici, 1840)
9. Oil-immersion objective (Wenham, 1870)
10. Compensating oculars (1886)
11. Apochromatic objectives (1886)
12. Iris diaphragm (Bausch and Lomb, 1887)
13. Abbé condenser (Abbé, 1888)
14. Double-objective binocular microscope (Greenough, 1892)
15. Ultramicroscope (dark-field) (Zsigmondy, 1900)
16. Electron microscope (Knoll and Ruska, 1931)
17. Phase-contrast microscope (Zernicke, 1935)
18. Reflecting microscope (Burch, 1943)
19. Fluorescence microscope (Coons, 1945)

The electron microscope, first developed in the 1930s, is a powerful tool for the study of cell structure because its resolving power greatly exceeds that of the light microscope. The source of ordinary light for the light microscope is replaced with a beam of electrons emitted by a tungsten filament; the glass lenses of the light microscope are replaced with magnets for shaping the electron beam; and the human eye is replaced with a fluorescent screen or a camera (Figs. 1-1 and 1-2). The system must be completely evacuated and samples must be dry, nonvolatile, and cut into extremely thin sections. Image formation depends on differences in electron scattering, which in turn depends upon the density of objects in the electron beam. Since the organic materials of which the cell is composed have rather uniform densities, the tissue must be treated with "electron stains" containing heavy metals such as osmium, lead, or uranium. These react selectively with cellular structures, so that areas where the metal has been deposited look dark on the photograph.

The resolution of the electron microscope has been about 5 Å*, although continued improvement in the magnetic lens field will doubtless lower this limit. In practice a resolution of 10 to 15 Å is common as compared with 2,500 Å for the best light microscope.

Other methods of preparing and visualizing specimens for electron microscopy have been useful. By evaporating a heavy metal in a vacuum chamber at an angle to the specimen, a **shadow casting** is created. The specimen becomes coated with the metal, depending upon the height of surface features; photographs of such preparations look like aerial views of the landscape seen in late afternoon (see Fig. 3-3). Recently, the **scanning electron microscope** has been introduced. Because of its great depth of field, this instrument has been used to obtain strikingly

*An angstrom (Å) = 1/10,000 micron (μ), or 1/100,000,000 cm.; 1 inch = 2.54 cm.

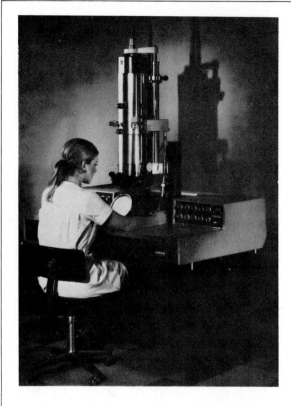

FIG. 1-1
A modern electron microscope. (Courtesy Philips
Electronic Instruments, Mount Vernon, N. Y.)

FIG. 1-2
Comparison of optical paths
of light and electron
microscopes. Note that to
facilitate comparison, the
schematic of the light
microscope has been
inverted from its usual
orientation with light source
below and image above.

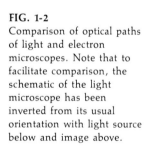

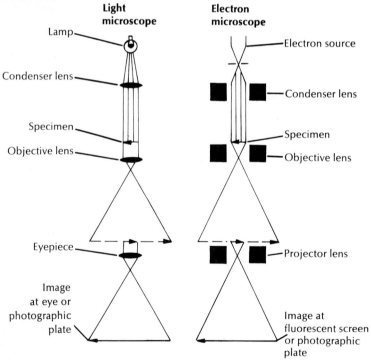

Light microscope

Lamp

Condenser lens

Specimen

Objective lens

Eyepiece

Image at eye or photographic plate

Electron microscope

Electron source

Condenser lens

Specimen

Objective lens

Projector lens

Image at fluorescent screen or photographic plate

beautiful photographs of the surfaces of animals and epithelial tissues. Examples are shown in Figs. 3-3 and 30-16.

Other useful methods are (1) isotopic tracers, in which a radioactive isotope of an element can be substituted in a chemical compound and can then be detected by some method (Geiger tube, scintillation counter (Fig. 1-3), photographic plate, etc.) so that the chemical pathway or other information can be acquired; (2) chromatography, which involves the separation of organic or inorganic components by allowing a solution of a mixture to flow, for example, over a porous surface such as filter paper, or through a porous gel (Figs. 1-4 and 1-5); (3) centrifugation, which separates materials of different densities from

each other and which can be employed for many analytical purposes; and (4) colorimetry and related methods, which measure and identify materials by determining the differential absorption of radiant energy of different concentrations.

The phase-contrast microscope has proved useful in studying living cells. If two cell components transmit the same amount of light, it is difficult with the light microscope to distinguish differences between them. If there is a difference in refractive index between the components, the phase-contrast microscope will show a difference in their brightness. This microscope is especially useful in the study of mitochondria and chromosomes in the living cell.

In biochemistry the spectrophotometer is widely

FIG. 1-3

Liquid scintillation counter used for measuring radioactive isotopes. Biologic samples containing a radioactive tracer (such as carbon 14) are combined with a liquid organic fluor in glass vials. The release of beta or gamma rays by the isotope produces minute light flashes (scintillations) in the liquid fluor, which are converted into electron pulses. These are automatically counted over a measured period of time and recorded.

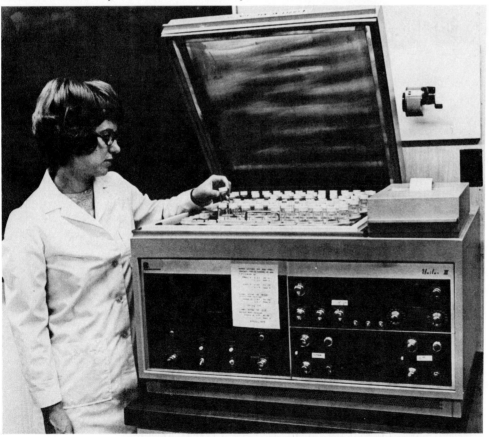

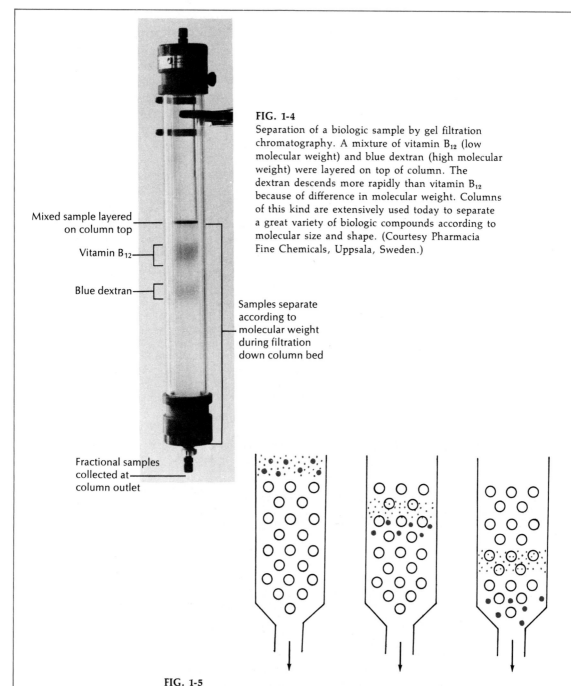

Mixed sample layered
on column top

Vitamin B$_{12}$

Blue dextran

Samples separate
according to
molecular weight
during filtration
down column bed

Fractional samples
collected at
column outlet

FIG. 1-4
Separation of a biologic sample by gel filtration chromatography. A mixture of vitamin B$_{12}$ (low molecular weight) and blue dextran (high molecular weight) were layered on top of column. The dextran descends more rapidly than vitamin B$_{12}$ because of difference in molecular weight. Columns of this kind are extensively used today to separate a great variety of biologic compounds according to molecular size and shape. (Courtesy Pharmacia Fine Chemicals, Uppsala, Sweden.)

FIG. 1-5
How a gel filtration column works. Column is filled with minute porous polysaccharide beads, which swell with water to form a gel. Large molecules (large red dots) cannot enter pores of beads and therefore pass through bed in fluid space surrounding beads. Smaller molecules (small red dots) penetrate pores of gel beads to varying extent depending on their size; they filter more slowly through bed. Since molecules migrate at different rates, they emerge from outlet in separated fractions.

used to determine different chemical substances. Ultraviolet light is absorbed by different chemicals of characteristic wavelengths. By determining the wavelength absorbed by an unknown substance and comparing it with a table of known absorbances, identification can be made. Variant forms of this mechanism are in use, but all are based on the selective powers of light absorption of components when traversed by a beam of light.

■ HOMEOSTASIS AS THE IDEAL ADAPTATION*

Homeostasis is one of the most important concepts in biology. It may be described as the tendency of an organism to maintain a constancy of the internal fluid environment that surrounds the cells and tissues of the body. The concept was first developed by the famous French physiologist Claude Bernard over a period of years, beginning about 1857. Bernard emphasized what he called the *milieu intérieure,* or extracellular fluid, which served as a medium for the exchange of foods and waste between the cells and the environment. He thought that all vital mechanisms within the body had only the one object of preserving constancy in the conditions of life in the internal environment within the range suitable for continued normal functioning. The concept has been broadened by many others such as W. B. Cannon to include regulatory devices and feedback mechanisms. Cannon stressed the coordinated physiologic processes that promote steady states.

Homeostasis in its broader sense is the most important adaptation of all organisms. It involves the self-regulatory mechanisms for functional stability that an organism must maintain to survive in a constantly changing external environment. Regulatory controls are found in all systems of the body. To appreciate the significance of the general concept, one has only to mention a few of these homeostatic control mechanisms, such as the acid-base balance, the maintenance of relatively constant carbon dioxide levels in the blood, the constant temperature controls of birds and mammals, the control of hormone functioning, the glucose balance in the blood, and the regulation of water by the kidneys. Indeed, it is difficult to think how our bodily processes could function without such regulatory mechanisms. In population studies in the ecosystem, homeostasis is implicit at

*Refer to p. 20, Principle 17.

all stages in the ecologic organization of the community.

In any feedback mechanism, homeostasis is particularly evident. For instance, sugar (glucose) is a major source of energy for body activities and is found as a normal constituent in blood, with the excess stored as glycogen in the liver. When the blood supply falls to a level below normal, the adrenal medulla is stimulated by the hypothalamus to secrete epinephrine (a hormone) into the blood. One of the effects of epinephrine is to convert glycogen into blood glucose in the liver by enzymatic action. More glucose in the blood stimulates the pancreas to secrete insulin, which indirectly causes more glucose to be stored as glycogen. The amount of glucose in the blood acts as a feedback mechanism in each case. A feedback is the output of a product of a system that, when fed back into the system, influences further production of that product. The above effect is called **negative feedback** and aims to produce a condition of equilibrium. **Positive feedback** can occur when there is a lack of normal control mechanisms; for example, a fever tends to increase metabolism and to bring about still higher temperatures, producing a vicious circle.

Homeostatic mechanisms show an evolutionary development. Regulatory mechanisms are much more precise in higher than in lower vertebrates. However, temperature control in a warm-blooded form such as a mammal, for instance, is a gradual development from the young to maturity. Young mammals fluctuate between the ectothermic (cold-blooded) and endothermic (warm-blooded) conditions.

■ CONCEPT OF THE ECOSYSTEM*

Every animal is part of a group of similar individuals that interbreed and share common factors of the environment. They are organized into populations, communities, and ecosystems. The individuals of a particular species make up a **population.** When a population coexists with other populations of different species but with similar requirements and all adapted to a certain complex of environmental conditions, a **community** is formed. A community is a localized aggregation of populations so organized that it is more or less independent of other communities and so constituted that it is self-sufficient in its ener-

*Refer to p. 22, Principle 33.

getics, or energy transformations. Within its organization are different strata of populations, each adapted for a particular role in the maintenance of the community. One stratum may be the producers, such as plants, which can convert solar energy to chemical energy in plant products. A second stratum is the consumers, which eat plants and other animals that live on plants. A third stratum contains the decomposers (bacteria and fungi chiefly), which decompose dead organisms and organic debris, etc. and release basic chemical substances to be used again by plants. The community and the nonliving environment interact to form an ecologic system, or **ecosystem.** An ecosystem therefore is the sum total of the physical and biologic factors operating within an area that includes living organisms and abiotic substances. The exchange of materials between the living and nonliving occurs in a circular path. Such a system usually consists of a few species with large populations and many species with small populations.

The ecosystem is the basic functional unit in ecology. It may be large or small, provided that it meets the requirements of major components that operate together to form a functional stability. Life cycles result from the interactions between the biotic and abiotic components of the system. A forest region, a meadow, a lake, a desert, or even an ocean are examples of ecosystems. If the system is well balanced, no supporting materials are ever exhausted. An ecosystem is characterized by self-regulation, balanced energetics, functional and interspecies diversity, and independence except for a solar energy source.

But in another sense, the ecosystem looms as man's greatest problem. The dominant role he plays in controlling the factors of the environment has placed an enormous responsibility on his shoulders. His failure to live up to this responsibility has produced serious disturbances in the homeostasis of his surroundings, and it has only been within recent decades that he has made much of an attempt to face this challenge. Despite warnings by ecologists and conservationists, man has largely ignored the basic principles of ecologic organization as he has expanded his technologic skills and has increased his numbers so rapidly. Man has suddenly been brought up with a jolt as he has begun to realize the enormity of his mistakes.

These problems are discussed in greater detail in Chapter 39.

■ CYCLE OF EXISTENCE*

Most animals have a more or less definite life-span of existence. On the other hand, some biologists consider the protozoans as immortal because the parent loses its identity when its substance passes directly to the daughter cells in binary fission. As all biologists know, the life-span of animals varies enormously. Why animals grow old and die is not known, although many facts about senescence and other aspects of aging have been studied in recent years. There may be a buildup of insoluble materials within the cells of an animal, or the wear and tear on the machinery of life cannot be replaced as quickly or as efficiently as in the younger periods of the life-span. However, many tissue cells in the body are constantly being shed (perhaps 1% to 2% daily in man) so that the material in an animal's body is constantly in a state of flux and transformation, even though the individuality may be the same. M. Rubner had a theory of aging that has not been wholly discredited. He considered that various mammals during their lifetime used up about the same number of calories per unit of body weight. A mouse, for instance, lives about 3 to 4 years and an elephant about 70 years. He showed that there is a more rapid tempo, or physiologic time, in the mouse than in the elephant. A mouse's heart beats from 520 to 780 times per minute and that of an elephant from 25 to 28. He concluded that the 3 years or so of a mouse's life corresponded, on a physiologic time scale, to the much longer life of the elephant. There are many obscure points in his theory, but it can be concluded that the attainable length of life of a mammal is dependent partly on its rate of energy dissipation, that is, the pace of its life.

There are a number of correlations between the length of the life-span and certain factors. In general, mammals with a longer gestation period have a longer life-span. Among homoiothermic animals the larger forms tend to live longer than the smaller ones. The smaller the animal, the greater its surface in proportion to its weight and thus the more rapid is its heat loss. To preserve a constancy of body temperature, small mammals and birds must have a higher unit rate of metabolism. The larger the animal, the longer it takes to grow to maturity; also, the larger it is, the more complex it is. Increased complexity requires time for the various divisions of labor to occur. When a bacterium divides, only half

*Refer to p. 20, Principle 22.

an animal needs to be regenerated, but most metazoans start with a single cell (as the zygote) and must build a complete, complex body from there.

Some animals seem to have more definite lifespans than others. There is far more uniformity of length of life in mammals, birds, and insects than in most other metazoans. However, length of life of various animals has often been acquired from zoological gardens, where animals living in captivity have constant diet, little or no exercise, little exposure to seasonal change, and lack of other conditions found in the wild. It is difficult to draw valid conclusions from such sources. Perhaps most animals living under natural conditions in the wild state have little or no opportunity to live a normal life-span because as they grow older they weaken and fall prey to predators.

Since animals that live more rapidly have a shorter life, cold-blooded forms that hibernate, or are very inactive, may have a longer life-span. Certain reptiles are known to live longer when the temperature is colder. Dietary restrictions often add to the life-span of experimental animals.

The complete life cycle, which includes development, adult equilibrium, and senescence, has a central position in the life sciences. It is the essential unit (rather than merely the adult animal). J. T. Bonner has emphasized the importance of the life cycle as the unit of evolution; not only adults are adapted to ecologic conditions, but all stages of the life cycle are also. He considers the life stages as steps in the chemical reactions that occur in definite sequence in time within living animals. All steps of the life cycle must be adaptive to ensure the survival of the animal. From an evolutionary viewpoint natural selection operates at each step of the life cycle. Innovations or variations are introduced during the early stages (especially in the nuclear genes of the zygote by mutation and recombination), and later unfavorable ones are eliminated by differential reproduction. Genes do not restrict their expressions to the formation of the adult but throughout the course of development.

■ SOME IMPORTANT BIOLOGIC PRINCIPLES AND CONCEPTS

A principle or generalization is a statement of fact that has a wide application and can be used to formulate other principles and concepts. The physical sciences would be difficult to master without clear-cut

formulas and rules. The biologic sciences do not lend themselves to the mathematical exactness of the physical sciences, and few generalizations of universal application can be made. However, certain well-formulated principles are indispensible for clear thinking; they help the student see important relationships and form basic conclusions. Principles in biology, as in other sciences, are based on observation and experimentation and have been tested by many workers over long periods of time. They are, of course, always subject to revision and new interpretation in the light of new knowledge.

The following list of basic concepts is not intended to be exhaustive. References to some of these principles have already been made. Further references will be made in later chapters where they will have significant application and will be better understood. These principles demonstrate the essential unity of the organism and the integration of all biologic systems.

1. All organisms are composed of protoplasm, which is the physical basis of life. Wherever there is life there is protoplasm. Not everything found in protoplasm, however, is alive, for many lifeless materials such as yolk granules, waste materials, etc. may be scattered through it. The elements in protoplasm are shared with nonliving matter, but protoplasm is highly selective and many of the elements in inanimate matter are not found in it. The mystery of protoplasm does not lie in its chemical elements but in the way they are linked together into compounds and are organized into the complex substance to which we ascribe the phenomenon of the life process. (Chapters 1 and 2.)

2. Many biologic phenomena are now explained at the molecular level. Ultimately it is believed that living processes can be understood by the interactions of molecules, atoms, and electrons. Molecular biology thus stresses the physicochemical aspects of life. By such means it has been possible to understand more about energy transformations and the enzymes responsible for such reactions. Throughout the biologic kingdom the basic metabolic processes are similar. The basic principles of this biochemical unity are also similar to those found in the inanimate world. (Chapter 2.)

3. The property of replication may be considered as the most unique life feature. The power to make exact copies of its patterns as well as innovations or mutations has made possible the evolution and development of the biologic units (organisms) as we now know them. This property, together with the directive forces of evolution (natural selection, etc.), has gradually led to higher and higher levels of organization, complexity, and adaptation. (Chapters 31 and 33.)

4. Most of the unique and distinctive features of protoplasm are caused by the presence and composition of certain large molecules (macromolecules). These macromolecules are the polysaccharides, the nucleic acids, the proteins, and the fats. Polysaccharides are made up of subunits (glucose) and are the chief sources of energy. Nucleic acids, deoxyribonucleic acid (DNA) and ribonucleic acid (RNA), form the hereditary duplicating mechanisms of the genes. Proteins form the framework of protoplasmic units (cells) and enzymes, which are basic for all the phenomena of the life process. (Chapters 2 and 33.)

5. Protoplasm has a unique physicochemical organization, which varies to produce the potentialities of each particular group of organisms. Superficially, protoplasm appears much the same in all organisms, but its submicroscopic organization and chemical characteristics vary widely with different species and with different cells in the same organism. This organization accounts for functional and morphologic differences between organisms. That of man, for instance, is not the same as that of the ameba. Because of these facts and others, protoplasm, as seen under the light microscope, reveals little of its intrinsic makeup. Its true fundamental nature seems to be ultramicroscopic and hence a fruitful source of investigation in the relatively new and immature field of molecular biology. (Chapters 2 and 3.)

6. In all protoplasmic systems the ground substance is differentiated into a superficial region (ectoplasm) and an inner region (endoplasm). Because of its position the ectoplasm serves as the boundary between the external environment and the inner part of the protoplasmic system. The ectoplasm is specialized to perform many roles, such as exchange between the environment and the protoplasmic system, conduction, respiration, differentiation of development, fertilization, and general integration of the system. Classification in protozoa is based on ectoplasmic differentiation, such as pseudopodia, cilia, and flagella. This distinction between ectoplasm and endoplasm may be seen best in some eggs and ameboid cells; body cells in general do not show it clearly. (Chapters 3 and 7.)

7. All protoplasm has certain general properties that are the basic expressions of life. These characteristics distinguish the living from the nonliving and are the criteria by which we measure the dynamic aspects of the protoplasmic system. They also represent the adaptive nature of living systems, as revealed in their mechanisms of survival. One fundamental property of protoplasm is its power to respond to its environment (irritability). Another is the ability to perform such essential physiologic functions as ingestion and digestion of food, absorption, circulation, excretion, and reproduction. Although all protoplasm has these properties, in multicellular organisms the protoplasm in cells or groups of cells (tissues) tends to be specialized to carry on these functions. Thus we have nerve cells for receiving and transmitting stimuli, muscle cells for contraction, etc. (Chapters 4 and 5.)

8. Protoplasmic systems are differentiated and organized into compartment units. This concept implies some kind of partition between the various units that go to make up living systems. Integration of whatever kind takes place through membranes, films, ectoplasmic-endoplasmic differentiations, etc. Diffusion and osmosis of bodily fluids and what they carry occur along concentration gradients through membranes. A single cell is made up of many unitary compartments, such as the nucleus, food vacuoles, and cytoplasmic inclusions. Even nervous transmission occurs through membranes from one compartment to another (synaptic junctions). Some compartments (blood vessels and gut) are large and extensive to facilitate internal transport and to overcome the limitations of diffusion and osmosis. (Chapters 3 and 4.)

9. Specialization and division of labor are correlated with the organization level of the organism. The evolutionary trend is toward more specialized organs and division of labor. In Protozoa, specialized cytoplasmic structures (organelles) illustrate division of labor. Among Metazoa there is a progressive sequence of specialization, ranging from the cell-tissue level of coelenterates to the organ-system level of the higher forms. Whenever structures become differentiated or specialized, they are always accompanied by physiologic divisions of labor. (Chapters 5 and 7.)

10. All organisms come from preexisting organisms. This concept means that all protoplasmic units have come from similar units. This is the principle of **biogenesis,** in contrast to the principle of **abiogen-**

esis (spontaneous generation), which has been discredited. The principle of biogenesis has been broadened to include the various units of protoplasmic systems, such as "all nuclei from previous nuclei," "all chromosomes from previous chromosomes," and "all centrosomes from previous centrosomes." Just how far this concept can be carried is a debatable point. It may be well to point out that biogenesis does not include the first beginning of life itself. If the animate has come from the inanimate, this fact would be difficult to reconcile with the principle. (Chapters 31 and 32.)

11. Reproduction involves the division of parental material to form offspring. All new organisms are formed at the expense of the old. Whatever the nature of reproduction, whether of one-celled or many-celled animals, there is division of the parent or parents. This division may be equal or unequal and may or may not involve the destruction of the parental organism. In one-celled and some multicellular animals, there is a simple division of the parent into two daughter cells or organisms. In sexual forms the egg and sperm are the real offspring, and fertilization can be considered as a restorative process for the new parental body. (Chapters 31 and 32.)

12. All organisms develop, within limits, a characteristic form. Within variable limits all organisms develop a particular and predictable size and shape. This generalization emphasizes the science of **morphogenesis,** which is concerned with the developmental concepts of hormones, axial gradients, organizers, and specific patterns of embryologic development. Many biologists have been interested in the interpretation of form in organisms on the basis of physical forces and causes. In some cases, shapes and forms of animals can be explained by physical factors. For instance, the form of some of the low organisms such as Protozoa can be explained on the basis of surface tension forces according to the principle of maxima and minima of surface films. Closely allied to the principle of form is that of organic symmetry. (Chapters 5 and 32.)

13. Animals have a diversity of body plans. These plans are not infinitely diverse, but most animals follow one or the other of a few major types that correspond in certain ways to the evolutionary unity of the ancestral form. A basic similarity of body plan is expected from common ancestors. Specialization, however, for a certain way of life has produced some modification in the evolutionary history of all major groups of animals. Chief among these major plans are the flatworm type, the tube-within-a-tube type, the segmented type, and the vertebrate type. (Chapters 5 and 35.)

14. The final form (morphology) of an animal is the result of the interactions between basic structural features and functions. In the early development of an animal structural features precede function. A certain genetic pattern may be inherited, but what this pattern actually becomes depends to a great extent on the functional experience of the animal. This functional modification is especially pronounced in the vertebrate body. The size of a muscle, the structure and nature of bone, tendons, and ligaments, the elaboration and distribution of blood vessels, and the detailed patterns of nervous systems are all dependent on functional modifications. (Chapters 5 and 32.)

15. All organisms have the capacity for growth. Growth can be accomplished in two ways—by cell division and by cell expansion. In all true growth the increase in size is an actual increase in protoplasm and not merely a swelling induced by water or some other agent. This involves a distinction between volume (wet weight) increase and dry weight increase. In early cleavage stages of embryologic development the cells become smaller with successive divisions so that there is no actual increase in the whole structure. For true growth to occur, new material must be built up into protoplasm. Mere cell division is a preliminary to actual growth, however, for such division increases the total surface area of cell membranes, which facilitates diffusion of the minerals, proteins, etc. that go into the makeup of protoplasm. Growth within the protoplasmic structure (intussusception) must also be distinguished from accretionary growth, which involves the addition of materials externally. Examples of accretion are the growth of crystals, the shells of clams and snails, teeth, etc. These substances cause an increase in size but not necessarily in living matter. Patterns of growth are inherited so that certain proportions are usually constant from generation to generation. Growth may be *allometric* if the growth rate differs from the mean rate of the body; it is positively allometric if the part grows faster, and negatively if slower. Other factors such as vitamins and hormones also play a part. (Chapters 3, 5, and 32.)

16. All patterns of life are rhythmic in nature. The activities of the living organism fluctuate around some mean that best promotes physicochemical equilibrium. Cycles and rhythms are involved in practically every phase of the life process. This principle ranges from the ordinary physiologic cyclic patterns of heartbeat, respiration, metabolism, reproduction, etc. to those larger cycles that include the external influences of light and temperature, such as the rhythms of day and night, sun and moon, and seasonal changes. The best-adapted organism is one that can adjust its own cycles of activity to those imposed by its surroundings. For such adjustments animals have evolved precise timing mechanisms (internal "clocks") that enable them to synchronize their own activities with those of their external environment. (Chapter 40.)

17. All organisms tend to maintain a constancy of conditions within their internal media. This is commonly known as the principle of **homeostasis.** It refers to the stabilization of internal conditions for which all organisms strive. External environmental factors tend to upset internal conditions in the organism; the organism is constantly trying to counteract these influences. There is an evolutionary progression in the development of this principle. Lower forms are more restricted in their activities because they have not evolved a control of stabilization to the same degree as higher ones. As an example, the temperature of warm-blooded animals is more easily regulated than that of cold-blooded animals so that the former enjoy greater freedom of activity under more varied environmental temperatures. Organisms must also maintain stable fluid conditions of salt, acids, alkalines, foods, and oxygen within the body. (Chapters 1 and 27.)

18. All organisms use energy in their living processes. All aspects of life require energy in some form. Energy is the capacity to do work and can be measured in calories. The energy potentialities of most food substances can be measured, although some required substances are not energy giving. The ultimate source of all energy is the sun. Plants utilize this energy directly by using the process of photosynthesis to form carbohydrates. Energy made and stored in plants is used by animals that eat plants. These animals may be eaten by other animals that get energy this way. According to their type of nutrition, organisms may be divided into holophytic, or those that carry on photosynthesis; holozoic, or those that ingest and digest organic materials; and

saprophytic, or those that absorb decayed organic matter through their body surface. Organisms that make their own food are also called autotrophic; those that depend on other organisms for their nutrition are called heterotrophic. (Chapters 2, 4, 28, and 38.)

19. All the basic activities of animals are mediated by enzymes. All metabolic processes involve series of enzymes, each of which controls a specific chemical reaction. In this process a substance undergoing a chemical reaction unites with an enzyme of a specific configuration to form a specific enzyme-substrate complex. By such means, enzymes can control the speed and particular nature of the reaction. Enzymes are large protein molecules with or without other chemical groups. (Chapter 4.)

20. All organisms are fundamentally alike in their basic requirements. Life processes are about the same everywhere in the animal kingdom. Animals carry on similar metabolic processes such as nutrition, digestion, respiration, and excretion. They must adjust to their environment and must develop adaptations for doing so. They must reproduce. This uniformity is to be expected, for all organisms are composed of protoplasm and the requirements of protoplasm are much the same wherever found. Organisms do show some differences in their basic requirements. Some can synthesize food elements that others cannot; some require different vitamins; some must even get along without oxygen and must get their energy by anaerobic processes. (Chapters 4 and 27 to 31.)

21. The parts of any one organism are so closely connected that the character of one part must receive its pattern from the character of all the rest. This may be referred to as correlation of growth. A single tooth may indicate whether the animal was carnivorous or herbivorous, whether it was a mammal or other vertebrate, etc. Certain structural features always coexist. This principle is helpful to paleontologists in the reconstruction of an organism from fossil parts. Thus, if a fossil lower jaw is strengthened inside by a shelf of bone, it belonged to an ape; if strengthened outside, it belonged to a human being. (Chapter 35.)

22. All organisms pass through a characteristic life history. This principle is concerned with the life cycle, which is more or less characteristic for each species of organisms. It involves the life-span and the various phases of the cycle, such as the period of development, the reproductive span, and the post-reproductive period. It also includes other factors, such as litter size, frequency of litters, age differential of reproductive capacity, and population relations. (Chapters 1, 31 and 32.)

23. Existing organisms have developed by a process of gradual change from previously existing organisms. This is the **evolutionary concept,** better known as organic evolution. It is based on the belief that present-day forms have descended with modifications from primitive forms that may have been radically different in structure and behavior. This principle is the key to our modern interpretation of animal origins and relationships. It gives us an explanation of phylogeny, or racial relationships, and helps in the taxonomic groupings of animals. Much fundamental evidence is still lacking because more specific ancestral forms are needed for understanding the exact relationships of animal groups. The idea that there has been direct evolutionary progression from simple to complex forms is only partly true, for it is not always possible to suggest what ancestral forms may have been like. This does not, however, invalidate the concept of evolution. (Chapters 35 and 36.)

24. Each major group of animals usually has a certain basic adaptive pattern that may have determined its evolutionary divergence. Such basic adaptive features are often obscure in a widespread group that has undergone considerable evolutionary diversity. Moreover, a basic adaptation may be retained by certain taxa of a group and may be lost by others. Good examples of basic adaptive features are shown by rodents (gnawing), by bats (flight), and by primates (arboreal). Such original adaptiveness may have arisen early in the evolution of a group and thereafter may have determined the degree of specialization and adaptive radiation. (Chapters 35 and 40.)

25. Animals that have many morphologic characters in common have a common descent. The more characters organisms have in common, the more closely they are related. This phylogenetic scheme forms the basis for modern classification of animals. A few common characters shared by two groups may have limited significance because of the possibility of convergent evolution; in other words, the characters may have originated independently. When these common characters are homologous, or similar in origin, the evidence for relationships is considered fundamental. (Chapters 5, 6, and 35.)

26. Organisms of higher levels may repeat in their embryonic development some of the corresponding stages of their ancestors. This is better known as the **biogenetic law,** which was formerly interpreted to mean that the embryonic stages of an animal are similar to the adult stages of its phylogenetic ancestors. This was the principle of **recapitulation,** or the idea that ontogeny (the life history) repeats phylogeny (the ancestral history). The modern interpretation is merely that some of the corresponding embryonic stages of the early ancestor are repeated. The earlier viewpoint would assume that all evolutionary advancements were added to the terminal stages of the life histories of organisms, but early embryonic stages have also undergone evolution. Evolution has produced many ontogenies in the phylogeny of an animal. Some biologists have suggested that **paleogenesis** is a better term to express the tendency for early developmental patterns to become more or less stabilized in successive ontogenies of later descendants. In early stages this tendency is more noticeable than it is in later stages because the developmental adjustments produced by the evolution of animals may involve embryonic adaptations **(caenogenesis)** and other changes that may appear in the terminal or adult stages. The principle must be considered generalized and not absolute, for some animal ontogenies do not repeat ancestral ontogenies at all. (Chapters 6, 32, and 35.)

27. All organisms inherit a certain pattern of structural and functional organization from their progenitors. This generalization involves the laws of **heredity** and applies to all living things. The highest and lowest types of animals have the capacity for reproducing their kind and transmitting their characteristics to their offspring. Hereditary transmission is much the same in all organisms. What is inherited by an offspring is not necessarily the exact traits as expressed by the parents because heredity is not as simple as this. What is inherited is a certain type of organization that, under the influence of developmental and environmental forces, gives rise to a certain visible appearance. Many potentialities may be inherited, but only one of these may express itself visibly. Heredity is the transmission of a sequence code (in DNA molecules) of amino acids for the formation of varied protein patterns or enzymes characteristic for each organism. (Chapter 33.)

28. Patterns of organization can be changed suddenly by mutation. Sudden changes in the appearance of an animal or plant different from anything inherited from the parents do appear in nature. Some of these changes are not transitory but are transmitted to the offspring. Mutations that occur in somatic (body) cells disappear with that generation. Mutations in the germ cells can be inherited by future generations. The latter usually involve changes in the genes. Mutations may be induced by artificial means, such as x-rays, radium, and mustard gas, but their natural causes are largely obscure. Although most mutations are considered harmful, some may be useful under favorable environmental conditions. Mutations play an important role in the evolutionary process. (Chapters 33 and 35.)

29. All organisms are sensitive to changes in their environment. No organism could survive long without mechanisms for responding to the environment. This is a basic reaction of all protoplasmic units. Specialization and division of labor have resulted in sensory organs that are especially sensitive to changes in the environment. Most of an animal's activities are directed toward finding a favorable ecologic environment and avoiding unpleasant stimuli. (Chapters 30 and 38.)

30. There is a definite gradient of physiologic activities in the body, from the anterior region of high activity to the posterior region of low activity. This is the principle of **axial gradients.** Metabolic rates vary within an animal, being greater at the anterior end and progressively smaller toward the posterior end. This rate is correlated with the regeneration of lost parts. The anterior end of a fragment of an animal may regenerate a new head, whereas the posterior end with a lower metabolism will form a tail. What a regenerating part becomes is determined mainly by its relation to the animal as a whole. Some forms such as flatworms and coelenterates seem to demonstrate the principle better than do most groups. (Chapter 5.)

31. All organisms are adapted in some way to their environment. To survive, an organism must adapt to the conditions imposed by its environment. Universal adaptations by which animals can adjust to all conditions are nonexistent. Adaptations are always special adjustments to particular conditions

and are always relative. Some animals are better adjusted to their environment than are others. Most animals have become specialized in their adaptive relations so that the more perfectly they are adapted to one environment, the less they are fitted for adjustment to a different environment. Adaptations are either inherited or acquired. Inherited adaptations are present from birth, such as the sense organs; acquired adaptations originate in response to definite stimuli, such as the formation of antibodies against a particular disease. (Chapters 35, 38, and 40.)

32. All organisms have some capacity to adjust themselves to changes in their environment. This is the principle of **acclimatization,** which refers to the process by which an organism, within the limits of its life history, is able to become inured to conditions that are normally harmful or injurious to it, such extremes of heat, cold, salinity of water medium, oxygen pressure, toxins, and many others. This is a physiologic process that must be distinguished from adjustments that are made over many generations, such as the accumulation of mutant genes that may favor the new adjustment in an organism over a long period of time. Acclimatization also does not refer to the routine and rapid adjustments that physiologic organs are able to make in their normal functioning, such as the ability to adapt to dim and strong light or the ability to detect differential sensitivity. (Chapters 37, 38, and 40.)

33. All organisms fit into a scheme of interrelationships between themselves and their environment. No animal can live apart from its environment. All animals are influenced by environmental forces, but the interrelationship is mutual, for each organism also influences its environment. The factors of the environment may be **biotic,** which includes interrelations between the animal and other animals within its range, or **physical,** which involves such forces as temperature, moisture, soil, air, light, and many others. Biotic factors may involve members of the same or different species. The basic environmental unit of any organism is the *ecosystem,* which includes the above factors. The interrelationships are often different in the two groups, because competition between members of the same species includes the search for the same food, shelters, and water; those of different species include such problems as food chains, population pressures, and other general community relations. Through the operation of these environmental factors a balance of nature is worked out. (Chapter 38.)

34. In all group organizations, individuals profit mutually from an unconscious cooperation. No animal lives to itself throughout its life history, for it either comes in contact with other members of the same species or with other species of the animal community to which it belongs. In animals with definite social organizations there are optimal population sizes that determine their success. Definite hazards appear when there are too few or too many organisms within a population. The rate of evolution appears to bear a definite relationship to an optimal-sized population. Many biologic processes are dependent on an optimum factor of numbers involved in any particular process. It is easy to see that overcompetition for mates, for food, and for shelter may result in a decrease rather than an increase. (Chapters 38 and 40.)

35. In metazoan forms the segregation of germ plasm and somaplasm represents the first specialization of cells. This principle stresses the separation of **somatic cells,** which take care of the general bodily functions of locomotion, nutrition, etc., from the **germinal cells,** which are responsible for reproduction. In general the principle holds true, but there are cases in which sex cells have come directly from soma cells in some animals. In colonial Protozoa, which are intermediate between the Protozoa and the Metazoa, the first differentiation is that between nutritive and reproductive cells. (Chapters 7, 31, and 32.)

36. Embryonic germ layers are the forerunners of adult organs and structures. The differentiation of the early embryo into three germ layers is an important event in the embryology of most metazoans, for these germ layers give rise to the future body structures. Some lower metazoans are diploblastic; that is, they have only two germ layers, **ectoderm** and **endoderm,** so that their capacity for developing complex organs is restricted. In higher metazoans (triploblastic animals) the third germ layer, **mesoderm,** which forms most of the body organs, is added. The importance of germ layers in animal development gave rise to the **germ layer theory,** which states that germ layers have been formed in much the same way throughout all metazoans and that each layer is destined to form certain specific organs. For example, the skin and nervous system are derived from

ectoderm and muscle and skeleton from mesoderm. Modern embryologists, however, have found many exceptions to this theory. Muscle usually comes from mesoderm, but lower animals with only the two germ layers also have muscle, which must come from ectoderm or endoderm. (Chapters 5 and 32.)

37. A body cavity of some form is characteristic of most bilateral animals. Body cavities are varied in form. Coelenterates and other radially symmetric forms have only a digestive cavity. In flatworms and some others the space between ectoderm and endoderm is filled with mesenchyme or its derivatives. The roundworms have a form of cavity known as the **pseudocoel.** The true **coelom,** a space that appears in the mesoderm, is characteristic of the higher phyla that have a tube-within-a-tube arrangement. The coelom encloses most of the internal organs, and its development has made possible the differentiation of many systems in the evolution of the animal. (Chapters 5, 10, 11, and 12.)

38. An organism is a biologic system whose parts are organized into a functional whole. An organism is not a mere summation of its constituent parts. It is a self-sufficient unit, and its parts, whether they are cells, tissues, or organs, cannot survive apart from the whole. The organism as a whole has properties that cannot be explained merely by considering the sum of the properties of its individual parts. It is impossible to understand the whole organism by analyzing its parts, for such a procedure destroys the organization that is the basic part of life. This, then, refutes much of the mechanistic interpretation of life; although physics and chemistry may be able to explain the parts, they cannot as yet explain the life process as a whole for the animal is a unit that loses its essential characters when it is divided. (Chapters 1 and 5.)

References
Suggested general readings

Arber, A. 1954. The mind and the eye. New York, Cambridge University Press. *This small work is a general analysis of the nature of biologic research.*

Bates, M. 1960. The forest and the sea. New York, Random House. *Written in a popular style, this revealing book will give the inquisitive student much to think about regarding man's relations to the world around him.*

Bonner, J. T. 1962. The ideas of biology. New York, Harper & Row, Publishers.

Bonner, J. T. 1965. Size and cycle: an essay on the structure of biology. Princeton, N. J., Princeton University

Press. *The author considers the whole life-span of animals rather than the adult animal as the fundamental unit in the understanding of the various disciplines concerned with biology.*

Calder, R. 1954. Science in our lives. East Lansing, Michigan State College Press. *The author stresses the fact that the essentials of a great scientific discovery depend upon three factors—the method, the man, and the moment. Is not this last factor mainly responsible for independent discovery by more than one worker?*

Cannon, W. B. 1945. The way of an investigator. ew York, W. W. Norton & Co., Inc.

Conant, J. B. 1951. Science and common sense, New Haven, Conn., Yale University Press. *This masterly treatise deals with all science, but Chapters 8 and 9 are devoted to the living organisms. The nature of the control experiment and the methods biologists have employed are explained. The history of the investigations on spontaneous generation is used as an example.*

Elsasser, W. M. 1966. Atom and organism. Princeton, N. J., Princeton University Press. *A theoretical physicist presents his views on the nature of life. He believes that the laws of biology are more or less unique and not altogether deducible from the laws of physics and chemistry. He presents arguments **against** both a mechanistic and vitalistic view of life.*

Gray, P. (editor). 1970. Encyclopedia of the biological sciences, ed. 2. New York, Reinhold Publishing Corp. *A reference work of great importance to all majors in the life sciences.*

Gray, P. 1967. The dictionary of the biological sciences. New York, Reinhold Publishing Corp. *A useful reference work for biology students. Genera and species are not found among the main entries.*

Grobstein, C. 1964. The strategy of life. San Francisco, W. H. Freeman & Co., Publishers. *An excellent paperback on the revolution of present-day biology.*

Haggis, G. H. (editor). 1964. Introduction to molecular biology. New York, John Wiley & Sons, Inc. *Traces the exciting discoveries in this discipline. The Appendix includes a concise version of the "Origin of Life."*

Hall, T. S. 1951. A source book in animal biology. New York, McGraw-Hill Book Co. *An excellent biologic anthology of great selections from the leading biologists of all times. Suitable for the beginning student.*

Hocking, B. 1972. Biology or oblivion: lessons from the ultimate science. Cambridge, Mass., Schenkman Publishing Co., Inc. *Wide-ranging, entertaining, and absorbing exploration of the field of biology.*

Jaeger, E. C 1955. A source-book of biological names and terms, ed. 3. Springfield, Ill., Charles C Thomas, Publisher. *This is a useful book for all students who are interested in the meaning and derivation of biologic terms. It is perhaps the best in the field.*

Johnson, W. H., and W. C. Steere (editors). 1962. This is life. New York, Holt, Rinehart & Winston, Inc. *This is an anthology of essays in modern biology.*

Kendrew, J. C. 1966. The thread of life. Cambridge, Mass., Harvard University Press. *A simple introductory account of the exciting development of molecular biology and its possibilities in the future.*

Lanham, U. 1968. The origins of modern biology. New York, Columbia University Press.

Pennak, R. W. 1964. Collegiate dictionary of zoology. New York, The Ronald Press Co. *An extremely useful reference work for all students of zoology.*

Selected *Scientific American* articles

Bronowski, J. 1958. The creative process. **199:**59-65 (Sept.).

Comfort, A. 1961. The life span of animals. **205:**108-119 (Aug.).

Crewe, A. V. 1971. A high-resolution scanning electron microscope. **224:**26-35 (April).

Everhart, T. E., and T. L. Hayes. 1972. The scanning electron microscope. **226:**54-69 (Jan.).

Gray, G. W. 1951. The ultracentrifuge. **184:**43-51 (June).

Terman, L. M. 1955. Are scientists different? **192:**25-29 (Jan.).

Wald, G. 1958. Innovation in biology. **199:**100-113 (Sept.).

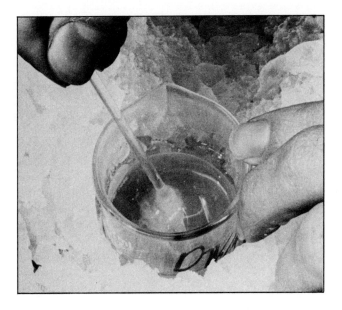

A DNA molecule is an extremely long helical chain, and its molecular architecture is reflected in its gross structure. This photograph shows the final step in a common DNA isolation scheme, in which many long, viscous strands of concentrated and purified DNA are being removed from an ice-cold alcohol suspension by winding them on a glass rod.

CHAPTER 2

MATTER AND LIFE

■ BASIC STRUCTURE OF MATTER
Nature of atoms and molecules

Although the ancient Greeks had certain conceptions of the composition of matter, such as the universe being composed of the four elements—fire, air, earth, and water—our present concepts of the nature of matter have originated within the past two centuries. A. L. Lavoisier, the great French scientist of the eighteenth century, compiled the first list of elements, a total of 28, and explained the precise nature of respiration (1778). J. Dalton, the English chemist, conceived (1808) that matter was composed of atoms, which combine in definite proportions to form chemical compounds. The Italian investigator A. Avogadro showed how many atoms of each kind make up each compounded atom and gave the concept of the molecule (1811). By 1869 no fewer than 92 kinds of atoms had been ascertained, and D. I.

Mendeléeff worked out his periodic table of the elements. In this table he arranged the elements into 8 groups on the basis that a relation of the chemical elements can be expressed by their properties, which are periodic functions of their atomic weights. If the elements are in a group of elements that have similar properties and relations, they follow a regular progression in the individual differences of their members.

For many years the atom was considered solid and indivisible. In 1911 Lord Rutherford showed that every atom consists of a positively charged nucleus surrounded by a negatively charged planetary system of electrons. The nucleus, containing most of the atom's mass, is made up of two kinds of particles, **protons** and **neutrons.** These two particles have about the same weight, each being about 1,800 times heavier than an electron. The protons bear positive charges, and the neutrons have no charges (neutral). Although there is the same number of protons in the nucleus as there are electrons revolving around the nucleus, the number of neutrons may vary. For every positively charged proton in the nucleus, there is a negatively charged electron. The total charge of the atom is thus neutral.

The number of protons in the nucleus is the **atomic number** of the atom. Thus hydrogen, helium, and lithium, containing respectively 1, 2, and 3 protons in their nuclei, have atomic numbers of 1, 2, and 3, respectively. The **mass number** of an atom is the total number of protons and neutrons in its nucleus. The nucleus of an oxygen atom contains 8 protons and 8 neutrons. It therefore has a mass number of 16. The heaviest natural element, uranium, has a nucleus of 92 protons (its atomic number is thus 92) and 146 neutrons, and so its mass number is 238. The elements are designated by convenient symbols that show the atomic number as a subscript, and the total number of protons and neutrons (mass number) as a superscript. Thus oxygen is designated $_8O^{16}$, hydrogen $_1H^1$ (1 proton, no neutrons), helium $_2He^4$ (2 protons, 2 neutrons), and so on (Fig. 2-1).

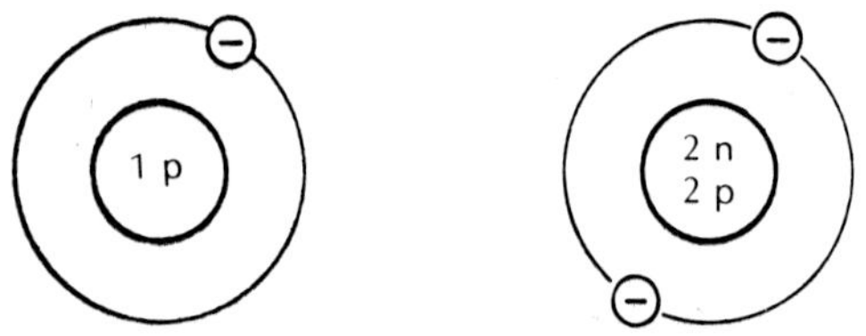

ATOMIC WEIGHT. The atomic weight of an atom is nearly the same as its mass number. However, a quick examination of a periodic table shows that none of the elements have atomic weights of exact integers. How are atomic weights derived? Obviously an atom weighs far too little to be weighed, or even to serve as a useful index for weight comparison. A hydrogen atom, for example, weighs 1.67×10^{-24} grams. Consequently, physicists have assigned a set of meaningful relative weights to the elements, using carbon as a base for comparison. Carbon, $_6C^{12}$, with 6 protons and 6 neutrons was assigned the integral value 12. By this scale, protons and neutrons have masses of 1.0073 and 1.0087, respectively—masses close to, but not exactly, one. (Electrons have an almost negligible mass of 0.00055.) Thus no element, except carbon 12, has an atomic weight of an exact integer: hydrogen weighs 1.0080; helium, 4.0026; lithium, 6.939; and so on.

ISOTOPES. It is possible for two atoms of the same element to have the same number of protons in their nuclei, but a different number of neutrons. For example hydrogen exists in nature primarily as $_1H^1$, that is, it contains 1 proton, but no neutron. However, there are also trace amounts of two other forms of hydrogen: $_1H^2$, which has 1 proton and 1 neutron and is called deuterium, and $_1H^3$, which has 1 proton and 2 neutrons and is called tritium. These three varieties differ only in the number of neutrons in the nucleus (Fig. 2-2). Such forms of an element, having the same charge but different atomic weights, are called **isotopes.**

RADIOACTIVE ISOTOPES. Although most of the naturally occurring elements are stable, all elements have at least one radioactive isotope. These isotopes undergo spontaneous disintegration with the emission of one or more of three types of parti-

Hydrogen
Atomic number 1
Mass number 1

Helium
Atomic number 2
Mass number 4

FIG. 2-1
Two lightest atoms. Since first shell closest to atomic nucleus can hold only 2 electrons, helium shell is closed so that helium is chemically inactive.

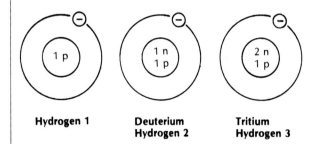

Hydrogen 1

Deuterium
Hydrogen 2

Tritium
Hydrogen 3

FIG. 2-2
Three isotopes of hydrogen. Of the three isotopes, hydrogen 1 makes up about 99.98% of all hydrogen, and deuterium makes up about 0.02%. Tritium is found only in traces in water. Numbers indicate approximate atomic weights. Most elements are mixtures of isotopes. Some elements (for example, tin) have as many as 10 isotopes.

cles, or rays—**gamma rays** (a form of electromagnetic radiation), **beta rays** (electrons), and **alpha rays** (positively charged helium nuclei stripped of their electrons). Most of the isotopes of greatest use in biologic tracer studies are beta and gamma emitters. Virtually all are prepared synthetically in nuclear reactors and cyclotrons. Among the commonly used radioisotopes are carbon 14 ($_6C^{14}$), tritium ($_1H^3$), and phosphorus 32 ($_{15}P^{32}$). Using radioisotopes, biologists are able to trace movements of elements and tagged compounds through organisms. Our present understanding of metabolic pathways in animals and plants is due in very large part to the recent use of this powerful analytical tool. Radioisotopes are also used to great advantage in the diagnosis of disease in man, such as cancer of the thyroid gland.

ELECTRON "SHELLS" OF ATOMS. According to Bohr's planetary model of the atom, the electrons revolve around the nucleus of an atom in precise orbits, or shells. This simplified picture of the atom has been greatly modified by recent experimental evidence.

According to the quantum theory the electrons surrounding the nucleus exist at discrete energy levels, called quantum levels. This theory replaces the older idea that electrons revolve around the nucleus in definite shells, or orbit patterns. A quantum level represents a discrete energy value. These energies, and hence the energies of electrons in these quantum levels, increase as the distance from the nucleus increases. Electrons tend to move as close to the nucleus as possible. However, there is a physical maximum to the number of electrons that can occupy each quantum level. Thus as the inner level becomes filled, additional electrons are forced into more distant quantum levels. These outer electrons are more excited and have a higher energy content.

Although the picture of the atom described by the quantum theory provides a much better basis for understanding the atom, the old planetary model is still useful in interpreting chemical phenomena. The number of concentric "shells," or the paths of the electrons in their orbits, varies with the element. Each shell can hold a maximum number of electrons. The first shell next to the atomic nucleus can hold a maximum of 2 electrons (hydrogen has only 1), and the second shell can hold 8; other shells also have a maximum number, but no atom can have more than 8 electrons in its outermost shell. Inner shells are filled first, and if there are not enough electrons to fill all the shells, the outer shell is left incomplete. Hydrogen has 1 proton and 1 electron in its single orbit but no neutron. Since its shell can hold 2 electrons, it has an incomplete shell. Helium has 2 electrons in its single shell, and its nucleus is made up of 2 protons and 2 neutrons. Since the 2-

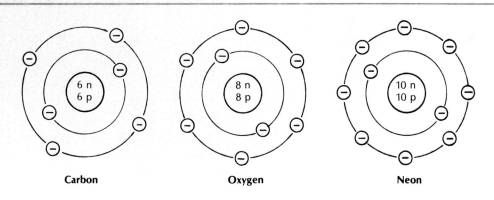

Carbon   **Oxygen**   **Neon**

FIG. 2-3

Electron shells of three common atoms. Since no atom can have more than 8 electrons in its outermost shell and 2 electrons in its innermost shell, neon is chemically inactive. However, the second shells of carbon and oxygen, with 4 and 6 electrons, respectively, are open so that these elements are electronically unstable and react chemically whenever appropriate atoms come into contact. Chemical properties of atoms are determined by their outermost electron shells.

electron arrangement in helium's shell is the maximum number for this shell, the shell is closed and precludes all chemical activity. There is no known compound of helium. Neon is another inert gas (chemically inactive [Fig. 2-3]) because its outer shell contains 8 electrons, the maximum number. However, stable compounds of xenon (an inert gas) with fluorine and oxygen are formed under special conditions. Oxygen has an atomic number of 8. Its 8 electrons are arranged with 2 in the first shell and 6 in the second shell (Fig. 2-3). It is active chemically, forming compounds with almost all the elements except the inert gases.

The number of electrons in the outer shell varies from 0 to 8. With either 0 or 8 in this shell, the element is chemically inactive. When there are fewer than 8 electrons in the outer shell, the atom will tend to lose or gain electrons to have an outer shell of 8, which will result in a charged ion. Atoms with 1 to 3 electrons in the outer shell tend to lose them to other atoms and to become positively charged ions because of the excess protons in the nucleus. Atoms with 5 to 7 electrons in the outer orbit tend to gain electrons from other atoms and to become negatively charged ions because of excess electrons over the protons. Positive and negative ions tend to unite.

A combination of two or more elements forms a **compound.** A **molecule** is the smallest particle of an element or compound that can exist by itself. Compounds always contain different elements, but the term molecule is often applied to single elements. Molecules of most elementary gases are made up of 2 atoms. Thus a molecule of oxygen is O_2 and that of hydrogen is H_2. In such cases the 2 atoms are always found together. Carbon dioxide (CO_2) and methane (CH_4) are true compounds because they consist of different elements, each of which is present in definite proportions.

GRAM ATOMIC AND GRAM MOLECULAR WEIGHTS. To appreciate the quantitative expression of compounds, the student should be familiar with the terms **gram atomic weight** and **gram molecular weight.** A gram atomic weight of an element is its atomic weight expressed in grams. The number of atoms in a gram atomic weight is 6.02×10^{23} (Avogadro's number). This means that the gram atomic weight of all elements has the same number of atoms. One gram of hydrogen (its gram atomic weight) has the same number of atoms as does 16 grams of oxygen or 12 grams of carbon, their gram atomic weights. A gram molecular weight, or **mole,** is the sum of the atomic weights of the atoms in a molecule. This also means that the same number of moles of all substances have the same number of molecules. Thus a mole (18 grams) of water (H_2O) has the same number of molecules as does a mole (32 grams) of oxygen (O_2) or a mole (28 grams) of carbon monoxide (CO).

IONS AND OXIDATION STATES. We have seen that every atom has a tendency to complete its outer shell to increase its stability. Let us examine how 2 atoms with incomplete outer shells, sodium and chloride, can interact to fill their outer shells. Sodium, with 11 electrons, has 2 electrons in its first shell, 8 in its second shell, and only 1 in the third shell. The third shell is highly incomplete; if this third-shell electron were lost, the second shell would be the outermost shell and produce a stable atom. Chlorine, with 17 electrons, has 2 in the first shell, 8 in the second and 7 in the incomplete third shell. Chlorine must gain an electron to fill the outer shell and become a stable atom. Clearly, the transfer of the third-shell sodium electron to the incomplete chlorine third shell would yield simultaneous stability to both atoms. But since electrons bear negative charges, the sodium-to-chlorine transfer will create two charged atoms, called **ions.** Both sodium and chlorine become ionized, sodium becoming electropositive (Na^+) and chlorine electronegative (Cl^-). Since unlike charges attract, a chemical bond is formed, called an **ionic bond** (Fig. 2-4). The ionic compound formed, sodium chloride, can be represented in electron dot notation ("fly-speck formulas") as:

$$Na\cdot \ + \ \cdot\ddot{\underset{..}{Cl}}: \ \longrightarrow \ Na^+ \ + \ (:\ddot{\underset{..}{Cl}}:)^-$$

Processes that involve the **loss of electrons** are **oxidation** reactions; those that involve the **gain of electrons** are called **reduction** reactions. Since oxidation and reduction always occur simultaneously, each of these processes is really a "half-reaction." The entire reaction is called an **oxidation-reduction** reaction, or simply **redox** reactions. The terminology is confusing because oxidation-reduction reactions involve electron transfers, rather than (necessarily) any reaction with oxygen. However, it is easier to learn the system than to try to change accepted usage.

We now need to introduce the concept of **oxida-**

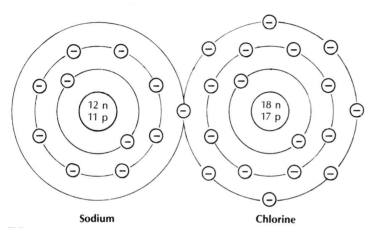

Sodium   **Chlorine**

FIG. 2-4

Ionic bond. When an atom of sodium and one of chlorine react to form a molecule, a single electron in outer shell of sodium is transferred to outer shell of chlorine. This causes outer or second shell (third shell is empty) of sodium to have 8 electrons and also chlorine to have 8 electrons in its outer or third shell. The compound thus formed is called sodium chloride (NaCl). By losing 1 electron, sodium becomes a positive ion, and by gaining 1 electron, chlorine (chloride) becomes a negative ion. This ionic bond is held together by a strong electrostatic force.

tion number. This term refers to the charge an atom would have if the bonding electrons were arbitrarily assigned to the more electronegative of two interacting elements. For example, in the sodium chloride reaction, we consider that the bonding electron has been transferred to the chlorine atom. Consequently sodium, having lost its electron, becomes electropositive and is said to have an oxidation number of +1. Chlorine, with its newly acquired electron, becomes electronegative, and takes an oxidation number of −1. Some elements always exist in compounds in the same oxidation number. For example, oxygen almost always has an oxidation number of −2; sodium, +1; magnesium, +2; potassium, +1; and calcium, +2. However, most metals commonly have two or more oxidation numbers. For example, iron may exist as +2 or +3, chromium as +3 or +6, manganese as +2, +4, or +7. Other elements, such as hydrogen, can take either positive or negative oxidation numbers.

COVALENT BONDS. Stability can also be achieved when 2 atoms **share** electrons. Let us again consider the chlorine atom, which, as we have seen, has an incomplete 7-electron outer shell. Stability is attained by gaining an electron. One way this can

be done is for 2 chlorine atoms to share one pair of electrons (Fig. 2-5). To do this, the 2 chlorine atoms must **overlap** their third shells, so that the electrons in these shells can now spread themselves over both orbits. Many other elements can form covalent (or electron-pair) bonds. For example: hydrogen (H_2)

$$H\cdot + H\cdot \ \rightarrow \ H\!:\!H$$

and oxygen (O_2)

$$\ddot{\underset{..}{O}}\!: + :\!\ddot{O} \ \rightarrow \ \ddot{\underset{..}{O}}\!:\!:\!\ddot{O}$$

In this case, oxygen with an oxidation number of −2 must share 2 pairs of electrons to achieve stability. Each atom now has 8 electrons available to its outer shell, the stable number.

Covalent bonds are of great significance to living systems, since the major elements of protoplasm (carbon, oxygen, nitrogen, hydrogen) almost always share electrons. Carbon, which usually has an oxidation number of either +4 or −4 (its outer shell contains 4 electrons), can share its electrons with hydrogen to form methane:

$$\cdot\dot{C}\cdot + 4\,H\cdot \ \rightarrow \ \begin{matrix} & H & \\ H\!:&\!\!\ddot{C}\!\!:&\!H \\ & H & \end{matrix}$$

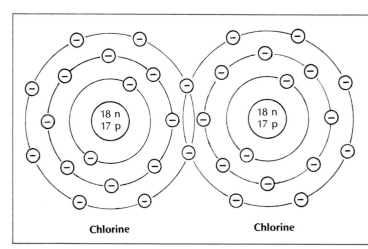

FIG. 2-5
Covalent bond. Each chlorine atom has 7 electrons in its outer shell, and by sharing one pair of electrons, each atom acquires a complete outer shell of 8 electrons, thus forming a molecule of chlorine (Cl_2). Such a reaction is called a molecular reaction, and such bonds are called covalent bonds.

Carbon now achieves stability with 8 electrons, and each hydrogen atom becomes stable with 2 electrons. Carbon also forms covalent bonds with oxygen:

$$\cdot \dot{C} \cdot + 2\, \ddot{O}\colon \; \rightarrow \; \ddot{O}\colon\colon C\colon\colon \ddot{O}$$

Carbon can also bond with itself (and hydrogen) to form, for example, ethane:

$$\begin{matrix} & H\;H \\ H\colon\!\ddot{C}\colon\!\ddot{C}\colon\!H \\ & \ddot{H}\;\ddot{H} \end{matrix} \quad \text{or} \quad \begin{matrix} & H\;\;\;H \\ H\!-\!C\!-\!C\!-\!H \\ & H\;\;\;H \end{matrix}$$

Carbon can join in "double bond" configuration:

$$\begin{matrix} H \quad\; H \\ \colon\!C\colon\colon\!\dot{C}\colon \\ H \quad\; H \end{matrix} \quad \text{or} \quad \begin{matrix} H \quad\;\;\; H \\ \diagdown \;\;\; \diagup \\ C\!=\!C \\ \diagup \;\;\; \diagdown \\ H \quad\;\;\; H \end{matrix}$$

Or even in "triple bond" configuration:

$$H\colon\!C\colon\colon\colon\!C\colon\!H \quad \text{or} \quad H\!-\!C\!\equiv\!C\!-\!H$$

These examples do not begin to adequately illustrate the amazing versatility of carbon, the element forming the backbone of all life.

Acids, bases, and salts

Every ionic compound is an acid, a base, or a salt. One commonly accepted definition of an **acid** is any compound that dissociates to yield hydrogen ions. An acid is classified as strong or weak, depending on the extent to which an ionic compound is dissociated. Those that dissociate completely in water (H_2SO_4, HNO_3, and HCl) are called strong acids.

Weak acids such as acetic acid (CH_3COOH) dissociate only weakly. A solution of acetic acid is mostly undissociated acetic acid molecules with only a small number of acetate and hydrogen ions present.

A **base** contains negative ions called hydroxyl ions and may be defined as a molecule or ion that will accept a proton. Bases are produced when compounds containing them are dissolved in water. NaOH (sodium hydroxide) is a strong base because it will dissociate completely in water into sodium (Na^+) and hydroxyl (OH^-) ions. Among the characteristics of a base is its ability to combine with a hydrogen ion. Like acids, bases vary in the extent to which they dissociate in aqueous solutions into hydroxyl ions.

A **salt** is a compound resulting from the chemical interaction of an acid and a base. Common salt, sodium chloride (NaCl), is formed by the interaction of hydrochloric acid (HCl) and sodium hydroxide (NaOH). In water the HCl is dissociated into H^+ and Cl^- ions. The hydrogen and hydroxyl ions combine to form water (H_2O), and the sodium and chloride ions combine to form salt:

$$\underset{\textbf{Acid}}{HCl} + \underset{\textbf{Base}}{NaOH} \; \rightarrow \; \underset{\textbf{Salt}}{NaCl} + H_2O$$

Organic acids are usually characterized by having in their molecule the carboxyl group (—COOH). The carboxyl group contains both a carbonyl group and a hydroxyl group on the same carbon atom:

$$\begin{matrix} R\!-\!C\!=\!O \\ | \\ O\!-\!H \end{matrix} \;\rightleftarrows\; \begin{matrix} R\!-\!C\!=\!O \\ | \\ O\!- \end{matrix} + H^+$$

R refers to an atomic grouping unique to the molecule. In water, the COO— group will behave as a weak acid. The common organic acids are acetic, citric, formic, lactic, and oxalic. The student will encounter many of these later in discussions of cellular metabolism.

Hydrogen ion concentration (pH)

Solutions are classified as acid, base, or neutral, according to the proportion of hydrogen (H^+) and hydroxyl (OH^-) ions they possess. In acid solutions there is an excess of hydrogen ions; in alkaline, or basic, solutions the hydroxyl ion is more common; and in neutral solutions both hydrogen and hydroxyl ions are present in equal numbers.

To express the acidity or alkalinity (or pH concentration) of a substance, a logarithmic scale, a type of mathematical shorthand, is employed that uses the numbers 1 to 14. In this scale, numbers below 7 indicate an acid range. The number 7 indicates neutrality, that is, the presence of equal numbers of H^+ and OH^- ions. According to this logarithmic scale, a pH of 3 is ten times more acid than one of 4; a pH of 9 is ten times more alkaline than one of 8.

In protoplasmic systems pH plays an important role, for, in general, slight deviations from the normal usually result in severe damage. Most substances and fluids in the body hover closely around the point of neutrality, that is, pH of around 7. Blood, for instance, has a pH of about 7.35, or just slightly on the alkaline side. Lymph is slightly more alkaline than blood. Saliva has a pH of 6.8, on the acid side. Gastric juice is the most acid substance in the body, about pH 1.6. The regulation of the pH in the body tissue fluids involves many important physiologic mechanisms; one of the most important is the buffer action of certain salts.

Buffer action

The hydrogen ion concentration in the extracellular fluids must be regulated so that metabolic reactions within the cell will not be adversely affected by a constantly changing hydrogen ion concentration. A change of only 0.2 pH unit from the normal blood pH of 7.35 can cause serious metabolic disturbances. To protect itself from sudden changes in acidity, the body has certain substances that resist any change in the pH when acids or alkalies are added to the body fluids. These are called **buffers.** The hydrogen ion concentration within the cells is probably greater (pH is lower) than the hydrogen ion concentration in the extracellular fluids because of the metabolic production of CO_2, which reacts with the cellular water to form carbonic acid (H_2CO_3). Certain phosphates, sulfates, and organic acid radicals also add to the acidic nature of the intracellular fluids. Within the cells the high content of protein serves as a buffer and thus tends to keep the pH from going too low.

The buffer function of blood is dependent on both plasma and red blood corpuscle buffer mechanisms. The chief buffer of plasma and tissue fluid is sodium bicarbonate ($NaHCO_3$). This salt dissociates into sodium ions (Na^+) and bicarbonate ions (HCO_3^-). When a strong acid (for example, HCl) is added to the fluid, the H^+ ions of the dissociated acid will react with the bicarbonate ion (HCO_3^-) to form a very weak acid, carbonic acid, which dissociates only slightly. Thus the H^+ ions from the HCl are removed and the pH is little altered.

Mixtures and their properties

Whenever masses of different kinds are thrown together, we have what is called a mixture. All the different states of matter (solids, liquids, gases) may be involved in these mixtures. The mixtures we are mainly interested in here are those in which water or other fluid is one of the states of matter. When something is mixed with a liquid, any one of three kinds of mixtures is formed.

MOLECULAR SOLUTIONS. If crystals of salts or sugars are added to water, the molecules or ions (in the case of salts) are uniformly dispersed through the water, forming a **true solution** (Fig. 2-6). Such solutions are transparent. In such a case the water is the **solvent** and the dissolved salt or sugar the **solute.** The freezing point of solutions is lower and the boiling point is higher than those of pure water.

SUSPENSIONS. If solids added to water remain in masses larger than molecules, the mixture is a suspension. Muddy water is a good example. When allowed to stand, the particles in suspension will settle out to the bottom. Suspensions have a turbid appearance and have the same boiling and freezing points as pure water.

COLLOIDS. Whenever the dispersed particles are intermediate in size between the molecular state and

FIG. 2-6
Three types of solutions or mixtures.

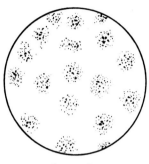

**Molecular solution
(transparent)**   **Colloid
(transparent or
slightly cloudy)**   **Suspension
(cloudy)**

the suspension, a third mixture is the result—the colloidal solution. Colloidal particles are rather arbitrarily considered to be between 1 and 500 millimicrons in size. If the particles are smaller, the solution is classified as a true solution; if larger, they are suspensions or emulsions. Colloids consist of two phases—a discontinuous, or dispersed, phase and a continuous, or dispersion, phase. These phases may be represented by the same states of matter or different ones. Some familiar examples are as follows:

Discontinuous phase	Continuous phase	Example
Solid	Liquid	Ink
Liquid	Liquid	Emulsion
Liquid	Solid	Gel
Solid	Solid	Stained glass
Gas	Solid	Foam, carbonated
	Liquid	water
Liquid		Fog
Solid	Gas	Smoke
	Gas	

A true colloidal solution is stable (that is, will not settle out), has about the same boiling point and freezing point as pure water, and is either transparent or somewhat cloudy.

Proteins, which are important constituents of protoplasm, form colloidal solutions because their large molecules are well within the size range of colloidal particles and behave like colloids. Since protein molecules also dissolve as molecules in solution, such solutions may also be called molecular.

One special form of colloidal solution is the **emulsion** in which both phases are immiscible liquids

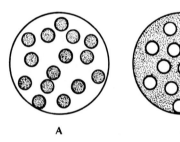

A   **B**

FIG. 2-7
Colloidal solution in which each phase is a liquid. **A,** Oil-in-water emulsion, with water being the continuous phase, oil the discontinuous phase. **B,** Water-in-oil emulsion, with water being the discontinuous phase, oil the continuous phase. Certain agents can bring about this phase reversal.

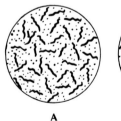

A   **B**

FIG. 2-8
Sol and gel. **A,** Sol condition in which gelatin particles are the discontinuous phase, water the continuous phase. **B,** Gel condition in which gelatin particles form continuous phase (network), enclosing water as discontinuous phase.

(Fig. 2-7). Cream is a good example. Here, droplets of oil, or fat, are dispersed in water. This type of colloidal solution has considerable significance in the makeup of protoplasm.

Colloidal emulsions also illustrate the property of some (but not all) colloids to reverse their phases. When gelatin is poured into hot water, the gelatin particles (discontinuous phase) are dispersed through the water (continuous phase) in a thin consistency that is freely shakable (Fig. 2-8). Such a condition is called a **sol.** When the solution cools, gelatin now becomes the continuous phase and the water is in the discontinuous phase. Moreover, the solution has stiffened and become semisolid and is called a **gel.** Heating the solution will cause it to become a sol again, and the phases are reversed. Some colloidal emulsions are not reversible. Heating egg white, for example, will change the egg albumin from a sol into an irreversible gel. In such cases the coagulated particles may collect into larger particles and settle out.

Why do colloids play such an important role in the structure of protoplasm? There are several reasons, among which may be mentioned the following:

1. Great surface exposure, which allows for many chemical reactions
2. The property of phase reversal, which helps explain how protoplasm can carry on diverse functions and change its appearance during metabolic activities
3. The property of undergoing gelatin or solation, which enables the protoplasm to contract, thus explaining such movements as ameboid movement
4. The inability of colloids to pass through membranes, which promotes the stability and organization of the cellular system, such as cell and nuclear membranes and cytoplasmic inclusions
5. The selective absorption or permeability of the cell membrane, which is largely dependent on the phase reversal of its colloidal structure.

■ BIOCHEMISTRY OF LIFE
Fitness of our planet for life

If the environment is an integral part of the life process, the physical surroundings of living substance must be uniquely suitable for life to exist. Organisms do not merely adapt to a particular environment, however; their own activities produce a change in the environment. It is thus seen that the relation between life and the environment is a reciprocal one. Many years ago (1913) the biochemist L. J. Henderson pointed out many facts about the uniqueness of the physical aspects of our planet for promoting the life process. In his classic book, *The Fitness of the Environment,* he shows how suitable the environment is for the process of organic evolution and the development of life.

The most abundant of all protoplasmic compounds is **water,** making up about 60% to 90% of most living organisms. All forms, terrestrial or aquatic, maintain an aqueous internal environment. Water therefore plays an important role in all protoplasmic systems. Some properties of this unique compound may be pointed out. It is the most versatile of all solvents. The English physiologist W. M. Bayliss once stated that practically all chemical substances will dissolve in water to some extent. It is evident that this property of water is of the utmost importance for the chemical reactions that must take place in the life process. Water also promotes the ionization of various compounds in the body, a chemical alteration that is essential for bodily processes (promotion of chemical reactions, ionic regulation of cells and organisms, the permeability of cell membranes, active transport of materials, etc.). Water is the only substance that occurs in nature in the three phases of solid, liquid, and vapor within the ordinary range of the earth's temperature. It has a high specific heat. The amount of heat required to raise the temperature of 1 gram of water from 15° to 16° C. is exactly 1 calorie; water thus has a specific heat of 1.0 calorie per gram. Most organic solvents have specific heats of about 0.5 calorie per gram; for iron the specific heat is less than 0.1. Thus a large amount of heat is required to raise the temperature of a given quantity of water. The presence of water has a great effect in moderating environmental temperature changes. Important from a biologic standpoint is the expansion of water before changing to ice instead of contraction (water is heaviest at 4° C.) so that aquatic organisms can live in the bottom of freshwater ponds and lakes in winter without being frozen in. The relatively high surface tension of water is caused by cohesion of water molecules at its surface; and the high latent heat of vaporization (requiring many hundreds of calories to change 1 gram of liquid water into water vapor) is evidence of the strong intramolecular forces in water.

Water possesses its many unique properties in large part because of its dipolar character. Water is a hydride of oxygen. The geometric shape of the molecule, in which the two covalent OH bonds form an angle of 104.5 degrees, produces a separation of electric charge of positive (the protons) and negative (cloud of electrons around the oxygen atom). A water molecule can be represented as:

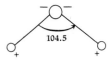

The dipolar character of water is responsible for the orienting effects water molecules have on each other and on other molecules dissolved in water. Because water molecules are charged, they can participate in **hydrogen bond** formation with the nitrogen and oxygen atoms of large macromolecules, especially proteins. Water can thus be bound to organic molecules and, in turn, serve as atomic bridges that hold numbers of large molecules together.

But water is not the only fitness factor for life on earth. One of these factors is the limited range of temperature extremes on earth. We do not experi-

ence on this planet the incredibly hot and cold temperatures that exist on other planets. Especially important in the fitness of earth for life is the abundance and availability of many key elements such as carbon, oxygen, and hydrogen that are crucial constituents in the organic compounds fundamental to life as we know it. Other elements (nitrogen, phosphorus, sulfur, etc.) also contribute to the unique properties of this planet for supporting living substance.

The chemical elements of life

Of the 92 naturally occurring elements, only 24 are considered essential for life. Of these, six have been selected by nature to play especially important roles in living systems. These major elements, shown in red in Fig. 2-9, are carbon (C), hydrogen (H), nitrogen (N), oxygen (O), phosphorus (P), and sulfur (S). Most organic molecules are built with these six elements. Another five essential elements found in less abundance in living systems and shown in light

FIG. 2-9

Periodic table of elements showing those essential to life. Six major elements in living systems are set in bright red; five essential minor elements are set in light red. Thirteen trace elements essential to life are set in boldface black. Other elements, notably nickel, aluminum, antimony, mercury, cadmium, lead, silver, and gold, are usually present in living systems in trace amounts but are not dietary essentials. The mass number is shown above, and the atomic weight (approximate) below, each element.

1 H 1.01																	2 He 4.00
3 Li 6.94	4 Be 9.01											5 B 10.8	6 C 12.0	7 N 14.0	8 O 16.0	9 F 19.0	10 Ne 20.2
11 Na 23.0	12 Mg 24.3											13 Al 27.0	14 Si 28.1	15 P 31.0	16 S 32.1	17 Cl 35.5	18 Ar 39.9
19 K 39.1	20 Ca 40.1	21 Sc 45.0	22 Ti 47.9	23 V 50.9	24 Cr 52.0	25 Mn 54.9	26 Fe 55.8	27 Co 58.9	28 Ni 59.7	29 Cu 63.5	30 Zn 65.4	31 Ga 69.7	32 Ge 72.6	33 As 74.9	34 Se 79.0	35 Br 79.9	36 Kr 83.8
37 Rb 85.5	38 Sr 87.6	39 Y 88.9	40 Zr 91.2	41 Nb 92.9	42 Mo 95.9	43 Tc (99)	44 Ru 101.	45 Rh 103.	46 Pd 106.	47 Ag 108.	48 Cd 112.	49 In 115.	50 Sn 119.	51 Sb 122.	52 Te 128.	53 I 127.	54 Xe 131.
55 Cs 133.	56 Ba 137.	57 La 139.	72 Hf 178.	73 Ta 181.	74 W 184.	75 Re 185.	76 Os 190.	77 Ir 192.	78 Pt 195.	79 Au 197.	80 Hg 201.	81 Tl 204.	82 Pb 207.	83 Bi 209.	84 Po (210)	85 At (210)	86 Rn (222)
87 Fr (223)	88 Ra (226)	89 Ac (227)															

58 Ce 140.	59 Pr 141.	60 Nd 144.	61 Pm (147)	62 Sm 150.	63 Eu 152.	64 Gd 157.	65 Tb 159.	66 Dy 162.	67 Ho 165.	68 Er 167.	69 Tm 169.	70 Yb 173.	71 Lu 175.
90 Th 232.	91 Pa (231)	92 U 238.	93 Np (237)	94 Pu (242)	95 Am (243)	96 Cm (247)	97 Bk (247)	98 Cf (249)	99 Es (254)	100 Fm (253)	101 Md (256)	102 No (256)	103 Lr (257)

red in Fig. 2-9 are calcium (Ca), potassium (K), sodium (Na), chlorine (Cl), and magnesium (Mg). Several other elements, called trace elements, are found in minute amounts in animals and plants, but are nevertheless indispensable for life. These are manganese (Mn), iron (Fe), iodine (I), molybdenum (Mo), cobalt (Co), zinc (Zn), selenium (Se), copper (Cu), chromium (Cr), tin (Sn), vanadium (V), silicon (Si), and fluorine (F). They are shown in black boldface type in Fig. 2-9. Other elements of the periodic table may be present in living things, often as contaminants, but none have been demonstrated essential for life. Many elements are extremely toxic to life in small amounts. Recently much emphasis has been given to their harmful effects. Mercury, for instance, is being found with increasing frequency in both fresh and salt waters. Mercury from paper mills, which until recently used phenyl mercuric acetate as a fungicide to prevent cellulose rot, and from spillage and leakage from plants engaged in chlorine formation, settles to the bottom of freshwater streams where it is converted by bacteria into soluble and highly toxic metallo-organic compounds. The half-life of these mercuric compounds in fish (which obtain it from the water and from invertebrate food organisms) may be as much as 1,000 days. In man who eats the fish the half-life is only about 70 days.

Most trace elements are necessary components of enzyme systems. Some trace elements are also important constituents of organic compounds, such as the iron of hemoglobin and myoglobin. Iodine, the heaviest essential trace element, is a vital part of the thyroid hormones, thyroxine and triiodothyronine. The discovery of necessary trace elements requires ingenious and elaborate dietary studies to determine whether or not a particular element promotes growth or prevents deficiency diseases in test animals. At the present time, tin, vanadium, silicon, and fluorine appear to be essential for the growth of young animals. It is not yet known just how our chemical milieu affects the structure of the animal body, or how necessary certain elements found in the body may be. Since living matter has come from the same elements that are found on this planet it is evident that the composition of the earth restricts what elements go into the composition of the organism. But evolution has been very selective in choosing certain elements as the building stones of life. It is clear from the listing above that only a select number of the

earth's elements are found in protoplasm. Their interactions pose many problems about which little is known.

Chemical complexity of living matter

The basis of biologic activities is chemical reactions. These reactions involve chemical elements that are found in all organisms as well as in the nonliving world. Life must have had its beginning in combinations and reactions of chemicals. At first, only a small number of combinations and reactions were necessary for the initiation of life. As time went on, more complex substances or compounds with successively higher levels of organization and reactions occurred. Despite the incredible complexity of living matter, all biologic phenomena operate, so far as is known, according to the physical laws of chemistry and physics.

Analysis of typical living matter reveals that it is composed of about 60% to 90% water (higher animals are usually about 60% to 70% water), 15% protein, 10% to 15% fats and lipids, 1% carbohydrates, and 5% inorganic ions of various kinds (Na^+, K^+, Cl^-, $SO_4^=$, etc.) (Fig. 2-10).

Many different categories of biomolecules are found in every cell of the organism. Most of them are dissolved or suspended in cellular water, either in ionized or nonionized form. Some organic sub-

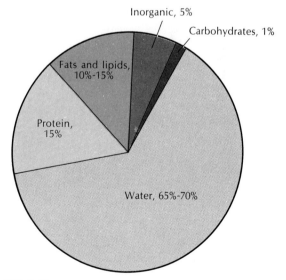

FIG. 2-10
Relative composition of major constituents of animal cells.

stances form a part of the hard substances (horn, claws, hoofs, keratin, etc.) of the body. Besides being in the cells and tissues of the body, organic components are also found in the environment, especially that of aquatic animals. Surprisingly, lake water may contain a far higher organic content than does ocean water. Some organic substances found in water are organic phosphorus, organic nitrogen, amino acids, carotenoid substances, and vitamins.

Inorganic matter actually outweighs the organic part of living matter. The inorganic portion includes (1) water, (2) mineral solids consisting of crystals, secreted precipitates, bone, and shells, and (3) cellular minerals, either free or combined with organic compounds. Most of the cellular minerals are in the form of ions (H^+, Ca^{++}, Na^+, and Mg^{++}, OH^-, $CO_3^=$, HCO_3^-, $PO_4^\equiv$, Cl^-, and $SO_4^=$ ions). Animals get these constituents from the water in which the minerals are dissolved from rocks and the soil.

Organic molecules

The term "organic compounds" has been applied to substances derived from plants and animals. All organic compounds contain carbon, but many also contain hydrogen and oxygen as well as nitrogen, sulfur, phosphorus, salts, and other elements. Organic compounds in a specific way are those carbon compounds in which the principal bonds are carbon-to-carbon and carbon-to-hydrogen. Carbon with its four valences has a great ability to bond with other carbon atoms in chains of varying lengths and configurations. More than a million organic compounds have been identified; more are being added daily. Carbon-to-carbon combinations introduce the possibility of enormous complexity and variety into molecular structure. As a chain it can combine with hydrogen to form an **aliphatic** compound:

Heptane

$$H-\overset{\overset{\displaystyle H}{|}}{\underset{\underset{\displaystyle H}{|}}{C}}-\overset{\overset{\displaystyle H}{|}}{\underset{\underset{\displaystyle H}{|}}{C}}-\overset{\overset{\displaystyle H}{|}}{\underset{\underset{\displaystyle H}{|}}{C}}-\overset{\overset{\displaystyle H}{|}}{\underset{\underset{\displaystyle H}{|}}{C}}-\overset{\overset{\displaystyle H}{|}}{\underset{\underset{\displaystyle H}{|}}{C}}-\overset{\overset{\displaystyle H}{|}}{\underset{\underset{\displaystyle H}{|}}{C}}-\overset{\overset{\displaystyle H}{|}}{\underset{\underset{\displaystyle H}{|}}{C}}-H$$

or a ring structure (**aromatic** compound)

Benzene

Other types of configurations include rings and chains joined to each other, multiple branches of chains, helix arrangements, etc. The diversity of carbon molecules has made possible the complex kinds of macromolecules that form the essence of life.

CARBOHYDRATES: NATURE'S MOST ABUNDANT ORGANIC SUBSTANCE. Carbohydrates are compounds made of carbon, hydrogen, and oxygen. They are usually present in the ratio CH_2O and are grouped as $H-C-OH$. Familiar examples of carbohydrates are sugars, starches, and cellulose. There is more cellulose, the woody structure of plants, than all other organic materials combined. Carbohydrates are made synthetically from water and carbon dioxide by green plants, with the aid of the sun's energy. This process, called **photosynthesis,** is a reaction on which all life depends, for it is the starting point in the formation of food.

Carbohydrates are usually divided into the following three classes: (1) **monosaccharides,** or simple sugars; (2) **disaccharides,** or double sugars; and (3) **polysaccharides,** or complex sugars. Simple sugars are composed of carbon chains containing 4 carbons (tetroses), 5 carbons (pentoses), or 6 carbons (hexoses). Other simple sugars have up to 10 carbons, but these are not biologically important. Simple sugars, such as glucose, galactose, and fructose, all contain a free sugar group,

in which the double-bonded O may be attached to the terminal C of a chain (an aldehyde) or to a nonterminal C (a ketone). The hexose **glucose** (also called dextrose) is the most important carbohydrate in the living world. Glucose is often shown as a straight-chain aldehyde, but in fact it tends to form a cyclic compound:

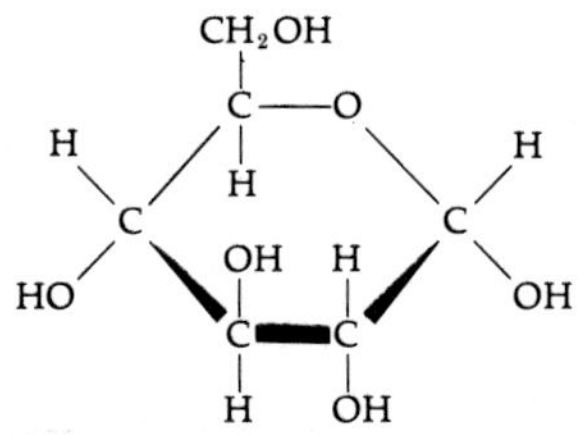

This formula shows the ring structure of glucose, but it is misleading because it obscures the three-dimensional form of the molecule. Rather than lying in a flat plane, the glucose molecule is "puckered," because the four atoms bonded to any single carbon molecule occupy the corners of a regular tetrahedron, four-cornered, pyramid-like figure. To better represent the structure of glucose than in the flat-plane ring structure above, organic chemists have devised a "chair" model to show the configuration of glucose and other hexose molecules:

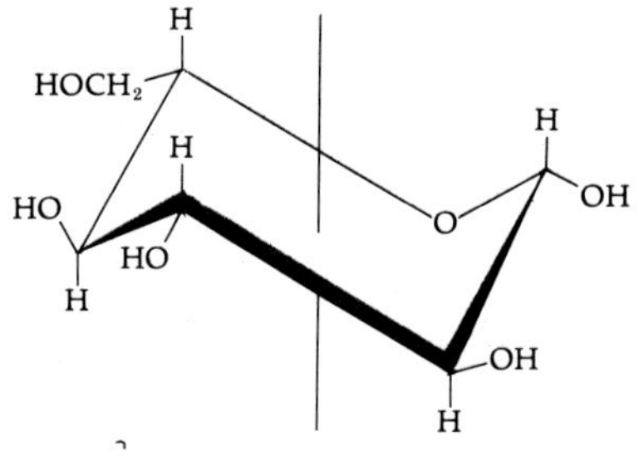

Although the three-dimensional chair conformation is the most accurate way to represent the simple sugars, we must remember that all forms of glucose, however represented, are the same molecule.

Other monosaccharides of biologic significance are galactose and fructose. Their straight-chain structure is compared with glucose in Fig. 2-11.

Disaccharides are double sugars formed by the bonding together of two simple sugars. An example is maltose (malt sugar) composed of two glucose molecules. As shown in Fig. 2-12 the two glucose molecules are condensed together with the removal of a molecule of water. This dehydration reaction, with the sharing of an oxygen atom by the two sugars, characterizes the formation of all disaccharides. Two other common disaccharides are sucrose (ordinary cane, or table, sugar), formed by the linkage of glucose and fructose, and lactose (milk sugar), comprised of glucose and galactose.

Polysaccharides are made up of many molecules of simple sugars (usually glucose) and are referred to by the chemist as polymers. Their empirical formula is usually written $(C_6H_{10}O_5)_n$, where n stands for the unknown number of simple sugar molecules of which they are composed. Starch is common in most plants and is an important food constituent. **Glycogen** (Fig. 2-13), or animal starch, is found mainly in liver and muscle cells. When needed, glycogen is converted into glucose and is delivered by the blood to the

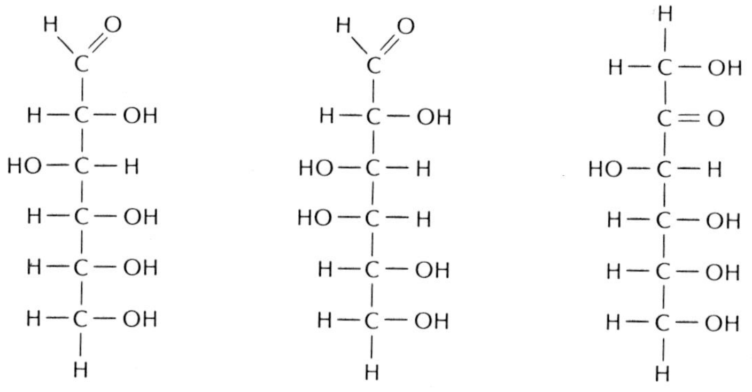

Glucose   **Galactose**   **Fructose**

FIG. 2-11
These three hexoses are the most common monosaccharides. Glucose and galactose are aldehyde sugars; fructose is ketone sugar.

tissues. Another polymer is **cellulose,** which is an important part of the cell walls of plants (Fig. 2-14). Cellulose cannot be digested by man, but some animals, such as the herbivores, with the aid of bacteria, and termites, with the aid of flagellates, can do so.

The main role of carbohydrates in protoplasm is to serve as a source of chemical energy. Glucose is the most important of these energy carbohydrates. Some carbohydrates become basic components of protoplasmic structure, such as the pentoses that form constituent groups of nucleic acids and of nucleotides.

PROTEINS: FOUNDATION SUBSTANCE OF PROTOPLASM. Proteins are large, complex molecules, characterized by a high nitrogen content. Proteins are characteristic for each form of life and for each tissue that composes an organism. The development of an egg into a complex, differentiated animal involves the formation of an enormous number of proteins that are specific for each tissue and organ. Proteins are composed of 20 commonly occurring amino acids (Fig. 2-15). Nearly all of the 20 amino acids are usually present in every protein. Since the

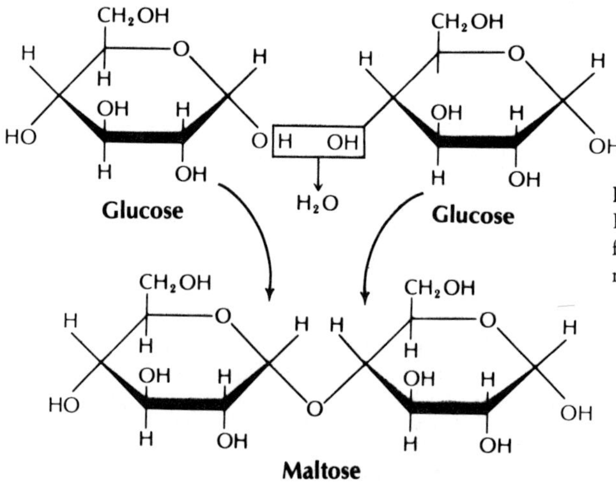

FIG. 2-12
Formation of a double sugar (disaccharide maltose) from two glucose molecules with the removal of a molecule of water.

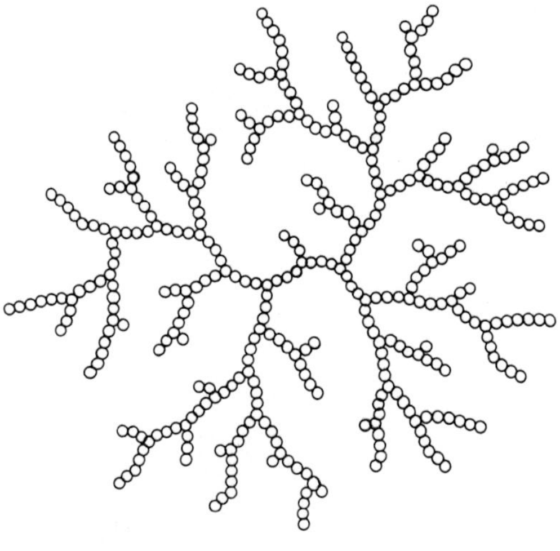

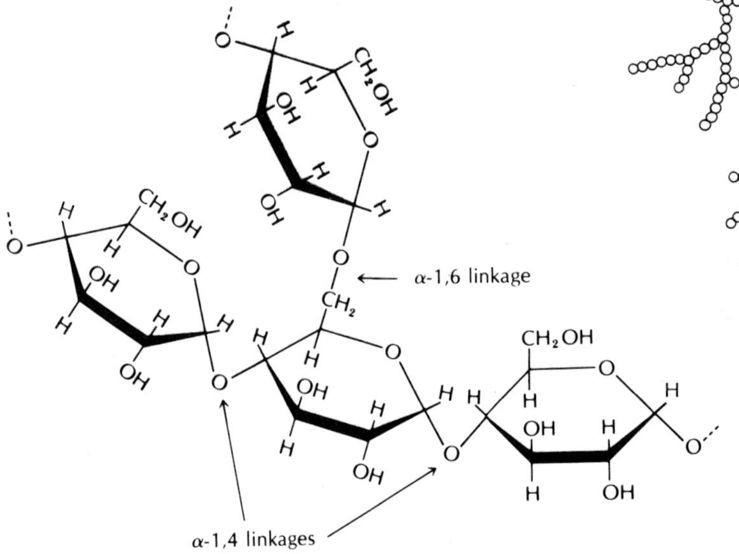

FIG. 2-13
Glycogen is a large, branched polysaccharide with a treelike structure. It is composed of linear chains of glucose molecules joined together by α-1,4 and α-1,6 linkages. A section of a glucose chain is shown enlarged at left.

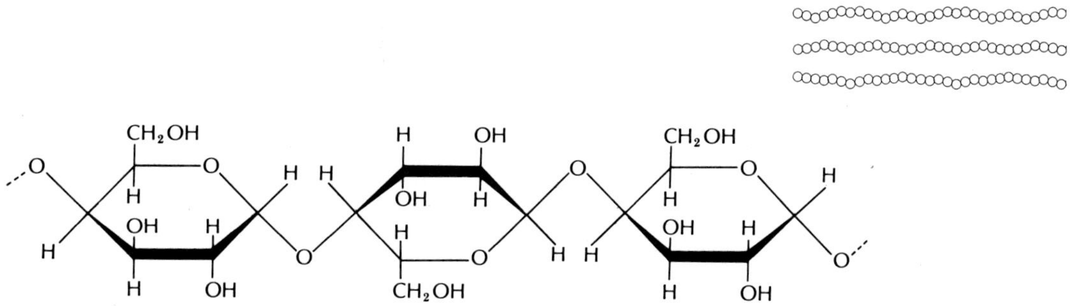

FIG. 2-14

Cellulose, the structural carbohydrate of plants, and the most abundant of all organic compounds. Cellulose differs chemically from glycogen in lacking molecular branching. Its glucose molecules are joined together in straight chains by β-1,4 linkages.

FIG. 2-15

Structural formulas of some important amino acids. Note the sulfur atom in cysteine, which is important in disulfide bond formation in proteins.

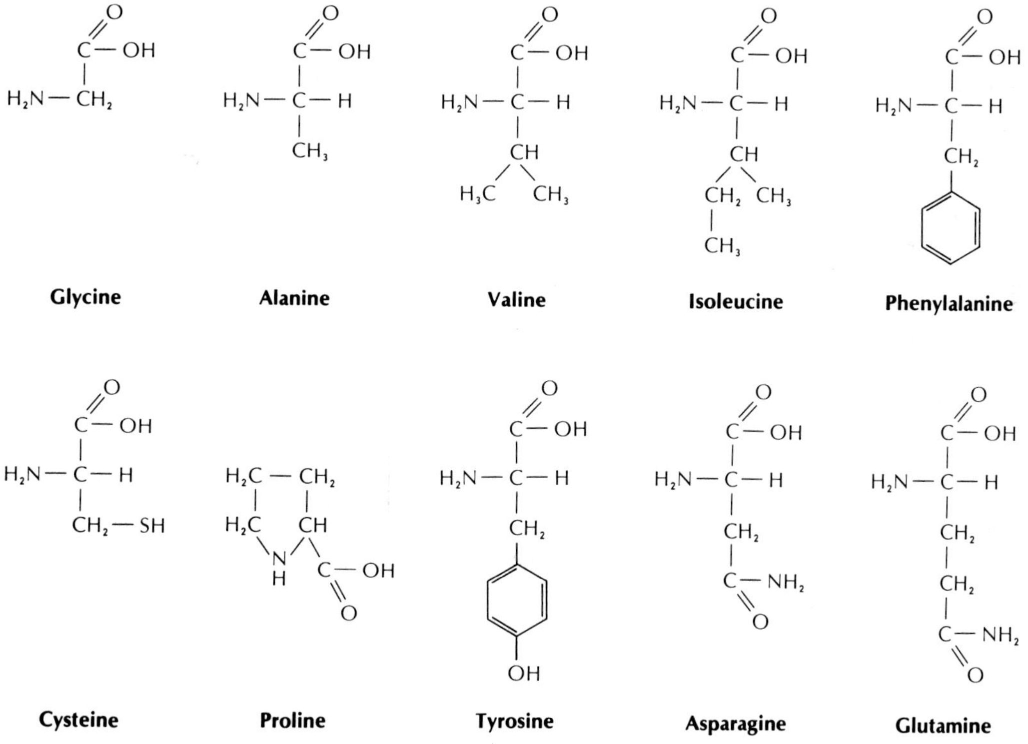

Glycine   **Alanine**   **Valine**   **Isoleucine**   **Phenylalanine**

Cysteine   **Proline**   **Tyrosine**   **Asparagine**   **Glutamine**

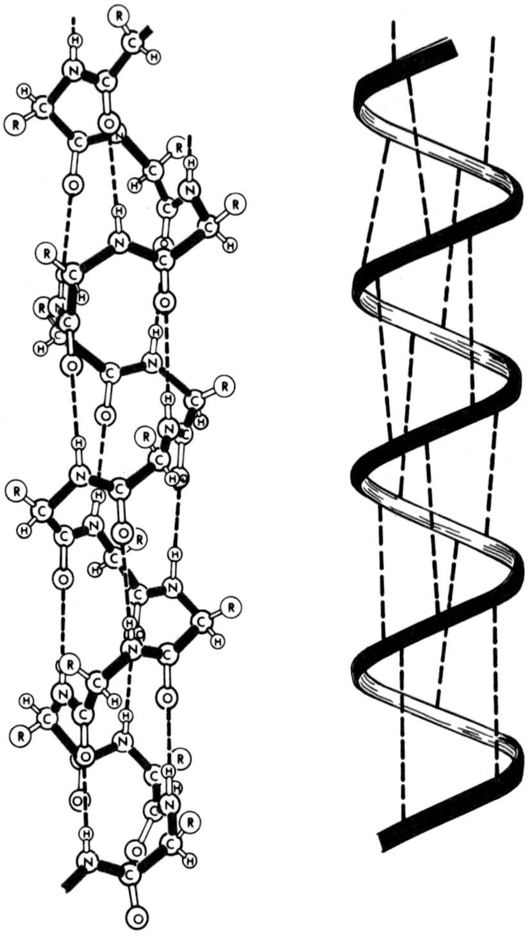

FIG. 2-16
Alpha-helix pattern of a polypeptide chain. *Dashed lines,* Hydrogen bonds that stabilize adjacent turns of the helix. *R,* Amino acid side chains. (Adapted from D. Green. 1956. Currents of biochemical research. New York, Interscience Publishers, Inc.)

amino acids can be, and are, arranged in all possible combinations, it is easy to see how an almost infinite variety of proteins can be produced.

Each of the 20 amino acids contains one amino group ($-NH_2$) and one carboxyl group ($-COOH$) attached to the same carbon atom. The remainder of the molecule is unique for each amino acid. The general formula for an amino acid is as follows:

$$H_2N-CH-COOH$$
$$|$$
$$R$$

Where the symbol R represents an atomic grouping unique for each acid.

One of the simplest amino acids is glycine:

$$H_2N-CH-COOH$$
$$|$$
$$H$$

R here represents the single hydrogen atom, but in other amino acids R could stand for a methyl group ($-CH_3$) or a variety of carbon radicals. Another amino acid, alanine, is represented thus:

$$H_2N-CH-COOH$$
$$|$$
$$CH_3$$

Here the group CH_3 is R.

In forming proteins, amino acids are linked together by a bond between the NH_2 group of one amino acid and the $COOH$ group of another to form a **peptide bond.**

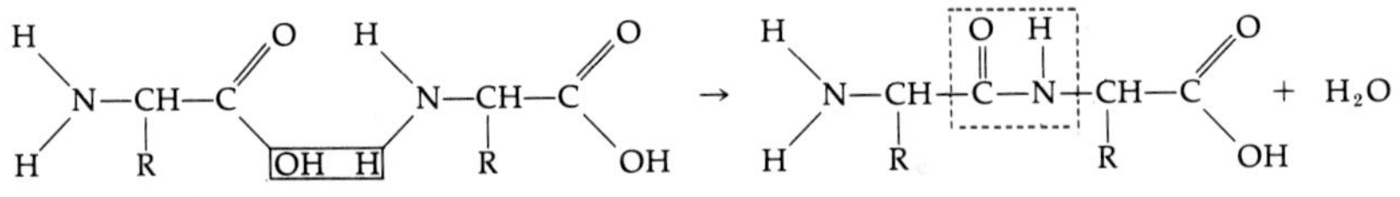

Peptide bond

The bonding of two amino acids, as shown above, forms a **dipeptide;** the addition of a third amino acid forms a **tripeptide.** In this manner—sequential addition of amino acids through peptide bonds—long **polypeptide** chains are build:

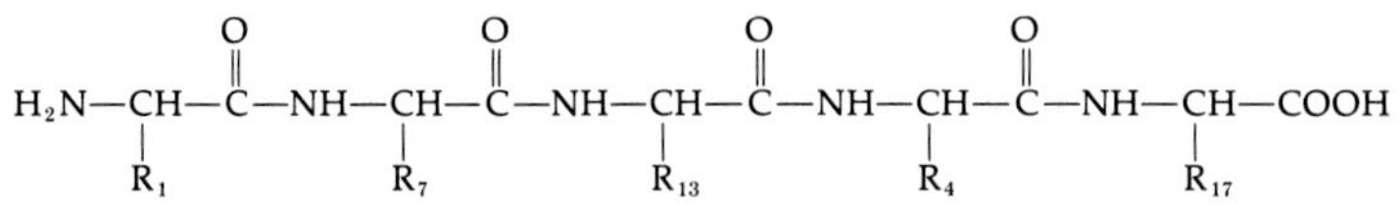

The polypeptide above consists of five different amino acids linked by peptide bonds. When the chain exceeds about 100 amino acids, the molecule is called a **protein** rather than a polypeptide.

A protein is not just a long string of amino acids, but is a highly organized molecule. For convenience, biochemists have recognized four levels of protein organization called primary, secondary, tertiary, and quaternary. The **primary** structure of a protein is determined by the kind and sequence of amino acids making up the polypeptide chain. The polypeptide chain or chains tend to spiral into a definite helical pattern, like the turns of a screw. This precise coiling, known as the **secondary** structure of the protein, most commonly takes a clockwise direction called an **alpha-helix** (Fig. 2-16). The spirals of the chains are stabilized by weak **hydrogen bonds,** usually between a hydrogen atom of one amino acid and the peptide-bond oxygen of another amino acid in an adjacent turn of the helix. Hydrogen bonds provide definite spacing to the helix: there is one amino acid every 1.5 Å along the axis, and there are an average 3.6 amino acids per turn.

The polypeptide chain (primary structure) not only spirals into helical configurations (secondary structure), but also the helices themselves bend and fold,

giving the protein its complex, yet stable, three-dimensional **tertiary** structure (Fig. 2-17). The folded chains are stabilized by the interactions between side groups of amino acids. One of these interactions is the **disulfide bond,** a strong covalent bond between pairs of cysteine (sis'tee-in) molecules that are brought together by folds in the polypeptide chain. Other kinds of bonds that help to stabilize the tertiary structure of proteins are hydrogen bonds, ionic bonds, and hydrophobic bonds.

The term **quaternary** structure describes those proteins that contain more than one polypeptide chain unit. For example, hemoglobin of higher vertebrates is composed of four polypeptide subunits nested together into a single protein molecule.

The complete structure of several proteins has now been worked out. This has proved to be a monumental task, for not only must the correct amino acid sequence be determined, but also the complete three-dimensional configuration, that is, the exact way the polypeptide chains are folded and bonded together. Insulin, the pancreatic hormone that governs glucose metabolism, was the first protein to have its amino acid sequence determined. Insulin is a small protein (mol. wt. 5,700), consisting of two polypeptide chains containing 51 amino acids. The amino acid

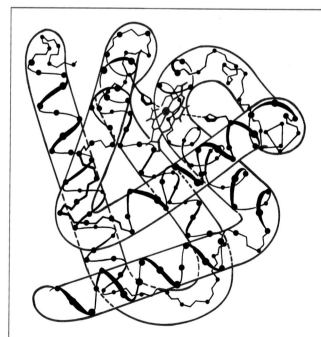

FIG. 2-17

Three-dimensional tertiary structure of the protein myoglobin. Adjacent folds of the polypeptide chain are held together by disulfide bonds that form between pairs of cysteine molecules. In the upper center of the molecule is the heme group, which combines with oxygen. (From H. Neurath. 1964. The proteins, ed. 2, vol. II. New York, Academic Press Inc.)

sequence was worked out by Frederick Sanger and his colleagues at Cambridge University in 1953. Sanger could not determine the three-dimensional configuration of insulin by the laborious techniques available at that time. By using x-ray diffraction, subsequent workers have constructed what are believed to be fairly accurate pictures of the shape of certain native proteins.

The molecular weights of protein molecules, classed as colloids, vary over a wide range; for example, insulin, 5,700; egg albumin, 40,000; hemoglobin, 68,000; serum globulin, 170,000; thyroglobulin, 660,000; snail blood hemocyanin, 6,600,000; tobacco mosaic virus protein, 60,000,000.

Proteins serve as the chief structural pattern of protoplasm and also form enzymes, hormones, chromosomes, and other cell components; they may release energy when utilized as food. The uniqueness of different cells is mainly attributable to the unique proteins they possess, and different species of organisms have certain proteins different from those of other species. The more closely two organisms are related, the more their proteins are alike; conversely, the more they are unlike, the more their proteins differ. Proteins thus serve as evidence for evolutionary relationship (species specificity). This generalization has practical application in grafting tissues from one animal to another because grafts are more likely to succeed in closely related animals that have similar protein patterns; in distantly related species these grafts will not "take" and so they degenerate.

NUCLEIC ACIDS: GENETIC APPARATUS OF THE CELL NUCLEUS. These complex substances of high molecular weight represent life at its most fundamental level. Genetic information is deposited in genes, which are built principally of deoxyribonucleic acid (DNA). DNA is a polymer built of repeated units called **nucleotides.** Each nucleotide contains three kinds of organic molecules—a **sugar,** a **phosphate group** and a **nitrogenous base.** The sugar in DNA is a pentose (5-carbon) sugar called **deoxyribose,** with the structural formula:

Deoxyribose sugar

The **phosphoric acid** has the structural formula:

Four **nitrogenous bases** are found in DNA. Two of them are organic compounds composed of nine-membered double rings, classed as **purines.** They are **adenine** and **guanine:**

Adenine   Guanine

The other two nitrogenous bases belong to a different class of organic compounds called **pyrimidines,** consisting of six-membered rings. The two pyrimidines found in DNA are **thymine** and **cytosine:**

Thymine   Cytosine

The sugar, phosphate group, and nitrogenous base are linked as shown in the generalized scheme below for a **nucleotide:**

Phosphate   Sugar   Nitrogenous base

In DNA, the backbone of the molecule is built of phosphoric acid and deoxyribose sugar; to this backbone are attached the nitrogenous bases (Fig. 2-18). However, one of the most interesting and important discoveries about the nucleic acids is that DNA is not a single polynucleotide chain but rather consists of **two** complimentary chains that are precisely crosslinked by specific hydrogen bonding of purine and pyrimidine bases. It was found that the number of adenines is equal to the number of thymines, and the number of guanines equals the number of cyto-

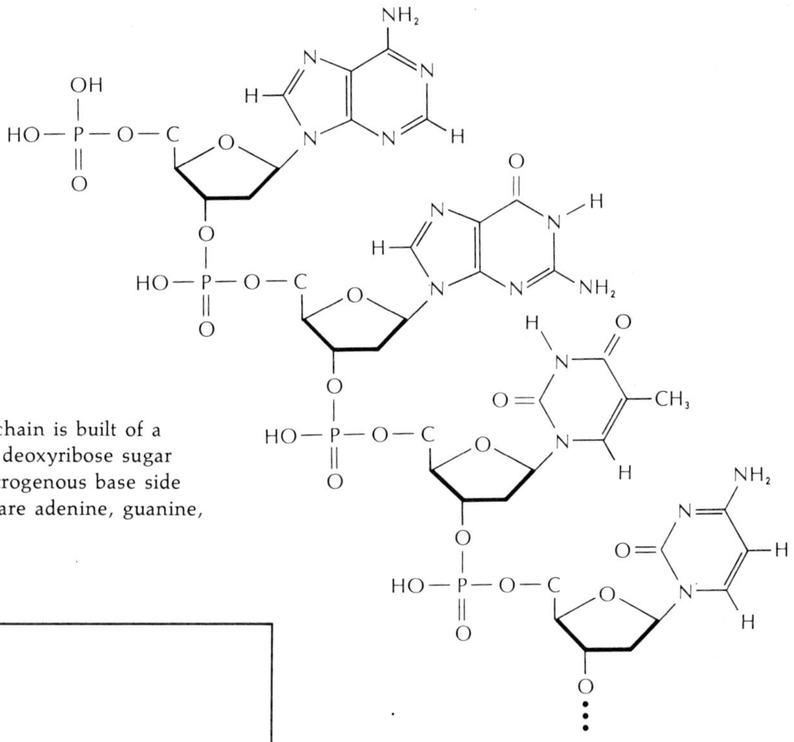

FIG. 2-18
Section of DNA. Polynucleotide chain is built of a
backbone of phosphoric acid and deoxyribose sugar
molecules. Each sugar holds a nitrogenous base side
arm. Shown from top to bottom are adenine, guanine,
thymine, and cytosine.

sines. This fact suggests a pairing of bases: adenine
with thymine (AT) and guanine with cytosine (GC).
The larger adenine (a purine) always attaches to the
smaller thymine (a pyrimidine) by two hydrogen
bonds; and the larger guanine (a purine) always
attaches to the smaller cytosine (a pyrimidine) by
three hydrogen bonds:

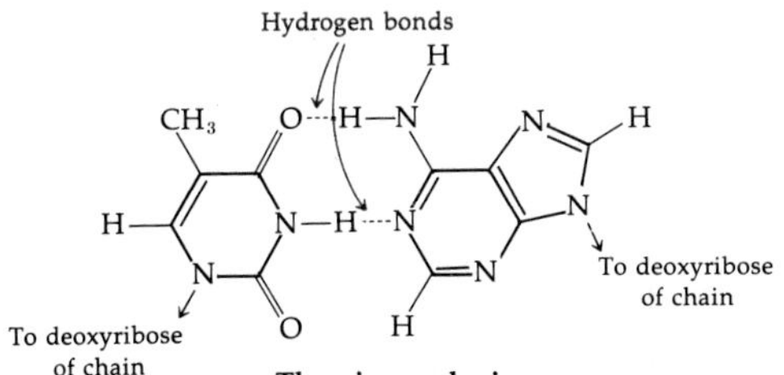

The result is a ladder structure. The uprights are
the sugar-phosphate backbones and the connecting
rungs are the paired nitrogenous bases, AT or GC.

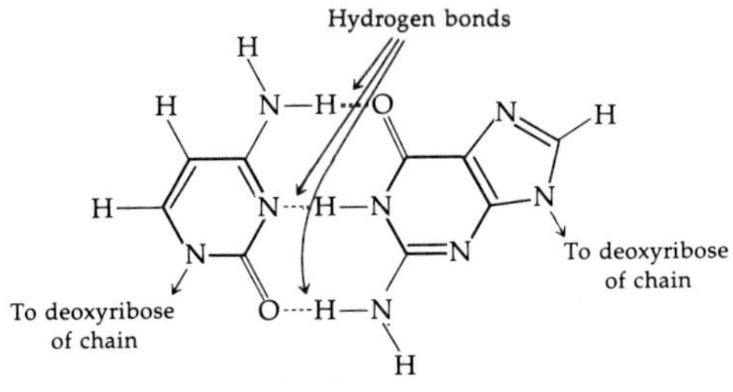

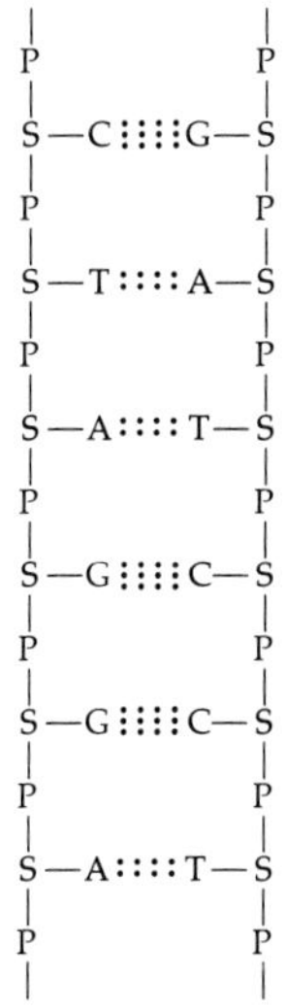

Thymine – adenine

Cytosine – guanine

However, the ladder is twisted into a **double helix,** with about 10 base pairs for each complete turn of the helix:

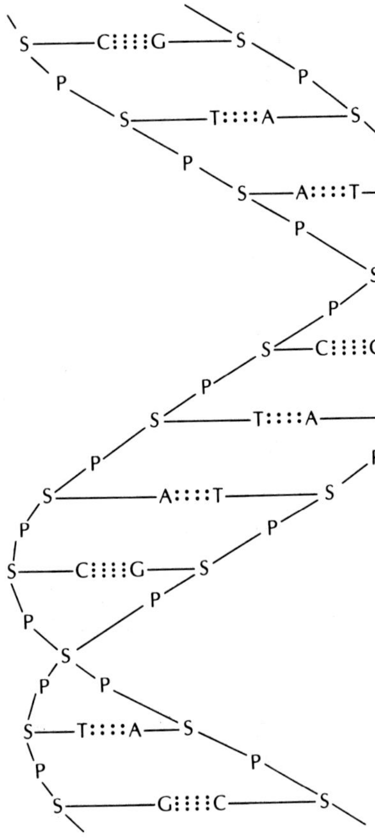

The determination of the structure of DNA was widely acclaimed as the single most important biologic discovery of this century. It was based on the x-ray diffraction studies of Maurice H. F. Wilkins and the ingenious proposals of Francis H. C. Crick and James D. Watson. Watson, Crick, and Wilkins were awarded the Nobel Prize for Medicine and Physiology in 1962 for their momentous work.

LIPIDS: FUEL STORAGE. Fats and fatlike substances are known as lipids. They include the true fats, oils, compound lipids, and steroids. The true fats, or simple lipids, are sometimes called the neutral fats. They consist of oxygen, carbon, and hydrogen and are formed by the combination of 3 fatty acid molecules and 1 glycerol molecule. True fats are therefore esters, that is, a combination of an alcohol (glycerol [Fig. 2-19]) and an acid. They also bear the term "triglyceride" because the glycerol radical is

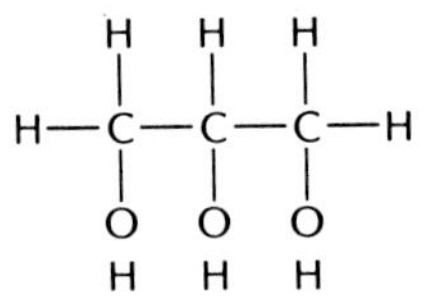

FIG. 2-19
Structural formula of glycerol (glycerin). This substance chemically is a type of alcohol and is obtained from fats and oils. True fats are formed by the combination of 1 molecule of glycerol with 3 fatty acid molecules and with elimination of 3 water molecules. The 3 fatty acid molecules may be the same, or they may be different.

combined with 3 radicals from fatty acid groups. A chemically pure fat such as stearin is an ester of glycerol and 3 molecules of a single fatty acid (stearic acid) (Fig. 2-20). Most natural fats, however, such as lard and butter are mixtures of chemically pure fats, for they usually have two or three different fatty acids attached to the 3 hydroxyl groups of glycerol. The production of a typical fat by the union of glycerol and stearic acid is shown by the following formula:

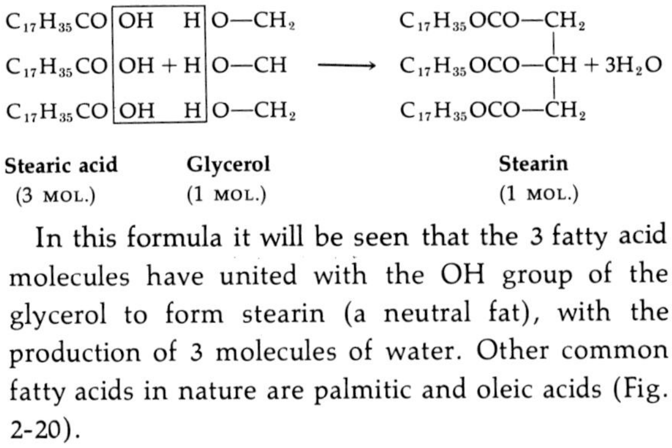

$C_{17}H_{35}CO$	OH	H	$O-CH_2$		$C_{17}H_{35}OCO-CH_2$
$C_{17}H_{35}CO$	OH $+$	H	$O-CH$	$\longrightarrow$	$C_{17}H_{35}OCO-CH + 3H_2O$
$C_{17}H_{35}CO$	OH	H	$O-CH_2$		$C_{17}H_{35}OCO-CH_2$

Stearic acid	**Glycerol**	**Stearin**
(3 MOL.)	(1 MOL.)	(1 MOL.)

In this formula it will be seen that the 3 fatty acid molecules have united with the OH group of the glycerol to form stearin (a neutral fat), with the production of 3 molecules of water. Other common fatty acids in nature are palmitic and oleic acids (Fig. 2-20).

Most true fats are solid at room temperatures, but plant oils (linseed, cottonseed, etc.) and animal oils (fish and whale) are liquid because of the nature of their fatty acids. Waxes such as beeswax differ from true fats in having an alcohol other than glycerol in their molecular structure.

Compound lipids are fatlike substances that, when broken down, will yield glycerol (or some other alcohol), fatty acids, and some other substances, such as a nitrogenous base (for example, choline), phosphoric acid, or a simple sugar. Among these lipids are the phospholipids (lecithin) found in egg yolk and prob-

A

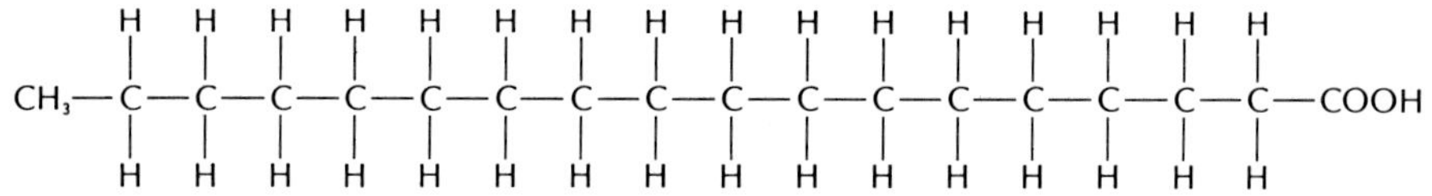

Stearic acid

B

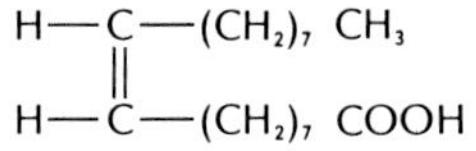

Oleic acid

FIG. 2-20

Saturated and unsaturated fatty acids. When all available bonds of carbon chain are filled with hydrogen ions, as in stearic acid, **A**, such fatty acids are called saturated. Unsaturated fats, such as oleic acid, **B**, have 1 or more double bonds ($C = C$) in their molecules because all available bonds of carbon are not filled with hydrogen atoms. Unsaturated fats tend to be oily liquids having lower melting points than more solid saturated fats.

ably every living cell and the cerebrosides (glycolipid) that are common in nervous tissue. The steroids, or solid alcohols, are not chemically related to fats but are included among the lipids because they have fatlike properties. Cholesterol ($C_{27}H_{45}OH$) is a common example of a steroid. Ergosterol, a plant steroid, becomes the hormone calciferol (commonly known as "vitamin D") when activated by ultraviolet rays. Male and female sex hormones and the adrenal gland hormones are other examples of steroids. Most of these steroids are derived from cholesterol.

Lipids have many functions in protoplasm. The true fats furnish a concentrated fuel of high-energy value and represent an economic form of storage reserves in the body. Excess carbohydrates can be transformed into fat, and to a limited extent fatty acids can be changed into glucose. Some phospholipids form part of the basic protoplasmic structure, such as lecithin, which gives a constant characteristic pattern to all cells. Phospholipids also share with proteins the basic structure of the plasma membrane and of the myelin sheaths or nerve fibers.

References
Suggested general readings

Allen, J. M. (editor). 1966. Molecular organization and biological function. New York, Harper & Row, Publishers. *A paperback by several specialists dealing with a number of aspects of molecular biology.*

Baker, J. J. W., and G. E. Allen. 1965. Matter, energy, and life. Palo Alto, Addison-Wesley Publishing Co., Inc. *Physical and chemical background to modern biology. Clearly written and illustrated paperback.*

Baldwin, E. 1962. The nature of biochemistry. New York, Cambridge University Press. *This little text is intended as a starting point for a basic understanding of the biochemical background of biologic study.*

Barry, J. M. 1964. Molecular biology: genes and the control of living cells. Englewood Cliffs, N. J., Prentice-Hall, Inc. *A paperback dealing with the structure of chemical molecules and the way they are formed.*

Bennett, T. P., and E. Frieden. 1966. Modern topics in biochemistry. New York, The Macmillan Co. *A paperback that will enable the beginning student to appreciate the role chemical compounds play in molecular biology.*

Rose, S. 1966. The chemistry of life. Baltimore, Penguin Books. *Biochemistry, from basic chemistry to cellular interactions. A paperback at undergraduate level.*

Watson, J. D. 1965. Molecular biology of the gene. New York, W. A. Benjamin, Inc. *Of the numerous works that have appeared on molecular biology, this is one of the clearest written and most comprehensive.*

Selected *Scientific American* articles

Doty, P. 1957. Proteins. **197**:173-184. (Sept).

Frieden, E. 1972. The chemical elements of life. **227**:52-60 (July).

Kendrew, J. C. 1961. The three-dimensional structure of a protein molecule. **205**:96-110 (Dec.).

Merrifield, R. B. 1968. The automatic synthesis of proteins. **218**:56-74 (March).

Perutz, M. F. 1964. The hemoglobin molecule. **211**:64-76 (Nov.).

Stein, W. H., and S. Moore. 1961. The chemical structure of proteins. **204**:81-92 (Feb.).

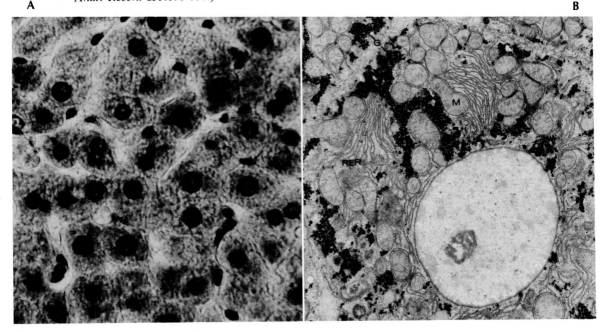

*Liver cells of rat. **A,** Magnified about 250 times. Note prominently stained nucleus in each polyhedral cell. **B,** Portion of single liver cell, magnified about 9,000 times. Single large nucleus dominates field; mitochondria (M), rough endoplasmic reticulum (RER), and glycogen granules (G) are also seen. (From Morgan, C. R., and Jersild, R. A. 1970. Anat. Record **166:**575-586.)*

CHAPTER 3

THE CELL AS THE UNIT OF LIFE

■ THE CELL CONCEPT

Although the idea that the cell represents the basic structural and functional unit of life is still often referred to as the cell theory, it has long passed the status of theory and should be known as the cell concept or doctrine. Perhaps no principle of biology is more accepted or is considered more important. It is virtually the chief cornerstone for biologic study and understanding.

As is the case with all important concepts, the cell concept has had an extensive background of development. The English scientist R. Hook (1665) is often credited with seeing and naming the first cells when he observed the small boxlike cavities in the surface of cork and leaves. The classic microscopist A. van Leeuwenhoek, the Dutch lens-maker, described many kinds of cells in addition to his famous protozoan discoveries (1675 to 1680). M. Malpighi, the

Italian microscopist who described capillary circulation among many other discoveries, no doubt observed cellular units. The French biologist R. Dutrochet gave some basic ideas about cells in 1824. In 1831 R. Brown discovered and described the nuclei of cells. J. Purkinje (1839) not only described cells as being the structural elements of plants and animals but also coined the term **protoplasm** for the living substance of cells. M. J. Schleiden and T. Schwann, German biologists, are often given credit for the cell theory formulation (1838) because of their rather extensive descriptions and diagrams, although they had erroneous ideas about how cells originate. In 1858 R. Virchow stressed the role of the cell in disease, or pathology, and stated that all cells came from preexisting cells. M. Schultz (1864) recognized the importance of the jellylike material, protoplasm, that makes up plant and animal cells. He concluded that protoplasm was the "physical basis of life," and defined a cell as a "mass of protoplasm containing a nucleus."

The cell doctrine, then, is a set of unifying concepts: all plants and animals are composed of cells or cell products, there is a basic unity of cellular construction, and all cells come from preexisting cells. The cell is thus *the* unit of life, and certainly much of our understanding of the life process can come from studies of cells and their components. Nevertheless the whole organism behaves as a complex multilevel system, and a full explanation of life as we see it in an organism must come from investigations pursued at all levels, from the "molecular" to the "organismic." The wide range of interests of contemporary biologists ensures that this will occur.

■ METHODS OF STUDYING CELLS

Since cells are small and mostly invisible, it follows that the microscope has been the tool of choice in studying them. But the microscope alone was unable to fulfill its function without the aid of staining methods. Fortunately the discovery and development of the aniline dyes by W. H. Perkin and others gave the investigators of the last half of the nineteenth century the opportunity to work out the details of cellular structures and cell division within the limits of the light microscope. It was at this time that cytology, the study of cells, developed into a flourishing science—a study that has greatly broadened under the impact of the electron microscope.

Cytologic techniques have constantly widened in every generation of investigators. Among these advances are the careful histologic techniques of fixing the tissues to preserve them as naturally as possible, the art of preparing and slicing tissues with a microtome, and proper staining methods for differential staining of cell constituents, or the selective affinity of the different cell components for the various stains. More precise physicochemical methods for locating specific entities within cells and for identifying them are constantly being sought. Ultraviolet light is employed because different chemical substances absorb rays of characteristic wavelengths. Some of the histochemical techniques for demonstrating inorganic or organic substances in cells and tissues are (1) the periodic acid–Schiff (PAS) reaction for showing carbohydrates, (2) the fluorescent antibody method that injects antibodies conjugated with a flu-

orescent substance into an animal, determining where the antibodies are localized in the cells, and (3) injecting tagged atoms that have been labeled with a radioactive isotope (tritium [^{3}H], iodine 131, and many others), and then photographing the desired specimen of tissue on a special photographic emulsion plate that will record the beta or other particles from the radioactive isotope.

■ ORGANIZATION OF THE CELL

Protoplasm is usually found in cells. An exception is the plasmodium of the noncellular slime molds (myxomycetes). Plasmodia are slow creeping, jelly-like structures with thousands of nuclei and without cell walls. As we have seen, the cell is the fundamental unit of biologic structure and function, and it is the minimum biologic unit capable of maintaining and propagating itself.

To see a living cell, gently scrape the inside of your cheek with a blunt instrument, put the scrapings on a slide in a drop of physiologic salt solution, and examine, unstained, with a microscope. The flat circular cells with small nuclei that you see are the squamous epithelial cells that line the mouth region.

Cells vary greatly in both size and form. Some of the smallest animal cells are certain parasites that may be 1 μ (1/25,000 inch) or less in diameter. At the other extreme we have the fertilized eggs of birds, some of which, including the extracellular material, are several inches in diameter. A red blood corpuscle in man has a diameter of about 7.5 μ. The longest cells are the nerve cells because the fibers, which are extensions of the cells, may be up to several feet long. Some striped muscle cells or fibers are several inches long.

The evolutionary pattern of life from the unicellular to the multicellular forms of life has increased in complexity as new qualities arise at each level. The various functions of the life process carried on at the unicellular stages tend to be allotted to specialized cells in multicellular organisms. Functional specialization is accompanied by structural specialization or division of labor, and the hierarchy of tissues, organs, and organ systems arise as a consequence in the evolutionary development of life. Although each cell is integrated with the functioning of the body as a whole, it nevertheless behaves as an independent metabolic unit. One cell of a group may

TABLE 3-1
Comparison of prokaryotic and eukaryotic cells

Characteristic	Prokaryotic cells	Eukaryotic cells
Chromosomes	Only nucleic acid and usually in a single piece	Nucleic acid plus protein with multiple chromosomes
Nuclear membrane	Absent	Well defined
Photosynthesis	Chlorophyll when present not in chloroplasts	Chlorophyll when present in chloroplasts
Flagella	9-2 fibrillar structure absent; most single stranded	9-2 fibrillar structure or modifications
Cytoplasmic inclusions (Golgi, mitochondria, endoplasmic reticulum, etc.)	Absent	Well defined
Cytoplasmic or amoeboid movement	Absent	Occurs
Cell wall	With muramic acid	When present, muramic acid absent

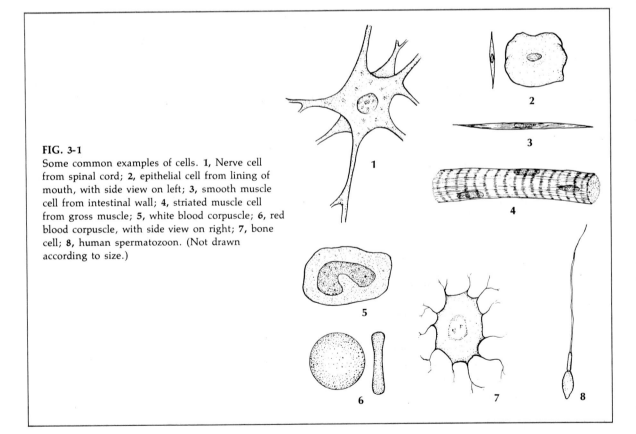

FIG. 3-1
Some common examples of cells. **1,** Nerve cell from spinal cord; **2,** epithelial cell from lining of mouth, with side view on left; **3,** smooth muscle cell from intestinal wall; **4,** striated muscle cell from gross muscle; **5,** white blood corpuscle; **6,** red blood corpuscle, with side view on right; **7,** bone cell; **8,** human spermatozoon. (Not drawn according to size.)

divide, secrete, or die, while adjacent cells may be in a different physiologic state.

A cell includes both its outer wall, or membranes, plus its contents. Typically, it is a semifluid mass of microscopic dimensions, completely enclosed within a thin, differentially permeable **plasma membrane.** It usually contains two distinct regions—the nucleus and cytoplasm. A few unicellular organisms, such as bacteria and blue-green algae, do not show a distinctive separation of nuclear and cytoplasmic constituents but have the chromatin material scattered through the cytoplasm. Cells without nuclei in these unicellular organisms are called prokaryotic in contrast to those with nuclei called eukaryotic (Table 3-1). The **nucleus** is enclosed by a **nuclear membrane** and contains the **chromatin** and one **nucleolus** or more. Within the **cytoplasm** are many **organelles,** such as mitochondria, Golgi complex, centrioles, and endoplasmic reticulum. Plant cells may contain in addition plastids, or chloroplasts.

The variety of different shapes assumed by cells is mostly correlated with their particular function (Fig. 3-1). Although many cells, because of surface tension forces, will assume a spherical shape when freed from restraining influences, there are others that retain their shape under most conditions because of their characteristic cytoskeleton, or framework.

The electron microscope reveals small cytoplasmic **microtubules** that may serve as cytoskeletal elements in maintaining the shape of cells. These tubules are straight, are of indefinite length, and have a diameter of about 200 to 270 Å. Their walls are made up of ten or more filamentous subunits. They appear to be most common near the cell center and may be closely related to the centriole. They form the spindle apparatus of dividing cells, the caudal sheath of spermatids, the marginal bands of nucleated erythrocytes, and the axoplasm of neurons. Their power to contract may be involved in the movements of the cytoplasm and the alterations in cell shape.

Components of the cell and their function*

All structures, or organelles, of the cell have separate, important functions. The **nucleus** (Figs. 3-2 and 3-3) has two important roles: (1) to store and carry hereditary information from generation to genera-

*Refer to p. 18, Principles 5 and 9.

tion of cells and individuals and (2) to translate genetic information into the kind of protein characteristic of a cell and thus determine the cells' specific role in the life process. A component of the nucleus, the **nucleolus,** is rich in ribonucleic acid (RNA), synthesizes ribosomal RNA, and probably acts as an intermediate between the genetic code of the chromosomes and the execution of the code in the cytoplasm. In the **mitochondria,** often called "powerhouses of the cell" (Fig. 3-2) the energy-yielding oxidations from the breakdown of complex organic compounds are localized. This energy is stored in high-energy phosphate bonds to be used in biologic activities as needed. (A T P)

The **centrioles** (Fig. 3-2) determine the orientation of the plane of cell division and probably supply the **basal granules,** or **kinetosomes,** which are concerned with the formation of motile fibrillar structures, such as cilia and flagella at the surface of cells. Another type of granules, **lysosomes,** contain hydrolytic enzymes. Often confused with the lysosomes are **peroxisomes,** membrane-bounded organelles with oxidative enzymes for the decomposition of hydrogen peroxide and other functions.

The Golgi complex (Figs. 3-2 and 3-3) is the primary site for the packaging of the secretory products that are synthesized on the ribosomes and migrate to the saccules, or stacks of flattened sacs, making up the Golgi complex. Here also carbohydrate molecules formed by the Golgi complex are added to the protein secretions to form glycoproteins before they are discharged for their various functions. **Plastids,** found in plants, serve as sites of synthesis of complex organic compounds from simpler substances, such as the formation of sugar from carbon dioxide and water.

PLASMA MEMBRANE. The electron microscope, together with biophysical and biochemical studies, has revealed much about the structure of the plasma membrane, or cytoplasmic membrane, as it is more properly called, that surrounds the cell. It is a sturdy envelope that encloses the cell and behaves as a selective "gatekeeper" that determines what can and what cannot enter or leave the cell. Membranes similar to, if not identical to, the cytoplasmic membrane also surround the organelles within the cell. Membranes therefore serve as partitions to subdivide the cell space into many self-contained compartments in which biochemical reactions may proceed. With the

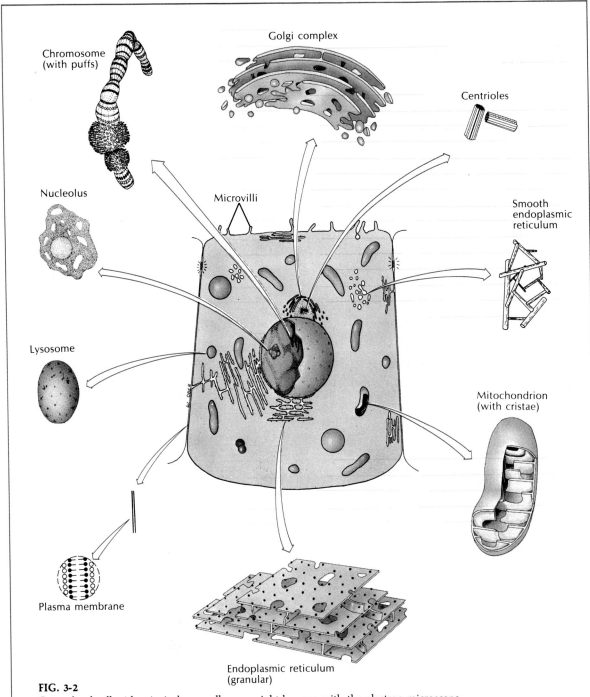

Chromosome
(with puffs)

Golgi complex

Centrioles

Nucleolus

Microvilli

Smooth
endoplasmic
reticulum

Lysosome

Mitochondrion
(with cristae)

Plasma membrane

Endoplasmic reticulum
(granular)

FIG. 3-2

Generalized cell with principal organelles, as might be seen with the electron microscope.
Each of the major organelles is shown enlarged. Membranes of organelles are believed to
be continuous with, or derived from, the plasma membrane by an infolding process.
Structure of other membranes (of nucleus, endoplasmic reticulum, mitochondria, etc.) is
probably similar to that of plasma membrane, shown enlarged at lower left.

A

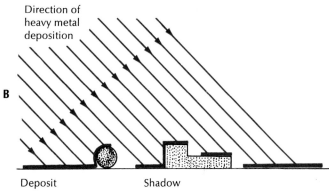

B

Direction of
heavy metal
deposition

Deposit   Shadow

FIG. 3-3
Fine structures of rat liver,
showing the three-dimensional
appearance of a cell surface
prepared by freeze-fracturing and
etching. The specimen is first
quickly frozen and then struck with
a cold microtome blade to make a
clean break. It is then put in a
vacuum and the surface shadowed
with carbon and platinum, as
shown in **B**. An electron
micrograph is then prepared. **A**
shows *N,* the nucleus; *P,* nuclear
pores; *rer,* rough endoplasmic
reticulum; *m,* fractured
mitochondria; *ser,* smooth
endoplasmic reticulum; and *Go,* a
Golgi complex. Note the
structureless area (*) where
glycogen is deposited. The
direction of shadow-cast is shown
by the encircled arrow. (×17,000.)
(**A**, From Orci, L. 1971. J.
Ultrastructural Research **35**:1-19,
Academic Press Inc.)

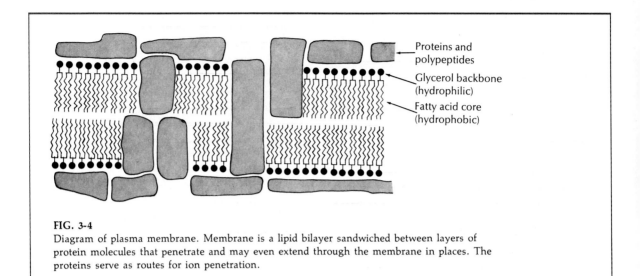

FIG. 3-4

Diagram of plasma membrane. Membrane is a lipid bilayer sandwiched between layers of protein molecules that penetrate and may even extend through the membrane in places. The proteins serve as routes for ion penetration.

electron microscope, the cell membrane appears as two dark lines, each about 25 to 30 Å thick, on each side of a light zone. The entire membrane is 75 to 100 Å thick (an angstrom is 1/100,000,000 cm.).

It is composed almost entirely of proteins and lipids. The proteins are enzymes that carry out specific functional roles. The lipids give the membrane its strength and provide other gross structural properties. Membrane lipids belong to a class of compounds called **phospholipids.** These are **amphipathic** molecules, that is, one end is insoluble in water (hydrophobic) while the opposite end is water soluble (hydrophilic) and polar, carrying an ionic charge. The nonpolar end consists of hydrocarbon chains of **fatty acids** and the polar (charged) end consists of **glycerol** attached to **phosphate** and other groups (Fig. 3-4).

The phospholipids are arranged in a bilayer sandwiched between two layers of protein. The proteins are not arranged in orderly array as originally thought; some penetrate into the lipid bilayer and may even extend all the way through it. The proteins are responsible for the **selective permeability** of living membranes, a subject that will be discussed in the next chapter.

ENDOPLASMIC RETICULUM. The double-layered endoplasmic reticulum is a system of channels and cavities interlacing the interior of a cell (Figs. 3-3, 3-5, and 3-6). This system has been closely associated with the storage and transport of products of cellular metabolism. The nuclear membrane is formed from parts of this membrane system, and it is so arranged that there is direct continuity between nucleus and cytoplasm by openings in the nuclear membrane (Fig. 3-5). The double-walled endoplasmic reticulum is a highly variable morphologic structure consisting of vesicles and tubules, and it often has the power to fragment and reform its structural features. There are two types: rough- and smooth-surfaced. The rough-surfaced type has on its outer surface small granules called **microsomes** (the dense granules are often called **ribosomes** [Fig. 3-6] because they contain ribonucleic acid). These ribosomes are important sites of protein synthesis. In some cases the granules can function without being attached to the membrane. The endoplasmic reticulum is also important as an intracellular transport system for the products synthesized by the ribosomes. Among the substances transported by this system are the zymogen bodies that give rise to digestive enzymes and that are carried to the smooth-surfaced Golgi vesicles (Figs. 3-2 and 3-3). Later these zymogen granules are discharged as enzymes through openings at the surface of the cell.

By this relationship of the endoplasmic reticulum to the surface membrane of the cell, the cell is a three-phase structure consisting of cytoplasm, cavities of the endoplasmic reticulum, and membranes separating the other two phases. The concept also includes the idea that there is really a one-membrane unit from the cell surface membrane to the system of interior membranes.

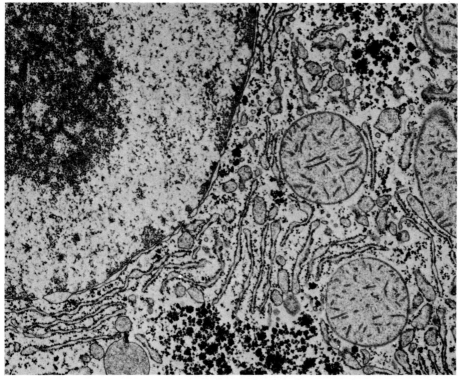

FIG. 3-5
Electron micrograph of part of hepatic cell of rat showing portion of nucleus (left) and surrounding cytoplasm. Endoplasmic reticulum and mitochondria are visible in cytoplasm, and pores are seen in nuclear membrane. (×14,000.) (Courtesy G. E. Palade, The Rockefeller University, New York.)

FIG. 3-6
Electron micrograph of portion of pancreatic exocrine cell from guinea pig showing endoplasmic reticulum with ribosomes (small dark granules). Oval body (upper left) is mitochondrion. (×66,000.) (Courtesy G. E. Palade, The Rockefeller University, New York.)

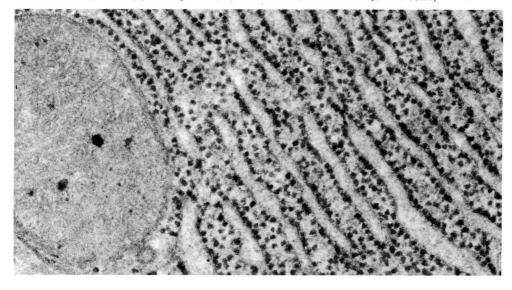

MITOCHONDRIA. Another important cytoplasmic organelle is the mitochondrion (Figs. 3-2 and 3-7). Mitochondria are found in all cells and can be detected by the light microscope. They show considerable diversity in shape, size, and number. Many of them are rodlike and are about 0.2 to 5 μ in greatest diameter. About two thirds of their structure is protein and one third is lipid. They are capable of altering their form in accordance with the physiologic condition of the cell. They may be scattered more

FIG. 3-7
Electron micrograph of elongated mitochondrion in pancreatic exocrine cell of guinea pig. (×50,000.) (Courtesy G. E. Palade, The Rockefeller University, New York.)

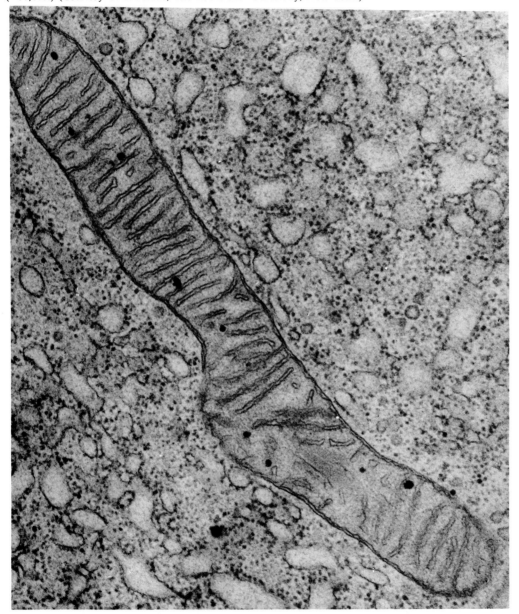

or less uniformly through the cytoplasm or they may be localized near cell surfaces and other regions where there is unusual metabolic activity. Investigation of the fine structure of the mitochondrion with the electron microscope and centrifuge microscope reveals that it is a double membrane system that may be formed from detached pockets of the cell membrane. The inner layer of the double membrane is much folded and forms projections (cristae) that extend into the interior fluid or matrix. There are thus two structural systems—the membrane system and the homogenous fluid matrix.

Research during the past few years shows that the mitochondria are the principal chemical sites for cellular respiration. They contain highly integrated systems of enzymes for providing energy in cell metabolism. Among these systems is the important Krebs cycle and its respiratory enzymes that produce the energy-rich adenosine triphosphate (ATP) so essential for many vital activities. The complete Krebs cycle (p. 76) is probably carried out in the mitochondria. DNA has been found recently in mitochondria.

Surfaces of cells and their specializations

Cell surfaces vary, depending on many factors. Several cell types are free and can move throughout the animal. These free cells have no direct junctional arrangements with other cells and include such types as leukocytes, red blood corpuscles, amebocytes, macrophages, and many others. Interstitial cells are undifferentiated, are located on epithelial structures, and often migrate to injured regions for repair. Pigment cells (chromatophores) have a certain amount of freedom to move about.

The surface cells of many types throughout the animal kingdom bear specialized structures of locomotion called cilia and flagella (Fig. 7-19). These are vibratile extensions of the cell surface, and their covering membrane is continuous with the plasma membrane. Internally cilia and flagella have the same structure—nine fibrils surrounding a pair of fibrils. (Exceptions have been noted having but a single central fibril.) At the base where they are anchored in the cell cytoplasm, each flagellum or cilium is attached to a granule, a **kinetosome.** The kinetosome is similar to the cilia and flagella, but has nine triplet filaments and no double central filaments. Flagella

and cilia have numerous functions such as propelling individual cells (protozoans), multicellular animals (planarians and ctenophores), or fluids or entities through tubular organs (sponges and gonoducts). Most animal sperm are provided with them for propulsion.

The surfaces of contiguous cells, or cells packed together, have junction complexes between them. There are several types of these specializations. The adjoining surfaces of cells are sealed only in restricted areas. Nearest the free surface, the two opposing membranes appear to fuse to form a **tight junction** (Fig. 3-8). Below this is a slightly widened **intermediate junction.** Next are desmosomes, small ellipsoidal disks scattered between the epithelial cells (Fig. 3-8). Desmosomes act as "spot welds" between ap-

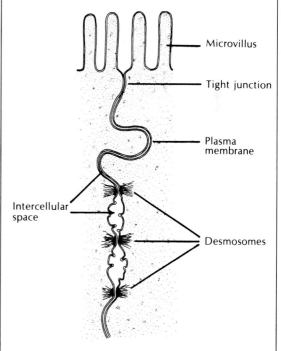

FIG. 3-8
Two opposing plasma membranes forming the boundary between two epithelial cells. Various kinds of junctional complexes are found. Tight junction is a firm, adhesive band completely encircling the cell. Desmosomes are isolated "spot welds" between cells that serve as sites of intercellular communication. Intercellular space may be greatly expanded in epithelial cells of some tissues.

Microvillus

Tight junction

Plasma membrane

Intercellular space

Desmosomes

posing plasma membranes. They measure about 250 to 410 millimicrons in their greatest diameter. From the cell cytoplasm tufts of fine filaments converge onto the desmosomes. Between the two apposed plates of a desmosome is a narrow intercellular space (200 to 240 Å wide). All of these special junctional complexes produce the **terminal bar,** found at the distal junctions of adjacent columnar epithelial cells. They form a complete beltlike junction just beyond the luminal or apical portion of the plasma membrane.

Other specializations of the cell surface are the interdigitations of confronted cell surfaces where the plasma membranes of the cells infold and interdigitate very much like a zipper. They are especially common in the epithelium of kidney tubules. The distal or apical boundaries of some epithelial cells, as seen with the electron microscope, show regularly arranged **microvilli.** They are small, fingerlike projections consisting of tubelike evaginations of the plasma membrane, with a core of cytoplasm (Figs. 3-2 and 3-8). They are well seen in the lining of the intestine where they greatly increase the absorptive or digestive surface. Such specializations are seen as **brush borders** by the light microscope. The spaces between the microvilli are continuous with tubules of the endoplasmic reticulum, which may facilitate the movement of materials into the cells.

■ CELL DIVISION (MITOSIS)

All cells of the body arise from the division of preexisting cells. Indeed, all the cells found in most multicellular organisms have originated from the division of a single cell, the **zygote,** formed from the union of an egg and a sperm (fertilization). Sperm and egg cells arise from a special kind of cell division called **meiosis,** which is described in Chapter 31. They contain only a single chromosome of each kind and are called **haploid** because each contains the **haploid number** of chromosomes (represented by the symbol n) characteristic for the species. At fertilization, the fusion of egg and sperm form a zygote now containing the **diploid number** of chromosomes ($2n$). Once diploidy is restored by fertilization, the zygote begins to reproduce by a complex process of cell division called **mitosis.** The purpose of mitosis is to produce **body** (somatic) **cells** each having at the start the same potentialities of the parent cell. Each of the two daughter cells produced by a cell division

must contain the same number and kind of chromosomes present in the parent cell before division. For this to happen, each chromosome must of course be duplicated. Mitosis, then, is principally a matter of chromosomal division. The rather complex mechanism of mitosis ensures the correct duplication and division of chromosomes.

Structure of chromosomes

The nucleus bears **chromatin** material, which in turn carries the **genes** responsible for hereditary qualities. During cell division this chromatin becomes arranged into definite **chromosomes** of varied shapes. Each animal within a species has the same number of chromosomes in each body cell. Man has 46 in each of his body cells (but not in his germ cells). The number of chromosomes a species possesses has no basic significance nor is there necessarily any relationship between two different species that have the same number. Both the guinea pig and the onion have 16 chromosomes. Since there are thousands of different species of animals, many species must of necessity have the same number.

CENTROMERE AND CHROMOMERES. The shape of the metaphase or anaphase chromosome, although constant for a specific species of animals, is of different types among animals, depending on the location of the centromere, length of arms, secondary constrictions, etc. Some chromosomes have the **centromere** at the midpoint, with equal limbs (metacentric); other chromosomes have the centromere closer to one end, with unequal arms (acrocentric). The ordinary light microscope may reveal only the outline of the metaphase chromosome, with two parallel strands (chromatids) united at one point (centromere) within it. With special techniques and the electron microscope, finer details are found, such as the presence of a coiled multiple filament **(chromonema)** along which are beadlike, dark-staining enlargements called **chromomeres.** The chromomeres, which may represent superimposed coils or DNA condensations, may contain aggregations of genes, or the genes may be located between them. The DNA is present in the chromosomes as greatly folded threads, which if fully stretched out and laid end to end would exceed a meter in length in man. The complex folding of the DNA is believed to be stabilized by histones, small basic proteins that are associated with DNA to form nucleoprotein.

HETEROCHROMATIN. Other parts of the chromosome include condensed and variable staining regions called **heterochromatin,** in contrast to the rest of the chromosome or **euchromatin.** Heterochromatin refers to chromosomes or regions of chromosomes that are condensed at interphase or prophase and do not unravel at telophase as do the other chromosomes. Two main types of heterochromatin are found in mammals: constitutive heterochromatin present in homologous chromosomes, and facultative heterochromatin, which results from the inaction of one of the two X chromosomes in the female. Centromeres and certain other regions consist of heterochromatin. Heterochromatin regions may not have functional genes, although DNA is present there and may code for microtubules. Investigations have thus far failed to reveal the function of heterochromatin, but there is the suggestion that it may represent genetic polymorphism that is expressed in different blood groups, tissue enzymes, etc.

PURPOSE OF MITOSIS. It will be seen that a biparental organism begins with the union of two gametes, each of which furnishes a **haploid** set of chromosomes (23 in man) to produce a somatic or **diploid** number of chromosomes (46 in man). The chromosomes of a haploid set are also called a **genome.** Thus, a fertilized egg (zygote) consists of a paternal genome and a maternal genome.

The purpose of mitosis is to ensure an equal distribution of each kind of chromosome to each daughter cell. A cell becomes highly abnormal in its reactions if it fails to receive its proper share of chromosomes, as the great German cytologist Boveri showed many years ago.

Stages in cell division

There are two distinct phases of mitosis: the division of the nuclear chromosomes (**karyokinesis**) and the division of the cytoplasm (**cytokinesis**). Mitosis, however, really refers to karyokinesis, the division of the nucleus (i.e., chromosomal segregation) which is certainly the most obvious and complex part of cell division, and that of greatest interest to the microscopist. Ordinarily karyokinesis and cytokinesis occur at the same time, although there are occasions when the nucleus may divide a number of times without a corresponding division of the cytoplasm. The result is a mass of protoplasm containing many nuclei and referred to as a multinucleate cell. Skeletal muscle is an example.

The process of mitosis is arbitrarily divided for convenience into four successive stages or phases, although one stage merges into the next without sharp lines of transition. These phases are prophase, metaphase, anaphase, and telophase. When the cell is not actively dividing, it is in the "resting" stage or **interphase** (Fig. 3-9). However, the cell is not really "resting" at this stage as it appeared to be to the early light microscopists. Subsequent research with radioactively labeled precursors (especially tritiated thymidine) has revealed that the cell is very active at this time, doubling the DNA content of the nucleus. Thus when the cell begins "active" mitosis, it already has a double set of chromosomes. These chromosomes are not visible at interphase; at this time the nucleus bears a deeply staining material called **chromatin** (Fig. 3-9), which appears granular, or in the form of thin, randomly coiled threads within the nuclear envelope.

Prophase. At the start of **prophase,** the chromosomes begin to become visible as elongate strands of chromatin. Each chromosome is composed of two highly coiled threads called **chromatids.** The two chromatids of each chromosome are strands of DNA coated with protein and joined at a small body, the **centromere.** As prophase proceeds, a mitotic apparatus begins to form. This consists of a spindle, asters, and centrioles. The **centrioles** are permanent, self-duplicating bodies, generally found in pairs, lying just outside the nuclear membrane. The centrioles are enclosed in a distinctive area of protoplasm; the whole complex is called a **centrosome.** The electron microscope has revealed that each centriole is a cylindrical body with walls composed of nine groups of tubulelike structures, each group containing three tubules. The structure of the centrioles is similar to that of the granules (kinetosomes) found at the bases of cilia and flagella. Each cell inherits one set of centrioles and produces another set.

During interphase the two centrioles lie at right angles to each other. These separate at the start of prophase and migrate to opposite sides of the nucleus. At the same time portions of the cytoplasm are attracted to the centrioles and transformed into fine filaments or fibers. Some of these fibers stretch across the cell between the two centrioles to form a spindle; other fibers radiate out from each centriole

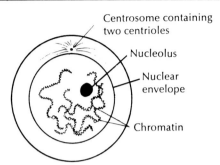

Centrosome containing
two centrioles

Nucleolus

Nuclear
envelope

Chromatin

1. Interphase

Chromatin material appears granular;
although not visible, each chromosome
reaches its maximum length and
minimum thickness; duplication of
chromosome occurs at this state

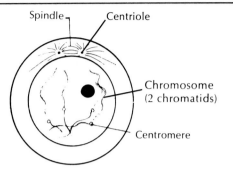

Spindle — Centriole

Chromosome
(2 chromatids)

Centromere

2. Early prophase

Each elongated chromosome now consists of
2 chromatids attached to single centromere;
double nature of chromosome not apparent;
centrosome divides and spindle starts
development

5. Prometaphase

Nuclear membrane disintegrates
or changes structure

6. Metaphase

Chromosomes arranged on equatorial
plate; centromeres (not yet divided)
anchored to equator of spindle

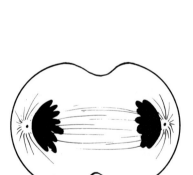

9. Early telophase

Chromosomes lie close together and form a
clump; nuclear reorganization begins

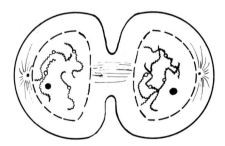

10. Late telophase

Chromosomes become longer and thinner;
chromosomes may lose identity; nuclear
membrane reappears and spindle-astral
fibers fade away; cell body divides into
2 daughter cells, each of which now enters
interphase

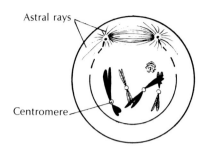

Astral rays

Centromere

3. Middle prophase
Each chromosome may be visible double; chromatids growing shorter and thicker; nuclear envelope disintegrates

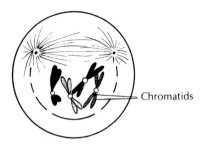

Chromatids

4. Late prophase
Double nature of short, thick chromosome more apparent; each chromosome made up of 2 half-chromosomes or sister chromatids; nucleolus usually disappears

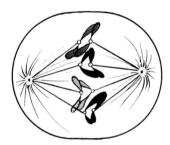

7. Early anaphase
By splitting of centromere each chromatid has its own centromere, which lies on equator and is attached to spindle fiber

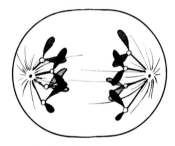

8. Anaphase
Chromatids, now called daughter chromosomes are in 2 distinct groups; daughter centromeres which may be attached at various points on different chromosomes, move apart and drag daughter chromosomes toward respective poles

FIG. 3-9
Diagrams of mitotic stages.

(the spindle poles) to form **asters** (Figs. 3-10 and 3-12). The whole structure—two centrioles surrounded by their starlike **asters** and connected by the spindle —is called the mitotic apparatus. It increases in size and prominence as the centrioles move farther apart (Fig. 3-10).

The spindle is shaped like two cones placed base to base. It was so named because to early cytologists it resembled the wooden spindle once used to twist threads together in spinning. Recent studies of the mitotic spindle have shown that the fibers are composed mostly of elongate protein molecules linked together in some way to form long, delicate, hollow microtubules. As many as 3,000 microtubules have

been counted in a single spindle. Late in prophase, the fibers of the spindle make connection with the centromere of each chromosome (recall that each of the half-chromosomes, or chromatids, of a chromosome are joined together at some point by the centromere). The chromosomes now begin to move, apparently under the control of the poles (centrioles) of the spindle. During this process, the nuclear membrane disappears, and the nucleolus disintegrates or becomes invisible.

Metaphase. During early metaphase the chromosomes quickly migrate into the *equatorial plate,* an imaginary plane formed by the bases of the spindle cones. The centromeres of the chromosomes line up

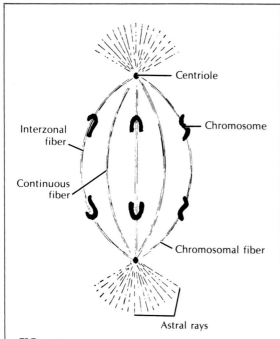

Centriole

Chromosome

Interzonal
fiber

Continuous
fiber

Chromosomal fiber

Astral rays

FIG. 3-10
Diagram of mitotic spindle (anaphase). It is not yet
clear how chromosomes are moved apart. Spindle
fibers may act as guides, along which
chromosomes move toward poles. Mechanism may
involve pushing produced by swelling between two
sets of separating chromosomes, or chromosomes
may be pulled to poles by folding of protein chains
of fibers that produce traction on attached
chromosomes. Fibers are tubes, not threads.

of the interzonal fibers passing between the separating centromeres (Fig. 3-10). They move at a speed
of about 1 micron/min. It has been suggested that
the spindle fibers contract in a way resembling the
contraction of muscle myofibrils. The high-energy
bond of ATP seems to be required. Another possibility is that fibers grow between the separating centromeres by the addition of protein molecules, while
shortening between centromeres and the spindle
poles by the loss of molecules. The idea is supported
by the observation that fibers maintain the same
thickness throughout anaphase, neither thickening
as they shorten, nor getting thinner as they length-
en. None of the several theories have yet been confirmed by experimentation, even though D. Mazia
and others have been able to isolate the delicate mitotic apparatus intact by digesting away the surrounding cytoplasm.

Telophase. When the daughter chromosomes
reach their respective poles, the telophase has
begun. The daughter chromosomes are crowded together and stain intensely. Two other events also
occur—the appearance of a **cleavage furrow** encircling the surface of the cell (Fig. 3-9) and a **cell plate**
(in plants) that originates from the central portions
of the interpolar spindle fibers. Eventually the cleavage furrow deepens and constricts the cell into two
daughter cells. Other changes that terminate the
telophase period are the disappearance of the spindle
fibrils that revert to a sol from a gel condition, the
gradual assumption of a chromatin network as the
chromosomes lose their identity, the formation of a
new nuclear membrane, and the manufacture of new
nucleoli by the chromosomes.

The cell cycle

The complete sequence of events leading to mitosis, the actual cell division, and the events following
are called the cell cycle. A cell prepares to divide long
before the actual division occurs. A human cell in tissue culture completes a cycle every 18 to 22 hours,
yet division occupies only about 1 hour of this
period. The most important preparation—replication
of DNA—occurs during the interphase, termed the
S period (period of synthesis) (Fig. 3-13). In man,
each chromosome contains about 175 million nucleotide pairs arranged in a double helix that makes one
turn for every 10 nucleotide pairs. Accordingly, there
are about 17.5 million turns in the DNA in each

precisely on the plate, with the arms of the chromosomes trailing off randomly in various directions.
Thus lined up, the centromeres form the *metaphase
plate* (Figs. 3-11 and 3-12).

Anaphase. The two chromatids of each double
chromosome thicken and separate. The single centromere that has held the two chromatids together
now splits so that two independent chromosomes
each with its own centromere are formed. The chromosomes part more, evidently pulled by the spindle
fibers attached to the centromeres. The arms of each
chromosome trail along behind as though the chromosome were being dragged through a resisting medium. To the cytologist watching this event through
the microscope, the chromosomes seem to be both
pulled by contraction of the chromosomal fibers
passing to the centrioles and *pushed* by elongation

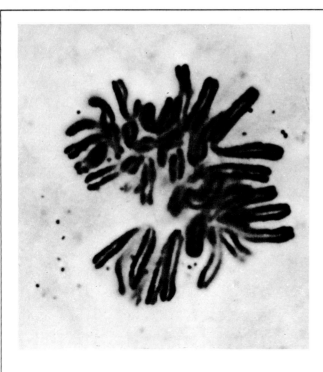

FIG. 3-11
Chromosome in tail epidermis of salamander *Ambystoma* shown arranged on equatorial plate in metaphase stage of mitosis. (Courtesy General Biological Supply House, Inc., Chicago.)

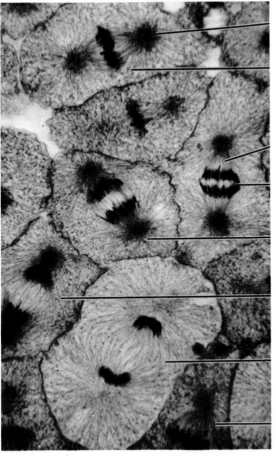

Aster

Metaphase

Spindle

Early anaphase

Late anaphase

Early telophase

Late telophase

Metaphase

FIG. 3-12
Stages of mitosis in whitefish. (Courtesy General Biological Supply House, Inc., Chicago.)

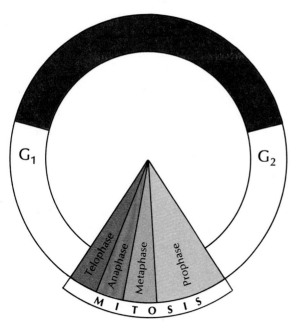

FIG. 3-13
The cell cycle (mitotic cycle) showing relative
duration of recognizable stages. **S,** Synthesis of
DNA. **G_1,** Presynthetic phase. **G_2,** Postsynthetic
phase. Actual duration times and relative duration
of phases vary considerably in different cells.

chromosome, all of which must somehow replicate
and untangle during the middle of the interphase (S
period). This period is preceded and succeeded by G_1
and G_2 periods respectively (*G* stands for "gap")
when no DNA synthesis is occurring. It is believed
that enzymes and substrates are being made ready
during the G_1 period for the DNA replication that
follows. During G_2, spindle, and aster proteins are
being synthesized. The energy demands of the cell
are high during the G_2 period.

The result of cell division is the formation of two
cells, each with an identical gene set, so that each
daughter cell is potentially the same as the mother
cell. Cell division is important for growth and replacement, wound healing, etc. Muscle cells rarely
divide, and nerve cells never divide after birth. The
more specialized the cell the less frequently it divides. However, some tissues continually divide because the body loses a percentage of its cells daily
and these must be replaced. Cell reproduction is faster in the embryonic state and slows down with age,
a condition that may be due to metabolic checks
brought on by larger cell populations.

■ THE CELL AS THE BASIC UNIT OF LIFE*

A cell is usually considered a combination of organelles and is organized completely enough to have
all the necessary materials and apparatus for performing metabolism and self-replication. A cell,
therefore, is considered to be the minimum unit that
manifests the vital phenomena of life. However, the
boundary between the living and the nonliving is not
as sharply drawn as formerly. Some particles smaller
than cells, for example, viruses, are regarded by
some as living. All viruses, however, are associated
with cells from which they derive their energy. Although viruses have the initial genetic mechanisms
and multiply, they do so at the expense of the host's
cells.

Many features or aspects of life can be produced
in the test tube without cellular organization. Genetic
investigations make use of strands of nucleic acid
that can synthesize duplications of DNA and proteins. Life must have originated from the nonliving
world in much simpler units than the complex cell.
Some present cells are simpler than others. Bacterial
cells as well as those of blue-green algae lack such
organelles as mitochondria, lysosomes, and a definite nucleus. Motile bacteria have a flagellum of a
single fibrous protein molecule instead of the nine-
fold symmetry of higher forms. There must have
been many precellular forms in the long evolution
of the cell because the properties of life did not arise
all at once. Many intermediate forms represent a
continuity from the nonliving to the living. Whether
or not some of these precellular units still exist has
never been satisfactorily settled, but the consensus
is that the whole spectrum of the origin of life from
inorganic matter could have occurred only once in
this planet's history. The cell as representing a combination of all the vital phenomena may logically be
considered the basic unit of life.

In the light of these factors a living organism must
consist of at lease one cell. If one considers all **unicellular** organisms—bacteria, fungi, algae, protozoans, etc.—it is probable that they constitute the
majority of all living forms. This level of complexity,
or of hierarchy, may still be considered the favorite
one of nature despite the attention we give to multicellular forms. Even in the simplest multicellular
types or cellular colonies, the cells of the aggregation
*Refer to pp. 18 and 23 for Principles 7, 8, and 38.

are very much alike, and therefore each ranks more or less an independent cell.

A technical distinction is thus made between a unicellular organism and a cell of a metazoan. A cell of a unicellular form may be considered homologous only with the multicellular organism as a whole and not with the individual cells of that organism. On this account, many biologists consider the unicellular organism as **acellular.** Since the unicellular organisms must carry on all the vital properties, it must have within the boundary of its limiting membrane organelles analogous to the organ systems of the multicellular forms. Differentiations and division of labor occur in protozoans, for instance, but some ciliates possess in their morphologic features and correlated functions the same general complexity of higher animals.

Cellular differentiation has many advantages over the unicellular condition. For example, it gives the large development of surfaces on which exchange of material occurs. It also promotes functional differentiation and size increase. In all higher organization the principle of cellular construction has been retained because nature has evolved no better method to meet the requirements of the environmental factors.

■ CELL NUMBER AND CELL CONSTANCY

Among multicellular animals, one of the chief causes (but not the only one) of variation in body size is difference in cell numbers. Large animals have many cells; smaller animals have fewer cells. Although most organisms have an indefinite number of cells, depending on the ultimate size the organism reaches at maturity, there are some examples of cell constancy **(eutely)** in the animal kingdom. This constancy in the number of cells may be characteristic of the entire animal or it may be restricted only to certain tissues or organs. Interest in eutely is focused upon the mechanism that terminates cell division with such precision at a definite point in development. Some protistan colonies have long been known to have a constant number of cells. *Pandorina morum* (a green alga) has 8 to 16 cells; *Pandorina charkowiensis* has 16 or 32 cells, and certain species of *Eudorina* have 64 cells. Many roundworms show the trait. *Epiphanes* (a rotifer) has 958 nuclei (syncytial patterns are common in the aschelminths). In this animal the nuclei or cells are distributed as follows: skin, 301; pharynx, 167; digestive system, 76; urogenital system, 43; muscles, 120; and nervous system, 247. Cell constancy does not apply to the gonads that form eggs and sperm continuously.

Even in higher forms, certain tissues undergo no increase in number of cells after birth. In man there appears to be no cell division in nervous and muscle tissue after the fetal stages. The number of glomeruli in the kidneys of some animals is fixed at birth.

■ FLUX OF CELLS

In many organisms, certain tissues continually shed their cells because of wear and tear or other causes. The epidermis of the skin, the lining of the alimentary canal, and the blood-forming tissues lose large numbers of cells daily. There must be a constant replacement of the cells that are lost, for there is no net loss or gain in the overall picture. In man it has been estimated that the number of cells shed daily is about 1% to 2% of all the body cells.

In its contact with the environment the organism is usually subject to a constant attrition of physical and chemical forces. Mechanical rubbing wears away the outer cells of the skin, and emotional stresses destroy many cells. Food in the alimentary canal rubs off lining cells, the restricted life cycle of blood corpuscles must involve a renewal of enormous numbers of replacements, and during active sex life many millions of sperm are produced each day. This loss is made up by a chain reaction of binary fission or mitosis.

At birth the child has about 2 trillion cells. This immense number has come from a single fertilized egg (zygote). Such a number of cells could be attained by a chain reaction in which the cell generations had divided about 42 times, with each cell dividing once about every 6 to 7 days. In about five more cell generations by the chain reaction, the cells have increased to 60 trillion at maturity (in an individual of 170 pounds). However, not all cells divide at the same rate and some cells (nervous and muscular), as we have seen, stop dividing altogether at birth. The growth of an organism is not merely an increase in number of cells but it also involves some molecular reproduction or increase in cell size.

The life-span of different cells varies with the tissue, the animal, and the conditions of existence. Nerve cells and muscle cells, to some extent, persist throughout the life of the higher animal. Red blood corpuscles live about 120 days. The normal process of metamorphosis found in many animals involves a great loss of cells. Many cells are removed in the shaping of organs during morphogenesis.

Cells undergo a senescence with aging. At some point in the life cycle of most cells there is a breakdown of cell substance, the formation of inert material, a slowing down of metabolic processes, and a decrease in the synthetic power of enzymes. These factors lead eventually to the death of the cell. In certain cases, parts of the cell such as scales, feathers, and bony structures, may persist after the death of the cell.

References
Suggested general readings

Bonner, J. T. 1955. Cells and societies. Princeton, N. J., Princeton University Press. *The author emphasizes the sameness of the basic biologic requirements of the organisms, but the methods of meeting these requirements are highly varied. It is an excellent review of some of the chief functions of organisms.*

Bonner, J. T. 1959. The cellular slime molds. Princeton, N. J., Princeton University Press.

De Robertis, E. D. P., W. W. Nowinski, and F. A. Saez. 1970. General cytology, ed. 5. Philadelphia, W. B. Saunders Co. *This concise and popular text contains a good description of mitosis.*

DuPraw, E. J. 1970. DNA and chromosomes. New York, Holt, Rinehart & Winston, Inc. *This paperback is packed with information about nucleic acids and chromosomes in all phases of activity.*

Jensen, W. A., and R. B. Park. 1967. Cell ultrastructure. Belmont, Calif., Wadsworth Publishing Co., Inc. *A collection of electron micrographs and illustrated cells and their components from both plants and animals.*

Levine, L. (editor). 1962. The cell in mitosis. New York, Academic Press, Inc. *The more complicated aspects of cell division are treated by various specialists in this field.*

Marsland, D. 1957. Temperature-pressure studies on the role of sol-gel relations in cell division. Washington, D. C., American Physiological Society, pp. 111-126.

Mercer, E. H. 1962. Cells: their structure and function. (Paperback.) New York, Doubleday & Co., Inc.

Novikoff, A. B., and E. Holtzman. 1970. Cells and organelles. New York, Holt, Rinehart & Winston, Inc. *Clearly written and well-illustrated description of structure of cells and cell components and their function.*

Porter, K. R., and M. A. Bonneville. 1963. An introduction to the fine structure of cells and tissues. Philadelphia, Lea & Febiger. *This work gives one an insight into what the electron microscope means to the modern study of histology.*

Ramsey, J. A., and V. B. Wigglesworth (editors). 1961. The cell and the organism. Cambridge, Cambridge University Press. *A great variety of biologic phenomena are treated in the light of recent advances in this revealing volume.*

Stern, H., and D. L. Nanney. 1965. The biology of cells. New York, John Wiley & Sons, Inc. *This introductory textbook in cell biology stresses the cell doctrine, the chromosome and gene, and the physicochemical basis of life.*

Swanson, C. P. 1969. The cell, ed. 3. Englewood Cliffs, N. J., Prentice-Hall, Inc. *One of a series of biologic monographs. A concise, well-illustrated account of the modern concept of the cell.*

Wyckoff, R. W. G. 1958. The world of the electron microscope. New Haven, Yale University Press. *This work explains the principles of this important tool of research and its practical application.*

Selected *Scientific American* articles

Baserga, R., and W. E. Kisieleski. 1963. Autobiographies of cells. **209:**103-110. (Aug.).

de Duve, C. 1963. The lysosome, **208:**64-72 (May).

Fox, C. F. 1972. The structure of cell membranes. **226:**31-38 (Feb.).

Green, D. C. 1964. The mitochondrion. **210:**63-74 (Jan.).

Hokin, L. E., and M. R. Hokin. 1965. The chemistry of cell membranes. **213:**78-86 (Oct.).

Mazia, D. 1961. How cells divide. **205:**100-120 (Sept.).

Neutra, M., and C. P. Leblond. 1969. The Golgi apparatus. **220:**100-107 (Feb.).

Racker, E. 1968. The membrane of the mitochondrion. **218:**32-39 (Feb.).

Zamecnik, P. C. 1958. The microsome. **198:**118-124 (March).

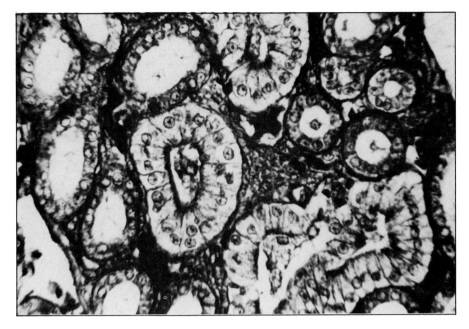

Section of kidney tissue showing cuboidal and columnar epithelial cells of the kidney tubules. In forming urine, these cells selectively modify the tubular fluid by reabsorbing some substances and secreting others. Both active and passive transport processes are required.

CHAPTER 4

PHYSIOLOGY OF THE CELL

◼ DIFFERENTIATION OF CELL FUNCTIONS

In the previous chapter, we described the generalized blueprint of an animal cell. We pointed out that a unicellular organism contains within its cell boundaries all the equipment required for life and its propagation. These basic functions are the metabolic production of energy, the biosynthesis of cellular constituents, the regulation of metabolism, the presence of a protective and selective membrane boundary, the ability to move, and the capacity to reproduce. As animals advanced evolutionarily to multicellular forms, the functional role of the body's cells changed. The advantages of multicellularity seem clear enough: organisms gained size, speed, and a certain independence from the often harsh physical forces of their immediate environment. But a price was paid. As the whole organism gradually gained

independence, its individual cells lost some of theirs. Systems evolved to take over specific functions such as respiration, digestion, and excretion; a circulatory system was required for internal transport and nervous and hormonal systems were added to coordinate everything. When cells specialized to perform these needs, they could no longer independently perform *all* the functional activities of an organism. Cell specialization encouraged, and eventually made necessary, further cell specialization.

Nevertheless, certain activities are performed by every animal cell, whether it is a liver cell, a nerve cell, or a muscle cell. All cells must produce energy, manufacture their own internal structures, control much of their own activity, and guard their boundaries. It is these basic shared activities that we will study in this chapter.

■ PHYSIOLOGY OF MEMBRANES

Every cell is surrounded by an incredibly thin, yet sturdy, envelope that plays a vital role in the life of cells. This is the **plasma membrane.** Once believed to be a rather static entity that defined cell boundaries and kept cell contents from spilling out, the plasma membrane has proved to be a dynamic structure having remarkable activity and selectivity. It is a permeability barrier that separates the internal and external environment of the cell, regulates the vital flow of molecular traffic into and out of the cell, and provides many of the unique functional properties of specialized cells. Membranes inside the cell—those surrounding the mitochondria, Golgi apparatus, endoplasmic reticulum, lysosomes, and other organelles—also play vital functional roles and share many of the structural features of the plasma membrane. Internal membranes provide the site for many, perhaps most, of the enzymatic reactions required by the cell. Proteins are formed on the membranes of ribosomes, and DNA is replicated on membrane sites. In fact, so many crucial cell functions are associated with membranes that research on cell membranes may reach the intensity in the 1970s that nucleic acid research reached in the 1960s.

PRESENT CONCEPT OF MEMBRANE STRUCTURE. Recent studies have largely confirmed the classic model of membrane structure proposed by Danielli and Davson in 1934, that of a lipid bilayer sandwiched between two layers of protein (see description of membrane composition on p. 49). This theory was remarkably visionary, especially since it was suggested long before there was experimental evidence to support it. Only in 1971 was the lipid bilayer positively confirmed at Cambridge University in England by M. Wilkins (Nobel Prize winner of DNA fame).

But the Danielli-Davson theory has required modification in respect to the proteins that were originally believed to rest in a static globular configuration on the lipid bilayer. In 1957, J. D. Robertson proposed the **unit membrane** hypothesis, which received widespread support from biologists. Robertson's hypothesis, based primarily on studies of the insulating myelin membrane of nerve fibers, retained the lipid bilayer of the Danielli-Davson model, but suggested that the surface proteins rested on the lipid bilayer as flat uninterrupted sheets. As the name of the theory suggested, the unit membrane structure was believed to be characteristic of all kinds of membranes. Unfortunately, the myelin membrane is not typical of most plasma membranes, and recent research has forced a reevaluation of the unit membrane hypothesis. New biophysical techniques such as optical rotary dispersion and freeze-etching (Fig. 3-3), have contributed important new information about membrane structure. The proteins are not arranged in orderly array upon the lipid bilayer, like a static crust. Some proteins penetrate into the lipid core of the bilayer, and others extend all the way through it (Fig. 3-4). Furthermore, the membrane appears to be remarkably restless and fluid, with proteins constantly moving and reorganizing their molecular configuration. With the debate about the lipid bilayer core resolved (the Danielli-Davson hypothesis firmly accepted after nearly 40 years of uncertainty), attention is presently directed to the proteins, and the roles they play in membrane transport mechanisms.

PASSAGE OF MATERIALS THROUGH MEMBRANES. The plasma membrane acts as a gatekeeper for the entrance and exit of the many substances involved in cell metabolism. Some substances can pass through with ease, others enter slowly and with difficulty, and still others cannot enter at all. This is what is called the **selective behavior** of the cell membrane. Because conditions outside the cell are different from and more variable than conditions within the cell, it is necessary that the passage of substances across the membrane be rigorously controlled. We presently recognize three principal ways that substances traverse the cell membrane. If substances move across driven by forces that arise outside the cell, the movement is called **passive transport.** If metabolic energy is required to transfer substances across the membrane, the movement is called **active transport.** A third transport mechanism is involved in moving larger molecules or particles into a cell. This is a kind of wholesale ingestion, called **endocytosis,** which requires that the cell membrane form a pocket or channel into which the material is enveloped.

PASSIVE TRANSPORT AND OSMOSIS. If some salt is dropped into a beaker of water, salt molecules and ions will spread through the water until the concentration of salt is uniform throughout. This happens because all molecules are in motion arising from their heat (kinetic) energy. A molecule (or ion) moves in a straight line until it meets another particle and then bounces off and takes a new direction. The effect of this constant activity is that every component in a solution will diffuse until it reaches equal concentration everywhere in the solution. This is an example of a system reaching a state of maximum disorder, or maximum **entropy** (a measure of order of a system). When a solute (say, a salt crystal) is first placed in a container of water, the system is highly ordered; it is also a highly improbable and unstable condition. As the solute dissolves and diffuses evenly throughout the water, the system approaches maximum randomness and disorder and also maximum stability and entropy.

Now if a living cell surrounded by a membrane is immersed in a solution having more solute molecules than the fluid inside the cell, a **concentration gradient** instantly exists between the two fluids. More solute molecules will strike the membrane from the outside than from the inside, creating a powerful driving force. The solute is pushed toward the inside, the side having the lower concentration. If the membrane is **permeable** to the solute, it will diffuse "downhill" across the membrane until its concentrations on each side are equal. This is passive transport because no energy is expended.

Most cell membranes are **semipermeable,** that is, permeable to water, but selectively permeable, or impermeable, to solutes. In passive transport it is this selectiveness that regulates molecular traffic. As a rule, gases (such as O_2 and CO_2) and lipids (such as hydrocarbons) are the only solutes that can penetrate biologic membranes with any degree of freedom. This happens because of the lipid nature of membranes. Most membranes are nearly impermeable to salts, sugars, and macromolecules. Such water-soluble molecules must pass through special pores in the membrane, but in most membranes these pores are too small to allow solutes to pass through by diffusion. If now such a membrane is placed between two unequal concentrations of a solution, water will start to flow through the membrane from the more dilute to the more concentrated solution. This is **osmosis.** To understand why this happens we must view the system from the standpoint of the concentration of water on each side. On the side of the stronger solution (higher concentration of solute) the water is present in **lower** concentration than it is on the side of the weaker solution.

Water therefore diffuses through the membrane from the side where it is most concentrated (weaker solution) to the side where it is least concentrated (stronger solution). Water, like any other material, tends to diffuse "downhill," that is, from a higher to a lower concentration. Osmosis differs from unrestricted diffusion in that only the water can diffuse; the solute is restricted by the selectively permeable membrane.

Another difference between osmosis and unrestricted diffusion is that the movement of water creates a volume change. This can be demonstrated by a familiar experiment in which a selectively permeable membrane, such as collodion membrane, is tied over the end of a funnel. The funnel is filled with a sugar solution and placed in a beaker of pure water so that the water levels inside and outside the funnel are equal. In a short time the water level in the glass tube will be seen to rise, indicating that water is passing through the collodion membrane into the sugar solution (Fig. 4-1). Inside the funnel are sugar molecules as well as water. In the beaker outside the funnel are only water molecules. Thus the concentration of water is greater on the outside because some of the space inside is taken up with sugar molecules. The water therefore will go from the greater concentration (outside) to the lesser (inside).

As the fluid rises in the tube against the force of gravity, it exerts a hydrostatic pressure on the collodion membrane and glass tubing (small arrows in Fig. 4-1). This hydrostatic pressure opposes the movement of water molecules into the funnel. Eventually, the hydrostatic pressure becomes so great that there is no further **net** movement of water from the beaker into the bag, and the fluid level in the glass tube stabilizes. We see, then, that osmosis can perform work. An **osmotic pressure** (large arrows in Fig. 4-1) drives water through the membrane into the solution and creates an opposing **hydrostatic pressure** head. When the hydrostatic pressure (measured by the height of the fluid column) equals the opposing osmotic pressure, no more water enters the osmometer. (Actually, water molecules continue to traverse the membrane, but the movement inward is matched by movement outward.) Osmotic pressure can thus be expressed in terms of the height of the fluid column, which in turn depends on the concentration of the sugar solution.

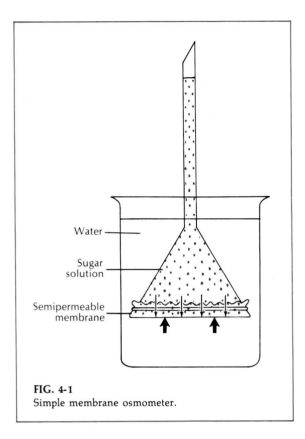

FIG. 4-1
Simple membrane osmometer.

Actually, the direct measurement of osmotic pressure in biologic solutions is seldom done today. This is because the osmotic pressures of most biologic solutions are so great that it would be impractical, if not impossible, to measure them with the simple membrane osmometer described. The osmotic pressure of human blood plasma would lift a fluid column over 250 feet—if one could construct such a long, vertical tube and find a membrane that would not rupture from the pressure. Indirect methods of measuring osmotic pressure are more practical. By far the most widely used measurement is the **freezing point depression.** This is a much faster and more accurate determination than is the direct measurement of osmotic pressure by the collodion membrane osmometer we described above. Pure water freezes at exactly 0° C. As solutes are added, the freezing point is lowered; the greater the concentration of solutes, the lower the freezing point. Human blood plasma will freeze at about −0.56° C.; seawater will freeze at about −1.80° C. Although the lowering of the freezing point of water by the presence of solutes is small,

great accuracy of measurement is possible because the instruments used by biologists can detect differences of as little as 0.001 centigrade degree.

ACTIVE TRANSPORT. We have seen that the cell membrane is a very effective barrier to the passive transfer of most solutes of biologic importance. Not only must such solutes cross the membrane, they must often be moved "uphill" against the forces of passive diffusion. For example, living cells must continuously transport potassium from outside to the inside where the potassium concentration is already 20 to 50 times higher than outside. Movement in such cases is by **active transport,** and always involves the expenditure of energy because materials are pumped against a "head." Although membrane transport mechanisms have been, and continue to be, a major area of research interest, it is still not understood how cell energy in the form of ATP is coupled to the transport process. It seems almost certain however that protein "carrier" molecules are required to breach the membrane barrier. The recent discovery that some proteins completely penetrate the membrane has helped us to understand how this may occur. The carriers are believed to be proteins positioned in the middle of the membrane with their margins exposed to both membrane surfaces. Behaving like a revolving shuttle, the protein captures a molecule on one side, rotates 180° to the opposite side, and discharges its fare. The carrier then rotates back, again presenting its attachment site for the pickup and transport of another molecule. In this way, molecules of all kinds and sizes can breach the membrane barrier. The transport is "active" because without metabolic energy the carrier will not rotate back to pick up another molecule. Protein carriers are usually quite specific, recognizing and transporting only one kind of molecule and ignoring all other kinds of molecules present in the fluid bathing the membrane.

The rate of active transport through a membrane is largely independent of the concentration gradient. Active transport is, however, very sensitive to temperature change, because it is powered by the cell's metabolic machinery. A rise of 10° C. will about double the rate of active transport. Diffusion, on the other hand, is proportional to the absolute temperature; a rise of 10° C. will increase diffusion rate by only 10/273, or a little less than 4%.

Active transport may occur across all cell membranes, but it has been most studied in nerve, muscle, and kidney cells. Active transport is also involved in the transport of certain nutrients across the epithelium of the digestive tract and of the digestive glands. The excitability of nervous and muscular tissues depends on precisely controlled differences in sodium and potassium concentration inside and outside cell membranes and on rapid shifts in membrane permeability to these ions.

ENDOCYTOSIS. The ingestion of solid or fluid material by cells was observed by microscopists nearly 100 years before phrases like "active transport" and "protein carrier mechanism" entered the biologists' vocabulary. Endocytosis is a collective term that describes two similar processes, **phagocytosis** and **pinocytosis.** Phagocytosis, which literally means "cell eating," is a common method of feeding among the Protozoa and lower Metazoa. It is also the way white blood cells (leukocytes) engulf cellular debris and uninvited microbes in the blood. By phagocytosis, the cell membrane forms a pocket that engulfs the solid material. The membrane-enclosed vesicle then detaches from the cell surface and moves into the cytoplasm where its contents are digested by intracellular enzymes. Pinocytosis, or "cell drinking" is similar to phagocytosis except that drops of fluid are sucked discontinuously through tubular channels into cells to form tiny vesicles. These may combine to form larger vacuoles. Both processes require metabolic energy and in this respect they are forms of active transport.

Biologists have been troubled by the apparent nonselectivity of phagocytosis and pinocytosis. Both seem to be bulk transport processes that deny any possibility of discrimination by the cell. Nevertheless, there is convincing physiologic evidence that pinocytosis *is* selective although the true explanation still evades experimental proof.

■ THE CENTRAL ROLE OF ENZYMES IN THE LIVING PROCESS

The whole life process involves numerous chemical reactions occurring within the cells. However, the chemical breakdown of large molecules and the release of energy for cellular activities would not proceed at any meaningful rate without enzymes. Enzymes are **biologic catalysts** that are required for almost every reaction in the body. As every chemist knows, catalysts are chemical substances that accel-

erate reactions without affecting the end products of the reactions and without being destroyed as a result of the reaction. Enzymes fit this definition, too. Enzymes are involved in every aspect of life phenomena. They control the reactions by which food is digested, absorbed, and metabolized. They promote the synthesis of structural materials to replace the wear and tear on the body. They determine the release of energy used in respiration, growth, muscle contraction, physical and mental activities, and a host of others. Little wonder that enzymes are absolutely fundamental to life.

NATURE OF ENZYMES. Enzymes are complex molecules varying in size from small, simple proteins with a molecular weight of 10,000 to highly complex molecules with weights up to 1 million. Some enzymes, such as the gastric enzyme pepsin, are pure proteins—delicately folded and interlinked chains of amino acids. Other enzymes contain special active nonprotein substances in addition to the protein portion of the molecule. Such an active and highly essential component is called a **prosthetic group** (working group). The prosthetic group is usually firmly attached to the protein. In some cases, however, biochemists have been able to detach the prosthetic group, yielding two molecules termed the **coenzyme** (prosthetic group) and the **apoenzyme** (protein part). Apoenzymes are inactive without the all-important coenzyme. Nearly all the vitamins have been shown to be essential parts of enzyme prosthetic groups. Most of the enzymes that have vitamin-containing coenzyme groups play crucial roles in cellular metabolism. Since a vitamin cannot be synthesized by the animal needing it, it is obvious why a dietary vitamin deficiency can be serious.

NAMING OF ENZYMES. Enzymes are named for the reactions they catalyze. Usually the suffix **-ase** is added to the root word of the substance, or substrate, on which the enzyme works. Thus sucrase acts on sucrose, lipase acts on lipids, and protease acts on proteins. Enzymes may also be named according to the nature of the reaction. For example, dehydrogenases catalyze dehydrogenations.

ACTION OF ENZYMES. An enzyme functions by combining in a highly specific way with the substance upon which it acts (the *substrate*). According to the classic *lock-and-key* theory, each enzyme contains an *active site,* which is a unique molecular configuration that is exactly complimentary to at least

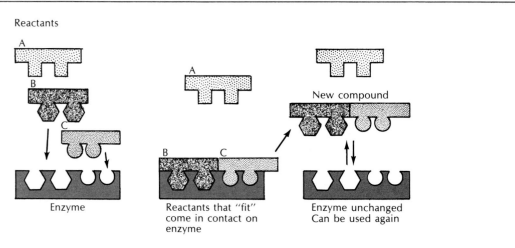

Reactants

A

B

C

Enzyme

A

Reactants that "fit" come in contact on enzyme

B   C

New compound

Enzyme unchanged Can be used again

FIG. 4-2
Enzyme action and specificity. Enzymes are believed to have surface configurations that "fit" specific substrates. Here molecules **B** and **C** fit into enzyme surface, but **A** does not. Reactions involving **B** and **C** are speeded up by the molecules' coming in contact briefly with enzyme. When reaction is complete, the enzyme, still unchanged, can dissociate from the substrate and is free to aid in further reactions. Molecule **A** and others not specific to this enzyme are unaffected by it.

a portion of the specific substrate molecule (Fig. 4-2). By fitting onto the substrate, the enzyme provides a unique chemical environment that somehow makes the substrate molecule more active. Enzyme and substrate combine to form an unstable enzyme-substrate complex.

$$E + S \rightleftharpoons ES \rightleftharpoons ES' \rightleftharpoons E + P$$

With the formation of ES, enough activation energy is introduced to form S′, an activated form of the substrate, which cleaves into its products P. These are liberated from the enzyme, which is restored to its original form.

The classic lock-and-key theory is still largely accepted by biochemists. However, the active site of the enzyme may be a flexible surface that infolds, and conforms to, the substrate, rather than a fixed and nonyielding template, as the original theory held. This newer **conformational theory** has not altered a firmly held principle of enzyme action: the enzyme and substrate must combine so that active groups on the enzyme come into precise alignment with reactive sites on the substrate. Only then can the substrate be altered chemically. The necessity of correct alignment explains the high specificity of enzymes (see below).

Enzymes that engage in important main-line sequences, for example, the crucial energy-providing reactions of the cell that go on constantly, seem to operate in enzyme sets rather than in isolation. Main-line enzymes are found in relatively high concentrations in the cell, and they may implement quite complex and highly integrated enzymatic sequences. One enzyme carries out one step, then passes the product to another enzyme that catalyzes another step, and so on.

SPECIFICITY OF ENZYMES. Most enzymes are highly specific. Such high specificity is a consequence of the exact molecular fit that is required between enzyme and substrate. However, there is some variation in degree of specificity. Some enzymes, such as succinic dehydrogenase, will catalyze the oxidation (dehydrogenation) of one substrate only, succinic acid. Others, such as proteases (for example, pepsin and trypsin), will act on almost any protein. Usually an enzyme will take on one substrate molecule at a time, catalyze its chemical change, release the product, and then repeat the process with another substrate molecule. The enzyme may repeat this process billions of times until it is finally worn out (a few hours to several years) and is broken down by scavenger enzymes in the

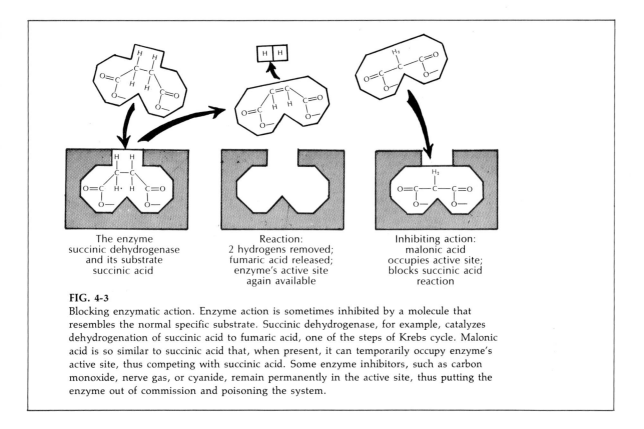

| The enzyme succinic dehydrogenase and its substrate succinic acid | Reaction: 2 hydrogens removed; fumaric acid released; enzyme's active site again available | Inhibiting action: malonic acid occupies active site; blocks succinic acid reaction |

FIG. 4-3

Blocking enzymatic action. Enzyme action is sometimes inhibited by a molecule that resembles the normal specific substrate. Succinic dehydrogenase, for example, catalyzes dehydrogenation of succinic acid to fumaric acid, one of the steps of Krebs cycle. Malonic acid is so similar to succinic acid that, when present, it can temporarily occupy enzyme's active site, thus competing with succinic acid. Some enzyme inhibitors, such as carbon monoxide, nerve gas, or cyanide, remain permanently in the active site, thus putting the enzyme out of commission and poisoning the system.

cell. Some enzymes are able to undergo successive catalytic cycles at dizzying speeds of up to a million cycles per minute; most operate at slower rates. The digestive enzymes, which are secreted into the digestive tract to degrade food materials, are "one-shot" enzymes. After breaking down their substrate, they are themselves digested and lost to the body. Despite the high specificity of enzymes, they can sometimes be fooled into accepting a molecule that resembles the normal substrate. Malonic acid is such a molecule, and when succinic dehydrogenase combines with it, the active site is blocked and the enzyme is inhibited or poisoned (Fig. 4-3). Malonic acid (which undergoes no chemical transformation) thus acts as a competitive inhibitor.

ENZYME-CATALYZED REACTIONS. Enzyme-catalyzed reactions are reversible. For example, succinic dehydrogenase catalyzes both the dehydrogenation of succinic acid to fumaric acid and the hydrogenation of fumaric acid to succinic acid.

$$\text{Succinic acid} + H_2O \rightleftharpoons \text{Fumaric acid}$$

Reversibility is signified by the double arrows. Enzymes vastly accelerate reactions in either direction. Many enzymes, however, catalyze reactions almost entirely in one direction. For example, the proteolytic enzyme pepsin can degrade proteins into amino acids, but it cannot accelerate the rebuilding of amino acids into any significant amount of protein. The same is true of most enzymes that catalyze the hydrolysis of large molecules such as nucleic acids, polysaccharides, lipids, and proteins. There is usually one set of reactions and enzymes that break them down, but they must be resynthesized by a different set of reactions and catalyzed by different enzymes. This apparent irreversibility exists because the chemical equilibrium usually favors the formation of the smaller degradation products. The net **direction** of any chemical reaction is dependent on the relative energy contents of the substances involved, so that other compounds (energy-rich compounds) must participate in the conversion of, for example, amino acids into proteins, thus making the synthetic process different from simply the reverse of the degradation reactions.

SENSITIVITY OF ENZYMES. Enzyme activity is sensitive to temperature and pH. As a general rule enzymes work faster with increasing temperature, but will do so only within certain limits. Moreover, this increase in velocity is not proportional to the rise in temperature. Usually the rate is doubled with each 10° C. rise, but a change from 20° to 30° C. has a greater effect than one from 30° to 40° C. The optimum temperature for animal enzymes is about body temperature. Above 40° to 50° C. most enzymes become unstable and may be inactivated altogether.

Each enzyme usually works best within a certain range of acidity or alkalinity. Pepsin of the acid gastric juice is most active at about pH 1.8; trypsin of the alkaline pancreatic juice is most active at about pH 8.2. Many work best when the pH is around neutrality. In strong acid or alkaline solutions, most enzymes irreversibly lose their catalytic power.

■ CELLULAR METABOLISM

The term "cellular metabolism" refers to the sum total of the chemical processes that are necessary for all the phenomena of life, such as the synthesis of new cell materials, the replacement of that which is destroyed during wear and tear, and whatever is needed by the cell to grow, reproduce, move, etc. For these processes cells require a continuous supply of nutrients obtained from the surrounding extracellular fluid. Living cells, like man-made machines, do work and consequently require fuel. This fuel is in the form of organic molecules, which, for animals, must be supplied in the diet. Animals, of course, are totally dependent on plants for ready-made fuels. Animal cells tap the stored energy of organic fuels (for example, simple sugars, fatty acids, amino acids) through

a series of controlled degradative steps. This process makes use of molecular oxygen from the atmosphere. In return the cells give off carbon dioxide as an end product, which is used by plant cells in making glucose and the more complex molecules. In this way the cellular energy cycle of life involves the harnessing of sunlight energy by green plants directly and by animal cells indirectly.

METABOLIC ENERGY. There are certain limited parallels between the combustion of fuel in a fire and metabolic combustion of fuel in a living cell. In both, oxygen is consumed and both are exergonic reactions in that energy is liberated as heat. If the fuel is burned in an internal combustion engine so that work is performed, the parallel is even better. But here the similarity ends. The burning of gasoline in a cylinder is an explosive event that promotes just one function, the rapid expansion of gas. Many chemical bonds are broken simultaneously, and much energy is lost as heat. In contrast, metabolic energy must be released gradually and coupled to a great variety of energy-consuming reactions. Although metabolic energy exchanges proceed with great efficiency, some heat is inevitably liberated. Heat can be put to use of course; higher vertebrates use it to elevate and maintain a constant internal body temperature, just as the heat of gasoline combustion is made to warm the occupants of a car. But for the most part, heat is a useless commodity to a cell, since it is a nonspecific form of energy that cannot be captured and redistributed to power metabolic processes. There is actually only one way in which the oxidative release of energy is made available for use by cells: it is coupled to the production of ATP (adenosine triphosphate) by addition of inorganic phosphate to ADP (adenosine diphosphate). The structure of ATP is given below:

Adenine   Ribose   Triphosphoric acid

The ATP molecule consists of a purine (adenine), a 5-carbon sugar (ribose), and 3 molecules of phosphoric acid linked together by two pyrophosphate bonds to form a triphosphate group. The pyrophosphate bonds are called **high-energy bonds** because they are repositories of a great deal of chemical energy. This energy has been transferred to ATP from other low-energy bonds in the respiratory process. Respiration, by the stepwise oxidation of fuel substrates, redistributes bond energies so that a few high-energy bonds are created and stored in ATP. Obviously this energy is gained at the expense of fuel energy; the end products of cellular respiration, CO_2 and H_2O, contain much less bond energy than do the fuel substrates (for example, glucose) that entered the oxidative pathway.

The high-energy pyrophosphate bonds of ADP and ATP are frequently designated by the symbol $\sim$. Thus a low-energy phosphate bond is shown as —P, and a high-energy one as $\sim$P. ADP can be represented as A—P $\sim$P and ATP as A—P $\sim$P $\sim$P.

The trapping of energy by ATP can be shown as follows:

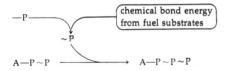

A low-energy phosphate group containing bond energy of 2,000 to 3,000 calories per mole is converted into a high-energy phosphate group containing bond energy of 8,000 to 10,000 calories per mole. Where does the low-energy phosphate group (—P) come from? Ultimately it comes from the diet. But in cellular respiration, ATP itself is the immediate source of the phosphate group necessary to start the oxidation process, donating the —P to the fuel substrate molecule (usually glucose).

ATP   ADP

Fuel substrate ⟶ Fuel substrate —P

The fuel molecule is phosphorylated with a low-energy phosphate group, and the phosphorylated fuel can then be oxidized to yield energy. Quite obviously, the fuel molecule must release more energy—actually it provides much more energy—than is loaned to it by ATP at the start.

This is a kind of initial deficit financing that is required for an ultimate energy return many times greater than the original energy investment. We will return later to this subject of energy budgeting.

The amount of ATP produced in respiration is totally dependent on its rate of utilization. In other words, ATP is produced by one set of reactions and immediately consumed by another. ATP is a great **energy-coupling** molecule, used to transfer energy from one reaction to another. For example, ATP formed from the oxidation of glucose is used to synthesize proteins or lipids, or provide power for some other process, such as the contraction of skeletal muscles. The point is, living organisms do not produce and put aside vast amounts of ATP, hoarded against some future energy need. What they do store is the fuel itself, in the form of carbohydrates and lipids especially. ATP is formed as it is needed, primarily by oxidative processes in the mitochondria. Oxygen is not consumed unless ADP and phosphate molecules are available, and these do not become available until ATP is hydrolized by some energy-consuming process. **Metabolism is therefore mostly self-regulating.**

There are minor exceptions to the rule that high-energy bonds cannot be stored in cells. Muscle cells contain a type of molecule especially adapted for energy storage. This is phosphocreatine:

$$
\begin{array}{c}
\text{H} \\
\quad \diagdown \\
\qquad \text{N} \sim \text{PO}_3^- \\
\diagup \\
\text{H}_2\text{N} - \text{C} \\
\diagdown \\
\qquad \text{N} - \text{CH}_3 \\
\qquad | \\
\qquad \text{CH}_2 \\
\qquad | \\
\qquad \text{COO}^-
\end{array}
$$

Phosphocreatine contains a high-energy bond that can provide instant power to the muscle contractile machinery, which often has sudden energy needs. Phosphocreatine is formed from creatine and ATP and is in direct chemical equilibrium with ATP. A short-term burst of activity will rapidly deplete the available ATP, making ADP available. High-energy phosphate is then transferred to ADP from phosphocreatine, providing more ATP for use in muscle contraction.

ATP GENERATED BY ELECTRON TRANSFER. Having seen that ATP is the one common energy denominator by which all cellular machines are energized, we are in a position to ask how is this energy captured from fuel substrates? The principal means is by the transfer of electrons from fuel substrates to molecular oxygen through a series of enzymes. This electron transfer system is made of a chain of large molecules localized in the inner mitochondrial membranes. The electron carriers are compounds that can be reduced by accepting electrons from the previous carrier, and then be oxidized again by passing electrons to the next carrier. Each successive carrier is a somewhat stronger oxidizing agent than the one before; that is, **successive carriers are increasingly stronger electron acceptors.** Finally, the electrons, as well as the hydrogen protons that accompany them, are transferred to molecular oxygen to form water:

$$\text{Fuel-}\widehat{(2H)} \quad A \quad B \quad C \quad D \quad \text{Oxygen} \rightarrow H_2O$$

It is conventional to represent electron transfer through the electron carrier system as the transfer of **hydrogen** atoms, although we must emphasize that it is the energized **electrons** and not the **protons** of the hydrogen that are the important energy packets. The proton of each hydrogen atom simply takes a free ride during this electron shuttle until, at the end, it bonds with reduced oxygen and forms water.

The whole function of the electron transport chain is the capture of energy from the original fuel substrate and the transformation of it into a form the cell can use. To do this, the large chemical potential of food molecules is drawn off in small steps (rather than in one explosive burst as in ordinary combustion) by successive electron carriers. At three points along the chain, ATP production takes place by the phosphorylation of ADP. This method of energy capture is called **oxidative phosphorylation** because the formation of high-energy phosphate is coupled to oxygen consumption, and this depends, as we have seen, on the demand for ATP by other metabolic activities within the cell. The actual **mechanism** of ATP formation by oxidative phosphorylation is not yet known; we can only say that the transfer of electrons does something that is translated into the production of high-energy phosphate bonds.

NATURE OF ELECTRON CARRIERS. Oxidative phosphorylation is much too complex to function efficiently, if at all, were the enzymes just floating freely in the cytoplasm of the cell. There is now abundant evidence that the oxidative enzymes and

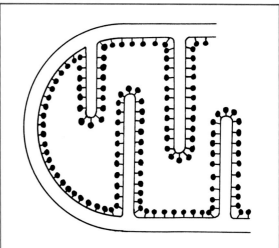

FIG. 4-4
Representation of section of mitochondrion as seen through high-resolution electron microscope showing the inner membrane spheres that bear enzymes of the respiratory chain. The density of the spheres is actually many times greater than that depicted in this diagram.

electron carriers are arranged in a highly ordered state on the membranes of the mitochondria.

It will be recalled that mitochondria are composed of two membranes. The outside membrane forms a smooth sac enclosing the inner membrane that is turned into numerous ridges called **cristae** (Fig. 3-2). The inner membrane is studded with enormous numbers of minute particles, which some investigators think are actually tiny stalked spheres. These particles, or spheres, bear the electron carriers of the respiratory chain responsible for oxidative phosphorylation. A section of a mitochondrion, as it might appear under the high-resolution electron microscope, is shown (highly diagrammatically) in Fig. 4-4.

Pairs of electrons, donated initially from food substrates, flow along the electron carriers in succession (Fig. 4-5). For most food substrates the initial electron acceptor is NAD (nicotinamide-adenine dinucleotide, a derivative of the vitamin niacin). The substrate is oxidized in the process (because it loses electrons) and NAD is reduced (because it gains electrons).

Next FAD (flavine-adenine dinucleotide, a riboflavin derivative) oxidizes the reduced NAD by accepting its electrons. FAD becomes reduced (having gained electrons) and NAD is returned to its original oxidized state. In the same way the pair of electrons is passed

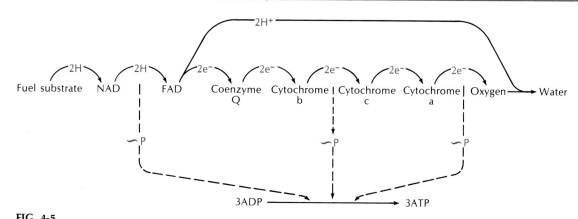

FIG. 4-5

Electron transport system. Electrons are transferred from one carrier to the next, terminating with molecular oxygen to form water. A carrier is reduced by accepting electrons and then is reoxidized by donating its electrons to the next carrier. ATP is generated at three points in the chain. These electron carriers are located on inner membrane spheres of mitochondria.

sequentially to coenzyme Q (chemically related to vitamin K) and then through a series of electron acceptors called **cytochromes.** The cytochromes are large molecules that belong to a class of proteins called chromoproteins because they contain colored prosthetic groups. The prosthetic groups of a cytochrome, like hemoglobin which is closely related to it, is an iron-bearing group that can be reversibly reduced.

$$Fe^{+++} + e^- \rightleftharpoons Fe^{++}$$

As the electrons are passed from cytochrome to cytochrome, each is successively reduced and then oxidized. Finally the electrons are passed to molecular oxygen. The transfer of electrons and the points of high-energy phosphate bond formation are shown in Fig. 4-5.

Thus for every pair of electrons moved along the carriers to oxygen, a total of 3 ATP molecules is formed.

ACETYL–COENZYME A: STRATEGIC INTERMEDIATE IN AEROBIC RESPIRATION. At this point the student may be forgiven if he thinks that the metabolic energy fixation process consists entirely of an electron transport chain that is capable of passing bond energy to ATP. Actually, we have only described the last (but especially crucial) step in **aerobic respiration.** The term "aerobic respiration" describes that kind of respiration requiring atmospheric oxygen, and this, of course, is the familiar sort of respiration practiced by the majority of animals. We have seen that ATP production is coupled to electron transfer, which in turn is completely dependent on oxygen, the final hydrogen and electron acceptor. Without oxygen the process stops because the electrons have nowhere to go. The components of the electron transfer chain would quickly become fully reduced and remain so, in the absence of the electron sink, molecular oxygen.

However, organisms do have a backup system enabling them to respire without oxygen. This is called **anaerobic respiration,** or **fermentation.** Although anaerobic respiration does support the lives of bacteria, yeasts, and a few other organisms, it is not nearly as efficient as aerobic respiration and consequently is not capable alone of maintaining life of higher animals. But it is useful, indeed essential, as a supplementary source of energy during rapid and intense muscular contraction. We will be satisfied now with just noting that animal cells possess the machinery for anaerobic respiration and will treat this subject in more detail later.

In aerobic respiration most fuel molecules are progressively stripped of their carbon atoms until those carbons are finally converted into molecules of CO_2. During this degradation the hydrogens and their electrons are removed and passed into the important energy-yielding electron transport chain we have already described. But to reach this final sequence, most carbon atoms appear in a 2-carbon group, called **acetyl–coenzyme A.** This is a critically important compound. Some two thirds of all the carbon atoms in foods eaten by animals appear as acetyl–coenzyme A at some stage. The strategic metabolic position of acetyl–coenzyme A is illustrated in Fig. 4-6. It is the final oxidation of acetyl–coenzyme A that provides the energized electrons used to generate ATP. Acetyl–coenzyme A is also the source of nearly all the carbon atoms found in the body's fats, as the reverse arrow in Fig. 4-6 indicates. The structure of acetyl–coenzyme A can be shown in abbreviated form as:

$$
\underset{\substack{\text{Acetyl} \\ \text{group}}}{CH_3-\overset{\displaystyle \overset{O}{\|}}{C}} \overset{}{\underset{\substack{\text{Coenzyme A} \\ \text{group}}}{-S-Co\ A}}
$$

The right hand side of the molecule is a coenzyme containing the vitamin **pantothenic acid,** another example of how vitamins play important structural roles in critical cellular functions.

OXIDATION OF ACETYL–COENZYME A. The breakdown (oxidation) of the 2-carbon acetyl group of acetyl–coenzyme A occurs in a sequence called the **citric acid cycle** (or **Krebs cycle,** after its British discoverer Sir Hans A. Krebs). The cycle is composed of a sequence of nine transformations and oxidations. To simplify an otherwise complex story, we have summarized the cycle in Fig. 4-7. The citric acid cycle begins with the condensation of acetyl–coenzyme A with oxaloacetate to form a 6-carbon compound citrate (citric acid). Citrate enters a series of reactions in which 2 molecules of CO_2 are produced from the original acetyl group. When the cycle is complete, the 4-carbon oxaloacetate is returned to its original form, ready to condense with another molecule of acetyl–coenzyme A. Oxaloacetate therefore acts as a carrier for the 2 carbons of the acetyl group; it is not itself used up in the cyclical process. As the acetyl group is oxidized carbon atom by carbon atom, four pairs of electrons and 4 protons (shown as four pairs of hydrogen atoms in Fig. 4-7) are transferred to the electron transfer chain (shown in the center of the cycle in Fig. 4-7). Three pairs of electrons are passed to NAD; the remaining pair is passed directly to FAD. Each pair of electrons then shuttles down the electron

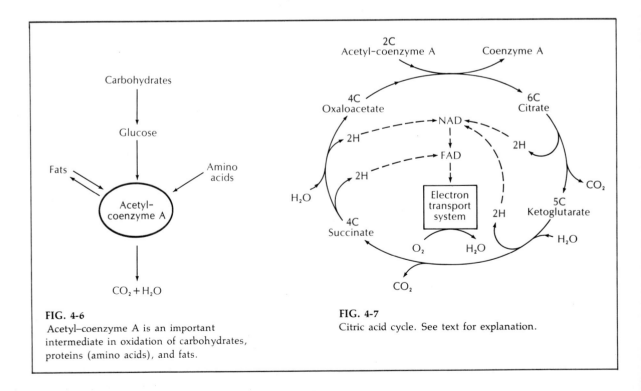

FIG. 4-6
Acetyl–coenzyme A is an important intermediate in oxidation of carbohydrates, proteins (amino acids), and fats.

FIG. 4-7
Citric acid cycle. See text for explanation.

transport chain to an atom of oxygen, as already described. Three molecules of ATP are generated for **each** molecule of NAD receiving electrons; this yields a total of nine ATP per acetyl group. Two more molecules of ATP are generated from the electrons passed directly to FAD. One more high-energy bond is generated at another point in the cycle; it forms a compound called GTP (guanosine-5′-triphosphate), which has the same energy yield as ATP, and for simplicity's sake, we will call it ATP. Thus the net yield is 12 molecules of ATP for the single acetyl group fed into the cycle. We must keep firmly in mind that 11 of these 12 high-energy phosphate bonds are generated by oxidative phosphorylation in the electron transport chain (Fig. 4-5) and not in the citric acid cycle itself. The citric acid cycle simply provides a means for the release of energized electrons during the oxidation of the acetyl group. All of these reactions occur in mitochondria. But electron release through the citric acid cycle is believed to occur in the **outer** membrane, whereas the electron carriers and the coupling to oxidative phosphorylation occurs in the **inner** mitochondrial membrane. Thus there is a spatial as well as functional separation of these processes.

GLUCOSE: MAJOR SOURCE OF ACETYL–COENZYME A. All the major fuels (glucose, fats, amino acids) serve as sources of acetyl–coenzyme A. Glucose, however, is a particularly important fuel for most tissues, especially the brain. Glucose is first converted to a 3-carbon compound called **pyruvate** (pyruvic acid) through a series of reactions that are called the Embden-Meyerhof pathway. Pyruvic acid is then enzymatically stripped of a carbon atom to form acetyl–coenzyme A. The general outline for this sequence is shown in Fig. 4-8. Again, we shall simplify a rather complex biochemical story by condensing this glucose metabolism pathway, which actually consists of ten consecutive enzymic reactions, into four major steps.

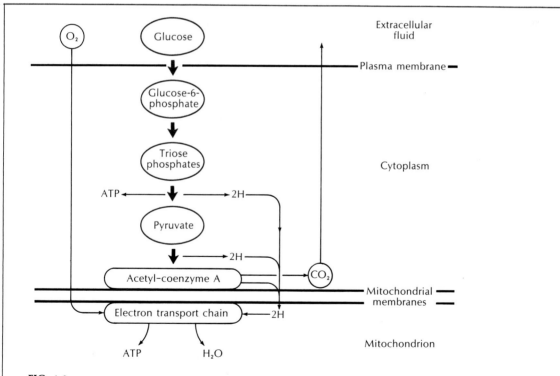

FIG. 4-8
Pathway for oxidation of glucose. Glucose is oxidized to acetyl–coenzyme A by the Embden-Meyerhof pathway (in red). Acetyl–coenzyme A then enters citric acid cycle (not shown) on outer mitochondrial membrane. Hydrogens removed in cycle are transferred to electron transport chain on inner mitochondrial membrane. See text for details.

The metabolism of glucose begins with its phosphorylation by ATP to form **glucose-6-phosphate** (see below).

$$\text{Glucose} + \text{ATP} \rightarrow \text{Glucose-6-phosphate} + \text{ADP} + \text{H}^+$$

Glucose-6-phosphate is an important intermediate because it is a "stem" compound that can lead into any of several metabolic pathways. However, the predominant metabolic fate for glucose-6-phosphate is entry into the Embden-Meyerhof sequence. Following still another phosphorylation, the 6-carbon glucose molecule is split into two 3-carbon sugars, called **triose phosphates.** Each triose phosphate is oxidized and rearranged to form **pyruvate** resulting in a yield of high-energy phosphate as ATP. A pair of hydrogen atoms and a molecule of CO_2 are then removed from pyruvate, forming acetyl–coenzyme A.

Let us now summarize the entire oxidation of glucose. Glucose first enters the cells of tissues, passing through the plasma membrane by a transport process that requires the presence of the pancreatic hormone **insulin.** Glucose is then phosphorylated and enters the Embden-Meyerhof pathway in the cytoplasm. This is shown in red in Fig. 4-8. Through this sequence it is split in the middle to form two 3-carbon sugars (triose phosphates) that are converted to pyruvate. Pyruvate is decarboxylated to form acetyl–coenzyme A. This sets the stage for entry into the citric acid cycle located on the outer mitochondrial membrane. After condensing with oxaloacetic acid to form citric acid, the 2-carbon acetyl fragment is oxidized—yielding 4 pairs of electrons and 4 protons that are passed along the electron transport chain located on the inner mitochondrial membrane. The electrons finally arrive at oxygen, the ultimate electron acceptor.

What has been accomplished? A molecule of glucose has been completely oxidized to CO_2 and H_2O. ATP has been generated at several points along the way. Let us now balance up the yield. First of all, 2 molecules of ATP were consumed in the initial phosphorylation of glucose. Then 14 molecules of ATP are generated by the transformations leading to the formation of acetyl–coenzyme A. (Remember that each glucose molecule is split into **two** 3-carbon sugars, each of which produces ATP when oxidized to acetyl–coenzyme A. Some ATP is produced directly, the rest results from oxidation followed by oxidative phosphorylation.) Our balance is now 12 ATP. Then we add the 12 ATP generated in the complete oxidation of **each** of the acetyl–coenzyme A molecules. This gives a total yield of 36 ATP. The whole sequence can be summarized as follows:

$$\text{Glucose } (C_6H_{12}O_6) + 6O_2 + 36ADP + 36\text{—P} \longrightarrow$$
$$6CO_2 + 6H_2O + 36ATP$$

EFFICIENCY OF OXIDATIVE PHOSPHORYLATION. It is probably obvious that no energy-transforming process can be 100% efficient, not even the remarkable cellular oxidative machinery produced by organic evolution. If we burn a mole of glucose (180 grams) in a bomb calorimeter, it releases about 686,000 calories. This is the **potential** energy for forming ATP. It has been determined that 8,000 to 12,000 calories are required to synthesize 1 mole of ATP. Consequently glucose theoretically **could** provide enough energy to generate 50 to 85 moles of ATP from ADP. It actually turns out 38 moles (36 + 2) of ATP. If we assume that each ATP mole represents an average energy equivalent of 10,000 calories, then 38 moles of ATP represents 380,000 calories. Thus the efficiency of glucose oxidation is 380:686, or about 55%. Engineers would be delighted if they could build machines that could do as well.

METABOLISM OF LIPIDS. Animal fats are triglycerides, molecules composed of glycerol and 3 molecules of fatty acids. These fuels are important sources of energy for many metabolic processes in

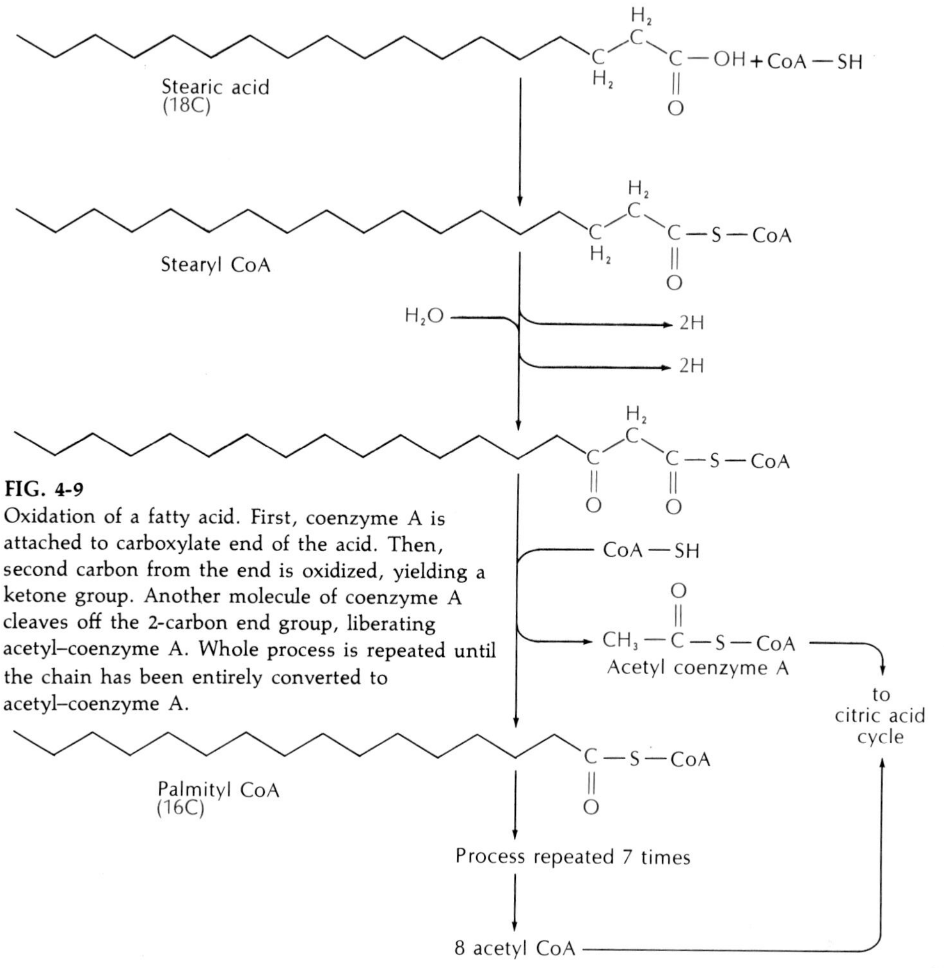

FIG. 4-9

Oxidation of a fatty acid. First, coenzyme A is attached to carboxylate end of the acid. Then, second carbon from the end is oxidized, yielding a ketone group. Another molecule of coenzyme A cleaves off the 2-carbon end group, liberating acetyl–coenzyme A. Whole process is repeated until the chain has been entirely converted to acetyl–coenzyme A.

all animals, not just obese victims of misplaced appetites. Most of fat is fatty acids, carboxylic acids with long hydrocarbon chains (Fig. 2-20). We know that fats enter the mitochondrial metabolic processes through acetyl–coenzyme A (Fig. 4-6). What happens in brief is that the long hydrocarbon chain of a fatty acid is sliced up by oxidation, two carbons at a time; these are released from the end of the molecule as acetyl–coenzyme A. The process is repeated until the entire chain has been reduced to several 2-carbon acetyl units. The oxidation of a fatty acid is diagrammed in Fig. 4-9, using a shorthand representation in which each jog in the chain symbolizes a saturated carbon ($-CH_2^-$) of the fatty acid **stearic acid,** one of the abundant naturally occurring fatty acids. First, the fatty acid is combined with coenzyme A. Then in a three-step process the third carbon from the end is oxidized (stripped of its hydrogens). Next, a molecule of acetyl–coenzyme A is sliced off the end, by **another** molecule of coenzyme A that then adds itself to the chain. Thus the hydrocarbon chain, now 2 carbons shorter, is left with coenzyme A on its end. The process is repeated until the whole chain has been chopped up into acetyl–coenzyme A. This material then enters the citric acid cycle to yield ATP in the manner described earlier.

How much ATP is gained from fatty acid oxidation? Five high-energy phosphate bonds are generated for each acetyl–coenzyme A unit split off. Then oxidation of each 2-carbon fragment produces 12 ~P. With allowance for the ATP expended to attach the first coenzyme A, it has been calculated that the complete oxidation of 18-carbon stearic acid will yield 147 ATP molecules. By comparison, 3 molecules of glucose (also totaling 18 carbons) yields 108 ATPs. Little wonder that fat is considered the king of animal fuels! Fats are more concentrated fuels than carbohydrates because fats are almost pure hydrocarbons; they contain more hydrogen per carbon atom than sugars do, and it is the energized electrons of hydrogen that generate high-energy bonds, when they are carried through the mitochondrial electron transport system.

GLYCOLYSIS: GENERATING ATP WITHOUT OXYGEN. Up to this point, we have been describing **oxidative,** or **aerobic,** metabolism, the kind of respiration that predominates in the majority of animals. It hardly needs emphasizing that the availability of oxygen is an obvious basic necessity of animal life. Nevertheless there are microorganisms, notably the yeasts and certain bacteria, that multiply happily with no oxygen at all. These organisms are called **an-aerobes.** They occupy important ecologic niches, some of the niches created by man. For example, oxygen-depleted streams are becoming regretably common appendages to our industrialized society. Anaerobic organisms use carbon compounds as fuel, breaking them down by a process commonly called **fermentation.** This term, meaning "cause to rise," was originally used to describe the action of yeasts that break down glucose into alcohol (ethanol) and CO_2. It is now applied to any microorganism that metabolizes foodstuffs without oxygen. The end products, which vary with the nature of the fermentive process, include butanol, acetone, lactic acid, and hydrogen gas.

Most higher organisms also have the capacity to ferment glucose, that is, break it down to produce high-energy phosphate in the absence of oxygen. The process is called **glycolysis.** It is used as a backup system for aerobic metabolism, providing a means for short-term generation of ATP during brief periods of heavy energy expenditure, when the slow rate of O_2 diffusion would be a limiting factor.

In glycolysis, glucose is split eventually into two molecules of **lactate** (lactic acid), yielding 2 molecules of ATP in the process. The glycolytic pathway is shown in Fig. 4-10. It will be seen that glycolysis utilizes the same Embden-Meyerhof pathway that, in oxidative metabolism, directs glucose into the citric acid cycle via acetyl–coenzyme A (compare Figs. 4-8 and 4-10). But in the absence of oxygen, both pyruvate and hydrogen accumulate in the cytoplasm because neither can proceed into their oxidative channels without oxygen. The problem is neatly solved by forming lactate. Pyruvate is converted into lactate that accepts the hydrogen, as shown below:

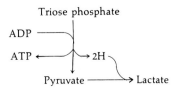

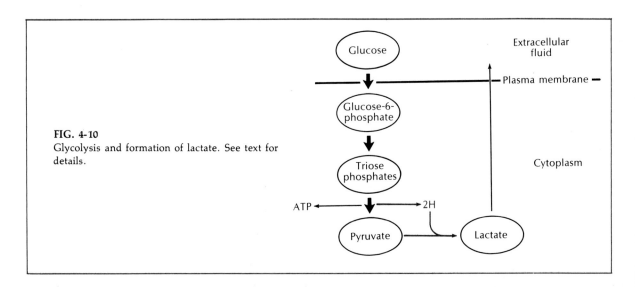

FIG. 4-10
Glycolysis and formation of lactate. See text for details.

Lactate then diffuses out into the blood, where it is later disposed of in the liver. Thus lactate formation prevents the cytoplasm from being swamped with pyruvate and allows **some** ATP formation. Of course, glycolysis is not an efficient producer of ATP; only 2 moles of ATP per mole glucose are generated by glycolysis as against 36 moles by oxidative phosphorylation. Nevertheless, the capacity to produce a little extra ATP during an emergency may mean the difference between life and death for an animal. Some animals rely heavily on glycolysis during normal activities. For example, diving birds and mammals fall back on glycolysis almost entirely to give them the needed energy to sustain a long dive. And salmon would never reach their spawning grounds were it not for glycolysis that provides almost all the ATP used in the powerful muscular bursts needed to carry them over falls and up rapids.

References

Suggested general readings

Barker, G. R. 1968. Understanding the chemistry of the cell. Institute of Biology's Studies in biology no. 13, New York, St. Martin's Press. *This paperback contains a concise account of cell chemistry, energy and metabolism, and research approaches.*

Brachet, J. 1957. Biochemical cytology. New York, Academic Press, Inc. *An advanced work on the morphology and biochemistry of the cell. The author summarizes in admirable fashion the great strides made in this field during the past few decades but also warns that many of our concepts will no doubt be radically changed in the future.*

Brachet, J., and A. E. Mirsky. 1959. The cell: biochemistry, physiology, morphology. New York, Academic Press, Inc.

Giese, A. C. 1968. Cell physiology, ed. 3. Philadelphia, W. B. Saunders Co. *An excellent cell physiology for students who have had good introductory courses in chemistry and physics.*

Guthe, K. F. 1968. The physiology of cells. New York, The Macmillan Co. *Paperback.*

Howland, J. L. 1968. Introduction to cell physiology: information and control. New York, The Macmillan Co. *Brief though moderately advanced treatment of biochemistry, enzyme action, and cell physiology.*

Lehninger, A. L. 1970. Biochemistry: the molecular basis of cell structure and function. New York, Worth Publishers. *A very lucidly written and amply illustrated undergraduate biochemistry text, particularly suitable for the student who leans more toward chemistry than biology.*

McElroy, W. D. 1964. Cell physiology and biochemistry, ed. 2. Englewood Cliffs, N. J., Prentice-Hall, Inc. *Paperback.*

McGilvery, R. W. 1970. Biochemistry: A functional approach. Philadelphia, W. B. Saunders Co. *Well-written mammalian biochemistry.*

Swanson, C. P. 1969. The cell, ed. 3. Englewood Cliffs, N. J., Prentice-Hall, Inc. *A good introduction to the concepts of molecular biology.*

Trumbore, R. H. 1966. The cell: chemistry and function. St. Louis, The C. V. Mosby Co. *A good basic, well-organized text, with summaries, on all aspects of the cell.*

Selected *Scientific American* articles

Changeux, J. R. 1965. The control of biochemical reactions. **212**:36-45 (April).

Frieden, E. 1959. The enzyme-substrate complex. **201**:119-125 (Aug.).

Green, D. C. 1960. The synthesis of fat. **202**:46-51 (Feb.).

Koshland, D. E., Jr. 1973. Protein shape and biological control. **229**:52-64 (Oct.).

Lehninger, A. L. 1960. Energy transformation. **202**:102-114 (May).

Loewenstein, W. R. 1970. Intercellular communication. **222**:79-86 (May).

Margaria, R. 1972. The sources of muscular energy. **226**:84-91 (March).

Neurath, H. 1964. Protein-digesting enzymes. **211**:68-79 (Dec.).

Phillips, D. C. 1966. The three-dimensional structure of an enzyme molecule. **215**:78-90 (Nov.).

Rustad, R. C. 1961. Pinocytosis. **204**:121-130 (April).

Solomon, A. K. 1960. Pores in the cell membrane. **203**:146-156 (Dec.).

Solomon, A. K. 1962. Pumps in the living cell. **207**:100-108 (Aug.).

Cerianthus membranaceus, *an organism exhibiting radial symmetry.*

CHAPTER 5

ARCHITECTURAL PATTERN OF AN ANIMAL

The architecture of most animals conforms to a well-defined plan. The basic uniformity of biologic organization derives from the supposed common ancestry of animals and from their basic cellular construction. Despite the vast differences of structural complexity of animals ranging from the simplest protozoa to man, all share an intrinsic material design and fundamental functional plan. In the last analysis, whatever unity we see among animal organization is explained by just one fact: all animals live on Earth, a planet bearing a unique set of physical properties that has molded the nature of life on it. Yet even as the biologist takes some comfort from his belief in the basic unity of life—and many areas of biologic endeavor, such as general physiology and molecular biology are grounded on this faith—we must admit that animals exhibit an incredible diversity of specific structural and functional adaptations. Animals are shaped by their particular habitat: its physical nature and its biotic community. Such adaptations are not *caused* by the environment; rather they arose when, by natural selection, beneficial variations were preserved. This concept will be clearer when evolution is discussed in a later chapter. In this chapter we deal with the fundamental uniformity of animal structure that is recognizable despite the numerous specializations, both subtle and prominent, that have modified the basic plan.

■ GRADES OF ANIMAL COMPLEXITY

How has increased animal complexity, so evident in animal phylogeny, arisen? The **acellular** or single-celled forms are complete organisms and carry on all the functions of higher forms. Within the confines of their cell, they often show complicated organization and division of labor, such as skeletal elements, locomotor devices, fibrils, and beginnings of sense organs. The **metazoan or multicellular animal**, on the other hand, has cells differentiated into tissues and organs that are specialized for specific functions. The metazoan cell is not the equivalent of a protozoan cell; it is only a specialized part of the whole organism and is incapable of independent existence.

Complexity has thus increased as animals became larger. It has increased as the result of specialization and division of labor within the body tissues. An ameba can move without muscles, digest food without an alimentary canal, and breathe without gills or lungs. More advanced animals have specialized organs for these functions. An alimentary canal is not a mere epithelial tube for secretion and absorption but has muscles to manipulate and nerves to control it. Specialization and division of labor have many advantages for adjustments to specific niches, but they require complicated machinery and more energy.

Does this mean that life is progressing toward higher and higher types, such as man? In the evolutionary picture the first animals were small and relatively simple, but there is no reason to believe that more recent animals are better adjusted to their environments than were their ancient ancestors. Nor is there any evidence that evolution has led in man's direction. We cannot successfully argue that man is better adapted to his environment than are a gypsy moth, a yellow perch, and a common starling to theirs.

■ ORGANIZATION OF THE BODY

The body consists of three different elements—body cells, extracellular structural elements, and body fluids. The body fluids of vertebrates include the blood, tissue or intercellular fluid between the cells, and lymph within the lymphatics. (One should note the potential confusion in the terms **intercellular,** meaning "**between** cells"; **intracellular,** meaning "**within** cells"; and **extracellular,** meaning "**outside** cells." Intercellular and extracellular are normally used synonymously.) An endothelial barrier separates the tissue fluid outside the lymphatics and the lymph within them. There is an exchange of materials between the blood within its cavities and channels and the intercellular (tissue) fluid between the cells.

Body fluids fill continuous spaces and are responsible for diffusion and convection.

Extracellular (or intercellular) substance is the material that lies between the cells. It affords mechanical stability, protection, storage, and exchange agents. It is mainly responsible for the firmness of tissues and gives support to the cells. Two types of intercellular tissue are recognized—formed and amorphous. The formed type includes collagen (white fibrous tissue). This is the most abundant protein in the animal kingdom and makes up the major part of the fibrous constituent of the skin, tendon, ligaments, cartilage, and bone. Elastin, which gives elasticity to the tissues, also belongs to the formed type. Amorphous intercellular substance (ground substance) is composed of mucopolysaccharides arranged in long chain polymers.

Principle of individuality

All organisms, however simple, are composed of **units** with coordinated interreactions. The smallest units capable of independent existence are the **cells.** Among some biologists the **gene** is considered the chief biologic unit; but many other units are recognized in both acellular and many-celled forms. Protozoans contain such units as the **contractile vacuole, nucleus,** and other organelles. In the metazoans units of different levels are **tissues, organs,** and **systems.** In some phyla **metamerism,** or the serially repeated division of the body into successive segments (as in the earthworm, for example) occurs.

In such forms each segment represents another kind of structural unit. In **polymorphism,** there is more than one form of the same species; these may consist of united individuals (Portuguese man-of-war) or they may be separate (certain ant colonies). Individuality is difficult to define because there are many gradations between separate organic entities and those of colonies whose members are attached together in some way. The organism is a historic entity that is made up of many stages in a life cycle, some of which may be very different, for example, the tadpole and frog, the caterpillar and butterfly, etc. Is each stage a separate individual or should the combined stages of a life cycle be considered an individual?

Within the cell itself are many other units. Whether there is an ultimate living unit, biologists do not know.

Grades of organization*

An animal is an organization of units differentiated and integrated for carrying on the life processes, but this organization goes from one level to another as we ascend the evolutionary path.

Protoplasmic grade of organization. This type is found in protozoans and other acellular forms. All activities of this level are confined to the one mass called the cell. Here the protoplasm is differentiated into specialized organelles that are capable of carrying on definite functions.

Cellular grade of organization. Here aggregations of cells are differentiated, involving division of labor. Some cells are concerned with reproduction and others with nutrition. The cells have little tendency to become organized into tissues. Some protozoan colonial forms having somatic and reproductive cells might be placed in this category. Many authorities also place the sponges at this level.

Cell-tissue grade of organization. A step beyond the preceding is the aggregation of similar cells into definite patterns of layers, thus becoming a tissue. Sponges are considered by some authorities to belong to this grade, although the jellyfish are usually referred to as the beginning of the tissue plan. Both groups are still largely of the cellular grade of organization because most of the cells are scattered and not organized into tissues. An excellent example of a tissue in coelenterates is the **nerve net,** in which the nerve cells and their processes form a definite tissue structure, with the function of coordination.

Tissue-organ grade of organization. The aggregation of tissues into organs is a further step in advancement. Organs are usually made up of more than one kind of tissue and have a more specialized function than tissues. The first appearance of this level is in the flatworms (Platyhelminthes), in which there are a number of well-defined organs such as eyespots, proboscis, and reproductive organs. In fact, the reproductive organs are well organized into a reproductive system.

Organ-system grade of organization. When organs work together to perform some function we have the highest level of organization—the organ system. The systems are associated with the basic bodily functions—circulation, respiration, digestion, etc. Typical of all the higher forms, this type of organization is first seen in the nemertean worms in which a com-

*Refer to p. 18, Principle 9.

plete digestive system, separate and distinct from the circulatory system, is present.

■ PRELIMINARY SURVEY OF ANIMAL EMBRYOLOGY

Embryology is the study of the progressive growth and differentiation that occurs during the early development of an organism. A brief summary of embryology is necessary for understanding the pattern of an animal and also for understanding some of the basic concepts used in describing animal groups and their classification. The developmental process and its control will be discussed in greater detail in Chapter 32.

General pattern of development

All animals have a characteristic life history. Many protozoans such as the ameba are potentially immortal and come from an ancestral line that has never experienced natural death from old age because their method of asexual binary reproduction is simply the dividing of the parent organism into two daughter cells, each essentially a continuation of the parent. Early in the life history of all metazoans, however, there occurs a differentiation of the germ cell from the body or soma cells. It is the uniting of the germ cells (male sperm and female ova) that gives rise to a new generation (sexual reproduction), while the body (soma) cells die. The real life history of a metazoan starts with the union of an ovum (egg) with a spermatozoan, a process called fertilization.

The fertilized egg, called a **zygote,** is really a one-celled organism, and from it develops a complete animal by the process of **differentiation.** How this occurs is only partly known. All the information necessary for development is contained within the fertilized egg, principally in the genes of the egg's nucleus. The actual blueprint for differentiation is coded within the DNA molecules of the genes. The heredity of the organism stabilizes the pattern of development; the variation that makes evolutionary changes possible is contributed by the gene segregation that occurs during the formation of gametes and recombination of maternal and paternal genes at fertilization.

During development of every species, certain basic characteristics of the phylum appear before the specific qualities of the species appear. Such basic qualities may be symmetry, a longitudinal axis, and if a vertebrate, a notochord, dorsal tubular nerve cord, three major pairs of sensory organs, paired pharyngeal pouches, a chambered heart, a liver, paired kidneys, paired pectoral and pelvic appendages, etc. There are overlappings of some of these characteristics in both vertebrates and invertebrates. As development continues, the individual acquires the morphologic characters of its lower taxa (class, order, family, and genera) and finally of his own species. This indicates that development proceeds from the general to the specific in gross morphologic characters. Species-specific characteristics as far as minor features are concerned, on the other hand, may appear quite early.

The stages of embryogenesis are as follows:

1. **Fertilization.** The activation of the egg by a sperm in biparental reproduction. The union of these male and female gametes forms a **zygote;** this is the starting point for development.

2. **Cleavage and blastulation.** The division of the zygote into smaller and smaller cells (cleavage) to form a hollow ball of tiny cells (blastula).

3. **Gastrulation.** The sorting out of cells of the blastula into layers (ectoderm, mesoderm, endoderm) that become **committed** to the formation of future body organs.

4. **Differentiation.** The formation of body organs and tissues, which take on their specialized functions. The basic body plan of the animal becomes established.

5. **Growth.** Increase in size of the animal by cell division or cell enlargement. Growth depends on the intake of food to supply material for the synthesis of protoplasm.

Although the embryonic development begins with fertilization, this event must obviously be preceded by the preparatory stages of egg and sperm development.

Types of eggs

Eggs may be classified as follows with respect to yolk distribution (Fig. 5-1):

The **isolecithal egg** (also called alecithal or homolecithal) is small, and the small amount of yolk (deutoplasm) and cytoplasm is uniformly distributed through the egg, with the nucleus near the center. Cleavage is usually **holoblastic** (total) and the cells (blastomeres) are nearly equal in size. Such eggs are found in the protochordates (such as sea squirts and

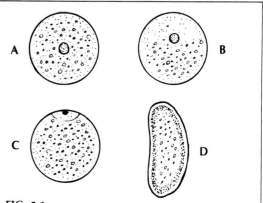

FIG. 5-1
Types of eggs. **A,** Isolecithal (echinoderms and amphioxus); small yolk evenly distributed. **B,** Telolecithal with holoblastic cleavage (amphibians and bony fishes); yolk concentrated at vegetal pole. **C,** Telolecithal with meroblastic cleavage (birds and reptiles); protoplasm concentrated in germinal disk at animal pole. **D,** Centrolecithal (insects and other arthropods); protoplasm centered but migrates out at cleavage, leaving yolk centered.

Amphioxus) and the echinoderms (sea stars and sea urchins).

In the **telolecithal egg** the large amount of yolk (50% to 90%) tends to be concentrated toward one pole (the vegetal pole) where metabolism is lower. The protoplasm and nucleus are found mainly at the opposite (animal) pole where metabolic activity is greater. There are two classes of this type: (1) Yolked eggs with **holoblastic** cleavage, in which the later cleavages produce unequal cells—small cells, or micromeres, at the animal pole and large cells, or macromeres, at the vegetal pole. Amphibians and bony fish have this kind of egg. (2) Yolked eggs (megalecithal) with **meroblastic** (partial or discoidal) cleavage, in which the small amount of protoplasm is concentrated at the animal pole in the germinal disk, or blastoderm, where cleavage occurs. These eggs are usually large, contain a great deal of albumin (egg white) derived from the oviducts, and have a hard or soft shell. Bird and reptile eggs are good examples.

In the **centrolecithal egg** the nucleus and the surrounding layer of protoplasm are at first in the center of the egg, but as cleavage occurs, most of the nucleated masses of cytoplasm migrate to the periphery and form a cellular layer (blastoderm), leaving the

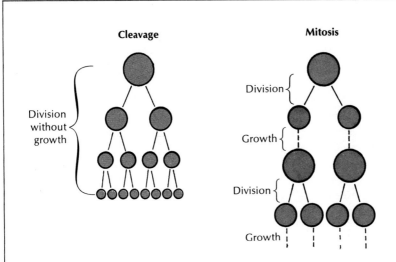

FIG. 5-2

Comparison of cell division in cleavage and mitosis. In cleavage, cells become progressively smaller with no interdivisional growth; in mitosis a period of growth follows each division so that daughter cells become as large as parent cell. Both types retain diploid hereditary constitution.

yolk in the center of the egg. From the blastoderm the embryo develops. These eggs are characteristic of the arthropods.

In addition to the plasma membrane universally present in eggs, there are in the various animal groups many kinds of protective membranes or envelopes around the eggs. These include the **vitelline,** or fertilization membrane, secreted by the egg, the **zona pellucida** formed by ovarian follicle cells of mammals, the **egg jelly** from the oviducts of bony fish and amphibians, and the **chitinous shell** (chorion) from the ovarian tubules of insects.

Fertilization and formation of the zygote

The fusion of the pronuclei of sperm and egg is really the starting point of embryonic development. Specifically, the process restores the diploid number of chromosomes, combines the maternal and paternal genetic traits, and activates the egg to develop. This will be treated in more detail in Chapter 32.

Cleavage and blastulation

The unicellular zygote now begins to divide, first into two cells, those two into four cells, those four into eight. Repeated again and again, these cell divi-

sions soon convert the zygote into a ball of cells. This process, called **cleavage,** occurs by mitosis. But unlike ordinary body-cell mitosis, there is no true growth and no increase in protoplasmic mass (Fig. 5-2). With each subsequent division, the cells are reduced in size by one half. The cleavage process converts a single, very large, unwieldy egg into many small, more maneuverable, ordinary-sized cells.

Cleavage patterns are much affected by the amount of yolk in the egg. In eggs with very little yolk, such as those of mammals, the cytoplasm is uniformly distributed through the egg, and the nucleus is in, or near, the egg center. In such eggs, cleavage is complete **(holoblastic),** and the daughter cells formed at each division are of approximately equal size. The eggs of frogs and other amphibians are richly supplied with yolk that tends to be massed in the so-called vegetal pole of the egg. The opposite, or animal, pole contains the egg cytoplasm and the nucleus. The early cleavage divisions tend to be displaced toward the animal pole because the mass of relatively inactive yolk in the vegetal pole retards the rate of cleavage in that region (Fig. 5-3). Birds and reptiles produce the largest eggs of all animals. Nearly all of this comparatively enormous size is storage food—the

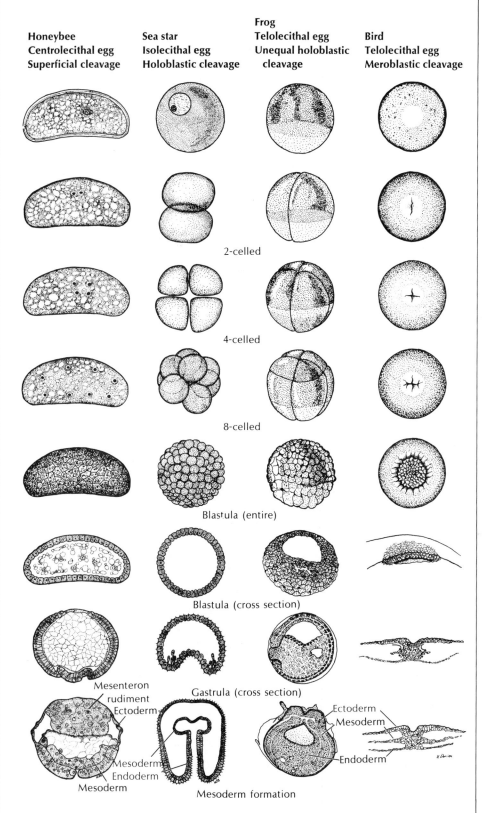

**Honeybee
Centrolecithal egg
Superficial cleavage**

**Sea star
Isolecithal egg
Holoblastic cleavage**

**Frog
Telolecithal egg
Unequal holoblastic cleavage**

**Bird
Telolecithal egg
Meroblastic cleavage**

2-celled

4-celled

8-celled

Blastula (entire)

Blastula (cross section)

Gastrula (cross section)

Mesenteron rudiment
Ectoderm
Mesoderm
Endoderm
Mesoderm

Ectoderm
Mesoderm
Endoderm

Mesoderm formation

FIG. 5-3
Examples of different types of cleavage.

part of the egg we commonly call yolk and the investment of albumin, or egg "white." The active cytoplasm, containing the nucleus, is but a tiny disk resting on top of the ball of yolk. Cleavage (called **meroblastic**) is confined to this area of cytoplasm, since the cell divisions cannot possibly cut through the vast bulk of inert yolk (Fig. 5-3).

Radial and spiral cleavage. Holoblastic cleavage is also classified on the basis of radial and spiral types. In **radial cleavage** the cleavage planes are symmetrical to the polar axis and produce tiers, or layers, of cells on top of each other. Radial cleavage is **indeterminate,** that is, there is no definite relation between the position of any blastomere and the specific tissue it will form in the embryo. Very early blastomeres, if separated, may each be capable of giving rise to a complete embryo. In **spiral cleavage** the cleavage planes are diagonal to the polar axis and produce alternate clockwise and counterclockwise quartets of unequal cells around the axis of polarity. Spiral cleavage is **determinate** and the fate of each blastomere can be foretold. All blastomeres must be present to form a whole embryo. With few exceptions, radial cleavage is found in the deuterostomes and spiral cleavage in the protostomes. See Chapter 32 for further discussion.

Cleavage, however modified by the presence of varying amounts of yolk, results in a cluster of cells, called a **blastula** (Fig. 5-3). In many animals, such as the amphibian and mammalian embryos, the cells rearrange themselves around a central fluid-filled cavity called the **blastocoel.** We have seen that cleavage has resulted in the proliferation of several thousand maneuverable cells poised for further development. There has been a great increase in total DNA content, since each of the many daughter cell nuclei, by chromosomal replication at mitosis, contains as much DNA as the original zygote nucleus.

Gastrulation

Gastrulation is a regrouping process in which new and important cell associations are formed. Up to this point the embryo has divided itself up into a multicellular complex; the cytoplasm of these numerous cells is nearly in the same position it was in the original undivided egg. In other words, there has been no significant movement or displacement of the cells from their place of origin. As gastrulation begins, the cells become rearranged in an orderly way by morphogenetic movements.

In **amphioxus** (Fig. 5-4) the blastoderm of the vegetal pole bends inward so that the whole embryo becomes converted into a double-walled, cup-shaped structure. The new cavity formed in the **archenteron** (primitive gut), with an opening, the **blastopore,** to the outside. In amphibians, the type of gastrulation that occurs in amphioxus is impossible. The cleavage divisions at the lower or vegetal pole are slowed down by the inert yolk so that the resulting blastula consists of many small cells at the animal pole and a few large cells at the vegetal pole. Cells on the surface begin to sink inward (invaginate) at one point, the **blastopore.** Through the curved groove of the blastopore, surface cells move as a sheet to the interior to form a two-layered embryo (Fig. 5-5). A rod-

FIG. 5-4
Holoblastic cleavage and gastrulation in isolecithal type of egg, such as that of amphioxus.

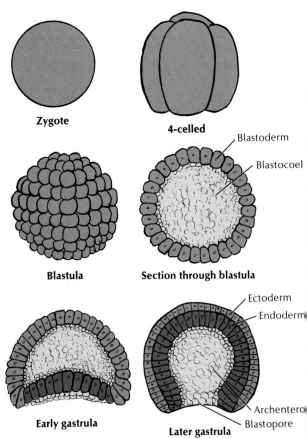

Zygote

4-celled

Blastoderm

Blastocoel

Blastula

Section through blastula

Ectoderm

Endoderm

Early gastrula

Later gastrula

Archenteron

Blastopore

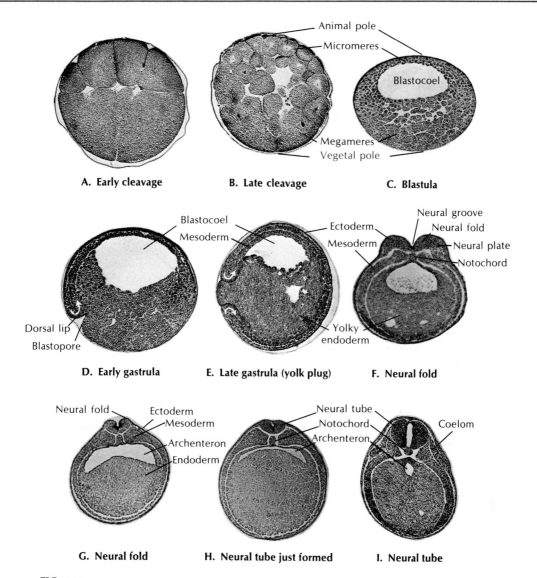

FIG. 5-5

Early embryology of frog. All views are sections through developing embryos. **A,** Beginning of fourth cleavage division, which will result in 16-cell stage. **B to C,** Unequal division of cells results in hollow blastula, small cells at animal pole, and yolk cells at vegetal pole. **D,** Gastrulation begins; multiplying animal cells overgrow vegetal cells and turn in, forming dorsal lip of blastopore. **E,** Involution continues until vegetal hemisphere is covered by ectodermal overgrowth (epiboly) leaving plug of yolk at blastopore. Mesoderm forms inside ectoderm, proliferated from dorsal lip of blastopore. Blastopore will become external opening of gut. **F,** Gastrulation completed. Germ layers are present, with gut forming in endoderm. Dorsal ectoderm thickens to form neural plate, which begins to fold. Notochord differentiates from mesoderm. **G to H,** Neural fold closes to form neural tube. **I,** Mesoderm spreads between ectoderm and endoderm and splits to form coelomic cavity. Triploblastic body plan is now established. Neural tube will become brain and spinal cord; other organs and systems will differentiate.

like **notochord** forms at this time, growing forward to run lengthwise along the dorsal side of the embryo. Continued rearrangements of cells form a third layer; these three layers, called **germ layers,** are the primary structural layers that play crucial roles in the further differentiation of the embryo. The outer layer, or **ectoderm,** will give rise to the nervous system and outer epithelium of the body. The middle layer, or **mesoderm,** will give rise to the circulatory, skeletal, and muscular structures. The inner layer, or **endoderm,** will develop into the digestive tube and its associated structures.

In certain simple metazoa, only two germ layers are formed, the endoderm and ectoderm. These animals are called **diploblastic.** In all higher forms, the mesoderm also appears, either from pouches of the archenteron, or from other cells. This three-layered condition is called **triploblastic.**

Formation of the coelom

The coelom, or true body cavity that contains the viscera, may be formed by one of two methods— **schizocoelous** or **enterocoelous** (Fig. 5-6). In schizocoelous formation the coelom arises from the splitting of mesodermal bands that originate from the blastopore region and grow between the ectoderm and endoderm; in enterocoelous formation the coelom comes from the fusion and expansion of outfolding pouches of the archenteron, or primitive gut.

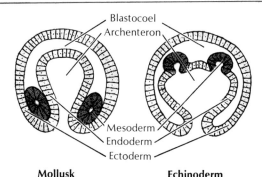

Blastocoel
Archenteron

Mesoderm
Endoderm
Ectoderm

Mollusk   **Echinoderm**
Schizocoelous   **Enterocoelous**

FIG. 5-6
Two types of mesoderm and coelom formation. Schizocoelous, in which mesoderm originates from wall of archenteron near lips of blastopore, and enterocoelous, in which mesoderm and coelom develop from endodermal pouches.

Since coelom formation occurs very early in embryonic development, the appearance of two quite different methods of formation among animals is believed to signal a fundamental division in metazoan evolution. The **deuterostome** division of the Metazoa (echinoderms, chaetognaths, protochordates, and chordates) follow the enterocoelous method of coelom formation. The **protostome** division (almost all other Metazoa) follow the schizocoelous method. There are, of course, other characteristics that distinguish these two phylogenetic divisions of bilateral animals, which are discussed later, but coelom formation is perhaps the most prominent difference to appear during early development.

Differentiation

With formation of the three primary germ layers, cells continue to regroup and rearrange themselves into primordial cell masses. These masses will continue to differentiate, leading ultimately to the formation of specific organs and tissues. During this process, cells become increasingly committed to specific directions of differentiation. Cells that previously had the potential to develop into a variety of structures, now lose this diverse potential and assume committments to become, for example, kidney cells, intestinal cells, or brain cells. Differentiation is discussed in more detail in Chapter 32.

Fate of germ layers

Following are some of the vertebrate structures that normally arise from the three germ layers:

Ectoderm
Epidermis of skin
Lining of mouth, anus, nostrils
Sweat and sebaceous glands
Epidermal coverings such as hair, nails, feathers, horns, epidermal scales, enamel of teeth
Nervous system, including sensory parts of eye, nose, ear

Endoderm
Lining of alimentary canal
Lining of respiratory passages and lungs
Secretory parts of liver and pancreas
Thyroid, parathyroid, thymus
Urinary bladder
Lining of urethra

Mesoderm
Skeleton and muscles
Dermis of skin
Dermal scales and dentin

Excretory and reproductive systems
Connective tissue
Blood and blood vessels
Mesenteries
Lining of coelomic cavity

The assignment of early embryonic layers to specific "germ layers" is for the convenience of embryologists and of no concern to the embryo. The idea that each germ layer can give rise to certain tissues and organs only, and to no others, is no longer held. It is now known that the interactions of cells play a part in determining their differentiation in vertebrate animals. The precise position of a cell with relation to other cells and tissues during early development often controls the real fate of that cell. Under some conditions a certain germ layer may give rise to structures normally arising from a different germ layer. Experiments have demonstrated that a presumptive ectodermal structure, when grafted into appropriate regions, will form organs that normally come from a different germ layer. The topographic position, therefore, of cells in their development must play a significant role in their final fate and destiny. Embryologic development must be considered quite flexible. The biologic system cannot be restricted to a definite pattern even though normally it appears to be. Are there any organs or structures that do not come from any germ layer? In a strict sense the germ cells do not originate from any of the three germ layers because they come directly from cells that were segregated in the early cleavage stages of the fertilized egg.

■ DIFFERENTIATION OF TISSUES, ORGANS, AND SYSTEMS

The different types of tissues originate from the basic properties of protoplasm. These properties, and the tissues that are the manifestations of these properties, are irritability and conductivity (nervous), contractility (muscle), supportive and adhesive (connective tissue), absorptive and secretion (epithelial), and fluidity and conductivity (vascular) (Fig. 5-7). This is a surprisingly short list of basic tissue types that are able to meet all requirements of the diverse morphologic patterns of all animals.

During embryonic development, the germ layers differentiate into the five major tissues by a process called **histogenesis.** Their origins are as follows: (1) epithelial tissue, from all three germ layers; (2) connective or supporting tissue, from mesoderm; (3) muscular or contractile tissue, from mesoderm; (4) nervous tissue, from ectoderm; and (5) vascular tissue, from mesoderm. A **tissue** is a group of similar cells (together with associated cell products) specialized for the performance of a common function. The study of tissues is called **histology.** All cells in metazoan animals take part in the formation of tissues. Sometimes the cells of a tissue may be of several kinds, and some tissues have a great many intercellular materials.

EPITHELIAL TISSUE. An **epithelium** is a tissue that covers an external or internal surface. It also includes hollow or solid derivatives from this tissue. Epithelial tissues (Figs. 5-8 and 5-9) are made up of

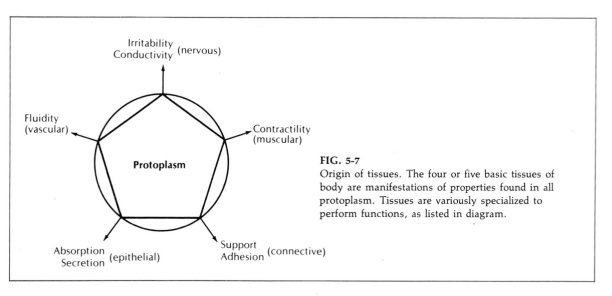

FIG. 5-7
Origin of tissues. The four or five basic tissues of body are manifestations of properties found in all protoplasm. Tissues are variously specialized to perform functions, as listed in diagram.

closely associated cells, with some intercellular material between the cells. Some cells are bound together by **intercellular bridges** of cytoplasm. Most of them have one surface free and the other surface lying on vascular connective tissue. A noncellular **basement membrane** is often attached to the basal cells. Epithelial cells are often modified to produce secretory glands that may be unicellular or multicellular. Some free surfaces (joint cavities, bursae, brain cavity) are not lined with typical epithelium.

Epithelia are classified on the basis of cell form and number of cell layers. **Simple epithelium** is one layer thick (Fig. 5-8), and its cells may be flat or **squamous** (endothelium of blood vessels), short prisms or **cuboidal** (glands and ducts), and tall or **columnar** (stomach and intestine). Any of these three forms of cells may occur in several layers as a **stratified epithelium** (skin, sweat glands, urethra) (Fig. 5-9). Some stratified epithelia can change the number of their cell layers by movement (**transitional**—bladder). Others have cells of different heights and give the appearance of stratified epithelia (**pseudostratified**—trachea). Many epithelia may be **ciliated** at their free surfaces (oviduct). Epithelia serve to protect, secrete, excrete, lubricate, etc.

CONNECTIVE TISSUE. Connective tissues bind together and support other structures. They are so common that the removal of all other tissues from the body would still leave the complete form of the body clearly apparent. Connective tissue is made up of **scattered cells,** a great deal of formed materials such as **fibers** and ground substance (**matrix**) secreted by the cells. There are three types of fibers; white or collagenous (collagen is the most common structural protein in the body), yellow or elastic, and branching or reticular. Connective tissue may be classified in various ways, but all the types fall under either **loose connective tissue** (reticular, areolar, adipose) or **dense connective tissue** (sheaths, ligaments, tendons, cartilage, bone) (Figs. 5-10 and 5-11). Adipose stresses cells, ligaments stress fibers, and cartilage stresses ground substance (matrix).

Connective tissues are derived from the **mesenchyme,** a generalized embryonic tissue that can differentiate also into vascular tissue and smooth muscle. Mesenchyme may also be considered the most primitive connective tissue. When its cells are closely

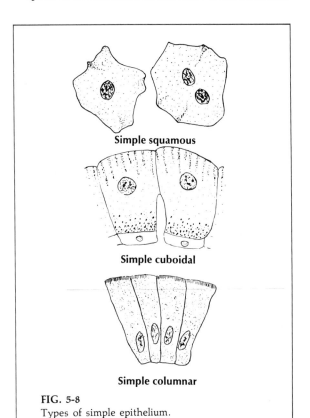

Simple squamous

Simple cuboidal

Simple columnar

FIG. 5-8
Types of simple epithelium.

FIG. 5-9
Types of stratified and transitional epithelial tissue.

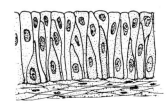

Pseudostratified

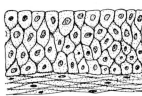

Stratified columnar

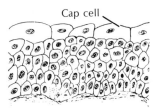

Cap cell

Transitional

packed together, it is called **parenchyma;** when loosely arranged with gelatinous material, it is called **collenchyma.**

MUSCULAR TISSUE. Muscle is the most common tissue in the body of most animals. It is made up of elongated cells or fibers specialized for contraction. It originates (with few exceptions) from the mesoderm, and its unit is the cell or **muscle fiber.** The unspecialized cytoplasm of muscles is called **sarcoplasm,** and the contractile elements within the fiber (cell) are the **myofibrils.** Functionally, muscles are either **voluntary** (under control of will) or **involuntary.** Structurally, they are either **smooth** (fibers unstriped) or **striated** (fibers cross-striped). The three kinds of muscular tissue (Figs. 5-12 and 5-13) are **smooth involuntary** (walls of viscera and walls of blood vessels), **striated involuntary** or **cardiac** (heart), and **striated voluntary** or **skeletal** (limb and trunk). Another type of muscular tissue is made up of the

myoepithelial cell (Fig. 5-14). Myoepithelial cells are found in sweat, salivary, and mammary glands between the epithelium and connective tissue. They extend branching processes around the secretory cells of the glands. Their function may be to squeeze secretions from the acini toward the surface openings of the larger ducts.

NERVOUS TISSUE. Nervous tissue is specialized for irritability and conductivity. The structural and functional unit of the nervous system is the **neuron** (Fig. 5-15), a nerve cell made up of a body containing the nucleus and its processes or fibers. It originates from an embryonic ectodermal cell called a **neuroblast.** (Part of the nervous system of echinoderms may be mesodermal in origin.) In most animals the bodies of nerve cells are restricted to the central nervous system and ganglia, but the fibers may be very long and ramify through the body. Neurons are arranged in chains, and the point of contact between

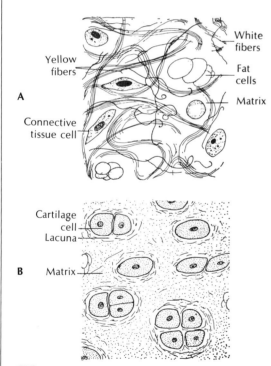

FIG. 5-10

A, Areolar, a type of loose connective tissue. **B,** Hyaline cartilage, most common form of cartilage in body and a type of dense connective tissue.

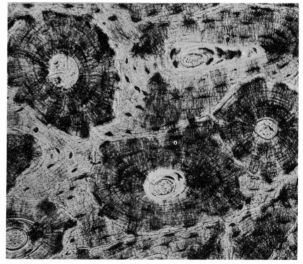

FIG. 5-11

Section of bone, a type of dense connective tissue, showing several cylindric haversian systems typical of bone.

Skeletal

FIG. 5-12
Three kinds of vertebrate muscle fibers, as they appear when viewed with the light microscope.

Cardiac

Smooth

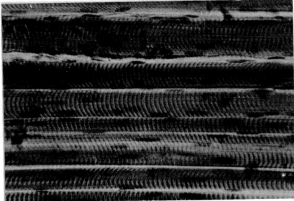

FIG. 5-13
Photomicrograph of skeletal muscle showing several striated fibers lying side by side. (Courtesy J. W. Bamberger, Los Angeles, Calif.)

FIG. 5-14
Two myoepithelial basket cells surrounding salivary secretory cells. Each myoepithelial cell is made up of a central body with long cytoplasmic processes and may be considered a fourth type of muscular tissue. Myoepithelial cells resemble smooth muscle cells.

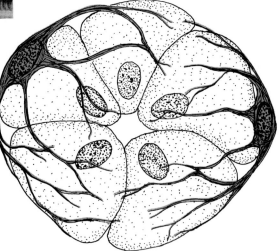

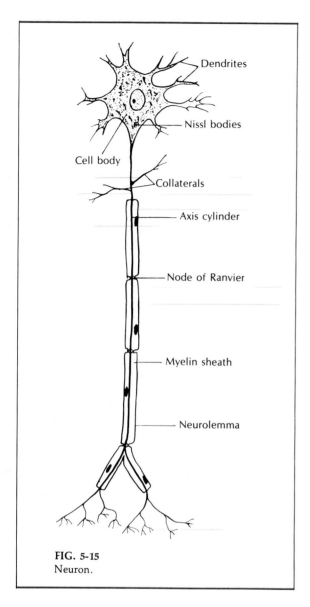

FIG. 5-15
Neuron.

Labels in figure: Dendrites, Nissl bodies, Cell body, Collaterals, Axis cylinder, Node of Ranvier, Myelin sheath, Neurolemma

neurons is a **synapse.** Some of the fibers bear a sheath (medullated or myelin); in others the sheath is absent (nonmedullated).

Sensory neurons are concerned with picking up impulses from sensory **receptors** in the skin or sense organs and transmitting them to nerve centers (brain or spinal cord). **Motor neurons** carry impulses from the nerve centers to muscles or glands (**effectors**) that are thus stimulated to act. **Association neurons** may form various connections between other neurons.

VASCULAR TISSUE. Vascular tissue is a fluid tissue composed of **white blood cells, red blood cells,**

platelets, and a liquid—plasma. Traveling through blood vessels, the blood carries to the tissue cells the materials necessary for their life processes. **Lymph** and tissue fluids, which arise from blood by filtration and serve in the exchange between cells and blood, also belong to vascular tissue.

Moist membranes

Important functional structures of the body are the moist mucous and serous membranes. These membranes are modified from epithelium and connective tissue and are kept moist by either thin watery secretions or thick mucous secretions. A **mucous membrane** is made up of a layer of epithelium (simple or stratified) resting upon a bed of connective tissue. Its surface is kept moist by goblet cells or multicellular glands. Mucous membranes have a wide distribution in the lining of hollow organs that communicate with the outside of the body, such as the alimentary canal, urinary and genital tracts, sinuses, and respiratory passageways. A **serous membrane** consists of a flat mesothelium (sometimes cuboidal or columnar cells in lower vertebrates) that is supported by a thin layer of connective tissue. This membrane is kept moist by a scanty fluid and contains various free cells from the mesothelium and the blood. Serous membranes are usually divided into a parietal portion, which lines the external walls of the cavities, and a visceral portion, which is reflected over the exposed surfaces of organs. The pericardium, pleura, and peritoneum are all serous membranes.

Both mucous and serous membranes perform important functions of lubrication, protection, support, and defense against bacterial infection.

■ ORGANS AND SYSTEMS*

DEFINITIONS. An **organ** is a group of tissues that performs a certain function. In higher forms many organs may have most of the various tissues in their makeup (Fig. 5-16). The heart has (Fig. 5-17) epithelial tissue for covering and lining, connective tissue for framework, muscular walls for contraction, nervous elements for coordination, and vascular tissue for transportation.

All organs have a characteristic structural plan. Usually one tissue carries the burden of the organ's chief function, as muscle does in the heart; the other

*Refer to p. 18, Principle 9.

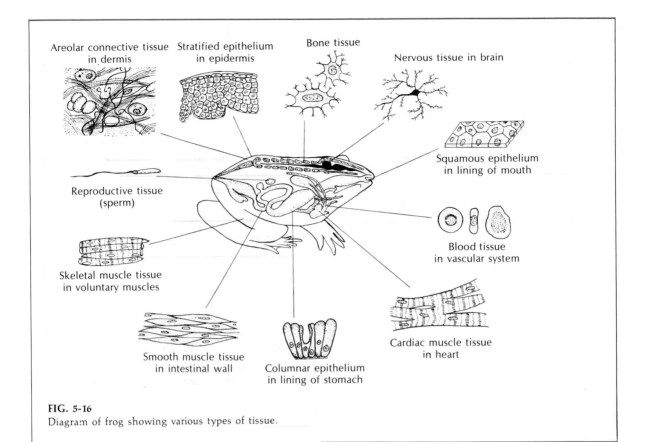

Areolar connective tissue in dermis

Stratified epithelium in epidermis

Bone tissue

Nervous tissue in brain

Squamous epithelium in lining of mouth

Reproductive tissue (sperm)

Blood tissue in vascular system

Skeletal muscle tissue in voluntary muscles

Smooth muscle tissue in intestinal wall

Columnar epithelium in lining of stomach

Cardiac muscle tissue in heart

FIG. 5-16
Diagram of frog showing various types of tissue.

FIG. 5-17
Heart showing various types of tissue in its structure.

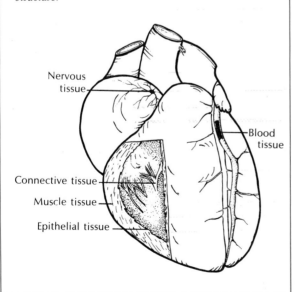

Nervous tissue

Blood tissue

Connective tissue

Muscle tissue

Epithelial tissue

tissues perform supportive roles. The chief functional cells of an organ are called its **parenchyma;** the supporting tissues are its **stroma.** For instance, in the pancreas the secreting cells are the parenchyma; the capsule and connective tissue framework represent the stroma.

Organs are, in turn, associated in groups to form **systems,** with each system concerned with one of the basic functions. The higher metazoans have 11 organ systems: skeletal, muscular, integumentary, digestive, respiratory, circulatory, excretory, nervous, special sensory, endocrine, and reproductive. However, all living organisms perform the same basic functions. The need for procuring and utilizing food and for movement, protection, perception, and reproduction are as important to an ameba, a clam, or an insect as to a man. Obviously, because of differences in size, structure, and environment, each must meet these problems in a different manner.

BODY CAVITY, OR COELOM

The coelom is the true body cavity. It is the space between the digestive tube and the outer body wall; it contains the visceral organs. Not all animals have a coelom, for example, the jellyfish and flatworms. Animals that do have a coelom have a "tube-within-a-tube" arrangement (Fig. 5-18). The outer tube is the body wall, the inner tube is the digestive tract, and the space between is the coelom. A true coelom develops between two layers of mesoderm—an outer somatic layer and an inner visceral layer—and is lined with mesodermal epithelium called the **peritoneum**.

The coelom is of great significance in animal evo-

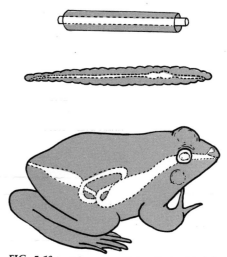

FIG. 5-18
Tube-within-a-tube arrangement.

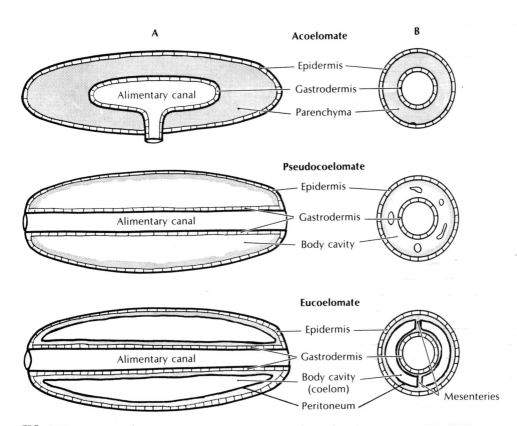

FIG. 5-19
Schematic drawings of types of body cavity organizations. **A,** Longitudinal sections; **B,** cross sections. In acoelomate type, space between epidermis and gastrodermis is filled with mesenchymal parenchyma, which may contain small spaces; in pseudocoelomate type, space between body wall and digestive tract is remnant of blastocoel and is not lined with mesodermal peritoneum; and in eucoelomate, or true coelomate, type, body cavity is lined with mesodermal peritoneum, which also covers digestive tract. Mesenteries are made up to two peritoneal layers.

lution because it provides spaces for visceral organs, permits greater size and complexity by exposing more cells to surface exchange, and contributes directly to the development of certain systems such as excretory, reproductive, and muscular. This fluid-filled cavity also serves as a hydrostatic fluid skeleton in some primitive forms, aiding them in movement, rapid change of shape, or burrowing. One has only to compare the slow, gliding locomotion of a planarian worm (without a coelom) with the rapid and nimble swimming movements of a polychete annelid worm (with a coelom) to see this advantage.

In roundworms and some others the body cavity is not lined with mesoderm and so is given the name **pseudocoel** (Fig. 5-19).

■ BODY PLAN AND SYMMETRY

Convenient terms for locating regions of an animal body are **anterior** for the head end, **posterior** for the opposite or tail end, **dorsal** for the back side, and **ventral** for the front or belly side. **Medial** refers to the midline of the body, **lateral** to the sides. **Distal** parts are farther from a point of reference; **proximal** parts are nearer. **Pectoral** refers to the chest region or the area supporting the forelegs, and **pelvic** refers to the hip region or the area supporting the hind legs.

Symmetry refers to balanced proportions, or the correspondence in size and shape of parts on opposite sides of a median plane (Fig. 5-20). A symmetrical body can be cut into two equivalent or mirrored halves. **Spherical symmetry** is found chiefly in protozoans. A sphere can be divided equally by any plane that passes through the center. These are usually floating or rolling forms. **Radial symmetry** applies to a few sponges, most coelenterates, and adult echinoderms that, like bottles or wheels, can be divided into similar halves by any plane passing through the longitudinal axis. In such forms one end of the longitudinal axis is usually the mouth or oral end and the other is the aboral end. A variant form is **biradial symmetry** in which, because of the presence of some part that is single or paired rather than radial, only one or two planes through the longitudinal axis divides the form into mirrored halves. Examples are the comb jellies, which are basically radial forms but have a bilateral pair of arms or tentacles, and sea stars, which have a sieve plate located between two of their radially arranged arms. Radial animals are usually

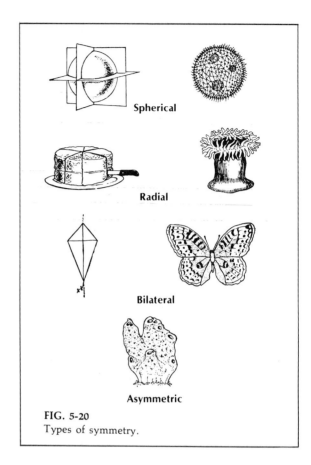

FIG. 5-20
Types of symmetry.

well suited to a sessile existence, for they respond to their environment equally well on all sides.

Most other animals have **bilaterally symmetrical bodies** (Fig. 5-21). In these types only a **sagittal plane** divides the animal into equivalent right and left halves. A sagittal plane passes through the anteroposterior axis and through the dorsoventral axis. A **frontal plane** divides a bilateral body into dorsal and ventral halves by running through the anteroposterior axis and the right-left axis at right angles to the sagittal plane. A **transverse plane** would cut through a dorsoventral and a right-left axis at right angles to both the sagittal and frontal planes and would result in anterior and posterior portions. Along with bilateral symmetry, we find differentiation of a head end, which has greater perception than the tail end. These forms are suited for forward movement, which is an asset in the search for food and protection

Most of the sponges lack any symmetry at all and are called **asymmetrical.**

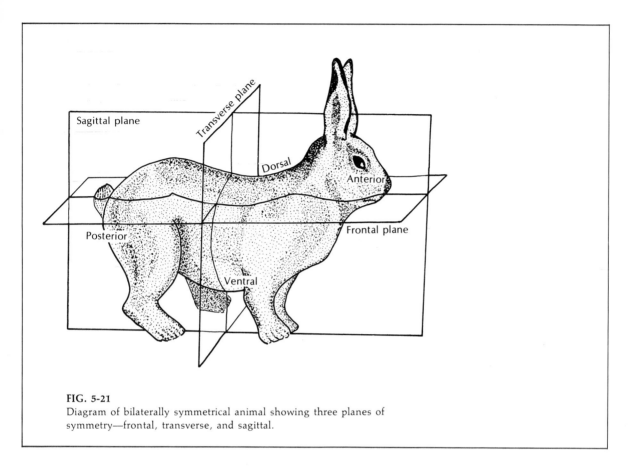

FIG. 5-21
Diagram of bilaterally symmetrical animal showing three planes of
symmetry—frontal, transverse, and sagittal.

■ CEPHALIZATION AND POLARITY

The differentiation of a definite head end is called
cephalization and is found chiefly in bilaterally symmetrical animals. The concentration of nervous tissue
and sense organs in the head bestows obvious advantages to an animal moving through its environment head first. This is the the most efficient positioning of instruments for sensing the environment
and responding to it. Usually the mouth of the animal
is located on the head as well, since so much of an
animal's activity is concerned with procuring food.
Cephalization is always accompanied by a differentiation along an anteroposterior axis (**polarity**). Polarity usually involves gradients of activities between
limits, such as between anterior and posterior ends
(see axial gradient theory on p. 22, Principle 30).

■ METAMERISM

Metamerism is the linear repetition of similar body
segments. Each segment is called a **metamere**, or
somite. In forms such as the earthworm and other
annelids, in which metamerism is best represented,
the segmental arrangement includes both external
and internal structures of several systems. There is
repetition of muscles, blood vessels, nerves, and the
setae of locomotion. Some other organs, such as
those of sex, are repeated in only a few somites. In
higher animals much of the segmental arrangement
has become obscure.

When the somites are similar, as in the earthworm,
the condition is called **homonomous metamerism;** if
they are dissimilar, as in the lobster and insect, it
is called **heteronomous metamerism.**

Segmentation often appears in embryonic stages
but becomes obscure in the adult. For example, muscles of vertebrate animals show a decided metamerism in the embryo but little in the adult. On the other
hand, the arrangement of the vertebrae is clearly
metameric in adult vertebrates.

True metamerism is found in only three phyla:

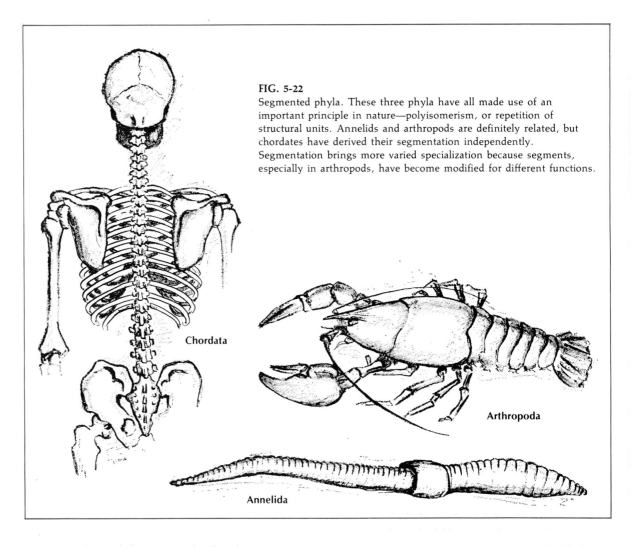

FIG. 5-22
Segmented phyla. These three phyla have all made use of an important principle in nature—polyisomerism, or repetition of structural units. Annelids and arthropods are definitely related, but chordates have derived their segmentation independently. Segmentation brings more varied specialization because segments, especially in arthropods, have become modified for different functions.

Chordata

Arthropoda

Annelida

Annelida, Arthropoda, and Chordata (Fig. 5-22), although superficial segmentation of the ectoderm and the body wall may be found among many diverse groups of animals.

TAGMATIZATION (TAGMOSIS). In the more advanced annelids and arthropods, the segments may be united into functional groups of two or more somites, each group being structurally separated from other groups and specialized to perform a certain function for the whole animal. The primitive annelids and arthropods however adhere to the primitive pattern of having a serial succession of identical somites. As will be seen later, tagmosis is more common among the arthropods than in the annelids because the soft-bodied annelid requires the numerous fluid-filled body somites in its locomotion. A good example

of tagmosis is the familiar insect that has three tagmata: head, thorax, and abdomen. Each tagma in this animal represents a grouping of a certain number of segments and each is specialized for a specific function or functions.

■ HOMOLOGY AND ANALOGY*

In comparative studies of animals the concepts of **homology** (similarity in origin) and **analogy** (similarity in function of appearance but not in origin) are frequently used to express relationship of animals, the basic patterns of morphology, and the way these patterns have varied. Two main types of homology are recognized by embryologists—that within a single

*Refer to p. 21, Principle 25.

individual, or **general homology,** and that dealing with structures of two different individuals, or **special homology.** Examples of general homology are the ribs or vertebrae, the pattern of which is often called serial homology, and the resemblance between the forelimb and hind limb or the left and right hand, bilateral homology. Special homology refers to correspondence between parts of different animals such as the arm of a man, the wing of a bird, the foreleg of a dog, and the pectoral fin of a fish—all of which are homologous in a broad meaning of the term.

However, homology is a relative term. If one always insists on criteria of similarity of structure *and* development for strict homology as is sometimes done, then many structures that are called homologous do not meet these requirements. For instance, the wing of a bird and the wing of a bat are homologous only insofar as both are derived from pentadactyl forelimbs. They are analogous, however, for they have the same function. In examining homologous structures it is best to keep in mind the question: To what extent or degree do they share a common origin? Of course, even this question contains a nebulous element, since all animals are related, having presumably descended from a common ancestor in the long distant past. Homology implies a close evolutionary kinship. Convergent evolution, or the independent origin of two similar structures, is caused by independent mutations that are favored under similar environments. Horns have appeared independently many times in mammals, but should they be considered homologous? Perhaps the best criterion for homology would be homologous genes, but this is impossibly strict at our present state of knowledge. Embryologists deal with phenotypes, or the visible expressions, which is not the same thing as the genotypes, or hereditary constitution.

Analogy denotes similarity of function. The wing of a bird is analogous to the wing of a butterfly because they have the same function, but they are not homologous because they have totally different origins. The term "analogy" in its usage may or may not denote similarity of origin.

The principle of homology, although difficult to define strictly, has a wide application in zoology. It is used as an argument for evolution because it is based on the idea of inheritance from common ancestors. Homology is also important in classifying animals. As more is known about animals, chemical identity may become an important aspect of homology among organisms. A plan of archetype with which the structures of animals are compared may thus be formulated, not only from gross morphologic structures, but also from their biochemical resemblances.

■ SIZE OF ORGANISM

Different species of animals show an enormous range in size, from the tiny protozoan weighing a fraction of a milligram to the whale weighing more than 100 tons. Many animals such as birds and mammals have definite age limits to growth, but in reptiles and fishes growth may continue throughout life, although at a reduced pace. Mammals vary from forms as small as shrews to those as large as whales. One may get the impression that evolution has progressed from the small to the large, but this has not always been the case.

It is a general principle that every organism is distinguished by a characteristic size. Of course, many factors may modify the dimensions of an individual animal, such as nutrition and other environmental factors.

When unicellular animals reach a certain size, they divide, for a size limitation is imposed upon the animal by its surface-volume ratio. Every cell depends on its surface membrane for the exchange of materials with the surrounding environment. The volume increases as the cube of the radius; the surface, as the square of the radius. Since the volume increases much faster than the surface, as the size of the cell increases, the surface membrane soon becomes inadequate to handle the increased demand for food, gas exchange, and waste removal. Protozoans can solve the problem by simply dividing when they reach a critical size. Increasing body size among animals is possible only by increasing the number of cells, which is the solution adopted by all the metazoams.

The ratio between the nucleus and cytoplasm may be a controlling factor in determining the size of cells, for these two units must be in a certain balance for the efficient functioning of the cell. The importance of cell size as a regulative force in metabolism is strikingly demonstrated by the fact that the volume is very much the same for any cell type and is independent of the animal's size. The cells found in a mouse are about the same size as those found in the largest mammals.

References

Suggested general readings

Arey, L. B. 1968. Human histology. A textbook in outline form, ed. 3. Philadelphia, W. B. Saunders Co. *An excellent summary of human histology.*

Balinski, B. I. 1970. An introduction to embryology, ed. 3. Philadelphia, W. B. Saunders Co. *One of the best embryology texts, although stressing mechanisms of development more than descriptive embryology.*

Bloom, W., and D. W. Fawcett. 1968. A textbook of histology, ed. 9. Philadelphia, W. B. Saunders Co. *Authoritative and superbly illustrated treatise on human histology. Advanced for the beginning student.*

Holmes, R. L. 1965. Living tissue. New York, Pergamon Press, Inc. *This little book is an introduction to functional histology.*

Montagna, W. 1956. The structure and function of skin. New York, Academic Press, Inc. *A good account of the most versatile organ in the body.*

Page, I. H. (editor). 1959. Connective tissue, thrombosis, and atherosclerosis. New York, Academic Press, Inc. *Although mainly clinical, there is much general information about the properties of connective tissue.*

Patt, D. L., and G. R. Patt. 1969. Comparative vertebrate histology. New York, Harper & Row, Publishers.

Patten, B. M. 1964. Foundations of embryology, ed. 2. New York, McGraw-Hill Book Co. *Well-written, clearly illustrated treatment of descriptive embryology.*

Romer, A. S. 1962. The vertebrate body, ed. 3. Philadelphia, W. B. Saunders Co. *An exemplary comparative anatomy text with good descriptions of body tissues and organs.*

Selected *Scientific American* articles

Fraser, R. D. B. 1969. Keratins. **221:**86-96 (Aug.).

Gross, J. 1961. Collagen. **204:**120-130 (May).

McLean, F. C. 1955. Bone. **192:**84-91 (Feb.).

Ross, R., and P. Bornstein. 1971. Elastic fibers in the body. **224:**44-52 (June).

PART TWO

THE DIVERSITY OF ANIMAL LIFE

■ The 1.5 million named species of animals are witness to the enormous diversity of animal life believed to have descended from a common ancestor by means of a single evolutionary process (discussed in Part IV). Classification of animals is based on evolutionary relationships—the only way the diversity of life can be understood. Through progressive adaptation to varied and changing environmental opportunities, animals have expanded into practically every environmental niche that can support life.

The following chapters deal with the methods used in classifying animals, the adaptive radiation that has occurred among them, the relationships of the various groups to each other, the evolutionary advances that have pointed the way toward higher phyla, and the distinctive characteristics of each of the types that make up the varied fauna of our earth.

CHAPTER 6
THE CLASSIFICATION OF ANIMALS

CHAPTER 7
THE ACELLULAR ANIMALS

CHAPTER 8
THE LOWEST METAZOANS

CHAPTER 9
THE RADIATE ANIMALS

CHAPTER 10
THE ACOELOMATE ANIMALS

CHAPTER 11
THE PSEUDOCOELOMATE ANIMALS

CHAPTER 12
THE MOLLUSKS

CHAPTER 13
THE SEGMENTED WORMS

CHAPTER 14
THE ARTHROPODS

CHAPTER 15
THE AQUATIC MANDIBULATES

CHAPTER 16
TERRESTRIAL MANDIBULATES—MYRIAPODS AND INSECTS

CHAPTER 17
THE LESSER PROTOSTOMES

CHAPTER 18
THE LOPHOPHORATE ANIMALS

CHAPTER 19
THE ECHINODERMS

CHAPTER 20
THE LESSER DEUTEROSTOMES

CHAPTER 21
THE CHORDATES: ANCESTRY AND EVOLUTION;
GENERAL CHARACTERISTICS

CHAPTER 22
THE FISHES

CHAPTER 23
THE AMPHIBIANS

CHAPTER 24
THE REPTILES

CHAPTER 25
THE BIRDS

CHAPTER 26
THE MAMMALS

Carolus Linnaeus, the great Swedish naturalist who founded our modern system of classification in the mideighteenth century. This statue of a youthful Linnaeus stands before his home in the old university town of Uppsala, Sweden.

CHAPTER 6

THE CLASSIFICATION OF ANIMALS*

Zoologists have named more than 1.5 million species and thousands more are added to the list each year. Yet some evolutionists believe that present species make up less than 1% of those that have existed in the past. Extinction has been, therefore, a major feature in the development of life. Why has this happened? Inability to adapt to changing environments, invasion of new diseases into a biotic community, failure of heredity to respond to selection pressures, and overspecialization may be possible causes for extinction. One group has often supplanted another group in the long evolution of life. Some lines, within a relatively short time, have given rise to numerous species; others have remained virtually unchanged for millions of years.

One no doubt wonders about this diversity of life. What produced it? Why do some groups have more

*Refer to p. 18, Principles 5 and 9.

variations than others? What has been the role of the environment in this diversity? If our concept about the origin of life is correct, why is not life the same the world over? These are not questions that have simple answers, but the student in his study of the many evolutionary patterns may arrive at some of the answers on his own.

■ CONCEPT OF SPECIES

Although the species concept is of the utmost importance in biology, most biologists do not agree on a single rigid definition that applies to all cases. In early concepts the species was considered a static unit, convenient merely for classification. Evolution later showed that these supposedly arbitrary divisions of living matter were ever changing and that they were in fact integrated on the basis of common ancestry and relationship. Genetics changed the concept further by stressing population equilibrium with the environment and the sharing of common gene (hereditary unit) pools. Morphologic characters have always been given high priority in the diagnosis of species, but careful analysis clearly shows that other characteristics, such as physiologic reactions, behavior patterns, chemical constituents, and ecologic requirements, also afford a basis for classifying organisms into species units.

The term "species characters" usually refers to those attributes that distinguish one species from all others. Criteria commonly used are morphologic characters, reproductive characteristics, range of population, ecologic segregation, genetic composition, and isolating mechanisms. The single property that maintains the integrity of a species more than any other property, especially among biparental organisms, is interbreeding. The members of a species can interbreed freely with each other, produce fertile offspring, and share in a common genetic pool. Interbreeding of different species is usually either physically impossible or produces sterile offspring. There are, of course, exceptions, but such cases are rare compared with the general population of species. Species are thus usually considered genetically closed systems, whereas **races** within a species are open systems and can exchange genes. **A species therefore may be defined as a group of organisms of interbreeding natural populations that are reproductively isolated from other groups and that share in common gene pools** (Mayr and Dobzhansky).

■ CRITERIA USED IN CLASSIFYING ANIMALS

Early taxonomists classified animals into groups according to anatomic and physiologic characters common to the group in question. A species was considered a type of primeval pattern or archetype divinely created. With the development of the evolutionary concept scientists realized that species were not fixed units but have evolved one from another and often flow into each other. Consequently, the criteria of taxonomy used by the systematist underwent a gradual change. At first each species was supposed to have a **type** that was used as a standard, and all individuals were fitted into that type which they most nearly resembled. Thus the **typologic specimen** was by the taxonomists duly labeled and deposited in some prestigious center such as a museum. Anyone classifying a particular group would

always take the pains to compare his specimens with the available typologic specimens. Since variations from the type specimen nearly always occurred, these differences were supposed to be attributable to imperfections during embryonic development and were considered of minor significance.

The typologist has evidently acquired his way of thinking from the propensity of primitive man to classify the diversity of nature into convenient categories (Mayr). From the typologist's viewpoint, evolution would have to occur in leaps and bounds; gradual evolution as we know it would be impossible. His viewpoint is the antithesis of those who think in terms of populations where every individual is unique and will change when placed in a different environment. For the typologist, only the particular *type* is real and any variations that other specimens possess are illusions. The student of populations on the other hand considers the type (average) as an abstraction, and the variations as real.

This typologic method of classifying persisted for a long period of time (and still does to some extent), although the broad basis of evolutionary descent as the foundation of classification was becoming more firmly established. The development of ecologic principles with the emphasis placed on population studies no doubt was responsible for the modern interpretation of the species. Anatomic and physiologic characters must be considered in their total distribution within the individuals of a population. No one pattern can be taken as truly representative nor can be thought of as an idealized abstraction of the characters of an individual. The species must be regarded as a **population** made up of individuals of common descent within a similar environment of a definite region and uniquely set apart from other species by a distinct evolutionary role. The test of determining the species status of an individual animal is to ascertain just where it fits into a definite population group.

■ PROBLEMS OF CLASSIFICATION

There are many complications in the taxonomy of animals, and only a few need be mentioned here. In the first place, the totality of the genes of a species is coadapted with each other, but its gene complex may be polluted by invasion of foreign genes. Such *hybridization* can produce striking changes in systematic relations that are often difficult to figure out. Such hybridization is common among plants and throws confusion into systematics. It is not known how common this process may be among animals, but it may be said that a complex of coadapted genes is a protection against pollution from other genes.

In polytypic species, that is, a species with several subspecies over an extensive geographic range, each population can breed successfully with an adjacent population, but those at the ends of the range are intersterile. The leopard frog, *Rana pipiens,* represents such a species. This species has an extensive range, covering a large part of North America. The frogs of Wisconsin can interbreed with those of Iowa or Nebraska, but not with those in Mexico. These two subspecies (from Wisconsin and Mexico) meet one of the fundamental requirements of a species, that is, inability to interbreed successfully. Are these subspecies actually true species?

Also, conclusions about common ancestries are not always valid because of lack of information about their fossil history, etc. Two structural patterns may have similar adaptive responses to survival requirements, but they may have developed these similar patterns independently without the benefit of common ancestry. Evolutionary knowledge is more precise with lower taxa, but with higher ones ancestral information is largely speculative, or it may be based on structural resemblances among the animals that now exist. This explains why classification of a group may vary from one generation of zoologists to another as new knowledge is acquired.

■ LINNAEUS AND THE EARLY HISTORY OF TAXONOMY

Although Aristotle, the great Greek philosopher and student of zoology, attempted to classify animals on the basis of their structural similarities, little was done about the grouping of animals until the English naturalist John Ray (1627-1705) brought forth his system of classification. He employed structural likenesses as the basis of his classification and worked out a number of groups. He seems to have been the first biologist to have a modern concept of species and paved the way for the work of Carolus Linnaeus (1707-1778), who gave us the modern scheme of classification.

Linnaeus was a Swedish botanist at the University of Uppsala. He had a great talent for collecting and classifying objects, especially flowers. Linnaeus worked out a fairly extensive system of classification

for both plants and animals. His scheme of classification was published in his great classic work *Systema Naturae,* which had gone through ten editions by 1758. Linnaeus emphasized structural features of plants and animals in his methods of classification. Actually his classification was largely arbitrary and artificial, and he believed strongly in the fixity of species. He divided the animal kingdom down to species, and according to his scheme each species was given a distinctive name. He recognized four classes of vertebrates and two classes of invertebrates. These classes were divided by him into orders, the orders into genera, and the genera into species. Since his knowledge of animals was limited, his lower groups, such as the genera, were very broad and included animals now placed in several orders or families. As a result, much of his classification has been drastically altered, yet the basic principle of his scheme is followed at the present time.

Although Linnaeus recognized four units, or taxa, in classification—class, order, genus, and species—since his time other major units of grouping have been added, so that the units now used are **phylum, class, order, family, genus,** and **species.** The major units can be subdivided into finer distinctions, such as subphylum, subclass, suborder, subfamily, subgenus, and subspecies.

■ BINOMIAL NOMENCLATURE AND THE NAMING OF ANIMALS

Linnaeus's great contribution was the introduction of two names for each kind of organism: the genus name and the species name. These words are from Latin or in Latinized form, because Latin was the language of scholars and universally understood. The generic name is usually a noun and the specific name an adjective. For instance, the scientific name of the common robin is *Turdus migratorius* (L. *turdus,* thrush; *migratorius,* of the migratory habit). This usage of two names to designate a species is called **binomial nomenclature.** There are times when a species is divided into subspecies, in which case a **trinomial nomenclature** is employed. Thus to distinguish the southern form of the robin from the eastern robin, the scientific term *Turdus migratorius achrustera* (duller color) is employed for the southern type. Taxa lower than subspecies are sometimes employed when four words are used in the scientific name, the last one usually standing for **variety.** In this latter case the nomenclature is **quadrinomial.**

The trinomial and quadrinomial nomenclatures are really additions to the linnaean system, which is basically binomial.

■ BASIS FOR FORMATION OF TAXONOMIC UNITS (TAXA)

Taxonomy, as already mentioned, emphasizes the natural relationships among the various animal types. Descent from a common ancestor makes for similarity in character, and the more recent this descent the closer the animals are grouped in taxonomic units. For instance, the genera of a particular family show less diversity than do the families of an order. Families must take more time to become diverse, and thus their common ancestor must have been more remote than that of the genera. The same principle applies to the higher categories, and therefore we should expect the common ancestors of the various phyla to be much older than those of classes. This principle, however, cannot be applied too rigidly to all groups of animals, for some have been much faster in their evolution than others. As we shall see later, some mollusks have taxa that have changed little in the course of millions of years. Some of their genera are actually much older than orders and classes in other groups.

As an illustration, one may use the following criteria to distinguish the phyla of the animal kingdom:

1. Unicellular or multicellular
2. Body saclike or tube-within-a-tube
3. Diploblastic or triploblastic
4. Segmented or nonsegmented
5. Presence or absence of a digestive system
6. Type of symmetry—asymmetry, bilateral, or radial
7. Presence or absence of appendages
8. Jointed or nonjointed appendages
9. Type of skeleton—exoskeleton or endoskeleton
10. Presence or absence of a notochord
11. Presence or absence of a coelom

■ WHY CLASSIFICATION VARIES AMONG DIFFERENT AUTHORITIES

Taxa are the outcome of changing concepts of classification and therefore are subject to man's diverse judgments.

1. It is very difficult to appraise all the fine distinctions among animals. The fact that two animals have similar characteristics does not establish their relationship. The similar characteristics may have developed entirely independently of each other, by convergent evolution, with no common ancestry involved. Taxonomists do not always agree about these lines of descent.

2. Many thousands of new species are named each year. Not all of these are well defined and more will have to be found out about them before they are firmly established in animal classification. For this reason much of the work of the taxonomist consists of revising what has already been described rather than describing new species.

3. There is some diversity of opinion among zoologists about subdivision of groups. Some are "splitters," inclined to much subdivision; others are "lumpers," preferring to lump together minor groups.

4. The **law of priority** also brings about frequent changes. The first name proposed for a taxonomic unit that is published and meets other proper specifications has priority over all subsequent names proposed. The rejected duplicate names are called **synonyms.** It is disturbing sometimes to find that a species that has been well established for years must undergo a change in terminology when some industrious systematist discovers that on the basis of priority, or for some other reason, the species is, according to this "law," misnamed.

■ RULES OF SCIENTIFIC NOMENCLATURE

To prevent confusion in the field of taxonomy and to lay down a uniform code of rules for the classification of animals, there was established in 1898 an International Commission on Zoological Nomenclature. This Commission meets from time to time to formulate rules and to make decisions in connection with taxonomic work.

The *Basic Rules of Nomenclature* laid down by the International Commission on Zoological Nomenclature are as follows:

1. The system of nomenclature adopted is the **binomial system** as described by Linnaeus in the tenth edition of his *Systema Naturae* (1758). This system is modified in some cases to include a trinomial nomenclature when a subspecific name is used.

2. Zoologic nomenclature is independent of botanical nomenclature and may employ the same names

for taxonomic units, but this procedure is not recommended.

3. The scientific names of animals must be either Latin or Latinized in form.

4. The genus name is a single word, nominative singular, and begins with a capital letter.

5. The species name may be a single or compound word, is printed with an initial lower case letter, and is usually an adjective in grammatical agreement with the generic name. In case the species name is derived from a personal name, it may be written with a capital initial letter. When a subspecies name is used, it also has an initial small letter.

6. The **author** of a scientific name is the one who first definitely published the name in connection with a description of the animal. The author's name should follow the species name and should rarely be abbreviated.

7. The **law of priority** states that the first published name in connection with a genus, species, or subspecies is the one recognized. All duplicate names are called synonyms.

8. When the genus name is not the one under which a species is placed by the original author, or if the generic name is changed, the original author's name is placed in parentheses. For instance, the name *Rana gryllus* was given by Le Conte to the common cricket frog, but since the generic name has been changed, it is now often written *Acris gryllus* (Le Conte).

9. A type specimen is the particular specimen or specimens on which the name of the species was established. It is customary for taxonomists to place such types in public museums or other places where they can be available to those who are interested. Such types must retain their original name even if the species is later divided. Whenever a new genus is described, one species is taken as the type of the genus and also retains the original name in case the genus is later divided into two or more genera. **No two genera of animals may have the same name.**

10. The name of a family is formed by adding *-idae* to the stem of the name of the type genus, and the name of a subfamily, by adding *-inae.*

■ RECENT TRENDS IN TAXONOMY

Since the concept of organic evolution, biologic units have been classified mainly on the basis of common descent. But it has not always been clear just what this evolutionary relationship is in all cases. Only the fossil record can supply convincing evidence, and that is often lacking. To determine the nature, properties, and genetic variations of species, data from many disciplines, such as genetics, cytology, ecology, behavior, and physiology, are brought to bear on the problem. The old concept of species as distinct and set apart by definite criteria is being replaced by the concept that a species is an intrabreeding, reproductively isolated gene pool, with the variations found within a dynamic population.

Numerical taxonomy

In systematics it is necessary to recognize the characters that have taxonomic value and to make a proper analysis of the data (characters) before assigning them to the proper taxon. Often the analysis of the data involves a statistical approach. In 1763 the Frenchman M. Adanson proposed a scheme of classification that involved the grouping of individuals into a particular species according to the number of shared characteristics. Thus each member of a species would have a majority of the total characteristics of the taxon, even though some of its characters are not shared with others of the same taxon. Such a classification has quantitative rather than qualitative significance. This classification lacks the phyletic relationship of evolutionary taxonomy (Adanson lived long before Darwin's time). However, this scheme has been revived (especially by plant taxonomists) in recent years and has given rise to **numerical taxonomy,** which makes use of the computer method for ascertaining calculations of similarity. Similar and dissimilar characteristics are simply fed into a computer and an analysis is made of its calculations for determining the taxon of a group.

Other approaches

Other techniques are now being developed in systematics that offer great promise in solving problems of phylogenetic relationships and evolution. Refinements of analytical methods are now being employed in electrophoresis, protein and genetic homology, blood groups, chromosomal analysis, etc. with this in view.

It has been shown that there are certain homologies among polynucleotide sequences in the DNA molecules of such different forms as fish and man. These sequences appear to be genes that have been retained with little change throughout vertebrate evolution. Possible phenotypic expressions of these homologous sequences are bilateral symmetry, notochord, hemoglobin, etc. By using a single strand of DNA from one species, short radioactive pieces of a DNA strand from another species, and mixing the strands together, it was found that some of the smaller strands paired with similar regions on the large strand, indicating that the paired parts had common genes.

Another recent technique involves the recognition of RNA codons by transfer RNA of another species. These new biochemical methods of classification are used to complement, rather than replace, the older, more established methods. In general, molecular evidences have not agreed very well with the fossil evidence. Nevertheless, the new molecular approach provides a potentially powerful tool for the systematist.

Some examples of scientific nomenclature

The examples listed in Table 6-1 will give you some idea of how animals are classified on the basis of relationship and likeness. Of all animals the anthropoid apes are generally agreed to be nearest man in relationship and structural features. In contrast to man and the gorilla are the frog, also a vertebrate like the others but diverging from them much earlier, and the little katydid, which is not a chordate but belongs to a lower phylum.

■ PHYLOGENY OF ANIMALS
The first animals*

The origin of life from nonliving beginnings may have taken more than 2 billion years. During this time there occurred the preliminary synthesis of organic compounds, and later of macromolecular systems that were the forerunners of the first living things. **Biopoiesis** therefore included the abiotic synthesis of macromolecular systems, followed by biogenesis, or the transformation of the macromolecular systems into the first living things.

The nature of the first life is purely speculative. It is generally agreed that such primitive forms could not now survive in the face of stronger competitors such as bacteria and fungi. This creature must have been very simple of construction, judged by bioge-

*Refer to p. 21, Principles 23 and 24.

netic standards of the present day. The earliest form of life must have had the basic characteristics of a living form, such as self-duplication and some capacity of metabolism and adaptive adjustments to its environment.

One naturally thinks of viruses as giving some insight into what early life must have been like. Viruses are the smallest biologic structures that have all the information needed for their own reproduction. They consist of a shell of protein enclosing a core of nucleic acid (DNA or RNA). The protein shell is not only protective but may serve for entrance into the walls of living cells that are attacked by the virus. Leaving the protein coat behind, the nucleic acid enters the cell, which it causes to produce large numbers of virus particles. In time the cell ruptures and the viruses are set free to attack other cells. Viruses require an electron microscope to be seen, for their size (100 to 2,000 Å) is smaller than one-half the wavelength of violet light, which is the minimum size for the light microscope. But viruses are neither cells nor organisms. Some viruses have the power to arise as mere fragments broken off from the nucleic acid material of a donor cell. These fragments have the power to direct the cell to form the protective mantle around the virus. Viruses are inactive in the free state and become reactivated when they enter a new cell.

Since plants and animals cannot be separated by biologic criteria, it is a moot question whether or not the earliest organisms were unicellular plants or animals. The first true animals were probably metazoans and heterotrophic by nature, although they could have been parasitic. They were undoubtedly simpler than the bacteria and were able to live an independent existence with the simplest of requirements for duplication, metabolism, and environmental adjustments.

Origin of Metazoa

The origin of the multicellular animals (Metazoa) has posed many problems to zoologists. It is generally believed that both plant and animal traits evolved gradually from the Protista. It has been customary to pick out flagellates as the ancestors of the many-celled animals. Some of the primitive flagellates had both ameboid and flagellate methods of locomotion, which could explain the types of gametes found among present metazoans. Many flagellates have

TABLE 6-1
Examples of classification of animals

Taxa	Man	Gorilla	Grass frog	Katydid
Phylum	Chordata	Chordata	Chordata	Arthropoda
Subphylum	Vertebrata	Vertebrata	Vertebrata	
Class	Mammalia	Mammalia	Amphibia	Insecta
Subclass	Eutheria	Eutheria		
Order	Primates	Primates	Salientia	Orthoptera
Suborder	Anthropoidea	Anthropoidea		
Family	Hominidae	Simiidae	Ranidae	Tettigoniidae
Subfamily			Raninae	
Genus	*Homo*	*Gorilla*	*Rana*	*Scudderia*
Species	*sapiens*	*gorilla*	*pipiens*	*furcata*
Subspecies			*pipiens*	*Brunner*

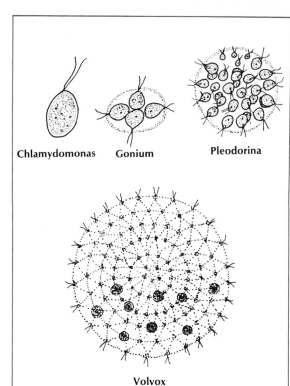

Chlamydomonas   Gonium   Pleodorina

Volvox

FIG. 6-1
Some flagellates live in clumps or colonies, joined at their outer surfaces or enclosed in gelatinous envelope. In *Gonium* all cells are alike and each may divide to form new colony. Some colonies have division of labor, possessing both somatic and reproductive cells. *Pleodorina illinoisensis* has 4 somatic and 28 reproductive cells; *Volvox* has hundreds of somatic and only a few reproductive cells. *Chlamydomonas* is solitary and does not form colonies.

tendencies to form colonies of few to many cells, and it is possible to pick out a progressive series of such aggregations of gradual complexity from simple to complex (Fig. 6-1). Thus *Chlamydomonas* (1 cell), *Gonium* (4 to 16 cells), *Pandorina* (16 cells), *Eudorina* (32 cells), *Pleodorina* (32 to 128 cells), and *Volvox* (many thousands of cells). The zooids, or cells, may be loosely connected or may be held together in a mucilaginous jelly. *Pleodorina* has its cells differentiated into a few somatic cells (sterile) and those capable of reproduction. This differentiation is carried further in *Volvox*, in which the majority of cells cannot reproduce. In all the others each zooid is capable of both asexual and sexual reproduction. These examples merely show phases of development as arbitrarily arranged but do not afford evidence that one gave rise to another. Some zoologists believe that it is unlikely flagellates evolved beyond the colony.

The Metazoa are often considered a diphyletic group, with sponges (Parazoa) as one separate line and the Eumetazoa as the other. Sponges probably arose from Protozoa and represent complex flagellate colonies at a differentiated cellular level, with a tendency toward a tissue level of construction. Their digestive collar cells are similar to the protozoan collar flagellates. Sponges also have many other differences in construction and embryonic development.

There is also a theory of metazoan origin from **syncytial ciliates.** This theory differs from the colonial theory just described in that the whole animal (a ciliate) is at first a syncytium (multinucleated) and later becomes multicellular by the development of internal cell boundaries. According to this theory, the first true metazoans were the flatworms resembling the acoel flatworms that are considered to be the most primitive bilateral animals. The chief advantage of the syncytial theory is that some syncytial ciliates have an established bilateral symmetry and an anteroposterior axis. But there are many objections to the theory because it implies that coelenterates with radial symmetry are derived from flatworms and not the reverse. This theory also fails to account for the flagellate sperm cell so characteristic of the metazoan pattern.

A third theory favors the origin of the Metazoa from plantlike protozoans (the **Metaphyta theory**). According to this theory, it is easy to explain the transformation of plantlike protozoans to cellular metaphytes because each group absorbs nutriment from each side equally. However, the metazoan has a mouth and a different method of nutrition; although some ciliates do have a mouth and gullet, such specialized requirements were later developments, and it is difficult to think that metazoans could be derived from such specialized forms.

Numerous **symbiotic relationships** of autotrophic algae are known to occur in many invertebrates, and this may be considered a fourth theory. All stages in the reduction of the cell wall of the symbiotic algae are represented among these relationships. Organelles such as mitochondria and chloroplasts may once have been free-living organisms. When symbionts were established, many of their functions were taken over by the nuclear DNA. Both mitochondria and the plastids of plants are derived from the prokaryotic symbiotes, but each has had a separate origin and differs from each other in structure and biochemical requirements. Higher plants may have developed from an early partnership between nonphotosynthesizing plants and algae. Algae may have provided food, while the host plant gave protection and support. So far as animals are concerned certain sea slugs (nudibranchs) retain intact in their digestive tracts the chloroplasts from algae they have eaten. Miotic division may have been a later advancement in evolution.

Phylogeny and ontogeny

As we have seen in the discussion on embryology earlier, all eumetazoans in their embryonic development pass through certain common stages. These are in succession the zygote, cleavage, blastula, gas-

trula, and pharyngula. The German evolutionist E. Haeckel of the last century believed that these successive stages of individual development corresponded to an adult form of an ancestral form. Thus the zygote would represent the protozoan or protistan stage; the blastula, the hollow colonial protozoans; and the gastrula, the adult coelenterate of the present. The human embryo with gill depressions in the neck was believed to represent the stage when our adult ancestors were fishes. On this basis he gave his generalization: ontogeny (individual development) repeats phylogeny (evolutionary descent).

This hypothesis of Haeckel is often called the **gastrea** hypothesis. It is based primarily on coelenterate development that occurs in a different way from what Haeckel supposed. For instance, he considered embolic gastrulation to be the primitive pattern, although it is now known that such gastrulation is rare in coelenterates. Moreover, he believed that the gastrea was the common ancestor of all eumetazoans.

K. E. von Baer, the embryologist, had noticed long before Haeckel the general similarity between the embryonic stages and the adults of certain animals, but arrived at a more correct interpretation. According to his view, the earlier stages of all embryos tend to look alike, but as development proceeds, the embryos become more dissimilar. There are exceptions to this generalization because the young of related species differ more than do the adults, for example, the feeding larvae (caterpillars) compared with the reproducing adults (butterflies).

The recapitulation of Haeckel would be correct only when changes in development were added on as new stages at the end of development, but the addition of new stages at ontogeny is not the usual course of development. When changes in ontogeny that have evolutionary significance do happen, they are more evident in later stages. Often there are special adaptations to embryonic life that are not found in the adult condition, such as the fetal membranes of the amniotes, and are of evolutionary importance because similar structures are absent in lower vertebrates. New development usually occurs by **developmental divergence,** by which a new path of embryonic development diverges away from a preexisting path. This is strikingly shown in neoteny, in which larval or embryonic characters are retained and the organism does not develop into the customary adult.

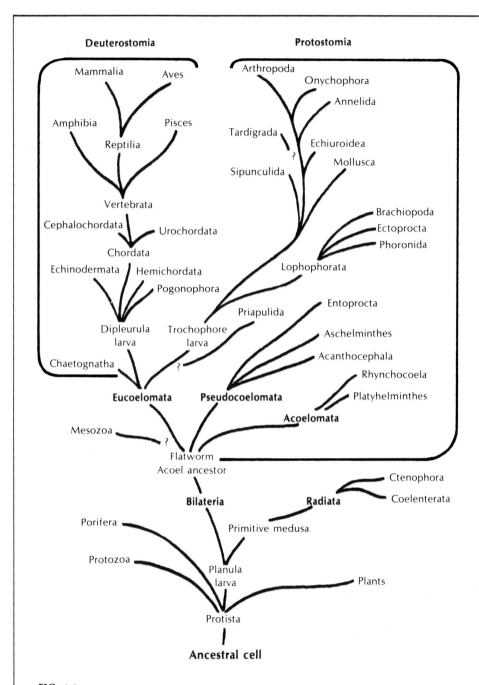

FIG. 6-2

Animal phylogeny (hypothetical). Basis for such a tree phylogeny is evolutionary relationship, or common ancestry, but such information is scanty. In general, time is represented by vertical levels—the higher the branch, the more recent the taxon—but this is also uncertain because fossil record is incomplete. Interpretations of relationship are based on common characters, similarity of embryologic development, basic structural patterns, fossils, etc.

Instead, it develops sex organs and functions as a new type of animal. It may be stated in conclusion that ontogeny repeats ontogeny, with variations, and that phylogeny consists of a series of ontogenies. In whatever ways animals have evolved, they have in general followed four important guidelines in their morphogenesis:

1. Development of germ layers from diploblastic to triploblastic conditions
2. Formation of a body cavity
3. Emergence of adaptations suitable for land and air existence from aquatic ancestral life
4. Grouping of higher metazoans into the great divisions of protostomes and deuterostomes

Much of our future discussion of the nature of the various animal groups will center around the adaptability of these guidelines.

Phylogeny of animals

Phylogeny is the science of ancestral history and racial relationships. Exact relationships of the members of the animal kingdom are often vague or nonexistent according to our present knowledge. This is especially the case with the large major groups (phyla), regarding which there is much disagreement among authorities. Within smaller taxonomic units (species, genera, orders, etc.) relationships have been more definitely established. The student should therefore remember that the sequence in which zoologists present the various groups does not indicate that each group has arisen directly from the one that has preceded it. Most existing forms are related indirectly to each other through common ancestors that are now extinct. Most common ancestors were sufficiently generalized in structure to give rise to many divergent groups, but such ancestors have either undergone evolution or else have become extinct because of their inability to adapt to a changing environment. The closing of the gaps in relationships is very much like supplying the missing parts of a jigsaw puzzle. If many parts are gone, the problem becomes complicated. However, if similarity of a structure and development mean anything in an evolutionary interpretation, then it is obvious that certain groups are closely connected because the evidence stands out clearly. It is a generalization widely accepted in biology that if two different organisms share many common traits, it is logical to assume that there is a relationship basis for this similarity

and that it has not been due to convergent or coincidental evolution.

With all the shortcomings any phylogenetic tree (Fig. 6-2) must possess, such a scheme of hypothetical relations has some value in visualizing the evolutionary picture, provided that it is not stressed too dogmatically. Many lines of evidence will be pointed out in the discussions of the various invertebrate phyla. The phylogenetic tree here presented on the basis of these evidences may serve to tie the phyla together in the evolutionary blueprint. A better way, perhaps, to represent phylogeny is by a fan-shaped scheme because there is little difference in the age of most phyla from a geologic viewpoint (Simpson).

Larger divisions of animal kingdom

Although the phylum is often considered to be the largest and most distinctive taxonomic unit, zoologists often find it convenient to combine phyla under a few large groups because of certain common embryologic and anatomic features. Such large divisions may have a logical basis, for the members of some of these arbitrary groups are not only united by common traits, but evidence also indicates some relationship in phylogenetic descent.

SUBKINGDOMS, BRANCHES, AND GRADES. Hyman has proposed a scheme for some of these larger groupings that should give the student a more comprehensive view of animal classification.

SUBKINGDOM PROTOZOA (ACELLULAR)—phylum Protozoa
SUBKINGDOM METAZOA (CELLULAR)—all other phyla
  BRANCH A (MESOZOA)—phylum Mesozoa
  BRANCH B (PARAZOA)—phylum Porifera
  BRANCH C (EUMETAZOA)—all other phyla
    GRADE I (RADIATA)—phyla Coelenterata (Cnidaria), Ctenophora
    GRADE II (BILATERIA)—all other phyla
      ACOELOMATA—phyla Platyhelminthes, Rhynchocoela
      PSEUDOCOELOMATA—phyla Acanthocephala, Aschelminthes, Entoprocta
      EUCOELOMATA—all other phyla

PROTOSTOME AND DEUTEROSTOME—DIVISIONS OF BILATERAL ANIMALS. The largest group of animals, the Bilateria, may be arranged into two major divisions—Protostomia and Deuterostomia. The characteristics and the phyla of each are as follows:

PROTOSTOMIA—Mouth usually formed from blastopore; schizocoelous formation of body cavity (coelom); mostly

TABLE 6-2
Diphyletic theory of phylogeny*

Protostomes	Deuterostomes
Mouth from, at, or near blastopore	New mouth from stomodaeum
Anus new formation	Anus from or near blastopore
Spiral cleavage (mostly) and involving all organ systems	Cleavage mostly radial and involves only dorsal myotomes
Mostly determinate embryonic development (mosaic)	Usually indeterminate embryology
Coelom forms as split in mesodermal bands; schizocoelous	Coelom from fusion of enterocoelous pouches
Mesoderm usually from blastomere 4d·	Mesoderm from various cells
Blood flows posteriorly in ventral vessels and anteriorly in dorsal vessels	Blood flows posteriorly in dorsal vessels and anteriorly in ventral vessels in vertebrates
Larva (when present) a trochophore type	Larva (when present) a pluteus type
Includes phyla Platyhelminthes, Rhynchocoela, Annelida, Mollusca, Arthropoda, minor phyla	Includes phyla Echinodermata, Hemichordata-Pogonophora, Chaetognatha, and Chordata

*Some variations from this table in each group.

spiral cleavage; determinate or mosaic pattern of egg cleavage; ciliated larva (when present) a trochophore or trochosphere type. Examples: phyla Platyhelminthes, Aschelminthes, Rhynchocoela, Annelida, Mollusca, Arthropoda, and others.

DEUTEROSTOMIA—Anus formed from blastopore; enterocoelous formation of coelom; mostly radial cleavage; indeterminate or equipotential pattern of egg cleavage; ciliated larva (when present) a pluteus type. Examples: phyla Echinodermata, Hemichordata, Pogonophora, and Chordata.

These divisions form the two main lines of evolutionary ascent in the animal kingdom and are often referred to as the diphyletic theory of phylogeny. (See Table 6-2.)

References
Selected general readings

Anfinson, C. B. 1959. The molecular basis of evolution. New York, John Wiley & Sons, Inc. *This book has made a great impact on all serious students of the life sciences. Chapter 1, although brief, will give the student a good phylogenetic orientation.*

De Beer, G. R. 1940. Embryos and ancestors. New York, Oxford University Press. *An excellent evaluation of the biogenetic law.*

Hadzi, J. 1963. The evolution of the Metazoa. New York, Pergamon Press, Inc.

Hoyer, B. H., B. J. McCarthy, and E. T. Bolton. 1964. A molecular approach in the systematics of higher organisms. Science, **144:**959-967.

Hyman, L. H. 1940-1967. The invertebrates, vols. 1-6. New York, McGraw-Hill Book Co. *Informative discussions on the phylogenies of most of the invertebrates are treated in this outstanding series of monographs.*

Jagersten, G. 1972. Evolution of metazoan life cycles. New York, Academic Press Inc.

Kerkut, G. A. 1960. The implications of evolution. New York, Pergamon Press, Inc. *Many long-established concepts are rather rudely upset in this treatise.*

Manville, R. H. 1952. The principles of taxonomy. Turtox News **30:** nos. 1 and 2. *A concise account of classification procedures.*

Marcus, E. 1958. On the evolution of animal phyla. Quart. Rev. Biol. **33:**24-58.

Mayr, E., E. G. Linsley, and R. C. Usinger. 1953. Methods and principles of systematic zoology. New York, McGraw-Hill Book Co. *The present status of the rules and regulations of taxonomy is well discussed.*

Mayr, E. (editor). 1957. The species problem. Washington, D. C., American Association for the Advancement of Science. *This is a symposium by many authorities on the problems of species.*

Moore, J. A. (editor). 1965. Ideas in modern biology. New York, The Natural History Press. *Includes papers of the*

Plenary Symposia of the XVI International Congress of Zoology held at Washington, D. C., August 20-27, 1963. Papers deal with genetics, cell biology, development, evolution, phylogeny, and behavior.

Morescalchi, A. 1970. Karyology and vertebrate phylogeny. Bull. Zool. **37**:1-20.

Raven, P. H., B. Berlin, and D. E. Breedlove. 1971. The origins of taxonomy. Science **174**:1210-1213.

Romer, A. S. 1968. The procession of life. Cleveland, The World Publishing Co., Inc. *An excellent summary of the evolutionary development of the structure, physiology, and habits of the major groups of animals.*

Schenk, E. T., and J. H. McMasters. 1948. Procedures in taxonomy, including a reprint of the International Rules of Zoological Nomenclature with summaries of opinions rendered, ed. 2. Stanford, Calif., Stanford University Press.

Simpson, G. G. 1945. The principles of classification and a classification of mammals. Bull. Amer. Museum Natural History, vol. 85. *A well-presented account of the bases of taxonomic procedures.*

Simpson, G. G. 1961. Principles of animal taxonomy. New York, Columbia University Press. *An outstanding contribution of the basic principles of systematics and morphologic diversity.*

Selected *Scientific American* article

Sokal, R. R. 1966. Numerical taxonomy. **215**:106-116 (Dec.).

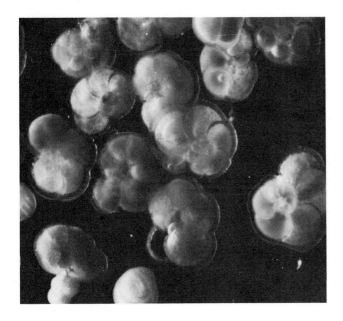

Living foraminiferans. These are ameboid marine protozoans that secrete a calcareous, many-chambered shell in which to live and then extrude protoplasm through pores to form a layer over the outside. The animal begins with one chamber, and as it grows, it secretes a succession of new and larger chambers, continuing this process throughout life. Foraminiferans are planktonic animals, and when they die, their shells are added to the ooze on the ocean's bottom. (Photo by Roman Vishniac, New York.)

CHAPTER 7

THE ACELLULAR ANIMALS

PHYLUM PROTOZOA*

■ PLACE IN ANIMAL KINGDOM

The protozoan is a complete organism in which all life activities are carried on within the limits of a single plasma membrane. As a protoplasmic mass not divisible into cells, a protozoan may be termed "acellular" and may be said to belong to a protoplasmic level of organization.

The term "Protista" is often used to include all acellular organisms, such as bacteria, slime molds, and protozoans; the term "Protozoa" is then reserved for those protistans that lack chlorophyll and whose nutrition is holozoic, saprozoic, or symbiotic.

■ BIOLOGIC CONTRIBUTIONS

1. **Protoplasmic specialization** (division of labor within the cell) in protozoans involves the organization of protoplasm into functional organelles.

*Pro-to-zo'a (Gr. *prōtos,* first, + *zōon,* animal).

2. The earliest indication of **division of labor between cells** is seen in certain colonial protozoans that have both somatic and reproductive zooids (individuals) in the colony.

3. **Asexual reproduction** by mitotic division is first developed in the protists.

4. The behavior of conjugant mates in certain protozoans may indicate the earliest differentiation of sex. **True sexual reproduction** with zygote formation is found in some protozoans.

5. The responses (taxes) of protozoans to stimuli represent the **beginning of reflexes and instincts** as we know them in metazoans.

6. The first appearance of an **exoskeleton** is indicated in certain shelled protozoans.

7. **All lines of nutrition** were developed in the protozoans—autotrophic, saprozoic, and holozoic.

117

PHYLOGENY AND ADAPTIVE RADIATION

PHYLOGENY. Protozoans are often placed close to the beginning of the genealogic tree. Metazoans may not have come directly from protozoans as we know them, but may have been derived from organisms similar to protozoans. Colonial protozoans, particularly among flagellates (Fig. 7-4), show various degrees of cell aggregation and differentiation that may be suggestive of how metazoans arose.

The flagellates are usually considered to be nearest the ancestral stem of both plant and animal kingdoms.

ADAPTIVE RADIATION. Protozoan evolution has been guided chiefly by the adaptive features of the outer layer, or **ectoplasm,** in which the various adaptations determine the shape, type of body covering, nutrition, food-getting apparatus, body movement, locomotion, and protective devices of the various species.

Within a single-cell body plan they have exploited a division of labor of morphologic structure (**organelles**) that has fitted them for most of the ecologic niches of the ecosystem. All types of nutrition are found among them, and no known source of energy supply is withheld from them. All modes of living, whether free-living or symbiotic, are represented among them.

THE PROTISTA

The term "Protista" is used by many biologists to refer to all organisms, whether plant or animal, in which specialization and differentiation is restricted to a single cell. Such a category includes the unicellular algae, the protozoans, the bacteria, yeasts, etc. On the other hand, the term "Protista" is often used to denote only those acellular organisms that have a diploid nucleus bounded by a nuclear membrane (eukaryotic), whereas the term "Monera" includes those acellular forms that have a haploid nucleus and do not have a nuclear membrane (prokaryotic) and applies to the bacteria and the blue-green algae.

The Monera lack a nuclear membrane, although they do have a nuclear area, and they lack an endoplasmic reticulum, Golgi apparatus, mitochondria, and lysosomes. Their walls contain murein, a polymer of amino acids and amino sugars, or their derivatives, particularly muramic acid, which does not occur in the walls of plants. Bacterial nuclei contain only one chromosome, which is composed exclusively of DNA. Most bacteria reproduce by binary fission, and each daughter cell receives a full set of genes in a complete chromosome.

There is still confusion as to what forms belong in the phylum Protozoa, since this has traditionally included both autotrophic and heterotrophic forms. Actually the autotrophic forms are more plantlike than animal-like, and some authorities prefer that they be included only with plants (Metaphyta).

THE PROTOZOANS

A protozoan is an organism that is made up of a mass of protoplasm that is not divided into cells and that carries on all the life processes within its cell membrane. In structure a protozoan may be likened to a cell from a multicellular animal, but functionally it is a complete organism that performs all the essential life processes of an animal.

It is erroneous to think of protozoans as simple animals. They have many complicated structures, and physiologically they are quite complex little animals. Within the cytoplasm of a protozoan there is specialization and division of labor. The specialized structures within the cytoplasm are called **organelles,** and each is fitted for a specific function similar to those of specialized organs or groups of cells of the metazoan (many-celled) animal. These organelles may perform as skeletons, sensory systems, conducting mechanisms, contractile systems, organs of locomotion, defense mechanisms, and so on.

Because the protozoans are not made up of cells they are termed **acellular animals** and they represent the **protoplasmic level of organization.** Many biologists place them close to the common ancestor of the many-celled forms. Some of the flagellates are quite close to the plants and may be considered as connecting links between plants and animals. Many flagellates are autotrophic (holophytic), that is, they contain chlorophyll and, like the plants, can manufacture their own carbohydrates by photosynthesis. Perhaps in the early evolution of animals some of these autotrophic flagellates may have lost their green chloroplasts to become colorless animals that must either absorb nutrients from their environment (saprozoic) or feed upon other plant or animal matter (heterotrophic, or holozoic).

Most protozoans are single, but many of them, particularly among the flagellates, live in distinct colonies of from several to hundreds of protozoan zooids. The distinctions between such protozoan colonies and metazoans made up of from several to millions of body cells is mainly a degree of division of labor. In metazoans the cells are dependent on each other for such functions as nutrition, movement, excretion, and reproduction, and certain cells become highly specialized for distinctive functions. In the protozoan colony, however, certain cells are specialized for reproduction, but all the rest can perform all other body functions independently of each other.

Size

Most Protozoa are small or microscopic, usually from 3 to 300 μ long. The largest are among the Foraminifera, some of which have shells 100 to 125 mm. in diameter. Certain amebas may be 4 to 5 mm. in diameter. Some of them are found in colonies in which each individual carries on its functions independent of the others, although in a few colonies there is a small amount of differentiation.

Number of species

The number of named species of Protozoa lies somewhere between 15,000 and 50,000, but this figure probably represents only a fraction of the total number of species. Some protozoologists think that there may be more protozoan species than all other species together because each species of the higher phyla may have its own unique protozoan parasites, and many protozoans bear parasites themselves. There are probably more parasitic species than free-living ones.

Ecologic relationships

Protozoans are found wherever life exists. Protozoans are highly adaptable and easily distributed from place to place. All species live in moist habitats, whether in seawater, ocean bottom, freshwater, foul water, soil, decaying organic matter, or plants, or as parasites in all kinds of animals from other protozoans to man. They may exist as free-living organisms or in symbiotic, parasitic, or other relationships. They include both solitary and colonial forms and may be sessile or free swimming. They form a large part of the floating plankton. The same species are often found widely separated in time as well as space. Some species may have spanned geologic eras of more than 100 million years.

Despite their wide distribution, many of them can live successfully only within narrow environmental ranges. Species adaptations vary greatly in protozoans, and successions of species frequently occur as environmental conditions change. Certain species can endure only in mild acidity; others can exist at greater acidities and may replace those that cannot as conditions change.

Protozoans have played an enormous role in the economy of nature. Their skeletons have formed gigantic ocean and soil deposits. As parasites they may give rise to diseases; as mutuals they may be indispensable to certain other forms of life, for example, the termites; and they may be important sources of water contamination.

Characteristics

1. **Acellular** (or one cell), some colonial
2. **Mostly microscopic,** although some large enough to be seen with the unaided eye
3. All symmetries represented in the group; shape variable or constant (oval, spherical, etc.)
4. **No germ layer present**
5. No organs or tissues, but **specialized organelles** found; nucleus single or multiple
6. Free living, mutualism, commensalism, parasitism all represented in the group
7. Locomotion by **pseudopodia, flagella, cilia,** and direct cell movements; some sessile
8. Some provided with a **simple protective exoskeleton,** but mostly naked
9. Nutrition includes all types: autotrophic (manufacturing own nutrients by photosynthesis), heterotrophic (depending on other plants or animals for food), saprozoic (using nutrients dissolved in the surrounding medium)
10. Habitat: aquatic, terrestrial, or parasitic, with or without locomotor organoids
11. Reproduction asexually by fission, budding, and cysts and sexually by conjugation of gametes

Brief classification

Four main groups of protozoans are commonly recognized—the flagellates, the ameboid, the spore-forming, and the ciliates. Traditionally, each of these groups has been represented as a taxonomic class under the phylum Protozoa. In recent years there have been many changes in the taxonomic arrangement of this group and some authorities assign subphylum or even phylum rank to each of the groups.

Subphylum Sarcomastigophora (sar''ko-mas-ti-gof'o-ra) (Gr. *sarkos,* flesh, + *mastix,* whip, + *phora,* pl. of bearing). With locomotor organelles of flagella or pseudopodia, or absent; monomorphic nuclei.

Superclass Mastigophora (mas-ti-gof'o-ra) (Gr. *mastix,* whip, + *phora,* bearing). Move by flagella or by pseudopodia; autotrophic or heterotrophic, or both.

Class Phytomastigophorea (fi''to-mas-ti-go-for'e-a) (Gr. *phyton,* plant, + *mastix,* whip, + *phora,* bearing). Plantlike flagellates usually bearing chromatophores. *Chilomonas, Euglena, Volvox, Ceratium, Peranema.*

Class Zoomastigophorea (zo''o-mas-ti-go-for'e-a) (Gr. *zōon,* animal, + *mastix,* whip, + *phora,* bearing). Flagellates without chromatophores. *Trichomonas, Trichonympha, Trypanosoma, Leishmania, Proterospongia.*

Superclass Opalinata (o''pa-lin-a'ta) (NF. *opaline,* like opal in appearance, + -*ata,* group suffix). Body covered with longitudinal rows of cilium-like organelles; parasitic; cytostome lacking; nuclei not differentiated into macronuclei and micronuclei. *Opalina, Protoopalina.*

Superclass Sarcodina (sar-ko-di'na) (Gr. *sarkos,* flesh, + -*ina,* belonging to). **(Rhizopoda).** Locomotion by pseudopodia; no definite pellicle; free-living or parasitic; uninucleate or multinucleate; mostly holozoic.

Class Actinopodea (ak''ti-no-po'de-a) (Gr. *aktis, aktinos,* ray, + *pous, podos,* foot). With pseudopodia radiating from a spherical body. *Actinosphaerium, Actinophrys, Thalassicolla.*

Class Rhizopodea (ri-zo-po'de-a) (Gr. *rhiza,* root, + *pous, podos,* foot). Variety of pseudopodia. *Amoeba, Entamoeba, Arcella, Globigerina.*

Class Piroplasmea (pir-o-plaz'me-a) (L. *pirum,* pear, + Gr. *plasma,* a thing molded). Rod-shaped or ameboid parasites. *Babesia.*

Subphylum Sporozoa (spor-o-zo'a) (Gr. *sporos,* seed, + *zōon,* animal). Typically spore forming; all parasitic; no cilia or flagella (except in flagellated microgametes).

Class Telosporea (tel-o-spor'e-a) (Gr. *telos,* end, + *sporos,* seed). Spore forming, with asexual and sexual reproduction; movement by gliding or body flexion (no pseudopodia). *Monocystis, Gregarina, Eimeria, Plasmodium.*

Class Toxoplasmea (tox-o-plaz'me-a) (Gr. *toxikon,* poison, + *plasma,* a thing molded). Spores lacking; asexual reproduction; locomotion by gliding or body flexion (no flagella or pseudopodia in any stage of life cycle). *Toxoplasma.*

Class Haplosporea (hap-lo-spor'e-a) (Gr. *haploos,* single, + *sporos,* seed). Spore forming; asexual reproduction by schizogony; no flagella, but pseudopodia may be present in some stages. *Haplosporidium.*

Subphylum Cnidospora (ni-do-spor'a) (Gr. *knidē,* nettle, + *sporos,* seed). Spores with polar filaments. *Nosema.*

Subphylum Ciliophora (sil-i-of'ora) (L. *cilium,* eyelid, + Gr. *phora,* bearing). Cilia or ciliary organelles in adult or young stages; tentacles in adults of some; usually two kinds

of nuclei (macronucleus and micronucleus); reproduction usually involves conjugation.

Class Ciliata (sil-i-aht'a) (L. *cilium,* eyelid, + -*ata,* group suffix). Characteristics of subphylum.

Subclass Holotrichia (ho-lo-trik'e-a) (Gr. *holos,* entire, + *thrix, trichos,* hair). Ciliation usually uniform over body; specialized buccal ciliature inconspicuous or absent. *Paramecium, Colpoda, Tetrahymena, Spirochona, Balantidium.*

Subclass Spirotrichia (spir-o-trik'e-a) (L. *spiro,* coil, + Gr. *thrix, trichos,* hair). Adoral zone of membranelles winding clockwise toward cytostome; cilia sparse; some with cirri on ventral surface. *Stentor, Blepharisma, Halteria, Epidinium, Euplotes.*

Subclass Peritrichia (per-i-trik'e-a) (Gr. *peri,* around, + *thrix, trichos,* hair). Conspicuous oral ciliature winding counterclockwise around apical end; body cilia usually lacking in adults; mostly attached stalked ciliates. *Vorticella, Carchesium, Trichodina.*

Subclass Suctoria (suk-tor'e-a) (L. *sugere, suctum,* to suck). Stalked ciliates with tentacles; no ciliature in adult. *Podophyra, Ephelota.*

Evolution

With the exception of certain shell-bearing Sarcodina, such as Foraminifera and Radiolaria, Protozoa have left no fossil records. Mastigophorans (flagellates) are considered to be the oldest of all protozoans and may have arisen from bacteria and spirochetes. That this group includes members that are chlorophyll bearing and resemble the plant algae may indicate a common origin for plants and animals. Evidence for the origin of Sarcodina from flagellates is shown by the fact that both flagellate and ameboid stages are found in some protozoans. However, the different orders of Sarcodina may have arisen independently from different kinds of flagellates. Sporozoa, which are all parasitic, are somewhat degenerate in structure and may have come from Sarcodina and Mastigophora. The origin of the ciliates is somewhat obscure, but they, too, have come from flagellates.

Protozoan fauna of plankton

Plankton is a general term for those organisms that passively float and drift with the wind, tides, and currents of both freshwater and marine water. It is composed mostly of microscopic animals and plants, of which protozoans form an important part. Sarcodina, particularly the foraminiferans and radiolarians, and Mastigophora make up most of the protozoan part of plankton, but the ciliates are also represented, and even some parasitic sporozoans are found there in other plankton animals.

Plankton is a very important source of food for other marine animals. Not only invertebrates feed upon plankton, but many fish and even the huge whalebone whales feed directly upon it. As the animals of the surface plankton die, they sink to deeper levels to serve as food for other animals there.

Symbiotic relationships

The term **symbiosis** refers to the intimate interrelationships between two organisms of different species for the purpose of deriving energy or for some other benefit. This special relationship may be beneficial to both species (mutualism) or beneficial to only one species but not harmful to the other (commensalism), or the relationship may be forced so that one receives benefit and the other furnishes all the energy and may actually be harmed (parasitism). These relationships are not always clear cut, and are not limited to protozoans, but protozoans are represented by all three major types of symbiosis.

MUTUALISM. *Paramecium bursaria* harbors green algae (zoochlorellae) that manufacture carbo-hydrates (by photosynthesis) for the benefit of the paramecium and receive a safe shelter in return. Zoochlorellae are also found in other protozoans, such as *Stentor,* certain amebas, and heliozoans. Zooxanthellae (yellow or brown algae) are found in certain ectoparasitic ciliates such as *Trichodina.*

Several species of flagellates live in the intestines of termites and wood-feeding roaches, where they secrete enzymes for digesting the cellulose that is thus made available to their hosts (Fig. 7-1). The flagellates cannot live outside the host; the termite or roach would starve without the flagellates. The protozoans ingest the wood after the insects have chewed it up into small bits. When termites lose their protozoan fauna (by high oxygen exposure, high temperatures, or prolonged starvation), they can survive only a short time even when fed an abundance of wood, unless they are reinfested with the flagellates.

COMMENSALISM. Commensal protozoans may live on the outside (ectocommensals) or, more commonly, on the inside (endocommensals) of another organism. Some *Vorticella* and suctorians are found attached to hydroids as ectocommensals. Endocom-

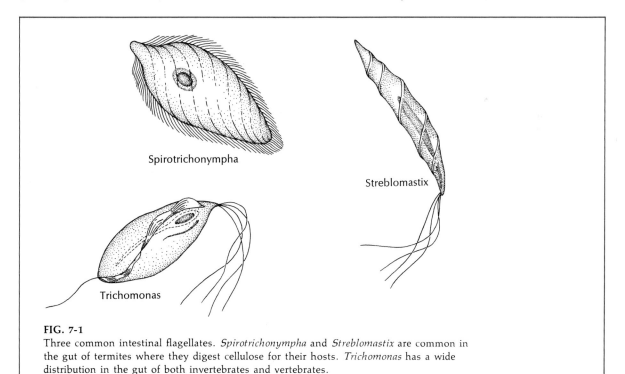

FIG. 7-1
Three common intestinal flagellates. *Spirotrichonympha* and *Streblomastix* are common in the gut of termites where they digest cellulose for their hosts. *Trichomonas* has a wide distribution in the gut of both invertebrates and vertebrates.

mensals are common in the digestive tubes of higher forms. Herbivorous mammals contain great numbers of ciliates and also a few flagellates and amebas. In ruminants such as cattle and sheep they are found in the first two stomach compartments, where they digest bacteria and foodstuffs in the host's food. Eventually they pass into the other compartments of the stomach and intestine, where they are destroyed and digested so that the host gets all the nutrition after all. Estimates are given of as many as 100,000 to 1 million ciliates per cubic millimeter of gut contents and a total number in a mature cow of 10 to 50 billion. Apparently Protozoa are not essential to the cattle, which can live and grow normally when the commensals are removed.

PARASITISM. Parasitism is the most common form of symbiosis among Protozoa. Ectoparasites live on the outside of the body and endoparasites live within the host. Most animals, especially higher ones, have one or more kinds of protozoan parasites. Protozoans themselves, even protozoan parasites, are often parasitized by other protozoans. For instance, the opalinid mastigophoran that lives in the frog's intestine is parasitized by a certain ameba.

Parasitic species are found among all classes of Protozoa; the class Sporozoa is entirely parasitic. Every vertebrate species probably harbors a parasitic or endocommensal ameba. Protozoan parasites have different ways of infesting the host. Some are transferred by contact of bodily parts *(Entamoeba histolytica);* by arthropod or other vectors *(Trypanosoma* and *Plasmodium);* by placenta (blood parasites); and by invasion of ovary or egg *(Babesia).*

Protozoan parasites may differ little in structure from free-living forms, but they undoubtedly have physiologic adaptations. Some protozoan parasites are adapted to a wide range of hosts and parasitize many species; others are restricted to a few species.

Contamination of water

Many protozoan parasites are transmitted in water and may give rise to serious diseases such as amebic dysentery. Protozoans also affect the taste and odor of water and often determine its drinking qualities. Water derived from surface sources and stored in large reservoirs is most likely to suffer. The chief cause of fishy, aromatic, or other odors is the production of aromatic oils by the disintegration of microscopic organisms. Certain flagellates, such as

Dinobryon and *Uroglena,* impart to water a pronounced fishy odor, not unlike cod-liver oil. *Synura,* another flagellate, gives a bitter and spicy taste to water. Of course other factors, such as algae and decomposing organic matter, also cause bad odors and taste.

Red tides and toxins

A "red tide" refers to the huge swarms of dinoflagellate algae in seawater that occasionally turn seawater dark yellow or brown and cause the destruction of large numbers of fish and other sea life. The Florida red tide is caused by the dinoflagellate *Gymnodinium brevis,* which produces a destructive neurotoxin while still alive. The organism associated with the California red tide is *Gonyaulax polyhedra* (Fig. 7-2), which is supposed to cause destruction by the depletion of oxygen in the water (from the dead algae). The California red tide is usually less severe than the Florida one, although recent investigation shows that it also produces a toxin released when the algae die. The chemical structure of the saxitoxin molecule produced by *Gonyaulax catenella* has been worked out. It is a potent nerve toxin that is stored

FIG. 7-2
Gonyaulax polyhedra, a dinoflagellate responsible for "red tides" along coast of southern California. This organism produces toxic alkaloidal substance that is very destructive to fish. Similar dinoflagellate, *Gymnodium brevis,* causes frequent "red tides" along Florida coast. Shellfish that feed on these organisms may be source of food poisoning in man.

in shellfish without harming them, but when other mammals, including man, eat the tainted shellfish they may be poisoned.

The California red tide occurs every 2 to 3 years and may be caused by organic runoff from the mountains, or the upwellings of nutrient-laden cold water along the coast, but the exact cause is not yet known.

Reproduction and life cycles
Reproduction

Reproduction in most Protozoa is primarily by cell division (asexual). It is comparable in some respects to cell division in the multicellular animals. The protozoan, however, has certain structural specializations (organelles), such as flagella, cilia, contractile vacuole, and gullet, that may be divided equally or unequally to the two daughter cells, so that a certain amount of differentiation or regeneration may be necessary to make the new animal complete. Some of these organelles are self-reproducing, but others are lost by resorption (dedifferentiation), then differentiated anew in each of the daughter organisms.

The method of reproduction varies. Some Protozoa simply undergo binary fission, budding, or sporulation. All of these are basically asexual processes. In others, however, asexual reproduction is often followed at certain periods by some form of sexual reproduction that may or may not be necessary for the continued existence of the organism.

BINARY FISSION. This process, the most common among Protozoa, involves the division of the organism, both nucleus and cytoplasm, into two essentially equal daughter organisms. Binary fission may be transverse (most ciliates) or longitudinal (Mastigophora). The nucleus divides by mitosis (Fig. 7-3). In many forms the chromosomes are similar in structure and behavior to metazoan chromosomes; in others the chromosomes are granular and highly atypical. Chromosome numbers appear to be constant for a species; for example, *Zelleriella intermedia* (Ciliata) has 24, *Entamoeba histolytica* (Sarcodina), 6; *Oxytricha fallax* (Ciliata), 24; and *Euglena viridis* (Mastigophora), 30.

BUDDING. Budding involves unequal cell division in which usually the parent organism retains its own identity while forming one or more small cells that assume the parent form after they become free. In

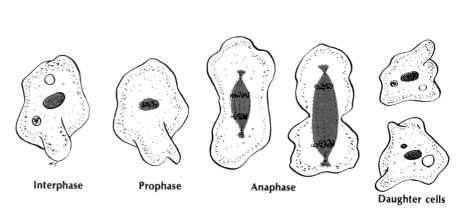

Interphase   **Prophase**   **Anaphase**   **Daughter cells**

FIG. 7-3

Mitosis in nucleus of *Amoeba*. There are many mitotic patterns among protozoans. In most cases nuclear membrane persists throughout mitosis, and division bodies of centrioles, centrosphere, and spindle are of nuclear rather than of cytoplasmic origin. Sometimes one of these division bodies may be absent. In many cases, chromosomes behave as in metazoans. In others, chromatin mass splits and passes to poles without forming chromosomes. Amitosis in protozoans is restricted mainly to macronucleus of ciliates.

some cases the bud may be as large as the parent. Budding may be either external (certain suctorians and ciliates) or internal (suctorians and sporozoans).

MULTIPLE DIVISION (SPORULATION). In multiple division the nucleus divides a number of times, followed by the division of the cytoplasm of the organism into as many parts as there are nuclei. It is a method of rapid multiplication and is characteristic of such parasitic forms as the Sporozoa. It is often found in protozoans with complicated life cycles, including asexual and sexual phases.

PROTOZOAN COLONIES. Protozoan colonies are formed when the daughter zooids remain associated together instead of moving apart and living a separate existence (Fig. 7-4). Protozoan colonies vary from individuals embedded together in a gelatinous substance to those that have protoplasmic connections among them. The arrangement of the individu-

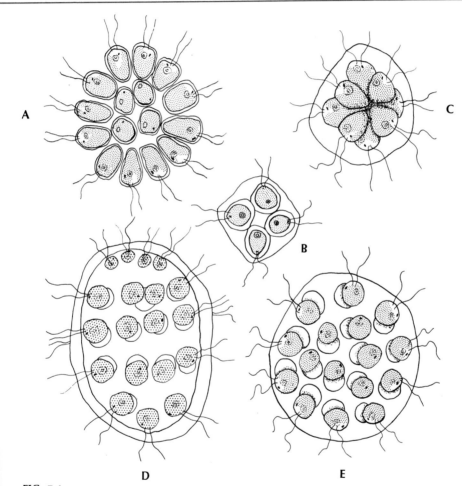

FIG. 7-4
Some colonial protozoans. Such colonies may have been the forerunners of the metazoans. **A,** *Gonium pectorale,* usually of 16 zooids. **B,** *G. sociale,* 4 zooids, each of which may give rise to a daughter colony by cell division or may serve as an isogamete. **C,** *Pandorina morum,* 16 zooids, each of which divides four times to form new colony. **D,** *Pleodorina illinoisensis,* 32 zooids, of which the four anterior are sterile and the others are capable of both sexual and asexual reproduction. **E,** *Eudorina elegans,* 32 zooids, each of which forms a new colony.

als results in certain types of colonies, such as **linear** (daughter cells attached endwise), **spherical** (grouped in a ball shape), **discoid** (platelike arrangement), and **arboroid** (treelike branches). Simple colonies may have only a few zooids *(Pandorina)* (Fig. 7-4), or they may have thousands of zooids *(Volvox)* (Fig. 7-17). Usually the individuals of a colony are structurally and physiologically the same, although there may be some division of labor such as differentiation of reproductive and somatic zooids. Division of labor, however, may be carried so far that it is difficult to distinguish between a protozoan colony and a metazoan individual.

SEXUAL PHENOMENA. Sex is found in certain protozoans but is absent in others. When sex is found, it may involve the formation of male and female gametes (similar or unlike in appearance) that unite to form a zygote (synkaryon), or there may be variant forms of this, such as the complete union of two mature sexual individuals that merge their cytoplasm and nuclei together to form the zygote. By division the zygote may give rise to many individuals or to a new colony. Other sexual phenomena have been described for Protozoa, such as **autogamy,** in which gametic nuclei arise and fuse to form a zygote in the same organism that produces the gametes; **endomixis,** which involves nuclear reorganization without fusion of micronuclei; **parthenogenesis,** or the development of an organism from a gamete without fertilization; and **conjugation,** in which there is an exchange of gametic nuclei of micronuclear origin between paired organisms (conjugants).

Some of these processes are described under the discussion of the paramecium.

Life cycles

Many Protozoa have very complex life cycles; others have simple ones. A simple life cycle may consist of an active phase and a cyst. In some cases the cyst may be lacking. *Amoeba* has a relatively simple life cycle. The more complex life cycles include two or more stages in the active phase and a reproductive phase that may include sexual as well as asexual phenomena. Some protozoans, for example, have both a ciliated and a nonciliated stage in the same organism; others have ameboid and flagellate stages; and still others, free-swimming and sessile stages, etc. The most complex protozoan life cycles are found in the parasitic class Sporozoa, a

good example of which is *Plasmodium* (Fig. 7-18), the malarial parasite, described in a later section.

Encystment is common among protozoans, helping them withstand drought and extreme weather. There is usually a complex series of events when free-living forms encyst. The organism becomes quiescent and many organelles (cilia, flagella, contractile vacuole, etc.) may disappear. A cyst wall is secreted over the surface so that the animal can withstand desiccation, temperature changes, and other harsh conditions. Reproductive cycles, such as budding, fission, and syngamy, may also occur in the encysted condition of some protozoans. The cysts of some protozoans may be viable for many years.

■ REPRESENTATIVE TYPES

A fairly representative member of each large group of protozoans will be described to give the student a basis for comparison of the groups. Forms such as *Amoeba* and *Paramecium,* however, although large and easy to obtain for study, are not wholly representative because their life histories are somewhat simple compared with other members of their respective groups.

■ SUBPHYLUM SARCOMASTIGOPHORA

The Sarcomastigophora includes both those protozoans that move by flagella (Mastigophora) and those that move by means of pseudopodia (Sarcodina). As a rule the nuclei are all of one type (monomorphic), although there may be more than one nucleus in an individual.

Superclass Sarcodina
Amoeba proteus

HABITAT. *Amoeba proteus* is widely distributed. It lives in slow streams and ponds of clear water, often in shallow water on the underside of lily pads and other aquatic vegetation, or on the sides of dams, in watering troughs, and in the sides of ledges where the water runs slowly from a brook or spring. They are rarely found free in water, for they require a substratum on which to glide.

STRUCTURE. The shape of the ameba is irregular and continuously changing because of its power to thrust out **pseudopodia,** or false feet, at any point on its body (Fig. 7-5). It is colorless and about 250 to 600 μ in its greatest diameter. Sometimes its shape is almost spherical when all its pseudopodia are with-

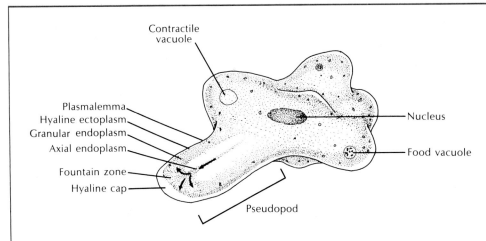

FIG. 7-5
Structure of *Amoeba* in active locomotion. Arrows indicate direction of streaming plasmasol. First sign of new pseudopodium is thickening of ectoplasm to form clear hyaline cap. Into this hyaline region flow granules from fluid endoplasm (plasmasol), forming a type of tube with walls of plasmagel and core of plasmasol. As plasmasol flows forward it fountains out and is converted into plasmagel. Substratum is necessary for ameboid movement, but only tips of pseudopodia touch it. See text for explanation of ameboid movement.

drawn. Although it possesses no cell wall, it has a thin delicate outer membrane called the **plasmalemma** (Fig. 7-5). Just beneath this is a nongranular layer, the **ectoplasm,** which encloses the granular **endoplasm.** The endoplasm is made up of an outer, relatively stiff **plasmagel** and a more fluid inner **plasmasol,** which exhibits flowing or streaming movements. In the gel layer the crystals and granules keep a fixed distance from each other; in the sol they bump into and move over each other.

A number of **organelles** are found within the endoplasm. The disk-shaped **nucleus** is granular and refractive to light. Another organelle is the **water expulsion vesicle (contractile vacuole),** a bubblelike body that grows to a maximum size and then contracts to expel its fluid contents to the outside. Scattered through the endoplasm are **food vacuoles,** which are drops of water enclosing food particles. There are also other vacuoles, **crystals,** and **granules** of various shapes and forms. Foreign substances such as sand and bits of debris may also be in the protoplasm, having been picked up accidentally.

METABOLISM. The ameba lives upon algae, protozoans, rotifers, and even other amebas. It shows some selection in its food, for it will not ingest every-thing that comes its way. Food may be taken in at any part of the body surface. When the ameba engulfs food, it thrusts out pseudopodia to enclose the food particle completely (phagocytosis) (Fig. 7-6). Along with the food, some water in which the food is suspended may also be taken in. These food vacuoles are carried around by the streaming movements of the endoplasm. A lysosome with enzymes fuses with each food vacuole, and digestion proceeds within the vacuole. As digestion proceeds, the vacuoles decrease in size because of loss of water and the passage of the digested material into the surrounding cytoplasm. Finally, indigestible material is eliminated by passing out through the plasmalemma as the animal flows away.

The ameba is able to live for many days without food but decreases in volume during this process. The time necessary for the digestion of a food vacuole varies with the kind of food, but is usually around 15 to 30 hours.

The ameba needs and utilizes energy like any other animal. It gets this energy by oxidation, which results in waste products such as carbon dioxide, water, and urea. Some of these waste substances are eliminated through the body surface, but some are

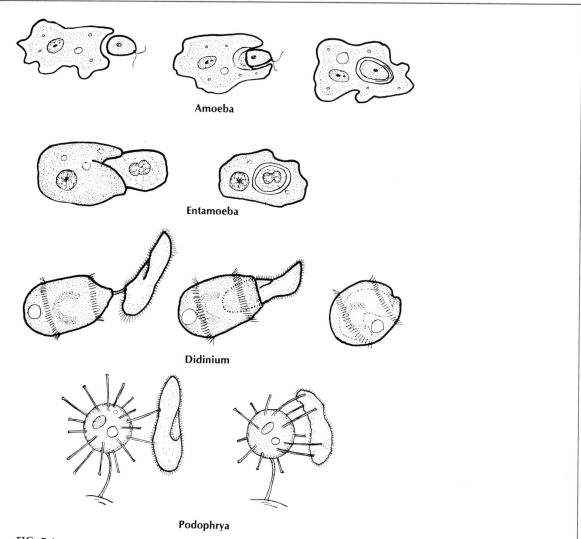

Amoeba

Entamoeba

Didinium

Podophrya

FIG. 7-6
Some typical methods of ingestion among protozoans. *Amoeba* ingests small flagellate.
Entamoeba, a parasite, engulfs leukocyte. *Didinium,* a holotrich, eats only paramecia. It
pierces prey before swallowing it whole. Suctorians such as *Podophrya* have protoplasmic
tentacles with funnel ends that suck protoplasm from prey.

discharged through the water expulsion vesicle, which also gets rid of excess water that the ameba is continually taking in. It is thus responsible for regulating the osmotic pressure of the body. The ameba has a certain amount of salt in its protoplasm that makes it hypertonic to the surrounding fresh water. Water will therefore enter the ameba by osmosis through its plasmalemma. It is interesting to note that marine amebas do not have contractile vacuoles because they are immersed in isotonic sea water (but when placed in fresh water they will form them).

Respiration occurs directly through the body surface by diffusion. Oxygen is dissolved in the water, which is everywhere in contact with the cell membrane so that the gas is easily accessible and diffuses into the ameba.

LOCOMOTION. Locomotion takes place by the formation of temporary locomotor structures, the pseudopodia, which are thrust out on any part of the body surface and into which the cytoplasm flows. This characteristic movement is called **ameboid movement** (Fig. 7-7). When a pseudopodium is beginning to form, a blunt, fingerlike projection called the **hyaline cap,** composed only of ectoplasm, first appears. A little later the granular plasmasol flows into this projection as it extends forward (Fig. 7-5). Usually the ameba forms several small pseudopodia at the start of the movement; one of these gradually becomes larger, while the others disappear.

Theories to account for ameboid movement are far from satisfactory. One theory (Mast's) is based upon a reversible transformation of the cytoplasm from a sol (fluid) to a gel (solid). According to this theory, the forward movement of the solated endoplasm is caused by the contraction of molecules of plasmagel in the ectoplasmic tube in the temporary posterior (tail) end of the ameba. This pressure pushes out the solated anterior end, forming a pseudopodium. As the fluid endoplasm reaches the tip of the advancing pseudopod, it fountains to the sides, undergoes gelation, and forms a plasmagel sleeve, or tube. Contraction of the posterior gel ectoplasm is immediately followed by solation to replenish the internal plasmasol.

This tail-contraction theory is now considered too simple by some investigators, who maintain that the ameba pulls itself forward by contraction of molecules of endoplasm at the anterior end (front-contraction theory). Evidence indicates that the endoplasmic axial core has a viscosity about the same as that of the ectoplasmic tube, instead of being much less, as in Mast's theory. The only regions with low viscosity and rapid movement of cytoplasm are the zone around the axial endoplasm and the region where the endoplasm is being formed from the posterior ectoplasm. Strands of free-moving cytoplasm stream forward in the axial endoplasm to the anterior fountain zone, where the molecules contract and fountain outward to join the ectoplasmic tube. The endoplasm anchored to the ectoplasmic tube contracts, everts, and pulls the axial endoplasm forward. Continuous streaming movements are kept up by the propagated contraction along the ectoplasm. The hyaline cap is formed from water squeezed from the endoplasm as the molecules contract. This water eventually moves backward in the channel just beneath the plasmalemma.

A necessary feature of ameboid movement is the

FIG. 7-7
Ameboid movement. Series photographed at intervals of about half a minute. *Right,*
Pseudopodium is extending toward escaping rotifer.

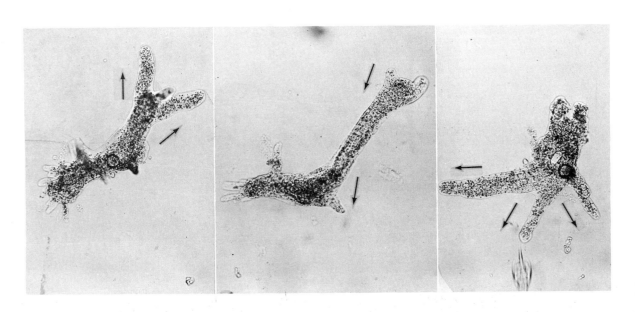

attachment of the ameba (at the tips of pseudopodia) to a substratum during the act of moving.

Ameboid movement is also found elsewhere in the animal kingdom, notably in the white corpuscles of blood, in amebocytes of sponges, etc.

REPRODUCTION. When the ameba reaches full size, it divides into two animals by the process of **binary fission.** Typical mitosis (Fig. 7-3) occurs with all the phases, taking about 30 minutes. During the process of division the shape of the ameba is spherical, with a number of small pseudopodia. The nuclear membrane disappears during the metaphase and the body elongates and separates by fission into two daughter cells. Under ordinary conditions the ameba attains a size for division about every 3 days.

Sporulation and budding have been reported to occur in the ameba, but binary fission seems to be the only regular method employed.

BEHAVIOR. The ameba reacts to stimuli just as any other animal does. Its reactions center around food getting, locomotion, changes in shape, avoidance of unfavorable environments, etc. Its responses to different forms of stimuli vary. In a positive reaction the ameba goes toward the stimulus; in a negative reaction it moves away. If touched with a needle it will draw back and move away, but when floating it will respond in a positive way to a solid object. It moves away from a strong light and may change its direction a number of times to avoid it, but it may react positively to a weak light. The ameba's rate of locomotion is lessened by colder temperatures and may cease entirely near the freezing point. Its rate increases up to 90° F., but it ceases to move at temperatures higher than this.

Its response to chemicals varies with the nature of the chemical. Although indifferent to most normal constituents in its medium, the ameba will react positively toward substances of a food character.

Other members of Sarcodina

RHIZOPODEA. The amebas belong to a class of sarcodines called Rhizopodea, all of which have pseudopodia similar to those of *Amoeba proteus.* There are many species of *Amoeba,* for example, *A. verrucosa* with short pseudopodia; *Pelomyxa carolinensis (Chaos chaos)* (Fig. 7-8), which is several times as large as *A. proteus;* and *A. radiosa* with many slender pseudopodia.

There are many entozoic amebas, most of which live in the intestine of man or other animals. Two common genera are *Endamoeba* and *Entamoeba. Endamoeba blattae* is an endocommensal in the in-

FIG. 7-8

Comparison in size of *Pelomyxa* (larger) and *Amoeba.* The former may attain length of 5 mm. Several paramecia also shown. (Courtesy Carolina Biological Supply Co., Burlington, N. C.)

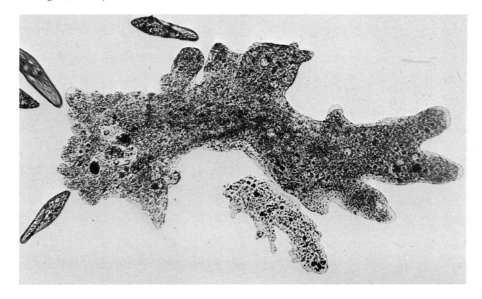

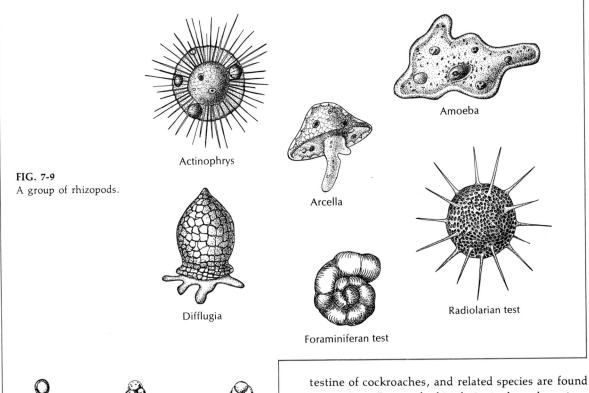

FIG. 7-9
A group of rhizopods.

Actinophrys

Arcella

Amoeba

Difflugia

Foraminiferan test

Radiolarian test

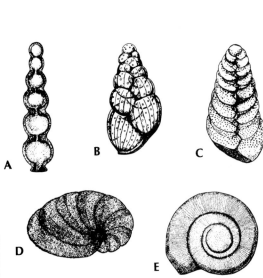

A

B

C

D

E

FIG. 7-10
Some foraminiferan fossil shells. Millions of square miles of thick ocean ooze is made up of tests of forams. **A,** *Nodosaria vigula* Brady (fossil, Brno, Moravia). **B,** *Textularia fucosa* Reiss (recent, Antilles Islands). **C,** *Bolivina punctata* d'Orb (fossil, Santa Monica, Calif., U. S. A.). **D,** *Monionina scapha* Ficht (recent, Samoa). **E,** *Spirillina vivipara* Ehrenberg (recent, Samoa). (Courtesy General Biological Supply House, Inc., Chicago.)

testine of cockroaches, and related species are found in termites. *Entamoeba histolytica* is the only serious rhizopod parasite of man. It lives in the intestinal wall, which it enters by secreting a substance that dissolves away the intestinal lining. It causes amebic dysentery and often produces severe lesions and abscesses that may spread to other organs. Some infected persons do not show severe symptoms but are carriers. Infection is spread by contaminated water or food containing cysts discharged in the feces.

Other species of *Entamoeba* found in man are *E. coli,* which causes intestinal disturbances, and *E. gingivalis,* which causes pyorrhea in the mouth by dissolving away the cement that holds the teeth to the bone.

Not all rhizopods are "naked" as are the amebas. Some have their delicate plasma membrane covered with a protective **test** or shell. *Arcella* and *Diffugia* (Fig. 7-9) are common sarcodines of Order Arcellinida (Testacida), which have a test of secreted siliceous material or pseudochitin reinforced with grains of sand. They move with pseudopodia that project from openings in the shell.

The **foraminiferans** (Order Foraminiferida) are an ancient group of shelled rhizopods found in all

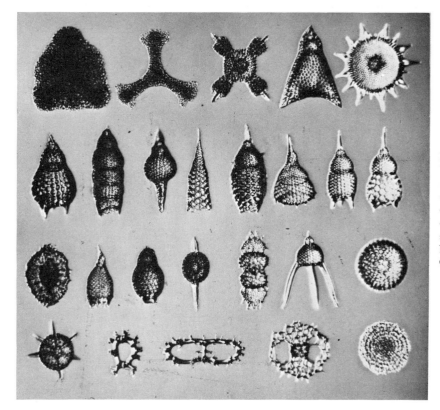

FIG. 7-11
Types of radiolarians. In his study of these beautiful forms collected on famous *Challenger* expedition, Haeckel worked out our present concepts on symmetry. (Courtesy General Biological Supply House, Inc., Chicago.)

oceans and a few in fresh and brackish water. They are mostly bottom living, but a few live in open water. Their tests are of numerous types (Figs. 7-10). Most tests are many-chambered and are made of calcium carbonate, although silica, silt, and other foreign materials are sometimes used. Slender pseudopodia extend through openings in the test, then branch and run together to form a protoplasmic net in which they ensnare their prey. Here the captured prey is digested and the digested products are carried into the interior by the flowing protoplasm. Their life cycles are complex, for they have multiple division and alternation of generations.

The **slime molds** also belong to the rhizopod group.

ACTINOPODEA. In this class of sarcodines the pseudopodia are slender and usually radiate out from the central test. Some forms have the slender pseudopods stiffened by an axoneme running down their center. These protozoans are beautiful little animals. Included in this group are the heliozoans, which are mostly freshwater forms. They include

Actinosphaerium, which is about a millimeter in diameter and can be seen with the naked eye, and *Actinophrys* (Fig. 7-9), only 50 μ in diameter; neither has a test. *Clathrulina* secretes a latticed test.

The radiolarians are the oldest known group of animals. They are all marine and nearly all pelagic (live in open water). Most of them are planktonic, though some live in deep water. Their highly specialized siliceous skeletons are intricate in form and of great beauty (Fig. 7-11). The body is divided by a central capsule that separates inner and outer zones of cytoplasm. The central capsule, which may be spherical, ovoid, or branched, is perforated to allow cytoplasmic continuity. The skeleton is made of silica or strontium sulfate and usually has a radial arrangement of spines that extend through the capsule from the center of the body. At the surface a shell may be fused with the spines. Around the capsule is a frothy mass of cytoplasm from which stiff pseudopodia arise. These are sticky for catching the prey that are carried by the streaming protoplasm to the central capsule to be digested. Radiolarians may have one or many nuclei. Their life history is not completely known, but binary fission, budding, and sporulation have been observed in them.

Role in building earth deposits

Two orders, Foraminifera and Radiolaria, which have existed since Precambrian times, have left excellent fossil records, for their hard shells have been preserved unaltered. Many of the extinct species are identical to present ones. They were especially abundant during the Cretaceous and Tertiary periods. Some of them were among the largest of protozoans, measuring up to 100 mm. or more in diameter.

For untold millions of years the tests of dead Foraminifera (Fig. 7-10) have been sinking to the bottom of the ocean, building up a characteristic ooze rich in lime and silica. Most of this ooze is made up of the shells of the genus *Globigerina*. About one third of all the sea bottom is covered with *Globigerina* ooze. This ooze is especially abundant in the Atlantic Ocean.

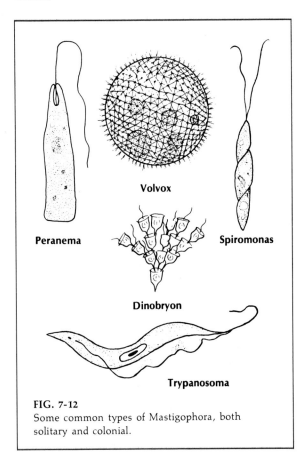

FIG. 7-12
Some common types of Mastigophora, both solitary and colonial.

The Radiolaria, with their less soluble siliceous shells (Fig. 7-11), are usually found at greater depths (15,000 to 20,000 feet), mainly in the Pacific and Indian Oceans. Radiolarian ooze probably covers about 2 to 3 million square miles. Under certain conditions, radiolarian ooze forms rocks (chert). Many fossil Radiolaria are found in the Tertiary rocks of California.

The thickness of these deep-sea sediments has been estimated to be from 2,000 to 12,000 feet. Although the average rate of sedimentation must vary greatly, it is always very slow. *Globigerina* ooze is probably increased 1 to 12.5 mm. in a thousand years. When one considers that as many as 50,000 shells of Foraminifera may be found in a single gram of sediment, one can form some idea of the magnitude of numbers of these microorganisms and the length of time it has taken them to form the sediment carpet on the ocean floor.

Of equal interest and of greater practical importance are the limestone and chalk deposits on land that were laid down when a deep sea covered the continents. Later, through a rise in the ocean floor and other geologic changes, this sedimentary rock emerged as dry land. The chalk deposits of many areas of England, including the White Cliffs of Dover, were laid down by the accumulation of these small microorganisms. The great pyramids of Egypt were made from limestone beds that were formed by a very large foraminiferan that flourished during the early Tertiary period. Since petroleum oil is of organic origin, the presence of fossil Foraminifera (and also Radiolaria) in oil-bearing rock strata may be significant to oil geologists.

Superclass Mastigophora

The superclass Mastigophora is made up of the flagellates, or the protozoans that move by means of flagella. They are usually considered the most primitive of the protozoan groups. This group is divided into the phytoflagellates (Phytomastigophorea), which have chlorophyll and are thus plantlike, and the zooflagellates (Zoomastigophorea), which do not have chlorophyll but are either holozoic or saprozoic and thus are animal-like.

The **phytoflagellates** usually have one or two flagella (sometimes four) and chromatophores (also called chromoplasts or chloroplasts), which contain the pigments used in photosynthesis. They are mostly free living and contain such familiar forms as *Euglena, Chlamydomonas, Peranema* (Fig. 7-12), *Volvox,* and the dinoflagellates that cause the red tides. *Chilomonas* is another common form that is the chief food of amebas. *Noctiluca,* a marine form, is luminescent and produces a striking greenish light at night. *Ceratium,* which is found in both freshwater and saltwater, has a body covering of cellulose plates and horns. Some of these flagellates are colonial (Fig. 7-4), living in characteristic groups, such as *Gonium* with 16 individuals (zooids), *Eudorina* with 32 zooids, and *Volvox* (Fig. 7-12) with thousands of zooids.

The **zooflagellates** are the colorless flagellates (because they lack chromatophores), with holozoic or saprozoic nutrition, and many are parasitic; some have pseudopodia as well as flagella. An example is *Proterospongia,* a colony of collared zooids (Fig. 7-13) that has been suggested as a forerunner of the sponges, in which collared cells are typical.

Some of the worst of the protozoan parasites are zooflagellates. Many of them belong to the genus *Trypanosoma* (Fig. 7-12) and live in the blood of fishes, amphibians, reptiles, birds, and mammals. Some are nonpathogenic, but those that infect the mammals produce severe diseases. *T. gambiense*

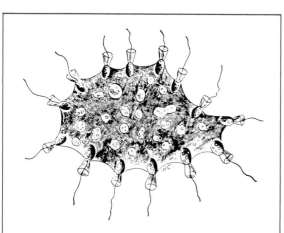

FIG. 7-13
Proterospongia, a colonial choanoflagellate. In a gelatinous mass, collared zooids are embedded on the outside and collarless ameboid zooids on the inside. Collared cells resemble choanocytes of sponges. Only choanoflagellates and sponges have these peculiar cells.

causes African sleeping sickness, and *T. rhodesiense* causes the more virulent Rhodesian sleeping sickness. Both are transmitted by the tsetse fly (*Glossina*). Their natural reservoirs (antelope and other wild mammals) are apparently not harmed by harboring these parasites. *T. cruzi* causes Chagas' disease in man in Central and South America. It is transmitted by the bite of a "kissing bug" (*Triatoma*). Three species of *Leishmania* cause severe diseases in man. One is a disease of the spleen and liver, another is a type of skin lesion, and a third is lesions in the mucous membranes of the nose and throat. These are transmitted by sandflies or by direct contact and are common in Africa, Asia, and Central and South America.

Several species of *Trichomonas* (Fig. 7-1) are commensals. *T. hominis* is found in the cecum and colon of man; *T. vaginalis,* sometimes found in the vagina of women, may cause vaginitis; and other species of *Trichomonas* are widely distributed through all classes of vertebrates and many invertebrates. *Giardia intestinalis* is often blamed for a severe diarrhea. It is transmitted through fecal contamination.

Euglena viridis, a phytoflagellate

HABITAT. The normal habitat of *Euglena viridis* (order Euglenida) (Fig. 7-14) is freshwater streams and ponds where there is considerable vegetation. They are sometimes so numerous as to give a distinctly greenish color to the water. Although light is necessary for their metabolism, they are often found at various depths below the surface of water, for they are fairly active forms.

STRUCTURE. The spindle-shaped body of *E. viridis* is about 60 μ (0.06 mm.) long, but some species are smaller and some larger (*E. oxyuris*) is 500 μ long. It is covered by a **pellicle** flexible enough to permit movement (Figs. 7-16). Inside the pellicle is a thin layer of clear **ectoplasm** surrounding the mass of **endoplasm.** From a flask-shaped **reservoir** in the anterior end a **flagellum** extends. It originates as the union of two delicate threads, or **axonemes,** each of which ends as a tiny granule, or **blepharoplast** (kinetosome), on the floor of the reservoir that seems to be essential for movement of the flagellum and that may function as a centriole in cell division. A tiny fibril, or **rhizoplast,** extending from one of the blepharoplasts to the **nucleus** near the center of the cell suggests that the flagellum is under nuclear control.

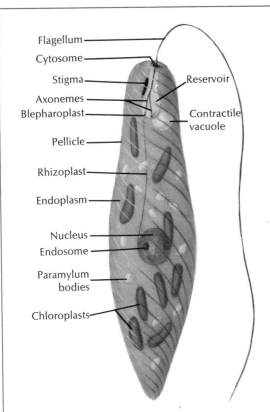

Flagellum
Cytosome
Stigma
Axonemes
Blepharoplast
Reservoir
Contractile vacuole
Pellicle
Rhizoplast
Endoplasm
Nucleus
Endosome
Paramylum bodies
Chloroplasts

FIG. 7-14
Euglena. Features shown are a combination of those visible in living and stained preparations.

FIG. 7-15
Cross section of flagella. Each flagellum has nine peripheral and two central fibrils, each made up of two microfibrils enclosed in a sheath. Cilia have the same construction. (Electron micrograph, courtesy I. R. Gibbons, Harvard University.)

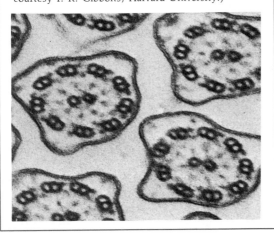

A swelling of the flagellum associated with the **eyespot** suggests a possible mechanism by which the organism reacts to light changes.

The electron microscope reveals that a flagellum consists of eleven fibrils arranged with nine of them in a circle and two in the center and all contained in an elastic outer sheath (Fig. 7-15). It is believed that contraction of fibrils on one side causes the flagellum to bend and that the elasticity of the outer layer causes it to return. Each of the fibrils is in turn made up of two microfibrils within a sheath. This same pattern is true of cilia and of the tail or flagellum of the spermatozoa of higher animals. This pattern appears to have a significant evolutionary sequence because the flagellum of bacteria has only two or three closely wound fibrils instead of eleven (Dillon).

A large **water expulsion vesicle (contractile vacuole),** which is formed by fusion of smaller vacuoles, empties wastes and excess water into the reservoir, the anterior opening of which is an exit.

Near the reservoir is a red **eyespot,** or **stigma** (Fig. 7-14). This is a shallow cup of pigment that allows light from only one direction to strike a light-sensitive receptor located as a swelling near the base of the flagellum. When the euglena is moving toward the light, the receptor is illuminated; when it changes direction, the shadow of the pigment falls on the receptor. Thus the animal, which depends upon sunlight for its photosynthesis, can orient itself toward the light. If given the choice, it will avoid shady areas and regions of bright light.

Within the cytoplasm are oval **chromatophores** (chloroplasts) that bear chlorophyll and give the euglena its greenish color. **Paramylum bodies** of various shapes are masses of starch, a means of food storage.

METABOLISM. The euglena derives its food mainly through **autotrophic (holophytic)** nutrition, which makes use of photosynthesis, a process that takes place within the chromatophores through the action of chlorophyll. This form also makes use of **saprozoic nutrition,** which is the absorption of dissolved nutrients through the body surface. It is doubtful whether the euglena ingests solid food particles through its mouth region **(holozoic nutrition),** although some flagellates such as *Peranema* ingest other organisms. **Respiration** and **excretion** occur by diffusion through the body wall.

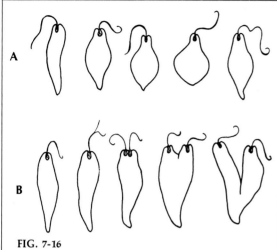

FIG. 7-16
A, Changes of shape in mastigophorans such as *Euglena* are called "euglenoid movement."
B, Reproduction in *Euglena* occurs by longitudinal fission, beginning at anterior end.

LOCOMOTION. The euglena swims freely by the movement of its flagellum, which moves in a whip-like manner or with rotary motion, with the undulation passing from base to tip. This motion pulls the animal forward in a straight course while the body rotates spirally. Flagellates can travel from a few tenths to 1 mm. per second, according to their size, with the speed increasing with the larger flagellates. Some flagellates use the flagella to push rather than to pull themselves along. *Euglena* can also change its shape by peristalsis-like "euglenoid" movement (Fig. 7-16, *A*).

REPRODUCTION. *Euglena* reproduces by longitudinal **binary** division. The nucleus undergoes mitotic division, while the body, beginning at the anterior end, divides lengthwise (Fig. 7-16, *B*). During inactive periods and for protection the euglena assumes a spherical shape surrounded with a gelatinous covering, thus becoming **encysted.** In this condition it can withstand drought and can become active when it is in water again. Encysted euglenas usually divide so that each cyst may contain two or more euglenas.

Volvox globator

The group to which *Volvox* belongs (order Volvocida) includes many freshwater flagellates, mostly green, that have a close resemblance to algae. The bodies of most species are enclosed in a cellulose membrane through which two short flagella project. Most of them are provided with green chromatophores. Many are colonial forms (Fig. 7-4).

Volvox (Fig. 7-17) is a green hollow sphere that may reach a diameter of 0.5 to 1 mm. It is a colony of many thousands of zooids (up to 50,000) embedded in the gelatinous surface of a jelly ball. Each cell is much like a euglena, with a nucleus, a pair of flagella, a large chloroplast (a type of chromatophore), and a red stigma. Adjacent cells are connected with each other by cytoplasmic strands. At one pole (usually in front as the colony moves), the stigmata are a little larger. Coordinated action of the flagella causes the colony to move by rolling over and over.

Here we have the beginning of division of labor to the extent that most of the zooids are somatic cells concerned with nutrition and locomotion, and a few germ cells located in the posterior half are responsible for reproduction. Reproduction is asexual or sexual. In either case only certain zooids located around the equator or posterior half take part.

Asexual reproduction in *Volvox* occurs by the repeated mitotic division of one of the germ cells, to form a hollow sphere of cells, with the flagellate ends of the cells inside. The sphere then invaginates, or turns itself inside out, to form a daughter colony like the parent colony. Several daughter colonies are formed inside the parent colony before they escape by rupture of the parent (Fig. 7-17).

In **sexual reproduction** some of the zooids differentiate into **macrogametes** (ova) or **microgametes** (sperm) (Fig. 7-17). The macrogametes are fewer and larger and are loaded with food for nourishment of the young colony. The microgametes, by repeated division, form bundles or balls of small flagellated sperm that, when mature, leave the mother colony and swim about to find a mature ovum. When a sperm fertilizes an egg, the zygote so formed secretes a hard, spiny, protective shell around itself. When released by the breaking up of the parent colony, the zygote remains quiescent during the winter. Within the shell the zygote undergoes repeated division until a small colony is produced that is released in the spring. A number of asexual generations may follow, during the summer, before sexual reproduction occurs again.

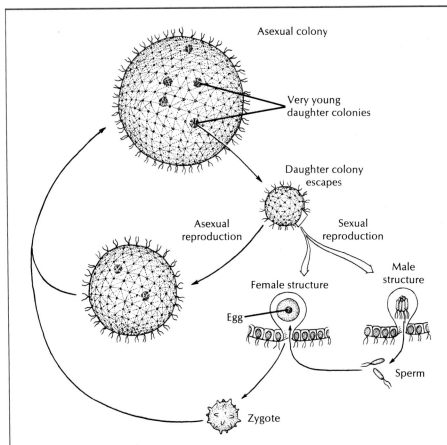

Asexual colony

Very young
daughter colonies

Daughter colony
escapes

Asexual
reproduction

Sexual
reproduction

Male
structure

Female structure

Egg

Sperm

Zygote

FIG. 7-17
Life cycle of *Volvox*. Asexual reproduction occurs in spring and summer when specialized
diploid reproductive cells divide to form young colonies that remain in the mother colony
until large enough to escape. Sexual reproduction occurs largely in autumn when haploid
sex cells develop. The fertilized ova may encyst and so survive the winter, developing into
mature asexual colonies in the spring. In some species the colonies have separate sexes;
in others both eggs and sperm are produced in the same colony.

■ SUBPHYLUM SPOROZOA

All sporozoans are endoparasites and their hosts
are found in all of the animal phyla. The adult stages
have no organelles of locomotion such as pseudopodia, flagella, or cilia. Sporozoans are a heterogeneous
group; the various taxa have little in common except
their parasitic habit and the fact that they bear spores
during some stage of their life cycle. Hosts are usually infected by means of spores enclosed in hard
walls, or sometimes the transmission is by naked
young. There is sometimes an intermediate host
such as mosquitoes, leeches, flies, or other vectors.

Although there are no locomotor organelles in the
adult stages, the reproductive cells are often flagellated, and some of the stages exhibit ameboid movement. Some forms have myonemes for contraction.

Nutrition is mostly saprozoic, or by absorption of
liquid nutrients through the body surface.

The life cycle usually includes both an asexual and
a sexual stage, although some, such as the greg-
arines have only a sexual stage. In many sporozoans
the complete life histories are not yet known.

Coccidia

The Coccidia are sporozoans whose life cycle involves both schizogony and sporogony and is usually

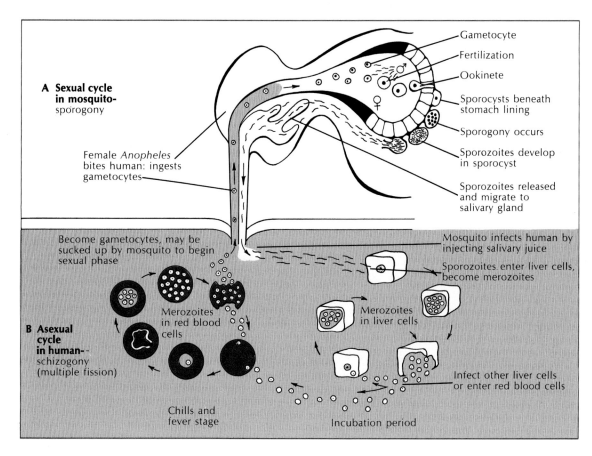

FIG. 7-18
Life cycle of *Plasmodium vivax,* the protozoan (class Sporozoa) that causes malaria in
man. **A,** Sexual cycle produces sporozoites in body of mosquito. **B,** Sporozoites infect man
and reproduce asexually, first in the cells of liver sinusoids and finally in red blood cells.
Malaria is spread by the *Anopheles* mosquito, which sucks up gametocytes along with
human blood and later, when biting another victim, leaves sporozoites in the new wound.

confined to one host. They are parasites that infect
epithelial tissues in both invertebrates (annelids,
arthropods, mollusks) and vertebrates. They are
found chiefly in the epithelial lining of the coelom,
alimentary canal, bile duct, blood vessels, etc. The
disease produced is called coccidiosis and may be
serious. The symptoms are usually severe diarrhea
or dysentery. Infection is by the ingestion of oocysts
and sometimes by separate sporozoites. *Eimeria* is a
common genus found in rabbits and chickens. *E.
magna* and *E. stiedae* infect the intestine and liver of
rabbits. *E. tenella* is often fatal to young fowl. Oocysts usually pass in the feces of the rabbit. Later

each zygote produces four sporocysts, which, when
ingested, repeat the life cycle.

Plasmodium

Probably the best known of the coccidians is *Plasmodium,* the parasite that causes malaria in man.
Malaria is one of the most widespread diseases in
the world. It is mainly a disease of tropical and
subtropical countries but is also common in the temperate zones. The vectors of the parasites are female
mosquitoes of the genus *Anopheles.* Four species of
Plasmodium are known to infect man. Each produces
its own peculiar clinical picture, although all malarial

parasites have similar cycles of development in the host (Fig. 7-18).

Man acquires malaria from the bite of the mosquito, which introduces the parasites from its salivary glands into the blood in the form of **sporozoites.** It was found in 1948 that the sporozoites first enter the cells of the liver. Here, as **cryptozoites,** they pass through a process of multiple division **(schizogony).** The products of this division, **merozoites,** then enter the red corpuscles. The period when the parasites are in the liver is called the incubation period. During this time antimalarial drugs may have little effect upon the parasites.

When the parasites enter the red blood corpuscles (usually only one to a cell), they become amebalike **trophozoites.** These feeding forms then develop into **schizonts,** which have granules of black pigment. Each schizont, by multiple fission (schizogony), divides into many daughter asexual merozoites (6 to 36 in number, according to the species of *Plasmodium*); these break out of the red cells to enter other red corpuscles and repeat the asexual cycle. In a few days the number of parasites is so great that the characteristic chills and fever occur; these symptoms are caused mainly by the toxins released by the parasites.

The time elapsing between the fever-chill stages of the cycle depends on the type of malaria. In *P. vivax* (benign tertian) the chills and fever occur every 48 hours; in *P. malariae* (quartan), every 72 hours; in *P. falciparum* (malignant tertian), usually every 24 to 48 hours, although it may be irregular; and in *P. ovale*, every 48 hours.

After this period of asexual reproduction, or schizogony, the merozoites become sexual forms, **gametocytes.** When these gametocytes are sucked up into the stomach of the mosquito, they become **microgametocytes** (male) and **macrogametocytes** (female). Zygotes formed by the union of gametes develop into ookinetes that penetrate into the stomach walls of the mosquito. Later the ookinetes enlarge to form **oocysts.** Each oocyst divides in a few days into thousands of sporozoites that rupture the cyst and migrate to the salivary glands, whence they are transferred to man by the bite of the mosquito.

The developmental cycle in the mosquito requires from 7 to 18 days but may be longer in cool weather. After being inoculated by the mosquito, man usually manifests the symptoms of the disease 10 to 14 days later.

Some latent forms of malaria may persist for some years without showing clinical symptoms, probably because of the small number of parasites in the blood. The body gradually acquires a resistance to the disease; however, this resistance does not prevent relapses.

The elimination of mosquitoes and their breeding places by insecticides, drainage, etc. has been effective in controlling malaria.

Other species of *Plasmodium* parasitize birds, reptiles, and mammals. Those of birds are transmitted chiefly by the *Culex* mosquito.

Gregarinida

The gregarine sporozoons are parasites that live mainly in the digestive tract and body cavity of certain invertebrates, especially insects and annelids. A familiar example is *Monocystis lumbrici*, which lives in the seminal vesicles of earthworms. They may cause sterility in the worm, for they destroy the sperm.

Earthworms are infected by ingesting spores, which contain sporozoites. Each sporozoite enters a bundle of immature sperm cells and becomes a trophozoite, which lives on the sperm cells. The trophozoites (gametocytes) produce gametes. Two gametes unite to form the zygote, which forms a hard case around itself. This is the spore case, or oocyst. The zygote nucleus divides to form 8 daughter nuclei, each of which becomes a sporozoite, and the cycle is ready to start over.

■ SUBPHYLUM CILIOPHORA
Class Ciliata

The ciliates are an interesting group, with a great variety of forms living in all types of freshwater and marine water. Ciliates are the most complex and diversely specialized of all the protozoans. Most of them are free living, but some are commensal and a few are parasitic. Most of them are solitary and motile, but some are sessile and some form colonies. There is great diversity of shape and size. In general they are larger than most other protozoans, but they range from very small (10 to 12 μ) up to 3 mm. long. All have cilia that beat in a coordinated rhythmic manner, though the arrangement of the cilia may vary.

Ciliates are always multinucleate, possessing at least one **macronucleus** and one **micronucleus,** but varying from one to many of either type. The macronuclei are apparently responsible for metabolic and

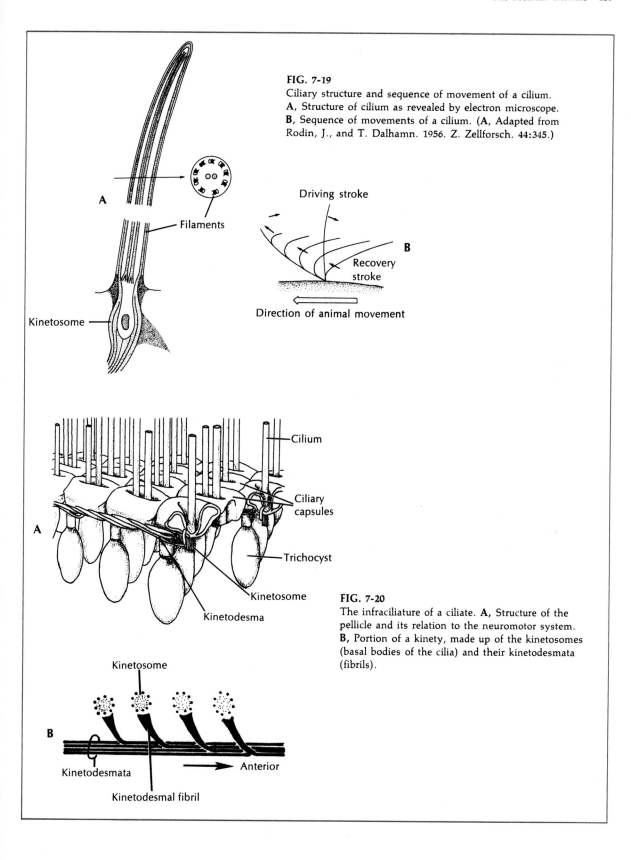

FIG. 7-19
Ciliary structure and sequence of movement of a cilium.
A, Structure of cilium as revealed by electron microscope.
B, Sequence of movements of a cilium. (**A,** Adapted from
Rodin, J., and T. Dalhamn. 1956. Z. Zellforsch. **44:**345.)

Driving stroke

Recovery stroke

B

Direction of animal movement

Filaments

Kinetosome

Cilium

Ciliary capsules

Trichocyst

Kinetosome

Kinetodesma

A

FIG. 7-20
The infraciliature of a ciliate. **A,** Structure of the
pellicle and its relation to the neuromotor system.
B, Portion of a kinety, made up of the kinetosomes
(basal bodies of the cilia) and their kinetodesmata
(fibrils).

Kinetosome

B

Kinetodesmata

Kinetodesmal fibril

Anterior

developmental functions and for maintaining all the visible traits, such as the pellicular apparatus. Macronuclei are varied in shape among the different species. The micronuclei control the sexual and reproductive processes and have a long-range control over the macronuclei. They divide mitotically; the macronuclei divide amitotically.

Ciliates are covered by a **pellicle,** which may be very thin or, in some species, form a thickened armor. The **cilia** are short and usually arranged in longitudinal or diagonal rows. Like flagella, they have two central and nine peripheral fibrils (Figs. 7-15 and 7-19). Cilia may cover the surface of the animal or may be restricted to the oral region or to certain bands. In some forms the cilia are fused into a sheet called an **undulating membrane,** or into smaller **membranelles,** both used to propel food into the **cytopharynx** (gullet). In other forms there may be fused cilia forming stiffened tufts called **cirri,** often used in locomotion by the creeping ciliates (Fig. 7-21).

The motor-coordinating system for ciliary movement apparently lies in the **intraciliature** just beneath the pellicle (Fig. 7-20, *A*). Each cilium terminates beneath the pellicle in a basal granule called a **kinetosome.** From each kinetosome a **kinetodesmal fibril** arises and passes along beneath the row of cilia, joining with the other fibrils of that row. The kinetosomes and fibrils (kinetodesmata) of that row make up what is known as a **kinety** (Fig. 7-20, *B*). All ciliates seem to have kinety systems, even those that lack cilia at some stage.

Many ciliates have contractile fibrils called **myonemes** that run in rows parallel with the rows of kinetosomes and permit extensive contraction and alteration in shape in the animal. In *Stentor* (Fig. 7-21) waves of contraction spread over the animal in both an anterior and a posterior direction, but electrical stimulus will cause contraction simultaneously in all areas of the body. The electron microscope reveals two systems of fibrils, the km fibers (composed of stacks of microtubules) and the M bands (bundles of microfilaments) lying beneath the km fibers. High-speed cinematic analysis (Newman, 1972) indicates that the M bands may be the ones responsible for contraction.

Most ciliates are holozoic. Most of them possess a cytostome (mouth) that in some forms is a simple opening and in others is connected to a gullet or ciliated groove. The mouth in some is strengthened with stiff, rodlike trichites for swallowing larger prey; in others ciliary water currents carry micro-

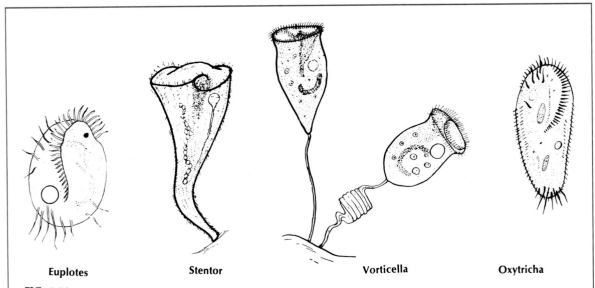

Euplotes   **Stentor**   **Vorticella**   **Oxytricha**

FIG. 7-21
Some representative ciliates. *Euplotes* and *Oxytricha* can use stiff cirri for crawling about.
Contractile myonemes in ectoplasm of *Stentor* and in stalks of *Vorticella* allow great expansion and contraction.

scopic food particles toward the mouth, as in the paramecia. *Didinium* has a proboscis for engulfing the paramecia it feeds upon (Fig. 7-6). Suctorians paralyze their prey and then suck out their contents through tubelike sucking tentacles (Fig. 7-6). In any case the food is digested within food vacuoles.

In one of the subclasses, Holotricha, small bodies called **trichocysts** are located in the ectoplasm (Figs. 7-20 and 7-23). When discharged, they form long threadlike filaments that pass through the pellicle and harden, except for the tips, which are sticky for attachment. In some ciliates the discharged trichocysts seem to help anchor the animal while feeding; in others they are apparently used to paralyze other small organisms for defense or capture of prey.

Among the more striking and familiar of the ciliates are *Stentor*, trumpet-shaped and solitary, with a bead-shaped macronucleus (Fig. 7-21); *Vorticella*, bell-shaped and attached by a contractile stalk (Fig. 7-21); *Euplotes* and *Stylonychia*, with flattened bodies and groups of fused cilia (cirri) that function as legs (Fig. 7-21); *Blepharisma*, slender and pink in color; and *Spirostomum*, very long and vermiform in appearance.

Paramecium, a representative ciliate

Paramecia are usually abundant in ponds or sluggish streams containing aquatic plants and decaying organic matter.

STRUCTURE. The paramecium is often described as slipper shaped. *Paramecium caudatum* is from 150 to 300 μ (0.15 to 0.3 mm.) in length and is blunt anteriorly and somewhat pointed posteriorly (Fig. 7-22). The animal has an asymmetric appearance because of the **oral groove**, a depression that runs obliquely backward on the ventral side.

The **pellicle** is a clear, elastic membrane divided into hexagonal areas by tiny elevated ridges (Fig. 7-20) and is covered over its entire surface with cilia arranged in lengthwise rows. Just below the pellicle is the thin clear **ectoplasm** that surrounds the larger mass of granular **endoplasm** (Fig. 7-23). Embedded in the ectoplasm just below the surface are the spindle-shaped **trichocysts** filled with a semifluid substance that may be discharged for attachment and defense. The trichocysts alternate with the bases of the cilia. The infraciliature can be seen only by special fixing and staining methods.

The **cytostome**, or **mouth**, at the end of the oral

groove leads into a tubular **cytopharynx**, or **gullet**. Along the gullet is an undulating membrane to keep food moving. Fecal material is discharged through an **anal pore (cytoproct)** posterior to the oral groove (Fig. 7-23). The endoplasm contains food vacuoles containing food in various stages of digestion. There are two water expulsion vesicles **(contractile vacuoles)**, each consisting of a central space surrounded by several **radiating canals** that collect fluid and empty it into the central vacuole.

P. caudatum has two nuclei: a large, kidney-shaped **macronucleus** and a smaller **micronucleus** fitted into the depression of the former. These can usually be seen only in stained specimens. The micronucleus is the reproductive nucleus and also gives rise to the macronucleus. The macronucleus is not essential to reproduction but is essential for normal metabolism. The number of micronuclei varies in different species. *P. multimicronucleatum* may have as many as seven.

METABOLISM. Paramecia are holozoic, living upon bacteria, algae, and other small organisms. They are selective in choosing their food, for some items are taken in and others rejected. The cilia in the oral groove sweep food particles in the water into

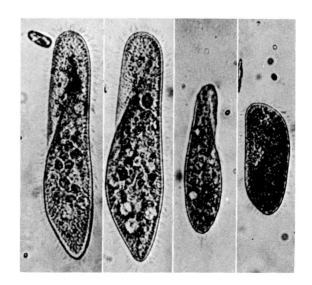

FIG. 7-22
Comparison of four common species of *Paramecium* photographed at same magnification. Left to right: *P. multimicronucleatum, P. caudatum, P. aurelia,* and *P. bursaria.* (Courtesy Carolina Biological Supply House, Burlington, N. C.)

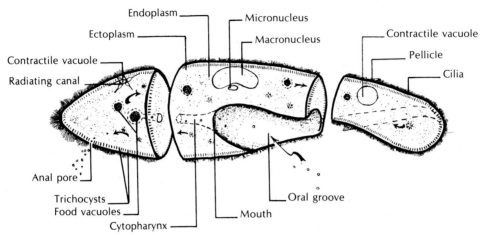

FIG. 7-23
General structure of *Paramecium* (shown as though sectioned).
Arrows indicate the direction of movement of food vacuoles
in the cytoplasm (cyclosis).

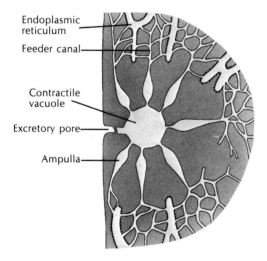

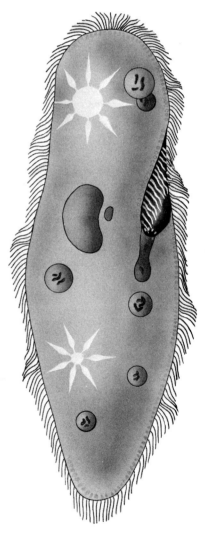

FIG 7-24
Enlarged section of a contractile vacuole (water expulsion vesicle) of
Paramecium. Water is believed to be collected by endoplasmic reticulum,
emptied into feeder canals and then into the vesicle. The vesicle contracts
to empty its contents to the outside, thus serving as an osmoregulatory
organelle. *Right, Paramecium,* showing gullet, food vacuoles, and nuclei.

the cytostome, whence they are carried into the cytopharynx by the undulating membrane. From the cytopharynx the food is collected into a food vacuole that is constricted off and dropped into the endoplasm. The food vacuoles circulate in a definite course through the protoplasm (cyclosis) while the food is being digested by enzymes from the endoplasm. The indigestible part of the food is ejected through the anal pore. Digestion in ciliates is fairly rapid. A *Didinium* can digest a whole paramecium in about 20 minutes.

Respiration takes place through the body surface by diffusion, oxygen dissolved in the surrounding water passing in and the waste, including carbon dioxide, passing out.

The two contractile water expulsion vesicles regulate the water content of the body and help eliminate nitrogenous waste. They lie close to the dorsal surface and drain fluid from the cytoplasm by means of radiating canals that connect the vesicle with the endoplasmic reticulum (Fig. 7-24). When the vacuole reaches a certain size, it discharges to the outside through a pore. The two vacuoles contract alternately. They contract more frequently at higher tempera-

tures and in the mature animal. Since the contents of the paramecium are hypertonic to the surrounding fresh water, osmotic pressure would cause water to diffuse into the cell, and thus one of the main functions of the vacuoles is to get rid of the excess water.

LOCOMOTION. The body of the paramecium is elastic, allowing it to bend and squeeze its way through narrow places. Its cilia can beat either forward or backward, so that the animal can swim in either direction. The cilia beat obliquely, thus causing the animal to rotate on its long axis. In the oral groove the cilia are longer and beat more vigorously than the others so that the anterior end swerves aborally. As a result of these factors, the animal follows a spiral path in order to move forward (Fig. 7-25, *A*). In swimming backward the beat and path of rotation are reversed.

BEHAVIOR.* When a ciliate, such as a paramecium, comes in contact with a barrier or a disturbing chemical stimulus, it reverses its cilia, backs up a short distance, and swerves the anterior end as it pivots on its posterior end. This is called an **avoiding reaction** (Fig. 7-25, *B*). While it is doing this, samples of the surrounding medium are brought into the oral groove. When the sample no longer contains the unfavorable stimulus, the animal moves forward. In this "trial-and-error" method the animal attempts many directions until it finds one that is favorable and then makes its escape from the injurious environment.

Paramecia do not always respond in the same manner to the same stimuli. Their physiologic states vary with conditions. A hungry animal will react in a different way from one that is well fed. In general its behavior is conditioned by factors that favor or hinder the normal life processes.

Automatic and fixed responses in orientation to particular stimuli are often called **tropisms** or **taxes**. Taxes refer more specifically to movement or locomotor responses to the stimuli. If the response is movement toward the stimulus, it is a positive response; an avoiding reaction is a negative response. With respect to the type of stimulus, a taxis or tropism might be classified as one of the following: thermotaxis (thermotropism), response to heat; phototaxis, response to light rays; thigmotaxis, response to contact; chemotaxis, response to chemical substances; rheotaxis, response to currents of air or

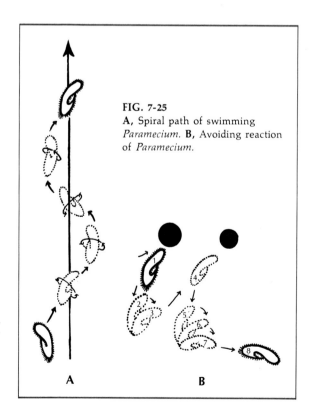

FIG. 7-25

A, Spiral path of swimming *Paramecium.* B, Avoiding reaction of *Paramecium.*

A   B

*Refer to p. 22, Principle 29.

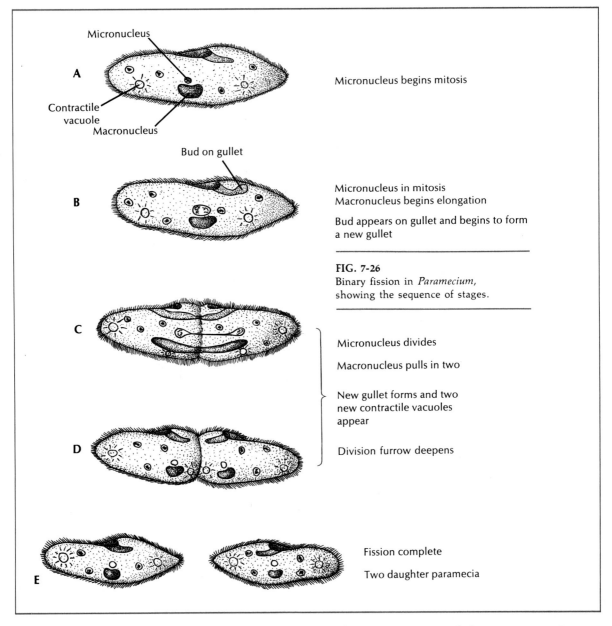

Micronucleus

A

Contractile
vacuole

Macronucleus

Micronucleus begins mitosis

Bud on gullet

B

Micronucleus in mitosis
Macronucleus begins elongation

Bud appears on gullet and begins to form
a new gullet

FIG. 7-26
Binary fission in *Paramecium*,
showing the sequence of stages.

C

Micronucleus divides

Macronucleus pulls in two

New gullet forms and two
new contractile vacuoles
appear

D

Division furrow deepens

E

Fission complete

Two daughter paramecia

water; galvanotaxis, response to constant electric current; or geotaxis, response to gravity.

Since no nervous system is found in protozoans (except perhaps the neuromotor system in ciliates), these responses must be attributable to the innate irritability of protoplasm. The complex responses of higher forms are believed to have developed from these simple mechanical responses.

REPRODUCTION. Paramecia reproduce only by **transverse binary fission** but have certain forms of nuclear reorganization called **conjugation** and **autogamy**.

In **binary fission** the micronucleus divides mitotically into 2 daughter micronuclei, which move to opposite ends of the cell (Fig. 7-26). The macronucleus elongates and divides amitotically. Another cytopharynx is budded off and two new contractile vacuoles appear. In the meantime a constriction furrow appears near the middle of the body and deepens until the cytoplasm is completely divided. The

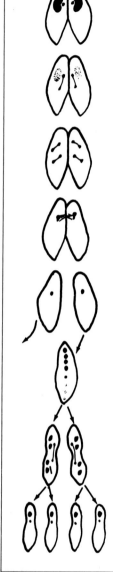

Mates attach by oral grooves

Macronuclei degenerate; micronuclei undergo first meiotic division

Following second meiotic division three haploid micronuclei degenerate

Remaining haploid micronucleus divides by mitosis to form "male" and "female" pronuclei. "Male" pronuclei are exchanged

Pronuclei fuse to form diploid zygote nucleus; mates separate

Zygote nucleus divides 3 times to form 4 macronuclei and 4 micronuclei; 3 micronuclei degenerate

Two mitotic divisions occur to produce 4 individuals, each with a macronucleus and a micronucleus

FIG. 7-27
Conjugation in *Paramecium caudatum.*

process of binary fission requires from $1/2$ to 2 hours.

The division rate usually varies from one to four times each day. About 600 generations can be produced in a year. If all descendants were to live and reproduce, the number of paramecia produced would soon equal the volume of the earth!

The process known as **conjugation** (Fig. 7-27) occurs in ciliates and a few other protozoans. This phenomenon happens only at intervals. It is the temporary union of two individuals that mutually ex-change micronuclear material. Conjugating individuals come together and attach by their oral surfaces, and a protoplasmic bridge forms between them. In thriving cultures one may see a number of these conjugating pairs swimming about.

A series of nuclear changes now occurs. The macronucleus starts to disintegrate and finally disappears. The micronucleus enlarges, forms a spindle, and divides by meiosis, resulting in 4 daughter nuclei, each with a haploid number of chromosomes.

Three of the daughter micronuclei degenerate. The remaining micronucleus divides unequally into 2 pronuclei, the smaller of which in each animal moves across the protoplasmic bridge into the other animal. Each of these exchanged (male) pronuclei fuses with the larger (female) pronucleus of the other animal, thus restoring the normal (diploid) number of chromosomes in the micronucleus of each animal.

The two paramecia now separate, and in each the fused micronucleus, which is comparable to a zygote in higher forms, divides by mitosis into 2, 4, and 8 micronuclei. Four of these enlarge and become macronuclei, and 3 of the other 4 disappear. Now the paramecium itself divides twice, resulting in four paramecia, each with 1 micronucleus and 1 macronucleus. After this complicated process, the animals may continue to reproduce by binary fission without the necessity of conjugation.

What is the meaning of this unique phenomenon? There is first of all an exchange of hereditary material so that each conjugant profits from a new hereditary constitution. It is not the same as the union of the gametes in higher forms (zygote formation), in which direct progeny is the result, for in conjugation the animals still continue asexual division. However, the result is similar because the nucleus of each exconjugant contains hereditary material from two individuals. The process does not seem necessary for rejuvenation; cultures can be maintained for many years without undergoing conjugation.

In 1937 it was discovered that not every paramecium would conjugate with any other paramecium of the same species. Sonneborn found that there were physiologic differences between individuals that set them off into **mating types**. Ordinarily conjugation will not occur between individuals of the same mating type but only with an individual of another (complementary) mating type. It was also found that within a single species there are a number of varieties,* each of which has mating types that conjugate among themselves but not with the mating types of other varieties. In *Paramecium aurelia*, for instance,

---

*Within each species of *Paramecium* the individuals exhibit morphologic and physiologic differences. Since these differences are usually more minor and more superficial than those that distinguish species, the groups within a species are referred to as strains, biotypes, or varieties. Most species of Protozoa can be divided into a number of these groups.

TABLE 7-1

Examples of conjugation in three varieties of *Paramecium aurelia*

Variety		1		2		3	
	Mating types	I	II	III	IV	V	VI
1	I	–	+	–	–	–	–
	II	+	–	–	–	–	–
2	III	–	–	–	+	–	–
	IV	–	–	+	–	–	–
3	V	–	–	–	–	–	+
	VI	–	–	–	–	+	–

+ indicates conjugation will occur.
– indicates conjugation will not occur.

---

each of six varieties has two mating types; conjugation, however, will occur only between members of opposite or complementary mating types within their own variety.

Mating types are usually designated by Roman numerals. Thus in variety 1 of *P. aurelia,* the mating types are called mating types I and II; in variety 2, mating types III and IV, etc. New varieties of this species have been described from time to time until 16 were known in 1957. Their wide and sporadic distribution pose interesting evolutionary problems. With few exceptions, each variety has only two interbreeding mating types. There is no morphologic basis for distinguishing mating types within a variety; such differences that exist must be physiologic. Some varieties, however, can be distinguished from each other morphologically.

Mating types are also found in other species of paramecia and in other ciliates.

As an example of how conjugation occurs, the mating types of three varieties of *P. aurelia* are shown in Table 7-1.

Autogamy refers to a process of self-fertilization. After the disintegration of the macronucleus and the division of the 2 micronuclei to form 8 micronuclei, 2 of the haploid gametic nuclei that result from this division enter a small bulge (paroral cone), fuse together, and restore the diploid number of chromosomes in the synkaryon, or zygote. The other 6 micronuclei degenerate, and the synkaryon divides twice to produce 2 macronuclei and 2 micronuclei. At

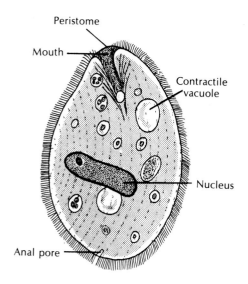

FIG. 7-28

Balantidium coli, a ciliate parasitic in man. This ciliate is common in hogs in which it does little damage; man becomes infected by food or water contaminated by hog feces. In man it produces intestinal ulcers and severe chronic dysentery. Infections are common in parts in Europe, Asia, and Africa, but rare in United States.

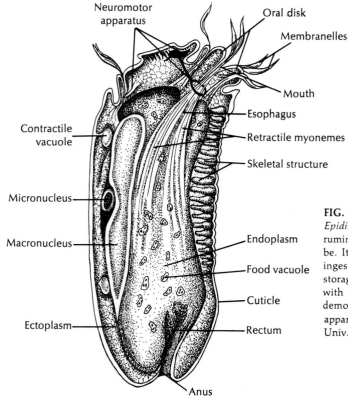

FIG. 7-29

Epidinium, which lives in the digestive tract of ruminants, shows how complex a protozoan can be. It has specialized organelles for coordination, ingestion and digestion, support, contraction, food storage, egestion, and hereditary continuity. It was with this ciliate that Sharp (1914) made his classic demonstration of the so-called neuromotor apparatus. (Modified from Sharp, R. G. 1914. Univ. Calif. Publ. Zool. **13**:43-122.)

the first binary fission each daughter cell will receive 1 of the macronuclei and by division of the micronuclei also 2 micronuclei. This process is similar to conjugation but does not involve two individuals.

Parasitic ciliates

Most parasitic ciliates are not very harmful. *Balantidium coli*, parasitic in man, is an exception (Fig. 7-28). It is often found in hogs, where it usually does no harm. Man becomes infected by water and food contaminated by cysts from the hog's feces. The parasite enters the intestinal submucosa and causes ulcers and severe and even fatal dysentery. It is not as common in America as it is in Europe, Asia, and Africa. Other similar species are found in cattle and horses—*Epidinium* (Fig. 7-29) in cattle, for example. Some ciliate parasites such as *Nyctotherus* also occur in the colon of frogs and toads. Tadpoles are infected when they eat the feces of frogs containing the cysts.

Suctorians

Suctorians are ciliates in which the young possess cilia and are free swimming and the adults grow a stalk for attachment, become sessile, and lose their cilia. They have no cytostome but have protoplasmic processes that serve as tentacles. Some of these have rounded knobs for capturing their prey—usually ciliates; some are sharp for piercing and for sucking up the protoplasm (Fig. 7-6).

One of the best places to find freshwater suctorians is in the algae that grows on the carapace of turtles. Common genera of suctorians found there are *Anarma* (without stalk or test) and *Squalorophrya* (with stalk and test). Other freshwater representatives are *Podophrya* (Fig. 7-8) and *Dendrosoma*. *Acinetopsis* and *Ephelota* are saltwater forms.

Suctorian parasites include the *Trichophrya*, which is parasitic on the gills of the small-mouthed black bass and may cause serious damage to the fish; *Allantosoma*, which occurs in the intestine of certain mammals; and *Sphaerophrya*, which is found in *Stentor*.

■ VALUE IN BIOLOGIC INVESTIGATION

Protozoans are widely studied by biologists everywhere. Geneticists have long given consideration to the problem of heredity and variation in Protozoa. The rapid multiplication of protozoans and the many generations one can produce in a short time make

these animals ideal for this study. The more some of these are studied, the more complex they turn out to be. The discovery of mating types in paramecia gives a suggestion of sex. Although individual paramecia cannot be labeled male or female, there is definitely a physiologic difference between individuals, just as there is in well-established sexual forms. In recent years much interest has centered upon the role *Paramecium aurelia* plays in explaining the phenomena of cellular transformations, that is, the changing of one type of cell into another type. Sonneborn made use of two hereditary strains of paramecia that he called "killers" and "sensitives." The killer strain contained in its cytoplasm certain visible particles (kappa particles) that liberated into the culture fluid a chemical substance (paramecin) that had the effect of destroying the other strain—the sensitive. Kappa particles, or killer substances, are transmitted directly by cytoplasmic genes (plasmagenes) from the cytoplasm of the parent cell to the daughter cells and not by nuclear genes as in ordinary heredity. This is often used as an example of cytoplasmic inheritance. Sonneborn was also able to change, experimentally, killers into sensitives and the reverse, thus producing a form of cellular transformation. Work by Beale and Gibson indicates that the killer substance is dependent on a special particle (metagon) that acts as an intermediary between the killer particles and nuclear genes. There are many problems in this interesting phenomenon still unsolved.

Many other studies of Protozoa of current interest involve their nutritional needs, population crowding, permeability, and problems of serology and immunity.

Derivation and meaning of names

Amoeba (Gr. *amoibē*, change). Genus of Sarcodina.

Arcella (L. diminutive of *arca*, box). Genus of Sarcodina; with a boxlike test.

Coccidia (Gr. *kokkos*, kernel or berry). Order of Sporozoa.

Difflugia (L. *diffluo*, flow apart). Genus of Sarcodina; refers to the flowing out of the pseudopodia.

Eimeria (after Eimer, German zoologist). Genus of the order Coccidia of Sporozoa.

Entamoeba (Gr. *entos*, within, + *amoibē*, change) **histolytica** (Gr. *histos*, tissue, + *lysis*, a loosing). Genus and species in Sarcodina.

Euglena (Gr. *eu*, true, good, + *glēnē*, eyeball or eye pupil). Refers to the stigma, or eyespot.

Euplotes (Gr. *eu*, true, good, + *plōtēr*, swimmer). Genus of Ciliata.

Foraminifera (L. *foramen*, hole, + *fero*, bear). The tests are frequently perforated.

Infusoria (L. *infusus*, poured in, as in infusions). So-called because of their abundance in a culture.

Leishmania (Leishman, who discovered it). Parasitic genus of Zoomastigophorea.

Monocystis (Gr. *monos*, single, + *kystis*, bladder). Parasitic genus of Sporozoa.

Noctiluca (L. the moon, from *nox*, night, + *lucere*, to shine). Genus of Mastigophora. Their luminescence is most apparent at night.

Opalina (L. *opalus*, opal). Genus of superclass Opalinata.

Paramecium (Gr. *paramēkēs,* oblong). Genus of Ciliata.

Pelomyxa (Gr. *pelos*, mud, + *myxa*, nasal mucus). Refers to the black and brown inclusions in body. Genus of Sarcodina.

Radiolaria (L. *radiolus*, small ray). Order of Sarcodina. The axopodia radiate in all directions.

Rhizopodea (Gr. *rhiza*, root, + *pous, podos*, foot). Many pseudopodia may be extended at one time.

Stentor (Grecian herald with loud voice). Genus of Ciliata. Shaped like a megaphone.

Trichomonas (Gr. *thrix*, hair, + *monas*, single). Genus of Mastigophora.

Trypanosoma (Gr. *trypanon*, auger, + *sōma*, body). Genus of Mastigophora. The body with undulating membrane is twisted.

Volvox (L. *volvere*, to roll). Genus of Mastigophora. Refers to their characteristic movement.

Vorticella (L. diminutive of *vortex*, a whirlpool). Genus of Ciliata. Movement of their cilia creates a spiral movement of water.

References

Suggested general readings

Allen, R. D. 1961. A new theory of ameboid movement and protoplasmic streaming. Exp. Cell. Res., Suppl. no. 8, 17-31.

Bacq, Z. M. 1947. L'acétyl choline et l'adrénaline chez les invertébrés. Biol. Rev. **22:**73-91.

Barrington, E. J. W. 1967. Invertebrate structure and function. Boston, Houghton Mifflin Co. *Stresses the importance of animal organization in the light of molecular and ultrastructural studies.*

Brieger, E. M. 1963. Structure and ultrastructure of microorganisms. New York, Academic Press, Inc. *Molecular biology has been focused for some time upon microorganisms in studying the basic organization of the cell at submicroscopic levels, but this treatise attempts to show that the bacterial cell is not a replica of the cell of higher organisms.*

Brown, H. P., and M. M. Jenkins. 1962. A protozoan (*Dileptus*) predatory upon Metazoa. Science **136:**710.

Chambers, R., and J. A. Dawson. 1925. Structure of undulating membrane, Biol. Bull. **48:**240-242.

Cheng, T. C. 1964. The biology of animal parasites. W. B. Saunders Co., Philadelphia. *A parasitology text containing a good account of protozoan parasites.*

Christianson, R. G., and J. M. Marshall. 1965. A study of phagocytosis in the ameba *Chaos chaos.* J. Cell Biol. **25:**443-457.

Corliss, J. O. 1961. The ciliated Protozoa: characterization, classification and guide to the literature. New York, Pergamon Press, Inc.

Cushman, J. A. 1948. Foraminifera, their classification and economic use, ed. 4. Cambridge, Mass., Harvard University Press.

Edmondson, W. T. (editor). 1959. Ward and Whipple's fresh-water biology, ed. 2. New York, John Wiley & Sons, Inc. *In this handbook there are useful keys of the major groups of Protozoa.*

Gojdics, Mary. 1953. The genus *Euglena.* Madison, University of Wisconsin Press. *A comprehensive account of this group. Much attention is given to taxonomy, but there are also good descriptions of morphology.*

Goldacre, R. J. 1961. The role of the cell membrane in the locomotion of *Amoeba.* Exp. Cell Res., Suppl. no. 8, 1-16.

Grasse, P. P. 1953. Traité de zoologie, vol. I. Paris, Masson & Cie, Editeurs.

Hall, R. P. 1953. Protozoology. Englewood Cliffs, N. J., Prentice-Hall, Inc. *A standard text on Protozoa. An excellent chapter on reproduction and life cycles.*

Hardy, A. C. 1956. The open sea, its natural history: the world of plankton. Boston, Houghton Mifflin Co. *Good descriptions of the Radiolaria and other protozoans in Chapter 6. Beautiful color illustrations.*

Hickman, C. P. 1972. Biology of the invertebrates, ed. 2. St. Louis, The C. V. Mosby Co.

Hitz, G. C. 1962. Nutritional requirements of ciliates: influence on population growth and on morphology. Am. Zool. **2:**416.

Hyman, L. H. 1940. The invertebrates: Protozoa through Ctenophora, vol. 1. New York, McGraw-Hill Book Co. *An extensive and exhaustive section is devoted to the morphology and physiology of protozoans.*

Jahn, T. L., and F. F. Jahn. 1949. How to know the Protozoa. Dubuque, Iowa, William C. Brown Co., Publishers. *A manual on the identification and description of protozoan forms.*

Jennings, H. S. 1906. Behavior of the lower organisms. New York, Columbia University Press. *A classic work on the tropisms (taxes) of protozoan forms. This treatise has had a profound influence on all subsequent investigations along this line.*

Kitching, J. A. 1938. Contractile vacuoles. Biol. Rev. **13:**403-448.

Papas, G. D., and P. W. Brandt. 1958. The fine structure of the contractile vacuole in *Amoeba.* J. Biophys. Biochem. Cytol. **4:**485-488.

Kudo, R. R. 1966. Protozoology, ed. 5. Charles C Thomas, Publisher, Springfield, Ill.

Mackinnon, D. L., and R. S. J. Hawes. 1961. An introduction to the study of Protozoa. Oxford, The Clarendon Press.

Manwell, R. D. 1961. Introduction to protozoology. New York, St. Martin's Press, Inc.

Marshall, J. M., and V. Nachmias. 1965. Cell surface and pinocytosis. J. Histochem. Cytochem. **13**:92-104.

Mast, S. O. 1947. The food vacuole of *Paramecium*. Biol. Bull. **92**:31-72.

Mayr, E. (editor). 1957. The species problem. Washington, D. C., American Association for the Advancement of Science. *This is a symposium by the authorities on the problems of species. The paper entitled "Breeding Systems, Reproductive Methods, and Species Problems in Protozoa" by Professor T. M. Sonneborn is a good analysis of mating types and the concept of syngen as applied to the varieties of Protozoa.*

Muller, M., and I. Toro. 1962. Studies on feeding and digestion in Protozoa. III. Acid phosphatase activity of food vacuoles of *Paramecium multimicronucleatum*. J. Protozool. **9**:98-102.

Noland, L. E. 1957. Protoplasmic streaming: a perennial problem. J. Protozool. **4**:1-6.

Pennak, R. W. 1953. Fresh-water invertebrates of the United States. New York, The Ronald Press Co. *A reference work with considerable attention devoted to Protozoa.*

Pettersson, H. 1954. The ocean floor, New Haven, Conn., Yale University Press. *The role the Foraminifera and Radiolaria have played in building up the sediment carpet of the ocean floor is vividly described in this little book.*

Pitelka, D. R. 1963. Electron-microscopic structure of Protozoa. New York, The Macmillan Co. *A fine account of the ultrastructure of Protozoa by one of the great authorities in the field.*

Roth, L. E. 1960. Electron microscopy of pinocytosis and food vacuoles in *Pelomyxa*. J. Protozool. **7**:176-185.

Sleigh, M. A. 1962. The form and beat of cilia of *Stentor* and *Opalina*. J. Exp. Biol. **37**:1-14.

Sleigh, M. A. 1962. The biology of cilia and flagella. New York, The Macmillan Co. *An excellent summary.*

Tartar, V. 1961. The biology of *Stentor*. New York, Pergamon Press, Inc.

Wichterman, R. 1953. The biology of *Paramecium*. New York, McGraw-Hill Book Co. *Those who think Protozoa are "simple animals" will be disillusioned by this well-written treatise on one genus of protozoans. A bibliography of more than 2,000 references gives an idea of the impressive amount of work performed on this animal.*

Yonge, C. M. 1928. Feeding mechanisms in invertebrates. Biol. Rev. **3**:21-76.

Selected *Scientific American* articles

Allen, R. D. 1962. Amoeboid movement. **206**:112-122 (Feb.).

Alvarado, C. A., and L. J. Bruce-Chevatt. 1962. Malaria. **206**:86-98 (May).

Bonner, J. T. 1949. The social amoebae. **180**:44-47 (June).

Bonner, J. T. 1950. *Volvox:* A colony of cells. **182**:52-55 (May).

Hawking, F. 1970. The clock of the malaria parasite. **222**:123-131 (June).

Hutner, S. H., and J. A. McLaughlin. 1958. Poisonous tides. **199**:92-98 (Aug.).

Sater, P. 1961. Cilia. **204**:108-116 (Feb.).

A large marine sponge surrounded by clams and calcareous algae, with a contracted sea anemone at upper right. Sponges have no organs and depend on flagellated cells to keep water flowing through canals to bring oxygen and food to the cells. (Photo by T. Lundälv, Kristinebergs Zoological Station, Sweden.)

CHAPTER 8

THE LOWEST METAZOANS

PHYLUM MESOZOA
PHYLUM PORIFERA

■ POSITION IN ANIMAL KINGDOM

The multicellular animals, or Metazoa, are typically divided into three branches—Mesozoa (a single phylum), Parazoa (phylum Porifera, the sponges), and Eumetazoa (all other phyla).

Although Mesozoa and Parazoa are multicellular, neither fits into the general plan of organization of the other phyla. Such cellular layers as they possess are not homologous to the germ layers of the Eumetazoa, and although mesozoan reproduction is complex, neither group has developmental patterns in line with the other metazoans. The poriferans are considered to be aberrant, that is, deviating widely from standard patterns. This is the reason for the name Parazoa, which means the "beside-animals."

■ BIOLOGIC CONTRIBUTIONS

Although the simplest in organization of all the metazoans, the mesozoans and sponges do comprise a higher level of morphologic and physiologic integration than that found in protozoan colonies. Both groups may be said to belong to a **cellular level of organization.**

The mesozoans, although composed simply of an outer layer of somatic cells and an inner layer of reproductive cells, nevertheless have a very complex reproductive cycle somewhat suggestive of that of the trematodes (flukes). Mesozoans are entirely parasitic.

The poriferans are more complex, with **several types of cells** differentiated for various functions, some of which are organized into **incipient tissues** of a low level of integration.

The developmental patterns of both phyla are different from those of other phyla and their embryonic layers are not homologous to the germ layers of other phyla.

The sponges have developed a unique **system of**

water currents on which they depend for food and oxygen.

▪ PHYLOGENY AND ADAPTIVE RADIATION

PHYLOGENY. Some authorities consider the Mesozoa to be degenerate flatworms, the affinity being based upon the resemblance between their life cycles and the fact that both produce ciliated larvae. Another view is that they are primitive, and intermediate between protozoans and other metazoans. If they are primitive, this may indicate that the first metazoans had a solid blastula instead of a hollow one.

They may have affinities with the protozoans, indicated by the widespread cilia in the group and the differentiation of only somatic and reproductive cells as in some colonial protozoans.

The origin of sponges is equally mysterious. The theory that they may have evolved from protozoan choanoflagellates such as *Proterospongia* (Fig. 7-23) lacks favor because sponges do not acquire collars until late in development. Their relation to protozoans is shown in their phagocytic method of nutrition and the resemblance of their flagellated larvae to colonial protozoans. Sponges have some resemblance to coelenterates in their calcareous skeletons, mesenchyme, and amebocytes as gamete-forming cells. They are not in the direct line of evolution of other animals, but an offshoot of the main line.

ADAPTIVE RADIATION. The mesozoans are adapted to a parasitic existence. Parasites often have complicated life histories, and this may explain some of their evolutionary development. The small size and small number of mesozoan species (50) would indicate a modest adaptive radiation.

The Porifera of many thousand species have been a highly successful group. In their evolutionary diversification their unique water-current system has been largely responsible for their degree of complexity from simple to the more complicated. The leuconoid pattern of sponge structure, for example, favors a larger size because it has more chambers for capturing food and for gaseous exchange than the asconoid and syconoid types.

▪ PHYLUM MESOZOA*

The minute, wormlike members of the phylum Mesozoa probably have the simplest structure of any metazoan form. They have an outer layer of 20 to 30 ciliated somatic cells (the number is constant within each species) surrounding a long slender axial cell in which are a number of reproductive cells. Mesozoans are all parasitic in other invertebrates and most of them range in size from 0.5 to 7 mm. in length. There are only two orders, the Dicyemida and the Orthonectida.

The dicyemids are endoparasitic in the kidneys of cephalopods. The **nematogen,** or **vermiform,** stage (Fig. 8-1) is most often seen clinging to the spongy tissue of the kidney by the cilia on their anterior ends. The germ cells in the long inner axial cell give rise asexually, by repeated division, to young **vermiform larvae,** much like the parent, which escape the parent, swim about for a time in the urine, then at-

*Mes-o-zo'a (Gr. *mesos,* middle, + *zōon,* animal).

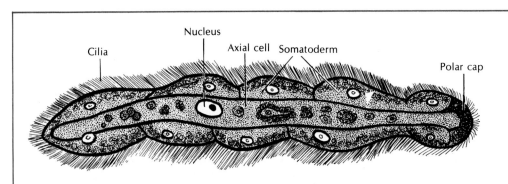

FIG. 8-1
Vermiform stage of *Dicyema* (a dicyemid mesozoan). These are common parasites of squids and octopuses and have both asexual and sexual stages in the life cycle.

tach to the kidney, and grow into mature nematogens. They apparently do little harm to the host.

When, after many generations, the population in the kidney becomes crowded, some of the members begin another type of reproduction by which dispersal of the species can occur. The germ cells in the axial cell now give rise to hermaphroditic gonads that produce eggs and sperm. The zygotes develop into minute (0.04 mm.) ciliated **infusoriform larvae** quite unlike the parent. Instead of remaining in the kidney, they are shed with the urine into the seawater. Formerly their fate was unknown, but recent investigation by Lapan and Morowitz (1972) indicates that they sink to the bottom where they are eaten by other cephalopods and so start a new population. According to these researchers there is no intermediate host.

Orthonectids parasitize a variety of invertebrates, such as bivalve mollusks, polychaetes and rhynchocoels. They have separate sexes, with the female being much larger than the male (Fig. 8-2). The zygotes develop into ciliated larvae that escape the host and infest new hosts.

There is still much to learn about these mysterious little parasites, but probably one of the most intriguing questions is their place in the evolutionary picture. Are they comparable to a blastula containing internal reproductive cells, or are they more nearly comparable to the gastrula-like planula larvae of coelenterates and protozoan colonies? Are they merely degenerate and simplified flatworms, or are they, as many investigators believe, truly primitive forms such as might have furnished the stepping stone from single-celled to many-celled animals?

■ PHYLUM PORIFERA*

Sponges belong to phylum Porifera, which means "pore bearing." The bodies of all sponges bear myriads of tiny pores and canals that are basic to their functional activity. It is upon the water currents carried through this unique system of pores and canals that the sponge depends for food and oxygen and for ridding itself of wastes.

Although sponges are multicellular, they show few of the characteristics of the Metazoa. There are no organ systems and their tissues are poorly defined. Organization goes little farther than the cellular level. The basic body form resembles a perforated sac, being composed of three layers: (1) an outer layer of flattened epidermal cells, (2) a gelatinous middle layer of ameboid cells, some of which secrete a skeleton of crystalline spicules or of proteinaceous spongin fibers, or both, and (3) an inner layer of flagellated cells.

The fossil record of sponges is incomplete. Sponges with organic skeletons of spongin are not preserved as fossils. The best fossils have been those of the glass sponges (class Hyalospongiae) that flourished from the Cambrian period to the present. The calcareous skeletons of the class Calcispongiae were poorly preserved. Less than a hundred years ago sponges were believed to be plants, for they cannot move about as adults.

*Po-rif'era (L. *porus,* pore, + *fera,* bearing).

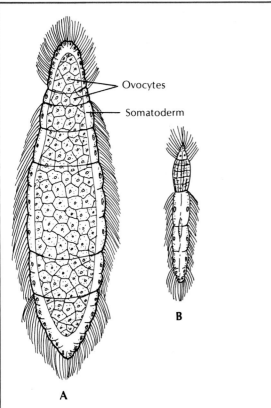

FIG. 8-2

A, Female and, **B,** male orthonectid *(Rhopalura).* This mesozoan parasitizes such forms as flatworms, mollusks, annelids, and brittle stars. Note that structure consists of a single layer of ciliated epithelial cells surrounding an inner mass of sex cells.

Ovocytes

Somatoderm

A

B

There is no evidence that any of the higher Metazoa arose from sponges; they seem to be "dead ends" in the evolutionary picture. In fact, because they vary so much from other metazoans, they are called Parazoa, which means the "beside-animals." It is interesting, though, that biochemical investigations show that sponges have about the same nucleic acids and amino acids as other metazoans. They have both phosphoarginine (an invertebrate phosphogen) and phosphocreatine (a chordate phosphate).

Characteristics

1. Mostly marine, although a few freshwater forms; all aquatic
2. All adult sponges **attached** and with a variety of body forms, such as **vaselike, globular,** and **many-branched**
3. Radial symmetry or none
4. **Multicellular;** body a loose aggregation of cells of mesenchymal origin; body surface, or dermal epithelium, simply a colloid with freely movable cells or a syncytium; mesenchyme usually with **skeletal spicules** or **horny fibers** and free ameboid cells
5. Body with many **pores (ostia), canals,** and **chambers** that serve for the passage of water
6. Most of the inner chambers and interior surfaces lined with **choanocytes,** or **flagellate collar cells**
7. Digestion intracellular and no excretory or respiratory organs; contractile vacuoles in cells of some freshwater sponges
8. No organs or definite tissues
9. **Skeleton usually of calcareous** or **siliceous crystalline spicules** or of **protein spongin**
10. Reactions to stimuli apparently local and independent; nervous system probably absent
11. Asexual reproduction by **buds** or **gemmules** and sexual reproduction by eggs and sperm; free-swimming, ciliated larva

Ecologic relationships

Most sponges are marine, but some of the horny sponges are found in freshwater. Few occur in brackish water. Many marine forms occur in the shallow water along shores, but glass and horny sponges are found chiefly at deeper depths; glass sponges may actually be attached to sea bottoms. Most of those with mineral skeletons are found in waters of the temperate zone; those with spongin fibers occur more frequently in warm waters. Sponges in calm waters may grow taller than those in rapidly moving waters. Adult sponges are always sessile and usually attached to rocks or other solid objects. A few live on soft sand or mud. Their external form is mostly attributable to environmental factors of currents and gravity.

Ecologically the biomass of sponges is only a small part of animal life, as they are restricted chiefly to localized regions.

Many animals (crabs, nudibranchs, mites, bryozoans) have associations with sponges, either as parasites or for protective purposes. Some crabs use pieces of sponge on their carapaces as protection and camouflage. Although most fishes tend to avoid sponges, some reef fishes are known to graze on shallow-water sponges.

Most sponges form colonies, some of which attain great size (1 to 2 meters in diameter); others are small (1 to 2 mm.). They vary in color, ranging from dull gray and brown to brilliant scarlet and orange.

Classes

There are three classes of sponges, classified mainly by the kinds of skeletons they possess.

Class Calcispongiae (cal-si-spun'je-e) (L. *calcis,* lime, + Gr. *spongos,* sponge) **(Calcarea).** Have spicules of carbonate of lime that often form a fringe around the osculum. Spicules are needle-shaped or three- or four-rayed. All three types of canal systems represented. All marine. Examples: *Scypha, Leucosolenia.*

Class Hyalospongiae (hy"a-lo-spun'je-e) (Gr. *hyalos,* glass, + *spongos,* sponge) **(Hexactinellida).** Have six-rayed siliceous spicules in three dimensions; often cylindric or funnel shaped. Choanocytes limited to certain chambers. Habitat mostly deep water; all marine. Examples: Venus's flower basket *(Euplectella), Hyalonema.*

Class Demospongiae (de-mo-spun'je-e) (Gr. *dēmos,* people, + *spongos,* sponge). Have siliceous spicules, spongin, or both. One family found in freshwater; all others marine. Examples: *Thenea, Cliona, Spongilla, Meyenia,* and all bath sponges.

Structure

Sponges vary enormously in their structure and other features. Most dry sponges we see consist only of skeletal framework. In the living condition many of them appear as slimy gelatinous masses. The common bath sponge has this appearance, and only when the protoplasmic mass is removed do we see the commercial sponge with its skeleton of fibers.

The surface of sponges possesses many small pores **(ostia)** for the inflow of water. The ostia open into **canals,** simple or complex, which may open directly into a central cavity, the **spongocoel,** or may be interrupted by small chambers before emptying into the spongocoel. The opening of the spongocoel to the outside is known as the **osculum.** Colonial sponges have many oscula. The sponge has no mouth and no organs.

The outer surface and the incurrent canals are covered with a thin layer of epithelium. The term "layer of cells" must be used with certain reservations, for the cells in sponges are loosely arranged in a gelatinous **mesoglea;** in some cases even cells may be lacking, and only a spongin sheet or syncytium is present.

In simple asconoid sponges the spongocoel is lined with characteristic flagellum-bearing cells, the **choanocytes,** commonly called "collar cells" because each has a little transparent collar that encloses a single flagellum. In more complex sponges the collar cells are confined to the radial canals and chambers and are not present in the spongocoel. In these sponges the spongocoel is lined with a thin epithelial layer.

Among the different sponges are a great variety of canal systems (Fig. 8-3). Most sponges fall into one or the other of three principal types.

1. *Asconoid type.* The canals pass directly from the ostia to the spongocoel, which is lined with collar cells. *Leucosolenia* has this type of canal.

2. *Syconoid type.* The incurrent canals (from the outside) lie alongside of the radial canals that empty into the spongocoel. Both types of canals end blindly in the body wall but are connected by minute pores. Only the radial canals are lined with collar cells. *Scypha* is an example of this type.

3. *Leuconoid or rhagon type.* The canals of this

FIG. 8-3

Three types of sponge structure. The degree of complexity from simple asconoid to complex leuconoid type has involved mainly the water and skeletal systems, accompanied by outfolding and branching of the collar cell layer. The leuconoid type is considered the major plan for sponges, for it permits greater size and more efficient water circulation.

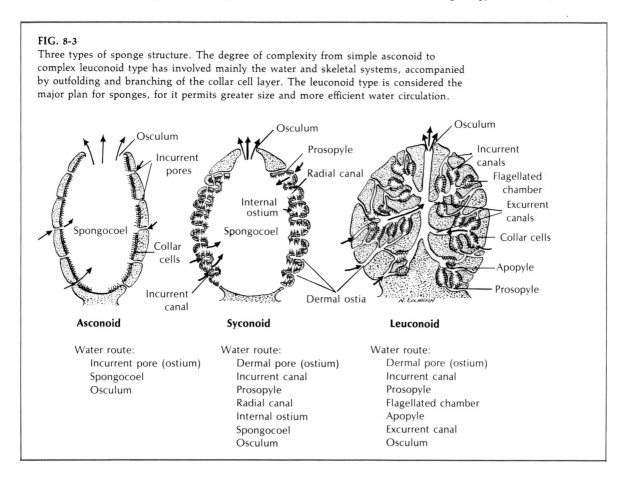

Asconoid

Water route:
Incurrent pore (ostium)
Spongocoel
Osculum

Syconoid

Water route:
Dermal pore (ostium)
Incurrent canal
Prosopyle
Radial canal
Internal ostium
Spongocoel
Osculum

Leuconoid

Water route:
Dermal pore (ostium)
Incurrent canal
Prosopyle
Flagellated chamber
Apopyle
Excurrent canal
Osculum

type are many branched and complex, with numerous chambers lined with collar cells. The spongocoel lacks collar cells. The larger sponges, including the bath sponge, are all this type.

These three types of canals are correlated with the evolution of sponges, from the simple to the complex forms. It has been mainly a matter of increasing the surface in proportion to the volume, so that there may be enough collar cells to meet the food demands. This problem has been met by the outpushing of the spongocoel of a simple sponge such as the asconoid type to form the radial canals (lined with choanocytes) of the syconoid type. The formation of incurrent canals between the blind outer ends of the radial canals completes this type. Further increase in the body wall foldings produces the complex canals and chambers (with collar cells) of the leuconoid type.

Types of cells

Although there are many types of cells in sponges (Fig. 8-4) None of them are actually arranged with the true regularity of tissues. Perhaps the nearest approach to tissues are the flat protective cells called **pinacocytes** that make up the outer and inner epithelial layers and the collar cells, or **choanocytes,** wherever they are found (Fig. 8-4). The pinacocytes are contractile cells that possess myonemes and help regulate the surface area of the sponge. Some of the pinacocytes are modified as **myocytes,** or muscle cells that form contractile circlets around the pores and oscula, and **porocytes** that form the ostia in

asconoid sponges. Choanocytes with their flagella create currents of food- and oxygen-laden water through the sponge. The collar of the choanocyte is composed of a circle of contractile protoplasmic tentacles with a long flagellum extending from the base. The collar traps small food particles and passes them down to the body of the cell where they are engulfed, ameboid fashion, and digested.

The middle layer, or mesoglea, is a gelatinous matrix that contains skeletal spicules and/or spongin fibers and several types of **amebocytes** (ameboid cells). These include scleroblasts, which form the spicules; spongioblasts, which form spongin; archeocytes, which have a variety of functions, such as the digestion of food and the formation of eggs and sperm; and collencytes (Fig. 8-4).

Sponges have great power of regeneration. H. V. Wilson discovered many years ago that when sponge cells were sifted through bolting cloth into small groups upon a surface under water, each group of cells would grow into a separate sponge.

Skeletons

The skeleton serves as the basis for classifying sponges (Fig. 8-5). In sponges such as *Scypha* the skeleton consists of **spicules** of calcium carbonate; glass sponge spicules are formed of siliceous material. These spicules are of many different forms and shapes. The straight ones are called the **monaxons;** those of three rays in one plane are **triradiates;** those of four rays in four planes are **tetraxons;** and those of many rays are **polyaxons.** Some sponges have the

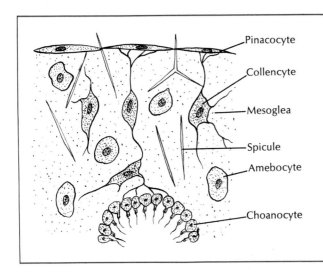

FIG. 8-4
Small section through sponge wall showing four types of sponge cells. Pinacocytes are protective and contractile; choanocytes create water currents and engulf food particles; amebocytes have a variety of functions; collencytes appear to be contractile and some authorities have suggested a nervous function for them.

Pinacocyte

Collencyte

Mesoglea

Spicule

Amebocyte

Choanocyte

spicules arranged in regular order, others in a haphazard arrangement. Spicules are formed by **scleroblasts,** special ameboid cells. In some sponges the larger spicules are called **megascleres;** the smaller ones are called **microscleres.**

Sponges, such as the bath sponge, freshwater sponges, and others, contain **spongin** (Fig. 8-5), a proteinlike substance. This type of skeleton is a branching, fibrous network that supports the soft, living cells of the sponge. Special cells from the mes-

enchyme, called **spongioblasts,** form this type of skeleton.

Metabolism

Metabolism in sponges is mainly a matter of individual cellular function. Their food consists largely of microscopic algae, bacteria, and organic debris drawn through the canal system by currents of water induced by the waving of the flagella on the choanocytes. As the flagella undulate spirally from base to

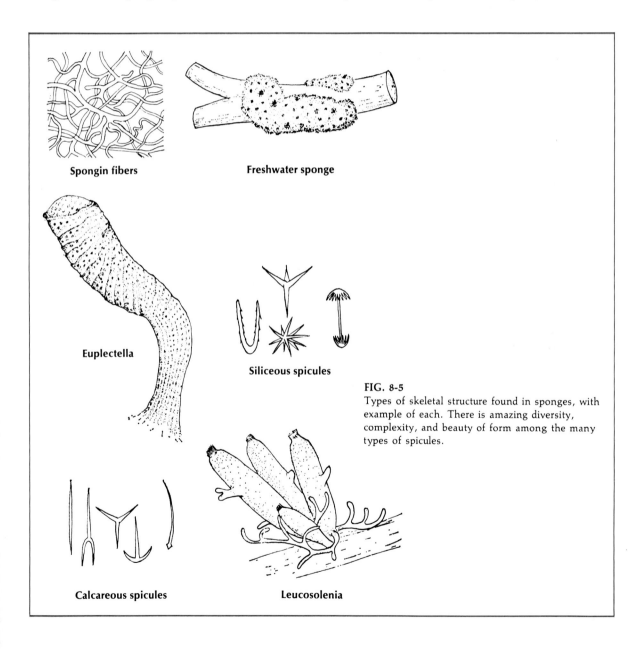

Spongin fibers

Freshwater sponge

Euplectella

Siliceous spicules

Calcareous spicules

Leucosolenia

FIG. 8-5
Types of skeletal structure found in sponges, with example of each. There is amazing diversity, complexity, and beauty of form among the many types of spicules.

tip, water currents are set up that bring particles of food to the outer surface of the collars, where the food adheres. Later the food reaches the base of the cells and passes into the cytoplasm, where food vacuoles are formed. Digestion is therefore **intracellular** in sponges. Amebocytes in the mesenchyme layer may aid in digestion and distribution of food from the choanocytes to other parts of the body. Absorption of food substances from cell to cell may also occur. Undigested food is ejected by amebocytes into outgoing currents.

Since water currents are of such primary importance to sponges, many studies have been made to determine the mechanics of these currents.

Most sponges are of the leuconoid type, containing large numbers of small chambers lined with flagellated choanocytes. In *Leuconia* it was estimated that there were some 81,000 incurrent canals bringing water in at a velocity of 0.1 cm. per second. There were, however, over 2 million flagellated chambers so that their total diameter was much greater than that of the incoming canals, reducing the velocity to only 0.001 cm. per second. This slow movement allows ample opportunity for capture of food particles by the collar cells. All of this water was expelled through a single osculum at a velocity of 8.5 cm. per second. This jet of water could carry waste products some distance away from the sponge. It is to the advantage of the sponge to discharge the water as far away from its oscula as possible to prevent reusing water containing its own waste and carbonic acid.

The size of the osculum in proportion to the volume of the sponge determines the pumping rate and speed at which water is expelled. The size of the oscula and pores can be regulated to some extent by the contractile myocytes that surround them. The amount of water passing through a sponge depends on its size; some large sponges have been found to filter more than 1,500 liters of water a day.

Only the finest food particles reach the choanocytes, because of the screening action of the prosopyles and the collars. Ostia average 50 μ in diameter and prosopyles about 5 μ. The tiny spaces between the cytoplasmic tentacles that make up a collar are only about 0.1 μ wide. Thus the choanocytes capture the smallest particles and slightly larger bits are engulfed by the archeocytes in the mesenchyme along the radial canals.

Excretion and respiration are taken care of by each individual cell by simple diffusion processes. Contractile vacuoles have been found in the amebocytes and choanocytes of freshwater sponges (*Spongilla* and *Ephydatia*).

Reproduction

Sponges reproduce both asexually and sexually. **Asexual** reproduction is mainly a matter of bud formation. After reaching a certain size, these buds may become detached, or they may remain to form colonies. Internal buds, or **gemmules,** are formed in freshwater and some marine sponges (Fig. 8-6). Here, archeocytes are collected together in the mesenchyme and become surrounded by a siliceous shell, or sometimes by a cluster of spicules. When the animal dies, the gemmules survive and preserve

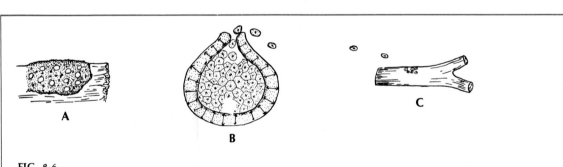

FIG. 8-6
Gemmule and colony formation in *Spongilla*, a freshwater sponge. **A,** Sponge in winter showing cluster of gemmules. **B,** Greatly enlarged cross section of a gemmule showing cells, some emerging from the pore. **C,** Cells from a gemmule that will develop into a new colony.

the life of the species during periods of severe drought or freezing. Later the cells in the gemmules escape through a special opening and develop into new sponges.

In **sexual** reproduction ova and sperm develop from archeocytes or from choanocytes. The ova are fertilized in the mesenchyme, develop there, and fi-

FIG. 8-7
Cluster of sponges, *Scypha*. About natural size.

nally break out into the spongocoel and then out the osculum. Some sponges are **monoecious** (having both male and female sex cells in one individual) and others are **dioecious** (having separate sexes).

During development the zygote undergoes cleavage and differentiation of cells in the mesenchyme of the sponge, and finally a flagellated **amphiblastula larva** (calcareous sponges) emerges (Fig. 8-9). This larval form is made up of flagellate cells at one end and nonflagellate at the other. In time the flagellate cells are invaginated into or overgrown by the nonflagellate group and become the choanocytes; in the parenchymula larva of Demospongiae the flagellate cells migrate into the interior. In both larvae what is ectoderm in other metazoans becomes internal cells. The larval form, after swimming around, soon settles down and becomes attached to some solid object, where it grows into an adult.

■ REPRESENTATIVE TYPES
Class Calcispongiae
Scypha

Scypha is a marine form, living in shallow water where it is usually attached to rocks. It may live singly or it may form a cluster or colony by budding.

The vase-shaped animal is about ¹/₂ to 1 inch long.

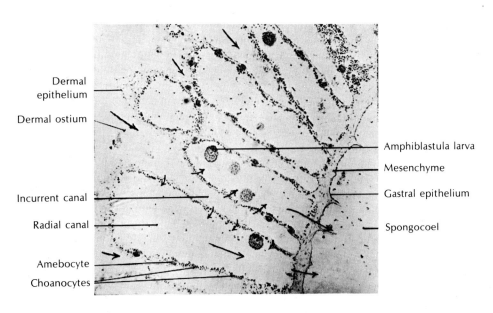

Dermal epithelium
Dermal ostium
Incurrent canal
Radial canal
Amebocyte
Choanocytes
Amphiblastula larva
Mesenchyme
Gastral epithelium
Spongocoel

FIG. 8-8
Cross section through wall of sponge *Scypha* showing canal system. Photomicrograph of stained slide.

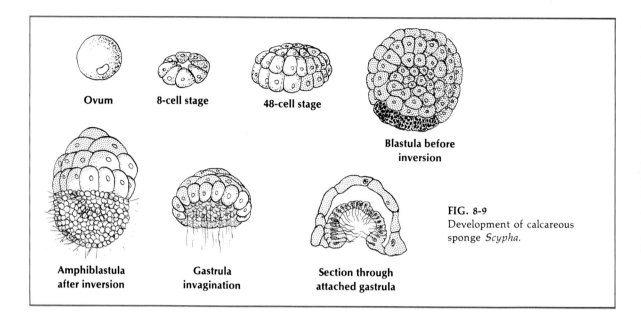

Ovum 8-cell stage 48-cell stage

Blastula before inversion

Amphiblastula after inversion **Gastrula invagination** **Section through attached gastrula**

FIG. 8-9
Development of calcareous sponge *Scypha.*

(Fig. 8-7). The **osculum** has a fringe of straight spicules that discourage small animals from entering, and spicules projecting from the body give the animal a bristly appearance. There is no outer covering or cortex in *Scypha,* the common form studied in America; a European genus called *Grantia,* often confused with *Scypha,* does have such a cortex. The entire outside of the body is full of tiny pores, or **ostia,** opening into the **incurrent canals** (Fig. 8-8) that end blindly near the **spongocoel.** From the spongocoel, flagellated **radial canals** run toward the outer surface and end blindly. Their openings into the spongocoel are called **apopyles,** or **internal ostia.** The incurrent and the radial canals are connected by small pores called **prosopyles.**

The incurrent canals are lined with spicules and pinacocytes, whereas radial canals are lined mainly with **choanocytes.** The spongocoel is lined by a thin epithelium.

The mesoglea of the body wall contains a large number of interlacing spicules of carbonate of lime that support and protect the soft parts of the body. About four types of spicules are recognized: **short monaxons, long monaxons, triradiates,** and **polyaxons.** All spicules originate from **scleroblasts. Ameboid** wandering cells in the mesoglea have a great variety of functions to perform, such as the digestion of food, the formation of skeleton, and the origin of the reproductive cells.

Water is drawn into the incurrent canals by the beating of the flagella and thence is passed through the prosopyles into the radial canals, which carry it to the spongocoel and finally discharge it through the osculum. The rate of flow is regulated by myocyte sphincters and by the action of the encircling pinacocytes.

Reproduction in *Scypha* is by two methods—asexual and sexual. The **asexual** method involves **budding;** that is, a small bud appears at the base of an adult sponge and grows into full size. It may adhere to the parent and thus help form a colony, or it may break free and form a new attachment.

Sexual reproduction occurs by the formation of eggs and sperm that develop from **archeocytes** or choanocytes. Both types of sex cells are formed in the same individual and thus *Scypha* is **monoecious.** The zygote, or fertilized egg, develops into an embryo while in position in the mesenchyme and later in a radial canal (Fig. 8-9). The flagellated amphiblastula larva finally escapes from the parent, swims around a while, settles down, becomes attached, and grows into an adult.

Class Hyalospongiae
Glass sponges

The glass sponges are nearly all deep-sea forms that are collected by dredging. Most of them are radially symmetric, with vase- or funnel-shaped bodies that are usually attached by stalks of root spicules to a substratum (Fig. 8-5). In size they range from 3 or 4 inches to more than 4 feet in length. Their distinguishing features are the skeleton of six-rayed siliceous spicules that are commonly bound together into a network forming a glasslike structure, and the **trabecular net** of living tissue produced by the fusion of the pseudopodia of many types of amebocytes. Within the trabecular net are elongated finger-shaped chambers lined with choanocytes and opening into the spongocoel. The osculum is unusually large and may be covered over by a sievelike plate of silica. There is no epidermis or gelatinous mesenchyme, and both the external surface and spongocoel are lined with the trabecular net. Because of the rigid skeleton and lack of contractility, muscular elements (myocytes) appear to be absent. The general arrangement of the chambers fits glass sponges into both syconoid and leuconoid types. Their structure is adapted to the slow constant currents of sea bottoms, for the channels and pores of the sponge wall are relatively large and uncomplicated and permit an easy flow of water. Little, however, is known about their physiology.

Glass sponges can reproduce asexually by buds and sexually by germ cells, which are developed from archeocytes. The fertilized egg develops into a flagellated larva that swims around and finally settles down into a sessile existence.

The latticelike network of spicules found in many glass sponges is of exquisite beauty, such as that of *Euplectella,* or Venus's flower basket (Fig. 8-5), a classic example of the Hyalospongiae. Many fossil sponges, especially of the Cretaceous and Jurassic periods, are members of this class.

Class Demospongiae

Class Demospongiae contains more than 80% of the sponge species, including the once common bath sponge and most of the familiar North American sponges. All of them are leuconoid and all are marine except the family Spongillidae, which are freshwater forms with wide distribution. The Demospongiae are quite varied in both color and shape. Some are encrusting, some are tall and fingerlike, and some are shaped like fans, vases, cushions, or balls (Fig. 8-10). Some are boring sponges that bore into mollusk shells and encrust. Some loggerhead sponges grow several meters in diameter.

Freshwater sponges are of special interest because they are the only living sponges most students ever see. They have a wide distribution and are not uncommon if one knows where to look. Most of them live in clear, well-oxygenated water of ponds, lakes, and slow streams where they may be found encrusting twigs, plant stems, old pieces of submerged wood, sluiceways, etc. In standing water they are usually found where there is some wave action. They look like a type of wrinkled, irregular scum, pitted with some pores (Fig. 8-5). They are yellowish or brownish, although some have a greenish tinge from the presence of the symbiotic algae (zoochlorellae). They may grow to several inches in diameter.

The most common genera of the family Spongillidae are *Spongilla* and *Myenia. S. lacustris,* the commonest freshwater sponge, develops fingerlike branches and is usually green; *S. fragilis* is unbranched. *Myenia* may be found in either still or running water. Freshwater sponges are most common in midsummer, although some are more easily found in the fall. They die and disintegrate in late autumn, leaving the gemmules (already described), which are able to survive drying and freezing for many months.

When examined closely, freshwater sponges reveal a thin dermis overlying large subdermal spaces (separated by columns of spicules) with many water channels in the interior. There are usually several oscula, each of which (at least in *Myenia*) is mounted on a small chimneylike tube. Their spiculation also includes a spongin network. Their cells include a variety of amebocytes.

Although the gemmule method of asexual reproduction is common among freshwater sponges, sexual reproduction also occurs.

The bath sponges belong to the family Spongiidae and the genera *Spongia* and *Hippospongia.* These are members of the group known as horny sponges that have only spongin skeletons. Sponges are collected by hooks, by dredging or trawling, or by divers. After their collection they are exposed out of water to kill them and are then placed in shallow water, where they are squeezed or treaded upon to remove the softened animal matter until only the horny, spongin skeleton remains. After being cleaned and bleached, they are trimmed and sorted for the market. A fungus disease has greatly depleted the sponge industry in the West Indies.

Sponges are often cultured by cutting out pieces

FIG. 8-10

Demospongiae. **A,** Unidentified sponge growing on underside of overhanging rock, surrounded by clusters of the hydrozoan *Tubularia.* Note the many oscula openings. **B,** Two specimens of a football-shaped sponge *Geodia baretti,* which occur in great numbers in Scandinavian waters below 60 meters. (Photos by T. Lundälv, Kristinebergs Zoological Station, Sweden.)

of the individual animals, fastening them to concrete or rocks, and dropping them into the proper water conditions. It takes many years for sponges to grow to market size.

Derivation and meaning of names

Dicyemida (Gr. *di-*, two, + *kyēma*, embryo). An order of the Mesozoa.

Leucosolenia (Gr. *leukos*, white, + *sōlēn*, pipe). Genus of sponges (Calcispongiae).

Orthonectida (Gr. *orthos*, straight, + *nēktos*, swimming). An order of the Mesozoa.

Parazoa (Gr. *para*, beside, + *zōon*, animal). Sponges are so called because they do not appear to be closely related to any group of the Metazoa.

Scypha (Gr. *skyphos*, cup). This genus is often incorrectly called *Grantia* or *Sycon*.

References

Bidder, G. P. 1923. The relation of the form of a sponge to its currents. Quart. J. Microsc. Sci. **67**:293-300.

deLaubenfels, M. W. 1936. Sponge fauna of the dry Tortugas with material for a revision of the families and orders of the Porifera. Carnegie Institute, Washington, D. C., Tortugas Laboratories, pub. 30. *An important work in resolving many of the difficulties of sponge classification.*

deLaubenfels, M. W. 1945. Sponge names. Science (n.s.) **101**:354-355. *The author points out the confusion resulting from the incorrect usage of the genus name Grantia, a European sponge, for Scypha, the form commonly used in American laboratories.*

Galstoff, P. E. 1925. The ameboid movement of dissociated sponge cells. Biol. Bull. **45**:153-161.

Gauguly, B. 1960. The differentiating capacity of dissociating sponge cells. Arch. Entwick. Org. **152**:23-24.

Gudgen, E. W. 1950. Fishes that live as inquilines (lodgers) in sponges. Zoologica N. Y. **35**:121-131.

Hyman, L. H. 1940. The invertebrates: Protozoa through Ctenophora. New York, McGraw-Hill Book Co.

Jewell, M. 1959. Porifera (section 11). In W. T. Edmondson (editor): Ward and Whipple's fresh-water biology, ed. 2. New York, John Wiley & Sons, Inc. *A taxonomic key to the freshwater sponges of the United States.*

Jones, W. C. 1962. Is there a nervous system in sponges? Biol. Rev. **37**:1-150. *The author refutes the idea that sponges have sensory and ganglionic cells.*

Jones, W. C. 1958. The effect of reversing the internal water current on the spicule orientation in *Leucosolenia variabilis*. Quart. J. Microsc. Sci. **99**:263-278.

Lapan, E. A., and H. J. Morowitz. 1972. The Mesozoa. Sci. Amer. **227**:94-101 (Dec.). *The author presents evidence that the Dicyemida (the Orthonectida may represent a different group because of their complexity) are not simplified flatworms but are at the simplest known level of multicellular organization and may help in understanding the processes of cell differentiation.*

McConnaughey, B. H. 1951. The life cycle of the dicyemid Mesozoa. Univ. Calif. Pub. Zool. **55**:295-336.

McConnaughey, B. H. 1963. The Mesozoa. In E. C. Dougherty (editor): The lower Metazoa: comparative biology and phylogeny. Berkeley, University of California Press. *A recent account of this enigmatic group by the foremost American student of the phylum. The author stresses the resemblance between the Mesozoa and the parasitic flatworms.*

Minchin, E. A. 1900. Porifera. In Lankester's treatise on zoology, part II. London, A. & C. Black Ltd. *A detailed account of the general morphology of sponges, invaluable to the student who wishes to know the basic structure of sponges.*

Rasmont, R., J. Bouillon, P. Castiaux, and G. Vandermeerssche. 1958. Ultrastructure of the choanocyte collar cells in fresh water sponges. Nature **181**:58-59.

Spiegel, M. 1954. The role of specific surface antigens in cell adhesion. I. The reaggregation of sponge cells. Biol. Bull. **107**:130-149.

Stunkard, H. W. 1954. The life history and systematic relations of the Mesozoa. Quart. Rev. Biol. **29**:230-244. *Presents evidence that the Mesozoa are degenerate flatworms.*

Thorson, G. 1950. Reproductive and larval ecology of some marine bottom invertebrates. Biol. Rev. **25**:1-45.

Van Beneden, E. 1876. Recherches sur les dicyémides. Bruxelles Acad. Roy. Belg. Bull. Cl. Sci. **41**:1160-1205, **42**:35-97. *One of the first investigations on the group that was discovered in 1839. He considered them intermediate between the Protozoa and the Metazoa.*

Vonk, H. J. 1955. Comparative physiology (nutrition, feeding, and digestion) Annu. Rev. Physiol. **17**:483-498.

Wilson, H. V. 1907. On some phenomena of coalescence and regeneration in sponges. J. Exp. Zool. **5**:245-258. *This classic experimental work on siliceous sponges first showed the phenomenon of regeneration after dissociation. A new sponge is formed by aggregation and fusion out of the cells of an old sponge, which have been separated by squeezing through a piece of gauze. This phenomenon also occurs in forms other than the Porifera.*

Yonge, C. M. 1928. Feeding mechanisms in invertebrates. Biol. Rev. **3**:21-76.

Sea anemones, the "flowers of the sea," are marine coelenterates and are radially symmetrical. The base of an anemone adheres to a substratum, with its mouth and surrounding rings of tentacles at the free end. They are carnivores, using myriads of tiny stinging cells on their tentacles to paralyze their prey and then bending the tentacles to carry the food to the mouth. (Courtesy R. C. Hermes, Homestead, Fla.)

CHAPTER 9
THE RADIATE ANIMALS
PHYLUM COELENTERATA (CNIDARIA)
PHYLUM CTENOPHORA

■ PLACE IN ANIMAL KINGDOM

These two phyla make up the radiate animals, which are characterized by **primary radial** or **biradial symmetry,** and represent the lowest of the eumetazoans. This type of symmetry is composed of an oral-aboral axis with the body parts arranged concentrically around it and is particularly suitable for sessile or sedentary animals. All other eumetazoans have bilateral symmetry. Biradial symmetry in some members is a step toward the bilateral symmetry of the higher metazoans.

Neither phylum has advanced beyond the **tissue grade of organization,** although incipient organs occur. In general the ctenophores are considered to have a higher structural grade than that of the coelenterates.

■ BIOLOGIC CONTRIBUTIONS

1. Both phyla have developed two well-defined **germ layers,** ectoderm and endoderm, with a third, or mesodermal, layer in some. The body plan is now saclike and the body wall is composed of two distinct layers, epidermis and gastrodermis, derived from the ectoderm and endoderm respectively. The gelatinous matrix, mesoglea, between these layers may be structureless, or may contain a few cells and fibers, or may be composed largely of mesodermal connective tissue and muscle fibers.

2. An internal body cavity, the **gastrovascular cavity,** is endodermal in origin and has a single opening, the mouth, which serves also as an anus.

3. **Extracellular digestion** occurs in the gastrovascular cavity, as well as intracellular digestion in the gastrodermis cells.

4. Most radiates have **tentacles,** or extensible projections around the oral end, that aid in food capture.

5. The first true **nerve cells** (protoneurons) occur in the radiates, with a **nerve net** arrangement that is the forerunner of the centralized nervous system of eumetazoans.

6. **Sense organs** appear first in the radiates and include well-developed statocysts (organs of equilibrium) and ocelli (photosensitive organs).

7. Locomotion in the free-moving forms is achieved either by **muscular contractions** (coelenterates) or **ciliary comb plates** (ctenophores). However both groups are still better adapted to floating or being carried by currents than to strong swimming.

8. **Biradial symmetry,** a forerunner of bilateral symmetry, is an advancement over true radial symmetry.

9. **Polymorphism** in the coelenterates has widened their ecologic possibilities. In many species the presence of both a polyp (sessile and attached) stage and a medusa (free-swimming) stage permits occupation of a benthic (bottom) and a pelagic (open-water) habitat by the same species.

10. Some unique features are found in these phyla, such as **nematocysts** (stinging organoids) in coelenterates and **colloblasts** (adhesive organoids) and **ciliary comb plates** in ctenophores.

■ PHYLOGENY AND ADAPTIVE RADIATION

PHYLOGENY. Although nearest to the ancestral stock of the eumetazoans, the origin of the radiate phyla is obscure. The coelenterates may have come from protozoans by way of a ciliated, free-swimming gastrulate animal similar to the planula larva found throughout the group. Such a form, pushed inward to produce a double-layered gastrula shape, would correspond roughly to a coelenterate with an outer ectoderm and an inner endoderm. This theory is supported by the fact that most forms in this group have such a larva somewhere in their life history.

The trachyline medusae (class Hydrozoa) are considered to be the most primitive of modern coelenterates because of their direct development from the planula and actinula larvae to the medusa (jellyfish) stages and because their ontogeny repeats the phylogenetic history of coelenterates better than does that of other members of the phylum.

In the light of their many resemblances to the coelenterates, the ctenophores should have originated from the same ancestral stock, such as the trachyline medusa, but evidence is not available. Biradial symmetry of the ctenophores could have derived from the tetramerous radial symmetry of the hydromedu-

sae. The ancestral form of this phylum was supposed to have been a more or less spherical animal with a concentration of cilia along eight meridional rows that later developed into the comb plates.

ADAPTIVE RADIATION. In their evolution neither phylum has deviated very far from its basic plan of structure. In polymorphic forms (coelenterates) both the polyp and medusa are constructed on the same scheme. Likewise, the ctenophores have adhered to the arrangement of the comb plates and their biradial symmetry. Each phylum has stuck to its pattern in whatever modification it has undergone in its adaptive radiation into different environments. To overcome the restriction imposed by a body structure that has little space for specialized organs, the coelenterates have to some extent compensated by the evolutionary development of different individuals (various types of polyps and medusae) specialized for performing different functions in their life cycle. This polymorphism has no doubt aided them in adapting to a greater variety of environments.

■ PHYLUM COELENTERATA (CNIDARIA)*

This large and interesting group of more than 9,000 species takes its name from the large cavity, or coelenteron, that serves as an intestine. The word "coelenteron" means "hollow intestine." The name "Cnidaria" is based on "nettle" and refers to the cells called cnidoblasts that produce the stinging organelles (nematocysts) so characteristic of this phylum.

The coelenterates may be regarded as the basic stock of the metazoan line, although their organization has a structural and functional simplicity not found in other Metazoa. Not totally restricted to a single plan, they have variety in their organization and have evolved different methods of integration. It is as though the group had been experimenting with different methods without settling upon a definite plan. However there is some unity of organization found among all the groups.

The coelenterates are all aquatic and include some of nature's strangest and loveliest creatures—the branching plantlike hydroids, the flowerlike sea

*Se-len"te-ra'ta (Gr. *koilos,* hollow, + *enteron,* gut, + L. *-ata,* pl. suffix meaning characterized by); ny-dar'e-a (Gr. *knidē,* nettle, + L. *-aria,* pl. suffix meaning like or connected with).

FIG. 9-1

Representative coelenterates. Class Hydrozoa: **1,** *Physalia;* **2,** *Gonionemus;* **3,** *Obelia.*
Class Scyphozoa: **4,** *Aurelia;* **5,** *Chrysaora.* Class Anthozoa: **6,** *Metridium;* **7,** *Astrangia;*
8, *Gorgonia;* **9,** staghorn coral *(Acropora).* (Not shown to scale.)

anemones, the jellyfishes, and those architects of the ocean floor, the corals, sea whips, sea fans, sea pansies, and all the hard corals whose eons of calcareous house-building have produced great reefs and coral islands (Fig. 9-1).

Coelenterates are radiate animals that have the parts arranged concentrically around an oral-aboral axis. The arrangement is often built around a definite number, such as four (tetramerous plan) or six (hexamerous plan) or a multiple of either. They have a saclike body plan of a solid body wall built around a central cavity, the **gastrovascular cavity,** that has one opening serving as both mouth and anus. This is in contrast to the sponges in which the outside opening is an exit, and it may indicate that the coelenterates have the same ancestry as higher forms.

Ecologic relationships

Most coelenterates occur in shallow marine water, usually in tropical or subtropical regions, although some are found in deep water and a few in freshwater. In general they have a wide distribution. There are no terrestrial species. Colonial hydroids are found usually in shallow coastal water attached to mollusk shells, rocks, wharves, and other animals, but occasionally they are found at great depths.

Floating and free-swimming medusae are found in open seas and lakes, often a long distance from the shore. Floating colonies such as the Portuguese man-of-war and *Velella* have floats or sails by which they are carried by the wind.

Reef-building corals are mostly found in warm, shallow water and are usually restricted to continental and island shores in tropical or subtropical regions. They flourish best at about 22° C. Cold currents along the coasts of otherwise warm countries will prevent their growth. In most cases they do not grow below 300 feet because of the decline in temperature. They seem to do best at temperatures near the upper limits of survival. Coral formation is also dependent on environmental conditions, such as light, temperature, and agitation of water.

Economic importance

As a group, coelenterates have little economic importance. Precious coral serves for jewelry and ornaments. Corals also are important in building the coral reefs and islands, some of which are used as habitations by man and other animals. In places where it is available, coral rock serves for building purposes.

Some mollusks and flatworms eat hydroids bearing nematocysts and utilize these stinging cells for their own defense. Some animals use coelenterates as food, although this is rarely done by man. Planktonic medusae may be of some importance as food for fish that are of commercial value; on the other hand, the reverse is true—the young of the fish fall prey to coelenterates.

Characteristics

1. Entirely aquatic, some in freshwater but mostly marine
2. **Radial symmetry** or biradial symmetry around a longitudinal axis with **oral** and **aboral ends;** no definite head
3. Two types of individuals—**attached polyps** and **free medusae**
4. Exoskeleton (perisarc) of chitin or lime in some
5. Body with two layers, epidermis and gastrodermis, with mesoglea between **(diploblastic)**; mesoglea with cells and connective tissue (ectomesoderm) in some **(triploblastic)**
6. **Gastrovascular cavity** (often branched or divided with septa) with a single opening that serves as both mouth and anus; extensible tentacles often encircling the mouth or oral region
7. Special stinging cell organoids called **nematocysts** in either or both epidermis and gastrodermis; nematocysts abundant on tentacles, where they may form batteries or rings
8. **Nerve net** of synaptic and nonsynaptic patterns; with some sensory organs; diffuse conduction
9. Muscular system (epitheliomuscular type) of an outer layer of longitudinal fibers at base of epidermis and an inner one of circular fibers at base of gastrodermis; modifications of this plan in higher coelenterates, such as separate bundles of independent fibers in the mesoglea
10. Reproduction by asexual budding (in polyps) or sexual reproduction by gametes (in all medusae and some polyps). Sexual forms monoecious or dioecious; **planula larva;** holoblastic cleavage; mouth from blastopore
11. No excretory or respiratory system
12. No coelomic cavity

Classes

Class Hydrozoa (hy-dro-zo'a) (Gr. *hydra,* water serpent, + *zōon,* animal). Solitary or colonial; asexual polyps and sexual medusae, although one type may be suppressed; hydranths with no mesenteries; medusae (when present) with a velum; both fresh water and marine. Examples: *Hydra, Obelia, Physalia, Tubularia.*

Class Scyphozoa (sy-fo-zo'a) (Gr. *skyphos,* cup, + *zōon,* animal). Solitary; polyp stage reduced or absent; bell-shaped medusae without velum; gelatinous mesoglea much enlarged; margin of bell or umbrella typically with eight notches that are provided with sense organs; all marine. Examples: *Aurelia, Cassiopeia, Rhizostoma.*

Class Anthozoa (an-tho-zo'a) (Gr. *anthos,* flower, + *zōon,* animal). All polyps; no medusae; solitary or colonial; enteron subdivided by at least eight mesenteries or septa with nematocysts; gonads endodermal; all marine.

Subclass Zoantharia. With simple unbranched tentacles. Sea anemones and hard corals. *Metridium, Adamsia, Astrangia, Cerianthus.*

Subclass Alcyonaria. With eight pinnate tentacles. Soft corals. *Tubipora, Alcyonium, Gorgonia, Renilla.*

Dimorphism in coelenterates

Coelenterates may be single or in colonies. Two morphologic types of individuals are recognized in the group.

1. **Polyps** have tubular bodies. A mouth surrounded by tentacles is located at one end; the other end is blind and usually attached by a pedal disk or other device to a substratum. Sometimes there is more than one type of polyp, each specialized for a certain function, such as feeding polyps and reproductive polyps.

2. **Medusae,** or free-swimming jellyfish, have bell- or umbrella-shaped bodies with a mouth located centrally on a projection of the concave side. Around the margin of the umbrella are the tentacles, which are provided with stinging cells.

Some species have both types of individuals in their life history *(Obelia* and other hydroids); others have only the polyp stage (hydra and Anthozoa); and still others have only the jellyfish, or medusa, stage (certain Scyphozoa). Many of the colonial hydroids are polymorphic, having a medusa stage and two or more types of polyps in the polyp stage.

Though polyps and medusae seem superficially to be different from each other, actually this difference is not pronounced. If a polyp form such as that of the hydra were inverted, broadened out laterally to shorten the oral-aboral axis, the hypostome lengthened to form a manubrium, and mesoglea greatly increased, the result would be a structure similar to a medusa, or jellyfish (Fig. 9-2). The great amount of mesoglea in the medusa makes it more buoyant so that it can float easily.

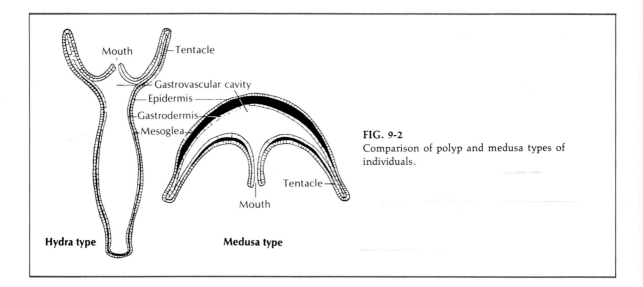

FIG. 9-2
Comparison of polyp and medusa types of
individuals.

In the evolution of the two types, polyps and medusae, probably the jellyfish represents the complete and typical coelenterate, whereas the polyp is a persistent larval, or juvenile, stage.

Nematocysts—the stinging organelles

One of the most characteristic structures in the entire coelenterate group is the stinging organoid called the **nematocyst** (Fig. 9-3). Seventeen different types of nematocysts have been described in the coelenterates so far; they are important in taxonomic determinations. The nematocyst is a tiny capsule composed of material similar to chitin and containing a coiled tubular "thread" or filament, which is a continuation of the narrowed end of the capsule. This end of the capsule is covered by a little lid, or **operculum.** The inside of the undischarged thread may bear little barbs, or spines.

The nematocyst is found in a modified interstitial cell called a **cnidoblast,** which is provided with a projecting triggerlike **cnidocil.** When the cnidocil is stimulated by food, prey, or enemies, the coiled thread turns inside out with explosive force, with the spines unfolding to the outside as the tube everts. Many chemicals (for example, weak acetic acid or methyl green) will cause discharge of nematocysts. Neither touch alone nor the presence of animal fluids (food) alone causes discharge, but touch combined with the presence of food does. The exploding force is probably pressure caused by the increase of water in the capsule, a result of osmotic changes, plus contraction of the cnidoblast that forces the operculum open.

The cnidoblasts may occur singly or in batteries consisting of one large and many small ones. They are found everywhere except on the basal disk but are especially abundant on the tentacles.

Nervous connections play no direct part in the discharge of nematocysts, for they are indirect effectors. However, nerves could affect the threshold of discharge: a full hydra ceases to discharge nematocysts at prey and so does an overstimulated one. When a nematocyst is discharged, the cnidoblast is digested and replaced from the interstitial cells.

The nematocysts of most coelenterates are not harmful to man, but the stings of the Portuguese man-of-war (Fig. 9-17) and certain large jellyfish such as *Cyanea* are quite painful and may even be dangerous to life. Some small freshwater worms that feed on hydrae digest all of the body except the nematocysts; they migrate to the surface of the predator, where they serve for defense.

The nerve net

The nerve net of the coelenterates is one of the best examples of a diffused nervous system in the animal kingdom. This plexus of nerve cells (protoneurons) connected with nerve fibers is found both at the base of the epidermis and at the base of the gastrodermis, because there are two interconnected nerve nets. In the hydra the net appears to be continuous. However, evidence exists that in some coe-

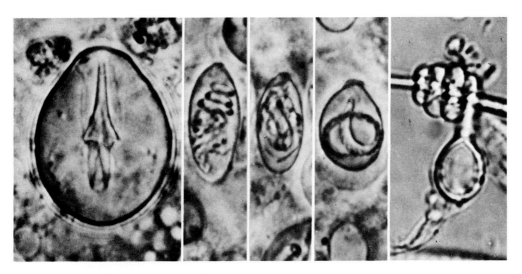

FIG. 9-3
Nematocysts of *Hydra littoralis,* swift-water hydra. Left to right: penetrant, streptoline glutinant, stereoline glutinant, volvent, and a discharged volvent attached to a copepod bristle. The largest and most familiar type is the penetrant, provided with barbed spines and long threadlike tube. Nematocysts are used to immobilize prey and to aid in locomotion. (Courtesy Carolina Biological Supply Co., Burlington, N. C.)

lenterates at least there are two kinds of nerve nets, one in which the protoneurons are continuous, or structurally united (Fig. 9-4), and one in which they are discontinuous, or separated by synaptic spaces. Impulses can pass in all directions over the net. Although the nerve net of coelenterates is generally unpolarized and is characterized by diffuse transmission, unrestricted spreading of excitation is found only in those parts specialized for conduction.

Mackie (1960) reported that in the siphonophore *Velella,* there are two nervous systems—one continuous and one synaptic. The continuous one would seem comparable to the syncytial giant nervous system in squids, annelids, and some arthropods, being specialized for rapid conduction to all parts for, say,

FIG. 9-4
Portion of syncytial giant fiber nervous system of *Velella. Velella* is provided with two nervous systems—one of syncytial (continuous) giant fibers and one of nonsyncytial (discontinuous) neurons. (Courtesy G. O. Mackie, University of Victoria, British Columbia.)

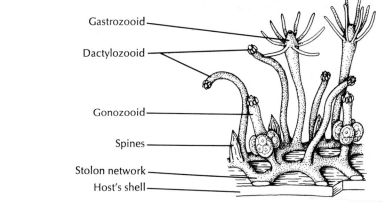

Gastrozooid

Dactylozooid

Gonozooid

Spines

Stolon network

Host's shell

FIG. 9-5

Hydractinia, a colonial hydrozoan. **A,** Hermit crab, *Pagurus floridanus,* with its coelenterate commensals. The crab lives in a snail shell (whose first owner had died) and on the shell two kinds of coelenterates have established their homes. *Hydractinia,* a tiny colonial hydroid, forms a velvety cover on the shell. *Adamsia,* a sea anemone, perches atop the shell. The hermit itself seeks out a suitable anemone, detaches it by gentle massage with its pincers, and then holds it against its shell home till the anemone attaches there. The hermit crab is camouflaged by its friends and protected by their nematocysts; the coelenterates get a free ride and bits of food from their host's meals. **B,** Portion of a colony of *Hydractinia,* showing the types of zooids and the stolon (hydrorhiza) from which they grow. (**A,** Photo by R. C. Hermes, Homestead, Fla.)

an escape response. The synaptic net conducts slower, spreading stimuli.

The nerve cells of the net are connected to slender sensory cells that receive external stimuli and to epitheliomuscular cells that react by contracting. The only localization of nervous function in the hydra is found around the hypostome and on the pedal disk where sensory and other nerve cells are more numerous. Separated bodily parts, when stimulated, often react just as they do in an intact animal.

Together with the contractile fibers of the epitheliomuscular cells, the sensory-nerve cell net combination is often referred to as a **neuromuscular system,** the first important landmark in the evolution of the nervous system. The nerve net is never com-

pletely lost from higher forms. Annelids have it in their digestive systems. In the human digestive system it is represented by the plexus of Auerbach and the plexus of Meissner. The rhythmic peristaltic movements of the stomach and intestine are coordinated by this counterpart of the coelenterate nerve ring.

Class Hydrozoa

The majority of hydrozoans are marine and colonial in form, with the typical life cycle composed of both the asexual polyp and the sexual medusa stages. The freshwater hydra, which has no medusoid stage, is, of course, an exception, as is the marine colonial *Hydractinia* (Fig. 9-5), which grows profusely on the

FIG. 9-6
Colony of *Tubularia larynx. Inset,* Single polyp showing the characteristic circlet of small tentacles surrounding the hypostome, the lower circlet of larger tentacles, and between them the gonophores (medusa buds that do not detach but shed their gametes while in place on the parent polyp). (Photos: colony, T. Lundälv, Kristinebergs Zoological Station, Sweden; *inset,* B. Tallmark, Uppsala University, Sweden.)

shells of certain hermit crabs and lacks a medusa stage. The colonial hydroid *Tubularia* develops medusa buds (gonophores) that never become free medusae, but shed their gametes while still attached to the parent polyp (Fig. 9-6).

The marine jellyfish *Liriope* has no hydroid stage; the larvae develop directly into medusae. Some medusae, such as *Sarsia,* not only reproduce sexually but bud young medusae from the manubrium or from the base of the tentacles.

The hydra, although exceptional, has, because of its size and ready availability, become a favorite as an introduction to this phylum. Combining its study with that of a representative colonial marine hydroid such as *Obelia* gives an excellent idea of the class Hydrozoa.

Hydra, a freshwater hydrozoan

The common freshwater hydra is a solitary polyp and one of the few coelenterates found in fresh water. Its normal habitat is the underside of aquatic leaves and lily pads in cool, clean fresh water of pools and streams. The hydra family is found throughout the world, with ten species occurring in the United States. Common species are the green hydra *(Chlorohydra viridissima),* which owes its color to symbiotic algae (zoochlorella) in its cells, and the brown hydra *(Pelmatohydra oligactis).*

THE BODY PLAN. The body of the hydra can ex-

tend to a length of 25 to 30 mm. or can contract to a tiny mass of jelly (Fig. 9-7). It is a cylindric tube with the lower (aboral) end drawn out into a slender stalk *(Pelmatohydra),* ending in a basal or pedal disk for attachment. This pedal disk is provided with gland cells to enable the hydra to adhere to a substratum and also to secrete a gas bubble for floating. In the center of the disk there may be an excretory pore. The **mouth,** located on a conical elevation, the **hypostome,** is encircled by six to ten hollow tentacles that, like the body, can be greatly extended when the animal is hungry.

The mouth opens into the **gastrovascular cavity,** which communicates with the cavities in the tentacles. In some individuals **buds** may project from the sides, each with a mouth and tentacles like the parent. Testes or ovaries, when present, appear as rounded projections on the surface of the body (Fig. 9-10).

THE BODY WALL. The body wall surrounding the gastrovascular cavity consists of an outer **epidermis** (ectodermal) and an inner **gastrodermis** (endodermal) with **mesoglea** between them (Fig. 9-8).

Epidermis. The epidermis is made up of small cubical cells and is covered with a delicate cuticle. This layer contains several types of cells—epitheliomuscular, interstitial, gland, cnidoblast, and sensory and nerve cells.

Epitheliomuscular cells make up most of the epi-

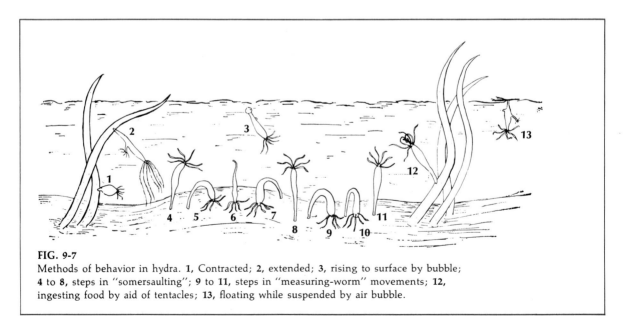

FIG. 9-7

Methods of behavior in hydra. **1,** Contracted; **2,** extended; **3,** rising to surface by bubble; **4** to **8,** steps in "somersaulting"; **9** to **11,** steps in "measuring-worm" movements; **12,** ingesting food by aid of tentacles; **13,** floating while suspended by air bubble.

dermis and serve both for covering and for muscular contraction. The bases of most of these cells are drawn out into longitudinal myofibrils (myonemes). Contraction of these fibrils shortens the body or tentacles.

Interstitial cells are undifferentiated cells among the bases of the epitheliomuscular cells. They can transform into cnidoblasts, sex cells, buds, etc. as needed.

Gland cells are tall cells around the pedal disk and mouth that secrete an adhesive substance for attachment and sometimes a gas bubble for floating.

Cnidoblasts are found throughout the epidermis, especially on the tentacles. They contain the cell organoid—the **nematocyst** or stinging cell—a rounded capsule enclosing a coiled tube that is continuous with the capsule wall. Four kinds of nematocysts are found in the hydra (Fig. 9-3): (1) The **penetrant** is long and threadlike with spines and thorns. When discharged, it is capable of entering the bodies of small animals that happen to touch the tentacles, paralyzing them with the druglike hypnotoxin that it secretes. The hydra may then seize its prey with its tentacles and draw it into the mouth. (2) The **streptoline glutinant** is a long barbed thread that usually coils when discharged. It produces an adhesive secretion used in locomotion and attachment. (3) The **volvent** is a short thread that coils in loops around the prey. (4) The **stereoline glutinant** is a straight unbarbed thread also used for attachment.

Sensory cells are scattered among the other epidermal cells, especially around the mouth, on the tentacles, and on the pedal disk. The free end of each sensory cell is flagellated. The other end branches into fine fibrils attached to the nerve plexus found in the epidermis, next to the mesoglea. The sensory cells are receptors for touch, temperature, and other stimuli.

Nerve cells of the epidermis are either bipolar with two processes or multipolar with many processes. These processes, or neurites, will conduct impulses in either direction. They lie in the epidermis near the level of the nuclei of the epidermal cells or adjacent to the mesoglea. There is no evidence that they lie in the mesoglea. Their processes connect with sensory cells, with the longitudinal fibers of the epitheliomuscular cells, and with other nerve cells. These connections are usually continuous in the hydra.

Gastrodermis. The gastrodermis, a layer of cells lining the gastrovascular cavity is made up chiefly of large, flagellated, columnar epithelial cells with irregular flat bases. The cells of the gastrodermis include nutritive-muscular, interstitial, and gland cells.

Nutritive-muscular cells are usually tall columnar cells and many have their bases drawn out into myonemes that run circularly around the body or tentacles. When the myonemes contract, they lengthen the body by decreasing its diameter. Some

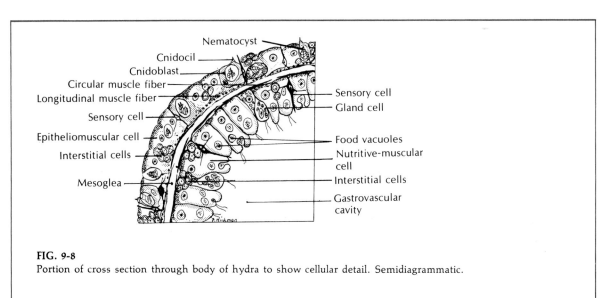

FIG. 9-8

Portion of cross section through body of hydra to show cellular detail. Semidiagrammatic.

serve as sphincters to close the mouth. The cells are highly vacuolated and often filled with food vacuoles. The free end of each cell usually bears two flagella, thus ensuring continual movement of food and fluids in the digestive cavity. Gastrodermal cells in the green hydra (*Chlorohydra*) bear green algae (zoochlorella), which give the hydras their color. This is probably a case of symbiotic mutualism, for the algae utilize the carbon dioxide and waste to form organic compounds useful to the host and receive shelter and other advantages in return. Nutritive-muscular cells may also secrete digestive enzymes into the coelenteron for the digestion of foods.

Interstitial cells are scattered among the bases of the nutritive cells. They may transform into other types of cells when the need arises.

Gland cells in the hypostome and in the column secrete digestive enzymes. Mucous glands about the mouth aid in ingestion.

Cnidoblasts are not found in the gastrodermis, for nematocysts are lacking in this layer.

Mesoglea. The mesoglea lies between the epidermis and gastrodermis and is attached to both layers. It is gelatinous, or jellylike, and has no fibers or cellular elements. It is a continuous layer that extends over both body and tentacles, thickest in the stalk portion and thinnest on the tentacles. This arrangement allows the pedal region to withstand great mechanical strain and gives the tentacles more flexibility. The mesoglea supports and gives rigidity to the body, acting as a type of elastic skeleton.

LOCOMOTION. A hydra has several ways of moving from one region to another (Fig. 9-7). One is gliding on the basal disk, aided by secretions from mucous glands. In a "measuring-worm" type of movement the hydra bends over and attaches its tentacles, slides its basal disk up close to the tentacles, and then releases its tentacles and straightens up. Another method is like a handspring, in which the animal attaches its tentacles, flips its basal disk completely over, and attaches it to a new position. The hydra may move from one place to another in an inverted position by using its tentacles as legs. To rise to the surface of the water it often forms a gas bubble on its basal disk and floats up.

FEEDING AND DIGESTION. The hydra feeds upon a variety of small crustaceans, insect larvae, and annelid worms. A hungry hydra waits for its food to come to it. It may, if necessary, shift to a more favorable location, but once attached to its chosen substratum, it waits with its tentacles fully extended. Any small organism that brushes against one of its tentacles is immediately stopped, harpooned by dozens of tiny nematocyst threads; some penetrate the prey's tissues to inject a poisonous paralyzing fluid, while some coil themselves about bristles, hairs, or spines for holding. The hapless prey may be many times larger than its captor. Now the tentacles, some of them attached to the prey, move slowly toward the hydra's mouth. The mouth slowly opens and the prey slides in. It is not swallowed by muscular action; the mouth simply widens and, well moistened with mucus, glides over and around the prey (Fig. 9-9).

The stimulus that actually causes the mouth to open is the chemical glutathione, which is found to some extent in all living cells. Glutathione is released from the prey through the wounds made by the nematocysts, but only those animals that release enough of the chemical are eaten by the hydra. This explains how a hydra distinguishes between *Daphnia,* which it relishes, and some other forms that it refuses. When commercial glutathione is placed in water containing hydras, each hydra will go through the motions of feeding even though no prey is present.

Inside the gastrovascular cavity contraction of the body wall forces the food downward. Gland cells in the gastrodermis discharge enzymes upon the food. The digestion started in the gastrovascular cavity is called **extracellular digestion,** but many of the food particles are drawn by pseudopodia into the nutritive-muscular cells of the gastrodermis, where **intracellular digestion** occurs. Indigestible particles are forced back out of the mouth, for there is no anus. Digested food products may be stored in the gastrodermis or distributed by diffusion to other cells, including the epidermis.

RESPIRATION AND EXCRETION. There are no special organs for gaseous exchange or excretion. These are apparently the function of individual cells. However, excess water that is taken in by osmotic pressure is expelled by periodically opening the mouth and contracting the body to force the water out of the mouth.

REPRODUCTION. The hydra uses both asexual and sexual methods of reproduction. **Asexual reproduction** is by **budding** (Fig. 9-10, *A*). Buds represent

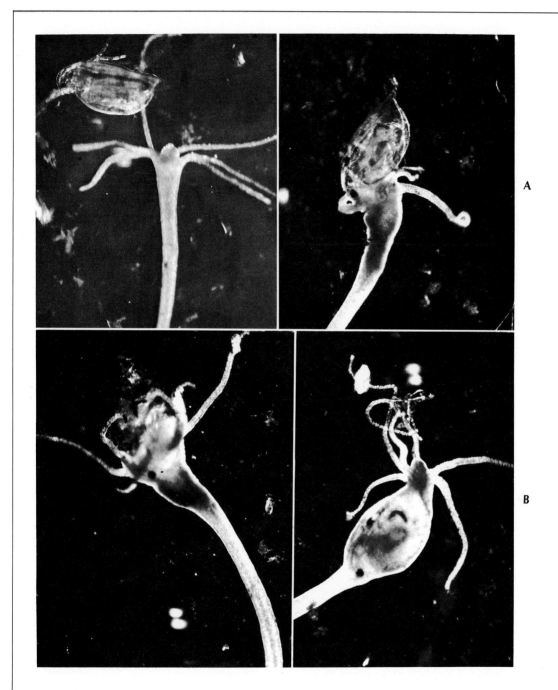

FIG. 9-9
Lunch for a hungry hydra. **A,** Unwary daphnid gets too close to waiting tentacle. Still kicking, it is drawn into widening mouth of captor. **B,** Swallowing process is slow but odds are against daphnid. Mission accomplished, hydra settles down to digest its lunch.

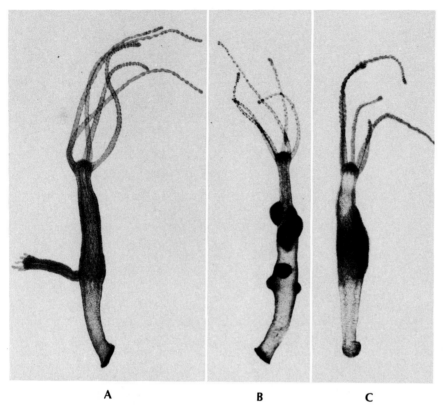

FIG. 9-10
Stained specimens of hydras in reproductive phases. **A,** Budding. The bud forms as an
outpocketing of the body wall, and its gastrovascular cavity is continuous with that of the
parent. **B,** Male hydra with testes. **C,** Female with ovary. Testes and ovaries develop from
the epithelial layer. (Courtesy Carolina Biological Supply Co., Burlington, N. C.)

outpocketings of the entire body wall, with the gastrovascular cavity of the bud being in communication
with the cavity of the parent. The bud acquires a
hypostome with a mouth and a ring of tentacles.
Eventually it constricts at its base and detaches to
lead a separate existence. Several buds may be
formed on the same animal.

In **sexual reproduction** most hydra species are
dioecious (have separate sexes); others are **monoecious** (hermaphroditic). Sexual reproduction involves
the formation of gonads, which are more common
in the autumn. Reduction of water temperature will
promote their formation, but work by Loomis has
shown that high pressures of free carbon dioxide and
reduced aeration in stagnant water may be a factor
responsible for sexuality in the hydra. Gonads (Fig.
9-10, *B* and *C*) are temporary structures formed from
interstitial cells that have accumulated at certain

points, multiply, and undergo all stages of gametogenesis. Haploid sperms are produced and set free
in the water. In the development of the egg one centrally located egg cell enlarges by the union of other
interstitial cells and eventually occupies most of the
space in the ovary. In some species the eggs ripen
one at a time, in succession; in others, several may
ripen at once (Fig. 9-11). The egg undergoes two
maturation divisions, producing two polar bodies,
and a haploid egg. *Pelmatohydra* has 15 chromosomes in its gametes.

While the zygote is still in the ovary, it undergoes
holoblastic **cleavage** and forms a hollow **blastula.**
The outer part of the blastula becomes the **ectoderm,**
which later forms the epidermis; the inner part of
the blastula delaminates to form an inner solid mass,
the **endoderm,** which forms the gastrodermis. The
coelenteron is later formed in this mass, and the

mesoglea is laid down between the ectoderm and endoderm. At about this time a shell, or cyst, is secreted about the embryo, which breaks loose from the parent. The embryo may pass the winter in the encysted condition. When weather conditions are more favorable, the embryo completes its development, the shell ruptures, and a young hydra with tentacles hatches out.

REGENERATION. The power of coelenterates to restore lost parts is pronounced. When the hydra is cut into several pieces, it is possible for each fragment to give rise to an entire animal (Fig. 9-12). A two-headed hydra can be produced by splitting one through the mouth. Parts of individuals of the same species (and sometimes of different species) may be grafted together, even though these fragments are too small to grow independently. The endodermal layer appears to be responsible for fusion, for cells of this layer project out ameboid processes that produce an interlacing effect.

Regeneration experiments have shown that the middle region of the hydra is in an indeterminate state and carries the potentialities of oral and aboral region simultaneously. A section of the middle region transplanted to the hypostome region will form a basal disk; transplanted to the posterior part of the body, it forms a hypostome at the aboral end.

Investigations by Brien and others on the growth patterns of the hydra indicated that the hydra could be considered immortal. When a stained graft was inserted at a "growth zone" just below the tentacles,

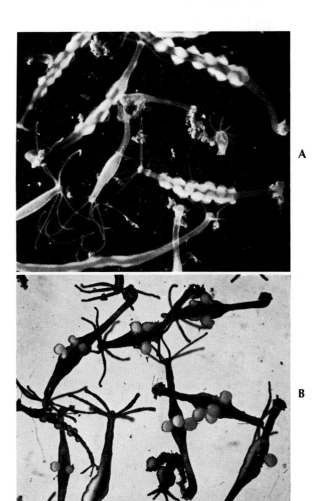

FIG. 9-11
Living hydras. **A,** Mostly males with testes. **B,** Mostly females with ovaries and developing embryos. (Courtesy General Biological Supply House, Inc., Chicago.)

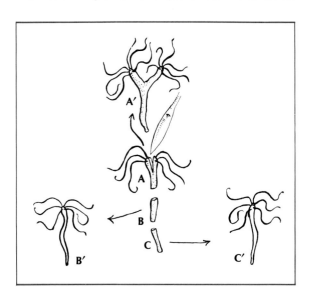

FIG. 9-12
Regeneration in hydra. Splitting hydra part way through mouth and hypostome, **A,** will give rise to two-headed polyp, **A'.** When pieces are cut off below head, **B** and **C,** each will give rise to whole individual, **B'** and **C'.**

the stained cells were seen to move gradually toward the base, where dead or worn out cells were shed. Thus cells arising at the growth zone occupy successively different levels of the hydra as they make their way toward the base (or toward the tip of the tentacles, where a similar renewing takes place). More recent investigations indicate that the so-called growth zone is not limited to one particular area but extends all along the trunk. The active renewal of all the cells in the hydra is believed to take about 45 days and appears to continue indefinitely. If the interstitial cells (which transform into the various types of cells) are destroyed by x-rays, the hydra will live for only a few days.

When a hydra is turned inside out, either by natural or artificial means, it was once believed (Trembley, about 1745) that the epidermal cells became gastrodermal cells and the gastrodermal cells became epidermal cells in their new positions. Many investigations, however, have shown that in some cases the hydras turn themselves right side out again and in other cases the layers are reversed by the migration of inside cells to the outside and of outside cells to the inside, thus restoring the original arrangement of the cells.

BEHAVIOR. The hydra will respond to various stimuli, both internal and external. Spontaneous movements of body and tentacles occur while the animal is attached. If the individual is well fed, its movements are slow, but they increase whenever it becomes hungry. These movements are produced by the contractile fibers in the wall when they are stimulated through the nerve net.

How the hydra reacts to stimuli depends on the intensity and kind of the stimuli and on the physiologic state of the hydra. A slight jar will cause the whole animal to contract rapidly. A localized stimulus, such as touching one of the tentacles with a sharp needle may produce the same effect. The explanation for this probably lies in the widespread transmission of the nerve impulse over all the nerve net. If such a localized stimulus is mild, there may be a more or less localized response, such as the contraction of a single tentacle or the pulling away of that part of the body touched.

A unique behavior pattern called the **contraction burst** has been described in some hydras under constant conditions (C. B. McCullough). At 5- to 10-minute intervals a hydra suddenly assumes a ball shape by contracting its longitudinal muscles. After a few seconds the individual usually extends itself again. It is believed that this behavior pattern enables the hydra to sample its environment intermittently, and the pattern may play a part in movement and light orientation. A neuronal pacemaker located in the subhypostome region may initiate the impulse for the action. Body contraction is also associated with the periodic expulsion of excess water through the mouth.

To light stimuli, hydras respond in an optimum way, tending to avoid very strong light but seeking moderately lighted regions. By trial and error they find the situation that best suits them. When subjected to a weak constant electric current, they orient the oral end toward the anode and the basal end toward the cathode.

Obelia and Gonionemus as examples of hydroid and medusa stages

Both *Obelia* and *Gonionemus* are marine forms that have polyp and medusa stages in their life histories. *Obelia* has a prominent hydroid (juvenile) stage but an inconspicuous medusa (adult) stage (Fig. 9-13). The reverse conditions are found in *Gonionemus,* in which the medusa is large and the hydroid small.

Obelia may be considered a typical colonial hydroid (Fig. 9-13). It is found on both the Atlantic and Pacific coasts. It attaches to stones and other objects by a rootlike base called **hydrorhiza,** from which arise branching stems (**hydrocauli**). On these stems are large numbers of polyps, which are of two types: **hydranths,** which are nutritive, and **gonangia,** which are reproductive. The hydranths furnish nutrition for the colony; the gonangia produce young medusae asexually by budding. The medusae are sexual, giving rise to sperm and eggs. When zygotes are formed, they develop through a series of stages, terminating in a polyp form, thus completing the life cycle (Fig. 9-13). In this way the polyps represent the asexual phase and the medusae represent the sexual phase.

The **hydrocaulus,** or stem that bears the polyps, is a hollow tube composed of a cellular body wall (**coenosarc**) surrounding the **gastrovascular cavity** and covered by a transparent chitinous **perisarc.** The coenosarc, like the body of the hydra, has an outer epidermis, an inner gastrodermis, and mesoglea be-

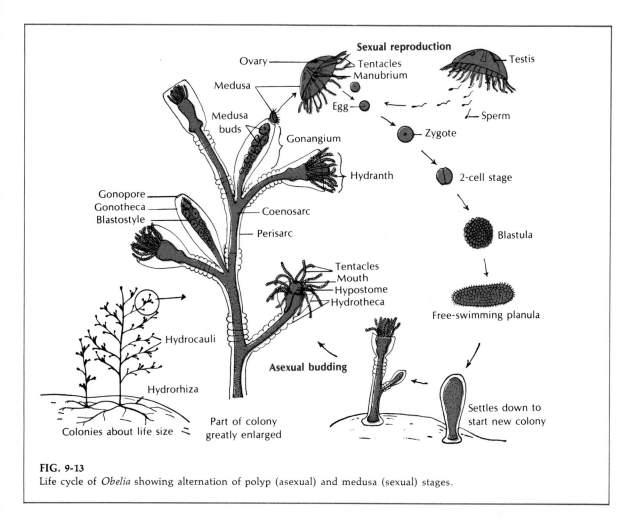

FIG. 9-13
Life cycle of *Obelia* showing alternation of polyp (asexual) and medusa (sexual) stages.

tween them. The gastrovascular cavity is continuous throughout the colony so that nourishment can be distributed from polyps to hydrorhiza. The protective perisarc is also continuous, being modified to cover the polyps.

The nutritive polyp, or **hydranth,** is much like a miniature hydra, with a **hypostome** and **mouth** surrounded by many **tentacles.** By means of the tentacles and **nematocysts,** these feeding polyps capture small crustaceans, worms, or insect larvae that come within their reach. Within the gastrovascular cavity food is reduced by digestive enzymes to a broth of small particles. This is driven throughout the colony by flagellary movement and by contractions of the hydranth. Cells of the gastrodermis pick up the food and complete digestion in food vacuoles; so digestion is both extracellular and intracellular. As far as is

known, starches, cellulose, and chitin are not digested by hydroids. The hydranth is protected by a cup-like **hydrotheca,** a continuation of the perisarc, into which the tentacles can contract.

In the reproductive **gonangium** the **medusae** develop as lateral buds called **gonophores.** The gonangium is surrounded by a protective sheath **(gonotheca)** with an opening through which young medusae escape (Fig. 9-13).

Not all hydroids have their polyps protected by this continuation of the perisarc as do *Obelia, Campanularia,* and others (known collectively as the Calyptoblastea). Many other hydroids, such as *Hydractinia* (Fig. 9-5), *Tubularia* (Fig. 9-6), and *Corymorpha* and *Eudendrium* (Fig. 9-14) (the Gymnoblastea) have naked polyps.

The cellular structure of the colony is much like

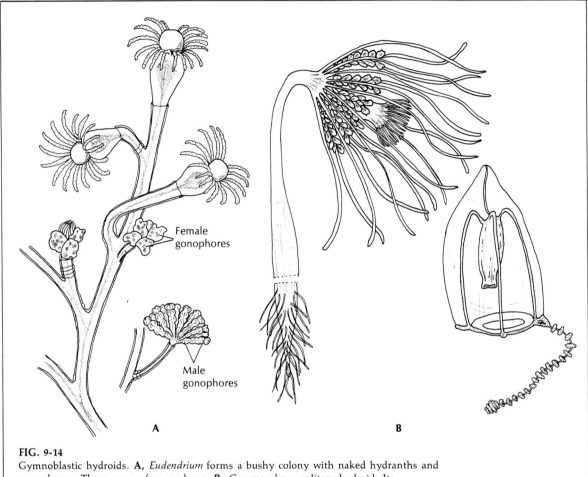

FIG. 9-14
Gymnoblastic hydroids. **A,** *Eudendrium* forms a bushy colony with naked hydranths and gonophores. There are no free medusae. **B,** *Corymorpha,* a solitary hydroid. Its gonophores produce free-swimming medusae, each with a single trailing tentacle.

that of the individual hydra. Myonemes from the epitheliomuscular and nutritive-muscular layers provide movement, stimulated through a **nerve net. Sensory cells** are most abundant around the mouth and tentacles.

Gonionemus is frequently studied as a type of jellyfish, since the medusa is much larger than that of *Obelia* (Fig. 9-15). It is fairly typical of the medusae of this class. The polyp of *Gonionemus* is extremely small. The medusa is bell shaped and 1 to 3 cm. in diameter. The convex, or aboral, side is called the **exumbrella,** whereas the concave, or oral, side is the **subumbrella.** Around the margin of the bell are a score or more of **tentacles** with nematocysts, each tentacle with a bend near the tip bearing an **adhesive pad.**

Inside the margin is a thin muscular membrane, the **velum,** which partly closes the open side of the bell. The velum distinguishes the hydrozoan from the scyphozoan jellyfish. It is used in swimming movements. Contractions in the velum and body wall bring about a pulsating movement that alternately fills and empties the subumbrellar cavity. As the animal contracts, forcing water out of the cavity, it is propelled forward, aboral side first, with a sort of "jet propulsion." The animal swims upward, turns over, and floats lazily downward, tentacles outspread to capture unwary prey. It rests while attached to vegetation by its adhesive pads.

Hanging down inside the bell is the **manubrium,** at the tip of which is the **mouth** surrounded by four

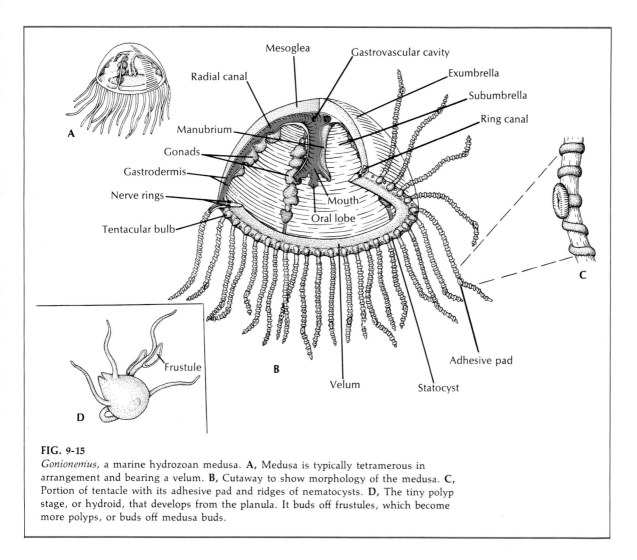

FIG. 9-15
Gonionemus, a marine hydrozoan medusa. **A,** Medusa is typically tetramerous in arrangement and bearing a velum. **B,** Cutaway to show morphology of the medusa. **C,** Portion of tentacle with its adhesive pad and ridges of nematocysts. **D,** The tiny polyp stage, or hydroid, that develops from the planula. It buds off frustules, which become more polyps, or buds off medusa buds.

oral lobes. From the mouth a **gullet** leads to the **stomach** at the base of the manubrium. Four **radial canals** lead out from the stomach to a **ring canal** around the margin, which connects with all the tentacles. The entire system from the gullet to the tips of the tentacles makes up the **gastrovascular cavity,** in which food is partly digested by enzymes and is distributed to all parts of the body. Digestion is completed in the cells of the gastrodermis. Worms, crustaceans, and small fish are favorite foods.

Medusae are dioecious. In *Gonionemus* the **gonads** are suspended under each of the radial canals (Fig. 9-15). The eggs or sperm are shed into the water outside. The fertilized egg develops into a ciliated **planula** larva, which swims about for a time, then settles down, attaches to some object, loses its cilia, and develops into a minute polyp. The cycle begins again with the young polyp budding off additonal polyps that finally produce new medusae by asexual budding.

Since free-swimming medusae are more active than the polyps, they need a little more elaborate nervous system. The **nerve net** is concentrated into two **nerve rings** at the base of the velum, one in the exumbrellar and one in the subumbrellar epithelium. **Statocysts** around the margin provide a sense of balance. Each statocyst is a small sac with a hard mass inside that moves about as the animal moves, acting as a stimulus to direct the movements. **Tentacular bulbs** are enlargements located at the base of the

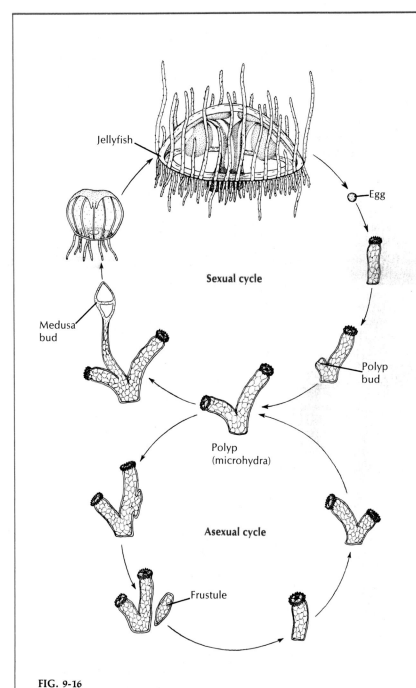

FIG. 9-16
Craspedacusta, a freshwater hydrozoan medusa—life cycle. The planula develops into a minute polyp, called microhydra *(center)*, which, by budding and by production of frustules, which give rise to more polyps, form a small colony (asexual cycle). The polyps may then bud off medusa buds, which become sexually mature medusae and produce eggs or sperm (sexual cycle). Fertilized eggs develop into planula larvae, which settle down and become tiny polyps (microhydra).

tentacles. Within the bulbs nematocysts are formed and migrate out to the batteries on the tentacles. The bulbs may also help in intracellular digestion. The entire animal seems to be photosensitive.

Freshwater medusae

The freshwater medusa *Craspedacusta sowerbyi* (Fig. 9-16) (class Hydrozoa, order Trachylina) may have evolved from marine ancestors in the Yangtze River of China. This interesting form has been found in many parts of Europe and all over the United States and in parts of Canada.

This animal has a hydroid phase that is tiny (2 mm.) and appears to be more or less degenerate, for it has no perisarc and no tentacles. It occurs in colonies of a few polyps. For a long time its relation to the medusa was not recognized, and thus the hydroid was given a name of its own, *Microhydra ryderi.* On the basis of its relationship to the jellyfish and the law of priority, both the hydroid (polyp) and the medusa should be called *Craspedacusta.* This hydroid gives rise to the medusae by budding.

This polyp has three methods of asexual reproduction: (1) by budding off new individuals, which may remain attached to the parent (colony formation); (2) by constricting off nonciliated planula-like larvae (frustules), which can move around and give rise to new polyps; and (3) by producing medusa buds, which develop into sexual jellyfish. Monosexual populations (that is, all male or all female) are characteristic of this species; both sexes of medusae are rarely found together.

The jellyfish, which may attain a diameter of 20 mm. when mature, has some odd features. The tentacles are numerous, are unequal in length, and are not provided with adhesive pads. Only one kind of nematocyst is found. The gonads are enlarged sacs that hang down inside the subumbrella, and the manubrium extends down almost to the level of the velum. Although the medusae are dioecious, usually all the jellyfish of a particular area are of the same sex.

Order Siphonophora

The siphonophorans are highly specialized hydrozoans. They are polymorphic swimming or floating colonies made up of several types of modified medusae and polyps. This is a very ancient group, judging by fossil and other evidence.

FIG. 9-17
Portuguese man-of-war *Physalia physalis* (order Siphonophora, class Hydrozoa) eating a fish. This colony of medusa and polyp types is integrated to act as one individual. As many as a thousand zooids may be found in one colony. They often drift upon southern ocean beaches, where they are a hazard to bathers. Although a drifter, the colony has restricted directional movement. Their stinging organoids secrete a powerful neurotoxin. (Courtesy New York Zoological Society.)

Physalia, or the Portuguese man-of-war (Fig. 9-17), is one such colony with a rainbow-hued float of blues and pinks that carries it along on the surface waters of the southern seas. Many are blown to shore on our eastern coast. The long graceful tentacles, actually zooids, are laden with nematocysts and capable of painful stings. The float, called a **pneumatophore,** is believed to have expanded from the original larval polyp. It contains an air sac arising from the body wall and filled with gas similar to air. The float acts as a type of nurse-carrier for future generations of individuals that bud from it and hang suspended in the water.

There are several types of polyp individuals. The

FIG. 9-18
Stephalia is a siphonophoran jellyfish with a circle of swimming bells around its reduced float, and a cluster of gastrozooids with their tentacles dangling below. Class Hydrozoa.

gastrozooids are feeding polyps with a single long tentacle arising from the base of each. Some of these long stinging tentacles become separated from the feeding polyp and are called **dactylozooids,** or fishing tentacles. These sting the prey and lift them to the lips of the feeding polyps. Among the modified medusoid individuals are the **gonophores,** which are little more than sacs containing either ovaries or testes.

An interesting symbiotic relationship exists between *Physalia* and a small fish called *Nomeus* that swims among the tentacles with perfect safety. Other larger fish trying to catch *Nomeus* are caught by the deadly tentacles. *Nomeus* probably feeds upon bits of *Physalia's* prey, or perhaps, according to one theory, it eats the zooids and nematocysts of the tentacles and so becomes immune to their poison. Another theory is that the fish releases an ectohormone that allows *Physalia* to recognize its commensal partner.

Some of the siphonophores, such as *Stephalia* (Fig. 9-18) and *Nectalia,* possess swimming bells as well as a float.

Class Scyphozoa

Class Scyphozoa contains most of the large jellyfish (Fig. 9-19). Some of these medusae, such as *Cyanea* (Fig. 9-19, *B*), are several feet in diameter, with tentacles more than 75 feet long. Others, however, are quite small. Most are found floating free in the open sea, but the members of one order are sessile and attach to seaweed, stones, and the like. The polypoid stage is lacking or is limited to a small larval stage.

The bell of the various species varies in depth from a shallow saucer shape to a deep helmet or goblet shape. The jelly (mesoglea) layers are unusally thick, giving the bell a fairly firm consistency. The jelly is 95% to 96% water. Unlike the hydromedusae, this layer in the scyphomedusae also contains ameboid

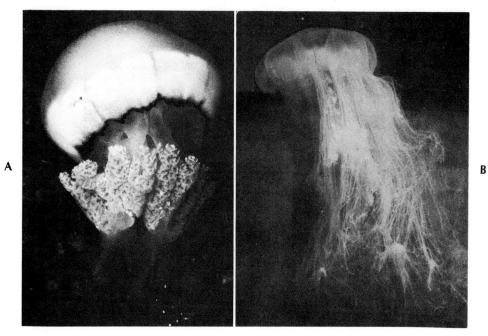

FIG. 9-19
Scyphozoan jellyfishes. **A,** *Rhizostoma pulmo,* the "lung medusa," from the north
Atlantic. It is large (50 cm. across), heavy bodied, and bluish white with violet edge
coloration. **B,** Large jellyfish *Cyanea.* Some Arctic forms attain diameter of more than 6
feet. Its many hundred tentacles may reach a length of 75 feet or more. (**A,** Photo by B.
Tallmark, Uppsala University, Sweden.)

cells and fibers, so that it is now called a **collenchyme.** Movement is by rhythmic pulsations of the
umbrella. There is no velum as in the hydromedusae. There may be many tentacles or few, and they
may be short as in *Aurelia* or long as in *Cyanea.*
Aurelia (Fig. 9-20) is a familiar species 3 or 4 inches
in diameter, commonly found in the waters off both
our east and west coasts, and is widely used as a type
for the study of Scyphozoa.

The margin of the umbrella is scalloped, usually
with each indentation bearing a pair of **lappets,** and
between them a sense organ called a **rhopalium** (tentaculocyst). *Aurelia* has 8 such notches. Some scyphozoans have 4, others 16. Each rhopalium is club-
shaped and contains a hollow statocyst for equilibrium and one or two sensory pits lined with sensory
epithelium. In some species the rhopalia also bear
ocelli.

The mouth is centered on the subumbrella side.
The manubrium is usually drawn out to form four

frilly **oral arms** that are used in capturing and ingesting prey.

The tentacles, manubrium, and often the entire
body surface are well supplied with nematocysts that
can give painful stings. In fact among the scyphozoans are the so-called stinging nettles, so dreaded
by swimmers, and the deadly sea wasp *Chironex
fleckeri,* the stings of which are considered quite dangerous and sometimes lethal. The nematocysts are
not primarily intended for man, however, but for
paralyzing small prey, which are then conveyed by
ciliated grooves in the lobes up to the mouth and digestive cavity. *Aurelia* and some others collect tiny
food organisms in mucus on the exumbrellar side
where they are carried by cilia to the tentacles
around the margin and then licked off by the oral
lobes. Flagella in the gastrodermis layer keep a current of water moving to bring food and oxygen into
the stomach and carry out wastes.

Cassiopeia, a jellyfish common to our Florida

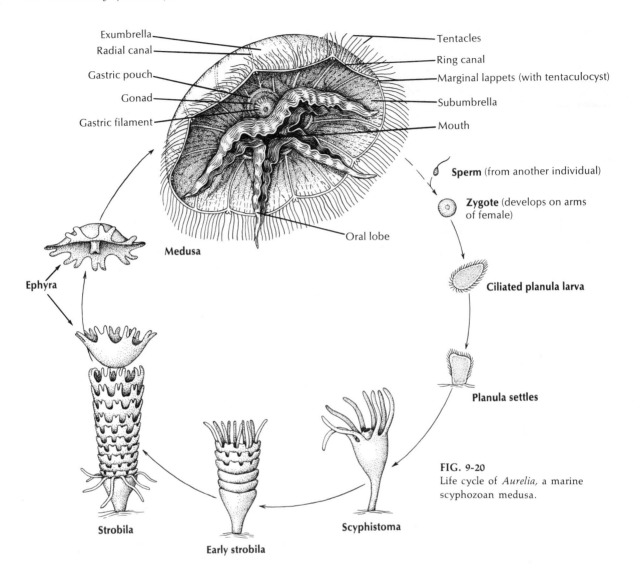

Exumbrella
Radial canal
Gastric pouch
Gonad
Gastric filament

Tentacles
Ring canal
Marginal lappets (with tentaculocyst)
Subumbrella
Mouth

Sperm (from another individual)

Zygote (develops on arms of female)

Oral lobe

Medusa

Ciliated planula larva

Ephyra

Planula settles

FIG. 9-20
Life cycle of *Aurelia,* a marine scyphozoan medusa.

Strobila

Scyphistoma

Early strobila

waters, and *Rhizostoma* (Fig. 9-19), which can also be found in colder waters, belong to a group that differs from *Aurelia* in their lack of tentacles and in the structure of the oral arms. The bases of the oral lobes branch to form a number of brachial arms, each enclosing a ciliated canal. The frilled edges of the lobes come together to form numerous ciliated grooves, or so-called mouths, that communicate with the brachial arms. *Cassiopeia* likes to lie on its back in shallow lagoons waiting for small fishes or crustaceans to come along, become entangled in mucus, and get swept into the "mouths" by ciliary action.

Internally, extending out from the stomach are four **gastric pouches** in which gastrodermis extends down in little tentacle-like projections called **gastric filaments.** These are covered with nematocysts to further quiet any prey that may still be struggling. Gastric filaments are lacking in the hydromedusae. A complex system of **radial canals** branch out from the pouches to a **ring canal** in the margin and make up a part of the gastrovascular cavity.

The **nervous system** in scyphozoans seems to be of a synaptic nerve net type, with a subumbrellar net that controls bell pulsations, and another more diffuse net that controls local reactions such as feeding.

The sexes are separate, with the gonads located in

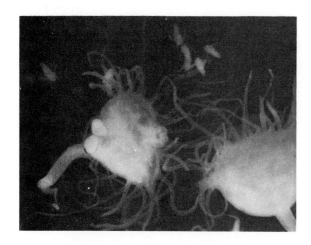

FIG. 9-21
Scyphistoma (polyp stage) of the large jellyfish *Cassiopeia.* The left individual shows the mouth at right and two buds at the left of the bell. The buds detach and develop into new scyphistomas. In the spring young ephyrae develop and bud off, one at a time, from the oral side to become medusae. Note the brine shrimp larvae captured by the tentacles, and the opened mouth, ready to feed.

the gastric pouches. Fertilization is internal, with the sperm being carried by ciliary currents into the gastric pouch of the female. The zygotes may develop in the seawater or may be brooded in folds of the oral arms. The ciliated planula larva becomes fixed and develops into a **scyphistoma,** a hydralike form. By a process of **strobilation** the scyphistoma of *Aurelia* forms a series of saucerlike buds, **ephyrae,** and is now called a **strobila** (Fig. 9-20). When the ephyrae break loose, they become inverted and grow into mature jellyfish.

In *Cassiopeia,* the scyphistoma does not form a strobila as in *Aurelia,* but buds off young medusae one at a time (monodisk strobilation). The scyphistoma can also constrict off small planula-like ciliated buds that swim about then settle down to become other scyphistomas (Fig. 9-21).

Class Anthozoa

The anthozoans, or "flower animals," are polyps with a flowerlike appearance. There is no medusa stage. Anthozoans are all marine and are found in both deep and shallow water and in polar seas as well as tropical seas. They vary greatly in size and may be single or colonial. Many of the forms are supported by skeletons.

The class is made up of two subclasses—Zoantharia, made up of the sea anemones and stony corals, and Alcyonaria, which includes the sea fans, sea plumes, sea pansies, and other soft corals. The zoantharians are built on a hexamerous plan (a plan of 6 or multiples of 6) and have simple tubular tentacles arranged in one or more circlets on the oral

disk. The alcyonarians are octamerous (built on a plan of 8) and always have 8 pinnately branched tentacles arranged around the margin of the oral disk.

The gastrovascular cavity is large and partitioned by mesenteries, or septa, that are inward extensions of the body wall. The walls and mesenteries contain both circular and longitudinal muscle fibers.

The mesoglea is a mesenchyme containing ameboid cells. There is a general tendency toward biradial symmetry in the septal arrangement and in the shape of the mouth and pharynx. There are no special organs for respiration or excretion, which occur by diffusion through the body walls and cells.

The sea anemones

Sea anemone polyps are larger and heavier than hydrozoan polyps (Fig. 9-22). Most of them range from $1/2$ inch or less to 2 inches in diameter, and from $1/2$ inch to several inches long, but some grow much larger. Some of them are quite colorful. Anemones are found in coastal areas all over the world, especially in the warmer waters, and they attach by means of their pedal disk to shells, rocks, timber, or whatever submerged substrata they can find. Some burrow in the bottom mud or sand.

Sea anemones are cylindric in form with a crown of tentacles arranged in one or more circles around the mouth on the flat **oral disk** (Fig. 9-23). The slit-shaped mouth leads into a **pharynx.** At one or both ends of the mouth is a ciliated groove called a **siphonoglyph,** that extends into the pharynx. The siphonoglyphs create water currents directed into the pharynx. The cilia elsewhere on the pharynx direct

FIG. 9-22
Sea anemone, *Metridium,* swallowing a stickleback it has captured. Poison from the
nematocysts of the tentacles can immobilize the fish in a few seconds, and it is swallowed within
15 minutes. (Photo by B. Tallmark, Uppsala University, Sweden.)

water outward. The currents thus created carry in oxygen and remove wastes, and they also help maintain an internal fluid pressure, or hydrostatic skeleton that serves in lieu of a true skeleton as a support for opposing muscles.

The pharynx leads into a large **gastrovascular cavity** that is divided into six radial chambers by means of six pairs of primary septa, or **mesenteries,** that extend vertically from the body wall to the pharynx (Fig. 9-23). These chambers communicate with each other and are open below the pharynx. Smaller mesenteries partially subdivide the larger chambers and provide a means of increasing the surface area of the gastrovascular cavity. The free edge of each incomplete septum forms a type of sinuous cord called a **septal filament** that is provided with nematocysts and with gland cells for digestion. In many anemones the lower ends of the septal filaments are prolonged into **acontia threads,** also provided with nematocysts and gland cells, that can be protruded through the mouth or through pores in the body wall to help overcome prey, or provide defense. The pores also aid in the rapid discharge of water from the body when the animal is endangered and contracts to a small size.

Sea anemones are carnivorous, feeding upon fish (Fig. 9-22) or almost any live animals of suitable size. Some species live on minute forms caught by ciliary currents.

Feeding behavior in many coelenterates is under chemical control. Some sea anemones may respond to reduced glutathione; others respond to two different compounds, where each of two phases is controlled by a different chemical activity. One phase in-

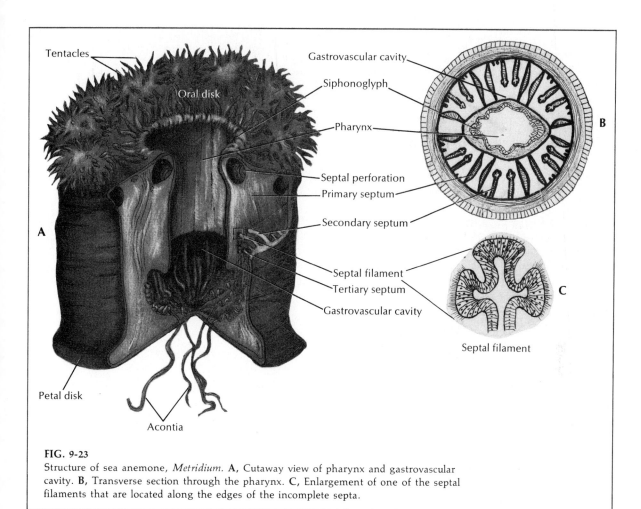

FIG. 9-23
Structure of sea anemone, *Metridium*. **A,** Cutaway view of pharynx and gastrovascular cavity. **B,** Transverse section through the pharynx. **C,** Enlargement of one of the septal filaments that are located along the edges of the incomplete septa.

volves asparagine, which causes a bending of the tentacles toward the mouth; in the second phase, reduced glutathione induces swallowing of the food. Asparagine is a feeding incitant, and reduced glutathione is a feeding stimulant for ingestion.

Sea anemones are muscular, having muscle fibers not only in the epidermis and gastrodermis but also in the collenchyme as well. There are definite muscle bands in the mesenteries. Anemones can glide along slowly on their pedal disks. They can expand and stretch out their tentacles in search of small vertebrates and invertebrates, which they overpower with tentacles and nematocysts and carry to the mouth.

When disturbed, sea anemones contract and draw in their tentacles and oral disks. Some anemones are able to swim, to a limited extent, by rhythmic bending movements, which may be an escape mechanism

from enemies such as sea stars and nudibranchs. *Stomphia,* for example, at the touch of a predatory sea star, will detach its pedal disk and make creeping or swimming movements to escape (Fig. 9-24). But this escape reaction can be caused not only by the touch of the star but by exposure to drippings exuded by the star, or to crude extracts made from its tissues. The sea star drippings contain steroid saponins that are toxic and irritating to most invertebrates. Extracts given off by nudibranchs can also provoke this reaction in sea anemones.

Anemones form some interesting commensal relationships with other animals. Some species habitually attach to the shells occupied by certain hermit crabs (Fig. 9-5). The hermit encourages the relationship and, finding its favorite species, which it recognizes by touch, it massages the anemone until it de-

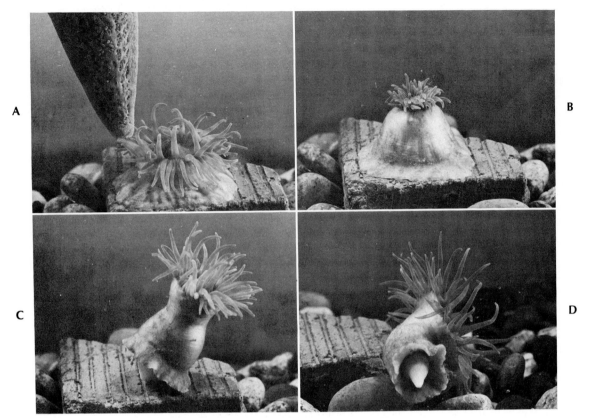

FIG. 9-24

Reaction of *Stomphia* to a predatory sea star *Dermasterias*. Upon contact with an arm of the star, **A**, the anemone contracts and, **B**, withdraws its tentacles; then, **C**, it loosens its pedal disk and, **D**, using a combination of tentacular movement and muscular body-wall movements enhanced by shifting of fluids in the gastrovascular cavity, it may roll, creep, or swim to a safer location. (Courtesy Dr. D. M. Ross; prior use in *Scientific American* acknowledged [cf. H. Feder, *Sci. Amer.* July 1972].)

taches and then it holds the anemone against its own shell until it is firmly attached. When the crab moves to a larger shell, it will move its anemone to the new shell, too. The hermit is not only immune to its passenger's stings, but is protected by them from its enemies. The anemone gets free transportation and particles of food dropped by the hermit crab.

Some species of crabs will hold small anemones in their pincers and use them for defense as well as steal food from their tentacles. Anemones also have commensal relationships with certain small fish that, in return for their immunity from stings, draw larger fish near enough to the tentacles for capture.

The sexes are separate in sea anemones, and the gonads are arranged on the margins of the mesenteries. The zygote develops into a ciliated larva.

Asexual reproduction sometimes occurs by budding and fragmentation.

Zoantharian corals

The zoantharian corals include the true stony corals (order Scleractinia) and the black corals (order Antipatharia). The stony corals might be described as miniature sea anemones that live in calcareous cups they themselves have secreted (Figs. 9-25 to 9-27). Like the anemones the coral polyp's gastrovascular cavity is subdivided by mesenteries, arranged in multiples of six, and its hollow tentacles surround the mouth, but there is no siphonoglyph.

Instead of a pedal disk, the epidermis at the base of the column and mesenteries secretes the limy skeletal cup, which takes on the same radial pattern

FIG. 9-25
Group of coral polyps, *Astrangia danae,* protruding
from their shallow cups. This is a common
zoantharian coral of the Atlantic coast. (Courtesy
General Biological Supply House, Inc., Chicago.)

FIG. 9-26
Stony coral *Oculina* showing cups in which polyps
once lived. A zoantharian coral. Note the pattern
of the polyp mesenteries in each cup.

FIG. 9-27
Caryophyllia smithii, a stony coral photographed at depth of 35 meters on west coast of Sweden. Septa
are plainly visible between the transparent hollow tentacles on the animal at left, and the mouth can be
seen in the right-hand animal. This is a solitary coral, not reef forming. Its cups may be 0.5 to 25 cm.
across. (Photo by B. Tallmark, Uppsala University, Sweden.)

as the polyp mesenteries (Fig. 9-26). The living polyp can contract into the safety of the cup when not feeding. Although some corals are solitary (Fig. 9-27), most are in colonies (Fig 9-25), which assume a great variety of branching or rounded shapes. The living polyps are found only on the surface layer of coral masses where they build new cups over the old ones of past generations. Thus over long periods of time large amounts of calcareous coral is built up.

The black, or thorny, corals usually form slender branching colonies several feet tall. Their polyps have six nonretractile tentacles.

Alcyonarian corals

Alcyonarian corals are the so-called soft corals. In this group the polyp has an octamerous arrangement of septa and eight pinnate tentacles (Fig. 9-28). The corals are all colonial, and the gastrovascular cavities of the polyps are interconnected by a system of canals within the mass of mesoglea in which the polyps are embedded. This mesogleal mass covers a secreted core of horny material or of limy spicules. There is a wide variety of shapes within the group, from the slender, branching sea whips (Fig. 9-29, *C*) with a core of horny skeleton to the sea fans (Fig. 9-29, *B*) with a skeleton of calcareous spicules and a tendency to branch in one plane only, and to the organ-pipe coral (Fig. 9-29, *A*) with its calcareous spicules fused into organlike tubes. *Alcyonium* is a fleshy coral whose spicules are scattered over the mesoglea (Fig. 9-28, *A*). *Renilla,* the sea pansy, is a colony shaped like a pansy leaf, with the polyps embedded in the fleshy upper side of the leaf, and a short stalk that supports the colony and is embedded in the sea floor (Fig. 9-28, *B*).

Few animals live on the abyssal plains of the ocean where the pressure is great, the water temperature is about 35° F., and there is total darkness except for

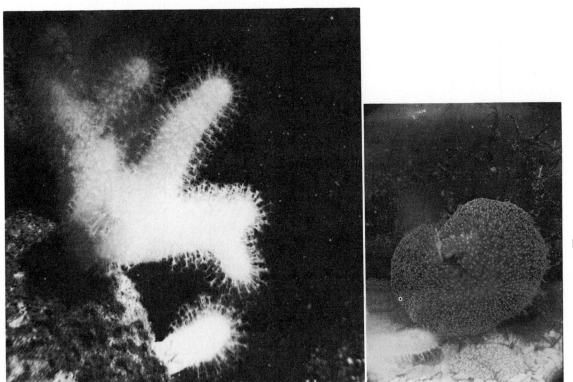

FIG. 9-28
A, *Alcyonium digitatum,* a colony-forming alcyonarian coral, abundant in Scandinavian waters. **B,** *Renilla,* the sea pansy, in which the polyps grow on the upper side of a leaflike rachis attached to the mud bottom by a stalk. (**A,** Photo by T. Lundälv, Kristinebergs Zoological Station, Sweden.)

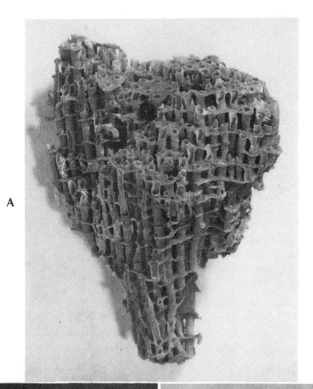

FIG. 9-29
Dried skeletons of alcyonarian corals. **A,** Organpipe coral *Tubipora.* These often build extensive reefs.
B and **C,** Gorgonian soft corals, dried. **B,** Sea fan, *Gorgonia.* **C,** Sea whip. These horny corals have
exoskeletons of calcareous spicules or hornlike gorgonin (an organic substance) and are often in fanlike
or featherlike colonies.

the luminescence given off by the few animals of that region. A deep-sea alcyonarian coral, *Umbellula,* has recently been photographed in its native habitat at a depth of nearly 16,000 feet. It has a 3-foot stalk anchored at one end to the bottom sediments and carrying at the other end a tuft of eight-tentacled polyps. There may be as many as 20 or 30 polyps in the cluster.

Coral reef formation

Most coral reefs are built by zoantharian corals of the order Scleractinia, which have the ability to construct massive skeletal structures of calcium carbonate. Some of the alcyonarian corals *(Tubipora, Heliopora)* and the hydrozoan *Millepora* also build minor reefs. Some other anthozoans and the calcareous algae also contribute to them. Reefs are characterized by high population density, intense calcium metabolism, and complex nutrient chains.

Reef-building corals are mainly restricted to the shallow waters of tropical seas, and flourish best in temperatures between 72° and 85° F. They are more common in the Indo-Pacific regions than elsewhere. Temperature seems to be the chief control, although depth of water, turbidity, and many other variables may also play a part.

Reefs tend to grow from 1 to 5 cm. per year, but some grow faster and others slower than this for there are many differences in calcification rates. Three kinds of coral reefs are commonly recognized, depending on how they are formed. The **fringing reef** may extend out to a distance of a quarter mile from the shore, with the most active zone of coral growth facing the sea. A **barrier reef** differs from a fringing reef in being separated from the shoreland by a lagoon of varying width and depth. The Great Barrier Reef off the northeast coast of Australia is more than 1,200 miles long and up to 90 miles from the shore. A third type of reef is the **atoll,** which is a reef that encircles a lagoon but not an island.

Of the many theories advanced to explain coral reef formation, none are entirely satisfactory. Darwin assumed that fringing reefs were first formed on a sloping shore, and as the sea floor subsided and the coral grew upward and outward, the fringing reefs became barrier reefs. In places an entire volcanic island may have become submerged, leaving only the coral atoll, which in time acquired a growth of vegetation. Another theory (Daly) stresses the lowering of the ocean level by the withdrawal of water for glacial formation. Wave action then produced flat areas. When the glaciers melted and the temperature became favorable, the corals began to grow on these surfaces, building higher as the ocean level rose. Borings to great depths on certain atolls confirm a belief that atolls developed on volcanic tops or mountains as they became submerged. Most reefs grow at the rate of 10 to 200 mm. each year. Most of the existing reefs could have been formed in 15,000 to 30,000 years.

Some solitary sea corals that do not form reefs have been found at depths of 4 or 5 miles, and a few have been collected from the waters of northern latitudes (Fig. 9-27).

Coral reefs have great economic importance, for they serve as habitats for a large variety of organisms, such as sponges, worms, echinoderms, mollusks, many kinds of fishes, and man. Nearly every phylum of marine life is found among their crevices and chasms.

SYMBIOTIC RELATIONSHIPS. Most corals (especially shallow-water and true reef corals) have an interesting symbiotic relationship with microscopic plants (zooxanthellae) that are found in the animal coral tissue. These photosynthetic plants are probably not used as food by the corals, but they render a service to the corals by removing carbon dioxide and nitrogenous and other wastes and make possible the close spacing of the polyps in reef communities. They also assist in the formation of the coral framework. T. Goreau found that light increased the rate of calcification because of its action on the symbiotic algae. However light has no effect on growth of corals that do not have algal symbionts.

CORALS AND THE CHRONOLOGY OF GEOLOGIC DEPOSITS. J. W. Wells has pointed out the importance of fossil corals in determining a relative chronology for geologic deposits and a time scale of events. In some corals it was possible to count the daily striations of calcium carbonate deposits within the annual bands of growth and to show that at the beginning of the Cambrian period (600-million years ago) a day was less than 21 hours long and there were 424 days in the year. Some Devonian corals had as many as 410 striations per year instead of 360 as in recent (Quaternary period) corals. These values are in agreement with the age of the Earth based upon the isotope dating of geophysics and the rotation-time variations of astronomy.

■ PHYLUM CTENOPHORA*
General relations

The ctenophores comprise a small group of fewer than 100 species, and they are strictly marine forms. They take their name from the eight rows of comb-like plates they bear for locomotion. Common names for ctenophores are "sea walnuts" and "comb jellies." With coelenterates, they represent the only two phyla with basic radial symmetry, in contrast to other metazoans, which have developed bilateral symmetry.

Nematocysts are lacking in ctenophores, except in one species *(Euchlora rubra)* that is provided with nematocysts on certain regions of its tentacles but lacks colloblasts. These nematocysts are a part of this ctenophore and are not obtained by eating hydroids.

In common with the coelenterates, ctenophores have not advanced beyond the tissue grade of organization. There are no definite organ systems in the strict meaning of the term.

Although they have some common characteristics, there is no convincing evidence that ctenophores were derived from coelenterates, although there may be some kinship. The ctenophores must be considered as a blind offshoot and not in direct evolutionary line with the higher forms.

Ecologic relationships

The ctenophores are strictly marine animals and are all free swimming except for a few creeping forms. They occur in all seas, but especially in warm waters.

Although feeble swimmers and more common in surface waters, ctenophores are sometimes found at considerable depths. They are often at the mercy of tides and strong currents, but they avoid storms by swimming downward in the water. In calm water they may rest vertically with little movement, but when moving, they use their ciliated comb plates to propel themselves mouth-end forward. Highly modified forms such as *Cestum* use sinuous body movements as well as their comb plates in locomotion.

The fragile, transparent bodies of ctenophores are

*Te-nof'o-ra (Gr. *kteis, ktenos,* comb, + *phora,* pl. of bearing).

easily seen at night because of their striking luminescence.

Characteristics

1. Symmetry **biradial;** arrangement of internal canals and the opposite position of the tentacles change the radial symmetry into a combination of the two **(radial + bilateral)**
2. Usually ellipsoidal or spherical in shape, **with eight rows of comb plates on the external surface**
3. Ectoderm, endoderm, and a mesoglea (ectomesoderm) with scattered cells and muscle fibers; ctenophores may be considered **triploblastic**
4. Nematocysts absent (except in one species) but **adhesive cells (colloblasts) present**
5. Digestive system consisting of a mouth, pharynx, stomach, and a series of canals
6. Nervous system consisting of an aboral sense organ **(statocyst),** with a subepidermal plexus arranged into eight strands beneath the eight comb plate rows
7. No polymorphism or attached stages
8. Reproduction monoecious; gonads (endodermal origin) on the walls of the digestive canals, which are under the rows of comb plates; cydippid larva

Comparison with Coelenterata

Ctenophores resemble the coelenterates in the following ways:

1. Form of radial symmetry; with the coelenterates, they form the group Radiata
2. Aboral-oral axis around which the parts are arranged
3. Well-developed gelatinous ectomesoderm (collenchyme)
4. No coelomic cavity
5. Diffuse nerve plexus
6. Lack of organ systems

They differ from the coelenterates in the following ways:

1. No nematocysts except in *Euchlora*
2. Development of muscle cells from mesenchyme
3. Presence of comb plates and colloblasts
4. Mosaic, or determinate, type of development
5. Presence of pharynx generally

Classes

Class Tentaculata (ten-tak-yu-la'ta) (L. *tentaculum,* feeler, + *-ata,* group suffix). With tentacles. Tentacles may or may not have sheaths into which they retract. Some types of this class flattened for creeping; others compressed to a bandlike form. In some the comb plates may be confined to the larval form. Examples: *Pleurobrachia, Cestum.*

Class Nuda (nu'da) (L. *nudus,* naked). Without tentacles; conical form; wide mouth and pharynx; gastrovascular canals much branched. Example: *Beroë.*

Class tentaculata
Representative type—Pleurobrachia

Pleurobrachia is often used as a type of this group of ctenophores. Its transparent body is less then an inch in diameter (Fig. 9-30). The oral pole bears the mouth opening, and the aboral pole has a sensory organ, the **statocyst.**

COMB PLATES. On the surface are eight equally spaced bands called **comb rows** that extend as meridians from the aboral pole, and end before reaching the oral pole (Fig. 9-30). Each band is made up of transverse plates of long fused cilia called **comb plates** (Fig. 9-31, *B*). Ctenophores are propelled by the beating of the cilia on the comb plates. The beat in each row starts at the aboral end and proceeds successively along the combs to the oral end. All eight rows normally beat in unison. The animal is thus driven forward with the mouth in advance. The animal can swim backward by reversing the direction of the wave.

TENTACLES. The two **tentacles** are long and solid and very extensible, and they can be retracted into a pair of **tentacle sheaths.** When completely extended, they may be 6 inches long. The surface of the tentacles bear **colloblasts,** or glue cells (Fig. 9-31, *A*), which secrete a sticky substance for catching and holding small animals.

BODY WALL. The cellular layers of ctenophores are similar to those of the coelenterates. Between the epidermis and gastrodermis is a jellylike **collenchyme,** which makes up most of the interior of the body and contains muscle fibers and ameboid cells.

DIGESTIVE SYSTEM AND FEEDING. The **gastrovascular system** consists of a mouth, a pharynx, a stomach, and a system of gastrovascular canals that branch through the jelly to run to the comb plates, tentacular sheaths, and elsewhere (Fig. 9-30). Two blind canals terminate near the mouth and an aboral canal passes near the statocyst and divides into two small **anal canals** through which undigested material is expelled.

Ctenophores live on small organisms such as marine eggs, crustaceans, and mollusks. The glue cells on the tentacles enable them to adhere to the small prey and carry it to the mouth. Digestion is both extracellular and intracellular.

RESPIRATION AND EXCRETION. Respiration and excretion occur through the body surface.

NERVOUS AND SENSORY SYSTEM. Ctenophores have a nervous system similar to that of the

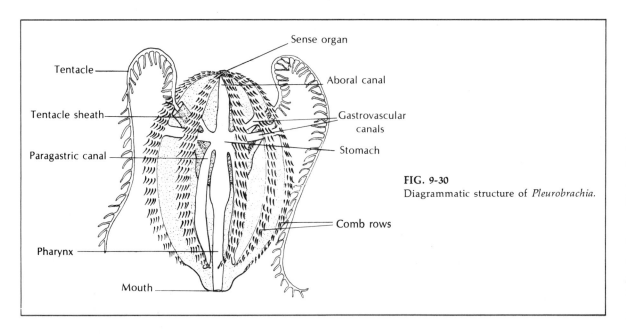

Tentacle

Tentacle sheath

Paragastric canal

Pharynx

Mouth

Sense organ

Aboral canal

Gastrovascular canals

Stomach

Comb rows

FIG. 9-30
Diagrammatic structure of *Pleurobrachia.*

coelenterates. It is made up of a subepidermal plexus of multipolar ganglion cells and neurites that may or may not anastomose. The nerve plexus is concentrated under each comb plate, but there is no central control as is found in higher animals.

The **sense organ** at the aboral pole consists of tufts of cilia that support a calcareous statolith, with the whole being enclosed in a bell-like container. This is an organ of equilibrium, for alterations in the position of the animal change the pressure on the tufts of cilia. The sense organ is also concerned in coordinating the beating of the comb rows, but does not trigger their beat.

The epidermis of ctenophores is abundantly supplied with sensory cells so that the animals are sensitive to chemical and other forms of stimuli. When a ctenophore comes in contact with an unfavorable stimulus, it often reverses the beat of its comb plates and backs up. The comb plates are very sensitive to touch, which often causes them to be withdrawn into the jelly.

REPRODUCTION. *Pleurobrachia,* in common with other ctenophores, is monoecious. The gonads are located on the lining of the gastrovascular canals

under the comb plates. Fertilized eggs are discharged through the epidermis into the water. Cleavage in ctenophores is determinate (mosaic), for the various parts of the animal are mapped out in the cleavage cells. If one of the cells is removed in the early stages, the resulting embryo will be deficient. This type of development is just the opposite of that of coelenterates. The free-swimming cydippid larva is superficially similar to the adult ctenophore and develops directly into an adult.

Other ctenophores

Ctenophores are fragile and beautiful creatures. Their transparent bodies glisten like fine glass, brilliantly iridescent during the day and luminescent at night.

One of the most striking ctenophores is *Beroë,* which may be more than 100 mm. in length and 50 mm. in breadth. Its shape is conical or ovoid, and it is provided with a large mouth but no tentacles. It is pink colored, and the body wall is covered with an extensive network of canals formed by the union of the paragastric and meridional canals. Venus's girdle *(Cestum)* is compressed and bandlike, may be

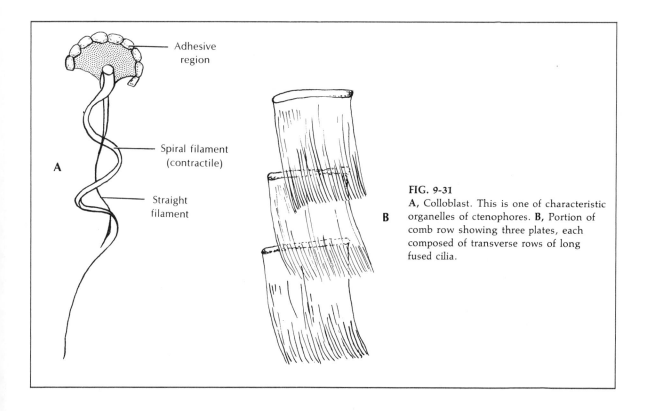

A

Adhesive region

Spiral filament (contractile)

Straight filament

B

FIG. 9-31
A, Colloblast. This is one of characteristic organelles of ctenophores. B, Portion of comb row showing three plates, each composed of transverse rows of long fused cilia.

more than a yard long, and presents a graceful appearance as it swims. The highly modified *Ctenoplana* and *Coeloplana* are very rare but are interesting because they have flattened disk-shaped bodies and are adapted for creeping rather than swimming. Both have unusually long tentacles. A common ctenophore along the Atlantic and Gulf coasts is *Mnemiopsis,* which has a laterally compressed body with two large oral lobes and unsheathed tentacles.

Nearly all ctenophores give off flashes of luminescence at night, especially such forms as *Mnemiopsis.* The vivid flashes of light seen at night in southern seas are often caused by members of this phylum.

Derivation and meaning of names

Aurelia (L. *aurum,* gold). Genus (Scyphozoa).

Calyptoblastea (Gr. *kalyptos,* covered, + *blastos,* bud, sprout). Order of Hydroida; also called Leptomedusae; having the gonophores in a gonotheca.

Cestum (Gr. *kestos,* girdle). Genus (Tentaculata). The body is compressed into a ribbonlike form.

Chlorohydra (Gr. *chlōros,* green, + hydra). Genus (Hydrozoa).

Craspedacusta (Gr. *kraspedon,* border, velum, + *kystis,* bladder). Genus (Hydrozoa).

Gonionemus (Gr. *gōnia,* angle, + *nēma,* thread). Genus (Hydrozoa).

Gymnoblastea (Gr. *gymnos,* naked, bare, + *blastos,* bud, sprout). Order of Hydroida; also called Anthomedusae; having naked medusa buds.

Hydra (Greek mythology, water serpent).

Metridium (Gr. *mētridios,* fruitful). Genus (Anthozoa).

Nomeus (Gr. herdsman). Refers to the habits of this commensal fish in inducing larger fish to chase it into the grasping tentacles of the Portuguese man-of-war.

Obelia (Gr. *obelias,* round cake). Genus (Hydrozoa).

Pelmatohydra (Gr. *pelma, pelmatos,* stalk or sole, + hydra). Genus (Hydrozoa). These hydras have a stalk at the basal end of body.

Physalia (Gr. *physallis,* bladder). Genus (Hydrozoa). Has a bladderlike gas-filled float.

Pleurobrachia (Gr. *pleuron,* side, + L. *brachia,* arms). Genus (Tentaculata). Refers to the tentacles, one on each side.

Siphonophora (Gr. siphon + *phora,* pl. of bearing). An order of Hydrozoa. Refers to the gastrozooid polyp with a single hollow tentacle in these polymorphic hydrozoan colonies.

Trachylina (Gr. *trachys,* rough, + L. *linum,* flax). An order of Hydrozoa. Refers to the appearance produced by the long-haired sensory cells and lithostyles or sense clubs.

References
Suggested general readings

Barnes, D. J. 1970. Coral skeletons: an explanation of their growth and structure. Science 170:1305-1308 (Dec. 18).

Bayer, F. M. 1971. Coral. Natural History **80**(3):42-47 (March). *Excellent color photographs of reef corals.*

Bayer, F. M., and H. B. Owre. 1968. The free-living lower invertebrates. New York, The Macmillan Co.

Bourne, G. C. 1900. Ctenophora, In E. R. Lankester: A treatise on zoology. London, A. & C. Black, Ltd. *Good, detailed descriptions of the morphology of ctenophores. Taxonomy only briefly considered.*

Brown, F. A., Jr. 1950. Selected invertebrate types. New York, John Wiley & Sons, Inc. *Good for certain representative marine forms.*

Buchsbaum, R. 1948. Animals without backbones. Chicago, University of Chicago Press. *Many excellent illustrations of coelenterates.*

Buck, J. 1973. Bioluminescent behavior in *Renilla.* Biol. Bull. **144**(1):19-37 (Feb.).

Bullough, W. S. 1950. Practical invertebrate anatomy. London, The Macmillan Co. *An excellent practical manual of certain selected types.*

Burnett, A. L. 1959. Hydra: an immortal's nature. Natural History **68**:498-507 (Nov.). *Describes among other aspects of the hydra's nature the experiments of P. Brien on the renewal of cells.*

Bushnell, J. H., Jr., and T. W. Porter. 1967. The occurrence and prey of *Craspedacusta sowerbyi* (particularly polyp stages) in Michigan. Trans. Amer. Microsc. Soc. **86**:22-27.

Chapman, G. 1953. Studies on the mesoglea of coelenterates. Quart. J. Microsc. Sci. **94**:155-168.

Compton, G. 1970. What is the world's deadliest animal? Science Dig. **68**(2):24-28 (Aug.).

Crowell, S. (editor). 1965. Behavioral physiology of coelenterates. Amer. Zool. **5**:335-389.

Darwin, C. 1842. The structure and distribution of coral reefs. Ed. 3 paperback, Berkley, University of California, in 1962. Ed. 3 first published by D. Appleton-Century Co. in 1896, in New York.

Dougherty, E. C. (editor). 1963. The lower Metazoa. Berkeley, University of California Press. *Deals with comparative biology and phylogeny of the lower metazoan phyla.*

Fraser, C. M. 1937. Hydroids of the Pacific coast of Canada and the United States. Toronto, University of Toronto Press. *This monograph and a similar one on the hydroids of the Atlantic coast are the most comprehensive taxonomic studies yet made on the American group.*

Goreau, T. 1961. Problems of growth and calcium deposition in reef corals. Endeavor **20:**32-39. *A summary of the recent concept involving the role of zooxanthellae in coral formation.*

Hardy, A. C. 1956. The open sea. Boston, Houghton Mifflin Co. *Many beautiful plates of medusae and other coelenterates in this outstanding treatise on sea life.*

Hickman, C. P. 1973. Biology of the invertebrates, ed. 2. St. Louis, The C. V. Mosby Co.

Hyman, L. H. 1940. The invertebrates: Protozoa through Ctenophora. New York, McGraw-Hill Book Co. *Extensive accounts are given of the coelenterates (Cnidaria) and the ctenophores in the last two chapters of this authoritative work.*

Jha, R. K., and G. O. Mackie. 1967. The recognition, distribution, and ultrastructure of hydrozoan nerve elements. J. Morphol. **123:**43-61.

Laubenfels, M. W. de. 1955. Are the coelenterates degenerate or primitive? System. Zool. **4:**43-48.

Lenhoff, H. M., and W. F. Loomis. 1961. The biology of hydra and some other coelentrates. Coral Gables, Fla., University of Miami Press.

Mackie, G. O. 1960. The structure of the nervous system in *Velella.* Quart. J. Microsc. Sci. **101:**119-131.

Macklin, M., T. Roma, and K. Drake. 1972. Water excretion by hydra. Science **179:**194-195 (Jan. 12).

McCullough, C. B. 1963. The contraction burst pacemaker system in hydras. Proc. Sixteenth Int. Congr. Zool. (Washington) **1:**25.

Morin, J. C., and J. W. Hastings, 1971. Energy transfer in a bioluminescent system. J. Cell. Physiol. **77:**313-318.

Pantin, C. F. A. 1950. Behavior patterns in lower invertebrates. Soc. Exper. Biol. Symp. **4:**175-182.

Pennak, R. W. 1953. Freshwater invertebrates of the United States. New York, The Ronald Press Co. *Chapter 4 is devoted to the freshwater coelenterates, including a good description of the rare freshwater jellyfish.*

Rees, W. J. (editor). 1966. The Cnidaria and their evolution. New York, Academic Press, Inc. *A symposium devoted entirely to the coelenterates of Great Britain.*

Russell, F. S. 1953. The medusae of the British Isles. Cambridge, Cambridge University Press. *In this fine monograph the many species of British medusae are described in text and by beautiful plates, many in color.*

Smith, F. G. W. 1948. Atlantic reef corals. Miami, University of Miami Press.

Stoddart, D. R. 1969. Ecology and morphology of recent coral reefs. Biol. Rev. **44:**433-498.

Tardent, P. 1963. Regeneration in the Hydrozoa. Biol. Rev. **38:**293-333.

Totton, A. K., and G. O. Mackie. 1960. Studies on *Physalia physalis* (L.). Discovery Reports **30:**301-407. Cambridge, Cambridge University Press. *This excellent monograph is an exhaustive treatment of the natural history and morphology (Totton) and the behavior and histology (Mackie) of the Portuguese man-of-war.*

Uchida, T. 1963. Two phylogenetic lines in the coelenterates from the viewpoint of their symmetry. Proceedings of the Sixteenth International Congress of Zoology (Washington), vol. 1, p. 24.

Warshofasky, F. 1966. Portuguese man-of-war. National Wildlife. **4**(4):37-40 (June-July).

Wells, J. W. 1963. Coral growth and geochronology. Nature (London) **197:**948-950.

Webster, G. 1971. Morphogenesis and pattern formation in hydroids. Biol. Rev. **46**(1):107.

Westfall, J. A. 1965. Nematocysts of the sea anemone, *Metridium.* Amer. Zool. **5:**373-393.

Yeatman, H. C. 1965. Ecological relationship of the ciliate *Kerona* to its host hydra. Turtox News **43:**226-227.

Yonge, C. M. 1949. The sea shore. London, William Collins Sons & Co., Ltd. *Many descriptions of coelenterates are scattered throughout this fascinating work.*

Selected *Scientific American* articles

Berrill, N. J. 1957. The indestructible hydra. **197:**118-125 (Dec.). *Describes the remarkable regenerative powers of this animal.*

Lane, C. E. 1960. The Portuguese Man-of-War. **202:**158-168 (March). *This beautiful but dangerous jellyish represents a remarkable colonial development.*

Loomis, W. F. 1959. The sex gas of hydra. **200:**145-156 (April). *The carbon dioxide tension of the water appears to play an important role in controlling sexual reproduction in hydras.*

Living specimens of
Haematoloechus. *These are flukes
that live as parasites in the lungs
of frogs.*

CHAPTER 10

THE ACOELOMATE ANIMALS

PHYLUM PLATYHELMINTHES
PHYLUM RHYNCHOCOELA

■ POSITION IN ANIMAL KINGDOM

1. The Platyhelminthes, or flatworms, and the Rhynchocoela, or ribbon worms, are the most primitive groups of animals to have **bilateral symmetry,** the type of symmetry assumed by all higher animals.

2. These phyla have only one internal space, the digestive cavity, with the region between the ectoderm and endoderm filled with **mesoderm** in the form of muscle fibers and mesenchyme (parenchyma). Since they lack a coelom of any kind, they are termed **acoelomate animals,** and because they have three well-defined germ layers, they are termed **triploblastic.**

3. Acoelomates show more specialization and division of labor than do the radiate animals because the mesoderm makes more elaborate organs possible. Thus the acoelomates are said to have reached the **organ-system level of organization.**

4. They both belong to the **protostome division** of

the Bilateria and have spiral and determinate cleavage.

■ BIOLOGIC CONTRIBUTIONS

1. The acoelomates have developed the basic **bilateral** plan of organization that has been widely exploited in the animal kingdom.

2. The **mesoderm** is now developed into a well-defined embryonic germ layer **(triploblastic),** thus making available a great source of tissues, organs, and systems and pointing the way to greater complexity of animal structure.

3. Along with bilateral symmetry, **cephalization** has been established. There is some centralization of the nervous system evident in the **ladder type of system** found in flatworms.

4. Along with the subepidermal musculature, there is also a mesenchymal system of muscle fibers.

5. An **excretory system** appears for the first time.

6. The rhynchocoels have developed the **first circulatory system** with blood and the **first one-way alimentary canal.** Although not stressed by zoologists, the rhynchocoel cavity in ribbon worms is actually a true coelom, but as merely a part of the proboscis mechanism, it is not of evolutionary significance.

7. Unique and specialized structures occur in both phyla. The parasitic habit of many flatworms has produced many modifications, such as organs of adhesion.

■ PHYLOGENY AND ADAPTIVE RADIATION

PHYLOGENY. Both phyla appear to be closely related, with the flatworms the more primitive. The order Acoela is thought to have originated from a planula-like ancestor and in turn gave rise to the other members of the phyla.

The Rhynchocoela could have arisen from the flatworms; the body construction of ciliated epidermis, muscles, and mesenchyme-filled spaces, etc. are similar in both groups. But the rhynchocoels are more advanced than the flatworms in having a complete digestive system, a vascular system, and a more highly organized nervous system.

ADAPTIVE RADIATION. The flatworm body plan with its creeping adaptation facilitated the development of bilateral symmetry, cephalization, ventrodorsal regions, and caudal differentiation. It was adapted in many ways for a parasitic existence into which its adaptive radiation led many of the more common parasites of man and beast. All its numerous species are variations on the same theme of body structure and functional behavior.

The ribbon worms have stressed the proboscis apparatus in their evolutionary diversity. Its use in capturing prey may have been secondarily evolved from its original function as a highly sensitive organ for exploring the environment. Being free living and active, the ribbon worms have advanced beyond the flatworms in their evolution. Perhaps the possession of a proboscis was deterrent to a parasitic habit but highly efficient as a predator tool.

■ PHYLUM PLATYHELMINTHES*
General relations

The term "worm" has been loosely applied to elongated, bilateral invertebrate animals without appendages. At one time zoologists considered worms

*Plat"y-hel-min'theez (Gr. *platys,* flat, + *helmins,* worm).

(Vermes) to be a group in their own right. Such a group included a highly diverse assortment of forms. Modern classification has broken up this group into phyla and reclassified them. By tradition, however, zoologists still refer to these animals as "flatworms," "ribbon worms," "roundworms," "segmented worms," etc.

The term Platyhelminthes, which means "flatworms," was first proposed by Gegenbaur (1859) and was applied to the animals now included under that heading. At first nemertines and some others were included but later were removed to other groups.

The flatworms are bilaterally symmetric, a basic plan for all animals that have advanced very far in complexity of organization. The flatworms may have evolved from a coelenterate-like ancestor that had acquired a creeping habit and a transformation of radial or biradial symmetry into a bilateral symmetry. Such a transformation would involve a number of body modifications, such as an oral-aboral flattening, with the oral end becoming the ventral surface, and the ventral surface adapting for locomotion with the aid of cilia and muscles. Along with the need for directional movement would come body elongation and a stressing of the anterior end, or cephalization. The advantages of cephalization would no doubt also have put a premium on the adaptive selection of suitable head types.

The small flatworms of the order Acoela (Fig. 10-1) seem to meet many of the requirements of an early ancestor of Platyhelminthes. They have many characteristics of the coelenterate planula larva, such as no epidermal basement membrane, no digestive system, a nerve plexus under the epidermis, and no distinct gonads.

The replacement of the bulky, jellylike mesoglea of coelenterates with a cellular mesodermal (ectomesodermal) mesenchyme (parenchyma) laid the basis for a higher organization, less bulk, and increased metabolic demands. From the mesenchyme the muscular, reproductive, and excretory systems arise.

Flatworms range in size from a millimeter or less to some of the tapeworms that are many feet in length. Their flattened bodies may be slender, broadly leaflike, or long and ribbonlike.

Ecologic relationships

The flatworms include both free-living and parasitic forms, but the free-living members are found ex-

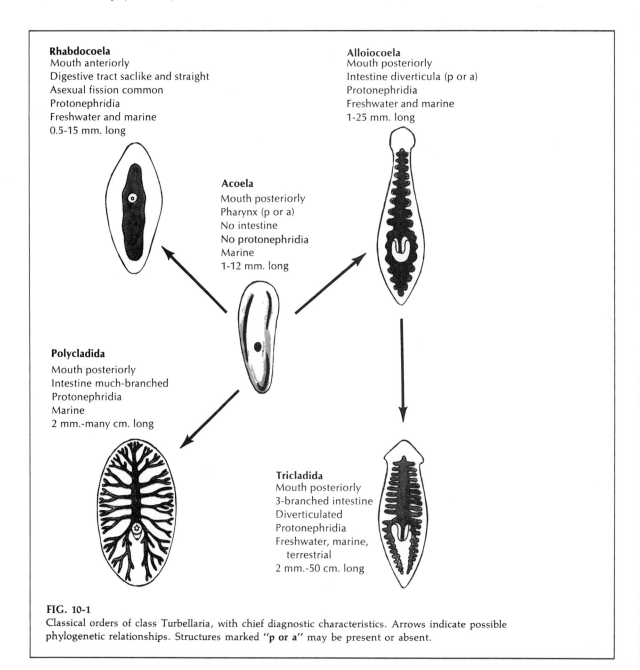

Rhabdocoela
Mouth anteriorly
Digestive tract saclike and straight
Asexual fission common
Protonephridia
Freshwater and marine
0.5-15 mm. long

Alloiocoela
Mouth posteriorly
Intestine diverticula (p or a)
Protonephridia
Freshwater and marine
1-25 mm. long

Acoela
Mouth posteriorly
Pharynx (p or a)
No intestine
No protonephridia
Marine
1-12 mm. long

Polycladida
Mouth posteriorly
Intestine much-branched
Protonephridia
Marine
2 mm.-many cm. long

Tricladida
Mouth posteriorly
3-branched intestine
Diverticulated
Protonephridia
Freshwater, marine,
  terrestrial
2 mm.-50 cm. long

FIG. 10-1
Classical orders of class Turbellaria, with chief diagnostic characteristics. Arrows indicate possible phylogenetic relationships. Structures marked "**p or a**" may be present or absent.

clusively in class Turbellaria. A few turbellarians are symbiotic or parasitic, but the majority are adapted as bottom dwellers in marine or fresh water, or live in moist places on land. Many, especially of the larger species, are found on the underside of stones and other hard objects in freshwater streams, or in the littoral zones of the ocean. Some may actually occur at considerable depths in the ocean. Pelagic existence is usually restricted to larval forms, although some adults are pelagic in tropical or subtropical seas. Those that live in the tidal zones of shores can adapt themselves to a wide range of temperature and salinity.

Relatively few turbellarians live in fresh water.

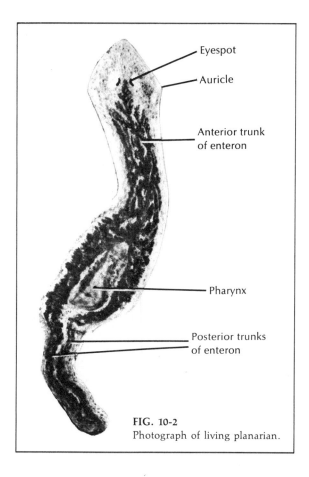

Eyespot

Auricle

Anterior trunk
of enteron

Pharynx

Posterior trunks
of enteron

FIG. 10-2
Photograph of living planarian.

Planarians (Fig. 10-2) and some others frequent streams and spring pools; others prefer flowing water of mountain streams. Some species occur in fairly hot springs.

Terrestrial turbellarians are found in fairly moist places under stones and logs. Land planarians, such as *Bipalium,* occur frequently in greenhouses where temperature conditions are favorable. Some of the turbellarians harbor symbiotic chlorellae within their mesenchyme. An example is the genus *Convoluta,* which may, at certain stages, digest their chlorellae.

The true parasitic flatworms include the classes Trematoda (the flukes) and Cestoda (the tapeworms). Some are ectoparasitic but most of them are endoparasitic. In species with an indirect life cycle many hosts may be involved. The first host of a trematode is usually a mollusk and the final host is a vertebrate. In trematodes, certain larval stages may be free living, but most cestodes are parasitic in all stages of their life cycle. Both flukes and tapeworms have undergone many adaptive changes that fit them for their parasitic existence. Such adaptations involve loss of sense organs, the appearance of adhesive organs or suckers, loss of digestive system, and excessive reproductive capacity.

Characteristics

1. Three germ layers **(triploblastic)**
2. **Bilateral symmetry;** definite polarity of anterior and posterior ends
3. **Body flattened dorsoventrally;** oral and genital apertures mostly on ventral surface
4. Body segmented in one class (Cestoda)
5. Epidermis may be cellular, syncytial, or absent (ciliated in some); **rhabdites** in epidermis or mesenchyme of some Turbellaria; thick cuticle with suckers or hooks in parasitic forms; cuticle, or **tegument,** a syncytial living tissue
6. Muscular system of a sheath form and of mesenchymal origin; layers of circular, longitudinal, and oblique fibers beneath the epidermis
7. No definite coelom **(acoelomate);** spaces between organs filled with **parenchyma,** a form of connective tissue or mesenchyme
8. The parenchyma with its fluid-filled spaces surrounded by the outer muscle layers produces an effective hydrostatic skeleton. The fluid in the interstices of the parenchyma also help in the transmission and transportation of metabolites, in the absence of a circulatory system.
9. Digestive system incomplete (gastrovascular type); absent in some
10. **Nervous system consisting of a pair of anterior ganglia with longitudinal nerve cords connected by transverse nerves and located in the mesenchyme** in most forms; similar to coelenterates in primitive forms
11. Simple sense organs; eyespots in some
12. Excretory system of two lateral canals with branches bearing **flame cells (protonephridia);** lacking in some primitive forms
13. Respiratory, circulatory, and skeletal systems lacking; lymph channels with free cells in some trematodes
14. Most forms monoecious; reproductive system complex with well-developed gonads, ducts, and accessory organs; internal fertilization; development direct; usually indirect in internal parasites

in which there may be a complicated life cycle often involving several hosts
15. Class Turbellaria mostly free living; classes Trematoda and Cestoda entirely parasitic

Classes

Class Turbellaria (tur"bel-lar'e-a) (L. *turbellae* [pl.], stir, bustle, + *-aria,* like or connected with). The turbellarians. Usually free-living forms with soft flattened bodies; covered with ciliated epidermis containing secreting cells and rodlike bodies (rhabdites); mouth usually on ventral surface, sometimes near center of body; no body cavity except intercellular lacunae in parenchyma; mostly hermaphroditic, but some have asexual fission. Examples: *Dugesia* (planaria), *Microstomum, Planocera.*

Class Trematoda (trem"a-to'da) (Gr. *trematōdes,* with holes, + *eidos,* form). The flukes. Body covered with thick living cuticle without cilia; leaflike or cylindric in shape; presence of suckers and sometimes hooks; alimentary canal usually with two main branches; nervous system similar to turbellarians; mostly monoecious; development direct in external parasites but usually indirect in case of internal parasites with alternation of hosts; all parasitic. Examples: *Fasciola, Opisthorchis (Clonorchis), Schistosoma.*

Class Cestoda (ses-to'da) (Gr. *kestos,* girdle, + *eidos,* form). The tapeworms. Body covered with thick, nonciliated, living cuticle; scolex with suckers or hooks and sometimes both for attachment; body divided into series of proglottids; no digestive or sense organs; general form of body tapelike; usually monoecious and self-fertilizing; many organs reduced; all parasitic, usually with alternate hosts. Examples: *Diphyllobothrium, Taenia, Echinococcus.*

Class Turbellaria

Turbellarians are small, mostly free-living worms that range in length from 5 mm. or less to 50 cm. Covered usually with ciliated epidermis, these are mostly creeping worms that combine muscular with ciliary movements to achieve locomotion. The mouth is on the ventral side. Unlike the trematodes and cestodes, they have simple life cycles.

Of the five orders of class Turbellaria, the order Acoela (Fig. 10-1) is the most primitive and probably has changed little from the ancestral form. Its members are small and have a mouth but no gastrovascular cavity or excretory system. Food is merely passed through the mouth or pharynx into temporary spaces that are surrounded by a syncytial mesenchyme where gastrodermal phagocytic cells digest the food intracellularly. The order has syncytial epidermis and a diffuse nervous system. An example of

the order is *Otocelis,* which lives as a commensal in the digestive system of sea urchins and sea cucumbers.

Order Rhabdocoela is characterized by a straight, unbranched gastrovascular cavity (Fig. 10-1). *Microstomum,* a representative of this order, has the habit of feeding on hydras and taking over the nematocysts for its own defense.

Order Tricladida has a three-branched digestive tract. The common planarians belong to this group (Fig. 10-2). Some members of the order are marine, such as *Bdelloura* (Fig. 10-5), an ectoparasite on the gills of *Limulus,* the horseshoe crab. *Bipalium* is a common large terrestrial form often found in greenhouses. This order is often divided into several smaller orders. Closely related with Tricladida is order Alloiocoela (Alloeocoela), with irregular saclike intestines. This group also is often subdivided.

Another order, Polycladida, is comprised of leaflike marine forms that have many intestinal branches from the central digestive cavity. *Planocera* is a polyclad ectoparasite that lives in the mouth of certain marine snails. One feature of some polyclads is the presence of a ciliated free-swimming larval form. One of these, called Müller's larva, is provided with ciliated projecting lobes and eyespots.

Representative type—*Dugesia,* the common brown planarian

HABITAT. A well-known representative of the triclad turbellarians is the common freshwater brown planarian (*Dugesia*) (Fig. 10-2). Several species of *Dugesia* are found on the underside of rocks and debris in brooks, ponds, and springs of the United States. They are small and flat and their dark mottled color blends perfectly with the rocks or plants to which they cling. Other common genera are *Dendrocelopsis* and *Cura,* both black planarians, and *Phagocata,* a white variety.

STRUCTURE. *Dugesia* is flat and slender and about $^3/_4$ inch or less in length. The head region is triangular, with two lateral lobes known as **auricles.** These are not ears but olfactory organs. Two **eyespots** on the dorsal side of the head near the midventral line give the animal a cross-eyed appearance. Its background color may range from a dark yellow to olive, or dark brown. The ventral side is lighter. Near the center of the ventral side is the **mouth,** through which the muscular **pharynx (proboscis)**

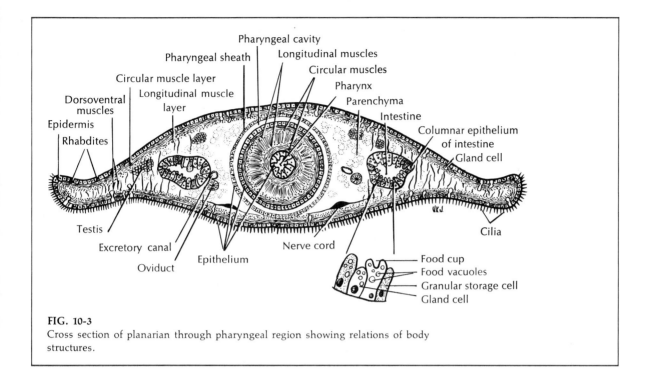

FIG. 10-3

Cross section of planarian through pharyngeal region showing relations of body structures.

can be extended for the capture of prey (Fig. 10-4, *C*). The **genital pore** opens posterior to the mouth.

The **skin** is ciliated epidermis resting on a basement membrane (Fig. 10-3). It contains rod-shaped **rhabdites** that, when discharged into water, swell and form a protective gelatinous sheath around the body. Single-cell mucous glands open on the surface of the epidermis. In the body wall below the basement membrane are layers of **muscle fibers** that run circularly, longitudinally, and diagonally. A meshwork of **parenchyma** cells, developed from mesoderm, fills the spaces between muscles and visceral organs.

LOCOMOTION. Freshwater planarians move by gliding, head slightly raised, over a slime tract secreted by its marginal adhesive glands. The beating of the epidermal cilia in the slime tract drives the animal along. Rhythmic muscular waves can also be seen passing backward from the head as it glides. A less common method is crawling. The worm lengthens, anchors its anterior end and by contracting its longitudinal muscles pulls up the rest of its body.

DIGESTIVE SYSTEM. The digestive system includes a mouth, pharynx, and **intestine.** The pharynx, enclosed in a **pharyngeal sheath** (Fig. 10-3), opens posteriorly just inside the mouth, through which it can extend (Fig. 10-4, *B*). The intestine has three many-branched trunks, one anterior and two posterior. The whole forms a **gastrovascular cavity** lined with columnar epithelium (Figs. 10-3 and 10-4, *B*).

Planarians are mainly carnivorous, feeding largely upon small crustaceans, nematodes, rotifers, and insects. They can detect food from some distance by means of chemoreceptors. They entangle their prey in mucous secretions from the mucous glands and rhabdites. The planarian grips its prey with its anterior end, wraps its body around the prey, extends its proboscis, and sucks up minute bits of the food. Intestinal secretions contain proteolytic enzymes for some **extracellular digestion.** Bits of food are sucked up into the intestine, where the phagocytic cells of the gastrodermis complete the digestion (**intracellular**). The gastrovascular cavity ramifies to most parts of the body, and food is absorbed through its walls into the body cells. Undigested food is egested through the pharynx. Planarians can go a long time without feeding, for they can draw food from the parenchyma cells back into the intestinal cells where it is digested.

EXCRETION. The excretory system consists of two longitudinal canals with a complex network of tubules that branch to all parts of the body and end in **flame cells** (protonephridia) (Fig. 10-4, *A*). The flame cell is hollow and contains a tuft of cilia that prevents stagnant layers of fluid on the inner surface of tubules. The two main canals open dorsally by excretory pores. The excretory system is involved largely with water regulation as well as excretion of organic wastes. Excretion of metabolic wastes occurs also through the epidermis and probably through the gastrodermis.

RESPIRATION. There are no respiratory organs. Exchange of gases takes place through the body surface.

NERVOUS SYSTEM. Two **cerebral ganglia** beneath the eyespots serve as the "brain." Two ventral **nerve cords** extending from the brain are connected by transverse nerves to form a "ladder type" of nervous system (Figs. 10-4, *B,* and 10-5). The more

primitive turbellarians may have three, four, or five pairs of longitudinal nerve trunks.

The **eyespots** (Fig. 10-6) are pigment cups into which retinal cells extend from the brain, with the photosensitive ends of the cells inside of the cup. They are sensitive to light intensities and the direction of a source, but can form no images. The pigment cups allow light to reach the light-sensitive ends of the cells through the openings of the cups only, thus revealing the exact direction of the light source. Planaria are negatively phototactic and are most active at night. The auricular sense organs on the side of the head are concerned with taste, smell, and touch.

REPRODUCTION. Triclad turbellarians reproduce both sexually and asexually. Asexually the animal merely constricts behind the pharyngeal region and separates into two animals. Each new animal regenerates its missing parts. Sexually the worm is **monoecious;** each individual is provided with both male and female organs (Fig. 10-4, *A*).

In the **male system** (Fig. 10-4, *A*) a row of **testes**

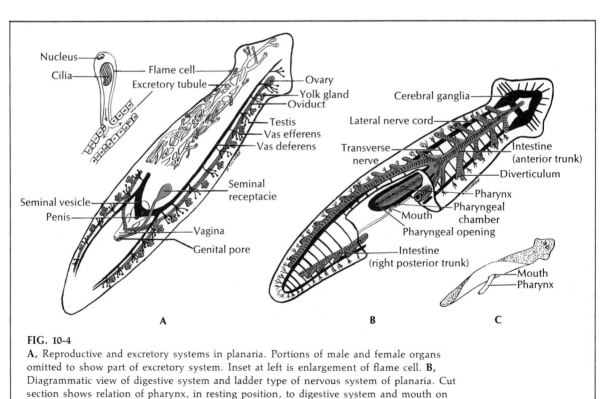

FIG. 10-4

A, Reproductive and excretory systems in planaria. Portions of male and female organs omitted to show part of excretory system. Inset at left is enlargement of flame cell. **B,** Diagrammatic view of digestive system and ladder type of nervous system of planaria. Cut section shows relation of pharynx, in resting position, to digestive system and mouth on ventral surface. **C,** Pharynx extended through ventral mouth.

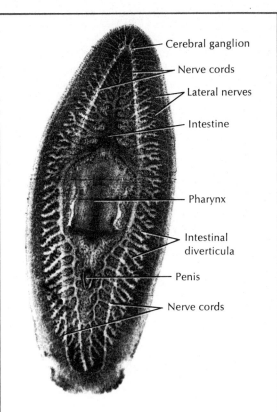

FIG. 10-5
Ladder type of nervous system shows up clearly in this photomicrograph of stained preparation of *Bdelloura,* a marine triclad.

FIG. 10-6
Ocelli, or eyes, of planarian. Pigment cup lets light enter open side, parallel to long axis of retinal (light-sensitive) cells. Planarian determines light direction from stimulation of light-sensitive cells.

on each side empty sperm by way of tiny tubes into a common duct, the **vas deferens,** which enlarges posteriorly to form a **seminal vesicle** where sperm are stored until discharged through the muscular **penis** (Fig. 10-5) during copulation (the mating act).

The **female system** contains two anterior **ovaries** that discharge eggs into tubular **oviducts,** which in turn join to form the **vagina.** Yolk ducts empty into the oviducts. Connected to the vagina is the seminal receptacle, which receives and stores sperm from the mating partner. Planarians have a reproductive system only during the breeding season, from early spring to late summer; at other times the worms reproduce asexually by fission.

Although hermaphroditic, turbellarians practice cross-fertilization, mutually exchanging sperm during copulation. At copulation each worm inserts its penis into the genital pore of the mate and in each worm sperm passes through the penis from the male seminal vesicle to the female seminal receptacle of the mate. After the worms separate sperm pass up the oviducts to fertilize eggs as they leave the ovaries. Several eggs together with yolk cells from the yolk glands become enclosed in small spherical egg capsules. The egg capsules, when laid, are attached by little stalks to the underside of stones or plants. Embryos finally emerge as little planarians (juveniles).

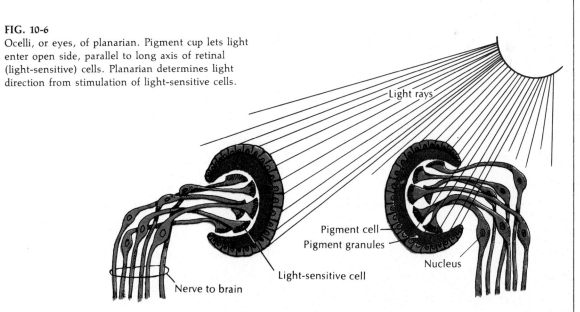

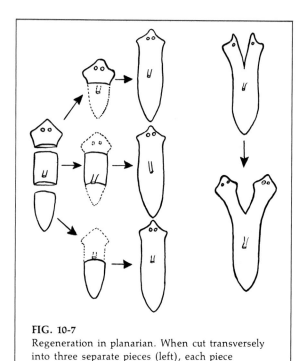

FIG. 10-7
Regeneration in planarian. When cut transversely
into three separate pieces (left), each piece
develops into planarian. If head is split (right), a
double-headed animal results.

REGENERATION.* Planarians have great power
to regenerate lost parts (Fig. 10-7). When an individual is cut in two, the anterior end will grow a new
tail and the posterior end a new head. Planarians
may also be grafted or cut in ways that produce
freakish designs such as two heads or two tails.

Experiments have shown that when a planarian is
cut across, free cells **(neoblasts)** from the mesenchyme migrate to the cut surface and aggregate there
to form a blastema, which develops into the new
part. X-radiation of a worm will destroy these neoblasts and no regeneration will occur. It has been
suggested that the neoblasts are attracted to the cut
region by chemical emanations, which cease when
the blastema is formed.

BEHAVIOR. Planarians respond to the same types
of stimuli as other animals do. Their ventral surfaces
are positively thigmotactic, whereas their dorsal surfaces are negatively thigmotactic. Planarians respond
in a positive way to the juices of foods. They avoid
strong light and will seek out dark or dimly lighted
regions. They are more active at night than during
*Refer to p. 22, Principle 30.

the day. The auricular lobes appear to be sensitive
to both water currents and chemical stimuli. In general their reactions to water currents depend on their
normal habitats. Those from flowing water are
usually positively rheotactic; those from still water
will not react or else are positively rheotactic to weak
currents only.

The behavior patterns of flatworms have been the
subject of numerous investigations all over the
world, for animal behaviorists have considered them
to be the lowest group that show any capacity for
learning in response to simple conditioned changes.

Class Trematoda

The flukes, or trematodes, are all parasitic (usually
in vertebrates), are chiefly leaflike in form, and as
adults, do not have cilia. Although parasitic, they
differ little structurally from the nonparasitic turbellarians. Some of their parasitic adaptations are the
lack of an epidermis, the presence of a living cuticle,
or tegument, and the development of special adhesive organs for clinging to their host, such as glandular or muscular disks, hooks, and true suckers.

In the Trematoda (and also Cestoda) the electron
microscope shows that the cuticle, or tegument, consists of a syncytial, living protoplasm that represents
extensions of deeper-lying nucleated cells (Fig. 10-8).
The outer surface of the tegument has many small
extensions (microtrichia) and pore canals. Trematodes retain other turbellarian characteristics, such as
a well-developed alimentary canal (but with the
mouth at the anterior, or cephalic, end) and similar
reproductive, excretory, and nervous systems, as
well as a musculature and mesenchyme that are only
slightly modified from those of the Turbellaria.
Sense organs are poorly developed in flukes and
occur only in larval stages and in a few adults
(eyespots in order Monogenea).

As in Turbellaria, the trematode body shows many
variations among the different groups. Although basically they all have the flattened form of flatworms,
some are round, some are elongated and oval, and
others are slender.

According to their parasitic habits, the class Trematoda is divided into three main orders or subclasses
—the Monogenea, which are chiefly ectoparasites
with simple life cycles (one host), the Aspidobothria,
which are chiefly endoparasites with simple life
cycles, and the Digenea, mostly endoparasites with

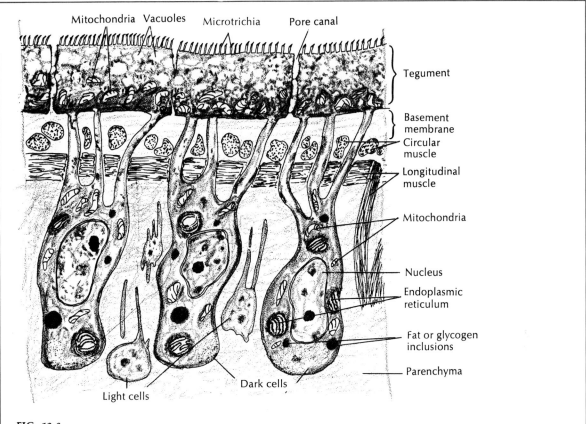

Mitochondria Vacuoles Microtrichia Pore canal

Tegument

Basement membrane

Circular muscle

Longitudinal muscle

Mitochondria

Nucleus

Endoplasmic reticulum

Fat or glycogen inclusions

Parenchyma

Dark cells

Light cells

FIG. 10-8
Section through tegument of cestode, as revealed by electron microscope, showing so-called "cuticle" composed of syncytial, living protoplasm rather than lifeless secretion, as formerly believed. Trematode tegument shows similar structure.

very complex life cycles in two or more hosts. Some authorities include the Aspidobothria under the Digenea.

MONOGENEA. Members of the order Monogenea, which require only one host in their life cycle, are mostly ectoparasites on the gills, skin, mouth, or urinary bladder of vertebrates and occasionally on crustaceans and cephalopods. Adhesive structures are found at both ends, but the posterior adhesive organ is usually a complex of hooks and suckers. The worms move by alternately attaching their anterior and posterior adhesive organs. Their eggs are few in number and only one egg is present at a time in the uterus. The egg hatches a ciliated larva that attaches to the host or swims around awhile before attachment. Common genera are *Polystoma* found in the urinary bladder of frogs, and *Gyrodactylus,* which

lives on the skin and gills of freshwater fish. Monogenetic flukes injure the fish by feeding on epithelial cells and blood and may do lethal damage.

ASPIDOBOTHRIA. The Aspidobothria have very large suckers that may occupy most of the ventral surface. The oral sucker may be absent. They resemble the Digenea in anatomy, but develop directly as in the Monogenea and require only one host. *Aspidogaster* lives in the pericardial cavity of freshwater clams. Other hosts are mollusks, fishes, and turtles.

DIGENEA. The order Digenea is almost exclusively endoparasitic, having from two to four hosts in the life cycle. They usually have an oral sucker around the mouth and a large ventroposterior sucker called an acetabulum. Their development involves a succession of forms, each of which occupies a different host. These different hosts have probably

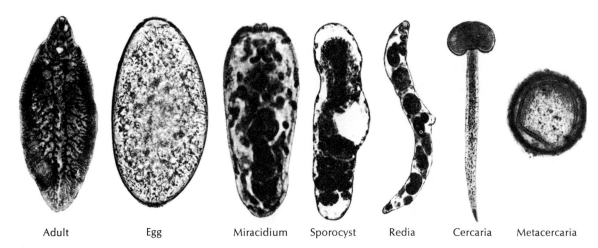

| Adult | Egg | Miracidium | Sporocyst | Redia | Cercaria | Metacercaria |

FIG. 10-9
Stages in life cycle of sheep liver fluke, *Fasciola hepatica* (class Trematoda, order Digenea).
(Not shown to scale.) (Courtesy Carolina Biological Supply Co., Burlington, N. C.)

been added to their life cycle as new groups of animals have emerged in evolution. When vertebrates arose and became definitive hosts, the trematodes still retained their early invertebrate hosts for the larval stages. The adults are found mainly in terrestrial, freshwater, and marine vertebrates and are usually specific for organs, such as the intestine, lungs, bile passages, kidneys, and urinary bladder.

Digenetic trematodes have the most complicated life histories in the animal kingdom. They usually have four larval stages (Fig. 10-9) **miricidium, sporocyst, redia,** and **cercaria.** Another stage, the **metacercaria** is considered a juvenile fluke. The miracidium nearly always enters a mollusk (bivalve or snail) as the first intermediate host, but there are exceptions. In 1944 the American investigator Martin found that sporocysts and cercariae of a certain fluke developed in a polychaete annelid. The larval stages may also be abbreviated, as when cercariae arise directly from sporocysts without the intervention of the redia stage.

Some of our most serious parasites belong to the Digenea. One of the first digenetic forms to be worked out was the sheep liver fluke *Fasciola hepatica* (Fig. 10-9). This fluke is responsible for "liver rot" in sheep and other ruminants, where the adult fluke lives in the liver and bile passageways. Eggs are shed in the feces and develop into ciliated **miracidia.** The

miracidium enters a snail and produces a baglike **sporocyst** of germ cells and one generation of **rediae.** The rediae give rise to **cercariae,** which encyst on aquatic vegetation as **metacercariae.** When the infested vegetation is eaten by the sheep or other ruminant, the cysts hatch into young flukes.

Opisthorchis (Clonorchis) sinensis—liver fluke of man

Opisthorchis (Fig. 10-10) is the most important liver fluke of man and is common in many regions of the Orient, especially in China, southern Asia, and Japan. Cats, dogs, and pigs are also often infected.

STRUCTURE. The worms vary from 10 to 20 mm. in length (Fig. 10-10). They have an oral sucker and a **ventral sucker** and are covered externally by a rough cuticle (tegument). The **digestive system** consists of a pharynx, a muscular esophagus, and two long, unbranched intestinal ceca. The **excretory system** consists of two protonephridial tubules, with branches provided with flame cells or bulbs. The two tubules unite to form a single median tubule that opens to the outside. The **nervous system,** like that of turbellarians, is made up of two cerebral ganglia connected to longitudinal cords that have transverse connectives. The **muscular system** also is of the planarian type.

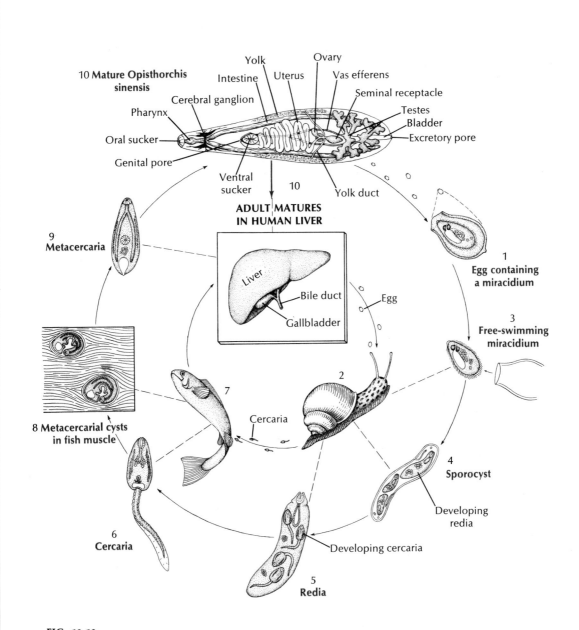

FIG. 10-10

Life cycle of human liver fluke, *Opisthorchis sinensis.* Egg, **1**, shed from adult trematode, **10**, in bile ducts of man, is carried out of body in feces and is ingested by snail (*Bythinia*), **2**, in which miracidium, **3**, hatches and becomes mother sporocyst. **4**, Young rediae are produced in sporocyst, grow, **5**, and in turn produce young cercariae. Cercariae now leave snail, **6**, find a fish host, **7**, and burrow under scales to encyst in muscle, **8**. When raw or improperly cooked fish containing cysts is eaten by man, metacercaria is released, **9**, and enters bile duct, where it matures, **10**, to shed eggs into feces, **1**, thus starting another cycle.

The **reproductive system** is hermaphroditic and complex. The **male system** is made up of two branched testes and two vasa efferentia that unite to form a single vas deferens, which widens into a seminal vesicle. To the seminal vesicle a protrusible penis, or cirrus sac, is attached within the genital opening. The **female system** contains a branched ovary with a short oviduct, which is joined by ducts from the seminal receptacle and from yolk glands. The oviduct is surrounded by Mehlis' gland, which presumably produces shells around the zygotes. From Mehlis' gland the much-convoluted uterus runs to the genital pore. Cross-fertilization is the usual method of fertilizing the eggs. Two flukes in copulation bring their genital pores in opposition and fertilize each other's ova. After the ova have received yolk and a shell, they pass through the genital pore.

LIFE CYCLE. The normal habitat of the adults is in the bile passageways of man (Fig. 10-10). The eggs, each containing a complete **miracidium,** are shed into the water with the feces but do not hatch until they are ingested by the snail *Bythinia* or related genera. The eggs, however, may live for some weeks in water. In the snail the miracidium enters the tissues and is transformed into the **sporocyst** (a baglike structure with embryonic germ cells), which produces one generation of **rediae.** The redia is elongated, with an alimentary canal, a nervous system, an excretory system, and many germ cells in the process of development. The rediae pass into the liver of the snail where, by a process of internal budding, they give rise to the tadpolelike **cercariae.** The cercariae escape into the water, swim about until they meet with fish of the family Cyprinidae, and then bore into the muscles or under the scales. Here the cercariae lose their tails and encyst as **metacercariae.** If man eats raw infested fish, the metacercarial cyst dissolves in the intestine, and the metacercariae are free to migrate up the bile duct, where they become adults. Here the flukes may live for 15 to 30 years. The effect of the flukes on man depends mainly on the extent of the infection. A heavy infection may cause a pronounced cirrhosis of the liver and result in death. Cases are diagnosed through fecal examinations. To avoid infection, all fish used as food should be thoroughly cooked. Destruction of the snails that carry larval stages would be a method of control.

Schistosoma—blood flukes

Three important species of blood flukes belong to genus *Schistosoma.* Infection by these flukes is called **schistosomiasis,** a disorder very common in Africa, China, southern Asia, and parts of South America. The old generic name was *Bilharzia,* and the infection was called **bilharziasis.** The blood flukes differ from most other flukes in being dioecious and having the two branches of the digestive tube united into a single tube in the posterior part of the body. The male is usually broader and encloses the very slender female (Fig. 10-11) in his gynecophoric canal, a ventral fold on the body.

The plan of the life history of blood flukes is similar in all species. Eggs are discharged in human feces or urine; if they get into water, they hatch out as ciliated **miracidia,** which must contact a certain kind of snail within 24 hours to survive. In the snail, they transform into **sporocysts,** which develop **cercariae** directly, without the formation of rediae. The cercariae escape from the snail and swim about until

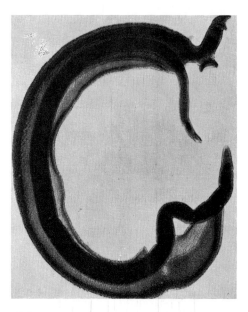

FIG. 10-11
Adult male and female *Schistosoma mansoni* in copulation. Male has long sex canal that holds female (the darkly stained individual) during insemination and oviposition. Man is usually host of adult parasites, found mainly in Africa but also in South America and elsewhere. Man becomes infected by wading or bathing in cercaria-infested waters. (AFIP No. 56-3334.)

they come in contact with the bare skin of a human being. They penetrate through the skin into a blood vessel, which they follow to the blood vessels of certain regions, depending on the kind of fluke. When they enter with drinking water, they may bore through the mucous membrane of the mouth or throat, but they do not survive in the stomach.

No encysted metacercarial stage is found, for the cercariae lose their tails as they enter the host, and transform into the adult condition. When the eggs are fertilized, the female leaves the male gynecophoric canal and goes into small blood vessels where she lays the eggs. The ova may be provided with spines that facilitate their penetration into the intestine or urinary bladder. By the time these ova reach the exterior, they have within them fully formed miracidia.

SCHISTOSOMA HAEMATOBIUM. This species is found chiefly in Africa and is one of the most dangerous of the blood flukes. The adults live in the blood vessels of the bladder and urinary tract. The sharp-spined eggs pass through the walls of the blood vessels into the urine causing lacerations of the mucous membrane of the bladder and bloody urine. Inflammation of the bladder may occur, and the ova may serve as nuclei for kidney stones. The eggs escape in the urine, and if the snail *Bulinus* is available, the miracidia will enter and start the life cycle. Poor hygienic conditions promote the spread of *Schistosoma.*

SCHISTOSOMA MANSONI. This species is common in the West Indies, parts of South America, and Africa. The adults live chiefly in branches of the portal and mesenteric veins. The spiny ova escape into the intestine and are discharged with the feces. Snails, including the common genus *Planorbis,* seem to be the intermediate host. Symptoms of this form of **schistosomiasis** are severe dysentery and anemia.

SCHISTOSOMA JAPONICUM. These flukes often cause trouble in Japan, China, and the Philippines. The life history is similar to that of the other blood flukes. Infection causes liver enlargement, formation of ulcers, and disturbances in the spleen. The chief intermediate host is the snail *Oncomelania.*

SCHISTOSOMA DERMATITIS (SWIMMER'S ITCH). Bathers in our northern lakes often suffer from a skin irritation caused by cercariae of *Schistosoma.* Soon after the bather leaves the water a prickling sensation is felt. After an hour or so the irritation subsides, to return as severe itching with edema and pustules. Infection is most common in July and August.

Several species of cercariae are known to cause these infections. Infections of man are purely accidental attempts of the flukes to penetrate man's skin and use him as a host, but it is a "dead end" for the cercariae, for none are known to survive. Many species of snails serve as intermediate hosts and aquatic birds are the final hosts. Wiping the body thoroughly and quickly after leaving the water has been recommended, for some authorities think the cercariae enter the skin as the water evaporates, but the efficacy of this preventive treatment has been disputed by some parasitologists.

Paragonimus westermani—lung flukes

Lung flukes are found in many parts of the Orient and to some extent in America. One of the most common is *Paragonimus westermani* (Fig. 10-12), which uses the carnivorous mammals, such as mink,

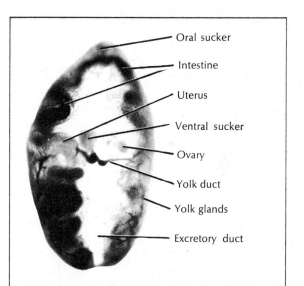

Oral sucker

Intestine

Uterus

Ventral sucker

Ovary

Yolk duct

Yolk glands

Excretory duct

FIG. 10-12
Pulmonary fluke, *Paragonimus westermani,* infects human lung, producing paragonimiasis. Adults are up to 2 cm. long. Eggs discharged in sputum or feces hatch into free-swimming miracidia that enter snails. Cercariae from snail enter freshwater crabs and encyst in soft tissues. Man is infected by eating poorly cooked crabs or by drinking water containing larvae freed from dead crabs. (AFIP No. 52-1862.)

cats, and dogs, as the definitive host. It is the only species of lung fluke known to infect man. In China, Korea, and Japan human infection may reach 40% to 50% in some regions. In the United States it appears to be most common in the Great Lakes area. The life history involves snails, freshwater crabs, and crayfish as intermediate hosts and man and other mammals as the final hosts.

Ova are coughed up in the sputum, and under moist conditions the miracidia develop in 3 weeks. After entering certain species of snails they develop a sporocyst and two generations of rediae. The cercariae, after leaving the snail, crawl about and encyst in freshwater crabs and crayfish. When man eats poorly cooked crustaceans, the ingested flukes pass through the intestinal walls into the abdominal cavity, then through the diaphragm into the lungs. Here they exist as adults, which may be about 20 mm. long.

Other trematodes

Fasciolopis buski (intestinal fluke of man) parasitizes man and pigs in India and China. Larval stages occur in snails (*Planorbis* and *Segmentina*), and the cercariae encyst on water chestnuts, an aquatic vegetation eaten raw by man and pigs.

Leucochloridium is noted for its remarkable sporocysts. Snails *(Succinea)* eat vegetation infected with egg capsules from bird droppings. The sporocysts that develop branch and enter the snail's head and tentacles. Here they enlarge greatly, become brightly striped with orange and green bands and pulsate at frequent intervals. Birds are attracted by the enlarged and pulsating tentacles, eat the snails, and so complete the life cycle.

Class Cestoda

The Cestoda, or tapeworms, differ in many respects from those of the preceding classes: their long flat bodies are usually made up of many sections, or **proglottids,** and there is a complete lack of a digestive system. They have no cilia, are covered with a modified epidermis, or tegument, and possess well-developed muscles. They have an excretory system and a nervous system somewhat similar to those of other flatworms. They are nearly all monoecious. They have no special sense organs but do have free sensory nerve endings. One of their most specialized structures is the **scolex,** or holdfast, which is the

organ of attachment. It is provided with a varying number of suckers and, in some cases, also with hooks.

All members of this class are endoparasites, and all, with a few exceptions, involve at least two hosts of different species. The adults are found only in vertebrates; other stages may be found in either vertebrates or invertebrates.

Trematodes and cestodes occupy different niches in the gut of animals. Trematodes can use undigested food, but cestodes, lacking a digestive system, must use previously digested food. As parasites, the cestodes have evolved a greater dependence on the host for their nutrition.

Class Cestoda is divided into two subclasses: Cestodaria and Eucestoda.

The Cestodaria includes forms that have undivided bodies and no scolices but are provided with some organ of attachment such as a rosette or proboscis. They have only one set of reproductive organs and give rise to 10-hooked larvae. Most of them are found as parasites in lower fish. *Amphilina foliacea* is a common type.

Members of Eucestoda have the body divided (rarely undivided) into proglottids and are provided with a scolex. Their larval forms have only six hooks, rather than 10 as in the Cestodaria. These are the typical tapeworms, of which the beef tapeworm, *Taenia saginata* (Fig. 10-13); the pork tapeworm, *T. solium;* and the dog tapeworm, *T. pisiformis* (Fig. 10-15), are examples.

Is the tapeworm a single animal with subdivided parts (proglottids), in other words is it a true segmented animal such as the annelids and arthropods, or it is a series of separate individuals loosely held together as a colony? Although the tapeworm forms its segments from the proliferation of the scolex, whereas in the true segmented animals new segments proliferate just in front of the anal segment, this difference is considered unimportant by some zoologists, who view cestodes as segmented. On the other hand the scolex might be regarded as the ancestral individual that gives rise by strobilation to daughter individuals, the proglottids. Zoologists are divided in their interpretation of these points.

The scolex, or holdfast organ (Fig. 10-14), is not a head specialized for perceiving or food handling, and some zoologists believe it is really the original posterior end that has been modified for attachment,

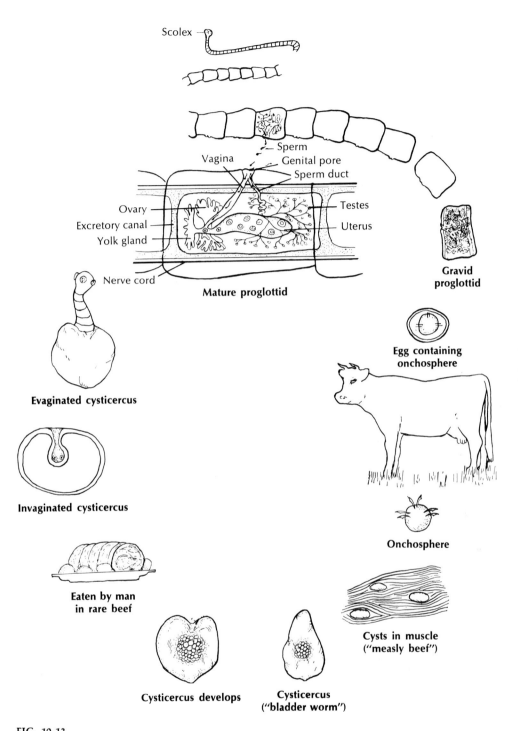

Scolex

Sperm
Vagina
Genital pore
Sperm duct
Ovary
Excretory canal
Yolk gland
Testes
Uterus
Nerve cord
Mature proglottid

Gravid proglottid

Egg containing onchosphere

Evaginated cysticercus

Invaginated cysticercus

Onchosphere

Eaten by man in rare beef

Cysts in muscle ("measly beef")

Cysticercus develops

Cysticercus ("bladder worm")

FIG. 10-13

Life cycle of beef tapeworm, *Taenia saginata.* Ripe proglottids break off in man's intestine, pass out in feces, and are ingested by cattle. Eggs hatch in cow's intestine, freeing onchospheres, which penetrate into muscles and encyst, developing into "bladder worms." Man eats infected rare beef and cysticercus is freed in intestine where it develops, forms a scolex, attaches to intestine wall, and matures.

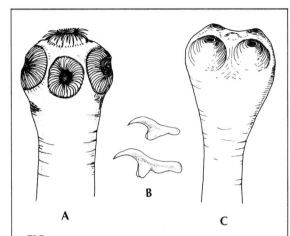

FIG. 10-14
Comparison of scolex (holdfast) of pork and beef tapeworms. **A,** Scolex of *Taenia solium* (pork tapeworm) with apical hooks and suckers. **B,** Hooks of *T. solium*. **C,** Scolex of *T. saginata* (beef tapeworm), with suckers only.

but embryology does not support this view. As long as the scolex is present, it is impossible to get rid of a tapeworm, for new proglottids will be formed as the old ones are shed. A proglottid (Fig. 10-15) is a sexually complete unit, for it is hermaphroditic. Its function is the production of ova and sperm, and its structure is specialized toward this end.

The most common tapeworms found in man are given in Table 10-1.

Taenia saginata—beef tapeworm

STRUCTURE. The beef tapeworm lives as an adult in the alimentary canal of man, whereas the larval form is found primarily in the intermuscular tissue of cattle. The mature adult may reach a length of 30 feet or more. Its **scolex** has four **suckers** for attachment to the intestinal wall, but no hooks (Fig. 10-14, *C*). A short neck connects the scolex to the body, or **strobila,** which may be made up of as many

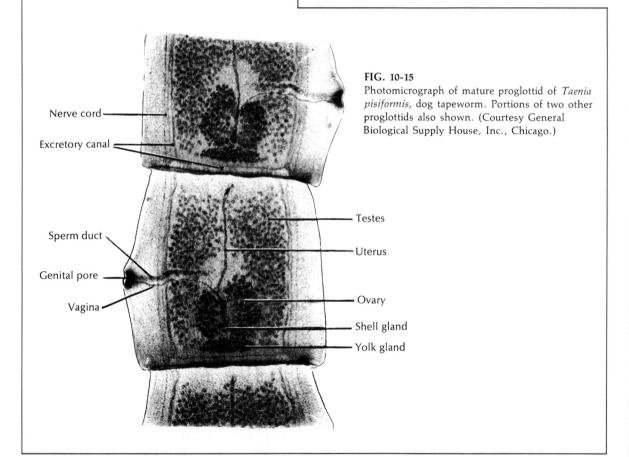

FIG. 10-15
Photomicrograph of mature proglottid of *Taenia pisiformis,* dog tapeworm. Portions of two other proglottids also shown. (Courtesy General Biological Supply House, Inc., Chicago.)

Nerve cord

Excretory canal

Sperm duct

Genital pore

Vagina

Testes

Uterus

Ovary

Shell gland

Yolk gland

TABLE 10-1
Common cestodes of man

Common and scientific name	Incidence in man
Beef tapeworm (*Taenia saginata*)	Eating rare beef; most common of all tapeworms in man
Pork tapeworm (*Taenia solium*)	Eating rare pork
Fish tapeworm (*Diphyllobothrium* [*Dibothriocephalus*] *latum*)	Eating rare or poorly cooked fish; fairly common in Great Lakes region
Dog tapeworm (*Dipylidium caninum*)	Unhygienic habits of children (larvae in flea and louse); not common
Dwarf tapeworm (*Hymenolepis nana*)	Larvae in flour beetles; not common
Hydatid worms (*Echinococcus granulosus*)	Larval cysts in man; infection by contact with dogs; 500 to 600 cases recorded In United States

as 2,000 **proglottids.** New proglottids are formed by **transverse budding** of the neck region. As they move backward, the proglottids increase in size, so that the proglottids are narrow near the scolex and broader and larger toward the posterior end. The terminal **gravid proglottids** (Fig. 10-18) are finally detached and shed in the feces.

The tapeworm shows some unity in its organization, for **excretory canals** in the scolex are also connected to the canals, two on each side, in the proglottids, and two **longitudinal nerve cords** from a **nerve ring** in the scolex run back into the proglottids also (Fig. 10-15). Attached to the excretory ducts are the **flame cells.** Each mature proglottid also contains **muscles** and **parenchyma** as well as a complete set of **male and female organs** similar to those of a trematode. Ova may be fertilized by the sperm of the same individual, or from other tapeworms if more than one should be present. When the **gravid proglottids** with their eggs break off and pass out with the feces, the proglottids disintegrate, and the eggs with the embryos may be scattered on the soil or grass, where they may be picked up by grazing cattle.

LIFE CYCLE. When cattle swallow the eggs, the eggshells dissolve, and the six-hooked larvae **(onco-spheres)** burrow through the intestinal wall into the blood or lymph vessels and finally reach voluntary muscle, where they encyst to become **bladder worms (cysticerci).** Here the larvae develop an invaginated scolex but remain quiescent. When infected "measly" meat (Fig. 10-16) is eaten by a suitable host, the

cyst wall dissolves, the scolex evaginates and attaches to the intestinal mucosa, and new proglottids begin to develop. It takes 2 or 3 weeks for a mature worm to form. When man is infected with one of these tapeworms, many gravid proglottids are expelled daily from his intestine. Usually only one tapeworm infects a host; an immunity against others is apparently established. Man usually becomes infected by eating rare roast beef, steaks, and barbecues. About 1% of American cattle are infected, and since a great deal of meat is consumed without government inspection, a certain amount of tapeworm infection must be expected, unless all meat is thoroughly cooked.

Other tapeworms

More than 1,000 species of tapeworms are known to parasitologists. Almost all vertebrates are infected. Nearly all tapeworms have an intermediate host and a final host that is infected by preying upon the former. Some of these tapeworms do considerable harm to the host by absorbing nourishment and by secreting toxic substances, but unless present in large numbers, they are rarely fatal.

TAENIA SOLIUM (PORK TAPEWORM). The adult lives in the small intestine of man, whereas the larvae live in the muscles of the pig. Adults may be 20 feet or longer. The scolex has both suckers and hooks arranged on its tip (Fig. 10-14, *A* and *B*), the **rostellum.** The life history of this tapeworm is similar to that of the beef tapeworm. Man becomes infected by eating improperly cooked pork. The inci-

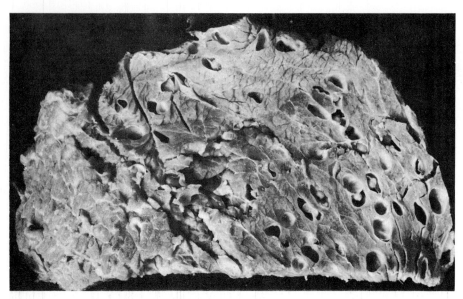

FIG. 10-16
"Measly" pork showing cysts of bladder worms, *Taenia solium.* Beef infected with beef tapeworm has similar appearance.

dence of infection is much lower than that of the beef tapeworm, probably because rare pork is less popular.

It is also possible for the larvae or bladder worms to develop in man, although this is not the normal procedure. If eggs or proglottids are ingested by man or broken-up proglottids from an adult tapeworm are carried back by reverse peristalsis to the stomach, the liberated embryos may migrate into any of several organs, including the brain. Disorders from these causes are called **cysticercosis.** One form of this may cause epilepsy.

DIPHYLLOBOTHRIUM LATUM (FISH TAPEWORM). The adult tapeworm is found in the intestine of man, dog, and cat; the immature stages are in crustaceans and fish. This tapeworm, often called the broad tapeworm of man, is the largest and most destructive of the cestodes that infect man. It sometimes reaches a length of 60 feet and may have more than 3,000 proglottids. Eggs discharged in water by the human host hatch and may be swallowed by the first intermediate host, a tiny crustacean *(Cyclops).* When the crustacean is eaten by a fish, the young larva migrates to the muscles where it grows to about an inch long. Usually it encysts in the muscle. When raw or poorly cooked fish is eaten by a suitable host, the larva is liberated and grows into adult form. It

has been known to live in man for many years. Broad tapeworm infections are found all over the world; in the United States infections are most common in the Great Lakes region.

DIPYLIDIUM CANINUM (DOG TAPEWORM). *Dipylidium* is common in pet dogs and cats and sometimes in children. It may be 1 foot or more in length and has about 200 proglottids. The larva is found in the louse and flea. The dog or cat becomes infected by licking or biting these ectoparasites. It takes about 2 weeks for the worm to mature.

HYMENOLEPIS NANA (DWARF TAPEWORM). The dwarf tapeworm is the smallest of human tapeworms and is common in the United States. No intermediate host is necessary. The adults are $1/2$ to 2 inches long and have 100 to 200 proglottids. After the eggs have been ingested, the larval forms are liberated and penetrate the intestinal mucosa, where they are transformed into cysticercoid larvae. After a few days they reenter the lumen of the intestine, evaginate their heads, become attached, and mature. Unsanitary toilet habits will cause superinfection. The tapeworm is also found in rats and other rodents.

ECHINOCOCCUS GRANULOSUS (HYDATID WORM). The adult (Fig. 10-17, *A*) is found in the dog, wolf, and a few others; the larvae are found in

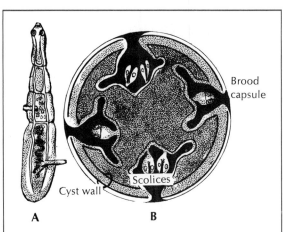

FIG. 10-17
Echinococcus granulosus, dog tapeworm, which may be dangerous to man. **A,** Adult tapeworm lives in intestine of dog or other carnivore. **B,** Early cyst or bladder worm stage found in cattle, sheep, hogs, and sometimes man, produces hydatid disease. Man acquires disease by unsanitary habits in association with dogs. When eggs are ingested, liberated larvae usually encyst in liver. Brood capsules containing scolices are formed from inner layer of each cyst. Cyst enlarges, developing other cysts with brood pouches. May grow for years, to size of orange, necessitating surgery.

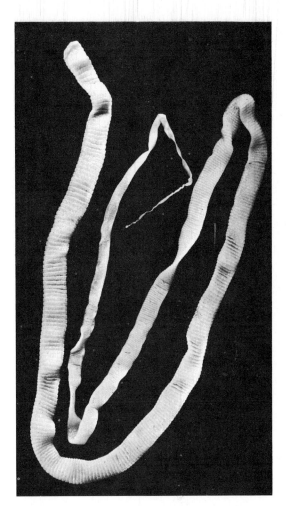

FIG. 10-18
Sheep tapeworm, *Moniezia expansa.* Note progressive increase in size. Young proglottids are budded from scolex and neck (center); oldest (gravid) proglottids shown at upper left.

more than 40 species of mammals, including man, monkeys, sheep, and cattle. Man thus serves as an intermediate host in the case of this tapeworm. The adults are only 5 to 10 mm. long and are composed of a scolex and four proglottids. The larval stages do most of the harm in the life cycle, for the cysticercus forms what is known as a **hydatid cyst,** usually in the liver but also in other organs (Fig. 10-17, *B*). Some of the cysts are 2 or 3 inches in diameter in man and may produce death by their pressure and other effects. A cyst enlarges by the continuous formation of brood capsules, each of which buds off many scolices internally. Many cases of hydatid cysts are reported each year in the United States. Man probably gets his infections from ingesting the eggs of a tapeworm eliminated by a dog.

MONIEZIA EXPANSA (SHEEP TAPEWORM). The adult is found in sheep and goats (Fig. 10-18); the larva grows in a small mite *(Galumna).*

TAENIA PISIFORMIS (DOG TAPEWORM). *T. pisiformis* is widely used as a type for study in the laboratory (Fig. 10-15). The larvae occur in the mesenteries and liver of rabbits; the adults are found in cats and dogs. Its occurrence in man is rare.

■ PHYLUM RHYNCHOCOELA (NEMERTINA)*
General relations

The rhynchocoels are often called the ribbon worms. Their name, which means "hollow beak,"

*Ring"ko-se'la (Gr. *rhynchos,* beak, + *koilos,* hollow). Nem"er-ti'na (Gr. *Nēmertēs,* one of the Nereids, unerring one, + *-ina,* belonging to).

refers to the proboscis, a long muscular tube that can be thrust out swiftly to grasp the prey. The phylum was formerly called Nemertea or Nemertina, with both names referring to the unerring aim of the proboscis.* These worms are still often spoken of as the nemertean or nemertine worms. They are thread-shaped or ribbon-shaped worms, nearly all of them marine. Some live in secreted gelatinous tubes. There are about 570 species in the group.

Some nemertine worms are less than an inch long, and others are several feet in length. *Lineus longissimus* is said to reach 30 meters. Their colors are often bright, though most are dull or pallid. In the odd genus *Gorgonorhynchus* (1931) the proboscis is divided into many proboscides, which appear as a mass of wormlike structures in the everted proboscis. This form may have arisen suddenly as a new genus, but this interpretation has been questioned.

With few exceptions, the general body plan of the nemertines is similar to that of Turbellaria. Like the latter, their epidermis is ciliated and has many gland cells. Another striking similarity is the presence of flame cells in the excretory system. Recently rhabdites have been found in several nemertines, including *Lineus.* However, they differ from flatworms in their reproductive system. They are mostly dioecious. In the marine forms there is a ciliated **pilidium larva.** This helmet-shaped larva has a ventral mouth but no anus—another flatworm characteristic. It also has some resemblance to the trochophore larva that is found in annelids and mollusks. Other flatworm characteristics are the presence of bilateral symmetry, mesoderm, and lack of coelom. All in all, the present evidence seems to indicate that the nemertines came from an ancestral form closely related to Platyhelminthes and Ctenophora.

The nemertines show some advances over the flatworms. One of these is the eversible **proboscis** and its sheath, for which there are no counterparts among Platyhelminthes. Another difference is the presence of an **anus** in the adult. These forms have a complete digestive system, the first to be found in the animal kingdom. They are also the simplest animals to have a **blood vascular** system.

Characteristics

1. Bilateral symmetry; highly contractile body that is cylindric anteriorly and flattened posteriorly

*See footnote on p. 219.

2. Three germ layers
3. Epidermis with cilia and gland cells; rhabdites in some
4. Body spaces with parenchyma, which is partly connective tissue and partly gelatinous
5. An **eversible proboscis,** which lies free in a cavity (rhynchocoel) above the alimentary canal
6. **Complete digestive system** (mouth to anus)
7. Body-wall musculature of outer circular and inner longitudinal layers with diagonal fibers between the two; sometimes another circular layer inside the longitudinal layer
8. **Blood vascular system with three longitudinal trunks**
9. No regular coelom; rhynchocoel may be considered true coelom
10. Nervous system usually a four-lobed brain connected to paired longitudinal nerve trunks or, in some, middorsal and midventral trunks
11. Excretory system of two coiled canals, which are branched with **flame cells**
12. Sexes separate with simple gonads; asexual reproduction by fragmentation; few hermaphrodites; pilidium larva in some
13. No respiratory system
14. Sensory **ciliated pits** or **head slits on each side of head,** which communicate between the outside and the brain; tactile organs and ocelli (in some)
15. In contrast to Platyhelminthes, few nemertines are parasitic

Ecologic relationships

A few of the nemertines are found in moist soil and fresh water, but by far the larger number are marine. At low tide they are often coiled up under stones. It seems probable that they are active at high tide and quiescent at low tide. Some nemertines such as *Cerebratulus* often live in empty mollusk shells. The small species live among seaweed, or they may be found swimming near the surface of the water. Nemertines are often secured by dredging at depths of 15 to 25 feet or deeper. A few are commensals or parasites. *Prostoma rubrum,* which is less than an inch long, is a well-known freshwater species.

Classes

Class Enopla (en'op-la). (Gr. *enoplos,* armed). Proboscis armed with stylets; muscular layer of outer circular and inner longitudinal muscles; no nerve plexus; intestinal ceca; mouth opens in front of brain. Example: *Amphiporus.*

Class Anopla (an'o-pla). (Gr. *anoplos,* unarmed). Proboscis lacks stylets; muscular layer of inner and outer longitudinal and middle circular muscles; nerve plexus present; mouth opens behind brain; intestinal pouches absent or rudimentary. Example: *Cerebratulus.*

Class Enopla

Most nemertines have a close resemblance to each other, although some are very long and difficult to study in the laboratory because their internal organs are not easily seen. Nemertines are slender worms and very fragile, with a great diversity in size. *Amphiporus,* which is taken here as a representative type, is one of the smaller ones.

Amphiporus ocraceus—a ribbon worm

Amphiporus (Fig. 10-19) is from 1 to 3 inches long and about 2.5 mm. wide. It is dorsoventrally flattened and has rounded ends. The body wall com-

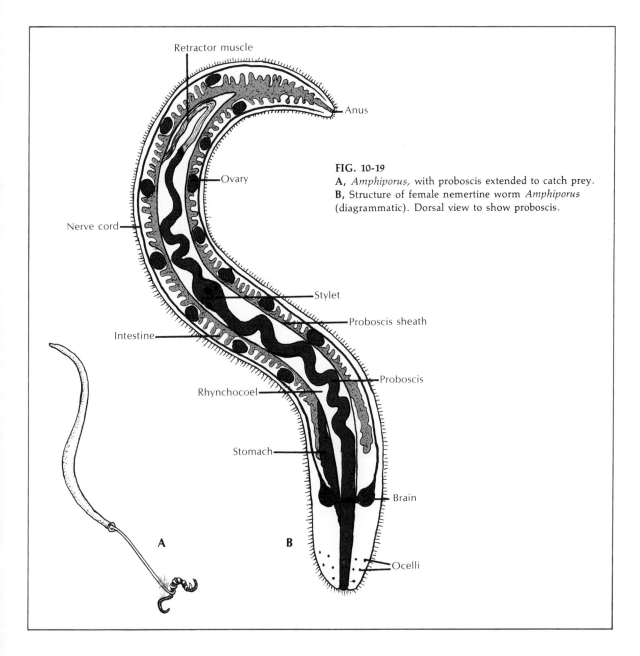

FIG. 10-19
A, *Amphiporus,* with proboscis extended to catch prey. **B,** Structure of female nemertine worm *Amphiporus* (diagrammatic). Dorsal view to show proboscis.

Retractor muscle
Anus
Ovary
Nerve cord
Stylet
Proboscis sheath
Intestine
Proboscis
Rhynchocoel
Stomach
Brain
Ocelli
A
B

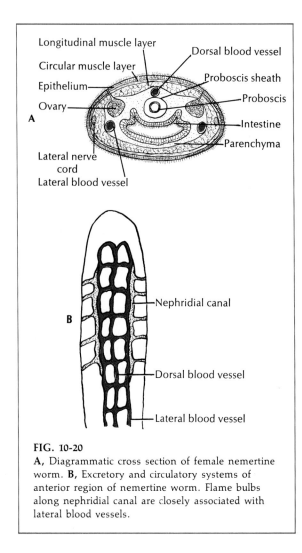

FIG. 10-20
A, Diagrammatic cross section of female nemertine worm. **B,** Excretory and circulatory systems of anterior region of nemertine worm. Flame bulbs along nephridial canal are closely associated with lateral blood vessels.

prises an epidermis of ciliated columnar cells and layers of circular and longitudinal muscles (Fig. 10-20). A partly gelatinous parenchyma fills the space around the visceral organs. Ocelli are located at the anterior end. The thick-lipped mouth is anteroventral, with the opening of the proboscis just above it.

The **proboscis** is not connected with the digestive tract, but is an eversible organ that can be protruded from its cavity, the **rhynchocoel,** and used for defense and catching prey (Fig. 10-19). It lies within a sheath to which it is attached by muscles. The rhynchocoel is filled with fluid, and by muscular pressure on this fluid the anterior part of the tubular proboscis is everted, or turned inside out. The proboscis apparatus is an invagination of the anterior

body wall, and its structure therefore duplicates that of the body wall. The retractor muscles attached at the end are used to retract the everted proboscis, much like everting the tip of a finger of a glove by a string attached to its tip. The proboscis is armed with a sharp-pointed stylet. A frontal gland also opens at the anterior end by a pore.

LOCOMOTION. *Amphiporus* can move with considerable rapidity by the combined action of its well-developed musculature and its cilia. It glides mainly against a substratum; some make use of muscular waves in crawling. Some nemertines have the interesting method of protruding the proboscis, attaching themselves by means of the stylet, and then drawing the body up to the attached position.

FEEDING AND DIGESTION. The nemertines are carnivorous and very voracious, eating either dead or living prey. In seizing their prey they thrust out the slime-covered proboscis, which quickly ensnares the prey by wrapping around it (Fig. 10-19, *A*). The stylet also pierces and holds the prey. Then by retracting the proboscis, the prey is drawn near the mouth and is engulfed by the esophagus that is thrust out to meet it.

The **digestive system** is complete and extends straight through the length of the body to the terminal **anus,** lying ventral to the proboscis sheath. The **esophagus** is straight and opens into a dilated part of the tract, the **stomach.** The blind anterior end of the intestine as well as the main intestine is provided with paired **lateral ceca.** The alimentary tract is lined with ciliated epithelium, and in the wall of the esophagus there are glandular cells.

Digestion is largely extracellular in the intestinal tube, and when the food is ready for absorption, it passes through the cellular lining of the intestinal tract into the blood vascular system. The indigestible material passes out the anus (Fig. 10-19, *B*), in contrast to Platyhelminthes, in which it leaves by the mouth.

CIRCULATION. The blood vascular system is simple and enclosed, with a single dorsal vessel and two lateral vessels (Fig. 10-20, *B*) connected by transverse vessels. All three longitudinal vessels join together anteriorly to form a type of collar. The blood is usually colorless, containing nucleated corpuscles, although in some nemertines the blood is red because of the presence of hemoglobin. There is no heart, and the blood is propelled by the muscular walls of the blood vessels and by bodily movements.

EXCRETION AND RESPIRATION. The excretory system contains a pair of lateral tubes with many branches and flame cells (Fig. 10-20, *B*). Each lateral tube opens to the outside by one or more pores. Waste is picked up from the parenchymal spaces and blood by the flame cells and carried by the excretory ducts to the outside.

Respiration occurs through the body surface.

NERVOUS SYSTEM. The nervous system includes a brain composed of four fused ganglia, one pair dorsal and one pair ventral, united by commissures. Five longitudinal nerves extend from the brain posteriorly—a large lateral trunk on each side of the body, paired dorsolateral trunks, and one middorsal trunk. These are connected by a network of nerve fibers. From the brain nerves run to the proboscis, to the ocelli and other sense organs, and to the mouth and esophagus. In addition to the ocelli, there are other sense organs, such as tactile papillae, sensory pits and grooves, and probably auditory organs.

REPRODUCTION AND DEVELOPMENT. The reproductive system in *Amphiporus* is dioecious. The gonads in either sex lie between the intestinal ceca. From each gonad a short duct (gonopore) runs to the dorsolateral body surface. Eggs and sperm are discharged into the water, where fertilization occurs. Egg production in the females is usually accompanied by degeneration of the other visceral organs.

Nemertines have spiral cleavage, mosaic blastomeres, and a hollow coeloblastula. Gastrulation is by embolic invagination and the mesoderm is partly from the endoderm and partly from the ectoderm.

A pilidium larva develops, which is helmet shaped and bears a dorsal spike of fused cilia and a pair of lateral lobes. The entire larva is covered with cilia and has a mouth and alimentary canal but no anus. In some nemertines the zygote develops directly without undergoing metamorphosis. The freshwater species, *Prostoma rubrum,* is hermaphroditic. A few nemertines are viviparous.

REGENERATION. Nemertines have great powers of regeneration. At certain seasons some of them fragment by autotomy, and from each fragment a new individual develops. This is especially noteworthy in the genus *Lineus.* Fragments from the anterior region will produce a new individual more quickly than will one from the posterior part, in ac-

cordance with the principle of the axial gradient. Sometimes the proboscis is shot out with such force that it is broken off from the body. In such a case a new proboscis is developed within a short time.

Derivation and meaning of names

Acoela (Gr. *a,* without, + *koilos,* hollow). Order of Turbellaria. These worms have no enteron.

Alloiocoela (Alloeocoela) (Gr. *alloios,* different, + *koilos,* hollow). Order of turbellarians.

Amphiporus (Gr. *amphi,* on both sides, + *poros,* pore). Genus (Enopla). Refers to the mouth and proboscis pores at the anterior end.

Cerebratulus (L. *cerebrum,* brain, + *ulus,* dim.) Genus (Anopla). Refers to the relatively prominent cerebral ganglia.

Digenea (Gr. *dis,* double, + *genos,* race). Subclass (Trematoda). These flukes require two or more hosts for their complete development.

Diphyllobothrium (Gr. *dis,* double, + *phyllon,* leaf, + *bothrion,* small hole). Genus (Cestoda). The scolex of this tapeworm has only two suckers instead of four, which is the number commonly found.

Dugesia (formerly called *Euplanaria* but changed by priority to *Dugesia* after Dugès, who first described the form in 1830). Genus (Turbellaria).

Echinococcus (Gr. *echinos,* spiny, + *coccus,* berry). Genus (Cestoda). The multiple scolices give a spiny and berrylike appearance to the dangerous tapeworm larval cysts.

Monogenea (Gr. *monas,* single, + *genos,* race). Subclass (Trematoda). Only one host required for development.

Opisthorchis (Gr. *opisthe,* behind, + *orchis,* testis). Genus (Trematoda). Testes are located in the posterior part of the body.

Paragonimus (Gr. *para,* beside, + *gonimos,* generative). Genus (Trematoda). Refers to the position of the reproductive organs. Testes lie side by side and the ovary lies opposite the uterus.

Polycladida (Gr. *poly,* many, + *klados,* branch). Order of turbellarians that have intestines of many branches.

Rhabdocoela (Gr. *rhabdos,* rod, + *koilos,* cavity). Order of turbellarians that have a straight intestine of smooth contour.

Schistosoma (Gr. *schistos,* divided, + *soma,* body). Genus (Trematoda). Male canal in which the female is held gives a split appearance to the body.

Taenia (Gr. *tainia,* band, ribbon). Genus (Cestoda).

Tricladida (Gr. *tri-,* three, + *klados,* branch). Order of turbellarians with 3-branched intestines.

References

Baer, J. C. 1952. Ecology of animal parasites. Urbana, University of Illinois Press.

Bils, R. F., and W. E. Martin. 1966. Fine structure and development of the trematode integument. Trans. Amer. Microsc. Soc. **85:**78-88.

Bogitsh, B. J. 1968. Cytochemical and ultrastructural observations on the tegument of the trematode *Megalodiscus temperatus*. Trans. Amer. Microsc. Soc. **87**:477-486.

Böhmig, L. 1929. Artikel. Nemertini. In W. Kükenthal and T. Krumbach: Handbuch der Zoologie, vol. 2, part 1, sec. 3. Berlin, Walter de Gruyter & Co.

Burger, O. 1890. Anatomie und Histologie des Nemertinen. Z. Wiss. Zool. **50**:1-279. *A classic study of the nemertines.*

Cheng, T. C. 1964. The biology of animal parasites. Philadelphia, W. B. Saunders Co.

Coe, W. R. 1943. Biology of the nemerteans of the Atlantic Coast of North America. Hartford, Transactions of Connecticut Academy of Arts and Science **35**:129. *This is only one of many valuable papers on Rhynchocoela by this author.*

Dawes, B. 1946. The Trematoda with special reference to British and other European forms. New York, Cambridge University Press. *A rather comprehensive review of the group. An excellent bibliography is included.*

Goodrich, E. S. 1945. The study of nephridia and genital ducts since 1895. Quart. J. Microsc. Sci. **86**:113-392.

Grassé, P.-P. (editor). 1961. Traité de zoologie, anatomie, systématique, biologie, vol. IV. Platyhelminthes, mésozoaires, acanthocéphales, némertiens (first fascicule). Paris, Masson & Cie. *An authoritative treatise on these groups.*

Hickman, C. P. 1973. Biology of the invertebrates, ed. 2. St. Louis, The C. V. Mosby Co.

Hyman, L. H. 1951. The invertebrates: Platyhelminthes and Rhynchocoela, vol. 2. New York, McGraw-Hill Book Co. *A comprehensive account of this group, with excellent figures and bibliography.*

Jenkins, M. M. 1966. Note on stalk formation in cocoons of *Dugesia dorotocephala* (Woodworth, 1897). Trans. Amer. Microsc. Soc. **85**:168.

Jenkins, M. M., and H. P. Brown. 1964. Copulation activity and behavior in the planarian *Dugesia dorotocephala* (Woodworth, 1897). Trans. Amer. Microsc. Soc. **83**:32-40.

Jennings, J. B. 1957. Studies on feeding, digestion, and food storage in free-living flatworms. Biol. Bull. **112**:571-581. *Detailed description of ingestion and digestion in certain turbellarians.*

Jennings, J. B. 1962. Further studies on feeding and digestion in triclad Turbellaria. Biol. Bull. **123**:571-597.

Jennings, J. B. 1963. Some aspects of nutrition in the Turbellaria, Trematoda, and Rhynchocoela. In E. C. Dougherty (editor). The lower Metazoa: Comparative biology and phylogeny. Berkeley, University of California Press.

Karling, T. G. 1963. Some evolutionary trends in turbellarian morphology. In E. C. Dougherty (editor): The lower Metazoa: Comparative biology and phylogeny. Berkeley, University of California Press.

Martin, W. E. 1952. Another annelid first intermediate host of a digenetic trematode. J. Parasit. **38**:1-4. *This report, together with an earlier one, describes the rare exceptions in which a group other than mollusks acts as the first intermediate host.*

Mueller, J. R. 1965. Helminth life cycles. Amer. Zool. **5**:131-139.

Olsen, O. W. 1962. Animal parasites: their biology and life cycles. Minneapolis, Minn., Burgess Publishing Co.

Oschman, J. L. 1967. Microtubules in the subepidermal glands of *Convoluta roscoffensis* (Acoela, Turbellaria). Trans. Amer. Microsc. Soc. **86**:159-162.

Pennak, R. W. 1953. Fresh-water invertebrates of the United States. New York, The Ronald Press Co. *Includes an excellent description of the structure and life history of the freshwater nemertine Prostoma rubrum.*

Poluhowich, J. J. 1968. Notes on the freshwater nemertean *Prostoma rubrum*. Turtox News **46**:2-6.

Rothman, A. H. 1963. Electron microscopic studies of tapeworms: the surface structures of *Hymenolepis diminuta* (Rudolphi, 1819) (Blanchard, 1891). Trans. Amer. Microsc. Soc. **82**:22-30.

Smyth, J. D. 1966. The physiology of trematodes. San Francisco, W. H. Freeman & Co.

Swellengrebel, N. H., and M. N. Sterman. 1961. Animal parasites in man. New York, D. Van Nostrand Co., Inc. *An English edition of a well-established Dutch text.*

Thomas, A. P. 1883. The life history of the liver fluke *(Fasciola hepatica)*. Quart. J. Microsc. Sci. (ser. 2) **23**:99-133. *This classic work represents the first life history of a digenetic trematode to be worked out. It gave a great impetus to work in the field of parasitology.*

Threadgold, L. T. 1963. The tegument and associated structures of *Fasciola hepatica*. Quart. J. Microsc. Sci. **104**:505-512.

Wardle, R. A., and J. A. McLeod. 1952. The zoology of tapeworms. Minneapolis, University of Minnesota Press. *A comprehensive account.*

Wells, M. 1968. Lower animals. New York, McGraw-Hill Book Co.

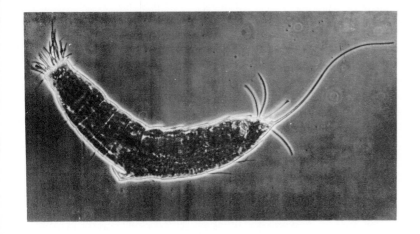

Cateria styx *is a minute worm (less than 1 mm. long) that has a segmented cuticle and lives in ocean mud or silt. It belongs to the phylum Kinorhyncha. A kinorhynch burrows by extending its head, anchoring with its spines, and drawing the body forward till the head is retracted into the neck and then repeating the process. (Courtesy R. P. Higgins. 1968. Trans. Amer. Microsc. Soc. 87:28.)*

CHAPTER 11

THE PSEUDOCOELOMATE ANIMALS

PHYLUM ROTIFERA
PHYLUM GASTROTRICHA
PHYLUM KINORHYNCHA
PHYLUM NEMATODA
PHYLUM NEMATOMORPHA
PHYLUM ACANTHOCEPHALA
PHYLUM ENTOPROCTA
PHYLUM GNATHOSTOMULIDA

■ PLACE IN ANIMAL KINGDOM

In the first seven phyla above, the original blastocoel of the embryo persists as a space, or body cavity, between the enteron and the body wall. Because this cavity lacks the peritoneal lining found in the true coelomates, it is called a **pseudocoel,** and the animals possessing it are called **pseudocoelomates.**

The Gnathostomulida are a new phylum whose evaluation is still incomplete. Although they lack pseudocoel spaces, they are placed with the pseudocoelomates because of other features.

Pseudocoelomates belong to the Protostomia division of the bilateral animals.

■ BIOLOGIC CONTRIBUTIONS

The pseudocoel is a distinct advancement over the solid body structure of the acoelomates. It may be filled with fluid or may contain a gelatinous substance with some mesenchyme cells. In either case

the pseudocoel (1) permits greater freedom of movement, (2) provides ample space for the development and differentiation of digestive, excretory, and reproductive systems, (3) provides a simple means of circulation or distribution of materials throughout the body, (4) acts as a storage place for waste products to be discharged to the outside by excretory ducts, and (5) often plays an important role as a hydrostatic organ.

The complete mouth-to-anus digestive tract is now well established and found in these phyla (except Gnathostomulida) and in all higher phyla.

■ PHYLOGENY AND ADAPTIVE RADIATION

PHYLOGENY. Affinities are difficult to establish within such a varied assemblage. Acanthocephala seem to be related to both the flatworms and the roundworms. The flattened shape of the body, the lack of a digestive system (as in cestodes), and a body wall of both circular and longitudinal muscles as well as some other features may indicate a relationship with the flatworms. The nuclear constancy and superficial segmentation, on the other hand, show roundworm affinities. Most of these phyla are believed to have been derived from rhabdocoel flatworms. The Entoprocta have often been included with the Ectoprocta, or moss animals, but the ectoprocts are true coelomate animals and the entoprocts are much simpler in their structure. Entoprocta may have come from an early offshoot of the same line that led to the Ectoprocta. The Gnathostomulida show some affinities to the Gastrotricha and Rotifera and some to the turbellarian flatworms. Among such diverse groups as these, affinities are very confusing and great caution must be used in taking a few similarities as evidence of relationship.

ADAPTIVE RADIATION. Each group of this vast assemblage of pseudocoelomates may have its own unique basic adaptive pattern in determining its evolutionary history. The most numerous group of all, the nematodes, have been able to adapt to almost every ecologic niche available to animal life. Their viability under the most harsh environmental conditions may be due to a wide range of physiologic response. The emphasis on longitudinal muscles has restricted their undulations, but their constant activity within the range of their restricted movement has enabled them to move into many niches. Being wholly parasitic, the Acanthocephala have undergone those modifications characteristic of parasitic forms. Their invaginable proboscis has evolved some changes in spine patterns and in a few other ways, but the size range from 1 or 2 mm. to almost 2 feet represents their chief evolutionary diversification. Variations in the evolutionary patterns of ciliary feeding and general structural organization are found in the Entoprocta.

◼ THE PSEUDOCOELOMATES

In contrast to the acoelomate phyla in which the space between the body wall musculature and the digestive tract is filled with a solid mesodermal parenchyma, the remaining phyla have a body cavity of some sort in which the organs are located. Such a cavity, in contrast to a solid parenchyma, provides a place in which organs are freely movable and promotes exchange of materials between the organs and the body wall. Vertebrates and higher invertebrates

have a true **coelom**, or peritoneal cavity, which is formed in the mesoderm during embryonic development and which is lined with peritoneum (a layer of mesodermal epithelium). A **pseudocoel** is a body cavity lying between the body wall and gut that is *not* lined with peritoneum. Seven distinct groups of animals belong to the pseudocoelomate category. These are Rotifera, Gastrotricha, Nematoda, Kinorhyncha, Nematomorpha, Acanthocephala, and Entoprocta. The first five of these groups have certain similarities that have lead many authorities to place them as classes in a phylum called Aschelminthes.* However, their organizational plan is different enough that other authorities prefer to consider them as separate phyla. Some group the five together loosely as individual phyla under a superphylum Aschelminthes. The Entoprocta has sometimes been grouped with the Ectoprocta and together called the Bryozoa (moss animals). However, because the ectoprocts have a true coelom, they are usually considered as a separate phylum, and the term "bryozoans" is usually restricted to the ectoprocts.

However one classifies them, the pseudocoelomates are a heterogeneous assemblage of animals that seem to have little in common except a pseudocoel. Most of them are small, some are microscopic, some are fairly large. Some, such as the nematodes, are found in freshwater, marine, terrestrial, and parasitic habitats; others, such as the Acanthocephala, are strictly parasitic. Some have unique characteristics such as the lacunar system of the acanthocephalans, the ciliary corona of the rotifers, or the zonites of the kinorhynchs.

Even in such a diversified grouping some characteristics are shared. In all there is a body wall of epidermis (often syncytial), a dermis, and muscles surrounding the pseudocoel. Except in the Gnathostomulida, the digestive tract is complete, and it, along with the gonads and excretory organs, are suspended in the pseudocoel and bathed in perivisceral fluid. The epidermis in many secretes a hard cuticle with some specializations such as bristles, spines, etc.

Constancy of cells or nuclei is common to several of the groups, and in most of them there is an emphasis upon the longitudinal muscle layer.

Characteristics of pseudocoelomates

1. Symmetry bilateral; unsegmented; triploblastic (three germ layers)

*As"kel-min'theez (Gr. *askos,* wineskin + *helmins,* worm).

2. Body cavity an unlined **pseudocoel**
3. Size mostly small; some microscopic; a few a meter or more in length
4. Body vermiform; body wall a **syncytial** or cellular **epidermis** with thickened cuticle, sometimes molted; muscular layers mostly of **longitudinal fibers; cilia** mostly absent
5. Digestive system complete with mouth, enteron, and anus; pharynx muscular and well developed; **tube-within-a-tube arrangement;** digestive tract usually only an epithelial tube with **no definite muscle layer**
6. Circulatory and respiratory organs lacking
7. Excretory system of canals and protonephridia in some; cloaca that receives excretory, reproductive, and digestive products may be present
8. Nervous system of cerebral ganglia or of a circumenteric nerve ring connected to anterior and posterior nerves; sense organs of **ciliated pits,** papillae, bristles, and some eyespots
9. Reproductive system of gonads and ducts that may be single or double; sexes nearly always separate, with the male usually smaller than the female; eggs microscopic with chitinous shell
10. Development may be direct or with a complicated life history; cleavage mostly determinate; **cell or nuclear constancy common**

■ PHYLUM ROTIFERA*

Rotifers derive their name from the characteristic ciliated crown, or **corona,** that, when beating, often gives the impression of rotating wheels. Rotifers range in size from 40 μ to 3 mm. in length, but most are between 100 and 500 μ long. The 1,500 to 2,000 species have a worldwide distribution. Some have beautiful colors, though most are transparent, and some have odd and bizarre shapes. Their shapes are often correlated with their mode of life. The floaters are usually globular and saclike; the creepers and swimmers are somewhat elongated and wormlike; and the sessile types are commonly vaselike, with a cuticular envelope (lorica). Some are colonial.

Ecologic relationships

Rotifers are a cosmopolitan group of about 2,000 species, some of which are found throughout the world. Most of the species are freshwater inhabitants, but a few are marine, some are terrestrial, and some are epizoic or parasitic.

*Ro-tif'e-ra (L. *rota,* wheel, + -*fera,* those that bear).

Rotifers are adapted to many kinds of ecologic conditions. Some can endure wide ranges of pH and temperature, though some are more specific in their requirements. Many are found only in acid waters with a pH of 4 to 6, but more are found in alkaline waters. Sessile rotifers are especially sensitive to chemical conditions.

Most species are benthic, occurring on the bottom or in vegetation of ponds or along the shores of large freshwater lakes where they swim or creep about the vegetation. Some terrestrial species frequent mosses and other plants where they often select a special kind of plant. A considerable part of the psammolittoral populations are rotifers, which live in the water film between sand grains off sandy beaches. Pelagic forms are common in the surface waters of freshwater lakes and ponds. These forms are provided with oil drops for buoyancy. Some planktonic rotifers may exhibit cyclomorphosis, that is, seasonal variations in body form.

Many species of rotifers can endure long periods of desiccation and resemble grains of sand. While in a desiccated condition, rotifers are very tolerant to temperature variations. This is especially true of moss-dwelling rotifers. True encystment in which a protective cyst is formed, occurs in only a few rotifers. Upon addition of water, desiccated rotifers resume their activity.

Strictly marine species of rotifers are rather few in number. Some of the littoral species of the sea may be freshwater ones that are able to adapt to saltwater.

Classification

Rotifers are usually divided into three orders (or classes):

SEISONACEA. Epizoic marine rotifers; elongated form; corona poorly developed; sexes similar in size and form. Example: *Seison.*

BDELLOIDEA. Swimming or creeping forms; anterior end retractile; corona usually with pair of trochal disks; no males; parthenogenesis. Examples: *Philodina* (Fig. 11-1), *Rotaria.*

MONOGONONTA. Swimming or sessile forms, with single ovary or testis; males reduced in size; eggs of three types (amictic, mictic, dormant). Examples: *Asplanchna, Epiphanes.*

Morphology and physiology

One of the best known genera is *Philodina* (Fig. 11-1), which is often used as a type for study.

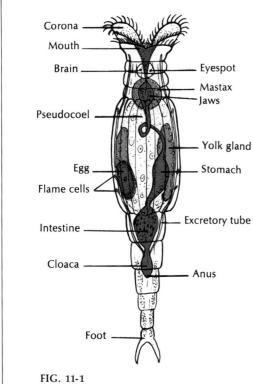

Corona
Mouth
Brain
Eyespot
Mastax
Jaws
Pseudocoel
Yolk gland
Egg
Stomach
Flame cells
Intestine
Excretory tube
Cloaca
Anus
Foot

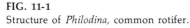

FIG. 11-1
Structure of *Philodina,* common rotifer.

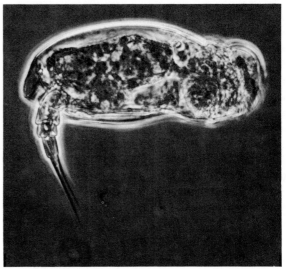

FIG. 11-2
Rotifer. Note the foot with its ringed joints and the two long toes. The toes contain pedal glands for attachment and, in swimming forms, the foot is often ventral and used as a rudder. (Courtesy R. P. Higgins.)

EXTERNAL FEATURES. The body of the rotifer comprises a head bearing a ciliated corona, a trunk, and a posterior tail, or foot. It is covered with a cuticle and is nonciliated except for the corona.

The ciliated **corona,** or crown, surrounds a nonciliated central area of the head called the **apical field,** which may bear sensory bristles or papillae. The appearance of the head end depends on which of the several types of corona it has—usually a circlet of some sort, or a pair of trochal disks. The cilia on the corona beat in succession, giving the appearance of a revolving wheel or pair of wheels, supplying the reason for the rotifers' name, which means "wheel-bearer." The **mouth** is located in the corona on the midventral side. The coronal cilia are used in both locomotion and feeding.

The **trunk** may be elongated as in *Philodina* (Fig. 11-1), or saccular in shape (Fig. 11-2). It contains the visceral organs and often bears sensory antennae. It is covered by a transparent cuticle that in *Philodina* and others is superficially ringed so as to simulate

segmentation, but in many other forms is much thickened to form an outer case or **lorica,** often arranged in plates or rings.

The **foot** is narrower and usually bears one to four toes. Its cuticle may be ringed so that it is telescopically retractile. It is tapered gradually in some forms (Fig. 11-1) and sharply set off in others (Fig. 11-2). The foot is an attachment organ and contains pedal glands that secrete an adhesive material used by both sessile and creeping forms. In swimming pelagic forms the foot is usually reduced. Rotifers move either by creeping with leechlike movements aided by the foot, or by swimming with the coronal cilia, or both.

INTERNAL FEATURES. Underneath the cuticle is the **syncytial epidermis,** which secretes the cuticle, and bands of **subepidermal muscles,** some circular, some longitudinal, and some running through the pseudocoel to the visceral organs. The **pseudocoel** is large, occupying the space between the body wall and the viscera. It is filled with fluid, some of the muscle bands and a network of mesenchymal ameboid cells.

The **digestive system** is complete. Some rotifers feed by sweeping minute organic particles or algae toward the mouth by the beating of the coronal cilia. The cilia are able to sort out and dispose of the larger unsuitable particles. The **pharynx** is fitted with a muscular portion **(mastax)** that is equipped with hard jaws **(trophi)** for sucking in and grinding up the food particles. The constantly chewing pharynx is often a distinguishing feature of these tiny animals. Carnivorous species feed on protozoans and small metazoans, which they capture by trapping or grasping. The trappers have a funnel-shaped area around the mouth. When small prey swim into the funnel, the lobes fold inward to capture and hold them till they are drawn into the mouth and pharynx. The hunters have trophi that can be projected and used like forceps to seize the prey, bring it back into the pharynx, and then pierce or break it up so the edible parts can be sucked out and the rest discarded. There are both salivary and gastric glands that are believed to secrete enzymes for extracellular digestion. Absorption occurs in the stomach.

The **excretory system** typically consists of a pair of **protonephridial tubules,** each with several **flame cells,** that empty into a common bladder. The bladder, by pulsating, empties into the **cloaca** into which the intestine and oviducts also empty. The fact that the rate of pulsation is fairly rapid—1 to 4 times per minute—would indicate that the protonephridia are important osmoregulatory organs, made necessary by the large amount of water swallowed in feeding.

The **nervous system** consists of a bilobed brain, dorsal to the mastax, that sends paired nerves to the sense organs, mastax, muscles, and viscera. Sensory organs include paired eyespots (in some species such as *Philodina),* sensory bristles and papillae, and ciliated pits and dorsal antennae.

REPRODUCTION. Rotifers are dioecious, but the males are usually smaller than the females. In the order Bdelloidea males are entirely unknown, and in Monogononta they seem to occur only for a few weeks of the year.

The female reproductive system consists of ovaries, yolk glands, and oviducts that open into the cloaca. In the Bdelloidea (*Philodina,* for example) all females are parthenogenetic and produce diploid eggs that hatch into diploid females. These reach maturity in a few days. In the order Seisonacea the females produce haploid eggs that must be fertilized and that develop into either males or females. In the Monogononta, however, females produce two kinds of eggs. During most of the year diploid females produce **amictic eggs** that are thin-shelled and diploid. These develop without fertilization into diploid amictic females. However such rotifers often live in temporary ponds or streams and are cyclic in their reproductive patterns. Just before a drastic environmental change the amictic eggs may develop into diploid mictic females that will produce thin-shelled haploid **mictic eggs.** If these eggs are not fertilized, they develop into haploid males. But if fertilized, the eggs develop a thick, resistant shell and become dormant. They survive as winter eggs or until environmental conditions are again suitable, at which time they hatch into diploid females. Winter eggs are often dispersed by wind or birds, a fact that may account for the peculiar distribution patterns of rotifers.

The male reproductive system includes a single testis and a ciliated sperm duct that runs to a genital pore (males usually lack a cloaca). The end of the sperm duct is specialized as a copulatory organ. Copulation is usually by hypodermic impregnation, that is, the penis can penetrate any part of the female body wall and inject the sperm directly into the pseudocoel.

Females hatch with adult features, needing only a few days growth to reach maturity. Males do not grow; they are sexually mature at hatching.

CELL OR NUCLEAR CONSTANCY. Most structures in rotifers are syncytial, but the nuclei in the various organs are said to show a remarkable constancy in numbers in any given species. For example, Martini reported that in one species of rotifer he always found 183 nuclei in the brain, 39 in the stomach, 172 in the corona epithelium, and so on. Not all zoologists are convinced that cell or nuclear constancy is always that absolute.

■ PHYLUM GASTROTRICHA

Gastrotrichs are small, ventrally flattened animals about 65 to 500 μ long, somewhat like rotifers but lacking the corona and mastax and having a characteristically bristly or scaly body. They are usually seen gliding on a substratum by means of ventral cilia or attaching their posterior end briefly by means of adhesive glands. They are assumed to be closely related to the rotifers, which they resemble in having cilia, protonephridia, and a similar muscle pattern. On the other hand, some characteristics are more similar to those of the nematodes.

Ecologic relationships

Gastrotrichs are found in both fresh and salt water. The 500 or so species are about equally divided between the two media. Freshwater forms have benthic habits and occur chiefly among the vegetation of ponds and lakes. They usually retain contact with a substratum of some sort. Many of their species are cosmopolitan, but only a few occur in both fresh water and the sea. Much is yet to be learned about their distribution. Gastrotrichs make up part of the psammolittoral populations found between the sand grains of sandy beaches. Experiments indicate that they can withstand low oxygen concentrations. Their heavy-shelled eggs can withstand harsh environmental conditions and may undergo dormancy for some years.

STRUCTURE. The gastrotrich (Fig. 11-3) is usually elongated, with a convex dorsal surface bearing a pattern of bristles, spines, or scales, and a flattened ciliated ventral surface. The head is often lobed and ciliated and the tail end may be forked.

A syncytial epidermis is found beneath the cuticle. Longitudinal muscles are better developed than are circular ones, and in most cases they are unstriped. Adhesive tubes secrete a substance for attachment. The pseudocoel is somewhat reduced and contains no amebocytes.

The digestive system is complete and is made up of a mouth, muscular pharynx, stomach-intestine, and anus (Fig. 11-3, *B*). Their food is largely algae, protozoans, and detritus, which is directed to the mouth by the head cilia. Digestion appears to be extracellular. Protonephridia are restricted to certain species.

The nervous system contains a brain near the pharynx and a pair of lateral nerve trunks. Sensory structures are similar to those in rotifers, except the eyespots are generally lacking.

Only females occur in freshwater species, and the eggs all develop parthenogenetically. The female reproductive system consists of one or two ovaries, a

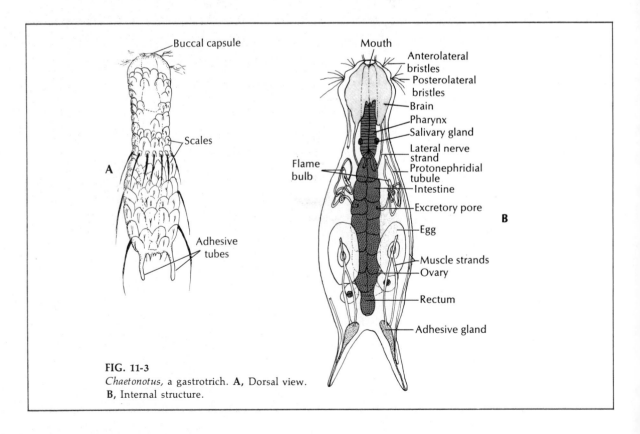

FIG. 11-3
Chaetonotus, a gastrotrich. **A,** Dorsal view. **B,** Internal structure.

uterus, an oviduct, and a gonopore, which may open anteriorly to, or in common with, the anus. Eggs are laid on some substratum such as weeds and hatch in a few days. Development is direct and the larvae have the same form as the adults. Species of *Chaetonotus* are common freshwater gastrotrichs (Fig. 11-3).

■ PHYLUM KINORHYNCHA*

The Kinorhyncha are marine worms a little larger than rotifers and gastrotrichs but usually not over 1 mm. long. Their name comes from the Greek words *kinein,* to move, + *rhynchos,* beak, and refers to their retractile proboscis.

Ecologic relationships

Kinorhynchs seem to be most abundant in shallow mud bottoms of ocean shores, but their general distribution is spotty. They have benthic habits in the slime and mud. Some live among algae where they feed on diatoms. The presence of algae in many species may account somewhat for their color. About a hundred species have been reported.

Phylogenetic relationships

The kinorhynchs share anatomic features with a number of other groups without being closely related to any. They are segmented as are the annelids and

*Kin"o-ring'ka (Gr. *kinein,* to move, + *rhynchos,* beak).

arthropods, but their segmentation is more superficial and secondary to the body plan (Fig. 11-4). They share with rotifers and gastrotrichs flame cells, spines, and retractile head ends; with the nematodes they share longitudinal cords, copulatory spicules, and pattern of nervous system, and like the Priapulida they have an eversible proboscis, spiny pharynx, and similar nervous system. Hymen considered them an offshoot from the same common stem that gave rise to the nematodes and gastrotrichs.

STRUCTURE. The body of the kinorhynch is divided into 13 or 14 rings (zonites), which bear spines, but they have no cilia (Fig. 11-4). The retractile head has a circlet of spines with a small retractile proboscis. The body is flat underneath and arched above. Their body wall is made up of a cuticle, a syncytial epidermis, and longitudinal epidermal cords much like those of nematodes. The arrangement of the muscles is correlated with the zonites, and circular, longitudinal, and diagonal muscle bands are all represented.

A kinorhynch cannot swim. In the silt and mud where it commonly lives it burrows by extending the head into the mud and anchoring it with spines. It then draws the body forward until the head is retracted into the body. When disturbed, the kinorhynch draws in the head and protects it with a closing apparatus of cuticular plates (p. 225).

The digestive system is complete, with a mouth at

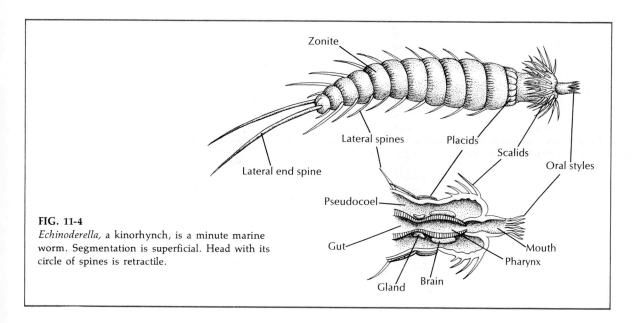

FIG. 11-4
Echinoderella, a kinorhynch, is a minute marine worm. Segmentation is superficial. Head with its circle of spines is retractile.

the tip of the proboscis (Fig. 11-4), a pharynx, an esophagus, a stomach-intestine, and an anus. Kinorhynchs feed on diatoms, or on organic material in the mud where they burrow.

The pseudocoel is filled with fluid-bearing amebocytes. The excretory system is made up of a multinucleated **solenocyte** (protonephridium) on each side of the eleventh zonite. Each solenocyte has one long and one short flagellum.

The nervous system is in contact with the epidermis, with a brain encircling the pharynx and a ventral ganglionated nerve cord extending throughout the body. Sense organs are represented by eyespots in some and by the sensory bristles.

Sexes are separate, with paired gonads and gonoducts. The young larvae have only the first three zonites, but as they grow new zonites are added to the others.

Among the most widely known of the genera of the Kinorhyncha are *Echinoderes, Echinoderella, Pycnophyes,* and *Trachydemus.*

■ PHYLUM NEMATODA*

The nematodes already number more than 12,000 species, and when all the species have been classified, they may even outnumber the arthropods. They are worldwide in distribution.

The distinctive characteristics of this extensive group of animals are their cylindric shape; their flexi-

*Nem"a-to'da (Gr. *nēma,* thread, + *eidos,* form).

ble but inelastic cuticle, which prevents them from changing length and thickness; and their unique manner of thrashing around, forming patterns of C's and S's. Other more or less unique features are (1) the pharynx, which is three-angled and highly muscular, and (2) the excretory system, consisting either of one or more large gland cells (the renette) opening by an excretory pore or a canal system without protonephridia, or both renette and canals together.

Most nematode worms are under 2 inches long, but many are microscopic, and some parasitic nematodes are over a yard long.

Ecologic relationships

Free-living nematodes occur in almost every conceivable kind of ecologic niche and habitat and are probably the most widespread of all metazoans. They have been found from the arctic regions to the tropics and occur in the sea, freshwater, and soil. They are present in high mountains, deserts, hot springs, and great ocean depths. Wherever found, their numbers may be enormous. Many thousands may be found in a single rotting apple. A fistful of soil may yield millions. Several billion per acre may occur in good farmland. Their general resistance probably exceeds that of any other group of animals. Vinegar eels, *Turbatrix aceti,* can withstand a pH of 1.5 and most marine species can endure a wide range of salinity. Even sandy beaches where nutrition is

TABLE 11-1
Common parasitic nematodes of man in the United States

Common and scientific name	Incidence in man
Hookworm (*Ancylostoma duodenale* and *Necator americanus*)	Larvae in soil burrow into skin (contact); common in Southern states
Pinworm (*Enterobius vermicularus*)	Infestation by inhalation of dust with ova and by contamination with fingers; very common in children
Intestinal roundworm (*Ascaris lumbricoides*)	Infestation by embryonated ova in contaminated food; common
Trichina worm (*Trichinella spiralis*)	Ingestion of infected pork muscle; most common parasite in United States
Whipworm (*Trichuris trichiura*)	Ingestion of contaminated food or by unhygienic habits; common in certain regions

scanty may yield hundreds of thousands per acre. Mucky ocean bottoms where food is more abundant will contain much greater populations. Many occur in specialized habitats where other animals are not found. Some species are cosmopolitan and can endure a wide range of conditions, whereas others require more specific conditions. Mosses and lichens have characteristic nematode species (Fig. 11-6) that can often endure extremes of temperature and desiccation by passing into a dormant (cryptobiotic) state and reviving when conditions are more favorable.

In addition to free-living species, many nematodes are parasites not only in animals but on nearly every kind of plant. Some of the common parasitic nematodes of man in this country are listed in Table 11-1.

Subclasses of nematodes

Whether Nematoda is considered a separate phylum, or a class under the Aschelminthes, the group is usually treated as a single class, with the following two subclasses:

Subclass Phasmidia (Secernentea). Body usually bears pair of minute sensory pouches (phasmids) near posterior tip; a pair of porelike sense organs (amphids) at anterior end; excretory system of lateral canals; both free-living and parasitic forms. Examples: *Rhabditis, Ascaris, Enterobius.*

Subclass Aphasmidia (Adenophorea). Phasmids lacking; spiral or disk-shaped amphids; excretory system of one or more renette cells (glandular); caudal glands present; mostly free-living, but includes some parasites. Examples: *Dioctophyme, Trichinella, Plectus.*

Ascaris lumbricoides—the intestinal roundworm

Because of the simplicity of its structure and life history as well as its availability, *Ascaris* is usually selected as a type for study in zoology. There are many species of this genus. One of the most common species, *A. megalocephala,* is found in the intestines of horses. The roundworm *A. lumbricoides suilla,* found in pigs, is morphologically similar to *A. lumbricoides* found in man, but it is rare for the larval form of the one found in man to grow to maturity in the pig and vice versa.

A. lumbricoides (Fig. 11-5) is one of the most common parasites found in man. Infestation normally occurs by swallowing embryonated ova. Unsanitary habits in which contaminated food and vegetables containing the ova are conveyed to the mouth repre-

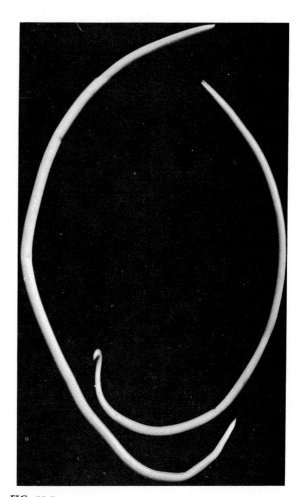

FIG. 11-5
Intestinal roundworm, *Ascaris lumbricoides,* male and female. Male (right) has characteristic sharp kink in end of tail.

sent one of the most frequent methods of infestation.

Morphology and physiology

The females of this species are about 20 to 30 cm. long. The whitish yellow worms are pointed at both ends, and four pale longitudinal lines caused by subcuticular thickenings extend the length of the body. Three lips, provided with papillae, surround the mouth. The male can be distinguished by its smaller size and by its sharply curved posterior end that bears two penial spicules in the genital pore.

The outer body covering is a thick, noncellular, many-layered **cuticle,** secreted by the epidermis. It

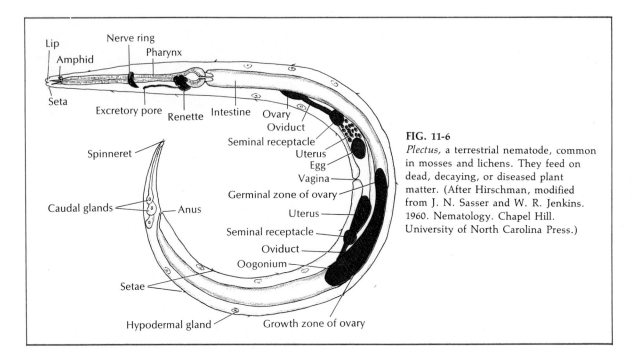

Lip
Nerve ring
Amphid
Pharynx
Seta
Excretory pore
Renette
Intestine
Ovary
Oviduct
Seminal receptacle
Uterus
Egg
Vagina
Germinal zone of ovary
Uterus
Seminal receptacle
Oviduct
Oogonium
Growth zone of ovary
Hypodermal gland
Setae
Anus
Caudal glands
Spinneret

FIG. 11-6
Plectus, a terrestrial nematode, common in mosses and lichens. They feed on dead, decaying, or diseased plant matter. (After Hirschman, modified from J. N. Sasser and W. R. Jenkins. 1960. Nematology. Chapel Hill. University of North Carolina Press.)

is permeable to water, respiratory gases, chloride, and certain other small ions, but protects the worm from many toxic compounds. It contains layers of fibers that give tough support to the body yet are elastic enough to permit the body to increase or decrease in volume without a change in the fluid pressure of the pseudocoel.

Underneath the cuticle are a **syncytial epidermis** and a layer of **longitudinal muscles.** There are no circular muscles. The muscles are arranged in four bands, or quadrants, marked off by four longitudinal cords of thickened epidermis. Each muscle cell is large and spindle shaped, with a protoplasmic process that runs to a nerve trunk in either the dorsal or ventral nerve cord. These processes provide the nerve supply to the muscles. The **pseudocoel** in which the organs lie is filled with fluid and contains fibers and giant cells (a nematode characteristic).

The **alimentary canal** of the nematode usually consists of a mouth (Fig. 11-6), a sucking pharynx, a long nonmuscular intestine lined with endodermal cells for absorption, a short rectum, and a terminal anus. An ascaris sucks up the partly digested, semifluid contents of its host's intestine. Its thick cuticle protects it from the host's digestive juices. Some plant parasites have an oral stylet with which to puncture plant cells. The pharynx, simple in the

ascaris, has various modifications in other types of nematodes. Most digestion is extracellular.

The **excretory system** of *Ascaris* consists of two lateral excretory canals, with a connecting transverse network, emptying through a pore just behind the mouth. There are no flame cells. Some nematodes have specialized renette cells for excretion (Fig. 11-6). Most nematodes excrete ammonia, but when water is scarce, *Ascaris* seems to excrete much of its nitrogen as urea. If this water-saving device is true of all nematodes, it may help explain their wide adaptability.

There are no special organs for circulation or respiration. Nematode parasites are predominately anaerobic. The Krebs cycle has been identified in some nematodes, but it is absent in *Ascaris.* They store glycogen from their food and then depend on the breakdown (glycolysis) of glycogen for their oxygen.

A **ring of nerve tissue** and ganglia around the pharynx gives rise to small nerves to the anterior end and to two nerve cords, one dorsal and one ventral. The chief sense organs are the **papillae** of the lips. Lateral chemoreceptors called **amphids** are characteristic of free-living nematodes (Fig. 11-6) but are reduced in *Ascaris.* Although cilia have been believed to be absent from nematodes, it is now known that the sensory processes in amphids are actually cilia.

Minute chemoreceptors called **phasmids** are found in many nematodes (Phasmidia).

Reproduction and life cycle

The **male reproductive system** consists of a long tubular testis, vas deferens, and seminal vesicle, an ejaculatory tube for discharging sperm, and penial spicules for attachment during copulation. The **female reproductive system** consists of ovaries, oviducts, and uterus and a vagina that empties through a genital pore.

A female ascaris may lay 200,000 eggs a day, which are eliminated in the host's feces. On the ground under suitable conditions, small worms may develop in the shells within 2 or 3 weeks. If ingested at this stage, the worms will mature in the host. When embryonated eggs are swallowed, they hatch into tiny larvae, which burrow through the intestinal wall into the veins or lymph vessels. In the blood they are carried through the heart to the lungs, move up the trachea, cross over into the esophagus, and then go down the alimentary canal to the intestine, where they grow to maturity in about 2 months. Here they copulate and the female begins her egg laying. Thus only one host is involved in the life cycle. They do their greatest damage to the host while the juvenile worms are migrating.

Other nematode parasites

Nearly all vertebrates and many invertebrates are parasitized by nematodes. Some of the more common nematode parasites of man are described below.

HOOKWORM. *Ancylostoma* and *Necator* are the common forms of hookworm that infest man. *Necator americanus* is the American form, although it was originally introduced from Africa. These worms are called hookworms because the male has a hook-shaped body; actually they have no hooks. The adults are 10 to 15 mm. long and have cutting plates in their mouths (Fig. 11-7) that cut holes in the intestinal mucosa of the host so that the worms can suck up blood and body fluids. They secrete an anticoagulant to prevent the blood from clotting while they are feeding. They often take in more blood than they can digest, and they leave a bleeding wound after feeding.

The life cycle is somewhat similar to that of the ascaris worm. Eggs are passed with the host's feces and hatch in about a day on warm, moist soil. They

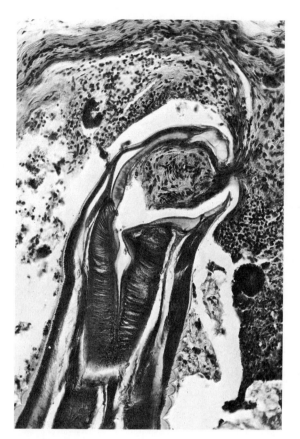

FIG. 11-7
Section through anterior end of hookworm attached to human intestine. Note cutting plates of mouth pinching off bit of mucosa from which thick muscular pharynx sucks blood. Mouth secretes anticoagulant to prevent blood clotting. (AFIP No. 33010.)

can live several weeks in the soil, feeding on bacteria and organic matter. If infested soil touches human skin, the larvae burrow through the skin into the blood. The bare foot is a common point of entry, and a mild irritation known as "ground itch" may occur there. Hookworms can also be acquired by swallowing the larvae. Hookworms can live for years in a host, and 25 to 50 worms are enough to cause anemia in the patient. Heavy infestations may result in retarded mental and physical growth and general loss of energy. Sanitary disposal of feces and the wearing of shoes are preventives. Hookworm disease can be treated with certain drugs.

TRICHINA WORM. *Trichinella spiralis* is the tiny nematode responsible for the serious disease **trichin-**

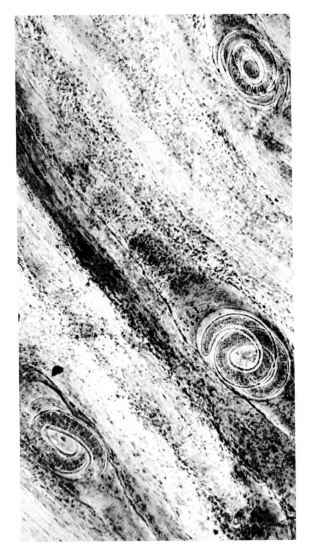

FIG. 11-8

Muscle infected with trichina worm, *Trichinella spiralis*. Larvae may live 10 to 20 years in these cysts. If eaten in poorly cooked meat, larvae are liberated in intestine. They quickly mature and release many larvae into blood of host.

osis. Adult worms burrow into the mucosa of the small intestine where the female produces living larvae. The larvae penetrate into blood vessels and are carried to the skeletal muscles where they coil up and form a cyst that becomes calcified (Fig. 11-8). The worms may live in the cysts for years if undisturbed. When meat containing live cysts is swallowed, the larvae are liberated into the intestine where they mature and produce living larvae.

Besides man trichina worms infest hogs, rats, cats, dogs, etc. Man acquires the parasite by eating improperly cooked pork. Hogs acquire them by eating garbage containing pork scraps with cysts, or by eating infested rats. Nearly 75% of all rats are said to be infested.

Heavy infestations, rare in the United States, cause **trichinosis,** a very serious disorder. Mild cases are fairly common. The simplest preventive measure is the thorough cooking of all pork. Cooking garbage before it is fed to hogs is required in many communities.

PINWORMS. The pinworm *Enterobius* is very common, especially in warm countries. In some communities 40% to 100% of the children are affected. The adults (Fig. 11-9, *A*) live in the large intestine and cecum. The larger females are about 12 mm. long. Females with eggs migrate to the anal region at night to lay their eggs (Fig. 11-9, *B*). Since this causes irritation, scratching at night is common, and fingers and bedding, and even the air, become contaminated, contributing to reinfestations. When ova are swallowed, they hatch in the duodenum and mature in the large intestine. No intermediate host is necessary. Each generation lasts 3 to 4 weeks. If reinfestation does not occur, they will die out. Careful sanitation and scrupulous cleanliness are the best preventives. Diagnosis can be made from perianal and perineal scrapings. An interesting fact is that incidence is higher among whites than blacks, and among children than adults. Pinworms in small numbers cause few symptoms, but heavy infestations are nearly always noted by intestinal disturbance and intense pruritus, as well as extreme nervousness and irritability.

WHIPWORMS. *Trichuris trichiura* are nematodes about an inch long with long whiplike anterior ends. They are parasitic in the intestines of children, largely in the tropics. Eggs passed in human feces embryonate in the soil and may be acquired by eating contaminated food or earth. Heavy infestation causes a wasting diarrhea or even anemia.

FILARIAL WORMS. Filarial worms *Wuchereria bancrofti* are tropical worms 2 to 4 inches long that live in the lymphatic glands and sometimes cause a condition called **elephantiasis.** The females discharge microscopic larvae, known as **microfilariae,** into the lymph and blood. The larvae develop no further unless sucked up by the right kind of mosquito.

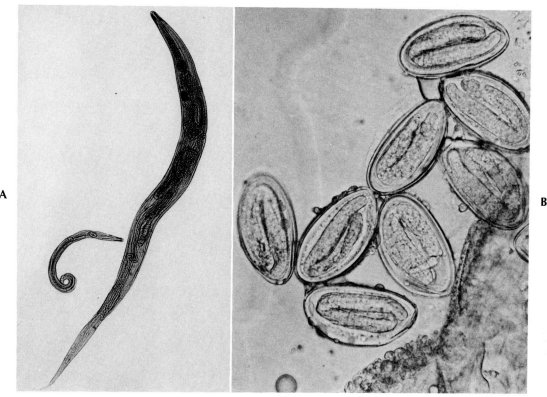

FIG. 11-9
Pinworms, *Enterobius vermicularis.* **A,** Adult pinworms; the male is much smaller. **B,**
Group of pinworm eggs, which are usually discharged at night around the anus of the
host, who, by scratching in his sleep, may get fingernails and clothing contaminated.
This may be the most common and widespread of all human helminth parasites.
(Courtesy Indiana University School of Medicine, Indianapolis.)

In the mosquito they metamorphose and then migrate to the proboscis, ready to enter human skin when the mosquito bites it. Returned to the human, the larvae migrate to a lymph gland where they coil up and mature. In most cases the filarial infection is not injurious to man, but repeated infestations cause obstruction in the flow of lymph, producing elephantiasis, a condition that involves excessive growth of connective tissue and enormous swelling of affected parts, such as the scrotum, legs, arms, or breast (Fig. 11-10). The last case reported in the United States was in South Carolina in 1912. Treatment is largely restricted to preventing new infestations of microfilariae in the blood by the use of the drug methylcarbamazine, which destroys the microfilariae.

GUINEA WORMS. *Dracunculus medinensis,* the guinea worm, is found largely in tropical areas. The male is small, but the female is from 2 to 4 feet long and lies just under the surface of the skin of its host, where it appears much like a varicose vein. The worm causes an ulcer in the skin through which it discharges living larvae upon contact with water. The larvae are picked up by the small aquatic crustacean *Cyclops,* where it undergoes further development (Fig. 11-11, *B*). Infections occur from drinking water containing the *Cyclops.* The worms can be removed surgically, but the time-honored method of removing the adult worm has been to wind it out on a small stick, a little each day (Fig. 11-11, *A*). It is interesting to note that the "fiery serpent" of Biblical times is believed to have been the guinea worm.

GIANT KIDNEY WORMS. *Dioctophyma renale* is a bright red nematode up to a meter long. It is parasitic in the abdominal cavity and kidneys of mammals. Eggs passed in the urine are swallowed by a

FIG. 11-10
Elephantiasis of leg caused by adult filarial worms of *Wuchereria bancrofti,* which live in lymph passages and block the flow of lymph. Larval microfilariae are transmitted by mosquitoes. (AFIP No. 44430-1.)

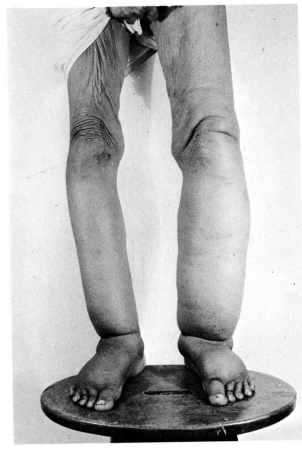

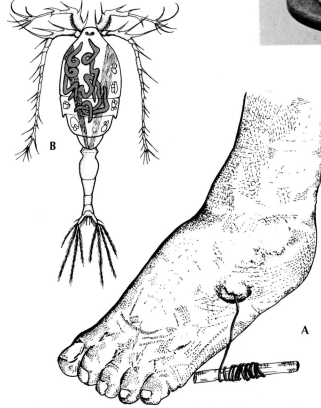

FIG. 11-11
A, Guinea worm, *Dracunculus,* being extracted from host by winding one turn of the stick each day to prevent rupturing worm and causing infection. A more modern method is surgery. **B,** *Cyclops,* a copepod, containing *Dracunculus* larvae in body cavity.

certain annelid worm (attached to a crayfish) where they hatch in 6 to 12 months and then encyst. The annelid must be eaten by a fish and the fish by a mammal to complete the cycle, which takes about 2 years in all. The adults digest away the kidney tissue of the mammalian host.

Nematology and human welfare

A student of nematodes once said that if the earth were to disappear, leaving only the nematode worms, the general contour of the earth's surface would be outlined by the worms. The widespread nature of nematode distribution and their varied roles in the economy of life have given an amazing uniqueness to this group. The fact that man and his domestic animals have always been plagued by intestinal roundworms has focused man's attention to methods of control that have been fairly successful. The group as a whole present many problems. Their taxonomy is only partially worked out and little is known of their phylogenetic relationships. Basic research has been restricted to recent years. Much needs to be learned about their ecologic relationships, about culturing them, about their position in the energy cycle, and about their basic adaptations to their habitats.

■ PHYLUM NEMATOMORPHA*

The popular name for the nematomorphs is "horsehair worms," based on an old superstition that the worms arise from horsehairs that happen to fall into the water. They were long included with the nematodes, with which they share the structure of the cuticle, presence of epidermal cords, longitudinal muscles only, and pattern of nervous system. However, the early larval form of some species has a striking resemblance to the Priapulida so that it is impossible to say to what group the nematomorphs are most closely related.

Ecologic relationships

The horsehair worms are free living as adults and parasitic in arthropods as juveniles. As a group they have worldwide distribution in every kind of aquatic habitat and may be found in both running and standing water. Adults do not feed but will live almost

---

*Nem"a-to-mor'fa (Gr. *nēma, nēmatos,* thread, + *morphē,* form).

anywhere in wet or moist surroundings if the oxygen is adequate. Juveniles cannot emerge from the arthropod host unless there is water nearby. Adults are often seen wriggling about in ponds or streams, with males being more active than females. The female discharges her eggs in water in long strings. Some juveniles, such as *Gordius,* a cosmopolitan genus, are believed to encyst on vegetation that may later serve as food for a grasshopper or other arthropod. The juveniles burrow into the arthropod hemocoel where they mature and reenter the water. In the marine form *Nectonema* juveniles occur in hermit crabs and other crabs.

Morphology

Horsehair worms are extremely long and slender, with a cylindric body. They may reach a length of 1.5 meters with a diameter of only 3 mm., but most are smaller with a diameter of not over 0.5 mm. The anterior ends are usually rounded, and the posterior ends are rounded or with two or three caudal lobes (Fig. 11-12). The males are usually shorter than the females, and the posterior ends are slightly curled as in nematodes. The body surface may bear small papillae *(areoles)* that give the body a rough appearance. The **body wall** is much like that of the nematodes—a fibrous cuticle, epidermis, and musculature of longitudinal muscles that may be arranged in bands by dorsal and ventral thickenings of the epidermis. A **pseudocoel** of some sort is found and may be filled with parenchyma.

The **digestive system** may be incomplete or degenerative. When present, there is an anterior mouth, pharynx, intestine, cloaca, and terminal anus. The larval forms absorb food from their arthropod hosts through the body wall. Even the adults take no food into the digestive tract but obtain their nutrition by absorption through the body surface.

Circulatory, respiratory, and excretory systems are lacking. There is a nerve ring around the pharynx and a midventral nerve cord. Sensory structures are limited to various sensory cells on the epidermis, and there are eyespots in some.

Each sex has a pair of gonads and a pair of gonoducts that empty into the cloaca. The female lays long strings of eggs that develop into larvae. These swim about looking for the proper aquatic arthropod host in which to complete their development. After several months in the body cavity of the host the

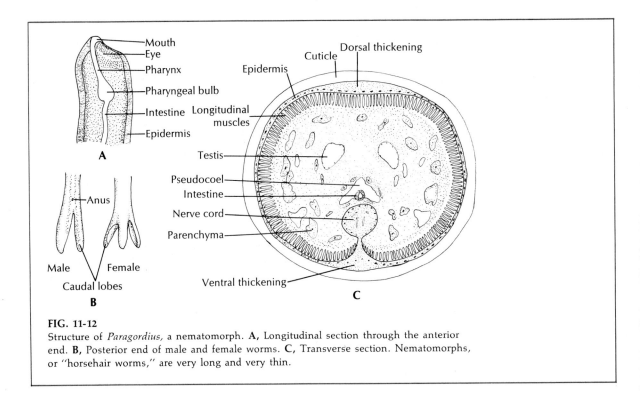

FIG. 11-12

Structure of *Paragordius,* a nematomorph. **A,** Longitudinal section through the anterior end. **B,** Posterior end of male and female worms. **C,** Transverse section. Nematomorphs, or "horsehair worms," are very long and very thin.

matured worm leaves its host and molts. It is now free living for the rest of its life.

■ PHYLUM ACANTHOCEPHALA*

The members of the phylum Acanthocephala are commonly known as "spiny-headed worms." The phylum derives its name from one of its most distinctive features, a cylindric invaginable proboscis bearing rows of recurved spines, by which it attaches itself to the intestine of its host. All acanthocephalans are endoparasitic, living as adults in the intestines of vertebrates.

The various species range in size from less than 2 mm. up to 650 mm. in length, with the females of a species being much larger than the males. Body shape also varies. In some forms it is long, slender, and cylindric; in others it may be laterally flattened or short and plump. The body surface may be smooth, but often it is wrinkled. Usually the body is capable of considerable extension and contraction because of the muscular arrangement. The color of the worms is often determined by the kind of food

*A-kan"tho-sef'a-la (Gr. *akantha,* spine or thorn, + *kephalē,* head).

they absorb from their hosts, ranging all the way from dirty brown to brighter colors.

Ecologic relationships

Acanthocephalans may be the most injurious of all wormlike parasites because of the damage produced by the proboscis hooks on the intestinal wall. About 500 species have been named, most of which parasitize fishes, birds, and mammals. They are worldwide in their distribution. As larvae the spiny-headed worms live in arthropods. Marine acanthocephalans pass their larval stages in amphipods, hermit crabs, and probably other crustaceans before maturing in fishes. The larvae of terrestrial forms live in insects, and the larvae of the worms that parasitize freshwater vertebrates usually live in crustaceans.

Phylogenetic relationships

Acanthocephalans possess several features that make it difficult to determine their relations to other animal groups. Their totally parasitic habits have no doubt been responsible for many of their distinctive features. There is no free-living stage, and there is no digestive system at any stage. They resemble the platyhelminths in their reproductive system and their method of development. They resemble the aschelminth phyla in having a pseudocoel and a syncytial epidermis and also a tendency toward cell or nuclear constancy. Van Cleve found in five different species of *Eorhynchus* the same number of nuclei in each of several organs. Increase in size of the worm is correlated with increase in cell size rather than in cell or nuclei number. Such constancy is considered by some to be a barrier to evolutionary development and progress.

Characteristics

1. Body cylindric; anterior end with **spiny retractile proboscis** and sheath
2. **Syncytial epidermis** covered with cuticle and containing **fluid-filled lacunae;** body wall with circular and longitudinal muscles
3. Fluid-filled pseudocoel
4. **No digestive,** respiratory, or circulatory systems
5. Excretory system (when present) with two branched ciliated protonephridia connected to a common excretory duct
6. Nervous system with a central ganglion on the proboscis sheath and nerves to proboscis and body; sensory papillae on proboscis and genital bursa
7. Separate sexes; larvae develop in body cavity of female; special **selector apparatus** in female system
8. **Parasitic** in intestine of vertebrates

Classification

The acanthocephalans have been classified in various ways by different authorities. Hyman divided the group into three orders, but no classes, basing the classification upon the arrangement of the proboscis spines and a few other characteristics.

Macracanthorhynchus hirudinaceus— intestinal spiny-headed worm of pigs

Macracanthorhynchus hirudinaceus occurs throughout the world in the small intestine of pigs and occasionally in other mammals. Its common occurrence and large size makes it suitable for study, and its life cycle has been known for a long time.

Morphology and physiology

The body of this parasite is cylindric and tapers posteriorly (Fig. 11-13, *C*). The female is 10 to 65

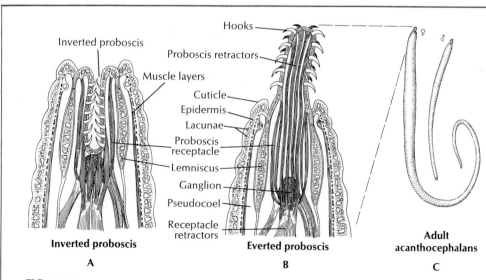

FIG. 11-13

Structure of a spiny-headed worm (phylum Acanthocephala). **A** and **B,** Eversible spiny proboscis by which the parasite attaches to the intestine of the host, often doing great damage. Lacking a digestive system, lacunae in the epidermis absorb and distribute food from host. **C,** Male is typically smaller than female.

cm. long and usually less than 1 cm. thick; the male is about one fourth the size of the female.

The body wall is made up of a thin cuticle, a syncytial epidermis, a thin dermis, and layers of circular and longitudinal muscles. The epidermis contains a **lacunar system** of ramifying fluid-filled canals (Fig. 11-13, *A* and *B*). This system, found only in acanthocephalans, absorbs and distributes food from the host. Attached to the neck region are two elongated **lemnisci** (extensions of the epidermis and lacunar system) that may act as reservoirs for the fluid of the neck region when the proboscis is invaginated.

The **proboscis,** bearing six rows of recurved hooks, is attached to the neck region (Fig. 11-13) and may be inverted into a **proboscis receptacle,** or sheath, by special muscles. When inverted, the proboscis spines point anteriorly; when everted, they point posteriorly. The proboscis is a forbidding organ and can cause serious damage to the intestine of its host.

A pair of **protonephridia,** much branched and with flame cells, unite to form a common tube that opens into the sperm duct or uterus. There is **no digestive system.** Acanthocephalans absorb their nourishment from the host through their thin, delicate cuticle. The actual digestion is done for them by the host. This power of absorption is strikingly shown when the worms are placed in water, for their bodies become swollen in a short time. After absorption into the body the food products are distributed by the fluid-filled lacunar canal system and the pseudocoel.

The **nervous system** consists of a ganglion on the proboscis (Fig. 11-13, *B*), a number of nerves to the proboscis and other parts of the body, and a few tactile sense organs on the proboscis and male bursa.

A pair of tubular **genital ligaments,** or ligament sacs, extend posteriorly from the end of the proboscis sheath. In the **male** are a pair of testes, each with a vas deferens, and a common ejaculatory duct that ends in a small penis. During copulation sperm are ejected into the vagina, travel up the genital duct and escape into the pseudocoel.

In the **female,** ligament-sac ovarian tissue breaks up into ovarian balls that float free in the ligament sacs. When the sacs rupture, the balls of ova are set free in the pseudocoel. One of the ligament sacs leads to a funnel-shaped **uterine bell** that receives the developing eggs and passes them on to the uterus (Fig. 11-14). An interesting and unique **selec-**

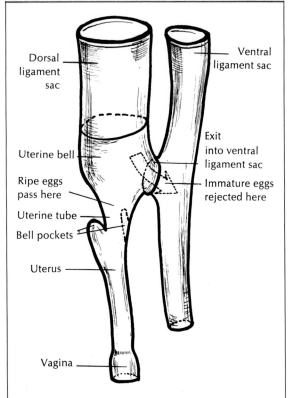

FIG. 11-14

Scheme of genital selective apparatus of female acanthocephalan. A unique device for separating immature from mature fertilized eggs. Eggs containing larvae enter uterine bell and pass on to uterus and exterior. Immature eggs are shunted into ventral ligament sac or into pseudocoel, to undergo further development.

tive apparatus operates here. Fertilized eggs escape from the ovarian balls and begin to develop in the pseudocoel or ligament sac. An opening in the uterine bell connected to the other ligament sac allows immature eggs or embryos to be sidetracked and returned to the pseudocoel or to one of the ligament sacs rather than passed on to the uterus. Only ripe eggs containing larvae are permitted to pass to the uterus, and on to the exterior.

The eggs, which are discharged in the feces of the vertebrate host, do not hatch until eaten by the intermediate host. For *Macracanthorhynchus* this is the larva, or grub, of the June beetle, *Phyllophaga.* Here the first larva (acanthor) burrows through the intes-

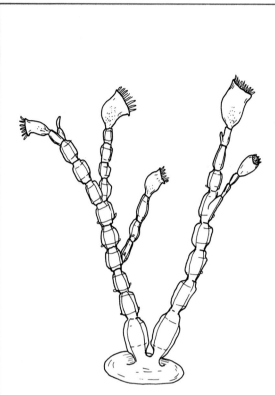

FIG. 11-15
Urnatella, the only freshwater entoproct, forms
small colonies of two or three stalks arising from a
basal plate. (Modified from C. Cori, 1936.
Kamptozoa. In H. G. Bronn (editor). Klassen und
Ordnungen des Tier-Reichs, vol. 4, part 2. Leipzig,
Akademische Verlagsgesellschaft.)

tine and encysts as a juvenile. Pigs become infested
by eating the grubs. Man rarely acquires this parasite. Acanthocephalans are very harmful, for great
damage is done by the spiny head. Multiple infestations may do considerable damage to the pig's intestine, and perforations may occur.

■ PHYLUM ENTOPROCTA*

The entoprocts are a small group of less than a
hundred species of tiny, sessile animals that, superficially, look much like hydroid coelenterates, but their
tentacles are ciliated and tend to roll inward. They
are the only pseudocoelomates that are not worm-
*En"to-prok'ta (Gr. *entos,* + *prōktos,* anus).

like. Most entoprocts are microscopic and none is
more than $1/4$ inch long. They are all stalked and sessile forms; some are colonial and some are solitary.
All are ciliary feeders.

Phylogenetic relationships

The entoprocts were once included with the phylum Ectoprocta in a phylum called Bryozoa, but the
ectoprocts are true coelomate animals and many
zoologists prefer to place them in a separate group.
Ectoprocts are still often referred to as bryozoans.
Probably the closest affinity to Entoprocta is the Ectoprocta, but there is little evidence of close relationship. The entoprocts may have arisen as an
early offshoot of the same line that led to the ectoprocts.

Ecologic relationships

With the exception of the genus *Urnatella,* all entoprocts are marine forms that have a wide distribution from the polar regions to the tropics. Most
marine species are restricted to coastal and brackish
waters and often grow on shells and algae. Some are
commensals on marine annelid worms. Freshwater
entoprocts occur on the underside of rocks in running water. *Urnatella gracilis* is the only common
freshwater species in North America (Fig. 11-15).

Morphology and physiology

The body, or **calyx,** of the entoproct is cup shaped,
bears a **crown,** or circle, of ciliated **tentacles,** and is
attached to a substratum by a **stalk** and an attachment disk with adhesive glands. Both tentacles and
stalk are continuations of the body wall. The 8 to
30 tentacles making up the crown are ciliated on
their lateral and inner surfaces, and each can move
individually. The tentacles can roll inward to cover
and protect the mouth and anus, but cannot be retracted into the calyx.

Movement is usually restricted in entoprocts, but
Loxosoma, which lives in the tubes of marine annelids, is said to be quite active, moving over the annelid and its tube freely.

The gut is U-shaped and ciliated, and both the
mouth and anus open within the circle of tentacles.
Entoprocts are **ciliary filter feeders.** Long cilia on the
sides of the tentacles keep a current of water containing protozoans, diatoms, and particles of detritus
moving in between the tentacles. Short cilia on the

inner surfaces of the tentacles capture the food and direct it downward toward the mouth.

The **body wall** consists of a cuticle, cellular epidermis, and longitudinal muscles. The **pseudocoel** is largely filled with a gelatinous parenchyma in which is embedded a pair of **protonephridia** and their ducts, which unite and empty near the mouth. There is a well-developed **nerve ganglion** on the ventral side of the stomach, and the body surface bears sensory bristles and pits. Circulatory and respiratory organs are absent. Exchange of gases occurs through the body surface, probably much of it through the tentacles.

Some species are monoecious, some dioecious, and some appear to be protandric—that is, the gonad at first produces sperm and later eggs. The gonoducts open within the circle of tentacles.

Fertilized eggs develop in a depression, or brood pouch, between the gonopore and the anus. Entoprocts have a modified spiral cleavage pattern with mosaic blastomeres. The coeloblastula is formed by emboly. The trochophore-like larva is ciliated and free swimming. It has an apical tuft of cilia at the anterior end and a ciliated girdle around the ventral margin of the body. Eventually the larva settles to the substratum and inverts to form the adult.

■ PHYLUM GNATHOSTOMULIDA*

The first gnathostomulid species was observed in 1928 in the Baltic but was not published until 1956. Since then they have been found in many parts of the world including our own Atlantic coast. Reidl (1969) reported 10 genera and 43 species, with the

---

*Gnathostomulida (nath"o-sto-myu'lid-a) (Gr. *gnathos,* jaw, + *stoma,* mouth, + L. *-ulus,* diminutive).

prediction that a great many more would be discovered.

Gnathostomulids are delicate wormlike animals, often a millimeter or less in length. They live in the interstitial spaces of very fine sandy coastal sediments and silt and can endure unusually stagnant conditions. They are often found in large numbers and frequently in association with gastrotrichs, nematodes, ciliates, tardigrades, and other small forms.

Lacking a coelom, a circulatory system, and an anus, the Gnathostomulida show some similarities to the Turbellaria and were at first included in that group. However their pharynx, armed with a pair of lateral jaws used to scrape fungi and bacteria off the substratum, is unlike the turbellarian pharynx, and although the epidermis is ciliated, each epidermal cell has but one cilium, a condition rarely found except in the sponges and coelenterates. There are no pseudocoelomate spaces. Their phylogeny is still obscure.

Gnathostomulids can glide, swim in loops and spirals, and bend the head from side to side. Sexual stages may include males, females, and hermaphrodites. Fertilization is internal.

Derivation and meaning of names

Ascaris (Gr. *askaris,* intestinal worm).

Dioctophyma (Gr. *diokto-* [irreg.] from *dionkoun,* to distend, + *phyma,* swelling). Refers to the papillae around the mouth, 6 to 18 in number.

Dracunculus (L. *draco,* dragon, + *-unculus,* small; from Gr. *drakontion,* guinea worm).

Enterobius (Gr. *enteron,* intestine, + *bios,* life).

Gordius (Greek mythologic king who tied an intricate knot).

Macracanthorhynchus (Gr. *makros,* long, + *akantha,* thorn, + *rhynchos,* beak).

Necator (L. *necator,* killer).

Oxyuris (Gr. *oxys*, sharp, + *oura*, tail).

Trichinella (Gr. *trichinos*, of hair, + *-ella*, diminutive).

References

Suggested general readings

Baer, J. G. 1952. Ecology of animal parasites. Urbana, University of Illinois Press. *Excellent account of the ways parasites have adapted themselves.*

Beauchamp, P. de. 1907. Morphologie et variations de l'appareil rotateur. Arch. Zool. Exp. Gen. **6:**1-29.

Behme, R., et al. 1972. Biology of nematodes. New York, MSS Information Corp.

Chitwood, B. G., and M. B. Chitwood. 1949. An introduction to nematology. Baltimore, Monumental Press. *An authoritative work on nematodes.*

Cori, C. 1936. Kamptozoa. In H. G. Bronn (editor). Klassen und Ordnungen des Tier-Reichs, vol. 4, part 2. Leipzig, Akademische Verlagsgesellschaft.

Donner, J. 1966. Rotifers, New York, Frederick Warne & Co., Inc.

Dougherty, E. C. (editor). 1963. The lower Metazoa. Berkeley, University of California Press. *This is a comparative phylogeny, morphology, and physiology of the invertebrates below the annelids. More than 30 specialists contributed monographs to this important work.*

Higgins, R. P. 1965. The homalorhagid Kinorhyncha of northeastern U. S. coastal waters. Trans. Amer. Microsc. Soc. **84:**65-72.

Hyman, L. H. 1951. The invertebrates: Acanthocephala, Aschelminthes, and Entoprocta, vol. 3. New York, McGraw-Hill Book Co. *Miss Hyman has described with great accuracy the many types of these phyla.*

Lee, D. L. 1965. The physiology of nematodes. San Francisco, W. H. Freeman & Co., Publishers.

Levine, N. D., et al. 1965. Recent advances in parasitology. Amer. Zool. **5:**73-163 (Feb.). *A symposium used as a refresher course in parasitology with special reference to useful techniques.*

Lapage, G. 1951. Parasitic animals. New York, Cambridge University Press. *Good descriptions of the life histories of many nematodes, including Ascaris and Trichina.*

Mariscal, R. N. 1965. The adult and larval morphology and life history of the entoproct *Barentsia gracilis* (M. Sars, 1835). J. Morph. **116:**311-388.

Meyer, A. 1933. Acanthocephala. In H. G. Bronn (editor): Klassen und Ordnungen des Tier-Reichs, vol. 4, part 2, sec. 2. Leipzig, Akademische Verlagsgesellschaft. *The definitive German account of acanthocephalans.*

Nickerson, W. 1901. On *Loxosoma davenporti*. J. Morph. **17:**357-376.

Pennak, R. W. 1953. Fresh-water invertebrates of the United States. New York, The Ronald Press Co. *This excellent work deals with the free-living, freshwater invertebrates and omits the parasitic forms. Many of the sections describe various types of Aschelminthes under the old classification.*

Remane, A. 1933. Rotatoria. In H. G. Bronn (editor): Klassen und Ordnungen des Tier-Reichs. **4:**1-4. Leipzig, Akademische Verlagsgesellschaft. *A good general account of the rotifers.*

Riedl, R. J. 1969. Gnathostomulida from America. Science **163:**445-452.

Roggen, D. R., D. J. Rask, and N. O. Jones. 1966. Cilia in nematode sensory organs. Science **152:**515-516.

Rogick, M. D. 1948. Studies on marine Bryozoa. Part II. *Barentsia laxa.* Biol. Bull. **94:**128-142. *The author considers the Entoprocta as a class under the Bryozoa.*

Sacks, M. 1964. Life history of an aquatic gastrotrich. Trans. Amer. Microsc. Soc. **83:**358-362.

Sasser, J. N., and W. R. Jenkins (editors). 1960. Nematology. Chapel Hill, University of North Carolina Press. *An important work by more than a score of eminent authorities in this difficult field.*

Van Cleve, H. J. 1941. Relationships of Acanthocephala. American Naturalist **75:**1-20. *In this and many other articles the most active American investigator of this group has attempted to appraise the position of the Acanthocephala in the animal kingdom. Read also his revealing article, "Expanding Horizons in the Recognition of a Phylum," Journal of Parasitology* **34:**1-20, 1948.

Selected *Scientific American* articles

Crowe, J. H., and A. F. Cooper, Jr. 1971. Cryptobiosis. **225:**30-36 (Dec.).

Edwards, C. A. 1969. Soil pollutants and soil animals. **220:**88-99 (April).

Hawking, F. 1958. Filariasis. **199:**94-100 (July).

Not all mollusks have shells as do the familiar snails and oysters. Nudibranchs, the marine sea slugs, are shell-less gastropods. The frilled sea slug, Tridachia, *is a dainty little green and white nudibranch familiar in Florida waters. (Courtesy R. C. Hermes, Homestead, Fla.)*

CHAPTER 12

THE MOLLUSKS

PHYLUM MOLLUSCA*
A MAJOR EUCOELOMATE PROTOSTOME GROUP

◼ PLACE IN ANIMAL KINGDOM

1. The mollusks are one of the major groups of true **coelomate** animals.

2. They belong to the **protostome** branch, or schizocoelomous coelomates, and have spiral and determinate cleavage.

3. **All the organ systems** are present and well developed.

4. Many mollusks have a **trochophore larva** similar to the trochophore larva of marine annelids, marine turbellareans, and some others.

◼ BIOLOGIC CONTRIBUTIONS

1. In mollusks gaseous exchange occurs not only through the body surface as in lower invertebrates, but also in specialized **respiratory organs** in the form of gills or lungs.

2. They have an open **circulatory system** with pumping **heart,** vessels, and blood sinuses.

3. The efficiency of their respiratory and circulatory systems has made greater body size possible. Invertebrates reach their largest size in some of the mollusks.

4. They introduce for the first time a fleshy **mantle** that, in most cases, secretes a shell.

5. Several features unique to the phylum are found, for example, the **radula** and the muscular **foot.**

6. The highly developed direct **eye** of higher mollusks is similar to the indirect eye of the vertebrates, but arises as a skin derivative in contrast to the brain eye of vertebrates (an example of convergent evolution).

*Mol-lus'ka (L. *molluscus,* soft).

■ PHYLOGENY AND ADAPTIVE RADIATION

PHYLOGENY. The trochophore larva of many marine mollusks resembles the trochophore larva of marine annelids. Both phyla share the same type of egg cleavage. This may indicate a relationship between the two groups. This view is supported by the discovery of living monoplacophorans that show segmentation (although their segmentation may be secondary rather than primitive as in the annelids).

Some mollusks have a ladder type of nervous system similar to that of turbellarian flatworms. Both the mollusks and the annelids may have come from a common platyhelminth origin.

Establishing relationship among the various classes of mollusks is difficult, although a generalized type of ancestral form has been hypothesized as having a muscular ventral foot, a mantle cavity formed from the soft dorsal integument, and a simple visceral mass, that might have diverged into the various mollusk forms.

ADAPTIVE RADIATION. The primitive molluskan plan was that of a ventral head-foot complex with bilateral symmetry and an anteroposterior axis. Later as the body became more bulky a mantle and visceral mass took shape above the foot, adding a dorsoventral axis with radial or biradial symmetry. Adaptive radiation has brought modifications of this basic plan, particularly in the gastropods with their asymmetric growth and torsion, and in cephalopods which have lengthened and in which the foot has moved forward and become transformed into prehensile tentacles around the mouth.

The versatile glandular **mantle** has probably shown more creative adaptive capacity than any other molluskan structure. It secretes the shell, when present; it encloses the mantle cavity into which the digestive system, nephridia, and gonads open; it is modified into gills, lungs, siphons, and apertures; and it functions in locomotion and even in the feeding processes. The **radula**, a tooth-bearing band in the mouth, is found in most mollusks (except bivalves) and is specialized for a variety of uses. The **shell**, too, has undergone endless evolutionary adaptations.

■ THE MOLLUSKS

Next to the arthropods the mollusks have the most named species in the animal kingdom—probably more than 100,000 living species. The name Mollusca, from the Latin *molluscus,* meaning soft, indicates one of their distinctive characteristics, a soft body. This very diverse group includes the chitons, tooth shells, snails, slugs, nudibranchs, sea butterflies, bivalves, squids, octopuses, and nautiluses. The group ranges from fairly simple organisms to some of the most complex invertebrates, and in size from almost microscopic to the giant squid that grows up to 50 feet long and is the largest of all the invertebrates.

The phylum is very old, with a continuous record since Cambrian time. Perhaps no group has left more or better fossils than mollusks because their shells facilitated fossil formation. It is somewhat difficult to establish relationship to other phyla. The trochophore larvae of many marine mollusks, as well as their type of egg cleavage, indicate an affinity with the annelids, and the ladderlike nervous system of some mollusks resembles that of the turbellarians. It is conceivable that a flatworm type of ancestor gave rise to both protostome groups—the segmentally arranged annelids and the nonsegmented mollusks. The only metamerism found in the segmented mollusk is in the adult *Neopilina,* the monoplacophoran discovered in 1952.

Zoologists have tried to reconstruct a hypothetical primitive mollusk with features that might be common to the ancestor of the various molluskan classes (Fig. 12-1). Such a generalized ancestor might have a body dorsally arched, calcareous shell, muscular foot on the ventral side, mantle cavity formed from the soft dorsal integument, simple complete digestive tube with a rasping radula, a pair of gills near the anus, and a pair of ganglia in the head, with two nerve trunks at different levels. Evolutionary diversity has resulted in modifications of the basic pattern. Amphineura (the chitons) is considered the most primitive class, although its fossil record dates from a period more recent than that of the other classes. Scaphopods and gastropods both have a univalve shell and a radula. The pelecypods (bivalves), with their pointed foot and lack of a head, eyes, or tentacles, have some resemblance to the scaphopods. The recent discovery of a bivalve gastropod *Tamanovalva limax,* off the coast of Japan may provide a "missing link" between the gastropods and the pelecypods, for this species has a slug-shaped head and creeping foot, but its shell is bivalved. Its early development is like that of the gastropods, but it soon changes into the bivalve form. The highly advanced

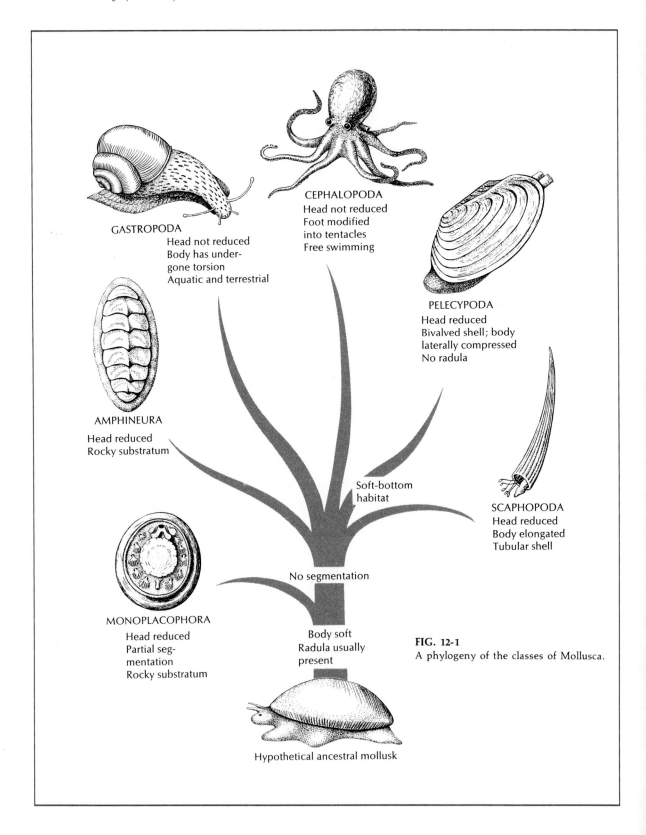

GASTROPODA
Head not reduced
Body has under-
gone torsion
Aquatic and terrestrial

CEPHALOPODA
Head not reduced
Foot modified
into tentacles
Free swimming

PELECYPODA
Head reduced
Bivalved shell; body
laterally compressed
No radula

AMPHINEURA
Head reduced
Rocky substratum

SCAPHOPODA
Head reduced
Body elongated
Tubular shell

Soft-bottom
habitat

No segmentation

MONOPLACOPHORA
Head reduced
Partial seg-
mentation
Rocky substratum

Body soft
Radula usually
present

Hypothetical ancestral mollusk

FIG. 12-1
A phylogeny of the classes of Mollusca.

cephalopods (squids and octopuses) seem isolated from the other classes but share some characteristics with the gastropods.

Among the distinctive structural features of the mollusks are the ventral muscular **foot,** the fleshy **mantle,** the **shell,** and a peculiar rasping organ called the **radula.**

THE FOOT. The foot is a highly muscular organ that often operates in conjunction with the hydrostatic pressure of the blood. In chitons and limpets the muscular foot acts as a powerful sucker bound to a substratum by mucus that enables the animal to withstand strong wave action, as well as move about to graze. Many small mollusks glide by means of cilia on the foot. In larger mollusks the rhythmic waves of muscular action in the foot combine with hydrostatic pressure to draw the animal forward. Burrowing forms can extend the foot into the mud by muscular action and then anchor it by enlargement with blood pressure while the body is being drawn forward. In pelagic forms the foot is modified

into parapodia or into thin mobile fins for swimming.

THE MANTLE. The mantle is a sheath of skin that surrounds the soft parts and hangs down as a free fold around the body, enclosing the mantle cavity space. The mantle secretes the shell and forms the siphons. The gills and lungs develop from the mantle, and the exposed surface of the mantle itself is often used for gaseous exchange. In cephalopods the muscular mantle is used for locomotion and for maintaining a flow of water over the gills.

THE SHELL. The molluskan shell, when present, is always secreted by the mantle and, whatever its shape, it is always underlaid by the mantle. The shell consists of three typical layers: (1) an outer horny **periostracum** of conchiolin, an organic substance, (2) a middle **prismatic layer** of crystalline calcium carbonate enmeshed in a protein network, and (3) an inner **nacreous layer** of iridescent nacre, or mother-of-pearl, which is formed of many thin layers of calcium carbonate (Fig. 12-2). The prismatic layer,

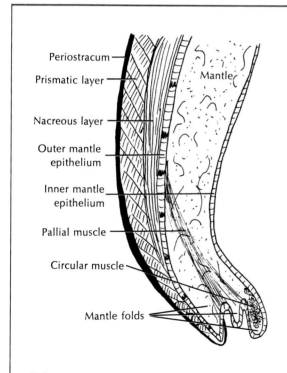

FIG. 12-2
Diagrammatic vertical section of shell and mantle of a bivalve.

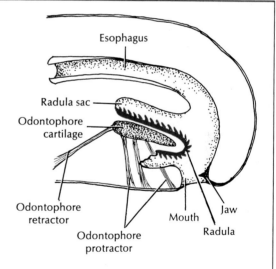

FIG. 12-3
Diagrammatic longitudinal section of gastropod head showing the radula and radula sac. The radula moves back and forth over the odontophore cartilage. As the animal grazes, the mouth opens, the odontophore is thrust forward, the radula gives a strong scrape backward bringing food into the pharynx, and the mouth closes. The sequence is repeated rhythmically. As the radula ribbon wears out anteriorly it is continually replaced posteriorly.

which makes up the bulk of the shell, is secreted by an area of the outer mantle epithelium near the mantle edge. The nacre is secreted by cells scattered over the entire outer surface of the mantle, and the periostracum is secreted in a groove along the edge of the mantle.

THE RADULA. The radula is a rasping, tongue-like organ found in all mollusks except the bivalves. The radula is a chitinous ribbon that bears many rows of fine teeth and is moved back and forth over a cartilaginous tongue (odontophore) by muscular action (Fig. 12-3). As the radula moves rhythmically back and forth, the whole odontophore apparatus can be protruded forward through the mouth. The function of the radula is twofold: to rasp off fine particles of food material and to serve as a conveyer belt for carrying the particles in a continuous stream toward the digestive tract. As the radula wears away anteriorly, it is continuously replaced by secretion at its posterior end.

Ecologic relationships

Mollusks are found in nearly all places that will support life, in all latitudes, and at altitudes up to 20,000 feet. Most are marine, living in shallow water along the seashore, in the open sea, or at great depths. But some are found in freshwater, and some have even been successful on land. The fossil evidence indicates that mollusks originated in the sea and much of their evolution probably took place along the shore because of the abundance of food and variety of habitats there. It is especially suitable for creeping and burrowing forms. The more mobile cephalopods are naturally pelagic (live in open water), but some live close to the bottom or near shore.

Only the gastropods and pelecypods left the sea and invaded fresh water and land. The filter-feeding habits of pelecypods restrict them to water. Many species are found in the estuaries of large rivers and are intermediate between the marine and freshwater forms. Such mollusks tolerate a salinity between that of the sea and fresh water. Burrowing mollusks had a better chance to invade and tolerate the conditions of estuaries. In time estuarine forms became tolerant of fresh water and some of them invaded the rivers.

Only gastropods have actually invaded land. According to fossil records pulmonate snails first invaded terrestrial habitats in the Carboniferous period. The vascular lining of the mantle cavity made possi-

ble the development of a lung. Terrestrial snails are usually restricted in their range by humidity, by shelter from heat and light, and by the presence of lime in the soil. Many pulmonates have evolved an operculum for closing the apertures of their shells during hibernation or estivation and can remain inactive for a long time.

It is in the tropics and subtropics that the most species are found, as well as the most strikingly beautiful color adaptations. Pelagic forms are usually colorless, but nudibranchs often adopt the color of forms on which they feed (green algae, anemones, hydroids).

Life cycles are generally short, although some mollusks live for several years. Nudibranchs probably live only a year; oysters may live up to 10 years; *Unio,* 12 years; freshwater snails, 4 to 5 years; squid *(Loligo)* up to 2 years; and giant squid 10 years (Spector).

Economic importance

A group as large as the mollusks would naturally affect man in some way. As food, oysters, clams, and snails are widely used. Pearl buttons are obtained from shells of bivalves. More than 40 species of mollusks are suitable for this purpose. The Missouri and Mississippi river basins furnish material for most of this industry in the United States. In some regions supplies are becoming so depleted that attempts are being made to propagate them artificially. Pearls, both natural and cultured, are produced in the shells of clams and oysters, most of them in a marine oyster, *Meleagrina,* found around eastern Asia.

Shells have been used as ornaments and also for utensils by primitive people in all ages. Shells of cowries (brightly colored snails of the genus *Cypraea*) were used as money in parts of the Far East, Africa, and the South Pacific. Purple snails *(Murex)* were used by the ancients as their chief source of dye.

Some mollusks are destructive. The burrowing shipworm *Teredo* (Fig. 12-20) does great damage to wooden ships and wharves. To prevent the ravages of shipworms, wharves must either be creosoted or built of cement. Snails and slugs are known to damage garden and other vegetation. In addition, snails often serve as intermediate hosts for serious parasites. The boring snail *Urosalpinx* (Fig. 12-7) is second only to the sea star in destroying oysters.

Characteristics

1. Body unsegmented (except in Monoplacophora)
2. Usually a definite **head** with mouth and sensory organs
3. The ventral body wall is specialized as a muscular **foot,** variously modified but used chiefly for locomotion
4. Dorsal body wall forms a pair of folds called the mantle, or **pallium,** which encloses the mantle cavity, is variously modified into gills or lungs, and secretes the **shell**
5. Body surface usually covered with ciliated epithelium bearing many mucous glands and sensory nerve endings
6. **Coelom** reduced and represented mainly by the pericardium, gonadal cavity, and kidney
7. Digestive system complete with digestive glands and liver; with a rasping organ **(radula)** usually present
8. Circulatory system of heart, pericardial space and blood vessels; blood mostly colorless, with erythrocruorin restricted and hemocyanin more common
9. Respiration by gills, by lungs, or direct
10. A pair of **metanephridia** (sometimes only one) close to the pericardial cavity and usually opening into the mantle cavity
11. Nervous system of four pairs of ganglia associated with nerve cords and subepidermal plexus; ganglia centralized in nerve ring in gastropods and cephalopods
12. Sensory organs of touch, smell, taste, and vision (in some), and statocysts for equilibrium
13. Cleavage is spiral and determinate, and the characteristic larva is the **trochophore** (veliger)

Classes of Mollusca

The classes of mollusks are based on such features as type of shell, type of foot, and shape of shell.

Class Monoplacophora (mon"o-pla-kof'o-ra) (Gr. *monos,* one, + *plax,* plate, + *phora,* pl. of bearing)—**segmented mollusks** (Fig. 12-1). Body bilaterally symmetric with a broad flat foot; a single limpetlike shell; mantle cavity with five or six pairs of gills; large coelomic cavities; radula present; internal segmentation only; six pairs of nephridia, two of which are gonoducts; separate sexes. Example: *Neopilina.*

Class Amphineura (am"fi-neu'ra) (Gr. *amphi,* on both sides, + *neuron,* nerve)—**chitons.** Elongated body with reduced head; bilaterally symmetric; radula present; row of eight dorsal plates usually; large, flat ventral foot, absent in some; nervous system a ring around the mouth with two pairs of ventral nerve cords; sexes usually separate, with a trochophore larva. Example: *Mopalia, Chaetopleura* (Fig. 12-4).

Class Scaphopoda (ska-fop'o-da) (Gr. *skaphē,* trough, boat, + *pous, podos,* foot)—**elephant tusk shells.** Body enclosed in a one-piece tubular shell open at both ends; conical foot; mouth with tentacles; head absent; mantle for respiration; sexes separate; trochophore larva. Example: *Dentalium* (Fig. 12-6).

Class Gastropoda (gas-trop'o-da) (Gr. *gastēr,* belly, + *pous, podos,* foot)—**snails and others.** Body usually asymmetric in a coiled shell (shell absent in some); head well developed, with radula; foot large and flat; mantle modified into a lung or gill; nervous system with cerebral, pleural, pedal, and visceral ganglia; dioecious or monoecious, with or without pelagic larva. Examples: *Littorina, Physa, Helix, Aplysia.*

Class Pelecypoda (pel-e-sip'o-da) (Gr. *pelekus,* hatchet, + *pous, podos,* foot)—**bivalves.** Body enclosed in a two-lobed mantle; shell of two lateral valves of variable size and form, with dorsal hinge; no head, but mouth with labial palps; no eyes (except a few) or radula; foot usually wedge shaped; gills platelike; sexes usually separate, with trochophore or glochidal larva. Examples: *Anodonta, Teredo, Venus.*

Class Cephalopoda (sef"a-lop'o-da) (Gr. *kephalē,* head, + *pous, podos,* foot)—**squids and octopuses.** Body with a shell, often reduced or absent; head well developed with eyes and a radula; foot modified into arms or tentacles; siphon present; nervous system of well-developed ganglia, centralized to form a brain; sexes separate, with direct development. Examples: *Loligo, Octopus, Sepia.*

Class Monoplacophora

The Monoplacophora (Fig. 12-1) is a relatively new class. In 1952 several specimens of a strange new mollusk were found off the west coast of Mexico at a depth of 2 miles. These new mollusks, *Neopilina galatheae,* have many molluskan characteristics and a superficial resemblance to the chitons and limpets but, unlike other mollusks, they are internally segmented. The low rounded shell is limpet shaped, and there is a broad flat foot. The five pairs of gills (six pairs in a more recent species *N. ewingi*) in the mantle cavity have a somewhat different fine structure from the typical mollusk gill. There is a complete digestive system with a radula and a crystalline style. Internally there are five pairs of nephridia and five pairs of gill hearts. The coelom is large and its paired dorsal compartments repeat the segmental pattern of the visceral organs. The embryonic development shows evidence of a relationship between the mollusks and the annelids. The segmental arrangement of this group brings up the question as to whether primitive mollusks were segmented—a question that has not as yet been answered.

Class Amphineura

The members of class Amphineura, made up of chitons and solenogasters, are strictly marine. Their name, meaning "nerves on both sides," refers to the two pairs of longitudinal nerve cords, pedal and pallial, that are connected to a ring around the mouth region. There are two groups, or orders, of amphi-

neurans—Polyplacophora, the chitons, and Aplacophora, the solenogasters.

The chitons (Fig. 12-4) are the most numerous of the Amphineura. They are somewhat flattened and have a convex dorsal surface that bears eight articulating limy **plates** that overlap posteriorly (Figs. 12-4 and 12-5, *A*). In early fossil forms the plates did not overlap. Most chitons are only 1 or 2 inches long, and the largest *(Cryptochiton)* rarely exceeds 8 to 10 inches.

A marginal **girdle** formed from the mantle sur-

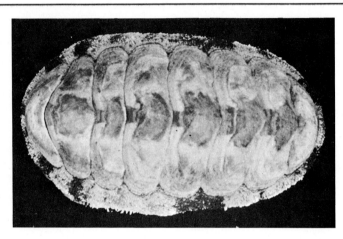

FIG. 12-4
Dorsal view of chiton. Note overlapping plat
and the surrounding marginal girdle.

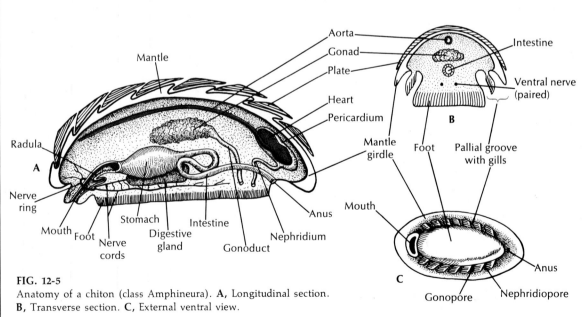

FIG. 12-5
Anatomy of a chiton (class Amphineura). **A,** Longitudinal section. **B,** Transverse section. **C,** External ventral view.

rounds or covers the plates. It may contain calcareous spines and scales that give it a shaggy appearance (Fig. 12-4). The **mantle,** which secretes the plates, covers the dorsal and lateral surfaces; a broad muscular **foot** very much like that of a snail covers most of the ventral surface (Fig. 12-5, *B* and *C*). The **pallial groove** between the foot and mantle surrounds the animal and contains two to 40 pairs of gills. The **head** is on the underside, separated from the foot by a narrow groove; it bears the mouth but no eyes or tentacles. The anus is found posteriorly in the pallial groove.

In general, chitons resemble the hypothetical ancestral mollusk described earlier. There are some differences, such as the gill arrangement, for the chiton usually has many gills and the ancestral prototype has only a single pair.

Chitons are sluggish and unless disturbed prefer to stay in one spot except for short foraging expeditions. They live upon seaweed or other plant life along the sea floor or shore. Their broad feet adhere to rocky surfaces along protected shores. A favorite depression in a rock that is exposed at low tide may be used for countless generations by chitons. Some attach to the underside of rocks. Chitons are mostly intertidal invertebrates, although some have been taken at great depths. Molested, they may roll up like pill bugs, because the plate joints are flexible. In the West Indies the natives cook them ("sea beef"). *Nuttallina* and *Cryptochiton* are familiar along the West coast; the tiny *Chaetopleura* is one of the common East coast forms.

The internal structure (Fig. 12-5, *A*) includes an alimentary canal with **radula;** an open circulatory system with a three-chambered heart surrounded by a pericardium in the posterodorsal region of the body; an excretory system of two long folded kidneys (metanephridia) that carry waste from the pericardial cavity to the exterior; a nervous system of two pairs of longitudinal nerve cords connected in the buccal region by a cerebral commissure; a sensory system of shell eyes (in some) on the surface of the shell, a pair of osphradia (sense organs for sampling water) near the anus, and scattered small sense organs; and a reproductive system of a single gonad (either male or female) with paired gonoducts to the outside. Females lay their eggs in masses or strings of jelly; fertilization is external.

The Aplacophora, or Solenogastres, are wormlike forms with no foot and no shell, except for minute spicules in the mantle. The body is completely enclosed in the mantle, the edges of which meet in a groove under the body. Some have a radula. The vascular and nervous systems resemble those of the chiton. Gills are restricted to a posterior cavity. Solenogastres are hermaphroditic; sex cells are discharged into the pericardium and carried out by the nephridia. Some authorities believe that the solenogasters are closer to the worms than to the mollusks. Others believe that they represent amphineurans whose development has been arrested (neoteny). They are usually found in fairly deep water and live on hydroids and corals. *Neomenia* and *Chaetoderma* are common species along the Atlantic coast.

Class Scaphopoda

The mollusks of class Scaphopoda are commonly called the tooth shells, or elephant tusk shells, because of their resemblance to those structures (Fig. 12-1). They have a slender, elongated body covered with the mantle, which secretes a tubular shell open at both ends. A burrowing foot protrudes through the larger end of the shell (Fig. 12-6). The mouth, which is near the foot, is provided with a radula and with contractile tentacles that are sensory and prehensile. Respiration takes place through the mantle, and the circulatory system consists merely of sinuses that are distributed among the different organs. Two saclike kidneys open near the anus. The sexes are separate, and the larva is a trochophore. After developing a mantle, the larva sinks to the bottom.

Members of this class are marine and live embedded in sand in shallow water, or sometimes at great depths, and they feed upon animals and plants of microscopic size (Fig. 12-6, *A*). *Dentalium* is a familiar example along our eastern seashore. Few living Scaphopoda are more than 3 inches long, but some fossil forms reached a length of 2 feet.

Class Gastropoda

Among the mollusks the class Gastropoda is by far the largest and most successful. It is made up of members of such diversity that there is no single general term in our language that can apply to them as a group. They include snails, limpets, slugs, whelks, conchs, periwinkles, sea slugs, sea hares, sea butterflies, etc. These animals are basically bilaterally symmetric, but by torsion the visceral mass

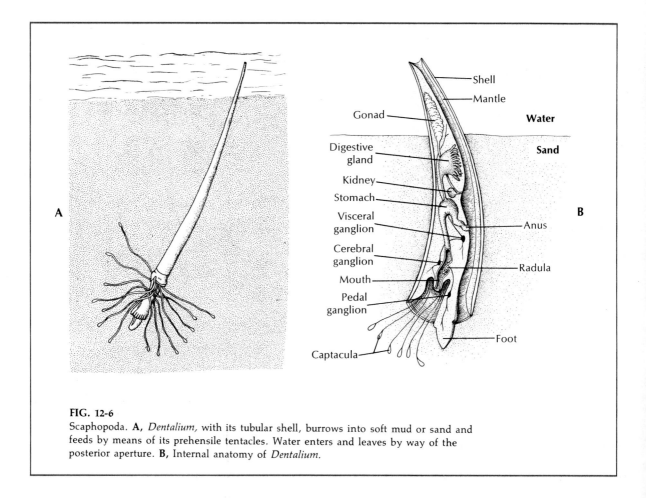

FIG. 12-6
Scaphopoda. **A,** *Dentalium,* with its tubular shell, burrows into soft mud or sand and feeds by means of its prehensile tentacles. Water enters and leaves by way of the posterior aperture. **B,** Internal anatomy of *Dentalium.*

has become asymmetric. The shell, when present, is always of one piece (univalve) and may be coiled or uncoiled.

Their range of habitat is large. In the sea gastropods are common both in the littoral zones and at great depths, and some are even pelagic (free swimming). Some are adapted to brackish water and some to freshwater. On land they are restricted by such factors as the mineral content of the soil and extremes of temperature, dryness, and acidity. Some have been found at great altitudes and even in polar regions. Snails have all kinds of habitats—small pools or large bodies of water, woodlands, pastures, under rocks, in mosses, on cliffs, in trees and underground.

Gastropods range from microscopic forms to giant marine snails that exceed 2 feet. Most of them are $1/2$ inch to 3 inches. Some fossil gastropods were 5

or 6 feet long. Their life-span is not well known, but some live from 5 to 15 years.

Gastropods are usually sluggish, sedentary animals because most of them have heavy shells and slow locomotor organs. Some are specialized for climbing, swimming, or burrowing. Shells are their chief defense, although they are also protected by coloration and by secretive habits. Some lack shells altogether. Some are distasteful to other animals, and a few such as *Strombus* can deal an active blow with the foot, which bears a sharp operculum. However, they are eaten by birds, beetles, small mammals, fish, and other predators. Serving as intermediate hosts for many kinds of parasites, snails are often harmed by the larval stages of the parasites.

Most gastropods live on plants or plant debris. Many marine forms live on seaweed and algae that they scrape from the rocks. A few live on detritus

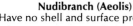

Oyster borer (Urosalpinx)
Can bore into oysters and suck
out their juices

Razor-shell clam (Ensis)
Rapid sand borer; can propel
itself long distances quickly

Rock borer (Pholas)
Can burrow into hard rock
with their abrading
teeth or ridges

Nudibranch (Aeolis)
Have no shell and surface projections (cerata)
often contain sting cells (nematocysts) which
slugs salvage from hydroids they eat

Scallop (Pecten)
Double edge of mantle contains row of steely
blue eyes; scallops move by clapping their valves

Long-neck clam (Mya)
United siphons highly retractile

FIG. 12-7
Some mollusks with unusual habits or structures. Nudibranchs and the oyster borer
belong to class Gastropoda; the others belong to class Pelecypoda.

strained from the water. Many are scavengers, living on dead and decayed flesh; others are carnivorous, preying upon clams, oysters, and worms. Some carnivores are provided with an extensible proboscis for drilling holes in other mollusks to eat the soft parts (Fig. 12-7, *Urosalpinx*).

To obtain and rasp food, snails have a **radula,** which is a long ribbon bearing a series of transverse rows of tiny chitinous teeth (Fig. 12-3). By means of muscles the radula moves rapidly back and forth like a handsaw over a cartilage in the upper part of the pharynx, thus tearing up the food. The arrangement, shape, and number of teeth are useful in taxonomic determinations.

Some of the sea butterflies secrete a free-floating mucus web, much like a butterfly net, about 2 meters in diameter in which small planktonic forms are caught. The webs with their collections are then drawn along lateral grooves in the proboscis into the mouth.

TORSION. Of all the mollusks, only gastropods undergo torsion. Torsion is a peculiar phenomenon that moves the mantle cavity to the front of the body and then twists the visceral organs and mantle cavity in a 180° rotation to produce the typical gastropod asymmetry. This occurs very early in the young embryo (veliger stage) and in some species the entire action may take only a a few moments. Before torsion occurs, the embryo is bilaterally symmetric with an anterior mouth and a posterior anus and mantle cavity (Fig. 12-8, *A*). Torsion is brought about by an uneven growth of the right and left muscles that attach the shell to the head-foot.

During the first stage of torsion there is a ventral flexure, bringing the anal region downward (Fig. 12-8, *B*) and then forward (Fig. 12-8, *C*), so that the

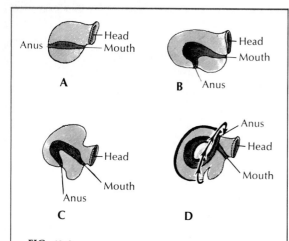

FIG. 12-8
Body torsion in snail. Early larva is symmetric, with mouth and anus at opposite ends of body, **A.** Most snails, however, undergo asymmetric growth in which anus shifts downward, **B,** then forward, and comes to lie near mouth, **C.** Later, anus and other parts rotate upward and lie above head, **D.** This rotation of visceral mass counterclockwise through angle of 180 degrees is caused by more rapid growth of left side of visceral mass.

anus opens anteriorly below (ventral to) the head and mouth. In the second phase the mantle cavity and associated viscera rotate 180°, so that the structures that were ventral shift up the right side to a dorsal position and the dorsal structures shift down the left side to a ventral position. This also shifts the left gill, kidney, etc. to the right side and the right organs to the left side. The mantle cavity with its anal opening faces forward and lies above the head and mouth (Fig. 12-8, *D*). Such a torsion allows the sensitive head end of the animal to be drawn into the protection of the mantle cavity, with the tougher foot forming a barrier to the outside.

There are varying degrees of torsion in the different groups of gastropods. In some gastropods, such as the sea hare *(Aplysia),* a reverse action, or **detorsion,** follows torsion, partially restoring the bilateral symmetry of the animal. The adaptive advantage of torsion to the animal is rather obscure, but it has been suggested that having the gills forward made better respiration possible.

COILING. Torsion must be distinguished from the coiling or spiral winding of the shell, as they are separate processes. The coiling of the shell and visceral mass may occur simultaneously with the torsion process, but fossil records indicate that coiling of the shell mainly occurred before torsion. The coiling may be caused by differential growth and muscular contraction so that a corkscrew-shaped cone is produced at right angles to the axis of the torsion. Coiling is not found in some snails or else is suppressed in some way (limpets). In nudibranchs a coiled shell is found in the embryo but is absent in the adult. The direction of coiling (right- or left-handed) of the shell is determined by the orientation of the early cleavage spindle, and its heredity is unique in that the character of the coiling is determined in the egg before fertilization, by the action of the mother's genes.

Major groups of gastropods

There are three major groups of gastropods, which some authorities call subclasses and others call orders. These are the Prosobranchia, Opisthobranchia, and Pulmonata.

PROSOBRANCHIA. The gills (ctenidia) are located anteriorly in front of the heart. Another name for this group is Streptoneura, which refers to the twisted nature of the nervous system (often in a figure eight). Prosobranchs have two tentacles, and the sexes are separate. An operculum is nearly always present. This is a horny plate on the foot used to seal the opening of the shell when the snail withdraws into the shell. This group contains most of the marine snails and a few of the freshwater ones. They range in size from the periwinkles and small limpets *(Patella* and *Fissurella)* to the giant conch *(Strombus),* the largest univalve in America. Familiar examples of prosobranchs are the abalone *(Haliotis),* which has an ear-shaped shell; the giant whelk *(Busycon)* (Fig. 12-10), which lays its eggs in double-edged, disk-shaped capsules attached to a cord a meter long; the common periwinkle *(Littorina);* the slipper or boot shell *(Crepidula);* the oyster borer *(Urosalpinx,* Fig. 12-7), which bores into oysters and sucks out their juices; the rock shell *(Murex),* of which a European species was used for making the royal purple of the ancient Romans; and the freshwater forms *(Goniobasis* and *Viviparus).*

OPISTHOBRANCHIA. The opisthobranchs are an odd assemblage of mollusks that include sea slugs, sea hares, sea butterflies, canoe shells, etc. In these the gill is displaced to the right side or rear

FIG. 12-9

Opisthobranchs. **A,** *Aplysia,* the sea hare, photographed on a glass surface. One small black eye is visible just beneath the second pair of tentacles. This species is much used in neurologic studies because of unusually large ganglia. **B,** *Dendrodoris,* a Pacific coast nudibranch, is a rich yellow and has a dorsal cluster of gills. (**A,** Photo by B. Tallmark, Uppsala University, Sweden.)

of the body. Two pairs of tentacles are usually found, and the shell is reduced or absent. All are monoecious. They are all marine, most of them shallow-water forms, hiding under stones and seaweed; a few are pelagic. There are two classical groups: Tectibranchia, with gill and shell usually present, and Nudibranchia, in which there is no shell or true gill, but adaptive gills are present around the anus.

Among the Tectibranchia are the sea hare *(Aplysia,* Fig. 12-9, *A)*, which grows to be more than 1 foot long and has large earlike anterior tentacles and a vestigial shell; and the pteropods, or sea butterflies (*Carolina* and *Clione*). In pteropods the foot is modified into fins for swimming; thus they are pelagic and form a part of plankton fauna.

The Nudibranchia are represented by the sea slugs, which are often brightly colored and carnivorous in their eating habits. The plumed sea slug *Aeolis,* which lives on sea anemones and hydroids, often draws the color of its prey into the elongated papillae (cerata), which cover its back (Fig. 12-7). It also salvages the nematocysts of the hydroids for its own use. The frilled sea slug, *Tridachia* (p. 246), is a lovely little green and white nudibranch common in Florida waters. *Dendrodoris* is one of the common West coast nudibranchs (Fig. 12-9, *B*).

PULMONATA. The pulmonates include the land and freshwater snails and slugs (and a few brackish or saltwater forms). They have a lung instead of a gill and single, monoecious gonads. The aquatic species have one pair of nonretractile tentacles, at the base of which are the eyes; land forms have two pairs of tentacles, with the posterior pair bearing the eyes. Among the thousands of land species some of the most familiar American forms are *Helix, Polygyra, Succinea, Anguispira, Zonitoides, Limax,* and *Agriolimax.* Aquatic forms are represented by *Helisoma, Viviparus, Campeloma, Lymnaea,* and *Physa. Physa* is a left-handed (sinistral) snail; that is, the shell coils to the left when viewed from the apex. Dextral shells (coiling to the right) are far more common among snails than are sinistral shells (Fig. 12-10). Genetic investigation has shown that the direction of coiling is always determined before the egg is fertilized.

Common land snail, a representative pulmonate

STRUCTURE. Some of our most common terrestrial snails belong to genus *Polygyra,* but an imported European snail, *Helix aspersa,* is often studied in this country because of its size and the ease with which it can be collected. In some of our southern states it is known as the garden snail.

The snail has a well-developed **head** that bears a **mouth** and two pairs of hollow retractile **tentacles.**

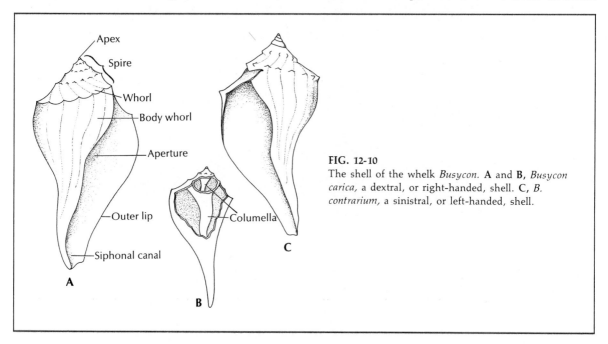

FIG. 12-10
The shell of the whelk *Busycon.* **A** and **B,** *Busycon carica,* a dextral, or right-handed, shell. **C,** *B. contrarium,* a sinistral, or left-handed, shell.

The longer pair is provided with olfactory organs and **eyes** that are adapted for light perception (Fig. 12-11).

The muscular **foot** is attached to the head. A slime gland located at the anterior end of the foot deposits a mucous film over which the ciliated foot glides with waves of muscular contraction. Movement of the foot is produced by changes in blood turgidity within hemocoelic spaces in the foot, along with an extensive system of muscles. Locomotion is slow—a "snail's pace"—and may not be more than 10 or 12 feet per hour, although some snails can travel faster. A snail moves forward by ripples of muscular action in the foot. Each wave starts at the back of the foot, by lifting a portion of the sole and placing it a little farther forward. The folds thus created are moved forward, with each fold being raised at its front end and relaxed at its hind end. As each fold reaches the anterior end of the foot, the snail moves forward. Extra speed is obtained by increasing the size of the ripples. In some cases of rapid movement the two sides of the foot sole may be stepping alternately.

The visceral mass forms a dorsal hump on top of which is the shell (Fig. 12-11). The visceral mass is surrounded by the **mantle,** which also secretes the shell. The mantle cavity is just beneath the mantle.

Where the soft parts are exposed, they are protected by a mucous membrane. The **genital pore** opens on the right side near the mouth, and the **anus** and **respiratory pore** are both located in the margin of the mantle at the edge of the shell. By means of the columella muscle, all soft parts can be withdrawn into the shell.

In the coiled **digestive system,** salivary glands open into the pharynx and a large digestive gland (liver) high in the spiral opens into the stomach (Fig. 12-11). A **radula** is used for grazing. A two-chambered **heart** and arteries deliver blood to the tissues. In some snails the blood contains the respiratory pigment erythrocruorin. For respiration the mantle cavity is modified into a **lung** provided with a network of blood vessels (Fig. 12-11). Air is drawn in through the respiratory pore. A **nephridium** removes waste from the pericardial cavity and empties it in the mantle cavity.

The **brain** consists of ganglia concentrated in the pharyngeal region. **Giant neurons** are found in mollusks as they are in annelids and some other phyla. The 30 or more giant neurons of the sea hare *Aplysia* may be as much a 1 mm. in diameter, probably the largest somatic cells in the animal kingdom. The nucleus is large; its DNA content may be thousands

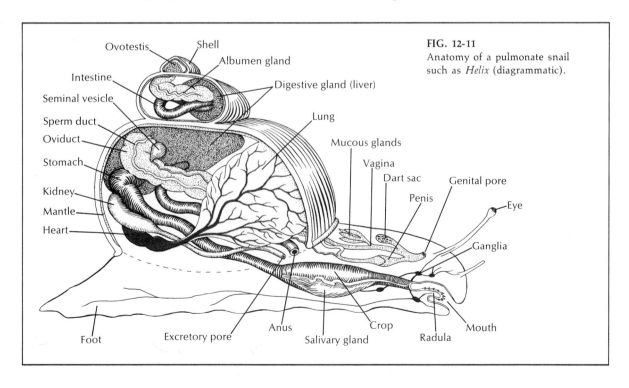

FIG. 12-11
Anatomy of a pulmonate snail such as *Helix* (diagrammatic).

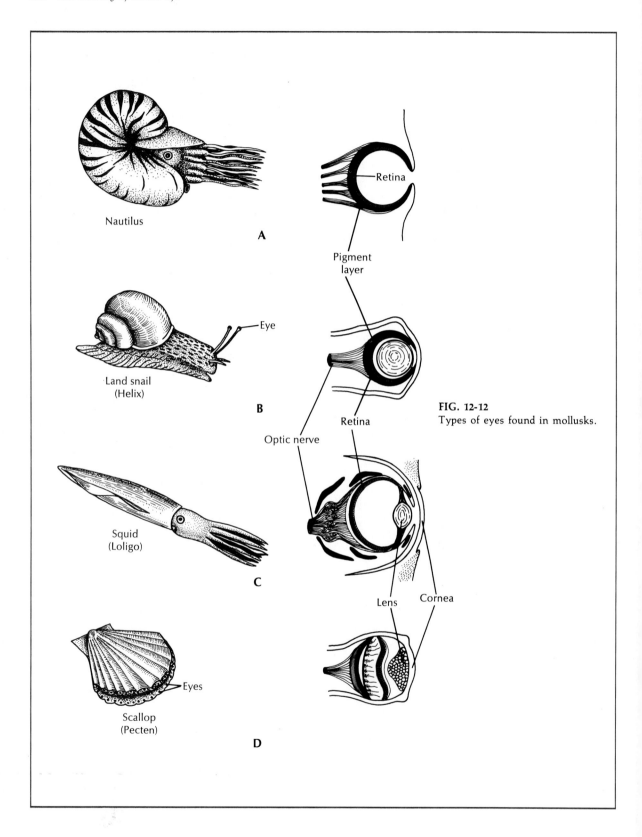

Nautilus

A

Land snail
(Helix)

B

Squid
(Loligo)

C

Scallop
(Pecten)

D

Eye

Retina

Pigment
layer

Optic nerve

Retina

Lens

Cornea

Eyes

FIG. 12-12
Types of eyes found in mollusks.

of times greater than the normal haploid value of that species, probably because of replication of chromosomes.

Sensory organs include the eyes and olfactory organs on the tentacles, a pair of statocysts for equilibrium near the brain, and tactile and chemical sense organs in the head and foot. The simplest type of gastropod eye is simply a cuplike indentation in the skin lined with pigmented photoreceptor cells. In many gastropods the eyecup contains a lens and is covered with a cornea (Fig. 12-12, *B*).

Land snails are monoecious, but cross-fertilization is the rule. An **ovotestis,** high in the spiral (Fig. 12-11), produces both eggs and sperm. A duct from the ovotestis divides into a **vas deferens,** which conducts sperm to the penis, and an **oviduct,** which leads to the vagina. Both empty through the common genital pore. The sperm are formed into spermatophores that are exchanged during copulation. Many terrestrial pulmonates eject a dart from a **dart sac** (Fig. 12-11) into the partner's body to heighten excitement. After copulation each partner deposits its eggs in shallow burrows in the ground, usually in damp places. Development is direct, and the young emerge as small snails.

BEHAVIOR AND NATURAL HISTORY. Snails are usually most active at night, when they glide about on their slime tracts foraging for food. Their food consists of green vegetation, which they rasp off by means of their radulae. They are partial to damp situations and by day are often found under patches of leaves or in burrows. When threatened with very dry weather, they form a temporary covering, or epiphragm, of mucus and limy secretions that cover the shell aperture. Some snails have a permanent **operculum,** which covers the aperture when the body is in the shell.

Class Pelecypoda

The pelecypods, or "hatchet-footed" animals, as their name implies, are the bivalved mollusks. They include over 7,000 species of mussels, clams, scallops, oysters, and shipworms, ranging from 1 mm. to 1 meter in length. Most of them are specialized for a sedentary type of life and have evolved a filter-feeding mechanism that depends on the gills for assistance in obtaining food by ciliary currents. They have no head, no radula, and little cephalization.

Most pelecypods are marine, but many live in brackish and fresh water. Because of their calcareous shells, they need to live in "hard" water, or water that is rich in limy salts.

There are many genera of clams in our freshwater streams and ponds, such as *Anodonta, Unio,* and *Lampsilis.* Some of them prefer a quiet pond or pool, but others are partial to moderately swift water. Those living in still water tend to have thinner shells. In our bigger lakes clams often have diurnal migratory habits, coming from the deeper water during the night to wander about the shallow water near shore and then returning to deeper water during the day.

THE SHELL. Bivalves are laterally compressed and covered by a pair of **valves** (shells) secreted by the mantle lobes and hinged together dorsally. An elastic **hinge ligament** holds the left and right valves together at the hinge and causes them to gape ventrally (Fig. 12-13, *A*). In many forms a tongue-and-groove arrangement in the valves prevents them from slipping. Powerful adductor muscles attached to the shell draw the valves together (Fig. 12-14).

The shell consists of the three typical layers (Fig. 12-2). The oldest and thickest part of the valve is the **umbo,** a rounded prominence near the hinge (Fig. 12-13, *A*) and surrounding it are successive concentric lines of growth.

THE MANTLE. The **mantle** is formed by folds of the body wall that hang down, one on each side of the soft body, adhering to the valves. The space between the mantle lobes is the **mantle cavity.** The body itself is made up of the visceral mass suspended from the dorsal midline, a muscular foot attached to the visceral mass anteroventrally, and a pair of **gills (ctenidia)** on each side (Fig. 12-14). Some pelecypods have a single gill rather than a double one on each side. Posteriorly the mantle is modified to form **excurrent** and **incurrent apertures** for regulating the intake and outgo of water (Fig. 12-14). Some of the marine bivalves have the mantle apertures enclosed and extended into long muscular siphons so that the clam can burrow into the mud and extend the siphons up to the water (Fig. 12-13). Cilia on the gills and inner surface of the mantle direct the flow of water over the gills.

THE FOOT. Pelecypods move by extending a slender muscular foot between the valves (Fig. 12-13). Blood swells the end of the foot to anchor it in mud or sand, and then longitudinal muscles contract

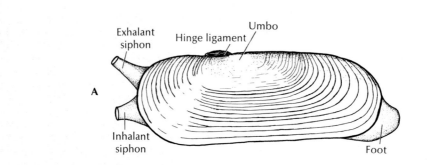

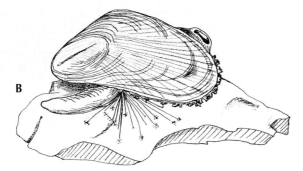

FIG. 12-13

Two marine pelecypods. **A,** *Tagelus gibbus,* the stubby razor clam, burrows into the muddy bottom with its foot and stretches its long siphons up to the surface for its inhalent water. **B,** *Mytilus,* a marine mussel, attaches itself to a rocky substrate by means of byssus threads secreted by a gland in the foot. The secretion flows down a groove in the foot, which places the thread in position, and then the foot moves to another location to place another thread. The secretion hardens upon contact with water.

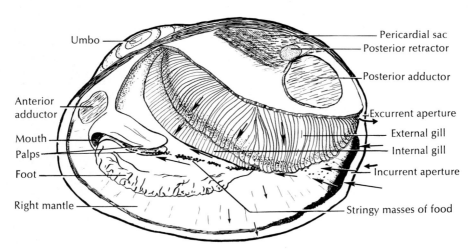

FIG. 12-14

Feeding mechanism of freshwater clam. Left valve and mantle are removed. Water enters mantle cavity posteriorly and is drawn forward by ciliary action to the gills and palps. As water enters the tiny openings of the gills, food particles are sieved out and caught up in strings of mucus that are carried by cilia to the palps and directed to the mouth. Sand and debris drop into the mantle cavity and are removed by cilia.

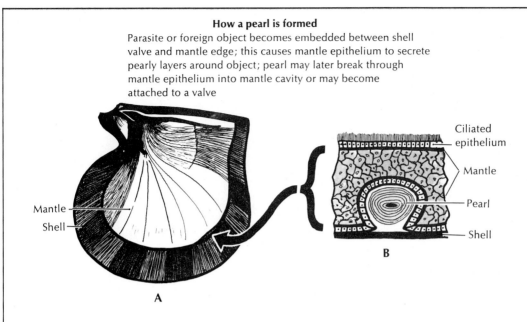

How a pearl is formed
Parasite or foreign object becomes embedded between shell valve and mantle edge; this causes mantle epithelium to secrete pearly layers around object; pearl may later break through mantle epithelium into mantle cavity or may become attached to a valve

FIG. 12-15
Pearl oyster *Margaritifera*. **A,** Interior of valve, with arrow pointing to site of pearl formation. **B,** Enlarged section of mantle, with pearl in position.

to shorten the foot and pull the animal forward. In most bivalves the foot is used for burrowing, such as in *Venus, Mya* (Fig. 12-7), and the razor clams (Figs. 12-7 and 12-13, *A*). A few even bore into wood (*Teredo,* Fig. 12-20) and rock (*Pholas,* Fig. 12-7). Probably the most primitive form of locomotion is creeping, represented in a few forms (*Solemya* and *Lepton*). Some, such as *Yoldia,* can leap over surfaces; *Kellia* is able to climb up a surface, and the scallops and file shells are able to swim jerkily by clapping their valves together to create a sort of jet propulsion.

Some forms are sessile as adults; for example, marine mussels such as *Mytilus* attach themselves by secreting a number of byssus threads that anchor them to a surface securely enough to withstand wave action (Fig. 12-13, *B*), and oysters cement one of their valves to a hard surface such as rocks or other oysters.

PEARL PRODUCTION. The production of a pearl is a protective device on the part of a bivalve. When a foreign substance such as a grain of sand or a parasite becomes enclosed between the mantle and the shell, the mantle secretes successive layers of nacre around the irritating substance (Fig. 12-15). *Meleagrina* is a pearl oyster used extensively in pearl culture by the Japanese. They culture pearls by opening the oyster, inserting a particle of nacre under the mantle, and placing the oyster in a wire cage to spend several more years in the ocean before being reopened.

FEEDING AND DIGESTION. Bivalves are filter feeders. A pair of ciliated **labial palps** on each side of the mouth directs microscopic food particles toward the mouth. These particles are brought into the mantle cavity by the respiratory current, are trapped in mucus secreted by the palps, and carried to the mouth by cilia on the surface of the palps (Fig. 12-14).

The floor of the stomach is folded into ciliary tracts for sorting the continuous stream of particles. A cylindric style sac opening into the stomach secretes a gelatinous rod called the **crystalline style,** which projects into the stomach and is kept whirling by means of cilia in the style sac (Fig. 12-16). The crystalline style is composed of mucoproteins and some enzymes (amylase for example). Surface layers of the rotating style dissolve in the stomach fluids, free-

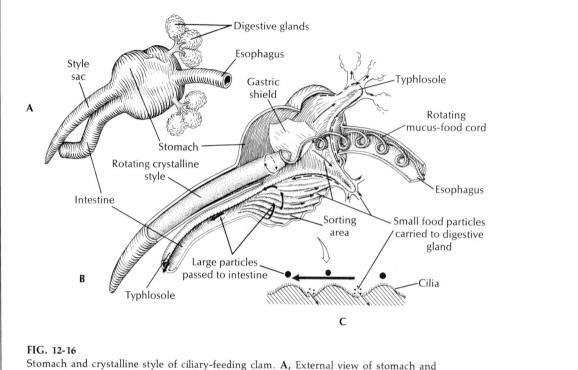

FIG. 12-16
Stomach and crystalline style of ciliary-feeding clam. **A,** External view of stomach and style sac. **B,** Transverse section showing direction of food movements. Food particles in incoming water are caught in a cord of mucus that is kept rotating by the crystalline style. Ridged sorting areas direct large particles to the intestine, and small food particles to digestive glands. **C,** Sorting action of cilia. (**B,** After J. E. Morton, 1967. Mollusca, ed. 4. London, Hutchinson & Co.)

ing enzymes for extracellular digestion. The end of the rotating style becomes attached to the mucus food mass, causing it to rotate too. As the mass spins, food particles detach from it and land on the ridged and ciliated sorting surface of the stomach floor. Here large or unsuitable particles are directed to the intestine for elimination. Smaller or partially digested nutritive particles are taken into the digestive gland or may be picked up by amebocytes. In either case digestion is completed intracellularly. Mollusks as a group have a great variety of enzymes, although there are some differences in the distribution of the enzymes among the various classes.

Some clams are deposit feeders. *Nucula,* for example, which is a burrowing form, has very long labial palps that extend out onto the mud to draw up organic deposits along ciliary grooves.

CIRCULATION. The circulatory system is an open one and consists of a heart, arteries, sinuses, and veins. The heart, which lies in the pericardial cavity (Fig. 12-14), is made up of two auricles and a ventricle (Fig. 12-17) and beats at the rate of about six times per minute. Blood is pumped through an anterior aorta to the foot and viscera and through a posterior aorta to the rectum and mantle. Part of the blood is oxygenated in the mantle and is returned to the ventricle through the auricles; the other part circulates through sinuses and passes in a vein to the kidneys, from there to the gills for oxygenation, and back to the auricles. Carbon dioxide and other wastes are carried to the gills and kidneys for elimination. The blood is colorless and contains nucleated ameboid corpuscles.

RESPIRATION. Respiration is carried on by both the mantle and the gills. In most bivalves each gill is formed of two walls (lamellae) joined together at their ventral margins. Each lamella is made up of

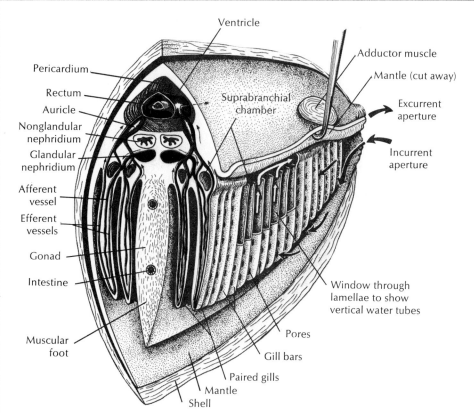

FIG. 12-17

Section through heart region of clam to show relation of circulatory and respiratory systems. *Blood circulation:* Ventricle pumps blood forward to sinuses of foot and viscera, and posteriorly to mantle sinuses. Blood returns from mantle to auricles; it returns from viscera to the kidney, and then goes to the gills, and finally to the auricles. *Respiratory water currents:* Water is drawn in by cilia, enters gill pores, and then passes up water tubes to suprabranchial chambers and out excurrent aperture. Blood in gills exchanges carbon dioxide for oxygen.

many vertical gill filaments strengthened by chitinous rods. Water enters the gills through innumerable small pores in the walls and is propelled by ciliary action. Partitions between the lamellae divide the gill internally into many vertical **water tubes** that carry water upward (dorsally) into a common **suprabranchial chamber** (Fig. 12-17) and through it to the excurrent aperture. Blood vessels or spaces in the interlamellar partitions are used for exchange of gases. The water tubes in the female often double as brood pouches for eggs and larvae during the breeding season. The various orders of Pelecypoda are classified on the basis of their gill types.

EXCRETION. A pair of U-shaped kidneys (nephridial tubules) lie just below the heart (Fig. 12-17). The glandular portion of each tubule opens into the pericardium; the bladder portion empties into the suprabranchial chamber. The kidneys remove waste from the blood that circulates through its network, and from the pericardial cavity.

NERVOUS AND SENSORY SYSTEM. The nervous system consists of three pairs of widely separated ganglia connected by commissures; and a system of nerves. The ganglia are the **cerebropleural ganglia** near the mouth, the **pedal ganglia** in the foot, and the **visceral ganglia** below the posterior adductor muscle. Neurosecretory cells that produce neurohumors such as acetylcholine and are comparable to

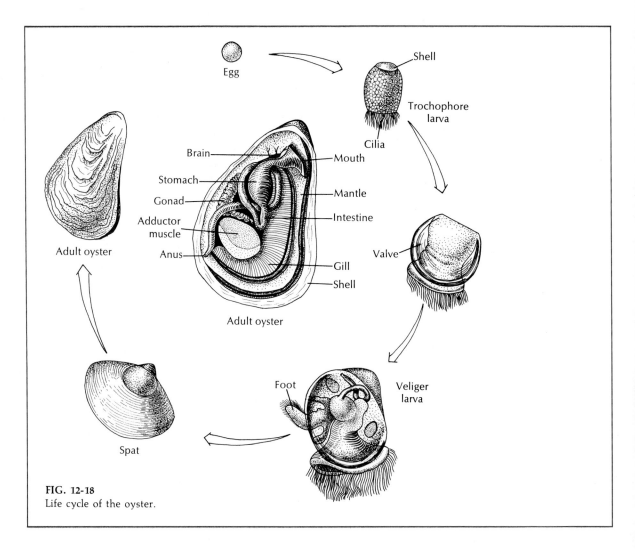

FIG. 12-18
Life cycle of the oyster.

those of arthropods and vertebrates, have been found in some mollusks.

Sense organs are poorly developed. They include a pair of **statocysts** in the foot for balance, a pair of **osphradia** in the incurrent area of the mantle cavity to test the quality of incoming water, and **tactile organs** and **ocelli** (in some forms) along the mantle margins. Ocelli are best represented in the scallop *Pecten,* where they are steel blue in color and located at intervals all around the mantle edge. They consist of cornea, lens, retina, and a pigmented layer (Fig. 12-12, *D*).

REPRODUCTION AND DEVELOPMENT. The sexes are usually separate. Reproductive organs are branched, lobate masses located around the intestinal coils of the visceral mass just above the foot (Fig.

12-17). The sperm ducts of the male or the oviducts of the female discharge their sperm or eggs into the suprabranchial chamber above the gills where they can be carried out with the outgoing current.

In most marine pelecypods fertilization occurs externally, although in some forms the larvae are incubated in the mantle cavity or gills. The embryo develops first into a free-swimming **trochophore larva,** followed by a short **veliger larval** stage, and then it soon develops into the young **spat** (Fig. 12-18). In oysters the spats swim about for 2 weeks or so before settling down to attach themselves permanently to a hard surface. Oyster spats are often netted and transported to desirable beds where they are provided with stones or cement for attachment. It takes about 4 years for an oyster to grow to commercial

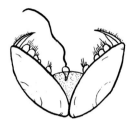

FIG. 12-19
Glochidium, or larval form of freshwater clam. When larva is released from brood pouch of mother, it may become attached to fish by clamping its valves closed. It remains as parasite on fish for several weeks. (Size, about 0.3 mm.)

size. Because of the many hazards in their life cycles, most mollusks produce an enormous number of eggs. It is said that the oyster, for example, may produce more than 50 million eggs in a single season.

In most freshwater clams fertilization is internal. Instead of leaving the body from the suprabranchial chamber, the eggs drop into the water tubes of the gills, which enlarge to form brood chambers. Sperm enter the gills with the incoming current to fertilize the eggs in the gills. Development proceeds within the brood chambers. There is no free-swimming larva. Instead, the veliger stage is a parasitic bivalved **glochidium larva** (Fig. 12-19). In some species, for example *Anodonta,* the valves bear hooks. When discharged into the water, the glochidia are carried by water currents or sink to the bottom. When they come into contact with the gills or body surface of a passing fish, they attach themselves by their hooks or by snapping their valves together. The larvae encyst and live as parasites on the fish for 8 to 12 weeks. During this time they develop into minature clams and the cysts break to release them. They then sink to the bottom to begin their existence as independent clams. This larval "hitchhiking" helps to distribute a form whose locomotion is very limited.

BORING BIVALVES. Many pelecypods are able to burrow into mud or sand, but some have evolved a mechanism for burrowing into much harder substances, such as wood or stone.

Teredo and *Bankia* are well-known wood-boring mollusks that can be very destructive to wooden ships and wharves. These strange little clams have a long, wormlike appearance, with a pair of slender siphons on the posterior end that keep water flowing over the gills, and a pair of small globular valves on the anterior end with which they burrow (Fig. 12-20, *A*). The valves have microscopic teeth so that they function as very effective wood rasps. The animals extend their burrows with an unceasing rasping motion of the valves. This sends a continuous flow of fine wood particles into the digestive tract where they are attacked intracellularly by the enzyme cellulase. In addition to wood as nourishment they also collect incoming plankton particles with their gills and palps. Fertilization is internal, with the zygotes becoming implanted in the gill tissue. The young *Teredo* are liberated as globular ciliated larvae $1/2$ μ in diameter that swim about looking for a likely spot to penetrate wood and metamorphose. Clinging to the new burrow with its foot and boring forward as it grows, the young *Teredo* may grow to a length of 100 to 125 mm. and a diameter of 5 mm., always extending its siphons backward toward the opening to pump in the oxygen- and food-bearing water. One species of *Bankia* makes burrows up to a meter long and 12 mm. in diameter.

Some clams bore into rock. The piddock *(Pholas)* bores into limestone, shale, sandstone, and sometimes wood or peat (Fig. 12-7). It has strong valves that bear spines by which it gradually cuts away the rock while anchoring itself by its foot. *Pholas* may grow to 15 cm. long and make rock burrows up to 30 cm. long.

Class Cephalopoda

Class Cephalopoda is far more advanced than any other class of Mollusca and in some respects is more advanced than any other invertebrate. All are marine, and they include the squids, octopuses, nautiluses, devilfish, and cuttlefish. They derive their name from Greek *kephalē,* head, and *pous,* foot. This derivation describes one of their most unusual features—the concentration of the foot in the head region. The edges of the foot are drawn out into arms and tentacles, which bear sucking disks for seizing prey. Part of the foot (epipodium) is modified to form the funnel for carrying water from the mantle cavity. The group goes back to the Cambrian period, with remarkable fossil records.

The largest invertebrate known is the giant squid

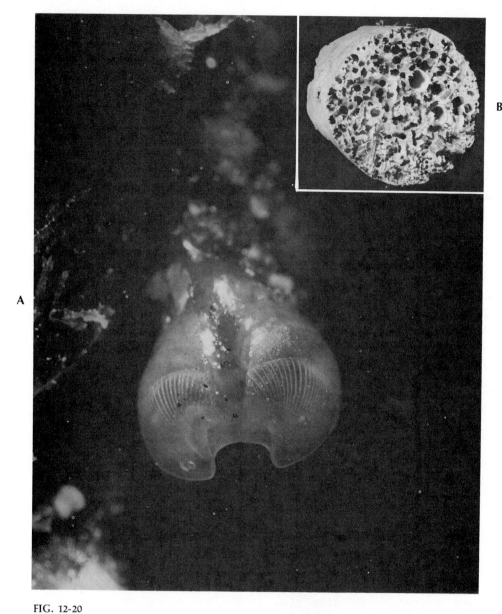

FIG. 12-20
A, Anterior end of the shipworm, *Teredo navalis,* a wood-boring bivalve. The bilobed
valves at the anterior end are modified to form a double rasping organ used to bore
tunnels in wood. The posterior end of the shipworm (not seen) is drawn out into siphons
that keep water flowing into the tunnel and over the gills. Shipworms can be very
destructive to ships and pilings. **B,** Piece of wood that has been riddled with tunnels by
shipworms. (**A,** Photo by R. Vishniac, New York.)

(Architeuthis), which may be up to 50 feet long. *Rossia,* a West coast squid, is only $1^1/_2$ to 3 inches long.

EVOLUTION. Mollusks appear to have differentiated along three lines of life habits. Bottom-dwelling filter feeders with a slow locomotion gave rise to pelecypods; herbivorous or carnivorous feeders with slow creeping movements evolved into gastropods; whereas a third group of active swimming predators became the cephalopods. As predators the cephalopods are not only swift, but they also have prehensile arms and tentacles for seizing and suckers for holding their prey. They can put out protective "smoke screens" from their ink sacs and can produce marvelous patterns of color changes.

Cephalopods probably originated from a form similar to the hypothetical mollusk ancestor. It is believed that the simple limpetlike shell became elongated as the visceral mass shifted away from the apex of the shell. Behind the visceral mass a septum was secreted at each growth period, so that the shell became a series of successive compartments. As the fossil record shows, the earliest shells were straight but later shells became coiled, producing the *Nautilus* type of shell (Fig. 12-21, *A*).

Although the earliest cephalopods bore heavy external shells, the shells were buoyant with gas chambers separated by septa and connected to the body by a tube, the **siphuncle** (Fig. 12-21, *B*). The evolution of the nonshelled or reduced-shell forms is obscure because of lack of fossils, but the evolu-

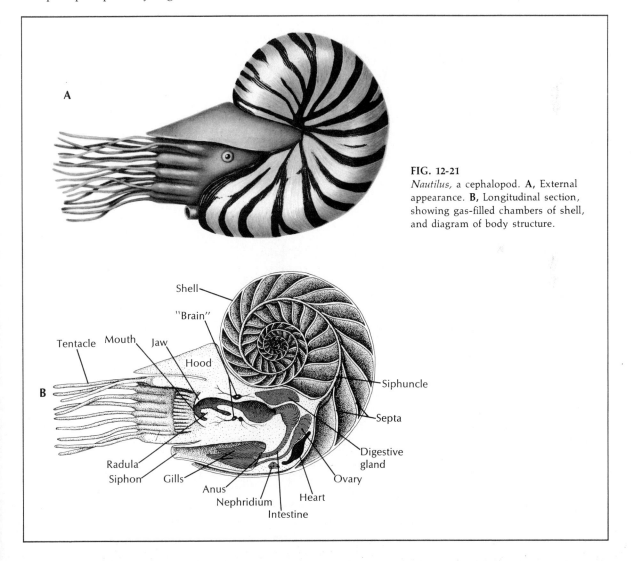

FIG. 12-21
Nautilus, a cephalopod. **A,** External appearance. **B,** Longitudinal section, showing gas-filled chambers of shell, and diagram of body structure.

FIG. 12-22

Octopuses (class Cephalopoda). **A,** Newly hatched octopus, less than 1 cm. long. Note the chromatophores in the skin by which quick color changes can occur. **B,** Adult octopus. The powerful suckers are used for crawling over rocks and seizing prey, mostly crabs. Its eyes are highly developed. (**A,** Photo by R. Hermes, Homestead, Fla.; **B,** From film *Marine Life,* courtesy Encyclopaedia Britannica Films, Inc.)

tionary tendency for a degenerate shell or none at all is evident.

ECOLOGY. The natural history of the cephalopods is known only in part. They are saltwater animals and appear sensitive to the degree of salinity. Few are found in the Baltic Sea, where the water has a low salt content. Cephalopods are found at various depths. The octopus is often seen in the intertidal zone, lurking among rocks and crevices, but occasionally is found at great depths. The more active squids are rarely found in shallow water, and some have been taken at 5,000 meters. *Nautilus* is usually taken from the ocean floor near islands (southwestern Pacific) where the water is several hundred meters deep.

COLOR CHANGES. Color changes of cephalopods are produced in the skin by the contraction and expansion of special pigment cells called **chromatophores** (Fig. 12-22, *A*), manipulated by tiny muscles attached to the edge of the cells. The pigment colors are black, brown, red, and yellow so

that a squid or octopus can assume a variety of colors when emotionally disturbed or for protection.

INK PRODUCTION. All cephalopods except *Nautilus* have ink sacs (Fig. 12-24) from which they expel ink when attacked. The ink contains melanin (black) pigment, which is formed by the oxidation of the amino acid tyrosine through the action of an enzyme. Some authorities think the discharged ink assumes the shape of a "dummy" to distract an enemy predator. Or the ink may paralyze the enemy's sense of smell.

FEEDING HABITS. All cephalopods are predaceous and carnivorous. Their chief food is small fish, mollusks, crustaceans, and worms. They are strong enough to pull clams apart; their horny beaks can quickly tear the flesh from a crustacean skeleton. They will even eat each other. Cephalopods are preyed upon by whales, seals, sea birds, and moray eels. The slender moray eel can go where the octopus hides and overcome it.

REPRODUCTION AND DEVELOPMENT. The sexes are separate in cephalopods. Females sometimes outnumber the males as much as 100 to 15. There is a certain amount of sexual dimorphism, too, so that it is possible to distinguish the sexes in some cases. One of the male arms is modified for transferring sperm to the female. The sperm are arranged in long tubes called **spermatophores,** which are formed in a special sac of the vas deferens. During mating, the male clasps the female and with his **hectocotylus** arm withdraws a bundle of spermatophores from his siphon and places them near her mouth or within her mantle cavity. In *Argonauta* and a few others, part of this arm is disengaged and left within the mantle chamber of the female; in others only the tip of the arm is detached. The hectocotylus arm is usually one of the fourth pair of arms in the squid; one of the third in the octopus. The squid lays her eggs in pencil-shaped masses of jelly and attaches them to some object. They are called "deadman's-fingers." The octopus lays eggs in long strings or bunches attached to a rock, where she remains to care for them.

Major groups of cephalopods

There are two orders of cephalopods: Tetrabranchia (four-gilled) and Dibranchia (two-gilled). Tetrabranchia is the more primitive. Its members populated the Paleozoic and Mesozoic seas but left only one genus, *Nautilus* (Fig. 12-22), of which there are three living species.

The order Dibranchia includes all living cephalopods except *Nautilus.* It has an internal shell

or none at all, a cylindrical body that may have a fin, a pair of gills, a pair of kidneys, eight to ten arms or tentacles with suckers, eyes with lenses, and a muscular mantle for locomotion. There are two suborders: Decapoda (squids), with ten arms and large coelom, and Octopoda (octopuses), with eight arms and reduced coelom.

Two of the decapod's arms are modified into tentacular arms for seizing prey. They can either be retracted into special pouches or doubled back upon themselves. They bear suckers only at the ends and are situated between the third and fourth arms on each side of the head. The suckers in squids are stalked (pedunculated), with horny rims bearing teeth; in octopuses the suckers are sessile and have no horny rims.

The squid—a representative cephalopod

ETERNAL STRUCTURES. The squid *Loligo* (Fig. 12-23) has an elongated torpedo-shaped body with a large head, which bears two large eyes and a mouth surrounded by ten arms provided with suckers. One pair of arms, the retractile tentacles, are longer than the others. Along each side of the body is a fleshy triangular fin. The mantle encloses the mantle cavity in which are the internal organs. This mantle ends just behind the head in a free margin, the **collar.** Under the collar there projects a conical structure, the **siphon** (Fig. 12-24), from which water can be forced by the contraction of the mantle. The shell, or **pen,** of *Loligo* is much reduced, consisting of a feather-shaped plate just beneath the skin of the back or anterior wall. It offers little protection

FIG. 12-23
School of young squids. As they swim along, each individual carefully maintains his position and distance with reference to others. If this pattern is disturbed, squids quickly revert to their original formation.

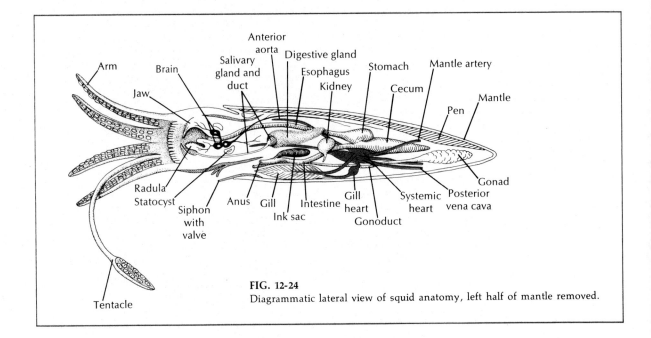

FIG. 12-24
Diagrammatic lateral view of squid anatomy, left half of mantle removed.

but does stiffen the body. There is also cartilage support for the neck region, the siphon, and the fins.

INTERNAL STRUCTURES (Fig. 12-24). The **digestive system** is made up of the usual divisions. The pharynx has a pair of horny **jaws,** a **radula,** and two pairs of salivary glands, and the stomach has the ducts of the **liver** and **pancreas** connected to it. The anus empties into the mantle cavity. A glandular **ink sac** opens into the mantle cavity near the anus.

The circulatory system (Fig. 12-24) is closed. **Branchial,** or gill, **hearts** pump blood through the gills, and a **systemic heart** forces blood to the various other organs. A pair of **kidneys** connect the pericardial space with the mantle cavity. **Respiration** is carried on by a pair of **gills** in the lower part of the mantle chamber.

The **nervous system** contains many pairs of ganglia concentrated mainly in the head region. This region is protected by a cartilaginous case. The **sensory** organs are fairly well developed. There are two complex **eyes** (Fig. 12-12, *C*). which contain cornea, lens, chambers, and direct retina with rods. They can form real images, as do the eyes of vertebrates, but because they are derived in a different manner, they are not homologous. Other sense organs are a pair of **statocysts** below the brain for equilibration and a pair of **olfactory organs.**

The sexes are separate, and from the **gonad** in each sex a duct leads forward to empty into the mantle cavity near the anus. The male is provided with a hectocotylus arm for transferring sperm to the female.

BEHAVIOR. Squids have an interesting method of locomotion that involves "jet propulsion." When going backward, which they can do with great speed, they take water into the mantle cavity, close the collar tightly around the neck, and eject the water forcibly in a jet from the siphon, which is directed toward the arms. In going forward, the siphon is directed backward. The two fins are used in steering and for swimming.

Squids also have two important devices for protecting themselves against their enemies. One of these is their remarkable power of changing color by contracting or expanding the various chromatophores in the skin. The other involves the ink sac. The squid secretes a jet of ink that serves as a decoy to confuse the enemy while the squid changes color and swims away.

Squids live upon fish, mollusks, and crustaceans, which they capture with their arms and specialized tentacles. The prey is drawn by the arms to the mouth, where the horny jaws bite out pieces that are then swallowed. The jaws and radula are operated by the highly muscular pharynx.

Many experiments on the behavior of cephalopods

have been made (J. Z. Young, M. J. Wells, and others). In general, cephalopods have nervous elements similar to those of vertebrates and their intelligence is perhaps the highest among the invertebrates. Their brain is the largest of the invertebrates, coordinated by millions of nerve cells. Each octopus behaves differently from other octopuses. Individual experience determines much of their behavior, for they have the capacity to remember their past experiences. With their suckers they can distinguish textural differences, although they cannot determine differences in weight and shape of objects. Their color sensitivity is amazing and their vision is very acute. They seem to be lacking the sense of hearing. Experimenters find it easy to modify their behavior patterns by devices of reward and punishment.

Other members of class Cephalopoda

The squid *Loligo* is little more than 1 foot long, but the giant squid *Architeuthis* may be more than 50 feet long, the largest of all invertebrates. It is found off the coast of Newfoundland. Another squid, called the cuttlefish *(Sepia),* produces sepia-colored ink, which is used by artists. The cuttlefish has a large pen or cuttlebone, which is often given canaries as a source of calcium.

The octopus, or devilfish, has no shell but possesses a large body and eight sucker-bearing tentacles (Fig. 12-22). Some of these animals are only a few inches in diameter and are harmless, but the giant octopus of the Pacific may reach an overall diameter of 30 feet and is very dangerous to divers.

The chambered nautilus (Fig. 12-21) is found on the bottom of the South Pacific seas. Its spiral shell in one plane is up to 10 inches in diameter and is made up of a series of compartments, each of which has been occupied in succession as the animal has grown larger. Thus each new chamber is larger than the preceding one. The animal occupies the outermost compartment. The unoccupied compartments are filled with gas, probably secreted by the **siphuncle,** a body mass of blood vessels and tissues enclosed in a tube that runs through the center of the partitions. The heads of these forms bear 60 to 90 arms but no suckers. The simple "pinhole camera" eye of the nautilus is shown in (Fig. 12-12, *C*).

The paper nautilus *(Argonauta)* has a delicately coiled shell formed from glands in its tentacles. This shell is adapted for housing its eggs and for a nursery for the young after they hatch.

Derivation and meaning of names

Anodonta (Gr. *an-*, without, + *odous, odontos,* tooth). A common genus of freshwater clams.

Busycon (Gr. *bous,* ox, + *sykon,* fig). Giant whelk.

Chiton (Gr. coat of mail, tunic). Refers to plates that cover this primitive mollusk.

Dentalium (L. *dentalis,* dental). Genus of Scaphopoda.

Helix (Gr. twisted; a spiral). Genus of land snails introduced from Europe.

Nautilus (Gr. *nautilos,* sailor). This sole survivor of the most ancient mollusks is an active swimmer.

Nudibranchia (L. *nudus,* naked, + Gr. *branchia,* gills). These gastropods have no shells.

Opisthobranchia (Gr. *opisthe,* behind, + *branchia,* gills). Gills are located posteriorly in these gastropods.

Physa (Gr. bellows, air stream). Refers to bubble of air carried by snail when it submerges.

Polygyra (Gr. *poly,* many, + *gyros,* circle). Large genus of land snails.

Polyplacophora (Gr. *poly,* many, + *plax ,* anything flat, + *phora,* bearing). An order of Amphineura with a shell of eight plates.

Prosobranchia (Gr. *prosō,* forward, + *branchia,* gills). Gastropods with gills located anteriorly.

Pulmonata (L. *pulmo,* lung, + *-ata,* group suffix).

Solenogastres (so-len-o-gas'treez) (Gr. *sōlēn,* pipe, + *gastēr,* belly). Synonym of Aplacophora.

Strombus (L. a kind of spiral snail). Giant conch.

References

Suggested general readings

Abbott, R. T. 1972. Kingdom of the seashell. New York, Crown Publishers, Inc.

Baker, F. C. 1898-1902. The Mollusca of the Chicago area. Part I: The Pelecypoda. Part II: The Gastropoda. Chicago, The Chicago Academy of Science. *An authoritative work by a specialist in this field. Taxonomy somewhat outdated.*

Buchsbaum, R. M., and L. J. Milne. 1960. The lower animals: Living invertebrates of the world. Garden City, N. Y., Doubleday & Co., Inc. *Superb photographs (many in color) and concise accounts of the invertebrate phyla.*

Cambridge Natural History. 1895. Mollusca (A. H. Cooke), vol. 3. London, Macmillan & Co., Ltd. *A standard work on mollusks.*

Edmondson, W. T. (editor). 1959. Ward and Whipple's fresh-water biology, ed. 2. New York, John Wiley & Sons, Inc. *A taxonomic key to the freshwater families of mollusks is included.*

Encyclopaedia Britannica. Articles on Mollusca and the various classes. Chicago, Encyclopaedia Britannica, Inc.

Galtsoff, P. S. 1961. Physiology of reproduction in molluscs. Amer. Zool. **1:**273-289.

George, W. C., and J. H. Ferguson. 1950. The blood of gastropod molluscs. J. Morph. **86:**315-336.

Grave, B. H. 1928. Natural history of the shipworm, *Teredo navalis.* Woods Hole, Mass. Biol. Bull. **55:**260-282.

Hyman, L. H. 1967. The invertebrates: Mollusca (vol. 6). New York, McGraw-Hill Book Co. *This volume covers four groups of the mollusks: Aplacophora, Polyplacophora, Monoplacophora, and Gastropoda. This volume upholds the fine traditions of the other volumes in this outstanding series.*

Keen, A. M. 1963. Marine molluscan genera of Western North America. Stanford, Calif., Stanford University Press.

Lankester, E. R. (editor). 1906. A treatise on zoology. Part V.: Mollusca (P. Pelseneer). London, A. & C. Black. *A classic monograph.*

Lemche, H. 1957. A new living deep-sea mollusk of the Cambro-Devonian class Monoplacophora. Nature **179:**413. *An account of Neopilina.*

MacGinitie, G. E., and N. MacGinitie. 1967. Natural history of marine animals, ed. 2. New York, McGraw-Hill Book Co. *A section is devoted to Mollusca with several excellent photographs of representative forms.*

Malek, E. A. 1962. Laboratory guide and notes for medical malacology. Minneapolis, Burgess Publishing Co. *Deals with those mollusks (mostly snails) that act as hosts for certain stages of parasites. Good illustrations of snail structure.*

Mead, A. R. 1961. The giant African snail: A problem in economic malacology. Chicago, University of Chicago Press.

Morton, J. E. 1967. Molluscs, ed. 4. London, Hutchinson & Co.

Pennak, R. W. 1953. Fresh-water invertebrates of the United States. New York, The Ronald Press Co.

Potts, W. T. W. 1967. Excretion in the mollusks. Biol. Rev. **42:**1-41.

Purchon, R. D. 1968. The biology of the Mollusca. New York, The Pergamon Press.

Russell-Hunter, W. D. 1968. A biology of the lower invertebrates. New York, The Macmillan Co.

Segal, E. 1961. Acclimation in mollusks. Amer. Zool. **1:**235-244.

Spector, W. S. (editor). 1956. Handbook of biological data, pp. 183-184. Philadelphia, W. B. Saunders Co.

Wells, M. J. 1962. Brain and behavior in cephalopods. Stanford, Calif., Stanford University Press.

Wilbur, K. M., and C. M. Yonge (editors). 1964, 1966. Physiology of Mollusca, vols. 1 and 2. New York, Academic Press Inc. *A pretentious monograph that summarizes much of the research work on mollusks.*

Young, J. Z. 1961. Learning and form discrimination by octopus. Biol. Reviews Cambridge Phil. Soc. **36:**31-96.

Selected *Scientific American* articles

Boycott, B. B. Learning in the octopus. **212:**42-50 (March).

Cadart, J. 1957. The edible snail. **197:**113-118 (Aug.). *Description of the cultivation of edible snails in France.*

Feder, H. M. 1972. Escape responses in marine invertebrates. **227:**92-100.

Korringa, P. 1953. Oysters. **189:**86-91 (Nov.). *Their life history is described.*

Lane, C. E. 1961. The teredo. **204:**132-142 (Feb.). *Biology of the shipworm.*

Steinbach, H. B. 1951. The squid. **184:**64-69 (April). *Facts about the largest of invertebrates and why it is of such importance in neurophysiologic studies.*

Willows, H. O. D. 1971. Giant brain cells in mollusks. **224:**68-75 (Feb.).

Some annelid worms are beautiful—such as the "feather-duster" worms. Spirographis *secretes a sturdy tube from which it can thrust its feathery spiral-shaped crown for feeding. Tiny organisms in the ocean water are caught in mucus on the radioles and directed by cilia to the mouth at the base of the crown. If disturbed, the worm can withdraw its crown to safety in the tube.*

CHAPTER 13

THE SEGMENTED WORMS

PHYLUM ANNELIDA*
A MAJOR EUCOELOMATE PROTOSTOME GROUP

■ POSITION IN ANIMAL KINGDOM

1. Annelids belong to the protostome branch, or schizocoelous coelomates, of the animal kingdom and have spiral and determinate cleavage.

2. Annelids as a group show a primitive metamerism with few differences between the different somites.

3. All organ systems are present and well developed.

■ BIOLOGIC CONTRIBUTIONS

1. The introduction of **metamerism** by the group represents the greatest advancement of this phylum and lays the groundwork for the more highly specialized metamerism of the arthropods.

2. A true coelomic cavity reaches a high stage of development in this group.

*An-nel'i-da (L. *annellus,* little ring, + *-ida,* pl. suffix).

3. Specialization of the head region into differentiated organs, such as the tentacles, palps, and eyespots of the polychaetes, is carried further in some annelids than in other invertebrates so far considered.

4. The tendency toward **centralization of the nervous system** is more developed, with cerebral ganglia (brain), two closely fused ventral nerve cords with unique giant fibers running the length of the body, and various ganglia with their lateral branches.

5. The circulatory system is much more complex than any we have so far considered. It is a closed system with muscular blood vessels and aortic arches ("hearts") for propelling the blood.

6. The appearance of the fleshy **parapodia,** with their respiratory function, introduces a suggestion of the paired appendages and specialized gills found in the more highly organized arthropods.

7. The well-developed **nephridia** in most of the somites have reached a differentiation that involves a removal of waste from the blood as well as from the coelom.

8. Annelids are the most highly organized animals capable of complete regeneration. However this ability varies greatly within the group.

■ PHYLOGENY AND ADAPTIVE RADIATION

PHYLOGENY. Annelid larval stages are similar to those of the flatworms, and the marine annelids have trochophore larvae in common with marine turbellarians, mollusks, and some of the minor phyla.

Annelids also have many arthropod characteristics, such as metamerism, a hypodermic-secreted cuticle, and a similar type of nervous system. There is some similarity of the parapodia of the annelid polychaetes to the appendages of certain arthropods. In their respiratory function the parapodia are also suggestive of the specialized gills of the aquatic arthropods. One phylum, the Onychophora, has both annelid and arthropod characters and is considered a classic example of an intermediate or transition form between two phyla.

In a genealogic tree the annelids are often placed between mollusks and arthropods, since all three have apparently come from a common ancestor.

ADAPTIVE RADIATION. A basic adaptive feature in the evolution of annelids is their septal arrangement resulting in fluid-filled coelomic compartments. Fluid pressure in these compartments is used as a hydrostatic skeleton in precise movements such as burrowing and swimming. Powerful circular and longitudinal muscles have been adapted for flexing, shortening, and lengthening the body.

There is a wide variation in feeding adaptations, from the sucking pharynx of the oligochaetes and the chitinous jaws of carnivorous polychaetes to the specialized tentacles and cirri of the ciliary feeders.

In polychaetes the parapodia have been adapted in many ways and for many functions, chiefly locomotion and respiration.

In leeches many of their adaptations are related to their bloodsucking habits.

■ THE ANNELIDS

The Annelida are worms whose bodies are divided into similar rings or segments. The origin of the name of the phylum describes this basic characteristic, for it comes from the Latin word *annellus,* meaning little ring. This biologic principle of body segmentation is commonly called **metamerism,** and the divisions are known as **segments, somites,** or **metameres.** Metamerism in Annelida is manifested not only in external body features but also in the internal arrangement of organs and systems. Circulatory, excretory, nervous, muscular, and reproductive organs all show a segmental arrangement, and there are internal partitions, or septa, between the somites (Figs. 13-3 and 13-4, *A*).

Significance of metamerism

No satisfactory reason can be given for the origin of metamerism. One theory is that chains of subzooids formed by asexual fission in flatworms, instead of separating into distinct individuals, may have held together and developed structural and functional unity with the passage of time. This theory would take into account the axial gradient idea that the anterior individuals would become the dominant part of the chain in determining the coordination of the whole. It is well known that in some platyhelminths the daughter individuals formed by asexual division cling together for some time before separating. Another theory stresses the secondary origin of metamerism by the repetition of body parts, such as muscles, nerves, nephridia, coelom, and blood vessels, in a single individual. Later, partitions were interposed to form definite segments. It is also possible that segmentation may have started in the musculature of an elongated, swimming animal, for the breaking up of the body into segments would facilitate swimming movement.

The locomotion produced by metamerism in conjunction with a fluid-filled coelomic cavity and powerful body-wall musculature represents an advancement over the ciliary and creeping methods of the lower Metazoa. The coordination of muscular action and the fluid-filled septal compartments have made possible efficient swimming and creeping over a substratum by undulatory body movements. Fluid-filled coelomic cavities also provide hydrostatic skeletons for burrowing. Differential turgor (especially if the coelom is subdivided by septa) can be effected by shifting the coelomic fluid from one part of the body to the other so that precise movements can occur.

Another advantage of metamerism is the possibility for each metamere to become specialized for particular functions. This would indicate that metamerism is not unlike the formation of cells in a metazoan body in which each cell or group of cells is differentiated for some definite purpose. This specialization, not appreciable in the Annelida but well developed in the Arthropoda, has made possible a rapid evolution of high organization in animals.

Although three phyla—Annelida, Arthropoda, and Chordata—are outstanding examples of metamerism, some other groups have tendencies toward a segmental arrangement, wholly or in part, such as one group (Monoplacophora) of the mollusks and the proglottid arrangement of the cestodes. True segmentation has arisen at least twice independently —in the Annelida-Arthropoda groups and in the Chordata. However, there may be many types of metamerism for different adaptive reasons, such as swimming (chordates), burrowing (annelids), reproduction (cestodes), and repetition of organs in elongated animals for effective control (rhynchocoels).

The whole evolutionary potential of the Annelida has been guided to a great extent by the morphologic organization of metamerism and its varied patterns of related structures.

■ ECOLOGIC RELATIONSHIPS

Annelids have a worldwide distribution and occur in marine waters, fresh water, and terrestrial soils. A few of the species may be called cosmopolitan, but the terrestrial soil species have poor powers of dispersal. In the frozen subsoils of the tundra terrestrial forms of annelids cannot exist.

Polychaetes are chiefly marine forms and make up about two-thirds of the annelid worms. Most of them live free in the open sea or in the littoral zones of the sea. They are usually divided into two groups —the sedentary polychaetes, or Sedentaria, and the free-moving polychaetes, or Errantia. Sedentary polychaetes are mainly tubicolous and spend all or much of their time in tubes or permanent burrows. Many of them, especially those that live in tubes, have elaborate devices for feeding and respiration. In sabellids and serpulids the cirri or tentacles around the mouth give rise to great featherlike "branchial crowns" that are involved in both feeding and respiration (p. 275). By means of ciliary currents water passes between the tentacles and carries food entangled in mucus to the mouth. The errant, or free-moving, polychaetes have various habitats; some are strictly pelagic and others live in crevices or under rocks or shells, never straying far in open water.

Many polychaetes are euryhaline and occur in brackish water. Freshwater polychaete fauna is more diversified in warmer regions than in the temperate zones.

Clitellates (Fig. 13-1) occur predominately in freshwater or terrestrial soils. Some freshwater species burrow in the bottom mud and sand and others among submerged vegetation. The swimming species usually have long setae, whereas the common earthworms, the most familiar example of the oligochaetes, have short setae for anchoring themselves as they move through the soil.

All the leeches (class Hirudinea) are predators, and many are specialized for piercing their prey and feeding upon blood or soft tissues. A few leeches are marine, but most of them live in freshwater or in damp regions. The greatest number of them occur in submerged vegetation or wood. Suckers are typically found at both ends of the body for attachment to their prey. Some are adapted for forcing their pharynx or proboscis into soft tissues as in the gills of fish. The most specialized leeches, however, have sawlike chitinous jaws with which they can cut through tough skin. Many leeches are not actually parasites, but live as carnivores upon small invertebrates.

Characteristics

1. Body **metamerically segmented;** symmetry bilateral; three germ layers
2. Body wall with outer circular and inner longitudinal muscle layers; transparent moist cuticle secreted by columnar epithelium (hypodermis) covers body
3. **Chitinous setae,** often present on fleshy appendages called **parapodia;** absent in some
4. Coelom (schizocoel) well developed in most and usually divided by septa; coelomic fluid for turgidity
5. **Blood system closed** and segmentally arranged; respiratory pigments (erythrocruorin and chlorocruorin) with amebocytes in blood plasma
6. Digestive system complete and not metamerically arranged
7. Respiration by skin or **gills**

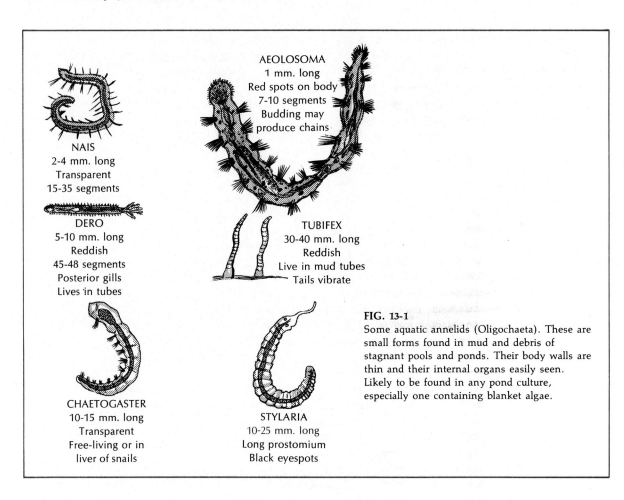

NAIS
2-4 mm. long
Transparent
15-35 segments

DERO
5-10 mm. long
Reddish
45-48 segments
Posterior gills
Lives in tubes

AEOLOSOMA
1 mm. long
Red spots on body
7-10 segments
Budding may
produce chains

TUBIFEX
30-40 mm. long
Reddish
Live in mud tubes
Tails vibrate

CHAETOGASTER
10-15 mm. long
Transparent
Free-living or in
liver of snails

STYLARIA
10-25 mm. long
Long prostomium
Black eyespots

FIG. 13-1
Some aquatic annelids (Oligochaeta). These are small forms found in mud and debris of stagnant pools and ponds. Their body walls are thin and their internal organs easily seen. Likely to be found in any pond culture, especially one containing blanket algae.

8. Excretory system typically a **pair of nephridia for each metamere**

9. Nervous system with a double ventral nerve cord and a pair of ganglia with lateral nerves in each metamere; brain a pair of dorsally located cerebral ganglia

10. Sensory system of tactile organs, taste buds, statocysts (in some), photoreceptor cells, and eyes with lenses (in some)

11. Hermaphroditic or separate sexes; larvae, if present, are trochophore type; asexual reproduction by budding in some; spiral and determinate cleavage

Classes

The annelids are classified primarily on the basis of the presence or absence of parapodia, setae, metameres, and other morphologic features.

Class Polychaeta* (pol"e-ke'ta) (Gr. *polys,* many, + *chaitē,* long hair). Body of numerous segments with lateral parapodia bearing many setae; head distinct, with eyes and tentacles; clitellum absent; sexes usually separate; gonads transitory; asexual budding in some; trochophore larva usually; mostly marine.

Subclass Errantia (er-ran'she-a) (L. *errare,* to wander, + *-ia,* pl. suffix). Body of many segments, usually similar except in head and anal regions; parapodia alike and with acicula; pharynx usually protrusible; head appendages usually present; free living, tube dwelling, pelagic, mostly marine. Examples: *Neanthes, Aphrodite, Glycera.*

Subclass Sedentaria (sed-en-ta're-a) (L. *sedere,* to sit, + *-aria,* like or connected with). Body with unlike segments and parapodia and with regional differentiation;

*Since order Oligochaeta and class Polychaeta are both provided with setae, they are sometimes placed together in a group called Chaetopoda (ke-top'o-da) (Gr. *chaitē,* long hair, + *pous, podos,* foot).

prostomium small or indistinct; head appendages modified or absent; pharynx without jaws and mostly nonprotrusible; parapodia reduced and without acicula; gills anterior or absent; tube dwelling or in burrows. Examples: *Arenicola, Chaetopterus, Amphitrite.*

Class Clitellata (cli"tel-la'ta) (L. *clitellae,* packsaddle, + *-ata,* group suffix). Body with clitellum; segmentation conspicuous; segments with or without annuli; segments definite or indefinite in number; parapodia absent; hermaphroditic; eggs usually in cocoons; mostly freshwater and terrestrial.

   Order Oligochaeta* (ol"i-go-ke'ta) (Gr. *oligos,* few, + *chaitē,* long hair). Body with conspicuous segmentation; setae few per metamere; head absent; coelom spacious and usually divided by intersegmental septa; development direct, no larva; chiefly terrestrial and freshwater. Examples: *Lumbricus, Allolobophora, Aeolosoma, Tubifex.*

   Order Hirudinea (hir"u-din'e-a) (L. *hirudo,* leech, + *-ea,* characterized by). Body with definite number of segments (33 or 34) with many annuli; body with anterior and posterior suckers usually; setae usually absent; coelom closely packed with connective tissue and muscle; terrestrial, freshwater, and marine. Examples: *Hirudo, Placobdella, Macrobdella.*

Class Clitellata

The Clitellata include those annelids that bear a **clitellum.** The clitellum is a thickened saddle-like portion of certain segments (Fig. 13-4, *B*) that is involved in copulation and in the production of cocoons. This group includes the oligochaetes, in which segmentation is usually conspicuous and the setae are present but few in number, and the leeches, Hirudinea, which do not have setae and in which the somites are marked by transverse grooves, called annuli that produce an appearance of more segments than there really are. Both groups are hermaphroditic.

Order Oligochaeta

The more than 2,000 species of oligochaetes are found in great variety of sizes and habitats. Most are terrestrial or freshwater forms, but some are parasitic. They also vary a great deal in bodily structures and organization.

Nearly everyone is familiar with the common earthworm, for it is almost worldwide in distribution. One cannot spade the soil without coming in contact

*Since order Oligochaeta and class Polychaeta are both provided with setae, they are sometimes placed together in a group called Chaetopoda (ke-top'o-da) (Gr. *chaitē,* long hair, + *pous, podos,* foot).

with these worms. Moreover, they crawl out on the sidewalks after a heavy rain and are easily seen there. The German name for them is *Regenwürmer* (rainworms).

Earthworms like moist, rich soil for their burrows. Golf courses are excellent places to see these holes because there the same burrow may be used for a long period of time and the castings of the worms are much in evidence. *Eisenia foetida,* the brandling, a smaller worm and often studied in zoology, is found in manure piles. Sandy, clay, and acid soils deficient in humus are unfavorable for earthworms.

At night earthworms emerge from their burrows to explore their surroundings, often keeping their tails in their burrows for quick withdrawal when disturbed. During very dry weather the earthworm may coil up in a slime-lined chamber several feet underground and pass into a state of dormancy.

Giant earthworms are found in South America, South Africa, Ceylon, and Australia. Many of them belong to the genus *Thamnodrilus* of the family Glossoscolecidae and are often found in humid mountain forests in rich dark soil. They mostly occur in branched and interconnected tunnels. In body structure the first segments may be broader than the others and may number from 150 to more than 250 somites. Giant worms may reach a length of 10 to 12 feet and their egg capsules may average 2 inches in length and more than an inch in thickness. Each capsule may contain more than one worm, and they are usually found in dead-end tunnels and not on the surface of the ground.

Some of the more common oligochaetes, other than earthworms, are the small freshwater forms such as *Aeolosoma* (Fig. 13-1)(1 mm. long), which contains red or green pigments, has bundles of setae (tiny bristles), and is often found in hay cultures; *Nais* (2 to 4 mm. long), which is brownish in color, with two or three bundles of setae on each segment; *Stylaria* (10 to 25 mm. long), which has two bunches of setae on each segment, with the prostomium extended into a long process, and black eyespots; *Dero* (5 to 10 mm. long), which is reddish in color, lives in tubes, and has three tail gills; *Tubifex* (30 to 40 mm. long), which is reddish in color and lives with its head in mud at the bottom of ponds and its tail waving in the water; *Chaetogaster* (10 to 15 mm. long), which has only ventral bundles of setae; and the Enchytraeidae, small whitish worms that live in both moist soil and water. Some oligochaetes such as *Aeolosoma* may form chains of zooids asexually by transverse fission (Fig. 13-1).

Lumbricus terrestris, the common earthworm—a representative oligochaete

EXTERNAL FEATURES. The body of the earthworm is cylindric, tapered at the ends and divided into 100 to 175 metameres (more in giant earthworms). Few metameres are added after hatching. The usual length of the earthworm is from 5 to 12 inches. The mouth at the anterior end is overhung by a fleshy **prostomium** (Fig. 13-4, *A*). The anus is on the last segment. Straddled over the back like a saddle (somites 31 to 37) is the swollen, glandular **clitellum** (Fig. 13-4,*B*), which is important in the reproductive process.

The outer surface of the worm is covered by a thin, transparent **cuticle** secreted by the **epidermis** beneath it. The basement membrane beneath the epidermis rests upon the muscle layers that make up most of the body wall. Certain pigments such as protoporphyrin are also found in the body wall.

SETAE. With the exception of the first and last somites, each segment bears four pairs of chitinous setae, which are located on the ventral and lateral surfaces. Each seta (Fig. 13-2) is a bristlelike rod set in a sac within the body wall. It is moved by retractor and protractor muscles attached to the sac. The setae project through small pores in the cuticle to the outside. When an earthworm is moving forward, the setae anchor the somites of a part of the body and prevent backward slipping. Setae are also used by the worm to hold fast in the burrow, as all robins well know. When a seta is lost, a new one is formed in a reserve follicle to replace it.

BODY PLAN. The earthworms have a tube-within-a-tube arrangement (Fig. 13-3). There is a body wall of external **circular** muscles and a thicker internal layer of **longitudinal** muscles lying just underneath the epidermis and its basement membrane. The body wall surrounds the **coelomic cavity.** The coelomic cavity is divided by **septa,** which mark the boundaries of the somites (Figs. 13-3 and 13-4, *A*). These septa are not always complete and may be absent between certain somites. The inner surface of the coelom as well as the outer surface of its organs is lined with **peritoneum.**

Coelomic fluid within the cavity gives rigidity to the body by maintaining turgor. The fluid contains

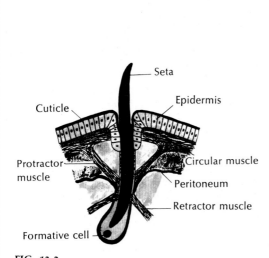

FIG. 13-2
Seta with its muscle attachments showing relation to adjacent structures. Setae lost by wear and tear are replaced by new ones, which develop from formative cell. (Modified from J. Stephenson, 1930, and others.)

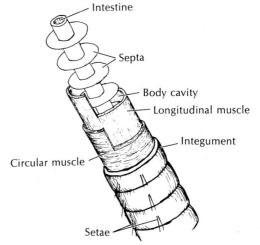

FIG. 13-3
Part of earthworm showing arrangement of muscle layers in body wall and septa. (Modified from Parry, D. A. 1960. Spider hydraulics. Endeavour **19:**156-162.)

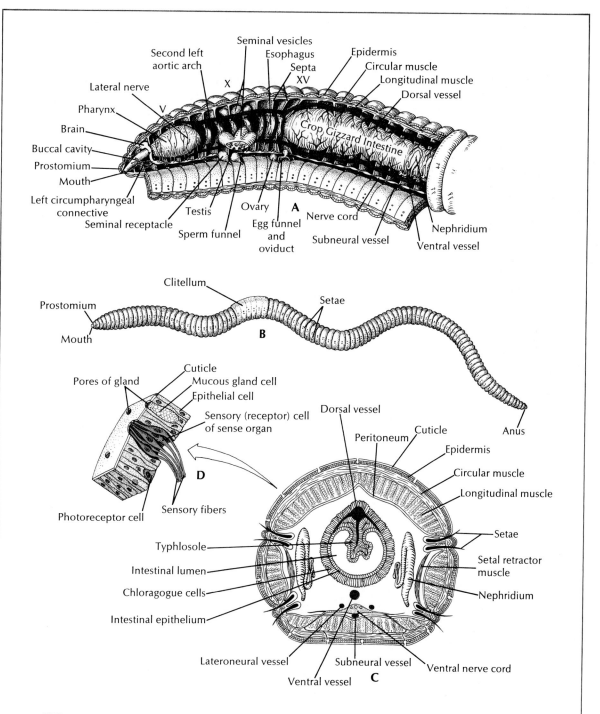

FIG. 13-4

Earthworm anatomy. **A,** Internal structure of anterior portion of worm. **B,** External features, lateral view. **C,** Generalized transverse section through region posterior to clitellum. **D,** Portion of epidermis showing sensory, glandular, and epithelial cells.

two main types of coelomic cells: **leukocytes,** which are phagocytic ameboid cells, and **eleocytes,** which come from the chloragogue cells of the digestive tract and carry nutritive granules to all parts of the body.

Through the center of the coelom runs the alimentary canal from the first to the last segment. The septa help hold it in place. The metameric arrangement is noticeable in the distribution of certain visceral organs; for instance, a pair of nephridia and a pair of nerve ganglia are located in each of the somites. Other visceral organs—reproductive apparatus and the main circulatory vessels—are closely located around the alimentary canal, which serves as a type of body axis.

LOCOMOTION. Earthworms move by a type of peristaltic movement. Contraction of circular muscles in the anterior end lengthen the body, thus pushing the anterior end forward where it is anchored by setae; contractions of longitudinal muscles then shorten the body, thus pulling the posterior end forward. As these waves of contraction pass along the entire body, it is gradually moved forward. Setae are rarely used as levers in crawling, but rather as anchors to prevent slipping. The fluid pressure in the individual coelomic compartments is a significant factor in the forward thrust of the body. Such pressure has been calculated to be equivalent to forces of $1\frac{1}{2}$ to 8 grams. When burrowing in soft soil, the earthworm thrusts its anterior end into crevices between soil particles and then enlarges its pharynx to push the soil aside. In firmer soil it burrows by literally eating its way through the soil, which accounts for the numerous castings.

Aquatic annelids swim by undulatory rather than peristaltic movements of the body.

DIGESTIVE SYSTEM. The alimentary canal of the earthworm (Fig. 13-4, *A*) is made up of (1) the **mouth** and **buccal cavity** in somites 1 to 3, (2) the muscular pharynx in somites 4 and 5, (3) the straight **esophagus,** with three pairs of **calciferous glands** in somites 6 to 14, (4) the thin-walled, enlarged **crop** in somites 15 to 16, (5) the thick, muscular **gizzard**

FIG. 13-5
Scheme of circulation in the earthworm. **A,** Somites 9, 10, and 11, showing three of the five aortic arches that receive blood from the dorsal vessel and pass it down to the ventral vessel to be carried posteriorly and distributed to the body tissues. **B,** Any somite posterior to the esophagus, showing distribution of blood from ventral vessel to intestine, nerve cord, and body wall and return of blood to dorsal vessel by the parietal and intestinal vessels. The dorsal vessel is the chief pumping organ.

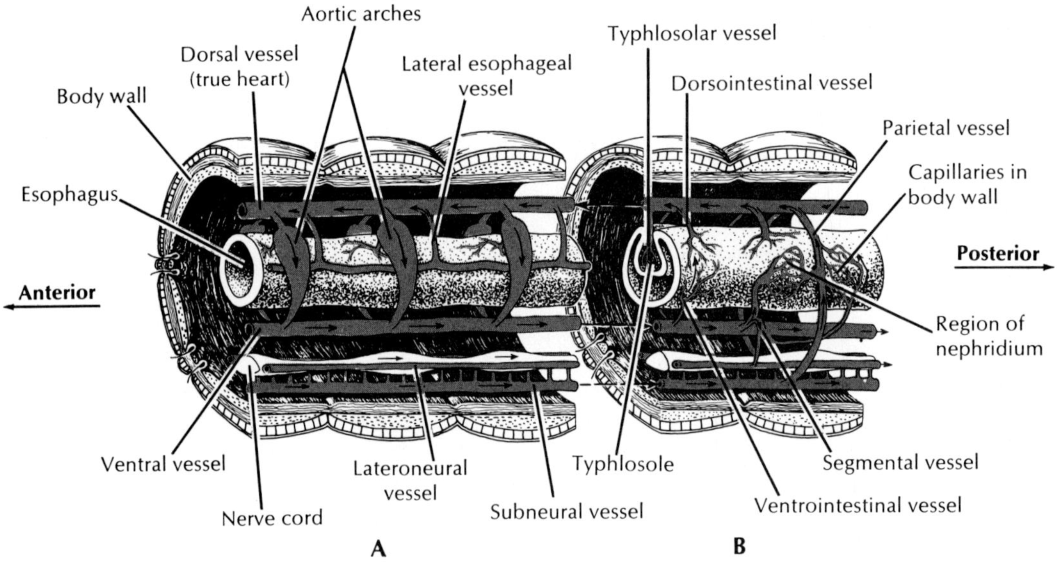

in somites 17 to 18, and (6) the long **intestine,** with slight bulges in each somite, from somite 19 to the last somite, where it ends in the **anus.**

A section of intestine reveals on its dorsal wall a peculiar infolded **typhlosole** (Figs. 13-4, *C,* and 13-5, *B*), which greatly increases the absorptive and digestive surface. The lining of the digestive system is made up of simple ciliated columnar epithelium. Longitudinal and circular muscles are found in the wall of the system, and curious yellow **chloragogue cells** (Fig. 13-4, *C*) from the peritoneum surround the digestive tract and fill much of the typhlosole. The chloragogue cells store foodstuffs and may convert protein into fat. When a cell is ripe (full of fat), its nucleus divides mitotically and the fat-containing portion of the cell constricts off as an **eleocyte.** The eleocytes, or wandering cells, apparently contribute their fat to other body cells. The chloragogue cells also function in excretion.

The food is mainly decayed organic matter, bits of leaves and vegetation, refuse, and animal matter. After being moistened by secretions from the mouth, food is drawn in by the sucking action of the muscular pharynx. The liplike prostomium aids in manipulating the food into position. The calciferous glands, by secreting calcium carbonate, may neutralize the acidity of the food, or they may simply excrete calcium as a metabolic product. Food is stored temporarily in the crop before being passed on into the gizzard, which grinds the food into small pieces. Digestion and absorption take place in the intestine. The digestive system secretes various enzymes to break down the food: pepsin, which acts upon protein; amylase, which acts upon carbohydrates; cellulase, which acts upon cellulose; and lipase, which acts upon fats. The indigestible residue is discharged through the anus. Naturally, earthworms take in a great deal of soil, sand, and other indigestible matter along with their food. In hard, firm soil, worms literally eat their way through the soil in their burrowing; this accounts for their numerous castings. The food products are absorbed into the blood, which carries them to the various parts of the body for assimilation. Some of the food is absorbed into the coelomic fluid, which also aids in food distribution.

CIRCULATORY SYSTEM. The circulatory system of the earthworm (Fig. 13-5) is a "closed system" consisting of a complicated pattern of blood vessels and capillaries that ramify to all parts of the body.

There are five main blood trunks, all running lengthwise through the body. These may be described as follows:

1. The **dorsal vessel** (single) runs above the alimentary canal from the pharynx to the anus. It is a pumping organ provided with valves and functions as the true heart. This vessel receives blood from vessels of the body wall and digestive tract and pumps it anteriorly into the five pairs of aortic arches. The chief function of the aortic arches is to maintain a steady pressure of blood into the ventral vessel.

2. The **ventral vessel** (single) lies between the alimentary canal and the nerve cord. This vessel serves as the aorta. It receives blood from the aortic arches and delivers it to the brain and rest of the body. As it passes backward, it gives off into each segment a pair of vessels to the walls and nephridia and a pair to the digestive tract.

3. The **lateral neural vessels** (paired) lie one on each side of the nerve cord. These receive the blood from the segmental vessels and carry it posteriorly with many branches to the nerve cord.

4. The **subneural vessel** (single), under the nerve cord, is the main vein. It receives blood from the nerve cord and passes it backward toward the tail and upward through paired **parietal vessels** in each segment (from somite 12 posteriorly). The parietal vessels also drain the blood from the nephridia and from the body wall and return it to the dorsal vessel.

In this scheme one can see that the ventral half of the alimentary tract is supplied from the ventral vessel and its dorsal half from the dorsal vessel. Nearly all the blood from the alimentary canal returns to the dorsal vessel by dorsointestinal vessels (two pairs per somite).

The propulsion necessary to force the blood along is provided by the peristaltic or milking action of the muscular walls of the blood vessels, particularly the dorsal vessel. Valves in the vessels prevent backflow.

The blood of the earthworm is made up of a **liquid plasma** in which are colorless ameboid cells, or **corpuscles.** Dissolved in the blood plasma is the pigment hemoglobin, of enormous molecular weight. This gives a red color to the blood and aids in the transportation of oxygen for respiration.

EXCRETION. The organs of excretion are the **nephridia,** a pair of which is found in each somite except the first three and the last one. Each one is

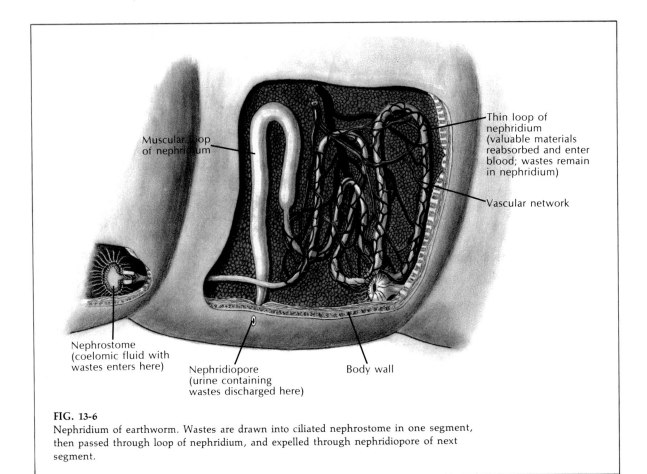

Muscular loop of nephridium

Thin loop of nephridium (valuable materials reabsorbed and enter blood; wastes remain in nephridium)

Vascular network

Nephrostome (coelomic fluid with wastes enters here)

Nephridiopore (urine containing wastes discharged here)

Body wall

FIG. 13-6
Nephridium of earthworm. Wastes are drawn into ciliated nephrostome in one segment, then passed through loop of nephridium, and expelled through nephridiopore of next segment.

found in parts of two successive somites (Fig. 13-6). A ciliated funnel, known as the **nephrostome,** is found just anterior to an intersegmental septum and leads by a small ciliated tubule through the septum into the somite behind, where it connects with the main part of the nephridium. This part of the nephridium is made up of several complex loops of increasing size, which finally terminate in a bladder-like structure leading to an aperture, the **nephridiopore;** this opens to the outside near the ventral row of setae. By means of cilia, wastes from the coelom are drawn into the nephrostome and tubule, where they are joined by organic wastes filtered from blood capillaries in the glandular part of the nephridium. All the waste is discharged to the outside through the nephridiopore.

Chloragogue cells may store waste temporarily before releasing it into the coelomic fluid. Some nitrog-

enous waste is also eliminated through the body surface.

Oligochaetes are largely freshwater animals, even such terrestrial forms as earthworms, which must exist in a moist environment. Osmoregulation is a function of both the body surface and the nephridia as well as the gut and the dorsal pores. *Lumbricus* will gain weight when placed in tap water and lose it when returned to the soil. Salts as well as water can pass across the integument, apparently by active transport.

RESPIRATION. The earthworm has no special respiratory organs, but the gaseous exchange is made in the moist skin, where oxygen is picked up and carbon dioxide given off. Blood capillaries are fairly numerous just below the cuticle, and the oxygen combines with the hemoglobin of the plasma and is carried to the various tissues.

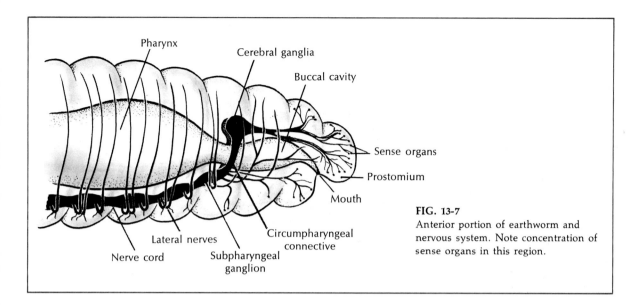

Pharynx

Cerebral ganglia

Buccal cavity

Sense organs

Prostomium

Mouth

Lateral nerves

Nerve cord

Circumpharyngeal connective

Subpharyngeal ganglion

FIG. 13-7
Anterior portion of earthworm and nervous system. Note concentration of sense organs in this region.

NERVOUS SYSTEM AND SENSE ORGANS. The nervous system in earthworms (Fig. 13-7) consists of a **central system** and **peripheral nerves.** The central system is made up of a pair of cerebral ganglia (the brain) above the pharynx and a pair of connectives passing around the pharynx connecting the brain with the first pair of ganglia in the nerve cord; a ventral nerve cord, really double, running along the floor of the coelom to the last somite; and a pair of fused ganglia on the nerve cord in each somite. Each pair of fused ganglia gives off to the body structures nerves containing both sensory and motor fibers. As in all higher forms, the sensory neurons carry impulses from special sensory cells in the epidermis to the nerve cord. Motor neurons run to muscles or glands.

Neurosecretory cells have been found in the brain and ganglia of annelids, both oligochaetes and polychaetes. They are endocrine in function and secrete neurohormones concerned with the regulation of reproduction, secondary sex characteristics, and regeneration.

For rapid escape movements the nerve cord of the earthworm is provided with three very large axons, commonly known as **giant fibers,** that run the length of the nerve cord (Fig. 13-8). Each of the giant fibers connects to one nerve cell body in each segment, with the cell bodies all lying in the midventral part of the cord. Examination with a fluorescent dye injection shows that the giant fibers are not made up of fused axons, as previously believed (B. Mulloney). Each giant fiber has three to five dendritic branches in each segment. The giant fibers carry impulses at a rate of 100 feet per second; in other nerves the speed is only 20 feet per second. The median fiber is concerned with sensory stimulation in the anterior part of the animal and transmits impulses to posterior effectors (muscles or glands). The lateral fibers (which are connected) are stimulated behind that level and send messages to anterior effectors. This mechanism explains the speed with which worms can withdraw into their burrows.

Sense organs are distributed all over the body and are particularly abundant at the anterior and posterior ends. Each sense organ consists of sensory cells surrounded by supporting epidermal cells (Fig. 13-4, *D*). The sensory cells are provided with small, hairlike tips, which project through pores in the cuticle; the bases of these projections are attached to sensory nerve fibers, which run to the central system. These sensory organs are somewhat similar to the taste buds of vertebrates. They are most numerous in the prostomium, where there may be as many as 700 per mm^2. There are also light-sensitive cells **(photoreceptors)** in the epidermis that pick up different degrees of light intensity (Fig. 13-4, *D*). In addition to these sensory organs, there are also free nerve endings between the cells of the epidermis, which are concerned with sense perception.

Earthworm behavior is largely, if not entirely, a

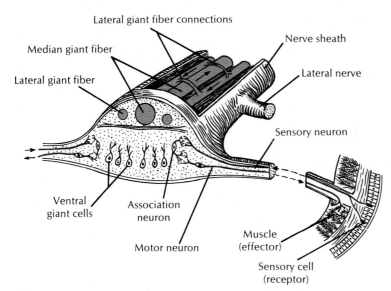

FIG. 13-8

Portion of nerve cord of earthworm showing arrangement of simple reflex arc *(in foreground)* and the three dorsal giant fibers *(in red)* that are adapted for rapid reflexes and escape movements. Ordinary crawling involves a succession of reflex acts, the stretching of one somite stimulating the next to stretch, and so on. Impulses are transmitted five times faster in giant fibers than in regular nerves so that all segments can contract simultaneously when quick withdrawal into a burrow is necessary.

matter of **reflex acts.** In the simplest reflex arc a sensory neuron makes direct contact with a motor neuron, and such arcs may occur in the nervous system of the earthworm. However, earthworms apparently have a more complicated kind of arc, which involves not only sensory and motor but also **association** neurons (Fig. 13-8). Stimuli are picked up by the **receptor,** or sensory neuron, and the impulse is carried into the central system; here, motor neurons may immediately receive it, or it first may be sent to the association neurons that relay it along to the motor. The motor neuron then carries the impulse to the **effector** (muscle or gland), and a reflex act is consummated. Reflex acts are usually more involved than this simple one, for many receptors are stimulated and many effectors act at the same time. The association neurons and fibers are mainly localized in the three **giant fibers** that are found in the dorsal side of the ventral nerve cord and connect the nerve cells in the ganglia with each other, thus facilitating widespread and rapid contractions of the worm's body.

REPRODUCTION AND DEVELOPMENT. Earthworms on monoecious (hermaphroditic); that is, both male and female organs are found in the same animal (Fig. 13-4, *A*). Two pairs of small testes lie in somites 10 and 11. A sperm funnel behind each testis is connected by a small tube (vas efferens) to one of a pair of sperm ducts (vas deferens), which pass posteriorly to the male openings on the ventral side of somite 15. The testes are surrounded by 3 pairs of large seminal vesicles in which immature sperm cells from the testes mature before being discharged during copulation through the sperm funnels and ducts.

A small pair of ovaries in somite 13 discharge mature eggs into the coelomic cavity where they are picked up by ciliated funnels and carried by oviducts to the outside through the female pores on somite 14. Two pairs of seminal receptacles in somites 9 and 10 receive and store sperm from another worm during copulation.

Reproduction in earthworms may occur at any season, but it is most common in warm moist weather

FIG. 13-9

Two earthworms in copulation. Anterior ends point in opposite directions as their ventral surfaces are held together by mucus bands secreted by clitellum. (Courtesy Dr. Guy Carter, Iowa City, Iowa.)

such as the spring of the year. The earthworm does not self-fertilize its eggs but receives sperm from another worm during **copulation,** which usually occurs at night. When two worms mate, they extend their anterior ends from their burrows and bring their ventral surfaces together with their anterior ends pointed in opposite directions (Fig. 13-9). This arrangement of the bodies places the seminal receptacle openings of one worm in opposition to the clitellum of the other worm (Fig. 13-10, *A*). The worms are held together by mucus bands and by special ventral setae, which penetrate each other's bodies in the regions of contact. Each worm secretes a slime tube about itself from somites 9 to 36. Sperm discharged from the sperm ducts of each worm travel by seminal grooves on the ventral surface to the openings of the seminal receptacles of the other worm (Fig. 13-10, *A*). After this reciprocal exchange

of sperm is made, the worms separate. This process of copulation requires about 2 hours.

Later each worm secretes a barrel-shaped cocoon about its clitellum within the posterior end of the slime tube. Eggs from the oviducts and albumin from the skin glands are passed into the cocoon while it still encircles the clitellum (Fig. 13-10, *B*). The worm then backs out, allowing the slime tube and cocoon to be slipped forward toward the head. As the cocoon passes the openings of the seminal receptacles sperm stored there from the mate are poured into it. Fertilization of the eggs now takes place within the cocoon (Fig. 13-10, *C*).

When the cocoon leaves the worm, its ends close, producing a lemon-shaped body. (Fig. 13-10, *E*). The size of the cocoon varies with different species of earthworms; those in *Lumbricus terrestris* are about 7 by 5 mm. In this form only one of several fertilized eggs develops into a worm, the others acting as nurse cells. Cocoons are commonly deposited in the earth, although they may be found at the surface near the entrance of burrows. Between copulations the earthworm continues to form cocoons so long as there are sperm in the seminal receptacles.

In their development the eggs are holoblastic, but the cleavage is unequal and spiral. The embryo passes through the blastula and gastrula stages and forms the three germ layers typical of the development of higher forms. The young worm is similar to the adult (Fig. 13-10, *F*) and escapes from the cocoon in 2 to 3 weeks. It does not develop a clitellum until it is sexually mature.

REGENERATION. Annelid worms are perhaps the most highly organized animals that have the power of complete regeneration. Not all annelids have this capacity, and in most of them there are limitations. Earthworms vary; some species can form two complete worms when cut in two, but other species cannot. In the common earthworm *(Lumbricus)* a posterior piece may regenerate a new head of three to five segments, and an anterior piece may form a new tail, with the level of the cut determining the number of segments regenerated. A cut at segment 50 will regenerate ten fewer segments than one made at the level of segment 40. Earthworms can also be grafted, and pieces of several worms have been grafted end to end to form long worms.

GENERAL BEHAVIOR. Earthworms are among the most defenseless of creatures, yet their abun-

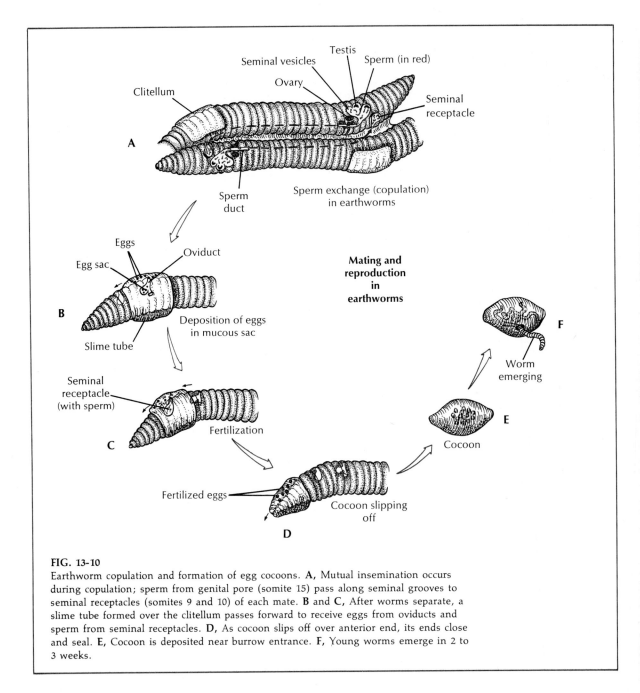

FIG. 13-10
Earthworm copulation and formation of egg cocoons. **A,** Mutual insemination occurs during copulation; sperm from genital pore (somite 15) pass along seminal grooves to seminal receptacles (somites 9 and 10) of each mate. **B** and **C,** After worms separate, a slime tube formed over the clitellum passes forward to receive eggs from oviducts and sperm from seminal receptacles. **D,** As cocoon slips off over anterior end, its ends close and seal. **E,** Cocoon is deposited near burrow entrance. **F,** Young worms emerge in 2 to 3 weeks.

dance and wide distribution indicate their ability to survive. Although they have no specialized sense organs, they are sensitive to many stimuli, such as **mechanical,** to which they are positive when it is moderate; **vibratory** (such as a footfall near them), which causes them to retire quickly into their burrows; and **light,** which they avoid unless it is very

weak. Chemical responses aid them in the choice of food.

Chemical responses as well as tactile are very important to the worm. It must be able to sample not only the organic content of the soil to find food, but must sense its texture, acidity, and calcium content. Earthworms can discriminate between different kinds of leaves and between acids, alkaloids, and sugars. *Lumbricus* will avoid soil with a pH below 4.1, for acid soils are deficient in calcium and some calcium is necessary for their well-being.

When irritated, earthworms eject coelomic fluid through their **dorsal pores,** one of which is located in each segment.

Experiments show that earthworms have some learning ability. They can be taught to avoid an electric shock, and thus an association reflex can be built up in them. Darwin credited earthworms with a great deal of intelligence in pulling leaves into their burrows, for they apparently seized the leaves by the narrow end, which is the easiest way for drawing such a shaped object into a small hole. Darwin assumed that the seizure of the leaves by the worms was not due to random handling or to chance but was purposeful in its mechanism. However, investigations since Darwin's time have shown that the process is mainly one of trial and error, for they often seize a leaf several times before attaining the right position.

EARTHWORM FARMING. Aristotle called earthworms the "intestines of the soil." Gilbert White, in his well-known *Natural History of Selborne,* spoke of their value in promoting the growth of vegetation by perforating and loosening the soil. He said a monograph on the value of earthworms to agriculture should be written some day.

About a century later, Charles Darwin fulfilled this need by writing his classic work *The Formation of Vegetable Mould Through the Action of Worms.* This work records the observations of the great naturalist on the habits of earthworms over a period of many years. He showed how worms brought the subsoil to the surface and mixed it with the topsoil. An earthworm can ingest its own weight in soil every 24 hours, and Darwin estimated that from 10 to 18 tons of dry earth per acre pass through their intestines annually and are brought to the surface, thus enriching the topsoil by bringing potassium and phosphorus up from the subsoil and also adding to the soil nitrogenous products from the worms' metabolism. All vegetable mold, he said, passes through their intestines many times. They expose the mold to the air and sift it into small particles. They drag leaves, twigs, and organic substances into their burrows, closer to the roots of plants.

Darwin doubted whether many other animals have played as important a part in the world's history as the lowly earthworms.

Many individuals have recommended culturing earthworms to build up the soil. Much work has been done on selective feeding and breeding with the idea of propagating the worms under favorable circumstances and then planting them and their capsules on soil to be improved. So far the results of such experiments are conflicting, and many scientists are not convinced that such methods are as effective as has been claimed.

Order Hirudinea (class Clitellata)

Leeches are mostly fluid feeders. Freshwater leeches inhabit ponds, lakes, quiet streams, and marshes. Marine forms are usually ectoparasitic and are found on crustaceans, turtles, fishes, and even dolphins.

Leeches have definite annelid characteristics, such as a ventral nerve cord with segmental ganglia, serial nephridia, and gonads in the coelom. The body is divided into somites (usually 34). However, these animals lack setae, which most other annelids possess, and they have copulatory organs and genital openings on the midventral line, where other annelids do not have them. Leeches appear to have more metameres than they really have because their somites are marked by transverse grooves (annuli) (Fig. 13-11).

Some leeches feed on tissue fluids and blood from open wounds, which they pump up by means of pharyngeal muscles. Others feed on vertebrate blood apparently obtained by discharging enzymes that digest the skin and blood vessel walls. Others live mainly on dead animals. Only the true bloodsuckers (certain genera of Hirudinidae including the so-called medicinal leech) have cutting plates, or "jaws," for actually cutting through tissues.

Tropical countries are plagued by more leeches than are temperate countries. Many of these attack human beings and are a nuisance. In Egypt and elsewhere in that region a small aquatic leech is often swallowed in drinking water and, by fastening on the pharynx and epiglottis, becomes a serious pest.

Hirudo medicinalis—the "medicinal leech"

STRUCTURAL CHARACTERISTICS. In this species and that of other true blood suckers the anterior sucker surrounds the mouth. The body is covered

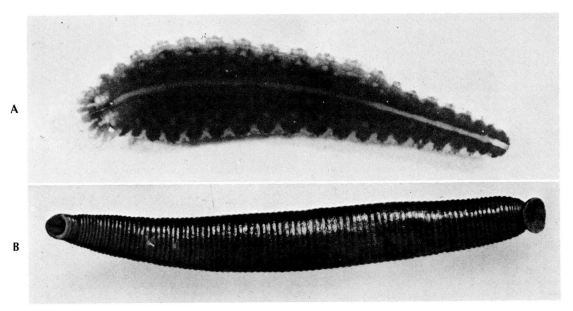

FIG. 13-11
A, *Placobdella,* common leech found on turtles. A living specimen, about 1 inch long. **B,**
Hirudo medicinalis, medicinal leech. This form was once widely used in bloodletting.
About 4 inches long but capable of great contraction and elongation.

with a cuticle, which is secreted by an epidermis beneath, and there are many mucous glands that open on the surface. A **clitellum** is found on segments 10 to 12 in the breeding season. Pigment and blood vessels are in the dermis. The muscular system is well developed, with circular, longitudinal, and oblique bands. Internally the **coelom** is reduced by mesenchyme and visceral organs to a system of sinuses.

The **alimentary canal** (Fig. 13-12, *A*) is made up of a mouth with three jaws and chitinous teeth, a muscular pharynx with salivary glands, a crop with eleven pairs of lateral ceca, a slender intestine, a rectum, and an anus; the latter opens anterior to the posterior sucker. The **circulatory system** has the typical annelid plan of dorsal, ventral, and lateral longitudinal vessels with many cross connections. **Respiration** takes place through the skin, and the excretory system has about 17 pairs of nephridia. The **nervous system** (Fig. 13-12, *B*) is not greatly different from that of the typical annelid plan, and there are sensory organs of taste, touch, and photoreception.

Leeches are hermaphroditic, but they have cross-fertilization. The **male organs** (Fig. 13-12, *B*) are the paired testes beneath the crop, a pair of vasa deferentia with some glands, and a median penis, which opens into the male pore. **Female organs** (Fig. 13-12, *B*) include a pair of ovaries with oviducts, an albumin gland, and a vagina, which opens near the male pore. The sperm is transferred in little packets (spermatophores) by the filiform penis, which penetrates the vagina of the mate in mutual copulation. The fertilized eggs are deposited in cocoons formed by glandular secretions. These cocoons may be attached to stones or other objects or even to the leech itself. Leeches have little or no regenerative ability.

BEHAVIOR. Leeches are found both on land and in water. They move in a manner similar to a measuring worm, that is, by looping movements of the body.

The bloodsucking leech attaches itself to the prey by a posterior sucker and then moves the anterior end about to locate a vulnerable spot to which to attach by the anterior sucker that surrounds the mouth and jaws. The jaws make a typical triradiate incision. Secretions from the leech's salivary glands contain an anticoagulant called **hirudin** that prevents clotting of the host's blood as it flows from the incision. The food is sucked up by the muscular pharynx and stored in the large crop. Blood suckers are said to be able to ingest as much as 3 to 4 times their body weight at a single feeding and then may not feed

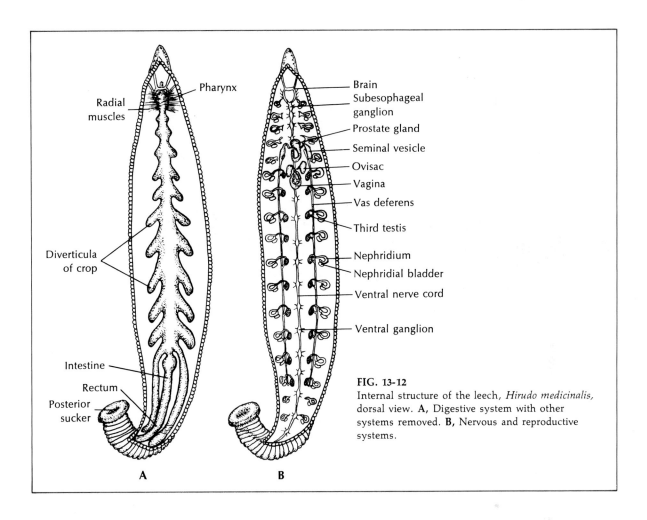

Radial muscles

Pharynx

Diverticula of crop

Intestine

Rectum

Posterior sucker

Brain
Subesophageal ganglion
Prostate gland
Seminal vesicle
Ovisac
Vagina
Vas deferens
Third testis
Nephridium
Nephridial bladder
Ventral nerve cord
Ventral ganglion

A   B

FIG. 13-12
Internal structure of the leech, *Hirudo medicinalis,*
dorsal view. **A,** Digestive system with other
systems removed. **B,** Nervous and reproductive
systems.

again for months. This is a useful adaptation, for leeches may not often have the chance to eat.

THE LEECH IN MEDICAL PRACTICE. For centuries the "medicinal leech" was employed in medical practice for bloodletting because of the mistaken idea that bodily disorders and fevers were caused by a plethora of blood. Since the leech is 4 or 5 inches long and can extend to a much greater length when distended with blood, the amount of blood it can suck out of a patient is considerable. Leech collecting and leech culture in ponds were practiced in Europe on a commercial scale during the nineteenth century. Wordsworth's interesting poem "The Leech-Gatherer" was based on this use of the leech.

Class Polychaeta

The polychaetes are the larger and older of the two classes of annelids, with more than 35,000 species,

most of them marine. Though the majority of them are from 5 to 10 cm. long, some are less than a millimeter and others may be as long as 3 meters. Some are brightly colored in reds and greens; others are dull or iridescent. Some are picturesque, such as the "feather-duster" worms (Figs. 13-13 and 13-14).

The name Polychaeta comes from a Greek word meaning "many setae," and herein lies one of the main differences between this class and the clitellates. Polychaetes have a well-differentiated head with sensory appendages and lateral parapodia with many setae, usually protruding in bundles from the parapodia. They show a pronounced differentiation of some body somites and a specialization of sensory organs practically unknown among clitellates. In contrast to clitellates, polychaetes have no permanent sex organs, possess no permanent ducts for their sex cells, and usually have separate sexes. Their

FIG. 13-13
A polychaete tubeworm, *Sabella pavonina.* The crown of tentacles is used in ciliary-mucus feeding. Photographed at depth of 35 meters off west coast of Norway. (Photo by T. Lundälv, Kristinebergs Zoological Station, Sweden.)

FIG. 13-14
Spirographis spallanzani, a polychaete tubeworm.

development is indirect, for they undergo a form of metamorphosis that involves a type of trochophore larva.

Ecologically polychaetes are divided into two groups. The Errantia include free-moving pelagic forms, active burrowers, crawlers, and tube worms that leave their tubes for feeding or breeding. The Sedentaria are sedentary worms that rarely expose more than the head end from the tubes or burrows in which they live.

Tube dwellers secrete many types of tubes. Some are parchmentlike (Figs. 13-13 and 13-17); some are firm, calcareous tubes attached to rocks or other surface (Fig. 13-14); and some are simply grains of sand or bits of shell or seaweed cemented together. Many burrowers in sand and mud flats line their burrows with mucus (Figs. 13-15 and 13-16).

Morphologically polychaetes may be classed as scale-bearing (the scaleworms), jaw-bearing (predatory forms), or crown-bearing (the ciliary-mucus feeders). There is typically a head or prostomium, which may or may not be retractile and which often bears eyes, antennae, and sensory palps. The first segment, the peristomium, surrounds the mouth and may bear setae or palps or chitinous jaws, or, as in the sabellid and serpulid worms, a tentacular crown. The trunk is segmented and most segments bear lateral fleshy expansions of the body called **parapodia,** which may have on them lobes, cirri, setae, etc. (Fig. 13-18, *C*). The parapodia are used in crawling, swim-

ming, or anchoring in tubes. They also serve as the chief respiratory organs, although some polychaetes also have gills. *Amphitrite,* for example, has three pairs of branched gills and long extensible tentacles (Fig. 13-15). *Arenicola,* the lugworm (Fig. 13-16), which burrows through the sand leaving characteristic castings at the entrance to its burrow, has paired gills on certain somites.

Most tube and burrow dwellers use mucus in some way to trap their food, often in combination with ciliary movement. Cilia on tentacles may create currents of water, filter out food particles, and move them toward the mouth. Or some worms make a mucus filter through which they pump water by rhythmic body contractions. Ciliary feeders may feed on plankton organisms or on the detritus that settles on a substratum. Terebellid polychaetes have long contractile tentacles that stretch out over a substrate to find food particles that are carried to the mouth in ciliated grooves (Fig. 13-15). In sabellids and serpulids, the tentacles form a stiff crown surrounding the head and their surface is increased by many fine pinnules (Fig. 13-13). The crown may be extended from the tube like an old fashioned feather duster. Here again food is trapped on the tentacles by mucus and carried to the mouth by cilia in grooves. Particles too large for the grooves are carried along the margins and cast off. Further sorting may occur near the mouth where only the smallest particles enter the mouth and larger ones are stored in a sac. Sand

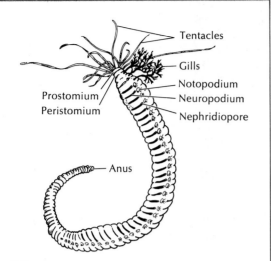

FIG. 13-15
Amphitrite ornata, a marine polychaete common to our eastern coast. It lives in mucus-lined tube buried in sand or mud near low-tide level. Its long extensible tentacles are feeding organs, each with ciliated groove that sweeps food particles toward mouth. Three pairs of gills are hollow, vascular outgrowths of body. Setae are located in the notopodia, and long, hooked seta, or uncinus, lies in each neuropodium.

grains from the sac are later used in enlarging the tube.

Chaetopterus is a polychaete that lives in a U-shaped parchment tube buried, except for the tapered ends, in sand or mud along the shore (Fig. 13-17). The worm attaches to the side of the tube by ventral suckers. Fans (especially modified parapodia) on segments 14, 15, and 16 pump water through the tube by rhythmic movements. A pair of enlarged parapodia in the tenth segment secrete mucus that forms a sheetlike film between the two. This extends back to form a long mucus bag that reaches a small food cup just in front of the fans. All the water passing through the tube is filtered through this mucus bag, the end of which is rolled up into a ball by cilia in the cup. When the ball is about the size of a BB shot, the fans stop beating and the ball of food and mucus is rolled forward to the mouth and swallowed.

Nereis (Neanthes) virens—clam worm

HABITAT. The clam worm (sandworm) is an errant polychaete that lives in or near the low-tide line of the seacoast. These animals often live in burrows that are lined with mucus from their bodies, and

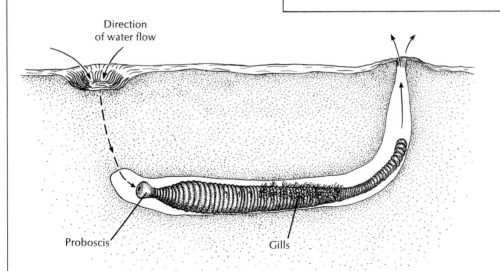

FIG. 13-16
Arenicola, the lugworm, lives in an L-shaped burrow in intertidal mud flats. It burrows by successive eversions and retractions of its proboscis. By peristaltic movements it keeps water filtering down through the sand and out the open end of the burrow. The worm then ingests the food-laden sand.

FIG. 13-17

Chaetopterus, a sedentery polychaete, in U tube and *Phascolosoma,* a sipunculid worm, in center. *Chaetopterus* lives in a parchment tube through which it pumps water with its three pistonlike fans. The fans beat 60 times per minute to keep water currents moving. The winglike notopodia of the twelfth segment continuously secrete a mucus net that strains out food particles. As the net fills with food, the food cup rolls it into a ball and, when the ball is large enough (about 3 mm.), the food cup bends forward and deposits the ball in a ciliated groove to be carried by cilia to the mouth and swallowed. (Courtesy The American Museum of Natural History, New York.)

sometimes they are found in temporary hiding places such as under stones where they stay, bodies covered and heads protruding. They are most active at night, when they wiggle out of their hiding places and swim about or crawl over the sand in search of food.

EXTERNAL STRUCTURE. The body is made up of about 200 somites and may be longer than 15 inches. The anterior somites are distinct from the others and form a definite **head,** which is divided into the **prostomium** and **peristomium** (Fig. 13-18, *A*). The prostomium has a pair of stubby **palps** (for touch and chemical sense), a pair of short **prostomial tentacles,** and two pairs of small dorsal **eyes.** The peristomium is made up of the ventral **mouth,** a pair of chitinous **jaws,** and four pairs of **peristomial ten-**

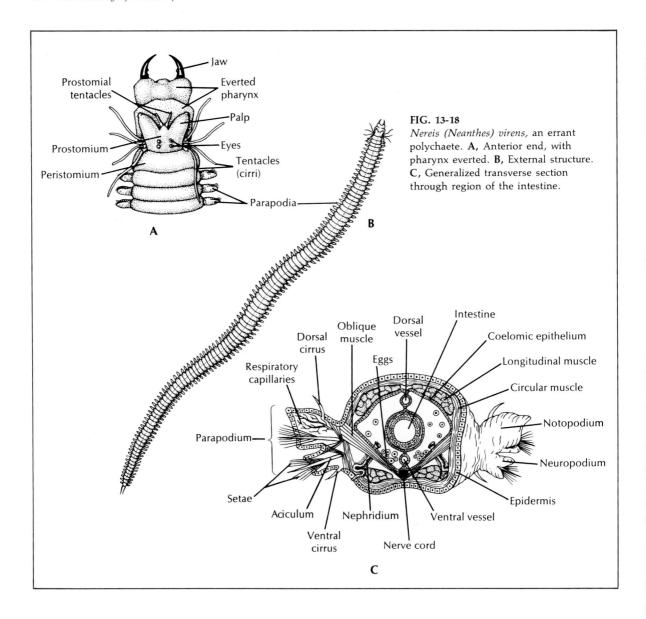

FIG. 13-18

Nereis (Neanthes) virens, an errant polychaete. **A,** Anterior end, with pharynx everted. **B,** External structure. **C,** Generalized transverse section through region of the intestine.

tacles on the dorsal side. The tentacles and eyes represent specialized organs for sensory perception; the tentacles are for touch, the palps are for taste and smell, and the eyes are for light perception.

Along the sides of the body are the fleshy **parapodia,** a pair to each somite except those forming the head. Each parapodium is formed of two lobes: a dorsal **notopodium** and a ventral **neuropodium** (Fig. 13-18, *C*). Each lobe is supported by one or more chitinous spines (acicula). The parapodia bear setae and are abundantly supplied with blood vessels.

Covering the body of the animal is a cuticle and epidermis, underneath which are circular and longitudinal muscles (Fig. 13-18, *C*).

LOCOMOTION. The worm moves by means of its circular and longitudinal muscles and by its parapodia. The latter are manipulated by oblique muscles that run from the midventral line to the parapodia in each somite. Parapodia are used both for creeping over the sand and for swimming. The animal swims by a lateral undulatory wriggling of the body unlike the peristaltic movement of the

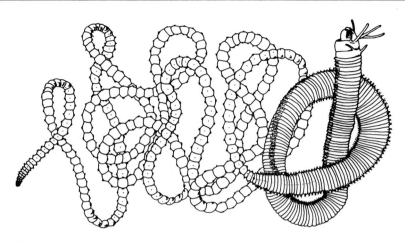

FIG. 13-19
Eunice viridis, the Samoan palolo worm. The posterior segments make up the epitokal region, consisting of segments packed with gametes and provided with an eyespot on the ventral side. Once a year the worms swarm and the epitokes detach, rise to the surface, and discharge their ripe gametes, leaving the water milky. By the next breeding season the epitokes are regenerated. (Modified from W. M. Woodworth, 1907.)

earthworm. It can dart through the water with considerable speed.

The worm will usually seek some kind of burrow if it can find one. When a worm is placed near a glass tube, it will wriggle in without hesitation. In its burrow it is able to suck or pump water in by dorsoventral undulatory movements that pass in waves from the anterior to the posterior end of the body.

INTERNAL STRUCTURE. The **coelomic cavity** is lined with peritoneum and is divided by septa between the somites. The **digestive system** contains a mouth, a protrusible pharynx with chitinous jaws, a short esophagus, a stomach-intestine, and an anus on the terminal segment. Digestive glands open into the esophagus.

The clam worm lives upon small animals, other worms, larval forms, etc. It seizes them with its chitinous jaws, which are protruded through the mouth when the pharynx is everted. By withdrawing the pharynx, the food is swallowed. Movement of the food through the alimentary canal is by peristalsis.

The circulatory, excretory, and nervous systems are basically similar to those of the oligochaetes. There are no special respiratory organs; this function is taken care of by the body wall and parapodia. Sensory organs include the **eyes** as photoreceptors, and

the palps and tentacles as organs of touch and taste.

Sexes are separate, but the reproductive organs are not permanent, for the sex cells are budded off from the coelomic lining and are carried to the outside by the nephridia and by bursting through the body wall. Fertilization is external, and the zygote develops into a free-swimming trochophore larva, which later transforms into a worm.

The palolo worm *Eunice viridis,* whose habitat is the South Pacific, lives in burrows among the coral reefs at the bottom of the sea. Just before swarming, their posterior somites become filled with enormous numbers of eggs or sperm (Fig. 13-19). These somites are cast off as one unit, called the epitoke, at the time of swarming and float to the surface, where they burst, releasing the sex cells. This swarming always occurs on the first day of the last quarter of the October-November moon and lasts for a few days. Usually, just before sunrise on that day, the surface of the sea is covered with these posterior units of the worms, which burst just as the sun rises. Fertilization of the eggs occurs at this time. The anterior part of the worm regenerates a new posterior part. A related form *(Leodice)* found in the Atlantic waters swarms in the third quarter of the June-July moon.

Derivation and meaning of names

Amphitrite (Greek mythology, sea goddess). A common genus of marine polychaete annelids.

Archiannelida (Gr. *archē,* origin, + L. *annellus,* little ring, + *-ida,* pl. suffix). A former class of simple, primitive worms now considered an order under Polychaeta.

Arenicola (L. *arena,* sand, + *colere,* to inhabit). A genus of polychaete worms.

Enchytraeidae (NL. *Enchytraeus,* from Gr., living in an earthen pot, + *-idae,* suffix for family). These small, white or reddish worms were often encountered by florists in potting plants.

Lumbricus (L. intestinal worm). Genus of common earthworm.

References

Barnes, R. D. 1965. Tube-building and feeding in chaetopterid polychaetes. Biol. Bull. **129:**217-233.

Barnes, R. D. 1968. Invertebrate zoology, ed. 2. Philadelphia. W. B. Saunders Co. *An excellent and comprehensive account of the Annelida.*

Barrett, T. J. 1947. Harnessing the earthworm. Boston, Bruce Humphries, Inc. *An explanation of earthworm farming and the possibilities it has. A popular account.*

Brown, C. H. 1950. Quinone tanning in the animal kingdom. Nature **165:**275.

Buchsbaum, R., and L. J. Milne. 1960. The lower animals. Garden City, N. Y., Doubleday & Co., Inc. *Excellent photographs (many in color) by two famous field naturalists.*

Cambridge Natural History. 1896. Annelida. London, Macmillan Co., Ltd. *Good account of morphology.*

Clark, L. B., and W. N. Hess. 1940. Swarming of the Atlantic palolo worm, *Leodice fucata.* Tortugas Lab. Papers **33:**21.

Clark, R. B., and U. Scully. 1964. Hormonal control of growth in *Nereis diversicolor.* Gen. Comp. Endocrinol. **4:**(1):82-90.

Dales, R. P. 1967. Annelids. New York, Hutchinson & Co. *A concise up-to-date account of the annelids.*

Darwin, C. R. 1911. The formation of vegetable mould through the action of worms. *A classic account of the way in which earthworms improve and transform the surface of the soil.*

Durchon, M. 1962. Neurosecretion and hormonal control of reproduction in Annelida. Gen. Comp. Endocrinol. 1(Suppl. 1):227-240.

Edmondson, W. T. (editor). 1959. Fresh-water biology, ed. 2. New York, John Wiley & Sons, Inc. *Good sections on aquatic Oligochaeta by C. J. Goodnight and on Polychaeta by O. Hartman. Keys and figures included.*

Fox, H. M. 1938. On the blood circulation and metabolism of sabellids. Proc. Roy. Soc. Bull. **125:**554-556.

Hanson, J. 1948. Formation and breakdown of serpulid tubes. Nature **161:**610.

Hanson, J. 1949. The histology of the blood system in Oligochaeta and Polychaeta. Biol. Rev. **24:**127-173.

Hess, W. N. 1925. Nervous system of the earthworm, *Lumbricus terrestris* L. J. Morph. Physiol. **40:**235-259. *A widely known investigation of great value to all students of the annelids.*

Hickman, C. P. 1973. Biology of the invertebrates, ed. 2. St. Louis, The C. V. Mosby Co.

Krivanek, J. O. 1956. Habit formation in the earthworm *Lumbricus terrestris.* Physiol. Zool. **29:**241-250. *An interesting study in animal behavior.*

Laverack, M. S. 1963. The physiology of earthworms. New York, The Macmillan Co.

Mann, K. H. 1962. Leeches *(Hirudinea),* their structure, physiology, ecology, and embryology. New York, Pergamon Press, Inc.

Manwell, C. 1960. Comparative physiology: blood pigments. Ann. Rev. Physiol. **22:**191-244.

Michaelsen, W. 1900. Oligochaeta. In Das Tierreich. Berlin, Friedlander & Sohn.

Moment, G. B. 1953. On the way a common earthworm, *Eisenia foetida,* grows in length. J. Morph. **93:**489-503.

Moore, J. P. 1956. Annelida. Encyclopaedia Britannica, Chicago, Encyclopaedia Britannica, Inc. *A comprehensive description of the group with many revealing illustrations.*

Parker, T. J., and W. A. Haswell. 1972. A textbook of zoology, vol. 1, ed. 7. New York, The Macmillan Co. *A comprehensive treatment of the phylum, with emphasis upon morphology.*

Russell-Hunter, W. D. 1969. A biology of the higher invertebrates. New York, The Macmillan Co. *A concise and up-to-date discussion of this great group.*

Scharre, E. 1962. Neurosecretion in *Lumbricus terrestris.* Gen. Comp. Endocrinol. **2**(1):1-3.

Stephenson, J. 1930. The oligochaetes, Oxford, Oxford University Press.

Wells, M. 1968. Lower animals. New York, McGraw-Hill Book Co.

Woodworth, W. M. 1907. The palolo worm, *Eunice viridis* (Gray). Bull. Museum Zool. Harvard Coll. **51:**(1):6-21.

Zappler, G. 1958. Darwin's worms. Natural History **67:**488-495. *An account of an experiment on the annelid intelligence by the great naturalist. Should be read by all beginning zoology students.*

Spiders are known for their careful architecture spun from silk from their own silk glands. This web, found on a dewy morning, was 2 to 3 feet across.

CHAPTER 14

THE ARTHROPODS

PHYLUM ARTHROPODA
A MAJOR PROTOSTOME GROUP
SUBPHYLUM TRILOBITA
SUBPHYLUM CHELICERATA

Position of arthropods in animal kingdom

1. Arthropods belong to the protostome branch, or schizocoelous coelomates, of the animal kingdom and have spiral and determinate cleavage.

2. Arthropods have much of the characteristic structure of higher forms—bilateral symmetry, triploblastic, coelomic cavity, and organ systems.

3. They share with the annelids the property of conspicuous segmentation, but in their somites they emphasize greater variety and more grouping for specialized purposes, and to the somites have been added the jointed appendages with pronounced division of labor, resulting in greater variety of action.

Biologic contributions

1. **Cephalization** makes additional advancements, with centralization of fused ganglia and sensory organs in the head.

2. The **somites** have gone beyond the sameness of the annelid type and are now **specialized** for a variety of purposes, forming functional groups of somites (tagmosis).

3. The presence of paired **jointed appendages** diversified for numerous uses makes for greater adaptability.

4. Locomotion is by extrinsic limb muscles, in contrast to the body musculature of annelids. **Striated muscles** are emphasized, thus ensuring rapidity of movement.

5. Although **chitin** is found in a few other forms below arthropods, its use is better developed in the arthropods.

6. The gills, and especially the **tracheae**, represent a breathing mechanism more efficient than that of most invertebrates.

7. The alimentary canal shows greater specializa-

tion by having chitinous teeth, compartments, and gastric ossicles.

8. Behavior patterns have advanced far beyond those of most invertebrates, with a higher development of primitive intelligence and **social** organization.

9. **Metamorphosis** is common in development.

10. Many arthropods have well-developed protective coloration and protective resemblances.

Phylogeny and adaptive radiation

PHYLOGENY. The evidence indicates that arthropods are more closely related to the annelids than to any other group. Although they cannot be said to come directly from annelids, they probably come from the same ancestors as did segmented worms such as the polychaetes.

Phylum Onychophora, with both annelid and arthropod characters, seems to represent a connecting link and may be considered a descendant from a line close to the primitive ancestor of annelids and arthropods (see p. 385).

Within the phylum, the branchiopods are probably the most primitive. The crustaceans, myriapods, and insects seem to have evolved along one line and the radically different chelicerates along another. The insects appear to have been derived from primitive Precambrian annelids and are more closely related to the centipedes, millipedes, and symphylids than to other arthropods.

ADAPTIVE RADIATION. In their evolutionary diversity arthropods seem to have been guided chiefly by the various modifications and specialization of their chitinous exoskeleton and jointed appendages. They are adapted to more types of habitats than are any other phylum. By their locomotory mechanisms and other adaptations they were one of the first groups to make the transition from water to land, where they have undergone amazing adaptive radiation. Being able to walk without dragging the body gave them a tremendous adaptive advantage over lower invertebrates.

■ THE ARTHROPODS

Phylum Arthropoda is the most extensive phylum in the animal kingdom, containing over three fourths of all known forms. Between 700,000 and 800,000 species have been recorded and probably as many more remain to be classified. The phylum includes spiders, scorpions, ticks, mites, crustaceans, millipedes, centipedes, and insects. In addition, many fossil forms are arthropods, for the phylum goes back to Precambrian times.

Arthropoda (ar-throp'o-da) means "joint footed," the name coming from Greek *arthros,* joint, and *pous,* foot. Arthropods are characterized by a chitinous exoskeleton and a linear series of somites, each with a pair of jointed appendages. Body organs and systems are well developed, for they represent, on the whole, an active and energetic group. They share with the nematodes an almost complete lack of cilia.

Arthropods are one of man's greatest competitors for food supplies, and they spread serious diseases. However, not all arthropods are harmful; many crustaceans serve as food, silkworms furnish silk, insects cross-pollinate plants, bees furnish honey and beeswax, and other insects yield useful drugs and dyes.

Ecologic relationships

The arthropods are more widely and more densely distributed throughout all regions of the earth than are members of any other phylum. They are found in all types of environment from low ocean depths to very high altitudes, and from the tropics far into both north and south polar regions. Different species are adapted for life in the air, on land, in fresh, brackish, and marine waters, and in or on the bodies of plants and other animals. Some species live in places where no other form could survive.

Although all types—carnivorous, omnivorous, and symbiotic—occur in this vast group, the majority are herbivorous. Most aquatic arthropods depend on algae for their nourishment, and the majority of land forms live chiefly on plants. In diversity of ecologic distribution the arthropods have no rivals.

Characteristics

1. Symmetry bilateral; triploblastic; body metameric
2. **Appendages jointed,** with one or two pairs to a somite and often modified for specialized functions
3. **Exoskeleton of chitin** secreted by the underlying epidermis and shed at intervals
4. Body often divided into **three regions:** the **head,** usually of six somites, the **thorax,** and the **abdomen,** the latter two divisions having a variable

number of somites; head and thorax often united into a cephalothorax

5. Muscles mostly **striated** and rapid in action; unstriated muscle in visceral organs

6. True coelom small in adult; most of body cavity a **hemocoel filled with blood**

7. Digestive system complete with mouth, enteron, and anus; **mouthparts modified from somites and adapted for different methods of feeding**

8. Circulatory system open, with dorsal heart, arteries, and mesenchymal blood cavities (sinuses)

9. **Cilia practically absent throughout group**

10. Respiration by body surface, **gills, air tubes (tracheae),** or **book lungs**

11. Excretory system by **green glands** or by a variable number of **malpighian tubules** opening into the digestive system

12. Nervous system of dorsal brain connected by a ring around the gullet to a double nerve chain of ventral ganglia

13. Sensory organs well developed and include eyes, antennae (tactile and chemical), balancing organs, auditory organs, and sensory bristles

14. Sexes nearly always separate, with paired reproductive organs and ducts; fertilization internal; oviparous or ovoviviparous; metamorphosis direct or indirect; parthenogenesis in a few forms

Comparison of Arthropoda with Annelida

Similarities between Arthropoda and Annelida are as follows:

1. External segmentation

2. Segmental arrangement of muscles

3. Metamerically arranged nervous system with dorsal cerebral ganglia

The most striking differences between the two phyla are that arthropods have the following:

1. Fixed number of segments

2. Usually a lack of intersegmental septa

3. Body segments usually grouped into three regions—head, thorax, and abdomen

4. Coelomic cavity reduced

5. Open (lacunar) circulatory system

6. Special mechanisms (gills, tracheae, book lungs) for respiration

7. Chitin for exoskeleton

8. Jointed appendages

9. Compound eye and other well-developed sense organs

10. Absence of cilia

11. Metamorphosis in many cases

Classification of arthropods

Subphylum Trilobita (try"lo-by'ta) (Gr. *tri-*, three, + *lobos*, lobe). All fossil forms; Cambrian to Permian; thorax and abdomen distinct; biramous appendages on all but last somite. Example: *Triarthrus*.

Subphylum Chelicerata (ke-lis"e-ra'ta) (Gr. *chēlē*, claw, + *keras*, horn, + *-ata*, group suffix). First pair of appendages modified to form chelicerae with claws; pair of pedipalps and four pairs of legs; no antennae; cephalothorax and abdomen usually unsegmented.

    Class Merostomata (mer-o-sto'ma-ta) (Gr. *mēros*, thigh, + *stoma*, mouth, + *-ata*, group suffix). Aquatic chelicerates; cephalothorax; compound lateral eyes; appendages with gills; sharp telson.

        Subclass Eurypterida (yu-rip-ter'i-da) (Gr. *eurys*, broad, + *pteryx*, wing or fin, + *-ida*, pl. suffix). Extinct; cephalothorax covered by dorsal carapace; abdomen with 12 segments and postanal telson; pair of simple ocelli and pair of compound eyes. Example: *Eurypterus*.

        Subclass Xiphosurida (zif-o-su'ri-da) (Gr. *xiphos*, sword, + *oura*, tail). Cephalothorax convex- and horseshoe-shaped abdomen unsegmented and terminated by long spine; three-jointed chelicerae and six-jointed walking legs; pair of simple eyes and pair of compound eyes; book gills. Example: *Limulus* (horseshoe, or king crab).

    Class Pycnogonida (pik-no-gon'i-da) (Gr. *pyknos*, compact, + *gony*, knee) **(Pantopoda).** Size usually small (3 to 4 mm.), but some reach 500 mm.; body chiefly cephalothorax; abdomen tiny; usually eight pairs of long walking legs, but some with 10 to 12 pairs; pair of subsidiary legs (ovigers) for egg bearing; mouth on long proboscis; four simple eyes; no respiratory or excretory system. Example: *Pycnogonum* (sea spider).

    Class Arachnida (ar-ak'ni-da) (Gr. *arachnē*, spider). Chelicerae with claws, pedipalps, and four pairs of legs; no antennae or true jaws; abdomen with or without appendages and generally distinct from cephalothorax; respiration by gills, tracheae, or book lungs; excretion by malpighian tubules or coxal glands; dorsal bilobed brain connected to ventral ganglionic mass with nerves; simple eyes; sexes separate; chiefly oviparous; no true metamorphosis. Examples: scorpions, spiders, mites, ticks, harvestmen.

Subphylum Mandibulata (man-dib"u-la'ta) (L. *mandibula*, mandible, + *-ata*, group suffix). One or two pairs of antennae form first two pairs of cephalic appendages, and functional jaws (mandibles) form third pair.

    Class Crustacea (crus-ta'she-a) (L. *crusta*, shell, + *-acea*, characterized by). Aquatic with gills; body with dorsal carapace: telson at posterior end; hard exoskeleton of chitin reinforced with limy salts; appendages biramous and modified for capturing food, walking, swimming, respiration, and reproduction; coelom reduced and hemocoel present; head of five segments with two pairs of antennae, a pair of jaws, and two pairs of maxillae; sexes usually separate; development with nauplius stage. See résumé of subclasses on pp. 333 to 339.

Class Diplopoda (di-plop′o-da) (Gr. *diploos,* double, + *pous, podos,* foot). Body subcylindric; head with short antennae and simple eyes; body with variable number of somites; short legs, usually two pairs to a somite; maxillae and jaws; separate sexes; oviparous; malpighian tubules for excretion; dorsal brain with double ventral nerve cord. Examples: *Julus, Spirobolus* (millipedes).

Class Chilopoda (ki-lop′o-da) (Gr. *cheilos,* lip, + *pous, podos,* foot). Form elongated and dorsoventrally flattened; pair each of long antennae, jaws, and maxillae; variable number of somites, each with a pair of legs; malpighian tubules for excretion; respiration with tracheae; separate sexes; oviparous; dorsal brain and double ventral nerve cord; pair of long antennae. Examples: *Cermatia, Lithobius, Geophilus* (centipedes).

Class Pauropoda (pau-rop′o-da) (Gr. *pauros,* small, + *pous, podos,* foot). Minute (1 to 1.5 mm.); cylindric body of double segments and bearing nine or ten pairs of legs; no eyes; genital openings near head end. Example: *Pauropus.*

Class Symphyla (sym′filla) (Gr. *syn,* together, + *phylon,* tribe). Slender (1 to 8 mm.) with long, filiform antennae; body of 15 to 22 segments with 10 to 12 pairs of legs; no eyes; genital openings near head end. Example: *Scutigerella* (garden centipede).

Class Insecta (in-sec′ta) (L. *insectus,* cut into). Body with head, thorax, and abdomen distinct and usually pronounced constriction between thorax and abdomen; pair of antennae; mouth parts modified for different food habits; head with six somites; thorax with three somites, and abdomen with variable number, usually 11 somites; thorax with two pairs of wings (sometimes one pair or none) and three pairs of jointed legs; respiration by branched tracheae; brain of fused ganglia and double ventral nerve cord; eyes both simple and compound; separate sexes; usually oviparous; metamorphosis gradual or abrupt. A brief description of insect orders is given on pp. 370 to 377.

Why have arthropods been so successful?

Some of the criteria used in judging the success of an animal group are the number and variety of its species, the variety of its habitats, how widely it is distributed, its powers of self-defense, the variety of its food habits, and its adaptability to changing conditions. Arthropods are so numerous and so diversified that they obviously have met most of these

conditions. Some of the structural and physiologic patterns that have been helpful to them are briefly summarized below.

CHITIN. Chitin is a nonliving and noncellular protein-carbohydrate compound secreted by the underlying epidermis. It is made up of an outer waxy layer, a middle horny layer, and an inner flexible layer. The chitin may be soft and permeable or it may form a veritable coat of armor. Between joints it is flexible and thin to permit free movements. In crustaceans it is harder because it is infiltrated with calcium salts. In general, it is admirably adapted for attachment of muscles, serving as levers and centers of movement, preventing the entrance and loss of water, and affording the maximum of protection without sacrificing mobility.

Chitin is also used for biting mouthparts, grinders in the stomach, lenses for the eyes, sound production, sensory organs, copulatory organs, and ornamental purposes.

Because a chitinous exoskeleton is nonliving, it cannot grow. An arthropod, as it grows, must shed its outer shell at intervals and grow a larger one—a process called **ecdysis,** or molting. Arthropods molt from four to seven times before reaching adulthood. Such an exoskeleton naturally limits the size of arthropods. Few of them exceed 2 feet in length and most are far below this limit. The largest is the Japanese crab *Macrocheira,* which has about an 11-foot span; the smallest is the parasitic mite *Demodex,* which is less than 0.1 mm. long.

SEGMENTATION AND APPENDAGES. Arthropods share with annelids and vertebrates the characteristic of segmentation. Typically each somite is provided with a pair of jointed appendages, but this arrangement is often modified, with both segments and appendages specialized for adaptive functions. This has made for greater efficiency and wider capacity for adjustment to different habitats.

RESPIRATORY DEVICES. Aquatic arthropods breathe mainly by some form of gill that is efficient; most land forms have the highly efficient tracheal system of air tubes that delivers oxygen directly to the tissue cells and makes high metabolism possible.

SENSORY ORGANS. Sensory organs are found in great variety, from the mosaic eye to those simple senses that have to do with touch, smell, hearing, balancing, chemical reception, etc. Arthropods are keenly alert to what goes on in their environment.

LOCOMOTION. The jointed appendages, equipped with striated muscles, have been modified and adapted into swift and efficient walking legs, swimming appendages, or wings.

REPRODUCTION AND METAMORPHOSIS. Arthropods lay large numbers of eggs, and many of them pass through metamorphic changes—larva, pupa, and adult stages, and some go through a series of nymphal stages preceding adulthood. The larval form is often adapted for eating a different kind of food from that of the adult, resulting in less competition within a species.

BEHAVIOR PATTERNS. Arthropods exceed most other invertebrates in the complexity and organization of their activities. Whether learning is involved to any great extent in their reactions to environmental stimuli is still open to question, but no one can deny the complex adaptability of the group. Fiddler crabs, for example, undergo a rhythmic color change of dark by day and light by night, regulated by means of an internal "time clock," which may work independently of external influences, can be set for different cycles, and is definitely inherited. As for social organization, no other group of invertebrates has carried this trait so far. The gregarious termites, ants, and bees have worked out marvelous systems of division of labor of great ingenuity and complexity.

Economic importance

Arthropods serve as an important source of food for man and other animals. Lobsters, crabs, shrimp, and crayfish are eaten all over the world. Plankton contains many crustaceans and is food for fish and other aquatic animals. Insects are important as food for many birds and animals, and some are relished by man in various areas of the world.

They are useful in other ways than as food. Some insects are predators that live on other insects, helping to keep the harmful ones in check. Arthropod products include shellac produced by scale insects, cochineal (a dye) from other scale insects, silk spun by silkworm larvae, honey and beeswax from bees, and so on. They are also useful in pollination of trees and other plants.

Many arthropods are harmful. Insects destroy millions of dollars worth of food each year. Arthropods carry devastating diseases: A mosquito carries malaria; copepods carry larval stages of the guinea worm and fish tapeworm; mites and ticks carry diseases and also live as ectoparasites; some spiders and scorpions deliver poisonous bites; barnacles foul ship bottoms; sow bugs and other insects damage gardens and greenhouse crops.

Subphylum Trilobita

The trilobites probably had their beginnings millions of years before the Cambrian period in which they flourished. They have been extinct some 200 million years. Their Precambrian ancestor probably also gave rise to all other arthropods as well. Their marine fossils were laid down when seas covered

FIG. 14-1
Trilobite (dorsal view) from plaster cast impression. All members of this class are now extinct. Some of the abundant fossils of this group may be the remains of molted exoskeletons.

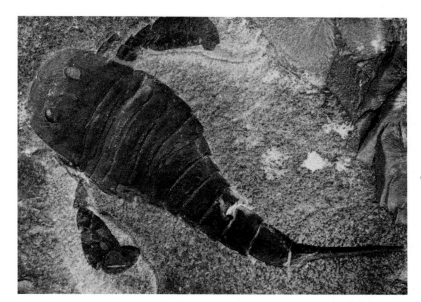

FIG. 14-2
Eurypterus, a fossil arthropod. Eurypterids flourished in Europe and North America from Ordovician to Permian periods.

what is now land and are now useful to geologists in the indexing of rock layers.

Their hard chitinous-calcareous shell was divided into three longitudinal lobes by furrows (Fig. 14-1). The head was one piece but showed signs of former segmentation; the thorax had a variable number of somites; and the somites of the pygidium, at the posterior end, were fused into a plate. The head bore a pair of antennae, compound eyes, mouth, and four pairs of biramous (two-branched) jointed append-ages. Each body somite except the last also bore a pair of biramous appendages with a fringe of fila-ments probably serving as gills.

Most trilobites could roll up like pill bugs and were from 2 to 27 inches long.

Subphylum Chelicerata

The chelicerate arthropods make up a very ancient group that includes the eurypterids (extinct), horse-shoe crabs, spiders, ticks and mites, scorpions, and sea spiders. They are characterized by having six pairs of appendages that include a pair of pedipalps, a pair of chelicerae, and four pairs of legs. They have no mandibles or antennae. They suck up liquid food from their prey.

Class Merostomata

This class is represented by the eurypterids, all now extinct, and by the xiphosurids, or horseshoe

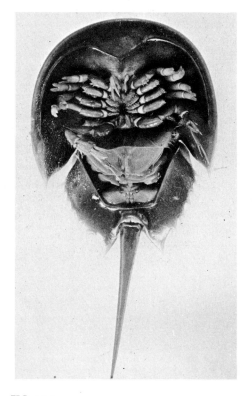

FIG. 14-3
Ventral view of horseshoe crab, *Limulus* (class Merostomata). They grow up to 15 to 18 inches long.

FIG. 14-4
Horseshoe crabs mate in shallow water at high tide. The larger female burrows into the
sand to lay her many eggs while the male covers them with sperm. As the tide recedes,
she covers her eggs with sand and they return to the sea. The eggs, warmed by the sun,
hatch in about 2 weeks. (Courtesy L. L. Rue, III, Blairstown, N. J.)

crabs, an ancient group sometimes referred to as
"living fossils."

SUBCLASS EURYPTERIDA. The eurypterids
(Fig. 14-2) were the largest of all fossil arthropods,
with some of them reaching a length of 9 feet. They
are found in rocks from the Ordovician to the Carboniferous periods. They had many resemblances to
the marine horseshoe crabs (Figs. 14-3 and 14-4) and
also to the scorpions, their land counterparts. The
head had six fused segments and bore both simple

and compound eyes and six pairs of appendages. The body was encased in a chitinous covering. The abdomen had 12 segments and a spikelike tail.

Theories of their early habitats differ. Some believe the eurypterids evolved mainly in fresh water, along with the ostracoderms; others hold that they arose in brackish lagoons.

SUBCLASS XIPHOSURIDA. The xiphosurids are an ancient group that goes back to the Cambrian period. Our common horseshoe crab *Limulus (Xiphosura)* (Fig. 14-3) goes back practically unchanged to the Triassic period. There are only three living genera (five species) living today: *Limulus,* which lives in shallow water along the North American Atlantic coast, *Carcinoscorpius* along the southern shore of Japan, and *Tachypleus* in the East Indies and along the coast of southern Asia. They usually live in shallow water.

Xiphosurids have an unsegmented, horseshoe-shaped **carapace** (hard dorsal shield) and a broad hexagonal abdomen, which has a long **telson,** or tailpiece. The cephalothorax bears five pairs of walking legs and a pair of chelicerae, whereas the abdomen has six pairs of broad thin appendages that are fused in the median line. On some of the abdominal appendages, **book gills** (flat, leaflike gills) are exposed. There are two compound and two simple eyes on the carapace. The horseshoe crab swims by means of its abdominal plates and can walk with its walking legs. It feeds at night on worms and small mollusks, which it seizes with its chelicerae.

During the mating season the horseshoe crabs come to shore at high tide to mate (Fig. 14-4). The female burrows into the sand where she lays her eggs, with the smaller male following her closely to add his sperm to the nest before she covers the eggs with sand. Here the eggs are warmed by the sun and protected from the waves until the young larvae hatch and return to the sea by another high tide. The larvae are segmented and are often called "trilobite larvae" because they resemble the trilobites, which are often considered to have been their ancestors.

Class Pycnogonida—the sea spiders

Some sea spiders are only a few millimeters long, but others are much larger. They have small, thin bodies and usually four pairs of long, thin legs. Some species are provided with chelicerae and palps. One unique feature is the subsidiary pair of egg-bearing legs (ovigers) on which the males carry the developing eggs. The mouth is located at the tip of a long suctorial proboscis used to suck juices from coelenterates and soft-bodied animals. Most of them have four simple eyes. They have a heart but no blood vessels. Excretory and respiratory systems are absent. The long, thin body and legs provide a large surface, in proportion to volume, that is evidently sufficient for diffusion of gases and wastes. Because of the small size of the body, the digestive system sends branches into the legs, and most of the gonads are also found there.

Sea spiders are found in all oceans, but they are most abundant in polar waters. They are probably more closely related to the Crustacea than to any other arthropods, although they do not have a nauplius larva. *Pycnogonum* is a common intertidal genus found on both the Atlantic and Pacific coasts.

Class Arachnida

The arachnids are not as closely knit a group as the insects. The name comes from the Greek word *arachnē,* spider. Besides spiders, the group includes scorpions, ticks, mites, harvestmen, and others. There are many differences among these with respect to form and appendages. Most of them are free living and are far more common in warm, dry regions than elsewhere. They are provided, as a rule, with claws, poison glands, fangs, and stingers. Most are highly predaceous and suck the fluids and soft tissues from the bodies of their prey. Some have interesting adaptations, such as the spinning glands of the spiders.

CHARACTERISTICS

1. Body of unsegmented cephalothorax and abdomen usually
2. Antennae and mandibles lacking
3. Six pairs of appendages usually, which include the chelicerae and pedipalps and four pairs of walking legs
4. Sucking mouthparts mainly
5. Poison glands in some
6. Respiration by book gills or by book lungs or tracheae
7. Excretion by malpighian tubules and coxal glands
8. Circulatory system typically of heart, arteries, veins, sinuses
9. Nervous system of arthropod plan, or more concentrated anteriorly
10. Eyes simple usually

11. Separate sexes

12. Development direct

ECONOMIC IMPORTANCE. Most arachnids are harmless and actually do much good by destroying injurious insects. A few, such as the black widow spider, can harm man by their bites. The sting of the scorpion may be quite painful. Some ticks and mites are vectors of diseases as well as causes of annoyance and painful irritations. Certain mites damage plants by sucking their juices or by damaging valuable fruits.

PHYLOGENETIC RELATIONS. Arachnids represent a very old group of invertebrates. It is believed that arthropods split into two branches probably as early as the Cambrian period. One branch became arachnids and the other developed into the other classes of arthropods. Some of the earliest arachnids were the large Paleozoic eurypterids. Spiders are closely related to scorpions, but the stem ancestor of spiders is unknown. Some authorities believe that spiders have come from some scorpion stock and by pedogenesis have developed and stressed their web-weaving habits. Their web-making may have originated from their habit of making silky cocoons.

Although the earliest arachnids were aquatic, the majority today are terrestrial.

THE SPIDERS—ORDER ARANEAE. The spiders are a very large group of more than 35,000 species

FIG. 14-5
Argiope aurantia, the common yellow and black garden spider, builds its orb web in gardens or tall grasses and then hangs there, head down, waiting for the arrival of fresh meat in the form of an unwary insect. (Photo by J. H. Gerard, Alton, Ill.)

Chelicerae
Right pedipalp

FIG. 14-6
A wolf spider strikes a defensive pose that shows his appendages, chelicerae, and eyes to good advantage.

distributed all over the world and found in most kinds of habitats. All are predaceous. Some chase their prey, others ambush them, but most of them spin a net in which to trap the forms they live upon. Many have poison glands in connection with their fangs, but very few have poisonous bites that are harmful to man. Most of them lie more or less concealed during the day.

The large garden spider *Argiope aurantia* (Fig. 14-5) is one of the forms often studied in the laboratory. Its body is compact and consists of a distinct cephalothorax and abdomen, both unsegmented. They are joined by a slender peduncle.

Up to eight **simple eyes** are borne on the dorsal side of the cephalothorax, and six pairs of appendages are on its ventral side. The first pair is the **chelicerae** (Fig. 14-6), which have terminal fangs (claws), each

provided with a duct from a poison gland (Fig. 14-7). The **pedipalps** (second pair) have basal parts with which they chew; they also have sensory functions and are used by the males to transfer the sperm. The four pairs of **walking legs** each have seven segments and terminate in two or three claws.

Arachnids use flexor muscles to flex the limbs at hinge joints, but they extend the limbs by means of fluid pressure within the limbs on the principle of a hydraulic system. There are no extensor muscles such as other arthropods have.

In the **digestive system** there is a sucking stomach provided with strong muscles and a digestive stomach with five pairs of digestive ceca (Fig. 14-7). The digestive gland (liver) is large and secretes its fluid into the intestine.

The **circulatory system** is made up of a dorsal heart

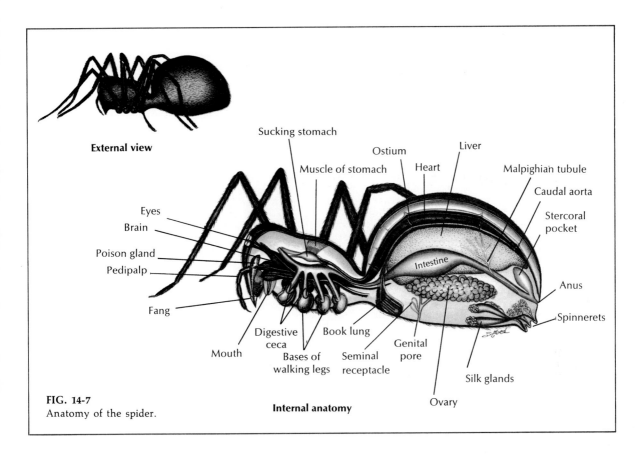

External view

FIG. 14-7
Anatomy of the spider.

Internal anatomy

Sucking stomach

Ostium

Liver

Muscle of stomach

Heart

Malpighian tubule

Caudal aorta

Stercoral pocket

Eyes

Brain

Poison gland

Pedipalp

Intestine

Anus

Fang

Spinnerets

Digestive ceca

Book lung

Genital pore

Mouth

Bases of walking legs

Seminal receptacle

Silk glands

Ovary

(Fig. 14-7), arteries, veins, and sinuses. In its course blood is carried to the book lungs where it is aerated.

Respiration is carried on by both tracheae, which play a minor part, and by book lungs. The **tracheae** make up a system of air tubes that carry air direct to the tissues and are described more fully under the insects. The **book lungs** are peculiar to arachnids and are usually paired (Fig. 14-7). Each consists of many parallel invaginated air pockets from a posterior chamber extending into a blood-filled anterior chamber. Air enters the posterior chamber by a slit in the body wall. Because these air pockets are flattened and leaflike, the whole structure is called a book lung. The blood carries the oxygen, for the tracheae are not as extensive as in the insects.

The **excretory system** consists of the malpighian tubules, which empty into the intestine, and a pair of **coxal glands** in the cephalothorax, which discharge through ducts between the legs. The malpighian tubules are discussed more fully in Chapter 16 on the insects. The **nervous system** is that of the arthropod

plan. It comprises a dorsal brain, a subpharyngeal ganglionic mass from which nerves run to the various organs, and a short ventral nerve cord. Spiders are sensitive to touch and smell and probably to other stimuli.

The **sense organs** are centered mainly in sensory hairs on the appendages and body, and there are usually eight **simple eyes.** The eyes are chiefly for the perception of moving objects, but some may form distinct images. Each eye is provided with a lens, optic rods, and a retina. Since vision is usually poor, a spider often depends on sensing the vibrations of its web to judge the size and activity of entangled prey. The female may also sense the approach of a prospective mate by the message he gently taps upon her web.

The **sexes** are separate, and the reproductive organs of each sex are located in the ventral part of the abdomen (Fig. 14-7). In the transfer of sperm the male spins a small web, deposits a drop of sperm upon it, and then picks up and stores the whole in the cavities of his pedipalps. When he mates, he inserts his pedipalps into the female genital opening, and the sperm are stored in the female's seminal receptacles. Before the mating act, spiders (usually the male) perform various courtship rituals, such as waving the appendages, assuming peculiar attitudes, touching the tip of the female's legs, and even offering her insect prey.

When the eggs are laid, they are fertilized by the sperm as they pass through the vagina. The eggs are usually laid in a silk cocoon; this may be carried around by the female, or it may be attached to a web

FIG. 14-8
A fishing spider, *Dolomedes sexpunctatus,* is sucking out the body juices of a minnow it has captured. Fishing spiders usually live near the water and are often seen running about over aquatic vegetation. They are hunters and do not use the web to ensnare prey.
(Photo by J. H. Gerard, Alton, Ill.)

or plant. A single cocoon may contain hundreds of eggs. About 2 weeks are necessary for hatching but the young remain in the sac for a few weeks longer and molt once before leaving the sac. Several molts are necessary before they become mature.

Web-spinning habits of spiders. No aspect of spider life is more interesting than the spinning of webs. Not all spiders spin webs, for some, such as the hunting, or wolf, spider (Fig. 14-6), simply chase and catch their prey, although some spin an anchor thread to warn them of the prey's approach. The majority of spiders, however, spin some kind of web in which to ensnare and wrap up their food. Here, entrapped forms are killed by the poison fangs. The spider punctures the body of its prey with its fangs, and then alternately injects digestive fluid through the puncture and sucks up the dissolved tissues until the prey has been sucked dry. Tarantulas, wolf spiders, and others may crush the prey with very strong jaws to facilitate the digestive process. Usually only small invertebrate forms are caught in these webs, but instances are known where mice have actually been caught. *Dolomedes,* a species common east of the Rocky Mountains, is known to kill small fish (Fig. 14-8).

The spinning organs of spiders consist of two or three pairs of **spinnerets,** fingerlike appendages containing many hundreds of microscopic tubes that run to special abdominal **silk glands.** A protein secretion from these glands passes through the tubes to the outside, where it hardens into silk thread on contact with the air.

Spiders use these threads for many purposes, such as lining their nests, forming cocoons, making balloons, and spinning their webs (Fig. 14-9). Two kinds of silk threads are used by spiders in making their nets, elastic and inelastic. The inelastic threads are generally used to make the framework of the net, which consists of threads radiating out from a center to an extensive periphery. The elastic type forms the spirals that run in concentric rows from the center outward and are supported by the radiating fibers. These spiral threads have viscid masses of sticky material for holding the prey when it becomes entangled in the web. The silk threads of spiders are stronger than steel threads of the same diameter.

Different species of spiders make different kinds of nets. Some are simple and primitive and consist merely of a few strands radiating out from a spider's

burrow or place of retreat. On the other hand, the orb-weaving spiders create beautiful geometric patterns (see p. 299). Cobwebs, which are usually irregular strands of silk, are formed by certain species of spiders. These untidy masses are often rendered more so by the dust that collects on them.

The earliest webs were probably formed of nonviscid silk, and the viscid type was a later innovation. Early ancestors of spiders may have had no silk apparatus for catching prey, and when silk first evolved, it was used for making egg sacs or covering eggs. The first web may have appeared when a spider

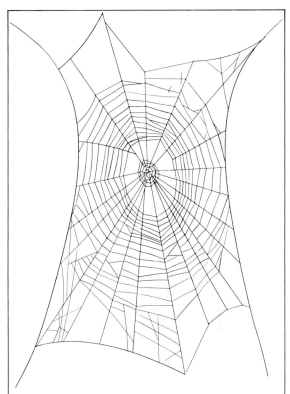

FIG. 14-9
Web of orb-weaving spider. From silk glands inside abdomen, liquid is forced out of spinnerets and hardens on contact with air. Each of four or five different kinds of glands secrete different kind of silk. Orb weaver follows a pattern in constructing web. Framework of boundary threads of dry silk is usually an irregular four- or five-sided figure. Radii, of dry silk are spun next and held in place at "hub" by platform of spiral dry silk. Functional spiral of sticky threads is spun last by spider, working inward from rim to hub, being careful to walk on nonsticky radii while doing so.

FIG. 14-10
Burrow of trap-door spider, with lid thrown open and its owner just retreating within.

hid in a crevice with an egg sac and formed for the shelter, from its drag line, a lining that became tubular in time. Some primitive spiders still build this type of net, whereas the orb web is a later development (B. J. Kaston).

The trap-door spider. Although spiders as a group have many interesting habits and adaptations, perhaps none is more fascinating than the trap-door spider. Several species of trap-door spiders, which are relatives of the tarantulas, are found in the southern and western parts of our country.

The common species in southern California is *Bothriocyrtum californica.* This spider exercises great architectural ability in the construction of her nest, which is a burrow in the ground that she excavates with her fangs, palps, and front legs. The burrow is 6 to 10 inches deep and a little more than 1 inch in diameter. Its walls are firm and smooth and are lined with a silken web. The burrow is waterproof. The semicircular trapdoor is hinged by means of the web that lines the door and the burrow (Fig. 14-10). When closed, the door fits snugly into the beveled entrance of the burrow flush with the surface of the surrounding ground. The door is carefully camouflaged with moss, lichen, and other vegetation so that it is difficult to detect.

On the inside surface of the trapdoor are little "arm holds" of silken web by which the spider, with her fangs, can keep the door closed against intruders. When watching for food, she holds the lid partly ajar, seizes her prey, and drags it into her burrow. Her food consists mainly of insects and small crustaceans. She spends her entire life in the burrow, where she raises her young. Some have been known to live more than 7 years. The small males have rarely been found.

Trap-door spiders have their enemies, as do most animals. The most deadly is a wasp of the Pompilidae family, which either slips in while the door is ajar or forces it open with her jaws. Inside, she overcomes the spider and paralyzes her by stinging. On the abdomen of the spider the wasp lays an egg that hatches in a few days into a larva that feeds on the paralyzed spider. Another pompilid wasp, often called the "tarantula hawk," can overcome and paralyze a tarantula many times its own size, drag it to a burrow already prepared, deposit her egg on it, and bury it. The spider is handicapped by being nearly blind and rarely wins the battle.

An African spider makes a trap-door burrow in the soft bark of trees. The burrow in this case is also lined with a silken web but is much smaller and more shallow than that found in America.

Our dangerous spiders. Nearly all spiders produce venom for killing their prey. However the bites of most spiders are harmless to man, and even the most poisonous will bite only when tormented, when defending their eggs or young, or when their web is violently disturbed. There are two spiders in the United States that can give severe or even fatal bites —the black widow spider and the brown recluse. Despite their large size and fearsome looks, the American tarantulas (Fig. 14-11) are *not* dangerous to man. They rarely bite, and their bite is not considered serious.

The black widow, *Latrodectus mactans* (Fig. 14-12), is not a large spider. It is shiny black and has a bright orange or red "hourglass" on the underside of the abdomen. The name "widow" was given to it because the female was reputed to eat her mate immediately after mating. Actually the male is seldom sacrificed in this way, although such a practice does occasionally occur among some spiders.

FIG. 14-11
A tarantula spider, *Dugesiella lentzi,* just captured. These interesting spiders are not dangerous; they make good pets and live for years. (Photo by J. H. Gerard, Alton, Ill.)

FIG. 14-12
This black widow spider *Latrodectus,* suspended on her web, has just eaten large cockroach. Note "hourglass" marking (orange colored) on ventral side of abdomen.

The black widow is worldwide in distribution and has been reported from all of our states, but is most abundant in the southern states and southern California. Their webs are very irregularly built, usually around rubbish, in dark cellars or corners, or under objects.

Black widow bites may cause acute pain or burning sensation, muscle spasm, vomiting, restlessness, and cyanosis. Patients may be incapacitated for several days. The venom is neurotoxic, that is, it acts upon the nervous system. About four or five out of each thousand bites reported have proved fatal.

The brown recluse spider, *Loxosceles reclusa,* native to Arkansas and Missouri but also found elsewhere in the eastern United States, is also poisonous to man. It is smaller than the black widow, is brown, and bears a fiddle-shaped dorsal stripe on its cephalothorax (Fig. 14-13). It seeks out small crevices and other protected places. The venom of the brown recluse is hemolytic rather than neurotoxic, producing a necrosis (death) of the tissue surrounding the bite. Its bite is serious and occasionally fatal.

SCORPIONS (ORDER SCORPIONIDA). Although scorpions (Fig. 14-14) are more common in tropical and subtropical countries, they have been

FIG. 14-13
The brown recluse, *Loxosceles reclusa,* is a small brown, venomous spider. Note the small violin-shaped marking on its cephalothorax. The venom is hemolytic and dangerous. (Photo by J. H. Gerard, Alton, Ill.)

FIG. 14-14
The striped scorpion, *Centruroides vittatus,* found in our southwestern states. Its sting is painful, but in this particular species is not considered lethal. (Photo by J. H. Gerard, Alton, Ill.)

FIG. 14-15
Pseudoscorpion, or false scorpion. Order Chelonethida
(Pseudoscorpionida). Most members of this group do
not exceed 5 to 8 mm. in length. They live under
stones, bark of trees, and sometimes between pages of
books. Their food is chiefly small insects and mites.
In winter they construct cocoons from silk glands,
which open on chelicerae. Note large pedipalps, which
resemble those of true scorpions. (Courtesy R. Weber
and W. Vesey.)

reported from at least 30 of our states. More than
600 species are known. Their habitat is in trash piles,
around dwellings, and in burrows of desert regions.
Most of them will burrow. They are most active at
night, and they feed mainly on insects and spiders.

The cephalothorax is made up of six segments and
is enclosed by a dorsal chitinous carapace, and by
various ventral plates. The carapace bears from two
to 12 eyes (depending on the species), which are
grouped into median and sometimes lateral eyes.
From the cephalothorax six pairs of appendages arise:
(1) the small paired chelicerae, each of three joints,
(2) the paired pedipalps, each of six joints and a
pincer, and (3) the four pairs of walking legs, each
of eight joints. The chelicerae and pedipalps, which
are provided with jaws and teeth, are used for seizing
and tearing their prey.

The abdomen consists of 12 segments, seven in the
preabdomen and five in the postabdomen, or "tail."
The postanal telson bears the terminal poison claw
or stinger (Fig. 14-14). Scorpions may or may not use
the stinger in overcoming their prey. On the ventral
side of the abdomen are curious comblike pectens,
which are tactile organs used for exploring the ground
and for sex recognition.

Of the more than 40 American species, only two

are said to be dangerous to man. The bite of most
scorpions produces only a painful swelling.

In mating, scorpions have an interesting courtship
ceremony in which the two mates seize each other's
claws and perform a dance. Scorpions bring forth
living young, which are carried on the back of the
mother, but they do not feed upon her tissues (a
popular superstition).

Pseudoscorpions, or false scorpions (Fig. 14-15),
belong to a different order. Their pedipalps are similar
to those of the true scorpions, but the pseudoscorpion
carries its poison glands within the pedipalps.

MITES AND TICKS (ORDER ACARINA). Mites
and ticks are arachnids in which the cephalothorax
and abdomen are fused into an unsegmented ovoid
with eight legs. The larvae have only six legs until
after the first molt.

Mites and ticks are found almost everywhere—in
both fresh and salt water, on vegetation, on the
ground, and parasitic in animals. There are about
15,000 known species, a great many of which are of
direct importance to humans.

Ticks, which are usually larger than mites, feed
upon the blood of vertebrates. They pierce the skin
and suck up blood until enormously distended. After
filling themselves, the ticks will drop off, digest the

FIG. 14-16
The wood tick *Dermacentor,* one species of which transmits Rocky Mountain spotted fever and tularemia, as well as producing tick paralysis in pets.

meal, undergo molting, and then climb a bush and wait for another victim. They can survive long periods between feedings.

The female tick lays her eggs on the ground. When hatched, the larvae climb up on vegetation to await a passing host.

Ticks are important vectors of disease. Texas cattle fever is caused by a protozoan parasite, *Babesia bige-*

mina, transmitted by the tick *Boophilus annulatus.* The wood tick *Dermacentor* (Fig. 14-16) is the vector for Rocky Mountain spotted fever, which is caused by a rickettsial organism carried in the salivary secretions of the tick. Tularemia (rabbit fever) is also carried by a species of *Dermacentor.*

Most mites are less than 2 mm. long, and they are quite varied in their habits and habitats. The spider mites, or red spiders, are destructive to plants. The mange mite, *Demodex,* is responsible for mange. Chiggers, or red bugs *(Eutrombicula),* lay their eggs on the ground. The larvae, when they find a suitable vertebrate host, attach to the skin and form, with the aid of digestive secretions, a tiny crater at the feeding site. The larva feeds on the partially digested tissues in the crater. Chiggers do not suck blood.

References

Selected general readings

Baerg, W. J. 1958. The tarantula. Lawrence, University of Kansas Press. *A monograph on the habits and natural history of a greatly misunderstood member of an animal group.*

Baker, E. W., and G. W. Wharton. 1952. An introduction to acarology. New York, The Macmillan Co. *A work on the much-neglected group of ticks and mites. Emphasizes taxonomy.*

Barrington, E. J. W. 1967. Invertebrate structure and function. New York, Houghton Mifflin Co. *Contains some excellent material on the arthropods.*

Cloudsley-Thompson, J. E. 1958. Spiders, scorpions, centipedes, and mites. New York, Pergamon Press, Inc. *This monograph emphasizes the behavior and ecology of the group.*

Collias, N. E. (editor). 1964. The evolution of external construction by animals. Amer. Zool. **4:**175-243 (May).

Discusses the evolution of bird-nest building, spider webs, caddisworm cases, termite nests, and bee nests.

Curtis, H. 1965. Spirals, spiders, and spinnerets. Amer. Sci. **53**:52-58.

Comstock, J. H., and W. J. Gertsch. 1940. The spider book, ed. 2. New York, Doubleday, Doran & Co., Inc. *A well-written account of spiders. Much emphasis is given to their habits and behavior.*

Fabre, J. H. 1919. The life of the spider. New York, Dodd, Mead & Co. *An interesting description of the adaptations of the spider. Especially good for the beginner in zoology.*

Hedgpeth, J. W. 1954. On the phylogeny of the Pycnogonida. Acta Zoologica **35**:193-213.

Herms, W. B. 1950. Medical entomology, ed. 4. New York, The Macmillan Co. *A section of this book is devoted to Arachnida. The role arachnids play in disease transmission is emphasized.*

Hickman, C. P. 1973. Biology of the invertebrates, ed. 2. St. Louis, The C. V. Mosby Co.

Kaston, B. J. 1966. Evolution of the web. Natural History **75**:26-33. (April).

Kaston, B. J., and E. Kaston. 1953. How to know the spiders. Dubuque, Iowa, William C. Brown Co. *An excellent taxonomic manual of the common spiders, fully illustrated. It contains much information about spiders in general.*

Kaston, B. J. 1965. Some little known aspects of spider behavior. Amer. Midland Naturalist **73**:336-356. *Should be read by all students of spiders.*

Levi, H. W. 1967. Adaptations of respiratory systems of spiders. Evolution **21**:571-575.

Levi, H. W., and L. R. Levi. 1969. A guide to spiders and their kin. New York, Golden Press, Inc.

Lougee, L. B. 1964. The web of the spider. Bloomfield Hills, Mich., Cranbrook Institute of Science. *A delightful, although brief, work on spiders. Many photographs of the various webs made by spiders, but text is kept at a minimum.*

McCook, H. C. 1889-1893. American spiders and their spinningwork. Philadelphia, Academy of Natural Science. *A great classic work that in many ways has never been surpassed in this field.*

Meglitsch, P. A. 1972. Invertebrate zoology, ed. 2. New York, Oxford University Press.

Moore, R. C., C. G. Lalicker, and A. G. Fisher. 1952. Invertebrate Fossils. McGraw-Hill Book Co., Inc. *Four chapters deal with the arthropods, two specifically with trilobites and chelicerates.*

Mullen, G. R. 1969. Morphology and histology of the silk glands in *Araneus sericatus* CL. Trans. Amer. Microsc. Soc. **88**:232-240.

Parry, D. W. 1960. Spider hydraulics. Endeavour **19**:156-158.

Savory, T. H. 1952. The spider's web. New York, Frederick Warne & Co., Ltd. *An interesting account of the way spiders spin and make use of their webs.*

Savory, T. H. 1964. Arachnida. New York, Academic Press, Inc.

Snodgrass, R. E. 1952. A textbook of arthropod anatomy. Ithaca, N. Y., Comstock Publishing Associates. *A fine comparative study of all groups in this vast field. The author also points out the inconsistencies of arthropod structure with theories of arthropod relationships.*

Vachon, M. 1953. The biology of scorpions. Endeavour **12**:80-86.

Wells, M. 1968. Lower animals. New York, McGraw-Hill Book Co., Inc.

Wilson, R. S. 1962. The control of dragline spinning in the garden spider. Quart. J. Microsc. Sci. **104**:557-571.

Witt, P. N. (editor). 1969. Web-building spiders. Amer. Zool. **9**:70-238 (Feb.).

Selected *Scientific American* articles

Savory, T. H. 1962. Daddy longlegs. **207**:119-128 (Oct.). *An interesting account of these familiar phalangids.*

Savory, T. H. 1968. Hidden lives. **219**:108-114 (July). *Describes the numerous invertebrates (cryptozoa) that are found living concealed near the surface of the ground under rocks and debris and how these forms enjoy certain advantages in living where they do.*

Marine hermit crab, Eupagurus bernhardus, *has chosen an empty snail shell* (Buccinum) *as its home. The thin cuticle on the crab's abdomen makes a protective home a necessity. As it grows, it will select larger shells. Because of the antics of hermit crabs, they might be called the clowns of the sea. (Courtesy B. Tallmark, Uppsala University, Sweden.)*

CHAPTER 15

THE AQUATIC MANDIBULATES

PHYLUM ARTHROPODA
CLASS CRUSTACEA

▪ THE MANDIBULATA

The mandibulates are those arthropods that possess mandibles (jaws). Antennae, mandibles, and maxillae make up the appendages of the head. These are sensory, mastigatory, and food-handling organs. The body may consist of a head and trunk, or it may have a head, thorax, and abdomen, or the head and thorax may be fused into a cephalothorax, so that the main divisions are cephalothorax and abdomen. The trunk appendages are mainly for walking or swimming, but in some groups they are highly specialized in function. Among the mandibulates are the crustaceans, which are very largely aquatic, and the myriapods (centipedes, millipedes, pauropods, and symphylans) and insects, all of which are very largely terrestrial animals. The myriapods and insects are dealt with in a separate chapter.

Class Crustacea

The members of class Crustacea get their name from the hard shells they bear (L. *crusta,* shell). The 30,000 or more species in this class include lobsters, crayfish, shrimp, crabs, water fleas, copepods, barnacles, and some others. It is the only arthropod class that is primarily aquatic. Many of them are marine, some live in fresh water, and others are found in moist soil. Most of them are free living, but a few are sessile, commensal, or parasitic. They differ from other arthopods chiefly by having gills for breathing, by having two pairs of antennae, one pair of mandibles, and two pairs of maxillae, by having the appendages modified for various functions, and by lacking malpighian tubules.

General nature of a crustacean

The crustacean body is made up of segments that vary in number among the different taxonomic groups. Most crustaceans have between 16 and 20, but some primitive forms have 60 or more segments. The more advanced crustaceans tend to have fewer segments as well as increased tagmatization.

There are usually three major body divisions, or

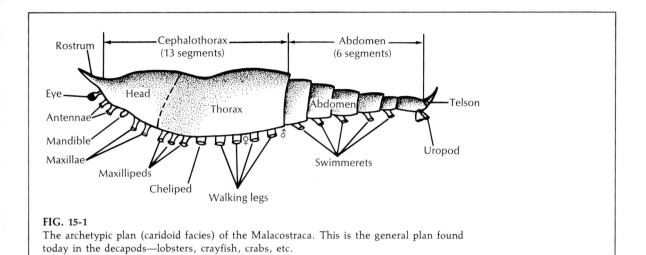

FIG. 15-1

The archetypic plan (caridoid facies) of the Malacostraca. This is the general plan found today in the decapods—lobsters, crayfish, crabs, etc.

tagmata: head, thorax, and abdomen. These three regions are not always distinct because of the fusion of metameres from different divisions. Thus the cephalothorax of the crayfish consists of both cephalic and thoracic segments.

The most advanced and by far the largest group of crustaceans is the subclass Malacostraca, which includes the lobsters, crayfish, crabs, shrimps, sow bugs, beach fleas, and many others. These all follow a generally similar body plan that is often referred to as the **caridoid facies** and is considered to be the primitive or ancestral plan of the group (Fig. 15-1). In this typical body plan the three body divisions or tagmata include a head with five or six segments, a thorax with eight segments, and an abdomen with six or eight segments. At the anterior end is the nonsegmented rostrum and at the posterior end is the nonsegmented telson, which, with the last abdominal segment and its uropods, form the tail fin in many forms. Primitively, each segment typically bears one pair of biramous (two-branched) appendages, but there are many modifications.

Relationships and origin of crustaceans

The relationship of the crustaceans to other arthropods has long been a puzzle. According to a widely held theory, the trilobites are considered the ancestors of the crustaceans, but some zoologists have proposed that both these groups evolved independently from different nonarthropod ancestors.

Some light was shed on the problem when the primitive crustacean *Hutchinsoniella macracantha* was discovered in 1954 in Long Island Sound. This form was assigned to a new subclass of its own (Cephalocarida) because of its biramous trunk limbs. It also has some features found only in the nauplius (larval) stages of other crustaceans, such as its ventral median eye. It is considered the most primitive of all known crustaceans. It shares with the trilobites limbs that are alike and, except for the first pair, are not specialized for a specific function but are used for locomotion and food getting. This little form (about 4 mm. long) may represent the primitive arthropod limb pattern that later became highly specialized in other crustaceans and may also indicate a definite trilobite relationship.

CRAYFISH—A REPRESENTATIVE CRUSTACEAN. The crayfish and lobsters have about the same plan of body structure. The lobster is a marine form, the crayfish a freshwater form. There are a number of genera, and more than 130 species have been described in the United States. Common genera in North America are *Orconectes* and *Cambarus,* which are found east of the Rocky Mountains, and *Astacus,* which occurs mainly west of the Rockies and also in Europe. The various species of crayfish and lobsters resemble each other closely except for minor details.

Habitat. The crayfish is one of the most common animals in fresh water. They are found in streams, ponds, and swamps over most of the world. Some

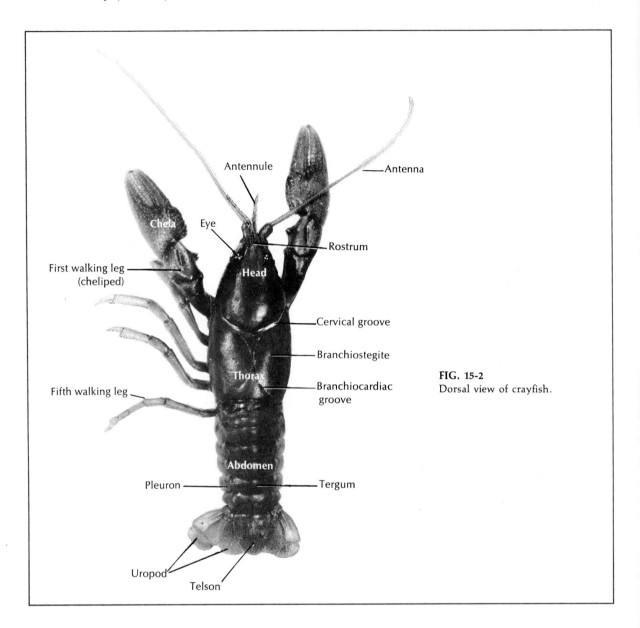

FIG. 15-2
Dorsal view of crayfish.

are found in slow-moving water, others prefer swift streams, and some blind ones dwell in caves. They are primarily bottom dwellers and spend the day under rocks and vegetation, coming out at night in search of food. Some are terrestrial in their habits and make characteristic chimneylike burrows in the soil if the water level is not too far under the surface.

Behavior. Endangered crayfish can dart off to a hiding place with considerable speed. In their burrows they face outward so that they can move out for food. While stationary, they keep currents of water moving by waving the bailer and swimmerets back and forth. They walk with most of the weight supported by the fourth pair of legs. When escaping from danger, they move rapidly backward by extending and flexing the abdomen, uropods, and telson, thus producing a series of backward darts.

With reference to orientation, crayfish are positively thigmotactic and try to get most of the body in contact with a surface. Many chemical substances, unless concentrated, will attract them. Most light sources will cause them to retreat, although they do

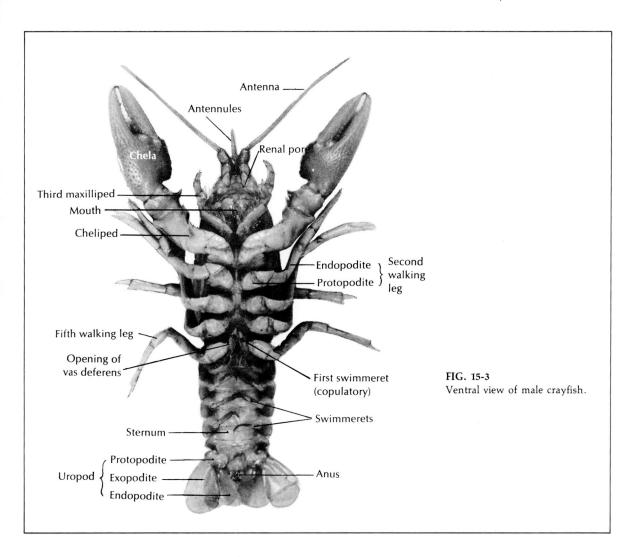

FIG. 15-3
Ventral view of male crayfish.

have a liking for red. They are mainly nocturnal animals.

The behavior of the crayfish is mostly instinctive, but they can be taught simple habits through experience.

External features. As with other crustaceans, the body of the crayfish is covered with an exoskeleton composed of chitin and lime. This hard protective covering is soft and thin at the joints between the somites, allowing flexibility of movement. The somites are grouped into two main regions: the **cephalothorax** (head and thorax) and the **abdomen** (Fig. 15-2). There are 19 somites in the body: five in the head, eight in the thorax, and six in the abdomen, each with a pair of appendages.

The cephalothorax, enclosed dorsally and laterally by a skeletal shell, the **carapace,** appears unsegmented except for a groove between the head and thorax (Fig. 15-2). The anterior tip of the carapace is called the **rostrum.** The **compound eyes,** which are stalked and movable, lie beneath and on each side of the rostrum.

In the abdomen each somite consists of a dorsal plate or **tergum** (Fig. 15-2), and a ventral transverse bar, or **sternum** (Fig. 15-3), joined together by a lateral **pleuron** on each side. The segmented abdomen terminates in the broad, flaplike **telson,** which is not considered a somite. On the ventral side of the telson is the **anus.**

The openings of the paired **vasa deferentia** are on

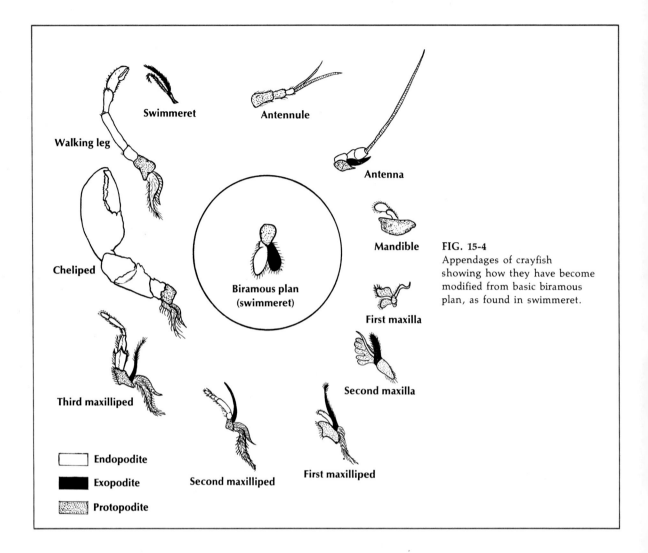

FIG. 15-4
Appendages of crayfish showing how they have become modified from basic biramous plan, as found in swimmeret.

the median side at the base of the fifth pair of walking legs, and those of the paired **oviducts** are at the base of the third pair. In the female the opening to the seminal receptacle is located in the midventral line between the fourth and fifth pairs of walking legs.

Appendages. The crayfish typically has one pair of jointed appendages on each somite (Fig. 15-3). These appendages differ from each other, depending on their functions. All, however, are variations of a common plan. This common plan is best illustrated by the two-branched, or **biramous** swimmerets of the abdomen (Fig. 15-4). It consists of a basal portion, the **protopodite,** which bears two branches, a lateral **exopodite** and a median **endopodite.** The

protopodite is made up of two joints, whereas the exopodite and endopodite have from one to several segments each. From this common biramous type of appendage have evolved all the different kinds of appendages in the crayfish.

Three kinds of appendages are recognized in the adult: (1) **foliaceous,** such as the maxillae, (2) **biramous,** such as the swimmerets and uropods, and (3) **uniramous,** such as the walking legs. All these appendages have been derived from the biramous type. This is shown in the embryonic crayfish, in which all the appendages arise as two-branched structures.

Structures that have a similar basic plan and have descended from a common form are said to be **homologous,** whether they have the same function or

TABLE 15-1
Crayfish appendages

Appendage	Protopodite	Endopodite	Exopodite	Function
Antennule	3 segments, statocyst in base	Many-jointed feeler	Many-jointed feeler	Touch, taste, equilibrium
Antenna	2 segments, excretory pore in base	Long, many-jointed feeler	Broad, thin, pointed squama	Touch, taste
Mandible	2 segments, heavy jaw and base of palp	2 distal segments of palp	Absent	Crushing food
First maxilla	2 thin medial lamellae	Small unjointed lamella	Absent	Food handling
Second maxilla	2 bilobed lamellae, extra plate, epipodite	1 small pointed segment	Doral plate, the scaphognathite (bailer)	Draws currents of water into gills
First maxilliped	2 medial plates and epipodite	2 small segments	1 basal segment plus many-jointed filament	Touch, taste, food handling
Second maxilliped	2 segments plus gill	5 short segments	2 slender segments	Touch, taste, food handling
Third maxilliped	2 segments plus gill	5 larger segments	2 slender segments	Touch, taste, food handling
First walking leg (cheliped)	2 segments plus gill	5 segments with heavy pincer	Absent	Offense and defense
Second walking leg	2 segments plus gill	5 segments plus small pincer	Absent	Walking and prehension
Third walking leg	2 segments plus gill; genital pore in female	5 segments plus small pincer	Absent	Walking and prehension
Fourth walking leg	2 segments plus gill	5 segments, no pincer	Absent	Walking
Fifth walking leg	2 segments; genital pore in male; no gill	5 segments, no pincer	Absent	Walking
First swimmeret	In female reduced or absent; in male fused with endopodite to form tube			In male, transfers sperm to female
Second swimmeret				
Male	Structure modified for transfer of sperm to female			
Female	2 segments	Jointed filament	Jointed filament	Creates water currents; carries eggs and young
Third, fourth, and fifth swimmerets	2 short segments	Jointed filament	Jointed filament	Create current of water; in female carry eggs and young
Uropod	1 short, broad segment	Flat, oval plate	Flat, oval plate; divided into 2 parts with hinge	Swimming; egg protection in female

not. Since the specialized walking legs, mouthparts, chelipeds, and swimmerets have all developed from a common type and have become modified to perform different functions, they are all homologous to each other (serially homologous). During this structural modification some branches have been reduced, some lost, some greatly altered, and some new parts added. The crayfish and its allies possess the best examples of **serial homology** in the animal kingdom.

Table 15-1 shows how the various appendages have become modified from the biramous plan to fit specific functions.

Regeneration and autotomy. Crayfish have the power to regenerate certain lost parts. In general, any of the appendages and the eyes will be renewed when removed, although regeneration is faster in young animals. A lost part is partially renewed at the next molting, and after several moltings it is completely restored. The new structure is not always identical to the one lost. If only part of the eye stalk is cut off, normal regeneration will occur, but if the entire eye stalk is removed, a structure similar to an antenna replaces it. Whenever the part regenerated is different from the lost part, such a regeneration is called **heteromorphosis.**

The power of self-amputation is called **autotomy.** It refers to the breaking off of an injured leg or chela at a definite preformed breakage plane by means of a reflex act. The breaking point is near the base of the legs and is marked by an encircling line on the basal segment of the chelae and at the third joint on the walking legs. If one of these appendages is injured, all parts terminal to the breaking point are cast off. The process is effected by a special autotomizer muscle, which contracts excessively to cause extreme flexion of the leg, putting pressure on the breakage plane until it ruptures (Fig. 15-5). Injury to the dactyl (claw) is unlikely to cause autotomy; the closer the injury is to the breakage plane, the greater is the chance of autotomy. Autotomy has the advantage of preventing excess loss of blood, for when the legs are broken off at the breaking point, the wound there closes more quickly. After a part is cast off in this manner, a replacement regenerates in the regular way.

Internal features. The crayfish has all the organs and systems found in the higher forms (Fig. 15-6). A few of the systems, such as the muscular and nervous, show the segmentation that is so obvious in

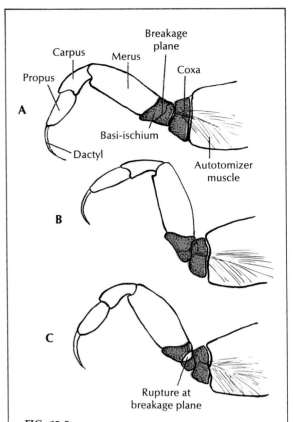

FIG. 15-5
Autotomy in the crab is a reflex act in which extreme flexion of the leg causes it to rupture at a preformed breakage plane. The stimulus is injury to the leg distal to the breakage plane. **A,** Leg in normal resting position showing location of preformed breakage plane. **B,** Extreme flexion of the leg from contraction of autotomizer muscle causes basi-ischium to press against rim of coxa. **C,** Continued contraction and pressure causes rupture at breakage plane. A double membrane at this location pinches off the nerve and blood vessels at autotomy and prevents blood loss and tissue damage at stump. Regeneration occurs faster at this location than elsewhere. (Modified from Wood, F. D., and H. E. Wood. 1932. J. Exp. Zool. **62:**1-55.)

the annelids, but most of them are modified from this plan. Most of the changes involve concentration of parts in a particular region or else reduction or complete loss of parts, for example, the intersepta.

Coelom. In contrast to annelids, arthropods have a much reduced **coelomic cavity,** which is divided into a number of separate spaces. One of these cavi-

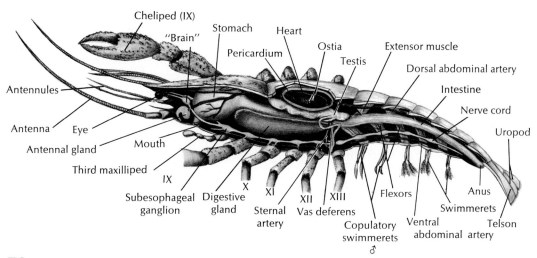

FIG. 15-6
Internal structure of male crayfish.

FIG. 15-7
Diagrammatic cross section through heart region of crayfish showing direction of blood flow in this "open" blood system. Blood is pumped from heart to body tissues through arteries, which empty into tissue sinuses. Returning blood enters sternal sinus, is carried to gills for gas exchange and then back to pericardial sinus by branchiocardiac canals. Note absence of veins.

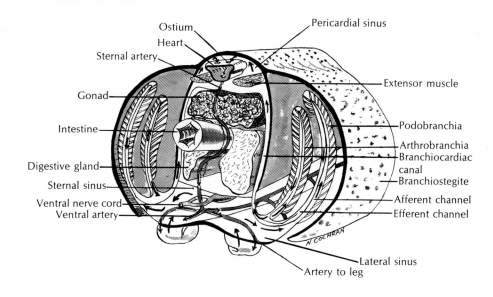

ties encloses the excretory glands and another the reproductive organs. The larger cavities around the alimentary canal are not true coelomic cavities; they contain blood and are known as **hemocoels.**

Muscular system. Striated muscles make up a considerable part of the body of a crayfish. It uses them for body movements and for manipulation of the appendages. The muscles are often arranged in opposite pairs. **Flexors** draw a part toward the body; **extensors** straighten it out. The abdomen has powerful flexors (Fig. 15-6), which are used when the animal swims backward—its best means of escape. Strong muscles on either side of the stomach manipulate the mandibles.

Respiratory system (Fig. 15-7). Crayfish breathe by means of **gills** which are delicate featherlike projections of the body wall and are located on each side of the thorax in the **gill chambers.** The gill chamber is covered by the lateral wall of the carapace and is open ventrally and at both ends. The "bailer," a part of the second maxilla, draws water over the gill filaments by moving back and forth. In *Cambarus* the gills are arranged in two rows; in some others there is a third row. Gills in the outermost row are attached to the bases of the seventh to twelfth pairs of appendages.

Feeding and digestion. Crayfish and lobsters will eat almost anything edible, dead or alive. Before reaching the mouth, the food is torn into small bites and kneaded into a usable size and shape by the maxillipeds, maxillae, and mandibles. Although the mandibles can crush, they do not really chew. Chewing is a function of the stomach. The mouth opens on the ventral side between the mandibles and leads to a short esophagus (Fig. 15-6). The stomach is large and thin-walled, contains two compartments, and is manipulated by a group of gastric muscles. When food reaches the anterior, or cardiac, portion of the stomach, it is chewed up by a **gastric mill** (composed of three movable chitinous teeth, one median and two ventral (Fig. 15-8). Then it passes through a filtering device in the posterior (pyloric) part of the stomach. A special sieve of fine bristles allows only the finest particles to enter a filter pouch through which the food moves in a fluid stream into the ducts of the large digestive glands. Digestion is completed in the tubules of the digestive glands, and the nutrients are absorbed into the blood, which carries them to the tissue cells to be utilized. Food particles too large for the filter pouch are funneled into the intestine for elimination, or else are regurgitated. The contents of the digestive tract are propelled by peristalsis. Digestive enzymes are similar to those of vertebrates, although there is no pepsinlike enzyme.

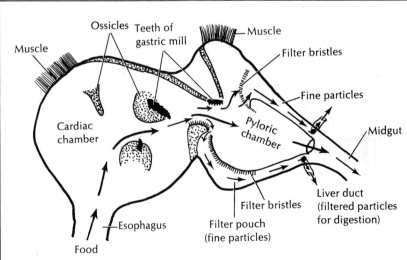

FIG. 15-8
Malacostracan stomach showing gastric "mill" and directions of food movements. Note that mill is provided with chitinous ridges, or teeth, and setae for churning and straining the food before it passes into the pyloric stomach.

The gastric mill enables these somewhat sedentary animals to swallow their food and chew it later while they are safely hidden from enemies.

The esophagus and hindgut are lined with chitin, which is periodically shed during ecdysis (molting). During the molting, calcium salvaged from the old exoskeleton may be stored in calcareous bodies called gastroliths in the walls of the stomach and then used again in the formation of the new exoskeleton.

Circulation. Crayfish and other malacostracans have an "open" or lacunar type of blood system. This means that there are no veins, but that the blood leaves the heart by way of arteries and returns to venous sinuses, or spaces, instead of veins before it reenters the heart. Recall that the annelids have a closed system, as do the vertebrates.

A dorsal heart is the chief propulsive organ. It is a single-chambered sac of striated muscle. Blood enters the heart from the surrounding pericardial sinus through three pairs of valves that prevent backflow into the sinus (Fig. 15-6). From the heart the blood enters a system of arteries and capillaries. Valves in the arteries prevent a backflow of blood. Small arteries or capillaries empty into the tissue sinuses, which in turn discharge into the large sternal sinus (Fig. 15-7).

From there afferent sinus channels carry blood to the gills where O_2 and CO_2 are exchanged. The blood is now returned to the pericardial sinus by efferent channels (Fig. 15-7).

Blood in arthropods is largely colorless. It includes ameboid blood cells of at least two types, and hemocyanin, a copper-containing respiratory pigment. Some of the smaller crustaceans have hemoglobin, an iron-containing pigment, in the blood instead of hemocyanin. The blood has the property of clotting, which prevents loss of blood in minor injuries. The blood corpuscles release a thrombinlike coagulant that precipitates the clotting.

Excretion. The excretory organs of crustaceans are usually a single pair of tubular structures located in the ventral part of the head anterior to the esophagus (Fig. 15-6). They are frequently called **antennal glands** because they open at the base of the antennae. The antennal glands, also called **green glands,** resemble nephridia in that they are long tubular structures surrounded by blood capillaries, but unlike nephridia they lack open nephrostomes.

The antennal gland consists of an internal end-sac or glandular portion bathed in blood, an excretory tubule or labyrinth, and a dorsal bladder, which opens to the exterior by a pore on the ventral surface of the basal antennal segment (Fig. 15-9). Wastes are probably removed from the blood by ultrafiltration (hydrostatic pressure) into the end-sac, and as the filtrate passes through the excretory tubule, it is modified by resorption of salts and water. Secretion of urinary components may also occur in the excretory tubule.

As with most aquatic invertebrates, the crustacean excretory organ functions principally to regulate the ionic and osmotic composition of the body fluids. Freshwater crustaceans such as the crayfish are constantly threatened with overdilution of water, which enters osmotically across the gills and other water-permeable surfaces of the body. The kidney, by forming a dilute, low-salt urine, acts as an effective flood-control device. In marine crustaceans such as lobsters and crabs the kidney functions to adjust the salt composition of the blood by selective modification of the salt content of the tubular urine. In these forms the urine remains isosmotic to the blood.

Nervous system. The nervous system of the crayfish and earthworm have much in common, although that of the crayfish is somewhat larger and has more fusion of ganglia (Fig. 15-6). The **brain** is a pair of **supraesophageal ganglia** that supply nerves to the eyes, antennules, and antennae and is joined by connectives to the **subesophageal ganglion,** a fusion of at least five pairs of ganglia that supply nerves to the mouth, appendages, esophagus, and

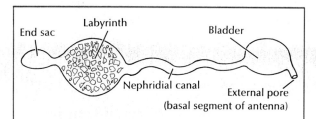

FIG. 15-9

Scheme of antennal gland (green gland) of crayfish. (In natural position organ is much folded.) Most selective resorption takes place in labyrinth, a complicated mass of tubules. Some crustaceans lack labyrinth, and excretory tubule (nephridial canal) is a much-coiled tube.

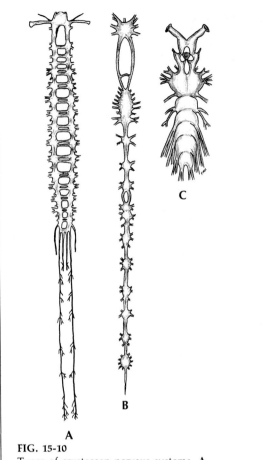

FIG. 15-10
Types of crustacean nervous systems. **A,**
Branchinecta (a fairy shrimp). **B,** *Astacus* (a
crayfish). **C,** *Argulus* (an ectoparasite). Note
tendency toward concentration and fusion of
ganglia. Ladderlike arrangement in **A** may be
considered a primitive condition, whereas that in
B is similar to nervous system of annelid type.

Sensory system. Crustaceans have better developed
sense organs than do the annelids. The largest sense
organs of the crayfish are the eyes and the statocysts. Tactile organs are widely distributed over the
body in the form of **tactile hairs,** delicate projections
of the cuticle, which are especially abundant on the
chelae, mouthparts, and telson. The chemical senses
of **taste** and **smell** are found in hairs on the antennules, antennae, mouthparts, and other places.

A saclike **statocyst** is found on the basal segment
of each antennule and opens to the surface by a dorsal pore. The statocyst contains a ridge that bears
sensory hairs formed from the chitinous lining and
grains of sand that serve as **statoliths.** Whenever the
animal changes its position, corresponding changes
in the position of the grains on the sensory hairs are
relayed as stimuli to the brain, and the animal can
adjust itself accordingly. The chitinous lining of the
statocyst is shed at each molting (ecdysis), and with
it the sand grains are also lost, but new grains are
picked up through the dorsal pore when the animal
renews its statocyst lining.

The eyes in crayfish are **compound** and are made
up of many units called **ommatidia** (Fig. 15-11). Covering the rounded surface of each eye is the transparent **cornea,** which is divided into some 2,500 small
squares known as **facets,** representing the outer ends
of the ommatidia. Each ommatidium, starting at the
surface, consists of a **corneal facet;** two **corneagen
cells,** which form the cornea; a **crystalline cone** of
four **cone cells** (vitrellae); a pair of **retinular cells**
around the crystalline cone; several retinular cells,
which form a central **rhabdome;** and black **pigment
cells,** which separate the retinulae of adjacent ommatidia. The inner ends of the retinular cells connect
with sensory nerve fibers that pass through optic
ganglia to form the optic nerve to the brain.

The movement of pigment in the arthropod compound eye makes possible two kinds of vision. In
each ommatidium are three sets of pigment (distal
retinal pigment cells, proximal retinal pigment cells,
reflecting pigment cells), and these are so arranged
that they can form a more or less complete collar or
sleeve around each ommatidium. For strong light or
day adaptation the distal retinal pigment moves inward and meets the outward moving proximal retinal
pigment so that a complete pigment sleeve is formed
around the ommatidium (Fig. 15-11). In this condition only those rays that strike the cornea directly

antennal glands. The double ventral nerve cord has
a pair of ganglia for each somite (from the eighth
to the ninteenth and gives off nerves to the appendages, muscles, etc. (Fig. 15-10, *B*). In addition to
this central system, there is also a sympathetic nervous system associated with the digestive tract.

In the primitive branchiopods the ventral nerve
cord has the ladder arrangement characteristic of
flatworms and annelids, for the two parts are separate and are connected by transverse commissures.
(Fig. 15-10, *A*).

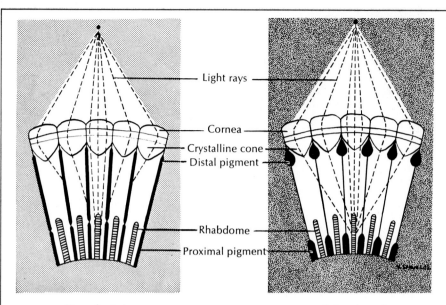

Light rays

Cornea

Crystalline cone

Distal pigment

Rhabdome

Proximal pigment

Day-adapted   **Night-adapted**

FIG. 15-11
Compound eye of arthropod showing migration of pigment in ommatidia for day and
night vision. Five ommatidia represented in each diagram. In daytime each ommatidium
is surrounded by a dark pigment collar so that each ommatidium is stimulated only by
light rays that enter its own cornea (mosaic vision); in nighttime, pigment forms
incomplete collars, and light rays can spread to adjacent ommatidia (continuous, or
superposition, image). (Redrawn from Moment, G. B. 1967. General zoology. Boston,
Houghton Mifflin Co.)

will reach the retinular cells, for each ommatidium
is shielded from the others. Thus each ommatidium
will see only a limited area of the field of vision (a
mosaic, or apposition, image). In dim light the distal
and proximal pigments separate so that the light
rays, with the aid of the reflecting pigment cells,
have a chance to spread to adjacent ommatidia and
to form a continuous, or superposition, image. This
second type of vision is less precise but takes advantage of the amount of light received. Compound eyes
also have the power to analyze polarized light, as has
been demonstrated in the honeybee.

Reproduction. The crayfish is dioecious and there
is some sexual dimorphism, for the female has a
broader abdomen than the male and lacks the modified swimmerets, which serve as copulatory organs
in the male. The **male organs** consist of paired **testes**
and a pair of **vasa deferentia,** which pass over the
digestive glands and down to **genital pores** located
at the base of the fifth walking legs (Fig. 15-6). The

female organs are paired **ovaries** and short **oviducts,**
which pass over the digestive glands and down to
the genital pores at the base of the third walking
legs.

Crayfish mate in the spring or early fall. If they
copulate in the fall, the eggs are usually not laid till
spring. In the process of copulation the male inverts
the female, holds her with his body, chelae, and telson, and transfers his sperm from the openings of
his vasa deferentia to her seminal receptacle by
means of the first two pairs of his swimmerets (Fig.
15-3). The seminal receptacle is a shallow cavity in
the midline between the fourth and fifth pairs of
walking legs. Here the sperm cells are retained until
the mature eggs pass out of the oviducts. Two to
three hundred eggs are discharged at one time, and
as they pass by in slimy strings, they are fertilized
by the sperm from the seminal receptacle.

Development. After fertilization, the masses of
eggs are attached to the swimmerets and remain

FIG. 15-12
Female crayfish with eggs attached to swimmerets. Crayfish carrying eggs are said to be "in berry." The young, when hatched, also cling to swimmerets, protected by tail fan.

FIG. 15-13
Life cycle and distribution *(in red)* of the Gulf shrimp *Pennaeus.* Pennaeids spawn at depths of 20 to 50 fathoms. The young larval forms make up part of the plankton fauna. Older shrimp spend their days hidden in the loose deposits on the bottom, coming up at night to feed.

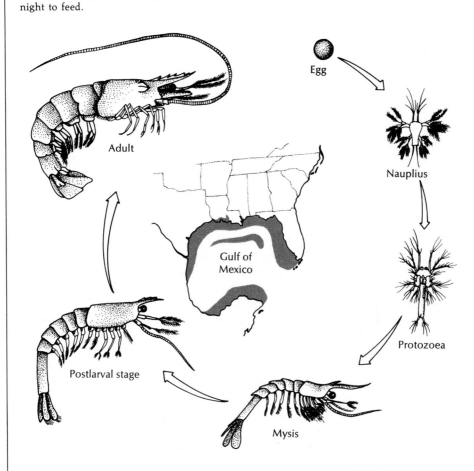

there during development (Fig. 15-12). It takes 5 to 6 weeks for hatching, and each embryo resembles the adult except for size. The young remain attached to the mother for several weeks, during which time they begin their molting. After leaving the mother, they undergo several more molts during the first season, reaching a length of up to 2 inches by fall. Crayfish have a life-span of 3 to 5 years.

Although the young of crayfish resemble the adult, this is not true of all members of this class. In many the larval stages are unlike the adult. Some of these larvae have a strong resemblance to types that are lower in the scale of life. One common larval form, called **nauplius,** is found in the life cycle of the shrimp *Penaeus* (Fig. 15-13) and many other species (Figs. 15-15, 15-16, and 15-20). The nauplius larva has an unsegmented body, frontal eye, and three pairs of biramous appendages.

With successive molts, the nauplius is transformed into a **metanauplius,** with six pairs of appendages; a **protozoea,** with seven pairs of appendages and developing somites; a **zoea,** with eight pairs of appendages and six more in early stages of development, as well as a distinct cephalothorax and abdomen. From the zoea stage develops the **mysis** larva, bearing 13 pairs of biramous appendages on the cephalothorax. Then the adult shrimp with 19 pairs of appendages comes from the mysis larva. Some of these larval stages resemble the adults of lower crustaceans, especially the mysis stage, which closely resembles the adult *Mysis.*

This correspondence between larval stages and the adults of types lower in the animal scale is called the **biogenetic law of recapitulation.** According to this principle, animals in their individual development pass through stages in the evolution of the race. Thus ontogeny repeats phylogeny. Although the biogenetic law in its original sense has been criticized, the description of the crustacean stages may at least give an insight into the course of evolutionary development of this group.

Ecdysis. Ecdysis, or molting, is absolutely necessary for the body to increase in size, because the exoskeleton is nonliving and does not grow as the animal grows. The molting process dominates the crustacean's whole way of life. Its behavior, its reproduction, its entire metabolic processes are affected directly or indirectly by the successive stages of the molting and intermolt cycle. The cycle is short in the young animal in which ecdysis must occur frequently to allow growth. The cycle lengthens as the animal grows to maturity, and in the adult it may occur only once or twice a year.

The cuticle, secreted by the epidermis, is composed of a thin outer **epicuticle** (mostly wax) and a thicker **endocuticle.** Just before each molt inorganic salts are withdrawn from the exoskeleton and stored in the gastroliths and elsewhere. At the same time a new cuticle is secreted beneath the old cuticle, while part of the old endocuticle is being digested away by enzymes in the molting fluid. By absorbing air and water, the crayfish ruptures the old cuticle, usually along the middorsal line, and backs out of the old exoskeleton, shedding even the lining of the digestive system and the cornea of the eyes as well as the gross external structures (Fig. 15-14). Then follows a rapid redeposition of the salvaged inorganic salts and other constituents to harden the new cuticle, together with tissue growth. During the period of molting the animal is defenseless and remains hidden away.

Hormonal control of molting. The signal for the preparations for ecdysis comes from the central nervous system. Externally certain conditions such as temperature, length of daylight hours, humidity (in the case of land crabs) may affect the length of the intermolt cycle. Internally the initiation of molting by the central nervous system depends on the storage of sufficient organic reserves. Starvation may inhibit and feeding promote it. Even the need for extensive regeneration of appendages may help initiate proecdysis.

The molting cycle is under hormonal control. In each eyestalk of the crayfish, lobster, crab, etc. is an **x-organ.** This is a group of neurosecretory cells in the medulla terminalis ganglion in the eyestalk. The axons of these cells terminate in a blood sinus called the **sinus gland** (which is probably not glandular in function, however). A **molt-inhibiting hormone** is formed in the neurosecretory cells and carried in their axons to the sinus gland where it is stored in the enlarged axon endings. The sole function of the molt-inhibiting hormone of the x-organ apparently is to control the crustacean molting glands, or **y-organs.**

The y-organs are located beneath the adductor muscles of the mandibles. They produce a **molt-accelerating hormone.** The y-organ is homologous to

FIG. 15-14
Molting sequence in the lobster, *Homarus americanus.* **A,** Membrane between carapace
and abdomen ruptures, and carapace begins slow elevation. This step may take up to 2
hours. **B** to **E,** Head, thorax, and finally abdomen are withdrawn. This process usually
takes no more than 15 minutes. **F,** Immediately after ecdysis, chelipeds are dessicated and
body very soft. Lobster now begins rapid absorption of water so that within 12 hours
body increases about 20% in length and 50% in weight. Tissue water will be replaced by
protein in succeeding weeks. (Courtesy D. E. Aiken, St. Andrews, New Brunswick.)

the prothoracic glands of insects, and its hormone
is similar to the insect hormone ecdysone. Removal
of the y-organs prevents ecdysis from occurring.
These organs are under the hormonal control of the
x-organs. Reduction in the blood-level of the molt-
inhibiting hormone permits the y-organs to secrete
and release their molt-accelerating hormone. Re-

moval of the eyestalks with their neurosecretory cells
will shorten the molting cycle, permitting ecdysis to
occur with greater than normal frequency.

Although the y-organ hormone initiates proecdysis, or the events that lead up to actual shedding of
the exoskeleton, once this process is initiated, the
cycle proceeds automatically without further action

of either of the controlling hormones. Thus the interaction of these two hormones is sequential, or cyclic. The release of the molt-inhibiting hormone is governed by the central nervous system; the molting hormone starts and integrates the molting cycle.

Evidence indicates that the molting hormone induces metamorphosis by increasing the synthesis of proteins through alterations in gene-activity patterns, in line with the modern concept of biochemical genetics.

Other endocrine functions. A number of hormones are produced by neurosecretory cells in or near the eyestalks. Some of them apparently control the spread of pigments in special branched cells (chromatophores) in the epidermis.

Crustacean pigments are of great variety, including yellow, red, orange, green, brown, and black. Most of the pigments are found in star-shaped chromatophores with many processes, but some are found in the tissues. Color changes by chromatophore action is common in crustaceans. Concentration of the pigment granules in the center of the cells causes a blanching, or lightening, effect. Dispersal of the pigment throughout the cells causes a darkening effect. In many marine crustaceans these color changes are diurnal. The pigment behavior is controlled by hormones from the neurosecretory system.

Migration of retinal pigment for light and dark adaptation in the eyes (Fig. 15-11) is apparently also under the control of eyestalk hormones.

In 1954 H. Charniaux-Cotton, a French investigator, discovered in an amphipod (*Orchestia,* the common beach flea) a pair of **androgenic** glands, which produce hormones that control the male sexual characteristics. These glands may be found in all male Malacostraca. When transplanted into an immature female, they produce masculinization of sex characters such as testes instead of ovaries. Their removal will cause spermatogenesis to cease in the testes. In isopods the glands are located in the testes; in all other malacostracans they are found between the muscles of the coxopodites of the last walking legs and are partly attached near the ends of the vasa deferentia.

Brief résumé of the crustaceans

The crustaceans are an extensive group with many subdivisions. There are many patterns of structure, habitat, and mode of living among them. Some are much larger than the crayfish; others are smaller, even microscopic. Some are highly developed and specialized; others have simpler organizations. The older classification divided the crustaceans into two subclasses—Entomostraca and Malacostraca. The entomostracans are simpler in structure and usually smaller than the malacostracans and do not have abdominal appendages. The term "Entomostraca" is largely being discontinued, for its members lack morphologic unity. The Malacostraca form a natural division, for its members have an eight-segmented thorax, a six- or seven-segmented abdomen, a gastric mill, and abdominal appendages.

Subclass Cephalocarida

Cephalocarida (sef"a-lo-kar'i-da) (Gr. *kephalē,* head, + *karis,* shrimp, + *-ida,* pl. suffix) is a small group of three or four species that have been found along both the Atlantic and Pacific coasts. Cephalocarids have many primitive characteristics. They are strictly marine (sometimes found in brackish water) and about 3 mm. long. They are provided with five pairs of biramous head appendages and nine pairs of small thoracic appendages. They are true hermaphrodites and are unique among arthropods in discharging both eggs and sperm through a common duct. *Hutchinsoniella* is the best known example.

Subclass Branchiopoda

Branchiopoda (bran"kee-op'o-da) (Gr. *branchia,* gills, + *pous, podos,* foot) are among the most primitive of all the crustaceans. One common example is the transparent fairy shrimp *(Eubranchipus)* (Figs. 15-15 and 15-16, *G*), which is found in temporary pools in pasture fields and elsewhere early in spring. They have 11 pairs of broad, leaflike trunk appendages, which are all basically alike. These appendages are used for locomotion, respiration, and egg bearing and bear chromatophores, which add to their beauty.

Males are not common in some species and parthenogenesis frequently occurs in the brine shrimp. The male of the fairy shrimp has a pair of eversible penes for the transfer of sperm to the female. The developing eggs are carried by the female in a ventral brood sac, where they are released in clutches at intervals of a few days. Released eggs are resistant to freezing and desiccation and remain viable after the ponds dry up. Both freezing and drying stimulate

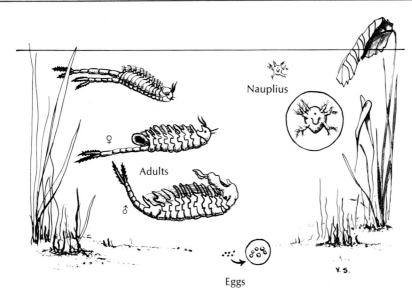

FIG. 15-15
Seasonal cycle of *Eubranchipus,* the fairy shrimp, found chiefly in temporary pools and ponds. Their appearance is sporadic and their distribution irregular. Eggs, in clutches of 10 to 250, can withstand drying during summer, usually hatch in winter and early spring, pass through a nauplius stage, and develop quickly into adults. Drying promotes hatching of eggs but is not always necessary. Twenty-seven species have been recorded for North America. Most species are around 1 inch in length. (Redrawn from Moment, G. B. 1967. General zoology. Boston, Houghton Mifflin Co.)

hatching but are not always necessary. In the following spring, usually between January and May in our northern states, these eggs hatch into free-swimming nauplius larvae with three pairs of appendages and a single eye. Their distribution is very sporadic, both geographically and in annual occurrence. The eggs may be distributed by the wind and on the feet of birds.

The subclass Branchiopoda is divided into four orders: Anostraca (fairy shrimps and brine shrimps), Notostraca (tadpole shrimps such as *Triops,* Fig. 15-16, *D*), Conchostraca (clam shrimps such as *Lynceus*), and Cladocera (water fleas such as *Daphnia,* Fig. 15-16, *F*).

Subclass Ostracoda

Ostracoda (os-trak'o-da) (Gr. *ostrakōdēs,* testaceous) are enclosed in bivalve shells and resemble tiny clams, 0.25 to 8 mm long. When they move, they thrust out their appendages through the open shell. In addition to the head appendages, there are usually two pairs of trunk appendages. Ostracods make up a part of the crustacean population of plankton. *Cypridina* and *Cypris* (Fig. 15-16, *C*) are common genera.

Subclass Copepoda

Copepoda (ko-pep'o-da) (Gr. *kōpē,* oar, + *pous, podos,* foot) are small crustaceans with elongated bodies and forked tails. One of the common forms is *Cyclops* (Fig. 15-16, *B*), which has a single median eye in the head region. Some copepods serve as intermediate hosts for certain parasites such as *Diphyllobothrium,* the broad tapeworm of fish. *Argulus,* a fish louse, is a common parasite on freshwater fish (Fig. 15-16, *E*).

Subclass Cirripedia

The Cirripedia (sir-ri-pe'di-a) (L. *cirrus,* curl, + *pes, pedis,* foot) include the barnacles, which are enclosed in a shell of calcareous plates. At one time they were mistaken for mollusks, but they have

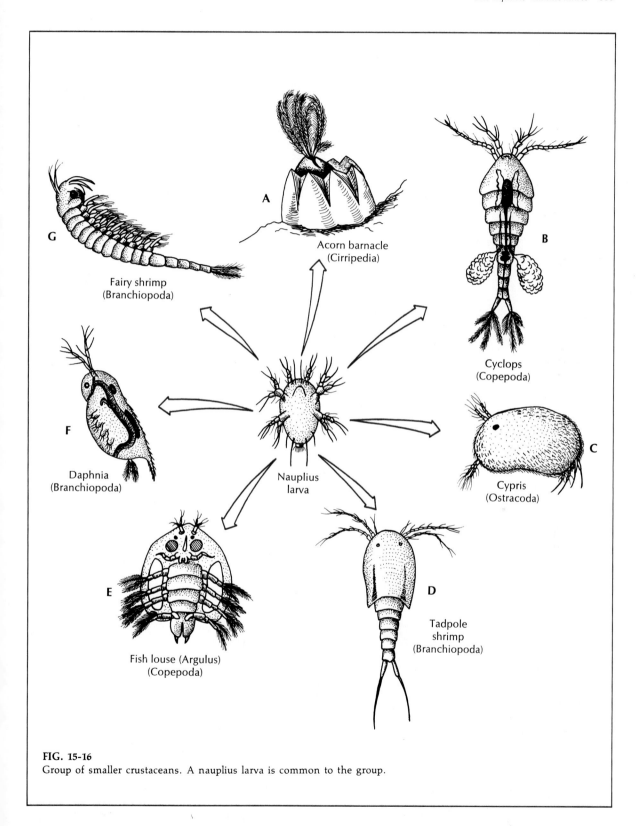

FIG. 15-16
Group of smaller crustaceans. A nauplius larva is common to the group.

jointed appendages, which they use for creating currents of water. In the larval stage they are free swimming but soon attach themselves to a firm surface, where they remain throughout their lives.

Sessile barnacles such as the familiar *Balanus* (Fig.

15-16, *A,* and 15-17) are enclosed in an immovable wall of overlapping plates with a movable operculum, or trapdoor, which is opened and closed by muscles when the animal feeds. When feeding, barnacles thrust out their feathery legs and scoop in plankton.

FIG. 15-17
Group of acorn barnacles, *Balanus,* and sea mussels, *Mytilus,* attached to rock in tidal zone.

FIG. 15-18
Gooseneck barnacles, *Lepas fascicularis,* attach by long stalks. The barnacle is enclosed in a soft mantle and protected by calcareous plates. It sweeps in food particles with its feathery jointed appendages. Adductor muscles can pull plates together when animal is not feeding. (Courtesy R. C. Hermes, Homestead, Fla.)

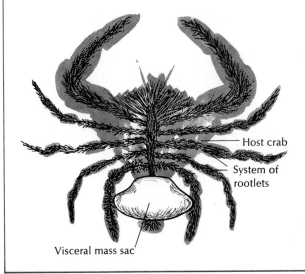

Host crab

System of rootlets

Visceral mass sac

FIG. 15-19
Sacculina, unusual crustacean (subclass Cirripedia) that parasitizes crabs and crayfish. *Sacculina* larva enters host through thin cuticle at one of the joints and discharges cells into crab's blood. At junction of crab's stomach and intestine, these cells become attached and form mass, which sends branching rootlets to all parts of host's body to suck up body fluids. Eventually *Sacculina* forms opening on ventral side of crustacean and extrudes a soft sac filled with eggs. Eggs develop parthenogenetically into nauplius larvae that infect other hosts. Parasite destroys crab's reproductive organs and alters its sex hormones so much that host assumes female form at next molt.

The aquatic mandibulates 337

They are cemented to stones or wharves or oyster beds usually just below high tide. When the tide goes out, they close up their shell houses and wait till the tide returns with another meal. Stalked barnacles have a long stalk or peduncle by which they attach (Fig. 15-18).

Barnacles frequently foul ship bottoms. So great may be their number that the speed of ships may be reduced 30% to 40%, necessitating drydocking the ship to clean them off.

Some barnacles are parasitic. *Sacculina* (Fig. 15-19) parasitizes crayfish and crabs, sending a system of threadlike structures throughout the host body and sucking out the host's body fluids.

Subclass Malacostraca

In the Malacostraca (mal-a-kos'tra-ka) (Gr. *malakos,* soft, + *ostrakon,* shell), the three most familiar orders are Isopoda, Amphipoda, and Decapoda.

ORDER ISOPODA. Isopoda (i-sop'o-da) (Gr. *isos,* equal, + *pous, podos,* foot) are found both in water and on land. They are flattened dorsoventrally and lack a carapace, and the legs are all similar except the anterior and posterior pair. The thorax and abdomen are usually fused. Common land forms are the sow bug, or pill bug (*Porcellio* and *Armadillidium*) (Fig. 15-20, *F*), which live under stones and in damp places. Although many are terrestrial, they breathe by gills, which is possible only under moist

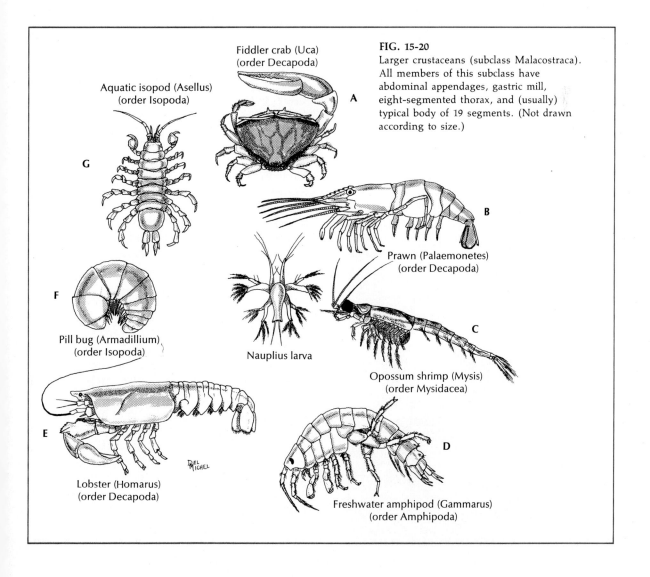

Aquatic isopod (Asellus)
(order Isopoda)

Fiddler crab (Uca)
(order Decapoda)

FIG. 15-20
Larger crustaceans (subclass Malacostraca). All members of this subclass have abdominal appendages, gastric mill, eight-segmented thorax, and (usually) typical body of 19 segments. (Not drawn according to size.)

Prawn (Palaemonetes)
(order Decapoda)

Nauplius larva

Opossum shrimp (Mysis)
(order Mysidacea)

Pill bug (Armadillium)
(order Isopoda)

Lobster (Homarus)
(order Decapoda)

Freshwater amphipod (Gammarus)
(order Amphipoda)

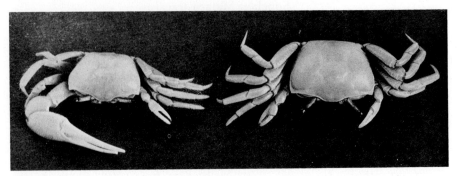

FIG. 15-21
Fiddler crab *Uca.* Male (left) waves its large chela back and forth in presence of female; hence the name "fiddler." Subclass Malacostraca.

FIG. 15-22
"Decorator crab," *Stenocionops furcata,* obtaining a sea anemone, *Calliactis tricolor,* to place upon its back. **A,** Crab prods and pinches anemone near its base until it relaxes and detaches its pedal disk. **B,** Crab then raises anemone toward its own carapace. Anemone usually attaches to carapace with tentacles first and then later turns and attaches its pedal disk. It took this crab 13 minutes to loosen anemone and get it firmly settled in place. Decorator crabs also attach sponges to their backs as part of their camouflage. (Courtesy D. M. Ross, from Cutress, C., D. M. Ross, and L. Sutton. 1970. Canad. J. Zool. **48** (2): 371-376.)

A

B

conditions. *Asellus* (Fig. 15-20, *G*) is a common freshwater form found under rocks and in aquatic plants. *Ligyda (Lygia)* is a common marine form that scurries about on beach or rocky shore, but always just above the water line, never in the water. Some isopods are parasites on fish and crustaceans.

ORDER AMPHIPODA. Amphipoda (am-fip'o-da) (Gr. *amphis,* on both sides, + *pous, podos,* foot) are laterally compressed and have no carapace. Thorax and abdomen are not sharply marked off from each other. Of the eight pairs of thoracic appendages, the first five are used in feeding and the others in crawling. The six pairs of abdominal appendages are employed in swimming and jumping. Amphipods include beach fleas *(Orchestia)* and the freshwater forms *Hyalella* and *Gammarus* (Fig. 15-20).

ORDER DECAPODA. Decapoda (de-cap'o-da) (Gr. *deka,* ten, + *pous, podos,* foot) have five pairs of walking legs, of which the first pair is modified to form pincers (chelae). The larger crustaceans belong to this group and include lobsters (Fig. 15-20, *E*), crayfish, shrimp (Fig. 15-13), and crabs (Fig. 15-20, *A*). This is an extensive group of many thousands of species.

The crabs, especially, exist in a great variety of forms. Although resembling the pattern of crayfish, they differ from the latter in having a broader cephalothorax and a much-reduced abdomen. Familiar examples along the seashore are the hermit crabs (p. 318), which live in snail shells because their abdomens are not protected by the same heavy exoskeleton as the anterior parts are; the fiddler crabs, *Uca* (Fig. 15-21), which burrow in the sand just below the high-tide level and come out to run about over the sand while the tide is out; and the spider crabs such as *Libinia* and the interesting decorator crabs *Stenocionops,* which cover their carapaces with sponges and sea anemones for protective camouflage (Fig. 15-22).

References

Adiyodi, K. G., and R. G. Adiyodi. 1970. Endocrine control of reproduction in decapod Crustacea. Biol. Rev. **45**:121-165.

Barnard, J. L., R. J. Menzies, and M. C. Bacescu. 1962. Abyssal Crustacea. New York, Columbia University Press. *A series of three papers based on the findings of the Columbia University research vessel Vema. Emphasis is on systematics, including new species.*

Barnwell, F. H. 1966. Daily and tidal patterns of activity in individuals of fiddler crab (genus *Uca*) from the Woods Hole region. Biol. Bull. **130**:1-17.

Bliss, D. E., and L. H. Mantel. 1968. Terrestrial adaptations in Crustacea. Amer. Zool. **8**:307-700.

Brown, F. A., Jr. 1954. Biological clocks and the fiddler crab. Sci. Amer. **190**:34-37 (April).

Buchsbaum, R., and L. J. Milne. 1960. The lower animals. Garden City, N.Y., Doubleday & Co., Inc.

Calman, W. T. 1909. Crustacea. In E. R. Lankester: A treatise on zoology, Part VII, third fascicle. London, A. & C. Black, Ltd. *Its taxonomy is outdated, but serious students of the arthropods will find a wealth of information in it.*

Caspari, E. (editor). 1964. Behavior genetics. Amer. Zool. **4**:97-173 (May).

Copeland, E. 1966. Salt transport organelle in *Artemia salenis* (brine shrimp). Science **151**:470-471.

Edmondson, W. T. (editor). 1959. Ward and Whipple's fresh-water biology, ed. 2. New York, John Wiley & Sons, Inc.

Green, J. 1961. A biology of Crustacea. London, H. F. & G. Witherby, Ltd. *A concise account of the structure and physiology of this group.*

Hickman, C. P. 1973. Biology of the invertebrates, ed. 2. St. Louis, The C. V. Mosby Co.

Huxley, T. H. 1880. The crayfish: An introduction to the study of zoology. London, Trubner & Co. *A classic study that is often used as a model of clear and simple biologic presentation.*

Jackson, R. M., and F. Raw. 1966. Life in the soil. (Paperback.) New York, St. Martin's Press, Inc. *Shows how organisms form an integral part of soil, and their importance in soil formation and plant growth.*

Passano, L. M. 1961. The regulation of crustacean metamorphosis. Amer. Zool. **1**:89-95.

Pennak, R. W. 1953. Fresh-water invertebrates of the United States. New York, The Ronald Press Co. *Unusually clear taxonomic keys and illustrative drawings.*

Russell-Hunter, W. D. 1969. A biology of the higher invertebrates. New York, The Macmillan Co. *(Paperback.)*

Schmitt, W. L. 1965. Crustaceans. Ann Arbor, The University of Michigan Press.

Smith, G., and W. F. R. Weldon. 1909. Crustacea. Cambridge Natural History, vol. 4. London, Macmillan & Co., Ltd. *A classic account of the general structure and organization of crustaceans.*

Southward, A. J. 1955. Feeding of barnacles. Nature **175**:1124-1125.

Waterman, T. H. (editor). 1960-1961. The physiology of Crustacea; metabolism and growth, vol. 1. Sense organs, integration, and behavior, vol. 2. New York, Academic Press, Inc. *This work represents the most recent and up-to-date monograph on this important group of arthropods.*

Wood, F. D., and H. E. Wood. 1932. Autotomy in decapod Crustacea. J. Exp. Zool. **62**:1-55.

Wulff, V. J. 1956. Physiology of the compound eye. Physiol. Rev. **36**:145-163.

The grasshopper, with its antennae, compound eyes, chitinous exoskeleton, wings, and three pairs of jointed legs, exhibits the typical insect characteristics. (Courtesy L. L. Rue, III, Blairstown, N. J.)

CHAPTER 16

TERRESTRIAL MANDIBULATES—THE MYRIAPODS AND INSECTS

PHYLUM ARTHROPODA
CLASSES CHILOPODA, DIPLOPODA, PAUROPODA, SYMPHYLA, INSECTA

Along with the arachnids, the insects and myriapods are primarily terrestrial arthropods. Only a few of them have returned to aquatic life, usually in fresh water.

The term "myriapod" is a common term for a group of several classes of mandibulate arthropods that have evolved a pattern of two tagmata—head and trunk—with paired appendages on all trunk somites except the last. The myriapods include the Chilopoda (centipedes), Diplopoda (millipedes), Pauropoda (pauropods), and Symphyla (symphylans).

The insects have evolved a pattern of three tagmata—head, thorax, and abdomen—with appendages on the head and thorax, but greatly reduced or absent from the abdomen. Insects may have arisen from an early myriapod or protomyriapod form.

The terrestrial mandibulates, unlike aquatic mandibulates, have no antennules and their appendages are unlike those of the crustaceans in being always uniramous, never biramous. And although some insect young are aquatic and have gills, the gills are not homologous to those of the crustaceans.

The insects and myriapods share with the onychophorans and some of the arachnids the use of tracheae for carrying the respiratory gases directly to and from all body cells.

Excretion is usually by malpighian tubules.

Class Chilopoda—the centipedes

The Chilopoda, or centipedes, are land forms with somewhat flattened bodies that may contain from a few to 177 somites (Fig. 16-1). Each somite, except the one behind the head and the two last ones, bears

340

FIG. 16-1
Centipede *Scolopendra.* Class Chilopoda. Most segments have one pair of appendages each. First segment bears pair of poison claws, which in some species can inflict serious wounds. Centipedes are carnivorous.

a pair of jointed appendages. The appendages of the first body segment are modified to form poison claws.

The head appendages are similar to those of an insect. There is a pair of antennae, a pair of mandibles, and one or two pairs of maxillae. A pair of eyes on the dorsal side of the head consist of groups of ocelli.

The digestive system is a straight tube into which salivary glands empty at the anterior end. Two pairs of malpighian tubules empty into the hind part of the intestine. There is an elongated heart with a pair of arteries to each somite. Respiration is by means of a tracheal system of branched air tubes that come from a pair of spiracles in each somite. The nervous system is typically arthropod, and there is also a visceral nervous system.

Sexes are separate, with unpaired gonads and paired ducts. Some centipedes lay eggs and others are viviparous. The young are similar to the adults.

Centipedes prefer moist places such as under logs, bark, and stones. They are very agile and are carnivorous in their eating habits, living upon earthworms, cockroaches, and other insects. They kill their prey with their poison claws and then chew it with their mandibles. The common house centipede, *Scutigera forceps* (Fig. 16-2), with 15 pairs of legs, is often seen scurrying around bathrooms and damp cellars, where they catch insects. Most species are harmless to man. Some of the tropical centipedes may reach a length of 1 foot.

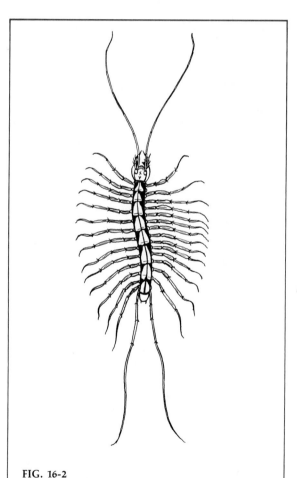

FIG. 16-2
Common house centipede, *Scutigera forceps.* Class Chilopoda. Often seen scurrying around the house, where it eats roaches, bedbugs, and other insects. Its bite is harmless to man.

FIG. 16-3
Millipede *Spirobolus.* Class Diplopoda. They have
two pairs of jointed appendages on each segment
except the four thoracic somites, each of which has
one pair. In contrast to centipedes, millipedes have
cylindric rather than flattened bodies and are
usually vegetarians. (Courtesy Carolina Biological
Supply Co., Burlington, N. C.)

Class Diplopoda

These forms are called millipedes, which literally
means "thousand feet" (Fig. 16-3). Even though
they do not have that many legs, they do have many
appendages since each abdominal somite has two
pairs of appendages, a condition that may have arisen from the fusion of pairs of somites. Their cylindric bodies are made up of 25 to 100 somites. The
short thorax consists of four somites, each bearing
one pair of legs.

The head bears two clumps of simple eyes, a pair
each of antennae, mandibles, and maxillae. The general body structures are similar to those of
centipedes, with a few variations here and there.
Two pairs of spiracles on each abdominal somite
open into air chambers, that give off the tracheal air
tubes. Two genital apertures are found toward the
anterior end.

In most millipedes the appendages of the seventh
somite are specialized for copulatory organs. After
copulation the eggs are laid in a nest and carefully
guarded by the mother. The larval forms have only
one pair of legs to each somite.

Millipedes are not so active as centipedes. They
walk with a slow, graceful motion, not wriggling as
the centipedes do. They prefer dark moist places
under logs or stones. They are herbivorous, feeding
upon decayed matter, whether plant or animal, al-

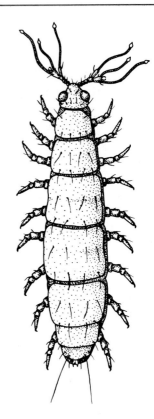

FIG. 16-4
A pauropod. Pauropods are minute, whitish
myriapods with three-branched antennae and nine
pairs of legs. They live in leaf litter, under stones,
etc.

though sometimes they eat living plants. When disturbed, they often roll up into a ball. Common examples of this class are *Spirobolus* and *Julus,* both of
which have wide distribution.

Class Pauropoda

The pauropods are a group of minute (2 mm. or
less), soft-bodied myriapods, of which only 60 or 70
species have been named so far. They have a small
head with branched antennae and are without eyes,
but have a pair of sense organs that have the appearance of eyes (Fig. 16-4). Their 12 trunk segments
usually bear nine pairs of legs (none on the first or
the last two segments). They have only one tergal
plate covering each two segments.

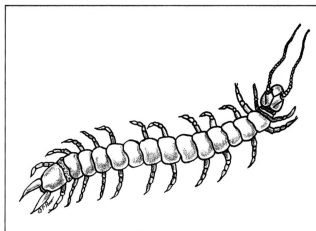

FIG. 16-5
Scutigerella, a symphylan, is a minute whitish myriapod that is sometimes a greenhouse pest. (Modified from Snodgrass, R. E. 1952.)

Tracheae, spiracles, and circulatory system are lacking. Pauropods are probably most closely related to the diplopods but are considered to be more primitive.

Although widely distributed, the pauropods are the least well-known of the myriapods. They live in moist soil, leaf litter, or decaying vegetation, and under bark and debris. Representative genera are *Pauropus* and *Allopauropus.*

Class Symphyla

Symphylans are small (2 to 10 mm.) and have centipede-like bodies (Fig. 16-5). They live in humus, leafy mold, and debris. Like the familiar garden centipede *Scutigerella,* they often become greenhouse pests. They are soft-bodied, with 14 segments, 12 of which bear legs and one a pair of spinnerets. The antennae are long and unbranched.

Symphylans are eyeless but have sensory pits at the bases of the antennae. The tracheal system is limited to a pair of spiracles on the head and tracheal tubes to the anterior segments only.

The mating behavior of *Scutigerella* is unusual. The male places a spermatophore at the end of a stalk. When the female finds it, she takes it into her mouth, storing the sperm in special buccal pouches. Then she removes the eggs from her gonopore with her mouth and attaches them to moss or lichen, or to the walls of crevices, smearing them during the handling with some of the semen and so fertilizing them. The young at first have only six or seven pairs of legs.

Class Insecta

The insects are the most successful biologically of all the groups of arthropods. There are more species of insects than of all the other classes of animals, combined. The recorded number of insect species has been estimated to be 800,000, with thousands of other species yet to be discovered and classified. There is also striking evidence that evolution is continuing among insects at the present time, even though the group as a whole is considered to be stable, according to the fossil record. Studies on *Drosophila* by Patterson, Dobzhansky, and others and on termites by Emerson afford some clear-cut cases of present-day evolutionary change.

It is difficult to visualize the significance of this extensive group and its role in the biologic pattern of animal life. The science of insects (**entomology**) occupies the time and resources of skilled men all over the world. The struggle between man and his insect competitors seems to be an endless one, for no sooner are they suppressed at one point than they break out at another. Yet paradoxically insects have so interwoven themselves into the economy of nature in so many useful roles that man would have a difficult time without them.

Distribution

Insects are among the most abundant and widespread of all land animals. They have spread into practically all habitats that will support life except that of the sea. Only an insignificant few are found there. The marine water striders *(Halobates)* are about the only insects that live on the open sea, but a considerable insect fauna is found in brackish water, in salt marshes, and on sandy beaches. They are found in fresh water, in soils, in forests, in plants, in deserts and wastelands, on mountain tops, and as parasites in and on the bodies of plants and animals.

Their wide distribution is made possible by their powers of flight and their highly adaptable nature. In most cases they can easily surmount barriers that are well nigh impassable to many other animals. Their small size allows them to be carried by currents of both wind and water to far regions. Their well-protected eggs can withstand rigorous conditions and can be carried long distances by birds and animals. Their agility and agressiveness enable them to fight for every possible niche in a location. No single pattern of biologic adaptation can be applied to them.

Size range

Insects range all the way from less than 1 mm. up to 20 to 25 cm. long. Some of the tropical moths have a wingspread of 20 to 30 cm. As a rule, the largest insects are found in tropical countries. Most insects, however, are rarely more than 2.5 cm. long, and a considerable number fall below this dimension. Some beetles are only 0.25 mm. long.

Relationships and origin

It is difficult to work out the ancestry of insects. Fossils give little help, although the fossil record indicates that the first insects were wingless and date back to the Devonian period. All of the few thousand species of fossil insects that have been found in amber and volcanic ash were winged, but wingless forms (order Collembola) have been found in older rocks. Both centipedes and insects have sharply marked off heads, provided with antennae and jaws, but insects have a thorax, which is wanting in centipedes, and lack the large digestive glands of the crustaceans. Some zoologists think they have come from a crustacean larval form, such as the zoea; others believe they are derived from polychaete worms. One theory holds that insects probably arose from the ancestral stock of the class Symphyla.

Characteristics

1. Body of three clearly defined regions—**head, thorax,** and **abdomen**

2. Head of six segments, with one pair of antennae and one pair of mandibles and two pairs of maxillae; **mouthparts adapted for sucking, chewing, and lapping**

3. Thorax of three segments, each with a pair of jointed walking legs; thorax may have two pairs, one pair, or no pair of **wings**

4. Abdomen of not more than 11 segments and **modified posteriorly as genitalia**

5. Respiration by a many-branched **tracheal system,** which communicates with the outside by **spiracles** on the abdomen

6. Digestive system of fore-, mid-, and hindgut and provided with salivary glands

7. Circulatory system of heart, aorta, and hemocoel; no capillaries or veins

8. Excretion by **malpighian tubules** that empty into the hindgut

9. Coelom very much reduced

10. Nervous system of a dorsal brain, subesophageal ganglia, and a double ventral nerve cord provided typically with a pair of ganglia to each somite

11. Sense organs consisting of simple and compound eyes, receptors for taste about the mouth, receptors for touch on various parts of body, and receptors for sound, proprioception, gravity, pressure, etc.

12. Reproduction by separate sexes; paired gonads with single duct in each sex; fertilization internal; few reproducing by parthenogenesis; most exhibiting **metamorphosis**

Adaptability

Insects have marvelous powers of distribution and the capacity to adjust themselves to new habitats. Most of their structural modifications center around the wings, legs, antennae, mouthparts, and alimentary canal. This wide diversity enables this vigorous group to take advantage of all available resources of food and shelter. Some are parasitic, some suck the sap of plants, some chew up the foliage of plants, some are predaceous, and some live upon the blood of various animals. Within these different groups, specialization occurs, so that a particular kind of insect will eat, for instance, the leaves of only one kind of plant. This specificity of eating habits lessens competition and to a great extent accounts for their biologic success.

Insects are fitted for a wide range of habitats,

especially in dry and desert regions. The hard and protective chitinous exoskeleton prevents evaporation, but the insects also extract the utmost in fluid from food and fecal material, as well as moisture from the water by-product of bodily metabolism.

The exoskeleton is made up of a complex system of plates known as sclerites connected to one another by concealed, flexible hinge joints. The muscles between the sclerites enable the insect to make precise movement. The rigidity of its exoskeleton is attributable to the unique scleroproteins and not to its chitin component, and its lightness makes flying possible. By contrast, the cuticle of crustaceans is stiffened by mineral matter and that of the arachnids by organic materials.

Food habits and mouthparts

The food habits of insects are determined to some extent by their mouthparts, which are usually biting or sucking in nature. The majority of insects feed on plant juices and plant tissues. Such a food habit is called **phytophagous.** Some insects feed on specific plants; others, such as grasshoppers, will eat almost any plant. The caterpillars of many moths and butterflies eat the foliage of only certain plants. Monarch butterflies (Fig. 16-16) are known to be poisonous to birds because their caterpillars assimilate cardiac glycosides from certain species of milkweed *(Asclepiadaceae).* This substance confers unpalatability on the butterflies after metamorphosis and induces vomiting in their predators.

Ants are known to have fungus gardens on which they subsist. Many beetles and the larvae of many insects live upon dead animals **(saprophagous).** Other insects are highly **predaceous,** catching and eating other insects as well as other types of animals. The so-called predaceous diving beetle *Cybister fimbriolatus,* however, has been found not to be as predaceous as supposed, but largely a scavenger.

Many insects, adults as well as larvae, are **parasitic.** Fleas, for instance, live on the blood of mammals, and the larvae of many varieties of wasps live upon spiders and caterpillars (Fig. 16-6). In turn, many are parasitized by other insects. Some of the latter are beneficial by controlling the numbers of injurious insects. When parasitic insects are themselves parasitized by other insects, the condition is known as **hyperparasitism,** which often becomes quite involved.

FIG. 16-6
The tomato sphinx, or southern hornworm, and parasites. This caterpillar stage of the sphinx moth, *Protoparce,* feeds on tomato leaves, tobacco, etc. A tiny wasp, *Apanteles,* may lay eggs inside the young caterpillar. When hatched, the wasp larvae feed on caterpillar's lymph and fatty tissue and then emerge and spin cocoons on the outside surface of the caterpillar. The young wasps emerge in 5 to 10 days, but the caterpillar usually dies. (Courtesy O. W. Olsen, Turtox News, August 1967.)

For each type of feeding, the mouthparts are adapted in a specialized way. The sucking mouthparts are usually arranged in the form of a tube and can pierce the tissues of plants or animals. This arrangement is well shown in the water scorpion *(Ranatra fusca),* a member of the order Hemiptera. This elongated, sticklike insect with a slender caudal respiratory tube has a beak in which are four piercing, needlelike stylets made up of two mandibles and two maxillae. These parts are fitted together to form two tubes, a salivary tube for injecting saliva into the prey and a food tube for drawing out the body fluid of the prey. The mosquito also combines piercing with needlelike stylets and sucking through a food channel (Fig. 16-7, *B*). In butterflies and moths the proboscis, usually coiled up when not in use, is fitted as a sucking tube for drawing nectar from flowers (Fig. 16-7, *C*). Biting mouthparts such as those of the grasshopper and many other herbivorous insects are adapted for seizing and crushing food (Fig. 16-7, *A*); those of most carnivorous insects are sharp and pointed for piercing their prey. Houseflies, blowflies, and fruitflies have sponging and lapping mouthparts (Fig. 16-7, *D*). At the apex of the labium are a pair

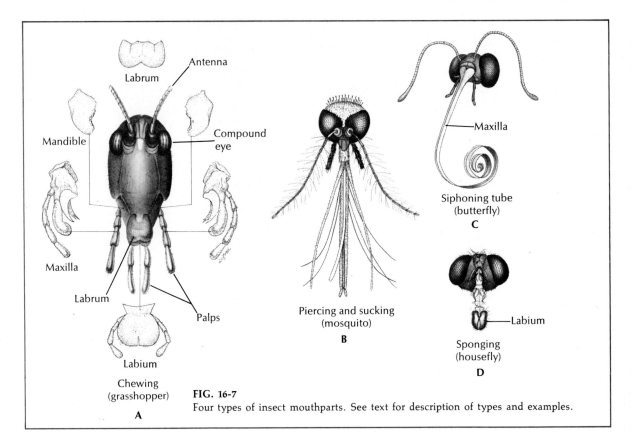

FIG. 16-7

Four types of insect mouthparts. See text for description of types and examples.

of large, soft lobes with grooves on the lower surface that serve as food channels. These flies lap up liquid food, or liquify it first with salivary secretions.

The kind of mouthparts an insect has determines the type of insecticide used in destroying it. For those that bite and chew their food the poison can be applied directly to the food; those that suck must be smothered with gaseous mixtures that interfere with their respiration.

Locomotion in insects

WALKING. Most insects, when walking, use a triangle of legs involving the first and last leg of one side together with the middle leg of the opposite side. In this way, insects keep three of their six legs on the ground, a tripod arrangement for stability. A slow-moving insect alternates its movement first on one side and then on the other. When it goes faster, the two phases tend to overlap and one side may begin before the other side has finished. In some insects all six legs may be used simultaneously and not alternately.

Some insects, such as the water strider *(Gerris),* are able to walk on the surface of water. The water strider has on its footpads nonwetting hairs that do not break but merely indent the surface film of water. As it skates along, *Gerris* uses only the two posterior pair of legs and steers with the anterior pair.

POWER OF FLIGHT. Insects share with birds and flying mammals (bats) the power of flight. Most insects have two pairs of wings, one on the mesothorax, the other on the metathorax. However, the wings of insects are not homologous to those of birds and bats, for the latter are derived in an entirely different manner. Wings of insects are extensions of the integument. They vary a great deal and are used for making distinctions in classification.

When only one pair of wings is present (Diptera), the missing pair (the metathoracic) is represented by a pair of clublike threads called **balancers,** or **halteres.** In flight these halteres vibrate rapidly in a fixed plane, producing an effect like that of a gyroscope.

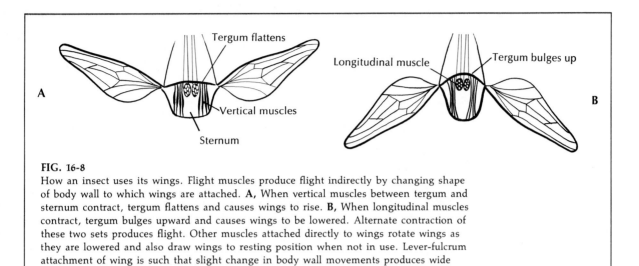

FIG. 16-8

How an insect uses its wings. Flight muscles produce flight indirectly by changing shape of body wall to which wings are attached. **A,** When vertical muscles between tergum and sternum contract, tergum flattens and causes wings to rise. **B,** When longitudinal muscles contract, tergum bulges upward and causes wings to be lowered. Alternate contraction of these two sets produces flight. Other muscles attached directly to wings rotate wings as they are lowered and also draw wings to resting position when not in use. Lever-fulcrum attachment of wing is such that slight change in body wall movements produces wide range in wing tips. In dragonflies wings are always in horizontal position and are controlled by direct muscles that produce slight but rapid up-and-down movements. (Modified from Snodgrass, R. E. 1952.)

Some wings are thin and membranous and are called **membranous wings.** Beetles and some others have the front pair of wings thickened and hardened into **horny wings (elytra),** which protect the more delicate flying wings behind. Grasshoppers and closely related forms have the front wings modified into flexible **leatherlike wings (tegmina).** Butterflies and moths have their wings covered with fine **scales,** which are easily shed when handled. The pattern of wing venation is more or less peculiar to families and orders and is useful in classification.

The wing beat, as well as the speed, of insects varies greatly with different insects. Many butterflies have very slow wing beats of only 5 or 6 per second; those of honeybees may be as many as 200 per second. Likewise, insects also vary greatly in their speed of flying. Some of the flies are reported by a number of observers to have great speeds, but when these flying records are carefully checked, they are usually found to be exaggerated. In a few orders of insects there are no wings at all, and others have generations without wings followed by generations with wings. For the mechanics of wing movement see Fig. 16-8.

Protection and coloration

Insects as a group display many colors. This is especially true of butterflies, moths, and beetles.

Even in the same species the color pattern may vary in a seasonal way, and there also may be color differences between males and females. Some of the color patterns in insects are probably highly adaptive, such as those for **protective coloration, warning coloration, mimicry** (Fig. 16-9), and others.

Besides color, insects have other methods of protecting themselves. The chitinous exoskeleton affords a good protection for many of them; others, such as stinkbugs, have repulsive odors and taste; and others protect themselves by a good offense, for many are very aggressive and can put up a good fight (for example, bees and ants); and still others are swift in running for cover when danger threatens. Since bats can detect their prey by echolocation, certain moths have evolved special ultrasonic ears (tympanic membranes) by which they can pick up the bat chirps and thus evade capture.

Neuromuscular coordination

Insects are active creatures and this implies good neuromuscular coordination. Their muscles are strong and numerous (more than 4,000 have been found in a caterpillar) and are mainly of the striated variety. Their strength is all out of proportion to their size, for a flea can hop a distance a hundred times its own length, and ants and bees can pull many times their own weight.

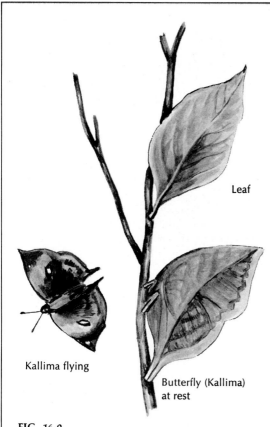

FIG. 16-9
Striking case of protective resemblance in butterfly *Kallima,* which mimics leaf when perched on twig. This butterfly is native of East Indies and was first described by famous English naturalist Alfred Russell Wallace.

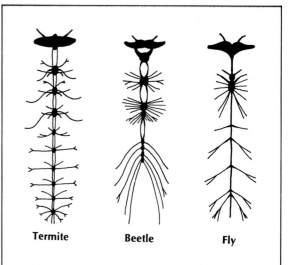

Termite   **Beetle**   **Fly**

FIG. 16-10
Nervous systems of three types of insects. Each has a supraesophageal ganglion (brain) and two ventral nerve cords lying close together or fused. In primitive termites there is a pair of ganglia for each segment of body. Number is reduced in higher forms (beetles and flies). Trend in higher insects is toward greater degree of centralization and cephalization.

The nervous system of insects is similar to that of the earthworm, with a ventral double nerve cord and paired ganglia in each somite. In some insects there is a tendency toward centralization and cephalization, with the ganglia being concentrated toward the anterior end of the body (Fig. 16-10). The skill and dexterity with which many insects can avoid danger are well known, for everyone realizes how quickly flies can avoid swats and dragonflies can evade a butterfly net.

Insect sense organs

Along with their neuromuscular coordination, the sensory perceptions of insects are unusually keen. Most of their sense organs are microscopic and locat-ed chiefly in the body wall. Each type usually responds to a specific stimulus. The various organs are receptive to mechanical, auditory, chemical, visual and other stimuli.

MECHANICAL SENSES. Mechanical stimuli, or those dealing with touch, pressure, vibration, etc. are picked up by sensilla that may be simply a seta, or hair, connected with a nerve cell process, or a nerve ending just under the cuticle and lacking a seta, or by a more complex organ (scolopophorous organ) consisting of sensory cells with their endings attached to the body wall. Such organs are widely distributed over the body, legs, and antennae (Fig. 16-11, *B*).

AUDITORY ORGANS. Airborne sounds may be detected by very sensitive hairs (hair sensilla) or by tympanal organs (scolopophores). In tympanal organs a number of sensory cells (ranging from a few to hundreds) extend to a very thin tympanic membrane that encloses an air space in which vibrations can be detected. Tympanal organs are found in certain Orthoptera (Fig. 16-20), Homoptera, and Lepi-

FIG. 16-11
A, Head of mosquito showing compound eyes, which cover most of head. At upper right
are bases of antennae and base of proboscis with maxillary palps. **B,** Portion of mosquito
antennae. Note the sensory hairs and the segmented nature of the antennae. (Courtesy
P. P. C. Graziadei, Florida State University, Tallahassee, Fla.)

doptera. Vibrations of the substrate are detected by organs usually located on the tibia.

CHEMORECEPTORS. Chemoreceptors (for taste or smell) are usually bundles of sensory cell processes that are often located in sensory pits. These are usually on mouthparts but in ants, bees, and wasps are also found on the antennae, and butterflies, moths, and flies also have them on the legs. The chemical sense is generally keen and some insects can detect certain odors for several miles.

VISUAL ORGANS. Insect eyes are of two types, simple and compound. Simple eyes are found in some nymphs and larvae and in many adults. Compound eyes (Fig. 16-11, *A*) are found in most adults and may cover most of the head and consist of thousands of ommatidia. The structure of the compound eye is similar to that of crustaceans (Fig. 15-11) and is described on pp. 328 and 329.

OTHER SENSES. Insects also have well-developed senses for temperature, especially on the antennae and legs, and for humidity, proprioception, gravity, etc.

Sound production

A sense of hearing is not present in all insects but appears to be well developed in those that produce sound. Everyone is familiar with many of the sounds of insects, such as the buzzing of bees, the chirping of crickets, and the humming of mosquitoes and flies. Sounds are produced in a variety of ways. Some are the result of the rubbing together of rough surfaces, such as those of the integument. Grasshoppers produce a sound effect by rubbing the femur of the last pair of legs over rough ridges on the forewings. Male crickets scrape their wing covers to produce their characteristic chirping. The hum of mosquitoes and bees is caused mainly by the rapid vibration of their wings. Some Hymenoptera produce their sound by a combination of means, such as the vibration of the wings and the special leaflike appendages in the tracheal system. The sound of the cicada is a long drawn-out one produced in a special chamber between the thorax and abdomen. Within this chamber is a membrane connected to a muscle whose rapid contractions cause the membrane to vibrate at different pitches. The cavity amplifies the sound by acting as a resonance chamber.

The humming of bees varies with the temperament of the hive. When excited, the more rapid vibration of the wings produces a difference in sound that is readily detected by those familiar with their ways. Sound may be a means of communication by which insects are able to warn of danger, to call their mates, etc.

Behavior

The keen sensory perceptions of insects enable them to respond to many stimuli. Most, if not all, of their responses are **reflexes.** In many cases these are clear cut and definite, such as the attraction of the moth for light, the detection of rotting flesh by carrion flies, and the avoidance of light by cockroaches. To attract their mates, many insects emit a delicate odor that is detected by the opposite sex. The female *Bombyx mori,* a moth, produces a powerful sex attractant (bombykol), which the male can detect with the sensory hairs on each of his two feathery antennae. Moths differ in the chemical composition of their sex attractants, but a moth will respond only to the sex attractant of his own species.

The taxis of contact **(thigmotaxis)** is illustrated by crickets and beetles, which are often found in narrow crevices; **rheotaxis,** or reaction to water currents, is shown by the caddis fly larvae, which live in rapids. It is now known that mosquitoes are attracted by lactic acid in sweat **(chemotaxis).**

Although insect behavior is mainly instinctive, their behavior patterns can be modified somewhat. Bees, for instance, can be taught to make simple associations between food and color, and ants will learn to make associations between certain odors and food supplies. Thus to some extent they have memory sufficient for the establishment of a conditioned reflex.

Reproduction

The sexes are always separate in insects, and fertilization is internal. Most are **oviparous,** but a few are **viviparous** and bring forth their young alive. **Parthenogenesis** occurs in aphids, gall wasps, and others. In a few (for example, *Miastor,* a type of fly) a process called **pedogenesis** is found. This involves parthenogenesis by larval stages rather than by the adults. Many larvae are produced, some of which pupate to become male and female adults.

Methods of attracting the opposite sex are often quite involved among insects. Some, like the female moth, give off a scent that can be detected by the males. Fireflies use flashes of light for this purpose, whereas many insects find each other by the sounds they make.

In most insects copulation occurs, sperm is deposited in the female vagina and fertilization is internal. However in the lower orders sperm are often encased in spermatophores that may be transferred to the female at copulation or may be deposited on the substratum to be found and picked up by the female who places it in her vagina. The silverfish places a spermatophore on the ground and then spins signal threads nearby to guide the female to it. During the evolutionary transition from aquatic to terrestrial life the use of spermatophores was widely employed, with copulation evolving much later. It is an adaptive advantage for the male to be able to deposit the sperm directly into the vagina.

Insects usually lay a great many eggs. Perhaps the greatest number of eggs is produced by the queen honeybee, which may lay more than a million eggs during her lifetime. At the other extreme, some viviparous flies bring forth a single young at a time. There seems to be a relation between the number of eggs produced and the care of the offspring. Forms that make no provision for their young bring forth many hundreds of eggs; those that have to provide for the larvae, such as the solitary wasps and bees, lay fewer eggs. Fewness of eggs in one generation may be offset by short life cycles. Houseflies and fruit flies require only about 10 days to complete their cycles of development and growth. On the other

FIG. 16-12
Tiger moth and eggs. She always lays her eggs upon the weed that serves as food for her larvae. (Courtesy C. G. Hampson, University of Alberta, Edmonton, Alberta.)

hand, the number of offspring produced bears no relation to the number of eggs laid, for in some chalcid flies each egg gives rise to more than a hundred embryos **(polyembryony).**

Insects reveal marvelous instincts in laying their eggs. Butterflies will lay their eggs only on the particular kind of plant on which the caterpillar feeds (Fig. 16-12), and the ichneumon fly, with unerring accuracy, seeks out a certain kind of larva on which her young are parasitic. Many, however, drop their eggs, which are often well protected, wherever they happen to be and make no further provision for them.

Metamorphosis and growth

When the young of animals undergo abrupt or pronounced changes in appearance during their development to the adult stage, the process of transition is called **metamorphosis.** Although this condition is not restricted to insects, they illustrate this biologic principle better than does any other group. The transformation, for instance, of the hickory horned devil caterpillar into the beautiful royal walnut moth represents an astonishing change in development. In insects metamorphosis is an outcome of the evolution of wings, which are restricted to the reproductive stage where they can be of the most benefit. Not all insects undergo metamorphosis, but most of them do in some form or other.

Complete metamorphosis (which nine tenths of all insects have) is of great adaptive value, for it separates the physiologic process of growth (larva), differentiation (pupa), and reproduction (adult) so that each stage of development can function most efficiently without hindrance from the others. Metamorphosis

FIG. 16-13
Ecdysis in the dog-day cicada, *Tibicen.* Before old cuticle is shed, a new one forms underneath. **A,** Old cuticle splits in dorsal midline as result of blood pressure and of air forced into thorax by muscle contraction. **B,** Emerging insect is pale and its new cuticle is soft. **C,** The wings begin to expand as blood is forced into the veins, and insect enlarges by taking in air. **D,** Within an hour or two the cuticle begins to darken and harden, and cicada is ready for flight. (Photos by J. H. Gerard, Alton, Ill.)

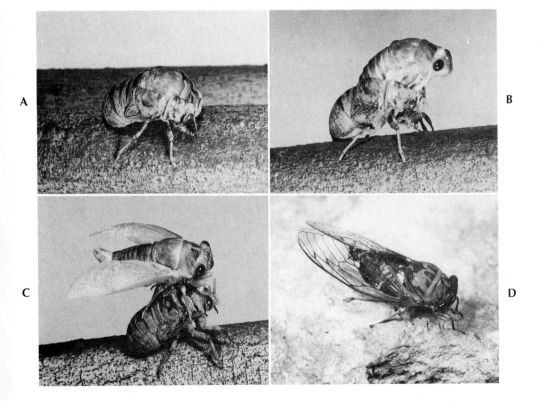

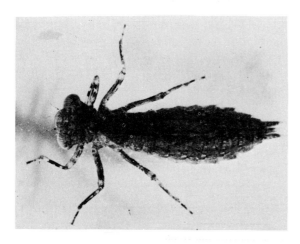

FIG. 16-14
The naiad, or aquatic larva, of a dragonfly. Order Odonata. Found in bottom of pools and streams.

speeds up the energy-transforming mechanisms of the insect life. It also gives a broader ecologic niche relationship, for the larvae often live in an entirely different environment from that of the adults and have different food habits. This specialization of development stages promotes also the evolution rate because of greater possibilities of mutations.

Since the exoskeleton acts as a restrictive armor that prevents expansion of the body, an insect can change its form or size by the process of **molting,** or **ecdysis.** Thus, when it becomes too large for its cuticle, a second epidermis is formed by the hypodermis, the old epidermis splits open along the back of the head and thorax, and the insect works its way out of the old exoskeleton (Fig. 16-13).

With regard to their growth and development, insects may be divided into four groups—those having no metamorphosis and those having incomplete, gradual, or complete metamorphosis.

FIG. 16-15
A, Young praying mantes (nymphs) emerging from their egg capsule. Egg capsules (oothecae) are glued to shrubbery and other objects in late summer and fall. When eggs hatch in spring, enormous swarm of wingless nymphs emerge from single capsule. **B,** Praying mantis, about life size. It gets this name from the way it holds its forelimbs but is far more interested in preying on other insects than in pious devotions. Order Orthoptera. (**B,** Courtesy J. W. Bamberger, Los Angeles, Calif.)

NO METAMORPHOSIS. A few insects called collectively the **ametabola** have no metamorphosis at all. The young hatch from the egg in a form similar to that of the adult and development consists of merely growing larger. Examples are the orders Thysanura and the Collembola. The stages in the life cycle are (1) egg, (2) juvenile, and (3) adult.

INCOMPLETE METAMORPHOSIS. In the **hemi-metabola** (Ephemeroptera, Odonata, Plecoptera) the eggs are laid in water and develop into aquatic naiads, which are quite different from the aerial adults (Fig. 16-14). The naiads have tracheal gills and other modifications for an aquatic life. They grow by successive molts, crawl out of water, and after the last molt become winged adults. The life cycle stages are (1) egg, (2) naiad, and (3) adult. The three orders that

FIG. 16-16

Metamorphosis of monarch butterfly, *Danaus plexippus.* **A,** Adult lays eggs on milkweed plant and hatched larvae feed on milkweed leaves. **B,** Larva hangs on milkweed as it prepares to pupate. At this stage wings develop internally but are not everted until last larval instar. They have chewing mouthparts but no compound eyes. **C,** Larva has transformed into chrysalis, or pupa, an inactive stage that does not feed and is covered by a cocoon or protective covering. **D,** Adult has emerged, with short, wrinkled wings. Wings will expand and harden and pigmentation will develop, and the butterfly will go on its way. (Photos by J. H. Gerard, Alton, Ill.)

have this type of metamorphosis are not closely related phylogenetically but have evolved this type of larval adaptation independently.

GRADUAL METAMORPHOSIS. In this type the newly hatched nymph resembles the adult in general but has no wings or genital appendages (Fig. 16-15). The body proportions of the nymph are also different from those of the adult. At each **instar** (growing stage) after each molt the nymph looks more like the adult until wings are developed. Both nymphs and adults have the same type of mouthparts and food habits. Insects with this type of metamorphosis are called **paurometabola** and include Orthoptera, Hemiptera, Homoptera, and many others. The stages in the life cycle are (1) egg, (2) nymph, and (3) adult.

COMPLETE METAMORPHOSIS. A large number of insect orders in their larval development to adults undergo changes that are referred to as complete metamorphosis (Fig. 16-16) and the group as the **holometabola.** About 88% of all insects experience this type of metamorphosis. The young emerge as wormlike segmented **larvae,** with little difference be-

tween the head, thorax, and abdomen. These wormlike forms are called by various names, such as caterpillars, maggots, bagworms, fuzzy worms, and grubs. The larva goes through several instar stages, increasing in size between molts, and then passes into a type of resting period, the **pupa (chrysalis).** It forms around its body a **case** from its outer body covering or a **cocoon** by spinning silk threads around itself. Within the cocoon or case the final metamorphosis occurs, and finally the adult, or **imago,** appears. When it emerges, the adult is fully grown and undergoes no further molting. In complete metamorphosis the stages of development are (1) egg, (2) larva, (3) pupa, and (4) adult.

PHYSIOLOGY OF METAMORPHOSIS. Metamorphosis in insects is controlled and regulated by hormones. There are three major endocrine organs involved in development through the larval stages to the pupa and eventually to the emergence of the adult. These organs are the **brain,** the **prothoracic glands,** and the **corpora allata** (Fig. 16-17).

The intercerebral part of the brain and the ganglia of the nerve cord contain several groups of neurose-

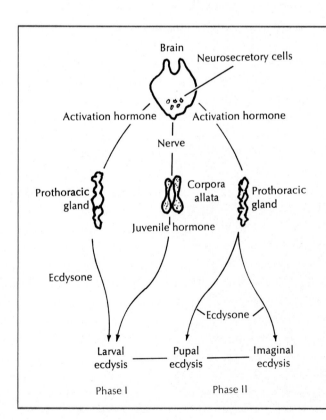

FIG. 16-17
Hormone control in complete metamorphosis of insect. Molting hormone (ecdysone) sets in motion the process of casting off skin (ecdysis) in larva. If juvenile hormone is present along with ecdysone (phase I), only simple molting of larva will occur, each ecdysis producing larger larva. When only ecdysone is secreted (phase II), larva will molt into pupa, and pupa will molt into mature form, or imago (metamorphosis).

cretory cells that produce an endocrine substance called the **activation hormone.** These neurosecretory cells may send their axons to another organ behind the brain, the **corpora cardiaca,** which serves as a storage place for the activation hormone. The corpora cardiaca are of nervous origin, similar to the neurohypophysis of vertebrates. In the blood the activation hormone is carried to the prothoracic gland, a glandular organ in the prothorax, which is stimulated to produce the **molting hormone, or ecdysone.** This hormone sets in motion certain processes that lead to the casting off of the old skin (ecdysis) by proliferation of the epidermal cells.

If the larval form is retained at the end of this process, it is called simple molting; if the insect undergoes changes into pupa or adult, it is called metamorphosis. Simple molting persists as long as a certain **juvenile hormone** (neotenine) is present in sufficient amounts, along with the molting hormone in the blood, and each molting simply produces a larger larva.

The juvenile hormone is produced by a pair of tiny glands **(corpora allata)** located near the corpora cardiaca. The kind of cuticle that is produced depends on the amount of juvenile hormone present. If only a small amount of this hormone is present, a pupal cuticle is the result. When the corpora allata cease to produce the juvenile hormone, the molting hormone alone is secreted into the blood and the adult emerges (metamorphosis). It is thus seen that the molting hormone is necessary for each molt but is modified by the juvenile hormone.

Experimental evidence shows that, when the corpora allata (and thus the juvenile hormone) are removed surgically, the following molt will result in metamorphosis into the adult. Conversely, if the corpora allata from a young larva are transplanted into an old larva, the latter can be converted into a giant larva, because no metamorphosis can occur. Many other experimental modifications on this theme have been performed. Some progress has also been made in determining the chemical nature of the hormones.

The mechanism of molting and metamorphosis just described is that found in insects with complete metamorphosis (holometabola), but the same factors also apply in general to the molting nymphal stages of the paurometabola or hemimetabola, in which there are no pupal stages. A recent method of insect control involves the use of compounds that mimic the juvenile hormones that prevent insects from becoming sexually competent when treated just before they become adults.

What factors initiate the sequence role of these three different hormones? How are they correlated with cyclic events in the life histories of insects? Experimentally it has been shown that low temperature activates the neurosecretory cells of the intercerebral gland of the brain, which then sets in motion the sequence of events already related. The chilling of the brain seems to be all important in the initiation of metamorphosis. Adults cannot molt and grow because they have no prothoracic glands. Many aspects of the control mechanism of these interesting processes have not yet been worked out.

Diapause

Diapause refers to a condition or state of physiologic dormancy or arrested development. Although the concept may apply to variant similar conditions in other animals, its original meaning has direct reference to insects. It is well known that there are periods in the life cycle of many insects when eggs, pupae, or even adults remain for a long time in a state of dormancy because external conditions of climate, moisture, etc. are too harsh or unfavorable for survival under states of normal activity. Because of diapause the insect egg has a mechanism for preventing evaporation from dry surroundings, the pupa can withstand extreme cold, and the adult can synchronize its life cycle with an abundance of food. Altogether, it is an important adaptation in the embryonic larval and pupal stages of most insects of the northern hemisphere. Diapause is that stage of the life cycle when the insect's morphogenesis is interrupted because of unfavorable environmental conditions and is resumed when climate, season, and food are favorable for development and survival. The evidence indicates that hormones are responsible for the control of diapause, for the latter occurs whenever the neurosecretory cells of the brain fail to secrete the molting hormone. Diapause always occurs at the end of an active growth stage of the molting cycle so that, when the diapause period is over, the insect is ready for another molt, or ecdysis.

Migration

A number of the larger and stronger winged insects apparently are able to make long flights, such as the monarch butterfly *(Danaus plexippus)* of North America. In early autumn immense swarms of these butterflies gather in the northern part of the United States and eastern Canada and make southward flights that may take them 2,000 miles or more to warmer regions, around the Gulf of Mexico and South America. Some of them journey as far as the Hawaiian Islands. Many observers have seen swarms of these butterflies far at sea. The actual flight of these forms is not as directional as that of birds; they are carried more by wind currents; this may account for their sporadic appearance in places in which they ordinarily do not resort. Most of the adult monarchs that drift southward in the fall have developed during the preceding summer. Those that go northward in the spring reproduce on milkweeds along the way and give rise to the fall migrants.

Ladybug beetles, which have been of great help in conquering the scale insect in California, are known to migrate to the mountains for a part of their life cycle.

Pheromones

Pheromones are substances that are secreted onto the outside of the body where they influence the behavior of other members of the same species. They are sometimes called the "social hormones." These include the odor trails made by ants, sex attractants, alarm substances, and territorial markers.

They do not include repellants against other species. They are chemical communication signals and are most numerous in the social insects.

Social instincts

Some insects such as bees and ants exhibit very complicated patterns of social instincts. It is true that most insects are more or less solitary and come together only for mating, whereas some are at times found together in large gregarious swarms. Others have worked out complex societies involving **division of labor.** In these societies the adults of one or both sexes live together with the young in a cooperative manner. The size and complexity of these insect organizations vary with the kind of insects. Among the bumblebees the groups are small and the groupings last only a season.

HONEYBEES. The honeybees have one of the most complex organizations in the insect world. Instead of lasting one season, their organization con-

tinues for a more or less indefinite period. As many as 60,000 to 70,000 honeybees may be found in a single hive. Of these, there are a single **queen,** a few hundred **drones** (males), and the rest **workers** (infertile females). The workers carry on all the activities of the hive except the laying of eggs. They gather the nectar from flowers, manufacture honey, collect pollen, secrete wax, take care of young, and ventilate and guard the hive. Each worker appears to do a specific task in all this multiplicity of duties. Their life-span is only a few weeks. One drone fertilizes the queen and stores sperm enough in her spermathecae to last her a lifetime. The life-span of drones is usually for the duration of the summer, for they are driven out or killed by the workers.

A new queen mates during a mating flight and then never again leaves her hive except to swarm. She may live as long as five seasons during which time she may lay a million eggs. According to N. Koenigen the queen bee can differentiate, by careful inspection with her head and front legs, between the slightly larger cells in which she lays her unfertilized eggs (which develop into males) and the worker cells in which she lays fertilized eggs. Whether a larva that is destined to become a female will develop into a worker or a queen will depend on the kind of food it is fed by the workers. Larvae that will become queens are fed "royal jelly," a secretion from the workers' salivary glands produced by the workers only when there is no queen or when an old queen fails to produce enough pheromones, or "queen substance," to prevent it.

The queen is responsible for keeping the hive going during the winter, and only one queen will be tolerated at one time. The queen secretes from her mandibular glands a "queen substance," a complex of at least three pheromones, which is licked off by the worker bees. This substance inhibits ovarian development in the workers and prevents them from becoming queens and also from working on queen cells. In the absence of a queen from the hive they quickly begin to rear a new one. In an overcrowded colony the queen substance may not be distributed to all workers so that reproductive individuals develop and swarming may occur.

Bees have developed a language expressed by a sort of dancing ritual by which the workers or scouts can inform the hive of the whereabouts, distance to, and type of food source they have discovered. This is described in more detail on pp. 920 and 921.

Bees collect nectar from many kinds of flowering plants, many of which give a distinctive flavor to the honey. But the bees make some changes in the nectar before making it into honey. Nectar has as its chief sugar the 12-carbon sugar, sucrose. With the enzyme invertase, bees convert sucrose into levulose and dextrose (6-carbon sugars); so honey contains little sucrose. To prevent premature fermentation, bees remove part of the moisture from the nectar by spreading it in the various cells of the honeycomb and fanning it with their wings to evaporate the excess water.

SOCIAL VESPIDS. The social vespids, or paper wasps, such as the bald-faced hornets (Fig. 16-18) and the yellow jackets (which build chiefly in the ground), also have a caste system of queens, workers, and drones. They construct a nest of papery material consisting of wood or foliage chewed up and elaborated by the wasp. There may be more than one queen in a vespid nest. The determination of queens and workers seems to occur in the larval stages. Males tend to be produced toward the end of short-lived colonies. Among the adults of *Polistes,* and to some extent *Vespula* and others, social hierarchies are established that determine which of the many queens becomes the active one, but hierarchies are less definite in large colonies.

ANTS AND TERMITES. Ants and termites also have complicated social lives. In both groups winged and fertile males and females are produced in large numbers at certain seasons. They leave the nest in swarms and engage in mating flights. After mating they shed their wings and start their colonies

In ant colonies the males die soon after mating and the queen either starts her own new colony, or joins some established colony, and does the egg laying. The sterile females are wingless and do the work of the colony—gather food, care for the young, and protect the colony. In many larger colonies there may be two or three types of individuals within each caste.

Ants have evolved some striking patterns of economic behavior, such as making slaves, farming fungi, herding "ant cows" (aphids), sewing their nests together with silk, and using tools.

In termite colonies, after mating, both queens and kings shed their wings (Fig. 16-19, *A*) and individual pairs will start new colonies. There may also be supplementary reproductives with shorter wings that reproduce within the nest to aid the queen in building up the population. The workers are nymphs and wingless, sterile adults of both sexes (Fig. 16-19, *B*). Soldiers are sterile adults with enlarged heads and

A

B

FIG. 16-18
A, Bald-faced hornet, *Vespula maculata* (order Hymenoptera), one of the paper wasps noted for their globular, papery nests, in which larvae are reared. **B,** Paper nest of bald-faced hornet, lower side removed to show tiers of cells. Cells are open on lower side while larvae are growing and sealed when larvae pupate. Nests are attached to bushes or trees and are composed of fibers of weatherworn wood.

A

FIG. 16-19

Termites. **A,** Reproductive adult. After the mating flight, adults shed their wings and then go in pairs to start a new colony. **B,** Workers are wingless sterile adults, which tend the nest, care for the young, etc. Termites are pale, soft bodied, and broad waisted in contrast to ants, which are dark, hard bodied, and narrow waisted. (Photos by J. H. Gerard. Alton, Ill.)

mandibles, whose duty is to attack intruders, plug up holes in the nest, and the like.

Termites (often called "white ants") differ from ants in being soft bodied and usually light colored and having a broad joint between the thorax and abdomen. Ants are dark and hard bodied and have a narrow constriction between the thorax and abdomen. At rest, termites hold their wings flat over the abdomen; ants usually hold them above the body.

Ants have a varied diet. In some the larvae are the real food digesters for the colony, as they can digest solid food and the adults feed upon liquid foods. By means of trophallaxis, a method of food exchange practiced between larvae and adults, nutrients are distributed among the members. Termites feed upon cellulose, but cannot digest it themselves, depending instead on myriads of flagellates in the digestive tract

to perform that function for them—an excellent example of mutualistic symbiosis.

Both ants and termites, like the bees, secrete various ectohormones, or pheromones, onto the body surface. Through mutual grooming and licking of members of the colony, and through trophallaxis, these pheromones are spread through the colony. Some are inhibitory ectohormones that prevent nymphs from becoming reproductive individuals or soldiers, thus becoming workers. Some external secretions may act as alarm substances, some stimulate group activity, and some provide odor trails that other individuals can follow.

The grasshopper—a representative type

Insects show a remarkable variety of morphologic characters. Some are more or less generalized in body

B

structure, some are highly specialized. The grasshopper, or locust, is a generalized type and is commonly studied in our laboratories for that reason. The big lubber grasshopper, *Romalea microptera,* is a favorite for study, but grasshoppers in general have a similar pattern.

HABITAT. Grasshoppers have a worldwide distribution and are found where there are open grasslands and abundant leafy vegetation. The prairies of the west have immense hordes of them, because there they have abundant food and ideal places to breed. Most American grasshoppers do not migrate to any extent. Some species have very short wings or no wings at all and thus are restricted in their range.

GRASSHOPPER PLAGUES. The early settlers of the great plains of our West often had to contend with great migratory swarms of grasshoppers, which came from arid regions, where food was scarce. So great were some of their swarms that railroad trains were unable to make their way through the teeming masses. One of the worst locust plagues ever reported occurred in Tunis and Algiers in 1908, when swarms of locusts darkened the sun for days as they flew in from the deserts and arid regions. They devastated hundreds of square miles. The insects still flair up in those regions, but better control measures have usually kept them in check.

EXTERNAL FEATURES. The body has the typical insect plan of **head, thorax,** and **abdomen** (Fig. 16-20). The head is made up of six fused somites; the thorax, with three somites, carries the legs and wings; and the abdomen has 11 somites.

Covering the body is the **cuticle** (exoskeleton) of chitin secreted by the epidermis underneath. It is

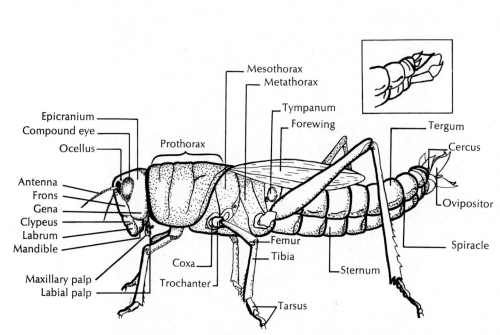

FIG. 16-20
External features of female grasshopper. Terminal segment of male shown in inset.

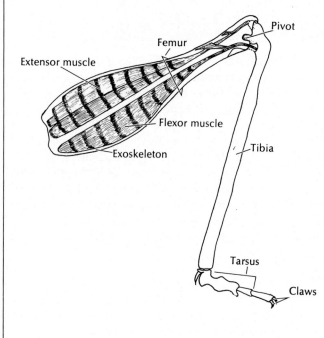

FIG. 16-21
Hind leg of grasshopper. Muscles that operate leg are found within hollow cylinder of exoskeleton. Here they are attached to internal wall, from which they manipulate segments of limb on principle of lever. Note pivot joint and attachment of tendons of extensor and flexor muscles, which act reciprocally to extend and flex limb.

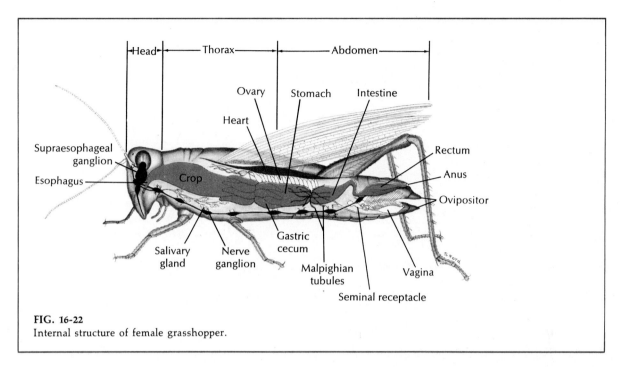

FIG. 16-22
Internal structure of female grasshopper.

divided into hard plates, or sclerites, that are separated by soft sutures, thus permitting the coat of armor to be moved freely.

The **head** bears a pair of compound eyes, three simple eyes (ocelli), and a pair of slender antennae. The **chewing mouthparts** consist of the upper lip, or labrum, attached to the clypeus; a tonguelike hypopharynx just back of the mouth; two heavy mandibles, toothed for chewing; two maxillae, each with several parts; and the lower lip, or labium, with its pair of labial palps (Figs. 16-7 and 16-20).

The **thorax** is made up of a prothorax, mesothorax, and metathorax, each bearing a pair of legs, and the mesothorax and metathorax each have a pair of wings also (Fig. 16-20). Each leg has seven segments and terminates with a pair of claws and a fleshy pulvillus used by the insect in clinging. The large muscular hind legs are adapted for leaping (Fig. 16-21). The membranous hindwings fold up like a fan under narrow leathery forewings. The wings are outgrowths of the epidermis and consist of a double membrane that contains the tracheae. The veins in the wings represent the thickened cuticle around the tracheae and serve to strengthen the wing. Although these veins vary in their patterns among the different species, they are constant within a species and serve as one means of classification.

On the abdomen 10 pairs of spiracles open into the respiratory tracheal system. The first segment of the abdomen bears on each side an oval tympanic membrane that covers the auditory sac of hearing. The terminal segments of the abdomen are modified in the two sexes for copulation and egg laying (Fig. 16-20).

INTERNAL FEATURES. The internal cavity of the grasshopper is a **hemocoel;** that is, it contains blood and is not a true coelomic cavity. Striated muscles are grouped for the movement of the mouthparts, wings, and legs. In the abdomen there are segmental muscles for respiratory and reproductive movements.

Digestion. The digestive system is the most conspicuous system in the body (Fig. 16-22). The ventral mouth is surrounded by the mouthparts and into it open salivary glands. There is a crop for storage, a muscular gizzard for grinding, a stomach with a series of gastric ceca, and an intestine and anus. Digestion and absorption is confined mainly to the stomach and ceca because the rest of the system is lined with chitin that is shed at each molt.

Circulation. The circulatory system is much reduced compared with that of many other arthropods. There is a tubular heart lying in the pericardial cavity, a dorsal aorta that extends from the heart to the head region, and a hemocoel made up of spaces between

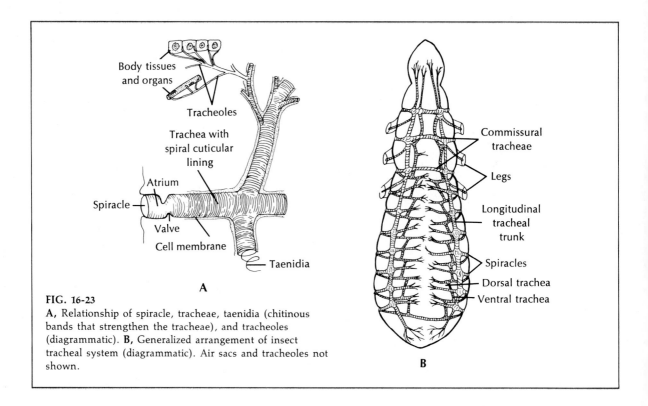

FIG. 16-23

A, Relationship of spiracle, tracheae, taenidia (chitinous bands that strengthen the tracheae), and tracheoles (diagrammatic). **B,** Generalized arrangement of insect tracheal system (diagrammatic). Air sacs and tracheoles not shown.

the internal organs. The system is an open one (lacunar) for there are no capillaries or veins. The blood apparently is not involved in the transportation of respiratory gases, for insects have a separate tracheal system for exchange of gases directly from the air.

Respiration. Respiration in insects is the function of an extensive network of tubes called tracheae that go everywhere in the body (Fig. 16-23).

The tracheal tubes consist of a single layer of cells and are lined with cuticle, that is shed at molting. The larger tubes are prevented from collapsing by spiral threads of chitin. The **spiracles** on each side of the body lead by branches into a longitudinal trunk that gives off the smaller tracheal tubes. The finer air tubes, called **tracheoles,** are connected directly to the body tissues to deliver oxygen and carry away carbon dioxide.

There are also several **air sacs** in the abdomen that pump air in and out of the tracheal system by the alternate contraction and expansion of the abdomen. The action of the spiracles is so synchronized that the first four pairs of spiracles are open at inspiration and closed at expiration, while the other six pairs are closed at inspiration and open at expiration.

Excretion. Insects and spiders have a unique excretory system consisting of **malpighian tubules** that operate in conjunction with specialized glands in the wall of the rectum. The malpighian tubules, variable in number, are thin, elastic, blind tubules attached to the juncture between the midgut and hindgut (Fig. 16-22 and 16-24, *A*). The free ends of the tubules lie free in the hemocoel and are bathed in blood (hemolymph).

Since the malpighian tubules are closed and lack an arterial supply, urine formation cannot be initiated by blood ultrafiltration as in the crustaceans and vertebrates. Instead potassium is actively secreted into the tubules (Fig. 16-24, *B*). This primary secretion of ions pulls water along with it by osmosis to produce a postassium-rich fluid. Other solutes and waste materials also are secreted or diffuse into the tubule. The fluid, or "urine," then drains from the tubules into the intestine. In the rectum, specialized rectal glands actively reabsorb most of the potassium and water, leaving behind wastes such as uric acid.

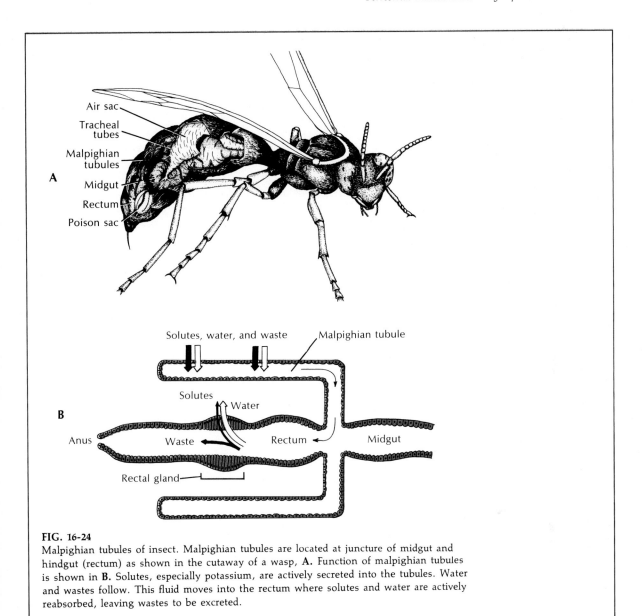

FIG. 16-24

Malpighian tubules of insect. Malpighian tubules are located at juncture of midgut and hindgut (rectum) as shown in the cutaway of a wasp, **A.** Function of malpighian tubules is shown in **B.** Solutes, especially potassium, are actively secreted into the tubules. Water and wastes follow. This fluid moves into the rectum where solutes and water are actively reabsorbed, leaving wastes to be excreted.

By cycling potassium and water in this way, insects living in dry environments may reabsorb nearly all water from the rectum, producing a nearly dry mixture of urine and feces. This is extremely important because terrestrial animals continually face the problem of conserving body fluids, and this may have contributed a good deal to the unusual success of insects on land.

Nervous and sensory systems. The nervous system consists of a brain (supraesophageal ganglion), two connectives, a subesophageal ganglion, and a ventral nerve cord with paired ganglia (Fig. 16-22). A pair of ganglia is found in each thoracic somite and five pairs are located in the abdomen, with nerves running to the visceral organs, legs, and wings. There is also an autonomic nervous system in two divisions. Nerves from this system supply the muscles of the digestive system, the spiracles, and the reproductive system.

Grasshoppers have all the major senses. There are

olfactory organs on the antennae; tactile hairs on the antennae, palps, cerci, and legs; taste organs on the mouthparts; compound eyes concerned with vision; ocelli for light perception; and auditory organs on the first abdominal somite that consist of a tympanic membrane within a circular chitinous ring. Grasshoppers produce sound by rubbing the rough surface of the hind tibias against the wings.

REPRODUCTION AND DEVELOPMENT. Sexes are separate in grasshoppers, and distinction between male and female can be determined by the posterior ends of the abdomen. In the male it is round; in the female it is pointed because of the **ovipositor** (Fig. 16-20).

In copulation the male inserts his copulatory organ into the vagina of the female and transfers his sperm. The sperm are stored in the seminal receptacles until the eggs are laid. The mature eggs, 3 to 5 mm. long, pass down the oviduct and pick up the yolk and shell before fertilization. A small opening in the egg, the **micropyle,** enables the sperm to enter and fertilize the egg. The female makes a tunnel in the ground with her ovipositor and then deposits her eggs. The eggs are usually laid in lots of 20 or so, and a single female may lay several lots. A few days later the adults die.

Development lasts about 3 weeks and ceases when cold weather comes. Growth begins in the spring when the temperature is warmer. The young **nymph** that hatches from the egg resembles the parent, but its head is disproportionately large and it lacks wings. As the young grasshopper grows, its chitinous exoskeleton is shed periodically, wings finally develop, and it reaches the adult form.

Some structural variations among insects

The grasshopper represents a generalized plan into which many other insects fit. However, some insects are highly specialized in their habits and reactions, and this is often accompanied by a corresponding difference in body structures. These differences are more pronounced in external characteristics than they are in internal features. Some of these variations are described here to get a view of their adaptive modifications. Others have already been mentioned in the first part of our discussion on insects.

BODY FORM. There are many patterns of body shape. Some insects are of the thick plump variety, such as beetles (Fig. 16-29); others have long slender bodies, such as the damselfly, crane fly, and walkingstick (Fig. 16-33). Many have bodies of a distinctly streamline form, which is represented by aquatic bugs and beetles (Fig. 16-37). Some insects are very much flattened (for instance, cockroaches), which is an adaptation for living in crevices. In the female of various species the ovipositor may be extremely long (ichneumon wasp, Fig. 16-41). Some also bear modifications of the cerci, such as the horny forceps of earwigs (Fig. 16-34) and those of stone flies and silverfish. Some insects such as moths have hairy coverings. Bees have many bristles for collecting pollen (Fig. 16-27).

ANTENNAE. Antennae may be long, as in cockroaches, some grasshoppers, and katydids, or short, as in dragonflies and most beetles. Some have plumed antennae, as in moths, and others have naked and club-shaped ones (Fig. 16-25). Butterflies have little knobs on the ends of their antennae.

LEGS. Legs of insects show modifications for special purposes. Terrestrial forms have walking legs with terminal pads and claws as in beetles. These pads may be sticky for walking upside down, as in houseflies. The mole cricket has the first pair of legs modified for burrowing in the ground (Fig. 16-26).

FIG. 16-25
Various types of insect antennae. *Left to right:* mosquito (plumose); May beetle (laminate); click beetle (serrate); tenebrionid beetle (moniliform); and water scavenger beetle (clavate).

Water bugs and many beetles have paddle-shaped appendages for swimming. For grasping its prey, the forelegs of the praying mantis are large and strong.

The **honeybee** is a good example of how an insect's legs are developed for special purposes (Fig. 16-27). The first pair of legs in this insect is suited to collect pollen by having a feathery **pollen brush** on the metatarsus and a fringe of hairs on the medial edge of the tibia for cleaning the compound eye. A semicircular indentation lined with teeth is found in the metatarsus, and this is covered over with a spine, the **velum** from the tibia. As the antenna is pulled through this notch, it is cleaned of pollen; hence this structure is called the **antennae cleaner.** The middle leg also has a **pollen brush** on the metatarsus and a **spur** on the tibia for removing wax from the wax glands on the abdomen. The hind limb is the most specialized of all, for it bears the **pollen basket,** the **pollen packer,** and **pollen brushes.** The pollen basket is a concavity on the tibia, with hairs along both edges that are kept moist with secretions from the mouth. The pollen packer consists of a row of stout bristles on the tibia and the auricle, a smooth plate

FIG. 16-26
Mole cricket, *Gryllotalpa.* Note how forelegs are adapted for digging. Order Orthoptera.

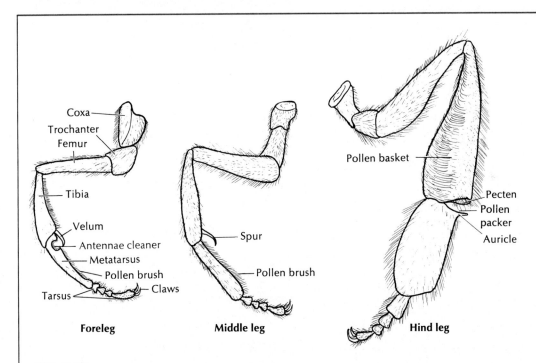

FIG. 16-27
Adaptive legs of honeybee (left side). In foreleg toothed indentation covered with velum is used to comb out antennae. Spur on middle leg removes wax from wax glands on abdomen. Pollen picked up on body hairs is combed off by pollen brushes on front and middle legs and deposited on pollen brushes of hind legs. Long hairs of pecten on hind leg remove pollen from brush of opposite leg; then auricle presses it into pollen basket when leg joint is flexed back. Bee carries load in both baskets to hive, pushes pollen into cell, to be cared for by other workers.

on the metatarsus. The pecten removes the pollen from the pollen brush of the opposite leg onto the auricle. When the leg is flexed, the auricle presses against the end of the tibia, compressing the pollen. Pollen brushes (combs) occur on the inner surface of the metatarsus and consist of rows of stout spines.

Bioluminescence

Some four or five orders of insects include forms that are self-luminous. Others may appear luminous, but this is probably caused by luminous bacteria. The best-known insects that produce their own light are the glowworm and the firefly. The glowworms may be the larvae of a fly (the New Zealand glowworm) or that of a beetle of the family Lampyridae. Some glowworms are wingless females.

The firefly (of the Lampyridae family of beetles) is famed all over the world for its display of light. The photogenic organs are located on the ventral surface of the last abdominal segment and consist of

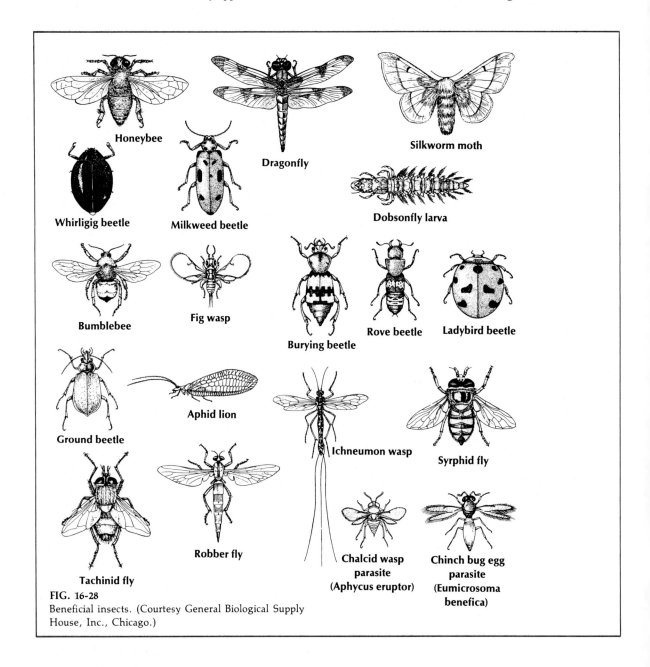

FIG. 16-28
Beneficial insects. (Courtesy General Biological Supply House, Inc., Chicago.)

two kinds of cells. Several layers of cells loaded with opaque uric acid crystals form a reflecting cup that encloses a mass of photogenic cells, with the opening of the cup directing the light outward. The photogenic cells are supplied with an extensive network of tracheae, ensuring them of a good supply of oxygen. In the firefly *Photinus* the male, while flying, flashes at regular intervals of about 5 seconds, each flash lasting 0.2 second. The female, on the ground, replies to his signal about 2 seconds after his flash, with her signal lasting about 0.4 second. Eventually the male, seeing her light flashing, flies to her for copulation.

For a discussion of the physiology of bioluminescence, see pp. 923 and 924.

Relation of insects to man's welfare

A large amount of space would be required to recount all the ways in which insects are harmful or beneficial to man's interests. Only a few can be mentioned here.

BENEFICIAL INSECTS (Fig. 16-28)

1. Many insects produce products useful to man. Among these are the honeybees, the culture of which is a multimillion dollar business. About 240 million pounds of honey and $4^1/_2$ million pounds of wax are produced annually in the United States alone. Several kinds of silkworms, especially *Bombyx mori,* are reared for the production of silk. Despite the growing use of synthetic fabrics, silk is still produced at the world rate of 65 to 75 million pounds annually. The lac insect (family Coccidae) in Indochina and the Philippines secretes so much wax that twigs of the host plant are encrusted $^1/_4$ to $^1/_2$ inch thick. This wax is collected and used in the making of shellac, of which the United States uses about 9 million dollars worth a year. Various dyes have been made from insects, and also certain drugs, such as cantharidin, which is made from a European blister beetle, the Spanish fly.

2. Insects are important in the cross-fertilization of

FIG. 16-29
Tumble bugs, or dung beetles, *Canthon pilularis,* chew off a bit of dung, roll it into a ball, and then roll it to where they wish to bury it in soil. One beetle pushes while other pulls. Eggs are laid in the ball and the larvae feed on the dung. Tumblebugs are black, a inch or less in length, and common in pasture fields. (Photo by J. H. Gerard, Alton, Ill.)

fruits and crops. The bees are indispensible in this respect. The Smyrna fig in California cannot be grown without a small fig wasp, *Blastophaga* (Fig. 16-28), which carries pollen from the male flower of the nonedible caprifig. The wasp can lay its egg only in the male flower, but it also tries the female flowers, thus pollinating them and making the edible fig possible.

Insects and higher plants, which diversified at the same time in the Cretaceous and Tertiary periods, evolved an intimate relationship of mutually advantageous adaptaions. Insects exploit flowers for food,

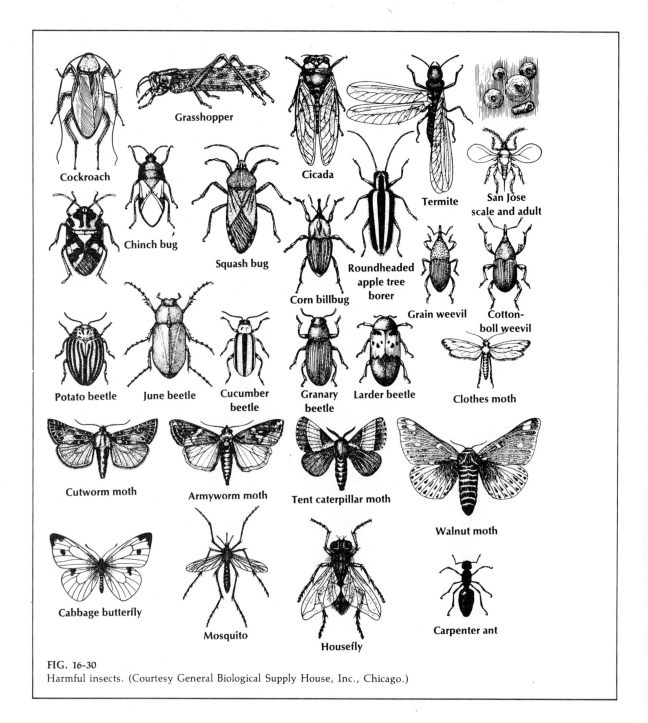

FIG. 16-30
Harmful insects. (Courtesy General Biological Supply House, Inc., Chicago.)

and flowers exploit insects for pollination. Each flower type or floral development of petal and sepal arrangement is correlated with the sensory adjustment of certain pollinating insects that are highly adapted for that kind of flower. The evolution of these mutual flower and host adaptations have resulted in amazing devices of allurement, traps, specialized structures, and precise timing (H. F. Becker).

3. Predaceous insects destroy a variety of harmful insects. Among these are tiger beetles, aphid lions, ant lions, praying mantids, ladybird beetles, wasps, and many others (Fig. 16-28). Some insects lay their eggs on the larvae of injurious insects, and the parasitic larvae hatched from these eggs devour their host (Fig. 16-6).

4. Many insects are scavengers. Many beetles and flies live on animal and plant refuse. Dead animals are eaten by the maggots of flies that lay their eggs in the carcasses. Tumblebugs roll up balls of dung in which they lay their eggs (Fig. 16-29); the developing larvae eat up the dung.

5. Insects are food for birds and for many other animals.

INJURIOUS INSECTS (Fig. 16-30)

1. Insects that eat and destroy plants and fruits include grasshoppers, chinch bugs, Hessian flies, corn borers, cottonboll weevils, San Jose scales, Mediterranean fruit flies, grain weevils, wireworms, and scores of others (Fig. 16-30), with the amount of damage running to more than a billion dollars each year. Every cultivated crop is bothered to some extent by insect pests.

2. Many insects anoy or harm man and other animals. Some are merely a nuisance. Some, like the bees and wasps, inject painful venoms. Biting flies inject their venom by biting; bees and wasps use their stingers. Many people are sensitive to bee and wasp stings, and probably more people die from their stings than from snake bites.

Many insects are parasitic. Chewing lice, sucking lice, fleas, and bedbugs are external parasites. The larvae of many insects are internal parasites of animals, for example, the ox warble flies, which burrow under the skin; the sheep botfly larvae, which burrow into the nasal passages; the horse botflies, in the stomachs of horses, and the screwworm larvae, which grow from eggs laid in wounds and feed upon the living tissues of their hosts.

3. Insects transmit diseases. Among the chief vectors of disease are the mosquitoes, which carry ma-

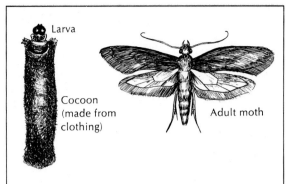

FIG. 16-31
Clothes moth *Tinea* and its larval case. Injury is inflicted by larvae, which build their cocoons from fabrics, usually woolen, upon which they feed. Adults lay eggs upon clothing or fabric not in daily use and kept more or less in the dark.

laria, yellow fever, and filariasis; houseflies, which carry typhoid fever, dysentery, and other diseases; tsetse flies, which carry African sleeping sickness; fleas, which carry bubonic plague; and body lice, which carry typhus fever.

4. Insects are destructive in the household. The ones that injure or damage food are weevils, cockroaches, and ants; those that damage clothing and furnishings are clothes moths (Fig. 16-31) and carpet beetles. Among these also are the termites (Fig. 16-19), which are highly destructive to buildings and all wooden structures.

Control of insects

The control of insects is one of the problems confronting man in his search for sound methods of ecologic management. Insects are interwoven into the ecologic system and serve many useful as well as destructive roles. The problem is to find ways of controlling destructive insects without destroying the rest. Insects must be contained but not eradicated. Two kinds of tactics are being emphasized—those involved in developing plants resistant to insect pests and those that strike directly at the insects in question. Methods that strike directly at the insects involve chemical, physical, or biologic methods.

Chemical methods of control destroy pests by poisons that act upon specific physiologic sites of the insect, such as nervous transmission, neuroendocrine secretions, reproductive processes, and inhibitors of energy metabolism or enzymes. The chief disadvantages of chemical control are the building up of resistance by the insects, and the side effects of the poisons on the environment. The use of chemical insecticides is becoming more restricted and many are now prohibited (DDT, etc.) because of pollution residues and because of their broad spectrum attacks on all kinds of insects, both good and bad.

Physical methods employed are mechanical exclusion and the use of heat, light, and sound to disrupt reproduction and behavior patterns. Pest insects may develop resistance to these methods.

Biologic controls make use of predators and parasites, development of resistant plants, the sterile-male techniques, the introduction of deleterious genes into insect populations, and so forth. One limitation of parasites and predators is that they may become too successful in destroying their food source and so turn to other sources. The chief advantage of biologic controls is that they are specific and are often in line with nature's own ways. They avoid environmental contamination and disruption of the ecology. They are expensive to produce and to get started, but are very effective. Most biologic controls fall into the following four main categories.

1. Viruses or pathogens that are natural enemies of insects, but that could be used in far more effective ways than found in nature. *Bacillus thuringiensis,* a spore-forming bacterium, has been used successfully against leaf-cutting lepidopterans injurious to lettuce crops. This bacterium forms a protein crystal that is a specific toxin. Many viruses are also being used against insect pests.

2. The use of natural predators that will attack and destroy certain harmful insects. The vedalia beetle (from Australia) counteracts the work of the cottony-cushion scale of citrus plants. Parasites from Europe control the alfalfa weevil. A beetle, *Agasicles,* has even been used successfully to control a weed pest— the alligator weed, which clogs up waterways in many Southern states. Both as adult and larva this beetle feeds on the leaves and other aerial parts of the weed.

3. The sterile-male approach that will interfere with metabolism or reproduction in insects. Males sterilized by irradiation have been very effective in eradicating screwworm flies, which attack livestock. Large numbers of sterile male flies are introduced into the natural population; females so mated lay infertile eggs.

4. Insect sex attractants that lure insects to traps.

Pheromones that serve as sex attractants are used to bait traps and the insects drawn to them are then destroyed. Such pheromones are highly specific and attract the target insect only.

Much is yet to be learned about insect control, but potential methods along the above lines seem to offer great possibilities.

Brief review of insect orders

The following is a classification of insects with a brief description of each order (Fig. 16-32). Some entomologists restrict the definition of an insect to those arthropods that have six legs and 14 postcephalic segments and do not add segments in postembryonic stages. By this definition the orders Protura and Collembola would be excluded as insects because their segmentation differs from that of the other orders.

Subclass Apterygota (ap-ter"i-go'ta) (Gr. *a,* not, + *pterygōtos,* winged). Primitive wingless insects that have not come from winged ancestors; with little or no metamorphosis; usually stylelike appendages on pregenital abdominal segments, in addition to cerci.

 Order 1. Protura (pro-tyu'ra) (Gr. *prōtos,* first, + *oura,* tail). These are considered the most primitive of insects, for they have no wings, no antennae, no compound eyes, and no metamorphosis. Appendages are present on abdomen as well as thorax. They are minute and are found under leaves, bark, and moss. Example: *Acerentulus.*

 Order 2. Collembola (col-lem'bo-la) (Gr. *kolla,* glue, + *embolos,* wedge)—**springtails** (Fig. 16-32). These have no wings, compound eyes, or tracheae (usually). They have a peculiar springing organ (furcula) on the ventral side of the fourth abdominal segment. They derive their name from the sticky secretion from a gland near the labium, by which they can adhere to objects. Most are under 5 mm. long. They are often found under leaves and bark. They are sometimes abundant in early spring on snowbanks. Example: *Achorutes.*

 Order 3. Thysanura (thy"sa-nu'ra) (Gr. *thysanos,* tassel, + *oura,* tail)—**bristletails** (Fig. 16-32). These are also wingless, with long antennae. The abdomen is provided with two or three long, jointed cerci. Some are quite small and others may be more than 1 inch long. A familiar form is the silverfish *(Lepisma),* which is often found in homes, where they eat the starch of book covers and clothing. Another one is the firebrat *(Thermobia),* often found about fireplaces.

Subclass Pterygota (ter"i-go'ta) (Gr. *pterygōtos,* winged). Usually winged, but if wingless, the condition is acquired; no abdominal appendages except cerci. This subclass includes 97% of all species of insects.

 Division 1. Exopterygota (ek"sop-ter-i-go'ta) (Gr. *exō,* out-

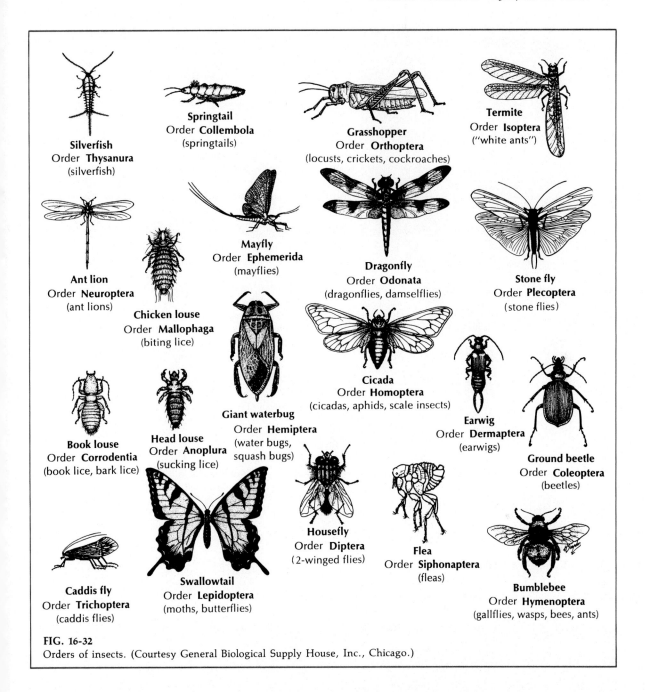

Silverfish
Order **Thysanura**
(silverfish)

Springtail
Order **Collembola**
(springtails)

Grasshopper
Order **Orthoptera**
(locusts, crickets, cockroaches)

Termite
Order **Isoptera**
("white ants")

Ant lion
Order **Neuroptera**
(ant lions)

Mayfly
Order **Ephemerida**
(mayflies)

Dragonfly
Order **Odonata**
(dragonflies, damselflies)

Stone fly
Order **Plecoptera**
(stone flies)

Chicken louse
Order **Mallophaga**
(biting lice)

Book louse
Order **Corrodentia**
(book lice, bark lice)

Head louse
Order **Anoplura**
(sucking lice)

Giant waterbug
Order **Hemiptera**
(water bugs,
squash bugs)

Cicada
Order **Homoptera**
(cicadas, aphids, scale insects)

Earwig
Order **Dermaptera**
(earwigs)

Ground beetle
Order **Coleoptera**
(beetles)

Caddis fly
Order **Trichoptera**
(caddis flies)

Swallowtail
Order **Lepidoptera**
(moths, butterflies)

Housefly
Order **Diptera**
(2-winged flies)

Flea
Order **Siphonaptera**
(fleas)

Bumblebee
Order **Hymenoptera**
(gallflies, wasps, bees, ants)

FIG. 16-32
Orders of insects. (Courtesy General Biological Supply House, Inc., Chicago.)

side, + *pterygōtos,* winged). With gradual metamorphosis.

Order 4. Orthoptera (or-thop'ter-a) (Gr. *orthos,* straight, + *pteron,* wing)—**grasshoppers (locusts), crickets** (Fig. 16-26), **cockroaches, walkingsticks** (Fig. 16-33), **praying mantids** (Fig. 16-15), **etc.** Two pairs of wings are found in this order. The forewings (tegmina) are thickened, and the hindwing is folded like a fan under the fore-

wing. These insects have chewing mouthparts and gradual metamorphosis. The group is very extensive. Most of them are destructive, but the praying mantis is useful in destroying other insects. Example: *Romalea.*

Order 5. Dermaptera (der-map'ter-a) (Gr. *derma,* skin, + *pteron,* wing)—**earwigs** (Figs. 16-32 and 16-34). The forewings are short, with large membranous hindwings. They have biting mouthparts and gradual meta-

FIG. 16-33
Walking stick. Note resemblance to twigs. (Shown slightly less than life size.)

morphosis. The tip of the abdomen bears a pair of forcepslike cerci. *Forficula* is a common example.

Order 6. Plecoptera (ple-kop'ter-a) (Gr. *plekein,* to fold, + *pteron,* wing)—**stone flies** (Fig. 16-32). The four wings are membranous and held pleated on the back when not in use. Mouthparts (not always present) are for chewing, and the metamorphosis is incomplete. The naiad is aquatic and bears tufts of tracheal gills. Example: *Pteronarcys.*

Order 7. Isoptera (i-sop'ter-a) (Gr. *isos,* equal, + *pteron,* wing)—**termites** (Fig. 16-32). These are often wrongly called white ants. They have chewing mouthparts and gradual metamorphosis. They can be distinguished from true ants by the broad union of the thorax to the abdomen. Sexual forms have four similar wings, which they shed after mating; workers and soldiers are wingless and blind. There are many subcastes. Termites are one of the best examples of a social insect, for they live in large colonies. Their diet is exclusively wood, and in tropical countries they are among the most destructive of insects. They are also fairly common in the temperate zones. To aid in their digestion of wood, termites have in their intestines flagellate protozoans that secrete enzymes for the breakdown of cellulose. The mounds of the colonies in the tropics are often imposing affairs. Example: *Reticulotermes.*

Order 8. Ephemerida (ef"e-mer'i-da) (Gr. *ephēmeros,* lasting but a day, + *-ida,* pl. suffix) **(Plectoptera)**—**mayflies** (Fig. 16-32). The wings are membranous, with the forewings larger than the hindwings. Adult mouthparts are vestigial, and the metamorphosis is incomplete. The naiads are aquatic, with lateral tracheal gills. *Ephemera* is a common form.

Order 9. Odonata (o-do-na'ta) (Gr. *odous, odontos,* tooth, + *-ata,* characterized by)—**dragonflies, damselflies**

FIG. 16-34
Earwig. Forcepslike cerci at posterior end are usually better developed in male and are used as organs for defense and offense. Order Dermaptera. (Stained preparation, greatly enlarged.)

(Fig. 16-32). This order gets its name from its toothlike biting mouthparts. These insects have two pairs of membranous wings, incomplete metamorphosis, and large compound eyes. They represent a beautiful group of insects that are often seen flying gracefully over ponds hawking for their food. The naiads are aquatic (Fig. 16-14), those of the dragonfly being provided with a long, hinged labium with which they capture their prey. The naiad has gills in its rectum and breathes by alternately drawing in and expelling water. *Gomphus* is a common example. The members of this order have nonflexible wings that cannot be folded flat over their bodies. These and the mayflies are often called the Palaeoptera (ancient-winged) because of this characteristic, in contrast to the Neoptera (new-winged), which can fold their wings.

Order 10. Corrodentia (cor"ro-den'che-a) (L. *corrodens,* gnawing, + *-ia,* pl. suffix)—**book lice** (Fig. 16-32). These are small insects with chewing mouthparts and four membranous wings (sometimes absent). Metamorphosis is gradual. They are sometimes found in books, since they have a fondness for the starch of the bindings, and also in bird's nests and under bark. *Troctes* is an example.

Order 11. Mallophaga (mal-lof'a-ga) (Gr. *mallos,* wool, + *phagein,* to eat)—**biting lice** (Fig. 16-32). These insects are less than $1/4$ inch long and are wingless. Their legs are adapted for clinging to the host, and their mouthparts are for chewing. Their metamorphosis is gradual. They live exclusively on birds and mammals, eating feathers, hairs, and skin debris. The common chicken louse is *Menopon.*

Order 12. Embioptera (em"bee-op'ter-a) (Gr. *embios,* lively, + *pteron,* wing)—**embiids.** These are small insects with elongated bodies, with wingless females and usually winged males. Their mouthparts are for chewing, and their metamorphosis is gradual. They make silk-lined channels in the soil and are colonial. They are mostly tropical forms. *Embia* is an example.

Order 13. Thysanoptera (thy"sa-nop'ter-a) (Gr. *thysanos,* tassel, + *pteron,* wing)—**thrips.** These are only a few millimeters long or smaller. Some are wingless, but others have four similar wings. They have sucking mouthparts and gradual metamorphosis. Parthenogenesis is common among them. They live by sucking the juices of plants. *Thrips* is an example.

Order 14. Anoplura (an"o-plu'ra) (Gr. *anoplos,* unarmed, + *oura,* tail)—**sucking lice** (Fig. 16-32). The bodies of these insects are small and depressed, and they are wingless. Their mouth is adapted for piercing and sucking, and they have no metamorphosis. These are the true lice, and three kinds have become pests to man: (1) the head louse *(Pediculus capitis),* which lives on the head hair and lays its eggs (nits) there; (2) the body louse *(Pediculus corporis),* sometimes called the "cootie," which lives on the body and head, lays its eggs in the clothing and hair, and is responsible for carrying typhus fever, trench fever, and other diseases; and (3) the crab louse *(Phthirius pubis),* which often gets in the pubic hair. Many other kinds are found on various mammals.

Order 15. Hemiptera (he-mip'ter-a) (Gr. *hēmi-,* half, + *pteron,* wing)—**true bugs** (Figs. 16-32 and 16-35). This is an extensive group of great economic importance. The front wings of these insects are thickened and leatherlike at the anterior half but membranous at the posterior half, whereas the hindwings are membranous and fold under the front ones. They have piercing and sucking mouthparts, and the metamorphosis is gradual. This order includes such groups as the water bugs, bedbugs, stinkbugs, chinch bugs, assassin bugs, and water striders. *Gerris* is the familiar water strider.

Order 16. Homoptera (ho-mop'ter-a) (Gr. *homos,* same, + *pteron,* wing)—**cicadas** (Fig. 16-13), **aphids, scale insects, leafhoppers.** These insects have two pairs of wings (absent in some) of uniform thickness and texture. The mouthparts are for piercing and sucking, and there is gradual metamorphosis. One of the most noted members of this order is the cicada, or 17-year locust *(Magicicada septemdecem).* Eggs are laid in trees, where they hatch into nymphs, which then drop to the ground. Then for 17 years they make their home in the soil, living on plant juices from the roots of trees. At the end of this time they crawl up a tree trunk, undergo their final molt, and emerge as adults. Some southern species require only 13 years for their cycle. Another interesting member of the order is the aphid *(Aphis mali),* or plant louse. Aphids have both sexual and parthenogenetic generations, and they are destruc-

FIG. 16-35

Box elder bug, *Leptocoris.* These often become a nuisance in the fall when they enter houses in swarms, seeking place to hibernate. However, they do no damage to house contents. Order Hemiptera.

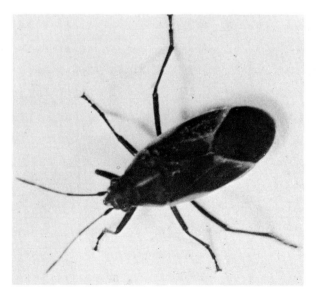

A 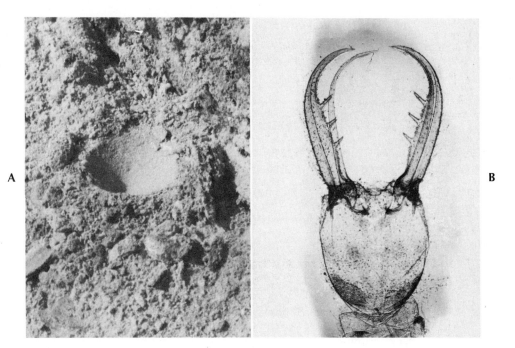 B

FIG. 16-36
A, Conical crater pit of ant lion larva, or "doodlebug," designed to trap ants. When ant starts to slide into sandy pit, ant lion, which is concealed in pit, helps by undermining sand beneath the ant. **B,** Head of ant lion larva *Myrmeleon* (greatly enlarged) showing large mandibles used for seizing prey. Order Neuroptera.

FIG. 16-37
Giant diving beetle *Dytiscus*. This beetle is more than 1 inch long, is very active in water, and is said to eat any prey it can overcome. Order Coleoptera. The larval form of *Dytiscus* is shown in Fig. 38-12, *C.*

FIG. 16-38
Common stag beetle, *Lucanus,* another coleopteran.

tive to plants. Scale insects are also destructive by sucking plant sap. Some are protected by a soft cottony covering.

Order 17. Zoraptera (zo-rap'ter-a) (Gr. *zōros,* pure, + *apteros,* wingless). These are small insects not exceeding 3 mm. in length. They have some resemblance to termites, for they occur in colonies and both winged and wingless forms are found despite their name. The winged forms have two pairs of wings, which they shed, as do termites. The wingless forms are blind. Of 16 species, only two occur in the United States. Unlike termites, they feed as predators or scavengers on small arthropods. They are commonly found under bark and in rotten logs. *Zorotypus* is a common genus.

DIVISION 2. ENDOPTERYGOTA (en"dop-teri-go'ta) (Gr. *endon,* inside, + *pterygōtos,* winged). With complete metamorphosis.

Order 18. Neuroptera (neu-rop'ter-a) (Gr. *neuron,* nerve, + *pteron,* wing)—**dobsonflies, ant lions** (Fig. 16-32), **lacewings.** This order takes its name from the many cross veins in the wings. The four wings are alike and are membranous. They have complete metamorphosis, with biting mouthparts. The huge larval form of *Corydalis,* the dobsonfly, looks formidable with its large mandibles but is harmless and is excellent bait for fish. The larva (doodlebug) of the ant lion (Fig. 16-36, *B*) has the interesting habit of making a conical crater in the sand and lying concealed in it until its prey accidentally falls into it (Fig. 16-36, *A*). With its very large jaws the larva quickly seizes the ant and makes a meal of it.

Order 19. Coleoptera (ko"le-op'ter-a) (Gr. *koleos,* sheath, + *pteron,* wing)—**beetles, weevils** (Figs. 16-32, 16-37, and 16-38). The beetles are the most extensive group of animals in the world. About one animal out of every three is a beetle. The forewings (elytra) are thick and leathery, whereas the hindwings are membranous and are folded under the forewings. Some beetles are wingless. They have chewing mouthparts and complete metamorphosis. This large group is divided into many families, each with thousands of species. Among the most familiar of these family groups are the ground beetles, tiger beetles, carrion beetles, whirligig beetles, click beetles, darkling beetles, stag beetles, fireflies, dung beetles (Fig. 16-29), and diving beetles. The other section of this important group is the weevil. Weevils have their jaws modified into snouts, and the most familiar one is the cotton-boll weevil *(Anthonomus grandis).*

Order 20. Strepsiptera (strep-sip'ter-a) (Gr. *strepsis,* a turning, + *pteron,* wing)—**stylops.** There is a distinct sexual dimorphism in these small forms, for the males have tiny or vestigial forewings and fan-shaped hindwings and the females no wings, eyes, or antennae. The mouthparts are for chewing, and the life cycle is complex **(hypermetamorphosis).** The females and larvae are wholly parasitic in bees, wasps, and other insects. There are relatively few species in the group. *Xenos* is a parasite in the wasp *Polistes.*

Order 21. Mecoptera (me-kop'ter-a) (Gr. *mēkos,* length, + *pteron,* wing)—**scorpion flies.** These have four narrow, membranous wings (some are wingless) and chewing mouthparts. Metamorphosis is complete. The male has a curious clasping organ at the tip of the abdomen, which resembles the sting of a scorpion;

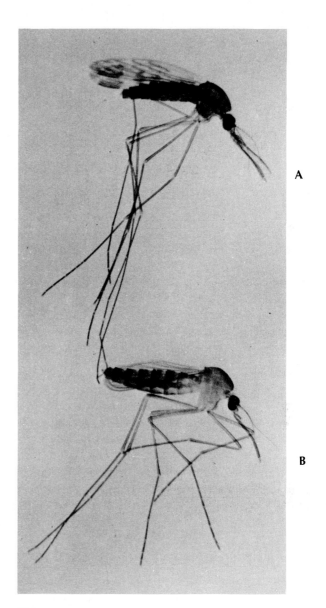

FIG. 16-39
Two common mosquitoes (females). **A,** *Anopheles,* the malaria carrier, bears dark blotches on its wings. **B,** *Culex.* In feeding, *Anopheles* holds its body at an angle to surface, whereas *Culex* holds its body horizontally. Order Diptera.

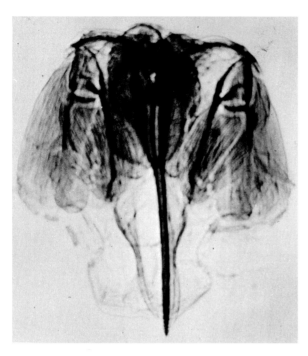

FIG. 16-40
Stinger of honeybee after dissection.
(Photomicrograph.)

hence the name of the order. *Boreus,* which is often found on snow in the winter, is one of the more familiar forms.

Order 22. Lepidoptera (lep"i-dop'ter-a) (Gr. *lepis, lepidos,* scale, + *pteron,* wing)—**butterflies, moths** (Figs. 16-12, 16-16, and 16-32). These insects are famed for their great beauty and are known the world over. Two pairs of wings are membranous and are covered with overlapping scales. The mouthparts are for sucking and are kept coiled under the head when not in use. The metamorphosis is complete, and the larval form is called a caterpillar (Fig. 19-16). The larval forms have glands for spinning their cocoons. Butterflies have a knob at the tip of each antenna; moths have plumed or feathered antennae. The major families are the tiger moths, regal moths, bagworm moths, swallow-tailed butterflies, sulfur butterflies, and gossamer butterflies. Examples: *Danaus, Callasamia.*

Order 23. Diptera (dip'ter-a) (Gr. *dis,* two, + *pteron,* wing)—**flies** (Figs. 16-32 and 16-39). These are the true flies. They are unique among insects in having only two wings, although some are wingless. In place of hindwings, they have **halteres** (hal-te'reez). Their metamorphosis is complete, and they have piercing and sucking mouthparts. Their larval forms are often known as maggots and those developing in water, as wigglers (or wrigglers). They are commonly separated into two great sections: the long-horned flies with antennae of more than five segments and the short-horned flies with

FIG. 16-41
Ichneumon wasp. By means of long ovipositor, female can bore deeply into tree and lay an egg near wood-boring beetle larva. There egg hatches into tiny larva, which parasitizes beetle larva. Ichneumon wasps are large insects; this specimen had an overall length of more than 6 inches. *Inset,* Typical compound insect eyes, antennae, and mouthparts of ichneumon head. Order Hymenoptera.

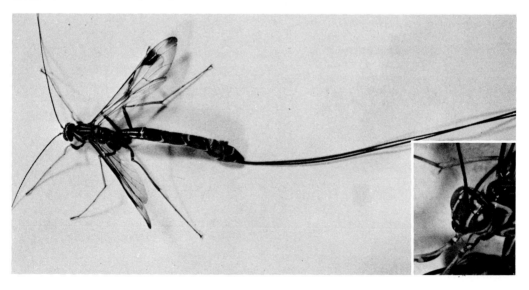

antennae of five or less joints. Among the long-horned flies are the crane flies, mosquitoes, moth flies, midges, gnats, and blackflies. Representatives of the short-horned flies are the fruit flies, flesh flies, botflies, houseflies, and bee flies. *Musca domestica* is the common housefly.

Order 24. Trichoptera (tri-kop'ter-a) (Gr. *thrix, trichos,* hair, + *pteron,* wing)—**caddis flies** (Fig. 16-32). These insects have two pairs of membranous wings with silky hairs. Metamorphosis is complete, and the mouthparts are vestigial. The larval forms have the interesting habit of living in fairly rapid waters in cases composed of sand and sticks bound together by their secretions. *Hydropsyche* is a common genus.

Order 25. Siphonaptera (sy"fo-nap'ter-a) (Gr. siphon + *a,* without, + *pteron,* wing)—**fleas** (Fig. 16-32). Fleas are wingless, with sucking mouthparts. Their bodies are laterally compressed, with legs adapted for leaping. Compound eyes are lacking, and simple eyes may be suppressed. They are ectoparasites on mammals and birds. There are different species of fleas, but they readily change hosts whenever the opportunity offers. Some of them are vectors for the bubonic plague and typhus fever. The one that is most annoying to man is *Pulex irritans.*

Order 26. Hymenoptera (hy"men-op'ter-a) (Gr. *hymen,* membrane, + *pteron,* wing)—**ants, bees, wasps** (Figs. 16-18 and 16-32). This order gets its name from the four membranous wings, which may be absent in some. They have complete metamorphosis and chewing or sucking mouthparts. Their wings have the peculiarity of being held together with hooks (hamuli). The ovipositor in the female is modified into a stinger (Fig. 16-40), piercer, or saw. Both social and solitary species are found. This group is a very large one and includes some of the most specialized members of the insect world (Fig. 16-41). Some are useful and others are destructive to man's interests. Among the bees the most familiar are the bumblebees, carpenter bees, honeybees, and mason bees. Mud-dauber wasps and bee wasps are among the most common of the wasps. The great family of ants has carried the social organization as far as any group of insects, and their marvelous instincts and reactions are among the most interesting in the animal kingdom. *Apis mellifica* is the common honeybee. *Vespula* is a common genus of hornets (Fig. 16-18).

References

Selected general references

Borror, D. J., and D. M. Delong. 1971. An introduction to the study of insects, ed. 3. New York, Rinehart & Co. *This up-to-date text emphasizes both the study and the identification of insects.*

Butler, C. G. 1955. The world of the honeybee. New York, The Macmillan Co. *A monograph on the organization and behavior of the honeybee community.*

Butler, C. G. 1961. The efficiency of the honeybee community. Endeavor **20**:5-10 (Jan.).

Butler, C. G. 1967. Insect pheromones. Biol. Rev. **42**:42-87.

Carpenter, F. M. 1953. The geological history and evolution of insects. Amer. Scientist **41**:256-270.

Cavill, G. W. K., and P. L. Robertson. 1965. Ant venoms, attractants, and repellents. Science **149**:1337-1345.

Cleveland, L. R. 1949. Hormone-induced sexual cycles of flagellates. I. Gametogenesis, fertilization and meiosis in *Trichonympha.* J. Morph. **85**:197-296. *Discusses the variety of sexual forms in roaches.*

Cleveland, L. R. 1928. Symbiosis between termites and their intestinal Protozoa. Biol. Bull. **54**:231-237. *A classical investigation.*

Collias, N. E., et al. 1964. The evolution of external construction by animals. Amer. Zool. **4**:175-243 (May). *Discusses the evolution of caddisworm cases, termite nests, bee nests, as well as spider webs, and bird nests.*

Dethier, V. G. 1963. The physiology of insect senses. New York, John Wiley & Sons, Inc. *This treatise shows how insects with few sense cells accomplish about the same functions as higher forms do with many sensory units.*

Eisner, T. E. 1966. Beetle's spray discourages predators. Natural History **75**:42-47 (Feb.).

Fisher, R. C. 1971. Aspects of the physiology of endoparasitic Hymenoptera. Biol. Rev. **46**:243-278.

Frisch, K. von. 1950. Bees: their vision, chemical senses, and language. Ithaca, N. Y., Cornell University Press. *An outstanding work on the way bees communicate with each other and reveal the sources of food supplies. A marvelous revelation of animal behavior.*

Goetsch, W. 1957. The ants. Ann Arbor, University of Michigan Press. *A concise and authoritative account of ants and their ways.*

Huffaker, C. B. 1970. Life against life—nature's pest control scheme. Environ. Reserve **3**:162-175.

Jaques, H. E. 1951. How to know the beetles. Dubuque, Iowa, William C. Brown Co. *A useful and compact manual for the coleopterist.*

Johannsen, O. A., and F. H. Butt. 1941. Embryology of the insects and myriapods. New York, McGraw-Hill Book Co. *A technical work on the development of insects and some of their allies.*

Kroeger, H., and M. Lezzi. 1966. Regulation of gene action in insect development. Ann. Rev. Entomol. **11**:1-22.

Lanham, U. 1964. The insects. New York, Columbia University Press. *A concise monograph on insects, dealing with their origin, evolution, ecology, and place in the order of nature.*

Lees, A. D. 1955. The physiology of diapause in arthropods. New York, Cambridge University Press.

Little, V. A. 1957. General and applied entomology. New York, Harper & Brothers. *Treats the subject from the viewpoints of anatomy, physiology, metamorphosis, and control.*

Menn, J. J., and B. Beroza (editors). 1972. Insect juvenile hormones: chemistry and action. New York, Academic Press Inc. *An appraisal of their structure and function.*

Miall, L. C. 1922. The natural history of aquatic insects. London, Macmillan & Co., Ltd. *Descriptions of both adult and larval forms of aquatic insects. Interesting to beginner.*

Neider, C. (editor). 1954. The fabulous insects. New York, Harper & Brothers. *An anthology of selections dealing with interesting insects and written by eminent authorities. The beginner will profit from reading this book.*

Nichols, D., and A. L. Cooke. 1971. The Oxford book of invertebrates. New York, Oxford University Press.

Pesson, P. 1959. The world of insects. Translated by R. B. Freeman, New York, McGraw-Hill Book Co. *A fascinating volume of illustrations (many in color) and pithy descriptions.*

Pierce, G. W. 1949. The songs of insects. Cambridge, Mass., Harvard University Press. *The author, a physicist, has applied the apparatus and methods of physics to an investigation of the sounds made by insects.*

Pringle, J. W. S. 1957. Insect flight. New York, Cambridge University Press. *This little monograph discusses the physiology, anatomy, and aerodynamics of insect flight.*

Richards, O. W. 1971. Biology of the social wasps (Hymenoptera, Vespidae). Biol. Rev. 46:483-528.

Roeder, K. D. 1963. Nerve cells and insect behavior. Cambridge, Mass., Harvard University Press.

Scheer, B. T. (editor). 1957. Recent advances in invertebrate physiology. Eugene, University of Oregon Publications. *Among the topics discussed are neuromuscular action, hormonal control, and patterns of rhythms.*

Segal, S. (editor). 1961. Metamorphosis in the animal kingdom. Amer. Zool. 1:1-171 (Feb). *A description of the common types of metamorphosis.*

Snodgrass, R. E. 1952. A textbook of arthropod anatomy. Ithaca, N. Y., Cornell University Press.

Ward, H. B., and G. C. Whipple. 1959. Freshwater biology. (Edited by W. T. Edmondson.) New York, John Wiley & Sons, Inc.

Wheeler, W. M. 1910. Ants. New York, Columbia University Press. *The classic work on these social insects.*

Wigglesworth, V. B. 1954. The physiology of insect metamorphosis. New York, Cambridge University Press. *The author believes that metamorphosis is merely another case of polymorphism, which is almost universal among animals. He adduces evidence that all levels of complexity are determined by the supply or deficiency of raw materials that may be produced within an endocrine gland and circulate as a hormone.*

Selected *Scientific American* articles

Batra, S. W. T., and L. R. Batra. 1967. The fungus gardens of insects. 217:112-120 (Nov.).

Bennet-Clark, H. C., and A. W. Ewing. 1970. The love song of the fruit fly. 223:85-92 (July).

Cambi, J. M. 1971. Flight orientation in locusts. 225:74-81 (Aug.). *The locust is equipped with elegant neural systems for controlling roll, pitch, and yaw during flight.*

Edwards, J. C. 1960. Insect assassins. 202:72-78 (June). *The assassin bug and certain other insects predigest their prey by injecting a venomous saliva.*

Evans, H. E. 1963. Predatory wasps. 208:144-154 (April). *Solitary wasps exhibit a highly specific behavior that suggests their course of evolution.*

Frisch, K. von. 1962. Dialects in the language of the bees. 207:78-86 (Aug.). *The specific bee dances are described in the article by this famous and much-honored behaviorist.*

Hinton, H. E. 1970. Insect eggshells. 223:84-91 (Aug.). *An account of the structural complexity of the shells as shown by the scanning electron microscope.*

Heinrich, B. 1973. The energetics of the bumblebee. 228:96-102 (April). *Traces the drastic economies of bumblebees' energy in their service of collecting nectar and pollinating flowers.*

Hocking, B. 1958. Insect flight. 199:92-98 (Dec.). *The flight machinery of insects is explained.*

Hölldobler, B. 1971. Communication between ants and their guests. 224:86-93 (March). *The chemical and mechanical language of ants and their arthropod guests is described in this well-illustrated article.*

Jacobson, M., and M. Beroza. 1964. Insect attractants. 211:20-27 (Aug.) *Chemists have isolated a number of insect sex attractants and are making synthetic ones as lures for insect control.*

Johnson, C. G. 1963. The aerial migration of insects. 209:132-138 (Dec.). *The author describes mass insect migrations and stresses that such migrations are more common than generally supposed.*

Jones, J. C. 1968. The sexual life of a mosquito. 218:108-116 (April).

Knipling, E. F. 1960. The eradication of the screw-worm fly. 203:54-61 (Oct.). *The sterile-male principle of insect control has been successfully applied for the elimination of this destructive pest of the southeastern United States.*

Lüscher, M. 1953. The termite and the cell. 188:74-78 (May). *An account of the influences that determine differentiation of termite nymphs into various forms.*

Lüscher, M. 1961. Air-conditioned termite nests. 205:138-145 (July).

Morse, R. A. 1972. Environmental control in the beehive. 226:93-98 (April).

Roeder, K. D. 1965. Moths and ultrasound. 212:94-102 (April).

Topoff, H. R. 1972. The social behavior of army ants. 227:71-79 (Nov.).

Wenner, A. M. 1964. Sound communication in honeybees. 210:116-124 (April).

Williams, C. M. 1950. The metamorphosis of insects. 182:24-28 (April).

Williams, C. M. 1967. Third-generation pesticides. 217:13-17 (July). *The author describes the development of highly selective insecticides.*

Wilson, D. M. 1968. The flight-control system of the locust. 218:83-90 (May). *The author describes his experiments on the neural control of locust flight.*

Wilson, E. O. 1958. The fire ant. 198:36-41 (March). *This imported species is now a serious pest in the South.*

Peripatus, *a caterpillar-like onychophoran with both annelid and arthropod characteristics. Does it represent a missing link between the annelids and arthropods, or an independent branch that has evolved from the common ancestors of all the segmented groups? (Courtesy Ward's Natural Science Establishment, Inc., Rochester, N. Y.)*

CHAPTER 17

THE LESSER PROTOSTOMES

PHYLUM SIPUNCULIDA
PHYLUM ECHIUROIDEA
PHYLUM PRIAPULIDA
PHYLUM PENTASTOMIDA
PHYLUM ONYCHOPHORA
PHYLUM TARDIGRADA

Place in animal kingdom

1. The phyla of the lesser protostomes are more or less distinct groups that present puzzling affinities to each other and to other groups. Some have been appended to other phyla in various classifications, but many zoologists at present are inclined to regard them as sufficiently different to be listed as independent phyla.

2. They also have much in common. All are coelomates and protostomes, although one or two have deuterostome characteristics in their embryologic development. Despite their lack of numbers and evolutionary diversity, they have survived in the competition with the more successful protostomes to which they are related in some degree.

Biologic contributions

Although the members of these phyla are assigned obscure roles in the animal kingdom, they take a fairly high rank in their morphologic and functional levels among the invertebrates. Organ systems, with few exceptions, are well represented. Respiratory systems in general are lacking, although the onychophorans have a tracheal system. The same may be said for the circulatory system, which is absent in some and present in others. Many of them burrow in the sand and mud and have ingenious methods for securing their food.

The group has many odd morphologic devices for performing their functions. For example, one may mention the introvert of the Sipunculida and the proboscis of the Echiuroidea for burrowing and feeding, the complex feeding apparatus of the Tardigrada, the stumpy, clawed legs of the Onychophora for locomotion, and the caudal appendages of the Priapulida of questionable function but perhaps for respiration.

Phylogeny and adaptive radiation

Phylogeny. The actual evolutionary development of these (and other) phyla is so hidden in the past that taxonomic divisions and relations are based largely upon embryologic and morphologic features. Such an odd assortment of phyla must have many puzzling phylogenetic affinities. Most have apparently developed from the annelid-arthropod stem line at different times.

The Priapulida had long been considered a class under the phylum Aschelminthes because of their supposed pseudocoelomate condition, and were given a close relationship to the class Kinorhyncha of that phylum. Present investigations show them to be true coelomates, although many zoologists believe that their exact status is debatable.

Sipunculida have a typical trochophore larva and appear to be related to the annelids, as do also the Echiuroidea, because of their similar embryologic development and for other reasons.

The three phyla—Pentastomida, Onychophora, and Tardigrada—have been often placed together as a group called Pararthropoda, or Oncopoda, because they have unjointed limbs with claws (at some stage) and a cuticle that undergoes molting.

Adaptive radiation. All these phyla have relatively minor economic and ecologic importance. All have undergone modest evolutionary diversification so that in most cases they are represented by few species. The small number of species indicates fairly stable ecologic conditions, low mutation rates, or other causes for few modifications.

In their evolutionary development the Sipunculida, the Priapulida, and the Echiuroidea have been guided mainly by the varied proboscis devices, which they have stressed in burrowing and food getting. The

Onychophora show the greatest structural diversity in the number of their stump legs, which vary from 14 to 44 pairs in the different species. Varied mechanisms of claws and feeding apparatus have evolved among the tardigrade species, whereas the pentastomidan adaptations have been chiefly those having to do with attachments to their hosts.

■ PHYLUM SIPUNCULIDA*

Sipunculids are benthic marine worms that live sedentary lives in burrows or tubes in mud or sand (Fig. 13-17), in borrowed shells, or among the rocks, coral, or vegetation of the intertidal zone. Some are tiny, slender worms, but some may reach a length of a foot or more. Some of them are commonly known as "peanut worms."

Sipunculids have no segmentation or setae. They are most easily recognized by a slender retractile introvert, or proboscis, that is continually and rapidly being run in and out the anterior end (Fig. 17-1). The trunk is more plump and its walls are muscular. When the introvert is everted, the mouth can be seen at its tip surrounded by tentacles. Undisturbed sipunculids usually extend the anterior end from the bur-

*Sigh-pun-kyu'li-da (L. *sipunculus,* little siphon, + *-ida,* pl. suffix), but sipunculid (sigh-pun'kyu-lid).

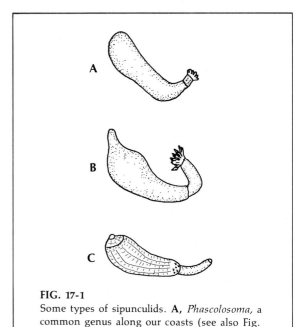

FIG. 17-1
Some types of sipunculids. **A,** *Phascolosoma,* a common genus along our coasts (see also Fig. 13-17). **B,** *Dendrostomum.* **C,** *Aspidosiphon.*

row or hiding place and stretch out the tentacles to explore and feed. They are deposit feeders living on organic matter collected by the tentacles and moved to the mouth by ciliary action. The introvert is extended by hydrostatic pressure produced by contraction of the body wall muscles against the coelomic fluid. It is retracted by special retractor muscles. Its surface is often rough because of surface spines, hooks or papillae.

There is a large fluid-filled coelom traversed by muscle and connective tissue fibers. The digestive tract is a long tube that doubles back on itself to end in the anus near the base of the introvert (Fig. 17-2). A pair of large nephridia open to the outside to expel waste-filled coelomic amebocytes; they also serve as gonoducts. Circulatory and respiratory systems are lacking, but the coelomic fluid contains red corpus-

cles that bear a respiratory pigment, hemerythrin, used in the transportation of oxygen. The nervous system has a bilobed cerebral ganglion just behind the tentacles and a ventral nerve cord extending the length of the body. The sexes are separate. Permanent gonads are lacking, and ovaries or testes develop seasonally in the connective tissue covering the origins of one or more of the retractor muscles. Sex cells are released through the nephridia. The larval form is usually a trochophore.

Of the 330 species, the best known genera are probably *Sipunculus*, *Dendrostomum*, *Phascolosoma*, and *Golfingia.*

Some authorities have placed the sipunculids and echiurids together as a class of Annelida, but the groups differ in many important respects and are now usually considered as separate phyla. Both, however, are no doubt related to the annelids.

■ PHYLUM ECHIURIDA*

Echiurids have some general resemblance to the annelids and some authorities have placed them with the annelids; however, they are entirely unsegmented and have other differences, and it seems best to assign them phylum rank. They have fewer species than the Sipunculida but are usually found in greater densities.

Echiurids are marine worms that burrow into mud or sand, or live in empty shells, sand dollar tests, rocky crevices, and so on. They are most common in the littoral zones of warm waters, but some have been dredged from waters as deep as 8,200 m.

They have a somewhat sausage-shaped cylindrical body with a muscular body wall, and anterior to the mouth is a flattened extensible proboscis that has given them the name of "spoon worms" because of its shape when contracted (Fig. 17-3). The proboscis cannot be retracted into the trunk. Echiurids are deposit feeders and the proboscis is normally used to extend out over the mud for exploration and feeding. A ciliated groove on the ventral side directs organic particles to the mouth (Fig. 17-4). The proboscis in some forms is short and in others long. *Bonellia,* which is only 8 cm. long, can extend its proboscis to a meter in length.

One common form, *Urechis,* secretes a funnel-shaped mucus net in its burrow through which it

*Ek-ee-yur'i-da (Gr. *echis,* adder, + *oura,* tail, + *-ida,* pl. suffix).

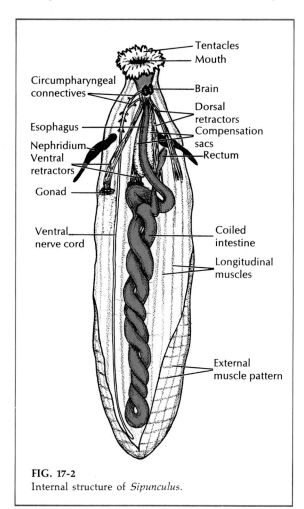

FIG. 17-2
Internal structure of *Sipunculus.*

Tentacles
Mouth
Circumpharyngeal connectives
Brain
Dorsal retractors
Compensation sacs
Esophagus
Nephridium
Ventral retractors
Rectum
Gonad
Ventral nerve cord
Coiled intestine
Longitudinal muscles
External muscle pattern

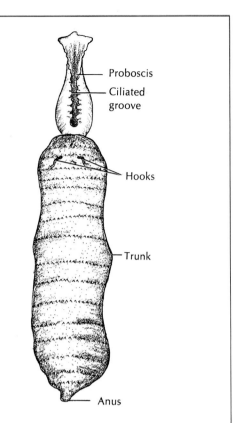

FIG. 17-3
Echiurus, an echiurid common on both Atlantic and Pacific coasts. The shape of the proboscis lends them the common name of "spoon worms."

pumps water, capturing bacteria and fine particulate material in the net. When loaded with food, the net is swallowed.

The muscular body wall is covered with cuticle and epithelium. There may be a pair of anterior setae, or a row of bristles around the posterior end. The coelom is large. The digestive tract is long and coiled and terminates at the posterior end. A pair of anal vesicles may have an excretory and osmoregulatory function. There is a closed circulatory system with a contractile dorsal vessel. Two or three pairs of nephridia serve mainly as gonoducts. A nerve ring runs around the pharynx and forward into the proboscis, and there is a ventral nerve cord. There are no specialized sense organs.

The sexes are separate, with a single gonad in each sex. The mature sex cells break loose from the gonads and leave the body cavity by way of the nephridia, and fertilization is usually outside. In some species sexual dimorphism is pronounced, with the female being much the larger of the two.

Bonellia is noteworthy both for its extreme sexual dimorphism and for the way sex is differentiated in this genus. At first most of the larvae are sexually indifferent and can develop either into males or females. Those larvae that come into contact with the female proboscis become tiny males that spend most of their lives in the nephridium of a female as para-

FIG. 17-4
Tatjanellia, an echiurid, is a detritus feeder. It extends its long proboscis to explore the bottom surface. Organic particles are picked up and carried along a ciliated food groove to the mouth. (After Zenkevitch; modified from Dawydoff, C. 1959.)

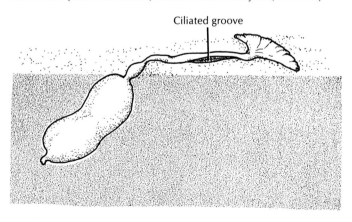

sites. Larvae that swim free and do not contact the female metamorphose into females. The stimulus for the male development appears to be a chemical one from the proboscis of the female. F. Baltzer (1925) was able to get various degrees of intersexuality by removing larvae at various stages of male transformation.

PHYLUM PRIAPULIDA*

The priapulids are a small group of marine worms found chiefly in the colder waters of both hemispheres. They have been reported along our Atlantic coast from Massachusetts to Greenland and on the Pacific from California to Alaska. They live in the bottom muck and sand of the sea floor and range from intertidal zones to depths of several thousand meters.

In the past they have been linked with various phyla, such as the echiurids and sipunculids, and they were placed in the phylum Aschelminthes by some authorities because of their resemblance to the Kinorhyncha and because they were considered to be pseudocoelomates. However a peritoneum has been identified (Shapeero, 1961), and they are now considered to be coelomates and are usually ranked as a separate phylum. Five genera have been named.

The small number of species indicate that either the priapulids have undergone little adaptive radiation or that only remnants remain of groups that were more widely distributed at one time. An eversible proboscis with many recurved teeth surrounding the mouth and lining the pharynx adapts them for predaceous habits. They are also adapted for burrowing into the mud, aided by the muscular body wall and fluid pressure. But although they can plow through the mud, priapulids prefer to remain quietly in a vertical position with the anterior end at the surface. They capture mostly slow-moving prey such as polychaetes, small crustaceans, and even other priapulids, which they swallow whole. Their stable ecologic niche has given them little opportunity or necessity for wide evolutionary diversification within the group. Only *Tubiluchus* is adapted to an interstitial habitat and is provided with a polythyridium (between pharynx and intestine) for filtering small particles of food.

External features. Priapulids are rarely more than 5 or 6 inches long. The cylindric body (Fig. 17-5) in-

*Pri"a-pyu'li-da (Gr. *priapos,* phallus, + *-ida,* pl. suffix).

cludes a proboscis, trunk, and one or two caudal appendages (lacking in *Halicryptus*). The trunk is annulated but not segmented, and has a warty appearance because of its many tubercles and spines. The large eversible proboscis bears rows of teeth that lead toward the mouth. It is used in sampling the sur-

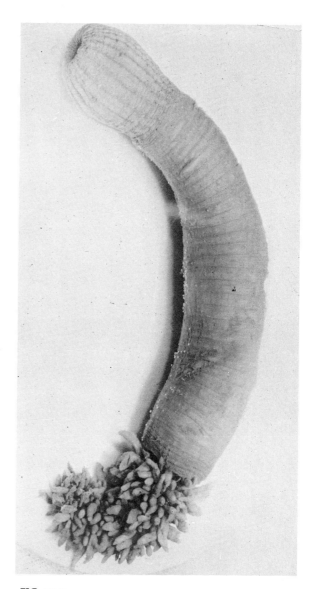

FIG. 17-5
Priapulus caudatus Lamarck. Proboscis (top) is partially withdrawn. Note rows of papillae on proboscis, superficial segmentation along trunk, and bushy caudal appendage. When fully expanded, some specimens may attain length of 145 mm. (Courtesy W. L. Shapeero, University of Washington, Seattle.)

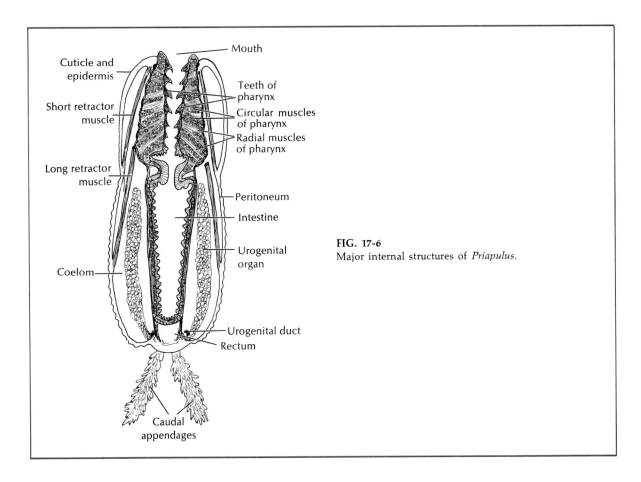

FIG. 17-6
Major internal structures of *Priapulus*.

roundings as well as in the capture of prey. When the proboscis is everted, the rows of teeth point forward. The anus and urogenital pores are located at the posterior end of the trunk. The caudal appendages are hollow stems containing many hollow vesicles that communicate with the coelom. They are believed to be respiratory in function and probably also chemoreceptive.

Internal features. The muscular body wall consists of a cuticle, a single layer of epidermis, layers of circular and longitudinal muscles, and a thin peritoneum. The chitinous cuticle is molted periodically throughout life.

The digestive system consists of the muscular, toothed pharynx, or proboscis, and a straight intestine and rectum (Fig. 17-6). There is a nerve ring around the pharynx and a midventral nerve cord.

The sexes are separate. The paired urogenital organs are each made up of a gonad and clusters of solenocytes and each has a urinary tube that carries both genital and urinary products to the outside. Gametes are shed into the sea where fertilization occurs. The larvae develop a lorica much like that of rotifers. They share the same habitat as the adults, and in about 2 years they molt and emerge as juveniles similar to the adults. The most common species is *Priapulus caudatus*, which has a single caudal appendage. *Priapulopsis bicaudatus* has two.

■ PHYLUM PENTASTOMIDA*

The wormlike Pentastomida are bloodsucking parasites found in the lungs and nasal passageways of carnivorous vertebrates. They occur mostly in tropical reptiles, most commonly in snakes and lizards, but they are also found in birds and mammals. Human infection is known in Africa and Europe, but so far none have been reported from North America.

Their life history usually includes an intermediate host for the larval stages, although they may com-

*Pen-ta-stom'i-da (Gr. *pente,* five, + *stoma,* mouth).

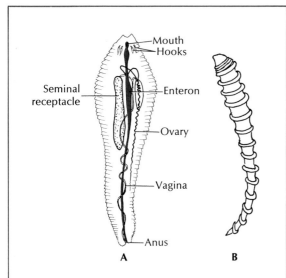

Seminal receptacle

Mouth
Hooks
Enteron
Ovary
Vagina
Anus

A **B**

FIG. 17-7

Two common pentastomids. **A,** *Pentastomum,* found in lungs of snakes and other vertebrates. Female is shown with some internal structures. **B,** Female *Armillifer,* a pentastomid with pronounced body rings. In Africa, man is parasitized by immature stages; adults (4 inches long or more) live in lungs of snakes. Human infection may occur from eating snakes or from contaminated food and water.

plete their entire life cycle in the same host. Their intermediate hosts are usually vertebrates that are eaten by the final host.

Linguatula taenioides lives in the nasal passageways of carnivorous mammals. Its eggs are discharged in mucus secretions, or feces, and hatch into larvae that look remarkably like four-legged tardigrades, having stumpy legs with claws on the ends. The larvae climb on vegetation that may be eaten by rabbits. In the rabbit the larvae encyst in the liver or another vital organ and undergo several molts in their development. If the rabbit is eaten, the larvae complete their development and find their way to the nasal passageways of the final host.

The adults are vermiform and usually vary in length from 1 to 5 inches. Transverse rings give them a segmented look (Fig. 17-7). Both larva and adult have hooks for attachment. Frontal glands secrete an anticoagulant for the blood of the host. The straight digestive system is adapted for sucking

blood. Their nervous system is of the general annelid type with three pairs of ganglia along the ventral nerve cord. They have no excretory or circulatory systems, and a definite respiratory system is lacking, except for breathing pores (stigmata) over the body surface. The sexes are separate, with the female much larger than the male. Fertilization is internal, and zygotes reach the external aquatic environment either by way of the mouth or by nasal secretions or sometimes by way of the feces.

Parasitic modifications have made phylogenetic affinities difficult. The worms show several arthropod characters, such as larval appendages, stigmata in the skin, and a molting cuticle, and the larvae resemble the tardigrades. R. Heymons believes that they may have evolved from the polychaetes because they show features of the Myzostomidae, a polychaete parasite or commensal group, and G. Osche suggests that they are offshoots of the myriapods.

◼ PHYLUM ONYCHOPHORA*

The onychophorans are the "walking worms," a small phylum of caterpillar-like animals, 2 to 5 inches long, that live in rain forests and other moist, leafy habitats in tropical and subtropical regions. The best known of the 7 genera is *Peripatus* and in common usage its name is sometimes used in speaking of the entire group.

Onychophorans are an ancient group whose fossil record shows that they have not changed substantially in 400 million years. They were originally marine animals and were probably far more common at one time than they are now. They show so many characteristics of both the annelids and the arthropods that they have been referred to as "missing links" between the two phyla. Although this concept may be questioned, they have nevertheless been of unusual interest to zoologists.

Onychophorans are nocturnal in habit and their food consists largely of insects and other small invertebrates that they capture by ejecting a stream of sticky material from glands on the oral papillae.

External features. The onychophoran body is more or less cylindric and shows no external segmentation, except for the paired appendages. The skin is soft and velvety, covered with a nonchitinous cuticle, and is studded with minute papillae, some of which have sensory bristles. The color may be gray,

*On"y-kof'o-ra (Gr. *onyx,* claw, + *pherein,* to bear).

FIG. 17-8
Peripatus in its natural habitat, usually moist places under logs, bark, or other rubbish. Phylum Onychophora. (Courtesy Ward's Natural Science Establishment, Inc., Rochester, N. Y.)

green, or reddish, and may be uniform or variegated. The head (p. 379) bears a pair each of antennae, annelid-like eyes, and oral papillae, and the mouth has a pair of chitinous jaws.

The unjointed legs are short and stocky and each bears a pair of claws. Some species have 17 pairs of legs; others have more (Fig. 17-8). Locomotion is slow. Waves of contractions affect both sides of the body, but only a few legs at a time are lifted and moved forward. They are the most primitive group to walk with the body raised upon legs.

Internal features. The body wall is muscular. The body cavity is a hemocoel imperfectly divided by muscular partitions into compartments. Slime glands on each side of the body cavity open on the oral papillae. Their secretions can be ejected for some distance, are sticky, and are used to capture prey.

The mouth is ventral and contains a pair of mandibles that work back and forth. There is a muscular pharynx and a straight digestive tract. Each segment contains a pair of **nephridia,** each nephridium with

a vesicle, ciliated funnel and duct, and an opening at the base of a leg. There is evidence that the chief excretory product is uric acid, a water-salvaging device used by birds, lizards, land snails, and most insects.

For respiration there is a **tracheal system** that ramifies to all parts of the body and communicates with the outside by many openings, or spiracles, scattered all over the body. The spiracles cannot be closed to prevent water loss; so although the tracheae are efficient, the animals are restricted to moist habitats. The tracheal system is somewhat different from that of arthropods and probably has evolved independently. In the open circulatory system the dorsal vessel serves as a pumping organ.

There are a pair of cerebral ganglia with connectives and a pair of widely separated nerve cords with connecting commissures. The brain gives off nerves to the antennae and head region and the cords send nerves to the legs. Sense organs include the pigment cup ocelli, taste spines around the mouth, tactile papillae on the integument, and hygroscopic receptors that orient the animal toward water vapor.

Onychophorans are dioecious, with paired reproductive organs. The males usually deposit their sperm in spermatophores in the female seminal receptacle, but in some species they are deposited on the body and the sperm penetrate the skin. Most species produce living young. *Peripatus* produces 30 or 40 young, each about $1/2$ inch long, but not all at the same time. In some species there is a placental attachment between mother and young (viviparous); in others the young develop in the uterus without attachment (ovoviviparous).

Onychophorans resemble the annelids with their segmentally arranged nephridia, muscular body wall, pigment-cup ocelli, and ciliated reproductive ducts. Arthropod characteristics are the tubular heart, presence of tracheae, hemocoel for a body cavity, and the large size of the brain. They differ from either phylum in their scanty metamerism, structure of the mandibles, and the separate arrangement of the nerve cords. They are more primitive than insects, and somewhat like the centipedes in the arrangement of internal metamerism. A fossil form, *Aysheaia,* discovered in the Burgess shale deposit of British Columbia and dating back to mid-Cambrian times, is very much like the modern onychophorans.

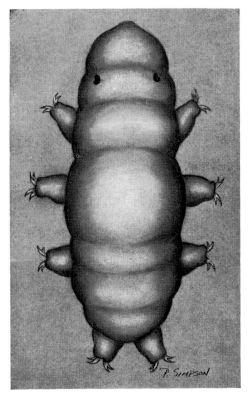

FIG. 17-9
Dorsal view of water bear (phylum Tardigrada).
They live in terrestrial mosses, pond debris,
blanket algae, and other places. Unique
characteristics are four pairs of unjointed legs and
complicated buccal apparatus for sucking liquid
food.

■ PHYLUM TARDIGRADA*

Water bears, or tardigrades (Fig. 17-9), are minute
forms, usually less than 1 mm. in length, and are
found in both fresh water and salt water, wet moss,
sand, damp soils, pond debris, liverworts, and lichens.

The body is elongated, cylindric, or a long oval
and is unsegmented. The head is merely the anterior
part of the trunk. It bears four pairs of short, stubby,
unjointed legs armed with four single or two double
claws, which may be unequal in size and are used
for clinging to its substrate. The last pair of appendages lies at the posterior end of the body. The body
is covered by a cuticle that is shed (ecdysis) four or
more times in its life history. Cilia are absent.

*Tar-dig′ra-da or tar-di-gray′da (L. *tardus,* slow, + *gradus,*
step).

The mouth is at the anterior end and opens into
a tubular pharynx that is adapted for sucking and has
two needle-like stylets that can be protruded through
the mouth. The stylets are used for piercing the cellulose walls of the plants they live upon, and the liquid contents are sucked in by the pharynx. Some tardigrades suck the body juices of nematodes, rotifers,
and other small animals. Glands empty into both the
pharynx and the digestive tract.

At the junction of the stomach and rectum, a pair
of malpighian tubes and a dorsal gland empty into
the digestive system and may be excretory in function.

Most of the body cavity is a hemocoel, with the
true coelom restricted to the gonadal cavity. There
are no circulatory or respiratory systems, as fluids
freely circulate through the body spaces and gaseous
exchange can occur through the body surface.

The muscular system consists of a number of long
muscle bands that have most of their origins and in-
sertions on the body wall. Circular muscles are absent, but the hydrostatic pressure of the body fluid
may act as a type of skeleton.

The brain is large and covers most of the dorsal
surface of the pharynx. It connects by way of circumpharyngeal connectives to the subpharyngeal ganglion, from which the double ventral nerve cord extends posteriorly as a chain of four ganglia. Sense
organs consist usually of a pair of pigmented
eyespots and various tactile organs.

Tardigrades are dioecious and have a single gonad.
Usually females are more common than males and
may make up an entire population. Parthenogenesis
is common. Both thin-shelled and thick-shelled
eggs, similar to the summer and winter eggs of rotifers, occur. Eggs are often found in the old cuticle
that has been shed (Fig. 17-10). The young are juveniles.

Tardigrades have great ability to withstand harsh
environmental conditions, such as desiccation and
freezing. Under such conditions the tardigrade loses
water, shrivels, and contracts into a more or less
rounded condition with low metabolism. This condition, called **cryptobiosis,** is found also in rotifers, nematodes, and some others. It is brought about when
there is water loss by evaporation or by osmotic
pressure, or when the surrounding oxygen level is
not sufficient to support oxidative metabolism. The
cryptobiote retains its structural integrity and can re-

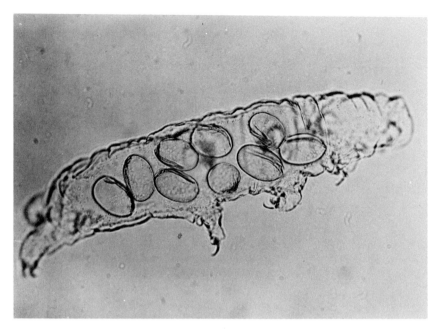

FIG. 17-10
Molted cuticle of a tardigrade containing a number of fertilized eggs. (From Sayre, R. M. 1969. Trans. Amer. Microsc. Soc. **88:**266-274.)

turn to normal life processes under certain conditions even after long periods. Cryptobiosis extends the normal life-span of the organisms in which it occurs.

Only 40 to 50 species of tardigrades have been found in North America, but 300 to 400 species occur in other parts of the world.

The affinities of tardigrades are among the most puzzling of all animal groups. They are coelomate animals and their mesoderm is enterocoelous like that of deuterostomes, although they may be highly modified protostomes. Various authorities have placed them near the arthropods, annelids, nematodes, or chaetognaths.

Common American genera of tardigrades are *Macrobiotus, Echiniscus,* and *Hypsibius.*

References

Baltzer, F. 1931. Echiurida. In W. Kükenthal and T. Krumbach (editors). Handbuch der Zoologie. Berlin, Walter de Gruyter & Co., vol. 2, part 2, section 9, pp. 62-168.

Barrington, E. J. W. 1967. Invertebrate structure and function. Boston, Houghton Mifflin Co. *The minor phyla are well considered in this work on the invertebrates.*

Chuang, S. H. 1963. Digestive enzymes of the echiuroid, *Ochetostoma erythrogrammon.* Biol. Bull. **125:**464-469.

Dawydoff, C. 1959. Echiuroidea. In P.-P. Grassé (editor). Traité de zoologie, vol. 5, part 1. Paris, Masson & Cie, pp. 855-907. *One of the best general treatments of the echiurids. Extensive bibliography.*

Grassé, P.-P. (editor). 1949. Traité de zoologie, vol. 4. Paris, Masson & Cie. *An excellent account of the onychophorans.*

Hickman, C. P. 1973. Biology of the invertebrates, (ed. 2). St. Louis, The C. V. Mosby Co. *General treatment and descriptions of all the minor coelomate phyla.*

Hill, H. R. 1960. Pentastomida. In McGraw-Hill encyclopedia of science and technology, vol. 9. New York, McGraw-Hill Book Co., pp. 623-624.

Hyman, L. H. 1959. The invertebrates. Vol. 5. Smaller coelomate groups. New York, McGraw-Hill Book Co.

Land, J. Van der. 1970. Systematics, zoogeography, and ecology of the Priapulida. No. 112. Zoologische verhandelingen uitgegeven door het Rijksmuseum van Natuurlijke Historie te Leiden. Leiden, E. J. Brill. *An up-to-date monograph.*

MacGinitie, G. E., and N. MacGinitie. 1967. Natural history of marine animals, ed. 2. New York, McGraw-Hill Book Co. *A section on the Echiurida, with descriptions of the physiology and behavior of Echiurus and Urechis.*

Manton, S. M. 1950. The locomotion of *Peripatus.* J. Linn. Soc. Zool. **41:**529-539.

Manton, S. M. 1952-1961. The evolution of arthropodan locomotory mechanisms (several papers). J. Linn. Soc. Zool., part 3.

Marcus, E. 1928. Zur vergleichenden Anatomie und Histologie der Tardigraden. Zool. Jahrb. Abt. Allg. Zool. **45**:99-192.

Pennak, R. W. 1953. Fresh-water invertebrates of the United States. New York, The Ronald Press Co. *A good general account of the tardigrades and keys to the common species.*

Pickford, G. 1947. Sipunculida. Encyclopaedia Britannica, vol. 20. *A concise, well-written account.*

Riggin, G. T., Jr. 1964. Tardigrades from the southern Appalachian mountains. Trans. Amer. Microsc. Soc. **83**:277-282.

Russell-Hunter, W. D. 1969. A biology of higher invertebrates. New York, The Macmillan Co.

Sayre, R. M. 1969. A method of culturing predaceous tardi-grades on the nematode *Panagrellus redivivus.* Trans. Amer. Microsc. Soc. **88**:266-274.

Shapeero, W. 1961. Phylogeny of Priapulida. Science **133**:879-880.

Shapeero, W. 1962. The epidermis and cuticle of *Priapulus caudatus* Lamarck. Trans. Amer. Microsc. Soc. **81**:352-355.

Shipley, A. E. 1920. Tardigrada and Pentastomida. In S. F. Harmer and A. E. Shipley (editors): The Cambridge natural history, vol. 4. London, Macmillan Co., Ltd.

Swedmark, B. 1964. The interstitial fauna of marine sand. Biol. Rev. **39**:1-42.

Tiegs, O. W., and S. M. Manton. 1958. The evolution of the Arthropoda. Biol. Rev. **33**:255-337.

Plumatella repens, *a freshwater bryozoan (phylum Ectoprocta). It grows in branching, threadlike colonies on the underside of rocks and vegetation in lakes, ponds, and streams. Safe in its protective case, the minute animal extends its tentacled lophophore to feed. Cilia on the tentacles create tiny currents that bring algae, protozoans, and detritus toward the mouth. If large inedible particles reach the mouth, some of the tentacles bend over and brush them off. (Courtesy R. Vishniac, New York.)*

CHAPTER 18

THE LOPHOPHORATE ANIMALS

PHYLUM PHORONIDA
PHYLUM ECTOPROCTA (BRYOZOA)
PHYLUM BRACHIOPODA

Position in animal kingdom

1. The lophophorate phyla possess a **true coelom,** that is, a body cavity that is lined with a layer of mesodermal epithelium called the peritoneum.

2. They belong to the **protostome** branch of the **bilateral** animals, but the Brachiopoda have some of the characteristics of the deuterostomes in their development.

3. The three phyla are usually grouped together because they all posess the crown of tentacles called a **lophophore,** which is specialized for sedentary filter feeding. The lophophore surrounds the mouth but not the anus, thus differing from the tentacular crown of Entoprocta.

Biologic contributions

1. The lophophore, a unique ridge that bears hollow, ciliated tentacles, is an efficient specialized filter feeding device, forming a ciliated route, or trough, for trapping and directing food particles to the mouth.

2. The brachiopods and phoronids possess **vascular systems** for circulation of food nutrients and other materials.

3. The blood in phoronids possesses red blood corpuscles that contain hemoglobin for carrying oxygen.

Phylogeny and adaptive radiation

Phylogeny. The three phyla apparently have a close relationship because of the common possession of a lophophore of similar construction. They occupy a unique position between the protostomes and the deuterostomes, and may be a connecting link between the two groups. Although classed as protostomes, one phylum (Brachiopoda) has an enterocoelous formation of the coelom, which is a deuterostome characteristic. Both lophophorates and deuterostomes have the same body regionalization of three divisions (protostome, mesosome, and metasome), except that the protostome is chiefly suppressed in the lophophorates. A common ancestry of the two groups is indicated by the evidence at hand. The lophophorates are considered to have a common trochophore larva, but it is highly modified in each of the phyla. The lack of a head in the lophophorates may be correlated with their ciliary method of feeding.

Adaptive radiation. Since all lophophorates are filter feeders, it may be supposed that their evolutionary diversification has been guided to a great extent by this function. Each phylum has had structural

modifications in this and other respects. Phoronids have varied their tubes according to the nature of their habitats.

Ectoprocts show a tendency toward an evolutionary development of a gelatinous zoecium from a chitinous one to allow more flexibility. Primitive forms had rigid calcareous skeletons, which have fossilized well. Some calcareous ectoprocts overcame rigidity by developing flexible, chitinous joints.

The brachiopods have been guided in their evolution by their shells and lophophores. The primitive form of the lophophore in these animals was a short tentacle-bearing ridge, which has gradually increased in length in its subsequent evolution. This increased length may take the form of lobulations, arm forma-

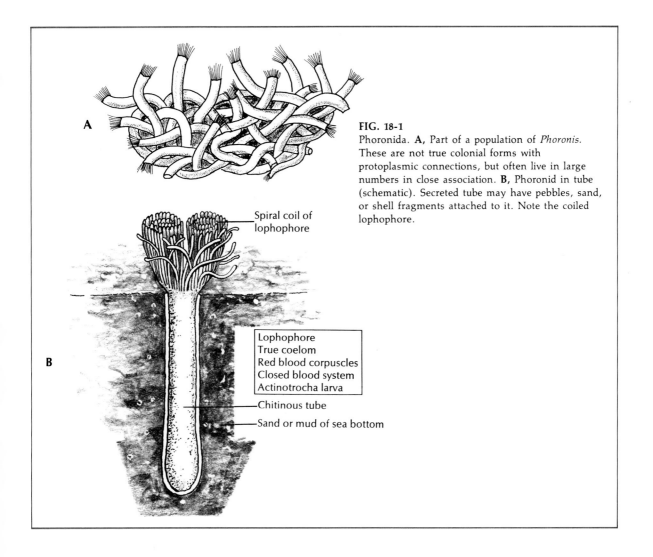

FIG. 18-1
Phoronida. **A,** Part of a population of *Phoronis.* These are not true colonial forms with protoplasmic connections, but often live in large numbers in close association. **B,** Phoronid in tube (schematic). Secreted tube may have pebbles, sand, or shell fragments attached to it. Note the coiled lophophore.

Spiral coil of lophophore

Lophophore
True coelom
Red blood corpuscles
Closed blood system
Actinotrocha larva

Chitinous tube
Sand or mud of sea bottom

tions, or spiral coils. Many brachiopods have undergone little change because their environment has remained fairly constant over long periods of time.

■ PHYLUM PHORONIDA*

Phylum Phoronida is made up of wormlike animals that live in tubes on the bottom of shallow seas (Fig. 18-1). Most are small (1 mm. or less), but some are more than 1 foot long. They have some resemblance to bryozoans and in the past were included in that group. Some species are more or less solitary, but other species are found in dense populations.

Each worm is enclosed in a leathery or chitinous tube that it has secreted and from which its tentacles are extended. The ciliated tentacles are borne on a lophophore, or ridge, that may be spirally coiled. When disturbed, the animal draws back into its tube with its tentacles completely hidden. The body wall is made of cuticle, epidermis, and both longitudinal and circular muscles. The coelomic cavity is a true coelom, being lined with peritoneal epithelium. It is subdivided by mesenteric partitions into compartments.

*Fo-ron'i-da (Gr. *phoros,* bearing, + L. *nidus,* nest).

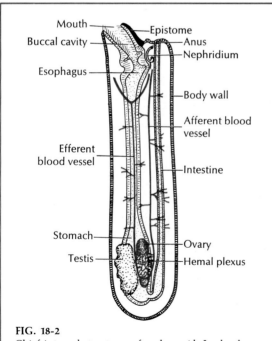

FIG. 18-2
Chief internal structures of a phoronid. Lophophore has been omitted.

Labels: Mouth, Buccal cavity, Esophagus, Efferent blood vessel, Stomach, Testis, Epistome, Anus, Nephridium, Body wall, Afferent blood vessel, Intestine, Ovary, Hemal plexus

The alimentary canal is U shaped and is ciliated most of its length. There is no respiratory system; respiratory gases diffuse through the body surface. There is a closed circulatory system of blood vessels that have contractile walls, but a heart is absent (Fig. 18-2). Nucleated red blood corpuscles are found also. The nervous system just below the epidermis consists of a nerve ring around the mouth, with nerves running to various parts of the body.

The members of this phylum are monoecious. When the sex cells are released, they pass through the paired nephridia to the space enclosed by the tentacles, where fertilization occurs. The larva, which is characterized by a large hoodlike lobe over the mouth, is called an actinotrocha and is commonly considered a type of the trochophore larva.

Two of the common species of the phylum are *Phoronopsis viridis,* a green phoronid with straight sandy tubes, found in tidal flats along the Pacific coast, and *Phoronis architecta,* whose sticklike tubes are often found buried in sea grass flats on our Atlantic and Gulf coasts.

■ PHYLUM ECTOPROCTA (BRYOZOA)*

The Ectoprocta have long been called bryozoans, or moss animals (from Gr. *bryon,* moss, and *zōon,* animal), a term that originally included the Entoprocta also. However, because the entoprocts are pseudocoelomates and have the anus located within the tentacular crown, they are no longer classed with the ectoprocts, which, like the other lophophorates, are eucoelomate and have the anus outside the circle of tentacles.

Ectoprocts are mostly minute animals that live in colonies, with each individual, or **zooid,** enclosed in a secreted case, or **zoecium,** that may be gelatinous or chitinous, or may be further stiffened with calcium. Most ectoprocts are marine, but some live in fresh water. Most colonies are attached to a substratum such as rocks, shells, plants, or the bodies of sea squirts. Some colonies are branching and plantlike, some form an encrustation (Fig. 18-3), some form interesting and individual shapes, and a few float free.

Ectoprocts have some resemblance to hydroids, but have a more advanced organization. More than 2,000 species are recognized.

*Ek-to-prok'ta (Gr. *ektos,* outside, + *prōktos,* anus). Bry-o-zo'a (Gr. *bryon,* moss, + *zōon,* animal).

Characteristics. Bilateral symmetry; unsegmented; form colonies by budding; each individual in a separate shell (zoecium); coelom; U-shaped alimentary canal; ciliated lophophore around mouth; anus outside lophophore; no excretory or vascular systems; nerve ganglion between mouth and anus; monoecious or dioecious reproduction; brood pouch (ooecium); a form of trochophore larva.

Structure and behavior. Ectoprocts are ordinarily found in shallow water. Their food consists of small plants and animals caught by the lophophore, which is also used for respiration. This ridge, the lophophore, which bears a circlet of ciliated tentacles, tends to be circular in marine forms (Fig. 18-4) and U shaped in freshwater Ectoprocta (Fig. 18-5 and p. 390). The entire crown can be drawn into the protective shell, or zoecium, by retractor muscles. The digestive system is U shaped with the mouth within the ring of tentacles and the anus outside of the lophophore. Digestion is extracellular. The true coelom is filled with a coelomic fluid, which distributes the food in the absence of a circulatory system.

Most members of the phylum go through a periodical renewal process, when the tentacles and internal organs degenerate into a compact mass called the "brown body." When new organs are regenerated from the body wall, the brown body is discharged through the anus. Since there is no excretory system, the brown body may have some relation to excretion.

The muscular system consists mainly of the body-wall muscles and the retractor muscles referred to earlier. The nervous system is composed of a single ganglion between the lophophore and anus and many nerves to the tentacles, digestive system, muscles, and a few other places. No sense organs have been detected.

Some of the individuals (zooids) (Fig. 18-4) are provided with an interesting structure, the **avicularium,** which resembles the beak of a bird and is used

FIG. 18-3
Skeletal remains of a colony of *Membranipora,* a marine encrusting form of Ectoprocta. Each little oblong zoecium is the calcareous former home of a tiny ectoproct. Some hydroids are visible at lower right. (Photo by B. Tallmark, Uppsala University, Sweden.)

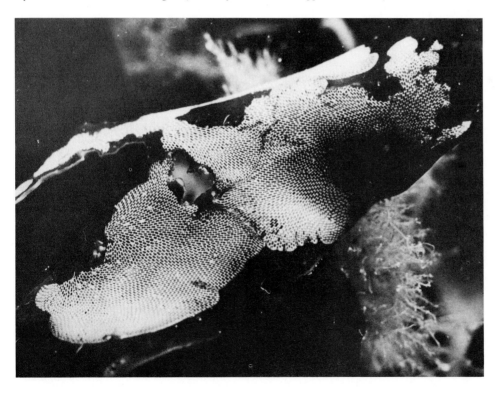

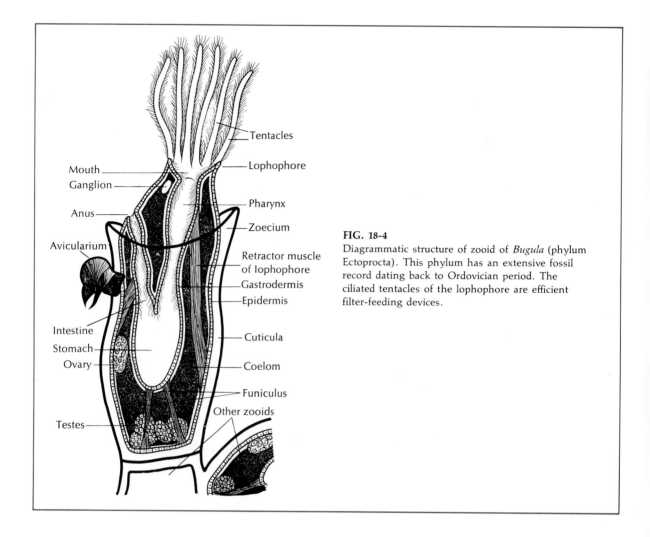

FIG. 18-4
Diagrammatic structure of zooid of *Bugula* (phylum Ectoprocta). This phylum has an extensive fossil record dating back to Ordovician period. The ciliated tentacles of the lophophore are efficient filter-feeding devices.

to keep other animals away as well as to remove debris that might settle on the colonies. When living zooids are observed, these avicularia are seen constantly opening and snapping closed.

Both sexual and asexual reproduction occur.

Freshwater ectoprocts form interesting mosslike colonies on the stems of plants or on rocks, usually in shallow water of ponds or pools. The colonies as a whole may have the power of sliding along the object on which they are supported. Freshwater ectoprocts also have a peculiar type of asexual reproduction by which internal buds known as **statoblasts** (Fig. 18-5, *B*) are formed. When the members of a colony die in late autumn, the statoblasts are released and in the spring can produce new colonies.

■ PHYLUM BRACHIOPODA*

The animals of phylum Brachiopoda resemble the mollusks in having two valves (shells), but in brachiopods the valves are dorsal and ventral instead of lateral, as in the mollusks. The ventral valve is larger than the dorsal one and is provided with a fleshy peduncle that is used for attaching the animal to the sea bottom or to some object. All members of the phylum are marine, benthic, and solitary.

Formerly these animals were classified with the Mollusca, but they differ from that phylum not only in the arrangement of the valves but also in internal details. Because brachiopods resemble the lamps of the ancients, they are often referred to as lamp

*Brak-i-op'o-da (Gr. *brachiōn,* arm, + *pous, podos,* foot).

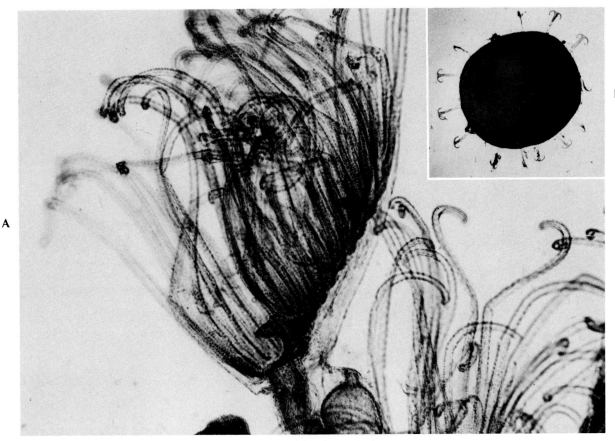

FIG. 18-5
A, Zooid of *Plumatella,* colonial freshwater ectoproct. **B,** Statoblast of *Pectinatella,* about 1 mm. in diameter and disk-shaped, is asexual reproductive bud that survives the winter.

shells. They represent a very ancient group, and in the past they were far more abundant than they are now. Some have undergone few changes since early geologic periods. They have left an excellent fossil record. About 200 living species are found in the group.

Characteristics. Bilateral symmetry; no segmentation; dorsal and ventral calcareous shells; peduncle as an attachment organ; true coelom; lophophore of two coiled ridges; heart and blood vessels; one or two pairs of nephridia; nerve ring; separate sexes; digestive system complete or incomplete; a form of trochophore larva.

Structure and behavior. Most brachiopod shells range between 5 and 80 mm. long, but many fossil forms are much larger. The valves may be held to-

gether by muscles as in *Lingula* and *Glottidia* (inarticulate brachiopods), or they may have an interlocking teeth and socket arrangement as in *Terebratella* (articulate brachiopods) (Fig. 18-6). Movement of the valves is by well-developed muscles.

The body occupies only the posterior part of the space between the valves (Fig. 18-7), and extensions of the body wall form mantle lobes that line and secrete the shell. The large horseshoe-shaped lophophore in the anterior mantle cavity bears long ciliated tentacles used in respiration and feeding. Ciliary water currents carry food particles between the gaping valves and over the lophophore. Food is caught in mucus on the tentacles and carried in a ciliated food groove along the arm of the lophophore to the mouth. Unwanted particles are carried down rejec-

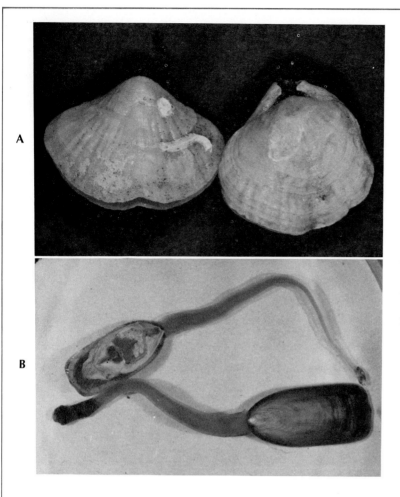

FIG. 18-6
A, Dorsal *(left)* and ventral *(right)* views of brachiopod *Terebratella.* The valves are articulated, and the stalk is attached to the dorsal valve. **B,** *Lingula,* brachiopod with nonarticulating valves and stalk emerging between the valves. Specimen on left has one valve removed.

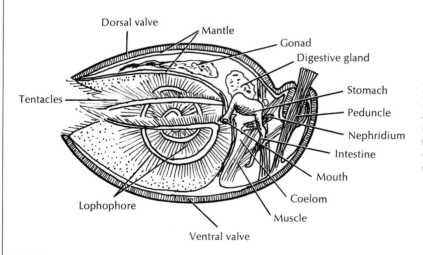

FIG. 18-7
Internal structure of articulate brachiopod (longitudinal section). Note that the lophophore occupies the anterior two thirds of the mantle cavity.

tion tracts to the mantle lobe and carried out in ciliary currents.

One or two pairs of nephridia open into the coelom and empty into the mantle cavity. There is a contractile heart and open circulatory system, and a nerve ring. The paired gonads in each sex discharge gametes through the nephridia.

The development of brachiopods is similar in some ways to the deuterostomes, with radial, mostly equal, holoblastic cleavage. The free-swimming larva resembles the trochophore. In the articulates, metamorphosis occurs after the larva has attached by a pedicle.

References

Atkins, D. 1932. The ciliary mechanism of the entoproct Polyzoa and a comparison with that of the ectoproct Polyzoa. Quart. J. Microsc. Sci. **75**:393-420.

Barnes, R. D. 1968. Invertebrate zoology, ed. 2. Philadelphia, W. B. Saunders Co.

Barrington, E. J. W. 1967. Invertebrate structure and function. New York, Houghton Mifflin Co. *An excellent up-to-date account.*

Bronn, H. G. (editor). 1939. Phoronidea. Klassen und Ordnungen des Tier-Reichs, vol. 4, part 4. Leipzig, Akademische Verlagsgesellschaft. *An authoritative monograph.*

Cannon, H. G. 1933. On the feeding mechanism of the Brachiopoda. Philosophical Trans. Roy. Soc. Bull. **222**:267-269.

Chapman, G. 1958. The hydraulic skeleton in the invertebrates. Biol. Rev. **33**:338-352.

Chuang, S. 1959. Structure and function of the alimentary canal in *Lingula unguis.* Proc. Zool. Soc. (London) **132**:293-311. *An account of the digestive system.*

Edmondson, W. T. (editor). 1959. Ward and Whipple's fresh-water biology, ed. 2. New York, John Wiley & Sons, Inc. *The section on freshwater Bryozoa (which includes both entoprocts and ectoprocts) is by the late M. D. Rogick, foremost authority on the group.*

Helmcke, J. G. 1939. Brachiopoda. In W. Kükenthal and T. Krumbach (editors). Handbuch der Zoologie, vol. 3, part 2, sect. 5. Berlin, Walter de Gruyter & Co.

Hickman, C. P. 1973. Biology of the invertebrates, ed. 2. St. Louis, The C. V. Mosby Co.

Hyman, L. H. 1959. The invertebrates: smaller coelomate groups, vol. 5. New York, McGraw-Hill Book Co. *Up-to-date accounts of Chaetognatha, Phoronida, Ectoprocta, Brachiopoda, and Sipunculida. This volume maintains the high traditions of the others in this outstanding modern treatise of zoology.*

Lynch, W. 1947. Behavior and metamorphosis of the larva of *Bugula neritina* (L). Biol. Bull. **92**:115-150.

Meglitsch, P. A. 1967. Invertebrate zoology. New York, Oxford University Press.

Rogick, M. D. 1935. Studies on fresh-water Bryozoa. II. The Bryozoa of Lake Erie. Trans. Amer. Microsc. Soc. **54**:245-263. *This and other papers on Bryozoa by the same author represent the best available accounts of American freshwater species.*

Russell-Hunter, W. D. 1969. A biology of higher invertebrates. (Paperback.) New York, The Macmillan Co. *With the author's companion volume, A Biology of Lower Invertebrates, the student can get a fine evaluation, within the limits of paperback editions, of the invertebrates.*

Tenney, W. R., and W. S. Woolcott. 1966. The occurrence and ecology of freshwater Bryozoans in the headwaters of the Tennessee, Savannah, and Saluda River systems. Trans. Amer. Microsc. Soc. **85**:241-245.

Sea urchins, the "pin cushions" of the sea, are bottom dwellers that attach themselves by hydraulically operated tube feet and feed by scraping the rocks with a peculiar five-toothed mechanism called "Aristotle's lantern." This is a group of Echinus acubus, *photographed at a depth of 25 meters off the west coast of Norway. (Courtesy T. Lundälv, Kristinebergs Zoological Station, Sweden.)*

CHAPTER 19

THE ECHINODERMS

PHYLUM ECHINODERMATA*
A MAJOR DEUTEROSTOME PHYLUM

Position in animal kingdom

1. Echinoderms belong to the deuterostome branch (enterocoelous coelomates) of the animal kingdom, which they share with the Chaetognatha, Hemichordata, Chordata, and a few minor phyla.

2. Echinoderms have radial and indeterminate cleavage.

3. The echinoderms share with the annelids, mollusks, and arthropods a very high type of invertebrate organization.

4. Unlike the other three phyla, this group has radial symmetry, but this type of symmetry has been secondarily acquired, for their larval forms are bilaterally symmetric.

5. They have no segmentation or well-defined head region.

*E-ki"no-der'ma-ta (Gr. *echinos,* sea urchin, hedgehog, + *derma,* skin, + *-ata,* characterized by).

Biologic contributions

1. They have a **mesodermal endoskeleton** of plates, which may be considered the first indication of the endoskeleton so well developed among vertebrates.

2. They have contributed a pattern of embryonic development similar to that of the highest group, the chordates. This pattern includes (a) an anus derived from the embryonic blastopore, (b) a mouth formed from a stomodaeum, which connects to the endodermal esophagus, (c) a mesoderm from evaginations of the archenteron (enterocoel), and (d) a nervous system in close contact with the ectoderm.

3. Most of the echinoderm characters are so out of line that few of them are copied by other phyla.

4. Some of their unique features are the **watervascular system, tube feet, pedicellariae, dermal branchiae,** and **calcareous endoskeleton.**

Phylogeny and adaptive radiation

Phylogeny

1. The echinoderms are a very ancient group of animals, which, according to fossil records, were differentiated back to Cambrian times. The fossil record gives no clues as to their origin.

2. The most primitive echinoderms were probably the stalked members, and from these the free forms arose. These early echinoderms were noncrinoid Pelmatozoa now wholly extinct.

3. From their larvae the evidence indicates that their ancestors were bilaterally symmetric and that their radial symmetry was secondarily acquired. Even in the adult condition certain aspects of bilateral symmetry can be discerned, since their symmetry is more biradial than truly radial.

4. Of all invertebrates, echinoderms are placed nearest to the chordates because of their similar embryonic development.

5. Another strong evidence of their chordate affinities is the similarity between the larval type of echinoderms and that of the prechordate acorn worm (*Balanoglossus*).

6. From the evidence at hand it appears that echinoderms and chordates may have originated from a common ancestor or at least from the same side of the phylogenetic tree.

7. Within the group the holothuroids are supposed to have come from a common stem with the crinoids, whereas the asteriods, echinoids, and ophiuroids have risen from a common pelmatozoan but noncrinoid ancestry.

Adaptive radiation. The evolution of the group has been guided and restricted by the hydraulic watervascular system of tube feet, which had primarily a food-catching function at first but which has been variously modified for other functions (locomotion, respiration, mucus production, burrowing, and sensory perception).

■ THE ECHINODERMS

The echinoderms are marine forms and include the sea stars, brittle stars, sea urchins, sea cucumbers, and sea lilies. They represent a bizarre group sharply distinguished from all other members of the animal kingdom. Their name is derived from their external spines or protuberances (Gr. *echinos,* sea urchin, hedgehog, + *derma,* skin). A calcareous endoskeleton is found in all members of the phylum, either in the form of plates or represented by scattered tiny ossicles.

The most noticeable characteristics of the echinoderms are (1) the spiny endoskeleton of plates, (2) the watervascular system, (3) the pedicellariae, (4) the dermal branchiae, and (5) radial or biradial symmetry. Radial symmetry is not limited to echinoderms, but no other group with such complex organ systems has radial symmetry.

They are an ancient group of animals extending back to the Cambrian period. An excellent fossil record gives no indication, however, of echinoderm ancestors. They have probably descended from bilateral ancestors despite their present type of symmetry, because their radial symmetry appears late in embryonic development and their early larvae are bilateral.

Of all invertebrates, the echinoderms are considered to be nearest in relation to the chordates. The larvae of echinoderms and of *Balanoglossus* (prechordate) are much alike. Although these two forms of larvae could have arisen independently by convergent evolution, evidence indicates that their similarity has real evolutionary meaning and that both echinoderms and chordates have come from a common ancestor. Chordate segmentation could have arisen after the two groups diverged.

The phylogeny of the echinoderms is obscure. The free-moving Eleutherozoa have probably arisen from the more ancient and primitive stem-bearing Pelmatozoa, which include the living crinoids and several extinct classes. Among the free-moving classes, the sea stars (Asteroidea) and brittle stars (Ophiuroidea) seem to be closely related, whereas the sea cucumber (Holothuroidea) no doubt arose from some crinoidlike ancestor.

Echinoderms belong to the Deuterostomia branch of the animal kingdom and are enterocoelous coelomates. The other phyla of this group are Chaetognatha, Hemichordata, Pogonophora and Chordata. It will be recalled that the deuterostomes have the following features in common: an anus from the blastopore, a coelom budded off from the archenteron (enterocoel), radial and indeterminate cleavage, a nervous system in close contact with the ectoderm, and a larva (when present) of the **dipleurula** type.

The primitive pattern of the echinoderms seems to have included radial symmetry, radiating grooves (ambulacra), and a tendency for the body openings (oral side) to face upward. Such an arrangement would be an advantage to a sessile animal; it could get its food from all directions. Living crinoids follow this primitive pattern. Modifications of the plan are found in free-moving forms, which have been favored by evolution. Attached forms, once plentiful, are now limited to the one class Crinoidea.

Zoologists have constructed a hypothetical ancestral larval form called a **dipleurula** (Fig. 19-1), from which existing larval forms may have arisen. The dipleurula is pictured as an elongated, bilaterally symmetric animal without a skeleton and with a complete digestive system and a coelom of three paired sacs. Through some degeneration or shifting of structures, this bilateral larva was transformed into the radial adult form. Although an attractive theory and widely accepted, many zoologists are doubtful about its usefulness. Perhaps a more satisfactory theory is the pentactula concept of five radial tentacles around the mouth, but this larval concept could be regarded as a later evolutionary development of the dipleurula.

The water-vascular system (Fig. 19-1), unique

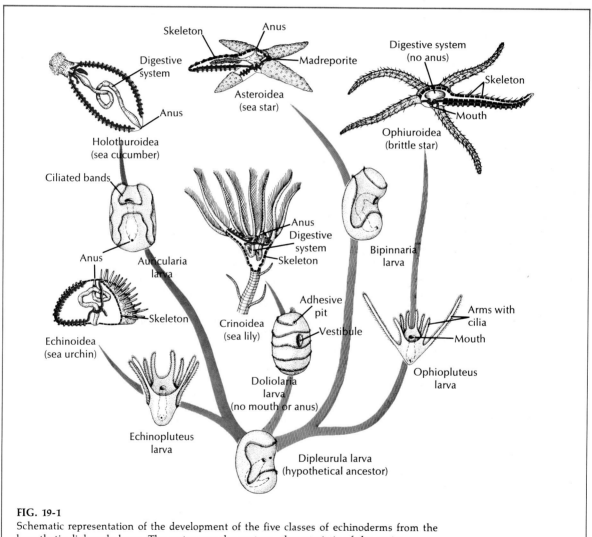

FIG. 19-1
Schematic representation of the development of the five classes of echinoderms from the hypothetic dipleurula larva. The water-vascular system, characteristic of the entire phylum, is shown in red.

in the echinoderms, sends out hollow protrusions, called **podia,** or **tube feet,** that probably originated as sensory or food-collecting structures. In sea stars, urchins, and sea cucumbers the podia have acquired suckers and are used in locomotion; food is no longer conveyed by ciliated grooves but is taken by mouth. The tube feet have also been modified for other functions. Some are pointed and without suckers for burrowing, many are mucus-producing for catching food and plastering burrow walls, and others are thin walled to serve for respiration.

The calcareous endoskeleton of most echinoderms is an important aspect of their evolution. Primitive echinoderms were heavily armored with complicated plates, but in existing forms the skeleton tends to be reduced for flexibility. In most groups it consists of closely fitting plates provided with spines and tubercles and held together by muscles and mesenchyme. Holothurians have almost no endoskeleton.

Ecologic relationships

Echinoderms occur in all depths of the ocean. All are typically marine forms and only a few are found in brackish water. Except for a few planktonic forms, they are chiefly benthic animals that live in various types of bottom habitats, mainly in the littoral zones. None are parasitic.

Ophiuroids (brittle stars, or serpent stars; Fig. 19-7) are by far the most active echinoderms and can move without the assistance of tube feet. Some species can swim.

The asteroids, or sea stars (Fig. 19-2), are commonly found in various types of bottom habitats, often on hard, rocky surface, but they are equally at home on sandy or soft bottom. They can crawl about at a very slow pace.

Holothurians, or sea cucumbers (Fig. 19-13), occur in all seas and at all depths from the intertidal zones to the deepest abysses. Many are found on sandy or mucky soil where they lie concealed.

FIG. 19-2
Group of *Asterias rubens* that has aggregated on a mussel bed to feed on mussels. Some of the mussels that were too close to the surface were washed down by heavy wave action to a rock shelf below, and the stars are feasting on them. (Photo by T. Lundälv, Kristinebergs Zoological Station, Sweden.)

Many crinoids use their cirri for attachment to stones or other hard substances. When detached from their hold, some crinoids can swim by elevating and lowering their arms. All comatulid crinoids move about while keeping their oral surfaces upward.

Echinoids, or sea urchins, are adapted for living on the ocean bottom and always keep their oral surfaces in contact with a substratum. They have a preference for hard bottoms, but may be found on sand. Sea urchins use their podia (tube feet) and spines in moving about (Fig. 19-10). When climbing a vertical surface, they make use of the adhesive power of the terminal disks of the podia.

There is a great variety of colors among the echinoderms, including oranges, reds, purples, blues, and browns. In the tissues of some echinoderms, such as sea urchins and sea stars, various types of blue and blue-green algae may be found and may be regarded as parasitic infections.

Characteristics

1. Body unsegmented with **radial symmetry;** body rounded, cylindric, or star shaped
2. **No head;** body often pentamerous (with divisions of five or more radiating areas, or ambulacra, alternating with spaces, or interambulacra)
3. **Endoskeleton of dermal calcareous ossicles with spines** or of calcareous spicules in epidermis; covered by an epidermis (ciliated in most); **pedicellariae** (in some)
4. A unique **water-vascular system** of coelomic origin, which pushes out the body surface as a series of tentacle-like projections (podia, or tube feet), that are protruded or retracted by alterations of fluid pressure within them; an external opening (madreporite or hydropore) usually present
5. Locomotion usually by **tube feet** (podia), which project from the ambulacral spaces
6. **Coelom extensive,** forming the perivisceral cavity and the cavity of the water-vascular system; coelom of enterocoelous type; coelomic fluid with amebocytes
7. Digestive system usually complete; axial or coiled; anus absent in ophiuroids
8. Vascular system reduced; hemal or lacunar system enclosed in coelomic channels
9. Respiration by **dermal branchiae,** by **tube feet,** by **respiratory tree** (holothuroids), and by **bursae** (ophiuroids)

10. Nervous system with circumoral ring and radial nerves; usually two or three systems of networks located at different levels
11. Sensory system (poorly developed) of tactile organs, chemoreceptors, podia, terminal tentacles, photoreceptors, and statocysts
12. Excretory organs absent
13. Sexes separate (a few hermaphroditic) with large gonads, single in holothuroids but multiple in most; simple ducts; fertilization usually external
14. **Development through specialized free-swimming larval stages** of many kinds, with metamorphosis
15. Autotomy and regeneration of lost parts conspicuous

Classification

There are about 6,000 living and 20,000 extinct or fossil species. Five classes of existing echinoderms are recognized and about that many extinct classes are known. Most of the extinct classes belong to the subphylum Pelmatozoa, which now has only one living class (Crinoidea).

Subphylum Eleutherozoa (e-lu"ther-o-zo'a) (Gr. *eleutheros,* free, + *zōon,* animal). Free-moving forms.

    Class Asteroidea (as"ter-oi'de-a) (Gr. *astēr,* star, + *eidos,* form, + *-ea,* characterized by)—**sea stars.** Star-shaped echinoderms, with the arms not sharply marked off from the central disk; ambulacral grooves with tube feet on oral side; tube feet with suckers; anus and madreporite aboral; pedicellariae present. Example: *Asterias* (Fig. 19-4).

    Class Ophiuroidea (o"fe-u-roi'de-a) (Gr. *ophis,* snake, + *oura,* tail, + *eidos,* form)—**brittle stars** and **basket stars.** Star shaped, with the arms sharply marked off from the central disk; ambulacral grooves absent or covered by ossicles; tube feet without suckers and not used for locomotion; pedicellariae absent; anus present. Examples: *Ophiothrix, Gorgonocephalus* (Fig. 19-7).

    Class Echinoidea (ek"i-noi'de-a) (Gr. *echinos,* sea urchin, hedgehog, + *eidos,* form)—**sea urchins, sea biscuits,** and **sand dollars.** More or less globular echinoderms with no arms; compact skeleton or test; movable spines; ambulacral grooves covered by ossicles; tube feet with suckers; pedicellariae present. Examples: *Arbacia, Strongylocentrotus, Lytechinus* (Fig. 19-10), *Mellita.*

    Class Holothuroidea (hol"o-thu-roi'de-a) (Gr. *holothourion,* sea cucumber, + *eidos,* form) **sea cucumbers.** Cucumber-shaped echinoderms with no arms; spines absent; microscopic ossicles in thick muscular wall; anus present; ambulacral grooves concealed; tube feet with suckers; circumoral tentacles (modified tube feet); pedicellariae absent; madreporite plate internal. Examples: *Thyone, Stichopus, Cucumaria* (Fig. 19-13).

Subphylum Pelmatozoa (pel″ma-to-zo′a) (Gr. *pelma,* stalk or sole, + *zōon,* animal). Chiefly stem-bearing forms.

Class Crinoidea (krin-noi′de-a) (Gr. *krinon,* lily, + *eidos,* form, + *-ea,* characterized by)—**sea lilies** and **feather stars.** Body attached during part or all of life by an aboral stalk of dermal ossicles; mouth and anus on oral surface; five arms branching at base and bearing pinnules; ciliated ambulacral groove on oral surface with tentacle-like tube feet for food collecting; spines, madreporite, and pedicellariae absent. Examples: *Antedon, Florometra* (Fig. 19-15).

Economic importance

Because of the spiny nature of their structure, echinoderms have a limited use as food for other animals. The eggs often serve as food, and the sea cucumber (trepang) is used by Orientals for soup. Sea urchins are considered a delicacy in many places. Where echinoderms are common along a seashore, they have been used for fertilizer.

Sea stars feed mainly on mollusks, crustaceans, and other invertebrates, but their chief damage is to clams and oysters. A single star may eat as many as a dozen oysters or clams in a day. To rid shellfish beds of these pests, rope nets in which the sea stars become entangled are sometimes dragged over oyster beds, and the collected sea stars destroyed. A more effective method is to distribute lime over areas where they abound. Lime causes the delicate epidermal membrane to disintegrate; lesions are then formed, which destroy the dermal branchiae and ultimately the animal itself.

The eggs of echinoderms are widely used in biologic investigations, for they are usually abundant and easy to collect and handle, and the investigator can follow their developmental stages with great accuracy. Artificial parthenogenesis was first discovered in sea urchin eggs when it was found that, by changing the chemical nature of sea water, development would take place without the presence of sperm.

Larval forms

The early development of the larval form is similar in all echinoderms (Fig. 19-1). The egg is usually fertilized in sea water or in brood pouches. It divides by total cleavage that is indeterminate. It passes through successive stages: a hollow single-layered blastula; a double-layered gastrula with an archenteron and a wide blastocoel; paired outpocketings of the archenteron, which form the coelom and watervascular system; mesenchyme formed in the blastocoel from ectodermal and endodermal cells; a mouth by the breaking through of a stomodaeum; and eventually a free, pelagic larva (in most), which shows similarities as well as differences among the various classes.

In most cases this early larva is bilateral and free swimming, and it lacks a calcareous skeleton but has calcareous spines, probably for support. The coelom consists of three pairs of pouches, and the future ventral side becomes concave. Some call this stage the **dipleurula.** Cilia that first covered most of the body are now restricted to a band around the concavity. This band, used in locomotion, is folded into loops, which vary in the different larval forms.

One of the most primitive of the echinoderm larvae is the **auricularia** (Holothuroidea). Its body is elongated, with a ciliated band partly around the preoral lobe and partly on body folds. In the **bipinnaria** (Asteroidea) larva, which is similar to the auricularia, the preoral ring of cilia is separated from the rest. In the other types of larvae the body lobes become elongated with supporting rods, and the ciliated bands become more elaborate. The transformation of the larvae into adults by metamorphosis is a complicated process involving shifting of the mouth position, development of the endoskeleton, torsion whereby the left side becomes the oral and the right side the aboral surface, development of a coelomic pouch into the water-vascular system, and assumption of radial symmetry.

Class Asteroidea—the sea stars

Sea stars are familiar along the shore lines where sometimes large numbers of them may aggregate on the rocks (Fig. 19-2). Sometimes they cling so tenaciously that they are difficult to dislodge without tearing off some of their tube feet. They also live in muddy or sandy bottoms and among coral reefs. They are often brightly colored. They range in size from a centimeter in greatest diameter to 2 or 3 feet. *Asterias* (Fig. 19-4) is one of the common East coast genera and is used in many zoology laboratories. *Pisaster* (Fig. 19-3, *D*) is common on the California coast, as is *Dermasterias,* the bat star (Fig. 19-3, *A*).

EXTERNAL STRUCTURE. Sea stars are composed of a central disk that merges gradually with the tapering arms (rays). They tend to be pentamerous and typically have 5 arms, but there may be

FIG. 19-3
Some West-Coast sea stars (class Asteroidea). **A,** *Dermasterias.* **B,** *Hippasteria.* **C,** *Pycnopodia.*
D, *Pisaster.* (**B,** Courtesy M. Newman, Vancouver Public Aquarium, British Columbia.)

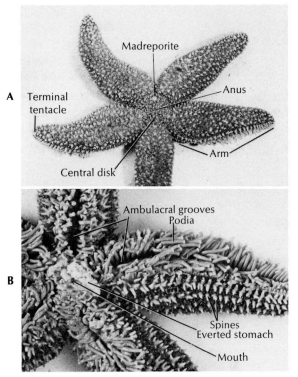

FIG. 19-4
Asterias. **A,** Aboral view. **B,** Oral view.

more (Fig. 19-3, *C*). The body is flattened and flexible, and covered with a ciliated pigmented epidermis. The mouth is centered on the under, or oral, side, surrounded by a soft peristomial membrane. An **ambulacral groove** along the oral side of each arm is bordered by moveable **spines** that protect the rows of **tube feet** (podia) projecting from the groove. Sea stars are the only eleutherozoans that have open ambulacral grooves (Fig. 19-4, *B*).

The aboral surface is usually rough and spiny. Around the bases of the spines are groups of minute pincerlike **pedicellariae** bearing tiny jaws manipulated by muscles that help keep the body surface free from debris, protect the skin gills, and aid in food capture (Fig. 19-5). The skin gills, or **dermal branchiae,** are soft delicate projections of the coelomic cavity, covered only with epidermis and peritoneum; they extend out through spaces between the ossicles and are concerned with respiration (Fig. 19-6). Also on the aboral side are the **anus** and the **madreporite** (Fig. 19-4, *A*), a calcareous sieve leading to the water-vascular system.

ENDOSKELETON. Beneath the epidermis of the sea star is a mesodermal endoskeleton of small cal-

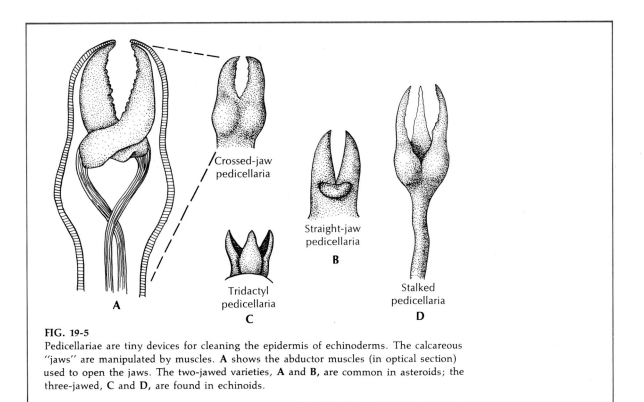

FIG. 19-5

Pedicellariae are tiny devices for cleaning the epidermis of echinoderms. The calcareous "jaws" are manipulated by muscles. **A** shows the abductor muscles (in optical section) used to open the jaws. The two-jawed varieties, **A** and **B,** are common in asteroids; the three-jawed, **C** and **D,** are found in echinoids.

careous plates, or **ossicles,** bound together with connective tissue. From these ossicles project the spines and tubercles that are responsible for the spiny surface. Muscles in the body wall move the rays and can partially close the ambulacral grooves by drawing their margins together.

COELOM. The coelom is large and filled with coelomic fluid containing amebocytes. The coelom projects into the little dermal branchiae, bringing the fluid close to the seawater for **gas exchange.** In **excretion** the amebocytes gather wastes and then escape through the walls of the dermal branchiae.

WATER-VASCULAR SYSTEM. The water-vascular system, which is found only in echinoderms, is a hydraulic system of canals and specialized tube feet (Fig. 19-6). In sea stars it serves as a means of locomotion and of food-getting. Water entering the porous madreporite on the aboral side passes through a **stone canal** to a **ring canal** around the mouth and then to the **radial canals,** one in each ray. A series of fine lateral canals, each with a valve, connect each radial canal with the tube feet, or **podia.** Nine small pouches in the ring canal are known as

Tiedemann's bodies and are believed to produce coelomocytes. Each podium is a muscular hollow tube, the inner end of which is a muscular sac, the **ampulla,** that lies within the coelomic cavity (Fig. 19-6, *A*), and the outer end of which usually bears a sucker. Some species lack the suckers. The podia pass through small pores between the ossicles in the ambulacral groove.

Used in locomotion, the water-vascular system operates on the principle of a hydraulic system. The valves in the lateral canals prevent backflow of fluid into the radial canals. When the ampulla muscles contract, fluid is forced into the podium. The extended podium can now be twisted in any direction. As the end touches substratum, it adheres by means of mucus secreted by the epidermis, and then becomes more firmly attached by the sucking action of the foot when the central portion is drawn back, creating a vacuum cup. The podium has only longitudinal muscles. Contraction of these muscles shortens the podium, forcing the fluid back into the ampulla and drawing the animal forward. It has been estimated that by combining mucous adhesion with suction a single podium can exert a pull equal to 25 to 30

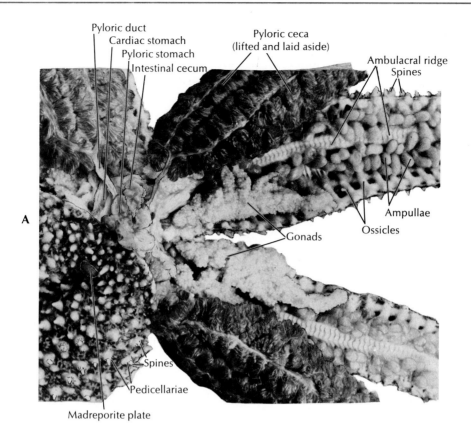

Pyloric duct
Cardiac stomach
Pyloric stomach
Intestinal cecum
Pyloric ceca
(lifted and laid aside)
Ambulacral ridge
Spines
A
Ampullae
Ossicles
Gonads
Spines
Pedicellariae
Madreporite plate

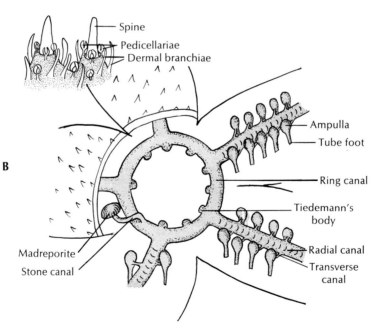

Spine
Pedicellariae
Dermal branchiae
Ampulla
Tube foot
Ring canal
Tiedemann's
body
B
Radial canal
Madreporite
Transverse
Stone canal
canal

FIG. 19-6

A, Sea star dissection, aboral view; at right, portions of two arms with digestive glands lifted aside to expose gonads and ampullae. At lower left, two undissected arms show madreporite and spines. **B,** Water-vascular system; upper left, detail of spines and skin gills.

grams. The coordinated effort of all or many tube feet is sufficient to draw the animal slowly up a vertical surface or over rocks.

On a soft surface, such as muck or sand, the suckers are ineffective, and so the tube feet are employed as legs. Locomotion now becomes mainly a stepping process involving a backward swinging of the middle portion of the podia, followed by a contraction, shoving the animal forward. Sea stars can move about 6 inches a minute. When inverted, the sea star twists its rays until some of its tube feet attach to the substratum as an anchor, and then it slowly rolls over.

Tube feet are innervated by the central nervous system rather than through the diffuse subepidermal plexus. Innervation of the tube is separate from that of the ampulla so that each acts independently of the other, but reciprocally with it.

FEEDING AND DIGESTIVE SYSTEM. Sea stars are carnivorous and feed on mollusks, crustaceans, tubeworms, etc. and are often very destructive to oyster beds. Some deep-sea forms are detritus feeders and use a filter feeding method.

The mouth on the oral side leads through a short esophagus to a large stomach in the central disk. The lower (cardiac) part of the stomach can be everted through the mouth during feeding (Fig. 19-4, *B*), but excessive eversion is prevented by gastric ligaments. The upper (pyloric) part is smaller and connects by ducts with a pair of large pyloric ceca (digestive glands) in each arm (Fig. 19-6). Digestion is mostly extracellular, although some intracellular digestion may occur in the ceca.

Some asteroids, particularly those with suckerless rays, swallow the prey whole. *Asterias* and other long-armed and suckered sea stars hold the prey and evert the stomach over it, digesting it first, then sucking up the broth. When feeding on a bivalve, such a sea star will hump over its prey, attaching its podia to the valves, and then exert a steady pull, using its feet in relays. A force of some 1,300 grams can thus be exerted. In half an hour or so the adductor muscles of the bivalve fatigue and relax. With a very small gap available, the star inserts its soft everted stomach into the space between the valves and wraps it around the soft parts of the shellfish. After feeding, the sea star draws in its stomach by contraction of the stomach muscles and relaxation of body wall muscles. Little indigestible material is ingested; so there is little fecal material ejected through the anal pore.

CIRCULATION. A **hemal system** is found in echinoderms but is much reduced in the sea stars. Its vessels are ill-defined sinus channels and are often referred to as hemal strands. The main vessels are a ring vessel around the mouth and a radial vessel in each arm, lying beneath the radial canal of the water-vascular system. From the ring vessel a sinus channel (axial sinus) extends up through the spongy mass of tissue (the axial gland) that extends along the stone canal. Coelomocytes are present in the hemal fluid. The hemal system may be useful in distributing digested products, but its general functions are not really known.

NERVOUS SYSTEM. The nervous system consists of three units placed at different levels in the disk and arm. The chief of these systems is the **oral** (ectoneural) system composed of a nerve ring around the mouth and a radial nerve into each arm, just beneath the radial canals. It appears to coordinate the tube feet. A **deep** (hyponeural) system lies aboral to the oral system, and an **aboral** system consists of a ring around the anus and radial nerves along the roof of the rays. An epidermal nerve plexus or nerve net freely connects these systems with the body wall and related structures. The epidermal plexus has recently been shown to coordinate the responses of the dermal branchiae to tactile stimulation—the only instance known in echinoderms where conduction occurs through a nerve net.

The sense organs are not well developed. There are tactile organs scattered over the surface and an ocellus at the tip of each arm. Their reactions are mainly to touch, temperature, chemicals, and differences in light intensity. Sea stars are usually more active at night.

REPRODUCTION. The sea stars have separate sexes. A pair of gonads lie in each interradial space (Fig. 19-6, *A*). Fertilization is external and occurs in early summer when the eggs and sperm are shed into the water. It has been shown that the maturation and shedding of sea star eggs are stimulated by a secretion from neurosecretory cells located on the radial nerves (H. Kastani).

REGENERATION AND AUTOTOMY. Echinoderms can regenerate lost parts. Sea star arms can regenerate readily, even if all are lost. Stars also have the power of autotomy and can cast off an injured

FIG. 19-7
A, Brittle star, *Ophiura albida.* **B,** Basket star, *Gorgonocephalus caryi.* Ophiuroids have sharply marked off arms and are very agile. They are bottom dwellers. (**A,** Photo by B. Tallmark, Uppsala University, Sweden; **B,** Courtesy M. Newman, Vancouver Public Aquarium, British Columbia.)

arm near the base. It may take months to regenerate a new arm.

Crown of thorns star

Since 1963 there have been numerous reports of increasing numbers of the crown of thorns sea star *(Acanthaster planci)* that were destroying the live polyps of the Pacific corals. Some of the suggested reasons for their increase were man's destruction of the giant triton, a gastropod that feeds upon sea stars, or that dredging, blasting, etc. had destroyed the creatures that feed upon the sea star larvae. Various attempts have been made to correct the problem. Indications at present are that the crown of thorns epidemic is tapering off and that recovery of the damaged reefs may be more rapid than was formerly believed.

Class Ophiuroidea—the brittle stars

In the class Ophiuroidea, as in Asteroidea, the members have a central disk with five (sometimes more) distinct arms (Fig. 19-7). The arms, however, are long and slender and are sharply marked off from the disk. They represent the largest class of echinoderms in number of species. Brittle stars are abundant wherever found. Basket stars have their rays branched in a complex fashion. Members of this class are found in shallow and deep water, and they have a wide distribution.

Some common ophiuroids along our Atlantic coast are *Amphipholis* (viviparous and hermaphroditic), *Ophioderma, Ophiothrix,* and *Ophiura* (Fig. 19-7, *A*). Along the Pacific coast are *Amphipholis, Amphiodia* with long delicate arms, the ultraspiny *Ophiothrix,* and the sand-colored *Ophioplocus* (viviparous). The basket star *Gorgonocephalus* (Fig. 19-7, *B*) is usually found at considerable depths. Some ophiuroids have variegated color patterns.

Brittle stars, unlike sea stars, have no pedicellariae, ambulacral grooves, or dermal branchiae. Their tube feet, called **tentacles,** are largely sensory in function and without suckers; they can pass food along the rays to the mouth and have a limited function in locomotion. Each of the jointed arms consists of a column of calcareous vertebrae connected by muscles and covered by plates. On the oral surface of the central disk are the **madreporite** and the **mouth,** with five movable plates that serve as jaws (Fig. 19-8).

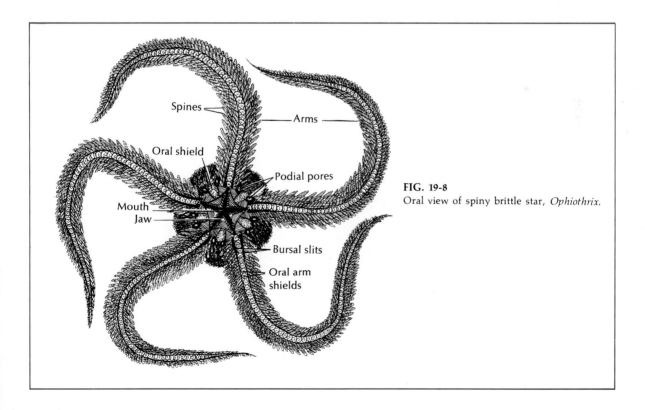

FIG. 19-8
Oral view of spiny brittle star, *Ophiothrix.*

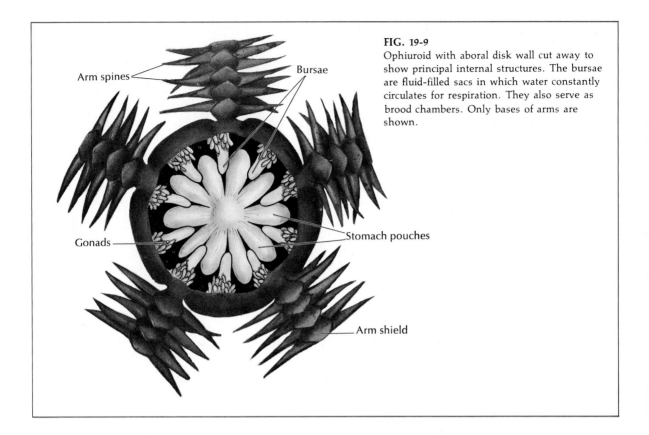

Arm spines

Bursae

Gonads

Stomach pouches

Arm shield

FIG. 19-9
Ophiuroid with aboral disk wall cut away to show principal internal structures. The bursae are fluid-filled sacs in which water constantly circulates for respiration. They also serve as brood chambers. Only bases of arms are shown.

There is no anus. The skin is leathery, with dermal plates (ossicles) and spines arranged in characteristic patterns. Cilia are mostly lacking.

The visceral organs are confined to the central disk; the rays are too slender to contain them (Fig. 19-9). The **stomach** is saclike and there is no intestine. Indigestible material is cast out of the mouth. Five pairs of **bursae** (peculiar to ophiuroids) open toward the oral surface by **genital slits** at the bases of the arms. Water circulates in and out of these sacs for exchange of gases. On the coelomic wall of each bursa are small **gonads** that discharge into the bursa their ripe sex cells, which pass through the genital slits into the water for fertilization. Sexes are usually separate; a few are hermaphroditic. Some brood their young in the bursae; the young escape through the genital slits or by rupturing the aboral disk. The larva, called the **ophiopluteus** (Fig. 19-1), metamorphoses into the adult. Water-vascular, nervous, and hemal systems are similar to those of the sea stars. In each arm there is a small **coelom,** a **nerve cord,** and a **radial canal** of the water-vascular system.

BEHAVIOR. Brittle stars are often found under stones and seaweed at low tide because they are negatively phototactic and positively thigmotactic. At high tide they are active, wandering about in search of small animals, which they capture with their rays. They move with a writhing, serpentlike motion of the arms. They can swim with their rays, but they often hold to objects with one or more rays while pushing with the others. Regeneration and autotomy are even more pronounced in brittle stars than in sea stars. Many individuals are regenerating parts most of the time, for they are very fragile.

Class Echinoidea—the sea urchins and sand dollars

The echinoids are the sea urchins, sand dollars, and heart urchins. They have a compact body enclosed in an endoskeletal test, or shell, made up of closely fitted plates. They lack arms, but their tests show the typical pentamerous plan of the echinoderms. Sea urchins are hemispherical and have medium to long spines; the disk-shaped sand dollars

FIG. 19-10
Sea urchin, *Lytechinus.* Note the slender suckered tube feet. They often attach to bits of shell, seaweed, etc., for camouflage. Stalked pedicellariae can be seen between the spines. This is a common Atlantic form from the Carolinas to the West Indies. (Courtesy R. O. Hermes, Homestead, Fla.)

and ovoid heart urchins have very short spines. *Athenosoma* has poison spines. Echinoids move about by means of their movable spines and tube feet. Some echinoids are colorful.

Echinoids have a wide distribution in all seas, largely in intertidal zones. Urchins often prefer rocky bottoms, but sand dollars and heart urchins like to burrow into a sandy substrate. *Arbacia* is a common East coast urchin; *Strongylocentrotus* is common along the Pacific coast; and *Lytechinus* (Fig. 19-10) is well-known in Florida waters.

STRUCTURE. The echinoid **test** is a compact skeleton of 10 double rows of plates that bear movable,

stiff spines (Fig. 19-11). The plates are firmly sutured. The five pairs of ambulacral rows are homologous to the five arms of the sea star and have pores (Fig. 19-11, *B*) through which the long tube feet extend. The plates bear small tubercles on which the round ends of the spines articulate as ball-and-socket joints. The spines are moved by small muscles around the bases.

There are several kinds of **pedicellariae**, the most common of which are three jawed and are mounted on long stalks (Figs. 19-5, *D,* and 19-10). Pedicellariae help keep the body clean and capture small organisms.

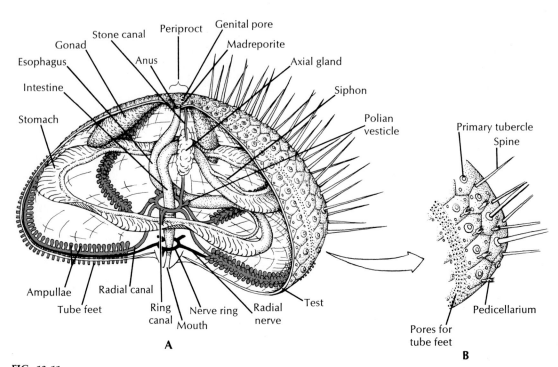

FIG. 19-11
A, Internal structure of the sea urchin; water-vascular system in red. **B,** Detail of portion of endoskeleton.

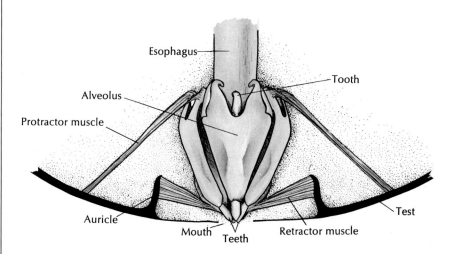

FIG. 19-12
Aristotle's lantern, complex mechanism used by sea urchin for masticating its food. Five pairs of retractor muscles draw lantern and teeth up into test; five pairs of protractors push lantern down and expose teeth. Other muscles produce variety of movements. Only major skeletal parts and muscles are shown in diagram.

The **mouth** is surrounded by five converging **teeth.** In some sea urchins branched gills (modified podia) encircle the peristome. The anus, genital openings, and madreporite are located aborally in the periproct (Fig. 19-11).

Inside the test (Fig. 19-11) is the coiled **digestive system** and a complex chewing mechanism called **Aristotle's lantern** (Fig. 19-12) to which the teeth are attached. A ciliated siphon connects the esophagus to the intestine and enables the water to bypass the stomach to concentrate the food for digestion in the intestine. Sea urchins eat algae and almost any organic material, which they graze or nibble off with their teeth. Sand dollars have short club-shaped spines that move the sand and its organic contents over the aboral surface and down the sides. Fine food particles drop down between the spines and are carried by ciliated tracts on the oral side to the mouth.

The **hemal** and **water-vascular systems** are similar to those of the asteroids (Fig. 19-11), with radial canals running upward along the ambulacral rows inside the test. The chief **nervous system** consists of a nerve ring around the mouth, five radial nerves that run along the radial canals, and a subepidermal plexus. There are few special sense organs, but the podia, spines, and pedicellariae are also sensory. Some echinoids have eyespots in the aboral epithelium. The gills, when present, and the podia provide areas for gas exchange. Amebocytes in the coelom have an excretory function, and there is a possibility that the axial gland may also have some excretory function.

Sexes are separate, and both eggs and sperm are shed into the sea for external fertilization. Some, such as the slate pencil urchins, brood their young in depressions between the spines. The **pluteus larvae** (Fig. 19-1) of nonbrooding echinoids may live a planktonic existence for several months and then metamorphose quickly into the young urchins.

Class Holothuroidea—sea cucumbers

In a phylum characterized by odd animals, class Holothuroidea contains members that both structurally and physiologically are among the strangest of all. These soft-bodied animals have a remarkable resemblance to the vegetable after which they are named (Fig. 19-13, *C*). In their evolution they appear to be more closely related to Echinoidea than to other echinoderms. They are bottom-dwelling

forms, living mostly in sand and mud. Low-tide pools with mucky bottoms are favorite places. Here they lie buried, with their tentacles sticking up into the clearer water.

Common species along the Eastern seacoast are *Cucumaria frondosa, Thyone briareus,* and the translucent *Leptosynapta.* Along the Pacific coast there are several species of *Cucumaria* (Fig. 19-13) and the striking reddish brown *Stichopus,* with very large papillae.

STRUCTURE. The bodies of sea cucumbers are elongated. The mouth and retractile tentacles are found at the oral end and the anus at the aboral end. They have leathery skins in which are embedded microscopic calcareous plates. The 10 to 30 retractile **tentacles** around the mouth correspond to the oral tube feet of other echinoderms (Fig. 19-13).

The body wall contains both circular and longitudinal muscles and is covered by a cuticle and nonciliated epidermis. The dorsal side bears two longitudinal zones of **tube feet;** the ventral side has three. However, podia are absent in some species. The ventral side often becomes flattened and is called the **sole.**

The **coelomic cavity** is large, filled with a fluid similar to seawater and contains many coelomocytes. The digestive system empties posteriorly into a muscular cloaca (Fig. 19-14). A **respiratory tree** composed of two long many-branched tubes also empties into the cloaca, which pumps seawater through it. The respiratory tree serves both for respiration and excretion. Water also passes through the walls of the respiratory tree into the coelom. Gaseous exchange also occurs in the skin and tube feet.

The **hemal system** is well developed in holothuroids. It consists of a ring around the esophagus, with radial vessels along the water canals and two main sinuses along the digestive system.

The **water-vascular system** (Fig. 19-14) consists of an internal madreporite, a ring canal around the esophagus, and five radial canals that connect to the tube feet. Opening into the ring canal are a number of **polian vesicles,** elongated sacs hanging in the coelom and serving as expansion chambers for the water-vascular system.

The main **nervous system** includes an oral nerve ring with five radial nerves. There appear to be sense organs for touch and light, and some species have statocysts for balance.

FIG. 19-13
The sea cucumber *Cucumaria frondosa,* with tentacles extended for feeding. In the background is the sea anemone *Tealia telina. Cucumaria* is a suspension feeder; minute plankton organisms adhere to the tentacles, which when loaded bend over and wipe off the food into the pharynx. (Photo by T. Lundälv, Kristinebergs Zoological Station, Sweden.)

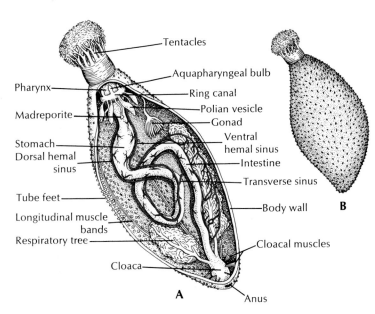

Tentacles
Aquapharyngeal bulb
Pharynx
Ring canal
Polian vesicle
Madreporite
Gonad
Ventral hemal sinus
Stomach
Dorsal hemal sinus
Intestine
Tube feet
Transverse sinus
Body wall
Longitudinal muscle bands
Respiratory tree
Cloacal muscles
Cloaca
Anus

A

B

FIG. 19-14
Anatomy of the sea cucumber *Thyone.*
A, Internal. **B,** External. *Red,* Hemal system.

The sexes are separate, but some holothuroids are hermaphroditic. There is one gonad in *Thyone* and *Cucumaria* (Fig. 19-14), two in *Stichopus*. Each gonad is composed of numerous tubules united at their base to form a tuft, which may be large at sexual maturity. A common gonoduct empties the sex cells through a **gonopore** to the outside, where fertilization occurs. The larval stage is called an **auricularia** (Fig. 19-1). Some species brood the young either inside the body or somewhere on the body surface.

BEHAVIOR. Sea cucumbers are sluggish, moving partly by means of their ventral tube feet and partly by waves of contraction in the muscular wall, which contains five powerful longitudinal muscle bands as well as circular muscle fibers. The dorsal tube feet are respiratory and tactile. The food consists of small organisms, which they entangle in the sticky mucus of their tentacles and suck into the mouth. Sea cucumbers have a peculiar power of self-mutilation. Some, when irritated, may cast out a part of their viscera by a strong muscular contraction that may either rupture the body wall or evert its contents through the anus. The lost parts are soon regenerated. There is an interesting commensal relationship between some sea cucumbers and a small fish, *Carapus,* that uses the cloaca and respiratory tree as shelter.

Class Crinoidea—sea lilies and feather stars

The crinoids are the most primitive of the echinoderms. As fossil records reveal, they were once far more numerous than now. They are essentially deepwater forms, although a few species live near shore. They differ from other echinoderms by being attached during part or all of their lives. Sea lilies have a flower-shaped body that is placed at the tip of an attached stalk (Fig. 19-16, *A*). The feather stars have long, many-branched arms, and the adults are free swimming (Fig. 19-15). At one stage of their life cycle they are attached to stalks, which they later absorb.

STRUCTURE. The body disk, or **calyx,** is covered with a leathery skin (tegmen) containing calcareous plates. Cuticle and epidermis are poorly developed. Five flexible arms branch to form many more arms, each with many lateral **pinnules** arranged like barbs on a feather (Fig. 19-16). Calyx and arms together

FIG. 19-15
The feather star, *Florometra serratissima,* a crinoid. Feather stars can crawl by holding on to objects with the adhesive ends of the pinnules and pulling along by contracting arms. They can also swim by raising and lowering alternate sets of arms. (Courtesy M. Newman, Vancouver Public Aquarium, British Columbia.)

are called the **crown.** Sessile forms have a long, jointed **stalk** attached to the aboral side of the body. This stalk is made up of plates, appears jointed, and may bear **cirri.** Madreporite, spines, and pedicellariae are absent. The upper (oral) surface bears the mouth, which opens into a short esophagus, from which the long **intestine** with diverticula proceeds aborally for a distance and then makes a complete turn to the **anus,** which may be on a raised cone. Crinoids feed upon small organisms that are caught in the ambulacral grooves with the aid of tube feet and mucus nets. Ciliated **ambulacral grooves** on the arms carry food to the mouth (Fig. 19-16, *B*). Tube feet in the form of tentacles are also found in the grooves. The **water-vascular system** has the echinoderm plan. The **nervous system** is made up of an **oral ring** and a **radial nerve,** which runs to each arm. The

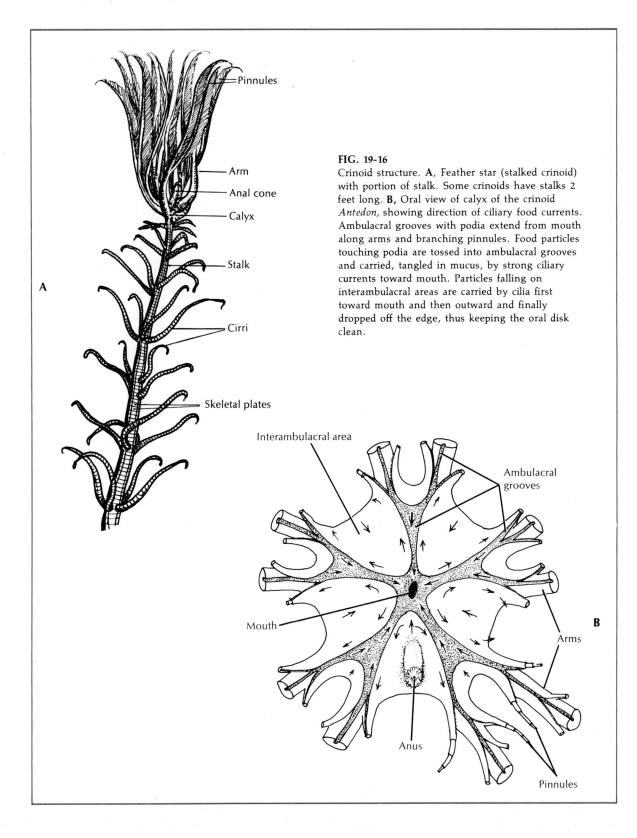

FIG. 19-16
Crinoid structure. **A,** Feather star (stalked crinoid) with portion of stalk. Some crinoids have stalks 2 feet long. **B,** Oral view of calyx of the crinoid *Antedon,* showing direction of ciliary food currents. Ambulacral grooves with podia extend from mouth along arms and branching pinnules. Food particles touching podia are tossed into ambulacral grooves and carried, tangled in mucus, by strong ciliary currents toward mouth. Particles falling on interambulacral areas are carried by cilia first toward mouth and then outward and finally dropped off the edge, thus keeping the oral disk clean.

Pinnules

Arm

Anal cone

Calyx

Stalk

Cirri

Skeletal plates

A

Interambulacral area

Ambulacral grooves

Mouth

Anus

Arms

Pinnules

B

aboral or entoneural system is the main one in crinoids, in contrast to most other echinoderms. Sense organs are scanty and primitive. The sexes are separate. The gonads are simply masses of cells in the genital cavity of the arms and pinnules. The gametes escape without ducts through a rupture in the pinnule wall. The **doliolaria larvae** (Fig. 19-1) are free swimming for a time before they become attached and metamorphose. Most crinoids are from 6 to 12 inches long.

References

Boolootian, R. A. (editor). 1966. Physiology of Echinodermata. New York, Interscience Publishers. *An up-to-date account of the physiology of this group.*

Boolootian, R. A., and A. C. Giese. 1959. Clotting of echinoderm coelomic fluid. J. Exp. Zool. **140**:207-229.

Buchsbaum, R., and L. Milne. 1960. The lower animals, living invertebrates of the world. Garden City, N. Y., Doubleday & Co., Inc. *Excellent color photographs of echinoderms.*

Burnett, A. L. 1960. The mechanism employed by the starfish *Asterias forbesi* to gain access to the anterior of the bivalve *Venus mercenaria.* Ecology **41**:583-584. *This investigator and others have shown that the starfish uses no narcotic agent to produce a small gap between the clam's valves through which the starfish can squeeze its stomach.*

Cambridge Natural History. 1906. Echinodermata (E. W. MacBride). London, Macmillan & Co., Ltd. *Authoritative and detailed treatment of general structures.*

Clark, A. M. 1962. Starfishes and their relations. British Museum, London.

Coe, W. R. 1972. Starfishes, serpent stars, sea urchins and sea cucumbers of the Northeast. New York, Dover Publications, Inc. *(Paperback.)*

Encyclopaedia Britannica. 1956. Echinoderma. Chicago, Encyclopaedia Britannica, Inc. *A fine, concise account of echinoderms with clear illustrations.*

Fontaine, A. R. 1965. Feeding mechanisms of the ophiuroid *Ophiocomina nigra.* J. Mar. Bio. Ass., U. K. **45**:373.

Goodbody, I. 1960. The feeding mechanism in the sand dollar *Mellita sexiesperforata.* Biol. Bull. **119**(1):80-86.

Grasse, P.-P. (editor). 1948. Traité de zoologie. XI. Echinodermes-Stomocordes-Protocordes (L. Cuénot). Paris, Masson & Cie.

Harvey, E. B. 1956. The American *Arbacia* and other sea urchins. Princeton, N. J., Princeton University Press. *A comprehensive monograph on this interesting group, which have furnished so many basic concepts in the field of cytology and development.*

Hyman, L. H. 1955. The invertebrates: Echinodermata, vol. 4. New York, McGraw-Hill Book Co. *The most comprehensive work yet published on the echinoderms.*

Jennings, H. S. 1907. Behavior of the starfish *Asterias forreri* De Loriol. Berkeley, Univ. Calif. Pub. in Zool. **4**:339-411. *An account of the reactions of the starfish, including the righting movement.*

Lankester, E. R. (editor). 1900. A treatise on zoology. Part III. The Echinoderma (F. A. Bather). London, A. & C. Black. *The technical descriptions of echinoderms in this work will always be of great value.*

MacGinitie, G. E., and N. MacGinitie. 1968. Natural history of marine animals. ed. 2, New York, McGraw-Hill Book Co. *The section on echinoderms includes some good descriptions and photographs of representative forms, especially those of the Pacific coast.*

Millott, N. (editor). 1967. Echinoderm biology. New York, Academic Press, Inc. *An interesting symposium on various aspects of a "noble group especially designed to puzzle the zoologist" (Hyman).*

Nicol, J. A. C. 1969. The biology of marine animals, ed. 2. New York, Pitman Publishing Corp.

Nichols, D. 1966. Echinoderms. London, Hutchinson & Co., Ltd.

Parker, T. J., and W. A. Haswell. 1972. A textbook of zoology, ed. 7, vol. 1. London, Macmillan & Co., Ltd. *Good general descriptions with considerable detail.*

Ricketts, E. F., and J. Calvin. 1968. Between Pacific tides, ed. 4 (revised by J. W. Hedgpeth). Stanford, Calif., Stanford University Press. *In many ways this is a unique book of seashore life. It stresses the habits and habitats of the Pacific coast invertebrates (including echinoderms), and the illustrations are revealing. It includes an excellent systematic index and annotated bibliography.*

Russell-Hunter, W. D. 1969. A biology of higher invertebrates. New York, The Macmillan Co.

Smith, R. I., and others (editors). 1957. Intertidal invertebrates of the central California coast. Berkeley, University of California Press. *This is a revision of S. F. Light's Laboratory and Field Text in Invertebrate Zoology. Consists mainly of taxonomic keys to the forms found in the intertidal zone.*

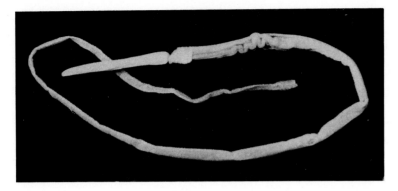

Saccoglossus is a fragile pinkish hemichordate that uses its proboscis for burrowing branching tunnels in ocean-bottom sediment, lining them with mucus. It may emerge at night to creep about over the eel grass. Cilia on the proboscis direct food particles to the mouth.

CHAPTER 20

THE LESSER DEUTEROSTOMES

PHYLUM CHAETOGNATHA
PHYLUM POGONOPHORA
PHYLUM HEMICHORDATA

The deuterostomes include, along with the Echinodermata, four other phyla—Chaetognatha, Pogonophora, Hemichordata, and Chordata. Two of the chordate subphyla—Urochordata and Cephalochordata—are also invertebrate groups.

The term lesser (or minor) deuterostomes is usually used in reference to the hemichordates, chaetognaths, and pogonophores, but only in the sense that each of these groups has a relatively small number of species. These groups are, however, widespread, and contain some commonly found invertebrate forms. Often smaller groups deserve much more attention than is usually given to them for they contribute much to our understanding of evolutionary diversity and relationships.

All three of these phyla have enterocoelous development of the coelom and some form of radial cleavage. The hemichordates show enough relationship to

the chordates that they were formerly included as a subphylum of the Chordata. The Pogonophora, with their tripartite body division, seem to have a close relationship with the hemichordates, but the Chaetognatha apparently are not closely related to any other group and are included here only because they represent a minor deuterostome group.

■ PHYLUM CHAETOGNATHA*

A common name for the chaetognaths is arrowworms. They are all marine animals and are considered by some to be related to the nematodes and by others to be related to the annelids. However, they actually seem to be aberrant and show no distinct relations to any other group. Only their embryology indicates their position as deuterostomes.

*Ke-tog'na-tha (Gr. *chaitē*, long flowing hair, + *gnathos*, jaw).

The name Chaetognatha means "hair-jawed" and refers to the sickle-shaped bristles on each side of the mouth. This is not a large group, for there are only some fifty known species. Their small, straight bodies resemble miniature torpedoes, or darts, ranging from 1 to 3 inches in length.

The arrowworms are all adapted for a planktonic existence, except for *Spadella*, a benthic genus. They usually swim to the surface at night and descend during the day. Much of the time they drift passively, but they can dart forward in swift spurts, using the caudal fin and longitudinal muscles, a fact that no doubt contributes to their success as planktonic predators. Horizontal fins bordering the trunk are used in flotation rather than in active swimming.

External features. The body of the arrowworm is unsegmented and is made up of head, trunk, and

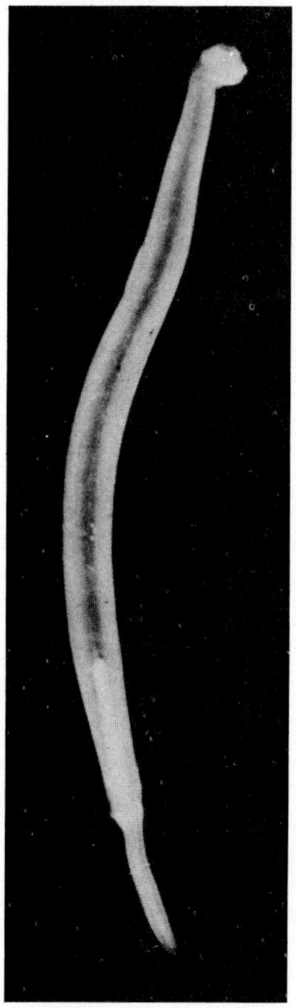

FIG. 20-1
Arrowworm, *Sagitta*. Head *(top)* is largely covered with hood formed from epidermis. When worm is engaged in catching its prey, hood is retracted to neck region. (Preserved specimen.)

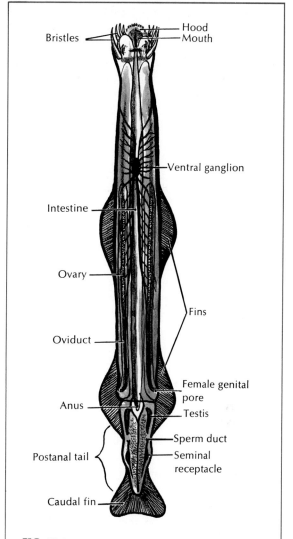

FIG. 20-2
Arrowworm, *Sagitta,* ventral view. These worms, rarely more than 2 or 3 inches long, form important part of marine plankton in both littoral and open sea waters. They have many resemblances to certain pseudocoelomates, and some authorities hesitate to call them coelomate animals, although they are enterocoelous and are placed in Deuterostomia. Among their features are postanal tail, hood, fins, and stratified epidermis.

postanal tail (Figs. 20-1 and 20-2). (They are the only group outside of the chordates that have a postanal tail.) On the underside of the head is a large vestibule leading to the mouth. There are teeth in the vestibule, and flanking it on both sides are curved chitinous spines used in seizing the prey. There is a pair of eyes on the dorsal side. A peculiar hood formed from a fold of the neck can be drawn forward over the head and spines. When the animal captures prey, the hood is retracted and the teeth and raptorial spines spread apart and then snap shut with startling speed. Arrowworms are voracious feeders, living on planktonic forms, especially copepods, and even small fish.

The body is covered with a thin cuticle and a layer of epidermis that is single-layered except along the sides of the body where it is stratified in a thick layer. These are the only invertebrates with a many-layered epidermis.

Internal features. Arrowworms are fairly advanced worms in that they have a complete digestive system, a well-developed coelom, and a nervous system with a nerve ring containing large dorsal and ventral ganglia and a number of lateral ganglia. Sense organs include the eyes, sensory bristles, and a U-shaped ciliary loop extending over the neck from the back of the head that is believed to detect water currents or their chemical nature. Vascular, respiratory, and excretory systems, however, are entirely lacking.

Arrowworms are hermaphroditic with either cross- or self-fertilization. The eggs of *Sagitta* are coated with jelly and are planktonic. Eggs of other arrowworms may be attached to the body and carried about for a time. The larvae develop directly without metamorphosis. Chaetognath embryology differs from that of other deuterostomes in that the coelom is formed by a backward extension from the archenteron rather than by pinched-off coelomic sacs. There is no true peritoneum lining the coelom. Cleavage is radial, complete and equal.

In some respects chaetognaths resemble the pseudocoelomates, especially the nematodes. Their lack of circular muscle, lack of peritoneum and a ring type of nervous system are nematode characteristics. However, their other characteristics show no affinity with the pseudocoelomates. Neither are there close affinities with any deuterostome group. They may have branched off very early from the base of the deuterostome line.

The best known species is *Sagitta,* the common arrowworm (Figs. 20-1 and 20-2).

■ PHYLUM POGONOPHORA*
Position in animal kingdom

Pogonophores belong to the Deuterostomia branch, with enterocoelous formation of the coelom and unequal holoblastic cleavage.

The body is divided into three regions, each with its own coelom, showing a relationship with the hemichordates. Because of their relationship to each other and to the chordates, the Pogonophora and Hemichordata are often called the **prechordates.**

Biologic contributions

1. Body is typically divided into three regions—protosome, mesosome, and metasome (trunk)—similar to the hemichordate pattern.

2. There is some indication of segmentation in the trunk.

3. A tentacular mass on the protosome forms a device for collecting food and probably for absorbing extracellularly digested food. This is the only group of free-living animals that has no digestive system.

Phylogeny and adaptive radiation

Phylogeny. This is a new phylum in the process of analysis. Belonging to the deuterostomes, pogonophores (beardworms) are in the same group with the echinoderms, hemichordates, and chordates. Their tripartite regions and coelom indicate a relationship to the hemichordates. However their embryology is unlike that of any other deuterostome, for they have determinate cleavage, no blastopore is formed, and the formation of germ layers is atypical.

Adaptive radiation. Although the general body plan is uniform throughout the phylum, there is much diversity among the species in the structure of the tentacular crown. The number of tentacles varies from one to more than 250. This might then be considered the basic adaptive feature of the group.

The pogonophores

The beardworms were entirely unknown before the twentieth century. The first specimens to be described were collected from deep-sea dredgings in 1900 off the coast of Indonesia. They have since been discovered in several seas including the western

---

*Po"go-nof'e-ra (Gr. *pōgōn,* beard, + *phora,* pl. of bearing).

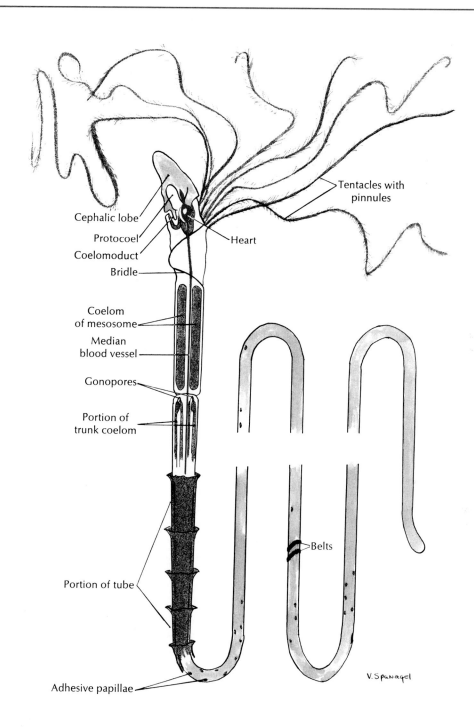

FIG. 20-3
Diagram of pogonophore showing principal external and internal structures. Portions of trunk have been omitted. Tentacles are often fused into cylinder. (Adapted from Ivanov, 1955, and others.)

Atlantic off the U. S. eastern coast. Only recently has sufficient material been available for an appraisal of the phylum. Some 80 species have been described so far.

These elongated tube-dwelling forms have left no known fossil record. Their closest affinities seem to be to the hemichordates.

Most pogonophores live in the bottom ooze on the ocean floor, mostly in deep water. This accounts for their delayed discovery. They are obtained only by dredging. The size varies widely from *Siboglinum pellucidum,* which is about 5.5 cm. long and 0.1 mm. in diameter to *Zenkevitchiana longissima* which may be up to 36 cm. long and only 0.8 mm. in breadth. They secrete very long tubes—three times their own length—of chitin and protein, in which they live, probably extending the anterior end only for feeding. The tubes may be uniform or arranged in rings (Fig. 20-3).

Characteristics

1. **Body elongated** and **tripartite** (protosome, mesosome, and metasome); body enclosed in a **secreted tube;** faint segmentation in trunk
2. Protosome small, with 1 to more than 250 **tentacles,** each provided with **pinnules**
3. Trunk with pair of raised girdles **(belts)** bearing chitinous adhesive organs
4. Coelom subdivided into a single protocoel in protosome and a pair of coelomic sacs each in the mesosome and metasome
5. **No digestive system;** digestion may be a function of the tentacles
6. **Closed circulatory system** of dorsal and ventral vessels; part of ventral vessel forms heart; blood with hemoglobin
7. No gills or special respiratory system; respiration may be a function of the tentacles
8. Excretory system of a **pair of nephridial coelomoducts,** connecting protocoel with outside
9. Nervous system of a ring-shaped or elongated brain in the cephalic lobe, with one or more median longitudinal cords in the epidermis
10. Sexes separate; paired gonads in the metacoel, with gonopores to the exterior; spermatophores present
11. Fertilization and brooding of eggs occurs in tube; cleavage holoblastic and unequal; gastrulation without blastopore; bilateral embryos

Classification of Pogonophora

Order Athecanephria (a-thek"a-nef're-a) (Gr. *a,* without, + *thēkē,* case, + *nephros,* kidney, + *-ia,* pl. suffix). Protosome-mesosome divided by constriction; coelomoducts with lateral nephridiopores; tentacles usually few and separate; postannular trunk with scattered adhesive papillae; fusiform spermatophores. Examples: *Oligobrachia, Siboglinum.*

Order Thecanephria (thek-a-nef're-a) (Gr. *thēkē,* case, + *nephros,* kidney). Protosome-mesosome usually without external constriction; coelomoducts with median nephridiopores; tentacles numerous and may be basically fused; postannular trunk with transverse rows of adhesive papillae; flat spermatophores. Examples: *Heptabrachia, Galathealinum.*

Structure and function

EXTERNAL FEATURES. The beardworm has a long cylindric body divided into a short anterior part that is subdivided into a **protosome** and a **mesosome,** and a very long slender trunk **(metasome)** separated by a diaphragm from the mesosome.

The "beard" of the beardworm consists of one to 268 long ciliated **tentacles** originating on the underside of the protosome (Fig. 20-3). The tentacles are hollow extensions of the coelom and bear minute pinnules. For a part or most of their length the tentacles lie parallel with each other, enclosing an intertentacular space between them into which the pinnules project (Fig. 20-4). Each tentacle contains an afferent and an efferent blood vessel and undoubtedly serves a respiratory function.

The tentacles also serve for food getting and, since there is no trace of a mouth or digestive system in a pogonophore, the tentacles are also believed to serve in digestion and absorption. Ivanov (1955) suggests that water is drawn from the tips of extended tentacles into the intertentacular cavity by ciliary action and then out between the tentacles, so that particles of detritus can be trapped in the network of pinnules. Digestive enzymes are probably secreted into the space when the basal portion of the tentacles is retracted into the tube and the food is digested extracellularly. Dissolved substances would then be absorbed by the pinnules and thence into the tentacular blood vessels for distribution to the body.

Jägersten (1957) has suggested that pogonophores may take up amino acids and micromolecules directly from the seawater. Considerable quantities of amino acids have been found in seawater near the sediment in which they live.

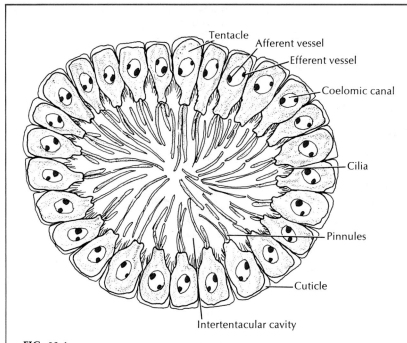

FIG. 20-4
Cross section of tentacular crown of pogonophore *Lamellisabella*. Tentacles arise from ventral side of protosome at base of cephalic lobe. Tentacles (which vary in number in different species) enclose a cylindric space, with the pinnules forming a kind of food-catching network. Food may be digested in this pinnular meshwork and absorbed into the blood supply of tentacles and pinnules.

The Pogonophora are the **only nonparasitic** animals that lack a digestive system and have external digestion.

A pair of ridges on the mesosome called the **bridle** probably aids the animal in clinging to its tube as it moves up and down. About midway down the trunk, or metasome, are two or three muscular rings called **belts,** or girdles, that serve to anchor the body in the tube when the anterior end is suddenly withdrawn. There are also numerous **adhesive papillae** on the trunk that may cling to the tube as the animal moves inside it. The arrangement of the papillae anterior to the belts strongly suggests metamerism.

INTERNAL FEATURES. The **coelom** consists of a single coelomic sac in the protosome and paired sacs in the mesosome and in the metasome (Fig. 20-3). Digestive and respiratory organs are absent. There is a well-developed **circulatory system** in which blood moves forward in a ventral vessel to the heart and then is pumped to the tentacles for oxygenation. From the tentacles the blood is carried to the head region and then posteriorly in a large dorsal vessel.

The **nervous system,** located in the epidermis, consists of a brain in the cephalic lobe and one or two longitudinal cords to the trunk region. The dorsal cord contains giant fibers probably associated with the need for "panic withdrawal" into the tube. Sense organs are limited to sense cells in the cephalic lobe; a ciliated band on the trunk may be a chemoreceptor.

Beardworms are dioecious, with a pair of gonads in the metasome. The sperm are contained in spermatophores, which Ivanov suggests may be transferred to the female by means of the tentacles. The female lays her clutch of eggs in her tube where they undergo embryonic development. The unequal but holoblastic cleavage produces a bilateral larva.

■ PHYLUM HEMICHORDATA—THE ACORN WORMS
Position in animal kingdom

1. Hemichordates belong to the deuterostome branch of the animal kingdom and are enterocoelous coelomates with radial cleavage.

2. Hemichordates have a combination of both invertebrate (echinoderm) and chordate characteristics.

3. A chordate plan of structure is suggested by gill slits and a restricted dorsal tubular nerve cord.

4. Because of their relationship to each other and to the chordates, Hemichordata and Pogonophora are called prechordates.

Biologic contributions

1. A **tubular dorsal nerve cord** (in collar zone) foreshadows the future condition in chordates, and a diffused net of nerve cells is similar to the uncentralized, subepithelial plexus of echinoderms.

2. The **gill slits** so characteristic of chordates are used primarily for filter feeding and only secondarily for breathing and are thus comparable to some of the protochordates.

3. The **stomochord** (**notochord** by tradition), an outpocketing forward from the roof of the mouth cavity, with its chitinous plate, is of debatable phylogenetic status. It is sometimes called a buccal diverticulum.

Phylogeny and adaptive radiation

Phylogeny. Hemichordate phylogeny has long been puzzling. They share characteristics with both the echinoderms and the chordates. They share with the chordates the gill slits, which are used primarily for filter feeding and secondarily for breathing as in some of the protochordates. A short dorsal, somewhat hollow nerve cord in the collar zone foreshadows the nerve cord of the chordates. A stomochord, or buccal diverticulum in the hemichordate mouth cavity, which was long believed to be a rudimentary notochord homologous with the notochord of chordates, may be of doubtful phylogenetic importance. The relationship to the echinoderms is striking. The early embryology is remarkably like that of echinoderms and the early tornaria larva is almost identical with the bipinnaria larva of asteroids. The similarity between the hydraulic action of the coelomic pouches and that of the water-vascular system in echinoderms, and the similarity in plan of the

subepithelial nerve plexus of the two groups are further evidence of their relations.

Within the phylum, the class Pterobranchia is considered more primitive than the class Enteropneusta and shows affinities with the Ectoprocta, Brachiopoda, and others because of the lophophore and sessile habits. Some believe that the pterobranchs may be similar to the common ancestors of both the hemichordates and the echinoderms.

Adaptive radiation. The pterobranchs, because of their sessile lives and being found largely in secreted tubes in ocean bottoms where conditions are fairly stable, have undergone little adaptive divergence. They have retained a tentacular type of ciliary feeding. The enteropneusts, on the other hand, although sluggish, are more active than the pterobranchs. Having lost the tentaculated arms, they use a proboscis to trap small organisms in mucus, or eat sand as they burrow and digest organic sediments from the sand. Their evolutionary divergence, though greater than that of the pterobranchs, is still modest.

The hemichordates

The hemichordates are marine animals that were formerly considered a subphylum of the chordates, based on their possession of gill slits and a rudimentary notochord. However it is now generally agreed that the hemichordate "notochord" is a stomochord and not homologous with the chordate notochord; so the hemichordates are given the rank of a separate phylum.

Hemichordates are vermiform bottom dwellers living usually in shallow waters. Some are colonial and some live in secreted tubes. Most are sedentary or sessile. Their distribution is fairly worldwide, but their secretive habits and fragile bodies make collecting difficult.

Members of class Enteropneusta (acorn worms) range from 20 mm. to 2.5 meters in length and 3 to 200 mm. in breadth. The Pterobranchia are smaller, usually from 5 to 14 mm., not including the stalk. About 70 species of enteropneusts and three small genera of pterobranchs are recognized.

Hemichordates have the typical tricoelomate structure of deuterostomes.

Characteristics

1. Soft-bodied; wormlike, or short and compact with stalk for attachment

2. Body divided into proboscis, collar, and trunk; probocis coelomic pouch single, but paired in other two; stomochord in posterior part of proboscis

3. Enteropneusta free living and of burrowing habits; pterobranchs sessile, with free-living and colonial members in tubes

4. Circulatory system of dorsal and ventral vessels and dorsal heart

5. Respiratory system of gill slits (few or none in pterobranchs) connecting the pharynx with outside as in chordates

6. No nephridia; a single glomerulus connected to blood vessels may have excretory function

7. A subepidermal nerve plexus thickened to form dorsal and ventral nerve cords, with a ring connective in the collar; dorsal nerve cord of collar hollow in some

8. Sexes separate in Enteropneusta, with gonads projecting into body cavity; in pterobranchs reproduction may be sexual, or else asexual (in some) by budding; tornaria larva in some Enteropneusta

Class Enteropneusta

The enteropneusts, or acorn worms, are sluggish wormlike animals that live in burrows or secreted tubes or under stones, usually in mud or sand flats of intertidal zones. *Balanoglossus* and *Saccoglossus* (Fig. 20-5) are common genera.

The mucus-covered body is divided into a tonguelike **proboscis**, a short **collar**, and a long **trunk** (protosome, mesosome, and metasome).

PROBOSCIS. The proboscis is the active part of the animal. It probes about in the mud, examining its surroundings and collecting food in mucus strands

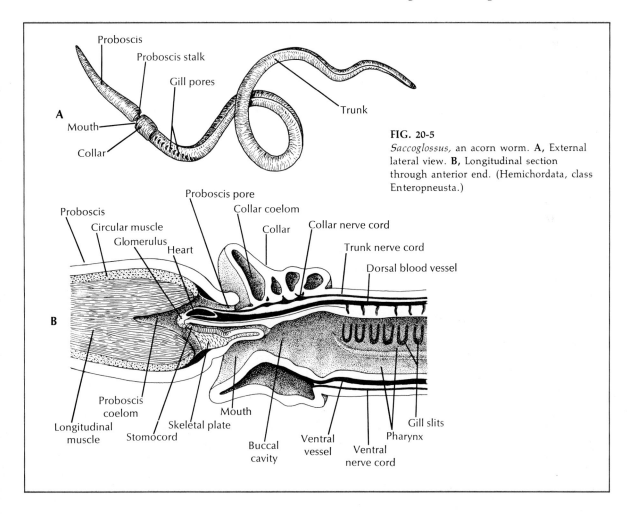

FIG. 20-5
Saccoglossus, an acorn worm. **A,** External lateral view. **B,** Longitudinal section through anterior end. (Hemichordata, class Enteropneusta.)

on its surface. These are carried by cilia to the groove at the edge of the collar, directed to the mouth on the underside, and swallowed. Large particles can be rejected by covering the mouth with the edge of the collar (Fig. 20-6).

Tube dwellers use the proboscis to excavate, thrusting it into the mud or sand and allowing cilia and mucus to move the sand backward. Or they may eat the sand and mud as they go, extracting from it its organic contents. They build U-shaped tubes with a back door and one or more front entrances. They can thrust the proboscis out the front for feeding. Defecation at the back door builds spiral fecal mounds that leave a telltale clue to the location of their tubes.

In the posterior end of the proboscis is a small coelomic sac (protocoel) into which extends the **stomochord**, a slender diverticulum of the gut that reaches forward into the buccal region and was formerly believed to be a notochord. A slender canal connects the protocoel with a **proboscis pore** to the outside (Fig. 20-5, *B*). The paired coelomic cavities in the collar also open by pores. By taking in water through the pores into the coelomic sacs the proboscis and collar can be stiffened to aid in burrowing. Contrac-

tion of the body musculature then forces the excess water out through the gill slits, reducing the hydrostatic pressure and allowing the animal to move forward.

BRANCHIAL SYSTEM. A row of **gill pores** are located dorsolaterally on each side of the trunk just back of the collar (Fig. 20-6, *A*). These open from a series of gill chambers that in turn connect with a series of **gill slits** in the sides of the pharynx. No gills are attached to the gill slits, but some respiratory gaseous exchange occurs in the vascular branchial epithelium, as well as in the body surface. Ciliary currents keep a fresh supply of water moving from the mouth through the pharynx and out the gill slits and branchial chambers to the outside.

FEEDING AND DIGESTION. Hemichordates are largely ciliary-mucus feeders. Back of the buccal cavity lies the large pharynx containing in its dorsal part the U-shaped gill slits (Fig. 20-5, *B*). Since there are no gills, it is assumed that the primary function of the branchial mechanism of the pharynx is food getting. Food particles caught in mucus and brought to the mouth by ciliary action on the proboscis and collar are strained out of the branchial water that leaves through the gill slits and are directed along the ven-

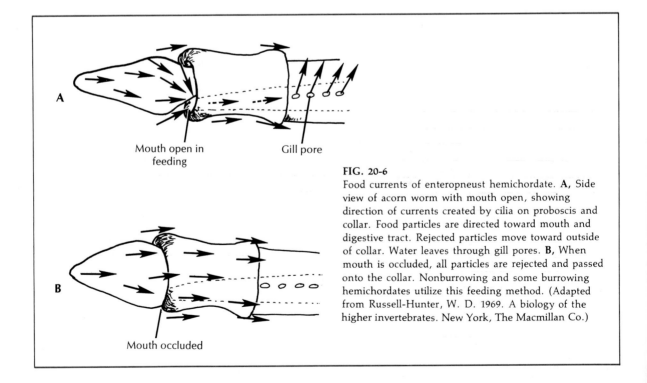

Mouth open in feeding

Gill pore

Mouth occluded

FIG. 20-6
Food currents of enteropneust hemichordate. **A,** Side view of acorn worm with mouth open, showing direction of currents created by cilia on proboscis and collar. Food particles are directed toward mouth and digestive tract. Rejected particles move toward outside of collar. Water leaves through gill pores. **B,** When mouth is occluded, all particles are rejected and passed onto the collar. Nonburrowing and some burrowing hemichordates utilize this feeding method. (Adapted from Russell-Hunter, W. D. 1969. A biology of the higher invertebrates. New York, The Macmillan Co.)

tral part of the pharynx and esophagus to the intestine where digestion and absorption occur (Fig. 20-6).

CIRCULATION AND EXCRETION. A middorsal vessel carries blood forward above the gut. In the collar the vessel expands into a sinus and a heart vesicle above the stomochord. Blood is then driven into a network of blood sinuses called the **glomerulus**, also above the stomochord, which is assumed to have an excretory function (Fig. 20-5, *B*). Blood travels posteriorly through a ventral vessel below the gut, passing through extensive sinuses to the gut and body wall. The blood is colorless.

NERVOUS AND SENSORY SYSTEM. The nervous system consists mostly of a subepithelial network, or plexus, of nerve cells and fibers to which processes of epithelial cells are attached. Thickenings of this net form dorsal and ventral nerve cords that are united posterior to the collar by a ring connective. The dorsal cord continues on into the collar and furnishes many fibers to the plexus of the proboscis. The collar cord contains giant nerve cells with processes running to the nerve trunks. This primitive nerve plexus system is highly reminiscent of that of the coelenterates and echinoderms.

Sensory receptors include neurosensory cells throughout the epidermis (especially in the proboscis, a preoral ciliary organ that may be chemoreceptive) and photoreceptor cells.

REPRODUCTION AND DEVELOPMENT. Sexes are separate in enteropneusts. Gonads are arranged in a dorsolateral row on each side of the anterior part of the trunk. Fertilization is external, and in some species a ciliated **tornaria** larva develops that at certain stages is so similar to the echinoderm bipinnaria that it was once believed to be an echinoderm larva. The familiar *Saccoglossus* of American waters does not have a tornaria stage.

Class Pterobranchia

The basic plan of this class is similar to that of the Enteropneusta, but certain structural differences are correlated with the sedentary mode of life of pterobranchs. The first pterobranch ever reported was obtained by the famed "Challenger" expedition of 1872-1876. Although first placed among the Polyzoa (Entoprocta and Ectoprocta), its affinities to the hemichordates were later recognized. Only two genera (*Cephalodiscus* and *Rhabdopleura*) are known in any detail.

Pterobranchs are small animals, usually within the range of 1 to 7 mm. in length, although the stalk may be longer. Many individuals of *Cephalodiscus* (Fig. 20-7) live together in gelatinous tubes, which

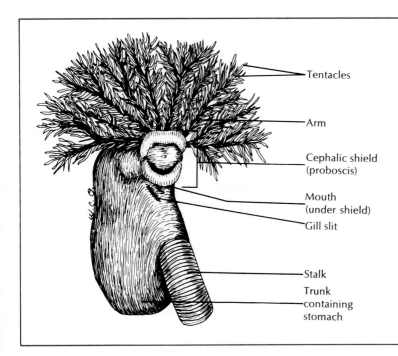

Tentacles

Arm

Cephalic shield
(proboscis)

Mouth
(under shield)

Gill slit

Stalk

Trunk
containing
stomach

FIG. 20-7
Cephalodiscus, a pterobranch hemichordate. These tiny (2 mm.) forms live in coenecium tubes in which they can move about. Ciliated tentacles and arms direct currents of food and water toward mouth. These deep-sea animals may be close to the ancestral stock of echinoderms and chordates.

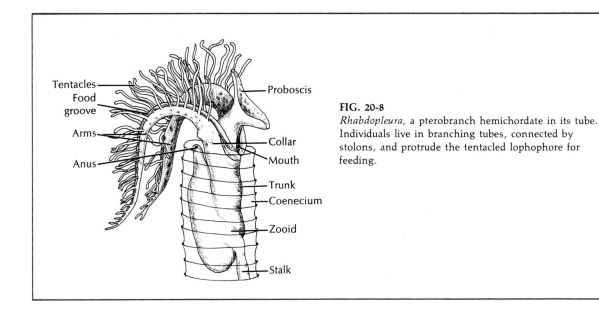

Tentacles
Food
groove
Arms
Anus
Proboscis
Collar
Mouth
Trunk
Coenecium
Zooid
Stalk

FIG. 20-8
Rhabdopleura, a pterobranch hemichordate in its tube. Individuals live in branching tubes, connected by stolons, and protrude the tentacled lophophore for feeding.

often form an anastomosing system. The zooids are not connected, however, and live independently in the tubes. Through apertures in these tubes, they extend their crown of tentacles. They are attached to the walls of the tubes by extensible stalks that can jerk the owners back when necessary.

The body of *Cephalodiscus* is divided into the three regions—proboscis, collar, and trunk—characteristic of the hemichordates. There is only one pair of gill slits, and the alimentary canal is U shaped, with the anus near the mouth. The proboscis is shield shaped. At the base of the proboscis there are five to nine pairs of branching arms with tentacles, somewhat similar to lophophores. Ciliated grooves on the tentacles and arms collect food. Some species are dioecious and others monoecious. Asexual reproduction by budding may also occur.

In *Rhabdopleura,* which is smaller than *Cephalodiscus,* the members remain together to form a colony of zooids connected by a stolon and enclosed in coenecium tubes (Fig. 20-8). The collar in these forms bears two branching arms or lophophores. No gill clefts or glomeruli are present. New individuals are reproduced by budding from a creeping basal stolon, which branches on a substratum. In none of the pterobranchs is there a tubular nerve cord in the collar, but otherwise their nervous system is similar to that of the Enteropneusta.

Atubaria, a little known genus, has no tube, but attaches its stalk to colonial hydroids.

The fossil graptolites of the middle Paleozoic era are often placed as an extinct class under Hemichordata. Their tubular chitinous skeleton and colonial habits indicate an affinity with *Rhabdopleura.* They are considered important index fossils of the Ordovician and Silurian geologic strata.

Derivation and meaning of names

Balanoglossus (Gr. *balanos,* acorn, + *glōssa,* tongue). Refers to the shape of the proboscis in this genus.

Saccoglossus (Gr. *sakkos,* sac, + *glōssa,* tongue). The most familiar genus of enteropneusts.

Sagitta (L. arrow). Refers to the shape of the body in this common genus of chaetognaths.

References

Barrington, E. 1965. The biology of Hemichordata and Protochordata. San Francisco, W. H. Freeman & Co., Publishers. *Concise account of behavior, physiology, and reproduction of hemichordates, urochordates, and cephalochordates.*

Bieri, R. 1959. The distribution of planktonic Chaetognatha in the Pacific and their relationship to the water masses. Limnol. Oceanogr. **4:**1-28.

Bullock, T. H. 1940. Functional organization of the nervous system of Enteropneusta. Biol. Bull. **79:**91-113.

Dahlgren, U. 1917. Production of light by Enteropneusta. J. Franklin Instit. **183:**735-754.

De Beer, G. 1955. The Pogonophora. Nature **176:**888. A *brief survey and appraisal.*

Edmondson, W. T. (editor). 1966. Marine biology III. New York, The New York Academy of Sciences.

Hartman, O. 1954. Pogonophora Johansson. Syst. Zool. **3:**183-185.

Hess, W. N. 1937. Nervous system of *Dolichoglossus kowalevskii.* J. Comp. Neurol. **68:**161-171.

Hyman, L. H. 1959. The invertebrates: smaller coelomate groups, vol. 5. New York, McGraw-Hill Book Co. *The best appraisal in English of this strange group.*

Ivanov, A. V. 1955. The main features of the organization of the Pogonophora. In Reports (doklady), Academy of Science, U.S.S.R., 100 (translated from Russian by A. Petrunkevitch in Syst. Zool. **4:**170).

Ivanov, A. V. 1963. Pogonophora (translated from Russian by D. B. Carlisle). New York, Consultants Bureau Enterprises, Inc. *The most recent appraisal of this relatively new group by the investigator who first studied the phylum in detail. Recently discovered species are also described.*

Kowalevsky, A. 1866. Anatomie des *Balanoglossus.* Mémoires de l'Académie impériale des sciences de St. Pé-

tersburg, ser. 7, vol. 10, no. 3. *A classic paper and the first accurate description of this group.*

Kuhl, W. 1938. Chaetognatha. In H. G. Bronn (editor). Klassen und Ordnungen des Tier-Reichs, vol. 4, sect. 4, part 1, book 2. Leipzig, Akademische Verlagsgesellschaft. *The best technical account.*

Manton, S. M. 1958. Embryology of Pogonophora and classification of animals. Nature (London) **181:**748-751.

Newell, G. E. 1951. The stomochord of Enteropneusta. Proc. Zool. Soc. **121:**741. *An appraisal of the status of the stomochord in comparison with a true notochord.*

Parry, D. A. 1944. Structure and function of the gut in *Spadella* and *Sagitta.* J. Marine Biol. Ass. U. K. **26:**16-36.

Van der Horst, C. J. 1932. Enteropneusta. In W. Kükenthal and T. Krumbach: Handbuch der Zoologie, vol. 3, part 2. *A detailed study of the phylum.*

Van der Horst, C. J. 1935. Hemichordata. In H. G. Bronn (editor). Klassen und Ordnungen des Tier-Reichs, vol. 4, part 4, book 2. Leipzig, Akademische Verlagsgesellschaft.

Yonge, C. M. 1928. Feeding mechanisms in the invertebrates. Biol. Rev. **3:**21-55. *An excellent paper.*

Two adult sea squirts. These bright orange, sessile, marine protochordates are ascidians, one of three classes of the subphylum Urochordata. Although adult ascidians are highly specialized and degenerative chordates, their free-swimming "tadpole" larvae bear all the right chordate hallmarks—notochord, gill slits, dorsal nerve cord, and postanal tail—and occupy an important position in theories of chordate ancestry.

CHAPTER 21

THE CHORDATES: Ancestry and evolution, general characteristics, protochordates

Position in animal kingdom

1. All available evidence indicates that chordates have evolved from the invertebrates, but it is impossible to establish the exact relationship.

2. Two possible lines of ancestry have been proposed in the phylogenetic background of the chordates. One of these is the annelid-arthropod-mollusk group; the other is the echinoderm-protochordate group.

3. The echinoderms as a group have certain characteristics that are shared with the chordates, such as **indeterminate cleavage, same type of mesoderm and coelom formation,** and **anus derivation from blastopore with mouth of secondary origin.** Thus the echinoderms appear to have a close kinship to the chordate phylum.

4. Taking the phylum as a whole, there is more fundamental unity of plan throughout all the organs and systems of this group than there is in any of the invertebrate phyla.

5. From gill filter-feeding ancestors to the highest vertebrates, the evolution of chordates has been guided by the specialized basic adaptation of the living endoskeleton, paired limbs, and nervous system.

Background

The great phylum chordata derives its name from one of the few common characteristics of this group —the **notochord** (Gr. *nōton,* back, + L. *chorda,* cord) (Fig. 21-1). This structure is possessed by all members of the phylum, either in the larval or embryonic stages or throughout life. The notochord is a rodlike, semirigid body of vacuolated cells, which extends, in most cases, the length of the body be-

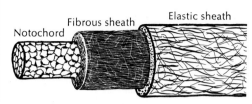

FIG. 21-1

Structure of notochord and its surrounding sheaths. Cells of notochord proper are thick walled, pressed together closely, and filled with semifluid. Stiffness caused mainly by turgidity of fluid-filled cells and surrounding connective tissue sheaths. This primitive type of endoskeleton is characteristic of all chordates at some stage of life cycle. Notochord provides longitudinal stiffening of main body axis, base for myomeric muscles, and axis around which vertebral column develops. In most vertebrates it is crowded out of existence. In man slight remnants are found in nuclei pulposi of intervertebral disks. Its method of formation is different in the various groups of animals. In amphioxus it originates from the endoderm; in birds and mammals it arises as an anterior outgrowth of the primitive streak.

tween the enteric canal and the central nervous system. Its primary purpose is to support and to stiffen the body, that is, to act as a skeletal axis.

The structural plan of chordates retains many of the features of invertebrate animals, such as bilateral symmetry, anteroposterior axis, coelom tube-within-a-tube arrangement, metamerism, and cephalization.

The animals most familiar to the student belong to the chordates. Man himself is a member and shares the common characteristics of this group. Ecologically the phylum has been very successful in the animal kingdom. They are among the most adaptable of organic forms and are able to occupy most kinds of habitat. From a purely biologic viewpoint, chordates are of primary interest because they illustrate so well the broad biologic principles of evolution, development, and relationship. They represent as a group the background of man himself.

Characteristics of chordates

1. Bilateral symmetry; segmented body; three germ layers; coelom well developed

2. **Notochord** (a skeletal rod) present at some stage in life cycle
3. **Nerve cord dorsal and tubular;** anterior end of cord usually enlarged to form brain
4. **Pharyngeal gill slits present at some stage in life cycle** and may or may not be functional
5. A **postanal tail** usually projecting beyond the anus at some stage and may or may not persist
6. **Heart ventral,** with dorsal and ventral blood vessels; closed blood system
7. Complete digestive system
8. Exoskeleton often present; well developed in some vertebrates
9. A cartilage or bony **endoskeleton** present in the majority of members (vertebrates)

The four distinctive characteristics that set chordates apart from all other phyla are the **notochord, dorsal tubular nerve cord, pharyngeal gill slits, and postanal tail.** These characteristics are always found in the early embryo, although they may be altered or may disappear altogether in later stages of the life cycle.

These four features are so important that each merits a short description of its own.

NOTOCHORD. This rodlike body develops in the embryo as a longitudinal outfolding of the dorsal side of the alimentary canal. It is endodermal in origin, although in some forms there is a possibility that the other germ layers have contributed to its formation. In most it is a rigid, yet flexible, rod extending the length of the body. It is the first part of the endoskeleton to appear in the embryo. As a rigid axis on which the muscles can act, it permits undulatory movements of the body. In most of the protochordates and in primitive vertebrates the notochord persists throughout life. In all vertebrates a series of cartilaginous or bony vertebrae are formed from the connective tissue sheath around the notochord and replace it as the chief mechanical axis of the body.

DORSAL TUBULAR NERVE CORD. In the invertebrate phyla the nerve cord (often paired) is ventral to the alimentary canal and is solid, but in the chordates the cord is dorsal to the alimentary canal and is formed as a tube. The anterior end of this tube in vertebrates becomes enlarged to form the brain. The hollow cord is produced by the infolding of ectodermal cells on the dorsal side of the body above the notochord. Among the vertebrates the nerve cord lies in the neural arches of the vertebrae, and the anterior brain is surrounded by a bony or cartilag-inous cranium.

PHARYNGEAL GILL SLITS. Pharyngeal gill slits are perforated slitlike openings that lead from the pharyngeal cavity to the outside. They are formed by the invagination of the outside ectoderm and the evagination of the endodermal lining of the pharynx. The two pockets break through when they meet, to form the slit. In higher vertebrates these pockets may not break through and only grooves are formed instead of slits; all traces of them usually disappear. In forms that use the slits for breathing, gills with blood vessels are attached to the margins of the slits and make the gaseous exchange with the water that enters the mouth and passes through the pharyngeal gill slits. The slits have in their walls supporting frameworks of gill bars. Primitive forms such as am-phioxus have a large number of slits, but only six or seven are the rule in the fish. The transitory ap-pearance of the slits in land vertebrates is often used as evidence for evolution.

POSTANAL TAIL. This chordate innovation, to-gether with somatic musculature and the stiffening notochord, provides the motility that larval tunicates and amphioxus need for their free-swimming exis-tence. As a structure added to the body behind the end of the digestive tract, it clearly has evolved specif-ically for propulsion in water. Its efficiency is later increased in fishes with the addition of fins.

What advancements do chordata show over other phyla?

One distinguishing characteristic of the chordates is the **endoskeleton,** which, as we have seen, is first found in the echinoderms. An endoskeleton is an in-ternal structure that provides support and serves as a framework for the body. Most chordates possess two types of endoskeletons in their life cycle. The first is the **notochord** (Fig. 21-1), possessed at some stage by all chordates. The second is the **vertebral column** and accessory structures such as the appen-dages. This second type of endoskeleton, which is more specialized and more adaptable for evolutionary growth, is possessed by only part, although the greater part, of the chordate phylum.

Even though the chordates have emphasized an endoskeleton, they have by no means cast aside the exoskeleton of the invertebrates. Many of the higher chordates (that is, the vertebrates) have keratinoid exoskeletons, although here the exoskeleton is main-ly for protection and not for attachment of muscles. Another noticeable distinction is that the endoskele-ton is a living tissue, whereas the exoskeleton is composed of dead noncellular material. The endo-skeleton has the advantage of allowing continuous growth, without the necessity of shedding. For this reason vertebrate animals can attain great size; some of them are the most massive in the animal king-dom. Endoskeletons provide much surface for muscle attachment, and size differences between animals re-sult mainly from the amount of muscle tissue they possess. More muscle tissue necessitates greater de-velopment of body systems, such as circulatory, di-gestive, respiratory, and excretory. Thus it is seen that the endoskeleton is the chief basic factor in the development and specialization of the higher ani-mals.

From an evolutionary viewpoint the function of the skeleton, as represented by the exoskeleton and the endoskeleton, has shifted more from a protective to a supportive one. The limy shells of clams and other mollusks and the chitinous armor of arthropods are excellent defensive armors, even though they also serve for attachment of muscles and support of bodi-ly structures. However, endoskeletons have their protective functions as well as supporting ones, as revealed by such excellent protective boxes as the cranium for the brain and the thorax for important visceral organs.

Another advantage chordates have is their method of **breathing.** An efficient respiratory system has gone hand in hand with a well-developed circulatory system. Certain arthropods, with their direct tracheal system, have evolved a very efficient respiration, but such a plan is fitted to animals of small size only. In either the gills of aquatic forms or the lungs of terrestrial forms, the blood circulates freely through the respiratory organs, ensuring rapid and efficient exchange of gases. Moreover, the blood system of chordates is admirably fitted to carry on so many functions that it has become a general factotum for bodily functions.

No single system in the body is more correlated with functional and structural advancement than is the **nervous system.** Throughout the invertebrate kingdom we have seen that there has been a more or less centralization of nervous systems. This ten-dency has reached its climax in the higher chordates in which we find the highly efficient tubular nervous system. Such a system allows the greatest possible utilization of space for the nervous units so necessary for well-integrated nervous patterns. Along with the advanced nervous system goes a better sensory system, which partly explains the power of this group to adapt itself to a varied environment.

■ ANCESTRY AND EVOLUTION

Since the early nineteenth century when the theory of organic evolution became the focal point for ferreting out relationships between groups of living organisms, zoologists have debated the question of vertebrate origins. Nearly every major invertebrate group has at one time or another been advanced as candidate for the vertebrate ancestral group, and all for one reason or another have failed to meet the qualification requirements for undisputed election. The great chasm between vertebrates and invertebrates seems nearly as wide today as it did 150 years ago when zoologists began serious speculation on the issue. The most serious obstacle to progress is the almost total absence of fossil material for reconstructing lines of descent. The earliest protovertebrates were in all probability soft-bodied forms that stood little chance of being preserved even under the most ideal conditions. Consequently, speculations must come from the study of living organisms, especially from an analysis of how they develop. But well over 600 million years have elapsed since the vertebrates had their beginnings, and even the highly conservative early developmental stages of animals cannot be expected to remain unchanged over such an enormous span of time. Nevertheless zoologists continue the attempt to reconstruct vertebrate origins by careful study of the available data, perhaps because, as one biologist puts it, it is an "interesting and enjoyable venture to speculate concerning the Cambrian and Precambrian happenings that have led to my own existence."*

EARLY THEORIES OF CHORDATE EVOLUTION. The earliest speculations about vertebrate origins understandably focused on the most successful, and in may respects most advanced, of invertebrate groups, the Arthropoda. It was quickly recognized that if one took an arthropod with its segmented

*Berrill, N. J. 1955. The origins of vertebrates. Oxford, Oxford University Press.

body, ventral nerve cord and dorsal heart and turned it over, one has the basic plan of a vertebrate. The idea was first proposed by St. Hilaire in 1818 and received detailed support by subsequent zoologists. Later, in 1875 Semper and Dohrn independently transferred the ancestral award to the Annelida, because this group shares a basic body plan with the arthropods and in addition has an excretory system that strikingly resembles that of primitive vertebrates. The annelid-vertebrate theory continued to receive support as late as 1922 when the Dutch biologist Delsman published his elaborate arguments in a book. The principal difficulty with the annelid theory was that an inverted annelid has its brain and mouth in the wrong relative positions. The annelid's nerve cord is ventral but connects to a dorsal brain via circumpharyngeal connectives through which the digestive tube passes (Fig. 16-4, *A*). When an annelid is inverted, as was required for the annelid-vertebrate theory, the mouth ends up on top of the head, and the brain below. Most proponents of the theory devoted their efforts to explaining away this discrepancy. The annelid theory was eventually discarded like the arthropod theory before it, when zoologists realized that they must base their speculations on developmental patterns of animals rather than on adult forms, which are the highly differentiated end products of development.

ECHINODERM THEORY. When, early in this century, further theorizing became rooted in developmental patterns of animals, it immediately became apparent that only the echinoderms deserved serious consideration as the vertebrate's ancestor. Echinoderms and chordates belong to the deuterostome branch of the animal kingdom in which the mouth is formed as a secondary opening and the blastopore of the gastrula becomes the anus. The coelom of these two phyla is primitively enterocoelous: it is budded off from the archenteron of the embryo. Furthermore, there is a great resemblance between the bipinnaria larvae of certain echinoderms and tornaria larvae of the hemichordates, a phylum bearing some chordate characteristics. Both have similar ciliated bands in loops, sensory cilia at the anterior end, and a complete digestive system of ventral mouth and posterior anus. Both echinoderm and chordate embryos show indeterminate cleavage; that is, each of the early blastomeres has equivalent potentiality for supporting full development of a complete embryo.

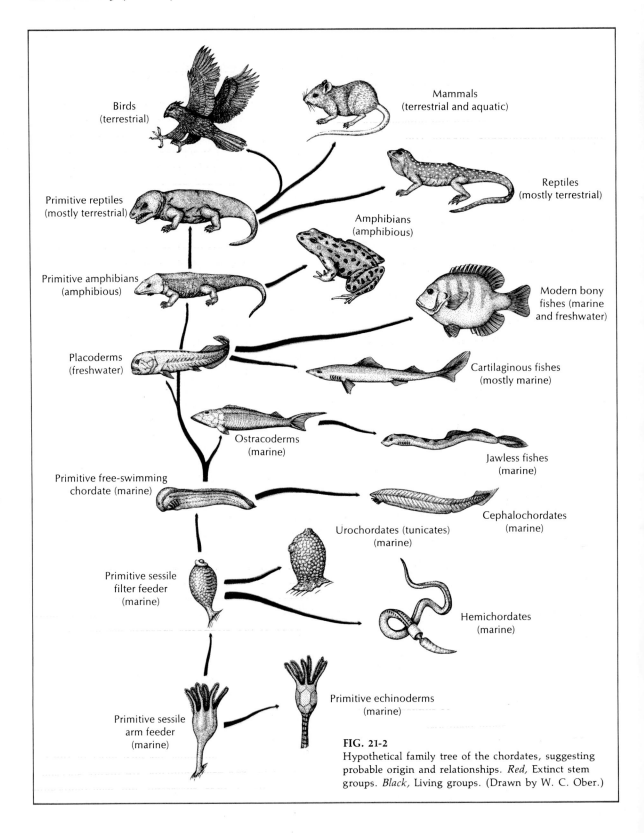

Birds
(terrestrial)

Mammals
(terrestrial and aquatic)

Primitive reptiles
(mostly terrestrial)

Reptiles
(mostly terrestrial)

Amphibians
(amphibious)

Primitive amphibians
(amphibious)

Modern bony
fishes (marine
and freshwater)

Placoderms
(freshwater)

Cartilaginous fishes
(mostly marine)

Ostracoderms
(marine)

Jawless fishes
(marine)

Primitive free-swimming
chordate (marine)

Cephalochordates
(marine)

Urochordates (tunicates)
(marine)

Primitive sessile
filter feeder
(marine)

Hemichordates
(marine)

Primitive echinoderms
(marine)

Primitive sessile
arm feeder
(marine)

FIG. 21-2
Hypothetical family tree of the chordates, suggesting
probable origin and relationships. *Red,* Extinct stem
groups. *Black,* Living groups. (Drawn by W. C. Ober.)

These characteristics are shared by brachiopods and pterobranchs (a hemichordate group) as well as by echinoderms, protochordates, amphioxus, and vertebrates. This is probably a natural grouping and almost certainly indicates interrelationships, although remote.

Biologists once believed that the phylogenetic distribution of phosphagens, organic compounds that act as energy reserves in muscle contraction, indicated an affinity between echinoderms and chordates. In 1931 Needham and others in England reported that most invertebrate muscle contains one type of phosphagen, arginine phosphate, whereas vertebrates have another, creatine phosphate. The protochordates and the echinoderms contained both, suggesting that of all invertebrates, these two groups alone showed a relationship to vertebrates. Much more recently French scientists reinvestigated the issue using chromatographic methods and more specific chemical tests than those available to Needham. Their findings that both kinds of phosphagens have a wide phylogenetic distribution among invertebrates made it clear that these compounds are of no value in suggesting animal relationships.

Efforts also have been made to relate primitive fossil echinoderms (stem-bearing forms similar to sea lilies today) with the vertebrate ostracoderms, a group of primitive, jawless fishes. Both groups have excellent fossil records and both coexisted during the Ordovician period, some 500 million years ago. But this approach again is an attempt to pin together affinities by looking at differentiated adult forms, and is the same flawed approach that misled earlier zoologists to relate annelids and vertebrates.

However the absence of biochemical support and the uncertainty of fossil evidence in no way invalidates the embryologic evidence for a remote kinship between echinoderms and chordates. It remains the best guess we have (Fig. 21-2). At the same time, we must acknowledge that there is no need to designate *any* invertebrate group as chordate ancestral stock. The vertebrates may have arisen independently from a simple unicellular protist stage, as the Canadian biologist J. R. Nursall has suggested, with all intermediate types having been lost.

Unable to narrow the search for a prechordate-chordate connecting link any further, zoologists are presently focusing on groups within the chordate phylum itself, which share enough anatomic and developmental features to make unraveling the evolutionary past a more profitable effort.

Candidates for vertebrate ancestral stock

All members of the phylum Chordata share four anatomic features at some time in their life histories: a notochord, forming a stiff mechanical axis for the body; a dorsal tubular nerve cord; pharyngeal gill slits used for breathing and feeding; and a postanal tail for larval motility. The three subphyla that bear these characteristics are the Cephalochordata (of which the lancelet amphioxus is its famous representative), the Urochordata (tunicates or sea squirts), and the Vertebrata. The vertebrates comprise by far the greatest number of the chordates so that the terms "vertebrate" and "chordate" are frequently used (somewhat incorrectly) as synonyms. The backboneless members of the phylum—cephalochordates and urochordates—are usually referred to as protochordates or prevertebrates and have long been considered good candidates for the vertebrate ancestral stock. The Hemichordata, described in the preceding chapter, previously were also assigned to the phylum Chordata since they possess a perforated pharynx (gill slits) and a middorsal thickening of nervous tissue, which may be a forerunner of the vertebrate tubular nerve cord. However the stomochord of the hemichordates (see Fig. 20-5) cannot be homologized to the notochord as was once believed. Furthermore the gill slits and dorsal nerve center may represent a convergent resemblance to similar chordate structures rather than a true homology. Consequently the hemichordates have been excluded from chordate membership although most biologists admit that a real, though remote, relationship exists. The other group formerly included with the chordates are the pogonophores. This is a highly aberrant and specialized group that possesses almost none of the diagnostic chordate features. Yet they are deuterostomes and their tripartite body organization and certain other characteristics indicates a relationship to the hemichordates. If we are to admit a remote affinity between hemichordates and chordates, we must do the same for the pogonophores.

POSITION OF AMPHIOXUS. The problems of sorting out lines of descent from this assemblage of candidates are enormous because most of the prechordates and protochordates are highly specialized remnants of once successful and well-represented groups. The first approach was to place the chordate and prechordate groups in an order of increasing morphologic complexity leading to the vertebrates. By this analysis, amphioxus becomes the logical structural ancestor because it possesses as an adult all four chordate characteristics mentioned above plus several vertebrate hallmarks: segmented musculature, the beginning of optic and olfactory sense organs, a liver diverticulum, beginnings of a ventral heart and separation of dorsal and ventral spinal roots in the vertebrate style (Figs. 21-8 and 21-9). Little wonder that amphioxus once attained a pinnacled position among zoologists searching for their vertebrate ancestor. But upon closer scrutiny, amphioxus failed to meet the qualifications for generalized ancestral type. Its notochord is overdeveloped into a forward extension for its specialized burrowing mode of life and it effectively prevents the development of a proper brain. Its kidney is a solenocyte type that bears little resemblance to vertebrate glomerular-tubular nephron. Its unique atrium has no vertebrate counterpart and there is a non–vertebrate-like proliferation of gill slits. Amphioxus today is usually regarded as a highly specialized and degenerative member of the chordate family: it lies as an offshoot, rather than in the main line, of chordate descent.

UROCHORDATA AND RECAPITULATION. Attention then became focused on the alternative protochordate group, the Urochordata (tunicates). This group is composed of three groups of which the ascidians (sea squirts) are the commonest and simplest. At first glance, more unlikely candidates for vertebrate ancestor could hardly be imagined. As adults, ascidians are virtually immobile forms surrounded by a tough, cellulose-containing tunic of variable color. Their adult life is spent in one spot attached to some submarine surface, filtering vast amounts of seawater from which they extract their planktonic food. As adults they lack notochord, tubular nerve cord, postanal tail, sense organs, and segmental musculature. Superficially they resemble sponges far more than they resemble any known vertebrate. Yet the chordate nature of ascidians is abundantly evident in their tadpole larvae. These tiny, active, site-seeking forms have all the right qualifications for membership in the prevertebrate club: notochord, hollow dorsal nerve cord, gill slits, postanal tail, brain, and sense organs (otolith balance organ and an eye com-

plete with lens). The discovery of this form in 1869 not only placed the urochordates squarely in the vertebrate camp but greatly influenced E. Haeckel in formulating his theory of recapitulation (biogenetic law; see Principle 26 on p. 21). According to this theory, adult stages of ancestors are repeated during the development of their descendants; in other words the development of an organism is an accurate record of past evolutionary history. We recognize now that this record is very slurred and telescoped and must be interpreted with caution. But at the time the true nature of the ascidian tadpole larva was first understood, it was considered to be a relic of an ancient free-swimming chordate ancestor of the ascidians. Adult ascidians then came to be regarded as degenerate, sessile descendants of the ancient chordate form.

GARSTANG'S THEORY OF CHORDATE LARVAL EVOLUTION. It remained for W. Garstang in England (1928) to introduce totally fresh thinking to the vertebrate ancestor debate. In effect, Garstang turned the sequence around: rather than the ancestral tadpole larva giving rise to a degenerative sessile ascidian adult, he suggested that the ancestral chordate stock was primarily a filter-feeding sessile marine group not unlike modern ascidians. The free-swimming tadpole larva was evolved from these attached, bottom-dwelling forms to meet the need for seeking out new habitats. Thus the tadpole larva was visualized as an ascidian creation, evolved within the group to enhance site-seeking capabilities. Garstang next suggested that at some point the tadpole larva became neotenous, that is, became capable of maturing gonads and reproducing in the larval stage. With continued larval evolution, a new group of free-swimming animals would appear. The best evidences for this theory are found in the living tunicates today, especially among the two planktonic groups, the thaliaceans and the larvaceans. In the latter group, the basic larval form is retained throughout life; they are in effect neotenous tunicates, although extremely specialized.

Garstang departed from previous thinking by suggesting evolution may occur in the larval stages of animals. This idea received slow acceptance by zoologists accustomed to thinking of developmental stages as being largely insulated from change, as embodied in the "biogenetic law." Yet, in all likelihood an evolutionary sequence similar to that

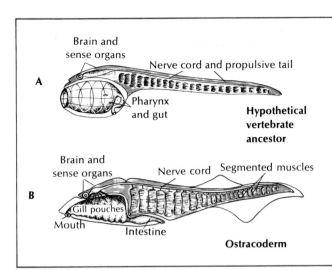

A

Brain and
sense organs

Nerve cord and propulsive tail

Pharynx
and gut

**Hypothetical
vertebrate
ancestor**

B

Brain and
sense organs

Nerve cord   Segmented muscles

Gill pouches

Mouth   Intestine

Ostracoderm

FIG. 21-3
Hypothetical vertebrate ancestor compared with an ostracoderm, earliest fossil vertebrate. According to the Garstang hypothesis, the common vertebrate ancestor was an ascidian tadpole larva that became neotenous and evolved into a new free-swimming species. Continued evolution led to the jawless ostracoderms and jawed placoderms. (Redrawn from Roe, A., and Simpson, G. G. 1958. Behavior and evolution. New Haven, Yale University Press.)

proposed by Garstang occurred. Garstang's theory has received detailed support from the Canadian zoologist N. J. Berrill, who however differs with Garstang in the way the evolutionary sequence proceeded.

The ascidian tadpole larva with its propulsive tail, stiffening notochord, and dorsal nerve cord that integrates sensory information and motor activity clearly suggests and foreshadows the vertebrate line. The resemblance of an early vertebrate ostracoderm to this hypothetical ancestor is suggested in Fig. 21-3.

Early habitat of the vertebrates

Until the beginning of this century it was assumed that the vertebrates had their origin in the sea. Then the discovery in Colorado of some vertebrate bony fragments led to a debate over the habitat of these earliest vertebrates that has taken more than 50 years to resolve. Geologists considered these particular fossils to be those of freshwater animals that had been swept downstream and deposited at river mouths. This interpretation has been vigorously supported by the American paleontologist A. S. Romer and was received with interest by zoologists who soon advanced several theories to explain certain vertebrate features as the outcome of a freshwater origin for the group.

If the early vertebrates lived in fresh water, they must have anatomic and physiologic defenses against the large osmotic gradient that existed between their salty body fluids inside and the virtually salt-free fresh water outside. The American kidney physiologist

Homer Smith argued that the heavy bony plates of the ostracoderm fishes (oldest fossil vertebrates yet discovered) developed as a barrier to the osmotic invasion of fresh water across the skin. He suggested further that the particular kind of kidney found in vertebrates evolved as a water excretory pump in the earliest vertebrates whose habitat was fresh water. Vertebrates have a filtration kidney composed of thousands of independent excretory units (nephrons), each of which begins with a tiny tuft of capillaries, the glomerulus, that functions as a filtration unit. The vertebrate kidney is ideally designed for excreting excess water that will osmotically enter a freshwater animal. Homer Smith's arguments offered much appeal and supported the claims of many paleontologists that their vertebrate fossil finds were of freshwater origin.

The theory, however, contained serious flaws. All freshwater animals must contend with the problem of osmotic invasion of water across the skin. Yet freshwater fishes today clothed with nothing more than a flexible coat of small scales and mucus have an extremely low permeability to water. Even if the ostracoderms lived in freshwater, it seems more likely that their heavy coat of dermal armor provided protection from contemporary invertebrate predators. Furthermore, as the English physiologist J. D. Robertson pointed out in 1957, the filtration type of kidney is neither uniquely vertebrate nor exclusively for freshwater. Kidneys of this basic functional pattern are widely distributed in nature: they are found in many arthropods and mollusks, as well as in the hagfishes, a group of primitive vertebrates that have had an exclusively marine history. These animals are isotonic in seawater and none have need of excreting large volumes of water. We must assume that the glomerular filtration unit evolved independently in different animal phyla because it is a very convenient way to put materials into a tubule where they can be selectively modified by downstream tubular activities. It is a highly flexible and successful excretory process in any environment. It is true that the vertebrate filtration kidney is an excellent water pump, and its presence in the early vertebrates would have been an important preadaptation when they did set about conquering the freshwater habitat. Robertson's views were supported by R. H. Denison, an American paleontologist who almost simultaneously (1956) published an exhaustive reevaluation of the fossil evidence. Denison showed that the great bulk of early vertebrate fossils are clearly and indisputably marine. Thus the habitat of the early vertebrates has been firmly returned to the sea, where it rested before this unintentionally mischievous controversy arose.

Jawless ostracoderms—earliest vertebrates

The earliest vertebrate fossils are fragments of bony armor discovered in Ordovician rock in Russia and in the United States. They were small, jawless crea- tures collectively called **ostracoderms** (os-trak'o-derm; Gr. *ostrakon,* shell, + *derma,* skin), which belong to the Cyclostomata division of the verte- brates. These earliest ostracoderms, called **heteros- tracans,** lacked paired lateral fins that subsequent fishes found so important for stability. Their swim- ming movements must have been inefficient and clumsy, although sufficient to propel them from one mud bed to another where they practiced their mud- grubbing search for nutrients in the form of organic debris on the sea bottom. Later in the Silurian and especially in the Devonian the heterostracan ostraco- derms underwent a major radiation, resulting in the appearance of several peculiar-looking forms varying in shape and length of the snout, dorsal spines, and dermal plates. Most continued their mud-loving exis- tence although at least one species became a surface feeder. Without ever evolving paired fins or jaws, the heterostracans dominated the early Devonian until eclipsed by another ostracoderm group, the **cepha- laspids.**

The cephalaspids improved the efficiency of a benthic life by evolving paired fins. These fins, locat- ed just behind the head shield, provided control over pitch and yaw that ensured well-directed forward movement. The best known genus in this group is *Cephalaspis* (Fig. 21-4), the subject of a brilliant and

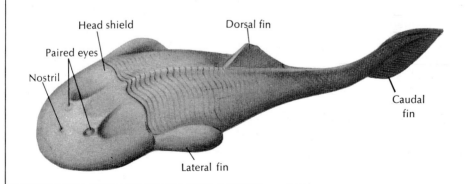

FIG. 21-4

Dorsolateral view of ostracoderm. This group represents oldest fossil vertebrate group yet discovered. They rarely exceeded 1 foot in length and were related to living cyclostomes (lampreys and hagfish). Like larval forms of living cyclostomes, ostracoderms were filter feeders. Many fine fossil specimens have been found in many parts of the world. Outstanding work of Stensiö has revealed amazing amount of detail about these primitive jawless vertebrates, most typical being genus *Cephalaspis.* They were all covered by bony armor, and some may have been provided with internal bony skeleton. They flourished from Ordovician to Devonian times, becoming extinct about 300 million years ago.

Head shield

Dorsal fin

Paired eyes

Nostril

Caudal fin

Lateral fin

classic series of studies by the Swedish paleozoologist E. A. Stensiö. *Cephalaspis* was a small animal, seldom exceeding 1 foot in length, and was covered by a well-developed armor, the head by a solid shield (rounded anteriorly), and the body by bony plates. It had no axial skeleton or vertebrae. The mouth was ventral and anterior, and it was jawless and toothless. Its paired eyes were located close to the middorsal line. A pineal eye was also present, in front of which was a single nasal opening. At the lateroposterior corners of the head shield were a pair of flaplike fins. The trunk and tail appeared to be adapted for active swimming. Between the margin of the head shield and the ventral plates there were ten gill openings on each side. They also had a lateral line system. They were adapted for filter feeding, which may explain the large expanded size of the head, made up as it is of a large pharyngeal gill-slit filtering apparatus.

A third major group of ostracoderms, the **anaspids,** were clearly related to the cephalaspids although they looked very different. Like the cephalaspids, they had paired fins, but these were peculiar, elongate, posterior stabilizers that formed a part of the caudal (tail) fin. The anaspids were much more streamlined than other ostracoderms and are believed to have practiced an active mode of life. Some paleontologists believe

that one of the two groups of living cyclostomes, the lampreys, may have evolved from the anaspids.

As a group, the ostracoderms were basically fitted for a simple, bottom-feeding life. Yet despite their anatomic limitations, they enjoyed a respectable radiation in the Silurian and Devonian periods. Their overall contribution was enormous, for they provided a blueprint for subsequent vertebrate evolution. But they could not survive the competition of the more advanced jawed fishes that began to dominate the Devonian, and in the end they disappeared.

Early jawed vertebrates

All jawed vertebrates, whether extinct or living, are collectively called gnathostomes ("jaw mouth") in contrast to jawless vertebrates, the cyclostomes ("circle mouth"). The latter are also often referred to as agnatha. The first jawed vertebrates to appear in the fossil record were the **placoderms** (plak'o-derm; Gr. *plax,* plate, + *derma,* skin). The advantages of jaws are obvious, since they allowed predation on large and active forms of food. The possessors of jaws would enjoy a tremendous advantage over jawless vertebrates, which were restricted to a wormlike existence of sifting out organic debris and small organisms in the bottom mud. Jaws arose through modifications of the first two of the serially-repeated carti-

FIG. 21-5
Climatus—one of first fish with jaws. This creature belonged to group called "acanthodians." One of their interesting characteristics was smaller paired fins between larger pectoral and pelvic fins. These smaller fins may have evolved from lateral fin folds, or they may have arisen independently. Acanthodians arose in early Devonian period and must have played important part in early evolution of vertebrates before they became extinct. (Modified from Romer, A. S. 1966. Vertebrate paleontology. Chicago, University of Chicago Press.)

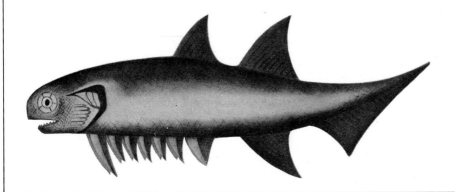

laginous gill arches. The beginnings of this trend can, in fact, be seen in some of the jawless ostracoderms where the mouth became bordered by strong dermal plates that could be manipulated somewhat as jaws with the gill arch musculature. The more anterior arches continued a gradual modification to permit more efficient seizing, and the skin surrounding the mouth was modified into teeth. Eventually the anterior gill arches became bent into the characteristic position of vertebrate jaws, as seen in the placoderms.

The early jawed fishes had well-developed lateral fins. These may have arisen as lateral folds in the body wall, which broke up into a series of fins, such as is found in the **acanthodians,** a primitive group contemporary with the placoderms (Fig. 21-5). In these, however, the anterior and posterior pairs were larger than the others and eventually became the pectoral and pelvic fins; the intermediate ones disappeared. Some paleontologists believe, however, that these numerous paired fins may have arisen independently as separate structures. Their internal skeleton so far as is known was composed of bone.

Placoderms evolved into a great variety of forms, some aberrant and grotesque in appearance. Some were quite large. They were armored fish covered with diamond-shaped scales or with large plates of bone. All became extinct by the end of the Paleozoic era.

Evolution of modern fishes and tetrapods

Reconstruction of the origins of the vast and varied assemblage of modern living vertebrates is, as we have seen, based in large part on fossil evidence. When a paleozoologist sits down with an array of fossil forms before him, many of them represented by more or less incomplete exoskeletal remains, he immediately encounters a conflict in his attempt to prove or disprove relationships between two distinct fossil groups. He sees on the one hand a number of structural features that are shared by both groups: this suggests relationship. He also sees features that are widely different for the two groups: this suggests early separation of lines of evolution. His task then is to decide whether shared features are **primitive** ones, that is, features that have been inherited from a common ancestor, or whether shared features have arisen independently because of similar habitats or some kind of retrogressive development. As for widely differing characteristics, he must decide whether these are latter-day specializations of little consequence or whether they represent fundamental and ancient departures in evolutionary trends. His problems obviously are complex, but presumably they are what keep paleontologists excited with their chosen profession. The outcome is that the paleontologist usually can only suggest group memberships; he seldom can make the decisive, indisputable judgments that students (and zoology professors too, for that matter!), who are not especially interested in the details, would like to have.

Modern textbooks, this one included, are fond of presenting animal relationships in the form of "family trees." The family tree often shows that group A gave rise to group B, group B proceeded to C, group C to D, and so on. Convenient as such presentations are, they are often misleading and sometimes downright wrong. The fact is that the earliest vertebrates tell us much less than we would like about subsequent trends in evolution. It is true that affinities become much easier to establish as the fossil record improves in more recent geologic times. For instance, the descent of birds and mammals from reptilian ancestors has been worked out in a highly convincing manner from the relatively abundant fossil record available. By contrast, the ancestry of modern fishes is shrouded in uncertainty. The Swedish paleontologist E. Jarvik has emphasized that the main vertebrate stem groups (such as cyclostomes, lungfishes, sharks, bony fishes, and stem tetrapods) became anatomically specialized some 400 to 500 million years ago and have changed relatively little since then. Thus main evolutionary lines as seen in the fossil record run back almost in parallel; if extended backwards to their illogical extreme, they would hardly ever meet. Obviously they must meet at some point in the distant past, but this exercise reveals that the crucial separations in vertebrate evolution occurred in the Cambrian, perhaps even the Precambrian, long before the fossil record became established for the convenience of paleozoologists.[*]

In the face of this, it would be a mistake for the student to conclude that all is chaos, for it is not. The vertebrates are clearly a natural, monophyletic group, distinguished by a great number of common characters. They have almost certainly descended

[*]Jarvik, E. 1967. Aspects of vertebrate phylogeny. In T. Ørvig (editor). Nobel Symposium 4, Stockholm, Almqvist & Wiksell.

from a common ancestor, the nature of which we have already discussed. Very early in their evolution, the vertebrates divided into two great stems, the cyclostomes and the gnathostomes. These two groups differ from each other in many fundamental ways, in addition to the obvious lack of jaws in the former group and their presence in the latter. Thus both groups are very old and of about the same age. On this basis we cannot say that cyclostomes are move "primitive" than gnathostomes even though the latter have continued on a marvelous evolutionary advance that produced most of the modern fishes, all of the tetrapods, and the reader of this book. Although the cyclostomes are represented today only by the hagfishes and the lampreys, these creatures too are successful in their own way.

In the chapters that follow, we will begin each with a summary of that group's origins and relationships within the group. The student is encouraged to make frequent reference to the geologic time scale on the inside back cover of this book. We have tried to base conclusions about group histories on recent paleontologic opinion. Although we recognize that family trees can give false impressions, we have nonetheless used them in the absence of a better alternative. They may be viewed as educated speculations and as such with a certain measure of skepticism; at the same time they are not science fiction! They are derived from close morphologic reasoning and a thorough understanding of general biologic principles by scientists who have devoted their lives to this form of detective work. Evolution, with its idea of life transforming itself through the ages, is supported by a vast wealth of fossil and living evidence. It is, after all, the framework of biology.

Summary of biologic contributions

1. A **living endoskeleton** is characteristic of the entire phylum. Two endoskeletons are present in the group as a whole. One of these is the rodlike **notochord,** which is present in all members of the phylum at some time; the other is the **vertebral column,** which largely replaces the notochord in higher chordates.

2. The endoskeleton does not interfere with **continuous growth,** for it can increase in size with the rest of the body. There is, therefore, no necessity for shedding it, as is the case with the nonliving exoskeleton of the invertebrate phyla. Moreover, the endoskeleton allows for almost indefinite growth so that many chordates are the largest of all animals.

3. The nature of the endoskeleton is such that it affords much surface for muscular attachment, and since the muscular and skeletal systems make up most of the bulk of animals, other bodily systems must also become specialized in both size and function to meet the metabolic requirements of these two great systems.

4. The endoskeleton in higher chordates consists of the **axial** and **appendicular** divisions. The axial is made up of the cranium, vertebral column, ribs, and sternum; the appendicular is composed of the pectoral and pelvic girdles and the skeleton of the appendages.

5. A **postanal** tail is a new additon to the animal kingdom and is present at some stage in most chordates.

6. A **ventral heart** is a new characteristic, and a closed blood system is better developed than it is in other phyla. Chordates have also developed a **hepatic portal system,** which is specialized for conveying food-laden blood from the digestive system to the liver.

7. **Perforated pharynx** (gill slits) are introduced for the first time. Gill slits or traces are present in the embryos of all chordates. Terrestrial chordates have developed lungs by modification of this same pharyngeal region.

8. A dorsal hollow nerve cord is universally present at some stage.

■ CHORDATE SUBPHYLA

There are three subphyla under phylum Chordata. Two of these subphyla are small, lack a vertebral column, and are of interest primarily as borderline or first chordates (protochordates). Since these subphyla lack a cranium, they are also referred to as Acrania. The third subphylum is provided with a vertebral column and is called Vertebrata. Since this phylum has a cranium, it is also called Craniata.

Protochordata (Acrania)
  Subphylum Urochordata (u"ro-kor-da'ta) (Gr. *oura,* tail, + L. *chorda,* cord, + *-ata,* characterized by) **(Tunicata).** Notochord and nerve cord only in free-swimming larva; adults sessile and encased in tunic. Example: *Molgula.*
  Subphylum Cephalochordata (sef"a-lo-kor-da'ta) (Gr. *kephalē,* head, + L. *chorda,* cord). Notochord and nerve cord found along entire length of body and persist throughout life; fishlike in form. Example: *Branchiostoma (Amphioxus).*

Craniata
  Subphylum Vertebrata (ver"te-bra'ta) (L. *vertebratus,* backboned). Bony or cartilaginous vertebrae surround spinal cord; notochord in all embryonic stages and persists in some of the fish. This subphylum may also be divided into two great groups (superclasses) according to whether they have jaws.
  Superclass Cyclostomata (si"klo-sto'ma-ta) (Gr. *kyklos,* circle, + *stoma,* mouth) **(Agnatha** [ag'na-tha]). Without true jaws or appendages. Example: *Petromyzon.*
  Superclass Gnathostomata (na"tho-sto'ma-ta) (Gr.

gnathos, jaw, + *stoma,* mouth). With jaws and (usually) paired appendages. Example: *Homo.*

Subphylum Urochordata (Tunicata)

The tunicates (Fig. 21-6) are widely distributed in all seas from near the shoreline to great depths. Most of them are sessile, at least as adults, although some are free living. The name "tunicate" is suggested by the nonliving tunic that surrounds them and contains cellulose. A common name for them is sea squirt because some of them discharge water through the excurrent siphon when irritated. They vary in size from microscopic forms to several inches in length.

As a group they may be considered as degenerative or specialized members of the chordates, for the adults lack many of the common characteristics of chordates. For a long time they were classified among the mollusks. In 1866 Kowalevsky, the Russian embryologist, worked out their true position.

Urochordata is divided into three classes—**Ascidiacea, Larvacea,** and **Thaliacea.** Of these, the members of **Ascidiacea,** commonly known as the ascidians, are by far the most common and are the best known. One of this group is described below as a representative tunicate. Ascidians may be solitary, colo-

nial, or compound. Each of the solitary and colonial forms has its own test, but among the compound forms many individuals may share the same test. In some of these compound ascidians each member has its own **incurrent siphon,** but the **excurrent opening** is common to the group. Most ascidians are monoecious, but some are dioecious, and they can also reproduce asexually by budding or gemmation. The larvae may develop outside or in the atrium of the parent. Many ascidians are among the most vividly colored of invertebrates.

The **Larvacea** and **Thaliacea** are pelagic forms of the open sea and are not often found in the intertidal zones where ascidians are common. The members of Larvacea are small, tadpolelike forms under 5 mm. in length and resemble ascidian tadpoles. They may represent persistent larval forms that have become neotenous. They secrete around themselves cellulose tunics and are filter feeders. *Oikopleura* is a larvacean form often collected in townet samplings of ocean plankton. The class Thaliacea is made up of members that may reach a length of 3 or 4 inches. Their transparent body is spindle shaped or cylindric and is surrounded by bands of circular muscles, with their incurrent and excurrent siphons at opposite ends.

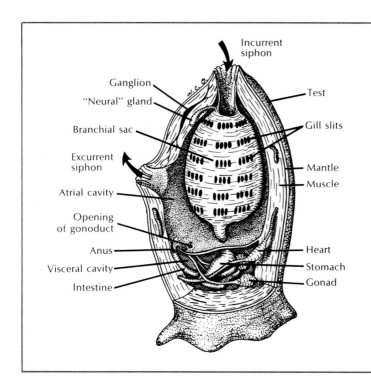

FIG. 21-6
Structure of adult solitary, simple ascidian. Arrows indicate direction of water currents.

They are mostly carried along by currents, although by contracting their circular muscle bands they can force water out of their excurrent siphons and can move by jet propulsion. Many are provided with luminous organs and give a brilliant light at night. Most of their body is hollow, with the viscera forming a compact mass on the ventral side. They appear to have come from attached ancestors like the ascidians. Some of them have complex life histories. In forms like *Doliolum* there is alternation of generations between sexual and asexual forms. After hatching from the egg, the larval tadpole changes into a barrel-shaped nurse or asexual stage, which produces small buds on a ventral stolon. These buds break free, become attached to another part of the parent, and develop into three kinds of individuals; one kind breaks free to become the sexual stage. *Salpa* also has alternation of generations and is a common form along the Atlantic coast.

Adult ascidian—*Molgula*

Molgula is globose in form and is attached by its base to piles and stones. Lining the test or tunic is a membrane or **mantle.** On the outside are two projections: the **incurrent** and **excurrent siphons** (Fig. 21-6). Water enters the incurrent siphon and passes into the pharynx (branchial sac) through the mouth. On the midventral side of the pharynx is a groove, the **endostyle,** which is ciliated and secretes mucus. Food material in the water is entangled by the mucus in this endostyle and carried into the esophagus and stomach. The intestine leads to the anus near the excurrent siphon. The water passes through the pharyngeal slits in the walls of the pharynx into the atrial cavity. As the water passes through the slits, respiration occurs.

The circulatory system contains a ventral **heart** near the stomach and two large vessels, one connected to each end of the heart. The action of the heart is peculiar in that it drives the blood first in one direction and then in the other. This reversal of blood flow is found in no other animal. The excretory system is a type of nephridium near the intestine. The nervous system is restricted to a nerve ganglion and a few nerves that lie on the dorsal side of the pharynx. A notochord is lacking. The animals are hermaphroditic, for both ovaries and testes are found in the same animal. Ducts lead from the gonads close to the intestines and empty near the anus. The germ cells are carried out the excurrent siphon into the surrounding water, where cross-fertilization occurs.

It will be seen that, of the four chief characteristics of chordates, adult tunicates have only one, the pharyngeal gill slits. However, the larval form gives away the secret of their true relationship.

Ascidian tadpole

The tadpole larvae (Fig. 21-7) among the different ascidians vary in certain details, but the basic plan is much the same in all. The development of the egg through the blastula and gastrula stages is somewhat similar to that of amphioxus. However, cleavage is determinate, and the mesoderm arises not from pouches but from clumps of cells of the archenteron. After a development of about 2 days the embryo hatches out into an elongated transparent larva about 1 to 5 mm. long. Its tail, four or five times longer than its trunk, is provided with a slender cuticular fin and contains the following structures: a **notochord** of vacuolated cells arranged in a single row; a hollow dorsal **nerve chord** extending from the tip of the tail to the sensory vesicle and made up of small cells; and a striated muscle band on each side of the notochord. Some mesenchymal cells are also found in the

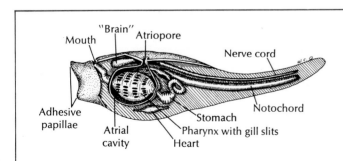

Mouth
"Brain"
Atriopore
Nerve cord
Adhesive papillae
Atrial cavity
Stomach
Pharynx with gill slits
Heart
Notochord

FIG. 21-7

Structure of tunicate (ascidian) tadpole larva. This larva is believed to closely resemble the ancestor of all vertebrates and shows all four principal chordate characteristics—notochord, dorsal nerve cord, pharyngeal gill slits, and postanal tail.

tail. In the larger head and trunk regions are found the three adhesive papillae; a digestive system of dorsal mouth, short esophagus, large pharynx with endostyle and **gill slits** (which open into the atrium), stomach, intestine, and anus (which opens into the atrium); the brain, which is a continuation of the nerve cord of the tail; a sensory vesicle containing an otolith for balance; and a dorsal median eye with lens and pigmented cup. A coelom and a circulatory system are present, but the heart is not formed until after metamorphosis. The larva does not feed but swims around for some hours, during which time it is at first positively phototactic and negatively geotactic but later becomes negatively phototatic and positively geotactic. By its adhesive papillae it now fastens itself vertically to some solid object and then undergoes retrograde metamorphosis to become an adult. In this process the tail is absorbed by phagocytes; the notochord, muscles, and nervous system (except a trunk ganglion) degenerate; the branchial sac enlarges with many gill slits; and the alimentary canal and circulatory system (with a heart) enlarge and develop. The body also undergoes a rotation so that the mouth and atrial openings (siphons) are shifted to the upper unattached end. Gonads and ducts arise in the mesoderm, and the whole animal becomes enclosed in a test or tunic.

The evolutionary significance of the ascidian tadpole has already been discussed.

Subphylum Cephalochordata
Amphioxus

Subphylum Cephalochordata is the most interesting of all the protochordates, for one of its members is the lancelet *Branchiostoma (Amphioxus)* (Fig. 21-8), one of the classic animals in zoology. This group is found mainly on the sandy beaches of southern waters, where they burrow in the sand, with the anterior end projecting out. One American species, *B. virginiae,* is found from Florida to the Chesapeake Bay. Altogether there are 28 species, of which 4 are North American, scattered over the world. They can swim in open water by swift lateral movements of the body.

Amphioxus is especially interesting, for it has the four distinctive characteristics of chordates in simple form, and in other ways it may be considered a blueprint of the phylum. It has a long, slender, laterally compressed body 2 to 3 inches long (Figs. 21-8 and 21-9), with both ends pointed. There is a long **dorsal fin,** which passes around the tail end to form the **caudal fin.** A short **ventral fin** is also found. These fins are reinforced by **fin rays** of connective tissue. The ventral side of the body is flattened and bears along each side a **metapleural fold.** There are three openings to the outside; the ventral anterior **mouth,** the **anus** near the base of the caudal fin, and the **atriopore** just anterior to the ventral fin.

The body is covered with a soft **epithelium** one

FIG. 21-8
Amphioxus. This interesting bottom-dwelling cephalochordate possesses the four distinctive chordate characteristics (notochord, dorsal nerve cord, pharyngeal gill slits, and postanal tail) that once made it the prime candidate for our vertebrate ancestor. However, because it also bears many specialized and degenerate features, zoologists now consider it a divergent offshoot from the main line of chordate evolution. (Photo by B. Tallmark.)

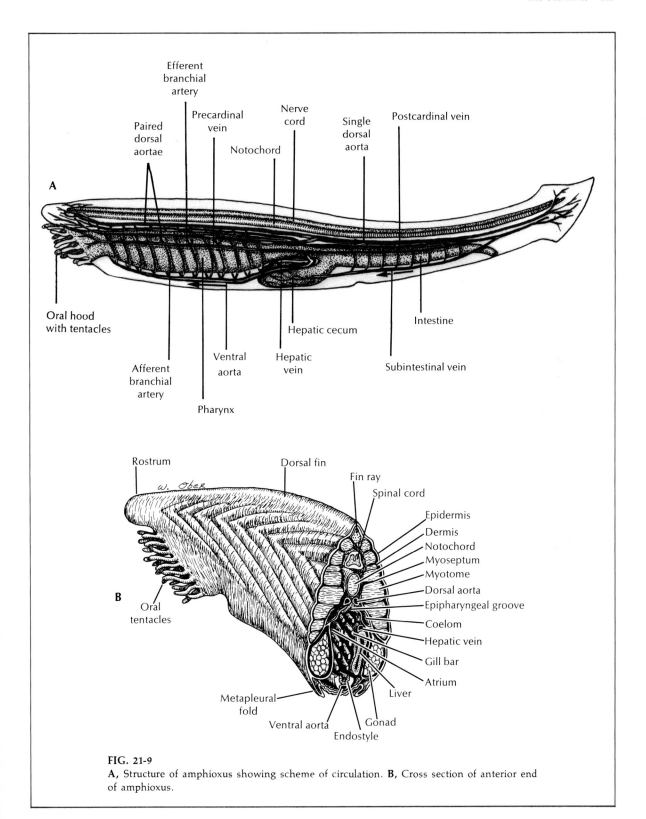

FIG. 21-9
A, Structure of amphioxus showing scheme of circulation. **B,** Cross section of anterior end of amphioxus.

layer thick resting upon some connective tissue. The **notochord,** which extends almost the entire length of the body, is made up of cells and gelatinous substance enclosed in a sheath of connective tissue. Above the notochord is the tubular dorsal **nerve cord,** with a slight dilation at the anterior end known as the **cerebral vesicle.** Along each side of the body and tail are the numerous <-shaped **myotomes,** or muscles, which have a metameric arrangement (Fig. 21-9). The myotomes are separated from each other by **myosepta** of connective tissue. The myotomes of the two sides alternate with each other. The anterior end of the body is called the **rostrum.** Just back of this and slightly below is a median opening surrounded by a membrane, the **oral hood,** which bears some twenty **oral tentacles (buccal cirri).** The oral hood encloses the chamber known as the **vestibule,** at the bottom of which lies the true **mouth** with a membrane, the **velum,** around it. Around the mouth are 12 **velar tentacles.** The cirri and tentacles serve to strain out large particles and have sensory functions. Ciliated patches on the walls of the buccal cavity in front of the velum produce a rotating effect and are called the **wheel organ,** which propels water currents. On the dorsal side of the oral hood is Hatschek's groove and pit, an embryonic relic from the first coelomic sac on the left. It may be homologous with the pituitary of vertebrates. Just behind the mouth is the large compressed pharynx with more than a hundred pairs of **gill slits,** which act as strainers in filter feeding as well as in respiration. From the pharynx the narrow tubular **intestine** extends backward to the anus. On the ventral side of the intestine is a large diverticulum, the **hepatic cecum.** The **coelom** is reduced and is confined to the region above the pharynx and around the intestine (Fig. 21-9). Connecting the coelom to the atrium are about a hundred pairs of ciliated **nephridia** of the solenocyte type, a modified kind of flame cell. The big cavity around the pharynx is the **atrium;** it is lined with ectoderm and is therefore not a coelom. The pharynx has a middorsal groove, the **hyperbranchial** (epipharyngeal) **groove,** and a midventral one, known as the **endostyle.** Both of these grooves are lined with cilia and gland cells (Fig. 21-9). Food is entangled by the mucus of the endostyle and is carried to the intestine; the water passes through the gill slits into the atrium and gives up oxygen to the blood vessels in the **gill bars.**

Although there is no heart, the **blood system** is similar to that of higher chordates (Fig. 21-9). The blood moves posteriorly in the **dorsal aorta** and anteriorly in the **ventral aorta;** a **hepatic portal** vein leads from the intestines to the liver. Blood is propelled by contractions of the ventral aorta and is carried to the dorsal aorta by vessels in the gill bars. The blood is almost colorless, with a few red corpuscles.

The **nervous system** is above the notochord and consists of a single dorsal **nerve cord,** which is hollow. This nerve cord gives off a pair of "nerves" alternately to each body segment, or myotome, as dorsal and ventral roots. Although this arrangement resembles the vertebrate spinal cord, the ventral root actually does not contain nerves but rather the drawn-out extensions of myotomic muscle fibers.

Ciliated cells with **sensory** functions are found in various parts of the body. Sexes are separate, and each sex has about 25 pairs of gonads located on the wall of the atrium. The sex cells are set free in the atrial cavity and pass out the atriopore to the outside, where fertilization occurs. Cleavage is total (holoblastic) and a gastrula is formed by invagination. The larva hatches soon after deposition and gradually assumes the shape of the adult.

No other chordate shows so well the basic diagnostic chordate characteristics as does the amphioxus. Not only are the four chief characters of chordates—**dorsal nerve cord, notochord, pharyngeal gill slits,** and **postanal tail**—well represented but also secondary characteristics, such as liver diverticulum, hepatic portal system, and the beginning of a ventral heart. Indicative also of the condition in vertebrates is the much thicker dorsal portion of the muscular layer. This is in contrast to the invertebrate phyla in which the muscular layer is about the same thickness around the body cavity. The metameric arrangement of the muscles suggests a similar plan in the embryos of vertebrates. Just where it is placed in the evolutionary blueprint of the chordates and the vertebrates is a controversial point. It is placed by many authorities near the primitive fish, ostracoderms, but whether it comes before or after these fish in the evolutionary line is not settled. Many regard the amphioxus as a highly specialized or degenerate member of the early chordates and believe that the overdeveloped notochord was developed in them as a correlation to their burrowing habits. The forward extension of the notochord into the tip of the snout may be one of the reasons for the small development of the brain of the amphioxus. Many authorities therefore assign amphioxus to a divergent side branch of some stage intermediate between the early filter-feeding prevertebrates and the vertebrates.

Subphylum Vertebrata

The third subphylum of the chordates, Vertebrata, has the same characteristics that distinguish the other two subphyla, but in addition it has a number of features that the others do not share. The characteristics that give the members of this group the name Vertebrata or Craniata are the presence of a braincase, or **cranium,** and a spinal column of vertebrae, which forms the chief skeletal axis of the body.

Characteristics

1. The chief diagnostic features of chordates—**notochord, dorsal nerve cord, pharyngeal gill slits,** and **postanal tail**—are all present at some stage of the life cycle.
2. They are covered with an **integument** basically of two divisions, an outer **epidermis** of stratified epithelium from the ectoderm and an inner **dermis,** or corium, of connective tissue derived from the mesoderm. This skin has many modifications among the various classes, such as glands, scales, feathers, claws, horns, and hair.
3. The notochord is more or less replaced by the spinal column of vertebrae composed of cartilage or bone or both. The vertebral column with the cranium, visceral arches, limb girdles, and two pairs of jointed appendages forms the distinctive **endoskeleton.**
4. **Many muscles** are attached to the skeleton to provide for movement.
5. The complete **digestive system** is ventral to the spinal column and is provided with large digestive glands, liver, and pancreas.
6. The circulatory system is made up of the **ventral heart** of two to four chambers; a closed blood vessel system of arteries, veins, and capillaries; and a blood fluid containing red blood corpuscles with hemoglobin, and white corpuscles. Paired aortic arches connect the ventral and dorsal aortae and give off branches to the gills among the aquatic vertebrates; in the terrestrial types the aortic arch plan is modified into pulmonary and systemic systems.
7. A **coelom** is well developed and is largely filled with the visceral systems.
8. The **excretory system** is made up of paired kidneys (mesonephric or metanephric types) provided with ducts to drain the waste to the cloaca or anal region.
9. The brain is typically divided into five vesicles.
10. Ten or 12 pairs of cranial nerves with both motor and sensory functions is the rule; a pair of spinal nerves supplies each primitive myotome; and an autonomic nervous system controls involuntary functions of internal organs.
11. An **endocrine system** of ductless glands scattered through the body is present.
12. The sexes are nearly always separate, and each sex contains paired gonads with ducts that discharge their products either into the cloaca or into special openings near the anus.
13. The **body plan** consists typically of **head, trunk,** and **postanal tail.** A **neck** may be present in some, especially terrestrial forms. Two pairs of appendages are the rule, although they are entirely absent in some. The coelom is divided into a pericardial space and a general body cavity; in addition, mammals have a thoracic cavity.

■ AMMOCOETE LARVA OF LAMPREY AS CHORDATE ARCHETYPE

The ammocoete larva of lampreys possesses many of the basic structures one would expect to find in a chordate archetype. Many of its structures are simple in form and similar to those in higher vertebrates. It has a heart, ear, eye, thyroid gland, and pituitary gland, which are characteristic of vertebrates but are lacking in amphioxus. This larva is so different from the adult lamprey that it was for a long time considered to be a separate species; not until it was shown to metamorphose into the adult lamprey was the exact relationship explained. This eel-like larva spends several years buried in the sand and mud of shallow streams, until it finally emerges as an adult that may continue to live in freshwater (freshwater lampreys) or else may migrate to the sea (marine lampreys).

Since Stensiö's important work on ostracoderms in 1927, the similarity of this ammocoete larva to the cephalaspids of that ancient group of fish has become more and more apparent, and many zoologists are substituting it for amphioxus as a basic ancestral type. It is true that the ammocoete has some degenerative specializations of its own, for it lacks the bony exoskeleton, an important feature in ostracoderms. Stensiö, Romer, and other paleontologists have emphasized that a hard or bony exoskeleton is characteristic of ancestral vertebrates and that cartilaginous structures in the adult represent a specialized embryonic condition that has been retained. Romer has shown that cartilage serves a real purpose in the embryo. Cartilage is not present in dermal bones such as certain skull bones that are laid down directly in membrane and have simple growth, but only in internal bones where it is necessary to maintain complicated relationships with blood vessels, muscles, and other bones throughout the entire growth period. Bone grows only by accretion and does not have the power to expand, which cartilage can do, and thus the latter represents an ideal embryonic material before the adult elements are fully formed.

Some of the generalized characteristics of the ammocoete larva will be pointed out in the following summary.

GENERAL CHORDATE FEATURES. The ammocoetes has a long, slender body, with the front end broader and blunter than the tail end (Fig. 21-10). A median membranous fin fold extends along most of the posterior dorsal border, passes around the caudal end, where the fin is broader, and then continues forward on the ventral side. The **notochord** is large and extends from the very tip of the tail to a region near the posterior end of the brain. The **dorsal nerve cord,** unlike that of the amphioxus, is enlarged anteriorly to form a complete brain. Instead of the numerous gill slits of the amphioxus, there are only seven pairs of gill pouches and slits in the ammocoetes (there are six pairs in shark embryos). Muscular segmentation is also found in the form of myotomes along the dorsal part of the body. The skeleton is meager and in a degenerate condition. Such parts as are found are entirely cartilaginous, for example, the gill bars of the branchial basket, the scattered plates of the braincase, and the small vertebrae near the notochord. The well-developed notochord is the chief supporting skeleton. There are no paired fins or jaws.

DIGESTIVE AND RESPIRATORY SYSTEMS. At the ventral anterior end of the larva is a cup-shaped **oral hood,** which encloses the **buccal cavity.** Numerous **oral papillae,** or branched projections, are attached to the sides and roof of the oral hood and surround the mouth cavity. They have a sensory function. Between the buccal cavity and the **pharynx** is the **velum,** consisting of a pair of flaps that help create currents of water entering the mouth. The expanded pharynx makes up a large part of the alimentary canal and bears in its lateral walls the seven pairs of **gill pouches.** Each of these pouches opens to the exterior by a **gill slit.** Each gill slit has a fold or **gill** on both its anterior surface and posterior sur-

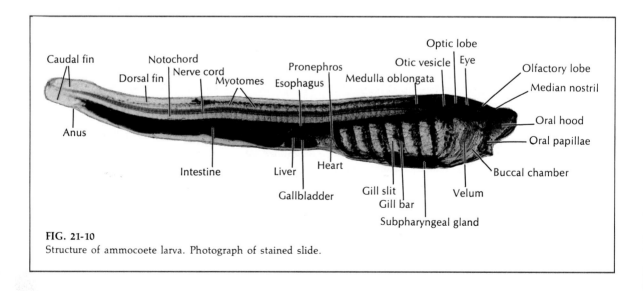

FIG. 21-10
Structure of ammocoete larva. Photograph of stained slide.

face, and the wall of the pharynx between adjacent gill slits contains supporting rods of cartilage (gill bars). The gills are richly supplied with blood capillaries and are bathed by currents of water that enter the mouth and pass to the exterior through the gill slits. Oxygen from the water enters the blood in the gills and carbon dioxide is given off by the blood in exchange. The endostyle (subpharyngeal gland), a closed furrow or tube extending for the length of four gill pouches, is found in the floor of the pharynx. It secretes mucus, which is discharged into the pharynx by a small duct. This sticky mucus entangles the food particles brought in by water currents produced by muscular contractions of the pharynx. Cords of food thus formed are carried into the intestine by the ciliated groove in the floor of the pharynx. During metamorphosis a portion of the endostyle is converted into the thyroid gland, which secretes the thyroid hormone containing iodine.

The pharynx narrows at its posterior end to form a short esophagus, which opens into the straight intestine. Opening into the intestine by the bile duct is the liver with which is associated a large and conspicuous gallbladder. Pancreatic cells are found in the wall of the anterior part of the intestine but do not form a distinct gland. The intestine opens posteriorly into the cloaca, which also receives the kidney ducts. The anus is found a short distance in front of the postanal tail.

The generalized vertebrate features of the ammocoetes are thus seen to be its jawless filter feeding, its relatively undifferentiated alimentary canal, its gill arrangement, and its endostyle characteristic of primitive feeding organisms. The development of the thyroid gland from part of the endostyle, as well as the muscular branchial movement, also represents a plan that higher vertebrates have followed.

CIRCULATORY SYSTEM. The hypothetical, primitive chordate plan of four major longitudinal blood vessels—dorsal aorta, subintestinal, right cardinal vein, and left cardinal vein—and two major connections between these blood vessels—right and left ducts of Cuvier and the aortic arches—is generally followed in the ammocoetes with certain modifications. The posterior end of the subintestinal vein has become modified to form the hepatic portal vein, the anterior portion to form the heart of one auricle (atrium) and one ventricle arranged in tandem. From the ventricle the short ventral aorta runs forward to

give off eight pairs of aortic arches to the gill pouches. Each arch is composed of an afferent branchial artery carrying blood to the capillaries of the gills and an efferent branchial artery carrying aerated blood from the gill capillaries to the dorsal aorta. The dorsal aorta gives off many branches to the body tissues, and a large posterior branch, the intestinal, to the intestine. The cardinal veins return blood from the tissues to the right and left ducts of Cuvier, which empty into the sinus venosus, a thin-walled chamber that empties into the atrium, and thence to the ventricle of the heart. Each cardinal vein is made up of an anterior cardinal and a posterior cardinal branch. The hepatic portal vein picks up blood laden with nutrients from the intestine and carries it to the liver. The hepatic vein carries blood from the liver to the sinus venosus, and so back to the heart.

EXCRETORY SYSTEM. The ancestral vertebrate kidney is supposed to have extended the length of the coelomic cavity and was made up of segmentally arranged uriniferous tubules. Each tubule opens at one end into the coelom by a nephrostome and at the other end into the common archinephric duct. Such a kidney has been called an archinephros, or holonephros, and is found in the embryos of hagfish (Fig. 21-11). Kidneys of higher vertebrates developed from this primitive plan. Embryologic evidence indicates that there are three generations of kidneys: pronephros, mesonephros, and metanephros. In all vertebrate embryos, the pronephros is the first and most primitive kidney to appear. As its name implies, it is located anteriorly in the body. It becomes the persistent kidney of adult hagfish. In all other vertebrates it degenerates during development and is replaced by a more centrally located and more structurally advanced kidney, the mesonephros. The mesonephros becomes the persistent kidney of adult fishes and amphibians. But in the developing embryos of amniotes (reptiles, birds, and mammals) the mesonephros is replaced in turn by the metanephros. The metanephros develops behind the mesonephros and is structurally and functionally the most advanced of the three kidney types. Thus three kidneys are formed in succession, each more advanced and each located more caudally than its predecessor.

In summary, we may state that the evolutionary sequence of adult vertebrate kidneys has been archinephros, mesonephros, and metanephros.

It is seen that the excretory system of the ammo-

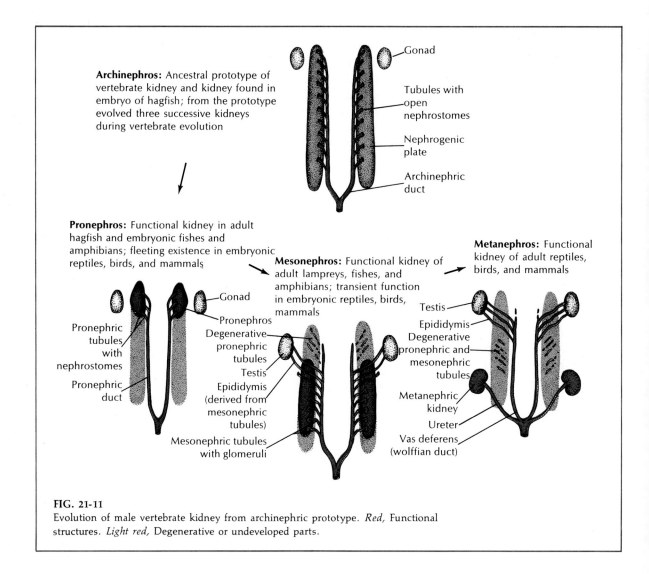

Archinephros: Ancestral prototype of vertebrate kidney and kidney found in embryo of hagfish; from the prototype evolved three successive kidneys during vertebrate evolution

Gonad

Tubules with open nephrostomes

Nephrogenic plate

Archinephric duct

Pronephros: Functional kidney in adult hagfish and embryonic fishes and amphibians; fleeting existence in embryonic reptiles, birds, and mammals

Mesonephros: Functional kidney of adult lampreys, fishes, and amphibians; transient function in embryonic reptiles, birds, mammals

Metanephros: Functional kidney of adult reptiles, birds, and mammals

Gonad

Pronephros

Degenerative pronephric tubules

Testis

Epididymis (derived from mesonephric tubules)

Mesonephric tubules with glomeruli

Pronephric tubules with nephrostomes

Pronephric duct

Testis

Epididymis

Degenerative pronephric and mesonephric tubules

Metanephric kidney

Ureter

Vas deferens (wolffian duct)

FIG. 21-11
Evolution of male vertebrate kidney from archinephric prototype. *Red,* Functional structures. *Light red,* Degenerative or undeveloped parts.

coetes conforms to the basic chordate plan, whereas the solenocyte type of flame cell found in amphioxus is altogether different.

REPRODUCTIVE SYSTEM. The **gonads** are paired ridgelike structures on the dorsal side of the coelom. Each gonad appears to have been formed by the fusion of a number of units. Since they lack genital ducts, the adult lampreys shed their gametes into the coelom, where an opening into the **mesonephric duct** allows the gametes to escape to the outside through the **urogenital papillae.**

NERVOUS AND SENSORY SYSTEM. Both brain and spinal cord conform to the basic chordate plan. The **brain** has the typical three divisions of **fore-brain, midbrain,** and **hindbrain.** Each of these divisions is associated with an important sense organ—olfaction, vision, and hearing, respectively. From the dorsal side of the forebrain are two outgrowths or stalks, each of which bears a vestigial **median eye,** the only instance among vertebrates of two median eyes. Other vertebrates may have two outgrowths, but the anterior one is the parietal body, which appears to have been a median eye, and the posterior one (epiphysis) is the pineal gland. In no living vertebrate does a median eye function as such. In some vertebrates only the pineal gland or body is present. A **pituitary gland,** formed from an evagination (infundibulum) of the forebrain and the **hypophysis** from the **nasohypophyseal canal** of the pharynx, is found on the ventral side of the brain. The functional **eyes** of the ammocoetes are small and develop from the forebrain. The spinal cord gives off a pair of **dorsal roots** (mainly sensory) and a pair of **ventral roots** (motor) in every muscle segment. The dorsal and ventral roots do not join as they do in higher vertebrates, nor do the nerves have myelin sheaths.

References

Barrington, E. J. W. 1965. The biology of Hemichordata and Protochordata. (Paperback.) San Francisco, W. H. Freeman & Co. *A synthesis of recent work on these deuterostomes most closely related to vertebrates. Discusses the possible homologies between the endostyle and the vertebrate thyroid gland.*

Beer, de, G. R. 1951. Vertebrate zoology, rev. ed. London, Sidgwick & Jackson, Ltd. *The student will glean a knowledge of representative types of vertebrates and how they fit into the evolutionary scheme from this excellent work, which should be read by all serious students of the zoologic sciences.*

Berrill, N. J. 1955. The origin of vertebrates. New York, Oxford University Press. *The author stresses the tunicates as the basic stock from which other protochordates and vertebrates arose. He believes that such a sessile filter feeder was really the most primitive animal and was not a mere degenerate side branch of chordate evolution.*

Colbert, E. H. 1955. Evolution of the vertebrates. New York, John Wiley & Sons, Inc. *A clear and well-written presentation.*

Halstead, L. B. 1968. The pattern of vertebrate evolution. San Francisco, W. H. Freeman & Co. *The author's thoughtful interpretation of fossil evidence is written in a relaxed style. He considers physiologic and ecologic aspects of evolution.*

Roe, A., and G. G. Simpson (editors). 1958. Behavior and evolution. New Haven, Conn., Yale University Press. *A masterpiece that integrates two great disciplines—evolution and behavior.*

Romer, A. S. 1959. The vertebrate story. Chicago, University of Chicago Press. *A comprehensive background of the evolutionary trends and relationships of the various vertebrate groups leading up to that of man himself.*

Romer, A. S. 1966. Vertebrate paleontology, ed. 3. Chicago, University of Chicago Press. *An authoritative work by a distinguished paleontologist.*

Smith, H. M. 1960. Evolution of chordate structure. New York, Holt, Rinehart & Winston, Inc. *This excellent introduction to comparative anatomy gives an appraisal of the basic structures of primitive chordates.*

Smith, H. W. 1953. From fish to philosopher. Boston, Little, Brown & Co. *An engagingly written evolutionary history of the kidney by a late authority in vertebrate kidney physiology.*

Two wolf eels (Anarhichas lupus) *gaze balefully at a skin-diving photographer from their rocky home 90 feet below the surface off the Norwegian coast. (Courtesy T. Lundälv, Sweden.)*

CHAPTER 22

THE FISHES

PHYLUM CHORDATA

CLASSES PETROMYZONTES, MYXINI, ELASMOBRANCHII, HOLOCEPHALI, AND OSTEICHTHYES

Fishes are the undisputed masters of the aquatic environment. Because fish live in a habitat that is basically hostile to man, we have not always found it easy to appreciate the incredible success of these vertebrates. Plato considered fish "senseless beings . . . which have received the most remote habitations as a punishment for their extreme ignorance." And the average North American today is probably unconscious of and uninformed about fish unless he happens to be a sports fisherman or tropical fish enthusiast. Nevertheless the world's fishes have enjoyed an adaptive radiation easily as spectacular as that of all the land vertebrates, with the possible exception of the mammals (the latter having succeeded in the air and in the sea as well as on land). Their numerous structural adaptations have produced a great variety of forms ranging from gracefully streamlined trout to grotesque creatures that dwell in the eternal

blackness of the ocean's abyssal depths. Considered either in numbers of species (more than 22,000 named species) or in numbers of individuals (countless billions), they easily outnumber the four terrestrial vertebrate classes combined.

Although fishes are the oldest vertebrate group, there is not the slightest evidence that, like the amphibians and reptile successors, they are declining from a period of earlier glory; there are indeed more bony fishes today than ever before and no other group threatens their domination of the seas. Their success can be attributed to one thing: they are perfectly adapted to their dense medium. A trout or pike can hang motionless in the water at any depth, varying its neutral buoyancy by adding or removing air from the swim bladder, or dart forward or at angles, using its fins as brakes and tilting rudders. Fish have excellent olfactory and visual senses and a unique lateral line system, which with its exquisite sensitivity to water currents and vibrations, provides a "distance touch" in water. Their gills are the most effective respiratory devices in the animal kingdom for extracting oxygen from water. With highly developed organs of salt and water exchange, bony fishes are excellent osmotic regulators capable of fine tuning their body fluid composition in their chosen freshwater or seawater environment. Fishes have evolved complex behavioral mechanisms for dealing with emergencies, and many have evolved elaborate reproductive behavior concerned with courtship, nest building, and care of the young. These are only a few of many such examples. The adaptations evident in this varied phylogenetic assemblage, which includes five of the nine vertebrate classes, are nearly as numerous as the number of species.

■ ORIGINS AND CLASSIFICATION OF MAJOR GROUPS OF FISHES

Theories of chordate and vertebrate origins were described in the preceding chapter. The fishes, like all other vertebrates take their origin from an unknown common ancestor that may have arisen from a free-swimming larval form of an ancient ascidian or ascidian-like stock. During the Cambrian, or perhaps even in the Precambrian, the earliest fishlike vertebrates branched into the jawless cyclostomes and the jawed gnathostomes (Fig. 22-1). All subsequent vertebrates descended from one or the other of these two great stems. The cyclostomes include the ostracoderms, which are now extinct, and the living hagfishes (Myxini) and lampreys (Petromyzontes). Biologists generally agree that the living cyclostomes are successors of the ostracoderms, but there is no agreement about which of the three ostracoderm groups (heterostracans, cephalaspids, anaspids) they are most closely related to. The weight of evidence points to an affinity between the anaspids and the living lampreys, but the heritage of the hagfishes is unknown. Both hagfishes and lampreys are highly specialized, and aside from both lacking jaws, they bear little resemblance to the ostracoderms. Moreover it has long been recognized that hagfishes and lampreys only superficially resemble each other. There are numerous anatomic and physiologic contrasts, as well as important differences in their mode of life and life cycles. This suggests that the hagfishes and lampreys have been phylogenetically independent for a very long time and should be considered as belonging to entirely separate classes, as ichthyologists now regard them.

All the rest of the fishes are gnathostomes that have descended from one or more early jawed ancestors. The placoderms have often been suggested, but the fossil evidence is actually so fragmentary that it is impossible to pick out ancestral groups with certainty. Whatever the early lines of descent, ichthyologists now recognize several natural groups of living jawed fishes, which are depicted in Fig. 22-1.

The sharks, skates, and rays comprise one natural, compact group, the class Elasmobranchii. These animals lost the heavy armor of the early placoderms, adopted cartilage instead of bone for the skeleton (a secondary degeneration), an active and predatory habit, and sharklike body form that has undergone only minor changes over the ages. The elasmobranchs flourished during the Devonian and Carboniferous, but declined dangerously close to extinction in the Permian. They staged a recovery in the early Mesozoic and radiated into the modest but thoroughly successful assemblage of modern sharks.

Obviously sharing a remote relationship to the elasmobranchs are the bizarre chimaeras of the class Holocephali. They appeared in the Jurassic, apparently as an offshoot of some now extinct elasmobranch stock. These peculiar yet strangely appealing fishes have many internal elasmobranch structures and have often been included with the latter under the name "Chondrichthyes."

We come now to the bony fishes, the dominant fishes today. This diverse assemblage offers perplexing classification problems. We can recognize three great stems of descent: first there are the actinopterygians, or ray-finned fishes, which radiated into the modern bony fishes; second are the crossopterygians, the lobe-finned fishes, from which the amphibians are descended; and third are the dipneusts, or lungfishes. Ichthyologists formerly classified the last two of these three groups together as Choanichthyes ("funnel-fish") in the belief that they shared internal nostrils, that is, nostrils that open into the mouth cavity in the style of higher vertebrates. But more thorough studies of fossil forms revealed that the internal nostril of the lungfish has a completely different origin; this and other differences in the skeleton has now set the lungfishes apart and excluded them from candidacy as ancestors to the

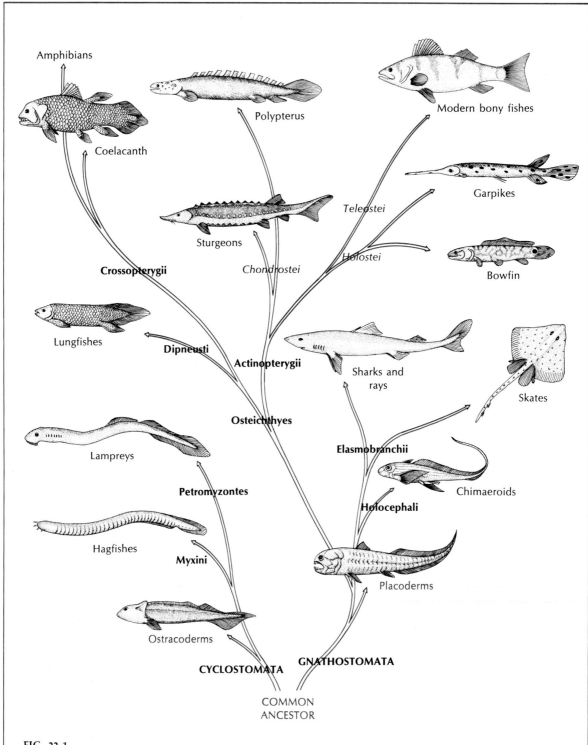

Amphibians

Coelacanth

Polypterus

Modern bony fishes

Garpikes

Crossopterygii

Teleostei

Sturgeons

Holostei

Chondrostei

Bowfin

Lungfishes

Dipneusti

Actinopterygii

Osteichthyes

Sharks and
rays

Skates

Lampreys

Elasmobranchii

Petromyzontes

Chimaeroids

Holocephali

Hagfishes

Myxini

Placoderms

Ostracoderms

CYCLOSTOMATA

GNATHOSTOMATA

COMMON
ANCESTOR

FIG. 22-1
Family tree of the fishes. Lines suggest probable relationships. *Red,* Extinct groups.

tetrapods. This distinction now rests with the crossopterygians, which are represented today by a single species, the coelacanth *Latimeria.* Both the crossopterygians, with one genus and the dipneusts, with three, are relic groups. These four survivors are meager evidence of stocks that flourished in the Devonian; the survivors have remained mostly unchanged over the subsequent 400 million years to present. The crossopterygians and the dipneusteans are now grouped with the actinopterygians under the class Osteichthyes (Teleostomi of some classifications), to produce a highly unbalanced living representation: one crossopterygian (the coelocanth), three genera of lungfishes, and some 18,000 actinopterygians. The latter are divided into three superorders as shown in Fig. 22-1. The evolution of these groups are discussed later in this chapter.

One further point of evolutionary importance must be mentioned. We pointed out in the previous chapter that the great weight of paleontologic opinion now favors a marine origin for the vertebrates. But at some later time, probably in the Silurian or Lower Devonian, most of the major groups of fishes penetrated into estuaries and then into freshwater rivers. This invasion was of momentous importance to subsequent vertebrate evolution. It compelled the development of physiologic mechanisms to maintain osmotic and ionic balance in the highly dilute freshwater environment. These Devonian invaders not only met the challenge but also established a body fluid concentration (about one third that of seawater) and ionic composition that fixed the body fluid pattern for all vertebrates to evolve later, whether aquatic, terrestrial, or aerial. Thus it was that when the early bony fishes (Osteichthyes) and elasmobranchs **returned** to the sea in the Triassic, they encountered a new osmotic challenge: they were now much more dilute than their surroundings. Rather than sliding back into osmotic equilibrium with seawater they chose to regulate osmotically and ionically. Bony fishes and sharks have solved this physiologic problem in different ways that are described later. Our interest at this point is that both the marine elasmobranchs and marine bony fishes are descendants of freshwater stocks: they have come full circle by returning to their ancestral home in the sea.

Classification of fishes

The following broad classification is a composite of schemes by several contemporary ichthyologists.

No one scheme is accepted by even the majority of ichthyologists. When we contemplate the incredible difficulty of ferreting out relationships among some 22,000 living species and a vast number of fossils of varying age, we can appreciate why fish classification has been, and will continue to be, undergoing continuous change.

Subphylum Vertebrata

Superclass Cyclostomata (sy"klo-sto'ma-ta) (Gr. *kyklos,* circle, + *stoma,* mouth) **(Agnatha).** No jaws; ventral fins absent; two semicircular canals; notochord persistent.

  Class Petromyzontes (pet"ro-my-zon'teez) (Gr. *petros,* stone, + *myzōn,* sucking)—**lampreys.** Mouth suctorial with horny teeth; nasal sac not connected to mouth; gill pouches, seven pairs. Examples: *Petromyzon, Lampetra.*

  Class Myxini (mik-sy'ny) (Gr. *myxa,* slime)—**hagfishes.** Mouth terminal with four pairs of tentacles; buccal funnel absent; nasal sac with duct to pharynx; gill pouches, 5 to 15 pairs; partially hermaphroditic. Examples: *Myxine, Bdellostoma.*

Superclass Gnathostomata (na"tho-sto'ma-ta) (Gr. *gnathos,* jaw, + *stoma,* mouth). Jaws present; usually paired limbs; three semicircular canals; notochord persistent or replaced by vertebral centra.

  Class Elasmobranchii (e-laz"mo-bran'kee-i) (Gr. *elasmos,* a metal plate, + *branchia,* gills)—**sharks, skates, and rays.** Endoskeleton cartilaginous, often calcified; placoid scales or no scales; 5 to 7 gill arches and gills in separate clefts along pharynx. Examples: *Squalus, Raja.*

  Class Holocephali (hol"o-sef'a-li) (Gr. *holos,* entire, + *kephalē,* head)—**chimaeras** or **ghostfish.** Gill slits covered with operculum; jaws with tooth plates; single nasal opening; without scales; accessory clasping organs in male; lateral line an open groove. Example: *Chimaera.*

  Class Osteichthyes (os"te-ik'thee-eez) (Gr. *osteon,* bone, + *ichthys,* a fish) **(Teleostomi)**—**bony fish.** Body primitively fusiform but variously modified; skeleton mostly ossified; single gill opening on each side covered with operculum; usually swim bladder or lung.

    Subclass Crossopterygii (cros-sop-terij'ee-i) (Gr. *krossoi,* fringe or tassels, + *pteryx,* fin, wing)—**lobed-finned fish.** Heavy bodied; paired fins lobed with internal skeleton of basic tetrapod type; premaxillae, maxillae present; scales large with tubercles and heavily overlapped; three-lobed diphycercal tail; skeleton with much cartilage; bony spines hollow; air bladder vestigial; gills hard with teeth; intestine with spiral valve; spiracle present. Example: *Latimeria.*

    Subclass Dipneusti (dip-nyu'sti) (Gr. *di-,* two, + *pneustikos,* of breathing)—**lungfish.** All median fins fused to form diphycercal tail; fins lobed or of filaments; scales of cycloid bony type; teeth of grinding plates; no premaxillae or maxillae; air bladder of single or paired lobes and specialized for breathing;

intestine with spiral valve; spiracle absent. Examples: *Neoceratodus, Protopterus, Lepidosiren.*

Subclass Actinopterygii (ak"ti-nop-te-rij'ee-i) (Gr. *aktis,* ray, + *pteryx,* fin, wing)—**ray-finned fish.** Paired fins supported by dermal rays and without basal lobed portions; nasal sacs open only to outside. Examples: *Salmo, Perca.*

■ SUPERCLASS CYCLOSTOMATA— LAMPREYS AND HAGFISHES

The living members of the cyclostomata are represented by some 50 species almost equally divided between two classes: Petromyzontes and Myxini (Fig. 22-2). They have in common the absence of jaws, internal ossification, scales, and paired fins, and both share porelike gill openings and an eel-like body form. At the same time there are so many important differences, some of which are indicated in the listing below, that they have been assigned to separate vertebrate classes.

Characteristics

1. Body slender, **eel-like,** rounded, with **soft skin** containing **mucous glands** but **no scales**
2. Median fins with cartilaginous fin rays, but **no paired appendages**
3. **Fibrous** and **cartilaginous** skeleton; notochord persistent
4. Suckerlike oral disk with well-developed teeth in lampreys; mouth with two rows of eversible teeth in hagfish
5. Heart with one auricle and one ventricle; aortic arches in gill region; blood with erythrocytes and leukocytes
6. Seven pairs of gills in lampreys; 5 to 15 pairs of gills in hagfish
7. Mesonephric kidney in lampreys; **pronephric kidney** anteriorly and mesonephric kidney posteriorly in hagfish
8. Dorsal nerve cord with differentiated brain; 8 to 10 pairs of cranial nerves
9. Digestive system without stomach; intestine with spiral fold in lampreys; spiral fold absent in hagfish
10. Sense organs of taste, smell, hearing; eyes moderately developed in lampreys but highly degenerate in hagfish
11. External fertilization; gonad single without duct; sexes separate and long larval stage in lampreys; hermaphroditic and direct development with no larval stage in hagfish

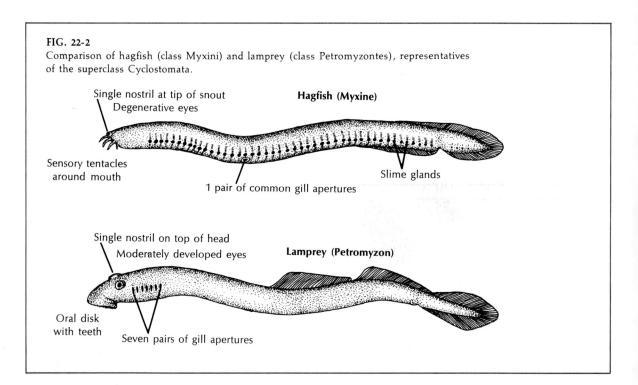

FIG. 22-2
Comparison of hagfish (class Myxini) and lamprey (class Petromyzontes), representatives of the superclass Cyclostomata.

Single nostril at tip of snout
Degenerative eyes
Hagfish (Myxine)
Sensory tentacles around mouth
1 pair of common gill apertures
Slime glands

Single nostril on top of head
Moderately developed eyes
Lamprey (Petromyzon)
Oral disk with teeth
Seven pairs of gill apertures

Class Petromyzontes—lampreys

All the lampreys of the northern hemisphere belong to the family Petromyzontidae. The destructive marine lamprey *Petromyzon marinus* is found on both sides of the Atlantic Ocean (America and Europe) and may attain a length of 3 feet. *Lampetra* also has a wide distribution in North America and Eurasia and ranges from 6 to 24 inches long. There are 19 species of lampreys in North America. About half of these belong to the nonparasitic brook type; the others are parasitic. The nonparasitic species have probably arisen from the parasitic forms by degeneration of the teeth, alimentary canal, etc. They may have done so by evolving from the parasitic species through pedomorphosis (reproduction by young, especially by parthenogenesis). The genus *Ichthyomyzon,* which contains three parasitic and three nonparasitic species, is restricted to eastern North America. On the west coast of North America the chief marine form is represented by *Lampetra (Entosphenus) tridentatus.*

All lampreys, marine as well as freshwater forms, spawn in the spring in North America (autumn in some European species) in shallow gravel and sand in freshwater streams. The males begin nest building and are joined later by females. Using their oral disks to lift stones and pebbles and vigorous body vibrations to sweep away light debris, they form an oval depression (Fig. 22-3). At spawning, with the female attached to a rock to maintain position over the nest, the male attaches to the dorsal side of her head. He curls the posterior part of his body tightly about that of the female such that the cloacal openings of both are together. As the eggs are shed into the nest, they are fertilized by the male. Only small numbers of eggs are shed at each mating, and the process may be repeated numerous times over a period of 1 to 2 days. The sticky eggs adhere to pebbles in the nest and soon become covered with sand. The adults die soon after spawning.

The eggs hatch in about 2 weeks, releasing small larvae (ammocoetes), which are so unlike their parents that early biologists were deceived into giving them distinct generic and specific names (such as *Ammocoetes branchioles*). The larva bears a remarkable resemblance to amphioxus and possesses the basic chordate characters in such simplified and easily visualized form that it has been considered a chordate archetype (see p. 447). After absorbing the remainder of its yolk supply, the young ammocoetes, now about 7 mm. long, leaves the nest gravel and drifts downstream to burrow in some suitable sandy, low-current area. Here it remains for an extraordinarily long time, 3 to 7 years, feeding on minute organisms and debris in the mud. During this time the ammocoetes grows and then in the fall rapidly metamorphoses into an adult. This change involves the development of larger eyes, the replacement of the hood by the oral disk with teeth, a shifting of the nostril to the top of the head, and the development of a rounder but shorter body.

Parasitic lampreys either migrate to the sea, if marine, or else remain in freshwater, where they attach themselves by their suckerlike mouth to fish and, with their sharp horny teeth, rasp away the flesh and suck out the blood (Fig. 22-4). To promote the flow of blood, the lamprey injects an anticoagulant into the wound. When gorged, the lamprey releases its hold but leaves the fish with a large gaping wound that may prove fatal. The parasitic freshwater adults live a year or more before spawning and then die; the marine forms may live longer.

The nonparasitic lampreys do not feed after emerging as adults, for their alimentary canal degenerates to a nonfunctional strand of tissue. Within a few months they also spawn and die.

The invasion of the Great Lakes by the landlocked sea lamprey *Petromyzon marinus* in this century has had a devastating effect on the fisheries there. No

FIG. 22-3
Brook lampreys, *Lampetra lamattei,* clearing pebbles for a spawning nest in a small creek. (Photo by J. W. Jordan, Jr.)

FIG. 22-4
Sea lampreys *(Petromyzon marinus)* attacking trout. **A,** Recently transformed sea lampreys, 6 to 7 inches long, attack 8-inch long brook trout *(Salvelinus fontinalis)* in experimental aquarium. **B,** Head of 15-inch long sea lamprey feeding on a rainbow trout *(Salmo gairdneri).* Note the single nostril on top of head and the eyes and gill aperatures. **C,** Lamprey detached from rainbow trout to show feeding wound that had penetrated body cavity and perforated gut. Trout died from wound. Note chitinous teeth on underside of lamprey head. (Courtesy United States Bureau of Sport Fisheries and Wildlife, Fish Control Laboratory, La Crosse, Wis.; **A,** by R. E. Lennon; **B** and **C,** by L. L. Marking.)

lampreys were present in the Great Lakes west of Niagara Falls until the Welland Ship Canal was built in 1829. Even then nearly 100 years elapsed before sea lampreys were first seen in Lake Erie. After that the spread was rapid and the sea lamprey was causing extraordinary damage in all the Great Lakes by the late 1940s. No fish species was immune from attack, but the lampreys preferred lake trout, and this multimillion dollar fishing industry was brought to total collapse in the early 1950s. Lampreys then turned to rainbow trout, whitefish, burbot, yellow perch, and lake herring, all important commercial species. These stocks were decimated in turn. The lampreys then began attacking chubs and suckers. Coincident with the decline in attacked species, the sea lampreys themselves began to decline after reaching a peak abundance in 1951 in Lakes Huron and Michigan and in 1961 in Lake Superior. The fall has been attributed both to depletion of food and to the effectiveness of control measures (electrical barriers and use of chemical larvicides in selected spawning streams). Lake trout are presently recovering slowly, but wounding rates are still high. Fishery organizations are experimenting with the introduction into the Great Lakes of species that appear to be more resistant to lamprey attack, such as kokanee salmon (landlocked Pacific sockeye salmon).

Class Myxini—hagfishes

The hagfishes are an entirely marine group that feed on dead or dying fishes, annelids, mollusks, and crustaceans. Thus they are neither parasitic like lampreys nor predaceous, but are scavengers. There are only 21 described species of hagfishes, of which the best known in North America are the Atlantic hagfish *Myxine glutinosa* (Fig. 22-2) and the Pacific hagfish *Bdellostoma stouti*.

Hagfishes have long been of interest to comparative physiologists. Unlike any other vertebrate, they are isosmotic to sea water like marine invertebrates. They are the only vertebrates to have both pronephric and mesonephric kidneys in the adult, although only the latter forms urine. They have no less than four sets of hearts positioned at different places in the body to boost blood flow through their low-pressure circulatory system. Despite other unique anatomic and physiologic features of interest to biologists, they probably have fewer human admirers than any other group of fishes. The sports fisherman who

catches one discovers that his hook is so deeply swallowed that retrieval is impossible. To cap the sportsman's misfortune, the animal secretes enormous quantities of slimy mucus from large and small mucous glands located all over the body and from special slime glands positioned along its sides (Fig. 22-2). A single hagfish is said to be capable of converting a bucket of water into a mass of whitish jelly in minutes. Their habit of biting into and entering the bodies of gill-netted fish has not endeared them to commercial fishermen either. Entering through the anus or gills, hagfishes set about eating out the contents of the body, leaving behind a loose sack of skin and bones. But as fishing methods have passed from the use of drift nets and set lines to large and efficient otter trawls, hagfishes have ceased to be an important pest.

The yolk-filled egg of the hagfish may be nearly 1 inch in diameter and is enclosed in a horny shell. There is no larval stage and growth is direct. They are hermaphroditic but can produce only one kind of gamete at a time; a single individual may produce sperm at one season and eggs the next.

Class Elasmobranchii (Chondrichthyes)—sharks, skates, and rays

There are more than 3,000 species in this ancient, compact, and highly developed group. Although a much smaller and less diverse assemblage than the bony fishes, their impressive combination of well-developed sense organs, powerful jaws and swimming musculature, and predaceous habits ensure them of a secure and lasting niche in the aquatic community. One of their distinctive features is their cartilaginous skeleton, which must be considered degenerate instead of primitive. Although there is some calcification here and there, bone is entirely absent throughout the class.

With the exception of whales, sharks are the largest living vertebrates. The larger sharks may reach 40 to 50 feet in length. The dogfish sharks so widely used in zoologic laboratories rarely exceed 3 feet.

Characteristics

1. **Body fusiform** or **spindle shaped,** with a **heterocercal** caudal fin (Fig. 22-10); paired pectoral and pelvic fins, two dorsal median fins; pelvic fins in male modified for **"claspers";** fin rays present

2. **Mouth ventral; two olfactory sacs that do not break into the mouth cavity;** jaws present

3. Skin with **placoid** scales and **mucous glands;** teeth are modified placoid scales

4. **Endoskeleton entirely cartilaginous;** notochord persistent; vertebrae complete and separate; appendicular, girdle, and visceral skeletons present

5. Digestive system with a J-shaped stomach and intestine with a spiral valve; liver, gallbladder, and pancreas present

6. Circulatory system of several pairs of aortic arches; dorsal and ventral aorta, capillary and venous systems, hepatic portal and renal portal systems; two-chambered heart

7. Respiration by means of five to seven pairs of gills with separate and exposed gill slits; **no operculum**

8. No swim bladder

9. Brain of two olfactory lobes, two cerebral hemispheres, two optic lobes, a cerebellum, and a medulla oblongata; ten pairs of cranial nerves

10. Sexes separate; gonads paired; reproductive ducts open into cloaca; oviparous, ovoviviparous, or viviparous; direct development; fertilization internal

11. Kidneys of mesonephros (opisthonephros) type

Classification of the Elasmobranchii

Class Elasmobranchii. Six living orders, three extinct orders

Subclass Selachii (se-lay'kee-i) (Gr. *selachos,* fish having cartilage instead of bones)—**modern elasmobranchs.** Of the six orders, the two most important are below:
Order Squaliformes—modern sharks and rays.
Order Rajiformes—skates.

Distinctive characteristics

The body of a shark such as a dogfish shark (Fig. 22-5) is fusiform or spindle shaped. In front of the ventral mouth is a pointed **rostrum;** at the posterior end the vertebral column turns up to form the **heterocercal** tail. The fins consist of the paired **pectoral** and **pelvic** fins supported by appendicular skeletons, two median **dorsal** fins (each with a spine in *Squalus*), and a median **caudal** fin. A median **anal** fin is present in the smooth dogfish *(Mustelus).* In the male the medial part of the pelvic fin is modified to form a **clasper,** which is used in copulation. The paired **nostrils** (blind pouches) are ventral and anterior to the mouth. The lateral eyes are lidless, and behind each eye is a spiracle (remnant of the first gill slit). Five gill slits are found anterior to each pectoral fin. The leathery skin is covered with placoid scales (dermal denticles), each of which consist of a wide basal plate of dermal dentin and a spine covered with vitrodentin, or a shiny enamel-like dentin. These scales are modified to form teeth in the mouth and are the remnants of the dermal plates of placoderms. A well-developed **lateral line system** serves as a "distance touch" in water for detecting and locating objects and moving animals (predators, prey, and social partners). It is composed of a canal system extending along the side of the body and over the head. The canal opens at intervals to the surface. Inside are special receptor organs **(neuromasts)** that are extremely sensitive to vibrations and currents in the water.

Internally the cartilaginous skeleton is made up of a **chondrocranium,** which houses the brain and auditory organs and partially surrounds the eyes and olfactory organs; a vertebral column; a visceral skel-

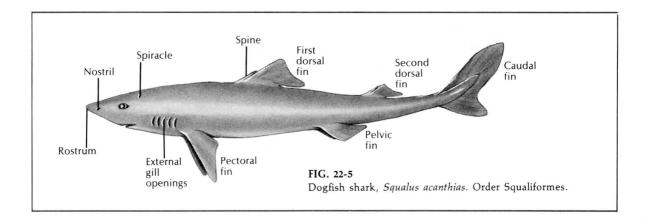

FIG. 22-5
Dogfish shark, *Squalus acanthias.* Order Squaliformes.

eton; and an appendicular skeleton. The jaws are suspended from the chondrocranium by ligaments and cartilages. Both the upper and the lower jaws are provided with many sharp, triangular teeth that, when lost, are replaced by other rows of teeth. Teeth serve to grasp the prey, which is usually swallowed whole. The muscles are segmentally arranged and are especially useful in the undulations of swimming.

The mouth cavity opens into the large **pharynx,** which contains openings to the separate gill slits and spiracles. A short, wide esophagus runs to the J-shaped stomach. A **liver** and **pancreas** open into the short, straight **intestine,** which contains the unique **spiral valve** that delays the passage of food and increases the absorptive surface (Fig. 22-6). Attached to the short rectum is the **rectal gland,** which secretes a colorless fluid containing a high concentration of sodium chloride. It assists the kidney in regulating the salt concentration of the blood. The chambers of the **heart** are arranged in tandem formation (Fig. 22-6) and the circulatory system is basically the same as that of the embryonic vertebrate and of the ammocoetes.

The **mesonephric kidneys,** are two long, slender

organs above the coelom and are drained by the **wolffian ducts,** which open into a single urogenital sinus at the **cloaca.** The wolffian ducts also carry the sperm from the testes of the male, which uses a clasper to deposit the sperm in the female oviduct. The müllerian duct, or oviduct (paired), carries the eggs from the **ovary** and coelom and is modified into a **uterus** in which a primitive placenta may attach the embryo shark until it is born. Such a relationship is actually **viviparous reproduction;** others simply retain the developing egg in the uterus without attachment to the mother's wall **(ovoviviparous reproduction).** Some sharks and rays deposit their fertilized eggs in a horny capsule called the "mermaid's purse," which is attached by tendrils to seaweed. Later the young shark emerges from this "cradle."

The nervous system is more advanced than that of ammocoete larval lampreys and is developed directly from the dorsal nerve cord of the embryo. The brain is typically made up of the three basic parts of the vertebrate brain—forebrain, midbrain, and hindbrain. These three parts form five subdivisions or regions—telencephalon, diencephalon, mesencephalon, metencephalon, and myelencephalon—

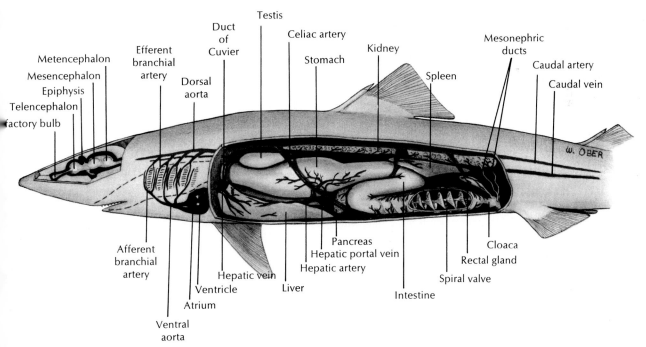

FIG. 22-6
Internal anatomy of dogfish shark, *Squalus acanthias.*

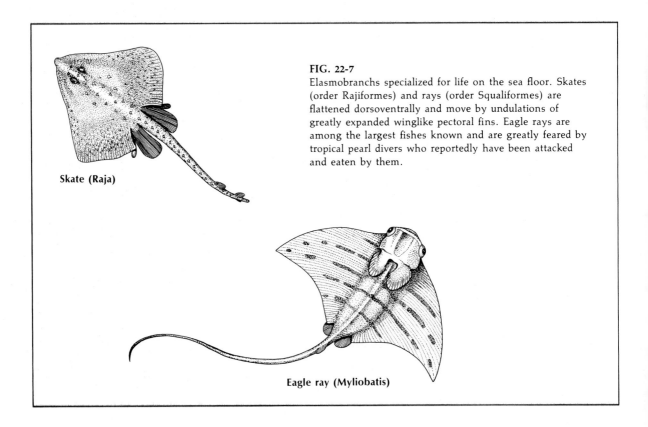

FIG. 22-7
Elasmobranchs specialized for life on the sea floor. Skates (order Rajiformes) and rays (order Squaliformes) are flattened dorsoventrally and move by undulations of greatly expanded winglike pectoral fins. Eagle rays are among the largest fishes known and are greatly feared by tropical pearl divers who reportedly have been attacked and eaten by them.

Skate (Raja)

Eagle ray (Myliobatis)

each with certain functions. There are ten pairs of cranial nerves that are distributed largely to the head regions. Surrounding the spinal cord are the neural arches of the vertebrae. Along the spinal cord a pair of spinal nerves, with united dorsal and ventral roots, is distributed to each body segment.

Elasmobranchs have developed an interesting solution to the physiologic problem of living in a hyperosmotic medium. Like the ancestors of the bony fishes, those of sharks and their kin lived in freshwater, only returning to the sea in the Triassic (about 230 million years ago) after becoming thoroughly adapted to freshwater existence. Their body fluid salt concentration was now much lower than the surrounding seawater; to prevent water from being drawn out of the body osmotically, the elasmobranchs have retained nitrogenous wastes, especially urea and trimethylamine oxide, in the blood. These solutes combined with the blood salts raised the blood solute concentration to slightly exceed that of seawater, eliminating an osmotic inequality between their bodies and the surrounding seawater.

The skates and rays (Fig. 22-7) are specialized for bottom dwelling. In these the pectoral fins are greatly enlarged and are used like wings in swimming. The gill openings are on the underside of the head, but the large spiracles are on top. Water for breathing is taken in through these spiracles to prevent clogging the gills, for their mouth is often buried in sand. Their teeth are adapted for crushing their prey—mollusks, crustaceans, and an occasional small fish. Two members of this group are of especial interest—the stingrays and the electric rays. In the stingrays the caudal and dorsal fins have disappeared and the tail is slender and whiplike. The tail is armed with one or more saw-edged spines, which can inflict very dangerous wounds. Such wounds may heal slowly and leave complications. Electric rays (Fig. 22-23) have smooth, naked skins and have certain dorsal muscles modified into powerful electric organs, which can give severe shocks and stun their prey. Stingrays also have electric organs in the tail.

Much has been written about the man-eating propensities of sharks, both by those exaggerating

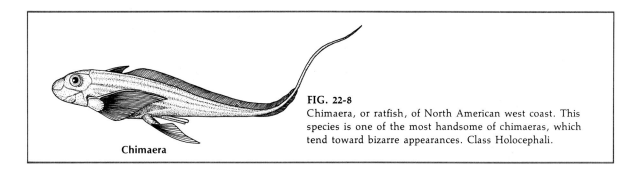

FIG. 22-8
Chimaera, or ratfish, of North American west coast. This species is one of the most handsome of chimaeras, which tend toward bizarre appearances. Class Holocephali.

their ferocious nature and by those seeking to write them off as harmless. It is true, as the latter group of writers argue, that sharks are by nature timid and cautious. But it also is a fact that certain of them are dangerous to man and deserve respect. There are numerous authenticated cases of shark attacks by *Carcharodon,* the great white shark (commonly reaching 20 feet and often larger); the mako shark, *Isurus;* the tiger shark, *Galeocerdo;* and the hammerhead, *Sphyrna.* More shark casualties have been reported from the tropical and temperate waters of the Australian region than from any other. During the Second World War there were several reports of mass shark attacks on the victims of ship sinkings in tropical waters.

Outside of North America, sharks and skates are much used for food. They make up about 1% of the present market for fish. In the United States disks of meat cut from skate "wings" are frequently sold in restaurants as scallops; the flavor and texture are similar to those of scallops, which the customer innocently supposes he is eating. There is a small market for shark leather, which has greater tensile strength and lasting qualities than mammalian leather.

Class Holocephali—chimaeras

The members of this small group, distinguished by such suggestive names as ratfish (Fig. 22-8), rabbitfish, spookfish and ghostfish, are remnants of an aberrant line that diverged from the elasmobranchs at least 300 million years ago (Carboniferous or Devonian). Fossil chimaeras (ky-meer'uz) were first found in the Jurassic, reached their zenith in the Cretaceous and early Tertiary (120 million to 50 million years ago), and have declined ever since. Anatomically they present an odd mixture of sharklike

and bony fish–like features. Instead of a toothed mouth, their jaws bear large flat plates. The upper jaw is completely fused to the cranium, a most unusual development in fishes. Their food is seaweed, mollusks, echinoderms, crustaceans, and fishes—all in all a surprisingly mixed diet for such a specialized grinding dentition. Chimaeras are not commercial species and are seldom caught. Despite their grotesque shape, they are beautifully colored with a pearly iridescence and have vivid emerald-green eyes that must be seen to be believed.

Class Osteichthyes (Teleostomi)— bony fish

In no other major animal group do we see better examples of adaptive radiation than among the bony fishes. Their adaptations have fitted them for every aquatic habitat except the most completely inhospitable. Body form alone is indicative of this diversity. Some have fusiform (streamlined) bodies and other adaptations for reducing friction. Predaceous, pelagic fish have trim, elongate bodies and powerful tail fins and other mechanical advantages for swift pursuit. Sluggish bottom-feeding forms have flattened bodies for movement and concealment on the ocean floor. The elongate body of the eel is an adaptation for wriggling through mud and reeds and into holes and crevices. Some, such as pipefishes, are so whiplike that they are easily mistaken for filaments of marine algae waving in the current (Fig. 22-21). Many other grotesque body forms are obviously cryptic or mimetic adaptations for concealment from predators or as predators. Such few examples cannot begin to express the amazing array of physiologic and anatomic specializations for defense and offense, food gathering, navigation, and reproduction in the diverse aquatic habitats to which bony fishes have adapted them-

FIG. 22-9

Osteolepis, primitive rhipidistian (crossopterygian) fish of middle Devonian times. This fish must be considered in direct line of descent between fish and amphibians because its type of skull was similar to that of primitive land vertebrates, and its lobe fin was of a pattern that could serve as beginning of tetrapod limb. This type of fin (archipterygium) consisted of median axial bones, with small bones radiating out from median ones. Some of its bones can be homologized with limb bones of tetrapods, such as humerus or femur and the ulna-radius or tibia-fibula elements. *Osteolepis* was covered by primitive cosmoid scales, not found in existing fish. These scales were of rhombic shape and consisted of basal bony layers, with a spongy layer of blood vessels covered with cosmine (dentin) and enamel. (Modified from Romer, A. S. 1966. Vertebrate paleontology. Chicago, University of Chicago Press.)

selves. Some of these adaptations are described in the pages that follow.

Origin, evolution, and diversity

The Osteichthyes are divided into three clearly distinct groups: Crossopterygii (lobe-finned fishes), Dipneusti (lungfishes), and Actinopterygii (ray-finned fishes). What are their origins? As pointed out earlier in this chapter, it seems safest to steer a middle course through current paleontologic debate and assign equal rank to all three groups within the class Osteichthyes. In other words, it is impossible to decide which one of these three groups, if any, might have served as ancestral stock for the other two. It is apparent from the fossil evidence that all three groups were distinct in the Devonian, some 400 million years ago. They are believed to have descended from an acanthodian of the Silurian, perhaps a creature similar to the fossil *Climatius,* illustrated in Fig. 21-5.

Subclass Crossopterygii

This group has had the least spectacular evolutionary radiation of all the Osteichthyes. It is by far the most important in an evolutionary sense because the tetrapods arose from one or more of their ancient members. The crossopterygians had nostrils that opened into the mouth (choanae), lungs as well as gills, and paired **lobed fins.** They first appeared in Devonian times, a capricious period of alternating droughts and floods when their lungs would have been a decided asset, if not absolutely essential, for their survival. They used their strong lobed fins as four legs to scuttle from one disappearing swamp to another that offered more promise for a continuing aquatic existence.

The crossopterygians are divided broadly into two groups. The **rhipidistians** (Fig. 22-9) appeared in the Devonian, flourished in the late Paleozoic, and then disappeared. This is the stock from which the amphibians have descended (Fig. 22-1). Among the primitive characteristics of the rhipidistians were the fusiform shape, two dorsal fins, and a heterocercal tail (Fig. 22-10). The paired fins bore sharp resemblances to a tetrapod limb, for they consisted of a basal arrangement of median or axial bones, with other bones radiating out from these median ones. Some of the proximal bones seem to correspond with the three chief bones of the tetrapod limb. The scales of these primitive fish were of the **cosmoid** type, a thick complex scale of dentinlike cosmine, enamel, and vascular pulp cavities. This type of scale is not found in modern fish but has been replaced by the bony cycloid type.

The other group of crossopterygians was the **coelacanths.** These also arose in the Devonian, radiated somewhat, and reached their evolutionary peak in the Mesozoic. At the end of the Mesozoic they also

Notochord or vertebral column

Protocercal (some larval fishes)

Vertebral column

**Heterocercal
(many primitive fishes, sturgeon, sharks)
(most ancestral type)**

**Hypocercal
(some primitive fishes, reptilian ichthyosaur)**

**Diphycercal
(lungfishes, crossopterygians, Polypterus)**

Homocercal (teleost fishes)

FIG. 22-10
Types of caudal fins among fish. Some functional
correlations may be seen among these different
types. Heterocercal tail, for example, is found in
fish without swim bladder, for it tends to
counteract gravity while swimming.

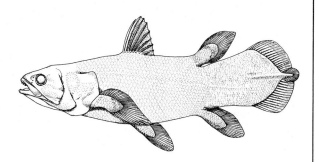

FIG. 22-11
Coelacanth, *Latimeria chalumnae.* This marine
surviving relic of the crossopterygians that flourished
some 350 million years ago has fleshy-based ("lobed")
fins with which its ancestors used to pull themselves
across land from pond to pond. (Courtesy Vancouver
Public Aquarium, British Columbia.)

disappeared but left one remarkable surviving species,
the living coelacanth *Latimeria chalumnae* (Fig. 22-
11). Since the last coelacanths were believed to have
become extinct 60 million years ago, the astonish-
ment of the scientific world can be imagined when
the remains of a coelacanth were found on a dredge
off the coast of South Africa in 1938. An intensive
search was begun in the Comoro Islands area near
Madagascar where, it was learned, native Comoran
fisherman occasionally caught them with hand lines
at great depths. By the end of 1971, a total of 69
specimens had been caught, many in excellent condition, although none have been kept alive after capture. The "modern" marine coelacanth is a descendant of the Devonian freshwater stock. The tail is
of the **diphycercal** type (Fig. 22-10) but possesses a
small lobe between the upper and lower caudal lobes,
producing a three-pronged structure (Fig. 22-11).
Coelacanths also show some degenerative features,
such as more cartilaginous parts and a swim bladder
that was either calcified or else persisted as a mere
vestige. They also lack the internal nostril so characteristic of crossopterygians, but this is probably a
secondary loss after the adoption of a deep-sea existence; obviously neither nostrils nor functional lungs
have any relevance for such a life habit.

Subclass Dipneusti

The lungfishes are considered by many paleontologists to be an offshoot of an early rhipidistian. This
view is contested by a minority of paleontologists,
among them the influential Swedish paleontologist

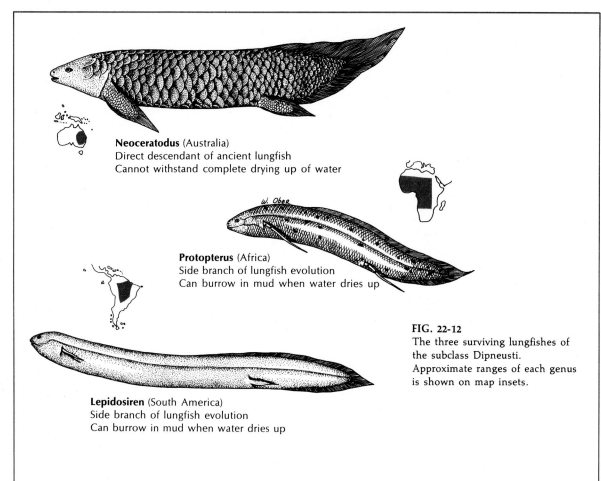

Neoceratodus (Australia)
Direct descendant of ancient lungfish
Cannot withstand complete drying up of water

Protopterus (Africa)
Side branch of lungfish evolution
Can burrow in mud when water dries up

Lepidosiren (South America)
Side branch of lungfish evolution
Can burrow in mud when water dries up

FIG. 22-12
The three surviving lungfishes of the subclass Dipneusti. Approximate ranges of each genus is shown on map insets.

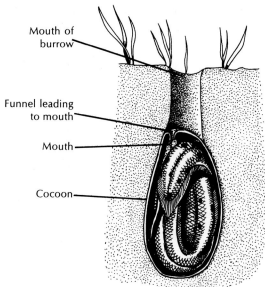

Mouth of burrow

Funnel leading to mouth

Mouth

Cocoon

FIG. 22-13
Estivating African lungfish *Protopterus.* As water disappears during dry season, *Protopterus* digs a burrow, coils tightly, and secretes mucus from skin and lips. Combined with mud, the mucus dries into a stiff, papery cocoon. A hollow tube of mucus extruded from mouth forms direct opening for air. Estivation may last several months during which nitrogenous wastes gradually accumulate in body tissues. When rains return, rising water level softens cocoon and *Protopterus* breaks out quickly, croaks several times with evident pleasure, and swims off. Subclass Dipneusti.

E. Jarvik, who prefers to leave their ancestry unassigned. The lungfishes were a distinct group in the Devonian. The earliest known lungfish, *Uranolophus,* shows a much closer relationship to the crossopterygians than do existing forms, for it had a heterocercal tail, two dorsal fins, cosmoid scales, and ossified shoulder girdle—structures that have undergone considerable modification in modern lungfishes. Even modern lungfishes resemble the crossopterygians with their lobe-shaped paired fins and lungs. At the same time all lungfishes, extinct or living, differ from crossopterygians in several significant skeletal features, including the totally different origin of the internal nostrils.

The three surviving genera of lungfishes (Fig. 22-12) have descended from Devonian ancestors such as *Uranolophus* or the somewhat later fossil *Dipterus.* The least specialized is *Neoceratodus,* the living Australian lungfish, which may attain a length of 5 feet. This lungfish is able to survive in stagnant, oxygen-poor water by coming to the surface and gulping air into its single lung, but it cannot live out of water. The South American lungfish *Lepidosiren* and the African lungfish *Protopterus* are evolutionary side branches of the Dipneusti, and they can live out of water for long periods of time. *Protopterus* lives in African streams and rivers that run completely dry during the dry season, with their mud beds baked hard by the hot tropical sun. The fish burrows down at the approach of the dry season and secretes a copious slime that is mixed with mud to form a hard

cocoon in which it estivates until the rains return (Fig. 22-13).

Subclass Actinopterygii

This huge assemblage contains all our familiar bony fishes. The fossil record reveals that the group had its beginnings in Devonian freshwater lakes and streams. The earliest actinopterygians were small fish with large eyes and extended mouths. Their tails were **heterocercal** (Fig. 22-10). They had a single dorsal fin and a single anal fin; paired fins were represented by the anterior pectoral fins and the posterior pelvic fins. Their skeletons were largely bone. Their trunks and tails were encased in an armor of heavy, rhombic scales **(ganoid).** Most of these early fish had functional lungs, as did all the Devonian fishes, but they lacked internal nostrils. All had gills (five pairs or less) and spiracles. These early actinopterygians belonged to the order Palaeonisciformes (pay"lee-o-nis-sif-for'meez). One common genus in the fossil record was *Cheirolepis,* a generalized type that had some resemblance to certain acanthodians. The palaeoniscids were distinctly different from their freshwater contemporaries, the crossopterygians and the lungfishes, with whom they shared the Devonian swamps and rivers. The palaeoniscids were rare in the Devonian, became a more secure group in the Carboniferous and Permian, but nearly disappeared in the Triassic when their descendants returned to the sea.

In their evolution, the actinopterygians have passed through three stages. The most primitive group are

FIG. 22-14
White sturgeon, *Acipenser.* Sturgeons, largest of all freshwater fish, were once abundant in North America but their size and very slow growth rate made them vulnerable to fishermen who destroyed them as nuisances. The largest North American sturgeons were in British Columbia's Fraser River system. Some caught early in this century reached 1,200 pounds, but sturgeons of this size will never be seen again on this continent. Even larger sturgeons are still collected in Russia, where the female's eggs (roe) are converted into gourmet caviar. Superorder Chondrostei.

FIG. 22-15
Long-nosed garpike, *Lepidosteus osseus.* This and other garpikes thrive in the rivers of the southern United States. They are voracious and solitary feeders that subsist on crayfish and small fishes. On sunny days they may often be seen floating on the river surface looking like drifting pieces of wood. Superorder Holostei.

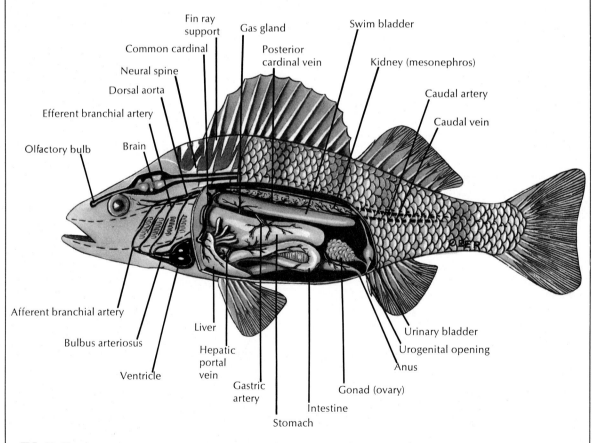

FIG. 22-16
Internal anatomy of the yellow perch, *Perca flavescens,* a freshwater teleost.

the Chondrostei, represented today by the freshwater and marine sturgeons (Fig. 22-14) and the bichir, *Polypterus,* of African rivers (Fig. 22-1). *Polypterus* is an interesting relic with a lunglike swim bladder and many other primitive characteristics; it resembles an ancient palaeoniscid more than any other living descendant. There is no satisfactory explanation for the survival to the present of certain fish such as this one and the coelacanth *Latimeria* when all of their kin perished millions of years ago.

A second actinopterygian group is the Holostei. There were several lines of descent within this group, which flourished during the Triassic and Jurassic. They declined toward the end of the Mesozoic as their successors, the teleosts, crowded them out. But they left two surviving lines, the bowfin, *Amia* (Fig. 22-1), of shallow, weedy waters of the Great Lakes and Mississippi valley, and the garpikes (or simply gars) of eastern North America (Fig. 22-15).

The third group is the Teleostei, the modern bony fishes (Fig. 22-16). No definite statements can be made about any criteria of success as they apply to modern teleosts. Diversity appeared early in teleost evolution, foreshadowing the truly incredible variety of body forms among teleosts today. The skeleton of primitive fish was largely ossified, but this condition regressed to a partly cartilaginous state among many of the Chondrostei and Holostei. Teleosts, however, have an internal skeleton almost completely ossified like the primitive members. The dermal investing bones of the skull (dermatocranium) and the chondrocranium (endocranium) around the brain and sense organs form a closer union among the teleosts than they did in the primitive bony fish. Other evolutionary changes among the teleosts were the movement of the pelvic fins forward to the head and thoracic region, the transformation of the lungs of primitive forms into air bladders (Fig. 22-16) with hydrostatic functions and without ducts, the changing of the heterocercal tail of primitive fish and of the intermediate superorders into a homocercal form, and the development of the thin cycloid and ctenoid scales from the thick ganoid type of early fish. Among other changes were the loss of the spiracles and the development of stout spines in the fins, especially in the pectoral, dorsal, and anal fins.

Characteristics of the Osteichthyes

1. **Skeleton more or less bony,** which represents the primitive skeleton; vertebrae numerous; noto-cord may persist in part; **tail usually homocercal**

2. Skin with mucous glands and with embedded dermal scales of three types: **ganoid, cycloid,** or **ctenoid;** some without scales; no placoid scales

3. Fins both median and paired, with **fin rays of cartilage or bone**

4. **Mouth terminal** with many teeth (some toothless); jaws present; olfactory sacs paired and may or may not open into mouth

5. Respiration by gills supported by bony gill arches and covered by a **common operculum**

6. **Swim bladder** often present with or without duct connected to pharynx

7. Circulation consisting of a two-chambered heart, arterial and venous systems, and four pairs of aortic arches; blood of nucleated red cells

8. Nervous system of a brain with small olfactory lobes and cerebrum and large optic lobes and cerebellum; ten pairs of cranial nerves

9. Sexes separate; gonads paired; fertilization usually external; larval forms may differ greatly from adults

Bony fish vary greatly in size. Some of the minnows are less than 1 inch long; other forms may exceed 10 feet in length. The swordfish is one of the largest and may attain a length of 12 to 14 feet. Most fish, however, are around 1 to 3 feet.

Classification of the Osteichthyes (Teleostomi)

Class Osteichthyes (os-te-ik′thee-eez)—**bony fish.** Three subclasses, 69 orders.

    Subclass Crossopterygii (cros-sop-te-rij′ee-i)—**lobe-finned fish.** Three extinct orders, one living order (Coelacanthoidei) containing one species, *Latimeria.*

    Subclass Dipneusti (dip-nyu′sti)—**lungfishes.** Four extinct orders, two living orders; these are Ceratodiformes containing the genus *Neoceratodus,* and Lepidosireniformes containing two genera, *Lepidosiren* and *Protopterus.*

    Subclass Actinopterygii (ak″ti-nop-te-rij′ee-i)—**ray-finned fishes.** Three superorders and 59 orders.

        Superorder Chondrostei (kon-dros′tee-i) (Gr. *chrondros,* cartilage, + *osteon,* bone)—**primitive ray-finned fish.** Eleven extinct orders, two living orders. These are Polypteriformes (the bichir, *Polypterus*) and Acipenseriformes (sturgeons and paddlefishes).

        Superorder Holostei (ho-los′tee-i) (Gr. *holos,* entire, + *osteon,* bone)—**intermediate ray-finned fish.** Four extinct orders, two living orders. These are Amiiformes (the bowfin, *Amia*) and Lepidosteiformes (gars, *Lepidosteus*).

        Superorder Teleostei (tel″e-os′tee-i) (Gr. *teleos,* complete, + *osteon,* bone)—**climax bony fish.** Body covered with thin scales without bony layer (cycloid or ctenoid) or scaleless; dermal and chondral parts of skull closely united; caudal fin mostly homocercal; mouth terminal; notochord a mere vestige; swim bladder mainly a hydrostatic organ and usually not opened to the esophagus; endoskeleton mostly bony. According to L. S. Berg there are 40 living orders containing 29 extinct families and 420 living families. Eight of the larger orders are below:

CLUPEIFORMES—54 families: tarpons, herrings, sardines, shad, alewifes, anchovies, smelts, whitefishes, pikes, mudminnows, salmon, trout.

CYPRINIFORMES—42 families: electric eel, characins, suckers, minnows, carp, goldfishes, loaches, catfishes.

ANGUILLIFORMES—29 families: freshwater eels, moray eels, conger eels, snipe eels.

GADIFORMES—4 families: codfishes and hakes.

GASTEROSTEIFORMES—4 families: sticklebacks.

SYNGNATHIFORMES—6 families: trumpet fishes, snipe fishes, pipefishes, seahorses.

PERCIFORMES—163 families: barracudas, mullets, perches, darters, sunfishes, grunters, croakers, moorish idols, damsel fishes, viviparous perches, wrasses, parrot fishes, trumpeters, sand perches, stargazers, blennies, wolf fishes, eel pouts, mackerels, tunas, swordfishes, and many others.

PLEURONECTIFORMES—6 families: flounders, flatfishes, soles, halibut.

■ STRUCTURAL AND FUNCTIONAL ADAPTATIONS OF FISH
Locomotion in water

To the human eye, fish appear capable of swimming at extremely high speeds. But our judgment is unconsciously tempered by our own experience that water is a highly resistant medium to move through. Most fishes, such as a trout or a minnow, can swim maximally about 10 body lengths per second, obviously an impressive performance by human standards. Yet when these speeds are translated into miles per hour it means that a foot long trout can swim only about 6.5 miles per hour. The larger the fish, the faster it can swim. A 2-foot long salmon can sprint 14 miles per hour and a 4-foot barracuda, the fastest fish measured, is capable of 27 miles per hour. Fish can swim this fast only for very brief

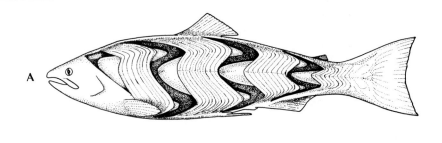

Path of tail through water

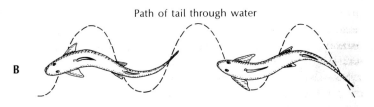

FIG. 22-17

A, Trunk musculature of a salmon. Segmental myotomes are **W** shaped when viewed from the surface. Musculature has been dissected away in four places to show internal anterior and posterior deflections of myotomes that improve muscular efficiency for swimming. **B,** Motion of swimming fish. Noncompressible water must be pushed aside by the forward motion of the head, driven by the snakelike stroke of the body. (**A** after Greene from Romer, A. 1970. The vertebrate body, ed. 4, Philadelphia, W. B. Saunders Co.; **B** modified from Marshall, P. T., and G. M. Hughes. 1967. The physiology of mammals and other vertebrates, New York, Cambridge University Press.)

periods during moments of stress; cruising speeds are much less.

The propulsive mechanism of a fish is its trunk and tail musculature. The axial, locomotory musculature is composed of zigzag muscle bands (myotomes) that on the surface take the shape of a W lying on its side (Fig. 22-17). Internally the muscle bands are deflected forward and backward in a complex fashion that apparently promotes efficiency of movement. The muscles are bound to broad sheets of tough connective tissue, which in turn tie to the highly flexible vertebral column. The way fishes swim is seen best in a relatively primitive fish such as a shark. When swimming, the shark body assumes the form of a sine wave. Waves of contraction begin on one side of the body at the front and proceed to the tail. When this wave has moved some distance, another wave is initiated at the front on the opposite side of the body. The process continues, with waves of contraction moving posteriorly, alternating from one side to the other. In higher bony fishes, the sweeping movement of the tail assumes a greater role.

Swimming is possible only because the density and noncompressibility of water offers great purchase for forward thrust. As a medium for locomotion, water offers another advantage: since the density of water is only slightly less than that of protoplasm, aquatic animals are almost perfectly supported and need expend no energy overcoming the force of gravity. Consequently, swimming is actually the most economic form of animal locomotion. For example, the energetic cost per kilogram body weight of traveling 1 kilometer is 0.39 kilocalories for a salmon (swimming), 1.45 for a gull (flying), and 5.43 for a ground squirrel (walking). However, the low energy cost of fish swimming is by no means fully understood. Relatively simple calculations show that a fish moves through water with only about one tenth the drag of a rigid model of the fish's body. The energy required to propel a submarine is many times greater than that consumed by a whale of similar size and moving at the same speed. Aquatic mammals and fishes create virtually no turbulence, a feat that man in his twentieth-century ingenuity is a long way from matching. The secret lies in the way aquatic animals bend their bodies and fins (or flukes) to swim, and in the textural properties of the body surface. It has recently been shown, for example, that the slimy surface of a fish reduces water

friction by at least 66%. Understanding the energetics of swimming remains part of the unfinished business of biology.

Neutral buoyancy and the swim bladder

All fishes are slightly heavier than water because their skeletons and other tissues contain heavy elements that are present only in trace amounts in natural waters. To keep from sinking, sharks must always keep moving forward in the water. The asymmetric (heterocercal) tail of a shark provides the necessary tail lift as it sweeps to and fro in the water and the broad head and flat pectoral fins (Fig. 22-5) act as planes to provide head lift. Sharks are also aided in their buoyancy problem by having very large livers containing a special fatty hydrocarbon called **squalene** that has a density of only 0.86. The liver thus acts like a large sack of buoyant oil that helps to compensate for the shark's heavy body.

By far the most efficient flotation device is a gas-filled space. The **swim bladder** (or gas bladder as it is often called) serves this purpose in the bony fishes. It arose from the paired lungs of the primitive Devonian bony fishes. Lungs were probably a ubiquitous feature of the Devonian freshwater bony fishes when, as we have seen, the alternating wet and dry climate probably made such an accessory respiratory structure essential for life. The primitive lung may have been similar to the lungs found in the existing *Polypterus,* the chondrostean fish of tropical Africa that so resembles the palaeoniscid ancestors of modern bony fish (Fig. 22-18). Swim bladders are present in all pelagic bony fishes but absent in most bottom dwellers such as flounders and sculpins.

By adjusting the volume of gas in the swim bladder, a fish can achieve neutral buoyancy and remain suspended indefinitely at any depth with no muscular effort. There are severe technical problems, however. If the fish descends to a greater depth, the swim bladder gas is compressed so that the fish becomes heavier and tends to sink. Gas must be added to the bladder to establish a new equilibrium buoyancy. If the fish swims up, the gas in the bladder expands, making the fish lighter. Unless gas is removed, the fish will rise with ever-increasing speed while the bladder continues to expand, until it pops helplessly out of the water. (This is a very real hazard for divers in helmeted diving suits who must carefully adjust the air pressure to prevent overinflation or underin-

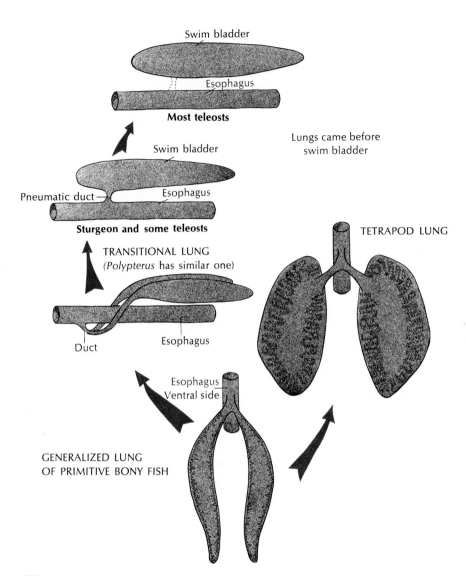

FIG. 22-18
Evolution of lungs and swim bladders. Fossil records of primitive bony fish indicate that most of them were provided with lungs that were adapted to climatic conditions existing during early evolution of fish. Lung originated on ventral side as double sac connected to throat by single duct. Embryologically, it may have started as gill pouches. From this generalized lung condition two lines of evolution occurred. (1) One line led to swim, or gas, bladder of modern teleost fish. Various transitional stages show that swim bladder and its duct (which is eventually lost) have shifted to dorsal position above esophagus to become structurally a buoyancy organ for flotation instead of for breathing. (2) Second line of evolution has led to tetrapod lung found in land forms. There has been extensive internal folding, but no radical change in lung position.

flation of their suits.) There are two ways fish adjust gas volume in the swim bladder. The less specialized fishes (trout, for example) have a **pneumatic duct** that connects the swim bladder to the esophagus (Fig. 22-18); these forms must come to the surface and gulp air to charge the bladder and obviously are restricted to relatively shallow depths. More specialized teleosts have lost the pneumatic duct (upper diagram in Fig. 22-18). Gas exchange depends on two highly specialized areas: a **gas gland** (Fig. 22-16) that secretes gas into the bladder and a **resorptive area,** or "oval," that can remove gas from the bladder. The gas gland contains a remarkable network of blood vessels **(rete mirabile)** arranged so that a vast number of arteries and veins in a tight bundle run in opposite directions

to each other. This is called **countercurrent flow,** and in some way, not yet understood, this arrangement makes possible a tremendous multiplication of gas concentration inside the swim bladder. The amazing effectiveness of this device is exemplified by a fish living at a depth of 8,000 feet. To keep the bladder inflated, the gas inside (mostly oxygen, but also variable amounts of nitrogen, carbon dioxide, carbon monoxide, and argon) must have a pressure exceeding 240 atmospheres. Yet the oxygen pressure in the fish's blood cannot exceed one-fifth atmosphere—equal to the oxygen pressure at the sea surface. Physiologists suggest that oxygen and other gases are actively transported within the gas gland's countercurrent exchange system; this would involve

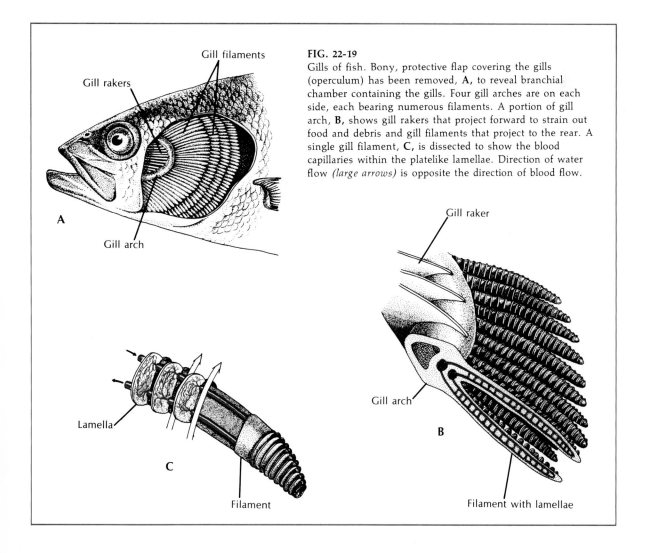

FIG. 22-19

Gills of fish. Bony, protective flap covering the gills (operculum) has been removed, **A,** to reveal branchial chamber containing the gills. Four gill arches are on each side, each bearing numerous filaments. A portion of gill arch, **B,** shows gill rakers that project forward to strain out food and debris and gill filaments that project to the rear. A single gill filament, **C,** is dissected to show the blood capillaries within the platelike lamellae. Direction of water flow *(large arrows)* is opposite the direction of blood flow.

Gill filaments

Gill rakers

Gill arch

A

Gill raker

Gill arch

Lamella

C

Filament

B

Filament with lamellae

a molecular carrier system not yet identified. Other theories have been proposed to explain swim bladder function, and this interesting organ is the object of considerable current research.

Respiration

Fish gills are composed of thin filaments covered with a thin epidermal membrane that is folded repeatedly into platelike **lamellae** (Fig. 22-19). These are richly supplied with blood vessels. The gills are located inside the pharyngeal cavity and covered with a movable flap, the **operculum.** This arrangement provides excellent protection to the delicate gill filaments, streamlines the body, and makes possible a pumping system for moving water through the mouth, across the gills, and out the operculum. Instead of opercular flaps as in bony fishes, the elasmobranchs have a series of **gill slits** out of which the water flows. In both elasmobranchs and bony fishes the branchial mechanism is arranged to pump water continuously and smoothly over the gills, even though to an observer it appears that fish breathing is pulsatile. The flow of water is opposite to the direction of blood flow (countercurrent flow), the best arrangement for extracting the greatest possible amount of oxygen from the water. Some bony fish can remove as much as 85% of the oxygen from the water passing over their gills. Very active fish, such as herring and mackerel, can obtain sufficient water for their high oxygen demands only by continually swimming forward to force water into the open mouth and across the gills. Such fish will be asphyxiated if placed in an aquarium that restricts free swimming movements, even though the water is saturated with oxygen.

A surprising number of fishes can live out of water for varying lengths of time by breathing air. Several devices are employed by different fishes. We have already described the lungs of the lungfishes, *Polypterus,* and the extinct crossopterygians. Freshwater eels often make overland excursions during rainy weather, using the skin as a major respiratory surface. The bowfin, *Amia,* has both gills and a lunglike swim bladder. At low temperatures it uses only its gills, but as the temperature and the fish's activity increases, it breathes mostly air with its swim bladder. The electric eel has degenerate gills and must supplement gill respiration by gulping air through its vascular mouth cavity. One of the best air breathers of all

is the Indian climbing perch that spends most of its time on land near the water's edge breathing air through special air chambers above the much-reduced gills.

Osmotic regulation

Freshwater is an extremely dilute medium with a salt concentration (0.001 to 0.005 gram moles per liter [M]) much below that of the blood of freshwater fishes (0.2 to 0.3 M). Water therefore tends to enter their bodies osmotically and salt is lost by diffusion outward. Although the scaled and mucus-covered body surface is almost totally impermeable to water, water gain and salt loss do occur across the thin membranes of the gills. Freshwater fish are **hyperosmotic regulators** that have several defenses against these problems (Fig. 22-20). First, the excess water is pumped out by the **mesonephric** kidney (see p. 450), which is capable of forming a very dilute urine Second, special **salt-absorbing cells** located in the gill epithelium are capable of actively moving salt ions, principally sodium and chloride, from the water to the blood. This, together with salt present in the fish's food, replaces diffusive salt loss. These mechanisms are so efficient that a freshwater fish devotes only a small part of its total energy expenditure to keeping itself in osmotic balance.

Marine bony fish are **hypoosmotic regulators** that encounter a completely different set of problems. Having a much lower blood salt concentration (0.3 to 0.4 M) than the seawater around them (about 1 M), they tend to lose water and gain salt. The marine teleost fish quite literally risks drying out, much like a desert mammal deprived of water. Again, marine bony fishes, like their freshwater counterparts, have evolved an appropriate set of defenses (Fig. 22-20). To compensate for water loss, the marine teleost drinks seawater. Although this behavior obviously brings needed water into the body, it is unfortunately accompanied by a great deal of unneeded salt. Unwanted salt is disposed of in two ways: (1) the major sea salt ions (sodium, chloride, and potassium) are carried by the blood to the gills where they are secreted outward by special **salt-secretory cells;** (2) the remaining ions, mostly the divalent ions (magnesium, sulfate, and calcium), are left in the intestine and voided with the feces. However, a small but significant fraction of these residual divalent salts in the intestine, some 10% to 20% of the total, pene-

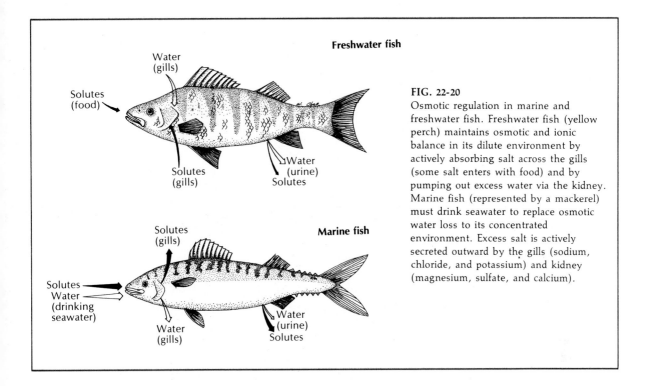

FIG. 22-20
Osmotic regulation in marine and freshwater fish. Freshwater fish (yellow perch) maintains osmotic and ionic balance in its dilute environment by actively absorbing salt across the gills (some salt enters with food) and by pumping out excess water via the kidney. Marine fish (represented by a mackerel) must drink seawater to replace osmotic water loss to its concentrated environment. Excess salt is actively secreted outward by the gills (sodium, chloride, and potassium) and kidney (magnesium, sulfate, and calcium).

trates the intestinal mucosa and enters the bloodstream. These ions are excreted by the kidney. Unlike the freshwater fish kidney, which forms its urine by the usual filtration-reabsorption sequence typical of most vertebrate kidneys (see p. 635), the marine fish's kidney excretes divalent ions by tubular secretion. Since very little if any filtrate is formed, the glomeruli have lost their importance and disappeared altogether in some marine teleosts. The pipefishes and the goosefish, shown in Figs. 22-21 and 22-22, are good examples of "aglomerular" marine fish.

Most bony fishes are restricted to either a freshwater or a seawater habitat. However some 10% of all teleosts can pass back and forth with ease between both habitats. Examples of these **euryhaline fishes,** (Gr. *eurys,* broad, + *hals,* salt), which must have highly adaptable osmoregulatory mechanisms, are salmon, steelhead trout, many flounders and sculpins, killifish, sticklebacks, and eels. Fishes that can tolerate only very narrow ranges of salt concentration—most freshwater and marine fishes—are said to be **stenohaline** (Gr. *stenos,* narrow, + *hals,* salt). Those fish that migrate from the sea to spawn in freshwater are **anadromous** (Gr. *ana,* up, + *dromos,* a running), such as salmon, shad, and marine lampreys. Freshwater forms that swim to the sea to spawn are **catadromous** (Gr. *kata,* down, + *dromos,* a running), such as the freshwater eel, *Anguilla.*

Coloration and concealment

Tropical coral reef fishes bear some of the most resplendent hues and strikingly brilliant color patterns in the animal kingdom. Viewed against the coral reef background of invertebrate and plant life that create a riot of color, the vivid markings and coloration of reef fishes attract relatively little attention. Their coloration is not always concealing, however, since tropical bony fishes tend toward vivid coloration even in areas of dull and somber backgrounds. Although conspicuous in these circumstances, they are protected by alertness and agility, or by their poisonous flesh. In this instance the coloration is an advertisement, warning would-be predators that they should seek their meal elsewhere.

Outside of coral reefs and other littoral habitats of the tropical seas, fishes, like most other animals, characteristically bear colors and patterns that serve to conceal them from enemies. Freshwater fishes wear subdued shades of green, brown, or blue above, grading to silver and yellow-white below. This is

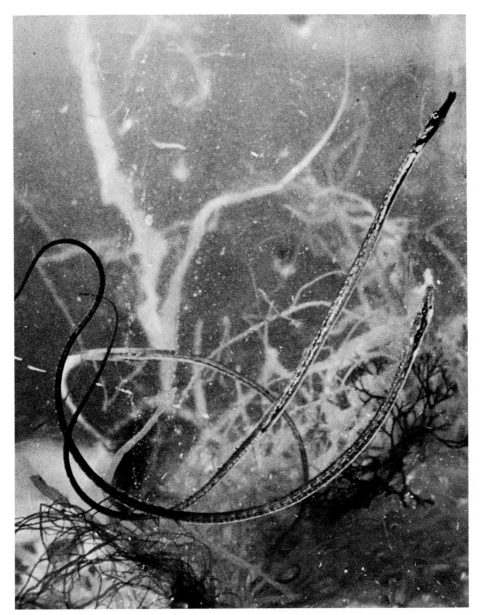

FIG. 22-21
Mimicry in pipefishes, *Entelurus aequoreus*. Their shape, coloration (which they change to match the background), and gentle swaying movements combine to make them almost impossible to see in their natural habitat of seaweed, eelgrass, and hydroids. Most pipefishes are only a few inches in length, although one species of the North American west coast may reach 18 inches. Pipefishes are also interesting because the male incubates the eggs in a special abdominal brood pouch; after mating and fertilization, the eggs are transferred there by the female. (Photo by B. Tallmark.)

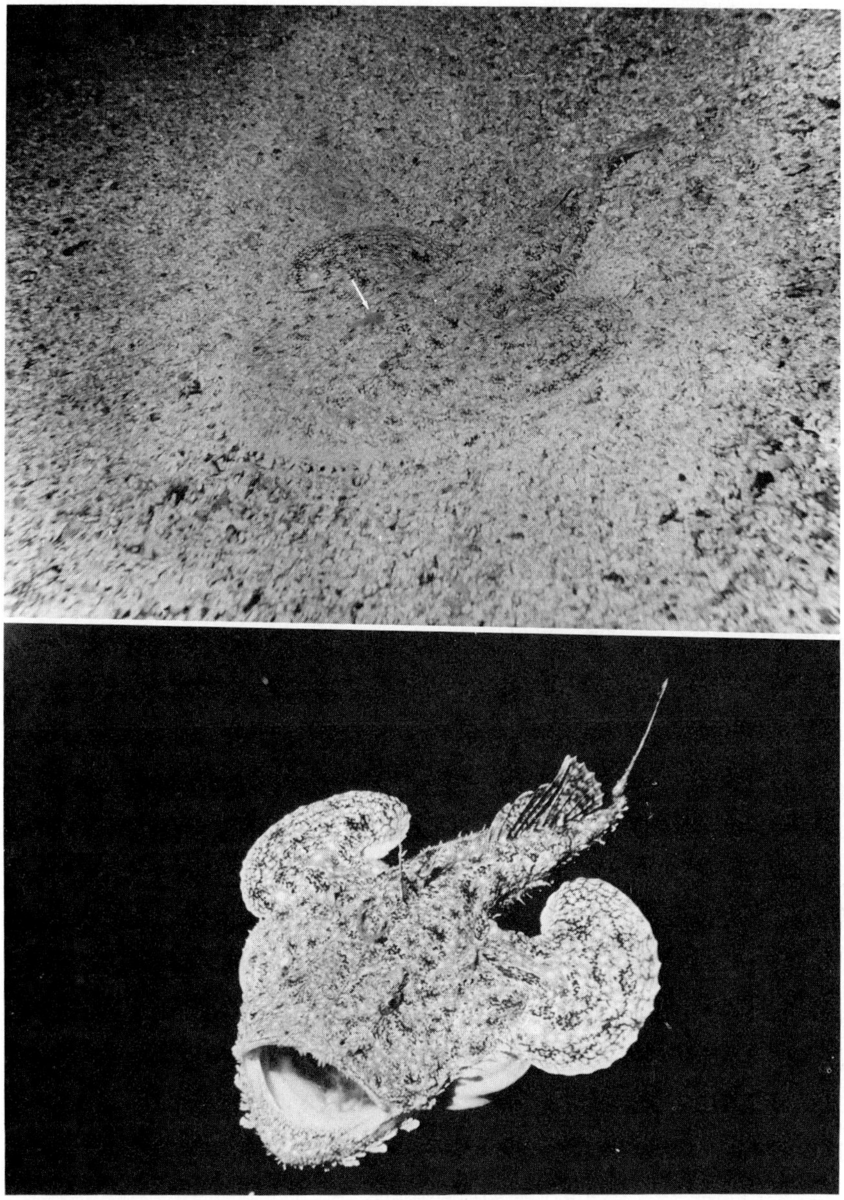

FIG. 22-22

A fish that fishes. **A,** Bedded down in the ocean bottom, beautifully concealed by protective coloration and fringed skin that breaks up the body outline, a goosefish, or angler, *Lophius piscatorius,* awaits its meal. Above its head swings a modified dorsal fin spine on the end of which is a fleshy tentacle *(arrow)* that contracts and expands in a convincing wormlike manner. When a fish approaches the alluring bait, the huge mouth opens suddenly, creating a strong current that sweeps the prey inside: in a split second all is over. Prodded out of its resting place by a skin diver, **B,** the goosefish reveals its true appearance. Photographed off the Norwegian coast at 80 feet depth. (Photos by T. Lundälv.)

obliterative shading. Seen from above against its normal background of water and stream bottom, the fish becomes almost invisible. Seen from beneath as it might be viewed by an aquatic predator, the pale belly of the fish blends with the water surface and sky above. The obliterative coloration is frequently enhanced by blotches, spots, and bars—conflicting patterns that like the camouflaged ships of the Second World War tend to break up the outline of the body. Some fishes, such as many flatfish species, can change their color to harmonize with the patterns of their background. Fish colors are chiefly the result of pigment within **chromatophores** in the dermal layer of the skin. The pigments are red, orange, yellow, and black and can be blended to produce other shades. Pigment dispersion within the branched chromatophores is controlled by the autonomic nervous system. Fish also bear **guanine** in their skin, a purine compound that gives many fish their silvery appearance.

Another form of concealment is protective resemblance, or **mimicry.** Pipefishes, for example, are of the shape and color of the seaweeds among which they live (Fig. 22-21). Pipefishes even sway slowly like seaweeds in a gentle current. There are many other examples of protective body form that appear to turn their owners into fragmented seaweed, or a leaf, or floating debris, or a weed-covered rock on the bottom. The goosefish, or angler, with its obliterative coloration and numerous fringes and branched appendages of skin, not only becomes nearly invisible when bedded down on the ocean floor, but sways a tempting bait above its huge mouth to attract prey (Fig. 22-22).

Electric fishes

The ability to produce strong electric shocks is confined to two groups of vertebrates: elasmobranchs (for example, electric ray) and teleosts (for example, electric eel, electric catfish). Best known and studied is the famous electric eel, *Electrophorus,* of the Amazon and Orinoco river systems of South America. This large, sluggish creature contains powerful electric organs that in 4- to 8-foot adults can produce

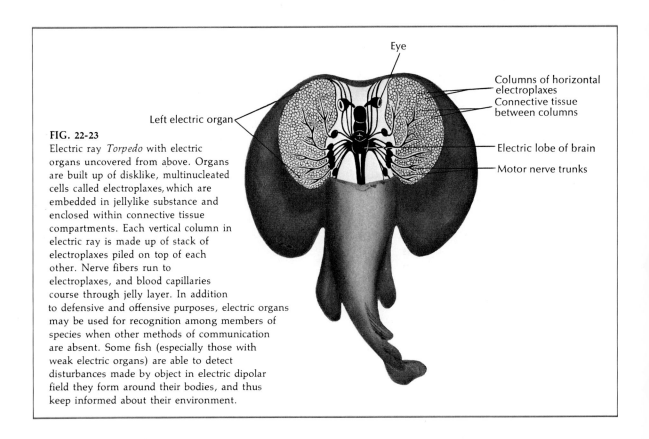

FIG. 22-23

Electric ray *Torpedo* with electric organs uncovered from above. Organs are built up of disklike, multinucleated cells called electroplaxes, which are embedded in jellylike substance and enclosed within connective tissue compartments. Each vertical column in electric ray is made up of stack of electroplaxes piled on top of each other. Nerve fibers run to electroplaxes, and blood capillaries course through jelly layer. In addition to defensive and offensive purposes, electric organs may be used for recognition among members of species when other methods of communication are absent. Some fish (especially those with weak electric organs) are able to detect disturbances made by object in electric dipolar field they form around their bodies, and thus keep informed about their environment.

Eye

Columns of horizontal electroplaxes
Connective tissue between columns

Left electric organ

Electric lobe of brain

Motor nerve trunks

paralyzing discharges exceeding 600 volts, quite ample to stun or kill its prey, or discourage potential enemies including large mammals. People who have accidentally met the electric eel under the latter's own terms report it is about as pleasant as contacting an uninsulated high-voltage wire. It is not necessary for the eel to actually touch its victims since the shocks stretch out in an electric field for many feet around the fish. Even small eels less than a foot long can discharge pulses exceeding 200 volts. The source of this impressive performance is the pair of electric organs lying on either side of the vertebral column and extending almost the entire length of the fish. Together they make up about 40% of the body weight. Each organ is composed of longitudinal columns of 6,000 to 10,000 thin, waferlike plates (electroplaxes) stacked one upon another like a long cylinder of coins. There are about 60 such columns in each organ. Each plate, or electroplax, is a modified muscle cell innervated by a nerve fiber and has very special electrical properties. One side of the plate is a nervous layer that will depolarize when stimulated; the opposite side is a nutritive layer that remains inactive. All the nervous sides of the plates face the tail of the animal. When the eel chooses to discharge its organ, motor impulses are sent out to the electric organs from a special neural center in the brain. These impulses travel to the numerous plates in a highly synchronized way so that all of the thousands of plates discharge simultaneously. Each plate develops a potential charge of about 150 millivolts and because the plates are arranged in series, the individual voltages summate like batteries arranged in series. Thus a high-voltage current flows from the tail to the head of the eel and completes the circuit in the water surrounding the eel. Electric eels must develop high voltages to overcome the high resistance of the freshwater in which they live.

Electric fishes living in the sea, such as the electric ray, *Torpedo* (Fig. 22-23), live in a low-resistance medium where high voltages are not required. They generate high amperages instead. In *Torpedo* each electric organ contains some 2,000 columns made up of about 1,000 electroplaxes arranged in vertical stacks. With fewer plates per column, the voltage produced is relatively low (about 50 volts), but the power output may be as much as 6 kilowatts! It is easy to understand why, when seasoned commercial fishermen happen to pick up one of these animals

in their trawl, they leave it strictly alone until it is dead and safe to handle.

Migrations

EEL. For centuries naturalists had been puzzled about the life history of the freshwater eel, *Anguilla* (an-gwil'la), a common and commercially important species of coastal streams of the North Atlantic. Each fall, large numbers of eels were seen swimming down the rivers toward the sea, but no adults ever returned. And each spring countless numbers of young eels, called "elvers," each about the size of a wooden matchstick, appeared in the coastal rivers and began swimming upstream. Beyond the assumption that eels must spawn somewhere at sea, the location of their breeding grounds was totally unknown. When the mystery was finally solved by the Danish scientist Johann Schmidt, the story was so fantastic that even today biologists occasionally pause to wonder if it can be true.

The first clue was provided by two Italian scientists, Grassi and Calandruccio, who in 1896 reported that elvers were not in fact larval eels; rather they were relatively advanced juveniles. The true larval eels, the Italians discovered, were tiny, leaf-shaped, completely transparent creatures that bore absolutely no resemblance to an eel. They had been called **leptocephali** by early naturalists who never suspected their true identity. In 1905 Johann Schmidt, supported by the Danish government, began a systematic study of eel biology that he continued until his death in 1933. Enlisting the cooperation of captains of commercial vessels plying the Atlantic, thousands of the leptocephali were caught in different areas of the Atlantic with the plankton nets Schmidt supplied them. By noting where in the ocean larvae in different stages of development were captured, Schmidt and his colleagues eventually reconstructed the spawning migrations.

When the adult eels leave the coastal rivers of Europe and North America, they swim steadily and apparently at great depth for 1 to 2 months until they reach the Sargasso Sea, a vast area of warm oceanic water southeast of Bermuda. Here, at depths of 1,000 feet or more, the eels spawn and die. The minute larvae then begin an incredible journey back to the coastal rivers of Europe. Drifting with the Gulf Stream and preyed upon constantly by numerous predators, they reach the middle of the Atlantic after 2 years.

By the end of the third year they reach the coastal waters of Europe where the leptocephali metamorphose into elvers, with an unmistakable eel-like body form. Here the males and females part company; the males remain in the brackish waters of coastal rivers and estuaries while the females continue up the rivers often penetrating hundreds of miles upstream. After 8 to 15 years of growth, the females, now 3 to 4 feet long, return to the sea to join the smaller males; both return to the ancestral breeding grounds thousands of miles away to complete the life cycle.

Schmidt found that the American eel *(Anguilla rostrata)* could be distinguished from the European eel *(A. vulgaris)* because it had fewer vertebrae—an average of 107 in the American eel as compared to an average 114 in the European species. Since the American eel is much closer to the North American coast line, it requires only about 8 months to make the journey. Recently it has been suggested that both American and European eels belong to the same species and that the adult European forms all die before reaching the spawning grounds. All eggs are laid by American eels (according to the theory), but when hatched, the larvae in the northern part of the spawning grounds will be carried by currents to Europe; those in the southern part will be carried toward the American coast. The larger number of vertebrae in the European eel could be explained on the basis of Jordan's law, which states that fish in colder waters have more vertebrae than those in warmer temperatures.

This theory remains to be confirmed. There are also several species of eels in the western Pacific and Indo-Pacific regions. They too undergo seaward spawning migrations, but nothing is known of the oceanic portion of their life histories. Many biologists suspect that once these species are studied with an investigation of the scale carried out by Schmidt, and by someone having his patience, there may well unfold an epic rivaling that of the common Atlantic eel.

HOMING SALMON. The life history of salmon is nearly as remarkable as that of the eel and certainly has received far more popular attention. Salmon are **anadromous**; that is, they spend their adult lives at sea but return to freshwater to spawn. The Atlantic salmon *(Salmo salar)* and the Pacific salmon (five species of the genus *Oncorhynchus* [on-ko-rink'us]) have this practice but there are important differences between the six species. The Atlantic salmon (as well as the closely related steelhead trout) make upstream spawning runs year after year. The five Pacific salmon species (king, sockeye, silver, humpback, and chum) each make a single spawning run, after which they die.

The virtually infallible homing instinct of the Pacific species is legend: after migrating downstream as a young smolt, a sockeye salmon ranges many hundreds of miles over the Pacific for nearly 4 years, grows to 5 to 10 pounds in weight, and then returns unerringly to spawn in the headwaters of its parent stream. Many years ago Canadian biologists marked and released nearly a half million young sockeyes born in a tributary of British Columbia's Fraser River. Eleven thousand of these were recovered 4 years later in the same parent stream; not one was discovered to have strayed into any of the dozens of other Fraser River tributaries. Experiments have shown that homing salmon are guided upstream by the characteristic odor of their parent stream. The salmon are apparently imprinted with the stream's odor while they are still unhatched embryos, since if the eggs are flown from the parent stream to another stream miles away, the adults still return to the parent stream after their residence at sea. The odor compound is a volatile organic substance, but its exact chemical nature remains unidentified; the embryonic salmon are probably conditioned to a mosaic of compounds released by the characteristic vegetation and soil in the watershed of the parent stream.

It is not yet understood how salmon find their way to the mouth of the river from the trackless miles of the open ocean. We cannot believe that they are capable of distinguishing the parent stream odor while still hundreds of miles at sea. Recent experiments suggest that adult salmon, like birds, are guided in the ocean by celestial cues (stars or azimuth position of the sun). To do this the salmon would require time-keeping abilities, certainly not an unreasonable possibility in view of the widespread presence of "biologic clocks" in many organisms from the simplest to the most advanced (p. 921). But the crucial experiment remains to be performed.

Reproduction

The teleosts show many types of sexual reproduction patterns. Although the hermaphroditic condition may occasionally occur abnormally in many species, only one or two families (for example, Serranidae) are truly hermaphroditic. In hermaphroditic forms the gonads are each divided into testicular and ovarian zones.

The fish *(Poecilia formosa)* discovered in Texas illustrates well the odd type of parthenogenesis called **pseudogamy,** or gynogenesis. This process involves the entrance and activation of an egg by a spermatozoon whose nucleus does not make genetic fusion with the nucleus of the egg. Sperm are furnished by males of related species, but the descendants are all like that of the female because they are genetically alike.

The **testes** are usually elongated, whitish organs divided into lobules that contain cysts of maturing germ cells. Within each cyst the maturing cells are always of the same stage of development. The lobules open into the spermatic duct (with secretory lining), which runs into the urogenital sinus. Males often become sexually mature before the females, and their testes may be active throughout the year; in others the testes have seasonal rhythms in reproductive activity. The **ovaries** may run the length of the abdominal cavity and are made up of many ovarian follicles supported by connective tissue. The ovary, with its membranous covering, may be continuous with the oviduct, or the ovaries may be naked and discharge their eggs into the peritoneal cavity, whence they are picked up by the oviducts (mullerian ducts). The paired oviducts may open through a common urogenital pore behind the anus, or they may open through genital pores. Some fish such as trout and salmon have no oviducts; others such as the freshwater eel have neither sperm ducts nor oviducts. Usually eggs are produced during a seasonal rhythm and the ovaries are quiescent at other times. A few (for example, hake) are known to have active ovaries at all times. Some fish such as the cod produce

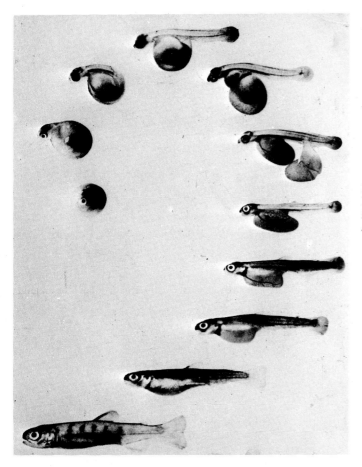

FIG. 22-24
Development of salmon from egg to fingerling. (Courtesy H. Kelly, U. S. Fish and Wildlife Service.)

enormous numbers of eggs (9 million have been found in the ovaries of a single female).

Most teleost fish are oviparous, laying eggs that are fertilized in the water by the sperm discharged by the male, usually in contact with the female. Gametes in water have limited viability unless they are united. Certain fish, however, are ovoviviparous (young born alive but nourished by egg yolk in mother) or viviparous (young with type of placental attachment to wall of uterus). Ovoviviparity or viviparity is found in the mosquito fish *(Gambusia)* and certain sea perch (Embiotocidae) as well as a few others.

Male fish have evolved many devices for transferring sperm into the female. Copulatory organs include the urogenital papillae, anal fins, and other specialized structures. Fertilization usually takes place while the egg is still in the ovarian follicle. In *Rhodeus* the female lays her eggs in the gill spaces of a freshwater mussel by means of a very long urogenital papilla. In fish that are ovoviviparous, the young develop within a cavity of the oviduct.

Soon after eggs are laid in water, they take up water and harden. Cleavage (meroblastic) occurs in the blastodisk of the zygote, and a blastoderm is formed. As cleavage continues, the blastoderm spreads over and encloses the yolk mass. The space between the blastoderm and yolk is the segmentation cavity, or blastocoel. Development proceeds, and eventually a larval form of fishlike appearance with a large yolk sac is hatched (Fig. 22-24). Temperature has a great effect on regulating the speed of hatching and subsequent development.

References

Suggested general readings

Berg, L. S. 1940. Classification of fishes, both recent and fossil. Ann Arbor, Mich., Edwards Brothers, Inc. *A comprehensive treatise of great value.*

Curtis, B. 1949. The life story of the fish. New York, Harcourt, Brace & Co., Inc.

Hardisty, M. W., and I. C. Potter. 1971-1972. The biology of lampreys, 2 vols. New York, Academic Press Inc. *Thorough presentation of systematics, life histories, ecology, behavior, physiology, and economic impact of this small but biologically interesting group.*

Hoar, W. S., and D. J. Randall (editors). 1969-1972. Fish physiology, 6 vols. New York, Academic Press Inc. *This series represents the most complete and authoritative treatise on the functional biology of fishes published. Technical and detailed.*

Lineaweaver, T. H., III, and R. H. Backus. 1970. The natural history of sharks. Philadelphia, J. B. Lippincott Co. *One of the best of many books dealing with these intriguing animals.*

Lagler, K. F. 1956. Freshwater fishery biology, ed. 2. Dubuque, Iowa, William C. Brown Co. *This is an excellent treatise on the principles and methods of freshwater fishery and research. Nearly every aspect of the subject is treated and the work is well documented throughout.*

Lanham, U. 1962. The fishes. New York, Columbia University Press. *A concise account of the evolution, structure, and function of fishes.*

Marshall, N. B. 1971. Explorations in the life of fishes. Cambridge, Harvard University Press. *Excellent general biology of fishes. Fairly technical.*

Norman, J. R. 1963. A history of fishes, ed. 2 (revised by P. H. Greenwood). New York, Hill & Wang, Inc. *A revision of a famous treatise that covers nearly all aspects of fish study.*

Romer, A. S. 1959. The vertebrate story. Chicago, University of Chicago Press.

Schultz, L. P., and E. M. Stern. 1948. The ways of fishes. New York, D. Van Nostrand Co., Inc. *A somewhat popular discussion of the behavior of fish. Especially good for the beginning student in zoology.*

Young, J. Z. 1963. The life of vertebrates, ed. 2. New York, Oxford University Press. *Many chapters in this excellent treatise are devoted to the structure, evolution, and adaptive radiation of fish.*

Selected *Scientific American* articles

Applegate, V. C., and J. W. Moffett. 1955. The sea lamprey. **192:**36-41 (April). *Life history and early control measures are described.*

Brett, J. R. 1965. The swimming energetics of salmon. **213:**

80-85 (Aug.). *Describes studies on the remarkable effi- ciency of swimming by fish.*

Carey, F. G. 1973. Fishes with warm bodies. **228**:36-44 (Feb.). *Some tuna and mackerel shark species employ a circulatory heat exchanger to conserve body heat and increase swimming power.*

Fuhrman, F. A. 1967. Tetrodotoxin. **217**:60-71 (Aug.) This powerful poison of puffer fish and newts is finding in- creasing use in neurophysiologic studies.

Gilbert, P. W. 1962. The behavior of sharks. **207**:60-68 (July).

Gray, J. 1957. How fishes swim. **197**:48-54 (Aug.). *This British biologists calculations of fish swimming efficiency have been referred to as "Gray's paradox."*

Grundfest, H. 1960. Electric fishes. **203**:115-124 (Oct.).

Hasler, A. D., and J. A. Larsen. 1955. The homing salmon. **193**:72-76 (Aug.) *Describe experiments that indicate salmon locate parent stream by using sense of smell.*

Jensen, D. 1966. The hagfish. **214**:82-90 (Feb.).

Johansen, K. 1968. Air-breathing fishes. **219**:102-111 (Oct.). *Not all fishes are water breathers. Many remarkable alternatives for air breathing evolved in fishes; several are described in this article.*

Leggett, W. C. 1973. The migrations of the shad. **228**:92-98 (Mar.).

Lissmann, H. W. 1963. Electric location by fishes. **208**:50- 59 (Mar.). *Certain fishes generate weak electric fields to sense their environment and locate prey.*

Lühling, K. H. 1963. The archer fish. **209**:100-108 (July). *This small fish of southeast Asia spouts a stream of water to down insects it sights on plants above the water.*

Millot, J. 1955. The coelacanth. **193**:34-39 (Dec.). *Its biolo- gy and evolutionary relationships are discussed.*

Ruud, J. T. 1965. The ice fish. **213**:108-114 (Nov.). *A fami- ly of transparent Antarctic fishes possess neither red blood cells nor hemoglobin. The physiologic conse- quences are described.*

Scholander, P. F. 1957. The wonderful net. **196**:96-107 (April). *Describes the arrangement of blood vessels in fish and other vertebrates to form countercurrent ex- change systems.*

Shaw, E. 1962. The schooling of fishes. **206**:128-138 (June).

Todd, J. H. 1971. The chemical languages of fishes. **224**: 98-108 (May). *Many fishes emit and sense chemical sig- nals that guide much of their behavior. Experiments with catfishes are described.*

American toad. This principally nocturnal yet familiar amphibian feeds upon large numbers of insect pests as well as snails and earthworms. The warty skin contains numerous poison glands that produce a surprisingly poisonous milky fluid, providing the toad with excellent protection from a variety of potential predators. (Photo by L. L. Rue, III, Blairstown, N. J.)

CHAPTER 23

THE AMPHIBIANS

PHYLUM CHORDATA
CLASS AMPHIBIA*

Amphibians were the first vertebrates to attempt the transition from water to land. Strictly speaking, their crossopterygian ancestors were the first vertebrates to make the attempt with any success—a feat that would have a poor chance now because present well-established competitors make it impossible for a poorly adapted transitional form to gain a foothold. Amphibians are not completely land adapted and hover between aquatic and land environments. This double life is expressed in their name. Structurally they are between the fish on the one hand and the reptiles on the other. Although more or less adapted for a terrestrial existence, few of them can stray far from moist conditions, although many have developed devices for keeping their eggs out of open water where the larvae would be exposed to enemies.

*Am-fib'e-a (Gr. *amphi,* both or double, + *bios,* life).

The more than 2,500 species of amphibians are grouped into three living orders: the newts and salamanders of the order Urodela, least specialized and most aquatic of all amphibians; the frogs and toads of the order Anura, largest and most successful group of amphibians and closest to the stock from which the higher tetrapods descended; and the highly specialized, secretive, earthworm-like caecilians of the order Gymnophiona.

■ THE MOVEMENT ONTO LAND

The movement from water to land is perhaps the most dramatic event in animal evolution, since it involves the invasion of a habitat that in many respects is less suitable for life. The origin of life was conceived in water, animals are mostly water in composition, and all cellular activities proceed in water. Nevertheless, animals eventually moved onto dry

484

land, carrying their watery composition with them. To survive and maintain this fluid matrix, various structural, functional, and behavioral changes had to evolve. Considering that almost every system in the body required some modification, it is remarkable that all vertebrates are basically alike in fundamental structural and functional pattern: whether aquatic or terrestrial, vertebrates are obviously descendants of the same evolutionary limb.

Amphibians were not the first to move onto land. Insects made the transition earlier and plants much earlier still. The pulmonate snails were experimenting with land as a suitable place to live about the same time the early amphibians were. Yet of all these, the amphibian story is of particular interest because their descendants became the most successful and advanced animals on earth.

Physical contrast between aquatic and land habitats

Beyond the obvious difference in water content of aquatic and terrestrial habitats—water is wet and land is dry—there are several sharp differences between the two environments of significance to animals attempting to move from water to land.

GREATER OXYGEN CONTENT OF AIR. Air contains at least 20 times more oxygen than water. Air has about 210 ml. of oxygen per liter; water contains 3 to 9 ml. per liter. Furthermore the diffusion rate of oxygen is low in water. Consequently aquatic animals must expend far more effort extracting oxygen from water than land animals expend removing oxygen from air.

GREATER DENSITY OF WATER. Water is about 1,000 times denser than air and about 100 times more viscous. Although water is a much more resistant medium to move through, its high density, about equal to that of animal protoplasm, buoys up the body. One of the major problems encountered by land animals was the need to develop strong limbs and remodel the skeleton to support their bodies in air.

CONSTANCY OF TEMPERATURE IN WATER. Natural bodies of water, containing a medium with tremendous thermal capacity, experience little fluctuation in temperature. The temperature of the oceans remains almost constant day after day. In contrast, both the range and the fluctuation in temperature are acute on land. Its harsh cycles of freezing, thawing,

drying, and flooding, often in unpredictable sequence, present severe thermal problems to terrestrial animals.

VARIETY OF LAND HABITATS. The variety of cover and shelter on land were great inducements for its colonization. The rich offerings of terrestrial habitats include coniferous and temperate forests, tropical forests, grasslands, deserts, mountains, oceanic islands, and polar regions. Even so, earth's hydrosphere (oceans, seas, lakes, rivers, and ice sheets), though offering a less diverse range of habitats, contain the greatest number and variety of living things on earth.

OPPORTUNITIES FOR BREEDING ON LAND. The provision of safe shelter for the protection of vulnerable eggs and young is much more readily accomplished on land than in water habitats.

■ EVOLUTION OF AMPHIBIANS

The amphibians are descended from the crossopterygians, a group of lobe-finned fishes. Their fossils bear so many structural resemblances to those of the earliest amphibians that there can be no question about the relationship.

APPEARANCE OF LUNGS. The lobe-finned fishes flourished in freshwater lakes and streams of the Devonian, a time of alternating droughts and floods. During dry periods, pools and streams began to dry up, water became foul, and the dissolved oxygen disappeared. The only fish that could survive such conditions were able to utilize the abundance of oxygen in the air above them. Gills were unsuitable because in air the filaments collapse together into clumps that soon dry out. Virtually all the survivors of this period had a kind of lung that developed as an outgrowth of the pharynx. It was a relatively simple matter to enhance the efficiency of this air-filled cavity by improving its vascularity with a rich capillary network and by supplying it with arterial blood from the last (sixth) pair of aortic arches. Oxygenated blood was returned directly to the heart by a pulmonary vein to form a complete pulmonary circuit. This was the origin of the **double circulation** characteristic of all tetrapods: a **systemic** circulation, which serves the body, and a **pulmonary** circulation, which serves the lungs. The crossopterygian lung was inflated by gulping a large bubble of air into the mouth and forcing it into the lung by compression of the buccopharyngeal cavity. One group of cros-

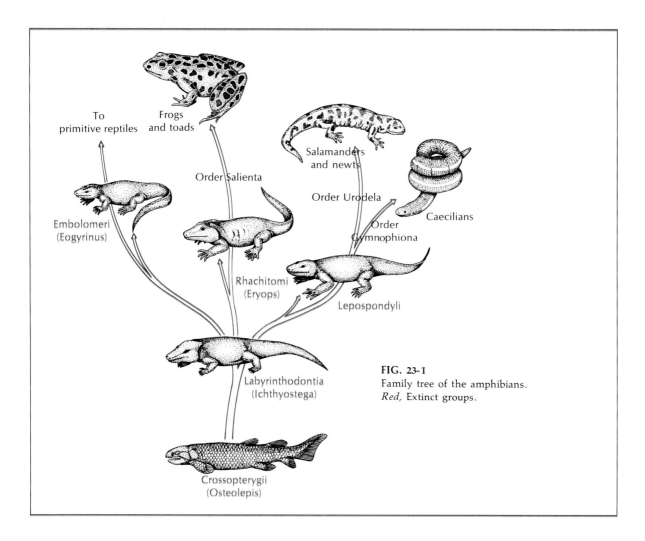

FIG. 23-1
Family tree of the amphibians.
Red, Extinct groups.

sopterygian fishes, the **rhipidistians,** had internal nostrils (nares) connecting the mouth to the excurrent nostril. This enabled them to breathe air with only the nose exposed above water. Since all tetrapods have retained this feature, the rhipidistians (of which *Osteolepis* was a primitive member; Fig. 23-1) are considered to be the ancestral tetrapod stock.

LIMBS FOR TRAVEL ON LAND. The evolution of limbs was also a product of difficult times during the Devonian period. When pools dried up altogether, fishes were forced to move to another pool that still contained water. The crossopterygians were equipped for the task, having strong lobed fins, used originally as swimming stabilizers, that could be adapted as paddles to lever their way across land in

search of water. The pectoral fins were especially well developed in these fishes. These had a series of skeletal elements in the fins and pectoral girdle that clearly foreshadowed the pentadactyl limb of tetrapods. We should note that the development of strong fins, and later, limbs, did not happen so that fish could colonize land, but **to permit them to find water and continue living like fish.** Land travel was simply and paradoxically a means for survival in water. But the evolution of lungs and limbs were fortunate and essential specializations that preadapted vertebrates for life on land.

EARLY AMPHIBIANS. The earliest amphibians of which we have adequate knowledge are represented by a 350 million-year-old fossil called *Ichthyostega* (Gr. *ichthys,* fish, + *stegos,* roof) (Fig. 23-1), collected from Upper Devonian strata in Greenland. This creature was clearly an intermediate between fish and amphibians. It had jointed, pentadactyl limbs for crawling on land, a more advanced ear structure for picking up airborne sounds, a foreshortening of the skull, and a lengthening of the snout that announced improved olfactory powers for detecting dilute airborne odors. Yet it was still fishlike in retaining a fishtail complete with fin rays and in having opercular (gill) bones. *Icthyostega* belonged to the Labyrinthodontia (Gr. *labyrinthos,* tortuous passage, + *odous, odontos,* tooth), the central stock from which all tetrapods arose. Unfortunately the subsequent evolution of the amphibians and the lines leading to the higher vertebrates have not been completely unraveled, and students of vertebrate evolution continue to debate varying theories vigorously. Without involving ourselves deeply in this complex controversy, we can say that according to one school (K. S. Thomson and others) all tetrapods have a monophyletic origin from some crossopterygian ancestor. This version is portrayed in Fig. 23-1. Another school (E. Jarvik) holds that the tetrapods have a diphyletic origin, with the urodeles and caecilians descending from one stock of crossopterygians and the anurans and all higher tetrapods from a separate crossopterygian stock. What we can safely say is that the amphibians originated during the Devonian and flourished during the subsequent Carboniferous period. Unlike the capricious Devonian, the Carboniferous was characterized by a warm, wet climate during which mosses and large ferns grew in profusion on a swampy landscape. Conditions were ideal for the amphibians. They radiated quickly into a great variety of species, feeding on the abundance of insects, insect larvae, and aquatic invertebrates available: this was the Age of Amphibians. But with water everywhere, there was little selective pressure to encourage movement onto land and many amphibians actually improved their adaptations for living in water. Their bodies became flatter for moving about in shallow water. Many of the urodeles (newts and salamanders), which may have descended from the lepospondyls (Fig. 23-1) developed weak limbs. The tail became better developed as a swimming organ. Even the anurans (frogs and toads), which are the most terrestrial of all amphibians, developed specialized hind limbs with webbed feet better suited for swimming than for movement on land. All

groups of amphibians use their porous skin as an accessory breathing organ. This specialization was encouraged by the swampy surroundings of the Carboniferous, but presented serious desiccation problems for life on land.

AMPHIBIANS' CONTRIBUTION TO VERTEBRATE EVOLUTION. Amphibians have met the problems of independent life on land only halfway. To be sure, they made several important contributions to the transition that required the evolution of their own descendants, the reptiles, to complete. Two changes of crucial importance were the change from gill to lung breathing and the development of limbs for locomotion on land. Amphibians also show strengthening changes in the skeleton so that the body can be supported in air. And a start was made toward shifting special sense priorities from the lateral line system of fish to the senses of smell and hearing. For this, both the olfactory epithelium and the ear required redesigning to improve sensitivities to airborne odors and sounds.

Despite these modifications the amphibians are basically aquatic animals. Their skin is thin, moist, and unprotected from desiccation in air. Most important, the amphibians remain chained to the aquatic environment by their mode of reproduction. Their eggs are shed directly into the water and are externally fertilized (with very few exceptions), and the larvae that hatch pass through an aquatic tadpole stage. It is true that many amphibians have developed ingenious devices for laying their eggs anywhere except in open water, to give their young a better chance for life. They may lay eggs under logs, under rocks, in the moist forest floor, in flooded tree holes, in pockets on the mother's back, or in folds of the body wall. However, it remained for the reptiles to complete the conquest of land with the development of an **amniotic egg** that finally freed the vertebrates from a reproductive attachment to water. With the appearance of reptiles at the end of the Paleozoic, the halcyon era for the amphibians began to fade. The reptiles captured rule of both water and land and removed most amphibians from both environments. From the survivors have descended the three modern orders of amphibians.

Characteristics

1. Skeleton mostly bony, with varying number of vertebrae; ribs present in some, absent in

others; notochord does not persist; **exoskeleton absent**

2. Body forms vary greatly from an elongated trunk with distinct head, neck, and tail to a compact, depressed body with fused head and trunk and no intervening neck
3. **Limbs, usually four (tetrapod),** although some are legless; forelimbs of some much smaller than hind limbs, in others all limbs small and inadequate; **webbed feet often present**
4. **Skin smooth and moist with many glands,** some of which may be poisonous; **pigment cells (chromatophores)** common, of considerable variety, and in a few capable of undergoing various patterns in accordance with different backgrounds; **no scales,** except concealed dermal ones in some
5. Mouth usually large with small teeth in upper or both jaws; **two nostrils open into anterior part of mouth cavity**
6. Respiration by gills, lungs, skin, and pharyngeal region either separately or in combination; external gills in the larval form and may persist throughout life in some
7. **Circulation with three-chambered heart,** two auricles and one ventricle, and a double circulation through the heart; skin abundantly supplied with blood vessels
8. Excretory system of paired mesonephric kidneys; urea main nitrogenous waste
9. Ten pairs of cranial nerves
10. Separate sexes; fertilization external or internal; metamorphosis usually present; **eggs with jelly-like membrane coverings**

Brief classification

Order Gymnophiona (jim"no-fy'o-na) (Gr. *gymnos,* naked, + *ophioneos,* of a snake) **(Apoda)—caecilians.** Body wormlike; limbs and limb girdle absent; mesodermal scales may be present in skin; tail short or absent; one family, 17 genera.

Order Urodela (yu"ro-dee'la) (Gr. *oura,* tail, + *dēlos,* visible) **(Caudata)—salamanders, newts.** Body with head, trunk, and tail; no scales; usually two pairs of equal limbs; eight families, 51 genera.

Order Salientia (say"lee-ench'e-a) (L. *saliens,* leaping, + *-ia,* pl. suffix) **(Anura)—frogs, toads.** Head and trunk fused; no tail; no scales; two pairs of limbs; mouth large; lungs; 10 vertebrae, including urostyle; 13 families, 100 genera.

■ STRUCTURE AND NATURAL HISTORY OF ORDERS
Gymnophiona (Apoda)

This little-known order contains some 75 species of burrowing, wormlike creatures commonly called **caecilians** (Fig. 23-1). They are distributed in tropical forests of South America (their principal home), Africa, and southeast Asia. They are characterized by their long, slender body, small scales in the skin, many vertebrae, long ribs, no limbs, and terminal anus. The eyes are small and most species are totally blind as adults (Fig. 23-2). Replacing eyes, which would be useless for a subterranean existence, are special sensory tentacles on the snout. Because they are almost all burrowing forms, they are seldom seen by man. Their food is mostly worms and small invertebrates, which they find underground. Fertilization is internal, and the male is provided with a protrusible cloaca by which he copulates with the female.

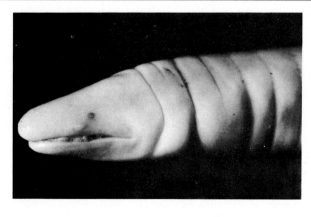

FIG. 23-2
Head and anterior region of caecilian. Order Gymnophiona (Apoda). These legless and wormlike amphibians may reach length of 18 inches and diameter of $3/4$ inch. Their body folds give them appearance of segmented worm. They have many sharply pointed teeth, a pair of tiny eyes mostly hidden beneath skin, a small tentacle between eye and nostril, and some forms have embedded mesodermal scales. (Courtesy General Biological Supply House, Inc., Chicago.)

The eggs are usually deposited in moist ground near the water; the larvae may be aquatic or the complete larval development may occur in the egg. In some species the eggs are carefully guarded in folds of the body during their development.

Urodela (Caudata)

The salamanders and newts of this order number about 230 species. These are the tailed amphibians, the least specialized in form and habit of all amphibians. Although urodeles are found in almost all temperate and tropical regions of the world, it is North America that contains most of the species. Urodeles are typically small; most of the common North American salamanders are less than 6 inches long. Some aquatic forms are considerably longer, and the carnivorous Japanese giant salamander may exceed 5 feet in length.

Urodeles have primitive limbs set at right angles to the body with the forelimbs and hind limbs about the same size. In some the limbs are rudimentary *(Amphiuma),* and in others *(Siren)* there are no hind limbs. Many aquatic amphibians never leave the water in their entire life cycle. Others assume a terrestrial life, living in moist places under stones and rotten logs, usually not far from the water. All have gills at some stage of their lives; some lose their gills when they become adults.

Although some species have lungs in place of gills, others have neither and breathe exclusively through the skin and pharyngeal region. This is true with the lungless salamanders of the family Plethodontidae, a common group in North America believed to have originated in swift mountain streams of the Appalachian mountains. Mountain brook water, which is cool and well oxygenated, is excellent for cutaneous breathing. The cold temperature of the water slows down metabolism and thus less oxygen is required. This family also has an interesting structural adaptation, the nasolabial groove, which helps clear the nostrils of water. The groove is flushed out by the secretions of a gland that empties into it and by cilia that carry the excess water away. Some plethodontids, such as *Aneides,* have large blood sinuses in the tips of the toes for digital respiration. It is believed that lungs are absent in some because they would provide too much buoyancy and would prevent their possessors from hiding quickly under aquatic covers.

In color, salamanders and newts are usually modest and unassuming but are occasionally strikingly brilliant. Color changes are not nearly so pronounced among them as among their relatives, the frogs.

Neither do the caudates show as much diversity of breeding habits as do Salientia. Both external and internal fertilization are found in the group, but some males deposit their sperm in capsules called **spermatophores** (Fig. 23-3). These are placed on leaves, sticks, and other objects, and the female picks them up with her cloacal lips and thus fertilizes her eggs. Aquatic species lay their eggs in clusters or stringy masses in the water; terrestrial forms may also lay their eggs in the water or in moist places. In *Salamandra,* a European form, the eggs are retained within the body of the female until the larvae have completed part of their development.

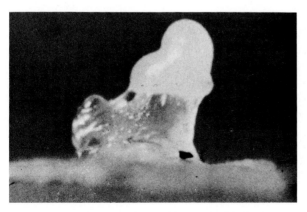

FIG. 23-3
Spermatophore of *Plethodon glutinosus,* consisting of jelly stalk and sperm cap. Males deposit these on leaves or sticks. During courtship display, male presses his mental hedonic gland against female skin, inducing her to detach sperm cap with cloacal lips and fertilize her eggs internally. (Courtesy J. A. Organ, Museum of Zoology, University of Michigan, Ann Arbor.)

Some salamanders show distinctive courtship behavior patterns. Plethodontid males often have a secondary sex character in the form of a hedonic or mental gland under the chin. Such glands are assumed to produce a secretion that excites the female during courtship.

By the use of radioactive tags it has been possible to record the movements of the salamander *Plethodon jordani* in its homing behavior. When displaced as far as 60 meters from their home areas both males and females made direct and rapid directions to their original home although males initiated homing movements sooner than females and occupied larger homing regions. Frequent climbing up on vegetation may indicate olfactory mechanisms of orientation to home-associated odors.

Although acoustic behavior is supposed to be scanty in salamanders, there is evidence that some can produce sound. Recently, it has been shown that *Siren intermedia* makes and responds to underwater sounds. These clicking sounds may be used for intraspecific communication because of the nocturnal habits and turbid habitat of these animals.

Blind salamanders are found in limestone caves in certain parts of the United States and Austria. One of these species is *Typhlotriton spelaeus,* which has functional eyes in the larval form and lives near the mouth of caves. As an adult, it withdraws deeper into the caves and the eyes degenerate. If the larval forms are kept in the light, they retain functional eyes when they mature; if they metamorphose in the dark, they lose the sight of their eyes. Blindness is not hereditary, for the larvae always have eyes.

Many urodeles never complete their development by metamorphosis to the definitive adult body form, but instead retain typical larval characters throughout life. These species continue to grow and become sexually mature, despite their larval body form; this phenomenon is called **neoteny.** Some species are permanent larvae, whereas others are only occasionally neotenous in nature, depending on conditions of the environment. Examples of **permanent larvae** are mud puppies of the genus *Necturus,* which live on bottoms of ponds and lakes and keep their external gills throughout life, and the congo eel *(Amphiuma means)* of the southeastern United States, which, with its useless, rudimentary legs, resembles an eel more than an amphibian. These species are fully and permanently neotenous, reproducing as permanent

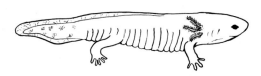

FIG. 23-4
An axolotl, larval form of tiger salamander, *Ambystoma tigrinum.* In Mexico and southwestern United States, larva does not metamorphose but breeds in this form.

larvae, and cannot be experimentally induced to metamorphose to adults. Examples of **occasional neoteny** are species of the genus *Ambystoma* and of the genus *Triturus.* The western axolotl, *Ambystoma tigrinum* (Fig. 23-4), widely distributed over Mexico and the southwestern United States, remains in the aquatic, gill-breathing, and fully reproductive larval form unless the water begins to dry up; then it will metamorphose to an adult, lose its gills, develop lungs, and assume the appearance of an ordinary salamander. Axolotls can be made to metamorphose by treating them with the thyroid hormone, thyroxin. Thyroxin is essential for normal metamorphosis in all amphibians. Recent research suggests that for some reason, the pituitary gland fails to become active in neotenous forms and does not release thyrotropin, which is required to stimulate the production of thyroxin by the thyroid gland.

Salamanders and newts live on worms, small arthropods, and small mollusks. Most of them will eat only things that are moving. Their food naturally is rich in proteins, and they do not usually store in their bodies great quantities of fat or glycogen. They are cold-blooded animals and have a low metabolism.

SOME COMMON AMERICAN SALAMANDERS. Most North American salamanders belong to the family Plethodontidae, which contains more species than all the other families combined (Fig. 23-5). The most common genus of this family in the eastern United States is *Desmognathus* (the dusky salamander), of which there are several species. On the West coast the more familiar genera of plethodontids are *Aneides* and *Batrachoseps.* The primitive family Salamandridae is represented in North America by the American newt *(Notophthalmus viridescens),*

FIG. 23-5

Common salamanders of the family Plethodontidae. **A,** Red-backed salamander, *Plethodon cinereus,* has dorsal reddish stripe with gray to black sides. **B,** Two-lined salamander, *Eurycea bislineata,* is yellow to brown with two dorsolateral black stripes. **C,** Long-tailed salamander, *Eurycea longicauda,* is yellow to orange with black spots that form vertical stripes on sides of tail. **D,** Slimy salamander, *Plethodon glutinosus,* is black with white spots.

which is found all over the eastern United States. This salamander usually passes through a land phase (eft) in its development before becoming an aquatic adult. The California newt *(Taricha torosa)* is common in the coast ranges of California. Another common family is Ambystomidae, which includes such familiar salamanders as the tiger salamander, spotted salamander *(Ambystoma maculatum),* and Jefferson salamander *(A. jeffersonianum).* Certain species of this family are also found on the Pacific coast. Family Cryptobranchidae is represented by the giant sala-

mander *Cryptobranchus,* which may reach a length of more than 2 feet. There are two species that are restricted to certain river systems in Pennsylvania and the Midwest. This salamander and those of Sirenidae are the only American salamanders that have external fertilization.

Salientia (Anura)

The more than 2,200 species of frogs and toads of this order are the most successful and most specialized of amphibians. They occupy a great variety of habitats, despite their aquatic mode of reproduction and water-permeable skin, which prevents them from wandering too far afield from sources of water, and their cold-bloodedness, which bars them from polar regions. When compared to the other two amphibian orders, the frogs and toads are conspicuous for their lack of tails as adults (although all pass through a tailed tadpole stage during development) and hence are collectively referred to as **anurans** (Gr. *an-*, without, + *oura,* tail). They are classified into 12 families, of which the best known to North Americans are the families Ranidae, containing most of our familiar frogs; Hylidae, the tree frogs; and Bufonidae, the toads.

Habitats and distribution

Probably the most abundant and successful of frogs are the 200 to 300 species of the genus *Rana,* found all over the temperate and tropical regions of the world except in New Zealand, the oceanic islands, and southern South America. They are usually found close to water, although some such as the wood frog *(Rana sylvatica)* spend most of their time on damp forest floors, often a distance from the nearest water. The wood frog probably returns to pools only for breeding in early spring. The larger frogs *R. catesbeiana* and *R. clamitans* (Fig. 23-6, *A* and *B*) are nearly always found in or near permanent water or swampy regions. The leopard frog *(R.*

FIG. 23-6
Some common North American frogs. **A,** Bullfrog *Rana catesbeiana* is largest of all American frogs. **B,** Green frog *Rana clamitans* is next to bullfrog in size. Body is usually green, especially around jaws; has dark bars on sides of legs. **C,** Leopard frog *Rana pipiens* has light-colored dorsolateral ridges and irregular spots. **D,** Spring peepers *Hyla crucifer,* the darlings of warm spring nights, when their characteristic peeping is so often heard. It is small (1 to 1½ inches) and light brown, with an "X" marked on back. (Photos by C. Alender.)

pipiens) (Fig. 23-6, *C*) has a wider variety of habitats and, with all its subspecies and phases, is perhaps the most widespread of all the North American frogs in its distribution. It has been found in some form in nearly every state but is sparingly represented along the extreme western part of the Pacific coast. It also extends far into northern Canada and as far south as Panama. The bullfrog *(R. catesbeiana)* is native east of the Rocky Mountains but has been introduced into most of the western states. *Rana clamitans* is confined mainly to the eastern half of the United States, although it has been introduced elsewhere. Within the range of any species of frogs, they are often restricted to certain habitats (for instance, to certain streams or pools) and may be absent or scarce in similar habitats of the range. The pickerel frog *(R. palustris)* is especially noteworthy this way, for it is known to be abundant only in certain localized regions.

Most of our larger frogs are solitary in their habits except during the breeding season. During the breeding period most of them, especially the males, are very noisy. Each male usually takes possession of a particular perch, where he may remain for a long time, trying to attract a female to that spot. At times frogs are mainly silent, and their presence is not detected until they are disturbed. When they enter the water, they dart about swiftly and reach the bottom of the pool, where they kick up a cloud of muddy water. In swimming, they hold the forelimbs near the body and kick backward with the webbed hind limbs, which propels them forward. When they come to the surface to breathe, only the head and foreparts are exposed, and as they usually take advantage of any protective vegetation, they are difficult to see.

During the winter months most frogs **hibernate** in the soft mud of the bottom of pools and streams. The wood frog hibernates under stones, logs, and stumps in the forest area. Naturally their life processes are at a very low ebb during their hibernation period, and such energy as they need is derived from the glycogen and fat stored in their bodies during the spring and summer months.

Adult frogs have numerous enemies, such as snakes, aquatic birds, turtles, raccoons, man, and many others; only a few tadpoles survive to maturity. Although usually defenseless, in the tropics and subtropics many frogs and toads are aggressive, jumping and biting at their potential enemies. Some defend themselves by feigning death. Most anurans can blow up their lungs so that they are difficult to swallow. When disturbed along the margin of a pond or brook, a frog will often remain quite still; when it thinks it is detected, it will jump, not always into the water where enemies may be lurking but

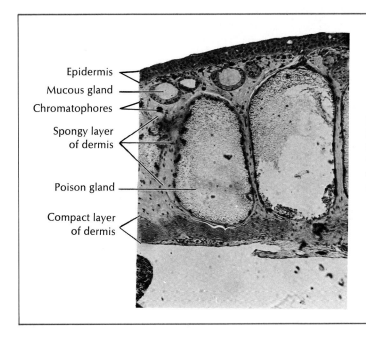

Epidermis
Mucous gland
Chromatophores
Spongy layer of dermis
Poison gland
Compact layer of dermis

FIG. 23-7
Histologic section of frog skin. Stratified epidermis is seen as dark layer at surface with thicker dermis below. Note small mucous glands and large poison glands in dermal layer.

into grassy cover on the bank. When held in the hand a frog may cease its struggles for an instant to put its captor off guard and then leap violently, at the same time voiding its urine. Their best protection is their ability to leap and their use of poison glands. Bullfrogs in captivity will not hesitate to snap at tormenters and are capable of inflicting painful bites.

The largest anuran is the West African *Rana goliath,* which is more than 1 foot long from tip of nose to anus. This giant will eat animals as big as rats and ducks. The smallest frog recorded is *Phyllobates limbatus,* which is only about $\frac{1}{2}$ inch long. This tiny frog, which is more than covered by a dime, is found in Cuba. Our largest American frog is the bullfrog *(Rana catesbeiana;* Fig. 23-6, *A),* which reaches a length of 8 or 9 inches. Most of our common species are only a few inches long, and the tree frogs are smaller than this.

Integument and coloration

The skin of the frog is thin and moist and is attached loosely to the body only at certain points. Histologically the skin is made up of two layers—an outer stratified **epidermis** and an inner spongy **dermis** (Fig. 23-7). The epidermis consists of a some-

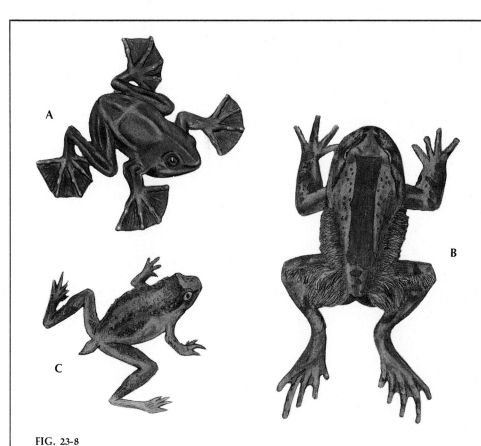

FIG. 23-8
Three frogs with unusual features. **A,** "Flying" frog of Borneo, *Polypedates nigropalmatus.* Large webs between digits of feet aid it in gliding from higher elevation to lower one. **B,** "Hairy frog" of Africa, *Astylosternus.* Hairlike filaments on groins and sides of male are actually cutaneous papillae, probably used for respiration. These filaments are unusually well developed during breeding season. **C,** Bell toad of northwestern Pacific coast, *Ascaphus truei.* Male's cloacal appendage serves as copulatory organ for fertilizing female's eggs. This frog and certain ovoviviparous frogs of Africa are only known salientians that have internal fertilization.

what horny outer layer of epithelium, which is periodically shed, and an inner layer of columnar cells, from which new cells are formed to replace those that are lost. The frog molts a number of times during its active months. In the process the outer layer of skin is split down the back and is worked off as one piece. The dermis is made up mostly of glands, pigment cells, and connective tissue. On its outer portion are the glands—small **mucous glands,** which secrete mucus for keeping the skin moist, and the larger **poison glands,** which secrete a whitish fluid that is highly irritating to enemies. The poison of *Dendrobates,* a South American frog, is used by Indian tribes to poison the points of their arrows.

The skin of both frogs and toads has many variations. Toads usually have warty skins, and frogs tend to have smooth, slimy skins. Whenever they live in or near the water, their skins are more or less smooth and slimy; in deserts or dry regions, however, their skins are rough and warty.

A skin modification occurs in the so-called hairy frog, *Astylosternus,* found in the Cameroons of Africa (Fig. 23-8, *B*). The males have fine cutaneous filaments on the thighs, groins, and sides, which have a remarkable resemblance to hair. These curious structures may have a respiratory function. The female Surinam toad, *Pipa,* has a modified skin on her back for carrying eggs and young (Fig. 23-9).

Skin color in the frog is produced by pigment granules scattered through the epidermis and by special pigment cells, **chromatophores,** located in the dermis (Figs. 23-7 and 23-10). Types of chromatophores include **guanophores,** which contain white crystals, **melanophores** with black and brown pigment, and **lipophores** (xanthophores) with red and yellow pigment. Of these the melanophores are most

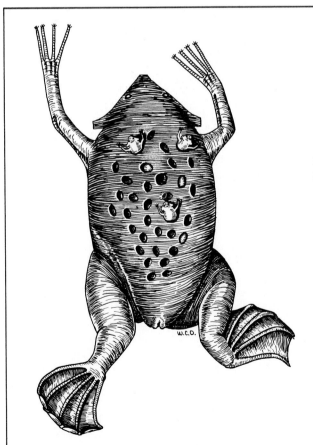

FIG. 23-9

Female Surinam toad, *Pipa pipa,* carrying young on back. As eggs are laid, male assists in positioning them on rough back skin of female. Skin swells, enclosing eggs. The approximately 60 young pass through the tadpole stage beneath skin and then emerge as small toads. Surinam toad is found mainly in Amazon and Orinoco river systems of equatorial South America.

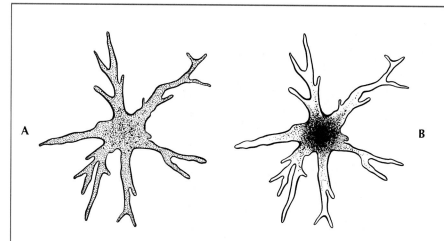

FIG. 23-10

Pigment cells (chromatophores). **A,** Pigment dispersed. **B,** Pigment concentrated. Pigment cell does not contract or expand; color effects are produced by streaming of cytoplasm, carrying pigment granules into cell branches for maximum color effect or to center of cell for minimum effect. Chromatophore of cephalopod, however, does change shape by muscular contraction. Control over dispersal or concentration of pigment is mostly through light stimuli to eye. Melanophore-stimulating hormone is known to influence activity of chromatophore (see text). Types of chromatophores depend on pigments they bear, for example, melanophores (brownish black), xanthophores (yellow or red), and guanophores (white).

important for adjusting coloration to blend with the background; some frogs and toads possess remarkably good powers of camouflage (Fig. 23-11). The degree of dispersion of the dark brown pigment granules of melanin within the melanophores is controlled by melanophore-stimulating hormone (MSH) of the pars intermedia of the pituitary gland (see Fig. 30-20). The release of MSH, which causes pigment dispersion, and thus darkening of the animal, is in turn controlled by the pattern of illumination on the retina of the eyes. When the animal is on a light background, both dorsal and ventral portions of the retina are illuminated by reflected light from the surface and by direct illumination from above. This inhibits the release of MSH, the pigment becomes concentrated in the centers of the melanophores, and the animal becomes pale in color. But on a dark, light-absorbing background, where only the ventral retina is illuminated, MSH is released from the pituitary and causes the melanophore pigment to disperse, darkening the skin. Both responses are concealing in their effect. Other factors, especially temperature, humidity, and activity, also influence the skin coloration.

Skeletal and muscular systems

In amphibians as in their fish ancestors, the well-developed **endoskeleton** of bone and cartilage provides a framework for the muscles in movement and protection for the viscera and nervous systems. But movement onto land and the necessity of transforming paddlelike fins into tetrapod legs capable of supporting the body's weight introduced a new set of stress and leverage problems. The changes are most noticeable in the anurans in which the entire musculoskeletal system is specialized for jumping and swimming by simultaneous extensor thrusts of the hind limbs. Since frogs and toads move with limbs instead of by swimming with serial contractions of myotomal trunk muscles, the vertebral column has lost the original flexibility characteristic of fishes and has, together with the enlarged pelvic girdle, become a rigid frame for transmitting force from the hind limbs to the body. The anurans are further specialized

FIG. 23-11
Cryptic coloration of the tree frog *(Hyla)*. Instead of the excellent protective markings borne by this harmless North American species, several of the dangerously poisonous South American tree frogs display vivid coloration to warn predators to keep away. All tree frogs have suction pads at ends of fingers and toes, enabling them to grip vertical surfaces. (Photo by L. L. Rue, III.)

by an extreme shortening of the body, more than in other more advanced tetrapods (Fig. 23-12). In these respects, urodeles represent much better than anurans the amphibians' transition state between aquatic and terrestrial forms. Nevertheless the limbs, girdles, and their muscles show a remarkable uniformity of pattern for all tetrapods, anurans included.

The **pectoral girdle,** consisting of **suprascapula, scapula, clavicle,** and **coricoid,** serves as support for

the forelimbs (Fig. 23-12). The forelimbs in frogs are used mainly to absorb the weight during landing after a jump. The pelvic girdle of frogs is much specialized, consisting of two **innominate** bones, each representing the fusion of a long **ilium,** a posterior **ischium,** and ventral **pubis** (Fig. 23-12). The pattern of bones and muscles in the limbs is the typical tetrapod type. There are three main joints in each limb (hip, knee, and ankle; or shoulder, elbow, and wrist). The hand

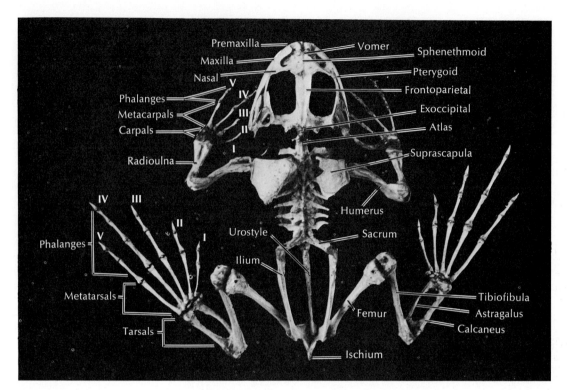

FIG. 23-12
Dorsal view of frog skeleton.

or foot is basically a five-rayed form with several joints in each of the digits. It is a repetitive system that can be plausibly derived from the crossopterygian fin. The bone structure of the fins of crossopterygian fishes is in fact distinctly suggestive of the amphibian limb; it is not difficult to imagine how selective pressures through millions of years remodeled the former into the latter. The muscles of the limbs are presumably derived from the radial muscles that moved the fins of fishes up and down, but the muscular arrangement has become so complex in the tetrapod limb that it is no longer possible to see parallels between this and fin musculature. Muscles are usually arranged in antagonistic groups so that movement is effected by the contraction of one group and the relaxation of the opposite group. Muscles may be classified according to their action.

abductor Moves the part away from the median axis of the body **(deltoid)**
adductor Moves the part toward the median axis of the body **(adductor magnus)**
flexor Bends one part on another part **(biceps brachii)**

extensor Straightens out a part **(triceps brachii)**
depressor lowers a part **(sternohyoid)**
levator Elevates a part **(masseter)**
rotator Produces a rotary movement **(gluteus)**

Despite the complexity of tetrapod limb musculature, it is possible to recognize two great groups of muscles on any limb: an anterior and ventral group that pulls the limb forward and toward the midline (protraction and adduction) and a second set of posterior and dorsal muscles that serves to draw the limb back and away from the body (retraction and abduction).

The trunk musculature, which in fish are segmentally organized into powerful muscular bands (myotomes) for locomotion by lateral flexion, was much modified during amphibian evolution. The dorsal (epaxial) muscles are arranged to support the head and brace the vertebral column. The ventral (hypaxial) muscles of the belly are more developed in amphibians than in fish since they must support the viscera in air without the buoying assistance of water. Anteriorly the hypaxial muscles form the hyoid musculature of the throat, used to raise and lower the floor of the mouth during breathing.

The frog skull is also vastly altered as compared to its crossopterygian ancestors; it is much lighter in weight, more flattened in profile, and contains fewer bones and less ossification. The front part of the skull, wherein are located the nose, eyes, and brain, is better developed, whereas the back of the skull, which contained the gill apparatus of fishes, is much reduced. Lightening of the skull was essential to mobility on land, and the other changes fitted the frog for its improved special senses and means for feeding and breathing.

Respiration and vocalization

Amphibians use three respiratory surfaces for gas exchange: the skin (cutaneous breathing), the mouth (buccal breathing), and the lungs. These surfaces are used to varying degrees by different groups and under different environmental conditions. The lungless salamanders of the family Plethodontidae breathe principally through the highly vascular skin (90% to 95% of total gas exchange) and to a much lesser extent through the vascular membranes of the mouth (5% to 10% of total gas exchange). The more terrestrial amphibians such as frogs and toads show a much greater dependence on lung breathing, since cutaneous breathing, simple and direct as it obviously is,

suffers from two disadvantages: (1) the skin must be kept thin and moist for gas exchange and consequently is too delicate for aerial life, and (2) the amount of gas exchange across the skin is mostly constant and cannot be varied to match changing demands of the body for more or less oxygen. Nevertheless the skin continues to serve as an important supplementary avenue for gas exchange in anurans, especially during hibernation in winter. Even under normal conditions when lung breathing predominates, most of the CO_2 is lost across the skin while most of the O_2 is taken up across the lungs.

The lungs are supplied by pulmonary arteries (derived from the sixth aortic arches) and blood is returned directly to the left atrium by the pulmonary veins. Frog lungs are ovoid, elastic sacs with their inner surfaces divided into a network of septa that are in turn subdivided into small terminal air chambers called **alveoli**. The alveoli of the frog lung are much larger than those of more advanced vertebrates, and consequently the frog lung has a smaller relative surface available for gas exchange: the respiratory surface of the common *Rana pipiens* is about 20 cm.^{2} per cm.^{3} of air contained, compared to 300 cm.^{2} for man. But the problem in lung evolution was not the development of a good internal vascular surface, but rather the problem of moving air into and

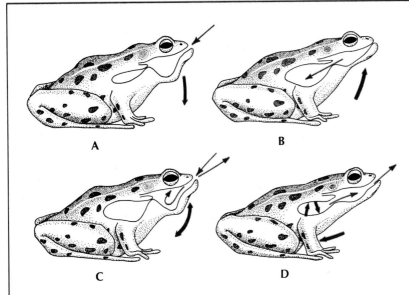

FIG. 23-13
Breathing in frog. The frog, a positive-pressure breather, fills its lungs by forcing air into them. **A,** Floor of mouth is lowered, drawing air in through nostrils. **B,** With nostrils closed and glottis open, the frog forces air into lungs by elevating floor of mouth. **C,** Mouth cavity rhythmically ventilated for a period. **D,** Lungs emptied by contraction of body wall musculature and by elastic recoil of lungs. (Adapted from Gordon, M. S., et al. 1968. Animal function: Principles and adaptations, New York, The Macmillan Co.)

out of it. A frog is a positive-pressure breather that fills its lungs by forcing air into them; this contrasts with the negative-pressure system of all the higher vertebrates. The sequence and explanation of breathing in a frog is shown in Fig. 23-13. One can easily follow this sequence in a living frog at rest: rhythmic throat movements of mouth breathing may continue some time before flank movements indicate that the lungs are being emptied.

Both male and female frogs have **vocal cords,** but those of the male are much better developed. They are located in the **larynx,** or voice box. Noise is produced by passing air back and forth over the vocal cords between the lungs and a large pair of sacs (vocal pouches) in the floor of the mouth. The latter also serve as effective resonators in the male. The chief function of the voice is to attract mates. Most species utter characteristic sounds that identify them. Nearly everyone is familiar with the welcome springtime calls of the spring peeper that produces a high-pitched sound surprisingly strident for such a tiny frog. Another sound familiar to residents of the more southern United States is the "jug-o-rum" call of the bullfrog. The bass notes of the green frog are banjolike, and those of the leopard frog are long and guttural. Many frog sounds are now available on phonograph records.

Circulation

As in fish, the amphibian circulation is a closed system of arteries and veins serving a vast peripheral network of capillaries through which blood is forced by the action of a single pressure pump, the heart.

The principal changes in circuitry involve the shift from gill to lung breathing. With the elimination of gills, a major obstacle to blood flow was removed from the arterial circuit. But two new problems arose. The first was to provide a blood circuit to the lungs. This was solved as we have seen by converting the sixth aortic arch into pulmonary arteries to serve the lungs and by developing new pulmonary veins for returning oxygenated blood to the heart. The second and evidently more difficult evolutionary problem was to separate the new pulmonary circulation from the rest of the body's circulation in such a way that oxygenated blood from the lungs would be selectively sent to the body and deoxygenated venous return from the body would be selectively sent to the lungs. In effect this meant creating a double circulation consisting of separate pulmonary and systemic circuits. This was eventually solved by placing a partition down the center of the heart, to create a double pump, one for each circuit. Amphibians and reptiles have made the separation to variable degrees, but the task was completed by the birds and mammals, which

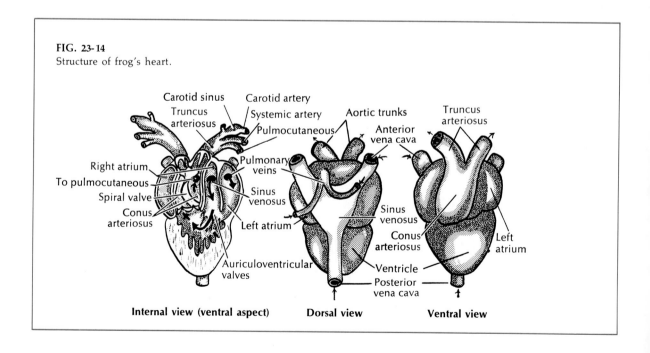

FIG. 23-14
Structure of frog's heart.

have a completely divided heart of two atria and two ventricles. The frog heart (Fig. 23-14) has two separate atria and a single undivided ventricle. Blood from the body (systemic circuit) first enters a large receiving chamber, the **sinus venosus,** which forces it into the **right atrium.** The **left atrium** receives freshly oxygenated blood from the lungs. Up to this point the deoxygenated blood from the body and oxygenated blood from the lungs are separated. But now both atria contract almost simultaneously, driving both right and left atrial blood into the single undivided **ventricle.** We should expect that complete admixture of the two circuits would happen here. In fact there is evidence that in at least some amphibians they remain mostly separated, so that when the ventricle contracts, oxygenated pulmonary blood is sent to the systemic circuit and deoxygenated systemic blood is sent to the pulmonary circuit. The **spiral valve** in the **conus arteriosus** (Fig. 23-14) may play an important role in maintaining selective distribution. The matter is controversial and has defied complete analysis despite the application of advanced techniques using radiopaque media and high-speed cineradiography.

Feeding and digestion

Frogs are carnivorous like most other adult amphibians and diet on insects, spiders, worms, slugs, snails, millipeds, or nearly anything else that moves and is small enough to swallow whole. They snap at moving prey with their protrusible tongue, which is attached to the **front** of the mouth and is free behind. The free end of the tongue is highly glandular and produces a sticky secretion, which adheres to the prey. The teeth on the premaxillae, maxillae, and vomers are used to prevent escape of prey, not for biting or chewing. The digestive tract is relatively short in adult amphibians, a characteristic of most carnivores, and it produces a variety of enzymes for breaking down proteins, carbohydrates, and fats.

The larval stages of anurans (tadpoles) are usually herbivorous, feeding on pond algae and other vegetable matter; they have a relatively long digestive tract, since their bulky food must be submitted to time-consuming fermentation before useful products can be absorbed.

Nervous system and special senses

The frog nervous system is of the basic vertebrate plan that was established in the early jawless fishes.

It is commonly divided into (1) the **central nervous system,** consisting of the brain and spinal cord; (2) the **peripheral nervous system,** consisting of the cranial and spinal nerves; and (3) the **autonomic nervous system,** composed of a chain of special ganglia on each side of the spinal column.

The three fundamental parts of the brain—forebrain (telencephalon), concerned with the sense of smell; midbrain (mesencephalon), concerned with vision; and hindbrain (rhombencephalon), concerned with hearing and balance—have undergone dramatic developmental trends as the vertebrates moved onto land and improved their environmental awareness. In general there is increasing cephalization with emphasis on information processing by the brain and a corresponding loss of independence of the spinal ganglia, which are capable of only stereotyped reflexive behavior. Nonetheless a headless frog preserves an amazing degree of purposive and highly coordinated behavior. With only the spinal cord intact, it maintains normal body posture and can with purposive accuracy raise its leg to wipe from its skin a piece of filter paper soaked in dilute acetic acid. It will even use the opposite leg if the closer leg is held.

The forebrain is divided into the **telencephalon** and **diencephalon** (Fig. 23-15). The telencephalon contains the olfactory center, which has assumed much increased importance for the detection of dilute airborne odors on land. The sense of smell is in fact one of the dominant special senses in frogs. The remainder of the telencephalon is the cerebrum, which is of little importance in amphibians and provides no hint of the magnificent development it is destined to attain in the higher mammals. The diencephalon consists of **thalamus, hypothalamus,** and **posterior pituitary,** which are important relay channels and regulative centers. The thalamus and hypothalamus are primitive integrative areas concerned with thirst, hunger, sexual drive, pleasure, and pain.

The most complex integrative activities of the frog brain are centered in the midbrain. The **optic lobe** (tectum) is a dorsal enlargement that integrates sensory information from the eyes as well as from other senses.

The hindbrain is divided into an anterior **metencephalon,** or cerebellum, and a posterior **myelencephalon,** or medulla. The cerebellum is concerned with equilibrium and movement coordination and is

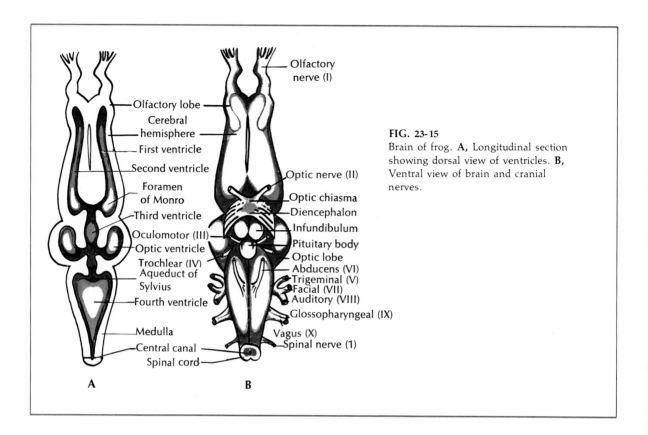

FIG. 23-15
Brain of frog. **A,** Longitudinal section showing dorsal view of ventricles. **B,** Ventral view of brain and cranial nerves.

not well developed in amphibians, which stick close to the ground and are not noted for dexterity of movement. The cerebellum becomes vastly developed in the fast-moving birds and mammals. The medulla is really the enlarged anterior end of the spinal cord through which pass all sensory neurons except those of vision and smell. Here are located centers for auditory reflexes, respiration, swallowing, and vasomotor control.

The evolution of a semiterrestrial life for the amphibians has necessitated a reordering of sensory receptor priorities on land. The pressure-sensitive lateral-line (acousticolateral) system of fish remains only in the aquatic larvae of amphibians and in a few strictly aquatic adult amphibian species. This system of course can serve no useful purpose on land, since it was designed to detect and localize objects in water by reflected pressure waves. Instead the task of detecting airborne sounds devolved upon the ear. The ear of a frog is by higher vertebrate standards a primitive structure: a middle ear closed externally by a large **tympanic membrane** (eardrum) and containing

a **stapes** (columella) that transmits vibrations to the inner ear. The latter contains the **utricle,** from which arise the semicircular canals and a **saccule** bearing a diverticulum, the **lagena.** The lagena is partly covered with a **tectorial membrane** that in its fine structure is not unlike that of the much more advanced mammalian cochlea. In most frogs this structure is sensitive to low-frequency sound energy not greater than 4,000 Hertz (cycles per second); in the bullfrog the main frequency response is in the 100 to 200 Hz range, which matches the energy of the male frog's low-pitched call. Although it has not been measured, we would expect the ear of the spring peeper to be sensitive to the high-frequency calls this species makes.

In most amphibians, the eye is the dominant special sense. The eye is basically of the fish type, but the lens is located farther from the cornea and provided with protractor muscles for limited accommodation (focusing on near or distant objects). The **retina** contains both **rods** and **cones,** providing frogs with color vision. The **iris** contains well-developed

FIG. 23-16
Pair of toads in amplexus. As eggs are
extruded from cloaca of female, male
fertilizes them with his sperm. (Photo by
C. Alender.)

circular and radial muscles and can rapidly expand
or contract the aperture (pupil) to adjust to changing
illumination. The upper lid of the eye is fixed, but
the lower is folded into a transparent **nictitating
membrane** capable of moving across the eye surface.
In all, frogs and toads possess good vision, a fact of
crucial importance to animals that rely upon quick
escape to avoid their numerous predators.

Other sensory receptors include tactile and chemical receptors in the skin, taste buds on the tongue
and palate, and a well-developed olfactory epithelium
lining the nasal cavity.

Reproduction and life cycle

Frogs and toads are cold-blooded (poikilothermous)
animals; their distribution and activities are therefore
controlled by seasonal changes and climatic conditions. Their activities are restricted to the warmer
seasons of the year, when they breed, feed, and grow.
During the winter months in colder climates they
spend their time in hibernation.

The time of spring emergency varies with different
species. One of their first interests after leaving their
dormant period is breeding. At this time the males
croak and call vociferously to attract females. The
breeding season usually extends for several weeks.
When their eggs are ripe, the females enter the water
and are mounted and clasped by the males in the
process called **amplexus** (Fig. 23-16). The male holds

the female by pressing the nuptial pads of his thumbs
against her breast just back of her forelegs. As the
female lays her eggs, the male discharges his seminal
fluid containing the sperm over the eggs. The sperm
penetrate the jelly layers of the eggs, assisted by
sperm lysins, which dissolve the membranes. When
the first spermatozoon touches the surface of the egg
proper, dramatic changes suddenly occur in the egg
cytoplasm, which are called the **cortical reaction.** A
fertilization membrane that prevents additional
sperm from entering the egg is immediately formed.
Thus only the first sperm penetrates the egg surface
(monospermy) and is drawn into the egg interior
where it swells into a **male pronucleus.** This soon
fuses with the **female** (egg) **pronucleus** to form a
zygote nucleus with the diploid number of chromosomes. After fertilization the jelly layers absorb water
and swell (Fig. 23-17). Eggs are laid in great masses,
which may include several thousands in the leopard
frog. The egg masses are usually anchored to vegetation or debris by the sticky jelly layers around the
egg. Not all the eggs have a chance to develop, for
some may not be fertilized and others are eaten by
turtles, insects, and other enemies.

Development of the fertilized egg (zygote) begins
about 2 to 3 hours after fertilization (Fig. 23-18). The
process involves cleavage, or segmentation of the
egg. Cleavage occurs more rapidly at the black, or
animal, pole, where there is more protoplasm. By

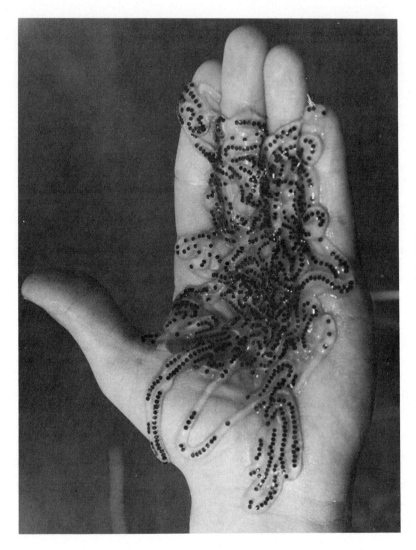

FIG. 23-17
Eggs of American toad. Toads lay eggs in strings; frogs lay eggs in clusters. (Photo by L. L. Rue, III.)

collecting masses of frog eggs early in the morning, it is possible to find eggs in various stages of development, such as 2-, 4-, and 8-cell stages. Generally one finds them still further along in development.

Eggs usually hatch into tadpoles within a period of 6 to 9 days depending on the temperature (Fig. 23-18). At the time of hatching, the tadpole has a distinct head and body with a compressed tail. The mouth is located on the ventral side of the head and is provided with horny jaws for scraping off vegetation from objects for food. Behind the mouth is a ventral adhesive disk for clinging to objects. In front of the mouth are two deep pits, which later develop into the nostrils. Swellings are found on each side of the head, and these later become external gills.

There are finally three pairs of external gills, which are later replaced by three pairs of internal gills within the gill slits. On the left side of the neck region is an opening, the **spiracle,** through which water flows after entering the mouth and passing the internal gills. The hind legs appear first, while the forelimbs are hidden for a time by the folds of the operculum. During metamorphosis the tail is resorbed, the intestine becomes much shorter, the mouth undergoes a transformation into the adult condition, lungs develop, and the gills are resorbed. The leopard frog usually completes its metamorphosis within a year or less; the bullfrog takes 2 or 3 years to complete the process.

Migration of frogs and toads is correlated with their

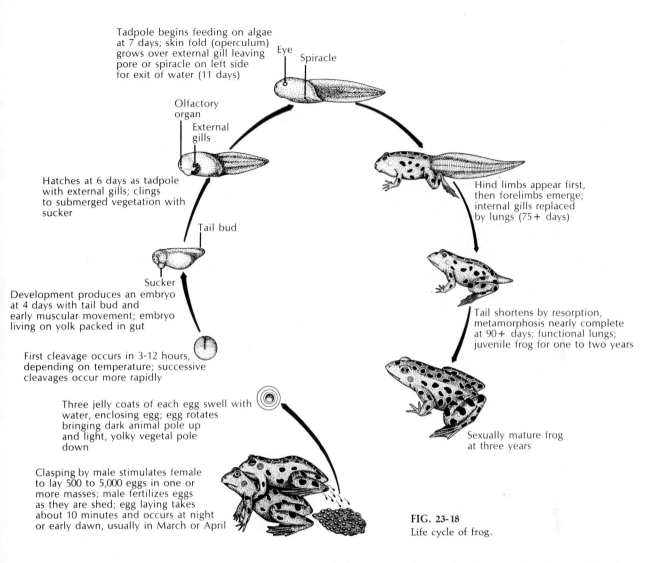

Tadpole begins feeding on algae at 7 days; skin fold (operculum) grows over external gill leaving pore or spiracle on left side for exit of water (11 days)

Eye Spiracle

Olfactory organ
External gills

Hatches at 6 days as tadpole with external gills; clings to submerged vegetation with sucker

Tail bud

Sucker

Development produces an embryo at 4 days with tail bud and early muscular movement; embryo living on yolk packed in gut

First cleavage occurs in 3-12 hours, depending on temperature; successive cleavages occur more rapidly

Three jelly coats of each egg swell with water, enclosing egg; egg rotates bringing dark animal pole up and light, yolky vegetal pole down

Clasping by male stimulates female to lay 500 to 5,000 eggs in one or more masses; male fertilizes eggs as they are shed; egg laying takes about 10 minutes and occurs at night or early dawn, usually in March or April

Hind limbs appear first, then forelimbs emerge; internal gills replaced by lungs (75+ days)

Tail shortens by resorption, metamorphosis nearly complete at 90+ days; functional lungs; juvenile frog for one to two years

Sexually mature frog at three years

FIG. 23-18
Life cycle of frog.

breeding habits. Males usually return to a pond or stream in advance of the females, whom they then attract by their calls. Some salamanders are also known to have a strong homing instinct, returning year after year to the same pool for reproduction guided by olfactory cues. The initial stimulus for migration in many cases is attributable to a seasonal cycle in the gonads plus hormonal changes that increase their sensitivity to temperature and humidity changes.

Unlike the salamanders, which mostly have internal fertilization, nearly all frogs and toads fertilize their eggs externally. A notable exception is the famed bell toad *(Ascaphus truei)* of the Pacific coast region (Fig. 23-8 *C*). This small toad (family Asca-

phidae), which is only about 2 inches long, is found in swift mountain streams of low temperature from British Columbia to northern California and has a conspicuous extension of cloaca, which serves as an intromittent or copulatory organ for fertilizing internally the eggs of the female. *Ascaphus* is the only American frog that has ribs in the adult condition. A related genus, *Liopelma,* is found in New Zealand. The only frog to bring forth its young alive is the ovoviviparous *Nectophrynoides* of Africa. This frog also has internal fertilization but has no external copulatory organ.

References

(See the end of Chapter 24.)

An American alligator of the Everglades eyes a hostile world while breathing comfortably through nostrils located high on the snout. Despite several anatomic advancements, such as complete separation of air and food passages and a completely divided heart, the crocodilians are a retrenched group dangerously near extinction. Only 25 species remain in the world. (Photo by L. L. Rue, III, Blairstown, N. J.)

CHAPTER 24

THE REPTILES

PHYLUM CHORDATA
CLASS REPTILIA*

Reptiles are the first truly terrestrial vertebrates. With some 6,000 species (about 300 species in the United States and Canada) occupying a great variety of aquatic and terrestrial habitats, they are clearly a successful group. Nevertheless, reptiles are perhaps remembered best for what they once were, rather than for what they presently are. The Age of Dinosaurs, which lasted 100 million years encompassing the Jurassic and Cretaceous periods, saw the appearance of a great radiation of reptiles, many of huge stature and awesome appearance, that completely dominated life on land. Then they suddenly declined. Out of the dozen or so principal groups of reptiles that evolved, four remain today. The most successful of these are the lizards and snakes of the order Squamata. A second group is the crocodilians;

*Rep-til'e-a (L. *repere,* to creep).

having survived for 200 million years, they may finally be made extinct by man. To a third group belong the tortoises and turtles of the order Testudines, an ancient group that has somehow survived and remained mostly unchanged from its early reptile ancestors. The last group is a relic stock represented today by a sole survivor, the tuatara of New Zealand.

Reptiles are easily distinguished from amphibians by their scaly hide, which resists desiccation. But more than any other single aspect, reptiles are set apart from amphibians by their method of reproduction. Reptiles lay eggs on land; amphibians must lay their eggs in water. This seemingly simple difference was in fact a remarkable evolutionary achievement that was to have a profound impact on subsequent vertebrate evolution. As we have seen, the amphibians are basically aquatic animals by virtue of their reproduction. To totally abandon an aquatic life, there evolved a sophisticated internally-fertilized egg containing a complete set of life-support systems. This **amniotic egg** could be laid on dry land. Within, the embryo floats and develops in an aquatic environment surrounded by a membrane (the amnion); it is provided with a yolk sac containing its food supply; another membrane serves as a surface for gas-exchange through the shell; and provision is made for storing waste products that accumulate during development (see p. 750). The early reptiles that developed this egg would certainly have enjoyed an immediate advantage over the amphibians. They could now hide their vulnerable eggs in a protected situation away from water—and away from the numerous creatures that fed freely on the eggs provided by amphibians each spring. With the evolution of this ultimate adaptation, conquest of land by the vertebrates was now possible.

■ ORIGIN AND ADAPTIVE RADIATION OF REPTILES

Biologists generally agree that reptiles arose from labyrinthodont amphibians sometime before the Permian period, which began about 270 million years ago. The oldest stem reptiles belonged to the order Cotylosaura (Fig. 24-1). Forming a transition between the labyrinthodont amphibians and stem reptiles is a lizardlike, partly aquatic animal, about 2 feet long, called *Seymouria.* It was found in 220 million-year-old Permian strata in Texas. It is considered a "stem" reptile since it shows characteristics of both groups and thus appears to have been a transition form. Despite its reptilian skeletal features, *Seymouria* was probably more amphibian than reptile, since there is evidence that it possessed a lateral-line system, typical of amphibians that have an aquatic larval stage. In other words, it is unlikely that *Seymouria* laid amniotic eggs as did the true terrestrial reptiles to follow. Nevertheless, *Seymouria* represents a nearly perfect transition form. From *Seymouria,* or a closely related form, arose the stem reptiles (cotylosaurs), showing more definite reptilian characteristics.

The adaptive radiation of reptiles, especially pronounced in the Triassic period (which followed the Permian), was correlated with the appearance of new ecologic niches. These were provided by the climatic and geologic changes that were taking place at that time, such as a variable climate from hot to cold,

mountain building and terrain transformations, and a varied assortment of plant life.

The Mesozoic was the age of the great ruling reptiles. Then suddenly they disappeared near the close of the Cretaceous some 65 to 80 million years ago. What caused their demise? Many changes were occurring during the Cretaceous: modern flowering plants were spreading rapidly, as were the aggressive and intelligent mammals. In general, the modern fauna and flora as we know it today was becoming well established and the dinosaurs were not sufficiently adaptable to survive. Their extinction probably resulted from the combined effect of climatic and ecologic factors, excessive specialization, and low reproductive potential. But this is speculation, and debate continues among paleontologists. Why did some reptiles survive against the fierce competition of the mammals? Turtles had their protective shells, snakes and lizards evolved in habitats of dense forests and rocks where they could meet the competition of any tetrapod, and crocodiles, because of their size, stealth, and aggressiveness, had few enemies in their aquatic habitats.

Characteristics

1. Body variable in shape, compact in some, elongated in others; **body covered with an exoskeleton of horny epidermal scales** with the addition sometimes of bony dermal plates; **integument with few glands**
2. **Limbs paired, usually with five toes,** and adapted for climbing, running, and paddling; absent in snakes
3. Skeleton well ossified; ribs with sternum forming a complete thoracic basket; **skull with one occipital condyle**
4. Respiration by lungs; **no gills;** cloaca used for respiration by some; branchial arches in embryonic life
5. **Three-chambered heart; crocodiles with four-chambered heart;** usually one pair of aortic arches
6. **Kidney a metanephros (paired);** uric acid main nitrogenous waste
7. Nervous system with the optic lobes on the dorsal side of brain; **12 pairs of cranial nerves** in addition to nervus terminalis
8. Sexes separate; **fertilization internal**
9. **Amniotic eggs, which are covered with leathery shells;** extraembryonic membranes including the

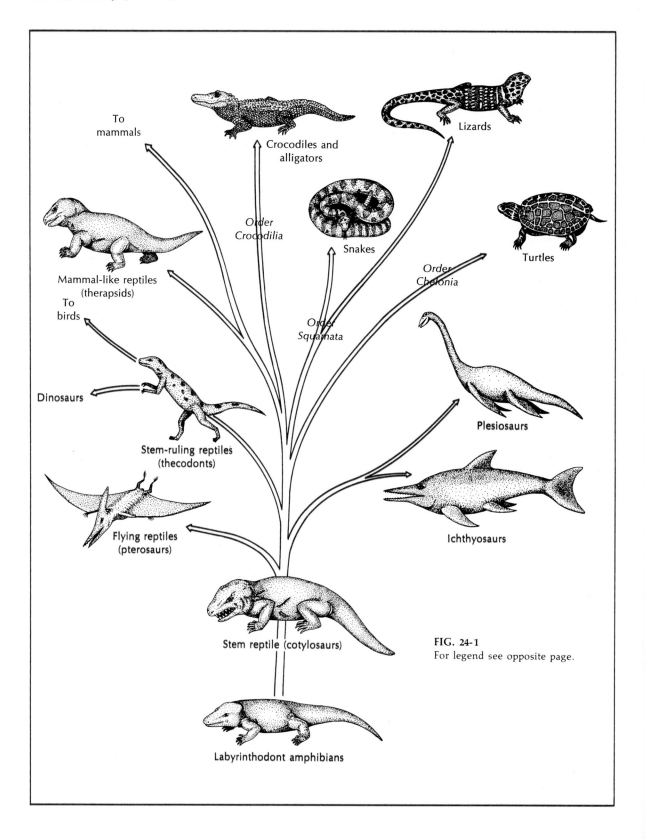

To
mammals

Crocodiles and
alligators

Lizards

*Order
Crocodilia*

Snakes

Mammal-like reptiles
(therapsids)

*Order
Chelonia*

Turtles

To
birds

*Order
Squamata*

Dinosaurs

Plesiosaurs

Stem-ruling reptiles
(thecodonts)

Flying reptiles
(pterosaurs)

Ichthyosaurs

Stem reptile (cotylosaurs)

FIG. 24-1
For legend see opposite page.

Labyrinthodont amphibians

amnion, chorion, yolk sac, and **allantois** present during embryonic life

Brief classification

Order Squamata (squa-ma'ta) (L. *squamatus,* scaly, + *-ata,* characterized by). Skin of horny epidermal scales or plates, which is shed; teeth attached to jaws; quadrate freely movable; vertebrae usually concave in front; anus a transverse slit. Examples: snakes (3,000 species), lizards (3,800 species), chameleons.

Order Testudines (tes-tu'din-eez) (L. *testudo,* tortoise) **(Chelonia).** Body in a bony case of dermal plates with dorsal carapace and ventral plastron; jaws without teeth but with horny sheaths; quadrate immovable; vertebrae and ribs fused to shell; anus a longitudinal slit. Examples: turtles and tortoises. 400 species.

Order Crocodilia (croc"o-dil'e-a) (L. *crocodilus,* crocodile, + *-ia,* pl. suffix) **(Loricata).** Four-chambered heart; vertebrae usually concave in front; forelimbs usually with five digits, hind limbs with four digits; quadrate immovable; anus a longitudinal slit. Examples: crocodiles and alligators. 25 species.

Order Rhynchocephalia (rin"ko-se-fay'le-a) (Gr. *rhynchos,* snout, + *kephalē,* head). Vertebrae biconcave; quadrate immovable; parietal eye fairly well developed and easily seen; anus a transverse slit. Example: *Sphenodon*—only species existing.

How Reptilia show advancement over Amphibia

1. Reptiles have developed some form of copulatory organ for internal fertilization.

2. The amniote eggs of reptiles are leathery and limy, which resist desiccation and they contain yolk and protective membranes to support embryonic development on land. Amphibian eggs have gelatinous covering.

3. Reptiles have dry, scaly skin, which is adapted to land life.

4. Reptiles also have a partial or complete separation of the ventricle, thus ensuring better oxygenation of the blood, but there is some mixing of venous and arterial blood through right and left arches.

5. Most reptiles have developed limbs well adapted for efficient locomotion on land.

Distinctive structures of reptilian body

In contrast to the soft, naked body of amphibians, reptiles have developed an **exoskeleton** that is largely waterproof. This exoskeleton consists of dead horny scales formed of keratin, a protein substance found in the epidermal layers of the skin. Reptilian scales are not homologous to fish scales, which are derived from the dermis and represent a kind of bone. Reptilian scales may be flat and fitted together like a mosaic, or they may be widely separated; in some cases they overlap like roof shingles. Beneath each scale is a dermal vascular papilla, which supplies nutriment to the scale. In some the outer layer of the scales is continually shed, either in small bits or sloughed off periodically in one piece. Turtles, however, add new layers of keratin under the old layers in the scales. Modified forms of scales are found in the reptile group. The scutes of a turtle shell are large, platelike scales. Also, in some reptiles the epidermal scales are reinforced by dermal plates (osteoderms). Scales are of diagnostic value in the identification of some reptiles, such as snakes.

The **skeleton** of the early reptiles (cotylosaurs) was similar to that of the labyrinthodont amphibians. Most changes in later reptiles involved a loss of skull elements (by fusion or otherwise) and adaptable transformations for better locomotion. The nostrils of an alligator or crocodile are on the dorsal side of

FIG. 24-1

Family tree of the reptiles. Transition from certain labyrinthodont amphibians to reptiles occurred in Carboniferous period to Mesozoic times. This transition was effected by development of amniote egg, which made land existence possible, although this egg may well have developed before oldest reptiles had ventured far on land. Explosive adaptation by reptiles may have been due partly to variety of ecologic niches into which they could move. Fossil record shows that lines arising from stem reptiles led to ichthyosaurs, plesiosaurs, and stem-ruling reptiles. Some of these returned to the sea. Later radiations led to mammal-like reptiles, turtles, flying reptiles, birds, dinosaurs, etc. Of this great assemblage, the only reptiles now in existence belong to four orders (Testudines, Crocodilia, Squamata, and Rhynchocephalia). The Rhynchocephalia, not shown in this diagram, is represented by only one living species, the tuatara *(Sphenodon)* of New Zealand. How the mighty have fallen!

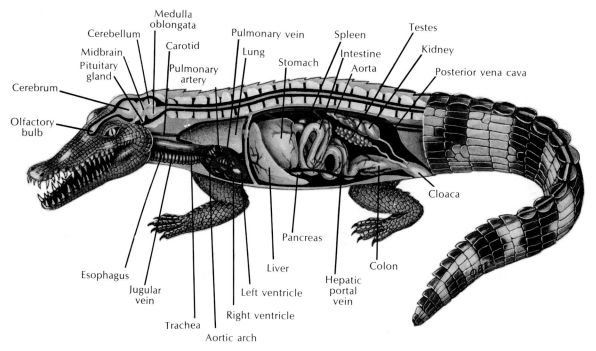

FIG. 24-2
Internal structures of male crocodile.

the head and the internal nares are at the back of the throat, which can be closed off by a fold. Thus the reptile can breathe while holding its prey submerged. The reptilian skull has developed a more flexible joint with the vertebral column, and more efficient girdles were evolved for supporting the body. The toes are provided with claws. Their peglike teeth vary somewhat but do not show the differentiation of the mammal; they may be set in sockets (thecodont) or fused to the surface of the bone (acrodont). In many the teeth are formed in two rows along the edges of the premaxillae and maxillae, in the upper jaw, and along the dentaries of the lower jaw. Teeth may also be present on some of the palate bones. They are absent in turtles, in which only a horny beak is present.

The **body cavity** of reptiles is mostly divided into sacs by mesenteries, ligaments, and peritoneal folds. The heart is always enclosed in a pericardial sac (Fig. 24-2). Among the turtles the lungs lie outside the peritoneal cavity. Lizards have a posthepatic septum that divides the peritoneal cavity into two divisions, and crocodiles have a similar one that contains muscle and may function in respiration. This partition, however, is not homologous to the mammalian diaphragm.

Most reptiles are carnivorous and their **digestive system** is adapted for such a diet. All are provided with a tongue. This is large, fleshy, and broad in crocodiles and turtles. In crocodiles a tongue fold with a similar fold of the palate can separate the air passage from the food passage. In some reptiles (chameleon) the tongue is highly protrusible and is used in catching their prey. When not in use, the anterior part of such a tongue is telescoped into the posterior portion. Buccal glands vary a great deal. In snakes the upper labial glands may be modified into poison glands. The stomach is usually spindle shaped and contains gastric glands. Pebbles or gastroliths are found in some reptiles (crocodiles) for grinding the food. A short duodenum receives the ducts from the liver and pancreas. The walls of the midgut are thrown into folds, but there are few glands. The **cloaca** (Fig. 24-2), which receives the rectum, ureters, and reproductive ducts, may be complicated. A urinary bladder is found in *Sphenodon,* turtles, and most lizards and opens into the ventral wall of the cloaca.

The **excretory system** is made up of the paired elongated or compact kidneys (metanephroi), composed of glomerular nephrons. Ureters carry the urine (fluid in turtles and crocodiles; semisolid with insoluble urates in the others) to the cloaca.

All reptiles have **lungs** for breathing. The glottis behind the tongue is closed by special muscles. The larynx contains arytenoid and cricoid cartilages but no thyroid cartilages. The trachea is provided with semicircular cartilage rings to keep it open. Lungs are mainly simple sacs in *Sphenodon* and snakes (in which one is reduced), but in turtles and crocodiles they are divided into irregular chambers, with the alveoli connected to branched series of bronchial tubes. In chameleons long, hollow processes of the lungs pass posteriorly among the viscera and represent forerunners of the air sacs of birds. Air is drawn into the lungs by the movements of the ribs and, in crocodiles, by the muscular diaphragm.

Since there is no branchial circulation in reptiles, on each side the fifth aortic arch is lost, the third becomes the carotid arch, the fourth the systemic arch, and the sixth the pulmonary arteries. The systemic arches are paired, in contrast to the single one of birds and mammals. In crocodiles there are two completely separated ventricles; in the other groups the ventricle is incompletely separated. **Crocodiles are thus the first animals with a four-chambered heart** (Fig. 24-2). The conus and truncus arteriosus are absent. Venous blood is returned to the sinus venosus of the heart through the paired precaval and the single postcaval veins. Renal and hepatic portal systems are present. The blood contains oval, nucleated corpuscles that are smaller than those of amphibians.

The male **reproductive system** consists of paired elongated testes that are connected to the vasa deferentia (the wolffian or mesonephric ducts) (Fig. 24-2). The latter carry sperm to the copulatory organ, which is an evagination of the cloacal wall and is used for internal fertilization. These copulatory organs are single in crocodiles and turtles but paired in lizards and snakes, in which they are called **hemipenes.** The female system is made up of large paired ovaries, and the eggs are carried to the cloaca by oviducts, which are provided with funnel-shaped ostia. The glandular walls of the oviducts secrete albumin and shells for the large amniote eggs. The embryonic membranes (amnion, yolk sac, allantois, and chorion), first appear in reptiles and are used for the nourishment, protection, and respiration of the embryo.

The reptilian **nervous system** shows many advancements over that of the amphibians. The cerebrum is enlarged, and there has been a transfer of some nervous integrative activities forward to the forebrain. Relatively, however, the brain is small, never exceeding 1% of the body weight. There are 12 pairs of cranial nerves in addition to the nervus terminalis. A better developed peripheral nervous system is associated with the more efficient limbs of reptiles. Sense organs vary among the different reptilian groups, but in general they are well developed. The lateral-line system, however, is entirely lost. The sense of hearing is poorly developed in most. All reptiles have a middle and inner ear, and crocodiles have an outer one as well. The middle ear contains the ear ossicle (stapes) and communicates with the pharynx by the eustachian tube. Jacobson's organ, a unique sense organ, is a separate part of the nasal sac and communicates with the mouth; it is especially well developed in snakes and lizards. It is innervated by a branch of the olfactory nerve and is used in smelling the food in the mouth cavity. In snakes, environmental odors trapped on the flickering tongue are carried into the mouth and to Jacobson's organ. This organ in some form is found in other groups, including the amphibians.

■ STRUCTURE AND NATURAL HISTORY OF ORDERS

SQUAMATA. The lizards and snakes of this order are the most recent products of reptile evolution, and they are by far the most successful of all living reptiles. They are covered with a flexible armor of overlapping scales that must be shed periodically as the outer layer dies. The order is divided into suborders Saura (Lacertilia), which includes the lizards, and Serpentes (Ophidia), which includes the snakes. Lizards have movable eyelids, external ear openings, and legs (usually); snakes lack these characteristics. Snakes are always legless, and unlike lizards they are adapted to swallow prey larger than themselves by virtue of a highly flexible lower jaw.

Lizards (suborder Saura). The lizards (Fig. 24-3) are an extremely diversified group, including terrestrial, burrowing, aquatic, arboreal, and aerial members (Fig. 24-3). *Draco,* a lizard found in India,

FIG. 24-3
Carolina anole, *Anolis*. These lizards are popularly called chameleons because they can
change their color, like true chameleons of the Old World. Family Iguanidae.

is able to volplane from tree to tree because of skin
extensions on the side. A few lizards such as the
glass lizards are limbless.

Many lizards live in the world's hot and arid regions, aided by several adaptations for desert life.
Since their skin lacks glands, water loss by this avenue is much reduced. They produce a semisolid urinary waste with a high content of crystalline urates.
Some, such as the Gila monster of the southwestern
United States deserts, store fat in their tails, which
they draw upon during drought to provide both energy and metabolic water. Especially interesting are
the techniques desert lizards use to maintain a relatively constant body temperature, using what physiologists term "behavioral thermoregulation."
Lizards, like other reptiles, are poikilothermic and ectothermic; if a lizard is placed under constant temperature conditions in the laboratory, its body temperature will soon become indistinguishable from
that of its surroundings. But in its natural environment where surrounding temperatures vary widely,
lizards modulate their body temperature by exploiting hour-to-hour changes in thermal flux from the
sun. In the early morning they emerge from their
burrows and bask in the sun with their bodies flattened to absorb heat. They then begin to actively
hunt insect prey. Toward midday they raise their
bodies from the hot sand and change body contour
to reduce the area exposed to the sun, a behavior

called **stilting**. During the hottest period of the day
they may retreat to their burrows. They emerge
again in midafternoon to stilt and later to bask as
the sun sinks lower and the air temperature drops.
These behavioral patterns help to maintain a relatively steady body temperature of 36° to 39° C. while
the air temperature is varying between 29° and 44°
C. Some lizards can tolerate intense midday heat
without shelter. The desert iguana of the southwestern United States prefers a body temperature of
42° C. when active and can tolerate a rise to 47° C.,
a temperature that is lethal to all birds and mammals
and most other lizards. The term "cold-blooded"
clearly does not apply to these animals!

The Gila monster *(Heloderma suspectum)* is one of
only two poisonous lizards in the world. This slow
moving species kills its prey with powerful jaws and
with the venom that is worked rather inefficiently
into the victim with chewing movements. It adapts
readily to captivity and requires little care. The well-
known horned toad *(Phrynosoma)* of the same region bears spiky scales that give it a grotesque appearance for frightening potential predators; in fact,
it has a placid temperament and makes a harmless
and interesting pet.

Chameleons are a group of 85 species of well-
adapted arboreal lizards, mostly African, that stalk
their insect prey slowly and deliberately, using their
prehensile tail and opposable digits to grasp

branches. Insects are caught with the sticky-tipped tongue that can be flicked rapidly and accurately to more than their own body length away. The tongue fits over the elongate hyoid bone like a wrinkled sleeve and is everted by the sudden contraction of circular muscles; longitudinal muscles retract the tongue. With their ability to change color in response to changes in temperature, light intensity, and emotion, and eyes that can swivel independently through 180 degrees, chameleons are among the most fascinating lizards to study.

Snakes (suborder Serpentes). The snakes arose from lizards, probably from burrowing forms. They are entirely limbless and lack both the pectoral and pelvic girdles (the latter persistent as vestiges in pythons). The numerous vertebrae of snakes, shorter and wider than those of tetrapods, promote quick lateral undulations through grass and over rough terrain. The ribs increase rigidity of the spinal cord, which provides more resistance to lateral stresses. The elevation of the neural spine gives the numerous muscles more leverage. Snakes also differ from lizards in having no movable eyelids (their eyes are permanently covered with a third eyelid) or external ear. Snakes are in fact totally deaf, although they are sensitive to low-frequency vibrations conducted through the ground.

Like lizards, snakes bear rows of scales that overlap as do shingles on a roof. In some snakes the scales are keeled and in others, smooth. Nearly all snakes have on their ventral surface from chin to anus a single row of transverse scales or scutes and one or two rows on the ventral surface of the tail. In moving, the snakes make use of these scutes by projecting their margins and using them for clinging against a surface while the body is driven forward by lateral undulations. On very smooth surfaces snakes cannot move forward easily. Snakes also move by alternately throwing the body into coils and then straightening it out. Most species are good swimmers; they swim by lateral convolutions of the body. Though persistent in folklore, it hardly needs to be said that there is no "hoop snake" that grasps its tail in its mouth and rolls like a hoop at high speed after its terrified victims. The story may have originated with the western sidewinder rattlesnake, which throws its body forward in successive contractions, traveling at an angle at surprising speed over loose sand.

Specialized as snakes obviously are in their shape, it perhaps comes as no surprise that many of them use a unique set of special senses to hunt down their prey. We alluded above to the deafness of snakes to airborne vibrations. Most also have relatively poor vision, with the tree-living snakes of the tropical forest being a conspicuous exception. In fact, the latter possess excellent binocular vision to help them track prey through the branches where scent trails would be impossible to follow. But most snakes live on the ground and rely on chemical senses to hunt food. In addition to the usual olfactory areas in the nose, which are not well developed, there are **Jacobson's organs**, a pair of pitlike organs in the roof of the mouth. These are lined with an olfactory epithelium and richly innervated. By flicking the forked tongue through the air, scent particles are picked up and conveyed to the mouth; the tongue is then drawn past Jacobson's organ or the tips of the forked tongue are inserted directly into the organs. Information is then transmitted to the brain where scents are identified.

Snakes of the subfamily Crotalinae within the family Viperidae are called **pit vipers** because of special heat-sensitive pits on their heads, between the nostrils and the eyes (Fig. 24-4). All of the best known North American poisonous snakes are pit vipers, such as the several species of rattlesnakes, water moccasin (Fig. 24-5), and the copperhead. The pits are supplied with a dense packing of free nerve endings from the fifth cranial nerve. They are exceedingly sensitive to radiant energy (long-wave infrared) and can distinguish temperature differences smaller than 0.2° C. from a radiating surface. Pit vipers use the pits to track warm-blooded prey and aim strikes, which they can do as effectively in total darkness as in daylight.

Nearly all pit vipers have a pair of teeth on the maxillary bones modified as fangs. These lie in a membrane sheath when the mouth is closed (Fig. 24-4). When the viper strikes, a special muscle and bone lever system erects the fangs when the mouth opens. The fangs are driven into the prey by the thrust, and venom is injected into the wound along a groove in the fangs. About 1,500 bites and 45 deaths from pit vipers are reported each year in the United States.

A pit viper immediately releases its prey after the bite and follows it until it is paralyzed or dies. Then it is swallowed whole. For this operation a snake is

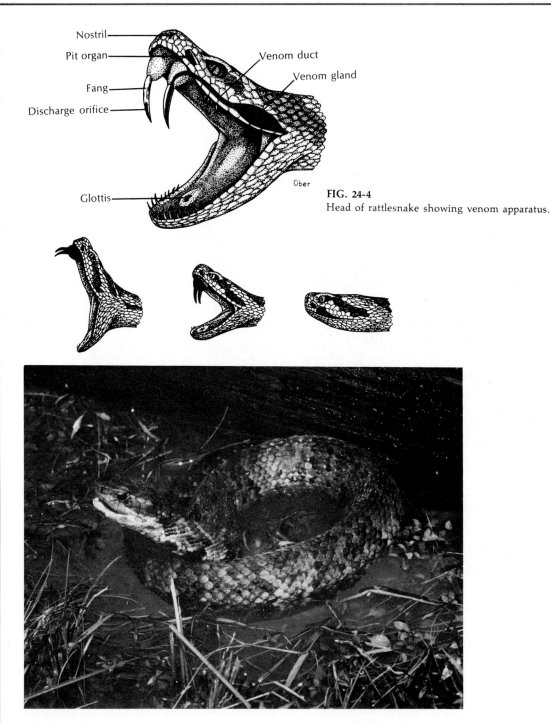

FIG. 24-4
Head of rattlesnake showing venom apparatus.

Nostril

Pit organ

Fang

Discharge orifice

Venom duct

Venom gland

Glottis

Ober

FIG. 24-5
Water moccasin *(Agkistrodon piscivorus),* also called cottonmouth from its habit of opening its mouth widely to expose white interior when threatened. This is a relatively common, large, and robust pit viper of the southern United States. It is mainly aquatic in habitat preference. There is great variety in coloration. It is omnivorous, eating frogs, snakes, fish, small mammals and birds, and large insects. Water moccasins are difficult to handle and can inflict painful bites. Family Viperidae. (Photo by L. L. Rue, III.)

FIG. 24-6
Rattlesnakes shedding skin. **A,** Eastern diamondback *(Crotalus adamanteus)* just before shedding. Note opaque eye, caused by air pocket beneath eye plate. Snake is blind at this time, remains hidden, and does not eat. **B,** Eastern timber rattlesnake *(Crotalus horridus)* shedding skin. A new segment is added to rattle each time snake sheds. Family Viperidae. (Photos by L. L. Rue, III.)

perfectly adapted. The jaw articulation is very loose and the two halves of the lower jaw are joined by an elastic ligament that allows them to come apart. The skin on the neck is highly elastic, and since snakes lack a sternum, the ribs are not joined ventrally so that they can come apart. This allows snakes to swallow prey much larger than themselves.

More people are bitten by the copperhead, *Agkistrodon mokasen,* than by any other species, although more deaths are caused by the Texas diamondback rattlesnake, *Crotalus atrox.* Rattlesnakes are characterized by the **rattle,** which consists of horny, ringlike segments held loosely together on the end of their tails. When aroused, the snake vibrates these

rapidly, producing a buzzing sound. Whenever the skin is shed, the posterior part remains behind to form another ring of the rattle (Fig. 24-6). The number of rattles is not an accurate indication of age, for the snake often sheds more than once a year, and rings of the rattle are frequently lost.

The tropical and subtropical countries are the homes of most species of snakes, both of the venomous and nonvenomous varieties. However, even here most of the members are nonpoisonous. Among warm countries only a few regions such as New Zealand are free from native snakes. Madagascar has no poisonous snakes. In India the death toll from snake bites averages about 25,000 annually. Many persons used to die from poisonous snakes in South America

FIG. 24-7
Boa constrictor *(Boa constrictor).* This specimen was kept at DePauw University for 23 years, until its death (1961) at nearly 30 years of age. During its vigorous years it ate from 15 to 30 rats or pigeons each summer (by first constricting them to death) but refused to eat in winter. During its last years it ate only 4 or 5 rats a year. It grew 2 to 3 feet in captivity; was nearly 9 feet long. Family Boidae.

FIG. 24-8
Coral snake, *Micrurus fulvius,* a very poisonous snake that inhabits southern United States and tropical countries. Only representative of cobra family in North America. Body bands black, red, and yellow. Family Elapidae. (Courtesy F. M. Uhler, U. S. Fish and Wildlife Service.)

until effective antivenom serums were developed. The tropics also furnish the enormous constrictor snakes (Fig. 24-7), some of which, like the regal python of the Malay Peninsula, attain a length of 30 feet; the anaconda of South America is another large snake. Their prey rarely exceeds the size of a large pig. The largest of the venomous snakes are the king cobra (hamadryad), which may reach 15 to 18 feet in length, and the bushmaster *(Lachesis)* of Central and South America, which, although never exceeding 10 feet in length, is thick and muscular. In the United States, the eastern diamondback of the Carolinas and Florida and the western diamondback of the southwestern canyons and deserts are the largest and most dangerous.

Poisonous snakes are usually divided into two groups based on the type of fangs. The vipers (family Viperidae) have tubular or grooved fangs at the front of the mouth; the group includes the American pit vipers mentioned above, and the Old World true vipers, which lack facial heat-sensing pits. Among

the latter are the common European adder and the African puff adder. The other family of poisonous snakes (family Elapidae) have short fangs, so that the venom must be injected by chewing. In this group are the cobras, mambas, coral snakes (Fig. 24-8), and sea snakes, all highly poisonous.

Predators that attack and swallow the highly venomous sea snakes regurgitate them quickly because of the internal bites from the snakes. The extreme toxicity of the snake's venom may have evolved so that predators would quickly regurgitate the snakes unharmed.

The mouth secretions (saliva) of all harmless snakes possess some toxic properties, and it is logical that evolution should have stressed this toxic tendency in certain species. Venom is frequently collected from snakes by "milking" the glands (Fig. 24-9). The venom is used to prepare antivenom serums to treat snakebite victims. Certain venoms have limited medicinal value as well. Snake poison is usually a thick, yellowish, slightly cloudy liquid that, on expo-

FIG. 24-9
"Milking" an eastern diamondback rattlesnake *(Crotalus adamanteus)* to collect venom for preparation of antiserum. (Photo by L. L. Rue, III.)

sure to the air, will crystallize into yellow crystals and may remain toxic for a long time.

There are two types of snake venom. One type acts mainly on the nervous systems (neurotoxic), affecting the optic nerves (causing blindness) or the phrenic nerve of the diaphragm (causing paralysis of respiration). The other type is hemolytic; that is, it breaks down the red blood corpuscles and blood vessels and produces extensive extravasation of blood into the tissue spaces. Many venoms have both neurotoxic and hemolytic properties. The toxicity of a venom is determined by the minimal lethal dose on laboratory animals. By this standard the venom of *Bothrops insularis*, a member of the fer-de-lance family in South America, appears to be the most deadly of poisons drop for drop.

Even the most deadly snakes have enemies. Man is the most important; he is far more dangerous to snakes than snakes are to him. In the United States the king snakes (*Lampropeltis;* Fig. 24-10) will eat other snakes, especially poisonous ones. The slender king snake will encircle even the big diamondback rattlesnake in its coils and squeeze it to death. The king snake is immune to the poison of the rattler. Where venomous snakes abound, natural enemies in some form are sure to act as a curb on their numbers. In India the mongoose (*Herpestes*), a little mammal, can attack and eat the hooded cobra and other poisonous snakes. Many snakes feed upon other snakes, which is the case with the king cobra of southeastern Asia, but the snakes it kills are mostly harmless. In South America the fer-de-lance has two deadly enemies, the skunk (*Conepatus*) and musurana (*Clelia*), a mild-appearing snake of the Brazilian forests. Snakes have many bird enemies, especially some of our large hawks.

Most snakes lay eggs (oviparous) (Fig. 24-10), but some bring forth the young alive. There is no evidence to show that snakes swallow their young to protect them.

TESTUDINES (CHELONIA). The tortoises and turtles are an ancient group that have plodded on

FIG. 24-10
Female king snake, *Lampropeltis getulus)* guarding her eggs. King snakes are highly beneficial constrictors, feeding on rodents and on poisonous snakes. Family Colubridae (Photo by L. L. Rue, III.)

from the Triassic to the present with very little change in their early basic morphology. They are enclosed in shells consisting of a dorsal **carapace** and a ventral **plastron**. Clumsy and unlikely as they appear to be within their protective shells, they are nonetheless a varied and successful group that seems able to accommodate to man's presence. The shell is so much a part of the animal that it is built in with the thoracic vertebrae and ribs. Into this shell the head and appendages can be retracted for protection. No sternum is found in these forms, and their jaws lack teeth but are covered by a horny sheath. The nasal opening is single. They have lungs, although aquatic forms have vascular sacs in the cloaca that serve for breathing when the animals are submerged. On their toes are horny claws for digging in the sand, where they lay their eggs. Some of the marine forms have paddle-shaped limbs for swimming. Fertilization is internal by means of a cloacal penis on the ventral wall of the male cloaca. All turtles are oviparous, and the eggs have firm, calcareous shells.

The great marine turtles, buoyed by their aquatic environment, may reach 6 feet in length and 1,000 pounds in weight, such as the leathery turtle; green sea turtles may reach 500 pounds (Fig. 24-11). Even some land tortoises may weigh several hundred pounds, such as the giant tortoises of the Galapagos Islands that so intrigued Darwin during his visit there in 1835. Most tortoises are rather slow moving; 1 hour of determined trudging will carry a large Galápagos tortoise about 1,000 feet. Their low metabolism probably explains their longevity, for some are believed to live more than 100 years.

Turtles eat both vegetable and animal products. Many marine forms capture fish and other vertebrates. Land tortoises live on insects, plants, and berries. The common box turtle *(Terrapene)* (Fig. 24-12) grows fat during the wild strawberry season.

The term "tortoise" is usually given to the land forms, whereas the term "turtle" is reserved for the aquatic forms. Among the former, the box tortoise *(Terrapene)* is one of the most familiar. It is about 6 inches long. It has a high arched carapace, with the front and rear margins curled up. The lower shell is hinged, with two movable parts so that it can be pulled up against the upper shell. The color markings vary, but usually the shell is a dark brown color with irregular yellow spots. Box tortoises are found in woods and fields and sometimes in marshes. Despite their slow movements, marked individuals have often been found a considerable distance away from the point of release. They lay their eggs in cavities dug out of loose soil and cover them over.

FIG. 24-11
Green sea turtle, *Chelonia.* Note that limbs are modified into flippers. Such turtles are strictly aquatic except when they lay their eggs on sandy shore. Some of these turtles may weigh as much as 400 to 500 pounds and are greatly prized for turtle soup. Family Cheloniidae.

FIG. 24-12
Common box tortoise, *Terrapene.* Family Testudinidae.

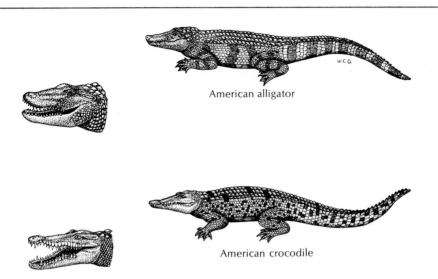

American alligator

American crocodile

FIG. 24-13

American crocodilians. The American alligator, *Alligator mississippiensis,* is found along the southeastern United States coast from central Texas to the Atlantic. The American crocodile, *Crocodylus acutus,* is now limited to the Everglades National Park in Florida. Both species are in danger of extinction. Enlargment of heads shows easily recognized differences between the two species. The crocodile has a more slender snout and the fourth tooth from the front of the lower jaw fits *outside* upper jaw and is visible when mouth is closed. It is not true, as commonly believed, that there is a difference in the way the jaws are hinged in the two species. Order Crocodilia.

FIG. 24-14

Tuatara, *Sphenodon punctatum,* the only living representative of order Rhynchocephalia. This "living fossil" reptile has well-developed parietal "eye" with retina and lens on top of head. Eye is covered with scales and is considered nonfunctional but may have been important sense organ in early reptiles. The tuatara is found only on islands of Cook Strait, New Zealand.

Parietal median eye
(covered with scales)

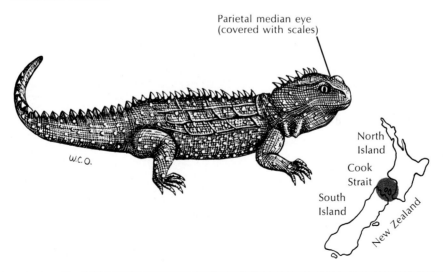

North
Island
Cook
Strait
South
Island
New Zealand

Snapping turtles *(Chelydra)* are found in nearly every pond or lake in eastern and central North America. They grow to be 12 to 14 inches in diameter and 20 to 40 pounds in weight. They are ferocious and are often referred to as the "tigers of the pond." They are entirely carnivorous, living on fish, frogs, waterfowl, or almost anything that comes within reach of their powerful jaws. They are wholly aquatic and come ashore only to lay their eggs.

CROCODILIA. The modern crocodiles are the largest living reptiles. They are what remains of a once abundant group in the Jurassic and Cretaceous. Having managed to survive virtually unchanged for some 160 million years, the modern crocodilians face a forbidding, and perhaps short, future in man's world. This order is divided into crocodiles and alligators. Crocodiles have relatively long slender snouts; alligators have short and broader snouts. With their powerful jaws and sharp teeth, they are formidable antagonists. Although all are carnivorous, many will not attack man. The "man-eating" members of the group are found mainly in Africa and Asia. The estuarine crocodile *(Crocodylus porosus)* found in southern Asia grows to a great size and is very much feared. It is swift and aggressive and will eat any bird or mammal it can drag from the shore to water where it is violently torn to pieces. Crocodiles are known to attack animals as large as cattle and deer.

Alligators are usually less aggressive than crocodiles. They are almost unique among reptiles in being able to make definite sounds. The male alligator can give loud bellows in the mating season. Vocal sacs are found on each side of the throat and are inflated when he calls. In the United States, *Alligator mississipiensis* is the only species of alligator; *Crocodylus americanus* is the only species of crocodile (Fig. 24-13).

Alligators and crocodiles are oviparous. Usually from 20 to 50 eggs are laid in a mass of dead vegetation. The eggs are about 3 inches long. The penis of the male is an outgrowth of the ventral cloaca.

RHYNCHOCEPHALIA. This order is represented by a single living species, the tuatara *(Sphenodon punctatum)* of New Zealand (Fig. 24-14). This animal is the sole survivor of a group of primitive reptiles that otherwise became extinct 100 million years ago. The tuatara was once widespread on the North Island of New Zealand but is now restricted to one or

two islands of Cook Strait where, under protection from the New Zealand government, it may recover. It is a lizardlike form 2 feet long or less that lives in burrows often shared with petrels. They are slow-growing animals with a long life; one is recorded to have lived 77 years.

The tuatara has captured the interest of biologists because of its numerous primitive features that are almost identical to those of Mesozoic fossils 200 million years old. These include a primitive skull structure of a type (diapsid, having two openings in the temporal bones) found in early Permian reptiles that were ancestors to the modern lizards. It also bears a well-developed parietal eye, complete with evidences of a retina (Fig. 24-14), and a complete palate. It lacks a copulatory organ. A specialized feature is the teeth, which are fused to the edge of the jaws, rather than being set in sockets. *Sphenodon* represents one of the slowest rates of evolution known among the vertebrates.

Classification

Body covered with horny (ectodermal) scales or plates; usually four limbs, each with five claws; skeleton ossified; one occipital condyle; lungs throughout life.

Order Testudines (Chelonia). Body enclosed in a shell of dorsal carapace and ventral plastron; jaws with horny sheaths; no teeth; quadrate bone immovable; vertebrae and ribs fused to shell usually.

    Suborder Cryptodira. Hidden-necked turtles; head withdrawn into shell by flexing neck ventrally. Eight families: Chelydridae (snapping turtles), Kinosternidae (musk and mud turtles), Testudinidae (tortoises and terrapins), Cheloniidae (sea turtles), Trionychidae (soft-shelled turtles), Dermochelyidae (leatherback turtles), Dermatemydidae (river turtles), Carettochelyoidea (pitted-shelled turtles).

    Suborder Pleurodira. Side-necked turtles; head withdrawn into shell by bending neck sideways. Two families: Chelidae (snake-necked turtles), Pelomedusidae (pelomedusid turtles).

Order Rhynchocephalia. Parietal organ (third "eye") present; scales granular; middorsal row of spines; quadrate bone immovable; vertebrae biconcave. Example: *Sphenodon.*

Order Squamata. Skin with horny epidermal scales; quadrate bone movable; vertebrae procoelus usually; copulatory organ (hemipenes) present.

    Suborder Sauria (Lacertilia)—lizards. Body slender, usually with four limbs; rami of lower jaw fused; eyelids movable; copulatory organs paired. Twenty families. The major ones are Gekkonidae (geckos), Iguanidae (New World lizards), Agamidae (Old World lizards), Chamaeliontidae (chameleons), Lacertidae (Old World lizards), Scincidae (skinks), Amphisbaenidae (worm lizards), Helodermatidae (poisonous lizards), Anguidae (plated lizards).

Suborder Serpentes (Ophidia)—snakes. Body elongated; limbs and ear openings absent; mandibles jointed anteriorly by ligaments; eyes lidless and immovable; tongue bifid and protrusible; teeth conical, and on jaws and roof of mouth. Eleven families. The major ones are Leptotyphlopidae (blind snakes), Boidae (boas and pythons), Colubridae (common snakes), Elapidae (poisonous fixed-fang snakes), Viperidae (vipers and pit vipers).

Order Crocodilia (Loricata). Body long; head large with long jaws; teeth many and conical; short limbs with clawed toes; tail long and heavy and bilaterally compressed; thick leathery skin with horny scutes; tongue nonprotrusible. Three families: Gavialidae (gavials), Alligatoridae (alligators and caimans), Crocodylidae (crocodiles).

References for amphibians and reptiles
Suggested general readings

Anderson, P. 1966. The reptiles of Missouri. Columbia, Mo., University of Missouri Press. *A good state survey of reptiles.*

Barbour, R. W. 1971. Amphibians and reptiles of Kentucky. Lexington, Ky., University of Kentucky Press. *Although this volume is primarily concerned with a state survey of amphibians and reptiles, it is in addition concerned with general distribution of species and contains a number of excellent photographs.*

Bellairs, A. 1970. The life of reptiles, 2 vols. New York, Universe Books. *An accurate well written treatise that is primarily concerned with anatomy and evolution. Perhaps a bit advanced for the beginning student.*

Bishop, S. C. 1943. Handbook of salamanders of the United States, of Canada and of lower California. Ithaca, N. Y., Comstock Publishing Co. *A volume that has now become a standard reference for the taxonomy and natural history of the salamanders.*

Blair, W. F., A. P. Blair, P. Brodkorb, F. R. Cagle, and G. A. Moore. 1968. Vertebrates of the United States, ed. 2. New York, McGraw-Hill Book Co. *An excellent source book that lists and discusses every vertebrate known to occur in the United States.*

Bucherl, W., E. E. Buckley, and V. Deulofeu (editors). 1968. Venomous animals and their venoms, 3 vols. New York, Academic Press Inc.

Carr, A. F., Jr. 1952. Handbook of turtles. Ithaca, N. Y., Cornell University Press. *One of the best handbooks on the subject.*

Cochran, D. M., and C. J. Goin. 1970. The new field book of reptiles and amphibians. New York, G. P. Putnam's Sons. *Probably the most useful field guide available for reptiles and amphibians. Most species descriptions are concise, well written, and clear.*

Conant, R. 1958. A field guide to reptiles and amphibians of eastern North America. Boston, Houghton Mifflin Co. *An extremely useful field guide. The fact that it is pocket-sized makes it very valuable for those doing research in the field.*

Gans, C. (editor). 1969. Biology of the Reptilia, 3 vols. New York, Academic Press Inc. *A work for the advanced student or researcher. The series will eventually consist of morphologic, embryologic and physiologic, and ecologic and behavioral portions. Specialists contribute chapters on topics that correspond to their research interests.*

Goin, C. J., and O. B. Goin. 1971. Introduction to herpetology, ed. 2. San Francisco, W. H. Freeman & Co. *A basic introductory text for the study of amphibians and reptiles.*

Klauber, L. M. 1956. Rattlesnakes, 2 vols. Berkeley and Los Angeles, University of California Press. *An excellent monograph on this group of reptiles.*

Minton, S. A., Jr., and M. R. Minton. 1969. Venomous reptiles. New York, Charles Scribner's Sons. *A semipopular account about poisonous snakes and the role they have played in human culture.*

Moore, J. (editor). 1964. Physiology of the amphibia. New York, Academic Press Inc. *Although for the advanced student, it is still a valuable reference on the amphibia.*

Oliver, J. A. 1955. The natural history of North American amphibians and reptiles. New York, Van Nostrand Co. *A good introduction to classification, distribution, and behavior of reptiles and amphibians.*

Schmidt, K. P. 1953. A checklist of North American amphibians and reptiles. American Society of Ichthyologists and Herpetologists, Chicago, University of Chicago Press. *A valuable taxonomic source that lists species and known distribution for all known amphibians and reptiles of the North American continent.*

Schmidt, K. P., and R. F. Inger. 1957. Living reptiles of the world. Garden City, N. Y., Hanover House. *A magnificent volume with excellent illustrations of the various families of living reptiles.*

Smith, H. M. 1946. Handbook of lizards of the United States and Canada. Ithaca, N. Y., Comstock Publishing Co. *A standard reference for those interested in the natural history and taxonomy of lizards.*

Stebbens, R. C. 1966. A field guide to western reptiles and amphibians. Boston, Houghton Mifflin Co. *A superbly illustrated and very well written field guide. Limited in that it covers only the western United States.*

Wright, A. H., and A. A. Wright. 1949. Handbook of frogs of the United States and Canada, ed. 3. Ithaca, N. Y., Comstock Publishing Co. *A volume that has, with time, become a standard reference work for the study of frogs.*

Wright, A. H., and A. A. Wright. 1957. Handbook of snakes of the United States and Canada, 2 vols. Ithaca, N. Y., Comstock Publishing Co. *These two volumes are indispensable and are probably the best available on snakes of the United States and Canada.*

Young, J. Z. 1962. The life of vertebrates, ed. 2. New York, Oxford University Press. *The author has attempted, with great effort and success, to combine the many facets of vertebrate life into a comprehensive volume. Excellent reading for advanced students.*

Selected *Scientific American* articles

Bogert, C. M. 1959. How reptiles regulate body temperature. **200:**105-120 (April). *Many reptiles, especially lizards, achieve a remarkable degree of temperature regulation by behavioral responses.*

Carr, A. 1965. The navigation of the green turtle. **212:**78-86 (May). *These marine turtles make regular migrations over vast ocean routes.*

Gamow, R. T., and J. E. Harris. 1973. The infrared receptors of snakes. **228:**94-100 (May). *The anatomy and physiology of the remarkable "sixth sense" of snakes is described.*

Gans, C. 1970. How snakes move. **222:**82-96 (June).

Minton, S. A., Jr. 1957. Snakebite. **196:**114-122 (Jan.). *A survey of the world's most poisonous snakes and their venoms.*

Muntz, W. R. A. 1964. Vision in frogs. **210:**110-119 (March). *Describes experiments on the selective response of frog's retina to the visual world.*

Riper, W. Van. 1953. How a rattlesnake strikes. **189:**100-102 (Oct.). *The rattlesnake strike is examined with high-speed photography.*

The eyes of this red-tailed hawk, with a visual acuity eight times better than man's, gather more detailed information about the environment than all other senses combined. (Photo by L. L. Rue, III, Blairstown, N. J.)

CHAPTER 25

THE BIRDS

PHYLUM CHORDATA
CLASS AVES*

Of the higher vertebrates, the birds are the most studied, the most observable, the most melodious and, many think, the most beautiful. With 8,600 species distributed over nearly the entire earth, birds far outnumber all other vertebrates except the fishes. Birds are found in forests and deserts, in mountains and prairies, and on all the oceans. Four species have even visited the North Pole, and one, a skua, was seen at the South Pole. Some birds live in total blackness in caves, finding their way about by echolocation, and others dive to depths greater than 45 meters to prey on aquatic life.

The single unique feature that distinguishs birds from all other animals is that they have feathers. If an animal has feathers, it is a bird; if it lacks feathers, it is not a bird. No other vertebrate group bears such an easily recognized and foolproof identification tag.

There is great uniformity of structure among birds. Despite some 130 million years of evolution during which they proliferated and adapted themselves to specialized ways of life, no one has difficulty recognizing a bird as a bird. In addition to feathers, all birds have forelimbs modified into wings (although they may not be used for flight); all have hindlimbs adapted for walking, swimming, or perching; all have horny beaks; and all lay eggs. Probably the reason for this great structural and functional uniformity is that birds evolved into flying machines. This fact greatly restricts diversity, so much more evident in other vertebrate classes. For example, birds do not begin to approach the diversity seen in their warm-

*Ay'veez (pl. of L. *avis*, bird).

blooded evolutionary peers, the mammals, where we find forms as unalike as a whale, a porcupine, a bat, and a giraffe.

Birds, however, do share with mammals the highest organ system development in the animal kingdom. But a bird's entire anatomy is designed around flight and its perfection. An airborne life for a large vertebrate is a highly demanding evolutionary challenge. A bird, must, of course, have wings for support and propulsion. Bones must be light and hollow yet serve as a rigid airframe. The respiratory system must be incredibly efficient to meet the intense metabolic demands of flight and serve also as a thermoregulatory device to maintain a constant body temperature. A bird must have a rapid and efficient digestion to process an energy-rich diet, it must have a high metabolic rate, and it must have a high-pressure circulatory system. Above all, birds must have a finely tuned nervous system and keen sense organs, especially superb vision, to handle the complex problems of headfirst, high-velocity flight over the landscape.

■ ORIGIN AND RELATIONSHIPS

About 150 million years ago, a flying animal drowned and settled to the bottom of a tropical freshwater lake in what is now Bavaria. It was rapidly covered with a fine silt and eventually fossilized. There it remained until discovered in 1861 by a workman splitting slate in a limestone quarry. The fossil was about the size of a crow, with a skull not unlike that of modern birds except that the beaklike jaws bore bony teeth set in sockets like those of reptiles (Fig. 25-1). The skeleton was decidedly reptilian with a long bony tail, clawed fingers, and abdominal ribs. It might have been classified as a reptile except that it carried the unmistakable imprint of **feathers,** those marvels of biologic engineering that only birds possess. The finding was dramatic because it settled once and for all the controversy over the ancestry of birds. The fossil was a "missing link" that proved beyond reasonable doubt that birds had evolved from reptiles. Two more skeletons were later found in the same limestone formation, in 1877 and in 1956.

Archaeopteryx (ar-kee-op'ter-ix, meaning "ancient wing"), as the fossil was named, was an especially fortunate discovery because the fossil record of birds is disappointingly meager. The bones of birds are lightweight and quickly disintegrate, so that only under the most favorable conditions will they fossilize. Nevertheless, there are certain localities where bird fossils are relatively abundant. One of these is the famous Rancho La Brea tarpits in Los Angeles where in one pit alone were found 30,000 fossil birds representing 81 species. By 1952 over 780 different fossil species had been recorded. Although most of these are relatively recent fossils, enough intermediate forms are known to provide a reasonable picture of bird evolution from the Jurassic when *Archaeopteryx* lived, to recent times (Fig. 25-2). Two well-known fossil birds in particular deserve mention. One was *Ichthyornis* (ik-thee-or'nis), a small ternlike sea bird that lived during the Cretaceous along the shores of North America's inland sea about 100 million years ago, 50 million years after *Archaeopteryx*. The other was *Hesperornis*, a flightless, loonlike diving bird (Fig. 25-2). Both were essentially modern birds in almost every way. By the close of the Cretaceous, about 63 million years ago, the characteristics of modern birds had been thoroughly molded. There remained only the emergence and proliferation of the modern orders of birds. Hundreds of thousands of bird species have appeared and nearly as many have disappeared, following *Archaeopteryx* to extinction. Only a minute fraction of these nameless species have been discovered as fossils.

Most paleontologists agree that the ancestors of both birds and dinosaurs were derived from a stem group of reptiles called thecodonts. Birds probably evolved from a single ancestor and thus have a monophyletic origin. However, existing birds are divided into two groups: (1) **ratite** (rat'ite, L. *ratitus,* marked like a raft, from *ratis,* raft), the flightless ostrichlike birds that have a flat sternum with poorly developed pectoral muscles, and (2) **carinate** (L. *carina,* keel), the flying birds that have a keeled sternum upon which the powerful flight muscles insert. This division originated from the view that the flightless birds (ostrich, emu, kiwi, rhea, etc.) represented a separate line of descent that never attained flight. This idea is now completely rejected. The flightless birds are descended from a flying ancestor but lost the use of their wings, which became unnecessary for their mode of life. Flightless forms are ground-living birds that can outrun predators, or they live where few carnivorous enemies are found.

A   **B**

FIG. 25-1
Archaeopteryx, the 150-million-year-old ancestor of modern birds. **A,** Cast of the second
and most perfect fossil of *Archaeopteryx,* which was discovered in 1877 in a Bavarian
stone quarry. **B,** Reconstruction of *Archaeopteryx.* (**A,** Courtesy of the American Museum
of Natural History, New York; **B,** Drawn by Sheila Ford.)

FIG. 25-2
Family tree of birds showing probable lines of descent and relationship. Thirteen of the
most familiar of the 27 recognized living orders of birds are pictured. Two extinct orders,
represented by *Archaeopteryx* and *Hesperornis,* are also pictured.

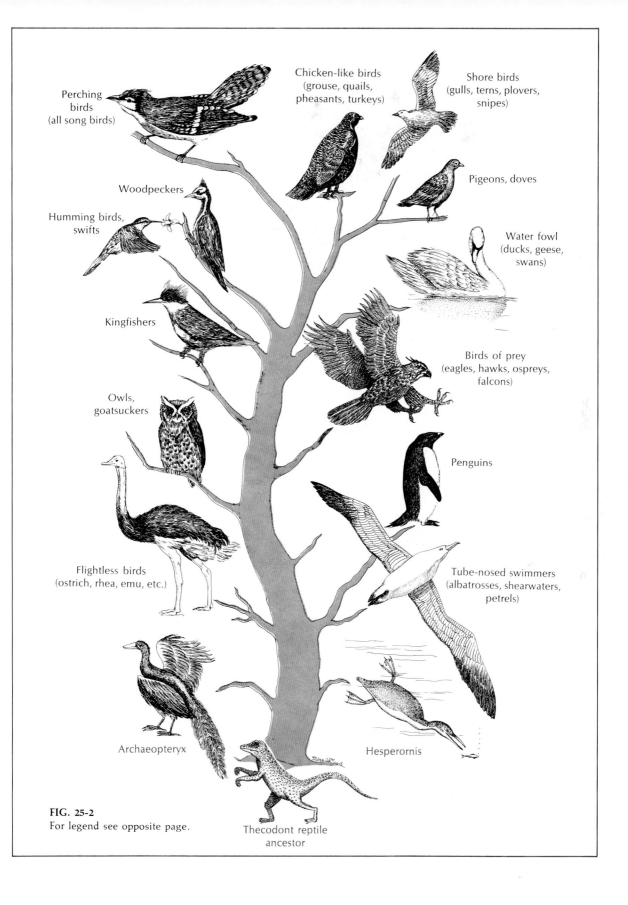

Perching birds
(all song birds)

Chicken-like birds
(grouse, quails,
pheasants, turkeys)

Shore birds
(gulls, terns, plovers,
snipes)

Pigeons, doves

Woodpeckers

Humming birds,
swifts

Water fowl
(ducks, geese,
swans)

Kingfishers

Birds of prey
(eagles, hawks, ospreys,
falcons)

Owls,
goatsuckers

Penguins

Tube-nosed swimmers
(albatrosses, shearwaters,
petrels)

Flightless birds
(ostrich, rhea, emu, etc.)

Archaeopteryx

Hesperornis

Thecodont reptile
ancestor

FIG. 25-2
For legend see opposite page.

Subclass Archaeornithes (ar"ke-or'ni-theez) (Gr. *archaios,* ancient, + *ornis, ornithos,* bird). Fossil. This included *Archaeopteryx* and possibly one or two other genera.
Subclass Neornithes (ne-or'ni-theez) (Gr. *neos,* new, + *ornis,* bird). Modern birds are placed in this group. Some extinct species with teeth are also included here because of their likeness to modern forms.

It may seem paradoxical that birds with their agile, warm-blooded, colorful, and melodious way of life should have descended from lethargic, cold-blooded, and silent reptiles. Yet the numerous anatomic affinities of the two groups is abundant evidence of close kinship and led the great English zoologist Thomas Henry Huxley to call birds merely "glorified reptiles." This unflattering description has never pleased bird lovers, who answer, "But how wonderously glorified!"

Characteristics

1. Body usually spindle shaped, with four divisions: head, neck, trunk, and tail; **neck disproportionately** long for balancing and food gathering
2. Limbs paired, with the **forelimbs usually adapted for flying;** posterior pair variously adapted for perching, walking, and swimming; foot with four toes (chiefly)
3. Epidermal **exoskeleton of feathers** and **leg scales;** thin integument of epidermis and dermis; no sweat glands; oil or preen gland at root of tail; **pinna of ear rudimentary**
4. **Skeleton fully ossified with air cavities or sacs;** skull bones fused with **one occipital condyle;** jaws covered with **horny beaks;** small ribs; vertebrae tend to fuse, especially the terminal ones; sternum well developed with keel or reduced with no keel; **no teeth**
5. Nervous system well developed, with brain and 12 pairs of cranial nerves
6. Circulatory system of **four-chambered heart,** with the **right aortic arch persisting;** reduced renal portal system; nucleated red blood cells
7. Respiration by slightly expansible lungs, with thin **air sacs** among the visceral organs and skeleton; **syrinx (voice box) near junction of trachea and bronchi**
8. Excretory system by metanephric kidney; ureters open into cloaca; **no bladder;** semisolid urine; uric acid main nitrogenous waste
9. Sexes separate; testes paired, with the vas deferens opening into the cloaca; **females with left ovary** and **oviduct;** copulatory organ in ducks, geese, ratites, and a few others
10. Fertilization internal; **amniotic eggs with much yolk** and **hard calcareous shells;** embryonic membranes in egg during development; **incubation external;** young active at hatching (**precocial**) or helpless and naked (**altricial**)

■ WHAT IS A FEATHER?

Feathers, more than any other single feature, distinguish a bird. A feather is almost weightless, yet possesses incredible toughness and tensile strength. A typical feather consists of a hollow **quill,** or calamus, thrust into the skin, and a **shaft,** or rachis, which is a continuation of the **quill** and bears numerous **barbs** (Fig. 25-3). The barbs are arranged in closely parallel fashion and spread diagonally outward from both sides of the central shaft, to form a flat, expansive, webbed surface, the **vane.** There may be several hundred barbs in each web. If the feather is examined with a microscope, each barb will appear like a miniature replica of the feather, with numerous parallel filaments, called **barbules,** set in each side of the barb, and spreading laterally from it. There may be 600 barbules on each side of a barb, which adds up to something over 1 million barbules for the feather. The barbules of one barb overlap the barbules of a neighboring barb in a herringbone pattern and are held together with great tenacity by tiny hooks. Should two adjoining barbs become separated—and considerable force is needed to pull the vane apart—they are instantly zipped together again by drawing the feather through the fingertips. The bird, of course, does it with its bill, and much of a bird's time is occupied with preening to keep its feathers in perfect condition.

There are different types of bird feathers for serving different functions. **Contour feathers** (Fig. 25-3, *E*) give the bird its outward form and are the type we have described above. Contour feathers that extend beyond the body and are used in flight are called **flight feathers. Down feathers** (Fig. 25-3, *H*) are soft tufts hidden beneath the contour feathers. They are soft because their barbules lack hooks. They are especially abundant on the breast and abdomen of water birds and on the young of game birds and function principally to conserve heat. **Filoplume feathers** (Fig. 25-3, *G*) are hairlike, degenerate

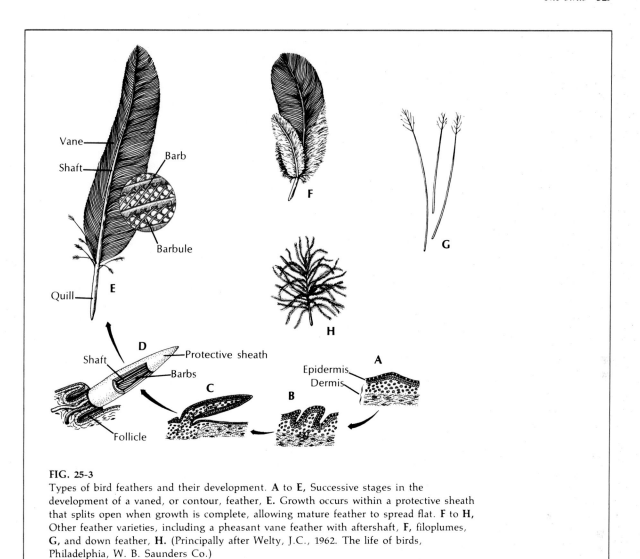

FIG. 25-3
Types of bird feathers and their development. **A** to **E,** Successive stages in the
development of a vaned, or contour, feather, **E.** Growth occurs within a protective sheath
that splits open when growth is complete, allowing mature feather to spread flat. **F** to **H,**
Other feather varieties, including a pheasant vane feather with aftershaft, **F,** filoplumes,
G, and down feather, **H.** (Principally after Welty, J.C., 1962. The life of birds,
Philadelphia, W. B. Saunders Co.)

feathers; each is a weak shaft with a tuft of short
barbs at the tip. They are the ''hairs'' of a plucked
fowl. They have no known function. The rictal bristles around the mouths of flycatchers and whippoorwills are probably modified filoplumes. A fourth type
of highly modified feather, called **powder-down
feathers,** are found on herons, bitterns, hawks, and
parrots. Their tips disintegrate as they grow, releasing a talclike powder that helps to waterproof the
feathers and give them metallic luster.

Feathers are epidermal structures that evolved
from the reptilian scale; indeed a developing feather
closely resembles a reptile scale when growth is just
beginning. One can imagine that in its evolution, the
scale elongated and its edges frayed outward until it
became the complex feather of birds. Strangely
enough, though modern birds possess both scales
(especially on their feet) and feathers, no intermediate stage between the two has been discovered on
either fossil or living forms.

Like a reptile's scale, a feather grows from a dermal papilla that pushes up against the overlying epidermis (Fig. 25-3, *A*). However, instead of flattening
like a scale, the feather rolls into a cylinder or feather
bud and is covered with epidermis. This feather bud
sinks in slightly at its base and comes to lie in a

feather follicle from which the feather will protrude. A layer of keratin is produced around the cylinder or bud and encloses the pulp cavity of blood vessels. This surface layer of keratin splits away from the deeper layer to form a sheath. The deeper layer now becomes frayed distally to form parallel ridges, the median one of which grows large to form the shaft (contour feathers) and the others the barbs. Then the sheath bursts and the barbs spread flat to form the vane (Fig. 25-3, *A* to *E*).

The pulp cavity of the quill dries up when growth is finished, leaving it hollow, with openings (umbilici) at its two ends. If the feather is to be a down feather, the sheath bursts and releases the barbs without the formation of a shaft or vane. Pigments (lipochromes and melanins) are added to the epidermal cells during growth in the follicle.

When fully grown, a feather is a dead structure. The shedding, or molting, of feathers is a highly orderly process. Except in penguins, which molt all at once, feathers are discarded gradually to avoid the appearance of bare spots. Flight and tail feathers are lost in exact pairs, one from each side, so that balance is maintained (Fig. 25-21). Replacements emerge before the next pair is lost and most birds can continue to fly unimpaired during the molting period; only ducks and geese are completely grounded during the molt. Nearly all birds molt at least once a year, usually in late summer after the nesting season. Many birds also undergo a second partial or complete molt just before breeding season, to equip them with their breeding finery, so important for courtship display.

■ ADAPTATIONS OF BIRD STRUCTURE AND FUNCTION AS REQUIREMENTS FOR FLIGHT

Just as an airplane must be designed and built to rigid aerodynamic specifications if it is to fly, so too must birds meet stringent structural requirements if they are to stay airborne. All the special adaptations found in flying birds come down to two things: more power and less weight. Flight by man became possible when he developed an internal combustion engine and learned how to reduce the weight-to-power ratio to a critical point. Birds did this millions of years ago. But birds must do much more than fly. They must feed themselves and convert food into high-energy fuel; they must escape predators; they

must be able to repair their own injuries; they must be able to air-condition themselves when overheated and heat themselves when too cool; and perhaps most important of all, they must reproduce themselves.

SKELETON. One of the major adaptations that allows a bird to fly is its light skeleton (Fig. 25-4). Bones are phenomenally light, delicate, and laced with air cavities, yet they are strong. The skeleton of a frigate bird with 7-foot wingspan weighs only 4 ounces (114 grams), less than the weight of all its feathers. A pigeon skull weighs only 0.21% of its body weight; the skull of a rat by comparison weighs 1.25%. The bird skull is mostly fused into one piece. The braincase and orbits are large to accommodate a bulging cranium and the large eyes needed for quick motor coordination and superior vision. The eyes of some birds are so large that they nearly touch in the middle of the skull. The anterior bones are elongated to form a beak. The lower mandible is a complex of several bones that hinge on two small movable bones, the quadrates. This provides a double-jointed action that permits the mouth to open widely. The upper jaw, consisting of premaxillae, maxillae, and other bones, is usually fused to the forehead, but in some birds, parrots, for instance, the upper jaw is hinged also. This adaptation allows greater flexibility of the beak in food manipulation and provides insect-catching species with a wider gap for successful feeding on the wing. The beaks of birds are strongly adapted to specialized food-habits —from generalized types, such as the strong, pointed beaks of crows to grotesque, highly specialized ones of flamingos, hornbills, and toucans (Fig. 25-5). The beak of a woodpecker is a straight, hard, chisel-like device (Fig. 25-6). Anchored to a tree trunk with its tail serving as a brace, the woodpecker delivers powerful, rapid blows to build nests or expose the burrows of wood-boring insects. It then uses its long, flexible, barbed tongue to seek out insects in their galleries. The woodpecker's skull is especially thick to absorb shock.

The bones of the pelvis (ilium, ischium, and pubis on each side) are fused with the lumbar and sacral vertebrae to form the **synsacrum,** which bears on each side a socket (**acetabulum**) for the head of the femur. Each leg is made up of the **femur,** the **tibiotarsus** (formed by the fusion of the tibia and proximal tarsals), the **tarsometatarsus** (formed by the fu-

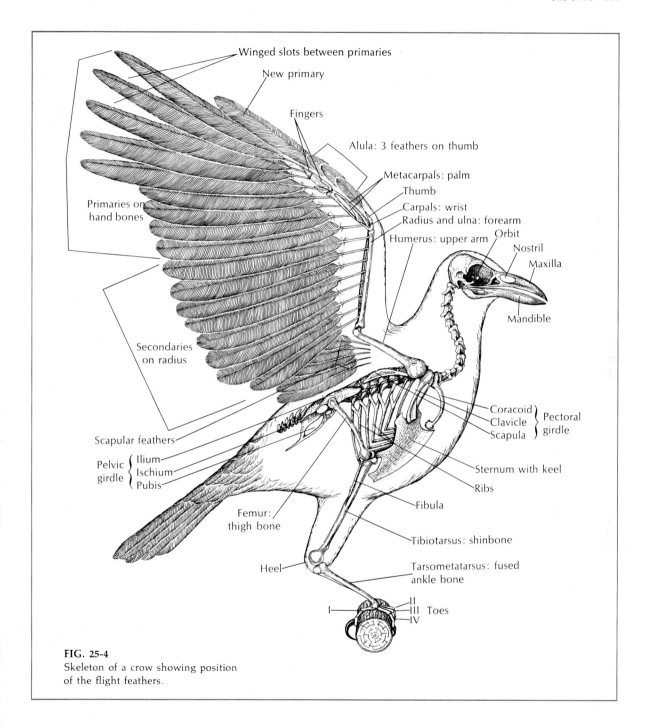

FIG. 25-4
Skeleton of a crow showing position
of the flight feathers.

sion of the distal tarsals and metatarsals), and the four **toes** (three in front and one behind). A sesamoid bone (the patella) is found at the knee joint. There are two to five phalanges in each toe. Woodpeckers and others have two front and two hind toes.

A few birds have only three toes, and the ostrich has two toes, unequal in size. Bird feet show a wide range of adaptations for walking, climbing, seizing, swimming, wading, etc. (Fig. 25-7).

The trunk is a rigid airframe, mainly because of

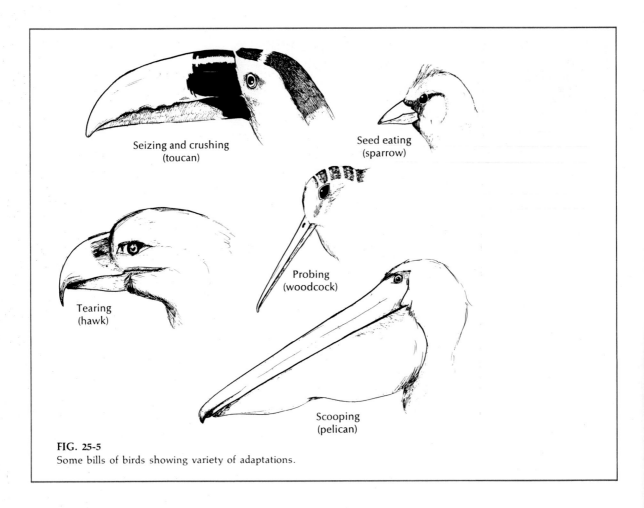

Seizing and crushing
(toucan)

Seed eating
(sparrow)

Probing
(woodcock)

Tearing
(hawk)

Scooping
(pelican)

FIG. 25-5
Some bills of birds showing variety of adaptations.

the fusion of the vertebrae and fusion of the ribs with the vertebrae and sternum. Special processes called **uncinate processes** form an additional brace by passing posteriorly from one rib over the one behind. This rigidity affords a firm point of attachment for the wings. To assist in this support and rigidity, the pectoral girdle of scapula, coracoid, and furcula (the latter also called the clavicle, or wishbone) are more or less firmly united and joined to the sternum. Where scapula and coracoid unite is a hollow depression, the glenoid cavity, into which the chief wing bone, the **humerus**, fits as a ball-and-socket joint. The neck of eight to 24 cervical vertebrae is extraordinarily flexible. In all flying birds the sternum (breast bone) is highly modified into a thin, flat keel for the insertion of large wing muscles. The forelimbs, or wing appendages, are the most highly modified of the paired appendages. Each consists of

a **humerus,** a **radius** and **ulna,** two **carpals,** and three **digits** (II, III, and IV). The other carpals are fused to the three metacarpals to form two long bones, the *carpometacarpus*. Of the digits, the middle one is longest and carries the large flight feathers. It consists of two phalanges. The second (alula) and fourth digits usually have only one (Fig. 25-4).

MUSCULAR SYSTEM. The muscles of birds have become highly adapted for flight. The locomotor muscles of the wings are relatively massive. The largest of these is the **pectoralis,** which depresses the wing in flight. Its antagonist is the **supracoracoideus** muscle, which raises the wing. Surprisingly perhaps, this latter muscle is not located on the backbone (anyone who has been served the back of the chicken knows it offers little meat) but is positioned under the pectoralis on the breast. It is attached by a tendon to the upper side of the humerus of the wing

FIG. 25-6
Adaptations for chopping holes in wood. This yellow-shafted flicker, *Colaptes auratus,* in common with other woodpeckers, has a heavy skull, chisel-edged beak, strong feet with two opposing toes (instead of the more common one) and stiff tail feathers that serve as a third, bracing leg. The black "mustache" mark beneath the eye of this male is the only visible difference between the sexes. Order Piciformes. (Photo by L. L. Rue, III.)

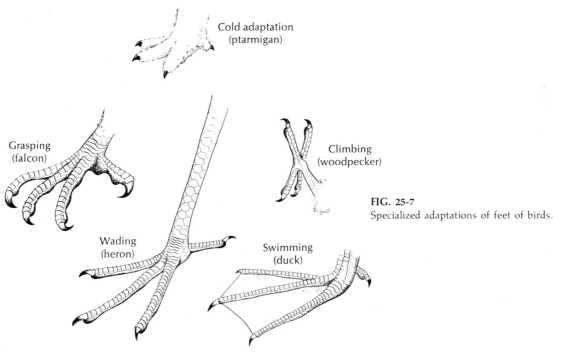

Cold adaptation
(ptarmigan)

Grasping
(falcon)

Climbing
(woodpecker)

FIG. 25-7
Specialized adaptations of feet of birds.

Wading
(heron)

Swimming
(duck)

so that it pulls from below by an ingenious rope and pulley arrangement. Both of these muscles are anchored to the keel. By thus placing the main muscle mass low in the body, aerodynamic stability is improved.

Leg muscles are not so highly modified as are flight muscles, since birds use their legs to stand, walk, and run much as their reptilian ancestors did. The main muscle mass is located in the thigh, surrounding the femur, and a smaller mass lies over the tibiotarsus (shank, or "drumstick"). Strong, but thin, tendons extend downward through sleevelike sheaths to the toes. Consequently the feet are nearly devoid of muscles, thus explaining the thin, delicate appearance of the bird leg. This arrangement places the main muscle mass near the bird's center of gravity, and at the same time confers great agility to the slender, lightweight feet. And since the feet are made up mostly of bone, tendon, and tough, scaly skin, they are highly resistant to damage from freezing. When a bird perches on a branch, an ingenious toe-locking mechanism (Fig. 25-8) is activated, which prevents a bird from falling off its perch when asleep. The same mechanism causes the talons of a hawk or owl to automatically sink deeply into its victim, as the legs bend under the impact of the strike. The powerful grip of a bird of prey was described by L. Brown,* "When an eagle grips in earnest, one's hand becomes numb, and it is quite impossible to tear it free, or to loosen the grip of the eagle's toes with the other hand. One just has to wait till the bird relents, and while waiting one has ample time to realize that an animal such as a rabbit would be quickly paralyzed, unable to draw breath, and perhaps pierced through and through by the talons in such a clutch."

Birds have lost the long reptilian tail, still fully evident in *Archaeopteryx,* and have substituted a pincushion-like muscle mound into which the tail feathers are rooted. It contains a bewildering array of tiny muscles, as many as 1,000 in some species, which control the crucial tail feathers. But the most complex muscular system of all is found in the neck of birds; the thin and stringy muscles, elaborately interwoven and subdivided, provide the bird's neck with the ultimate in vertebrate flexibility.

*Brown, L. 1970. Eagles, New York, Arco Publishing Co.

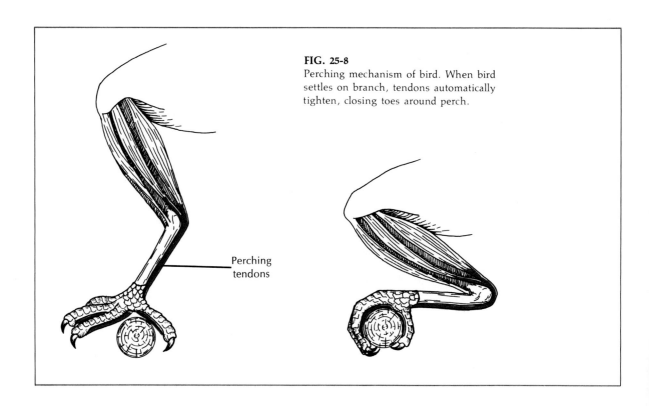

FIG. 25-8
Perching mechanism of bird. When bird settles on branch, tendons automatically tighten, closing toes around perch.

Perching tendons

DIGESTIVE SYSTEM. Birds eat an energy-rich diet in large amounts and process it rapidly with efficient digestive equipment. A shrike can digest a mouse in 3 hours and berries will pass completely through the digestive tract of a thrush in just 30 minutes. Furthermore, birds utilize a very high percentage of the food they eat. There are no teeth in the mouth, and the poorly developed salivary glands mainly secrete mucus for lubricating the food and the slender, horn-covered **tongue.** There are few taste buds. Hummingbirds and some others have sticky tongues, and woodpeckers have tongues that are barbed at the end. From the short **pharynx** a relatively long, muscular, elastic **esophagus** extends to the **stomach.** In many birds there is an enlargement **(crop)** at the lower end of the esophagus that serves as a storage chamber (Fig. 25-9).

In pigeons and some parrots the crop not only stores food, but produces milk by the breakdown of epithelial cells of the lining. This "bird milk" is regurgitated by both male and female into the mouth of the young squabs. It has a much higher fat content than cow's milk.

The stomach proper consists of a **proventriculus,** which secretes gastric juice, and the muscular **gizzard,** which is lined with horny plates that serve as millstones for grinding the food. To assist in the grinding process, birds swallow coarse, gritty objects or pebbles, which lodge in the gizzard. Certain birds of prey such as owls form pellets of indigestible materials, for example, bones and fur, in the proventriculus and eject them through the mouth. At the junction of the intestine with the rectum there are paired **ceca,** which may be well developed in some birds. Two **bile ducts** from the **gallbladder** or liver and two or three **pancreatic ducts** empty into the duodenum, or first part of the intestine. The **liver** is relatively large and bilobed. The terminal part of the digestive system is the **cloaca,** which also receives the genital ducts and ureters; in young birds the dorsal wall of the cloaca bears the bursa of Fabricius, which functions to produce immune bodies.

CIRCULATORY SYSTEM. The general plan of bird circulation is not greatly different from that of mammals. The four-chambered heart is large, with strong ventricular walls; thus birds share with mammals a complete separation of the respiratory and systemic circulations. However, the right aortic arch,

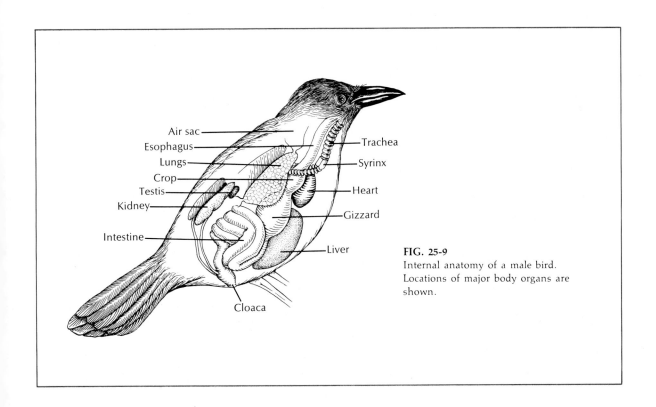

FIG. 25-9
Internal anatomy of a male bird. Locations of major body organs are shown.

instead of the left as in the mammals, leads to the dorsal aorta. The two jugular veins in the neck are connected by a cross vein, an adaptation for shunting the blood from one jugular to the other as the head is turned around. The brachial and pectoral arteries to the wings and breast are unusually large. The heartbeat is extremely fast, but, as in mammals, there is an inverse relationship between heart rate and body weight. For example, a turkey has a heart rate at rest of about 93 beats per minute, a chicken has 250 beats per minute, and a black-capped chickadee has a rate of 500 beats per minute when asleep, which may increase to a phenomenal 1,000 beats per minute during exercise. Blood pressure in birds is roughly equivalent to mammals of similar size. Bird's blood contains nucleated, biconcave red corpuscles that are somewhat larger than those of mammals. The phagocytes, or mobile ameboid cells, of the blood are unusually active and efficient in birds in the repair of wounds and in destroying microbes.

The body temperature of birds is high, ranging between 40° and 42° C. as compared to 36° to 39° C. for mammals. Some thrushes operate at 43.5° C. (110.5° F.), a temperature well past the lethal limit for most mammals, and only about 2.5° C. below the upper lethal temperature for these birds. Small birds tend to have rather more variable body temperatures than do large birds; for example, that of a house wren may fluctuate 8 centigrade degrees over 24 hours. Just as their homothermic counterparts, the mammals, so too must birds maintain relative constancy of their body temperature. If they begin to overheat, heat loss is accelerated by dilating blood vessels in the skin (to increase radiant heat loss) and by increasing the breathing rate (to increase evaporative cooling). Many species pant in extreme heat. In cold weather, birds ruffle their feathers to form a blanket of warm, insulative air next to the skin. They vasoconstrict peripheral blood vessels to reduce radiant heat loss. And if these physical mechanisms are not sufficient to prevent a drop in body temperature in very cold weather, a bird will shiver. As in mammals, this muscular movement creates needed heat. But it also increases food and oxygen consumption. A sparrow will consume twice as much oxygen and eat twice as much food at 0° C. (32° F.) as at 37° C. (98° F.). What person living in a northern climate has not marveled at the ability of the tiny chickadee, a minute furnace of cheerful activity, to survive direct exposure to the coldest winter weather?

RESPIRATORY SYSTEM. The respiratory system of birds differs radically from the lungs of reptiles and mammals and is marvelously adapted for meeting the high metabolic demands of flight. The lungs, which are relatively inexpansible because of their direct attachment to the body wall, are filled with numerous tiny **air capillaries** instead of alveoli of the mammalian type. Most unique, however, is the extensive system of nine interconnecting **air sacs** that are located in pairs in the thorax and abdomen and even extend by tiny tubes into the centers of the long bones (Fig. 25-10). The air sacs are connected to the lungs in such a way that perhaps 75% of the inspired air bypasses the lungs and flows directly into the air sacs, which serve as reservoirs for fresh air. On expiration, some of this fully oxygenated air is shunted through the lung, while the rest passes directly out. The advantage of such a system is obvious—the lungs receive fresh air during both inspiration and expiration. Rather than locating the respiratory exchange surface deep within blind sacs, which are difficult to ventilate as in mammals, birds have arranged to pass a continuous stream of fully oxygenated air through a system of richly vascularized air capillaries (Fig. 25-10, *B*). Although many details of the bird's respiratory system are not yet understood, it is clearly the most efficient of any vertebrate.

In addition to performing its principal respiratory function, the air sac system helps to cool the bird during vigorous exercise. A pigeon, for example, produces about 27 times more heat when flying than when at rest. The air sacs have numerous diverticula that extend inside the larger pneumatic bones of the pectoral and pelvic girdles, wings, and legs. Because they contain warmed air, they provide considerable buoyancy to the bird.

EXCRETORY SYSTEM. The relatively large paired metanephric kidneys are attached to the dorsal wall in a depression against the sacral vertebrae and pelvis. Urine passes by way of **ureters** to the **cloaca.** There is no urinary bladder. The kidney is composed of many thousands of **nephrons,** each consisting of a renal corpuscle and a nephric tubule. Urine is formed in the usual way by glomerular filtration followed by the selected modification of the filtrate in the nephric tubule. However, the urine becomes progressively concentrated by the absorption

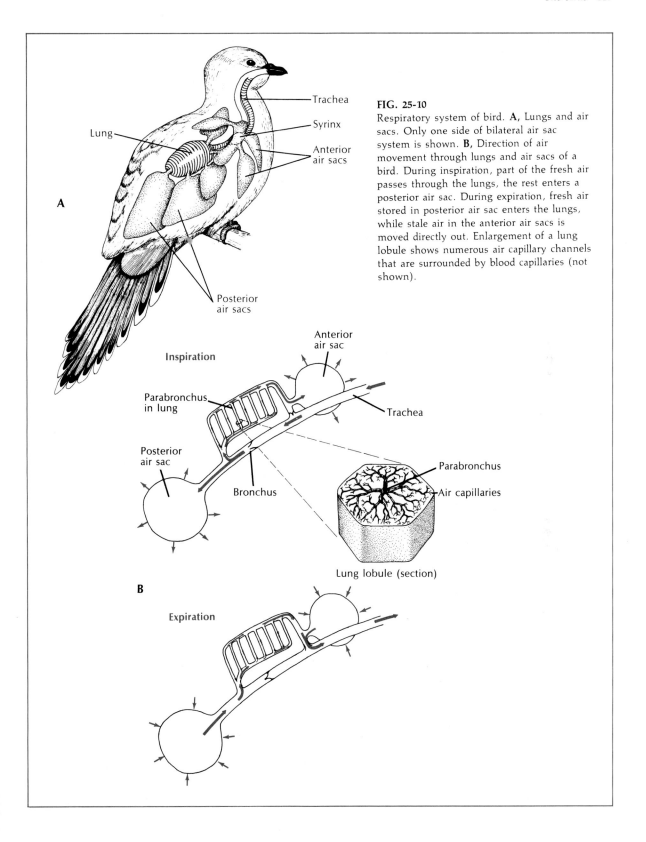

FIG. 25-10

Respiratory system of bird. **A,** Lungs and air sacs. Only one side of bilateral air sac system is shown. **B,** Direction of air movement through lungs and air sacs of a bird. During inspiration, part of the fresh air passes through the lungs, the rest enters a posterior air sac. During expiration, fresh air stored in posterior air sac enters the lungs, while stale air in the anterior air sacs is moved directly out. Enlargement of a lung lobule shows numerous air capillary channels that are surrounded by blood capillaries (not shown).

Secretory tubule
Central canal

FIG. 25-11
Salt glands of a marine bird (gull). One salt gland is located above each eye. Each gland consists of several lobes arranged in parallel. One lobe is shown in cross section, much enlarged. Salt is secreted into many radially arranged tubules, then flows into a central canal that leads into the nose.

of water by the tubules and the cloaca. The urine contains a high concentration of uric acid and creatine; when combined with fecal material, it becomes a white paste. There is a great advantage to excreting nitrogenous waste as uric acid instead of urea. Uric acid is nearly insoluble so that once excreted, water can be almost completely reabsorbed. The uric acid precipitates out and therefore does not contribute to the osmotic pressure of the urine. Birds thus excrete a "concentrated" urine with a relatively simple kidney. As we will see, a highly sophisticated urine-concentrating device has evolved in the mammalian kidney, enabling mammals to excrete their nitrogenous waste as soluble urea rather than insoluble uric acid.

Marine birds (also marine turtles) have evolved a unique solution for excreting the large loads of salt eaten with their food and in the sea water they drink. Sea water contains about 3% salt and is three times saltier than a bird's body fluids. Yet the bird kidney cannot concentrate salt in urine above about 0.3%. The problem is solved by special **salt glands,** one located above each eye (Fig. 25-11). These glands are capable of excreting a highly concentrated solution of sodium chloride—up to twice the concentration of sea water. The salt solution runs out the internal or external nostrils, giving gulls, petrels, and other sea birds a perpetual runny nose. The develop-

ment of the salt gland in some birds depends on how much salt the bird takes in its diet. For example, a race of mallard ducks living a semimarine life in Greenland have salt glands 10 times larger than those of ordinary freshwater mallards.

NERVOUS AND SENSORY SYSTEM. A bird's nervous and sensory system accurately reflects the complex problems of flight, and a highly visible existence in which it must gather food, mate, defend territory, incubate and rear young, and at the same time correctly distinguishing friend from foe. The brain of a bird has well developed **cerebral hemispheres, cerebellum,** and **midbrain tectum** (optic lobes). The **cerebral cortex**—the portion in mammals that becomes the chief coordinating center—is thin, unfissured, and poorly developed in birds. But the core of the cerebrum, the **corpus striatum,** has enlarged into the principal integrative center of the brain, where are controlled such activities as eating, singing, flying, and all the complex instinctive reproductive activities. Relatively intelligent birds, such as crows and parrots, have larger cerebral hemispheres than do less intelligent birds, such as chickens and pigeons. The **cerebellum** is a crucial coordinating center where muscle position sense, equilibrium sense, and visual cues are all assembled and used to coordinate movement and balance. The **optic lobes,** laterally bulging structures of the midbrain, form a

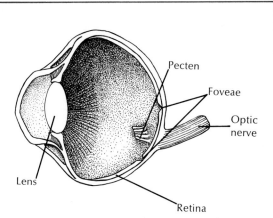

FIG. 25-12
Hawk eye has all the structural components of mammalian eye, plus a peculiar pleated structure, the pecten, believed to provide nourishment to the retina. The extraordinary keen vision of the hawk, which has a visual acuity eight times better than that of man, is attributed to the extreme density of cone cells in the foveae: 1.5 million per fovea compared to 0.2 million for man. Each hawk eye has two foveae as opposed to man's one, meaning that each hawk eye focuses on two objects simultaneously—the better to see its next meal!

visual-association apparatus comparable to the visual cortex of mammals.

Except in flightless birds and in ducks, the senses of smell and taste are poorly developed in birds. This lack however is more than compensated by good hearing and superb vision, the keenest in the animal kingdom. As in mammals, the bird ear consists of three regions: (1) the **external ear,** a sound-conducting canal extending to the **ear drum,** (2) a **middle ear,** containing a rodlike **columella** that transmits vibrations to the (3) **inner ear,** where the organ of hearing, the **cochlea,** is located. The bird cochlea is much shorter than the coiled mammalian cochlea, yet birds can hear roughly the same range of sound frequencies as man. Actually the bird ear far surpasses man's in capacity to distinguish differences in intensities and to respond to rapid fluctuations in song.

The bird eye resembles those of other vertebrates in gross structure, but is relatively larger, less spherical, and almost immobile; instead of turning their eyes, birds turn their heads with their long and flexi-

ble necks to scan the visual field. The light sensitive **retina** (Fig. 25-12) is elaborately equipped with rods (for dim light vision) and cones (for color vision). Cones predominate in day birds and rods are more numerous in nocturnal birds. A distinctive feature of the bird eye is the **pecten,** a highly vascularized organ attached to the retina near the optic nerve and jutting out into the vitreous humor (Fig. 25-12). It is believed to boost the retina's nourishment. The position of a bird's eyes in its head is correlated with its life habits. Vegetarians that must avoid predators have eyes placed laterally to give a wide view of the world; predaceous birds such as hawks and owls have eyes directed to the front. The fovea, or region of keenest vision on the retina, is placed (in birds of prey and some others) in a deep pit, which makes it necessary for the bird to focus exactly on the source. Many birds, moreover, have two sensitive spots (foveas) on the retina (Fig. 25-12)—the central one for sharp monocular views and the posterior one for binocular vision. Woodcocks can probably see binocularly both forward and backward. Bitterns, in their freezing stance of bill pointing up, can also see binocularly (Fig. 25-13). The visual acuity of a hawk is believed to be eight times that of man (enabling it to clearly see a crouching rabbit more than 1 mile away), and an owl's ability to see in dim light is more than 10 times that of the human eye (Fig. 25-14). Birds have good color vision, especially toward the red end of the spectrum.

REPRODUCTIVE SYSTEM. In the male the paired **testes** and accessory ducts are similar to those in many other forms. From the **testes** the **vasa deferentia** run to the cloaca. Before being discharged, the sperm are stored in the **seminal vesicle,** the enlarged distal end of the vas deferens. This seminal vesicle may become so large with stored sperm during the breeding season that it causes a cloacal protuberance. The high body temperature, which tends to inhibit spermatogenesis in the testes, is probably counteracted by the cooling effect of the abdominal air sacs. The testes of birds undergo a great enlargement at the breeding season, as much as 300 times, and then shrink to tiny bodies afterwards. Some birds, including ducks and geese, have a large, well-developed **copulatory organ** (penis), provided with a groove on its dorsal side for the transfer of sperm. However, in the more advanced birds, copulation is a matter of bringing the cloacal surfaces into contact, usually

FIG. 25-13
Concealing behavior of American bittern, *Botaurus lentiginosus.* When startled, bittern points bill skyward, directing downward-facing eyes toward danger. Its streaked breast, with coloration matching marsh vegetation, makes the bird exceedingly difficult to see. Order Ciconiiformes. (Photo by L. L. Rue, III.)

while the male stands on the back of the female. Some swifts copulate in flight.

In the female of most birds, only the left ovary and oviduct develop; those on the right dwindle to vestigal structures. The ovary is close to the left kidney. Eggs discharged from the ovary are picked up by the expanded end of the oviduct, the **ostium** (Fig. 25-15). The oviduct runs posteriorly to the cloaca. While the eggs are passing down the oviduct, **albu-**

min, or egg white, from special glands is added to them; farther down the oviduct, the shell membrane, shell, and shell pigments are also secreted about the egg. Fertilization takes place in the upper oviduct several hours before the layers of albumin, shell membranes, and shell are added. Sperm remain alive in the female oviduct for many days after a single mating. Hen eggs show good fertility for 5 or 6 days after mating, but then fertility drops rapidly. How-

FIG. 25-14

Screech owl, *Otus asio,* during brief interlude after successful forage. Owls possess eyes incredibly sensitive to light, enabling them to see prey in light one hundreth to one tenth the intensity required by man (0.000,000,73 footcandle). But this particular ultrasensitivity is traded off for relatively poor visual acuity, narrow visual field, and weak accommodation (ability to focus on near objects). Order Strigiformes. (Photo by L. L. Rue, III.)

ever the occasional egg will be fertile as long as 30 days after separation of the hen from the rooster. Some birds are determinate layers and lay only a fixed number (clutch) of eggs in a season. If any of the eggs of a set are removed, the deficit is not made up by additional laying (herring gull). Indeterminate layers, however, will continue to lay additional eggs for a long time if some of the first-laid eggs are continually removed (flickers, ducks, domestic poultry).

Most birds are probably determinate layers. Many birds such as songbirds lay an egg a day until the clutch is completed; others stagger their egg laying and lay every other day or so. Domestic geese usually lay every other day, which is probably the pattern for the large birds of prey. The European cuckoo (a parasitic bird that lays its eggs in the nest of others) tends to lay eggs similar in size and pattern to those of the host birds.

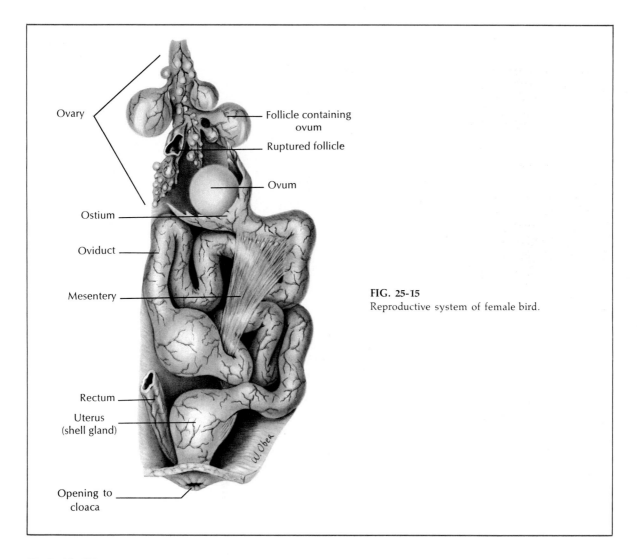

Ovary

Follicle containing ovum

Ruptured follicle

Ovum

Ostium

Oviduct

Mesentery

Rectum

Uterus (shell gland)

Opening to cloaca

FIG. 25-15
Reproductive system of female bird.

■ FLIGHT

What prompted the evolution of flight in birds, the ability to rise free of earth-bound concerns, as almost every human being has dreamed of doing? Much as we may envy the birds in their conquest of the air, we also recognize that evolution of flight was the pragmatic result of complex adaptive pressures; certainly the forerunners of birds did not take up flight just to enjoy a new experience. The air was a relatively unexploited habitat stocked with flying insect food. Flight also offered escape from terrestrial predators and opportunity to travel rapidly and widely to establish new breeding areas and to benefit from year-round favorable climate by migrating north and south with the seasons.

Birds unquestionably evolved from reptilian ancestors, but not directly from the membrane-winged flying reptiles (pterodactyls) as one might innocently suppose. They appear to have descended instead from a small group of thecodont reptiles (Pseudosuchia) that gave rise to both dinosaurs and birds. The fossil evidence is far too meager to give us a recorded history of the origins of bird flight. We can only theorize that the proavian reptiles passed through some sequence of swift running, flying leaps, tree climbing, parachute gliding—all leading toward an arboreal existence. Many adaptations of birds, such as the perching mechanism and active climbing habits, indicate such an arboreal apprenticeship. Natural selection favored the swifter and more agile,

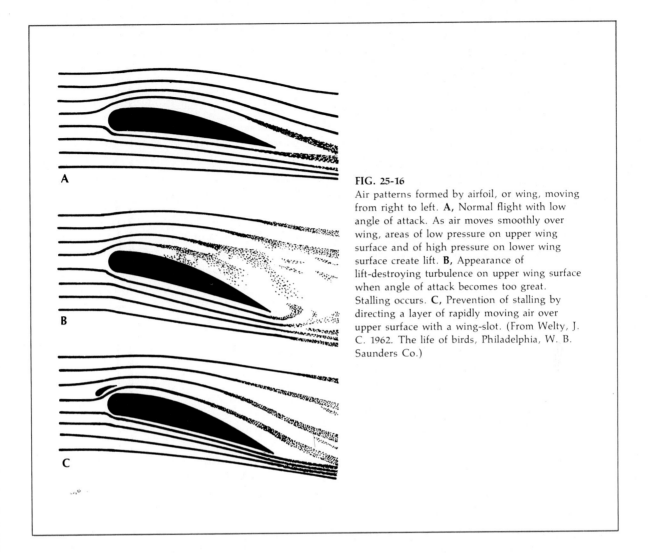

FIG. 25-16
Air patterns formed by airfoil, or wing, moving from right to left. **A,** Normal flight with low angle of attack. As air moves smoothly over wing, areas of low pressure on upper wing surface and of high pressure on lower wing surface create lift. **B,** Appearance of lift-destroying turbulence on upper wing surface when angle of attack becomes too great. Stalling occurs. **C,** Prevention of stalling by directing a layer of rapidly moving air over upper surface with a wing-slot. (From Welty, J. C. 1962. The life of birds, Philadelphia, W. B. Saunders Co.)

and eventually wings became strong enough to support the bird in air. One thing seems certain: feathers were an absolute requirement for flight and the evolution of flight and of feathers must have progressed together. There is absolutely no support for the idea that bird ancestors were originally membrane-winged flyers, like bats, that later developed feathers.

The bird wing as a lift device

Bird flight, especially the familiar flapping flight of birds, is complex. Despite careful analysis by conventional aerodynamic techniques and high-speed photography, there is much about it that is not understood. Nevertheless, the bird wing is an airfoil that is subject to recognized laws of aerodynamics. It is adapted for high lift at low speeds, and, not suprisingly perhaps, it resembles the wings of early low-speed aircraft. The bird wing is streamlined in cross-section, with a slightly concave lower surface (cambered) and with small tight-fitting feathers where the leading edge meets the air (Fig. 25-16). Air slips efficiently over the wing, creating lift with minimum drag. Some lift is produced by positive pressure against the under surface of the wing. But on the upper side, where the airstream must travel farther and faster over the convex surface, a negative pressure is created that provides more than two-thirds of the total lift.

The lift-to-drag ratio of an airfoil is determined by the angle of tilt (angle of attack) and the airspeed

(Fig. 25-16). A wing carrying a given load can pass through the air at high speed and small angle of attack, or at low speed and larger angle of attack. But as speed decreases, a point is reached where the angle of attack becomes too steep, turbulence appears on the upper surface, lift is destroyed, and stalling occurs. Stalling can be delayed or prevented by placing a **wing slot** along the leading edge so that a layer of rapidly moving air is directed across the upper wing surface. Wing slots were, and still are, used in aircraft when traveling at low speed. In birds, two kinds of wing slots have developed: (1) the **alula,** or group of small feathers on the thumb (Fig. 25-4), provides a midwing slot, and (2) **slotting between primary feathers** provides a wing-tip slot. In many songbirds, these together provide stall-preventing slots for nearly the entire outer (and aerodynamically most important) half of the wing.

Basic forms of bird wings

Bird wings vary in size and form because the successful exploitation of different habitats has imposed special aerodynamic requirements. The following four types of bird wings are easily recognized (D. B. O. Savile).

Elliptical wings. Birds that must maneuver in forested habitats such as sparrows, warblers, doves, woodpeckers, and magpies (Fig. 25-17) have elliptical wings. This type has a low **aspect ratio** (ratio of length to width). The outline of a sparrow wing is almost identical to that of the British Spitfire fighter plane of Second World War fame—also a highly maneuverable flyer. Elliptical wings are highly slotted between the primary feathers (Fig. 25-17), which is a kind that helps to prevent stalling during sharp turns, low-speed flight, and frequent landing and takeoff. Each separated primary behaves as a narrow wing

FIG. 25-17

Elliptical wing of the black-billed magpie, *Pica pica,* in slow flight. Note well-developed alula and separation (slotting) between primaries, typical of wings adapted for high maneuverability and low-speed flight. This member of the crow family is a familiar feature of the western North American landscape. Order Passeriformes. (Photo by C. G. Hampson, University of Alberta.)

with a high angle of attack, providing high lift at low speed. The high maneuverability of the elliptical wing is exemplified by the tiny chickadee, which if frightened, can change course within 0.03 second.

High-speed wings. Birds that feed on the wing, such as swallows, hummingbirds, and swifts, or that make long migrations, such as plovers, sandpipers, and terns (Fig. 25-18), have wings that sweep back and taper to a slender tip. They are rather flat in section, have a moderately high aspect ratio, and lack the wing-tip slotting characteristic of the preceding group. Sweepback and wide separation of the wing tips reduces "tip vortex," the drag-creating turbulence that tends to develop at wing tips. The fastest birds alive, such as falcons and sandpipers clocked at 177 and 110 miles per hour, respectively, belong to this group.

Soaring wing. The oceanic soaring birds have high-aspect-ratio wings resembling those of sailplanes. This group includes albatrosses, frigate-birds, gannets (Fig. 25-19), and gulls. Such long, narrow wings lack wing slots and are adapted for high speed, high lift, and dynamic soaring. They have the highest aerodynamic efficiency of all wings, but are less maneuverable than the wide, slotted wings of land soarers. Dynamic soarers have learned how to exploit the highly reliable sea winds, using adjacent air currents of different velocities.

High-lift wing. Vultures, hawks, eagles, owls (Fig. 25-20), and ospreys (Fig. 25-21)—predators that carry heavy loads—have wings with slotting, alulas, and pronounced camber, all of which promote high lift at low speed. Many of these birds are land soarers where broad, slotted wings provide the sensitive response and maneuverability required for static soaring in the capricious air currents over land.

FIG. 25-18

High-speed wing of arctic tern, *Sterna paradisaea.* Sweepback, taper, and absence of wing slotting adapts tern for high speeds during its incredibly long migrations, from arctic to antarctic and back each year, with some logging as much as 24,000 miles. This female was photographed beside its nest on island off east coast of Sweden. Arctic terns nest so close to water's edge that nests are often flooded during storms. Order Charadriiformes. (Photo by B. Tallmark.)

FIG. 25-19
Soaring wing of gannet, *Sula bassana.* Long, high-aspect wing of gannet, like that of a sailplane, has great aerodynamic efficiency and lift but is relatively weak and lacking in maneuverability. For 10 months of the year, gannets are out of sight of land. When one spots a fish, it folds its wings and drops like a stone from heights of 100 meters, hitting the water with great speed. It seldom misses its intended victim. Order Pelecaniformes. (Photo by B. Tallmark.)

FIG. 25-20
High-lift wing of snowy owl, *Nyctea scandiaca.* Wing is deeply emarginated with square slots between primary feathers, a highly efficient pattern for carrying heavy loads (prey) at low speed. Broad tail provides purchase on the air for quick maneuvering and assists in load-carrying. Order Strigiformes. (Photo by C. G. Hampson, University of Alberta.)

FIG. 25-21
Osprey, *Pandion haliaetus,* landing on nest. Note alulas *(top arrows)* and new primary feathers *(side arrows).* Feathers are molted in sequence in exact pairs so that balance is maintained during flight. These fish-eating birds have suffered a severe population decline in the United States in recent years because of illegal hunting and poor nesting success. Pesticides concentrated in fish eaten by ospreys causes egg shells to thin and burst during incubation. Order Falconiformes. (Photo by B. Tallmark.)

FIG. 25-20
For legend see opposite page.

FIG. 25-21
For legend see opposite page.

■ MIGRATION AND NAVIGATION

Perhaps it was inevitable that birds, having mastered the art of flight, should use this power to make the long and arduous seasonal migrations that have captured man's wonder and curiosity. The term **migration** refers to the regular, extensive, seasonal movements birds make between their summer breeding regions and their wintering regions. The chief advantage seems obvious: it enables birds to live in an optimal climate all the time, where abundant and unfailing sources of food are available to sustain their intense metabolism. Migrations also provide optimal conditions for rearing young when demands for food are especially great. Broods are largest in the far north where the long summer days and the abundance of insects combine to provide parents with ample food-gathering opportunity. Predators are relatively rare in the north and the brief once-a-year appearance of vulnerable young birds does not encourage the build-up of predator populations. Migration also vastly increases the amount of space available for breeding and reduces aggressive territorial behavior. And of course migration favors homeostasis by allowing birds to avoid climatic extremes.

Origin of migration

Migration has become so firmly established in the behavior of birds that it has long since become a hereditary instinct. But what prompted the origin of migration? Undoubtedly, it was ecologic pressures. We must suppose that birds, like other animals, will move only when compelled by some necessity. One theory of migration suggests that birds spread over the northern hemisphere when the latter was warm and food conditions were favorable all through the year. When the glacial era came and forced the birds to go south for survival, they came back in the spring when the ice age receded, only to be forced south again in winter because of the sharp establishment of the winter and summer seasons. This led in time to the firm establishment of the habit. Another theory centers around the view that the ancestral home of birds was in the tropics and some went north to avoid congestion and competition during the breeding season. After raising their young, they then returned.

Migration routes

Most migratory birds have well established routes trending north and south. And since most birds (and other animals) live in the northern hemisphere, where most of the earth's land mass is concentrated, most birds are south-in-winter and north-in-summer migrants. Of the 4,000 or more species of migrant birds (a little less than one half the total bird species), most breed in the more northern latitudes of the hemisphere; the percentage of migrants in Canada is far higher than is the percentage of migrants in Mexico, for example. Some use different routes in the fall and spring. Some, especially certain aquatic species, complete their migratory routes in a very short time. Others, however, make the trip in a leisurely manner, often stopping here and there to feed. Some of the warblers are known to take 50 to 60 days to migrate from their winter quarters in Central America to their summer breeding areas in Canada.

Not all members of a species perform their migrations at the same time; there is a great deal of straggling so that some members do not reach the summer breeding grounds until after others are well along with their nesting. Other birds, such as the purple martins and catbirds, return to a certain locality on almost the same day of the month each season. Records of catbirds kept in a certain eastern state reveal that the birds arrived in the particular locality about the middle of April and did not vary more than a day in a period of 5 years. Many observations also revealed that the same individual bird returns not only to the same locality but also to the same territory that it occupied in previous seasons.

Many of the smaller species migrate at night and feed by day; others migrate chiefly in the daytime; and many swimming and wading birds, either by day or night. The height at which they fly varies greatly. Some apparently keep fairly close to the earth, and others are known to fly as high as 4,000 to 5,000 feet. Many birds are known to follow landmarks, such as rivers and coastlines; but others do not hesitate to fly directly over large bodies of water in their routes. The routes of any two species rarely coincide, for there is almost infinite variety in the routes covered. Some birds have very wide migration lanes, and others, such as certain sandpipers, are restricted to very narrow ones, keeping well to the coastlines because of their food requirements.

Some species are known for their long-distance migrations. The arctic tern (Fig. 25-18), greatest globe spanner of all, breeds north of the Arctic Circle and

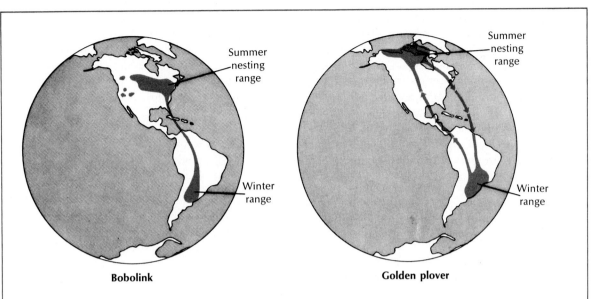

FIG. 25-22

Migrations of the bobolink and golden plover. The bobolink commutes 14,000 miles each year between nesting sites in North America and its wintering range in Argentina, a phenomenal feat for such a small bird. Although the breeding range has extended to colonies in western areas, these birds take no shortcuts but adhere to the ancestral eastern seaboard route.

The golden plover flies a loop migration, striking out across the Atlantic in its southward autumnal migration but returning in the spring via Central America and the Mississippi valley because ecologic conditions are more favorable at that time.

in winter is found in the antarctic regions, 11,000 miles away. This species is also known to take a circuitous route in migrations from North America, passing over to the coastlines of Europe and Africa and thence to their winter quarters. Other birds that breed in Alaska follow a more direct line down the Pacific coast of North and South America.

Many small songbirds also make great migration treks (Fig. 25-22). Africa is a favorite wintering ground for European birds, and many fly there from Central Asia as well.

Stimulus for migration

Although the Chinese have known for centuries that they could make caged songbirds sing in winter by artificially lengthening their days with candlelight, it was not until 1929 that W. Rowan in Canada showed the scientific world that the seasonal change in daylength is the principal timing factor for bird reproduction and migration. Subsequently many

other researchers have shown unequivocally that lengthening days in winter and spring stimulate the development of the gonads and the accumulation of fat—both important internal changes that predispose birds to migrate northward. There is evidence that increasing daylength stimulates the anterior lobe of the pituitary into activity. The release of pituitary gonadotropic hormone in turn sets in motion a complex series of physiologic and behavioral changes, leading to gonadal growth, fat deposition, migration, courtship and mating behavior, and care of the young.

Direction-finding in migration

Numerous experiments suggest that birds navigate chiefly by sight. Birds recognize topographic landmarks and follow familiar migratory routes—a behavior assisted by flock migration where navigational resources and experience of older birds can be pooled. But in addition to visual navigation, birds make use of other orientation cues at their disposal. Birds have an innate time sense, a kind of built-in clock of great accuracy; they have an innate sense of direction; and very recent work adds much credence to an old, much-debated theory that birds can detect, and navigate by, the earth's field of gravity. All of these resources are inborn and instinctive, although a bird's navigational abilities may improve with experience. Recently, experiments by G. Kramer and F. Sauer have demonstrated quite convincingly that birds can navigate by celestial cues—the sun by day and the stars by night. Kramer placed starlings in specially covered cages provided with six windows. The birds were trained to find food in a definite compass direction at a certain time of day. When tested at another time of day when the sun's position had changed, they compensated for the sun's motion and returned to the correct window. In other words, the birds possessed a built-in time sense that enabled them to maintain compass direction by referring to the sun, regardless of time of day. This is called **sun-azimuth orientation** (azimuth = compass bearing of the sun). Sauer's ingenious planetarium experiments strongly suggest that some birds, probably many, possess an innate knowledge of the constellations, and use them to navigate at night. But some of the remarkable feats of bird navigation still defy rational explanation. Most birds undoubtedly use a combination of environmental and innate cues to migrate. Migration is a rigorous undertaking, the target is often small, and natural selection relentlessly prunes off errors in migration, leaving only the best navigators to propagate the species.

■ BIRD POPULATIONS

Many censuses have been taken to ascertain bird populations within a particular area. The National Audubon Society and the Federal Fish and Wildlife

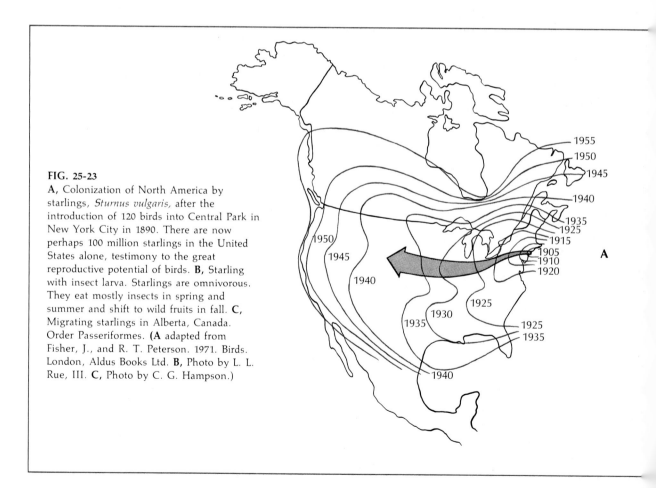

FIG. 25-23

A, Colonization of North America by starlings, *Sturnus vulgaris,* after the introduction of 120 birds into Central Park in New York City in 1890. There are now perhaps 100 million starlings in the United States alone, testimony to the great reproductive potential of birds. **B,** Starling with insect larva. Starlings are omnivorous. They eat mostly insects in spring and summer and shift to wild fruits in fall. **C,** Migrating starlings in Alberta, Canada. Order Passeriformes. (**A** adapted from Fisher, J., and R. T. Peterson. 1971. Birds. London, Aldus Books Ltd. **B,** Photo by L. L. Rue, III. **C,** Photo by C. G. Hampson.)

B

C

FIG. 25-23, cont'd
For legend see
opposite page.

Service have sponsored many such counts. Some are concerned with game birds and species on the verge of extinction. Emphasis is often placed upon breeding birds, making use of the territorial singing of the males. Some birds lend themselves to more accurate counts than do others, such as those that nest in colonies or have particular nesting habitats. Students of birds use various techniques for identifying birds,

FIG. 25-24
"Martha," the last passenger pigeon, *Ectopistes migratorius,* died in the Cincinnati Zoo in 1914. Passenger pigeons traveled and roosted in huge masses, with nesting areas exceeding 30 square miles. In 1869, hunters shot 7.5 million pigeons from a single nesting site. Hunted relentlessly for market, the population eventually dropped too low to sustain colonial breeding. The senior author of this text, when he was a boy, saw Martha some years before she died. (Courtesy Smithsonian Institution, Washington, D. C.)

such as sight, song, type of flying, and call notes. In 1914 a bird census made in the northeastern United States revealed about 125 pairs of birds per 100 acres (open farms) and 199 pairs per 100 acres of woodland. In 1949 a survey made in a spruce-fir forest in Maine gave a count of 370 pairs of breeding birds per 100 acres. Another count in the same region in 1950 showed 385 pairs of breeding birds per 100 acres.

Bird populations, like those of other animal groups, vary in size from year to year. Snowy owls (Fig. 25-20), for example, are subject to population cycles that closely follow cycles in their food crop, mainly rodents. Voles, mice, and lemmings in the north have a fairly regular four-year cycle of abundance; at population peaks, predator populations of foxes, weasels, buzzards, as well as snowy owls, increase because there is abundant food for rearing their young. After a crash in the rodent population, snowy owls move south, seeking alternate food supplies. They occasionally appear in large numbers in southern Canada and northern United States, where their total absence of fear of man makes them easy targets for thoughtless hunters.

Occasionally, the activities of man bring about spectacular changes in bird distribution. Both starlings (Fig. 25-23) and house sparrows have been accidentally or deliberately introduced into numerous countries, where they have become the two most abundant bird species on earth with the exception of the domestic fowl. Man also is responsible for the extinction of many bird species. More than 80 species of birds have followed the last dodo to extinction in 1681. Many died naturally, victims of changes in their habitat or competition with better-adapted species. But several have been hunted to extinction, among them the passenger pigeon, which only a century ago darkened the skies over North America in incredible numbers estimated in the billions (Fig. 25-24). Hunters kill millions of game birds annually as well as many nongame birds that happen to make convenient targets. An even greater number of game birds die indirectly as the result of eating lead pellets (which they mistake for seeds) or from the crippling effects of embedded pellets. One survey in Wisconsin revealed that the average hunter required 36 shots to down one goose. Of the Canada geese that survived the barrage to fly as far south as the Mississippi Valley, 44% contained embedded lead shot. But man's most destructive effects on birds are usually unintentional. The draining of marshes—more than 99% of Iowa's once-extensive marshland is now farmland—has destroyed waterfowl nesting. Deforestation has likewise had great impact on tree-nesting species. The vertical appendages of civilization, such as television towers, monuments, tall buildings, and electric transmission towers and lines take a fearful toll during bird migration in bad weather. Tree spraying programs have virtually eradicated songbirds from certain areas. Most birds, through their impressive reproductive potential, can replace in numbers those that become victims to man's activities. Someone has calculated that a single pair of robins, producing 2 broods of 4 young a season, will leave 19,500,000 descendants in 10 years, should all survive at least that long. But although some birds like robins, house sparrows, and starlings thrive on man's heavy-handed influence on his environment, most birds find the changes adverse, and to some species it is lethal.

■ BIRD SOCIETY

An old adage suggests that "birds of a feather flock together." Birds are indeed highly social creatures. Especially during the breeding seasons sea birds gather in often enormous colonies to nest and rear young. Land birds, with some conspicuous exceptions (such as starlings and rooks), tend to be less gregarious than sea birds during breeding and seek isolation for rearing their brood. But these same species that covet separation from their kind during breeding may aggregate for migration or feeding (Fig. 25-25). Togetherness offers advantages: mutual protection from enemies, greater ease in finding mates, less opportunity for individual straying during migration, and mass huddling for protection against low night temperatures during migration. Certain species may use highly organized cooperative behavior to feed, such as the pelicans, as shown in Fig. 25-26. At no time are the highly organized social interactions of birds more evident than during the breeding season, as they stake out territorial claims, select mates, build nests, incubate and hatch their eggs, and rear their young.

SELECTION OF TERRITORIES AND MATES. A pair of birds will usually select a territory on which to raise a brood. This territory is selected in the spring by the male, who jealously guards it against all other males of the same species. The male sings a great

FIG. 25-25
Flamingos *(Phoeniconaias minor)* on Lake Nakuru in Kenya, Africa, feeding on blue-green algae. There may be 2 million flamingos on the lake at times. They remove an estimated 70,000 tons of algae from this productive, alkaline lake each year. The lake is threatened by pollution from rapidly growing human population along shores. Order Ciconiiformes. (Photo by B. Tallmark.)

deal to help him establish priority on his domain. Eventually he attracts a female, and the pair start mating and nest building. The female apparently wanders from one territory to another until she settles down with a male. How large a territory a pair takes over depends on location, abundance of food, natural barriers, etc. In the case of robins a house may serve as the dividing line between two adjacent domains, and each pair will usually stay close to the lawn on its particular side of the house. When members of another species trespass, they are usually ignored; competition is greatest among the members of the same species. Song sparrows, however, try to keep off members of other species as well as their own. Birds may also defend their territories against other species because of environmental limitations and changes, where competition for food or other factors between different species (usually closely related) may occur.

This concept of territory claims by birds was greatly developed by H. E. Howard, an English ornithologist, in 1920. Since that time other competent students have verified and extended the concept. One of these was Mrs. M. M. Nice, whose work on the song sparrow is now classic; another was Lack, an English ornithologist, who worked with the English robin.

Elaborate courtship rituals are found in many birds, such as the prairie chicken, sage grouse, bowerbirds, and great crested grebes (Figs. 25-27 and 25-28). Songbirds have simpler rituals, consisting mostly of male displays and songs. Many modified aspects of territory are found. Some birds restrict their territories to nesting regions and share feeding grounds with others. Among hummingbirds the female has a separate nesting site, which she defends herself. Territories are usually deserted at the close of the nesting season and new ones are staked out the following spring. Song sparrows, however, keep their

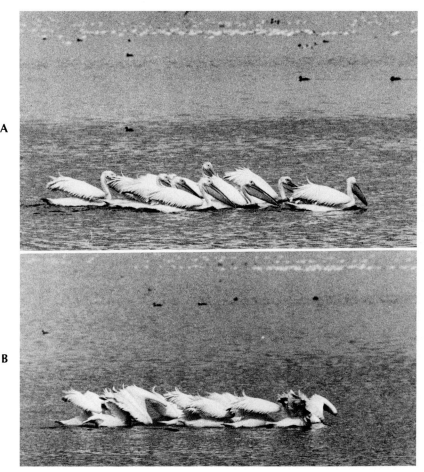

FIG. 25-26
Cooperative feeding behavior by the white pelican, *Pelecanus onocrotalus,* **A,** Pelicans on Lake Nakuru, East Africa, form a horseshoe to drive fish together, **B;** then they plunge simultaneously to scoop up fish in their huge bills. Pelicans were attracted to lake in mid-1960s to feed on fish *(Tilapia grahami)* introduced to control malaria by eating mosquito larvae. The pictures were taken 2 seconds apart. Order Pelecaniformes. (Photos by B. Tallmark.)

territories the year round. Territories are not absolutely fixed areas, but vary as the economic pressures vary.

NESTING AND CARE OF YOUNG. To produce offspring, all birds lay eggs that must be incubated by one or both parents. Cowbird eggs require only 9 to 10 days for hatching; most songbirds, about 14 days; the hen, 21 days; and ducks and geese, at least 28 days. Most of the duties of incubation fall upon the female, although in many instances both parents share in the task, and occasionally only the male performs this work.

Most birds build some form of nest in which to rear their young. These nests vary from depressions on the ground to huge and elaborate affairs. Some birds simply lay their eggs on the bare ground or rocks and make no pretense of nest building. Some of the most striking nests are the pendant nests constructed by orioles, the neat lichen-covered nests of hummingbirds (Fig. 25-29) and flycatchers, the chimney-shaped mud nests of cliff swallows, the floating nest of the red-necked grebe (Fig. 25-30), and the huge brush pile nests of the Australian brush turkey. Most birds take considerable pains to conceal

FIG. 25-27
Territorial defense by male black grouse, *Lyrurus tetrix,* of northern Eurasia. By posturing and displaying, each male advertises its chosen territorial boundaries. In this sequence, **A,** male on left chases out male on right, which has entered its territory. **B,** Now just as suddenly roles will reverse and the right-hand male will indignantly rout the male on left. Such seesaw battles, with neither bird actually harming the other, continue for hours. Black grouse display most of the year, even in the depth of winter. Order Galliformes. (Photos by B. Tallmark.)

their nests from enemies. Woodpeckers, chickadees, bluebirds, and many others place their nests in tree hollows or other cavities; kingfishers excavate tunnels in the banks of streams for their nests; and birds of prey build high in lofty trees or on inaccessible cliffs. A few birds such as the American cowbird and the European cuckoo build no nests at all but simply lay their eggs in the nests of birds smaller than themselves. When the eggs hatch, the young are taken care of by their foster parents. Most of our songbirds lay from three to six eggs, but the number of eggs laid in a clutch varies from one or two (some hawks and pigeons) to 18 or 20 (quail).

Nesting success is very low with many birds, espe-

FIG. 25-28
Wild turkey, *Meleagris gallapavo,* displaying. This species was once abundant over the entire eastern United States. It was disappearing until a few remnant flocks in Pennsylvania were given full protection, enabling the population to increase and spread. Turkey hunting is now a popular regulated sport in most eastern states. Many would agree with Benjamin Franklin that this handsome, vigorous, and completely native bird would make a more fitting national emblem than the bald eagle, a fish-eating scavenger found also in Canada and Asia. Order Galliformes. (Photo by L. L. Rue, III.)

cially in altricial species (see below). Surveys vary among investigators, but one investigation (V. Noland, Jr.) of 170 altricial bird nests reports only 21% as producing at least one young. Of the many causes of nest failures, predation by snakes, skunks, chipmunks, blue jays, crows, etc. is by far the chief factor.

Birds of prey probably have a much higher percentage of reproductive success.

When birds hatch, they are of two types; **precocial** or **altricial** (Fig. 25-31). The precocial young, such as quail, fowl, ducks, and most water birds, are covered with down when hatched and can run or swim

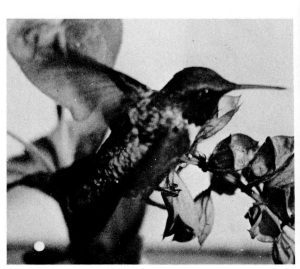
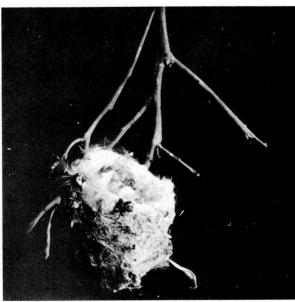

FIG. 25-29
Ruby-throated hummingbird *(Archilochus colubris)* and nest. Of 14 species of hummingbirds in United States, ruby-throated bird is the only one found east of the Mississippi. Order Apodiformes.

FIG. 25-30
Floating nest of red-necked grebe. Nests are rafts of plant fragments usually anchored to aquatic vegetation, but occasionally floating free. Nest becomes waterlogged during incubation and gradually sinks so that newly-hatched young are introduced immediately and unceremoniously to water. Order Podicipediformes. (Photo by L. L. Rue, III.)

Altricial
One-day-old meadow lark

Precocial
One-day-old ruffed grouse

FIG. 25-31
Comparison of 1-day-old altricial and precocial young. The altricial meadowlark at left is born nearly naked, blind, and helpless. The precocial ruffed grouse at right is covered with down, alert, legs strong, and able to feed itself.

FIG. 25-32
Nest of skylark, *Alauda arvensis,* with four 2-day old young. Light-reflecting edges and small white spots inside gaping mouths of young below parent are believed to be "targets" that stimulate feeding behavior by parent. The sparkling song and flight displays of skylarks in spring are well known and loved by Europeans. Skylarks nest on ground. Although young are altricial, they grow amazingly fast and begin some flying within a week. Note obliterative markings on breast of parent. Order Passeriformes. (Photo by B. Tallmark.)

FIG. 25-33

Family of Canada geese, *Branta canadensis.* Parents mate for life and young remain with parents in strong family unit during migrations, a characteristic that promotes breeding isolation and race separation within the species. Order Anseriformes. (Photo by L. L. Rue, III.)

as soon as their plumage is dry. The altricial ones, on the other hand, are naked and helpless at birth and remain in the nest for a week or more. The young of both types require care from the parents for some time after hatching. They must be fed, guarded, and protected against rain and the sun. The parents of altricial species must carry food to their children almost constantly, for most young birds will eat more than their weight each day. This enormous food consumption explains the rapid growth of the young and their quick exit from the nest (Fig. 25-32). The food of the young, depending on the species, includes worms, insects, seeds, and fruit. Pigeons are peculiar in feeding their young with "pigeon milk," the sloughed-off epithelial lining of the crop.

Many birds, such as the eagle, Canada goose (Fig. 25-33), and some of the songbirds, are known to mate for life. Others mate only for the rearing of a single brood. There are also cases in which one female mates with several males (**polyandry**), as illustrated by the European cuckoo; in other cases one male

mates with several females (**polygyny**), such as the ostrich. In most bird populations there are usually many sexually mature individuals that have no mates at all.

Résumé of living orders

Class Aves (birds) is made up of about 27 orders of living birds and a few fossil orders. More than 8,600 species and many subspecies have been described. Probably only a relatively few species remain to be discovered and named, but many subspecies are added yearly. With their powers of flight and wide distribution, most species of birds are more easily detected than are many animals. Only those that are solitary, shy, and restricted to remote regions have a chance of remaining undiscovered for any length of time. Altogether, the species are grouped into 170 families. Of the 27 recognized orders, 20 are represented by North American species.

The first four orders in the following list make up the group of **ratite,** or flightless, birds; the remainder are the **carinate,** or flying, birds.

Order Struthioniformes (stroo"thi-on-i-for'meez) (LL. *struthio*, ostrich, + L. *forma*, form)—**ostriches**. The ostrich (*Struthio camelus*) is the largest of living birds, with some specimens being 8 feet tall and weighing 300 pounds. These birds cannot fly. The feet are provided with only two toes of unequal size covered with pads, which enable the birds to travel rapidly through sandy country. The ostrich is found in the desert country of Africa and Arabia. Ostrich feathers are highly prized, and ostrich farming is extensive in South Africa, California, and elsewhere. There are 4 species, of which *Struthio camelus* is the largest.

Order Rheiformes (re"i-for'meez) (Greek mythology, *Rhea*, mother of Zeus, + form)—**rheas**. These flightless birds are restricted to South America and are often called the American ostrich.

Order Casuariiformes (kazh"u-ar"ee-i-for'meez) (NL. *Casuarius*, type genus, + form)—**cassowaries, emus**. This is a group of flightless birds found in Australia, New Guinea, and a few other islands. Some specimens may reach a height of 5 feet.

Order Apterygiformes (ap"te-rij"i-for'meez) (Gr. *a*, not, + *pteryx*, wing, + form)—**kiwis**. Kiwis are flightless birds about the size of the domestic fowl, found only in New Zealand. They all belong to the genus *Apteryx*, of which there are three species. Only the merest vestige of a wing is present. The egg is extremely large for the size of the bird.

Order Tinamiformes (tin-am"i-for'meez) (NL. *Tinamus*, type genus, + form)—**tinamous**.These are flying birds found in South America and Mexico. They resemble the ruffed grouse and are classed as game birds. There are more than 60 species in this order.

Order Sphenisciformes (sfe-nis"i-for'meez) (Gr. *sphēniskos*, dim. of *sphēn*, wedge, from the shortness of the wings, + form)—**penguins**. Penguins are found in the southern seas, especially in Antarctica. Although carinate birds, they use their wings as paddles rather than for flight. The largest penguin is the emperor penguin (*Aptenodytes forsteri*) of the Antarctic, which breeds in enormous rookeries on the shores of that region.

Order Gaviiformes (gay"vee-i-for'meez) (L. *gavia*, bird, probably sea mew, + form)—**loons**. Remarkable swimmers and divers, they live exclusively on fish and small aquatic forms. The familiar great northern diver (*Gavia immer*) is found mainly in northern waters.

Order Podicipediformes (pod"i-si-ped"i-for'meez) (L. *podex*, rump, + *pes, pedis*, foot)—**grebes**. The pied-billed grebe, or dabchick (*Podilymbus podiceps*), is a familiar example of this order. These birds are found on ponds and lakes all over the eastern half of the United States. They are shy, secretive birds and dive quickly when disturbed. Grebes are most common in old ponds where there are extensive growths of cattails, rushes, and water flags, of which they build their raftlike nests that float on the surface of the water.

Order Procellariiformes (pro-sel-lar"ee-i-for'meez) (L. *procella*, tempest, + form)—**albatrosses, petrels, fulmars, shearwaters**. As far as wingspan is concerned (more than 12 feet in some), albatrosses are the largest of flying birds. *Diomedea* is a common genus of albatrosses.

Order Pelecaniformes (pel"e-can-i-for'meez) (Gr. *pelekan*, pelican, + form)—**pelicans, cormorants, gannets, boobies, etc.** These are fish eaters with throat pouch and all four toes of each foot included within the web.

Order Ciconiiformes (si-ko"nee-i-for'meez) (L. *ciconia*, stork, + form)—**herons, bitterns, storks, ibises, spoonbills, flamingos**. A familiar wading bird of eastern North America is the great blue heron (*Ardea herodias*), which frequents marshes and ponds.

Order Anseriformes (an"ser-i-for'meez) (L. *anser*, goose, + form)—**swans, geese, ducks**. The members of this order have broad bills with filtering ridges at their margins, the foot web is restricted to the front toes, and they have a long breastbone with a low keel. The common domestic mallard duck is *Anas platyrhynchos*.

Order Falconiformes (fal"ko-ni-for'meez) (LL. *falco*, falcon, + form)—**eagles, hawks, vultures, falcons, condors, buzzards**. These are the great birds of prey (except owls) and are represented by many species. There are two American species of eagles, the American golden eagle (*Aquila chrysaëtos*) and the bald eagle (*Haliaeetus leucocephalus*).

Order Galliformes (gal"li-for'meez) (L. *gallus*, cock, + form)—**quail, grouse, pheasants, ptarmigan, turkeys, domestic fowl** (*Gallus domesticus*). Some of the most desirable game birds are in this order. The bobwhite quail (*Colinus virginianus*) is found all over the eastern half of the United States. The ruffed grouse (*Bonasa umbellus*), or partridge, is found in about the same region, but in the woods instead of the open pastures and grain fields, which bobwhite frequents.

Order Gruiformes (groo"i-for'meez) (L. *grus*, crane, + form)—**cranes, rails, coots, gallinules**. These are prairie and marsh breeders.

Order Charadriiformes (ka-rad"ree-i-for'meez) (NL. *Charadrius*, genus of plovers, + form)—**shore birds**, such as **gulls, oyster catchers, plovers, sandpipers, terns, woodcocks**.

Order Columbiformes (co-lum"bi-for'meez) (L. *columba*, dove, + form)—**pigeons, doves**.

Order Psittaciformes (sit"ta-si-for'meez) (L. *psittacus*, parrot, + form)—**parrots, parakeets**.

Order Cuculiformes (ka-koo"li-for'meez) (L. *cuculus*, cuckoo, + form)—**cuckoos, roadrunners**. The common cuckoo (*Cuculus canorus*) of Europe lays its eggs in the nests of smaller birds, who rear the young cuckoos. The incubation period of the cuckoo's egg is about 12 days, often less than that of the eggs of the foster parents. When hatched, the young cuckoo works its way under its companions and backs them out of the nest one by one. The American cuckoos, black billed and yellow billed, rear their own young.

Order Strigiformes (strij"i-for'meez) (L. *strix*, screech owl, + form)—**owls**. Owls are chiefly nocturnal birds and have probably the keenest eyes and ears in the animal kingdom.

Order Caprimulgiformes (kap"ri-mul"ji-for'meez) (L. *caprimulgus*, goatsucker, + form)—**goatsuckers, nighthawks, poorwills**.The birds of this group are most active at night and in twilight. They have small, weak legs, wide mouths, and short, delicate bills. The mouth is fringed with bristles in most species. The whippoorwills *(Antrostomus vociferus)* are common in the woods of the Eastern states, and the nighthawk *(Chordeiles minor)* is often seen and heard in the evening flying around city buildings.

Order Apodiformes (up-pod"i-for'meez) (Gr. *apous*, sandmartin; footless, + form)—**swifts, hummingbirds.** The swifts get their name from their speed on the wing. The familiar chimney swift *(Chaetura pelagica)* fastens its nest in chimneys by means of saliva. A swift found in China *(Collocalia)* builds a nest of saliva that is used by the Chinese for soup making. Most species of hummingbirds are found in the tropics, but there are 14 species in the United States, of which only one, the ruby-throated hummingbird, is found in the eastern part of the country. Most hummingbirds live upon nectar, which they suck up with their highly adaptable tongue, although some catch insects also.

Order Coliiformes (ka-lye"i-for'meez) (Gr. *kolios*, green woodpecker, + form)—**mousebirds.** Africa.

Order Trogoniformes (tro-gon"i-for'meez) (Gr. *trōgōn*, gnawing, + form)—**trogons.** Tropical.

Order Coraciiformes (ka-ray"see-i-for'meez or kor"uh-sigh"uh-for'meez) (NL. Coracii from Gr. *korakias*, a kind of chough [akin to *korax*, raven or crow], + form)—**kingfishers, hornbills, etc.** In the eastern half of the United States the belted kingfisher *(Megaceryle alcyon)* is common among most waterways of any size. It makes a nest in a burrow in a high bank or cliff along a water course.

Order Piciformes (pis"i-for'meez) (L. *picus*, woodpecker, + form)—**woodpeckers, toucans, puffbirds, etc.** Woodpeckers are adapted for climbing, with stiff tail feathers and toes with sharp claws. Two of the toes extend forward and two backward. There are many species of woodpeckers in North America, the more common of which are the flickers and the downy, hairy, red-bellied, redheaded, and yellow-bellied woodpeckers. The largest is the pileated woodpecker, which is usually found in deep and remote woods.

Order Passeriformes (pas"er-i-for'meez) (L. *passer*, sparrow, + form)—**perching birds.** This is the largest order of birds and is made up of 69 families. To this order belong the songbirds found in all parts of the world. Among these are the skylark, nightingale, hermit thrush, mockingbird, meadow lark, robin, and hosts of others. Others of this order, such as the swallow, magpie, starling, crow, raven, jay, nuthatch, and creeper, have no songs worthy of the name.

References
Suggested general readings

Bent, A. C. 1919. Life histories of North American birds. U. S. National Museum Bulletins. *More than a score of monographs in this outstanding series provide a wealth of information on birds.*

Broun, M. 1949. Hawks aloft. New York, Dodd, Mead & Co., Inc. *An account of the migration of hawks as observed from Hawk Mountain, Pennsylvania, with observations on the problem of protecting desirable birds of prey.*

Fisher, J., and R. T. Peterson. 1971. Birds. An introduction to general ornithology. London, Aldus Books. *Beautifully illustrated general biology of birds by two famous ornithologists.*

Gilliard, E. T. 1958. Living birds of the world. New York, Doubleday & Co., Inc. *A superb book of birds with illustrations (many in color).*

Griffin, D. R. 1964. Bird migration. Garden City, N. Y., The Natural History Press. *Interesting semipopular account of this subject; less detailed treatment and more engaging in style than Matthews' book.*

Headstrom, R. 1949. Birds' nests. New York, Ives Washburn, Inc. *Good photographs of nests with descriptions. Nests are grouped according to a scheme of nest construction.*

Kendeigh, S. C. 1952. Parental care and its evolution in birds. Urbana, University of Illinois Press. *The author believes that parental care has evolved independently in the different phyla, and that the complex patterns for the care of the young are definitely correlated with the development of the nervous and sensory systems.*

Krutch, J. W., and P. S. Eriksson. 1962. A treasury of birdlore. New York, Paul S. Eriksson, Inc. *Excerpts from the writings of famous ornithologists.*

Lorenz, K. Z. 1952. King Solomon's ring. Thomas Y. Crowell Co., New York.

Marshall, A. J. 1960-1961. Biology and comparative physiology of birds. 2 vols. New York, Academic Press, Inc. *A thorough, extensive, and technical treatment of these subjects.*

Matthews, G. V. T. 1968. Bird migration, ed. 2. Cambridge, The University Press. *An authoritative and comprehensive résumé of this subject, stressing the author's extensive homing experiments.*

Newton, A. 1896. A dictionary of birds. London, A. & C.

Black. *Despite its age, this famous work contains a wealth of valuable material on many aspects of bird anatomy, physiology, and general biology.*

Peterson, R. T. 1947. Field guide to the birds, ed. 2. Boston, Houghton Mifflin Co. *One of the best field manuals for ready identification.*

Pettingill, O. S., Jr. 1947. Silent wings. Madison, Wisconsin Society for Ornithology, Inc. *A pathetic account of the extinction of the passenger pigeon.*

Sparks, J., and T. Soper. 1970. Owls. Their natural and unnatural history. New York, Taplinger Publishing Co., Inc. *An engaging, well-illustrated account of owls in fact and fancy.*

Savile, D. B. O. 1957. Adaptive evolution in the avian wing. Evolution **11**:212-224. *The author presents a very useful classification of bird wings into four flight types according to habitat and function.*

Sturkie, P. D. 1954. Avian physiology. Ithaca, N. Y., Cornell University Press. *Dealing with the specialized physiology of birds and intended primarily for research workers, it includes data on nearly all aspects of bird organ systems and their physiology. The book is well documented.*

Tinbergen, N. 1953. Social behavior in animals. New York, John Wiley & Sons, Inc. *Good introduction to bird behavior.*

Tinbergen, N. 1960. The herring gull's world. New York, Basic Books, Inc., Publishers. *A superb account of bird behavior by a renowned ethologist. Excellent illustrations by the author.*

Van Tyne, J., and A. J. Berger. 1959. Fundamentals of ornithology. New York, John Wiley & Sons, Inc. *An up-to-date account of the structure, physiology, distribution, and taxonomy of one of the most interesting groups of animals. The chapter on migration is only one of many fine accounts about behavior patterns of birds.*

Welty, J. C. 1962. The life of birds. Philadelphia, W. B. Saunders Co. *Among the best of the ornithology texts; lucid style and excellent illustrations.*

Wolfson, A. (editor). 1955. Recent studies in avian biology. Urbana, University of Illinois Press. *This summary by various specialists is important to all serious students of birds, for it includes an evaluation of the concepts and problems of evolution, systematics, anatomy, migration, breeding behavior, diseases, etc. Excellent bibliographies are included.*

Selected *Scientific American* articles

Cone, C. D. Jr. 1962. The soaring flight of birds. **206**:130-140 (April). *The nature of air currents and patterns of soaring flight are analysed.*

Eklund, C. R. 1964. The Antarctic skua. **210**:94-100 (Feb.). *Discusses the biology of this large, aggressive, and cold-adapted bird.*

Emlen, J. T., and R. L. Penny. 1966. The navigation of penguins. **215**:104-113 (Oct.). *Penguins depend on an innate biologic clock and the sun's direction to guide them across hundreds of miles of featureless Antarctic landscape.*

Frings, H., and M. Frings. 1959. The language of crows. **201**:119-131 (Nov.).

Greenewalt, C. H. 1969. How birds sing. **221**:126-319 (Nov.). *The mechanism of bird song is quite different from that of musical instruments or the human voice.*

Lack, D., and E. Lack. 1954. The home life of the swift. **181**:60-64 (July). *A well-known British ornithologist and his wife describe the biology of this interesting bird.*

Peakall, D. B. 1970. Pesticides and the reproduction of birds. **222**:72-78 (April). *Pesticides threaten the survival of several species of birds of prey. The reasons are explained.*

Sauer, E. G. F. 1958. Celestial navigation by birds. **199**:42-47 (Aug.). *The author describes his ingenious planetarium experiments that demonstrate that migratory birds navigate by the stars.*

Schmidt-Nielsen, K. 1959. Salt glands. **200**:109-116 (Jan.). *The anatomy and physiology of this special salt-excretory organ of marine birds is described.*

Schmidt-Nielsen, K. 1971. How birds breathe. **225**:72-79 (Dec.).

Sladen, W. J. L. 1957. Penguins. **197**:44-51 (Dec.). *These interesting birds have several adaptions that fit them for life in a harsh environment.*

Stettner, L. J., and K. A. Matyniak. 1968. The brain of birds. **218**:64-76 (June).

Taylor, T. G. 1970. How an eggshell is made. **222**:88-95 (March).

Tickell, W. L. N. 1970. The great albatrosses. **223**:84-93 (Nov.). *Describes the reproductive behavior and movements of the largest of oceanic birds.*

Tucker, V. A. 1969. The energetics of bird flight. **220**:70-78 (May).

Welty, C. 1955. Birds as flying machines. **192**:88-96 (March). *A description of the remarkable adaptations that fit birds for flight.*

A coyote on the western Canadian prairie. Though shot, trapped, and poisoned at every opportunity by ranchers who consider coyotes pests despite their important role in rodent control, and hunted relentlessly for sport, this adaptable carnivore maintains a vigorous population in the West. (Photo by C. G. Hampson, University of Alberta, Edmonton, Alberta.)

CHAPTER 26

THE MAMMALS

PHYLUM CHORDATA
CLASS MAMMALIA*

Mammals, with their highly developed nervous system and numerous ingenious adaptations, occupy almost every environment on earth that will support life. Despite their relatively small numbers (4,500 species as compared to 8,600 species of birds, well over 20,000 fishes and 900,000 insects), they are overall the most biologically successful group in the animal kingdom, with the possible exception of the insects. Many potentialities that dwelled more or less latently in other vertebrates are highly developed in mammals. Mammals are enormously diverse in size, shape, form, and function. They range in size from the diminutive pigmy shrew, with a body length of less than $1^{1}/_{2}$ inches and weight of only a fraction of an ounce, to whales, exceeding 100 tons in weight. Commenting on the unequalled adaptive ra-

diation of this group, one biologist has compared the evolution of mammals to a biologic application of Parkinson's law: they expand (like work) to fill all the available space (time). As a climax to mammalian evolution there is man, a creature that we continue to hope will be an evolutionary success!

Yet despite their adaptability, and in some instances because of it, mammals have been influenced by man's heavy-handed presence more than any other group of animals. He has domesticated numerous mammals for food and clothing, as beasts of burden and as pets. He uses millions of mammals each year in biomedical research. He has introduced alien mammals into new habitats, occasionally with benign results, but more frequently with unexpected disaster. Although history provides us with numerous warnings, we continue to overcrop valuable wild stocks of mammals. The immensely valuable

*Mam-may'lee-a (L. *mamma*, breast).

564

whale industry is threatened with total collapse by exterminating its own resource—a classic example of self-destruction in the modern world, in which each segment of the industry is intent only on reaping all it can before bankruptcy enfolds it. In some cases destruction of a valuable mammalian resource has been deliberate, such as the officially sanctioned (and tragically successful) policy during the Indian wars of exterminating the bison to drive the plains Indians into starvation. Although commercial hunting by man has declined, the ever-increasing human population with the accompanying destruction of wild habitats has harassed and disfigured the mammalian fauna. We are becoming increasingly aware that our presence on this planet as the most powerful product of organic evolution makes us totally responsible for the character of our natural environment. Aware that man's welfare has been, and continues to be, closely related to that of the other mammals, it is clearly in our interest to preserve the natural environment of which all mammals, ourselves included, are a part. We need to remember that nature can do without man, but man cannot exist without nature.

■ ORIGIN AND RELATIONSHIPS

Long before the great dinosaurs had reached the peak of their evolutionary success, a group of late Paleozoic reptiles called the **pelycosaurs** appeared and flourished over a period of 40 million years and then nearly disappeared. The pelycosaurs were clearly reptilian, with sprawling gait and undifferentiated teeth. But their descendants, the **therapsids**, were a group of reptiles of the Triassic (early Mesozoic) with so many mammal-like characteristics that they are considered to be the direct ancestors of the living mammals (Fig. 26-1).

The evolution of the therapsids and their descendants was accompanied by several structural changes that brought them ever closer to full mammalian status. The clumsy limbs of the reptile that stuck out laterally were replaced by straight legs held close to the body, which provided speed and efficiency for hunting. Since reptilian stability was sacrificed by thus raising the animal from the ground, the muscular coordination center of the brain, the cerebellum, took on a greatly expanded role. Among the many changes in the bony structure of the head was the separation of air and food passages in the mouth. This enabled the animal to hold prey in its mouth

while breathing. It also made possible prolonged chewing and some predigestion of the food. At some point the premammals acquired warm-bloodedness —a physiologic and biochemical trick that bestowed tremendous benefits—and those two most characteristic of all mammalian identification tags: hair and mammary glands. Although the fossil sequence provides a good record of skeletal evolution, it is largely silent on the evolution of hair, glands, and warm-bloodedness, and of course, until these characteristics evolved, the premammals were not mammals. Despite the difficulty of selecting a sharp reptile-mammal boundary, it is certain that several groups of premammals had achieved full mammalian status by the end of the Triassic, some 200 million years ago.

Most of the living mammals belong to the subclass Theria and have descended from a common ancestor of the Jurassic, some 150 million years ago. However, the monotremes (subclass Prototheria), the egg-laying mammals of Australia, Tasmania, and New Guinea, are so different from the others and possess so many reptilian characters, that they are believed to have descended from an entirely different mammal-like reptile. The separation of Prototheria and Theria probably occurred some 50 million years earlier in the Triassic period. The geologic record during the following Jurassic and Cretaceous periods is fragmentary, in large part because the mammals of these periods were small creatures the size of a rat or smaller, with fragile bones that fossilized only under the most ideal circumstances.

When the dinosaurs vanished near the beginning of the Cenozoic era, the mammals suddenly expanded. This was partly attributable to the numerous ecologic niches vacated by the reptiles, into which the mammals could move as their divergent adaptations fitted them. There were other reasons for their success. Mammals were agile, warm blooded, and insulated with hair; they had developed placental reproduction and suckled their young, thus dispensing with vulnerable eggs and nests; and they were more intelligent than any other animal alive. During the Eocene and Oligocene epochs of the Tertiary (55 million to 30 million years ago), the mammals flourished and reached their peak. In terms of number of species, this was the golden age of mammals. They have declined in numbers ever since; only 932 of the 2,864 known mammalian genera

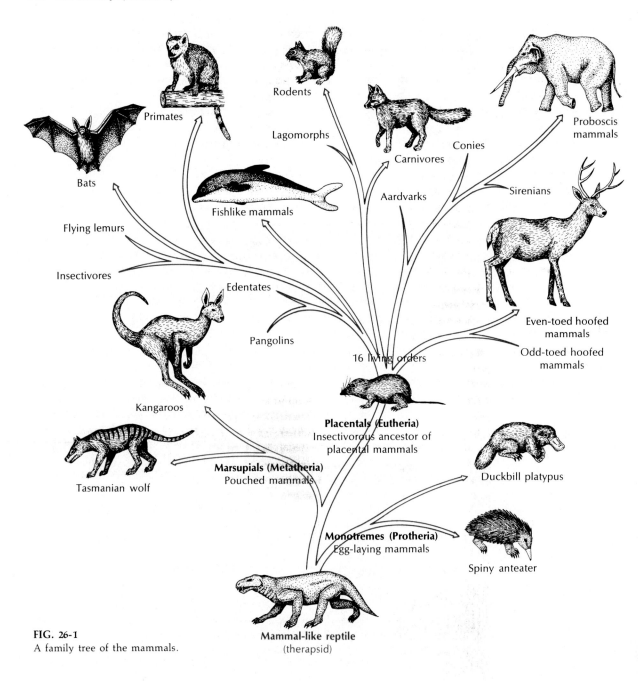

FIG. 26-1
A family tree of the mammals.

(33%) are still living. However, extinction of species within a group is a natural and expected consequence of changing conditions and does not necessarily portend group extinction. Rather than disappearing, mammals dominate the land environment as thoroughly now as they did 50 million years ago.

Characteristics

Since mammals and birds both evolved from reptiles, we can expect to find, and do find, many structural similarities among the three groups. It is in fact much easier to point to numerous resemblances between the mammals and the reptiles from which they descended, than to point to characteristics that are unique and diagnostic for mammals. Hair is the most obvious mammalian characteristic, although it is vastly reduced in some (such as whales) and although reptilian scales, from which hair is derived, may persist (such as on tails of rat and beaver). A second unique characteristic of mammals is the method of nourishing their young with milk-secreting glands; reptiles have nothing remotely similar. The mammalian skull contains the following several features that distinguish the group:

1. Mammals have two occipital condyles instead of one, as in reptiles, and a larger cranium.

2. The lower jaw in mammals is composed of one bone; the reptiles have more.

3. In mammals the lower jaw articulates directly with the skull and not through a quadrate bone, which is found in reptiles.

4. Mammals have a chain of three bones in the middle ear (incus, malleus, stapes); reptiles have only one columella (stapes) in the ear and retain the other two at the angle of the jaw.

5. The deciduous and permanent teeth of mammals have replaced the polyphyodont teeth of reptiles.

The single most important factor contributing to the success of mammals is the great development of the neocerebrum, permitting a level of adaptive behavior, learning, curiosity, and intellectual activity far beyond the capacity of even the brightest reptile.

We may summarize the mammalian characteristics as follows:

1. **Body covered with hair,** but reduced in some
2. **Integument with sweat, sebacous,** and **mammary glands**
3. Skeletal features of skull with **two occipital condyles, seven cervical vertebrae** (usually), and often an elongated tail
4. Mouth with teeth on both jaws
5. **Movable eyelids** and **fleshy external ears**
6. Four limbs (reduced or absent in some) adapted for many forms of locomotion
7. Circulatory system of a four-chambered heart, **persistent left aorta,** and **nonnucleated red blood corpuscles**
8. Respiratory system of lungs and a voice box
9. **Muscular partition between thorax** and **abdomen**
10. Excretory system of metanephros kidneys and ureters that usually open into a bladder
11. Nervous systems of a well-developed brain and 12 pairs of cranial nerves
12. **Warm blooded**
13. Cloaca present only in monotremes
14. Separate sexes; reproductive organs of a penis, **testes (usually in a scrotum),** ovaries, oviducts, and vagina
15. Internal fertilization; **eggs develop in a uterus** with **placental attachment** (except monotremes); **fetal membranes (amnion, chorion, allantois)**
16. Young nourished by **milk from mammary glands**

■ STRUCTURAL AND FUNCTIONAL ADAPTATIONS OF MAMMALS
Integument and derivatives

The mammalian skin and its modifications especially distinguish mammals as a group. As the interface between the animal and its environment, the skin is strongly molded by the animal's way of life. In general the skin is thicker in mammals than in other classes of vertebrates, although it is made up of the two typical divisions—epidermis and corium (dermis) (see Fig. 29-1, *B*). Among the mammals the corium becomes much thicker than the epidermis. The epidermis varies in thickness. It is relatively thin where it is well protected by hair, but in places subject to much contact and use, such as the palms or soles, its outer layers become thick and cornified with keratin.

HAIR. Hair is especially characteristic in mammals, although man himself is not a very hairy creature and in the whales it is reduced to only a few sensory bristles on the snout. The hair follicle from which a hair grows is an epidermal structure even though it lies mostly in the dermis and subdermal (subcutaneous) tissues (Fig. 29-1, *B*). The hair grows continuously by rapid proliferation of cells in the base of the follicle. As the hair shaft is pushed upward, new cells are carried away from their source of nourishment and turn into the dense type of keratin **(hard keratin)** that constitutes nails, claws, hooves, and feathers as well as hair. On a weight basis, hair is by far the strongest material in the body. It has a tensile strength comparable to rolled aluminum, which is nearly twice as strong, weight for weight, as the strongest bone.

Hair is more than just a cylinder of keratin. It consists of three layers: the medulla or pith in the center of the hair, the cortex with pigment granules next to the medulla, and the outer cuticle composed of imbricated scales. The hair of different mammals shows a considerable range of structure. It may be deficient in cortex, such as the brittle hair of the deer, or it may be deficient in medulla, such as the hollow, air-filled hairs of the wolverine, so favored by northerners for trimming the hoods of parkas because it resists frost accumulation. The hairs of rabbits and some others are scaled to interlock when pressed to-

FIG. 26-2

Beaver *(Castor fiber)* feeding on bark of trembling aspen. Although coarse guard hair is wet, the thick layer of fine underfur is nearly dry. Beavers feed mostly on the inner bark of higher, more tender branches of aspen, willow, and birch. Beavers are equipped with powerful jaws and chisel-sharp incisors for felling trees, which are used both for food and to provide building material for dams. Beavers are valuable conservationists and under protection are recovering rapidly from near-extermination during the nineteenth century. They are regularly moved from lowlands, where they are a nuisance to farmers, to mountain areas where their dams control floods and create marshes for waterfowl. As ponds silt up, they provide rich soil for vegetation, eventually becoming meadows and deciduous forest. Order Rodentia, family Castoridae. (Photo by L. L. Rue, III.)

gether. Curly hair, such as that of sheep, grows from curved follicles.

Each hair follicle is provided with a small strip of muscle that, when contracted, pulls the hair upright. These erector muscles are under the control of the sympathetic nervous system. During certain emotional states (fear and excitement), many animals erect their hair, particularly that on the neck and between the shoulders, thus increasing the apparent size of the body. In man, contraction of the erector muscles after excitement or cold stimulation causes the hair to stand up and the skin to dimple in above the muscle attachment. The result is "gooseflesh."

Mammals characteristically have two kinds of hair forming the **pelage**: (1) dense and soft **underhair** for insulation and (2) coarse and longer **guard hair** for protection against wear and to provide coloration. The underhair traps a layer of insulating air; in aquatic animals such as the fur seal, otter, and beaver, it is so dense that it is almost impossible to wet it. In water the guard hairs wet and mat down over the underhair, forming a protective blanket (Fig. 26-2). A quick shake when the animal emerges flings off the water and leaves the outer guard hair almost dry.

When a hair reaches a certain length, it stops growing. In rare instances, such as the mane of a horse, it may persist as a mature hair throughout the life of the animal. Normally however, it remains in the follicle only until a new growth starts, whereupon it falls out. In man, hair is shed and replaced throughout life. But in most mammals, there are periodical molts of the entire coat. In the simplest cases, such as in foxes and seals, the coat is shed once each year, during the summer months. In the fox, molting begins on the legs and hindquarters and progresses forward, with the new coat appearing as soon as the old is lost. Most mammals have two annual molts, one in the spring and one in the fall. The summer coat is always much thinner than the winter and is usually a different color. Several of the mustelid carnivores (such as the weasel) have white winter coats and colored summer coats. It was once believed that the white winter pelage of arctic ani-

FIG. 26-3
Arctic hare *(Lepus arcticus)* of the high arctic tundra in winter white coat. Its thick fur enables it to survive continuous exposure to cold. It feeds on the sparse vegetation exposed on ridges by the wind. Like its more southerly relative of the northern forests (the varying hare or snowshoe rabbit), its large paws provide support for locomotion on snow. Order Lagomorpha. (Photo by C. G. Hampson, University of Alberta.)

mals served to conserve body heat by reducing radiation loss, but recent research has shown that dark and white pelages radiate heat equally well. The winter white of arctic animals is simply camouflage in a land of snow (Fig. 26-3). The varying hare of North America have three annual molts: the white winter coat is replaced by a brownish gray summer coat, and this is replaced in autumn by a grayer coat, which is soon shed to reveal the winter white coat beneath.

Hairs are said to be homologous with the scales and feathers of reptiles and birds. However, hair probably originated from the tactile sensory pits (prototriches) of fish and amphibians, from apical bristles that at first were of sensory function. The distribution was the same as that of scales, since reptilian sensory pits were located on the apices of epidermal scales. This primitive pattern of distribution has been lost in advanced mammals.

Mammals in their inventiveness have modified hair variously to serve many purposes. The bristles of hogs, vibrissae on the snouts of most mammals, and the spines of porcupines and their kin are examples. Vibrissae, commonly and incorrectly called "whiskers," are really sensory hairs that provide an additional special sense to many mammals. The bulb at the base of each follicle is provided with a large sensory nerve. The slightest movement of a vibrissa generates impulses in the nerve endings that travel to a special sensory area in the brain. The vibrissae

FIG. 26-4
Nine-banded armadillo *(Dasypus novemcinctus)* in defensive position. Pelvic and pectoral bony shields are separated by nine bony movable bands. This species is abundant in South America and has invaded the southern United States where it is prospering on a diet of insects, scorpions, and worms. Reproduction is unique. Mating occurs in the summer. The single egg is immediately fertilized but does not implant in the uterus for 3 to 4 months. Then it suddenly splits into four cells, which implant separately and develop into four genetically identical embryos. Quadruplets are born in the spring and soon become independent. Order Edendata. (Photo by L. L. Rue, III.)

are especially long in nocturnal and burrowing animals. In seals they apparently serve as a "distance touch" sensitive to pressure waves and turbulence in the water caused by objects or passing fish. Vision is of little use to seals hunting in turbid water, where they are frequently found, and investigators have noted that blind seals, having lost their sight in some accident, remain just as fat and healthy as normal seals.

Porcupines, hedgehogs, the echidna, and a few other mammals have developed an effective and dangerous spiny armor; the spines of the common North American porcupine break off at the bases when struck and work deeply into their victim, aided by backward-pointing hooks on the tips. To assist slow learners like dogs in understanding what they are dealing with, porcupines rattle the spines and prominently display the white markings on the quills toward their tormentors.

Of quite different origin is the armadillo's shell. The scales are small bones of dermal origin and covered with a tough, horny epidermis. As revealed in the photograph of an armadillo in its defensive posture (Fig. 26-4), hair grows out between the scales and on the unscaled underside of the body.

COLORATION. We referred above to the cryptic white coloration of arctic mammals, an obviously useful adaptation for an all-white environment. Outside of the arctic, most mammals wear somber colors for protective purposes. Often the species is marked

FIG. 26-5
Zebras *(Equus burchelli)* at waterhole in East Africa. Animal in foreground is on alert for predatory lions while others drink. If danger is spotted, it will emit a short bark to set the herd running. Zebras travel in large herds of family units, each consisting of a stallion, several mares, and their foals. Order Perissodactyla. (Photo by C. G. Hampson, University of Alberta).

with "salt-and-pepper" coloration or a disruptive pattern that helps to make it inconspicuous in its natural surroundings. Examples are the leopard's spots, the stripes of the tiger, and the spots of fawns. Zoologists have long wondered what adaptive purpose, if any, is served by the clearly defined black and white pattern of zebras (Fig. 26-5). Although the zebra would appear to be a conspicuous target for predators, naturalists in Africa report that its stripes tend to blur the outline of the animal when it moves, making it difficult to distinguish from the background and from the more uniformly colored species of the African plains. Other mammals, for example, skunks, advertise their presence with conspicuous warning coloration.

An interesting aspect of color is the pair of rump patches of the pronghorn antelope, which are composed of long white hairs erected by special muscles. When alarmed, the animal can flash these patches in a manner visible for a long distance. They may be used as a warning signal to other members of the herd. The well-known "flag" of the Virginia white-tailed deer serves a similar purpose.

Mammalian coloration is principally caused by pigmentation in the hair, although in a few, bare sur-faces of skin may be found with bright hues, such as in the cheeks and sternal callosities of the mandrill (a species of baboon), which may be attributable to pigment or to blood capillaries in the skin. At least two types of chromatophores are found in mammals —melanophores (black and brown pigment) and xanthophores (red and yellow pigment). Pigment granules may also lie outside the regular pigment cells. Although the color of hair may fade to some extent, any noticeable change in the color of a mammal's fur coat must be brought about by molting.

Albinism, or a lack of pigment, may happen in most kinds of mammals, as also may **melanism,** or an excess of black pigment.

HORNS AND ANTLERS. Three kinds of horns or hornlike substances are found in mammals (Fig. 26-6). **True horns** found in ruminants, for example, sheep and cattle, are hollow sheaths of keratinized epidermis that embrace a core of bone arising from the skull. Horns are not normally shed, are not branched (although they may be greatly curved), and are found in both sexes. The horns of North American pronghorn antelope are unique in that they are shed each year after the breeding season. But unlike the shedding of deer antlers, the new horn replaces

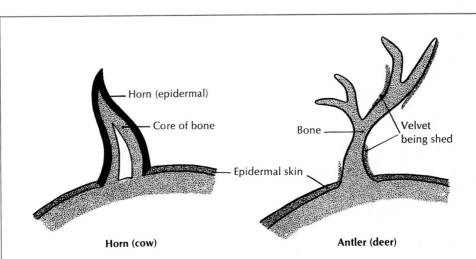

FIG. 26-6
Chief differences between horns and antlers. Bone, a dermal (mesoderm) derivative, forms basic part of each type, and when epidermal velvet with its hair is shed, bone forms all of antlers. Antlers are shed annually (in winter) when zone of constriction below burr appears near skull. Horns do not branch and are not shed. Injection of testosterone will prevent shedding of antlers.

the old by growing up inside and pushing it off. **Antlers** of the deer family are entirely bone when mature. During their annual growth, antlers develop beneath a covering of highly vascular soft skin called "velvet." When growth of the antlers is complete just prior to the breeding season, the blood vessels constrict and the stag tears off the velvet by rubbing the antlers against trees (Fig. 26-7). The antlers are dropped after the breeding season. New buds appear a few months later to herald the next set of antlers. For several years each new pair of antlers is larger and more elaborate than was the previous set. The

FIG. 26-7
Bull moose *(Alces alces)* shedding velvet. Antlers begin their growth each spring, stimulated by gonadotropin from the pituitary and progressively rising level of testosterone from the testes. When growth is complete in the late summer, vessels at base (burr) of antlers constrict. The skin, called velvet, dies, shrivels, and sloughs off, revealing bony antlers beneath. In late winter after breeding season, bone is resorbed at base, weakening joint, and antlers drop off. Moose are solitary in their habits; bulls travel alone, whereas cows travel accompanied by calves. Because of large amount of vegetable food needed to exist, winter is frequently a struggle for survival. Moose, called "elk" in Eurasia, are so abundant in Scandinavia that they are game cropped and marketed commercially. Order Artiodactyla. (Photo by L. L. Rue, III.)

FIG. 26-8

White rhinoceros *(Diceros simus)* of east African savanna. Unlike true horns, rhinoceros horn is a dense structure of closely packed fibers, derived from hair. In Asia, where the horn is believed to have aphrodisiac qualities, all three species of Asian rhinoceroses (Indian, Sumatran, Javan) have been hunted close to extinction. Only 50 Java rhinoceroses remain. The two African species (black and white) have been slaughtered for the same reason but are presently under protection in some areas. Rhinoceroses are grazers (white rhinoceros) and browsers (black rhinoceros) with poor eyesight and a wary, nervous disposition. Order Perissodactyla. (Photo by B. Tallmark.)

annual growth of antlers places a strain on the mineral metabolism since during the growing season a large moose or elk must accumulate 50 or more pounds of calcium salts from its vegetable diet. A **third kind of horn** is that of the rhinoceros. Hairlike horny fibers arise from dermal papillae and are cemented together to form a single horn (Fig. 26-8).

GLANDS. Mammals have a great variety of integument glands. Whatever the type of gland, they all appear to fall into one of three classes: eccrine, apocrine, and holocrine.

The **eccrine glands** (sweat glands) (Fig. 26-9) are found only in hairless regions (foot pads, etc.) in most mammals, although in some apes and in man they are scattered all over the body and are important devices for heat regulation. These glands have developed by the time of birth and are true secretory, or merocrine, glands; that is, the cell remains intact or is not destroyed in the process of secretion. Their secretory coils are restricted to the dermal region.

Sweat glands are used mainly to regulate the body temperature. They are common on such animals as the horse and man but are greatly reduced on the carnivores (cats) and are entirely lacking in shrews, whales, and others. Dogs are now known to have sweat glands all over the body. In human beings, racial differences are pronounced. Blacks, who have more than whites, can withstand warmer weather. **Lacrimal,** or tear, glands keep the surface of the eye moist and clean.

Apocrine glands (Fig. 26-9) are larger than eccrine glands and have longer and more winding ducts.

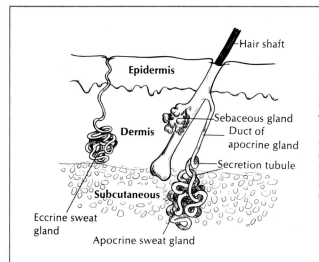

Eccrine sweat gland
Apocrine sweat gland
Epidermis
Dermis
Subcutaneous
Hair shaft
Sebaceous gland
Duct of apocrine gland
Secretion tubule

FIG. 26-9
Sweat glands (eccrine and apocrine). Phylogenetically apocrine glands are older. Eccrine glands, best developed in primates, develop from epidermis and play an important role in temperature regulation. Most glands in dog, pig, cow, horse, etc. are apocrine, but these have declined in man, along with hair, for they develop from follicular epithelium. Apocrine glands are not involved in temperature regulation, but their odorous secretions play a part in sexual attraction.

Their secretory coil is in the subdermis. They always open into the follicle of a hair or where a hair has been. Phylogenetically they are much older than the eccrine gland and are found in all mammals, some of which have only this kind of gland. Blacks have more apocrine glands than do whites, and women have twice as many as men. They develop about the time of sexual puberty and are restricted (in man) to the axillae, mons pubis, breasts, external auditory canals, prepuce, scrotum, and a few other places. Their secretion is not watery like ordinary sweat (eccrine gland) but is a milky, whitish or yellow secretion that dries on the skin to form a plastic-like film. Only the tip of the secretory cell is destroyed in the process of secretion. Their secretion is not involved in heat regulation, but their activity is known to be correlated with certain aspects of the sex cycle, among other possible functions.

The most common apocrine glands include the **scent** glands, found in all terrestrial species. Their location and function vary greatly. Some are defensive in nature, others convey information to members of the same species, and still others are involved in the mating process (Fig. 26-10). These glands are often located in the preorbital, metatarsal, and interdigital regions (deer); preputial region on the penis (muskrats, beavers, canine family, etc.); base of tail (wolves and foxes); and anal region (skunks, minks, weasels). These last, the most odoriferous of all glands, open by ducts into the anus and can dis-

charge their secretions forcefully for several feet. During the mating season many mammals give off strong scents for attracting the opposite sex. Man also is endowed with scent glands. But civilization has taught us to dislike our own scent, a concern that has stimulated a lucrative deodorant industry to produce an endless output of soaps and odor-masking compounds.

Mammary glands, which provide the name for mammals, are probably modified apocrine glands, although recent studies suggest that they may have derived from sebaceous glands (see below). Whatever their evolutionary origin, they occur on all female mammals and on most, if not all, male mammals; on the latter they are often covered by hair. They develop by the thickening of the epidermis to form a milk line along each side of the abdomen in the embryo. On certain parts of these lines the mammae appear, while the intervening parts of the ridge disappear. They secrete milk for the nourishment of the young. Milk varies in composition; in rapidly growing young it may contain 40% fat (seals, walruses). In the human female, the mammary glands begin at puberty to increase in size because of fat accumulation and reach their maximum development in about the twentieth year. The breasts (or mammae) undergo additional development during pregnancy. In other mammals, the breasts are swollen only periodically when they are distended with milk during pregnancy and subsequent nursing of the young.

FIG. 26-10
Olfactory exploration during courtship in thirteen-lined ground squirrels *(Citellus tridecemlineatus).* Scent is by far the most important means of communication in mammals. Each produces an odor characteristic of the species; the same species may vary its scent for different communicative purposes. Anal glands, as in these two ground squirrels, are especially common. This species lives on the prairies of western Canada and United States and feeds on grains, other vegetation, and insects. Order Rodentia, family Sciuridae. (Photo by C. G. Hampson.)

FIG. 26-11
Types of mammalian nipples. Monotremes have no nipples and young lick up milk from ridged depression as milk exudes from mother's skin. All other mammals (marsupials and placentals) have nipples. Mammary glands are assumed to have originated from apocrine sweat glands. Primitive arrangement of nipples (which vary in number) consists of two series, or milk lines, along abdomen. Composition of milk varies among different species of mammals. Whale milk, for instance, contains four times more protein and ten times more fat than cow milk, but it lacks sugar.

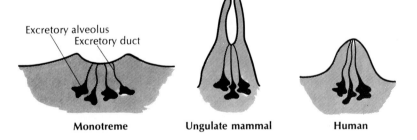

Excretory alveolus
Excretory duct

Monotreme   **Ungulate mammal**   **Human**

The outlets of the gland are by elevated nipples (absent in monotremes) (Fig. 26-11). The glands are located on the thorax of primates, bats, and a few others but on the abdomen or inguinal region in other mammals. Nipples vary in number from two in the human being, horse, bat, etc., to 25 in the opossum. The number is not always constant in the same species.

The third type of gland **(holocrine)** is one in which the entire cell is discharged in the secretory process and must be renewed for further secretion. Most of them open into hair follicles, but some are free and

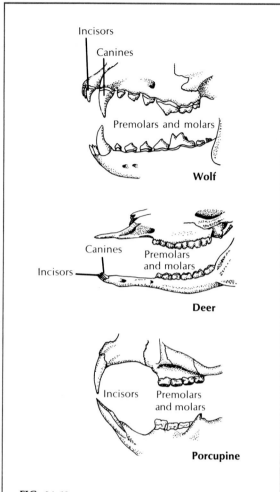

FIG. 26-12
Adaptations of mammal tooth patterns for different kinds of diet. Sharp canines of wolf are designed for stabbing, and premolars and molars are for cutting rather than grinding. Browsing deer has predominantly grinding teeth; lower incisors and canines bite against a horny pad in the upper jaw. Porcupine has no canines; self-sharpening incisors are used for gnawing. (Modified from Carrington, R. 1968. The mammals, New York, Life Nature Library.)

open directly onto the surface. The **sebaceous gland** is the most common example. In most mammals sebaceous glands are found all over the body; in man they are most numerous in the scalp, forehead, and face. Sebaceous glands that open into the hair follicles keep the skin and hair soft and glossy. A modi-

fied sebaceous gland, known as the meibomian gland, is found at the edge of each eyelid and provides an oily film between the eyelids and eyeball.

TEETH. All mammals, with the exception of certain species of whales, monotremes, and anteaters, have teeth. The character of the dentition is correlated with the food habits of the mammal (Fig. 26-12). Typically, mammals have two sets of teeth: deciduous and permanent. They are set in sockets in the jaw, and the various types are molars, premolars, incisors, and canines. The rounded or pointed eminence on the masticating surface is called a **cusp**. Each type of tooth is specialized for some aspect of food getting or mastication. Incisors with simple crowns and slightly sharp edges are mainly for snipping or biting; canines with long conical crowns are specialized for piercing; premolars with compressed crowns and one or two cusps are suited for shearing and slicing; and the molars with large bodies and variable cusp arrangement are for crushing and mastication. Molars always belong to the permanent set.

Each tooth is made up of **dentin** from the corium and the covering **enamel** from epithelium. **Cementum** from the corium lies around the **root** of the tooth, within which is the **pulp cavity** of nerves and blood vessels (see Fig. 28-2). Certain insectivores cut only milk teeth. This condition is called **monophyodont**; in contrast is the **diphyodont** dentition of both deciduous and permanent teeth. Teeth that are alike are called **homodont**, which is characteristic of lower vertebrates; when they are differentiated to serve a variety of purposes, they are called **heterodont**. Teeth were already differentiated into these four types in the higher mammal-like reptiles. Carnivorous animals have teeth with sharp edges for tearing and piercing. They have well-developed canines, but some of the molars are poorly developed. In the herbivores the canines are suppressed, whereas the molars are broad, with enamel ridges for grinding. Such teeth are also usually high crowned, in contrast to the low crowns of carnivores.

The incisors of rodents have enamel only on the anterior surface so that the softer dentin behind wears away faster, resulting in chisel-shaped teeth that are always sharp. Moreover, rodent incisors grow throughout life and must be worn away to keep pace with the growth. If two opposing incisors fail to meet, the growth of the incisors continues and re-

FIG. 26-13
Malocclusion in the woodchuck *(Marmota monax).* Teeth of rodents are two pairs of
deeply implanted incisors that grow throughout life to keep pace with wear. Should
incisors not meet correctly, continued growth prevents feeding and animal starves to
death, as did this unfortunate one. Natural selection often works harshly to remove
genetic defects! Woodchucks (also called "groundhogs") are hibernators. Order Rodentia,
family Sciuridae. (Photo by L. L. Rue, III.)

sults in serious consequences to the animal (Fig. 26-13).

The **tusks** of the elephant and the wild boar are modifications of teeth. The elephant tusk is a modified upper incisor and may be present in both males and females; in the wild boar the tusk is a modified canine present only in the male. Both are formidable weapons.

The number and arrangement of permanent teeth are expressed by a **dental formula.** The figures above the horizontal line represent the number of incisors, canines, premolars, and molars on one half of the upper jaw; the figures below the line indicate the corresponding teeth in one half of the lower jaw.

Man	Dog
2-1-2-3	3-1-4-2
2-1-2-3	3-1-4-3

Food and feeding

On the basis of food habits, animals may be divided into herbivores, carnivores, omnivores, and insectivores.

Herbivorous animals that feed upon grasses and other vegetation form two main groups: **browsers** or **grazers,** such as the ungulates (horses, swine, deer, antelope, cattle, sheep, and goats), and the **gnawers** and **nibblers,** such as the rodents and rabbits. Herbivorous mammals have a number of interesting adaptations for dealing with their massive diet of plant food. Cellulose, the structural carbohydrate of plants, is a potentially nutritious foodstuff, being comprised of long chains of glucose. However, the glucose molecules in cellulose are linked by a type of chemical bond that few enzymes can attack. No vertebrates synthesize cellulose-splitting enzymes. Instead the herbivorous vertebrates harbor a microflora of anaerobic bacteria in huge fermentation chambers in the gut. These bacteria break down the cellulose, releasing a variety of fatty acids, sugars, and starches that the host animal can absorb and utilize. In some herbivores, such as horse and rabbit, the gut has a capacious sidepocket, or diverticulum, called a **cecum,** which serves as a fermentation chamber and absorptive area. Hares and rabbits often eat their fecal pellets, giving the food a second pass through the fermenting action of the intestinal bacteria. The ruminants (cattle, sheep, antelope, deer, giraffe, and other hooved mammals) have a huge four-chambered stomach (Fig. 26-14). When a ruminant feeds, grass passes down the esophagus to the rumen, where it is broken down by the rich microflora and then formed into small balls of cud. At its leisure the ruminant returns the cud to its mouth where it is deliberately chewed at length to crush the fiber. Swallowed again, the food returns to the rumen where it is digested by the cellulolytic bacteria. Finally, the pulp passes to the reticulum, then to the omasum, and finally to the abomasum ("true" stomach) where proteolytic enzymes are secreted and normal digestion takes place. Herbivores in general have large and long digestive tracts and must eat a large amount of plant food to survive. A large African elephant weighing 6 tons must consume between 300 and 400 pounds of rough fodder each day to obtain sufficient nourishment for life.

Carnivorous mammals feed mainly on herbivores. This group includes foxes, weasels, cats, dogs, wolverines, fishers, lions, and tigers. Carnivores are well equipped with biting and piercing teeth and powerful clawed limbs for killing their prey. Since their protein diet is much more easily digested than is the woody food of herbivores, their digestive tract is shorter and the cecum small or absent. Carnivores eat separate meals and have much more leisure time for play and exploration. In general, carnivores lead more active—and by man's standards more interesting—lives than do the herbivores. Since a carnivore

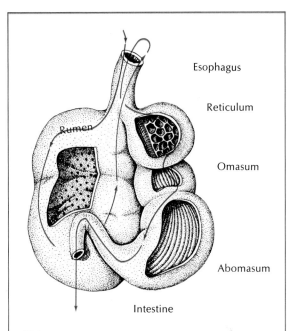

FIG. 26-14
Ruminant's stomach. Food passes first to rumen (sometimes via reticulum) and then is returned to mouth for chewing (chewing the cud, or rumination) *(black arrow).* After reswallowing, food passes to reticulum, omasum, and abomasum for final digestion *(red arrow).* See text for further explanation.

must find and catch its prey, there is a premium on intelligence; many carnivores, the cats for example, are noted for their stealth and cunning in hunting prey (Fig. 26-15). Although evolution seems to have favored the carnivores, their very success has lead to a selection of herbivores capable of either defending themselves or of detecting and escaping carnivores. Thus for the herbivores, there has been a premium on keen senses and agility (Fig. 26-16). Some herbivores, however, survive by virtue of their sheer size (for example, elephants) or by defensive group behavior (for example, muskox).

But man has changed the rules in the carnivore-herbivore contest. Carnivores, despite their superior intelligence, have suffered much from man's presence and have been virtually exterminated in some areas. Herbivores on the other hand, especially the rodents with their potent reproductive potential, have consistently defeated man's most ingenious ef-

FIG. 26-15
Lioness *(Panthera leo)* with Thompson gazelle she killed. Although lions live in groups called "prides," they typically hunt alone at dawn and dusk or at night. Lions stalk prey and then charge suddenly to surprise victim. They lack stamina for long chase. Prey is killed by biting throat to suffocate, or in case of small antelope such as this one, by breaking the animal's back with force of charge. Lions gorge themselves with kill and then sleep and rest for periods as long as 1 week before eating again. Order Carnivora, family Felidae. (Photo by L. L. Rue, III.)

forts to banish them from his environment. Indeed the problem of rodent pests in agriculture has been intensified; man has removed carnivores, which served as the herbivores' natural population control, but has not been able to devise a suitable substitute.

Omnivorous mammals live on both plant food and animals. Examples are pigs, raccoons, rats, bears, man, and most other primates (Fig. 26-17). Many carnivorous forms also eat fruits, berries, and grasses when hard pressed. The fox, which usually feeds upon mice, small rodents, and birds, will eat frozen apples, beechnuts, and corn when its normal sources are scarce.

Insectivorous mammals are those that subsist chiefly on insects and grubs. Examples are bats, moles, and shrews. The insectivorous category is not a well-distinguished one however, because many omnivores, carnivores, and even some herbivores will eat insects on occasion, for example, bears, raccoons, mice, baboons (Fig. 26-17), and ground squirrels.

The smaller the mammal, the more it must eat relative to its body size. This happens because the metabolic rate of an animal—and therefore the amount of food it must eat to sustain this metabolic rate—varies in rough proportion to the surface area, rather than to the body weight. Surface area is proportional to a 0.7 power of body weight. Putting it another way, the amount of food a mammal (or bird) eats is proportional to a 0.7 power of its body weight. This means that as the size of animals gets smaller, their metabolism becomes more intense. A 3-gram mouse will consume **per gram** 5 times more food than does a 10 kg. dog and about 30 times more food than does a 50,000 kg. elephant. One can easily see why small mammals (shrews, bats, mice) must spend much more time hunting and eating food than do large mammals. The smallest shrews weighing only 2 grams may eat more than their body weight each day and will starve to death in a few hours if deprived of food (Fig. 26-18). In contrast, a large carnivore can remain fat and healthy with only one

FIG. 26-16
Impala *(Aepyceros melampus)* of the African savanna. **A,** Adult male impala. **B,** Harem of impala. Dominant male with long recurved horns can be seen near center of herd. The others are females and young males with small horns. The many species of African antelopes have different food preferences, a fact that greatly reduces competition between species. They coexist in a harmonious community with high population density, taking advantage of the great variety of habitats the savanna offers, which includes grassland, parkland, dense bush, riverside forest, flooded plain, and swamp. Impalas live in wooded savanna. Order Artiodactyla. (Photos by B. Tallmark.)

FIG. 26-17
Olive baboons *(Papio anubis)* feeding on large ants. Baboons are omnivorous; they eat
seeds, fruits, bulbs, grasses, small mammals and birds, eggs, and insects. Here baboons
have torn open ant nest and pick up disturbed ants as they emerge. Baboons live in
troops of variable size, which may include several adult males. They search for food
during the day and sleep in trees at night. Order Primates, suborder Anthropoidea,
superfamily Cercopithecoidea. (Photo by B. Tallmark.)

FIG. 26-18
The masked (common) shrew *(Sorex cincereus)*
feeding on deer mouse it has killed. Note small size of
shrew relative to much larger mouse. This tiny but
fierce mammal, with a prodigious appetite for insects,
mice, snails, and worms, spends most of its time
underground and so is seldom seen by man. Shrews
are a primitive group believed to closely resemble the
insectivorous ancestors of placental mammals. Order
Insectivora. (Photo by C. G. Hampson.)

meal every few days. The mountain lion is known
to kill an average of one deer a week, although it
will kill more frequently when game is abundant.

For most mammals, searching for food and eating
is the main business of living. Seasonal changes in
food supplies are considerable in temperate zones.
Living may be easy in the summer when food is
abundant, but in winter many carnivores must range
far and wide to eke out a narrow existence (Fig. 26-
19). Some migrate to regions where food is more
abundant (see below). Others hibernate and sleep
the winter months away. But there are many provi-
dent mammals that build up food stores during
periods of plenty. This habit is most pronounced in
many of our rodents, such as squirrels, chipmunks,
gophers, and certain mice. All the tree squirrels—
red, fox, and gray—collect nuts, conifer seeds, and
fungi and bury these in caches for winter use. Often
each item is hidden in a different place (scatter
hoarding) and scent marked to assist relocation in
the future. The chipmunk is one of the greatest pro-
viders, for it spends the autumn months in collecting
nuts and seeds. Some of its caches may exceed a
bushel. Pikas lay in large hoards of grass and thistles
to carry them over the winter (Fig. 26-20).

FIG. 26-19
Coyote *(Canis latrans)* leaping to break snow crust, while hunting for mice, voles, and shrews. These small mammals, which comprise the principal food of coyotes during the difficult winter months, live on the frozen ground surface, protected by the snow cover above. The resilient coyote is still rather common in North America, despite relentless harassment by man. Order Carnivora. (Photo by C. G. Hampson.)

FIG. 26-20
American pika *(Ochotona princeps)* making hay. This member of the rabbit order lives in colonies in the Rocky Mountains where it excavates extensive tunnel systems among boulders and rocky ledges. They do not hibernate but prepare for the winter by storing grasses, twigs, thistles, and gooseberries in a haystack in special storerooms after carefully drying them first under the autumn sun. Order Lagomorpha. (Photo by L. L. Rue, III.)

Body temperature regulation

Temperature profoundly affects chemical (and biochemical) processes: rates of most biochemical reactions approximately double with each 10-centigrade-degree increase in temperature. Since biochemical reactions form the basis of the functional responses of animals, it follows that an animal's body temperature is the major factor that determines its level of activity. The "cold-blooded," or poikilothermic, animals have variable body temperatures. Although under some circumstances they may be able to maintain some uniformity of body temperature by behavioral responses, they are unable to regulate their body temperature physiologically.

Mammals share with birds the ability to maintain their body temperature well above that of their surroundings. They are homoiothermic, or "warm-blooded" (the terms "warm-blooded" and "cold-blooded" are hopelessly subjective and nonspecific but are so firmly entrenched in our vocabulary that most biologists find it easier to accept the usage than to try to change people). Homoiothermy has allowed mammals to stabilize their internal temperature so that biochemical processes and nervous function can proceed at steady high levels of activity. Thus they can remain active in winter and exploit habitats unavailable to the poikilotherms. Most mammals have body temperature between 36° and 38° C. (somewhat lower than that of birds, ranging between 40° and 42° C.). This constant temperature is maintained by a delicate balance between heat production and heat loss—not a simple matter when mammals are constantly alternating between periods of rest and bursts of activity. Heat is produced by the animal's metabolism, which includes the oxidation of foodstuffs, basal cellular metabolism, and muscular contraction. Heat is lost by radiation and conduction to a cooler environment and by the evaporation of water. The mammal can control both processes of heat production and heat loss within rather wide limits. If it becomes too cool, it can increase heat production by increasing muscular activity (exercise or shivering) and by decreasing heat loss by increasing its insulation. If the mammal becomes too warm, it decreases heat production and increases heat loss. We will examine these processes in the examples that follow.

ADAPTATIONS FOR HOT ENVIRONMENTS. Despite the harsh conditions of deserts—intense heat during the day; cold at night; scarcity of water, vegetation, and cover—many kinds of animals live there successfully. The smaller desert mammals are mostly fossorial (fitted for digging burrows) and nocturnal. The lower temperature and higher humidity of burrows helps to reduce water loss by evaporation. Water loss is replaced by free water in their food, or by drinking water if it is available. Water is also formed in the cells by the metabolic oxidation of foods. This gain from **oxidation water,** as it is called, can be very significant, since water is not always available for drinking. In fact some desert mammals, such as the kangaroo rats and ground squirrels of American deserts, sand rats of the Sahara Desert, and gerbils of Old World deserts can, if necessary, derive all the water they need from their dry food, drinking no water at all. Such animals can produce a highly concentrated urine and form nearly solid feces.

The large desert ungulates obviously cannot escape the desert heat by living in burrows. Animals such as camels and the desert antelopes (gazelle, oryx, and eland) possess a number of adaptations for coping with heat and dehydration. Those of the eland are shown in Fig. 26-21. The mechanisms for controlling water loss and preventing overheating are closely linked together. The eland, like other desert antelopes has a glossy, pallid color that reflects direct sunlight. The fur is an excellent insulation that works to keep heat out of desert animals, just as it serves to keep heat in arctic dwellers. The fur is not uniformly distributed over the body, however. Beneath the animal and on the axillae, groin, and scrotum or mammary glands, the pelage is very thin. These are provided with a rich capillary network and serve as thermal "windows" from which heat can be lost from the blood by convection and conduction. Heat is also lost by convection from the horns, which are well vascularized. The large ears of many mammals living in warm areas serve a similar purpose as heat radiators (such as those of the jackrabbit of the American Southwest, and African elephant (Fig. 26-22). Fat tissue of the eland, an essential food reserve, is concentrated in a single hump on the back, instead of being uniformly distributed under the skin where it would impair heat loss by radiation. The eland avoids evaporative water loss—the only device an animal has for cooling itself when the environmental temperature is higher than that of the body—by permitting its body temperature to

A

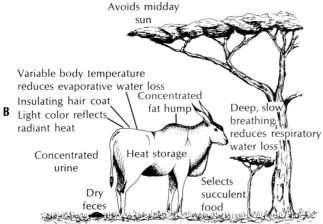

B

Avoids midday
sun

Variable body temperature
reduces evaporative water loss

Insulating hair coat

Light color reflects
radiant heat

Concentrated
fat hump

Deep, slow
breathing
reduces respiratory
water loss

Concentrated
urine

Heat storage

Dry
feces

Selects
succulent
food

FIG. 26-21
The common eland *(Taurotragus oryx),* inhabitant of the arid, open savanna of central
Africa. It is one of 72 species of African antelopes that occupy a variety of habitats that
include open savanna, bush savanna, marshes, and flooded grassland. Special food
preferences reduce competition between different species. The drawing shows physiologic
and behavioral adaptations of the eland for maintaining a constant body temperature in a
hot environment. See text for explanation. Order Artiodactyla. (Photo by C. G. Hampson,
University of Alberta. Drawing from Tallmark, B. 1972. Fauna och Flora [Stockholm],
67:163-175; after Taylor, C. P. 1969. Sci. Amer. **220:**88-95.)

FIG. 26-22
African elephant bull *(Loxodonta africana)* showing large ears, which serve as thermal
windows. The ears of a large elephant measure 6 feet long and 5 feet across and are
fanned back and forth to accelerate heat loss. The blood may cool as much as 5
centigrade degrees as it circulates through. Elephants also bathe themselves and wallow in
mud when overheated. Note cattle egrets *(Bubulcus ibis)* on elephant, which feed on
insects routed from brush by elephant's movements. Order Proboscidea. (Photo by B.
Tallmark.)

vary. At night the body cools down to less than 34°
C. by radiating heat to the cool surroundings. During the day the body temperature slowly rises to
more than 41° C. as the body stores heat. Only then
must the eland prevent further rise through evaporative cooling by sweating and panting. The large body
of the eland serves as a "heat sink," thus conserving
water and also reducing heat input, as the rising
body temperature approaches that of the hot environment. The eland also conserves water by concentrating its urine and forming dry feces. All these adaptations are also found developed to a similar or
even greater degree in camels, the most perfectly
adapted of all large desert mammals.

ADAPTATIONS FOR COLD ENVIRONMENTS.
In cold environments, mammals use two major
mechanisms to maintain homoiothermy, which are
(1) **decreased conductance,** that is, reduction of heat
loss by increasing the effectiveness of the insulation
and (2) **increased heat production.** The excellent insulation of the thick pelage of arctic animals is familiar to everyone. All mammals living in cold regions
of the earth increase the thickness of their fur in
winter, some by as much as 50%. As described earlier, the thick underhair is the major insulating layer,
whereas the longer and more visible guard hair
serves as protection against wear and for protective
coloration. But the body extremities (legs, tail, ears,
nose) of arctic mammals cannot be insulated as well
as can the thorax. To prevent these parts from becoming major avenues of heat loss, they are allowed
to cool to low temperatures, often approaching the
freezing point. This **regional heterothermy,** as it is
called, is achieved by a special geometric arrangement of the blood vessels supplying the extremities.
The artery and vein serving a leg, for example, are
in contact with each other so that **countercurrent
heat exchange** can occur between the warm arterial

blood passing into the leg, and the cold venous blood returning from the leg. In this way, heat is shunted directly from artery to vein and carried back to the core of the body, without ever reaching the peripheral areas. Without such a device, the blood would lose its heat through the poorly-insulated distal regions of the leg; cold blood would then return directly to the core where it would have to be reheated at a heavy energy cost. A consequence of this peripheral heat exchange system is that the legs and feet must operate at low temperatures. The feet of the arctic fox and barren-ground caribou are held just above the freezing point; in fact, the temperature may be below 0° C. in the footpads and hooves. To keep feet supple and flexible at such low temperatures, the fats have very low melting points, perhaps 30 centigrade degrees lower than the ordinary body fats. Furthermore, the nerves serving the legs continue to conduct impulses at temperatures far below those that cause ordinary nerves to cold-block.

Once a homoiotherm has lowered its heat loss to a minimum with pelage insulation and effective regional heterothermy, it has left only one response to prevent excessive cooling: it must produce more heat. This is done mainly by augmented **muscular activity,** through exercise or shivering. We are all familiar with the effectiveness of both activities. A man can increase his heat production as much as eighteenfold within 12 minutes by violent shivering if suddenly immersed in water at 4° C. Many other mammals can doubtless do as well when under great cold stress. Another source of heat is the increased oxidation of foodstuffs, especially brown fat stores. This mechanism is called **nonshivering thermogenesis.**

Small mammals the size of lemmings, voles, and mice meet the challenge of cold environments in a different way. Their very smallness is a disadvantage because the ratio of surface area to body volume is greatest in the smallest mammals. In effect, they have a relatively much greater surface area exposed for heat loss than do large mammals. Moreover, small mammals cannot insulate themselves as well as can large mammals because there is an obvious practical limit to how much pelage a mouse, for example, can carry before it becomes an immobile bundle of fur. Consequently, these forms have successfully exploited the excellent insulating qualities of snow by living under it in runways on the forest floor, where, incidentally, their food is also located. In this **subnivean environment** the temperature seldom drops below −5° C. even though the air above may fall to −50° C. The snow insulation decreases thermal conductance from small mammals in the same way that pelage does for large mammals. Living beneath the snow is really a kind of avoidance response to cold. Discussed below are two additional ways that mammals may survive low temperatures: migration and hibernation.

Migration

Migration is a much more difficult undertaking for mammals than for birds, and not surprisingly, few mammals make regular seasonal migrations, preferring instead to center their activities in a defined and limited territory. Nevertheless, there are some striking examples of mammalian migrations. More migrators are found in North America than on any other continent. The mammal to attract the most attention is the barren-ground caribou of Canada, which undergoes direct and purposeful mass migrations spanning 100 to 700 miles twice annually (Figs. 26-23 and 26-24). From winter ranges in the boreal forests (taiga) they migrate rapidly in late winter and spring to calving ranges on the barren grounds (tundra). The calves are born in mid-June. Harassed by warble and nostril flies that bore into their flesh, they move southward in July and August feeding little along the way. In September they reach the forest, feeding there almost continuously on low ground vegetation. Mating (rut) occurs in October. The caribou have suffered a drastic decline in numbers. Since primitive times when there were several million, they have dropped to less than 200,000 in 1958. The decline has been caused by excessive hunting by man, low calf crops, and destruction of the vulnerable forested wintering areas by fires accidentally started by man during recent exploration and exploitation activities in the North. Since 1958 the population has increased slowly (J. P. Kelsall).

The plains bison, before its deliberate near-extinction by man, made huge circular migrations to separate summer and winter ranges.

The largest mammal migrations of all are made by the oceanic seals and whales. One of the most remarkable migrations is that of the fur seal that breeds on the Pribilof Islands about 200 miles off the coast of Alaska and north of the Aleutian Islands.

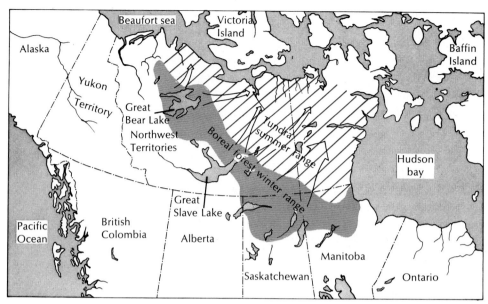

FIG. 26-23
Summer and winter ranges of the barren-ground caribou of Canada. The principal spring
migration routes are indicated by arrows; routes vary considerably from year to year.
(Adapted from Kelsall, J. P. 1968. The migratory barren-ground caribou of Canada.
Ottawa, Canadian Wildlife Service, Queens Printers.)

From wintering grounds off southern California the
females journey 3,000 miles across open ocean, arriving at the Pribilofs in the spring where they congre-
gate in enormous numbers. The young are born
within a few hours or days after arrival of the cows.
Then the bulls, having already arrived and established territories, collect harems of cows, which they
guard with vigilance. After the calves have been
nursed for about 3 months, cows and juveniles leave
for their long migration southward. The bulls do not
follow but remain in the Gulf of Alaska during the
winter.

Although we might anticipate that the only
winged mammals, bats, would use their gift to migrate, few of them do. Most spend the winter in hibernation. The three species of American bats that
do migrate, the red bat, the silvery-haired bat, and
the hoary bat, spend their summers in the northern
states and their winters in the south. Their migrations may take them far out over the Atlantic Ocean.

Hibernation

Many small and medium-sized mammals in north-
temperate regions solve the problem of winter scar-

city of food and low temperature by entering a prolonged and controlled state of dormancy. True hibernators, such as ground squirrels, woodchucks, mar-
mots, and jumping mice, prepare for hibernation by
building up large amounts of body fat. Some, such
as the marmot, also lay in stores of food in their den.
Entry into hibernation is gradual. After a series of
"test drops" during which body temperature drops
a few degrees and then returns to normal, the animal
cools to within a degree or less of the ambient temperature. Metabolism decreases to a fraction of normal. In the ground squirrel (Fig. 26-25), for example,
the respiratory rate drops from a normal rate of 200
per minute to 4 or 5 per minute, and the heart rate
from 150 to 5. In most, body temperature is moni-
tored so that if it drops dangerously close to the
freezing point, the animal will awaken. Even under
stable temperature conditions, the hibernator awak-
ens at irregular intervals to eliminate wastes and
then goes back to sleep. During arousal, the hibernator shivers violently and employs nonshivering thermogenesis to produce heat.

Some mammals (such as bears, badgers, raccoons,
and opossums) enter a state of prolonged sleep in

FIG. 26-24

Migrating barren-ground caribou *(Ranger tarandus groenlandicus)*. **A,** Caribou moving southeast in autumn toward forest winter range. Two males in foreground in autumn pelage and velvet-covered antlers. Adult females in background. Cows and younger animals normally lead the herd; males follow. Caribou feed while moving, grazing on low ground vegetation, principally though not exclusively lichens. **B,** Herd of caribou crossing lake in winter. Trails and beds in snow are evident in foreground. (**A,** Courtesy D. Thomas, Canadian Wildlife Service, Ottawa; **B,** from Kelsall, J. P. 1968. The migratory barren-ground caribou of Canada. Ottawa, Canadian Wildlife Service, Queens Printers.)

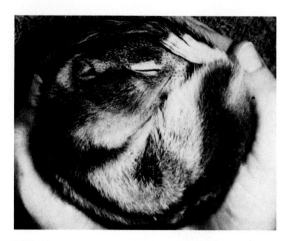

FIG. 26-25
Hibernating golden-mantled ground squirrel *(Citellus lateralis)* of the Rocky Mountains. Like other hibernators, this species rolls into tight ball with clenched front paws under chin, eyes and mouth closed tight, ears folded back, and tail wrapped over head. Body feels decidedly cold to touch and is so rigid it can be rolled or tossed like a ball. Experiments with ground squirrels have revealed that onset and end of hibernation is determined by an internal rhythm; even under conditions of constant temperature and light and with unlimited food available, they hibernate at the normal season. Order Rodentia. (Photo by C. G. Hampson.)

winter with little or no drop in body temperature. This is not true hibernation. Bears of the northern forest den-up for several months. Heart rate may drop from 40 to 10 beats per minute, but body temperature remains normal and the bear is awakened if sufficiently disturbed. One intrepid but reckless biologist learned how lightly a bear sleeps when he crawled into one's den and attempted to measure its rectal temperature with a thermometer!

Flight and echolocation

Mammals have not exploited the skies to the same extent that they have the terrestrial and aquatic environments. However, many mammals scamper about in trees with amazing agility; some can glide from tree to tree, and one group, the bats, is capable of full flight. Gliding and flying evolved independently in several groups of mammals, including the marsupials, rodents, flying lemurs, and bats. And anyone who has watched a gibbon perform in a zoo

realizes he is seeing something akin to flight in this primate, too. Among the arboreal squirrels, all of which are nimble acrobats, by far the most efficient is the flying squirrel (Fig. 26-26). These forms actually glide rather than fly, using the gliding skin that extends out from the sides of the body.

Bats, the only group of flying mammals, are nocturnal insectivores and thus occupy a niche left vacant by birds (Fig. 26-27). Their outstanding success is attributed to two things: one is, of course, flight; the other is the capacity to navigate by echolocation. Together these adaptations enable bats to fly and avoid obstacles in absolute darkness, to locate and catch insects with precision, and to find their way deep into caves (another habitat largely ignored by both mammals and birds) to sleep away the daytime hours. Most research has been concentrated on members of the family Vespertilionidae, to which most of our common North American bats belong. When they are in flight, these bats emit short pulses 5 to 10 milliseconds in duration in a narrow directed beam from the mouth. Each pulse is frequency modulated, that is, it is highest at the beginning, up to 100,000 Hertz (cycles per second) and drops to perhaps 30,000 Hz. at the end. Sounds of this frequency are ultrasonic to the human ear, which has an upper limit of about 20,000 Hz. The pulses are produced at a rate of 30 to 40 a second, increasing to perhaps 50 a second as the bat nears an object. Furthermore, the pulses are spaced so that the echo of each is received before the next pulse is emitted, an adaptation that prevents jamming. Since the transmission-to-reception time decreases as the bat approaches an object, it can increase the pulse frequency to obtain more information about the object. The pulse length is also shortened as it nears the object. The external ears of bats are large, like hearing trumpets, and shaped variously in different species. Less is known about the bat's inner ear, but it obviously is capable of receiving the ultrasonic sounds emitted. Bat navigation is so refined that biologists believe the bat builds up a mental image of its surroundings from echo scanning, virtually as complete as the visual image from eyes of diurnal animals.

Bats have undergone some adaptive radiation, yet for reasons not fully understood, all are nocturnal, even the fruit-eating bats that use vision and olfaction to find their food instead of sonar. The tropics have many kinds of bats, including the famed vam-

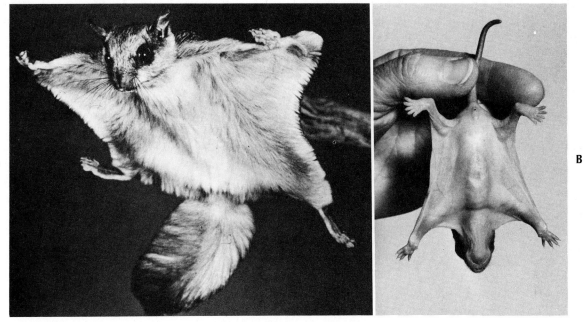

FIG. 26-26
Flying squirrels *(Glaucomys sabrinus).* **A,** In full flight. Area of undersurface is nearly trebled when gliding skin is spread. Glides of 40 to 50 yards are possible; good maneuverability is achieved by adjusting position of gliding skin during flight with special muscles. Flying squirrels are nocturnal and have superb night vision. They feed on nuts, seeds, and insects and hoard food for winter. They do not hibernate, but activity is much reduced in winter. **B,** A 3-week old youngster reflexly spreads gliding skin when picked up. Order Rodentia, family Sciuridae. (Photos by C. G. Hampson.)

FIG. 26-27
Large brown bat *(Eptesicus fuscus)* often found in old barns and church steeples. They frequently hang head down as this one is doing. Single young is usually born in June. Order Chiroptera.

pire bat. This species is provided with razor-sharp incisors used to shave away the epidermis to expose underlying capillaries. After infusing an anticoagulant to keep the blood flowing, it laps up its meal and stores it in a specially modified stomach. It is said that dogs can hear an approaching vampire's sonar and thus awaken and escape.

Territory and home range

Virtually all mammals, with aquatic mammals as perhaps the only exception, have territories: areas from which individuals of the **same** species are excluded. In fact, most wild mammals, like many people, are basically unfriendly to their own kind, especially so to their own sex during the breeding season. If the mammal dwells in a burrow or den, this forms the center of its territory. If it has no fixed address, the territory is marked out, usually with the highly

developed scent glands described earlier in this chapter. Territories vary greatly in size of course, depending on the size of the animal and its feeding habits. The grizzly bear has a territory of several square miles that it guards zealously against all other grizzlies. Mammals usually use natural features of their surroundings in staking their claims. These are marked with secretions from the scent glands, or by urinating or defecating. When an intruder knowingly enters another's marked territory, it is immediately placed at a psychologic disadvantage. Should a challenge follow, the intruder almost invariably breaks off the encounter in a submissive display characteristic for the species. An interesting exception to the territorial nature of most mammals is the prairie dog, which lives in large, friendly communities called prairie-dog "towns" (Fig. 26-28). When a new litter has been reared, the adults relinquish the old home

FIG. 26-28

Family of prairie dogs *(Cynomys ludovicianus)*. These highly social prairie dwellers are plant eaters that comprise an important source of food to many animals. They live in elaborate tunnel systems so closely interwoven that they form "towns" of as many as 1,000 individuals. Towns are subdivided into wards, in turn divided into coteries, the basic family unit, containing one or two adult males, several females and their litters. Although prairie dogs display ownership of burrows with territorial calls, they are friendly with inhabitants of adjacent burrows. The name "prairie dogs" derives from the sharp, doglike bark it makes when danger threatens. Western cattle and sheep ranchers have nearly eradicated prairie dogs in some areas by mass poisoning programs with disastrous results. The tunnel systems, which served as a natural sponge to prevent flash floods, filled in and serious erosion followed. Order Rodentia. (Photo by L. L. Rue, III.)

to the young and move to the edge of the community to establish a new home. Such a practice is totally antithetic to the behavior of most mammals, which drive off the young when they are self-sufficient.

The **home range** of a mammal is a much larger foraging area surrounding a defended territory. Home ranges are not defended in the same way a territory is; home ranges may in fact overlap, producing a neutral zone used by the owners of several territories, for seeking food.

Population

POPULATION SIZE. A population of animals includes all the animals of a species that interbreed. By this definition we note that there may be several distinct populations of the same species in a biome (an ecologic entity of plants and animals in an area), but they do not interbreed because they are separated by topographic or climatic barriers, or for some other reason.

Many surveys have been made to determine the number and kind of mammals found within a given area. These surveys, for which a certain amount of error must be allowed, are made by trapping, observation, tracks, signs, and other devices. Surveys are of great practical importance in conservation programs and ecologic studies. Several different techniques are employed. Actual counts of all the individuals of a species' population may be made on a given area, such as Darling did with various herds of red deer. Also used are indirect methods of marking captured specimens, releasing them, and from the ratio of recaptures to the marked numbers, estimating the number in the whole population of the species in the area being studied. Another common method is the sampling method; individuals are counted on a specific part of a large area and then an estimate of the total on the large area is made from the number on the sample area.

POPULATION CYCLES. No animal population lives in isolation. All mammals live in a community of numerous populations of different species, all of which form the biomass, or the sum of all living matter in an area. Each species is affected by the activities of other species and by the changes, especially climatic, that occur. Thus populations are always changing in size. Populations of small mammals are lowest before the breeding season, and greatest just after the addition of new members. Beyond these ex-

pected changes in population size, animal populations may fluctuate from other causes. Irregular fluctuations are commonly produced by variations in food supply or by disease. These are **density-independent** causes, since they affect a population whether it is crowded or dispersed. However, the most spectacular fluctuations are **density dependent;** that is, they are correlated with population crowding. Cycles of abundance are common among many rodent species. The population peaks and mass migrations of the Scandinavian and arctic North American lemmings are well known. Lemmings breed all year round, although more in the summer than in winter. The gestation period is only 21 days; young born at the beginning of the summer are weaned in 14 days and are themselves capable of reproducing by the end of the summer. Lemmings experience a 4-year cycle in abundance. Their numbers gradually increase from a density of 1 animal in 10 acres to 40 to 70 per acre at the peak of the fourth year. At this density, having devastated the vegetation by tunneling and grazing, they begin long, mass migrations to find new undamaged habitats for food and space. They swim across streams and small lakes as they go, but cannot distinguish these from large lakes and rivers and the sea, in which they drown. Since lemmings are the main diet of many carnivorous mammals and birds, any change in lemming population density affects all their predators as well.

The varying hare (snowshoe rabbit) of North America shows 10-year cycles in abundance. The well-known fecundity of rabbits enables them to produce litters of 3 or 4 young up to five times per year. The density may increase to 4,000 hares competing for food in each square mile of northern forest. Predators (owls, minks, foxes, and especially lynxes) also increase (Fig. 26-29). Then the population crashes precipitously, for reasons which have long been a puzzle to scientists. Rabbits die in great numbers, not from lack of food, or from an epidemic disease (as was once believed), but evidently from some density-dependent psychogenic cause. As crowding increases, hares become more aggressive, show signs of fear and defense, and stop breeding. The entire population reveals symptoms of pituitary-adrenal gland exhaustion, an endocrine imbalance called "shock disease" that leads to death. But there is much about these dramatic crashes that is not understood. Whatever the causes, population crashes that

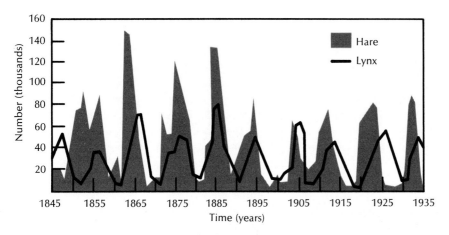

FIG. 26-29

Changes in population of varying hare and lynx in Canada as indicated by pelts received by the Hudson's Bay Company. The abundance of lynx (predator) follows that of the hare (prey). (After Odum, E. P. 1959. Fundamentals of ecology. Philadelphia, W. B. Saunders Co.)

follow superabundance are clearly advantageous to the species, since the survivors have a much better chance for successful breeding.

Reproduction

Fertilization is always internal, and all mammals are viviparous, except the monotremes, which lay eggs. Most mammals have definite mating seasons, usually in the winter or spring and timed to coincide with the most favorable time of the year for rearing the young after birth. Many males are capable of fertile copulation at any time, but the female mating function is restricted to a periodic cycle, known as the **estrous cycle.** The female will receive the male only during a relatively brief period known as **estrus,** or heat. The estrous cycle is divided into stages marked by characteristic changes in the ovary, uterus, and vagina. During **proestrus,** or period of preparation, new ovarian follicles grow and the uterus becomes distended with fluid. This is followed by **estrus,** when mating occurs. Almost simultaneously the ovarian follicles burst, releasing the eggs **(ovulation),** which are fertilized. Implantation of the fertilized egg and **pregnancy** follows. However, should mating and fertilization **not** occur, estrus is followed by **metestrus,** a period of repair. Each emptied ovarian follicle develops into a **corpus luteum,** or so-called yellow body. This stage is followed by **diestrus,** during which the corpora lutea regress and

the uterus becomes small and anemic. Then the cycle repeats itself, beginning with proestrus. Animals vary regarding the time of ova discharge from the ovary, but usually it occurs during the estrus period or shortly thereafter. How often females are in heat also varies greatly among the different mammals. Unless she has fruitfully mated, the female rat comes in estrus about every 4 days; the female dog, about every 6 months; the female house mouse, every 4 to 6 days; and the cow, every 21 days. Female rabbits are believed to breed at any time. Those animals that have only a single estrus during the breeding season are called **monestrus;** those that have a recurrence of estrus during the breeding season are called **polyestrus.** Dogs, foxes, and bats belong to the first group; field mice and squirrels are all polyestrus as are many mammals living in the more tropical regions of the earth (Fig. 26-30). The Old World monkeys and man have a somewhat different cycle, in which the postovulation period is terminated by **menstruation,** during which the lining of the uterus, or endometrium, collapses and is discharged with some blood. This is called a **menstrual cycle** and is described in Chapter 31.

Gestation, or period of pregnancy, varies greatly among the mammals. Mice and rats have about 21 days; rabbits and hares, 30 to 36 days; cats and dogs, 60 days; cows, 280 days; and elephants, 22 months. The marsupials (opossum) have a very short gesta-

FIG. 26-30

African lions *(Panthera leo)* mating. Lions breed at any season, although predominantly in spring and summer. During the short period a female is receptive, she may mate repeatedly. Three or four cubs are born after gestation of 100 days. Once the mother introduces the cubs into the pride, they are treated with affection by both adult males and females. Cubs go through an 18- to 24-month apprenticeship learning how to hunt and then are frequently driven from the pride to manage for themselves. Order Carnivora, family Felidae. (Photo by L. L. Rue, III.)

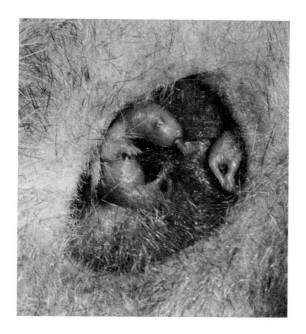

FIG. 26-31

Baby opossums *(Didelphis marsupialis)* about 2 weeks old in mother's pouch. After the mating in spring and a brief 13-day gestation, bee-sized embryos with enlarged forelegs and sharp hooked claws emerge from cloaca and squirm unaided to pouch and take nipples in their mouths. Pouch is provided with 13 nipples, much less than the 20 to 30 embryos born. The first 13 born are the lucky ones, the remainder are lost. Young remain in pouch 3 months and then are weaned by autumn and leave family group. In opossums, as in other marsupials, young remain in uterus for such a brief time that only a rudimentary placenta is formed. Order Marsupialia, family Didelphidae. (Photo by L. L. Rue, III.)

tion period of 13 days; at the end of that time the tiny young leave the vaginal orifice and make their way to the marsupial pouch where they attach themselves to nipples. Here they remain for more than 2 months before emerging (Fig. 26-31).

The number of young produced by mammals in a season depends on many factors. Usually the larger the animal, the smaller the number of young in a litter. Perhaps one of the greatest factors involved is the number of enemies a species has. Small rodents that serve as prey for so many carnivores produce, as a rule, more than one litter of several young each season. Field mice are known to produce as many as 17 litters of four to nine young in a year. Most carnivores have but one litter of three to five young a year. Large mammals, such as elephants and horses, have only one young.

The condition of the young at birth also varies. Those young born with hair and open eyes and ability to move around are called precocial (ungulates and jackrabbits); those that are naked, blind, and helpless (carnivores and rodents) are known as **altricial**. Mammals exhibit a great deal of parental care and will fiercely fight in defense of their young. When disturbed, they will often carry their young to more secure places.

■ MAN AND MAMMALS

Some 10,000 years ago, at the time man developed agricultural methods, he also began the domestication of mammals. Dogs were certainly among the first to be domesticated, probably entering voluntarily into their dependence on man. The dog is an extremely adaptable and genetically plastic species derived from wolves. Much less genetically variable and certainly less social than dogs is the domestic cat, probably derived from an African race of wildcat. Wildcats look like oversized domestic cats and are still widespread in Africa and Eurasia. The domestication of cattle, buffalos, sheep, and pigs probably came much later. It is believed that the beasts of burden—horses, camels, oxen, and llamas—probably were subdued by early nomadic peoples. It is of interest that certain domestic species no longer exist as wild animals, for example, the one-humped Arabian camel, and the llama and the alpaca of South America. All of the truly domestic animals breed in captivity and have become totally dependent on man; many have been molded by selective breeding

to yield characteristics that are desirable for man's purposes.

Some mammals hold special positions as "domestic" animals. The elephant has never been truly domesticated because it will not breed in captivity. In Asia, adults are captured and submit to a life of toil with astonishing docility. The reindeer of northern Scandinavia are domesticated only in the sense that they are "owned" by nomadic peoples who continue to follow them in their seasonal migrations. The eland of Africa (Fig. 26-21) is presently undergoing experimental domestication in several places. It is placid and gentle, immune to native diseases, and produces an excellent meat. Finally, we should not leave the subject of domestication without mentioning the albino rat, a domesticated brown rat. It has been suggested that the gentle nature of the albino, which has contributed so much to medical and psychologic research, is the result of a small defect in the amygdaloid nucleus of the brain. Fierce and intractable wild rats can be converted to docile, easily-handled, tame rats by destroying a small part of the amygdala in an operation.

In the introduction to this chapter we alluded to the senseless exploitation by man of the great marine whales, the largest animals that have ever lived. Despite 50 years of careful scientific study that has shown conclusively that several species of whales are now in real danger of total extinction, ambitious hunting continues by the two remaining whaling nations, the Soviet Union and Japan. In a recent conference (Stockholm, 1972), a recommended 10-year moratorium was rejected by the whaling nations and replaced with a quota system, the efficacy of which remains dubious. The whale tragedy is just one example of our inability to reconcile progress with the preservation of wildlife. The extermination of a species for commercial gain is so totally indefensible that no debate is required. Once a species is extinct, no amount of scientific or technical ingenuity will bring it back. What has taken millions of years to evolve, man can destroy in a decade of thoughtless exploitation. Many people are concerned with the awesome impact we have on wildlife, and there is more determination today to reverse a regretable trend than ever before. If given half a chance, mammals will usually make spectacular recoveries from man's depredations, as have the sea otter and the saiga antelope, both once in danger

FIG. 26-32
Brown rat *(Rattus norvegicus)*. Originally from the tropical forests of Asia, this species
and the less pugnacious tree-living black rat have spread all over the world. Living all too
successfully beside man in his habitations, the brown rat not only causes great damage to
man's food stores, but also spreads disease, including bubonic plague (carried through
infected fleas and a disease that greatly influenced human history in medieval Europe),
typhus, infectious jaundice, *Salmonella* food poisoning, and rabies. Order Rodentia, family
Murridae. (Photo by L. L. Rue, III.)

of extinction and now numerous. Paradoxically, in
Africa where conservationists wage a seesaw battle
with opposing interests, it appears that the commercial gain of tourism and game cropping will do more
to save the fauna than the outraged concern of naturalists.

Mammals can, of course, be enemies of man. Rodents and rabbits are capable of inflicting staggering
damage to growing crops and stored food (Fig. 26-
32). Man has provided an inviting forage for rodents
with his agriculture, and convenienced them further
by removing their natural predators. Rodents also
carry various diseases. Bubonic plague and typhus
are carried by house rats. Tularemia or rabbit fever,
is transmitted to man by the wood tick carried by
rabbits, woodchucks, muskrats and other rodents.
Rocky Mountain spotted fever is carried by ticks on
ground squirrels. Trichina worms and tapeworms are

acquired by man through hogs, cattle, and other
mammals.

Résumé of more important orders

Many authorities think that when all mammals are
classified there will be some 20,000 species and
subspecies. Classification is based upon major differences, such as the character of the teeth, modifications of the limbs or digits, the presence or absence
of claws and hoofs, and the complexity of the nervous
system. According to G. G. Simpson's classification
of mammals, there are 18 living and 14 extinct orders
of mammals. Of the living orders, many of their
families and genera are also extinct. Class Mammalia
is divided into two subclasses as follows: Subclass
Prototheria includes the monotremes, or egg-laying
mammals. Subclass **Theria** includes two infraclasses,
the **Metatheria** with one order, the marsupials, and

the **Eutheria** with the rest of the orders, all of which are placental mammals (Fig. 26-1).

SUBCLASS PROTOTHERIA (pro"to-thir'e-a) (Gr. *prōtos,* first, + *thēr,* wild animal). The egg-laying mammals.

Order Monotremata (mon"o-tre'mah-tah) (Gr. *monos,* single, + *trēma,* hole)—**egg-laying mammals, e.g., duck-billed platypus, spiny anteater.** This order is represented by the duckbills and spiny anteaters of Australia, Tasmania, and New Guinea. The most noted member of the order is the duck-billed platypus *(Ornithorhynchus anatinus).* The spiny anteater *(Tachyglossus),* about 17 inches long, is covered with coarse hair and spines. It has a long, narrow snout adapted for feeding on ants, its chief food. Monotremes represent the only order that is oviparous, and there is no known group of extinct mammals from which they can be derived. Their fossils date from the Pleistocene epoch.

SUBCLASS THERIA (thir'e-a) (Gr. *thēr,* wild animal)

Infraclass Metatheria (met"a-thir'e-a) (Gr. *meta,* after, + *thēr,* wild animal). The marsupial mammals.

Order Marsupialia (mar-su"pe-ay'le-a) (Gr. *marsypion,* little pouch)—**pouched mammals, e.g., opossums, kangaroos, koala.** These are primitive mammals characterized by an abdominal pouch, the **marsupium,** in which they rear their young. Although the young are nourished in the uterus for a short time, there is rarely a placenta present. This order is represented by the opossum, the kangaroo, the koala, the Tasmanian wolf, the wombat, and many others. Only the opossum is found in the Americas, but the order is the dominant group of mammals in Australia.

Infraclass Eutheria (yu-thir'e-a) (Gr. *eu,* true, + *thēr,* wild animal). The placental mammals.

Order Insectivora (in-sec-tiv'o-ra) (L. *insectum,* an insect, + *vorare,* to devour)—**insect-eating mammals, e.g., shrews, hedgehogs, moles.** The principal food of animals in this order is insects. The most primitive of placental mammals, they are widely distributed over the world except Australia. Placental mammals and marsupials are believed to have arisen independently from common ancestors during the Cretaceous period, but in time the placentals became dominant in most parts of the world because of their superior intelligence. Insectivora are small, sharp-snouted animals that spend a great part of their lives underground. The shrews are the smallest of the group; some of them are the smallest mammals known.

Order Chiroptera (ky-rop'ter-a) (Gr. *cheir,* hand, + *pteron,* wing)—**flying mammals.** The bats are in some respects the oddest of all mammals, for they are provided with wings. The wings are modified forelimbs in which the second to fifth digits are elongated to support a thin integumental membrane for flying. The first digit (thumb) is short with a claw. There are many families and species of bats the world over. The common North American forms are the little brown bat *(Myotis),* the free-tailed bat *(Tadarida),* which lives in the Carlsbad Caverns, and the large brown bat *(Eptesicus).* In the Old World tropics the "flying foxes" *(Pteropus)* are the largest of bats, with a wingspread of 4 to 5 feet, and live chiefly on fruits.

Order Dermoptera (der-mop'ter-a) (Gr. *derma,* skin, + *pteron,* wing)—**flying lemurs.** These are related to the true bats and consist of the single genus *Galeopithecus.* They are found in the Malay peninsula and the East Indies. They cannot fly in the strict sense of the word but glide like flying squirrels.

Order Carnivora (car-niv'o-ra) (L. *caro,* flesh, + *vorare,* to devour)—**flesh-eating mammals, e.g., dogs, wolves, cats, bears, weasels.** To this extensive order belong some of the most intelligent and strongest of animals. They all have predatory habits, and their teeth are especially adapted for tearing flesh. In most of them the canines are used for killing their prey. They are divided among two suborders: Fissipedia, whose feet contain toes, and Pinnipedia, with limbs modified for aquatic life. **Suborder Fissipedia** consists of the well-known carnivores—wolves, tigers, dogs, cats, foxes, weasels, skunks, and many others. They vary in size from certain tiny weasels to the mammoth Alaskan bear and Bengal tiger. They are distributed all over the world except in the Australian and antarctic regions, where there are no native forms. This suborder is divided into certain familiar families, among which are **Canidae** (the dog family), consisting of dogs, wolves, foxes, and coyotes; **Felidae** (the cat family), whose members include the domestic cats, tigers, lions, cougars, and lynxes; **Ursidae** (the bear family), made up of bears; and **Mustelidae** (the fur-bearing family), containing the martens, skunks, weasels, otters, badgers, minks, and wolverines. **Suborder Pinnipedia** includes the aquatic carnivores, sea lions, seals, sea elephants, and walruses. Their limbs have been modified as flippers for swimming. They are all saltwater forms and their food is mostly fish.

Order Tubulidentata (tu"byu-li-den-ta'ta) (L. *tubulus,* tube, + *dens,* tooth)—**aardvarks.** The aardvark is the Dutch name for earth pig, a peculiar animal with a piglike body found in Africa. The order is represented by only one genus *(Orycteropus)* with three or four species.

Order Rodentia (ro-den'che-a) (L. *rodere,* to gnaw)—**gnawing mammals, e.g., squirrels, rats, woodchucks.** The rodents are the most numerous of all mammals. Most of them are small. They are found on all continents and many of the large islands. They have no canine teeth, but their chisel-like incisors (never more than four) grow continually. Their basic adaptive feature, therefore, is gnawing. This adaptation has been largely responsible for the evolution of a very active and diversified group. Some rodents are useful for their fur. The beaver *(Castor),* the largest rodent in the United States, has a valuable pelt. Many rodents are utilized as food by carnivores and by man. The common families of this order are **Sciuridae** (squirrels and woodchucks), **Muridae** (rats and mice), **Castoridae** (beavers), **Erethizontidae** (porcupines), and **Geomyidae** (pocket gophers). The rodents are the most successful order of mammals because of their diverse adaptive radiation, which has enabled them to occupy so many ecologic niches. The most primitive living rodent is the tailless sewellel *(Aplodontia),* or mountain beaver, found in a restricted region of the Pacific coast.

Order Pholidota (fol"i-do'ta) (Gr. *pholis,* horny scale)—

pangolins. In this order there is one genus *(Manis)* with seven species. They are an odd group of animals whose body is covered with overlapping horny scales that have arisen from fused bundles of hair. Their home is in tropical Asia and Africa.

Order Lagomorpha (lag"o-mor'fa) (Gr. *lagōs,* hare, + *morphē,* form)—**rabbits, hares, pikas.** The chief difference between this order and Rodentia is the presence of four upper incisors, one pair of which is small, and the other large, with enamel on the posterior as well as anterior surface of the tooth.

Order Edentata (ee"den-ta'ta) (L. *edentatus,* toothless)—**toothless mammals, e.g., sloths, anteaters, armadillos.** These forms are either toothless or else have degenerate teeth without enamel. The group includes the anteaters, sloths, and armadillos. Most of them live in South America, although the nine-banded armadillo *(Dasypus novemcinctus)* extends into Texas and Florida. The sloths are very sluggish animals that have the queer habit of hanging upside down on branches. The hairs of sloths have tiny pits in which green algae grow and render them invisible against a background of mosses and lichens. The extinct ground sloths were represented by some members as large as small elephants.

Order Cetacea (see-tay'she-a) (L. *cetus,* whale)—**fishlike mammals, e.g., whales, dolphins, porpoises.** This order is well adapted for aquatic life. Their anterior limbs are modified into broad flippers; the posterior limbs are absent. Some have a fleshy dorsal fin and the tail is divided into transverse fleshy flukes. The nostrils are represented by a single or double blowhole on top of the head. Teeth may be absent, but when present they are all alike and lack enamel. They have no hair except a few on the muzzle, no skin glands except the mammary and those of the eye, and no external ear, and their eyes are small. The order is divided into two suborders: Odontoceti and Mysticeti. **Suborder Odontoceti** is made up of toothed members and is represented by the sperm whales, porpoises, and dolphins. The killer whale *(Orcinus),* the most savage member, does not hesitate to attack the larger whales. It is also destructive to seal rookeries. The sperm whale *(Physeter)* is the source of sperm oil, which is obtained from the head. A peculiar substance, ambergris, is sometimes formed in its stomach and is used in perfumes. This substance may be derived from squids, which form the principle food of sperm whales. Porpoises and dolphins, which also belong to this suborder, are only 6 or 7 feet long and feed mainly on gregarious fish. Another member of this group is the narwhal *(Monodon),* which has one of its two teeth modified into a twisted tusk 8 to 10 feet long and projecting forward like a pike. The other suborder, **Mysticeti,** or whalebone whales, includes many species, some of which are gigantic. Instead of teeth, they have a peculiar straining device of whalebone (baleen) attached to the palate. They live on the microscopic animals at the surface (plankton), which they strain out of the water with the whalebone. The largest of the whales is the blue whale *(Balaenoptera),* which has a dorsal fin and may grow 100 feet long and weigh 125 tons. When whales emerge to blow air through the blowhole, the warm air from the lungs is condensed by the cool sea air to form the familiar spout.

Order Proboscidea (pro"ba-sid'e-a) (Gr. *proboskis,* elephant's trunk, from *pro,* before, + *boskein,* to feed)—**proboscis mammals, e.g., elephants.** This order includes only the elephants, the largest of living land animals. They have large heads, massive ears, and thick skins (pachyderm). Hair is confined mainly to the tip of the tail. The two upper incisors are elongated as tusks, and the molar teeth are well developed. There are two genera of elephants: the Indian *(Elephas maximus),* with relatively small ears, and the African *(Loxodonta africana),* with large ears. There is also a small African form, the pigmy elephant *(Elephas cyclotis),* which is found in West Africa. The larger elephants may attain a height of 10 to 11 feet and a weight of 6 to 7 tons. The Asiatic or Indian elephant has long been domesticated and is trained to do heavy work. The taming of the African elephant is more difficult but was extensively done by the ancient Carthaginians and Romans, who employed them in their armies. Barnum's famous "Jumbo" was an African elephant.

Order Hyracoidea (hy"ra-coi'de-a) (Gr. *hyrax,* shrew)—**hyraxes, e.g., conies.** Conies are restricted to Africa and Syria. They have some resemblance to short-eared rabbits but have teeth like rhinoceroses, with hoofs on their toes and pads on their feet. They have four toes on the front and three toes on the back feet. They are herbivorous in their food habits and live among rocks or in trees.

Order Sirenia (sy-re'ne-a) (Gr. *seirēn,* sea nymph)—**sea cows, e.g., manatees.** Sea cows, or manatees, are large, clumsy aquatic animals. They have a blunt muzzle covered with coarse bristles, the only hairs these queer animals possess. They have no hind limbs, and their forelimbs are modified into swimming flippers. The tail is broad with flukes but is not divided. They live in the bays and rivers along the coasts of tropical and subtropical seas. There are only two genera living at present: *Trichechus,* found in the rivers of Florida, West Indies, Brazil, and Africa, and *Halicore,* the dugong of India and Australia.

Order Perissodactyla (pe-ris"so-dak'ti-la) (Gr. *perissos,* odd, + *dactylos,* toe)—**odd-toed hoofed mammals.** The odd-toed hoofed mammals have an odd number (one or three) of toes, each with a cornified hoof. Both the Perissodactyla and the Artiodactyla are often referred to as **ungulates,** or hoofed mammals, with teeth adapted for chewing. They include the horses, the zebras, the tapirs, and the rhinoceroses. The horse family (Equidae), which also includes asses and zebras, has only one functional toe. Tapirs have a short proboscis formed from the upper lip and nose. The rhinoceros *(Rhinoceros)* includes several species found in Africa and southeastern Asia. Their most striking character is the horn (one or two) on top of the snout. All are herbivorous.

Order Artiodactyla (ar"te-o-dak'ti-la) (Gr. *artios,* even, + *daktylos,* toe)—**even-toed hoofed mammals.** The even-toed ungulates include swine, camels, deer, hippopotamuses, antelopes, cattle, sheep, and goats. Most of them have two toes, although the hippopotamus and some others have four. Each toe is sheathed in a cornified hoof. Many such as the cow, deer, and sheep, have horns. Many of them are ruminants, that is, animals that chew the cud. Like Perissodactyla, they are strictly herbivorous. The group is divided into nine living families and many extinct ones and includes some of the most valuable domestic animals. This extensive order is commonly divided into three suborders: the **Suina** (pigs, peccaries, hippopotamuses), the **Tylopoda** (camels), and the **Ruminantia** (deer, giraffes, sheep, cattle, etc.). In the odd-toed ungulates the middle, or third, digit is stressed (second and fourth also in some), and the main axis of weight passes through this. In the even-toed ungulates the third and fourth toes (sometimes also the second and fifth) are stressed, and the main axis of the leg passes between the third and fourth toes so that they bear equally the weight of the animal.

Order Primates (pry-may'teez) (L. *prima,* first)—**highest mammals, e.g., lemurs, monkeys, apes, man.** This order stands first in the animal kingdom in brain development, although other structural features may be equaled or excelled by lower mammals. Most of the species are arboreal, apparently derived from tree-dwelling insectivores. The primates represent the end product of a line that branched off early from other mammals and have retained many primitive characteristics. It is believed that their tree-dwelling habits of agility in capturing food or avoiding enemies were largely responsible for their advances in brain structure. The brain of primates is so well developed that the cerebral hemispheres cover the rest of the brain, especially in the higher primates. As a group, they are generalized, with five digits (usually provided with flat nails) on both forelimbs and hind limbs. All have their bodies covered with hair except man. Forelimbs are often adapted for grasping, as are the hind limbs sometimes. The group is singularly lacking in claws, scales, horns, and hoofs. There are three suborders:

Suborder Lemuroidea (lem"yu-roi'de-a) (L. *lemures,* ghosts). These are primitive arboreal primates, with their second toe provided with a claw and a long nonprehensile tail. They look like a cross between squirrels and monkeys. They are found in the forests of Madagascar, Africa, and the Malay peninsula. Their food is mostly plants and small animals.

Suborder Tarsioidea (tar"se-oi'de-a) (Gr. *tarsos,* ankle). There is only one genus *(Tarsius)* in this group. Tarsiers are small, solitary primates that live in the Philippines and adjacent islands.

Suborder Anthropoidea (an"thro-poi'de-a) (Gr. *anthrōpos,* man). This suborder consists of monkeys, apes, and man. There are three superfamilies:

1. **Superfamily Ceboidea** (se-boi'de-a) (Gr. *kēbos,* long-tailed monkey)—**(Platyrhinii).** These are New World monkeys, characterized by the broad flat nasal septum and by the absence of ischial callosities and cheek pouches. Their thumbs are nonopposable and their tails are prehensile. Familiar members of this superfamily are the capuchin monkey *(Cebus)* of the organ grinder, the spider monkey *(Ateles),* and the howler monkey *(Alouatta).*

2. **Superfamily Cercopithecoidea** (sur"ko-pith"e-koi'de-a) (Gr. *kerkos,* tail, + *pithēkos,* monkey)—**(Catarrhinii).** These Old World monkeys have the external nares close together, and many have internal cheek pouches. They never have prehensile tails, there are calloused ischial tuberosities on their buttocks, and their thumbs are opposable. Examples are the savage mandrill *(Cynocephalus),* the rhesus monkey *(Macacus)* widely used in biologic investigation, and the proboscis monkey *(Nasalis).*

3. **Superfamily Hominoidea** (hom"i-noi'de-a) (L. *homo, hominis,* man). The higher (anthropoid) apes and man make up this superfamily. Their chief characteristics are lack of a tail and lack of cheek pouches. There are two families: Pongidae and Hominidae. The Pongidae family includes the higher apes, gibbon *(Hylobates),* orangutan *(Simia),* chimpanzee *(Pan),* and the gorilla *(Gorilla).* The other family, Hominidae, is represented by a single genus and species *(Homo sapiens),* modern man. Man differs from the members of family Pongidae in being more erect, in having shorter arms and larger thumbs, and in having lighter jaws with smaller front teeth. Most of the apes also have much more prominent supraorbital ridges over the eyes. Many of man's differences from the anthropoid apes are associated with his higher intelligence, his speech centers in the brain, and the fact that he is no longer an arboreal animal.

References
Suggested general readings

Andersen, H. T. (editor). 1969. The biology of marine mammals. New York, Academic Press Inc. *Despite its name, this authoritative book is restricted to physiology of marine mammals. Graduate level.*

Anthony, H. E. 1928. Field book of North American mammals. New York, G. P. Putnam's Sons. *An excellent manual on the classification, distribution, and characteristics of mammals.*

Blair, W. F., A. P. Blair, P. Brodkorr, F. R. Cagle, and G. A. Moore. 1968. Vertebrates of the United States, ed. 2. New York, McGraw-Hill Book Co. *This work is of great importance to all students of vertebrates, especially to those who are interested in taxonomy. The taxonomic keys are illustrated and identify all vertebrates down to species.*

Burns, E. 1953. The sex life of wild animals. New York, Rinehart & Co. *An interesting account of the mating behavior of many mammals.*

Burton, M. 1962. University dictionary of mammals of the world. New York, Thomas Y. Crowell Co. *A handy volume, giving essential information on nearly every family and species of mammals. Paperback.*

Davis, E. E., and F. B. Golley. 1964. Principles of mammalogy. New York, Reinhold Publishing Corporation. *An excellent introductory text of the mammals, including their classification, adaptations, evolution, distribution, populations, and behavior.*

Eimerl, S., and I. DeVore. 1965. The primates. Life Nature Library. New York, Time Inc. *Imaginatively written and beautifully illustrated semipopular treatment. Highly recommended.*

Grassé, P.-P. (editor). 1955. Mammifères. In Traité de zoologie, vol. 17, 2 parts. Paris, Masson & Cie. *A technical treatise in this extensive series.*

Hall, E. R., and K. R. Kelson. 1959. The mammals of North America, 2 vols. New York, The Ronald Press Co. *Full descriptions of species and subspecies with distribution maps. Taxonomic keys, records, and revealing line drawings of skull characteristics are included in this authoritative work. An extensive index and bibliography add much to the usefulness of this work.*

Lawick-Goodall, J. van, and H. van Lawick. 1971. Innocent killers. Boston, Houghton Mifflin. *An engrossing description of the biology of hyenas, jackals, and wild dogs by the author of the acclaimed field study of chimpanzees called* In the Shadow of Man *and her wildlife photographer husband.*

Matthews, L. H. 1969. The life of mammals. London, Weidenfeld & Nicolson. *A well-written general account of mammals: what they are and what they do.*

Mech, D. L. 1970. The wolf: the ecology and behavior of an endangered species. Garden City, N. Y., The Natural History Press. *A thorough, illustrated account; detailed, yet interestingly written.*

Reader's Digest Association. 1970. The living world of animals. London & Montreal, The Reader's Digest Association. *Surprising as it may be, this publishing firm has produced a truly fine account of animals, stressing their adaptations and ecology. Fine illustrations. Highly recommended.*

Sanderson, I. T. 1955. Living mammals of the world. New York, Garden City Books (Hanover House). *A beautiful and informative work of many photographs and concise text material. It is a delight to any zoologist regardless of his specialized interest.*

Scheffer, V. B. 1958. Seals, sea lions, and walruses. A review of the Pinnipedia. Stanford, Calif., Stanford University Press. *Treats the group from the standpoint of their evolution, characteristics, and classification. Many fine photographs and a good bibliography are included.*

Scheffer, V. B. 1969. The year of the whale. New York, Charles Scribner's Sons. *Beautifully written and engaging combination of fiction and fact about a baby sperm whale's first year of life.*

Schmidt-Nielsen, K. 1964. Desert animals. New York, Oxford University Press. *Deals with the problems desert-dwelling forms (including man) must meet and solve to survive.*

Seton, E. T. 1925-1928. Lives of game animals, 4 vols. New York, Doubleday, Doran & Co. *A classic.*

Simpson, G. G. 1945. The principles of classification and a classification of mammals. Bulletin of the American Museum of Natural History 85:1-350.

Schaller, G. B. 1963. The mountain gorilla. Chicago, University of Chicago Press. *A thorough and definitive study of the ecology and behavior of this fascinating primate.*

Walker, E. P. 1968. Mammals of the world. 3 vols. Baltimore, The Johns Hopkins Press. *The only single compendium of information on all known and living mammalian genera. A valuable reference work.*

Young, J. Z. 1957. The life of mammals. New York, Oxford University Press. *This is a well-known work about mammals, especially their anatomy, histology, physiology, and embryology. Classification is not treated.*

Selected *Scientific American* articles

Barnett, S. A. 1967. Rats. **216**:78-85 (Jan.). *Different species of rats are compared and their social behavior described.*

Bartholomew, G. A., and J. W. Hudson. 1961. Desert ground squirrels. **205**:107-116 (Nov.). *Two species living in California's Mojave Desert have developed interesting adaptations for survival in desert heat and aridity.*

Drinker, C. K. 1949. The physiology of whales. **181**:52-55 (July). *Adaptations for diving and comparisons of diving abilities among whales are described in this brief article.*

Flyger, V., and M. R. Townsend. 1968. The migration of polar bears. **218**:108-116 (Feb.). *The habitat, biology, and wide migrations of these large arctic carnivores are described.*

Griffin, D. R. 1958. More about bat "radar." **199**:40-44 (July).

Irving, L. 1966. Adaptations to cold. **214**:94-101 (Jan.). *The homeothermic birds and mammals have evolved several adaptations that permit them to survive in cold environments.*

King, J. A. 1959. The social behavior of prairie dogs. **201**:128-140 (Oct.).

Kooyman, G. L. 1969. The Weddell seal. **221**:100-106 (Aug.). *This Antarctic seal swims for miles under shelf ice on one breath of air, returning unerringly to its breathing hole.*

McVay, S. 1966. The last of the great whales. **215**:13-21 (Aug.). *Most of the 12 commercially hunted whale species have been nearly exterminated. The indifference and unrestricted fishing that has characterized the whaling industry is recounted in this article.*

Modell, W. 1969. Horns and antlers. **220**:114-122 (April). *Their differences, growth, and structure are described.*

Montagna, W. 1965. The skin. **212**:56-66 (Feb.). *The structure and diverse functions of mammalian skin are described.*

Mrosovsky, N. 1968. The adjustable brain of hibernators. **218**:110-118 (March). *Hibernation is preceded by a remarkable series of changes and a resetting of the hypothalamic thermostat.*

Mykytowycz, R. 1968. Territorial marking by rabbits. **218**:116-126 (May). *Describes the use of odor-producing glands by Australian colonial rabbits to mark territories.*

Pearson, O. P. 1954. Shrews. **191**:66-70 (Aug.). *This group of tiny, voracious, burrow-dwelling, and elusive forms are among the most fascinating of mammals.*

Pruitt, W. O., Jr. 1960. Animals in the snow. **202**:60-68 (Jan.). *Describes the adaptations of the many homeotherms that live where snow persists for more than half the year.*

Schmidt-Nielsen, K., and B. Schmidt-Nielsen. 1953. The desert rat. **189**:73-78 (July). *The kangaroo rat possesses several adaptations that enable it to live in hot, arid regions, eating only dry food and drinking no water at all.*

Schmidt-Nielsen, K. 1959. The physiology of the camel. **201:**140-151 (Dec.).

Scholander, P. F. 1963. The master switch of life. **209:**92-106 (Dec.). *Diving vertebrates are obviously specialized for making prolonged dives without breathing. One of the most important adaptations is the capacity to grossly redistribute the circulation.*

Taylor, C. R. 1969. The eland and the oryx. **220:**88-95 (Jan.). *These large African antelopes thrive in desert or near-desert regions without drinking water. Their adaptations for heat and aridity are described.*

Wimsatt, W. A. 1957. Bats. **197:**105-114 (Nov.). *Describes their variation and biology.*

PART THREE

ACTIVITY AND CONTINUITY
OF LIFE

■ Motility, nutrition, respiration, internal transport, excretion, irritability, and integration are necessary activities of animal life. These and the metabolic processes that provide energy for them are functional activities inherent in the simplest to the most complex animals. Animals have evolved various ways to perform these common tasks. Not all animals have specialized systems for carrying out specific functions; specialization of organ systems seems to be correlated with increase in body size and overall complexity. Higher forms exercise finer control over physiologic processes and exhibit greater internal constancy. But regardless of the degree of specialization, all animals possess integrative systems for coordinating internal activities and for relating the animal to its external environment.

Since all living things are mortal, species perpetuation is another important activity in the lives of animals. To comprehend the continuity of life, it is necessary to understand how hereditary units are transmitted from parent to offspring, how such units direct the structural and functional differentiation of a fertilized egg into an adult organism, and how the capacity for evolutionary diversity has been built into the hereditary mechanism.

CHAPTER 27
INTERNAL FLUIDS: CIRCULATION, RESPIRATION, AND EXCRETION

CHAPTER 28
DIGESTION AND NUTRITION

CHAPTER 29
SUPPORT, PROTECTION, AND MOVEMENT

CHAPTER 30
COORDINATION: NERVOUS SYSTEM, SENSE ORGANS, AND ENDOCRINE SYSTEM

CHAPTER 31
THE REPRODUCTIVE PROCESS

CHAPTER 32
PRINCIPLES OF DEVELOPMENT

CHAPTER 33
PRINCIPLES OF INHERITANCE

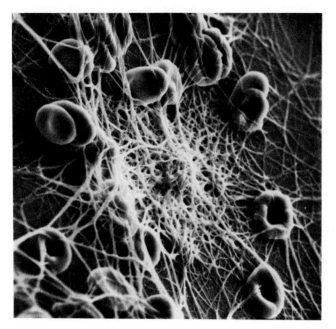

Human red blood cells entrapped in fibrin clot. Clotting is initiated after tissue damage by the disintegration of platelets in the blood, leading to a complex series of intravascular reactions that end with the conversion of a plasma protein, fibrinogen, into long, tough, insoluble polymers of fibrin. Fibrin and entangled erythrocytes form the blood clot, which arrests bleeding. An aggregation of platelets probably underlies the raised mass of fibrin in center. (Scanning electron micrograph, ×5180; courtesy N. F. Rodman, University of Iowa, Iowa City, Iowa.)

CHAPTER 27

INTERNAL FLUIDS: CIRCULATION, RESPIRATION, AND EXCRETION

Single-celled organisms live a contact existence with their environment. Nutrients and oxygen are obtained, and wastes are released, directly across the cell surface. These animals are so small that no special internal transport system, beyond the normal streaming movements of the cytoplasm, is required. Even some primitive multicellular forms, such as sponges, coelenterates and flatworms, have such a simple internal organization and low rate of metabolism that no circulatory system is needed. Most of the more advanced multicellular organisms, because of their size, activity, and complexity, require a specialized circulatory, or vascular, system to transport nutrients and respiratory gases to and from all tissues of the body. In addition to serving these primary transport needs, circulatory systems have acquired additional functions; hormones are moved about, finding their way to target organs where they assist the nervous system to integrate body function. Water, electrolytes, and the many other constituents of the body fluids are distributed and exchanged between different organs and tissues. An effective response to disease and injury is vastly accelerated by an efficient circulatory system. The warm-blooded birds and mammals depend heavily on the blood circulation to conserve or dissipate heat as required for the maintenance of constant body temperature.

■ INTERNAL FLUID ENVIRONMENT

The body fluid of a single-celled animal is the cellular cytoplasm, a fluid substance in which the various membrane systems and organelles of the cell are suspended. In multicellular animals the body fluids are divided into two main phases, the **intracellular** and the **extracellular.** The intracellular phase (also called intracellular fluid) is the fluid inside all

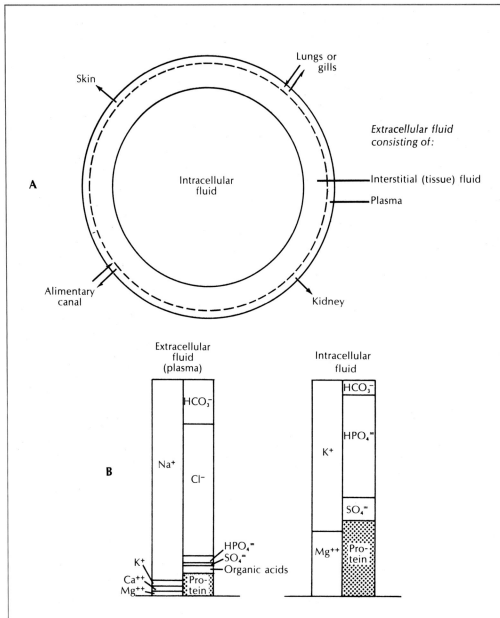

FIG. 27-1

Fluid compartments of body. **A,** All body cells can be represented as belonging to a single large fluid compartment that is completely surrounded and protected by extracellular fluid *(milieu intérieur).* This fluid is further subdivided into plasma and interstitial fluid. All exchanges with the environment occur across the plasma compartment. **B,** Electrolyte composition of extracellular and intracellular fluids. Total equivalent concentration of each major constituent is shown. Equal amounts of anions (negatively charged ions) and cations (positively charged ions) are in each fluid compartment. Note that sodium and chloride, major plasma electrolytes, are virtually absent from intracellular fluid (actually they are present in low concentration). Note the much higher concentration of protein inside the cells.

the body's cells. The extracellular phase (or fluid) is the fluid outside and surrounding the cells (Fig. 27-1, *A*). The significance of this extracellular fluid as a protective environment of the cells was recognized over a century ago by the great French physiologist, Claude Bernard. Bernard called this extracellular fluid the **milieu intérieur,** meaning the body's internal environment immediately surrounding the cells, as opposed to the external environment, or outside world. The environment outside the animal he called the **milieu extérieur.** Thus the cells, the sites of the body's crucial metabolic activities, are bathed by their own aqueous environment, the milieu intérieur, which buffers them from the often harsh physical and chemical changes occurring outside the body. Even today, English-speaking biologists frequently use the French phrase in referring to the extracellular fluid.

In animals having closed circulatory systems (vertebrates, annelids, and a few other invertebrate groups) the extracellular fluid is further subdivided into blood **plasma** and **interstitial** fluid (Fig. 27-1, *A*). The blood plasma is contained within the blood vessels, while the interstitial fluid, or tissue fluid as it is sometimes called, occupies the space immediately around the cells. Nutrients and gases passing between the vascular plasma and the cells must traverse this narrow fluid separation. The interstitial fluid is constantly formed from the plasma by filtration through the capillary walls.

COMPOSITION OF THE BODY FLUIDS. All these fluid spaces—plasma, interstitial, and intracellular—differ from each other in solute composition, but all have one feature in common—they are mostly water. Despite their firm appearance, animals are 70% to 90% water. Man for example is about 70% water by weight: of this, 50% is cell water, 15% is interstitial fluid water, and the remaining 5% is in the blood plasma. As Fig. 27-1, *A*, shows, it is the plasma space that serves as the pathway of exchange between the cells of the body and the outside world. This exchange of respiratory gases, nutrients, and wastes is accomplished by specialized organs (kidney, lungs, gill, alimentary canal), as well as by the integument.

The body fluids contain many inorganic and organic substances in solution. Principal among these are the inorganic electrolytes and proteins. Fig. 27-1, *B*, shows that **sodium, chloride,** and **bicarbonate** are the chief extracellular electrolytes, whereas **potassium, magnesium, phosphate, sulfate,** and **proteins** are the major intracellular electrolytes. These differences are dramatic; they are always maintained despite the continuous flow of materials into and out of the cells of the body. The two subdivisions of the extracellular fluid—plasma and interstitial fluid—have similar compositions except that the plasma has more proteins which are too large to filter through the capillary wall into the interstitial fluid.

COMPOSITION OF BLOOD. Among the lower invertebrates that lack a circulatory system (such as flatworms and coelenterates) it is not possible to distinguish a true "blood." These forms possess a clear, watery tissue fluid containing some primitive phagocytic cells, a little protein, and a mixture of salts similar to sea water. All invertebrates with closed circulatory systems maintain a clear separation between blood contained within blood vessels, and tissue (interstitial) fluid surrounding the vessels.

In vertebrates, blood is a complex liquid tissue composed of plasma and formed elements, mostly corpuscles, suspended in the plasma. When the red blood corpuscles and other formed elements are spun down in a centrifuge, the blood is found to be about 55% plasma and 45% formed elements.

The composition of mammalian blood is as follows:

Plasma
1. Water 90%
2. Dissolved solids, consisting of the plasma proteins (albumin, globulins, fibrinogen), glucose, amino acids, electrolytes, various enzymes, antibodies, hormones, metabolic wastes, and traces of many other organic and inorganic materials
3. Dissolved gases, especially oxygen, carbon dioxide, and nitrogen

Formed elements (Fig. 27-4)
1. Red blood corpuscles (erythrocytes), for the transport of oxygen and carbon dioxide
2. White blood corpuscles (leukocytes), serving as scavengers and as immunizing agents
3. Platelets (thrombocytes), functioning in blood coagulation

The plasma proteins are a diverse group of large and small proteins that perform numerous functions. They may be separated by classic electrophoretic techniques into six major groups (Fig. 27-2) although more modern chromatographic and immunoelectrophoretic methods show that there are probably

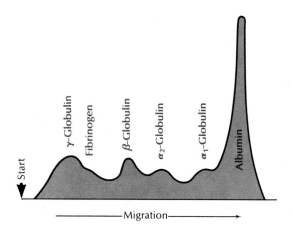

FIG. 27-2

Electrophoretic pattern of major plasma proteins. When blood plasma is placed in an electric field under the right pH conditions, proteins in solution will migrate at different rates according to their net charge. The profile is called a Tiselius pattern, after the Swedish biochemist who developed the technique.

hundreds of individual proteins in the plasma. The major protein groups are (1) **albumin,** the most abundant plasma protein, which constitutes 60% of the total; (2) the **globulins** (α_1, α_2, β, and γ), a diverse group of high-molecular weight proteins (35% of total) that includes immunoglobulins and various metal-binding proteins; and (3) **fibrinogen,** a very large protein that functions in blood coagulation.

Red blood cells, or **erythrocytes,** are present in enormous numbers in the blood, about 5.4 million per cubic millimeter in an adult man and 4.8 million in women. They are formed continuously from large nucleated **erythroblasts** in the red bone marrow. Here, hemoglobin is synthesized and the cells divide several times. In mammals the nucleus shrinks during development to a small remnant and eventually disappears altogether. Almost all other characteristics of a typical cell also are lost: ribosomes, mitochondria, and most enzyme systems. What is left is a biconcave disk consisting of a baglike membrane, the **stroma,** packed with the blood-transport-

FIG. 27-3

Mammalian and amphibian red blood corpuscles. The erythrocytes of a gerbil, **A,** are biconcave disks containing hemoglobin and surrounded by a tough stroma. The frog erythrocytes, **B,** are convex disks, each containing a nucleus, which is plainly visible in the scanning electron micrographs as a bulge in the center of each cell. (Mammalian erythrocytes, × 6,300; frog erythrocytes, × 2,400; courtesy P. P. C. Graziadei, Florida State University.)

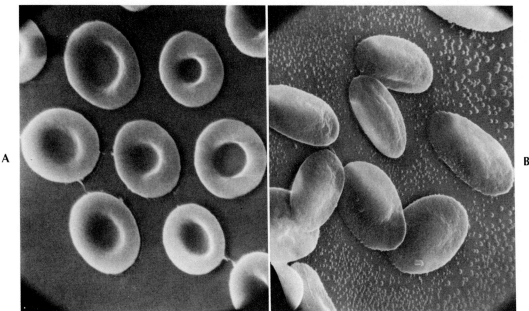

ing pigment **hemoglobin.** About 33% of the erythrocyte by weight is hemoglobin. The biconcave shape (Fig. 27-3, *A*) is a mammalian innovation that provides a much larger surface for gas diffusion than would a flat or spherical shape. All other vertebrates have nucleated erythrocytes that are usually ellipsoidal, rather than round, disks (Fig. 27-3, *B*).

The erythrocyte enters the circulation for an average life-span of about 4 months. During this time it may journey 700 miles, squeezing repeatedly through the capillaries, which are sometimes so narrow that the erythrocyte must bend to get through. At last it fragments and is quickly engulfed by large scavenger cells called **macrophages** located in the liver, bone marrow, and spleen. The iron from the hemoglobin is salvaged to be used again; the rest of

the heme is converted to **bilirubin,** a bile pigment. It is estimated that 10 million erythrocytes are born, and another 10 million destroyed every second. The white blood cells or **leukocytes,** form a wandering system of protection for the body. In adults they number only about 7,500 per cubic millimeter, a ratio of 1 white cell to 700 red cells. There are several kinds of white blood cells: **granulocytes** (subdivided into neutrophils, basophils, and eosinophils), **lymphocytes,** and **monocytes** (Fig. 27-4). All have the capacity to pass through the wall of capillaries and wander by ameboid movement through the tissue spaces. Monocytes and granulocytes have great power to engulf and digest bacteria and other foreign particulate matter, a process called **phagocytosis.** They also clean up and digest the debris of the

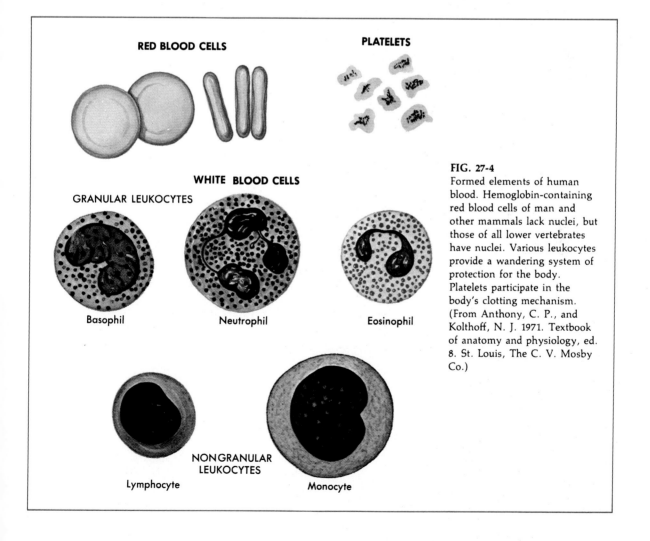

RED BLOOD CELLS

PLATELETS

WHITE BLOOD CELLS

GRANULAR LEUKOCYTES

Basophil   Neutrophil   Eosinophil

NONGRANULAR LEUKOCYTES

Lymphocyte   Monocyte

FIG. 27-4
Formed elements of human blood. Hemoglobin-containing red blood cells of man and other mammals lack nuclei, but those of all lower vertebrates have nuclei. Various leukocytes provide a wandering system of protection for the body. Platelets participate in the body's clotting mechanism. (From Anthony, C. P., and Kolthoff, N. J. 1971. Textbook of anatomy and physiology, ed. 8. St. Louis, The C. V. Mosby Co.)

body's own tissues, such as fragments of worn-out red blood cells, blood clots, or the remains of wounds and disease repair. Lymphocytes are especially important in producing gamma globulins, which act as immune bodies **(antibodies)** that destroy or neutralize toxic molecules **(antigens)**.

The platelets, or thrombocytes, are minute, colorless bodies about one third the diameter of red blood cells. They initiate the coagulation of blood. When blood spills from a vessel, as in a wound, the platelets rapidly disintegrate to release factors that start the formation of a clot. Platelets also readily clump together and plug torn vessels by entangling white blood cells (see photograph, p. 605).

COAGULATION OF BLOOD. It is essential that animals have ways of preventing the rapid loss of body fluids after an injury. Since blood is flowing and is under considerable hydrostatic pressure, it is especially vulnerable to hemorrhagic loss. The most primitive means of preventing hemorrhage, and one used by many soft-bodied invertebrates, is spasmic contraction of body musculature and blood vessels. Vessels in the wound are thus narrowed off or even shut tight, stopping the flow. But most firm-bodied animals (vertebrates and higher invertebrates), and especially those having high blood pressures, have special cellular elements and proteins in the blood capable of forming plugs, or clots, at the site of injury. When tissue cells are damaged, a substance called **thromboplastin** is released from the injured tissue as well as from the blood platelets (see above). This substance, in the presence of calcium, converts a normally inactive protein, **prothrombin**, into an active enzyme, **thrombin**. Thrombin then converts **fibrinogen**, a very large plasma protein, into an insoluble threadlike protein, **fibrin**. These fibrin threads form a stringy, tangled network that entangles white and red blood cells, forming a gel-like clot. In time the fibrin threads shrink and squeeze out a faintly yellow fluid called **serum.** The difference between **plasma** and **serum** is that the latter lacks fibrinogen and is incapable of clotting.

The conversion of prothrombin into thrombin is a critical step in the clotting reaction. In addition to thromboplastin, many other coagulation factors are required. A recent estimate listed no less than 35 compounds that participate in some way in coagulation. A deficiency of a single factor can delay or prevent the clotting process. Why has such a complex

clotting mechanism evolved? Probably it is necessary to provide a fail-safe system capable of responding to any kind of internal or external hemorrhage that might occur, yet not be activated into forming dangerous intravascular clots when no injury has occurred.

Several kinds of clotting abnormalities in man are known. Of these, **hemophilia** is perhaps best known. Hemophilia is a condition characterized by the failure of the blood to clot, so that even insignificant wounds can cause continuous severe bleeding. Called the "disease of kings," it once ran through the royal families of Europe, notably those of Queen Victoria of England and Alfonso XIII, the last king of Spain. Hemophilia is caused by an inherited lack of antihemophilic factor. The disorder is transmitted through females, but almost invariably appears only in males.

BLOOD GROUPS. In blood transfusions the donor's blood is checked against the blood of the recipient. Blood differs chemically from person to person, and when two different (incompatible) bloods are mixed, **agglutination** (clumping together) results. The basis of these chemical differences is the presence in the red blood corpuscle of **agglutinogens (antigens)** A and B, and, in the plasma, **agglutinins (antibodies)** a and b. According to the way these antigens and antibodies are distributed, there are four main blood groups: **A, B, AB,** and **O (Table 27-1).** Group **A** blood cells have **A** antigens on them, but the serum of a group **A** person has no **a** antibody because if it did, it would destroy its own blood cells. It does contain group **b** antibodies, however. It is therefore imcompatible with either group **B** or group **AB** blood because these contain **B** antigens. Similarly group **B** blood contains **B** antigens and lacks **b** antibodies. But since it contains **a** antibodies it is incompatible with any blood (group **A** or **AB**) that contains **A** antigens. Group **O** blood contains both antibodies **a** and **b** but no antigens. Group **AB** contains both **A** and **B** antigens but no antibodies. We see then that the blood group names identify their *antigen* content. Persons with type **O** blood are called universal donors because, lacking antigens, their blood can be infused into a person with any blood type. Even though it contains **a** and **b** antibodies, these are so diluted during transfusion that they do not react with **A** or **B** antigens in a recipient's blood. In practice, however, clinicians insist on matching

TABLE 27-1
Major blood groups

Blood group	Antigens in red corpuscles	Antibodies in serum	Can give blood to	Can receive blood from	Frequency in United States (%)		
					Whites	Blacks	Chinese
O	None	a, b	All	O	45	38	46
A	A	b	A, AB	O, A	41	27	28
B	B	a	B, AB	O, B	10	21	23
AB	AB	None	AB	All	4	4	13

blood types to prevent any possibility of incompatibility.

Rh FACTOR. It is difficult to think of another area of physiology as totally linked with the name of a single man as blood grouping is with the name of Karl Landsteiner. This Austrian—later American—physician discovered the ABO blood groups in 1900. The great importance of his work became abundantly clear during World War I when blood transfusion was first attempted on a large scale. In 1927 Landsteiner in collaboration with Philip Levine in the United States discovered the MN blood group, also present in all mankind. This classification is not important in transfusions but may be crucial in determining relationship in paternity cases. Then in 1940, 10 years after receiving the Nobel Prize in recognition of contributions that were more than adequate for the lifetime of any scientist, Landsteiner made still another famous discovery. This new group was called the Rh factor, named after the Rhesus monkey, in which it was first found. About 85% of individuals have the factor (positive) and the other 15% do not (negative). He also found that Rh-positive and Rh-negative bloods are incompatible; shock and even death may follow their mixing when Rh-positive blood is introduced into an Rh-negative person who has been sensitized by an earlier transfusion of Rh-positive blood. The Rh factor is inherited as a dominant; this accounts for a peculiar and often fatal form of anemia of newborn infants called **erythroblastosis fetalis.** Although the fetal and maternal bloods are separated by the placenta, this separation is not perfect. Some admixture of fetal and maternal bloods usually occurs, especially right after birth when the placenta ("afterbirth") separates from the uterine wall. This admixture of blood, normally of

no consequence, can be serious **if** the father is Rh positive, the mother Rh negative, and the fetus Rh positive (by inheriting the factor from the father). The fetal blood, containing the Rh antigen, can stimulate the formation of Rh-positive antibodies in the blood of the mother. The mother is permanently immunized against the Rh factor. During the second pregnancy these antibodies may diffuse back into the fetal circulation and produce agglutination and destruction of the fetal red blood cells. Because the mother is usually sensitized at the end of the first pregnancy, subsequent babies will be more severely threatened than is the first.

Erythroblastosis fetalis can now be prevented by giving an Rh-negative mother anti-Rh antibodies just after the birth of her first child. These antibodies remain long enough to neutralize any Rh-positive fetal blood cells that may enter her circulation, thus preventing her own antibody machinery from being stimulated to produce the Rh-positive antibodies. Active, permanent immunity is blocked.

■ CIRCULATION

The circulatory system of vertebrates is made up of a system of tubes, the **blood vessels,** and a propulsive organ, the **heart.** This is a **closed circulation** because the circulating medium, the **blood,** is confined to vessels throughout its journey from the heart to the tissues and back again. Many invertebrates have an **open circulation;** the blood is pumped from the heart into blood vessels that open into tissue spaces. The blood circulates freely in direct contact with the cells and then reenters open blood vessels to be propelled forward again. In invertebrates having open circulatory systems, there is no clear separation of the extracellular fluid into plasma and interstitial fluids, as there is in closed systems. Closed systems are more suitable for large and active animals because the blood can be moved rapidly to the tissues needing it. In addition, flow to various organs can be readjusted to meet changing needs by varying the diameters of the blood vessel.

Closed circulatory systems work in parallel with a cooperative system, the **lymphatic system.** This is a fluid "pick-up" system. It recollects tissue fluid (lymph) that has been squeezed out through the walls of the capillaries and returns it to the blood circulation. In a sense "closed" circulatory systems are not absolutely closed because fluid is constantly leak-

ing out into the tissue spaces. However, this leakage is but a small fraction of the total blood flow.

Although it seems obvious to us today that blood flows in a circuit, the first correct description of blood flow by the English physician William Harvey, initially received vigorous opposition when published in 1628. Centuries before, Galen had taught that air enters the heart from the windpipe and that blood was able to pass from one ventricle to the other through "pores" in the interventricular septum. He also believed that blood first flowed out of the heart in all vessels, arteries, and veins alike and then returned to the heart by these same vessels—an idea

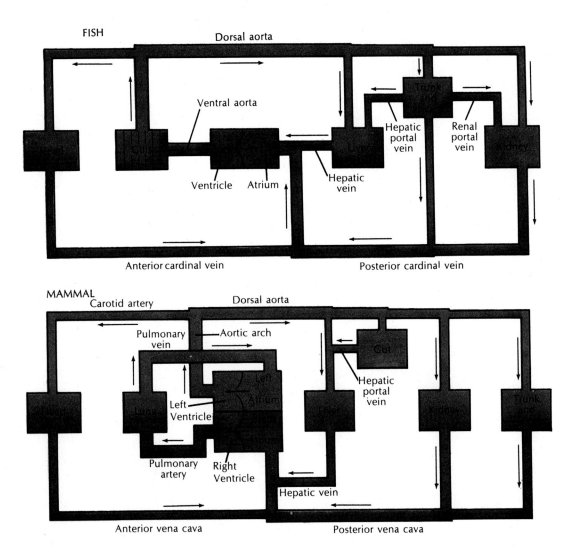

FIG. 27-5
Plan of circulatory system of fish *(above)* and mammals *(below)*.

of ebb and flow of the blood. Even though there was almost nothing right about this theory, it was still doggedly trusted at the time of Harvey's publication. Harvey's conclusions were based on sound experimental evidence. He made use of a variety of animals for his experiments, including the little snake found in English meadows. By tying ligatures on arteries, he noticed that the region between the heart and ligature swelled up. When veins were tied off, the swelling occurred beyond the ligature. When blood vessels were cut, blood flowed in arteries from the cut end nearest the heart; the reverse happened in veins. By means of such experiments, Harvey worked out a correct scheme of blood circulation, even though he could not see the capillaries that connected the arterial and venous flows.

PLAN OF THE CIRCULATORY SYSTEM. All vertebrate vascular systems have certain features in common. A **heart** pumps the blood into **arteries** that branch and narrow into **arterioles** and then into a vast system of **capillaries.** Blood leaving the capillaries enters **venules** and then **veins** that return the blood to the heart. Fig. 27-5 compares the circulatory systems of gill-breathing (fish) and lung-breathing (mammal) vertebrates. The principal differences in circulation involve the heart in the transformation from gill to lung breathing. The fish heart contains two main chambers, the **atrium** (or **auricle**) and the **ventricle.** Although there are also two subsidiary chambers, the **sinus venosus** and **conus arteriosus** (not shown in Fig. 27-5), we still refer to the fish heart as a "two-chambered" heart. Blood makes a single circuit through the fish's vascular system; it is pumped from the heart to the gills, where it is oxygenated, and then flows into the dorsal aorta to be distributed to the body organs. After passing through the capillaries of the body organs and musculature, it returns by veins to the heart. In this circuit the heart must provide sufficient pressure to push the blood through two sequential capillary systems, one in the gills and the other in the organ tissues. The principal disadvantage of the single-circuit system is that the gill capillaries offer so much resistance to blood flow that the pressure drops considerably before entering the dorsal aorta. This system can never provide high and continuous blood pressure to the body organs.

Evolving land forms with lungs and their need for highly efficient blood delivery had to solve this prob-

lem by introducing a **double** circulation. One **systemic** circuit with its own pump provides oxygenated blood to the capillary beds of the body organs; another **pulmonary** circuit with its own pump sends deoxygenated blood to the lungs. Rather than actually developing two separate hearts, the existing two-chambered heart was divided down the center into four chambers—really two two-chambered hearts lying side-by-side. Needless to say such a great change in the vertebrate circulatory plan, involving not only the heart but the attendant plumbing as well, took many millions of years to evolve (see p. 500). The partial division of the atrium and ventricle began with the ancestors of present-day lungfish. Amphibians accomplished the complete separation of the atrium, but the ventricle is still undivided in this group. In some reptiles the ventricle is completely divided, and the four-chambered heart appears for the first time. All birds and mammals have the four-chambered heart and two separate circuits—one through the lungs (pulmonary) and the other through the body (systemic). The course of the blood through this double circuit is shown in Fig. 27-5.

THE HEART. The vertebrate heart is a muscular organ located in the thorax and covered by a tough, fibrous sac, the **pericardium** (Fig. 27-6). As we have seen, the higher vertebrates have a four-chambered heart. Each half consists of a thin-walled atrium and a thick-walled ventricle. Heart (cardiac) muscle is a unique type of muscle found nowhere else in the body. It resembles striated muscle, but the cells are branched, and dense end-to-end attachments between the cells are called intercalated disks. There are four sets of valves. **Atrioventricular valves** (A-V valves) separate the cavities of the atrium and ventricle in each half of the heart. These permit blood to flow from atrium to ventricle but prevent backflow. Where the great arteries, the **pulmonary** from the right ventricle and the **aorta** from the left ventricle, leave the heart, **semilunar valves** prevent backflow.

The contraction of the heart is called **systole** (sis'to-lee), and the relaxation, **diastole** (dy-as'to-lee). The rate of the heartbeat depends on age, sex, and especially, exercise. Exercise may increase the **cardiac output** (volume of blood forced from either ventricle each minute) more than fivefold. Both the heart **rate** and the **stroke volume** increase. Heart rates among vertebrates vary with the general level

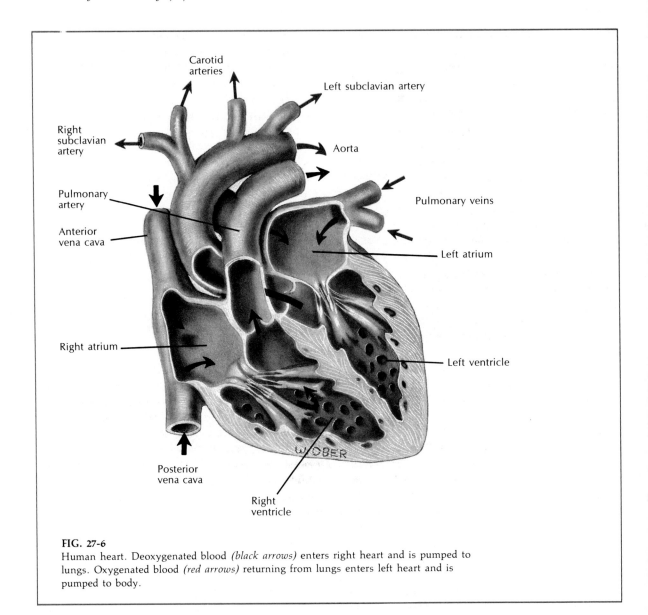

FIG. 27-6
Human heart. Deoxygenated blood *(black arrows)* enters right heart and is pumped to lungs. Oxygenated blood *(red arrows)* returning from lungs enters left heart and is pumped to body.

of metabolism and the body size. The cold-blooded codfish has a heart rate of about 30 beats per minute; a warm-blooded rabbit of about the same weight has a rate of 200 beats per minute. Small animals have higher heart rates than do large animals. The heart rate in an elephant is 25 beats per minute, in a man 70 per minute, in a cat 125 per minute, in a mouse 400 per minute, and in the tiny 4-gram shrew, the smallest mammal, the heart rate approaches a prodigious 800 beats per minute. We must marvel that the shrew's heart can sustain this frantic pace through-

out this animal's life, brief as it is. The only rest a heart enjoys is the short interval between contractions. The mammalian heart does an amazing amount of work during a lifetime. Someone has calculated that the heart of a man approaching the end of his life has beat some 2.5 billion times and pumped 300,000 tons of blood!

EXCITATION OF THE HEART. The heart beat originates in a specialized muscle tissue, called the **sinoatrial node,** located in the right atrium near the entrance of the caval veins (Fig. 27-7). This tissue

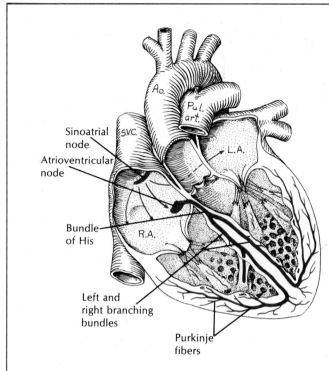

FIG. 27-7

Neuromuscular mechanisms controlling beat of the heart. Arrows indicate spread of excitation from the sinoatrial node (S-A node), across the atria, to the A-V node (atrioventricular node). Wave of excitation is then conducted very rapidly to ventricular muscle over the specialized bundle of His and Purkinje fiber system.

serves as the **pacemaker** of the heart. The contraction originates in the pacemaker and spreads across the two atria to the **atrioventricular (A-V) node.** At this point the electrical activity is conducted very rapidly to the apex of the ventricle through specialized fibers (bundle of His and Purkinje fiber system) and then spreads more slowly up the walls of the ventricles. This arrangement allows the contraction to begin at the apex or "tip" of the ventricles and spread upward to squeeze out the blood in the most efficient way; it ensures that both ventricles will contract simultaneously. Although the vertebrate heart can beat spontaneously—the excised fish or amphibian heart will beat for hours in a balanced salt solution—the heart rate is normally under nervous control. The control (cardiac) center is located in the medulla and sends out two sets of motor nerves. Impulses sent along one set, the **vagus** (parasympathetic) nerves, apply a brake action to the heart rate, and impulses sent along the other set, the **accelerator** (sympathetic) nerves, speed it up. Both sets of nerves terminate in the sinoatrial node, thus guiding the activity of the pacemaker. The cardiac center in

turn receives sensory information about a variety of stimuli. Pressure receptors (sensitive to blood pressure) and chemical receptors (sensitive to carbon dioxide and pH) are located at strategic points in the vascular system. This information is used by the cardiac center to increase or reduce the heart rate and cardiac output in response to activity or changes in body position. The heart is thus controlled by a series of feedback mechanisms that keep its activity constantly attuned to body needs.

CORONARY CIRCULATION. It is no surprise that an organ as active as the heart needs a very good blood supply of its own. The heart muscle of the frog and other amphibians is so thoroughly channeled with spaces between the muscle fibers that sufficient oxygenated blood is squeezed through by the heart's own pumping action. In birds and mammals, however, the heart muscle is very thick and has such a high rate of metabolism that it must have its own vascular **(coronary)** circulation. The coronary arteries break up into an extensive capillary network surrounding the muscle fibers and provide them with oxygen and nutrients. Heart muscle has an extremely high oxygen demand, removing 80% of the oxygen from the blood, in contrast to most other body tissues, which remove only about 30%.

ARTERIES. All vessels leaving the heart are called arteries whether they carry oxygenated blood (aorta) or deoxygenated blood (pulmonary artery). To withstand high, pounding pressures, arteries are invested with layers of both elastic and tough, inelastic connective tissue fibers. The elasticity of the arteries allows them to yield to the surge of blood leaving the heart during systole and then to squeeze down on the fluid column during diastole. This smooths out the blood pressure. Thus the arterial pressure in man varies only between a high of 120 mm. Hg (systole) and a low of 80 mm. Hg (diastole), rather than dropping to zero during diastole as we might expect in a fluid system with an intermittent pump. As the arteries branch and narrow into **arterioles,** the walls become mostly smooth muscle (Fig. 27-8). Contraction of this muscle narrows the arterioles and reduces the flow of blood. The arterioles thus control the blood flow to body organs, diverting it to where it is needed most. The blood must be given a hydrostatic pressure sufficient to overcome the resistance of the narrow passages through which the blood must flow. Consequently large animals tend to have higher blood pressure than do small animals.

Blood pressure was first measured in 1733 by Stephen Hales, an English clergyman with unusual inventiveness and curiosity. He tied his mare "to have been killed as unfit for service" on her back and exposed the femoral artery. This he cannulated with a brass tube, connecting it to a tall glass tube with the windpipe of a goose. The use of the windpipe was both imaginative and practical; it gave the apparatus flexibility "to avoid inconveniences that might arise if the mare struggled." The blood rose 8 feet in the glass tube and bobbed up and down with the systolic and diastolic beats of the heart. The weight of the 8-foot column of blood was equal to the blood pressure. We now express this as the height of a column of mercury, which is 13.6 times heavier than water. Hale's figures, expressed in millimeters of mercury, indicate he measured a blood pressure of 180 to 200 mm. Hg, about normal for a horse. Today, blood pressure can be measured with great accuracy with a sensitive pressure transducer; the electronic signal from this instrument is displayed on a graphic recorder.

CAPILLARIES. The Italian Marcello Malpighi was the first to describe the capillaries in 1661, thus confirming the existence of the minute links between the arterial and venous systems that Harvey knew must be there but could not see. Malpighi studied the capillaries of the living frog's lung, which incidentally is still one of the simplest and most vivid preparations for demonstrating capillary blood flow.

The capillaries are present in enormous numbers, forming extensive networks in nearly all tissues. In muscle there are more than 2,000 per square millimeter (1,250,000 per square inch), but not all are open at once. Indeed, perhaps less than 1% are open in resting skeletal muscle. But when the muscle is active, all the capillaries may open to bring oxygen and nutrients to the working muscle fibers and to carry away metabolic wastes.

Capillaries are extremely narrow, averaging less than 10 μ in diameter in mammals, which is hardly any wider than the red blood cells that must pass through them. Their walls are formed of a single layer of thin **endothelial** cells, held together by a delicate basement membrane and connective tissue fibers. Capillaries have a built-in leakiness that allows water and most dissolved substances in the blood plasma to filter through into the interstitial space. The capillary wall is **selectively permeable,** however, which means that it filters some dissolved materials and retains others. In this case the plasma proteins, which are the largest dissolved molecules in the plasma, are held back. These proteins, espe-

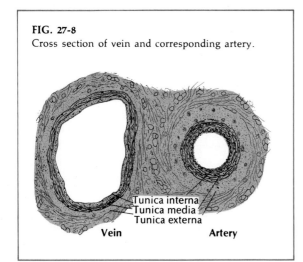

FIG. 27-8
Cross section of vein and corresponding artery.

Tunica interna
Tunica media
Tunica externa
Vein   **Artery**

cially the albumins, contribute an **osmotic pressure** of about 25 mm. Hg in mammals (Fig. 27-9). Although small, this protein osmotic pressure is of great importance to fluid balance in the tissues. At the arteriole end of the capillaries the blood pressure is about 40 mm. Hg (in man). This **filtration pressure** forces water and dissolved materials through the capillary endothelium into the tissue space where they circulate freely around the cells. As the blood proceeds through the narrow capillary, the blood pressure drops steadily to perhaps 15 mm. Hg. At this point the hydrostatic pressure is less than the osmotic pressure of the plasma proteins, still about 25 mm. Hg. Water now is drawn back into the capillaries. Thus it is the balance between hydrostatic pressure and protein osmotic pressure that determines the direction of capillary fluid shift. Normally water is forced out of the capillary at the arteriole end, where hydrostatic pressure exceeds osmotic pressure, and drawn back into the capillary at the venule end where osmotic pressure exceeds hydrostatic pressure. Any fluid left behind is picked up and removed by the **lymph capillaries.**

VEINS. The venules and veins into which the capillary blood drains for its return journey to the heart are thinner walled, less elastic, and of considerably larger diameter than their corresponding arteries and arterioles (Fig. 27-8). Blood pressure in the venous system is low, from about 10 mm. Hg where capillaries drain into venules to about zero in the right atrium. Because pressure is so low, the venous return gets assists from valves in the veins, from muscles surrounding the veins, and from the rhythmic pumping action of the lungs. If it were not for these mechanisms, the blood might pool in the lower extremities of a standing animal—a very real problem for people who must stand for long periods. The veins that lift blood from the extremities to the heart contain valves that serve to divide the long column of blood into segments. When the muscles around the veins contract, as in even slight activity, the blood column is squeezed upward and cannot slip back because of the valves. The well-known risk of fainting while standing at stiff attention in hot weather can usually be prevented by deliberately pumping the leg muscles. The negative pressure created in the thorax by the inspiratory movement of the lungs also speeds the venous return by sucking the blood up the large vena cava into the heart.

LYMPHATIC SYSTEM. Gasparo Aselli, an Italian anatomist, first discovered the nature of lacteals in 1627. In a dog that had recently been fed and cut open, he noticed white cordlike bodies in the mesenteries of the intestine that he first mistook for nerves. When he pricked these cords with a scalpel, a milky fluid gushed out. It is now known that this fluid is largely fat that is carried after digestion to the thoracic duct. The thoracic duct and its relations to the lacteals were discovered by the Frenchman Jean Pecquet in 1647. These vessels are part of the complete lymphatic system demonstrated almost simultaneously but independently by O. Rudbeck in Sweden (1651) and T. Bartholin in Denmark (1653), using dogs and executed criminals.

The lymphatic system (Fig. 27-10) is an accessory drainage system for the body. As we have seen, the

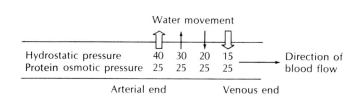

FIG. 27-9
Fluid movement across the wall of a capillary. At arterial end of the capillary, hydrostatic (blood) pressure exceeds protein osmotic pressure contributed by the plasma proteins, and a plasma filtrate (shown as "water movement") is forced out. At venous end, protein osmotic pressure exceeds the hydrostatic pressure, and fluid is drawn back in. In this way plasma nutrients are carried out into the interstitial space where they can enter cells, and metabolic end products from the cells are drawn back into the plasma and carried away.

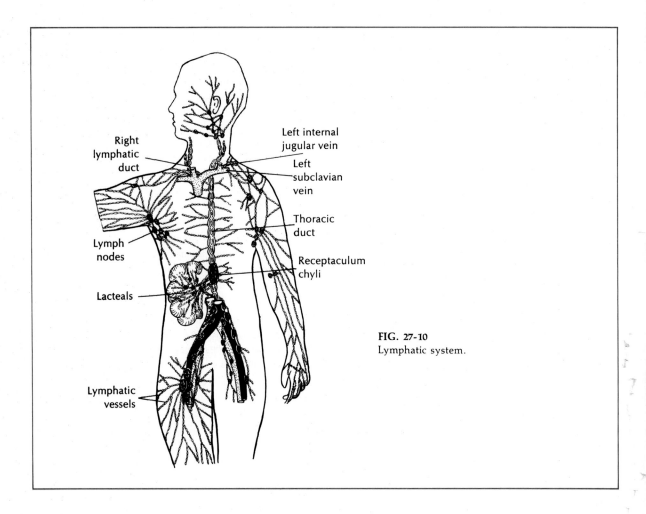

Right
lymphatic
duct

Left internal
jugular vein

Left
subclavian
vein

Thoracic
duct

Lymph
nodes

Receptaculum
chyli

Lacteals

Lymphatic
vessels

FIG. 27-10
Lymphatic system.

blood pressure in the arteriole end of the capillaries forces a plasma filtrate through the capillary walls and into the interstitial space. This tissue fluid bathing the cells is **lymph,** a clear, nearly colorless liquid. Lymph and plasma are nearly identical except that lymph contains very little protein, which was screened out as the plasma was squeezed through the capillary walls. Most of the lymph returns to the vascular system at the venous end of the capillaries by the capillary fluid-shift mechanism described earlier. Usually, however, outflow from the capillaries slightly exceeds backflow. This difference is gathered up and returned to the circulatory system by lymphatic vessels. The system begins with tiny, highly permeable lymph capillaries. These lead into larger lymph vessels, which in turn drain into the large **thoracic duct.** This enters the left subclavian vein in the neck region. Lymph flow is very low, a minute fraction of the blood flow.

Located at strategic intervals along the lymph vessels are **lymph nodes** (Fig. 27-10) that have several defense-related functions. They are effective filters that remove foreign particles, especially bacteria, that might otherwise enter the general circulation. They are also germinal centers for **lymphocytes** (p. 609) and they produce gamma globulin **antibodies**—both essential components of the body's defense mechanisms.

■ RESPIRATION

The energy bound up in food must be released by oxidative processes. As oxygen is used by the body cells, carbon dioxide is produced; this process is called **respiration.** Most animals are **aerobic,** meaning that they require and receive the necessary oxygen directly from their environment. A few animals, called **anaerobic,** are able to live in the absence of oxygen. Forms such as worms and arthropods dwelling in the oxygen-depleted mud of lakes, and parasites living anaerobically in the intestine, derive the necessary oxygen from the metabolism of carbohydrates and fats. However, anaerobic metabolism often occurs in the muscles of basically aerobic animals during vigorous muscle contraction.

Small aquatic animals such as the one-celled protozoans obtain what oxygen they need by direct diffusion from the environment. Carbon dioxide, the gaseous waste of metabolism, is also lost by diffusion to the environment. Such a simple solution to the problem of gas exchange is really only possible for very small animals (less than 1 mm. in diameter) or those having very low rates of metabolism. As animals became larger and evolved a waterproof covering, specialized devices such as lungs and gills developed that greatly increased the effective surface for gas exchange. But because gases diffuse so slowly through protoplasm, a circulatory system was necessary to distribute the gases to and from the deep tissues of the body. Even these adaptations were inadequate for advanced animals, with their high rates of cellular respiration. The solubility of oxygen in the blood plasma is so low that plasma alone could not carry enough to satisfy metabolic demands. So special oxygen-transporting blood proteins such as hemoglobin evolved, greatly increasing the oxygen-carrying capacity of the blood. Thus, what began as a simple and easily satisfied requirement led to the evolution of several complex and essential respiratory and circulatory adaptations.

Problems of aquatic and aerial breathing

How an animal respires is largely determined by the nature of its environment. The two great arenas of animal evolution—water and land—are vastly different in their physical characteristics. The most obvious difference is that air contains far more oxygen —at least 20 times more—than does water. Atmospheric air contains oxygen (about 21%), nitrogen (about 79%), carbon dioxide (0.03%), a variable amount of water vapor, and very small amounts of inert gases (helium, argon, neon, etc.). These gases are variably soluble in water. The amount of oxygen dissolved depends on the concentration of oxygen in

the air and on the water temperature. Water at 5° C. fully saturated with air contains about 9 ml. of oxygen per liter. (Note that by comparison, air contains about 210 ml. of oxygen per liter.) The solubility of oxygen in water decreases as the temperature rises. For example, water at 15° C. contains about 7 ml. of oxygen per liter, and at 35°C., only 5 ml. of oxygen per liter. The relatively low concentration of oxygen dissolved in water is the greatest respiratory problem facing aquatic animals. Unfortunately it is not the only one. Oxygen diffuses much more slowly in water than in air, and water is much denser and more viscous than air. All of this means that successful aquatic animals must have evolved very efficient ways of removing oxygen from water. Yet even the most advanced fishes with highly efficient gills and pumping mechanisms may use as much as 20% of their energy just extracting oxygen from water. By comparison, a mammal uses only 1% to 2% of its resting metabolism to breathe.

It is essential that respiratory surfaces be kept thin and always wet to allow diffusion of gases between the environment and the underlying circulation. This is hardly a problem for aquatic animals, immersed as they are in water, but it is a very real problem for air breathers. To keep the respiratory membranes moist and protected from injury, air breathers have in general developed invaginations of the body surface and then added pumping mechanisms to move air in and out. The lung is the best example of a successful solution to breathing on land. In general, **evaginations** of the body surface, such as gills, are most suitable for aquatic respiration and **invaginations,** such as lungs, are best for air breathing. We will now consider the specific kinds of respiratory organs employed by animals.

CUTANEOUS RESPIRATION. Protozoa, sponges, coelenterates, and many worms respire by direct diffusion of gases between the organism and the environment. We have noted that this kind of **integumentary respiration** is not adequate when the mass of living protoplasm exceeds about 1 mm. in diameter. But by greatly increasing the surface of the body relative to the mass, many multicellular animals respire in this way. Integumentary respiration frequently supplements gill or lung breathing in larger animals such as amphibians and fishes. For example, an eel can exchange 60% of its oxygen and carbon dioxide through its highly vascular skin. During their

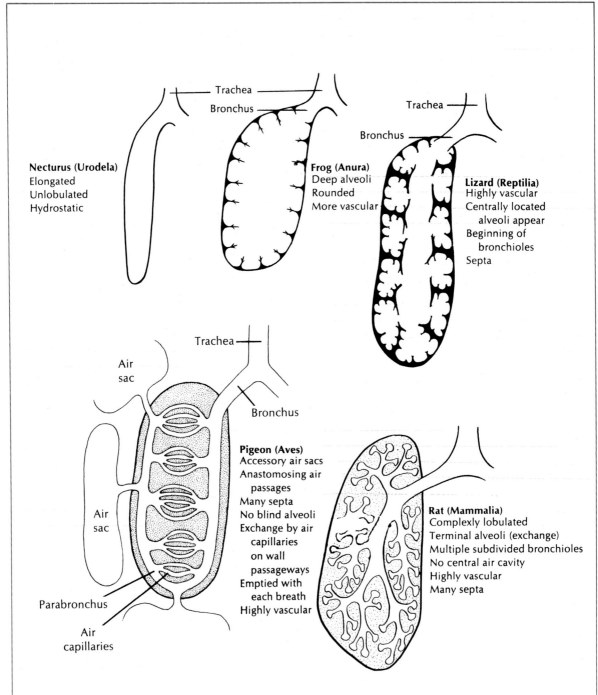

FIG. 27-11
Internal structures of lungs among vertebrate groups. In general, evolutionary trend has been from simple sacs with little exchange surface between blood and air spaces to complex, lobulated structures of complex divisions and extensive exchange surfaces.

winter hibernation, frogs exchange all their respiratory gases through the skin while submerged in ponds or springs.

GILLS. Gills are unquestionably the most effective respiratory device for life in water. Gills may be simple **external** extensions of the body surface, such as the **dermal branchiae** of starfish or the **branchial tufts** of marine worms and aquatic amphibians. Most efficient are the **internal** gills of fishes (described on p. 474) and arthropods. Fish gills are thin filamentous structures, richly supplied with blood vessels arranged so that blood flow is opposite to the flow of water across the gills. This arrangement, called **countercurrent flow,** provides for the greatest possible extraction of oxygen from water. Water flows over the gills in a steady stream, pushed and pulled by an efficient branchial pump, and often assisted by the fish's forward movement through the water (see Fig. 22-19).

LUNGS. Gills are unsuitable for life in air because when removed from the buoying water medium, the gill filaments collapse and stick together; a fish out of water rapidly asphyxiates despite the abundance of oxygen around it. Consequently air-breathing vertebrates possess lungs, highly vascularized internal cavities. Lungs of a sort are found in certain invertebrates (pulmonate snails, scorpions, some spiders, some small crustaceans) but these structures cannot be ventilated and consequently are not very efficient. Lungs that can be ventilated efficiently are characteristic of the terrestrial vertebrates. The most primitive vertebrate lungs are those of lungfishes **(Dipneusti),** which use them to supplement, or even replace, gill respiration during periods of drought. Although of simple construction, the lungfish lung is supplied with a capillary network in its largely unfurrowed walls, a tubelike connection to the pharynx, and a primitive ventilating system for moving air in and out of the lung. Amphibians also have simple baglike lungs, whereas in higher forms the inner surface area is vastly increased by numerous lobulations and folds (Fig. 27-11). This increase is greatest in the mammalian lung, which is complexly divided into many millions of small sacs **(alveoli),** each veiled by a rich vascular network. It has been estimated that the human lungs have a total surface area of from 50 to 90 m.^{2}—50 times the area of the skin surface—and contain 1,000 miles of capillaries.

Moving air into and out of lungs has been an evolutionary design problem that was, of course, solved although one wonders if an imaginative biologic engineer, given the proper resources, couldn't have come up with a better design. Unlike the efficient one-way flow of water across fish gills, air must enter and exit a lung at the same point. Furthermore, a tube of some length—the bronchi, trachea, and mouth cavity—connects the lungs to the outside. This is a "dead-air space" containing a volume of air that shuttles back and forth with each breath, adding to the difficulty of properly ventilating the lungs. In fact, lung ventilation is so inefficient that in normal breathing only about one-sixth of the air in the lungs is replenished with each inspiration. One group of vertebrates, the birds, vastly improved lung efficiency by adding an extensive system of air sacs (Fig. 27-11) that serve as air reservoirs during ventilation. Upon inspiration, some 75% of the incoming air bypasses the lungs to enter the air sacs. At expiration, some of this fresh air passes directly through the lung passages. Thus the air capillaries receive nearly fresh air during both inspiration and expiration (see Fig. 25-10). The beautifully designed bird lung is the result of selective pressures during the evolution of flight and its high metabolic demands.

Frogs force air into the lungs by first lowering the floor of the mouth to draw air into the mouth through the external nares (nostrils); then, by closing the nares and raising the floor of the mouth, air is driven into the lungs. Much of the time, however, frogs rhythmically ventilate only the mouth cavity, which serves as a kind of auxiliary "lung" (see Fig. 23-13). Amphibians, therefore employ a **positive pressure** action to fill their lungs, unlike most reptiles, birds, and mammals, which breathe by sucking air into the lungs (**negative pressure** action).

TRACHEAE. Insects and certain other terrestrial arthropods (centipedes, millipedes, and some spiders) have a highly specialized type of respiratory system; in many respects it is the simplest, most direct and most efficient respiratory system found in active animals. It consists of a system of tubes **(tracheae)** that branch repeatedly and extend to all parts of the body. The smallest end channels **(air capillaries),** less than 1μ in diameter, sink into the plasma membranes of the body cells. Oxygen enters the tracheal system through valvelike openings **(spiracles)** on each side of the body and diffuses directly to all cells of the body. Carbon dioxide diffuses out in the opposite direction. Some insects can ventilate the tracheal

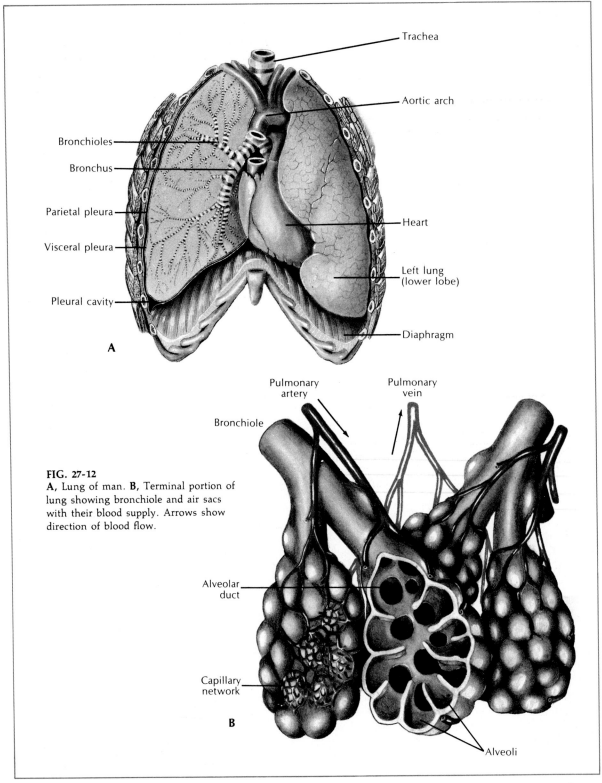

FIG. 27-12
A, Lung of man. **B,** Terminal portion of
lung showing bronchiole and air sacs
with their blood supply. Arrows show
direction of blood flow.

system with body movements; the familiar telescoping movement of the bee abdomen is an example. The tracheal system is simple because blood is not needed to transport the respiratory gases; the cells have a direct pipeline to the outside.

Respiration in man

In mammals the respiratory system is made up of the following: the nostrils (external nares); the **nasal chamber,** lined with mucus-secreting epithelium; the **posterior nares,** which connect to the **pharynx,** where the pathways of digestion and respiration cross; the **epiglottis,** a flap that folds over the **glottis** (the opening to the larynx) to prevent food from going the wrong way in swallowing; the **larynx,** or voice box; the **trachea,** or windpipe; and the two **bronchi,** one to each lung (Fig. 27-12). Within the lungs each bronchus divides and subdivides into smaller tubes **(bronchioles)** that lead to the air sacs **(alveoli).** The walls of the alveoli are thin and moist to facilitate the exchange of gases between the air sacs and the

adjacent blood capillaries (Fig. 27-13). Air passageways are lined with mucus-secreting ciliated epithelium and play an important role in conditioning the air before it reaches the alveoli. There are partial cartilage rings in the walls of the tracheae, bronchi, and even some of the bronchioles to prevent those structures from collapsing.

In its passage to the air sacs the air undergoes three important changes: (1) it is filtered free from most dust and other foreign substances, (2) it is warmed to body temperature, and (3) it is saturated with moisture.

The lungs consist of a great deal of elastic connective tissue and some muscle. They are covered by a thin layer of tough epithelium known as the **visceral pleura.** A similar layer, the **parietal pleura,** lines the inner surface of the walls of the chest (Fig. 27-12). The two layers of the pleura are in contact and slide over one another as the lungs expand and contract. The "space" between the pleura, called the **pleural cavity,** contains a partial vacuum. Actually, no real

FIG. 27-13

Appearance of inside of terminal bronchus of rat lung showing alveoli and alveolar ducts leading to deeper alveoli. This scanning electron micrograph was made after the inflated lung was quick-frozen in liquid nitrogen to preserve the appearance of the living lung. (× 180.) (From Kuhn, C., III, and E. H. Finke. 1972. J. Ultrastructural Res. **38:**161-173.)

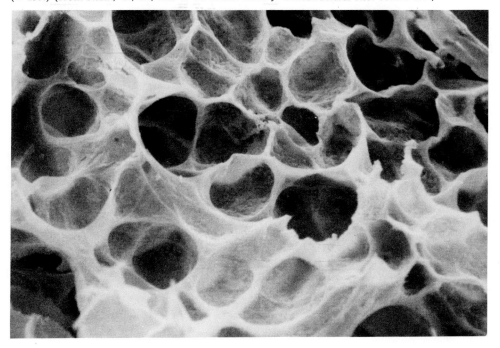

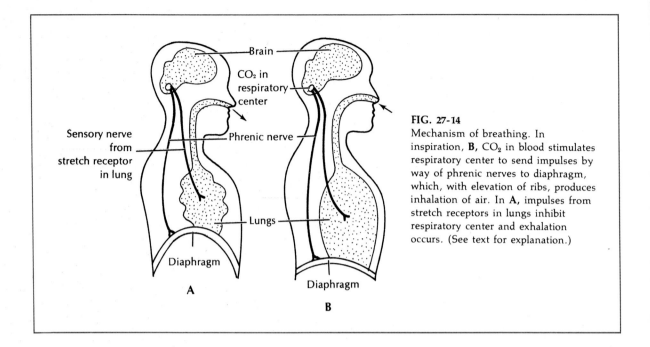

FIG. 27-14
Mechanism of breathing. In inspiration, **B**, CO_2 in blood stimulates respiratory center to send impulses by way of phrenic nerves to diaphragm, which, with elevation of ribs, produces inhalation of air. In **A**, impulses from stretch receptors in lungs inhibit respiratory center and exhalation occurs. (See text for explanation.)

pleural space exists; the two pleura rub together, lubricated by lymph. The chest cavity is bounded by the spine, ribs, and breastbone, and floored by the **diaphragm,** a dome-shaped, muscular partition between chest cavity and abdomen.

MECHANISM OF BREATHING. The chest cavity is an air-tight chamber. In **inspiration** the ribs are elevated, the diaphragm contracted and flattened, and the chest cavity is enlarged. The resultant increase in volume of chest cavity and lungs causes the air pressure in the lungs to fall below atmospheric pressure; air rushes in through the air passageways to equalize the pressure. **Expiration** is a less active process than inspiration. When their muscles relax, the ribs and diaphragm return to their original position and the chest cavity size decreases. The elastic lungs then contract and force the air out.

CONTROL OF BREATHING. Respiration must adjust itself to the varying needs of the body for oxygen. Respiration is normally involuntary and automatic but may come under voluntary control. The rhythmic inspiratory and expiratory movements are controlled by a nervous mechanism centered in the **medulla oblongata** of the brain (Fig. 27-14). By placing tiny electrodes in various parts of the medulla of experimental animals, neurophysiologists located separate **inspiratory** and **expiratory neurons** that act

reciprocally to stimulate the inspiratory and expiratory muscles of the diaphragm and rib cage (intercostal) muscles. The rate of breathing is determined by the amount of carbon dioxide in the blood: a slight rise in the blood CO_2 stimulates respiration; a fall will decrease breathing.

COMPOSITION OF INSPIRED, EXPIRED, AND ALVEOLAR AIRS. The composition of expired and alveolar airs is not identical. Air in the alveoli contains less oxygen and more carbon dioxide than does the air that leaves the lungs. Inspired air has the composition of atmospheric air. Expired air is really a mixture of alveolar and inspired airs. The variations in the three kinds of air are shown in Table 27-2. The water given off in expired air depends on the relative

TABLE 27-2
Variation in respired air

	Inspired air (vol. %)	Expired air (vol. %)	Alveolar air (vol. %)
Oxygen	20.96	16	14.0
Carbon dioxide	0.04	4	5.5
Nitrogen	79.00	80	80.5

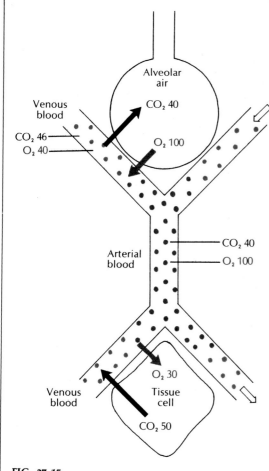

FIG. 27-15
Exchange of respiratory gases in lungs and tissue cells. Numbers present partial pressures in millimeters of mercury.

pressure at sea level is equivalent to 760 mm. of mercury (Hg), the partial pressure of O_2 will be 21% (percentage of O_2 in air) of 760, or 159 mm. Hg. The partial pressure of oxygen in the lung alveoli is greater (100 mm. Hg pressure) than it is in venous blood of lung capillaries (40 mm. Hg pressure) (Fig. 27-15). Oxygen then naturally diffuses into the capillaries. In a similar manner the carbon dioxide in the blood of the lung capillaries has a higher concentration (46 mm. Hg) than has this same gas in the lung alveoli (40 mm. Hg), so that carbon dioxide diffuses from the blood into the alveoli.

In the tissues respiratory gases also move according to their concentration gradients (Fig. 27-15). Here the concentration of oxygen in the blood (100 mm. Hg pressure) is greater than in the tissues (0 to 30 mm. Hg pressure), and the carbon dioxide concentration in the tissues (45 to 68 mm. Hg pressure) is greater than that in blood (40 mm. Hg pressure). The gases in each case will go from a high to a low concentration.

TRANSPORT OF GASES IN BLOOD. In some invertebrates the respiratory gases are simply carried dissolved in the body fluids. However, the solubility of oxygen is so low in water that this means is adequate only for animals having low rates of metabolism. For example, only about 1% of man's oxygen requirement can be transported in this way. Consequently in just about all the advanced invertebrates and the vertebrates, nearly all the oxygen and a significant amount of the carbon dioxide are transported by special colored proteins, or **respiratory pigments,** in the blood. In most animals (all vertebrates) these respiratory pigments are packaged into blood corpuscles. This is necessary because if this amount of respiratory pigment were free in blood, the blood would have the viscosity of syrup and would barely flow through the blood vessels, if at all.

Respiratory pigments. The two most widespread respiratory pigments are **hemoglobin,** a red, iron-containing protein present in all vertebrates and many invertebrates, and **hemocyanin,** a blue, copper-containing protein present in the crustaceans and cephalopod mollusks. Among other pigments are **chlorocruorin** (klor-a-kroo'o-rin), a green-colored, iron-containing pigment found in four families of polychaete tube worms. Its structure and oxygen-carrying capacity are very similar to those of hemoglobin, but it is carried free in the plasma rather than

humidity of the external air and the activity of the person. At ordinary room temperature and with a relative humidity of about 50%, an individual in performing light work will lose about 350 ml. of water from the lungs each day.

GASEOUS EXCHANGE IN LUNGS. The diffusion of gases, both in internal as well as external respiration, takes place in accordance with the laws of physical diffusion; that is, the gases pass from regions of high pressure to those of low pressure. The pressure of a gas refers to the partial pressure that that gas exerts in a mixture of gases. If the atmospheric

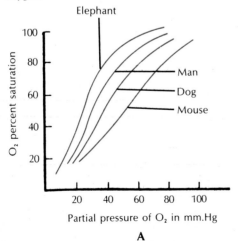

FIG. 27-16
Chemical structure of heme, an iron porphyrin compound composed of four pyrrole rings joined together with methane groups. There are four heme units in each hemoglobin molecule, each capable of carrying one oxygen molecule; thus each hemoglobin molecule when fully loaded is transporting four O_2 molecules.

being enclosed in blood corpuscles. Some polychaete worms have both chlorocruorin and hemoglobin present in their blood. **Hemerythrin** is a red pigment found in some polychaete worms. Although it contains iron, this metal is not present in a heme group (despite its name!) and its oxygen-carrying capacity is poor.

Hemoglobin is a complex protein. Each molecule is made up of 5% **heme,** an iron-containing compound giving the red color to blood (Fig. 27-16), and 95% **globin,** a colorless protein. The heme portion of the hemoglobin has a great affinity for oxygen; each gram of hemoglobin (there are about 15 grams of hemoglobin in each 100 ml. of human blood) can carry a maximum of approximately 1.3 ml. of oxygen; each 100 ml. of fully oxygenated blood contains about 20 ml. of oxygen. Of course, for hemoglobin to be of value to the body it must hold oxygen in a loose, reversible chemical combination so that it can be released to the tissues. The actual amount of oxygen bound to hemoglobin depends on the oxygen partial pressure surrounding the blood corpuscles, a rela-

FIG. 27-17
Oxygen dissociation curves. Curves show how the amount of oxygen bound to hemoglobin (oxyhemoglobin) is related to oxygen pressure. **A,** Small animals have blood that gives up oxygen more readily than does the blood of large animals. **B,** Oxyhemoglobin is sensitive to carbon dioxide pressure; as carbon dioxide enters blood from the tissues, it shifts the curve to the right, decreasing affinity of hemoglobin for oxygen.

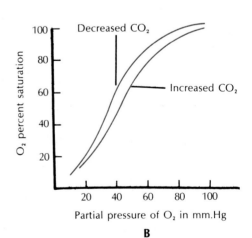

tionship expressed in the oxygen dissociation curve in Fig. 27-17. When the oxygen tension is high, as it is in the capillaries of the lung alveoli, hemoglobin becomes almost fully saturated to form oxyhemoglobin. Then, when the oxygenated blood leaves the lung and is distributed to the systemic capillaries in the body tissues, it enters regions of low oxygen partial pressure because oxygen is continuously consumed by cellular oxidative processes. The oxyhemoglobin now releases its bound oxygen, which diffuses into the cells. As the oxygen dissociation curve shows (Fig. 27-17), the lower the surrounding oxygen tension, the greater the quantity of oxygen released. This is an important characteristic because it allows more oxygen to be released to those tissues that need it most (have the lowest oxygen pressure). Another characteristic facilitating the release of oxygen to the tissues is the sensitivity of oxyhemoglobin to carbon dioxide. Carbon dioxide shifts the oxygen dissociation curve to the right (Fig. 27-17, *B*), a phenomenon that has been called the **Bohr effect** after the Danish scientist who first described it. Therefore as carbon dioxide enters the blood from the respiring tissues, it encourages the release of additional oxygen from the hemoglobin. The opposite event occurs in the lungs; as carbon dioxide diffuses from the venous blood into the alveolar space, the oxygen dissociation curve shifts back to the left, allowing more oxygen to be loaded onto the hemoglobin.

Unfortunately for man and other higher animals, hemoglobin has even a greater affinity for carbon monoxide (CO) than it has for oxygen—in fact, the affinity is about 200 times greater for CO than for O_2. Carbon monoxide is becoming an atmospheric contaminant of ever-increasing proportions as the world's population and industrialization continues rapidly upward. This odorless and invisible gas displaces oxygen from hemoglobin to form a stable compound called **carboxyhemoglobin.** Air containing only 0.2% CO may be fatal. Children and small animals are poisoned more rapidly than adults because of their higher respiratory rate.

Transport of carbon dioxide by the blood. The same blood that transports oxygen to the tissues from the lungs must carry carbon dioxide back to the lungs on its return trip. However, unlike oxygen that is transported almost exclusively in combination with hemoglobin, carbon dioxide is transported in three major forms.

1. Most of the carbon dioxide, about 67%, is converted in the red blood cells into bicarbonate and hydrogen ions, by undergoing the following series of reactions:

$$CO_2 + H_2O \rightleftharpoons H_2CO_3$$

Carbonic
acid

This reaction would normally proceed very slowly, but an enzyme in the red blood cells, **carbonic anhydrase,** catalyzes the reaction to proceed almost instantly. As soon as carbonic acid forms, it instantly and almost completely ionizes as follows:

$$H_2CO_3 \rightleftharpoons HCO_3^- + H^+$$

Carbonic   Bicarbonate   Hydrogen
acid   ion   ion

The hydrogen ion is buffered by several buffer systems in the blood, thus preventing a severe drop in blood pH. The bicarbonate ion remains in solution in the plasma and red blood cell water, since unlike carbon dioxide bicarbonate is extremely soluble.

2. Another fraction of the carbon dioxide, about 25%, combines reversibly with hemoglobin. It is carried to the lungs, where the hemoglobin releases it in exchange for oxygen.

3. A third small fraction of the carbon dioxide, about 8%, is carried as the physically dissolved gas in the plasma and red blood cells.

■ EXCRETION AND HOMEOSTASIS

At the beginning of this chapter we described the double-layered environment of the body's cells: the extracellular fluid *(milieu intérieur)* that immediately surrounds the cells, and the external environment *(milieu extérieur)* of the outside world. The life-supporting metabolic activities that occur within the body's cells can proceed only as long as they are bathed by a protective extracellular fluid environment of relatively constant composition. Yet there are many activities that threaten to throw the system out of balance. It is apparent that body fluid composition can be altered either by metabolic events occurring within the cells and tissues or by events occurring across the surface of the body. In other words, a living system is "open at both ends." On the inside, metabolic activities within the cell require a steady supply of materials, and these activities turn out a continuous flow of products and wastes. On the outside, materials are constantly being exchanged between the plasma and the external environment. Water, which makes up about two thirds of the body weight of animals, is always entering and leaving the body. Water is also formed within the cells as a by-product of oxidative processes. Ionized inorganic and organic salts are continually moving between the cells and body fluids and also between the animal and its environment. Protein is constantly being formed, transported, and broken down again within the tissues, yielding nitrogenous wastes that must be excreted. Obviously, body composition is a dynamic rather than a static thing. It is often described as operating as a **dynamic steady state.** This means that constancy of composition is maintained despite the continuous shifting of components of the system. This kind of internal regulation is **homeostasis.**

Homeostasis is maintained by the coordinated activities of numerous body systems, such as the nervous system, endocrine system, and, especially, the organs that serve as sites of exchange with the external environment. The last includes the kidneys, lungs or gills, alimentary canal, and skin. Through these organs enter oxygen, foodstuffs, minerals, and other constituents of the body fluids; here water is exchanged and metabolic wastes are eliminated. The kidney is the chief regulator of the body fluids. We tend to regard the kidney strictly as an organ of excretion that serves to rid the body of metabolic wastes. But in fact it is as much a regulatory organ as an excretory organ. It is responsible for individually monitoring and regulating the concentrations of body water, of the salt ions in the blood, and of other major and minor body fluid constituents. In its task of fine-tuning the composition of the internal environment, the kidney is assisted by the other organs of exchange such as the lungs, skin, and digestive tract, as well as by many internal mechanisms. Several other specialized structures have evolved among the vertebrates that assist in body fluid regulation in various environments; for example, the salt-secreting cells of fish gills and the salt glands of birds and reptiles.

How aquatic animals meet problems of salt and water balance

MARINE INVERTEBRATES. Most marine invertebrates are in osmotic equilibrium with their seawater environment. They have body surfaces that are permeable to salts and water so that their body fluid concentration rises or falls in conformity with changes in concentrations of seawater. Because such animals are incapable of regulating their body fluid osmotic pressure, they are referred to as **osmotic conformers.** Invertebrates living in the open sea are seldom exposed to osmotic fluctuations because the ocean is a highly stable environment. Oceanic invertebrates have, in fact, very limited abilities to withstand osmotic change. If they should be exposed to dilute seawater, they are quickly killed because their body cells cannot tolerate dilution and are helpless to prevent it. These animals are restricted to living in a narrow salinity range, and are said to be **stenohaline** (Gr. *stenos,* narrow, + *hals,* salt). An example is the marine spider crab, represented in Fig. 27-18.

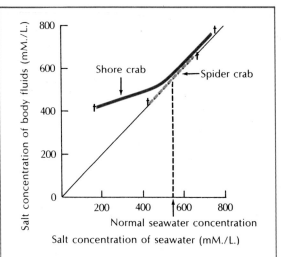

FIG. 27-18

Salt concentration of body fluids of two crabs as affected by variations in the seawater concentration. The 45-degree line represents equal concentration between body fluids and seawater. Since the spider crab cannot regulate its body-fluid salt concentration, it conforms to whatever changes happen in the external seawater environment. The shore crab, however, can regulate osmotic concentration of its body fluids to some degree because in dilute seawater the shore crab can hold its body-fluid concentration above the seawater concentration. For example, when the seawater is 200 mM. per liter, the shore crab's body fluids are about 430 mM. per liter. Crosses at ends of lines indicate tolerance limits of each species.

Conditions along the coasts and in estuaries and river mouths are much less constant than those of the open ocean. Here animals must be able to withstand large, and often abrupt, salinity changes as the tides move in and out and mix with freshwater draining from rivers. These animals are referred to as **euryhaline** (Gr. *eurys,* broad, + *hals,* salt), which means that they can survive a wide range of salinity change. Most coastal invertebrates also show varying powers of **osmotic regulation.** For example, the brackish water shore crab can resist body fluid dilution by dilute (brackish) seawater (Fig. 27-18). Although the body fluid concentration falls, it does so less rapidly than the fall in seawater concentration. This crab is a **hyperosmotic regulator** because in a dilute environment it can maintain the concentration of its blood above that of the surrounding water.

What is the advantage of hyperosmotic regulation over osmotic conformity, and how is this regulation accomplished? The advantage is that by regulating against excessive dilution, thus protecting the body cells from extreme changes, these crabs can successfully live in the physically unstable but **biologically rich** coastal environment. Their powers of regulation are limited however, since if the water is highly diluted, their regulation fails and they die. To understand how the brackish-water shore crab and other coastal invertebrates achieve hyperosmotic regulation, let us examine the problems they face. First, the salt concentration of the internal fluids is greater than in the dilute seawater outside. This causes a steady osmotic influx of water. As with the membrane osmometer placed in a sugar solution (p. 67), water diffuses inward because it is more concentrated outside than inside. The shore crab is not nearly as permeable as a membrane osmometer—most of its shelled body surface is in fact almost impermeable to water—but the thin respiratory surfaces of the gills are highly permeable. Obviously the crab cannot insulate its gills with an impermeable hide and still breathe. The problem is solved by removing the excess water through the action of the kidney (the antennal gland located in the crab's thorax). The second problem is salt loss. Again, because the animal is saltier than its environment, it cannot avoid loss of ions by outward diffusion across the gills. Salt is also lost in the urine. This problem is solved by special salt-secreting cells in the gills that can actively remove ions from the dilute seawater and move them

into the blood, thus maintaining the internal osmotic concentration. This is an **active transport** process that requires energy because ions must be transported against a concentration gradient, that is, from a lower salt concentration (in the dilute seawater) to an already higher one (in the blood).

INVASION OF FRESHWATER. Some 400 million years ago, during the Silurian and Lower Devonian, the major groups of jawed fishes began to penetrate into brackish-water estuaries and then gradually into freshwater rivers. Before them lay a new unexploited habitat already stocked with food, in the form of insects and other invertebrates, which had preceded them into freshwater. However, the advantages of this new habitat were traded off for a tough physiologic challenge: the necessity of developing effective osmotic regulation. Freshwater animals must keep the salt concentration of their body fluids higher than that of the water. Water therefore enters their bodies osmotically and salt is lost by diffusion outward. Their problems are similar to the brackish-water shore crab, but more severe and unremitting. Freshwater is much more dilute than are coastal estuaries, and there is no retreat, no salty sanctuary into which the freshwater animal can retire for osmotic relief. He must, and has, become a permanent and highly efficient hyperosmotic regulator. The scaled and mucus-covered body surface of a fish is about as waterproof as any flexible surface can be. The water that inevitably enters across the gills is pumped out by the kidney. Even though the kidney is able to make a very dilute urine some salt is lost; this is replaced by salt in food and by active absorption of salt (primarily sodium and chloride) across the gills. The bony fishes that inhabit our lakes and streams today are so well adapted to their dilute surroundings that they need expend very little energy to regulate themselves osmotically. Osmotic regulation in fishes is described on p. 474 and illustrated in Fig. 21-20.

Crayfish, aquatic insect larvae, mussels, and other freshwater animals are also hyperosmotic regulators and face the same hazards as freshwater fish; they tend to gain too much water and lose too much salt. Not surprisingly, all of these forms solved these problems in the same direct way that fish did. They excrete the excess water as urine, and they actively absorb salt from the water by some salt-transporting mechanism on the body surface.

Amphibians, when they are living in water, also

must compensate for salt loss by absorbing salt from the water. They use their skin for this purpose. Physiologists learned some years ago that pieces of frog skin will continue to actively transport sodium and chloride for hours when removed and placed in a specially balanced salt solution. Fortunately for biologists, but unfortunately for frogs, these animals are so easily collected and maintained in the laboratory that frog skin has become a favorite membrane system for studies of ion-transport phenomena.

MARINE FISHES. The great families of bony fishes that inhabit the seas today maintain the salt concentration of their body fluids at about one third that of seawater (body fluids = 0.3 to 0.4 M.; seawater = 1 M.). Obviously they must be osmotic regulators. Bony fishes living in the oceans today are descendants of earlier freshwater bony fishes that moved back into the sea during the Triassic about 200 million years ago. The return to their ancestral sea was probably prompted by unfavorable climatic conditions on land and the deterioration of freshwater habitats, but we can only guess at the reasons. During the many millions of years that the freshwater fishes were adapting themselves so well to their environment, they established a body fluid concentration equivalent to about one-third seawater, thus setting the pattern for all the vertebrates that were to evolve later, whether aquatic, terrestrial, or aerial. The ionic composition of vertebrate body fluid is remarkably like dilute seawater too, a fact that is doubtless related to their marine heritage.

When some of the freshwater bony fishes of the Triassic ventured back to the sea, they encountered a new set of problems. Having a much lower internal osmotic concentration than the seawater around them, they lost water and gained salt. Indeed, the marine bony fish quite literally risks drying out, much like a desert mammal deprived of water. The way the marine bony fishes osmotically regulate was described on p. 474 and therefore will not be detailed here. In brief, to compensate for water loss, the marine teleost drinks seawater. This is absorbed from the intestine, and the major sea salt, sodium chloride, is carried by the blood to the gills, where specialized salt-secreting cells transport it back into the surrounding sea. The ions remaining in the intestinal residue, especially magnesium, sulfate, and calcium, are voided with the feces or excreted by the kidney. In this roundabout way, marine fishes rid themselves of the excess sea salts they have drunk, resulting in a net gain of water, which replaces the water lost by osmosis. Samuel Taylor Coleridge's ancient mariner, surrounded by "water, water, everywhere" would doubtless have been tormented even more had he known of the marine fishes' simple solution for thirst. A marine fish carefully regulates the amount of seawater it drinks, consuming only enough to replace water loss and no more.

The cartilaginous sharks and rays (elasmobranchs) solved their water balance problems in a completely different way. This primitive group is almost totally marine. The salt composition of shark's blood is similar to the bony fishes, but also contains a large amount of organic compounds, especially urea and trimethylamine oxide. Urea is, of course, a metabolic waste that most animals quickly excrete in the urine. The shark kidney, however, conserves urea, causing it to accumulate in the blood. The blood urea, added to the usual blood electrolytes, raises the blood osmotic pressure to slightly exceed that of seawater. In this way the sharks and their kin turn an otherwise useless waste material into an asset, eliminating the osmotic problem encountered by the marine bony fishes.

How terrestrial animals maintain salt and water balance

The problems of living in an aquatic environment seem small indeed compared to the problems of life on land. Remembering that our bodies are mostly water, that all metabolic activities proceed in water, and that the origins of life itself were conceived in water, it seems obvious that animals were meant to stay in water. Yet many animals, like the plants preceding them, inevitably moved onto land, carrying their watery composition with them. Once on land, the terrestrial animals continued their adaptive radiation, undaunted by the threat of desiccation, until they became abundant even in some of the most arid parts of the earth. Terrestrial animals lose water by evaporation from the lungs and body surface, by excretion in the urine, and by elimination in the feces. Such losses are replaced by water in the food, by drinking water if it is available, and by forming metabolic water in the cells by the oxidation of foodstuffs. In some desert rodents, metabolic water may constitute most of the animals' water gain. Particularly revealing is a comparison (Table 27-3) of water bal-

TABLE 27-3
Water balance in man and kangaroo rat,
a desert rodent*

	Man (%)	Kangaroo rat (%)
Gains		
drinking	48	0
free water in food	40	10
metabolic water	12	90
Losses		
urine	60	25
evaporation (lungs and skin)	34	70
feces	6	5

*Partly from Schmidt-Nielsen, K. 1972. How animals work, Cambridge, Cambridge University Press.

ance in man, a nondesert mammal that drinks water, with that of the kangaroo rat, a desert rodent that may drink no water all.

The excretion of wastes presents a special problem in water conservation. The primary end-product of protein catabolism is ammonia, a highly toxic material. Fishes can easily excrete ammonia across their gills, since there is an abundance of water to wash it away. The terrestrial insects, reptiles, and birds have no convenient way to rid themselves of toxic ammonia; so instead they convert it into uric acid, a nontoxic, almost insoluble compound. This enables them to excrete a semisolid urine with little water loss. The use of uric acid has another important benefit. All of these animals lay eggs that are impermeable bags enclosing the embryos, their stores of food and water, and whatever wastes that accumulate during development. By converting ammonia to uric acid, the developing embryo's waste can be precipitated into solid crystals, which are stored harmlessly within the egg until hatching.

Marine birds and turtles have evolved a unique solution for excreting the large loads of salt eaten with their food. Located above each eye is a special **salt gland** capable of excreting a highly concentrated solution of sodium chloride—up to twice the concentration of seawater. In birds the salt solution runs out the nares (see p. 538). Marine turtles and lizards shed their salt gland secretion as salty tears. Salt glands are important accessory organs of salt excretion to

these animals, since their kidney cannot produce a concentrated urine, as can the mammalian kidney.

Invertebrate excretory structures

In such a large and varied group as the invertebrates it is hardly surprising that there is a great variety of morphologic structures serving as excretory organs. Many Protozoa and some freshwater sponges have special excretory organelles called contractile vacuoles. The more advanced invertebrates have excretory organs that are basically tubular structures that form urine by first producing an ultrafiltrate or fluid secretion of the blood. This enters the proximal end of the tubule and is modified continuously as it flows down the tubule. The final product, urine, frequently has a composition very different from the blood from which it originated.

CONTRACTILE VACUOLE. This tiny spherical intracellular vacuole of protozoans and freshwater sponges is not a true excretory organ, since ammonia and other nitrogenous wastes of metabolism readily leave the cell by direct diffusion across the cell membrane into the surrounding water. The contractile vacuole is really an organ of water balance. Because the cytoplasm of freshwater Protozoa is considerably saltier than their freshwater environment, they tend to draw water into themselves by osmosis. In *Paramecium* (Fig. 7-24) this excess water is collected by minute canals within the cytoplasm and conveyed to the contractile vacuole. This grows larger as water accumulates within it. Finally the vacuole is emptied through a pore on the surface, and the cycle is rhythmically repeated. Although the contractile vacuole has been carefully studied, it is not yet known how this system is able to pump out pure water while retaining valuable salts within the animal. Contractile vacuoles are common in freshwater Protozoa but rare or absent from marine Protozoa, which are isosmotic with seawater and consequently neither lose nor gain too much water.

NEPHRIDIA. The nephridium is the most common type of invertebrate excretory organ. All nephridia are tubular structures, but there are large differences in degree of complexity. One of the simplest arrangements is the flame cell system (or **protonephridia**) of the flatworm. In *Planaria* and other flatworms this takes the form of two highly branched systems of tubules distributed throughout the body (Fig. 10-4). Fluid enters the system through specialized "flame" cells, moves slowly into and down the tubules, and is excreted through pores that open at intervals on the body surface. It is believed that fluid containing wastes enters the flame bulb from the surrounding tissues by **pinocytosis** (cell drinking). In this process fluid-filled vesicles are formed just under the outer cell surface. The vesicles are transported across the cell and set free into the ciliated lumen of the flame cell. Here, the rhythmic flame beat creates a negative fluid pressure that drives the fluid into the tubular portion of the system. It is probable that as the fluid passes down the tubules, the tubular epithelial cells add certain waste materials to the tubular fluid (secretion) and withdraw valuable materials from it (reabsorption) to complete the formation of urine. The flame cell system, like the contractile vacuole of protozoans, is primarily a water balance system, since it is best developed in free-living freshwater forms. Branched flame cell systems are typical of primitive invertebrates that lack circulatory systems. Since there is no circulation to carry wastes to a compact excretory organ such as the kidney of higher invertebrates and vertebrates, the flame cell system must be distributed to reach the cells directly.

The protonephridium just described is a **"closed"** system, that is, the urine is formed from a fluid that must first enter the tubule by being transported across the flame cells. Another type of nephridium, typical of many annelids (segmented worms), is the **open,** or "true," nephridium. In the earthworm *Lumbricus* there are paired nephridia in every segment of the body, except the first three and the last one (Fig. 13-6). Each nephridium occupies parts of two successive segments. In the earthworm each nephridium is a tiny, self-contained "kidney" that independently drains to the outside through pores (**nephridiopores**) in the body wall.

Coelomic fluid containing wastes to be excreted is swept into a ciliated, funnellike opening (**nephrostome**) of the nephridium and carried through a long, twisted tubule of increasing diameter. It then enters a bladder and is finally expelled through a nephridiopore to the outside. The nephridial tubule is surrounded by an extensive network of blood vessels. Solutes, especially sodium and chloride, are reabsorbed from the formative urine during its travel through the tubule. The addition of the blood vascular network to the annelid nephridium makes it a much more versatile and effective system than the flame

cell system. However, the basic process of urine formation is the same: fluid flows continuously through a tubule while materials are added here and taken away there, until urine is formed.

ARTHROPOD KIDNEYS. The **antennal glands** of crustaceans form a single, paired tubular structure located in the ventral part of the head. Their structure and function was described on p. 327. These excretory devices are an advanced design of the basic nephridial organ. However, they lack open nephrostomes. Instead, a protein-free filtrate of the blood (ultrafiltrate) is formed in the end sac by the hydrostatic pressure of the blood. In the tubular portion of the gland, the filtrate is modified by the selective reabsorption of certain salts and the active secretion of others. Thus, crustaceans have excretory organs that are basically vertebrate-like in the functional sequence of urine formation.

Insects and spiders have a unique excretory system consisting of **malpighian tubules** that operate in conjunction with specialized glands in the wall of the rectum (see p. 362). The thin, elastic, blind malpighian tubules are closed and lack an arterial supply. Consequently urine formation cannot be initiated by blood ultrafiltration as in the crustaceans and vertebrates. Instead salts, largely potassium, are actively secreted into the tubules. This primary secretion of ions creates an osmotic drag that pulls water, solutes, and waste materials into the tubule. The fluid, or "urine," then drains from the tubules into the intestine, where specialized rectal glands actively reabsorb most of the potassium and water, leaving behind wastes such as uric acid. This unique excretory system is ideally suited for life in dry environments. We must assume that it has contributed to the great success of this most abundant and widespread group of land animals.

Vertebrate kidney function

The kidneys of man and other vertebrates play a critical role in the body's economy. As vital organs their failure means death; in this respect they are neither more nor less important than are the heart, lungs, or liver. The kidney is part of many interlocking mechanisms that maintain **homeostasis** —constancy of the internal environment. However, the kidney's share in this regulatory council is an especially large one. It must, and does, individually monitor and regulate most of the major constituents of the blood and several minor constituents as well. In addition it silently labors to remove a variety of potentially harmful substances that animals deliberately or unconsciously eat, drink, or inhale.

Perhaps even more remarkable than the job the kidney does is the way in which it does it. These small organs, which in man weigh less than 0.5% of the body's weight, receive nearly 25% of the total blood flow (cardiac output), amounting to about 2,000 liters of blood per day. This vast blood flow is channeled to approximately 2 million nephrons, which comprise the bulk of the two human kidneys. Each nephron is a tiny excretory unit consisting of

a pressure filter (**glomerulus**) and a long **nephric tubule.** Urine formation begins in the glomerulus where an ultrafiltrate of the blood is squeezed into the nephric tubule by the hydrostatic blood pressure. The ultrafiltrate then flows steadily down the twisted tubule. During its travel some substances are added to, and others are subtracted from, the ultrafiltrate. The final product of this process is urine.

All mammalian kidneys are paired structures that lie embedded in fat, anchored against the dorsal abdominal wall. Each kidney contains a medial indentation, the **hilus,** which is the site of entry and exit of blood vessels and which leads to the **ureter.** The

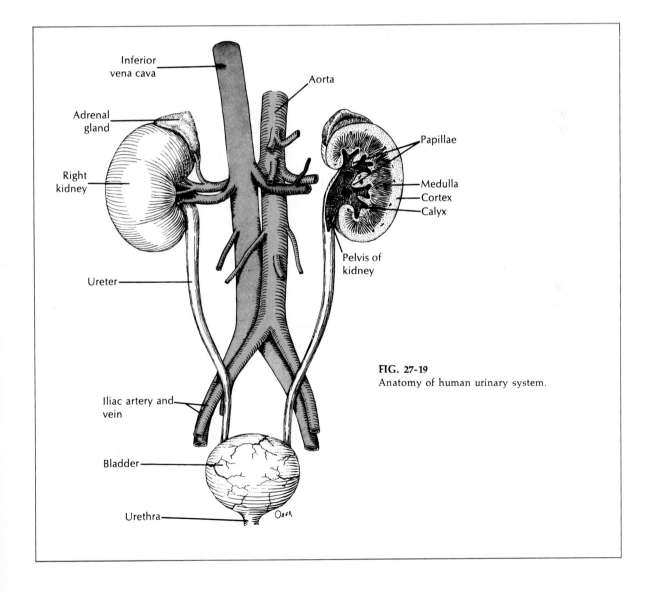

FIG. 27-19
Anatomy of human urinary system.

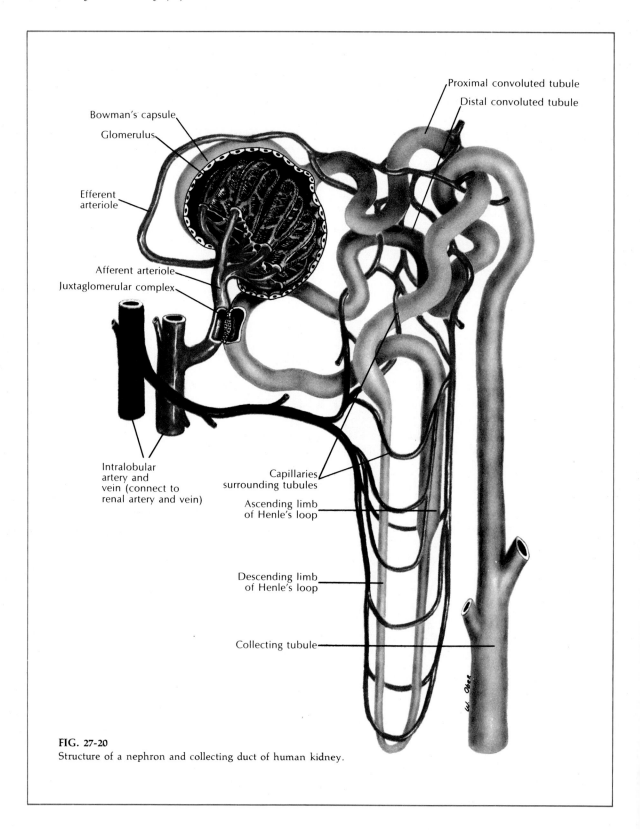

FIG. 27-20
Structure of a nephron and collecting duct of human kidney.

two ureters, 10 to 12 inches long in man, extend to the dorsal surface of the urinary bladder. Urine is discharged from the bladder by way of the single **urethra** (Fig. 27-19). In the male the urethra is the terminal portion of the reproductive system as well as of the excretory system. In the female the urethra is solely excretory in function, opening to the outside just anterior to the vagina.

Urine formed by the kidney nephrons drains into the renal pelvis and then into the muscular ureter, which conveys it to the urinary bladder by rhythmic muscular contractions. Once urine has entered the bladder, a small flap of epithelial tissue acts as a valve to prevent backflow into the urethra. The urinary bladder is a highly expansible muscular bag. As the bladder becomes distended, stretch receptors located in the muscular walls are stimulated, producing nerve impulses that travel to the brain to signal the sensation of fullness. The release of urine from the bladder is regulated by a muscular sphincter located at the juncture of bladder and urethra. Voiding of urine is involuntary in infants but comes under voluntary control during early childhood.

Since each of the thousands of nephrons in the kidney forms urine independently, each is, in a way, a tiny, self-contained kidney that produces a miniscule amount of urine—perhaps only a few nanoliters per hour. This amount, multiplied by the number of nephrons in the kidney, produces the total urine flow. The kidney is an "in parallel" system of independent units. However, as we will see later, these "independent" nephrons actually work together to create large osmotic gradients in the kidney medulla. This makes it possible for the mammalian kidney to concentrate urine well above the salt concentration of the blood.

As indicated above, the nephron, with its pressure filter and tubule, is intimately associated with the blood circulation (Fig. 27-20). Blood from the aorta is delivered to the kidney by way of the large **renal artery,** which breaks up into a branching system of smaller arteries. The arterial blood flows to each nephron through an **afferent arteriole** to the **glomerulus** (glo-mer'yoo-lus), which is a tuft of blood capillaries enclosed within a thin, cuplike **Bowman's capsule.** Blood leaves the glomerulus via the **efferent arteriole.** This vessel immediately breaks up again into an extensive system of capillaries, the **peritubular capillaries,** which completely surround the

nephric tubules. Finally, the blood from these many capillaries is collected by veins that unite to form the **renal vein.** This vein returns the blood to the vena cava.

GLOMERULAR FILTRATION. Let us now return to the glomerulus, where the process of urine formation begins. The glomerulus acts as a specialized mechanical filter in which a protein-free filtrate resembling plasma is driven by the blood pressure across the capillary walls and into the fluid-filled space of Bowman's capsule. As shown in Fig. 27-21, the net filtration pressure of about 20 mm. Hg is the difference between the blood pressure in the glomerular capillaries, believed to be about 65 mm. Hg, and the opposing colloid osmotic and hydrostatic back pressures. Most important of these negative pressures is the colloid (protein) osmotic pressure, which is created because the proteins are too large to pass the glomerular membrane. The ultrafiltrate that is formed now begins to flow down the nephric tubule. This consists of several segments. The first segment, the **proximal convoluted tubule,** leads into a long, thinwalled, hairpin loop called the **loop of Henle** (Fig. 27-20). This loop drops deep into the medulla of the kidney and then returns to the cortex to join the third segment, the **distal convoluted tubule.** The collecting duct empties into the kidney **pelvis,** a cavity that collects the urine before it passes into the **ureter,** on its way to the **urinary bladder** (Fig. 27-19).

TUBULAR REABSORPTION. The ultrafiltrate that enters this complex tubular system must undergo extensive modification before it becomes urine. About 200 liters of filtrate are formed each day by the average person's kidneys. Obviously the loss of this volume of body water, not to mention the many other valuable materials present in the filtrate, cannot be tolerated. How does tubular action convert the plasma filtrate into urine? To answer this question researchers developed tubular micropuncture techniques. After exposing the kidneys of anesthetized amphibians or mammals, tiny glass pipettes with sharpened tips only a few microns in diameter were used to withdraw samples of formative urine from different segments of the tubule. The samples were chemically analyzed with specially developed ultramicroanalytical techniques. Such experiments showed that most of the water and solutes were reabsorbed by the proximal tubule. Some vital materials such as glucose and amino acids were comple-

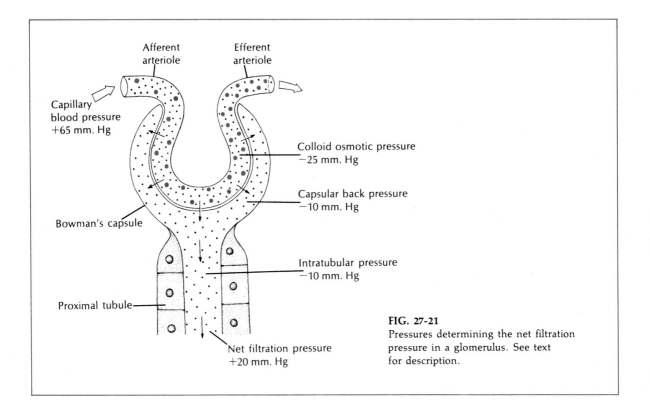

Afferent arteriole

Efferent arteriole

Capillary blood pressure +65 mm. Hg

Colloid osmotic pressure −25 mm. Hg

Capsular back pressure −10 mm. Hg

Bowman's capsule

Intratubular pressure −10 mm. Hg

Proximal tubule

Net filtration pressure +20 mm. Hg

FIG. 27-21
Pressures determining the net filtration pressure in a glomerulus. See text for description.

ly reabsorbed. Others such as sodium, chloride, and most other minerals underwent variable reabsorption. That is, some were strongly reabsorbed, others weakly reabsorbed, depending on the body's need to conserve each mineral. Much of this reabsorption is by **active transport,** in which cellular energy is used to transport materials from the tubular fluid, across the cell, and into the peritubular blood that will return them to the general circulation.

The reabsorption of sodium, the dominant cation in the plasma, illustrates the flexibility of the reabsorption process. About 600 grams of sodium are filtered by the human kidney every 24 hours. Nearly all of this is reabsorbed, but the exact amount is precisely matched to sodium intake. With a normal sodium intake of 4 grams per day, the kidney excretes 4 grams, and reabsorbs 596 grams each day. A person on a low-salt diet of 0.3 gram of sodium per day still maintains salt balance because only 0.3 gram escapes reabsorption. On a very high salt intake, the kidney can excrete up to about 10 grams of sodium per day, but not more. It may seem odd that the maximum sodium excretion is only 10 grams per day

when about 600 grams are filtered. One can logically ask: "Why can't more sodium be excreted by simply allowing more to escape tubular reabsorption?" The reason is that the filtration-reabsorption sequence has a built-in restriction in flexibility. Sodium (and other ions) are reabsorbed in both the proximal and distal portions of the convoluted tubule. Some 85% of the salt and water is reabsorbed in the proximal tubule; this is an **obligatory reabsorption** because it is governed entirely by physical processes (the osmotic pressure of the solutes) and cannot be controlled physiologically. In the distal tubule, however, sodium reabsorption can be controlled. This is called **facultative reabsorption,** meaning the reabsorption can be adjusted physiologically according to need. We can say that proximal reabsorption is involuntary and distal reabsorption is voluntary, although of course we are not aware of the adjustments the kidney is performing on our behalf. The flexibility of distal reabsorption varies considerably in different animals: it is restricted in man but very broad in many rodents. These differences have appeared because of selective pressures during evolution that fitted rodents for dry environments where they must conserve water and at the same time excrete considerable sodium.

Sodium reabsorption in the distal tubule is controlled by **aldosterone,** a steroid hormone of the adrenal cortex. The release of aldosterone is governed by a complex series of reactions that begin in the kidney. Although the precise signal for release is not known, a **decrease** in sodium intake causes the enzyme **renin** (ree'nin) to be released from the **juxtaglomerular complex,** located at a point in the nephron where the distal tubule touches the afferent arteriole (see Fig. 27-20). Renin enters the blood and converts a plasma globulin into **angiotensin I.** An enzyme in the plasma then converts angiotensin I into **angiotensin II,** which stimulates the adrenal cortex to release aldosterone. Aldosterone then increases the distal tubular reabsorption of sodium. If sodium intake is **increased,** the renin output drops, aldosterone is not released from the adrenal cortex, and more sodium is excreted by the kidney.

For most substances there is an upper limit to the amount of substance that can be reabsorbed. This upper limit is termed the **transport maximum** for that substance. For example, glucose is normally completely reabsorbed by the kidney because the transport maximum for the glucose reabsorptive mechanism is poised well above the amount of glucose normally present in the plasma filtrate (about 100 mg. per 100 ml. of filtrate). If the plasma glucose level rises above normal, a condition called hyperglycemia, the concentration in the filtrate will rise accordingly and a greater amount of glucose will be presented to the proximal tubule for reabsorption. As the level rises, eventually a point is reached (about 300 mg. per 100 ml. of plasma) where the reabsorptive capacity of the tubular cells is saturated. If the plasma level continues to rise, glucose will begin to appear in the urine (glycosuria). This condition happens in the untreated disease diabetes mellitus.

TUBULAR SECRETION. In addition to reabsorbing large amounts of materials from the plasma filtrate, the kidney tubules are able to secrete certain substances into the tubular fluid. This process, which is the reverse of tubular reabsorption, enables the kidney to build up the urine concentrations of materials to be excreted, such as hydrogen and potassium ions, drugs, and various foreign organic materials. The distal tubule is the site of most tubular secretion.

In the kidney of bony marine fishes, reptiles, and birds, tubular secretion is a much more highly developed process than it is in mammalian kidneys. Marine bony fishes actively secrete large amounts of magnesium and sulfate, which are by-products of their mode of osmotic regulation, which was described earlier (p. 474). Reptiles and birds excrete uric acid instead of urea as their major nitrogenous waste. This material is actively secreted by the tubular epithelium. Since uric acid is nearly insoluble, it forms crystals in the urine and requires little water for excretion. Thus the excretion of uric acid is an important adaptation for water conservation.

WATER EXCRETION. The total osmotic pressure of the blood is carefully regulated by the kidney. When fluid intake is high, the kidney excretes a dilute urine, saving salts and excreting water. When fluid intake is low, the kidney conserves water by forming a concentrated urine. A dehydrated man can concentrate his urine to four times his blood concentration.

Understanding how the mammalian kidney can adjust water excretion according to intake, and especially how it produces a concentrated urine, has been a relatively recent development in kidney research. Until 1951, the function of the long, hairpinlike loop of Henle was unknown. In that year Hargitay and Kuhn advanced that the loops of Henle formed a countercurrent system to concentrate urine. It was discovered that the osmotic pressure of the kidney tissue fluid increased progressively from the cortex to the urinary papillae, where the urine is discharged into the kidney pelvis. This increase is suggested by the color gradation in Fig. 27-22. The long loops of Henle, which reach deep into the medulla of the kidney, function as a **countercurrent multiplier system** (Fig. 27-22). Sodium is actively reabsorbed from the water-impermeable ascending limb of Henle's loop, enters the surrounding tissue fluid, and diffuses into the descending loop. By cycling sodium between the two opposing limbs, the concentration of urine becomes multiplied in the bottom of the loop. A tissue fluid osmotic concentration is established that is greatest at the bottom of the loop deep in the medulla and lowest at the top of the loop in the cortex. The actual concentrating of the urine, however, does not occur in the loops of Henle, but in the collecting ducts that lie parallel to the loops of Henle. As the urine flows down the collecting duct into regions of

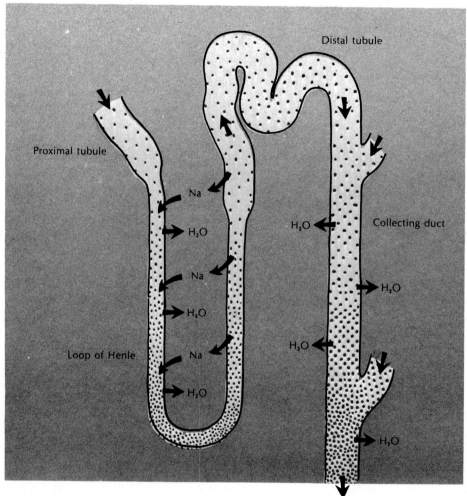

FIG. 27-22

Mechanism of countercurrent multiplier system of mammalian kidney. Relative concentration of sodium and other solutes is represented by density of red dots inside tubule, and intensity of color outside tubule. Concentration increases from top (cortex) to bottom (medulla). Final water absorption occurs in collecting duct. ∕ = active transport of sodium; ∕ = passive transport of sodium; ⟵ = passive transport of water.

increasing sodium concentration, water is osmotically withdrawn from the urine. The amount of water saved, and the final concentration of the urine, depends on the permeability of the walls of the collecting duct. This is controlled by the **antidiuretic hormone** (ADH, or vasopressin), which is released by the posterior pituitary gland (neurohypophysis). The release of this hormone is governed in turn by special receptors in the brain that constantly sense the osmotic pressure of the blood. When the blood

osmotic pressure drops, as during dehydration, more antidiuretic hormone is released, collecting duct permeability is increased, more water is withdrawn from the urine, and the urine becomes concentrated. An opposing sequence of events occurs during overhydration.

The varying ability of different mammals to form a concentrated urine is closely correlated with the length of the loops of Henle. The beaver, which has no need to conserve water in its aquatic environment, has short loops and can concentrate its urine to only about twice that of the blood plasma. Man with relatively longer loops can concentrate his urine 4.2 times that of the blood. As we would anticipate, desert mammals have much greater urine concentrating powers. The camel can produce a urine 8 times the plasma concentration, the gerbil 14 times, and the Australian hopping mouse 22 times. In this creature, greatest urine concentrator of all, the loops of Henle extend to the tip of a long renal papilla that pushes out into the mouth of the ureter.

References

General references for Part Three

Barrington, E. J. W. 1968. The chemical basis of physiological regulation. Glenview, Ill., Scott, Foresman & Co. *A selection of topics in comparative physiology, with emphasis on experimental approach. Clearly written.*

Eckstein, G. 1969. The body has a head. New York, Harper & Row, Publishers. *Written in an unusually appealing nontechnical style, this book presents a unique view of animal physiology.*

Florey, E. 1966. General and comparative physiology. Philadelphia, W. B. Saunders Co. *A detailed graduate level text, numerous illustrations, good invertebrate-vertebrate balance.*

Gordon, M. S. (editor). 1972. Animal function: principles and adaptations, ed. 2. New York, The Macmillan Co. *Graduate level vertebrate physiology.*

Hoar, W. S. 1966. General and comparative physiology. Englewood Cliffs, N. J., Prentice-Hall, Inc. *The best balanced of college comparative physiology texts.*

Prosser, C. L., and F. A. Brown, Jr. 1961. Comparative animal physiology. Philadelphia, W. B. Saunders Co. *This is a later edition of the well-known 1950 edition. Indispensable to all serious students of biology.*

Ramsay, J. A. 1952. A physiological approach to the lower animals. New York, Cambridge University Press.

Schmidt-Nielsen, K. 1964. Animal physiology, ed. 2. Englewood Cliffs, N. J., Prentice-Hall, Inc. *One of the best backgrounds for beginning the study of physiology before taking up the deeper physicochemical aspects of the subject.*

Schottelius, B. A., and D. D. Schottelius. 1973. Textbook of physiology, ed. 17. St. Louis, The C. V. Mosby Co. *Intermediate-level college human physiology text. Clearly written and illustrated.*

Vertebrate adaptations. Readings from Scientific American, with introductions by N. K. Wessels, 1968. San Francisco, W. H. Freeman & Co., Publishers. *Excellent collection of articles on gas exchange, vascular system, and water balance. The editors introductions helpfully summarize and round out the content. Highly recommended supplementary reading.*

Wilson, J. A. 1972. Principles of animal physiology. New York, The Macmillan Co. *A clearly written general and comparative physiology at the advanced undergraduate level.*

Wood, D. W. 1970. Principles of animal physiology. New York, American Elsevier Publishing Co., Inc. *Intermediate-level general physiology.*

Suggested general readings for Chapter 27

Brooks, S. M. 1960. Basic facts of body water and ions. New York, Springer Publishing Co., Inc. *An excellent elementary account of the fluid and electrolyte balance in the body.*

Chapman, G. 1967. The body fluids and their function. Institute of Biology's Studies in Biology no. 8. New York, St. Martin's Press. *Brief, comparative treatment of body fluids, their transport, and regulation.*

Cott, H. B. 1957. Adaptive coloration in animals. London, Methuen & Co., Ltd. *This monograph deals with the role of color adaptation in the animal kingdom.*

Edney, E. B. 1957. The water relations of terrestrial arthropods. New York, Cambridge University Press.

Graubard, M. 1964. Circulation and respiration. The evolution of an idea. New York, Harcourt, Brace & World, Inc. *Historic development of blood flow concepts. Selected writings from Aristotle, Galen, Vesalius, Fabricius, Harvey, Malpighi, Boyle, and others.*

Potts, W. T. W., and G. Parry. 1964. Osmotic and ionic regulation in animals. New York, The Macmillan Co. *An excellent treatise on concepts of homeostatic mechanisms, with special emphasis on osmoregulation and fluid balance.*

Schmidt-Nielsen, K. 1972. How animals work. Cambridge, The University Press. *This paperback deals especially with respiration, temperature regulation, and energy cost of locomotion.*

Snively, W. D., Jr. 1960. Sea within: the story of our body fluid. Philadelphia, J. B. Lippincott Co. *A popular, interesting treatise on the "interior sea" and its importance in our bodies in health and disease.*

Selected *Scientific American* articles

Adolph, E. F. 1967. The heart's pacemaker. **216**:32-37 (March).

Aird, R. B. 1956. Barriers in the brain. **194**:101-106 (Feb.). *Blood vessels in the brain regulate the passage of materials needed for the brain's metabolism.*

Clarke, C. A. 1968. The prevention of "rhesus" babies. **217**:46-52 (Nov.).

Clements, J. A. 1962. Surface tension in the lungs. **207**:120-130 (Dec.). *The surface of the lungs is coated with a surfactant that lowers surface tension and prevents collapse of the lungs.*

Comroe, J. H., Jr. 1966. The lung. **214**:56-68 (Feb.). *Physiology of the human lung.*

Fenn, W. O. 1960. The mechanism of breathing. **202**:138-148 (Jan.). *The human respiratory system is described.*

Fox, H. M. 1950. Blood pigments. **182**:20-22 (March). *Compares the characteristics of the major respiratory pigments: hemoglobin, hemocyanin, and chlorocruorin.*

Hock, R. J. 1970. The physiology of high altitude. **222**:52-62

(Feb.). *People living at high altitudes adapt physiologically to chronic low oxygen.*

Laki, K. 1962. The clotting of fibrinogen. **206**:60-66 (March).

Mayerson, H. S. 1963. The lymphatic system. **208**:80-90 (June). *The lymphatics are a crucial "second" circulatory system that picks up fluids leaking from the bloodstream.*

McKusick, V. A. 1965. The royal hemophilia. **213**:88-95 (Aug.). *Story of a defective gene that plagued European royalty for three generations.*

Smith, H. W. 1953. The kidney. **188**:40-48 (Jan.). *The structure, physiology, and evolution of the vertebrate kidney.*

Solomon, A. K. 1962. Pumps in the living cell. **207**:100-108 (Aug.). *Active transport processes in kidney tubules are described.*

Wiggers, C. J. 1957. The heart. **196**:74-87 (May). *The structure and physiology of a remarkable organ.*

Williams, C. M. 1953. Insect breathing. **188**:28-32 (Feb.). *How the tracheal system of insects is constructed and functions.*

Wolf, A. V. 1958. Body water. **199**:125-132 (Nov.).

Wood, J. E. 1968. The venous system. **218**:86-96 (Jan.).

Zucker, M. B. 1961. Blood platelets. **204**:58-64 (Feb.). *These minute elements of the blood plasma play important roles in stopping hemorrhage.*

Zweifach, B. W. 1959. The microcirculation of the blood. **200**:54-60 (Jan.). *Describes the anatomy and function of the capillary bed that serves the body's tissues.*

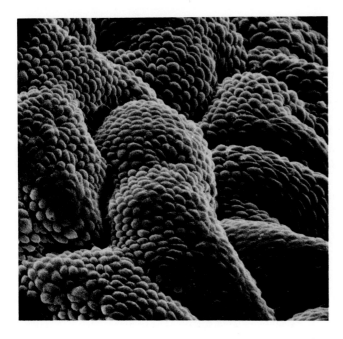

Epithelium of frog stomach. Epithelial cells, resembling cobblestone paving, cover the stomach surface and line gastric pits in which they secrete hydrochloric acid and pepsinogen, which flows out onto the surface. The stomach epithelium is protected from self-digestion by mucus secretions and rapid cellular renewal. (× 500.) (Scanning electron micrograph, courtesy P. P. C. Graziadei, Florida State University.)

CHAPTER 28

DIGESTION AND NUTRITION

All organisms require energy to maintain their highly ordered and complex structure. This energy is chemical energy that is released by transforming complex compounds acquired from the organism's environment into simpler ones. Obviously, if living organisms must depend on the breakdown of complex foodstuffs to build and maintain their own complexity, these foodstuffs must somehow be synthesized in the first place. Most of the energy for this synthesis is provided by the powerful radiations of the sun, the one great source of energy reaching our planet that is otherwise a virtually isolated system. Organisms capable of capturing the sun's energy by the process of photosynthesis are, or course, the green plants. Green plants are **autotrophic organisms** capable of synthesizing all the essential organic compounds needed for life. Autotrophic organisms need only inorganic compounds absorbed from their surround-

ings to provide the raw materials for synthesis and growth. Most autotrophic organisms are the chlorophyll-bearing **phototrophs,** although some, the chemosynthetic bacteria, are **chemotrophs,** gaining energy from inorganic chemical reactions.

Almost all animals are **heterotrophic organisms** that depend on already synthesized organic compounds for their nutritional needs. Animals, with their limited capacities to perform organic synthesis, must feed on plants and other animals to obtain the materials they will use for growth, maintenance, and the reproduction of their kind. The foods of animals, usually the complex tissues of other organisms, can seldom be utilized directly. Food is usually too large to be absorbed by the body cells and may contain material of no nutritional value as well. Consequently, food must be broken down, or digested, into soluble molecules sufficiently small to be utilized. One important difference, then, between autotrophs and heterotrophs is that the latter must have digestive systems.

Animals may be divided into a number of categories on the basis of dietary habits. **Herbivorous** animals feed mainly on plant life. **Carnivorous** animals feed mainly on herbivores and other carnivores. **Omnivorous** forms eat both plants and animals. A fourth category is sometimes distinguished, the **insectivorous** animals, which are those birds and mammals that subsist chiefly on insects.

The ingestion of foods and their simplification by digestion are only initial steps in nutrition. Foods reduced by digestion to soluble, molecular form are **absorbed** into the circulatory system and **transported** to the tissues of the body. There they are **assimilated** into the protoplasm of the cells. Oxygen is also transported by the blood to the tissues, where food products are **oxidized,** or burned to yield energy and heat. Much food is not immediately utilized but is **stored** for future use. Then the wastes produced by oxidation must be **excreted.** Food products unsuitable for digestion are rejected by the digestive system and are **egested** in the form of feces.

The sum total of all these nutritional processes is called **metabolism.** Metabolism includes both constructive and tearing-down processes. When substances are built into new tissues, or stored in some form for later use, the process is called **anabolism.** The breaking down of complex materials to simpler ones for the release of energy is called **catabolism.** Both processes occur simultaneously in all living cells.

Feeding mechanisms

Only a few animals can absorb nutrients directly from their external environment. Blood and intestinal parasites may derive all their nourishment as primary organic molecules by surface absorption; some aquatic invertebrates may soak up part of their nutritional needs directly from the water. For most animals, however, working for their meals is the main business of living, and the specializations that have evolved for food procurement are almost as numerous as species of animals. In this brief discussion we will consider some of the major food-gathering devices.

FEEDING ON PARTICULATE MATTER. Drifting microscopic particles fill the upper few hundred feet of the ocean. Most of this uncountable multitude is **plankton,** plant and animal microorganisms too small

to do anything but drift with the ocean's currents. The rest is organic debris, the disintegrating remains of dead plants and animals. Altogether this oceanic swarm forms the richest life domain on earth. It is preyed on by numerous larger animals, invertebrates and vertebrates, using a variety of feeding mechanisms. Some protozoans, such as the ameboid sarcodines, ingest particulate food by a process called **phagocytosis.** The animal, stimulated by the proximity of food, pushes out armlike extensions of the plasmalemma (cell membrane) and engulfs the particle into a food vacuole, in which it is digested. Other protozoans have specialized openings, called **cytostomes,** through which the food passes to be enclosed in a food vacuole.

By far the most important method to have evolved for particle feeding is **filter feeding** (Fig. 28-1). It is a primitive, but immensely successful and widely employed mechanism. The majority of filter feeders employ ciliated surfaces to produce currents that draw drifting food particles into their mouths. Most filter-feeding invertebrates, such as the tube-dwelling worms and bivalve mollusks, entrap the particulate food in mucus sheets that convey the food into the digestive tract. Filter feeding is characteristic of a sessile way of life, the ciliary currents serving to bring the food to the immobile or slow-moving animal. However, active feeders such as tiny copepod crustaceans and herring are also filter feeders, as are immense baleen whales. The vital importance of one component of the plankton, the diatoms, in supporting a great pyramid of filter-feeding animals is stressed by N. J. Berrill[*]: "A humpback whale . . . needs a ton of herring in its stomach to feel comfortably full—as many as five thousand individual fish. Each herring, in turn, may well have 6,000 or 7,000 small crustaceans in its own stomach, each of which contains as many as 130,000 diatoms. In other words, some 400 billion yellow-green diatoms sustain a single medium-sized whale for a few hours at most." Filter feeding utilizes the abundance and extravagance of life in the sea.

Filter feeders are as a rule nonselective and omnivorous. Sessile filter feeders take what they can get, having only the options of continuing or ceasing to filter. Active filter feeders, however, such as fish and

[*] Berrill, N. J. 1958. You and the universe, New York, Dodd, Mead & Co.

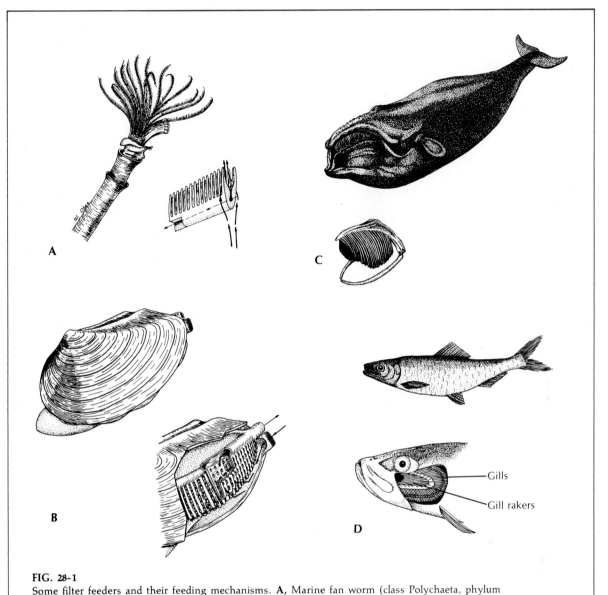

Gills

Gill rakers

FIG. 28-1
Some filter feeders and their feeding mechanisms. **A,** Marine fan worm (class Polychaeta, phylum Annelida). **B,** Marine clam (class Pelecypoda, phylum Mollusca). **C,** Right whale (class Mammalia, phylum Chordata). **D,** Herring (class Osteichthyes, phylum Chordata). **A,** Fan worms have a crown of tentacles. Numerous cilia on the edges of the tentacles draw water *(solid arrows)* between pinnules where food particles are entrapped in mucus; particles are then carried down a "gutter" in the center of the tentacle to the mouth *(broken arrows)*. **B,** Bivalve mollusks use their gills as feeding devices as well as for respiration. Water currents created by cilia on the gills carry food particles into the inhalant siphon and between slits in the gills where they are entangled in a mucus sheet covering the gill surface. Particles are then transported by ciliated food grooves to the mouth (not shown). Arrows indicate direction of water movement. **C,** Whalebone whales filter out plankton, principally large copepods called "krill," with whalebone, or baleen. Water enters the swimming whale's open mouth by the force of the animal's forward motion and is strained out through the more than 300 horny baleen plates that hang down like a curtain from the roof of the mouth. Krill and other plankters caught in the baleen are periodically wiped off with the huge tongue and swallowed. **D,** Herring and other filter-feeding fishes use gill rakers, which project forward from the gill bars into the pharyngeal cavity to strain off plankters. Herring swim almost constantly, forcing water and suspended food into the mouth; food is strained out by the gill rakers and the water passes on out the gill openings.

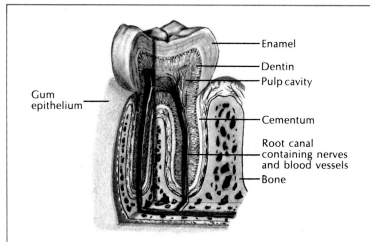

FIG. 28-2

Structure of human molar tooth. Tooth is built of three layers of calcified tissue covering: enamel, which is 98% mineral, and hardest material in the body; dentin, which composes the mass of the tooth, and is about 75% mineral; cementum, which forms a thin covering over the dentin in the root of the tooth, and is very similar to dense bone in composition. Pulp cavity contains loose connective tissue, blood vessels, nerves, and tooth-building cells. Roots of the tooth are anchored to the wall of the socket by a fibrous connective tissue layer called the "periodontal membrane." (After Netter, F. H. 1959. The Ciba collection of medical illustrations, vol. 3. Summit, N. J., Ciba Pharmaceutical Products, Inc.)

baleen whales, are much more selective in their feeding.

FEEDING ON FOOD MASSES. Some of the most interesting animal adaptations are those that have evolved for procuring and manipulating solid food. Such adaptations, and the animals bearing them, are partly shaped by what the animal eats.

Predators must be able to locate prey, capture it, hold it, and swallow it. Most animals use teeth for this purpose (Fig. 28-2). Although teeth are variable in size, shape, and arrangement, vertebrates as different as fish and mammals sometimes have remarkably similar tooth arrangements for seizing the prey and cutting it into pieces small enough to swallow. Mammals characteristically have four different types of teeth, each adapted for specific functions. **Incisors** are for biting and cutting; **canines** are designed for seizing, piercing, and tearing; **premolars** and **molars,** at the back of the jaws, are for grinding and crushing (Fig. 28-3). This basic pattern is often greatly modified (see Fig. 26-12). Herbivores have suppressed canines but have well-developed molars with enamel

ridges for grinding. Such teeth are usually high crowned, in contrast to the low-crowned teeth of carnivores. The well-developed, self-sharpening incisors of rodents grow throughout life and must be worn away by gnawing to keep pace with growth (see Fig. 26-13). Some teeth have become so highly modified that they are no longer useful for biting or chewing food. An elephant's tusk is a modified upper incisor used for defense, attack, and rooting, whereas the male wild boar has modified canines used as weapons.

Many carnivores among the fishes, amphibians, and reptiles swallow their prey whole. Snakes and some fishes can swallow enormous meals. This, together with the absence of limbs, is associated with some striking feeding adaptations in these groups—recurved teeth for seizing and holding the prey, and distensible jaws and stomachs to accommodate their large and infrequent meals.

Teeth are not vertebrate innovations; biting, scraping, and gnawing devices are common in the invertebrates. Insects, for example, have three pairs

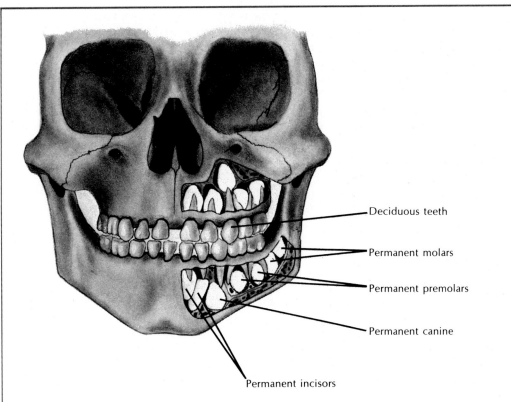

Deciduous teeth

Permanent molars

Permanent premolars

Permanent canine

Permanent incisors

FIG. 28-3
Human deciduous and permanent teeth. Partly dissected skull of a 5-year-old child, showing milk (deciduous) teeth and permanent teeth. Milk teeth begin to erupt at 6 months and are gradually replaced by the permanent teeth beginning at about 6 years of age. There are 20 deciduous teeth, 5 on each side of each jaw, and 32 permanent teeth, 8 on each side of each jaw. These 8 are arranged as follows: 2 incisors, 1 canine (also called cuspid), 2 premolars (bicuspids), 3 molars. The last molar, known as the wisdom tooth, erupts between ages of 17 and 25 or not at all. Upper permanent molars are not seen in this frontal view. (Adapted from Arey, L. B. 1965. Developmental anatomy. Philadelphia, W. B. Saunders Co.)

of appendages on their heads that serve variously as jaws, teeth, chisels, tongues, or sucking tubes. Usually the first pair is crushing teeth; the second, grasping jaws; and the third, a probing and tasting tongue.

Herbivorous, or plant-eating animals, whether vertebrate or invertebrate, have evolved special devices for crushing and cutting plant material. Despite its abundance on earth, the woody cellulose that encloses plant cells is to many animals an indigestible and useless material; herbivores, however, make use of intestinal microorganisms to digest cellulose, once it is ground up. Certain invertebrates such as snails have rasplike, scraping mouthparts. Insects such as locusts have grinding and cutting mandibles; herbivorous mammals such as horses and cattle use wide, corrugated molars for grinding. All these mechanisms serve to disrupt the tough cellulose cell wall, to accelerate its digestion by intestinal microorganisms, as well as to release the cell contents for direct enzymatic breakdown.

FEEDING ON FLUIDS. Fluid feeding is especially characteristic of parasites, but is certainly practiced among free-living forms as well. Most internal parasites (endoparasites) simply absorb the nutrient surrounding them, unwittingly provided by the host.

External parasites (ectoparasites) such as leeches, lampreys, parasitic crustaceans, and insects use a variety of efficient piercing and sucking mouthparts to feed on blood or other body fluid. Unfortunately for man and other warm-blooded animals, the ubiquitous mosquito excels in its blood-sucking habit. Alighting gently, the mosquito sets about puncturing its prey with an array of six needlelike mouthparts (see Fig. 16-7, *B*). One of these is used to inject an anticoagulant saliva (responsible for the irritating itch that follows the "bite," and serving as vector for microorganisms causing malaria, yellow fever, encephalitis, and other diseases); another mouthpart is a channel through which the blood is sucked. It is of little comfort that only the female of the species dines on blood. Far less troublesome to man are the free-living butterflies, moths, and aphids that suck up plant fluids with long, tubelike mouthparts.

Digestion

In the process of digestion, which means literally "carrying asunder," organic foods are mechanically and chemically broken down into small units for absorption. Animal foods vary enormously, having in common only their organic composition. Even though food solids consist principally of carbohydrates, proteins, and fats, the very components that make up the body of the consumer, these components must nevertheless be reduced to their simplest molecular units before they can be utilized. Each animal reassembles some of these digested and absorbed units into organic compounds of the animal's own unique pattern. Cannibals enjoy no special metabolic

benefit from eating their own kind; they digest their victims just as thoroughly as they do food of another species!

The digestive tract is actually an extension of the outside environment into, or through, the animal. Since most animals eat all manner of organic and inorganic materials, the gut's lining must be something like the protective skin; yet it must be permeable so that foodstuffs can be absorbed. Once in the gut the foods are digested and absorbed as they are slowly moved through it. Movement is either by **cilia** or by **musculature.** In general, the filter feeders that use cilia to feed, such as bivalve mollusks, also use cilia to propel the food through the gut. Animals feeding on bulky foods rely upon well-developed gut musculature. As a rule the gut is lined with two opposing layers of muscle—a longitudinal layer, in which the smooth muscle fibers run parallel with the length of the gut, and a circular layer, in which the muscle fibers embrace the circumference of the gut. This arrangement is ideal for mixing and propelling foods. The most characteristic gut movement is **peristalsis** (Fig. 28-4). In this movement a wave of circular muscle contraction sweeps down the gut for some distance, pushing the food along before it. The peristaltic waves may start at any point and move for variable distances. Also characteristic of the gut are **segmentation** movements that divide and mix the food.

Intracellular versus extracellular digestion

Man and other vertebrates and the higher invertebrates digest their food **extracellularly** by secreting

FIG. 28-4
Peristalsis. Food is pushed along before a wave of circular muscle contraction. (From Schottelius, B. A., and D. D. Schottelius. 1973. A textbook of physiology, ed. 17. St. Louis, The C. V. Mosby Co.)

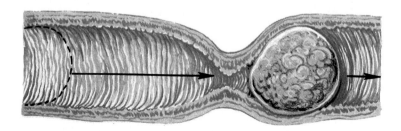

digestive juices into the intestinal lumen. There, foodstuffs are enzymatically split into molecular units small enough to be selectively absorbed by the intestinal epithelium, transported by the circulation, and utilized by all body cells. Digestion, then, occurs outside the body's tissues. **Intracellular** digestion is a primitive process typical of the lower invertebrates. This type of digestion is best illustrated by the single-celled Protozoa, which capture food particles by phagocytosis, enclose these particles within food vacuoles, and then digest them. Obviously the big limitation to intracellular digestion is that only small particles of food can be handled. Nevertheless, many multicellular invertebrates have intracellular digestion. Intracellular digestion is typical of filter-feeding marine animals such as brachiopods, rotifers, bivalves, and cephalochordates, as well as the coelenterates and flatworms. In all of these forms the food particle, phagocytized by the cell, is enclosed within a membrane as a food vacuole (Fig. 28-5). Digestive enzymes are then added. The products of digestion, the simple sugars, amino acids, and other molecules, are absorbed into the cell cytoplasm where they may be utilized directly or may be transferred to other cells. The inevitable food wastes are extruded from the cell.

It is believed that cellular enzymes are packaged into membrane-bound vacuoles called **lysosomes** (Fig. 28-5). These somehow join with, and discharge their enzymes into, the food vacuoles. All cells seem to contain lysosomes, even those of higher animals that do not practice intracellular digestion. Lysosomes have been called "suicide-bags" because they rupture spontaneously in dying or useless cells, digesting the cell contents. Lysosomes also play a role in the lives of healthy cells in cleaning up residues left by growth processes.

It is probable that the obvious limitations of intracellular digestion were responsible for shaping the evolution of extracellular digestion. Extracellular digestion offers several advantages: bulky foods may be ingested; the digestive tract can be smaller, more specialized, and more efficient; and food wastes are more easily discarded. Only with extracellular digestion could the enormous variation in feeding methods of the higher animals have evolved.

The vertebrate digestive plan is similar to that of the higher invertebrates. Both have a highly differentiated alimentary canal with devices for increasing the surface area, such as increased length, inside folds, and diverticula. The more primitive fishes (lampreys and sharks) have longitudinal or spiral folds in their intestine. Higher vertebrates have developed elaborate folds and small fingerlike projections **(villi).** Also, the electron microscope reveals that each cell lining the intestinal cavity is bordered

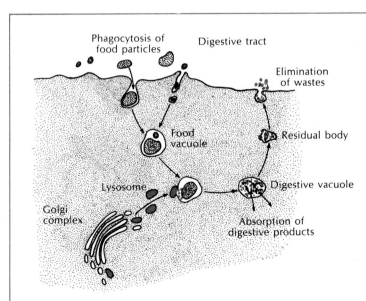

FIG. 28-5

Intracellular digestion. Lysosomes containing digestive enzymes (lysozymes) are produced within the cell, possibly by the Golgi complex. Lysosomes fuse with food vacuoles and release enzymes that digest the enclosed food. Usable products of digestion are absorbed into the cytoplasm and indigestible wastes are expelled to the outside.

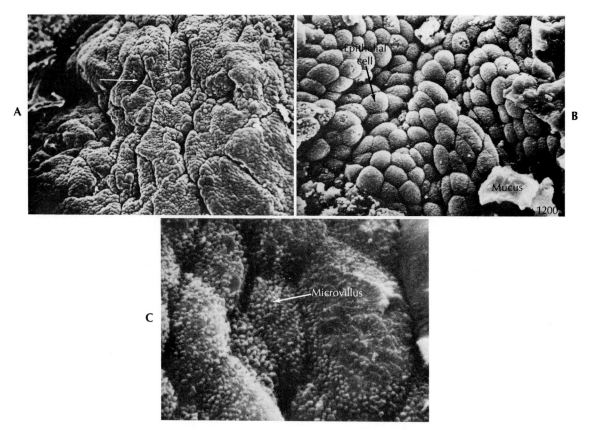

FIG. 28-6

Microscopic structure of human stomach. **A,** At low magnification (× 230) the epithelial cells look like paving stones. Numerous openings to gastric glands are evident; one is indicated by white arrow. **B,** A higher magnification (× 1,200) shows individual epithelial cells. Microvilli give cell surfaces a fuzzy appearance. Degenerating epithelial cells can be seen at left and at top. **C,** At much higher magnification (× 6,500) the individual microvilli are evident on the rounded surfaces of epithelial cells. These scanning electron micrographs were made after the tissue was preserved in a fixative, vacuum dehydrated, and shadowed with a thin layer of gold-palladium (see inset in Fig. 3-3). (From Pfeiffer, C. J. 1970. J. Ultrastructural Res. **33:**252.)

by hundreds of short, delicate processes called **microvilli** (Fig. 28-6). These processes, together with larger villi and intestinal folds, may increase the internal surface of the intestine more than a million times compared to a smooth cylinder of the same diameter. The absorption of food molecules is enormously facilitated as a result.

Digestion in man, an omnivore

ANATOMY OF MAN'S DIGESTIVE SYSTEM. The structural plan of the digestive system of man is shown in Fig. 28-7. The **mouth** is provided with teeth and a tongue for grasping, masticating, manipulating, and swallowing the food. Three pairs of salivary glands lubricate the food and, in man at least, perform limited digestion. In man and other mammals two sets of teeth are formed during life—the temporary, or "milk," teeth (also called deciduous teeth) and the permanent teeth (Fig. 28-3).

The **pharynx** is the throat cavity that serves for the passage of food. It is actually a complex reception chamber, receiving openings from (1) the nasal cavity, (2) the mouth, (3) the middle ear by way of two

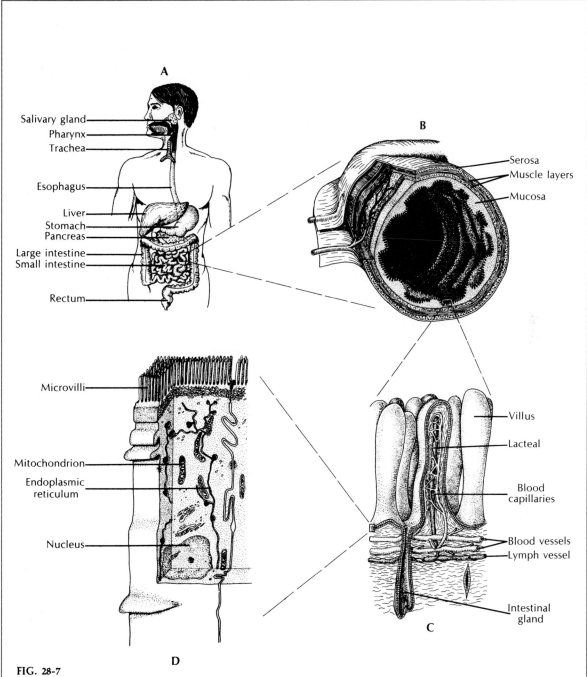

A

Salivary gland
Pharynx
Trachea

Esophagus

Liver
Stomach
Pancreas
Large intestine
Small intestine

Rectum

B

Serosa
Muscle layers
Mucosa

Microvilli

Mitochondrion

Endoplasmic
reticulum

Nucleus

D

Villus

Lacteal

Blood
capillaries

Blood vessels
Lymph vessel

Intestinal
gland

C

FIG. 28-7

A, Digestive system of man. **B,** Portion of small intestine. **C,** Portion of mucosa lining of intestine, showing fingerlike villi. **D,** Optical section of single lining cell, as shown by electron microscope.

eustachian tubes, (4) the esophagus, and (5) the trachea via the glottis.

The **esophagus** is a muscular tube connecting pharynx and stomach. It opens into the stomach by the cardiac opening.

The **stomach** is an enlargement of the gut between esophagus and intestine. In man it is divided into the **cardiac** region (adjacent to esophagus), **fundus** (central region), and **pyloric antrum** region (adjacent to intestine). The stomach is principally a storage organ but aids in digestion.

The **small intestine** is the principal digestive and absorptive area of the gut. It is divided grossly into three regions—**duodenum, jejunum,** and **ilium.** Two large digestive glands, the **liver** and **pancreas,** empty into the duodenum by the **common bile duct.**

The **large intestine (colon)** in man is divided into **ascending, transverse,** and **descending** portions, with the posterior end terminating in the **rectum** and **anus.** At the junction of the large and small intestines is the **colic cecum** and its vestigial **vermiform appendix.** The large intestine lacks villi but contains glands for lubrication.

The stomach, small intestine, and large intestine are all suspended by **mesenteries,** thin sheets of tissue that are modified from the **peritoneum,** or lining, of the coelom and the abdominal organs. Organs such as liver, spleen, and pancreas are also held in place by mesenteries, which carry blood and lymph vessels as well as nerves to the various abdominal organs.

ACTION OF DIGESTIVE ENZYMES. We have already pointed out that digestion involves both mechanical and chemical alterations of food. Mechanical processes of cutting and grinding by teeth and muscular mixing by the intestinal tract are important in digestion. However, the reduction of foods to small absorbable units relies principally on chemical breakdown by **enzymes.** Enzymes, the highly specific organic catalysts essential to the orderly progression of virtually all life processes, have been discussed earlier (p. 69).

It is well to state that although digestive enzymes are probably the best known and most sudied of all enzymes, they represent but a small fraction of the numerous, perhaps thousands, of enzymes that ultimately regulate all processes in the body. The digestive enzymes are **hydrolytic** enzymes **(hydrolases),** so called because food molecules are split by the

process of **hydrolysis,** that is, the breaking of a chemical bond by adding the components of water across it:

$$R\text{—}R \ + \ H_2O \ \xrightarrow[\text{enzyme}]{\text{Digestive}} R\text{—}OH \ + \ H\text{—}R$$

In this general enzymatic reaction, R—R represents a food molecule that is split into two products, R—OH and R—H. Usually these reaction products must in turn be split repeatedly before the original molecule has been reduced to its numerous subunits. Proteins, for example, are composed of hundreds, or even thousands, of interlinked amino acids, which must be completely separated before the individual amino acids can be absorbed. Similarly, carbohydrates must be reduced to simple sugars. Fats (lipids) are reduced to molecules of glycerol and fatty acids, although some fats, unlike proteins and carbohydrates, may be absorbed without being completely hydrolyzed first. There are specific enzymes for each class of organic compounds. These enzymes are located in various regions of the alimentary canal in a sort of "enzyme chain," in which one enzyme may complete what another has started, the product moving along posteriorly for still further hydrolysis.

DIGESTION IN THE MOUTH. In the mouth, food is broken down mechanically by the teeth and is moistened with saliva from the salivary glands. In addition to **mucin,** which helps to lubricate the food for swallowing, saliva contains the enzyme **amylase.** Salivary amylase is a carbohydrate-splitting enzyme that begins the hydrolysis of plant and animal starches (Fig. 28-8). Starches, as we have seen previously (p. 37), are long polymers of glucose. Salivary amylase does not completely hydrolyse starch, but breaks it down mostly into 2-glucose fragments called **maltose.** Some free glucose as well as longer fragments of starch are also produced. When the food mass (bolus) is swallowed, salivary amylase continues to act for some time, digesting perhaps half of the starch before the enzyme is inactivated by the acid environment of the stomach. Further starch digestion resumes beyond the stomach in the intestine.

SWALLOWING. Swallowing is a reflex process involving both voluntary and involuntary components. Swallowing begins with the tongue pushing the moistened food bolus toward the pharynx. The nasal cavity is reflexly closed by raising the soft palate. As the food slides into the pharynx, the epiglottis is

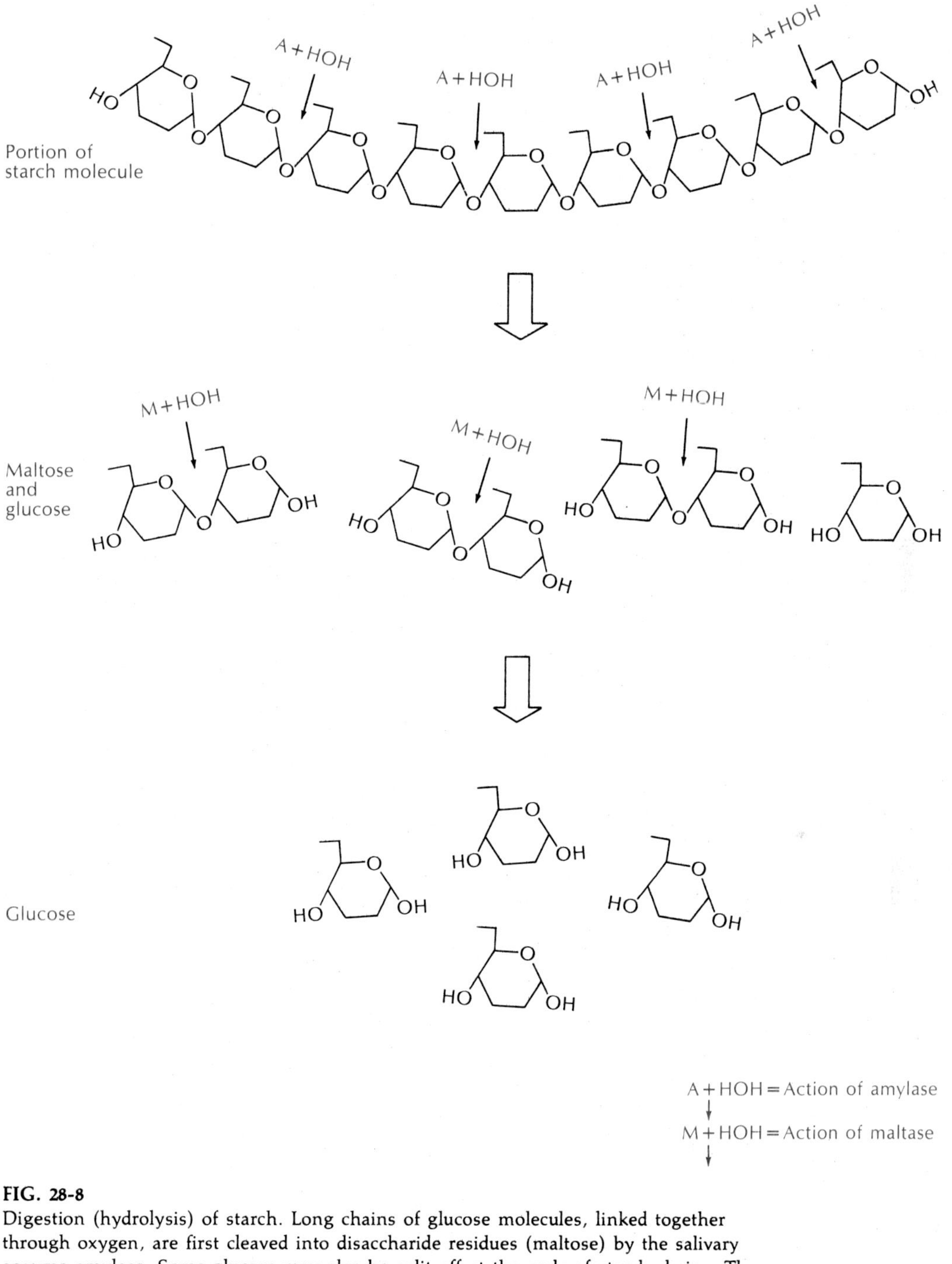

FIG. 28-8
Digestion (hydrolysis) of starch. Long chains of glucose molecules, linked together
through oxygen, are first cleaved into disaccharide residues (maltose) by the salivary
enzyme amylase. Some glucose may also be split off at the ends of starch chains. The
intestinal enzyme maltase then completes the hydrolysis by cleaving the maltose molecules
into glucose. A molecule of water is inserted into each enzymatically split bond.

tipped down over the windpipe, nearly closing it. Some particles of food may enter the opening of the windpipe, but are prevented from going further by contraction of laryngeal muscles. Once in the esophagus, the bolus is forced smoothly toward the stomach by peristaltic contraction of the esophageal muscles.

DIGESTION IN THE STOMACH. When food reaches the stomach, the **cardiac sphincter** opens reflexly to allow entry of the food, then closes to prevent regurgitation back into the esophagus. The stomach is a combination storage, mixing, digestion, and release center. Large peristaltic waves pass over the stomach at the rate of about three each minute; churning is most vigorous at the intestinal end where food is steadily released into the duodenum. About 3 liters of **gastric juice** are secreted each day by deep, tubular glands in the stomach wall. Two types of cells line these glands: (1) **chief cells** secrete an enzyme precursor called **pepsinogen** and (2) **parietal cells** secrete **hydrochloric acid.** Pepsinogen is an inactive form of enzyme that is converted into the active enzyme **pepsin** by hydrochloric acid and by other pepsin already present in the stomach. Pepsin is a **protease** (protein-splitting enzyme) that acts only in an acid medium—pH 1.6 to 2.4. It is a highly specific enzyme that splits large proteins by preferentially breaking down certain peptide bonds scattered along the peptide chain of the protein molecule. Although pepsin, because of its specificity, cannot completely degrade proteins, it effectively breaks them up into a number of small polypeptides. Protein digestion is completed in the intestine by other proteases that can together split all peptide bonds.

Rennin is another enzyme found in the stomachs of the suckling newborn of many mammals, although not of man. Rennin is a milk-curdling enzyme that transforms the proteins of milk into a finely flocculent form that is more readily attacked by pepsin. Rennin extracted from the stomachs of calves is used in cheese-making. Human infants, lacking rennin, digest milk proteins with acidic pepsin, the same as adults do.

The unique ability of the stomach to secrete a strong acid is still an unsolved problem in biology, in part because it has not been possible to collect pure parietal cell secretion or to isolate these cells in culture for study. It is well known that acid solutions can readily destroy organic matter. Since the stomach contains not only a strong acid but also a powerful proteolytic enzyme, it seems remarkable that the stomach mucosa is not digested by its own secretions. That it is not is due to another protective gastric secretion **mucin,** a highly viscous organic compound that coats and protects the mucosa from both chemical and mechanical injury. Sometimes, however, the protective mucus coating fails, allowing the gastric juices to begin digesting the stomach. The result is a peptic ulcer. We should note that despite the popular misconception that "acid stomach" is unhealthy, a notion that is carefully nourished by the makers of patent medicine, stomach acidity is normal and essential.

The secretion of the gastric juices is intermittent. Although a small volume of gastric juice is secreted continuously, even during prolonged periods of starvation, secretion is normally increased by the sight and smell of food, by the presence of food in the stomach, and by emotional states such as anger and hostility. The most unique and classical investigation in the field of digestion was made by the U. S. Army surgeon, William Beaumont, during the years 1825 to 1833. His subject was a young, hard-living French Canadian voyageur, named Alexis St. Martin, who in 1822 had accidentally shot himself in the abdomen with a musket, the blast "blowing off integuments and muscles of the size of a man's hand, fracturing and carrying away the anterior half of the sixth rib, fracturing the fifth, lacerating the lower portion of the left lobe of the lung and the diaphragm, and perforating the stomach." Miraculously, the wound healed, but a permanent opening, or fistula, was formed which permitted Beaumont to see directly into the stomach. St. Martin became a permanent, although temperamental, patient in Beaumont's care, which included food and housing. Over a period of 8 years, Beaumont was able to observe and record how the lining of the stomach changed under different psychic and physiologic conditions, how foods changed during digestion, the effect of emotional states on stomach motility, and many other facts about the digestive processes of his famous patient.

DIGESTION IN THE SMALL INTESTINE. The major part of digestion occurs in the small intestine. Three secretions are poured into this region—**pancreatic juice, intestinal juice,** and **bile.** All of these secretions have a high bicarbonate content, especially the pancreatic juice, which effectively neutralizes the gastric acid, raising the pH of the liquefied food mass, now called **chyme,** from 1.5 to 7 as it enters the duodenum. This change in pH is essential because all the intestinal enzymes are effective only in a neutral or slightly alkaline medium.

About 2 liters of **pancreatic juice** are secreted each day. The pancreatic juice contains several enzymes of major importance in digestion. Two powerful proteases, **trypsin** and **chymotrypsin,** are secreted in inactive form as **trypsinogen** and **chymotrypsinogen.** Trypsinogen is activated in the duodenum by **enterokinase,** an enzyme present in the intestinal juice. Chymotrypsinogen is activated by trypsin. These two proteases continue the enzymatic digestion of proteins begun by pepsin, but now inactivated by the alkalinity of the intestine. Trypsin and chymotrypsin, like pepsin, are highly specific proteases that split apart peptide bonds deep inside the protein molecule. Pancreatic juice also contains **carboxypeptidase,** which splits amino acids off the ends of polypeptides; **pancreatic lipase,** which hydrolyzes fats into fatty acids and glycerol; and **pancreatic amylase, which is a starch-splitting enzyme identical to salivary amylase in its action.**

Intestinal juice from the glands of the mucosal lining furnishes several enzymes. **Aminopeptidase** splits off terminal amino acids from polypeptides; its action is similar to the pancreatic enzyme carboxypeptidase. Three other enzymes are present that complete the hydrolysis of carbohydrates: **maltase** converts maltose to glucose (Fig. 28-8); **sucrase** splits sucrose to glucose and fructose; and **lactase** breaks down lactose (milk sugar) into glucose and galactose.

Bile is secreted by the cells of the **liver** into the **bile duct,** which drains into the upper intestine (duodenum). Between meals the bile is collected into the **gallbladder,** an expansible storage sac that releases the bile when stimulated by the presence of fatty food in the duodenum. Bile contains no enzymes. It is made up of water, bile salts, and pigments. The bile salts (sodium taurocholate and sodium glycocholate) are essential for the complete absorption of fats, which, because of their tendency to remain in large, water-resistant globules, are especially resistant to enzymatic digestion. **Bile salts** reduce the surface tension of fats, so that they are broken up into small droplets by the churning movements of the intestine. This greatly increases the total surface exposure of fat particles, giving the fat-splitting lipases a chance

to reduce them. The characteristic golden yellow of bile is produced by the **bile pigments** that are breakdown products of hemoglobin from worn-out red blood cells. The bile pigments also color the feces.

It is well to emphasize the great versatility of the liver. Bile production is only one of the liver's many functions, which include the following: storehouse for glycogen, production center for the plasma proteins, site of protein synthesis and detoxification of protein wastes, destruction of worn-out red blood cells, center for metabolism of fat and carbohydrate, and many others.

DIGESTION IN THE LARGE INTESTINE. The liquefied material, now called **chyle,** reaching the large intestine, or **colon,** is low in nutrients, since most important food materials have already been absorbed into the bloodstream from the small intestine. The main function of the colon is the absorption of water and some minerals from the intestinal chyle that enters. In removing more than half the water from the chyle, the colon forms a semisolid feces consisting of undigested food residue, bile pigments, secreted heavy metals, and bacteria. The feces are eliminated from the rectum by the process of **defecation,** a coordinated muscular action that is part voluntary and part involuntary.

The colon contains enormous numbers of bacteria that enter the sterile colon of the newborn infant early in life. In the adult about one third of the dry weight of feces is bacteria; these include both harmless bacilli as well as cocci that can cause serious illness if they should escape into the abdomen or bloodstream. Normally the body's defenses prevent invasion of such bacteria. The bacteria break down organic wastes in the feces and provide some nutritional benefit by synthesizing certain vitamins (biotin and vitamin K), which are absorbed by the body.

ABSORPTION. Most digested foodstuffs are absorbed from the small intestine, where the numerous finger-shaped **villi** provide an enormous surface area through which materials can pass from intestinal lumen into the circulation. Little food is absorbed in the stomach because digestion is still incomplete and because of the limited surface exposure. Some materials, however, such as drugs and alcohol are absorbed in part there, which explains their rapid action.

The villi (Fig. 28-7) contain a network of blood and lymph capillaries. The absorbable food products (amino acids, simple sugars, fatty acids, glycerol, and triglycerides as well as minerals, vitamins, and water) pass first across the epithelial cells of the intestinal mucosa and then into either the blood capillaries or the lymph vessel. Small molecules can enter either system, but since blood flow is several hundred times greater than lymph flow, it is unlikely that the lymph system carries much absorbed material out of the intestine. However, any materials that do enter the lymph system will eventually get into the blood by way of the thoracic duct.

Carbohydrates are absorbed almost exclusively as simple sugars (for example, glucose, fructose, and galactose) because the intestine is virtually impermeable to polysaccharides. Proteins, too, are absorbed principally as their subunits, amino acids, although it is believed that very small amounts of small proteins or protein fragments may sometimes be absorbed. Simple sugars and amino acids are transferred across the intestinal epithelium by both passive and active processes.

Immediately after a meal these materials are in such high concentration in the gut that they readily diffuse into the blood, where their concentration is initially lower. However, if absorption were passive only, we would expect transfer to cease as soon as the concentrations of a substance became equal on both sides of the intestinal epithelium. This would leave much valuable foodstuff to be lost in the feces. In fact, very little is lost because passive transfer is supplemented by an **active transport** mechanism located in the epithelial cells, which pick up the food molecules and transfer them into the blood. Materials are thus moved **against** their concentrated gradient, a process that requires the expenditure of energy. Although not all food products are actively transported, those which are, such as glucose, galactose, and most of the amino acids, are handled by transport mechanisms that are specific for each kind of molecule.

Nutritional requirements

The food of animals must include **carbohydrates, proteins, fats, water, mineral salts,** and **vitamins.** Carbohydrates and fats are required as fuels for energy demands of the body and for the synthesis of various substances and structures. Proteins, or actually the amino acids of which they are composed, are needed for the synthesis of the body's specific pro-

teins and other nitrogen-containing compounds. Water is required as the solvent for the body's chemistry and as the major component of all the body fluids. The inorganic salts are required as the anions and cations of body fluids and tissues and form important structural and physiologic components throughout the body. The vitamins are accessory food factors that are frequently built into the structure of many of the enzymes of the body.

All animals require these broad classes of nutrients, although there are differences in the amounts and kinds of food required. The student should note that of the basic food classes listed above, some nutrients are used principally as fuels (carbohydrates and lipids), whereas others are required principally as structural and functional components (proteins, minerals, and vitamins). Any of the basic foods (proteins, carbohydrates, fats) can serve as fuel to supply energy requirements, but, conversely, no animal can thrive on fuels alone. A **balanced diet** must satisfy all metabolic requirements of the body—requirements for energy, growth, maintenance, reproduction, and physiologic regulation. The recognition many years ago that many diseases of man and his domesticated animals were caused by, or associated with, dietary deficiencies led biologists

TABLE 28-1
Nutrients required by man*

Amino acids	Elements	Vitamins
Established as essential		
Isoleucine	Calcium	Ascorbic acid (C)
Leucine	Chlorine	Choline
Lysine	Copper	Folic acid
Methionine	Iodine	Niacin
Phenylalanine	Iron	Pyridoxine (B$_6$)
Threonine	Magnesium	Riboflavin (B$_2$)
Tryptophan	Manganese	Thiamine (B$_1$)
Valine	Phosphorus	Vitamin B$_{12}$
	Potassium	Vitamins A,D,E,
	Sodium	and K
Probably essential		
Arginine	Fluorine	Biotin
Histidine	Molybdenum	Pantothenic acid
	Selenium	Polyunsaturated
	Zinc	fatty acids

*From White, A., P. Handler., and E. L. Smith. 1968. Principles of biochemistry, New York, 1968, McGraw-Hill Book Co., Inc.

to search for specific nutrients that would prevent such diseases. These studies eventually yielded a list of **"essential" nutrients** for man and other animal species studied. The essential nutrients are those that are needed for normal growth and maintenance and that **must** be supplied in the diet. In other words, it is "essential" that these nutrients be in the diet because the animal cannot synthesize them from other dietary constituents. For man more than 20 organic compounds (amino acids and vitamins) and more than 10 elements have been established as essential (Table 28-1). Considering that the body contains thousands of different organic compounds, the list in Table 28-1 is remarkably short. Animal cells have marvelous powers of synthesis enabling them to build compounds of enormous variety and complexity from a small, select group of raw materials.

In the average diet of Americans and Canadians, about 50% of the total calories (energy content) comes from carbohydrates and 40% comes from lipids. Proteins, essential as they are for structural needs, supply only a little more than 10% of the total calories of the average North American's diet. Carbohydrates are widely consumed because they are more abundant and cheaper than proteins or lipids. Actually, man and many other animals can subsist on diets devoid of carbohydrates, provided sufficient total calories and the essential nutrients are present. Eskimos, for example, live on a diet that is high in fat and protein and very low in carbohydrate.

Lipids are needed principally to provide energy. In recent years much interest and research has been devoted to lipids in our diets because of the association between fatty diets and the disease **arteriosclerosis** (hardening and narrowing of the arteries). The matter is complex, but evidence suggests that arteriosclerosis may occur when the diet is high in saturated lipids but low in polyunsaturated lipids. For unkown reasons such diets, which are typical of middle-class and affluent North Americans, promote a high blood level of cholesterol, which may deposit in platelike formations in the lining of the major arteries. For this reason the polyunsaturated fatty acids are often considered essential nutrients for man. Generally speaking, animal fat is more saturated, whereas fat from plants is more unsaturated.

Proteins are expensive foods and restricted in the diet. Proteins, of course, are themselves not the essential nutrients, but rather they contain essential amino acids. Of the 20 amino acids commonly found in proteins, eight and possibly 10 are essential to man (Table 28-1). The rest can be synthesized. One must keep in mind that the terms "essential" and "nonessential" relate only to dietary requirements and to which amino acids can and cannot be synthesized by the body. All 20 amino acids are essential for the various cellular functions of the body. In fact, some of the so-called nonessential amino acids participate in more crucial metabolic activities than the essential amino acids. In general, animal proteins have more of the essential amino acids than do proteins of plant origin. An adult man would require about 67 grams of whole wheat bread each day to meet his amino acid requirement, but only 19 grams of beefsteak to meet these same requirements. This is because proteins are less concentrated in plants and contain relatively less of the essential amino acids. We should note, however, that beefsteak, because it is also high in saturated lipids, is a much less desirable food for man than fish or chicken.

Because animal proteins are so nutritious, they are in great demand by all countries. North Americans eat far more animal proteins than do Asians and Africans; on the average a North American eats 66 grams of animal protein a day, supplemented by milk, eggs, and cereals. In the Middle East, the individual consumption of protein is 14 grams, in Africa 11 grams, and in Asia 8 grams. Protein deficiency is a vital factor in the 10,000 deaths estimated by the United Nations as occurring daily from malnutrition. A protein and calorie deficiency disease of children, **kwashiorkor,** is the world's major health problem today and is growing more serious daily. The name kwashiorkor literally means "golden boy" because of characteristic changes in pigmentation of skin and hair. It occurs especially in nursing infants displaced from the breast by a newborn sibling. The disease is characterized by retarded growth, anemia, liver and pancreas degeneration, renal lesions, acute diarrhea, and a mortality of 30% to 90%. Because animal proteins are relatively scarce and expensive, there is presently a great effort by scientists to find cheap, plentiful sources of plant proteins. With the world's population expected to double to over 7 billion in just 30 years, at the present unchecked rate of growth, the search for protein takes on a desperate urgency. It is altogether fitting that the 1970 Nobel Peace Prize should go to the man, Dr. Norman Borlaug, who developed a dwarf wheat variety that has vastly increased the wheat yield in India, West Pakistan, and Mexico, producing what is called the "green revolution." However, Dr. Borlaug himself emphasizes that the green revolution only defers for a few years the mass famine that seems inevitable unless the human population is stabilized.

VITAMINS. Vitamins are relatively simple organic compounds that are required in small amounts in the diet for specific cellular functions. They are not sources of energy, but are often associated with the activity of important enzymes that have vital metabolic roles. Plants and many microorganisms synthesize all the organic compounds they need; animals, however, have lost certain synthetic abilities during their long evolution and depend ultimately on plants to supply these compounds. Vitamins, therefore, represent synthetic gaps in the metabolic machinery of animals. We have seen that several amino acids are also dietary essentials. These are not considered vitamins, however, because they usually enter into the actual **structure** of tissues or proteinaceous tissue secretions. Vitamins, on the other hand, are essential **functional** components of enzyme catalytic systems. Nevertheless, the distinction between certain vitamins (A, D, E, and K) and other dietary essentials not classified as vitamins is not so clear as it was once believed to be.

Vitamins are usually classified as fat-soluble (soluble in fat solvents such as ether) or water-soluble. The water-soluble ones include the B complex and vitamin C. The family of B vitamins, so grouped because the original B vitamin was subsequently found to consist of several distinct molecules, tends to be found together in nature. Almost all animals, vertebrate and invertebrate, require the B vitamins; they are "universal" vitamins. The dietary need for vitamin C and the fat-soluble vitamins A, D, E, and K tends to be restricted to the vertebrates, although some are required by certain invertebrates.

Lipid-soluble vitamins. *Vitamin A* exists in two forms, as A_1 and A_2, which differ only slightly in chemical structure. Vitamin A consists of one half of a β-carotene molecule; carotenes are a class of organic compounds synthesized only by plants. Animals convert the plant carotenoid to vitamin A; so many animal foods, especially butter, eggs, and milk, serve as sources of the vitamin precursor, as well as green leafy and yellow vegetables. Vitamin A functions in visual photochemistry and is also necessary for normal cell growth. A deficiency causes night blindness and growth retardation.

Vitamin D is derived from eggs, fish oils, and beef fat. The two forms of vitamin D are vitamin D_2 (ergocalciferol) and vitamin D_3 (cholecalciferol). The latter is synthesized in the skin by irradiation with ultraviolet rays of sunlight. Vitamin D promotes absorption of calcium in the intestine; its deficiency causes rickets, a disease characterized by defective bone formation.

Vitamin E is a poorly understood group of tocopherols found in grains, meat, milk, and green leafy vegetables. It is necessary for the integrity of biologic membranes. Deficiency causes sterility, defective embryonic growth, muscular weakness, and anemia. Its current popularity with food faddists as a panacea for nearly all ailments is largely wishful thinking.

Vitamin K is necessary for the synthesis of prothrombin in the liver. Its deficiency causes failure of blood to clot. It is scarce in the normal diet but is synthesized by bacteria in the colon.

B-complex (water-soluble) vitamins. *Thiamine (vitamin B_1)* is an essential coenzyme in pyruvate metabolism. Thiamine is synthesized by plants and is found in beans, grain, yeast, and roots. It is also present in eggs and lean meat. A deficiency causes beriberi, nervous system degeneration, and cessation of growth.

Riboflavin (vitamin B_2) is a part of the oxidative flavoprotein coenzyme (FAD). It is present in many foods such as green beans, eggs, liver, and milk. A deficiency, rare in man, causes nonspecific digestive disturbances and certain kinds of dermatitis.

Nicotinic acid (niacin) is converted in the body to hydrogen donor-acceptor coenzymes (NAD and NADP). It is found in green leaves, egg yolk, wheat germ, and liver. Oxidative metabolism is depressed with its deficiency, producing a complex of symptoms called pellagra.

Pyridoxine (vitamin B_6) is a coenzyme that functions in amino acid metabolism. It is widely distributed in foods, especially meat, eggs, yeast, and cereals. A dietary deficiency is rare in man but causes a variety of symptoms in experimental animals (dermatitis, growth retardation, anemia).

Pantothenic acid is part of coenzyme A that functions in carbohydrate and lipid metabolism. It is present in many foods and a deficiency is not known in man. In experimental animals its lack causes dermatitis, growth retardation, graying of hair, and other abnormalities.

Folic acid is essential for growth and blood cell formation. It is found in green leaves, soybeans, yeast, and egg yolk. A deficiency causes anemia and kidney hemorrhage.

Biotin (vitamin H) functions in fatty acid oxidation and synthesis. It is found in liver, egg yolk, meat, and many other sources. A deficiency is very rare in man. In experimental animals a lack causes dermatitis.

Cyanocobalamin (vitamin B_{12}) is essential for red blood cell formation. It is a unique vitamin in that it is synthesized only by bacteria. Chief sources are milk, egg yolk, liver, and oysters. A deficiency causes pernicious anemia.

Vitamin C. Vitamin C is necessary for the formation of intercellular material. It is found in citrus fruits, tomatoes, and other fresh vegetables. A deficiency causes scurvy, perhaps the most famous deficiency disease of man. Scurvy was the scourge of sailors and soldiers who subsisted for long periods on dried meats and grains. Scurvy is characterized by capillary bleeding and defective wound healing.

References

Suggested general readings
(See also general references for Part Three, p. 639.)

Brooks, F. P. 1970. Control of gastrointestinal function. New York, The Macmillan Co. *A concise, well-illustrated summary of human digestive physiology.*

Jennings, J. B. 1965. Feeding, digestion and assimilation in animals. Oxford, Pergamon Press, Inc. *A general, comparative approach. Excellent account of feeding mechanisms in animals.*

Morton, G. 1967. Guts. The form and function of the digestive system. Institute of Biology's Studies in Biology, no. 7, New York, St. Martin's Press.

Selected *Scientific American* articles

Davenport, H. W. 1972. Why the stomach does not digest itself. **226:**86-93 (Jan.).

Kretchmer, N. 1972. Lactose and lactase. **227:**70-78 (Oct.). *Most adult mammals, man included, lack lactase, the enzyme that breaks down lactose, or milk sugar.*

Neurath, H. 1964. Protein-digesting enzymes. **211:**68-79 (Dec.). *Describes studies of enzyme structure and how these digestive enzymes exert their catalytic effect.*

Rogers, T. A. 1958. The metabolism of ruminants. **198:**34-38 (Feb.). *How cellulose is digested in the four-chambered stomach of cows and other ruminants.*

Spain, D. M. 1966. Atherosclerosis. **215:**48-56 (Aug.). *How dietary factors contribute to increasing prevalence of this disease.*

Trowell, H. C. 1954. Kwashiorkor. **191:**46-50 (Dec.). *The most severe and widespread nutritional disease known to medical science is described.*

Young, V. R., and N. S. Schrimshaw. 1971. The physiology of starvation. **225:**14-21 (Oct.). *Studies with human volunteers have helped to clarify the body's nutritional needs.*

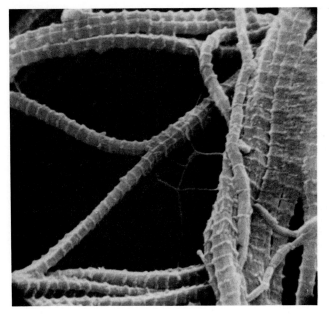

Skeletal muscle of frog. In this teased preparation motor nerve fibers (center) leading to myoneural junctions on the muscle fibers are plainly visible. Cross striations, actually Z-lines between successive functional units (sarcomeres), give muscle fibers a segmented appearance. (× 5,200.) (Scanning electron micrograph, courtesy P. P. C. Graziadei, Florida State University.)

CHAPTER 29

SUPPORT, PROTECTION, AND MOVEMENT

■ INTEGUMENT AMONG VARIOUS GROUPS OF ANIMALS

The integument is the outer covering of the body, a protective wrapping that includes the skin and all structures that are derived from, or associated with, the skin, such as hair, setae, scales, feathers, and horns. In most animals it is rough and pliable, providing mechanical protection against abrasion and puncture and forming an effective barrier against the invasion of bacteria. It provides moisture-proofing against fluid loss or gain. The skin protects the underlying cells against the damaging action of the ultraviolet rays of the sun. But in addition to being a protective cover, the skin serves a variety of other important functions. For example, in warm-blooded (homeothermic) animals, it is vitally concerned with temperature regulation, since most of the body's heat is lost through the skin. The skin contains re-

ceptors of many senses that provide essential information about the immediate environment. It has excretory functions and, in some forms, respiratory functions as well. Through skin pigmentation the organism can make itself more or less conspicuous. Skin secretions can make the animal attractive or repugnant or provide olfactory cues that influence behavioral interactions between individuals.

Invertebrate integument

Many protozoans have only the delicate cell or plasma membranes for external coverings; others, such as *Paramecium*, have developed a protective pellicle. Most multicellular invertebrates, however, have more complex tissue coverings. The principal covering is a single-layered **epidermis**. Some invertebrates have added a secreted noncellular **cuticle** over the epidermis for additional protection; some groups,

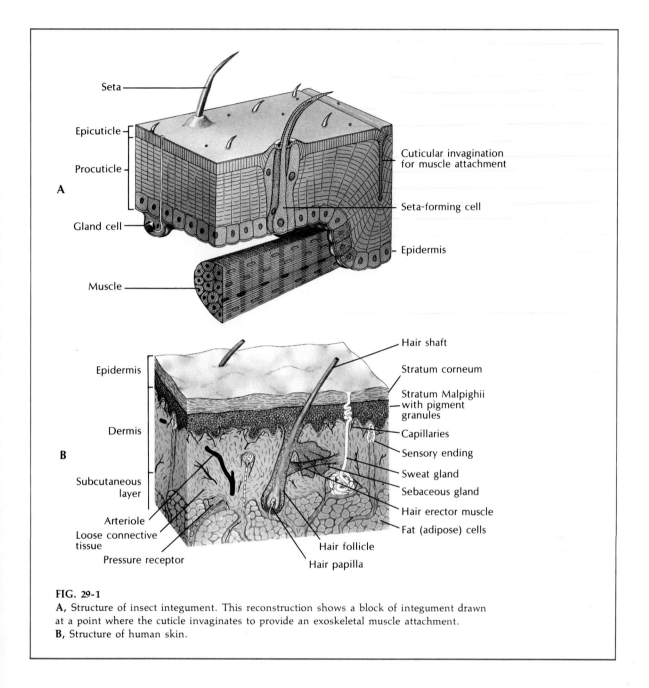

FIG. 29-1

A, Structure of insect integument. This reconstruction shows a block of integument drawn at a point where the cuticle invaginates to provide an exoskeletal muscle attachment. **B,** Structure of human skin.

such as many parasitic worms, have only a thick resistant cuticle and lack a cellular epidermis.

The molluskan epidermis is delicate and soft and contains mucous glands, some of which secrete the calcium carbonate of the shell. The cephalopod has developed a more complex integument, consisting of a cuticle, a simple epidermis, a layer of connective tissue, a layer of reflecting cells (iridocytes), and, finally, a thicker layer of connective tissue.

Arthropods have the most complex of invertebrate integuments, providing not only protection but also skeletal support. The development of a firm exoskeleton and jointed appendages suitable for the attachment of muscles has been a key feature in the great evolutionary success of this largest of animal groups. The arthropod integument consists of a single-layered **epidermis** (also called **hypodermis**), which secretes a complex cuticle of two zones (Fig. 29-1, *A*). The inner zone, the **procuticle,** is composed of protein and chitin. Chitin is a polysaccharide that resembles plant cellulose. The chitin is laid down in molecular sheets like the veneers of plywood, thus providing great strength to the procuticle layer. The outer zone of cuticle, lying above the procuticle, is the thin **epicuticle.** The epicuticle is nonchitinous, consisting instead of a complex of proteins and lipids. The epicuticle is significant in providing a protective moisture-proofing barrier to the integument.

The arthropod cuticle may remain as a tough but soft and flexible layer, or it may be hardened by one of two ways. In the decapod crustaceans, for example, crabs and lobsters, the cuticle is stiffened by **calcification,** the deposition of calcium carbonate. In insects hardening is achieved by a process called **sclerotization,** in which the protein molecules of the chitin form stabilizing cross-linkages. Arthropod chitin is one of the toughest materials synthesized by animals; it is strongly resistant to pressure and tearing and can withstand boiling in concentrated alkali, yet it is light, having a specific weight of only 1.3.

When arthropods molt, the epidermal cells first divide by mitosis. Enzymes secreted by the epidermis dissolve most of the procuticle; the digested materials are then absorbed and consequently not lost to the body. Then, in the space beneath the old cuticle, a new epicuticle and procuticle are formed. After the old cuticle is shed, the new cuticle is thickened and calcified or sclerotized.

Vertebrate integument

The basic plan of the vertebrate integument, as exemplified by human skin (Fig. 29-1, *B*), includes a thin, outer stratified epithelial layer, the **epidermis,** derived from ectoderm and an inner, thicker layer, the **dermis** (corium), which is of mesodermal origin and is made up of nerves, blood vessels, connective tissue, pigment, etc. Only in cyclostomes is a noncellular dead cuticle found. The epidermis consists usually of several layers of cells. The basal part is made up of columnar cells that undergo frequent mitosis to renew the layers that lie above. Thus the outer layers, mostly cornified and dead, are sloughed off constantly. As the squamous cells degenerate,

their cytoplasm is transformed into granules of **keratin.** In many vertebrates (fish and amphibians) the epidermis is provided with mucous glands that lubricate the exterior surface of the body. In fish the keratin is not abundant in the outer cells that are molted before much keratin can be formed. Epidermal cells are slowly replaced, since they are sloughed off separately when the layer of mucus is worn away. Granular poison glands are also found in some amphibians.

Land forms such as reptiles, birds, and mammals have a thicker epidermis, the outer layer of which is much cornified with keratin. This is for added protection to prevent drying out. Mammals are well provided with glands (Fig. 29-2). Indeed, the entire epidermal cutaneous system of the amniotes may be considered a glandular system and, with the exception of the sweat glands, is mostly holocrine in nature. The keratinized outer layers, secreted by the basal epidermal cells, are constantly being shed in small fragments (mammals) or in a single piece (some reptiles). Hair, feathers, keratin, and sebum (all dead cells when shed) may be considered as a secretion of the epidermis.

The **color** of the skin is often attributable to special pigment that may be in the form of granules scattered through the layers of the epidermis (mammals) and to special pigment cells, **chromatophores,** that are found chiefly in the dermis (fish and amphibians). The hormonal control of the expansion and contraction of pigment within amphibian chromatophores is discussed on p. 496.

The skin is variously modified in different vertebrates to form the so-called skin derivatives. These include the bony and horny structures, such as scales, claws, nails, horns, and antlers, in addition to glands, hair, and feathers. True bony structures develop in the dermis, and bony plates were very common in such primitive forms as ostracoderms and placoderms. Certain bony plates of the head were modified to form the dermatocranium of the skull. Fish scales (Fig. 29-3) are bony dermal plates that are covered with live epidermis bearing a superficial layer of dead cells that are constantly being replaced. Amphibians have moist naked skins without scales (except the tiny dermal scales of the order Gymnophiona). The superficial layer of their epidermis contains keratin, which is replaced when lost. In strictly land tetrapods, keratinized epithelial struc-

Goblet mucous cell
(vertebrate intestine)

Unicellular glands
(skin of invertebrate)

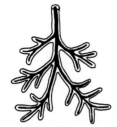

Coiled tubular gland
(sweat gland)

Compound tubular gland
(certain gastric glands and liver)

Compound alveolar gland
(salivary glands)

FIG. 29-2

Types of exocrine glands. Exocrine glandular products are either secretions (formation of useful products from raw materials of blood) or excretions (waste products from blood). They are carried outside cells that form them, either to interior of certain organs or to surface. Secretory function is usually displayed by epithelial tissue but sometimes also by nervous and connective tissue. Invertebrates often have glands that resemble those of vertebrates but invertebrates usually have greater variety of unicellular glands. Most glands form secretions in three ways: (1) merocrine, in which secretion is formed by cell; (2) apocrine, in which part of cell forms secretion; and (3) holocrine, in which entire cell is discharged in secretion.

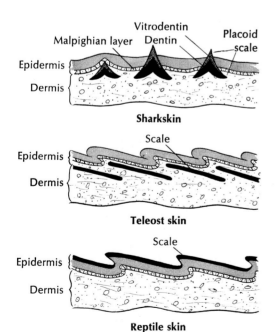

FIG. 29-3

Integument in classes Elasmobranchii, Osteichthyes, and Reptilia showing different types of scales. Placoid scales of sharks are derived from dermis and have given rise to teeth in all higher vertebrates. Teleost fish have bony scales from dermis, and reptiles have horny scales from epidermis. Only dermal scales are retained throughout life; epidermal scales are shed.

tures have largely replaced the bony plates. Reptiles have horny scales (of epidermal origin) that prevent loss of water (Fig. 29-3). These scales are also found on the legs and feet of birds and on the tails of certain mammals. In the armadillo (Fig. 26-4) the entire body is covered with large horny scales. In crocodiles and some turtles the scales form horny plates that overlie the bony dermal plates of the back and belly. Some lizards also have bony plates as well as horny scales. Horny scales are formed by the folding of the epidermis over mesodermal papillae, the upper surface of which becomes the cornified scale.

Some of the many marvelous skin derivatives found in mammals have been described in Chapter 26. That most mammalian characteristic, hair, serves many functions including protection, temperature regulation, and protective coloration, as well as being modified variously into sensory vibrissae, bristles, spines, and even the rhinoceros horn (p. 574). True horns and antlers are described on p. 572. Mammals also inherited the reptilian claws and modified them into nails and hoofs. A claw is shaped to cover the sides, top, and tip of a terminal joint; a nail is flattened and covers the dorsal surface of the distal phalange; and a hoof extends across the end of the digit and covers the plantar surface also. In the horse the hoof, which is developed from the claw of one toe, is the only part of the foot touching the ground. Other hoofed animals may have spongy pads or other parts of the foot on which to walk.

■ SKELETAL SYSTEMS

Skeletons are supportive systems that provide rigidity to the body, surfaces for muscle attachment, and protection for vulnerable body organs. The familiar bone of the vertebrate skeleton is only one of several kinds of supportive and connective tissues, serving various binding and supportive functions, which we will discuss in this section.

EXOSKELETON AND ENDOSKELETON. Although animal supportive and protective structures take many forms, there are two principal types of skeletons: the **exoskeleton,** typical of mollusks and arthropods, and the **endoskeleton,** characteristic of vertebrates. The invertebrate exoskeleton is mainly protective in function and may take the form of shells, spicules, and calcareous or chitinous plates. It may be rigid, as in mollusks, or jointed and movable as in arthropods. Unlike the endoskeleton, which grows with the animal, the exoskeleton is often a limiting coat of armor, which must be periodically shed (molted) to make way for an enlarged replacement. Some invertebrate exoskeletons, such as the shells of snails and bivalves, grow with the animal. Vertebrates, too, have traces of exoskeleton that serve to remind us of our invertebrate heritage. These are, for example, scales and plates of fishes, fingernails and claws, hair, feathers, and other cornified integumentary structures.

The vertebrate endoskeleton is formed inside the body and is composed of bone and cartilage surrounded by soft tissues. It not only supports and protects, but it is also the major body reservoir for calcium and phosphorus. In the higher vertebrates the red blood cells and certain white blood cells are formed in the bone marrow.

Cartilage

Cartilage and bone are the characteristic vertebrate supportive tissues. The **notochord,** the semirigid axial rod of protochordates and vertebrate larvae and embryos, is also a primitive vertebrate supportive tissue. Except in the most primitive vertebrates, for example, amphioxus and the cyclostomes, the notochord is surrounded or replaced by the backbone during embryonic development. The notochord is composed of large, vacuolated cells and is surrounded by layers of elastic and fibrous sheaths. It is a stiffening device, preserving body shape during locomotion (p. 431).

Vertebrate cartilage is the major skeletal element of primitive vertebrates. Cyclostomes and elasmobranchs have purely cartilaginous skeletons. In contrast, higher vertebrates have principally bony skeletons as adults, with some cartilage interspersed. Cartilage is a soft, pliable, characteristically deep-lying tissue. Unlike connective tissue, which is quite variable in form, cartilage is basically the same wherever it is found. The basic form, **hyaline cartilage** (Fig. 4-10, *B*), has a clear, glassy appearance. It is composed of cartilage cells **(chondrocytes)** surrounded by firm complex protein gel interlaced with a meshwork of collagenous fibers. Blood vessels are virtually absent. In addition to forming the cartilagenous skeleton of the primitive vertebrates and that of all vertebrate embryos, hyaline cartilage makes up the articulating surfaces of many bone joints of higher adult vertebrates and the supporting tracheal, laryngeal, and bronchial rings. The basic cartilage has several variants. Among these is **calcified cartilage,** where calcium salt deposits produce a bonelike structure. **Fibrocartilage,** resembling connective tissue, and **elastic cartilage,** containing many elastic fibers, are other variations of basic hyaline cartilage found among the vertebrates.

Bone

Bone differs from other connective and supportive tissues by having significant deposits of inorganic calcium salts laid down in an extracellular matrix. Its structural organization is such that bone has nearly the tensile strength of cast iron, yet is only one third as heavy. Most bones develop from cartilage **(endochondral bone)** by a complex replacement of embryonic cartilage with bone tissue. A second type of bone is **membrane bone** that develops directly from sheets of embryonic cells. In higher vertebrates membrane bone is restricted to bones of the face and cranium; the remainder of the skeleton is endochondral bone. Despite differences in origin, endochondral bone and membrane bone are not distinguishable histologically. In the fishes the dermal scales and plates that may cover most of the body are formed from membrane bone.

Two kinds of bone structure are distinguishable—**spongy** (or **cancellous**) and **compact** (Fig. 29-4). Spongy bone consists of an open, interlacing framework of bony tissue, oriented to give maximum strength under the normal stresses and strains that the bone receives. Compact bone is dense, appearing absolutely solid to the naked eye. Both structural kinds of bone are found in the typical long bones of the body such as the humerus (upper arm bone) (Fig. 29-4).

FIG. 29-4
Section of proximal end of human humerus, showing appearance of spongy and compact bone. (From Bloom, W., and D. W. Fawcett. 1968. A textbook of histology, Philadelphia, W. B. Saunders Co.)

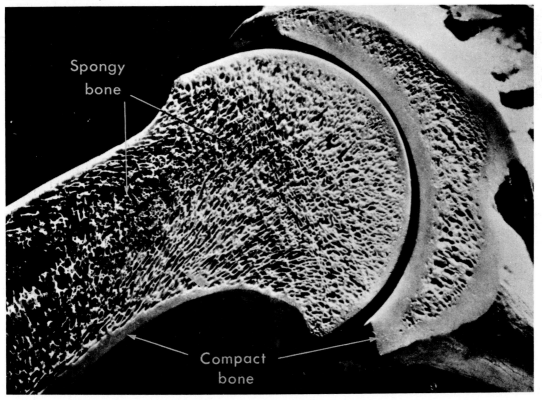

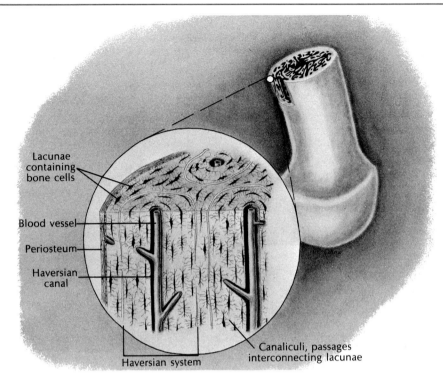

Lacunae
containing
bone cells

Blood vessel

Periosteum

Haversian
canal

Haversian system

Canaliculi, passages
interconnecting lacunae

FIG. 29-5
Structure of bone, showing the dense calcified matrix and bone cells arranged into
haversian systems. Bone cells are entrapped within the cell-like lacunae, but receive
nutrients from the circulatory system via tiny canaliculi that interlace the calcified matrix.
Bone cells were known as osteoblasts when they were building bone, but in mature bone
shown here, they become resting osteocytes. Bone is covered with a compact connective
tissue called "periosteum."

Microscopic structure of bone

Compact bone is composed of a calcified bone matrix arranged in concentric rings. The rings contain
cavities **(lacunae)** filled with bone cells (osteocytes)
that are interconnected by many minute passages
(canaliculi). These serve to distribute nutrients
throughout the bone. This entire organization of lacunae and canaliculi is arranged into an elongated
cylinder called a **haversian system** (Fig. 29-5). Bone
consists of bundles of haversian systems cemented
together and interconnected with blood vessels.
Bone growth is a complex restructuring process, involving both its destruction internally by bone-resorbing cells **(osteoclasts)** and its deposition externally by bone-building cells **(osteoblasts).** Both pro-

cesses occur simultaneously so that the marrow cavity inside grows larger by bone resorption while new
bone is laid down outside by bone deposition. Bone
growth responds to several hormones, in particular
parathyroid hormone, which stimulates bone resorption, and **calcitonin,** which inhibits bone resorption.
These two hormones are responsible for maintaining
a constant level of calcium in the blood (p. 713).

Plan of the vertebrate skeleton

The vertebrate skeleton has undergone a great
transformation in the course of evolution. This is
hardly surprising. The move from water to land
forced dramatic changes in respiration and body
form. The most pronounced differences are found in

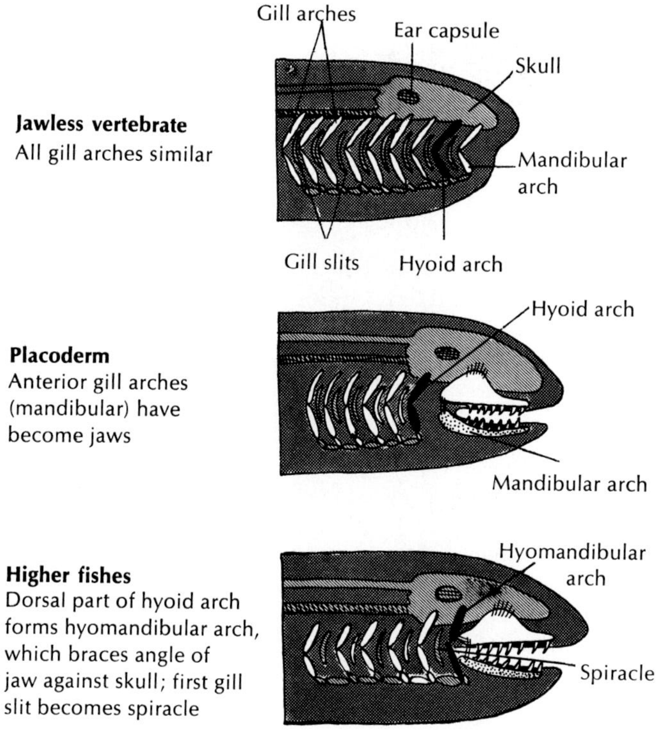

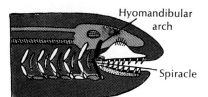

Jawless vertebrate
All gill arches similar

Gill arches
Ear capsule
Skull
Mandibular arch
Gill slits
Hyoid arch

Placoderm
Anterior gill arches (mandibular) have become jaws

Hyoid arch
Mandibular arch

Higher fishes
Dorsal part of hyoid arch forms hyomandibular arch, which braces angle of jaw against skull; first gill slit becomes spiracle

Hyomandibular arch
Spiracle

FIG. 29-6
How vertebrate got its jaws. First two arches in order are mandibular and hyoid; mandibular forms jaws, and hyoid forms supporting accessory structure. Jaws at first are separate from cranium, but later the two are consolidated.

the bones of the gill apparatus, called the **visceral skeleton,** and bones of the skull. The early vertebrates tend to have a larger number of skull bones than more recently evolved forms. Some fish may have 180 skull bones; amphibia and reptiles, 50 to 95; and mammals, 35 or fewer. Man has 29.

The evolutionary history of the skull shows that it is derived from three sources: (1) the **endocranium,** or neurocranium, which is the original skull that surrounds the brain and is best seen in its primitive basic plan in the sharks; (2) the **dermocranium,** which represents the outer membranous bony cap that originated from fused dermal scales of the head and overlies the endocranium (best seen in the bowfin, *Amia*); and (3) the **splanchnocranium,** which is the endoskeletal part of the visceral skeleton that supports the gills and is also represented in a fairly primitive plan in sharks. The jaws are formed by certain visceral or gill arches that are a part of the splanchnocranium, as shown in Fig. 29-6. In the lower vertebrates these three skull components are

more or less separated from each other; in higher forms they are all fused together or incorporated into a single unit—the vertebrate skull. There is a basic plan of homology in the skull elements of vertebrates from fish to man; evolution has meant reduction in numbers of bones through loss and fusion in accordance with size and functional changes. The best known diagnostic character of the mammalian skeleton is the lower jaw, which consists on each side of only one bone, the dentary.

The vertebral column varies greatly with different animals and with different regions of the vertebral column in the same animal. In fish it is differentiated only into trunk and caudal vertebrae; the column in many of the other vertebrates is differentiated into **cervical** (neck), **thoracic** (chest), **lumbar** (back), **sacral** (pelvic), and **caudal** (tail) vertebrae. In birds and also in man the caudal vertebrae are reduced in number and size, and the sacral vertebrae are fused. The number of vertebrae varies among the different animals. The python seems to lead the list with 435.

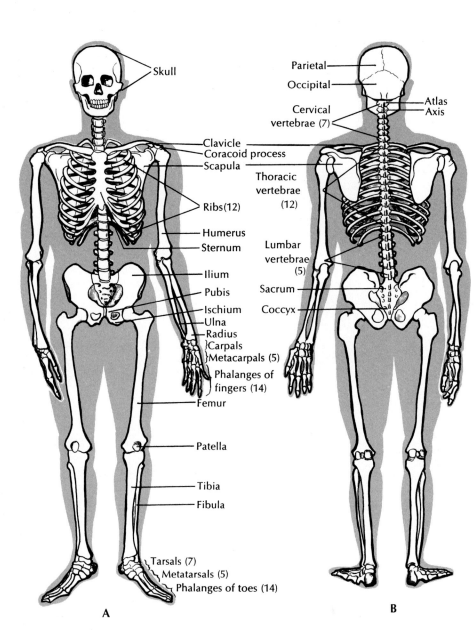

FIG. 29-7

Human skeleton. **A,** Ventral view. **B,** Dorsal view. Numbers in parentheses indicate number of bones in that unit. In comparison with other mammals, man's skeleton is a patchwork of primitive and specialized parts. Erect posture brought about by specialized changes in legs and pelvis enabled primitive arrangement of arms and hands (arboreal adaptation of man's ancestors) to be used for manipulation of tools. Development of skull and brain followed as consequence of premium natural selection put upon dexterity, better senses, and ability to appraise environment.

In man (Fig. 29-7) there are 33 in the child, but in the adult 5 are fused to form the **sacrum** and 4 to form the **coccyx.** Besides the sacrum and coccyx, man has 7 cervical, 12 thoracic, and 5 lumbar vertebrae. The first cervical vertebra is modified for articulation with the skull and is called the **atlas.** The number of cervical vertebrae (7) is constant in nearly all mammals.

Ribs show many variations among the vertebrates. Primitive forms have a pair of ribs for each vertebra from head to tail, but higher forms tend to have fewer ribs. Certain fish have two ventral ribs for each vertebra, and in some fish there are dorsal and ventral (pleural) ribs on the same vertebra. In tetrapods the single type of rib is supposed to correspond to the dorsal one of fish. The ribs of many land vertebrates are joined to the sternum. The sternum is lacking in snakes. The ribs of vertebrate animals are not all homologous, for they do not all arise in the same way. Ribs, in fact, are not even universal among vertebrates; many, including the leopard frog, do not have them at all. Others, such as the elasmobranchs and some amphibians, have very short ribs. Man has 12 pairs of ribs, although evidence indicates that his ancestors had more. The ribs together form the thoracic basket, which supports the chest wall and keeps it from collapsing.

Most vertebrate animals have paired appendages. None are found in cyclostomes, but both the cartilaginous and bony fishes have pectoral and pelvic fins that are supported by the pectoral and pelvic girdles, respectively. Forms above the fish (except snakes) have two pairs of appendages, also supported by girdles. The basic plan of the land vertebrate limb (tetrapod) is called **pentadactyl,** terminating in five digits. Among the various vertebrates there are

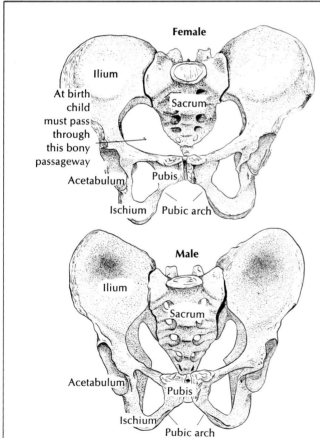

FIG. 29-8

Chief difference between male and female skeletons is structure of pelvis. Female pelvis has less depth with broader, less sloping ilia, more circular bony ring (pelvic canal), wider and more rounded pubic arch, and shorter and wider sacrum. Most structures of female pelvis are correlated with childbearing functions. In evolution of human skeleton, pelvis has changed more than any other part because it has to support weight of erect body. (Anterior view.)

many modifications in the girdles, limbs, and digits that enable animals to meet specific modes of life. For instance, some amphibians have only three or four toes on each foot, and the horse has only one. Also, the bones of the limbs may be separate or they may be fused in various ways. Whatever the modification, the girdles and appendages in forms above fish are all built on the same plan and their component bones can be homologized. In man the pectoral girdle is made up of 2 scapulae and 2 clavicles; the arm is made up of humerus, ulna, radius, 8 carpals, 5 metacarpals, and 14 phalanges. The pelvic girdle (Fig. 29-8) consists of 2 innominate bones, each of which is composed of 3 fused bones—ilium, ischium, and pubis; the leg is made up of femur, patella, tibia, fibula, 7 tarsals, 5 metatarsals, and 14 phalanges. Each bone of the leg has its counterpart in the arm, with the exception of the patella. This kind of correspondence between anterior and posterior parts is called **serial homology.**

Some bones (heterotopic) are not associated with the skeleton. These include the **os cordis** in the interventricular septum of the heart (ox, sheep, and goat), the **baculum** (os priapi or os penis) in the spongy bodies of the penis (rodents, carnivores, lower primates, and others), **sclerotic plates** of the eye (birds), and many others. Wherever stress occurs, the mesoderm has the potential to form bone.

■ ANIMAL MOVEMENT

Movement is a unique characteristic of animals. Plants may show movement, but this usually results from changes in turgor pressure or growth rather than from specialized contractile proteins as in animals. Movement appears in many forms in animal tissues, ranging from barely discernible streaming of cytoplasm, the swelling of mitochondria, or movement of the mitotic spindle during cell division, to frank movements of powerful striated muscles that carry a mammal at high speed across the landscape. Recently it has become evident that virtually all animal movement depends on a single fundamental mechanism: **contractile proteins,** which can change their form to elongate or contract. This contractile machinery is always composed of ultrafine fibrils— fine filaments, striated fibrils, or tubular fibrils (microtubules)—arranged to contract when powered by **ATP** (adenosine triphosphate). By far the most im-

portant protein contractile system is the **actomyosin system,** composed of two proteins, **actin** and **myosin.** This is an almost universal biomechanical system found in protozoans through vertebrates and performs a long list of diverse functional roles. In this section we will examine the three principal kinds of animal movement: ameboid, ciliary, and muscle.

Ameboid movement

Ameboid movement is a form of movement especially characteristic of the freshwater ameba and other sarcodine protozoans; it is also found in many wandering cells of higher animals, such as white blood cells, embryonic mesenchyme, and numerous other mobile cells that move among the tissue spaces. Ameboid cells constantly change their shape by sending out and withdrawing **pseudopodia** (false feet) from any point on the cell surface. Such cells are surrounded by a delicate, highly flexible, membrane called **plasmalemma** (Fig. 29-9). Beneath this lies a nongranular layer, the **hyaline ectoplasm,** which encloses the **granular ectoplasm.** Optical studies of an ameba in movement suggest that the outer layer of cytoplasm **(ectoplasm)** actively contracts in the **fountain zone** at the tip of the pseudopod to pull a central core of rather rigid endoplasm forward. The latter is then converted into ectoplasm, which slips posteriorly under the plasmalemma and joins the endoplasm at the rear to begin another cycle. There are other theories of ameboid movement, in particular one that favors the posterior end of the animal as the locus of contraction. According to this theory, an ameba is pushed rather than pulled forward. Although no completely satisfactory analysis exists, it is certain that ameboid movement is based on the same fundamental contractile system that powers vertebrate muscles: an actomyosin machinery driven by ATP.

Ciliary movement

Cilia are minute hairlike motile processes that occur on the surfaces of the cells of many animals; they are a particularly distinctive feature of ciliate protozoans. Except for the nematodes, in which cilia are absent, and the arthropods, in which they are rare, cilia are found in all major groups of animals. Cilia perform many roles, either in moving small animals through their aquatic environment (such as protozoans) or in propelling fluids and materials across the epithelial surfaces of larger animals. Cilia play prom-

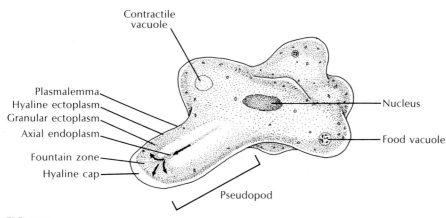

FIG. 29-9

Structure of *Amoeba* in active locomotion. Animal is moving in the direction of the advancing pseudopod. Although there is no entirely satisfactory explanation of ameboid movement, scheme shown is based on a recent analysis of cytoplasmic flow.

FIG. 29-10

Ciliary structure and movement. **A,** Section of the pellicle of *Paramecium,* showing arrangement of cilia, basal granules, and interconnection fibers (so-called neuromotor system). **B,** Structure of a cilium as revealed by the electron microscope. **C,** Sequence of movements of a cilium of *Paramecium.* Power stroke is to the right and recovery to the left. (**B** adapted from Rhodin, J., and T. Dalhamn. 1956. Z. Zellforsch. **44:**345.)

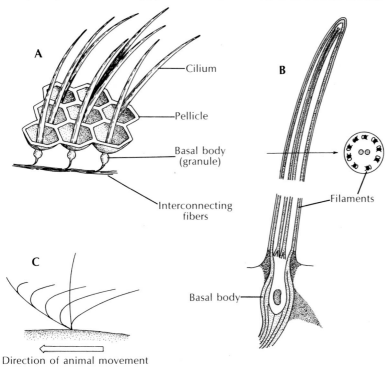

inent roles in filter feeding as described in the previous chapter (Fig. 28-1). In some animals, cilia may be modified to form undulating membranes (such as in *Trypanosoma,* p. 132) or grouped in macrocilia each consisting of 2,000 to 3,000 cilia bundled together (such as ctenophores, p. 196). Cilia are of remarkably uniform diameter (0.1 to 0.5 μ) wherever they are found. The electron microscope has shown that each cilium contains a peripheral circle of nine double filaments and an additional two filaments in the center (Fig. 29-10). (Exceptions to the 9 + 2 arrangement have been noted; certain sperm tails have but one central fibril.) A **flagellum** is a whiplike structure, larger than a cilium, and usually present singly at one end of a cell. They are found in members of flagellate protozoans, in animal spermatozoa, and in sponges.

According to one currently favored theory of ciliary movement, the fibrils behave as "sliding filaments" that move past one another much like the sliding filaments of vertebrate skeletal muscle that is described in the next section. During contraction, fibrils on the concave side slide outward past fibrils on the convex side to increase curvature of the cilium; during the recovery stroke, fibrils on the opposite side slide outward to bring the cilium back to its starting position. For such a system to work, the fibrils must be interconnected by molecular bridges, which in fact cannot be seen with the electron microscope.

Cilia contract in a highly coordinated way, the rhythmic waves of contraction moving across a ciliated epithelium like windwaves across a field of grain. The columns of cilia are coordinated by an interconnected fiber system through which the excitation wave passes with each stroke.

Muscular movement

Contractile tissue reaches its highest development in muscle cells called **fibers.** Although muscle fibers themselves can only shorten, they can be arranged in so many different configurations and combinations that almost any movement is possible.

TYPES OF VERTEBRATE MUSCLE. Vertebrate muscle is broadly classified on the basis of the appearance of muscle cells (fibers) when viewed with a light microscope (Fig. 5-12). **Striated muscle** appears transversely striped (striated), with alternating dark and light bands. We can recognize two types of striated muscle: **skeletal** and **cardiac muscle.** A third kind of vertebrate muscle is **smooth** (or visceral)

muscle, which lacks the characteristic alternating bands of the striated type.

Skeletal muscle is typically organized into sturdy, compact bundles or bands. It is called skeletal muscle because it is attached to skeletal elements and is responsible for movements of the trunk, appendages, the respiratory organs, eyes, mouthparts, and so on. Skeletal muscle fibers are extremely long, cylindric, multinucleate cells, which may reach from one end of the muscle to the other. They are packed into bundles called **fascicles,** which are enclosed by tough connective tissue. The fascicles are in turn grouped into a discrete **muscle** surrounded by a thin connective tissue layer. Most skeletal muscles taper at their ends, where they connect by tendons to bones. Other muscles, such as the ventral abdominal muscles, are flattened sheets.

In most fish, amphibians, and, to some extent, reptiles, there is a segmented organization of muscles alternating with the vertebrae. The skeletal muscles of higher vertebrates, by splitting, by fusion, and by shifting, have developed into specialized muscles best suited for manipulating the jointed appendages that have evolved for locomotion on land. Skeletal muscle contracts powerfully and quickly but fatigues more rapidly than does smooth muscle. Skeletal muscle is sometimes called **voluntary muscle** because it is innervated by motor fibers and is under conscious cerebral control.

Smooth muscle lacks the striations typical of skeletal muscte (Fig. 5-12). The cells are long, tapering strands, each containing a single nucleus. Smooth muscle cells are organized into sheets of muscle circling the walls of the alimentary canal, blood vessels, respiratory passages, and urinary and genital ducts. Smooth muscle is typically slow acting. It is under the control of the autonomic nervous system; thus, unlike skeletal muscle, its contractions are involuntary and unconscious. The principal functions of smooth muscles are to push the contents of a tube, such as the intestine, along its way by active contractions or to regulate the diameter of a tube, such as a blood vessel, by sustained contraction.

Cardiac muscle, the seemingly tireless muscle of the vertebrate heart, combines certain characteristics of both skeletal and smooth muscle. It is fast acting and striated like skeletal muscle, but contraction is under involuntary autonomic control like smooth muscle. Actually, the autonomic nerves serving the heart can only speed up or slow down the rate of contraction; the heartbeat originates within specialized cardiac muscle and the heart will continue to beat even after all autonomic nerves are severed. Until very recently, cardiac muscle was believed to be one large unseparated mass **(syncytium)** of branching, anastomosing fibers. Many histologists, their understanding vastly increased by the electron microscope, now consider cardiac muscle to be comprised of closely opposed, but separate, uninucleate cell fibers.

TYPES OF INVERTEBRATE MUSCLE. Smooth and striated muscles are also characteristic of invertebrate animals, but there are many variations of both types and even instances where the structural and functional features of vertebrate smooth and striated muscle are combined in the invertebrates. Striated muscle appears in invertebrate groups as diverse as the primitive coelenterates and the advanced arthropods. The thickest muscle fibers known, about 3 mm. in diameter and 6 cm. long, and easily seen with the unaided eye, are those of giant barnacles and Alaska king crabs living along the Pacific coast of North America. These cells are so large that they can be readily cannulated for physiologic studies and are understandably popular with muscle physiologists.

It is not possible in this short space to describe adequately the tremendous diversity of muscle structure and function in the vast assemblage of invertebrates. We will mention only two functional extremes.

Bivalve mollusk muscles contain fibers of two types. One kind can contract rapidly, enabling the bivalve to snap shut its valves when disturbed. Scallops use these "fast" muscle fibers to swim in their awkward manner. The second muscle type is capable of slow, long-lasting contractions. Using these fibers, a bivalve can keep its valves tightly shut for days or even months. Obviously these are no ordinary muscle fibers! It has been discovered that such retractor muscles use very little metabolic energy and receive remarkably few nerve impulses to maintain the activated state. The contracted state has been likened to a "catch mechanism" involving some kind of stable cross-linkage between the contractile proteins within the fiber. However, despite considerable research, no completely satisfactory explanation for this retractor mechanism exists.

Insect flight muscles are virtually the functional antithesis of the slow, holding muscles of bivalves. The wings of some of the small flies operate at frequencies greater than 1,000 per second. The so-called **fibrillar muscle,** which contracts at these incredible frequencies—far greater than even the most active of vertebrate muscles—show unique characteristics. They have very limited extensibility; that is, the wing leverage system is arranged so that the muscles shorten hardly at all during each downbeat of the wings. Furthermore, the muscles and wings operate as a rapidly oscillating system in an elastic thorax (see Fig. 16-8). Since the muscles rebound elastically during flight, they receive impulses only periodically rather than one impulse per contraction; one reinforcement impulse for every 20 or 30 contractions is enough to keep the system active.

STRUCTURE OF STRIATED MUSCLE. In recent years the electron microscope and advanced biochemical methods have been focused on the fine structure and function of the striated muscle fiber. These efforts have been so successful that more has been learned of muscle physiology in the last decade than in the previous century. The discussion that follows will be limited to the striated muscle, since its physiology is presently much better understood than is that of smooth muscle.

As we earlier pointed out, striated muscle is so named because of the periodic bands, plainly visible under the light microscope, which pass across the widths of the muscle cells. Each cell, or **fiber,** contains numerous **myofibrils** packed together and invested by the cell membrane, the **sarcolemma** (Figs. 29-11 and 29-15). Also present in each fiber are several hundred nuclei usually located along the edge of the fiber, numerous mitochondria (sometimes called **sarcosomes**), a network of tubules called the **sarcoplasmic reticulum** (to be discussed later), and other cell inclusions typical of any living cell. Most of the fiber, however, is packed with the unique **myofibrils,** each 1 to 2 μ in diameter.

The characteristic banding of the muscle fiber represents the fine structure of the myofibrils that make up the fiber. In the resting fiber are alternating light- and dark-staining bands called the **I bands** and **A bands,** respectively (Fig. 29-11). The functional unit of the myofibril, the **sarcomere,** extends between successive Z lines. The myofibril is actually an aggregate of much smaller parallel units called **myofilaments.** These are of two kinds—thick filaments, 110

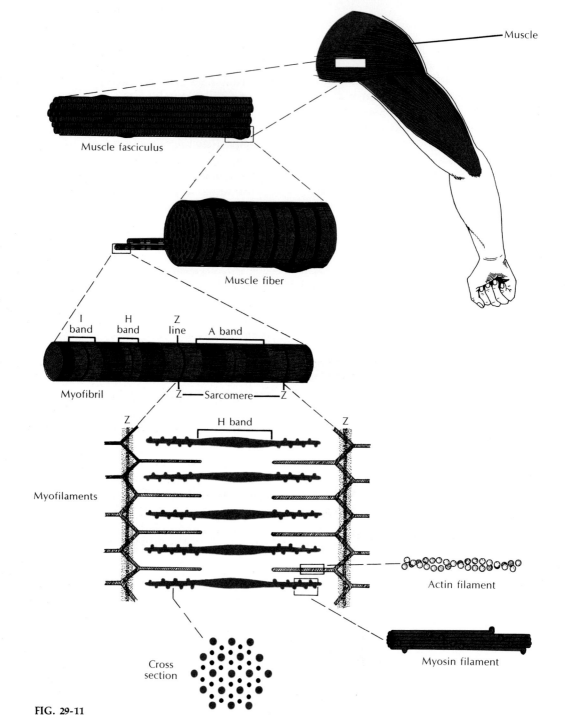

FIG. 29-11

Organization of vertebrate skeletal muscle from gross to molecular level. Actin (thin) and myosin (thick) filaments are enlarged to show supposed shapes of individual molecules and probable positioning of the cross bridges (shown as knobs) on myosin molecules, which serve to link thick and thin filaments during contraction. Cross section shows that each thick filament is surrounded by six thin filaments and that each thin filament is surrounded by three thick filaments. (Modified from Bloom, W., and D. W. Fawcett. 1968. A textbook of histology, Philadelphia, W. B. Saunders Co.)

Å in diameter composed of the protein **myosin**, and thin filaments, 50 Å in diameter composed of the protein **actin** (Fig. 29-11). These are the actual contractile proteins of muscle. The thick myosin filaments are confined to the A band region. The thin actin filaments are located mainly in the light I bands but extend some distance into the A band as well. In the relaxed muscle, they do not quite meet in the center of the A band. The Z line is a dense protein, different from either actin or myosin, which serves as the attachment plane for the thin filaments and keeps them in register. These relationships are diagrammed in Fig. 29-11.

CONTRACTION OF STRIATED MUSCLE. The thick and thin filaments are spatially arranged in a highly symmetric pattern, so that each thick filament is surrounded by six thin filaments; conversely, each thin filament lies among three thick filaments (see cross section at bottom of Fig. 29-11). The two kinds of filaments are linked together by molecular bridges which, it is believed, extend outward from the thick filaments to hook onto active sites on the thin filaments. During contraction the cross-bridges swing rapidly back and forth, alternately attaching and releasing the active sites in succession in a kind of ratchet action (Fig. 29-12).

In 1950, the English physiologists A. F. Huxley and H. E. Huxley independently proposed a **sliding filament model** to explain striated muscle contraction. This widely accepted theory, now firmly based on abundant experimental and electron microscopic evidences, is one of the most important contributions in animal physiology to have appeared in the last 20 years. The model proposed that the thick and thin filaments slide past one another. Both kinds of filament maintain their original length, but the thin actin filaments now extend farther into the A band as shown in Fig. 29-13. As contraction continues, the Z lines are drawn closer together. During very strong contraction the thin filaments touch and crumple in the center of the A band. Striated muscle contracts so rapidly that each cross-bridge may attach and release 50 to 100 times per second.

The contractile machinery has been most thoroughly studied in mammals (most of the electron microscopists used a specific thigh muscle, the psoas muscle, of the rabbit), but recent comparative studies indicate a remarkable uniformity of the sliding-filament mechanism throughout the animal kingdom. Even the contractile proteins myosin and actin are biochemically similar in all animals. The actomyosin contractile system evidently appeared very early in animal evolution and proved so flawless that no significant changes occurred thereafter.

Energy for contraction. Muscles perform work when they contract and, of course, require energy to do so. Resting muscles use little energy but consume large amounts during vigorous exercise. Muscles use only 20% of the energy value of food molecules when contracting; the remainder is released as heat. This is a rapid source of body heat as everyone knows; exercising is the quickest way to warm up when one is cold.

The immediate source of energy for muscular con-

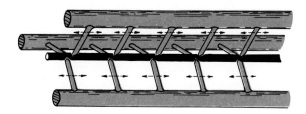

FIG. 29-12
Ratchetlike action of cross bridges between thick and thin filaments of skeletal muscle fibers. Cross bridges swing from site to site, pulling thin filaments past the thick. Each thick filament is actually surrounded by six thin filaments and is linked by six sets of cross bridges. For simplicity, this diagram shows only one set of cross bridges on each thick filament.

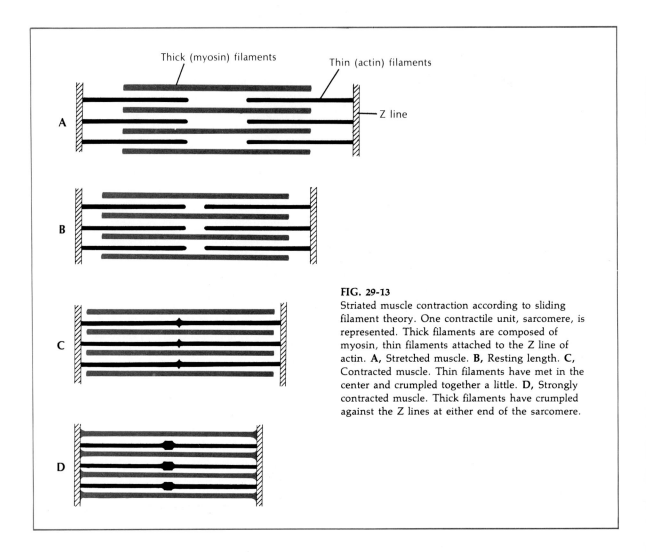

Thick (myosin) filaments

Thin (actin) filaments

Z line

A

B

C

D

FIG. 29-13
Striated muscle contraction according to sliding
filament theory. One contractile unit, sarcomere, is
represented. Thick filaments are composed of
myosin, thin filaments attached to the Z line of
actin. **A,** Stretched muscle. **B,** Resting length. **C,**
Contracted muscle. Thin filaments have met in the
center and crumpled together a little. **D,** Strongly
contracted muscle. Thick filaments have crumpled
against the Z lines at either end of the sarcomere.

traction is ATP. When muscle is stimulated to contract, the energy released by ATP powers the ratchetlike mechanism between actin and myosin, causing
the filaments to telescope.

Although the ATP stored in muscle supplies the
immediate energy for contraction, the supply is limited and quickly exhausted. However, muscle contains a much larger energy storage form, **creatine
phosphate,** which can rapidly transfer energy for the
resynthesis of ATP. Eventually even this reserve is
used up and must be restored by the breakdown of
carbohydrate. Carbohydrate is available from two
sources—from **glycogen** stored in the muscle and
from **glucose** entering the muscle from the bloodstream. If muscular contraction is not too vigorous

or too prolonged, glucose can be completely oxidized
to carbon dioxide and water by **aerobic glycolysis.**
But during prolonged or heavy exercise, the blood
flow to the muscles, although greatly increased above
the resting level, is not sufficient to supply oxygen
as rapidly as required for the complete oxidation of
glucose. When this happens, the contractile machinery receives its energy largely by **anaerobic glycolysis,** a process that does not require oxygen (see
p. 80). The presence of this anaerobic pathway,
although not nearly as efficient as the aerobic one,
is of great importance; without it, all forms of heavy
muscular exertion such as running would be impossible.

During anaerobic glycolysis, glucose is degraded to

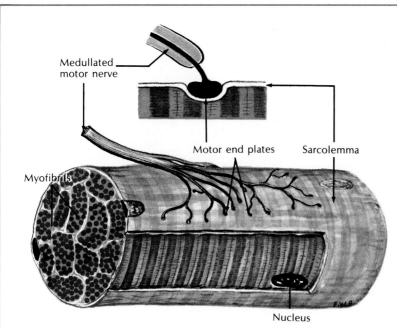

FIG. 29-14
Motor end plates (myoneural junctions) of a motor nerve on a single muscle fiber. (From Schottelius, B. A., and D. D. Schottelius. 1973. A textbook of physiology, ed. 17. St. Louis, The C. V. Mosby Co.)

lactic acid with the release of energy. This is used to resynthesize creatine phosphate, which, in turn, passes the energy to ADP for the resynthesis of ATP. Lactic acid accumulates in the muscle and diffuses rapidly into the general circulation. If the muscular exertion continues, the build up of lactic acid causes enzyme inhibition and fatigue. Thus the anaerobic pathway is a self-limiting one, since continued heavy exertion leads to exhaustion. The muscles incur an **oxygen debt** because the accumulated lactic acid must be oxidized by extra oxygen. After the period of exertion, oxygen consumption remains elevated until all of the lactic acid has been oxidized, or resynthesized to glucose.

To summarize, the sequence of chemical sources of energy can be expressed in abridged form as follows:

$$ATP \rightleftharpoons ADP + H_3PO_4 + \text{Energy for contraction}$$

$$\text{Creatine phosphate} \rightleftharpoons \text{Creatine} + H_3PO_4 +$$
Energy for resynthesis of ATP (anaerobic)

$$Glucose \underset{\text{(anaerobic)}}{\rightleftharpoons} \text{Lactic acid} +$$
Energy for resynthesis of creatine phosphate

$$Glucose + O_2 \xrightarrow{\text{(aerobic)}} CO_2 + H_2O +$$
Energy for resynthesis of creatine phosphate

Stimulation of contraction. Skeletal muscle must, of course, be stimulated to contract. If the nerve supply to a muscle is severed, the muscle will **atrophy,** or waste away. Skeletal muscle fibers are arranged in groups of approximately 100, each group under the control of a single motor nerve fiber. Such a group is called a **motor unit.** As the nerve fiber approaches the muscle fibers, it splays out into many terminal branches. Each branch attaches to a muscle fiber by a special structure, called a **synapse,** or **myoneural junction** (Fig. 29-14). At the synapse is a tiny gap, or cleft, that thinly separates nerve fiber and muscle fiber. In the synapse is stored a chemical, **acetylcholine,** which is released when a nerve impulse reaches

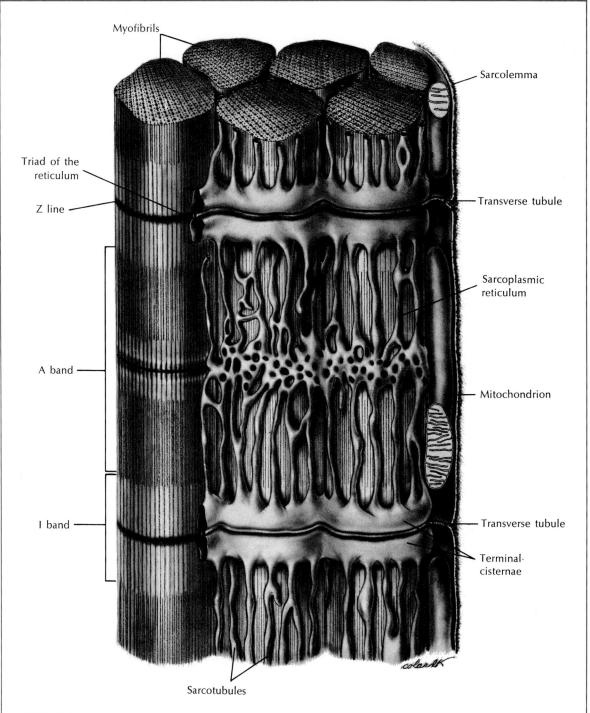

Myofibrils

Sarcolemma

Triad of the
reticulum

Z line

Transverse tubule

A band

Sarcoplasmic
reticulum

Mitochondrion

I band

Transverse tubule

Terminal
cisternae

Sarcotubules

FIG. 29-15

Three-dimensional representation of vertebrate striated muscle showing distribution of
sarcoplasmic reticulum and connecting transverse tubules (T system). In the frog muscle
shown here, the transverse tubules are positioned at the Z regions where they serve to
conduct electrical depolarizations and energy-rich supplies to the myofibrils via the
sarcoplasmic reticulum. (From Bloom, W., and D. W. Fawcett, 1968. A textbook of
histology, Philadelphia, W. B. Saunders Co.)

the synapse. This substance is a chemical mediator that diffuses across the narrow junction and acts on the muscle fiber membrane to generate an electrical depolarization. The potential spreads rapidly through the muscle fiber, causing it to contract. Thus the synapse is a special chemical bridge that couples together the electrical potentials of nerve and muscle fibers.

Coupling of excitation and contraction. For a long time physiologists were puzzled as to how the electrical potential at the myoneural junction could spread quickly enough through the fiber to cause simultaneous contraction of all the densely packed filaments within. Recently it was discovered that vertebrate skeletal muscle contains an elaborate communication system that performs just this function. This is the endoplasmic reticulum (called the **sarcoplasmic reticulum** in muscle). As shown in the three-dimensional diagram (Fig. 29-15), the sarcoplasmic reticulum is a system of fluid-filled channels running parallel to the myofilaments and connecting at the Z lines to **transverse tubules** (T system), which communicate with the sarcolemma that surrounds the fiber. The sarcoplasmic reticulum and T system are ideally arranged for speeding the electrical depolarization from the myoneural junction to the myofilament within. It also serves as a distribution network for glucose, oxygen, minerals, and other supplies needed for muscle contraction.

References
Suggested general readings
(See also general references for Part Three, p. 639.)

Bendall, J. R. 1969. Muscles, molecules and movement. American Elsevier Publishing Co., Inc. *Contains a wealth of well-organized information on muscle structure and physiology, pitched at the advanced undergraduate level.*

Montagna, W. 1956. The structure and function of the skin. New York, Academic Press, Inc. *An account of this important organ.*

Montagna, W. 1959. Comparative anatomy. New York, John Wiley & Sons, Inc. *A clearly written presentation of the comparative anatomy of the skin, skeleton, and muscles.*

Morton, D. J. 1952. Human locomotion and body form. Baltimore, The Williams & Wilkins Co. *An account of the mechanics of movement with relation to body form.*

Rothman, S. 1954. Physiology and biochemistry of the skin. Chicago, University of Chicago Press. *A highly technical account of the skin, its physical properties, its physiologic functions, and its chemical constituents. Especially for the advanced student.*

Waring, H. 1963. Color change mechanisms of cold-blooded vertebrates. New York, Academic Press, Inc. *Color change mechanisms considered from the specialist's level.*

Selected *Scientific American* articles

Chapman, C. B., and J. H. Mitchell, 1965. The physiology of exercise. **212:**88-96 (May). *Muscular activity mobilizes several adaptive nervous, respiratory, circulatory, and metabolic responses.*

Hayashi, T. 1961. How cells move. **205:**184-204 (Sept.). *The actomyosin contractile mechanism and its ubiquitous presence throughout the animal kingdom is described.*

Hoyle, G. 1958. The leap of the grasshopper. **198:**30-35 (Jan.). *The powerful muscle system of the grasshopper's hindleg can propel the animal 20 times its body length.*

Hoyle, G. 1970. How is muscle turned on and off? **222:**84-93 (April). *Calcium plays a crucial role in muscle contraction.*

Huxley, H. E. 1965. The mechanism of muscular contraction. **213:**18-27 (Dec.). *The sliding filament theory is described.*

McLean, F. C. 1955. Bone. **192:**84-91 (Feb.). *Structure and physiology of bone.*

Merton, P. A. 1972. How we control the contraction of our muscles. **226:**30-37 (May.) *Voluntary movements of skeletal muscle are controlled by a sensitive feedback mechanism.*

Smith, D. S. 1965. The flight muscles of insects. **212:**76-88 (June). *Although structurally similar to vertebrate skeletal muscle, insect flight muscle can contract much more rapidly and carry far more weight.*

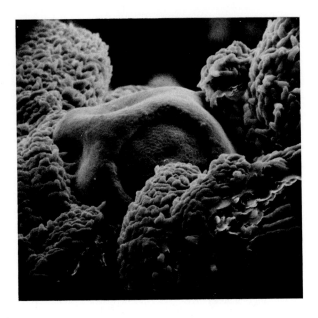

A scallop eye peering out from among folds of the mantle. Scallops (Pecten) bear dozens of eyes, resembling small, bright blue pearls, clearly visible when the valves of the shell are open. Each eye is a remarkably advanced structure with cornea, cellular lens, a double retina, and a tapetum, or reflective layer, containing crystals of guanine. (Scanning electron micrograph, × 275; courtesy of P. P. C. Graziadei, Florida State University.)

CHAPTER 30

COORDINATION: NERVOUS SYSTEM, SENSE ORGANS, AND ENDOCRINE SYSTEM

■ NERVOUS SYSTEM

The origin of the nervous system is based on one of the fundamental principles of protoplasm—irritability. Each cell responds to stimulation in a manner characteristic of that type of cell. But certain cells have become highly specialized for receiving stimuli and for conducting impulses to various parts of the body. Through evolutionary changes, these cells have become the most complex of all body systems—the nervous system. The endocrine system is also important in coordination, but the nervous system has a wider and more direct control of body functions than does the endocrine system.

The evolution of the nervous system has been correlated with the development of bilateral symmetry and cephalization. Along with this development, animals acquired exteroceptors and associated ganglia. The basic plan of the nervous system is to code the sensory information, internally or externally, and transmit it to regions of the central nervous system where it is processed into appropriate action. This action may be of several types, such as simple reflexes, automatic behavior patterns, conscious perception, or learning processes.

Nervous systems of invertebrates

Nervous systems as such are mostly lacking in protozoans, which depend upon primitive irritability and its conduction across the cell surface to respond to stimuli. Nevertheless there are instances of remarkable neural development in certain protozoans. The relatively complex **neuromotor apparatus** of ciliates coordinates the beat of the cilia, and certain species, such as *Epidinium* (see Fig. 7-29), have neurofibrils passing from an anterior motor mass—the beginnings of a central nervous system!

The metazoa show a progressive increase in nervous system complexity that we believe recapitulates to some extent the evolution of the nervous system. The coelenterates have a **nerve net** (see Fig. 9-4) containing bipolar and multipolar cells (protoneurons). These may be separated from each other by synaptic junctions, but they form an extensive network that is found in and under the ectoderm over all the body. An impulse starting in one part of this net will be conducted in all directions, since the synapses do not restrict transmission to one-way movement, as they do in higher animals. There are no differentiated sensory, motor, or connector components in the strict meaning of those terms. Branches of the nerve net connect to receptors in the epidermis and to the epitheliomuscular cells. Most responses tend to be generalized, yet many are astonishingly complex for so simple a nervous system (such as the swimming anemone in Fig. 9-24). Such a type of nervous system is retained among higher animals in the form of nerve plexuses in which such generalized movements as peristalsis are involved.

Flatworms are provided with two anterior **ganglia** of nerve cells from which two main nerve trunks run posteriorly, with lateral branches extending to the various parts of the body (Fig. 10-4). This is the true beginning of a differentiation into a **peripheral nervous system,** extending to all parts of the body, and a **central nervous system,** which coordinates everything. It is also the first appearance of the **linear** type of nervous system, which is more developed in higher invertebrates. Higher invertebrates have a more centralized nervous system, with the two longitudinal nerve cords fused (although still recognizable) and many ganglia present. The annelids have a well-developed nervous system consisting of distinctive **afferent** (sensory) and **efferent** (motor) neurons (Fig. 13-8). At the anterior end, the ventral nerve cord divides and passes upward around the digestive tract to join the bilobed brain. In each segment the double nerve cord bears a double ganglion, each with two pairs of nerves. Arthropods have a system similar to that of earthworms, except that the ganglia are larger and the sense organs better developed.

Mollusks have a system of three pairs of ganglia; one pair is near the mouth, another pair at the base of the foot, and one pair in the viscera. The ganglia are joined by connectives. The mollusks also have a number of sense organs, especially well developed in the cephalopods. Among the echinoderms the nervous system is radially arranged.

The nerve cord in all invertebrates is ventral to the alimentary canal and is solid. This arrangement is in pronounced contrast to the nerve cord of vertebrates, which is dorsal to the digestive system, single, and hollow.

Nervous system of vertebrates

Vertebrates have, as a rule, a brain much larger than the spinal cord. In lower vertebrates this dif-

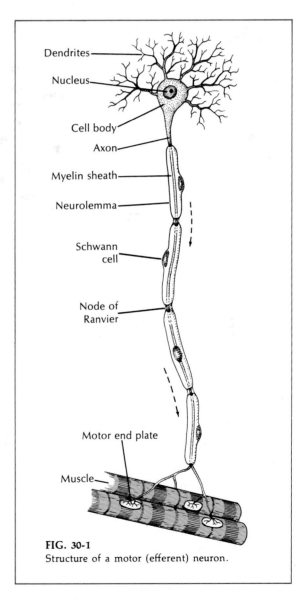

FIG. 30-1
Structure of a motor (efferent) neuron.

Dendrites

Nucleus

Cell body

Axon

Myelin sheath

Neurolemma

Schwann cell

Node of Ranvier

Motor end plate

Muscle

ference is not significant, but higher in the vertebrate kingdom the brain increases in size, reaching its maximum in mammals, especially man. Along with this enlargement has come an increase in complexity, bringing better patterns of coordination, integration, and intelligence. The nervous system is commonly divided into central and peripheral parts; the central division is chiefly concerned with integrative activity and the peripheral part with the conduction of sensory and motor information to all parts of the body.

The neuron: structural and functional unit of the nervous system

The neuron is a cell body and all its processes. Although extremely varied in form, a typical type is shown in Fig. 30-1. From the nucleated cell body extends an **axon,** which carries impulses *away* from the cell. Typically several branching **dendrites** surround the cell body. These carry impulses *toward* the cell body. Axons are usually covered with a soft, white lipid-containing material called **myelin.** This insulating material is often laid down in concentric rings by specialized **Schwann cells** to form a myelin sheath. This is enclosed by an outer membrane called the **neurolemma.**

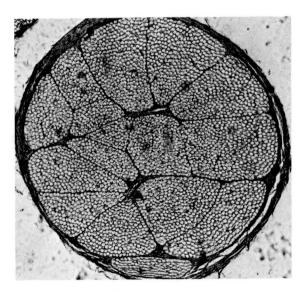

FIG. 30-2
Cross section of nerve showing cut ends of nerve fibers (small white circles). Such a trunk may contain thousands of fibers. Both afferent and efferent fibers are present.

Neurons are commonly divided into three types—**motor, sensory,** and **association** or **connector.** The dendrites of sensory neurons are connected to a **receptor,** and their axons are connected to other neurons; associators are connected only to other neurons; and motor neurons are connected by their axons to an **effector.** Nerves are actually made up of many nerve processes—axons or dendrites or both—bound together with connective tissue (Fig. 30-2). The cell bodies of these bundles of nerves are located either in ganglia or somewhere in the central nervous system (brain or spinal cord).

Nature of the nerve impulse

The nerve impulse is the chemical-electrical message of nerves, the common functional denominator of all nervous system activity. Despite the incredible complexity of the nervous system of advanced animals, nerve impulses are basically alike in all nerves and in all animals. It is an **all-or-none** phenomenon; either the axon is conducting an impulse, or it is not. All impulses are alike, and the only way an axon can vary its effect on the tissue it innervates is by changing the **frequency** of impulse conduction. Frequency change is the language of a nerve fiber. A fiber may conduct no impulses at all, or a very few per second up to a maximum approaching 1,000 per second. The higher the frequency (or rate) of conduction, the greater is the level of excitation. The same general rule applies to sense organs as well; that is, the more a sense organ is excited by a stimulus, the greater is the frequency of impulses sent out over the axons of the sensory nerves.

RESTING POTENTIAL. To understand what happens when an impulse is conducted down a fiber, we need to know something about the resting, undisturbed fiber. Nerve cell membranes, like all cell membranes, have special permeability properties that create ionic imbalances. Sodium and chloride predominate on the outside, whereas potassium ions are more common inside (Fig. 30-3). These differences are quite dramatic; there is about 10 times more sodium outside than in and 25 to 30 times more potassium inside than out. However, the nerve cell membrane is much more permeable to potassium than to sodium. The result is that potassium ions tend to leak outward much more rapidly—about 25 times more rapidly—than do sodium ions, which tend to leak inward. Since both potassium and sodium ions are positively charged, they repel each other. Thus the outward movement of potassium is checked by the sodium, causing the outside of the membrane to become positively charged (Fig. 30-3). The potassium ions do not drift away from the membrane, but are held there by equal numbers of negatively charged ions in the intracellular fluid. Thus the extracellular fluid outside and the intracellular fluid inside are both electrically neutral solutions, but the potassium diffusion front creates enough charge separation to form a bioelectrical potential **across** the membrane. This, then, is the origin of the resting transmembrane potential, which is positive outside, negative inside. This potential difference called the **resting potential** is usually about −70 mV., inside negative.

ACTION POTENTIAL. The nerve impulse is a rapidly moving change in electrical potential called the **action potential** (Fig. 30-4). This is a very rapid and brief depolarization of the axon membrane; in fact, not only is the resting potential abolished, but in most nerves the potential actually reverses for an instant so that the outside becomes negative as compared to the inside. Then, as the action potential moves ahead, the membrane returns to its normal resting potential, ready to conduct another impulse. The entire event occupies only a fraction of a millisecond. Perhaps the most significant property of the nerve impulse is that it is **self-propagating;** that is, once started, the impulse moves ahead automatically, much like the burning of a fuse.

What causes the reversal of polarity in the cell membrane during passage of an action potential? Careful studies have shown that when the action potential arrives at a given point, the cell membrane suddenly becomes much more permeable to sodium ions than before. Sodium rushes in. Actually only an extremely small amount of sodium traverses the membrane in that instant—less than one one-millionth of the sodium outside—but this brief shift of positive ions inward causes the membrane potential to disappear, even reverse. An electrical "hole" is created. Potassium, now finding its electrical barrier gone, begins to move out. Then, as the action potential passes on, the membrane quickly regains

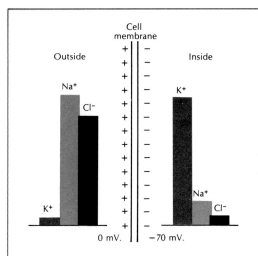

FIG. 30-3
Ionic composition inside and outside a resting nerve cell. An active sodium pump located in the cell membrane drives sodium to the outside, keeping its concentration low inside. Potassium concentration is high inside, and although the membrane is "leaky" to potassium, this ion is held inside by repelling positive charge outside the membrane.

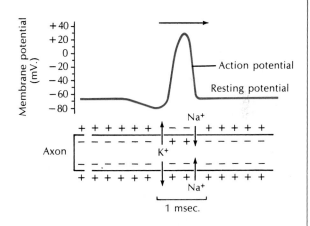

FIG. 30-4
Action potential of nerve impulse. The electrical event, moving from left to right, is associated with rapid changes in membrane permeability to sodium and potassium ions. When the impulse arrives at a point, sodium ions suddenly rush in, making the axon positive inside and negative outside. Then the sodium holes close and potassium holes open up. This restores the normal negative resting potential.

its resting properties. It becomes once again practically impermeable to sodium, and the outward movement of potassium is checked. The rising phase of the action potential is associated with the rapid influx (inward movement) of sodium (Fig. 30-4). When the action potential reaches its peak, the somewhat slower outflux (outward movement) of potassium causes the potential to reverse again, and the action potential falls toward the resting level.

SODIUM PUMP. The resting cell membrane has a very low permeability to sodium. Nevertheless, some sodium ions leak across, even in the resting condition. When the axon is active, sodium flows inward with each passing impulse, and although the amount is very small, it is obvious that the ionic gradient would eventually disappear if the sodium ions were not moved back out again. This is done by a **"sodium pump"** located in the axon plasma membrane. Although no one has ever actually seen the "pump," we do know quite a bit about it because it has been the object of intense biochemical and biophysical studies. The sodium pump is an active transport device capable of combining with sodium on the inside surface of the membrane, then moving to the outside surface where the sodium is released. It is probably composed of phosphate-containing protein molecules. The sodium pump requires energy, since it is moving sodium "uphill" against the

FIG. 30-5

Transmission of impulses across nerve synapses. **A,** Cell body of a motor nerve is shown covered with the terminations of association neurons. Each termination ends in a synaptic knob; hundreds of synaptic knobs may be on a single nerve cell body and its dendrites. **B,** Synaptic knob enlarged 60 times more than **A.** An impulse traveling down the axon will cause some synaptic vesicles to move down to the synaptic cleft and rupture, releasing transmitter molecules into the cleft. **C,** Synaptic cleft as it might appear under a high-resolution electron microscope. Transmitter molecules from a ruptured synaptic vesicle move quickly across the gap to produce an electrical potential change in the postsynaptic membrane.

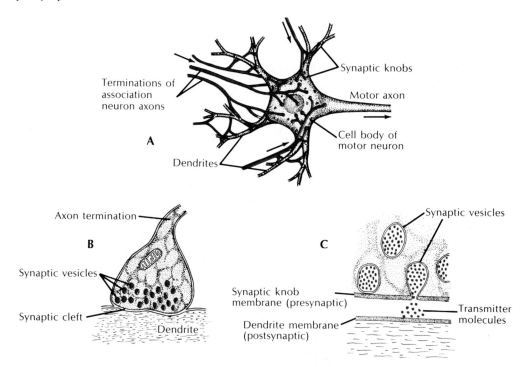

sodium electrical and concentration gradient. This energy is supplied by ATP through cellular metabolic processes. There is evidence that in some cells sodium transport outward is linked to potassium transport inward; the same carrier molecule may act as a two-way shuttle, carrying ions on both trips across the membrane. This kind of pump is called a sodium-potassium pump. Active transport was discussed earlier on p. 68.

Synapses: junction points between nerves

A synapse is found at the end of a nerve axon, where it connects to the dendrite or cell body of the next neuron. At this point the membranes are separated by a narrow gap having a very uniform width of about 20 millimicrons. The synapse is of great functional importance because it acts as a one-way valve that allows nerve impulses to move in one direction only. It is also through the many synapses that information is modulated from one nerve to the next.

The axon of most nerves divides at its end into many branches, each of which bears a synaptic knob that sits on the cell body of the next nerve (Figs. 30-5 and 30-6). The axon terminations and knobs of several nerves may almost cover a nerve cell body and its dendrites. An impulse coming down a nerve axon sprays out into the many branches and synaptic endings on the next nerve cell. Many impulses therefore converge at the cell body at one moment. But these may not excite the cell body enough to fire off an impulse. A neuron requires much prompting to fire. Usually many impulses must arrive at the cell body simultaneously, or within a very brief interval, to

FIG. 30-6
Synaptic knobs in the abdominal ganglion of the marine nudibranch mollusk *Aplysia*. The knobs are strongly attached to rounded surface in left center, which is probably the receptive surface of a postsynaptic neuron. Note firm attachment point of knob in upper center. Note also dividing fibers at left, which form direct connections between fibers. (Scanning electron micrograph, × 6,000; From Lewis, E. R., T. E. Everhart, and Y. Y. Zeevi. 1969. Science **165:**1140-1143.)

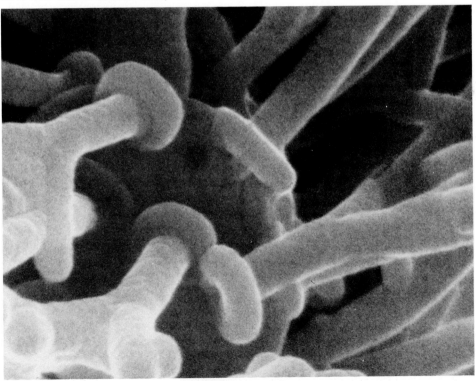

raise the cell body to its firing threshold level. This is called **summation;** it is the cumulative excitatory effect of many incoming impulses that pushes a nerve cell up to firing threshold.

The synapse is a kind of chemical bridge. The electron microscope shows that each synaptic knob contains numerous synaptic vesicles. These are filled with molecules of **chemical transmitter** (Fig. 30-5, *B*). For most synapses this transmitter substance is either **acetylcholine** or **norepinephrine.** When an impulse arrives at the knob, it induces some of the vesicles to move to the base of the knob and release their contents of transmitter molecules into the synaptic cleft (Fig. 30-5, *C*). These move rapidly across the narrow cleft to act on the nerve-cell membrane below. A small potential is produced in the postsynaptic membrane. If reinforced by the arrival of more impulses and by the release of more packets of transmitter molecules, either at the same or adjacent synapses, the small potential may be built up into a large one, sufficient to fire off an impulse in a nerve cell.

Synapses, then, are critical determinants in nervous system function. Although a nerve impulse is an all-or-none event, the synapses act like variable gates that may or may not allow impulses to proceed from one neuron to the next.

Reflex arc as functional unit

Neurons work in groups called **reflex arcs** (Fig. 30-7). There must be at least two neurons in a reflex arc, but usually there are more. The parts of a typical reflex arc consist of (1) a **receptor,** a sense organ in the skin, muscle, or other organ; (2) an **afferent** or sensory neuron, which carries the impulse toward the central nervous system; (3) a **nerve center,** where synaptic junctions are made between the sensory neurons and the association neurons; (4) the **efferent** or motor neuron, which makes synaptic junction with the association neuron and carries impulses out from the central nervous system; and (5) the **effector,** by which the animal responds to its environmental changes. Examples of effectors are muscles, glands, cilia, nematocysts of coelenterates, electric organs of fish, and chromatophores.

A reflex arc at its simplest consists of only two neurons—a sensory (afferent) neuron and a motor (efferent) neuron. Usually, however, association neurons are interposed (Fig. 30-7). Association neurons may connect afferent and efferent neurons on the same side of the spinal cord, connect them on opposite sides of the cord, or connect them on different levels of the spinal cord, either on the same or opposite sides. In almost any reflex act a number

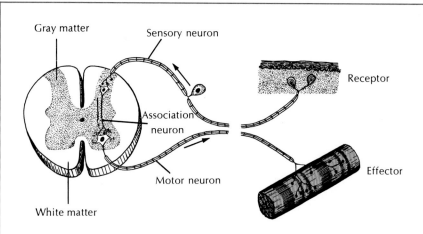

FIG. 30-7
Reflex arc. Impulse generated in the receptor is conducted over a sensory nerve to the spinal cord, relayed by an association neuron to a motor nerve cell body, and by the motor axon to an effector.

of reflex arcs are involved. For instance, a single afferent neuron may make synaptic junctions with many efferent neurons. In a similar way an efferent neuron may receive impulses from many afferent neurons. In this latter case the efferent neuron is referred to as the **final common path.**

A **reflex act** is the response to a stimulus carried over a reflex arc. It is **involuntary** and may involve the cerebrospinal or the autonomic nervous divisions of the nervous system. Many of the vital processes of the body such as control of breathing, heartbeat, diameter of blood vessels, sweat gland secretion, and others are reflex actions. Some reflex acts are inherited and innate; others are acquired through learning processes (conditioned).

Organization of the nervous system

The basic plan of the vertebrate nervous system is a dorsal longitudinal hollow nerve cord that runs from head to tail. During early embryonic development, the central nervous system begins as an ectodermal **neural groove,** which by folding and enlarging becomes a long, hollow, **neural tube.** The cephalic end enlarges into the brain vesicles and the rest becomes the spinal cord. The spinal nerves (31 pairs in man) have a dual origin. The **spinal ganglia** (dorsal root ganglia in Fig. 30-8), containing the sensory neurons, differentiate from specialized cells, called **neural crest cells,** that pinch off from the edges of the neural groove as it closes to form a tube. The ventral roots contain motor fibers that originate in the spinal cord. Both dorsal (sensory) and ventral (motor) roots meet some distance beyond the cord to form a mixed **spinal nerve** (Fig. 30-8).

Central nervous system

The central nervous system is composed of the brain and spinal cord.

SPINAL CORD. The cord is enclosed by the vertebral canal and additionally protected by three layers, the **meninges** (men-in'jeez): a tough outer **dura mater,** a thin spider web–like **arachnoid,** and a delicate innermost sheath, the **pia mater.** Between the arachnoid and the pia mater is a space containing

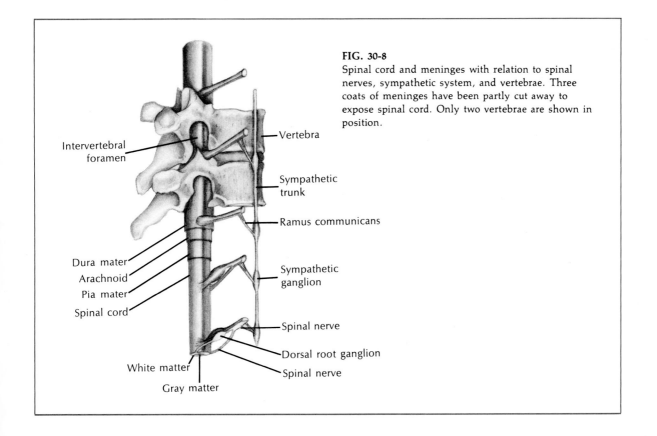

FIG. 30-8
Spinal cord and meninges with relation to spinal nerves, sympathetic system, and vertebrae. Three coats of meninges have been partly cut away to expose spinal cord. Only two vertebrae are shown in position.

Intervertebral foramen

Vertebra

Sympathetic trunk

Ramus communicans

Dura mater
Arachnoid
Pia mater
Spinal cord

Sympathetic ganglion

Spinal nerve

Dorsal root ganglion

White matter

Spinal nerve

Gray matter

cerebrospinal fluid, a secreted fluid forming a protective cushion and thermal insulation for the cord. The meninges and cerebrospinal fluid blanket are continuous with those covering the brain.

In cross section the cord shows two zones—an inner H-shaped zone of gray matter, made up of nerve cell bodies, and an outer zone of white matter, made up of nerve bundles of axons and dendrites (Fig. 30-7). The gray matter contains association neurons and the cell bodies of motor neurons. The white matter of the cord consists of longitudinal fibers linking different levels of the cord with each

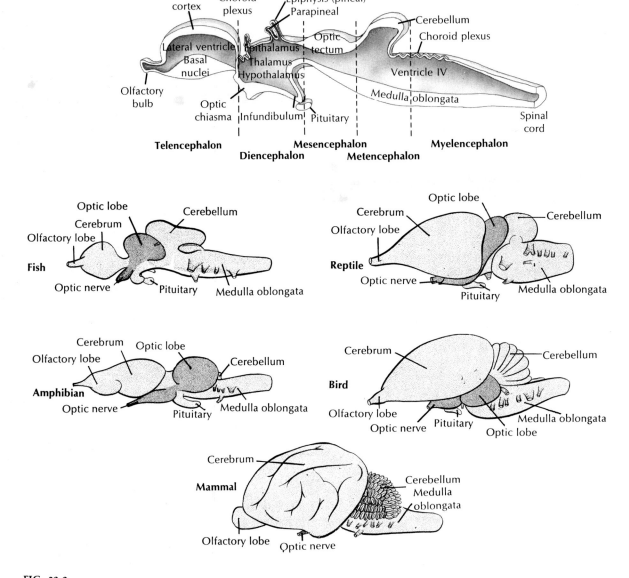

FIG. 30-9

General topography *(top drawing)* and comparative structure of vertebrate brains, showing the principal brain divisions and their development in different vertebrate groups. (*Top,* Adapted from Romer, A. S. 1949. The vertebrate body, Philadelphia, W. B. Saunders Co.)

other and with the brain. The fibers are bundled into **ascending tracts,** carrying impulses to the brain, and **descending tracts** carrying impulses away from the brain. The sensory (ascending) tracts are located mainly in the dorsal part of the cord; the motor (descending) tracts are found ventrally and laterally in the cord. Fibers also cross over from one side of the cord to the other, with the sensory fibers crossing at a higher level than the motor fibers. Although the different tracts cannot be distinguished in a sectioned cord, even with a microscope, their position is known from painstaking mapping experiments.

BRAIN. The brain in vertebrates shows an evolution from the linear arrangement in lower forms (fish and amphibians) to the much-folded and enlarged brain found in higher vertebrates (birds and mammals) (Fig. 30-9). The brain is really the enlarged anterior end of the spinal cord. The ratio between the weight of the brain and spinal cord affords a fair criterion of an animal's intelligence. In fish and amphibians this ratio is about 1:1, in man the ratio is 55:1—in other words, the brain is 55 times heavier than the spinal cord. Although man's brain is not the largest (the elephant's brain is four times heavier) nor the most convoluted (that of the porpoise is even more wrinkled), it is by all odds the best. Indeed the human brain is the most complex structure known to man. It has no parallel in the living or nonliving world. This "great ravelled knot," as the British physiologist Sir Charles Sherrington

called man's brain, may in fact be so complex that it will never be able to understand its own function.

The primitive three-part brain is made up of prosencephalon, mesencephalon, and rhombencephalon (forebrain, midbrain, and hindbrain) (Fig. 30-9; Table 30-1). The prosencephalon and rhombencephalon each divide again to form the five-part brain characteristic of the adults of all vertebrates. The five-part brain includes the telencephalon, diencephalon, mesencephalon, metencephalon, and myelencephalon. From these divisions the different functional brain structures arise.

The impressive evolutionary improvement of the vertebrate brain has accompanied the increased powers of locomotion of the more advanced vertebrates and greater environmental awareness. In the primitive vertebrate brain each of the three parts was concerned with one or more special senses: the prosencephalon with the sense of smell, the mesencephalon with vision, and the rhombencephalon with hearing and balance. These primitive but very fundamental concerns of the brain have been in some instances amplified, and in others reduced or overshadowed, during continued evolution as sensory priorities were shaped by the animal's habitat and way of life.

The brain is made up of both white and gray matter, with the gray matter on the outside (in contrast to the spinal cord in which the gray matter is inside). The gray matter of the brain is mostly in the convoluted **cortex.** In the deeper parts of the brain the white matter of nerve fibers connects the cortex with lower centers of the brain and spinal cord or connects one part of the cortex with another. Also in deeper portions of the brain are collections of nerve cell bodies (gray matter) that provide synaptic junctions between the neurons of higher centers and those of lower centers.

There are also nonnervous elements in the nervous system such as connective, supporting, and capsule cells and the **neuroglia.** Neuroglia cells greatly outnumber the neurons and play various vital roles in the functioning of the neurons. One of their functions is to bind together the nervous tissue proper. Another is their activity in pathologic processes of regeneration. Neuroglia cells are unfortunately the chief source of tumors of the central nervous system.

The main divisions of the brain are given in Table

TABLE 30-1
Divisions of the vertebrate brain

Embryonic vesicles		Main components in adults
Forebrain	Telencephalon	Olfactory bulbs
		Cerebrum
		Lateral ventricles
	Diencephalon	Thalamus
		Hypothalamus
		Infundibulum
		Third ventricle
Midbrain	Mesencephalon	Optic lobes (tectum)
		Cerebral peduncles
		Red nucleus
		Aqueduct of Sylvius
Hindbrain	Metencephalon	Cerebellum
		Pons
		Fourth ventricle
	Myelencephalon	Medulla

30-1. The **medulla,** the most posterior division of the brain, is really a conical continuation of the spinal cord. The medulla together with the more anterior midbrain constitute the "brainstem," an area in which numerous vital, and largely subconscious, activities are controlled, such as heartbeat, respiration, vasomotor tone, and swallowing. It contains the roots of all the cranial nerves except the first and is traversed by many sensory and motor fiber tracts. Although it is small in size and largely hidden from view by the much enlarged "higher" centers, it is in fact the most vital brain area; whereas damage to higher centers may result in severely debilitating loss of sensory or motor function, damage to the brainstem usually results in death.

The **pons,** between the medulla and the midbrain, is made up of a thick bundle of fibers that carries impulses from one side of the cerebellum to the other.

The **cerebellum,** lying above the medulla, is concerned with equilibrium, posture, and movement. Its development is directly correlated with the animal's mode of locomotion, agility of limb movement, and balance. It is usually weakly developed in amphibians and reptiles, which are relatively clumsy forms that stick close to the ground, and well developed in the more agile bony fishes. It reaches its apogee in birds and mammals in which it is greatly expanded and folded. The cerebellum does not initiate movements, but operates as a precision error-control center, or servomechanism, that programs a movement initiated somewhere else, such as in the motor cortex. Primates, and especially man, which possess a manual dexterity far surpassing that of other animals, have the most complex cerebellum of all, since hand and finger movements may involve the simultaneous contraction and relaxation of hundreds of individual muscles.

Between the medulla and diencephalon is the **midbrain.** This is the anterior portion of the brainstem. The white matter of the midbrain consists of ascending and descending tracts that go to the thalamus and cerebrum. On the upper side of the midbrain are the rounded **optic lobes,** serving as centers for visual and auditory reflexes. The midbrain has undergone little evolutionary change in size among vertebrates but has changed in function. It is responsible for the most complex behavior of fish and amphibians; the midbrain serves the higher integrative functions in these lower vertebrates that the cerebrum serves in higher vertebrates.

The **thalamus,** above the midbrain, contains masses of gray matter surrounded by the cerebral hemispheres on each side. This is the relay center for the sensory tracts from the spinal cord. Centers for the sensations of pain, temperature, and touch are supposedly located in the thalamus. In the **hypothalamus** are centers that regulate body temperature, water balance, sleep, and a few other body functions. The hypothalamus also has neurosecretory cells that produce pituitary-regulating neurohormones. These pass down fiber tracts to the anterior and posterior pituitary where the hormones are released into the circulation.

The anterior region of the brain, the **cerebrum,** can be divided into two anatomically distinct areas, the **paleocortex** and the **neocortex.** As its name implies, the paleocortex is the ancient telencephalon. Originally concerned with smell, it became well developed in the advanced fishes and early terrestrial vertebrates, which depend on this special sense. In mammals, and especially in primates, the paleocortex is a deep lying area called the rhinencephalon ("nose brain"), which actually has little to do with the sense of smell. Instead it seems to have acquired a variety of ill-defined functions concerned with consciousness, sleep, memory, emotional control, and sex. Together with a portion of the midbrain it is often called the **limbic-midbrain system.**

Though a late arrival in vertebrate evolution, the neocortex completely overshadows the paleocortex and has become so expanded that it envelopes the diencephalon and midbrain (Fig. 30-10). Almost all the integrative activities primitively assigned to the midbrain were transferred to the neocortex, or cerebral cortex as it is usually called. The cerebral cortex is incompletely divided into two hemispheres by a deep longitudinal fissure. For some reason, one hemisphere, almost always the left, becomes dominant over the other. Since each hemisphere controls the contralateral body side, left hemispheric dominance would seem to explain the great preponderance of right-handedness in man, although many left-handed persons have dominant left cerebral hemispheres as well. Hemispheric dominance is best expressed in speech, interpretive functions, and learned behavior and less for motor function. Thus even extensive damage to the right hemisphere may cause varying degrees of left-sided paralysis but have little effect on intellect. Conversely damage to the left hemisphere usually has disastrous effects on intellect.

It has been possible to localize function in the cerebrum by direct stimulation of exposed brains of people and experimental animals, by postmortem examination of people suffering from various lesions, and by surgical removal of specific brain areas in ex-

perimental animals. The cortex contains discrete motor and sensory areas (Fig. 30-10) as well as large "silent" regions, called **association areas,** concerned with memory, judgment, reasoning, and other integrative functions. These regions are not directly connected to sense organs or muscles. Only forms that have well-developed cerebral cortices are able to learn and modify their behavior by experience.

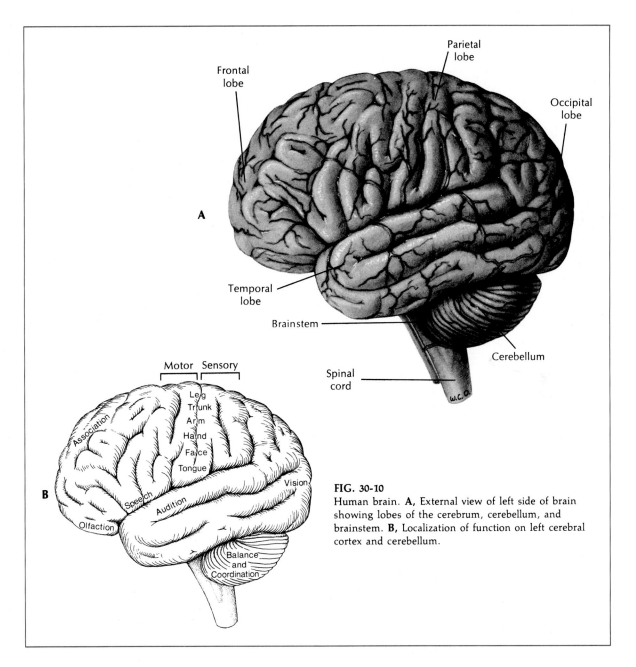

FIG. 30-10
Human brain. **A,** External view of left side of brain showing lobes of the cerebrum, cerebellum, and brainstem. **B,** Localization of function on left cerebral cortex and cerebellum.

Peripheral nervous system

This system is made up of the paired cranial nerves that run to and from the brain, and the paired spinal nerves that run to and from the spinal cord. They consist of bundles of axons and dendrites and connect with the receptors and effectors in the body.

CRANIAL NERVES. In the higher vertebrates, including man, there are 12 pairs of cranial nerves. They are primarily concerned with the sense organs, glands, and muscles of the head and are more specialized than spinal nerves. Some are purely sensory (olfactory, optic, auditory); some are mainly, if not entirely, motor (oculomotor, trochlear, abducens, spinal accessory, hypoglossal); and the others are mixed with both sensory and motor neurons (trigeminal, facial, glossopharyngeal, vagus). The majority of the cranial nerves arise from or near the medulla. Some bear autonomic nerve fibers, especially the facial and vagus. For convenience the various cranial nerves are also designated by the Roman numerals I to XII, as well as by specific names. The numbering begins at the anterior end of the brain and proceeds to the posterior end of the brain.

SPINAL NERVES. The spinal nerves contain both sensory and motor components in approximately equal numbers. In higher vertebrates and man there are 31 pairs: cervical, 8 pairs; thoracic, 12 pairs; lumbar, 5 pairs; sacral, 5 pairs; and caudal, 1 pair. Each nerve has two roots by which it is connected to the spinal cord (Fig. 30-8). All the sensory fibers enter the cord by the dorsal root, and all the motor fibers leave the cord by the ventral root. The nerve cell bodies of motor neurons are located in the ventral horns of the gray zone of the spinal cord; the sensory nerve cell bodies are in the dorsal spinal ganglia just outside the cord. Near the junction of the two roots, the spinal nerve divides into a small dorsal branch (ramus) that supplies structures in the back, a larger ventral branch that supplies structures in the sides and front of the trunk and in the appendages, and an autonomic branch that supplies structures in the viscera. To supply a large area of the body, the ventral rami of several spinal nerves may join to form a network (plexus). These are the **cervical, brachial,** and **lumbosacral plexuses.**

Autonomic nervous system

The autonomic nerves govern the involuntary functions of the body that do not ordinarily affect consciousness. The cerebrum has no direct control over these nerves, thus one cannot by volition stimulate or inhibit their action. Autonomic nerves control the movements of the alimentary canal and heart, the contraction of the smooth muscle of the blood vessels, urinary bladder, iris of eye, etc., and the secretions of various glands.

Subdivisions of the autonomic system are the **parasympathetic** and the **sympathetic.** Most organs in the body are innervated by both sympathetic and parasympathetic fibers, and their actions are antagonistic (Fig. 30-11). If one speeds up an activity, the other will slow it down. However, neither kind of nerve is exclusively excitatory or inhibitory. For example, parasympathetic fibers inhibit heartbeat but will excite peristaltic movements of the intestine; sympathetic fibers will increase heartbeat but slow down peristaltic movement.

The **parasympathetic** system consists of motor nerves, some of which emerge from the brain by certain cranial nerves and others from the pelvic region of the spinal cord by certain spinal nerves. Parasympathetic fibers **excite** the stomach and intestine, urinary bladder, bronchi, constrictor of iris, salivary glands, and coronary arteries. They **inhibit** the heart, intestinal sphincters, and sphincter of the urinary bladder.

In the **sympathetic** division the nerve cell bodies are located in the thoracic and upper lumbar areas of the spinal cord. Their fibers pass out through the ventral roots of the spinal nerves, separate from these, and go to the sympathetic ganglia, which are paired and form a chain on each side of the spinal column. From these ganglia some of the fibers run through spinal nerves to the limbs and body wall, where they innervate the blood vessels of the skin, the smooth muscles of the hair, the sweat glands, etc.; and some run to the abdominal organs as the splanchnic nerves. Sympathetic fibers **excite** the heart, blood vessels, sphincters of the intestines, urinary bladder, dilator muscles of the iris, and others. They **inhibit** the stomach, intestine, bronchial muscles, and coronary arterioles.

All preganglionic fibers, whether sympathetic or parasympathetic, release **acetylcholine** at the synapse for stimulating the ganglion cells. The terminations of the parasympathetic and sympathetic nervous systems release different types of chemical transmitter substances. The parasympathetic fibers

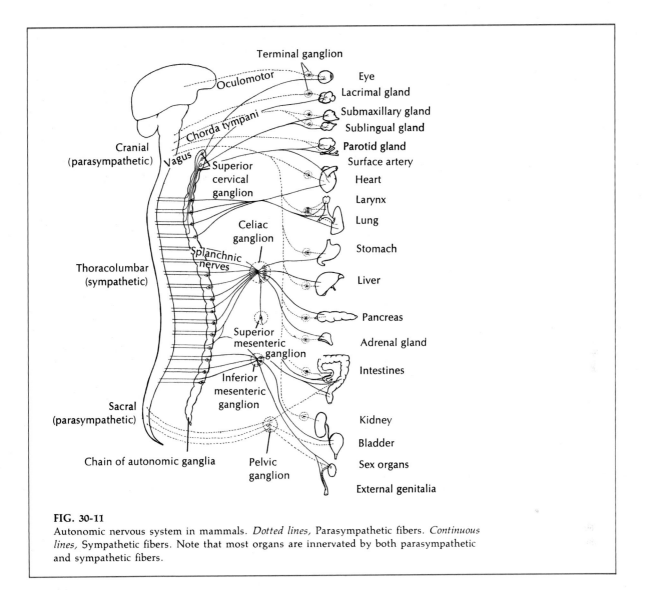

FIG. 30-11

Autonomic nervous system in mammals. *Dotted lines,* Parasympathetic fibers. *Continuous lines,* Sympathetic fibers. Note that most organs are innervated by both parasympathetic and sympathetic fibers.

release **acetylcholine** at their endings whereas the sympathetic fibers release **norepinephrine** (also called noradrenaline). These chemical substances produce characteristic physiologic reactions. Since there is some physiologic overlapping of sympathetic and parasympathetic fibers, it is now customary to describe nerve fibers as either adrenergic (norepinephrine effect) or cholinergic (acetylcholine effect).

■ SENSE ORGANS

Animals require a constant inflow of information from the environment to regulate their lives. Sense organs are specialized receptors designed for detecting environmental status and change. An animal's sense organs are its first level of environmental perception; they are data input channels for the brain.

A **stimulus** is some form of energy—electrical, mechanical, chemical, or radiant. The task of the sense organ is to transform the energy form of the stimulus it receives into nerve impulses, the common language of the nervous system. In a very real sense, then, sense organs are **biologic transducers.** A microphone, for example, is a man-made transducer that converts mechanical (sound) energy into electrical energy. And like the microphone that is sensitive only to sound, sense organs are, as a rule, quite specific for one kind, or **modality,** of stimulus energy. Thus eyes respond only to light, ears to sound, pressure receptors to pressure, and chemoreceptors to chemical molecules. But, again, all of these different forms of energy are converted into nerve impulses. Since all nerve impulses are qualitatively alike, how do animals perceive and distinguish the different **sensations** of varying stimuli? The answer is that the real perception of sensation is done in localized regions of the brain, where each sense organ has its own hookup. Impulses arriving at a particular sensory area of the brain can be interpreted in only one way. This is why pressure on the eye causes us to see "stars" or other visual patterns; the mechanical distortion of the eye initiates impulses in the optic nerve fibers that are perceived as light sensations. Although the operation hopefully could never be done, the deliberate surgical switching of optic and auditory nerves would cause the recipient to quite literally see thunder and hear lightning!

Classification of receptors

Receptors are classified on the basis of their location. Those near the external surface are called **exteroceptors** and are stimulated by changes in the external environment. Internal parts of the body are provided with **interoceptors,** which pick up stimuli from the internal organs. Muscles, tendons, and joints have **proprioceptors,** which are sensitive to changes in the tension of muscles and provide the organism with a sense of position.

Another way of classifying receptors is on the basis of the energy form used to stimulate them, such as **chemical, mechanical, photo,** or **thermal.**

Chemoreception

Chemoreception is the most primitive and most universal sense in the animal kingdom. It probably guides the behavior of animals more than any other sense. The most primitive animals, protozoans, use **contact chemical receptors** to locate food and adequately oxygenated water and to avoid harmful substances. These receptors elicit a simple trial-and-error behavior, called **chemotaxis.** More advanced animals have specialized **distance chemical receptors.** These are often developed to a truly amazing degree of sensitivity and are responsible for complex

behavioral activities. Distance chemoreception, usually referred to as sense of smell, or olfactory sense, guides feeding behavior, location and selection of sexual mates, territorial and trail marking, and alarm reactions of numerous animals. The social insects produce species-specific odors, called **pheromones,** which comprise a highly developed chemical language. Pheromones are a diverse group of organic compounds released by epithelial glands that serve either to initiate specific patterns of behavior, such as attracting mates or marking trails (releaser pheromones), or to trigger some internal physiologic change such as metamorphosis (primer pheromones). Insects have a variety of chemoreceptors on the body surface for sensing specific pheromones as well as other nonspecific odors.

In all vertebrates and in insects as well, the senses of **taste** and **smell** are clearly distinguishable. Although there are similarities between taste and smell receptors, in general the sense of taste is more restricted in response and is less sensitive than the sense of smell. Taste and smell centers are also located in different parts of the brain. In higher forms, **taste buds** are found on the tongue and in the mouth cavity (Fig. 30-12). A taste bud consists of a few sensitive cells surrounded by supporting cells and is provided with a small external pore through which the slender tips of the sensory cells project. The basal ends of the sensory cells contact nerve endings from cranial nerves. Taste bud cells in vertebrates have a short life of about 10 days and are continually being replaced.

The four basic taste sensations of man—sour, salt, bitter, and sweet—are each attributable to a different kind of taste bud. The tastes for salt and sweet are found mainly at the tip of the tongue, bitter at the base of the tongue, and sour along the sides of the tongue. Taste buds are more numerous in ruminants (mammals that chew the cud) than in man. They tend to degenerate with age; the child has more buds widely distributed over the mouth.

Sense organs of **smell** (olfaction) are found in a specialized epithelium located either in the nasal cavity (terrestrial vertebrates) or in pouches on the snout (aquatic vertebrates). The sense of smell is much more complex than that of taste. There are millions of olfactory receptor cells in the nasal epithelium, and as many as a thousand of these may converge on a single neuron. This allows great sum-

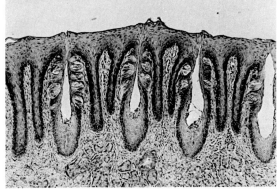

FIG. 30-12

Taste buds. **A,** Scanning electron micrograph of circumvallate papillae on surface of tongue of puppy dog. Taste buds (not visible) are located in walls of circular trench surrounding papillae. The numerous filiform papillae surrounding the two circumvallate papillae lack taste buds. (× 55.) **B,** Light micrograph of section of rabbit's tongue. Taste buds are little oval bodies on sides of slitlike recesses. (× 400.) (**A,** Courtesy P. P. A. Graziadei, Florida State University, Tallahassee, Fla.; **B,** courtesy J. B. Bamberger, Los Angeles, Calif.)

mation and vastly improves sensitivity. Some people can detect many thousands of different odors, and it is obvious that many other vertebrates can easily outdo man. Gases must be dissolved in a fluid to be smelled; therefore the nasal cavity must be moist. The sensory cells with projecting hairs are scattered singly through the olfactory epithelium. Their basal ends are connected to fibers of the olfactory cranial nerve that runs to one of the olfactory lobes. The

sensitivity to certain odors approaches the theoretical maximum for the chemical sense. The human nose can detect 1/25 millionth of 1 mg. of mercaptan, the odoriferous principal of the skunk. This averages out to about 1 molecule per sensory ending. Since taste and smell are stimulated by chemicals in solution, their sensations may be confused. The taste of food is dependent to a great extent on odors that reach the olfactory membrane through the throat. All the various "tastes" other than the four basic ones (sweet, sour, bitter, salt) are really the result of the flavors' reaching the sense of smell in this manner. The sense of smell is the least understood sense. Of the numerous theories that have been proposed, the favored ones today postulate some kind of **physical interaction** between the odor molecule and a protein receptor site on a cell membrane. This interaction somehow alters membrane permeability and leads to depolarization in the receptor cell, which triggers a nerve impulse. One theory (J. E. Amoore) proposes that odor molecules have specific stereochemical shapes and that the range of detectable odors is attributable to differences in the way the molecule smelled fits the receptor site.

Mechanoreception

Mechanoreceptors are sensitive to quantitative forces such as touch, pressure, stretching, sound, and gravity. Many receptors in and on the body constantly monitor information about conditions within the body (muscle position, body equilibrium, blood pressure, pain, etc.) and conditions in the environment (sound and other vibrations such as water currents).

TOUCH AND PAIN. Although superficial touch receptors are distributed over all the body, they tend to be concentrated in the few areas especially important in exploring and interpreting the environment. In most animals these areas are on the face and limb extremities. Of the more than half a million separate sensitive spots on man's body surface, most are found on his lips, tongue, and fingertips. Many touch receptors are bare nerve-fiber terminals, but there is an assortment of other kinds of receptors of varying shapes and sizes. Each hair follicle is crowded with receptors that are sensitive to touch.

The sensation of deep touch and pressure is registered by relatively large receptors called **pacinian**

corpuscles. They are common in deep layers of skin (Fig. 29-1, *B*), in connective tissue surrounding muscles and tendons, and in the abdominal mesenteries. Each corpuscle, easily visible to the naked eye, is built of numerous layers like an onion. Any kind of mechanical deformation of the pacinian corpuscles is converted into nerve impulses that are sent to sensory areas of the brain.

Pain receptors are relatively unspecialized nerve fiber endings that respond to a variety of stimuli that

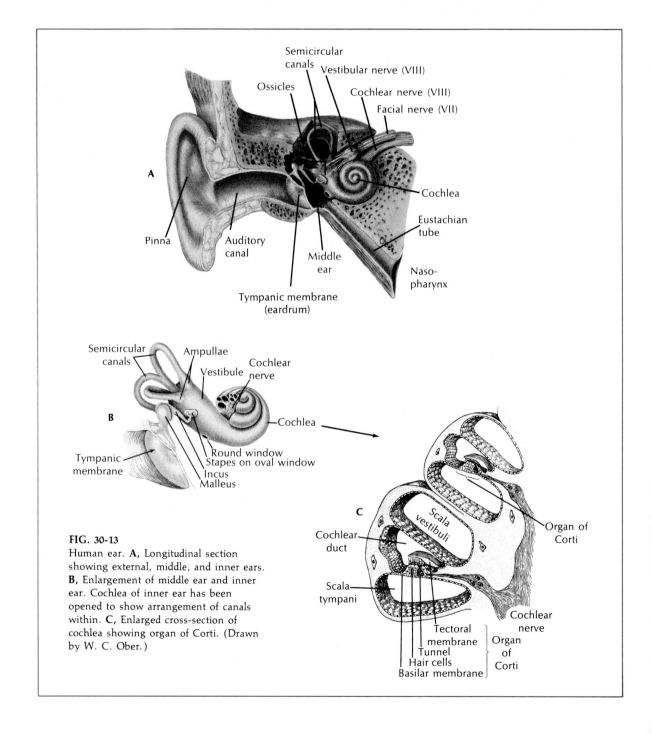

FIG. 30-13
Human ear. **A,** Longitudinal section showing external, middle, and inner ears. **B,** Enlargement of middle ear and inner ear. Cochlea of inner ear has been opened to show arrangement of canals within. **C,** Enlarged cross-section of cochlea showing organ of Corti. (Drawn by W. C. Ober.)

signal possible or real tissue damage. It is still uncertain whether pain fibers respond directly to injury or indirectly to some substance, such as histamine, which is released by damaged cells.

LATERAL LINE SYSTEM OF FISH. The lateral line is a distant touch reception system for detecting wave vibrations and currents in water. The receptor cells, called **neuromasts,** are located free on the body surface in primitive fishes and aquatic amphibians, but in the advanced fishes they are located within canals running beneath the epidermis; these open at intervals to the surface. Each neuromast is a collection of hair cells with the sensory hairs embedded in a gelatinous, wedge-shaped mass, known as a **cupula.** This projects into the center of the lateral line canal so that it will bend in response to any disturbance of water on the body surface. The lateral line system is one of the principal sensory systems that guide fish in their movements.

HEARING. The ear is a specialized receptor for detecting sound waves in the surrounding air (Fig. 30-13). Another sense, equilibrium, is also associated with the ears of all vertebrate animals. Among the invertebrates, only certain insects have true sound receptors. In its evolution the ear was at first associated more with equilibrium than with hearing. Hearing sense is found only in the internal ear, which is the only part of the ear present in many of the lower vertebrates; the middle and the external ears were added in later evolutionary developments. The internal ear is considered to be a development of part of the lateral line system of fish. Some fish apparently can transmit sound from their swim bladders by the Weberian ossicles (series of small bones) to some part of the inner ear, since they lack a cochlea.

The ear found in higher vertebrates is made up of three parts: (1) the **inner ear,** which contains the essential organs of hearing and equilibrium and is present in all vertebrates; (2) the **middle ear,** an air-filled chamber with one or more ossicles for conducting sound waves to the inner ear, present in amphibians and higher vertebrates only (Fig. 30-14); and (3) the **outer ear,** which collects the sound waves and conducts them to the tympanic membrane lying next to the middle ear, present only in reptiles, birds, and mammals but most highly developed in the latter.

Outer ear. The outer, or external, ear of higher vertebrates is made up of two parts: (1) the **pinna,** or skin-covered flap of elastic cartilage and muscles, and (2) the **auditory canal** (Fig. 30-13, *A*). In many mammals, such as the rabbit and cat, the pinna is freely movable and is effective in collecting sound waves. The auditory canal condenses the waves and passes them to the eardrum. The walls of the auditory canal are lined with hair and wax-secreting glands as a protection against the entrance of foreign objects.

Middle ear. The middle ear is separated from the external ear by the eardrum, or tympanic membrane, which consists of a tightly stretched connective membrane. Within the air-filled middle ear a remarkable chain of three tiny bones, **malleus** (hammer), **incus** (anvil), and **stapes** (stirrup), conduct the sound waves across the middle ear (Fig. 30-13, *B*). This bridge of bones is so arranged that the force of sound waves pushing against the eardrum is amplified as many as 90 times where the stapes contacts the oval window of the inner ear. Muscles attached to the middle ear bones contract when the ear receives very loud noises, thus protecting the inner ear from damage. However, these muscles cannot contract quickly enough to protect the inner ear from the damaging effects of a sudden blast. The middle ear communicates with the pharynx by means of the eustachian tube, which acts as a safety device to equalize pressure on both sides of the eardrum.

Inner ear. The inner ear consists essentially of two labyrinths, one within the other. The inner one is called the **membranous labyrinth** and is a closed ectodermal sac filled with the fluid, **endolymph.** The part involved with hearing (cochlea) is coiled like a snail's shell, making two and a half turns in man (Fig. 30-13, *B*). Surrounding the membranous labyrinth is the **bony labyrinth,** which is a hollowed-out part of the temporal bone and conforms to the shape and contours of the membranous labyrinth. In the space between the two labyrinths, **perilymph,** a fluid similar to endolymph, is found.

The cochlea is divided into three longitudinal canals that are separated from each other by thin membranes (Fig. 30-13, *B* and *C*). These canals become progressively smaller from the base of the cochlea to the apex. One of these canals is called the **vestibular canal** (scala vestibuli); its base is closed by the oval window. The **tympanic canal** (scala tympani), which is in communication with the vestibular

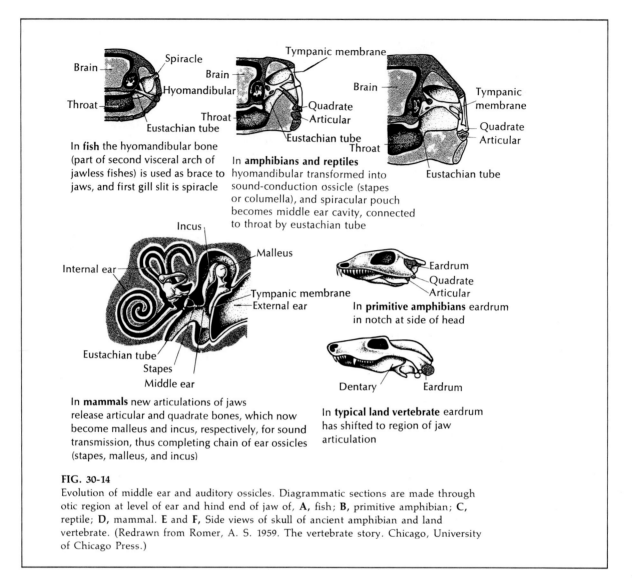

In **fish** the hyomandibular bone (part of second visceral arch of jawless fishes) is used as brace to jaws, and first gill slit is spiracle

In **amphibians and reptiles** hyomandibular transformed into sound-conduction ossicle (stapes or columella), and spiracular pouch becomes middle ear cavity, connected to throat by eustachian tube

In **primitive amphibians** eardrum in notch at side of head

In **mammals** new articulations of jaws release articular and quadrate bones, which now become malleus and incus, respectively, for sound transmission, thus completing chain of ear ossicles (stapes, malleus, and incus)

In **typical land vertebrate** eardrum has shifted to region of jaw articulation

FIG. 30-14
Evolution of middle ear and auditory ossicles. Diagrammatic sections are made through otic region at level of ear and hind end of jaw of, **A**, fish; **B**, primitive amphibian; **C**, reptile; **D**, mammal. **E** and **F**, Side views of skull of ancient amphibian and land vertebrate. (Redrawn from Romer, A. S. 1959. The vertebrate story. Chicago, University of Chicago Press.)

canal at the tip of the cochlea, has its base closed by the round window. Between these two canals is the **cochlear duct,** which contains the organ of hearing, the **organ of Corti** (Fig. 30-13, *C*). The latter organ is made up of fine rows of hair cells that run lengthwise from the base to the tip of the cochlea. There are at least 24,000 of these hair cells in the human ear, each cell with many hairs projecting into the endolymph of the cochlear canal and each connected with neurons of the auditory nerve. The hair cells rest on the **basilar membrane,** which separates the tympanic canal and cochlear duct, and are covered over by the **tectorial membrane** found directly above them.

Sound waves picked up by the external ear are transmitted through the auditory canal to the eardrum, which is caused to vibrate. These vibrations are conducted by the chain of ear bones to the oval window, which transmits the vibrations to the fluid in the vestibular and tympanic canals. The vibrations of the endolymph cause the basilar membrane with its hair cells to vibrate, so that the latter are bent against the tectorial membrane. This stimulation of the hair cells causes them to initiate nerve impulses in the fibers of the auditory nerve, with which they are connected. According to the **place theory** of pitch discrimination it is stated that when sound waves strike the inner ear the entire basilar membrane is set in vibration by a traveling wave of displacement, which increases in amplitude from the oval window toward the apex of the cochlea. This displacement wave reaches a maximum at the region of the basilar membrane, where the natural frequency of the membrane corresponds to the sound frequency. Here, the membrane vibrates with such ease that the energy of the traveling wave is completely dissipated. Hair cells in that region will be stimulated and the impulses conveyed to the fibers of the auditory nerve. Those impulses that are carried by certain fibers of the auditory nerve are interpreted by the hearing center as particular tones. The **loudness** of a tone depends on the number of hair cells stimulated, whereas the **timbre,** or quality, of a tone is produced by the pattern of the hair cells stimulated by sympathetic vibration. This latter characteristic of tone enables one to distinguish between different human voices and different musical instruments, even though the notes in each case may be of the same pitch and loudness.

SENSE OF EQUILIBRIUM. Closely connected to the inner ear and forming a part of it are two small sacs, the **saccule** and **utricle,** and three **semicircular canals.** Like the cochlea, the sacs and canals are filled with endolymph. They are concerned with the sense of balance and rotation. They are well developed in all vertebrates, and in some lower forms they represent about all there is of the internal ear, for the cochlea is absent in fish. They are innervated by the nonacoustic branch of the auditory nerve. The utricle and saccule are hollow sacs lined with sensitive hairs on which are deposited a mass of minute calcium carbonate crystals called **otoconia.** In bony fishes, these crystals are formed into compact stonelike structures called **otoliths.** Similar stony accretions are found within statocysts, the balance organs of many invertebrates. Although the anatomic nature of these static balance organs varies in different groups, they all function in the same basic way: the weight of the stony accretion presses on the hair cells to give information about the position of the head (or entire body) relative to the force of gravity. As the head is tilted in one direction or another, different groups of hair cells are stimulated; conveyed

to the brain, this information is interpreted with reference to position.

The semicircular canals of vertebrates are designed to detect changes in movement: acceleration or deceleration. The three semicircular canals are at right angles to each other, one in each plane of space. They are filled with fluid, and at the opening of each canal into the utricle there is a bulblike enlargement, the **ampulla,** which contains hair cells but no otoconia. Whenever the fluid moves, these hair cells are stimulated. Rotating the head will cause a lag, because of inertia, in certain of these ampullae. This lag produces consciousness of movement. Since the three canals of each internal ear are in different planes, any kind of movement will stimulate at least one of the ampullae.

Vision

Light sensitive receptors are called **photoreceptors.** These receptors range all the way from simple light-sensitive cells scattered randomly on the body surface of the lowest invertebrates (dermal light sense) to the exquisitely developed vertebrate eye. Although dermal light receptors contain little photochemical substance and are far less sensitive than optical receptors, they are important in locomotory orientation, pigment distribution in chromatophores, photoperiodic adjustment of reproductive cycles, and other behavioral changes in many lower invertebrates. The arthropods, however, have **compound eyes** composed of many independent visual units called **ommatidia.** The eye of a bee contains about 15,000 of these units, each of which views a separate narrow sector of the visual field. Such eyes form a mosaic of images from the separate units. The compound eye probably does not produce a very distinct image of the visual field, but it is extremely well suited to picking up motion, as anyone knows who has tried to swat a fly.

The vertebrate eye is built like a camera—or rather we should say a camera is modeled somewhat after the vertebrate eye. It contains a light-tight chamber with a lens system in front that focuses an image of the visual field on a light-sensitive surface (the retina) in back (Fig. 30-15). Because eyes and cameras are based on the same laws of optics, we can wear glasses to correct optical defects in our eyes. But here the similarity between eye and camera ends. The human eye is actually replete with optical shortcom-

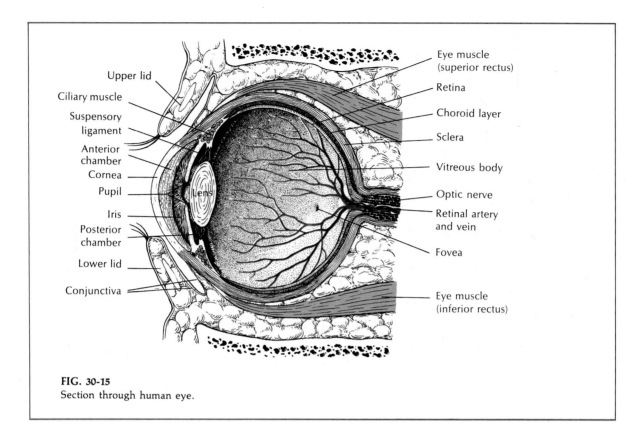

FIG. 30-15
Section through human eye.

ings; projected on the retina of the normal eye are more colored fringe halos, apparitions, and distortions than would be produced by even the cheapest camera lens. Yet the human brain corrects for this "poor design" so completely that we perceive a perfect image of the visual field. It is in the retina and the optic center of the brain that the marvel of vertebrate vision can be understood.

The spherical eyeball is built of three layers: (1) a tough outer white **sclerotic** coat (sclera) serving for support and protection, (2) the middle **choroid** coat containing blood vessels for nourishment, and (3) the light-sensitive **retinal** coat (Fig. 30-15). The **cornea** is a transparent modification of the sclera. A circular curtain, the **iris,** regulates the size of the light opening, the **pupil.** Just behind the iris is the **lens,** a transparent, elastic ball that bends the rays and focuses them on the retina. In land vertebrates the cornea actually does most of the bending of light rays, while the lens adjusts the focus for near and far objects. Between the cornea and the lens is the outer chamber filled with the watery **aqueous humor;** be-

tween the lens and the retina is the much larger inner chamber, filled with the viscous **vitreous humor.** Surrounding the margin of the lens and holding it in place is the **suspensory ligament.** This, together with the **ciliary muscle,** a ring of radiating muscle fibers attached to the suspensory ligament, makes possible the stretching and relaxing of the lens for close or distant vision (accommodation).

The **retina** is composed of photoreceptors, the **rods** and **cones** (Fig. 30-16). Approximately 125 million rods and 7 million cones are present in each human eye. Cones are primarily concerned with color vision in ample light; rods, with colorless vision in dim light. The retina is actually made up of three sets of neurons in series with each other: (1) photoreceptors (rods and cones), (2) intermediate neurons, and (3) ganglionic neurons whose axons form the optic nerve.

The **fovea centralis,** the region of keenest vision, is located in the center of the retina, in direct line with the center of the lens and cornea. It contains only cones. The acuity of an animal's eyes depends

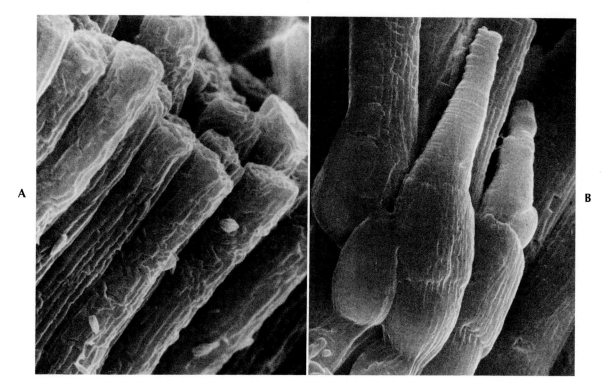

FIG. 30-16

Rods and cones of vertebrate retina. **A,** Outer segments of rods of bullfrog eye. Dendrites of nerve fibers pass up vertical fissures in rods. **B,** Cones and rods of mud puppy *(Necturus)* eye. Cone in center consists of conical outer segment and bulb-shaped inner segment giving rise to nerve fiber at base. Cones are much less sensitive to light than rods, and function in color vision. (Scanning electron micrographs, × 5,000; **A,** Courtesy E. R. Lewis, University of California, Berkeley, Calif.; **B,** from Lewis, E. R., Y. Y. Zeevi, and F. S. Werblin. 1969. Brain Res. **15:**559-562.)

on the density of cones in the fovea. The human fovea and that of a lion contain about 150,000 cones per square millimeter. But many water and field birds have up to 1 million cones per square millimeter. Their eyes are as good as man's eyes would be if aided by eight-power binoculars.

At the peripheral parts of the retina only rods are found. This is why we can see better at night by looking out of the corners of our eyes because the rods, adapted for high sensitivity with dim light, are brought into use.

CHEMISTRY OF VISION. Each rod contains a light-sensitive pigment known as **rhodopsin.** Each rhodopsin molecule consists of a large, colorless protein, **opsin,** and a small carotenoid molecule, **retinal** (formerly called retinene), a derivative of vitamin A. When a quantum of light strikes a rod and is ab-

sorbed by the rhodopsin molecule, the latter undergoes a chemical bleaching process that causes it to split into separate opsin and retinal molecules. In some way not yet understood this change triggers the discharge of a nerve impulse in the receptor cell. The impulse is relayed to the optic center of the brain. Rhodopsin is then enzymatically resynthesized so that it can respond to a subsequent light signal. The amount of intact rhodopsin in the retina depends on the intensity of light reaching the eye. The dark-adapted eye contains much rhodopsin and is very sensitive to weak light. Conversely most of the rhodopsin is broken down in the light-adapted eye. It takes about half an hour for the light-adapted eye to accommodate to darkness, while the rhodopsin level is gradually built up. The remarkable ability of the eye to dark- and light-adapt vastly increases the versatility of the eye; it enables us to see by starlight as well as by the noonday sun, 10 billion times brighter.

The light-sensitive pigment of cones is called **iodopsin.** It is similar to rhodopsin, containing **retinal** combined with a special protein, **cone opsin.** Cones function to perceive color and require 50 to 100 times more light for stimulation than do rods. Consequently night vision is almost totally rod vision; this is why the landscape illuminated by moonlight appears in shades of black and white only. Unlike man, who has both day and night vision, some vertebrates specialize for one or the other. Strictly nocturnal animals such as bats and owls have pure rod retinas. Purely diurnal forms such as the common gray squirrel and some birds have only cones. They are, of course, virtually blind at night.

COLOR VISION. How does the eye see colors? According to the trichromatic theory of color vision, there are three different types of cones that react most strongly to red, green, and violet light. Colors are perceived by comparing the levels of excitation of the three different kinds of cones. This comparison is made both in nerve circuits in the retina and in the visual cortex of the brain. Color vision is present in all vertebrate groups with the possible exception of the amphibians. Bony fish and birds have particularly good color vision. Surprisingly, most mammals are color blind; exceptions are primates and a very few other species such as squirrels.

■ ENDOCRINE SYSTEM

The endocrine system is the second great integrative system controlling the body's activities. Endocrine glands, or specialized tissues, secrete **hormones** (from the Greek root meaning "to excite") that are transported by the blood for variable distances to some part of the body where they produce definite physiologic effects. Hormones are effective in minute quantities; some are active when diluted several billion times in the blood. The endocrine system is a slow-acting integrative system as compared to the nervous system, and, in general, hormonal effects are long lasting. Some hormones are excitatory, others inhibitory. Many physiologic processes are governed by antagonistic hormones (one that stimulates, the other that inhibits the process). Such combinations are very effective in maintaining homeostatic conditions.

Endocrinology is a young field. Its birthdate is usually given as 1902, the year two English physiologists, W. H. Bayliss and E. H. Starling, demonstrated the action of an internal secretion. They were interested in determining how the pancreas secreted its digestive juice into the small intestine at the proper time of the digestive process. In an anesthetized dog they tied off a section of the small intestine beyond the duodenum (the part of the intestine next to the stomach) and removed all nerves leading to this tied-off loop, but left its blood vessels intact. Bayliss and Starling found that the injection of hydrochloric acid into the blood serving this intestinal loop had no effect upon the secretion of pancreatic juice, but when they introduced 0.4% hydrochloric acid directly inside the intestinal loop, a pronounced flow of pancreatic juice into the duodenum occurred through the pancreatic duct. Now when they scraped off some of the mucous membrane lining of the intestine and mixed it with acid, they found that the injection of this extract into the blood caused an abundant flow of pancreatic juice. They concluded that when the partly digested and slightly acid food from the stomach arrives in the small intestine, the hydrochloric acid reacts with something in the mucous lining to produce an internal secretion, or chemical messenger, which is conveyed by the bloodstream to the pancreas, causing it to secrete pancreatic digestive juices. They called this messenger **secretin.** In a 1905 Croonian lecture at the Royal College of Physicians, Starling first used the word "hormone," a general term to describe all such chemical messengers, since he correctly surmised that secretin was only the first of many hormones that remained to be described.

Since hormones are transported in the blood, they reach virtually all body tissues. This makes it possible for certain hormones, such as the growth hormone of the pituitary gland, to have a very widespread action, affecting most, if not all cells during growth of an animal. However, most hormones, despite their ubiquitous distribution, are highly specific in their action. Usually only certain cells will respond to the presence of a given hormone. For example, only the pancreatic cells respond to secretin, even though secretin is carried throughout the body by the circulation. All other cells simply ignore its presence. The cells that respond to a particular hormone are called **target-organ cells.**

Recently it has become evident that the classic definition of a hormone as the bloodborne product of a discrete ductless gland no longer perfectly applies to the heterogeneous endocrine system recognized today. Some cells secrete hormones that may diffuse only a short distance to neighboring cells to exert their effect without ever entering the bloodstream. These are the "local hormones"—prostaglandins and kinins. **Prostaglandins,** for example, are a family of 20-carbon fatty acids, synthesized in numerous body tissues from polyunsaturated fatty acids in the diet, that have a diverse list of physiologic effects. These include regulation of smooth muscle contraction and stimulation of specific metabolic processes. Prostaglandins are unfortunately misnamed since the source of their high concentration in semen, in which they were first discovered, is the seminal vesicles rather than the prostate gland. They are among the most potent hormones known. They produce their effect extremely rapidly and then are just as quickly metabolized and inactivated. Prostaglandins have ascended to recent prominence because they are being used very successfully to facilitate labor contractions and birth; they also can be used to promote abortion and thus show promise as birth control agents.

Even before local hormones were discovered, it had become evident that certain nerve cells are capable of secreting hormones. Such specialized nerve cells are called **neurosecretory cells** and their secreted products are called **neurosecretory hormones.** Subsequent studies demonstrated that these neurosecretory cells are crucial links between the body's two great integrative systems. Such knowledge made it possible to understand how, for example, increasing day length in the spring stimulates the breeding cycle of birds: increasing amounts of light received by the eyes are relayed via nerve tracts in the brain to neurosecretory cells that release hormones, which set the reproductive cycle into motion.

Neurosecretory cells are now known to be the main source—in some instances the only source—of hormones of many invertebrate groups. Neurosecretion is also a widespread phenomenon among the vertebrates. Because neurosecretion is obviously a very ancient physiologic activity and because it serves as a crucial link between the nervous system and the ductless gland system, we believe that hormones first evolved as nerve cell secretions. Later, nonnervous endocrine glands appeared in other parts of the body. These remote glands remained chemically linked to the nervous system, however, by the neurosecretory hormones. The vertebrate hypothalamic-hypophyseal complex mentioned above in connection with the regulation of breeding in birds, is a much studied example.

Mechanisms of hormone action

How do hormones exert their effects? This question has been the object of intense research in recent years. The student can readily appreciate that it is much easier to observe the physiologic effect of a hormone than to determine how the specific hormone acts to produce the effect. Although we have known for years that insulin lowers the blood glucose level, we are still uncertain as to *how* insulin does this.

It now seems that there may be no more than two basic mechanisms of hormone action.

1. *Stimulation of protein synthesis.* Several hormones, such as the thyroid hormones and the insect molting hormone, ecdysone, stimulate the synthesis of specific enzymes and other proteins, by causing the transcription of particular kinds of messenger RNA. These hormones therefore act directly on specific genes. It is possible that the hormone may activate genes by somehow antagonizing the repressor on ribonucleic acid (RNA) polymerase, according to the Jacob-Monod model for gene action (see p. 774). With the repressor removed by the hormone, RNA polymerase then begins building the enzymes (proteins) that will set in motion the observed action of the hormone.

2. *Activation of a "second messenger," cyclic AMP (adenosine 3′,5′-monophosphate).* **Cyclic AMP** is formed from ATP by the action of a special enzyme **adenyl cyclase** (also called adenylyl cyclase) located in the cell membrane (Fig. 30-17). There is rapidly accumulating evidence that the mechanism of action for many, if not most, vertebrate hormones is as follows: When a hormone (the "first messenger") reaches a target cell, it binds to receptor sites on the cell membrane. Such sites are highly specific and will recognize only one hormone of the many circulating in the bloodstream. Hormone binding in some way increases adenyl cyclase activity in the membrane, which in turn transforms ATP into cyclic AMP.

ATP → (Adenyl cyclase) → Cyclic AMP

Cyclic AMP then diffuses into the cell where it acts as a "second messenger" to alter (usually stimulate) cellular processes. Since cyclic AMP is such a powerful regulator (it is 1,000 times less abundant than ATP), it is rapidly degraded into inert AMP by another enzyme, phosphodiesterase, so that its effect is short lived.

Endocrinologists have long searched for a single, fundamental mechanism through which all hormones act. The second messenger concept has been shown to apply to many, but not all, hormonal actions. Some hormones seem to act by means of both mechanisms described above; others by one or the other. Still other hormonal actions may be exerted through yet unidentified intermediates. One has the feeling that we have read only the first few chapters of a detective story in which most of the central characters have made their appearance but none has yet revealed his full part in the intrigue. Certainly the unveiling of cyclic AMP in the 1960s has had far-reaching consequences on subsequent endocrinologic research. But as E. W. Sutherland, discoverer of cyclic AMP, has said: "Our present understanding of the biological role of cyclic AMP is probably very small compared to what it will be in the future."*

Invertebrate hormones

Over the last 40 years physiologists have shown that the invertebrates have endocrine integrative systems that approach the complexity of the vertebrate endocrine system. Not surprisingly, however, there

*Sutherland, E. W. 1972. Studies on the mechanism of hormone action. Science **177**:401-408.

are few, if any, homologies between invertebrate and vertebrate hormones. Invertebrates have different functional systems, different growth patterns, and different reproductive processes than do vertebrates, as well as being separated phylogenetically for a vast span of time. Most studies have been concentrated in the huge phylum Arthropoda, especially the insects and crustaceans. However, recent research has revealed hormonal systems, especially neurosecretory systems, in most of the other invertebrate phyla, too.

Neurosecretions are known to influence growth, asexual reproduction, and regeneration of hydra (phylum Coelenterata). Neurosecretory hormones also regulate regeneration, training, reproduction, and other aspects of flatworm physiology. We have known for some time that mollusks have neurosecretory hormones, especially among the gastropods and pelecypods. In the polychaete annelids, amputation of a portion of part of the worm body causes neurosecretory cells in the cerebral ganglia to secrete hormones that trigger regeneration. The cerebral ganglia of young worms produces a "juvenile hormone" that has a braking effect on metamorphosis; if the brain is removed the worms become sexually premature.

The chromatophores (pigment cells) of shrimp and crabs are controlled by hormones from the **sinus gland** in the eyestalk or in regions close to the brain. Many crustaceans are capable of remarkably beautiful color patterns that change adaptively in relation to their environment; these changes are governed by an elaborate system of endocrine glands and hormones.

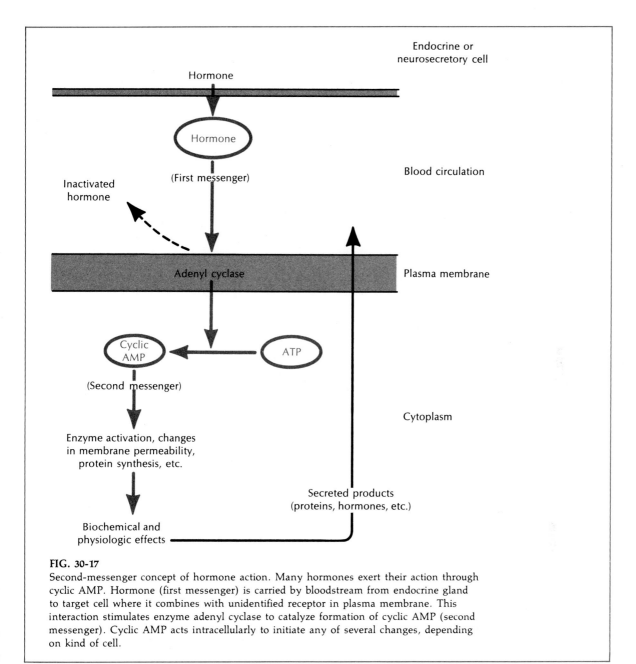

FIG. 30-17

Second-messenger concept of hormone action. Many hormones exert their action through cyclic AMP. Hormone (first messenger) is carried by bloodstream from endocrine gland to target cell where it combines with unidentified receptor in plasma membrane. This interaction stimulates enzyme adenyl cyclase to catalyze formation of cyclic AMP (second messenger). Cyclic AMP acts intracellularly to initiate any of several changes, depending on kind of cell.

Growth and metamorphosis of arthropods are under endocrine control. As described earlier (p. 351), growth of an arthropod is a series of steps in which the rigid, nonexpansible exoskeleton is periodically discarded and replaced with a new larger one. This process is especially dramatic in insects. In the type of development called **holometabolous**

seen in many insect orders (for example, butterflies, moths, ants, bees, wasps, and beetles), there is a series of wormlike larval stages, each requiring the formation of a new exoskeleton; each stage ends with a molt. The last larval stage enters a state of quiescence (pupa) during which the internal tissues are dissolved and rearranged into adult structures

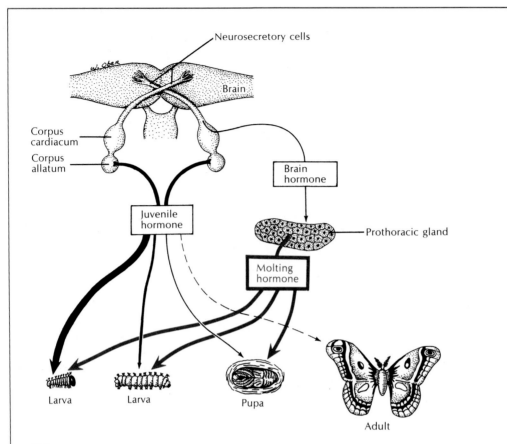

FIG. 30-18

Endocrine control of molting in a butterfly. Butterflies mate in the spring or summer, and eggs soon hatch into the first of several larval stages (called instars). After the final larval molt, the last and largest larva (caterpillar) spins a cocoon in which it pupates. The pupa, or chrysalis, overwinters and an adult emerges in the spring to start a new generation. Two hormones interact to control molting and pupation. The molting hormone, produced by the prothoracic gland and stimulated by a separate brain hormone, favors molting and the formation of adult structures. These effects are inhibited, however, by the juvenile hormone, produced by the corpora allata. Juvenile hormone output declines with successive molts, and the larva undergoes adult differentiation.

(metamorphosis). Finally the transformed adult emerges. Insect physiologists have discovered that molting and metamorphosis are controlled by the interaction of two hormones, one favoring growth and the differentiation of adult structures, the other favoring the retention of larval structures. These two hormones are the **molting hormone** (also referred to as **ecdysone** [ek'duh-sone]) produced by the corpora cardiaca and the **juvenile hormone,** produced by the corpora allata (Fig. 30-18). The structure of both hormones has recently been determined. It required the extraction of 1,000 kg. (about 4 tons) of silkworm pupae to show that the molting hormone is a steroid.

Molting hormone (α-ecdysone) of silkworm

The juvenile hormone has an entirely different structure:

$$CH_3-CH_2-\underset{\underset{CH_3}{|}}{\overset{\overset{O}{||}}{C}}-CH-CH_2-CH_2-\underset{\underset{|}{}}{\overset{\overset{CH_2-CH_3}{|}}{C}}=CH-CH_2-CH_2-\underset{\underset{COOCH_3}{|}}{\overset{\overset{CH_3}{|}}{C}}=CH$$

Juvenile hormone of silkworm

The molting hormone (ecdysone) is under the control of a neurosecretory hormone from the brain, called **brain hormone** (or ecdysiotropin). At intervals during larval growth, brain hormone is released into the blood and stimulates the release of molting hormone. Molting hormone appears to act directly on the chromosomes to set in motion the changes leading to a molt. The molting hormone favors the formation of a pupa and the development of adult structures. It is held in check, however, by the juvenile hormone, which favors the development of larval characteristics. During larval life the juvenile hormone predominates, and each molt yields another larger larva. Finally, the output of juvenile hormone decreases and the final pupal molt occurs.

Chemists have synthesized several potent analogs of the juvenile hormone, which hold great promise as insecticides. Minute quantities of these synthetic analogs induce abnormal final molts, or prolong or block larval development. Unlike the usual chemical insecticides, they are highly specific and do not contaminate the environment.

The pattern of endocrine regulation in crustaceans shows certain parallels to that in insects. Some crustaceans, such as the crayfish, reach a definite adult size after a final molt; others, such as the lobster, molt continuously and keep growing ever larger until death. Molting is controlled by a neurosecretory hormone, called **molt-inhibiting hormone,** produced by the X-organ in the eyestalk, and by a **molting hormone** (ecdysone) produced by the Y-organ located in the thorax. During intermolt, the molt-inhibiting hormone inhibits activity of the Y-organ. But at intervals, just before the molt, the output of molt-inhibiting hormone drops; this permits the Y-organ to secrete molting hormone, which stimulates molting.

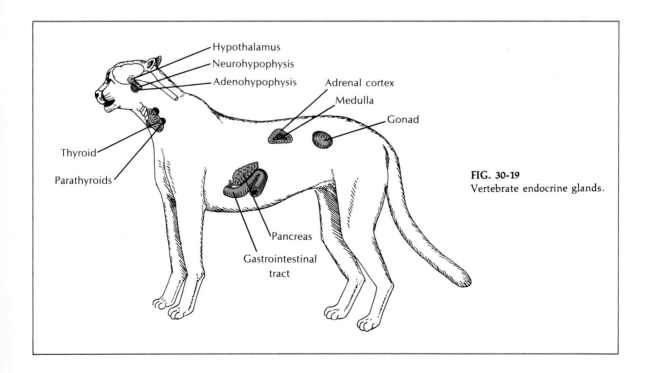

FIG. 30-19
Vertebrate endocrine glands.

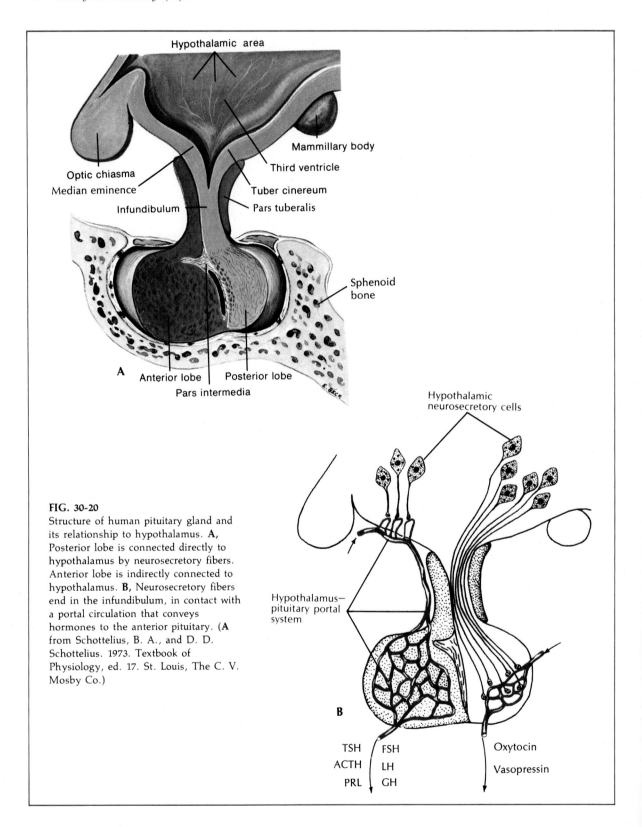

FIG. 30-20
Structure of human pituitary gland and its relationship to hypothalamus. **A,** Posterior lobe is connected directly to hypothalamus by neurosecretory fibers. Anterior lobe is indirectly connected to hypothalamus. **B,** Neurosecretory fibers end in the infundibulum, in contact with a portal circulation that conveys hormones to the anterior pituitary. (**A** from Schottelius, B. A., and D. D. Schottelius. 1973. Textbook of Physiology, ed. 17. St. Louis, The C. V. Mosby Co.)

Vertebrate endocrine organs

The vertebrate endocrine glands are small, well-vascularized organs located in certain parts of the body (Fig. 30-19). **Endocrine** glands are ductless groups of cells arranged in cords or plates; their hormonal secretions enter the bloodstream and are carried throughout the body. **Exocrine** glands, in contrast, are provided with ducts for discharging their secretions onto a free surface. Examples of exocrine glands are sweat glands and sebaceous glands of skin, salivary glands, and the various enzyme-secreting glands lining the wall of the stomach and intestine. Since the endocrine glands have no ducts, their only connection with the rest of the body is by the bloodstream; they must capture their raw materials from the blood and secrete their finished hormonal products into it. Consequently it is not surprising that the endocrine glands receive enormous blood flows. The thyroid gland is said to have the highest blood flow per unit of tissue weight of any organ in the body.

In the remainder of this section we will describe some of the best understood and most important of the vertebrate hormones. The hormones of reproduction are discussed in the next chapter. Space does not permit us to deal with all the hormones and hormonelike substances that have been discovered. The mammalian hormonal mechanisms are the best understood, since laboratory mammals and man have always been the objects of the most intensive research. Research with the lower vertebrates has revealed that all vertebrates share similar endocrine organs. All vertebrates have a pituitary gland, for example, and all have thyroid glands, adrenal glands (or the special cells of which they are composed), and gonads. There are some important differences, nevertheless, as we will seek to point out.

Hormones of the pituitary gland and hypothalamus

The pituitary gland, or **hypophysis,** is a small gland (0.5 gram in man) lying in a well-protected position between the roof of the mouth and the floor of the brain (Fig. 30-20). It is a two-part gland having a double embryologic origin. The **anterior pituitary** (adenohypophysis) is derived embryologically from the roof of the mouth. The **posterior pituitary** (neurohypophysis) arises from a ventral portion of the brain, the **hypothalamus,** and is connected to it

by a stalk, the **infundibulum** (Fig. 30-20, *A*). Although the anterior pituitary lacks any *anatomic* connection to the brain, it is nonetheless *functionally* connected to it by a special portal circulatory system (Fig. 30-20, *B*).

The anterior pituitary consists of an anterior lobe (pars distalis) and an intermediate lobe (pars intermedia) as shown in Fig. 30-20. The anterior lobe, despite its minute dimensions, produces at least six protein hormones. All but one of these six are **tropic hormones,** that is, they regulate other endocrine glands (Fig. 30-21 and Table 30-2). Because of the strategic importance of the pituitary in influencing most of the hormonal activities in the body, the pituitary has been called the body's "master gland." This name is misleading, however, since the tropic hormones are themselves regulated by neurosecretory hormones from the hypothalamus, as well as by hormones from the target glands they stimulate. The **thyrotropic hormone** (TSH) regulates the production of thyroid hormones by the thyroid gland. The **adrenocorticotropic hormone** (ACTH) stimulates the adrenal cortex. Two of the tropic hormones are commonly called **gonadotropins** because they act on the gonads (ovary of the female, testis of the male). These are the **follicle stimulating hormone** (FSH) and the **luteinizing hormone** (LH). The fifth tropic hormone is **prolactin,** which stimulates milk production by the female mammary glands and has a variety of other effects in the lower vertebrates. The functions of the two gonadotropins and prolactin are discussed in the next chapter in connection with the hormonal control of reproduction.

The sixth hormone of the anterior pituitary is the **growth hormone** (also called somatotropic hormone). This hormone performs a vital role in governing body growth through its stimulatory effect on cellular mitosis and protein synthesis, especially in new tissue of young animals. If produced in excess, the growth hormone causes giantism. A deficiency of hormone in the child or young animal causes dwarfism. Growth hormone and prolactin are so closely related chemically that growth hormone contains considerable prolactin activity; that is, it tends to stimulate milk production (like prolactin) as well as promote growth.

The intermediate lobe of the pituitary (Fig. 30-20) produces **intermedin** (also called melanophore stimulating hormone [MSH]), which controls the disper-

TABLE 30-2
Hormones of the vertebrate pituitary—chemical nature and actions

	Hormone	Chemical nature	Action
Adenohypophysis			
Pars distalis (distal lobe)	Thyrotropin (TSH)	Glycoprotein	Stimulates thyroid to secrete thyroid hormones
	Adrenocorticotropin (ACTH)	Polypeptide	Stimulates adrenal cortex to secrete steroid hormones
	Gonadotropins (two)		
	1. Follicle stimulating hormone (FSH)	Glycoprotein	Stimulates gamete production and secretion of sex hormones
	2. Luteinizing hormone (LH, ICSH)	Glycoprotein	Stimulates sex hormone secretion and ovulation
	Prolactin (LTH)	Protein	Stimulates mammary gland growth and secretion in mammals; various reproductive and nonreproductive functions in lower vertebrates
	Growth hormone (GH)	Protein	Stimulates growth
Pars intermedia (intermediate lobe)	Intermedin (MSH)	Polypeptide	Pigment dispersion in amphibian melanophores
Neurohypophysis			
Pars nervosa	Vasopressin (ADH)	Octapeptide	Antidiuretic effect on kidney
	Oxytocin	Octapeptide	Stimulates milk ejection and uterine contraction
	Vasotocin	Octapeptide	Antidiuretic activity
	Isotocin and others in lower vertebrates	Octapeptide	Function unknown
Median eminence	Thyrotropin releasing factor (TRF) Corticotropin releasing factor (CFR) Follicle stimulating hormone–releasing factor (FRF) Luteinizing hormone–releasing factor (LRF) Prolactin inhibiting factor (PIF) Growth hormone–releasing factor (GFR)	Probably all polypeptides	Control release of distal lobe hormones

sion of melanin within the melanophores of amphibians. Its action is described in Chapter 23. In other vertebrates, intermedin appears to perform no important physiologic role, even though it will cause darkening of the skin in man if injected into the circulation. Intermedin and ACTH are chemically very similar and ACTH will also cause skin darkening when it is secreted in abnormally large amounts.

As pointed out above, the pituitary gland is not the top director of the body's system of endocrine glands, as endocrinologists once believed. The pituitary serves higher masters, the neurosecretory centers of the hypothalamus; and the hypothalamus is itself under the ultimate control of the brain. The hypothalamus contains groups of neurosecretory cells, which are specialized giant nerve cells. Polypeptide hormones are manufactured in the cell bodies and then travel down the nerve fibers to their endings where the hormones are stored until released into the blood. The discharge of neurosecretory hormones may occur when a nerve impulse travels down the same neurosecretory fiber (these

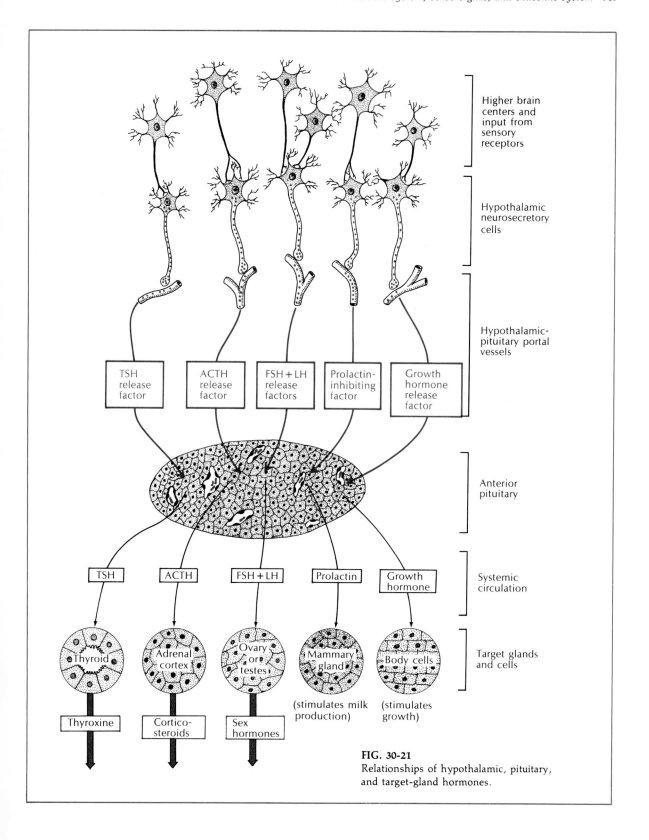

FIG. 30-21

Relationships of hypothalamic, pituitary, and target-gland hormones.

specialized nerve cells can, in most cases, still perform their original impulse-conducting function) or the release may be activated by ordinary fibers traveling alongside them.

Both the anterior and posterior lobes of the pituitary are under hypothalamic control, but in different ways. Neurosecretory fibers serving the posterior lobe travel down the infundibular stalk and into the posterior lobe, ending in proximity to blood capillaries, into which the hormones enter when released. In a sense the posterior lobe is not a true endocrine gland, but a storage-release center for hormones manufactured entirely in the hypothalamus.

The anterior pituitary's relationship to the hypothalamus is quite different. Neurosecretory fibers do not travel to the anterior lobe, but end some distance above it, in the **median eminence,** at the base of the infundibular stalk (Fig. 30-20). Neurosecretory hormones released here enter a capillary network and complete their journey to the anterior lobe via a short, but crucial, pituitary portal system. These hormones are called **releasing factors** because they govern the release of the anterior pituitary hormones (Table 30-2). There appears to be a specific releasing factor for each pituitary tropic hormone. Two of the releasing factors, TRF (thyrotropin releasing factor) and LRF (luteinizing hormone releasing factor), have recently been isolated and characterized chemically. Both are polypeptides. The structure of TRF, a tripeptide is as follows:

Pyroglutamic acid–histidine–proline NH₂
(thyrotropin releasing factor)

The hormones of the posterior lobe are chemically similar polypeptides consisting of eight amino acids (referred to as octapeptides). All vertebrates, except the most primitive fishes, secrete two posterior lobe octapeptides. However, their chemical structure has changed slightly in the course of evolution. The two posterior lobe hormones secreted, for example, by

fish are not identical to those secreted by mammals. Altogether, seven different posterior lobe hormones have been identified from the different vertebrate groups (Table 30-2).

Vasotocin is found in all vertebrate classes except mammals. It is a water balance hormone in amphibians, especially toads, in which it acts to conserve water by increasing permeability of the skin and the urinary bladder and by decreasing urine flow. The action of vasotocin is best understood in amphibians, but it appears to play some water conserving role in birds and reptiles as well.

The two mammalian posterior-lobe hormones are **oxytocin** and **vasopressin** (Fig. 30-22). They are formed, as we have seen, in the cell bodies of neurosecretory cells in the hypothalamus and then transported down the nerve cell axons to the posterior lobe. These hormones are among the fastest-acting hormones in the body, since they are capable of producing a response within seconds of their release from the posterior lobe.

Oxytocin has two important specialized reproductive functions in adult female mammals. It causes contraction of uterine smooth muscles during parturition (birth of the young). Doctors sometimes use oxytocin clinically to induce labor and facilitate delivery and to prevent uterine hemorrhage after birth. The second and most important action of oxytocin is that of milk ejection by the mammary glands in response to suckling.

Vasopressin, the second posterior lobe hormone, acts on the kidney to restrict urine flow, as already described on p. 638. It is therefore often referred to as the **antidiuretic hormone** (ADH). Vasopressin has a second, weaker effect of increasing the blood pressure through its generalized constrictor effect on the smooth muscles of the arterioles. Although the name "vasopressin" unfortunately suggests that the vasoconstrictor action is the hormone's major effect, it is probably of little physiologic importance, except perhaps to help sustain the blood pressure during a severe hemorrhage.

Hormones of metabolism

Many hormones act to adjust the delicate balance of metabolic activities in the body. Metabolism includes the **anabolic** activities of tissue synthesis, building up of energy reserves, and maintenance of tissue organization and the **catabolic** activities of en-

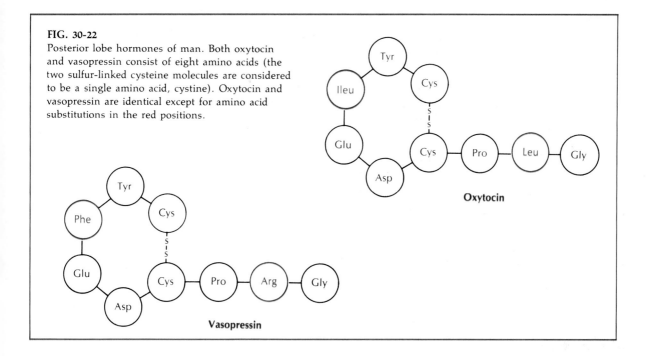

FIG. 30-22
Posterior lobe hormones of man. Both oxytocin and vasopressin consist of eight amino acids (the two sulfur-linked cysteine molecules are considered to be a single amino acid, cystine). Oxytocin and vasopressin are identical except for amino acid substitutions in the red positions.

ergy release and tissue destruction. Such activities are mediated almost entirely by enzymes. The numerous enzymatic reactions proceeding within cells are complex, but each step in a sequence is in large part self-regulating, as long as the equilibrium between substrate, enzyme, and product remains stable. However, hormones may alter the activity of crucial enzymes in a metabolic process, thus accelerating or inhibiting the entire process. We must emphasize that hormones never initiate enzymatic processes. They simply alter their rate, speeding them up or slowing them down. The most important hormones of metabolism are those of the thyroid, parathyroid, adrenal, and pancreas.

THYROID HORMONES. The two thyroid hormones **thyroxine** and **triiodothyronine** are secreted by the thyroid gland. This largest of endocrine glands is located in the neck region of all vertebrates; in many animals, including man, it is a bilobed structure. The thyroid is made up of thousands of tiny spheres, called **follicles;** each follicle is composed of a single layer of epithelial cells enclosing a hollow, fluid-filled center. This fluid contains stored thyroid hormone that is released into the bloodstream as it is needed.

One of the unique characteristics of the thyroid is its high concentration of iodine; in most animals this single gland contains well over half the body store of iodine. The epithelial cells actively trap iodine from the blood and combine it with the amino acid tyrosine, creating the two thyroid hormones. Each molecule of thyroxine contains four atoms of iodine as indicated by the following structural formula:

$$HO \underset{I}{\overset{I}{\bigcirc}} O \underset{I}{\overset{I}{\bigcirc}} CH_2-CH-COOH$$
$$\underset{NH_2}{}$$

Thyroxine

Triiodothyronine is identical to thyroxine except that it has three instead of four iodine atoms. Thyroxine is formed in much greater amounts than triiodothyronine, but both hormones have two important similar effects. One is to promote the normal growth and development of growing animals. The other is to stimulate the metabolic rate.

We do not know exactly how the thyroid hormones promote growth, although there is evidence that they stimulate protein synthesis through their effect on messenger RNA. Certainly the undersecretion of thyroid hormone dramatically impairs growth, especially of the nervous system. The human **cretin,** a mentally retarded dwarf, is the tragic product of thyroid malfunction from a very early age. Conversely the oversecretion of thyroid hormones causes precocious development, particularly in lower vertebrates. In one of the earliest demonstrations of hormone action, Gudernatsch in 1912 induced precocious metamorphosis of frog tadpoles by feeding them bits of horse thyroid. The tadpoles quickly resorbed their tails, grew limbs, changed from gill to lung respiration, and became froglets about one-third normal size. The result of a similar experiment is shown in Fig. 30-23.

The control of oxygen consumption and heat production in birds and mammals is the best known ac-

tion of the thyroid hormones. The thyroid enables warm-blooded animals to adapt to cold by increasing their heat production. Acting as a kind of biologic thermostat, the thyroid senses and controls the body temperature by releasing more or less thyroxine, as required. Thyroxine in some way causes cells to produce more heat and store less chemical energy (ATP); in other words, thyroxine **reduces** the efficiency of the cellular oxidative phosphorylation system (p. 74). This is why cold-adapted animals eat more food than do warm-adapted ones, even though they are doing no more work; the food is being converted directly to heat, thus keeping the body warm.

The synthesis and release of thyroxine and triio-

FIG. 30-23
Precocious metamorphosis of frog tadpoles *(Rana pipiens)* caused by adding thyroid hormone to the water. When frog tadpoles had developed hindlimb buds, small amounts of thyroxine were added to aquarium water. In only 3 weeks tadpoles metamorphosed to normal, but miniature, adults about one third the size of the mother. (From Turner, C. D., and J. T. Bagnara. 1971. General endocrinology, ed. 5, Philadelphia, W. B. Saunders Co.)

dothyronine is governed by **thyrotropic hormone** (TSH) from the anterior pituitary gland (Fig. 30-21). Thyrotropic hormone controls the thyroid through a **negative feedback mechanism.** If the thyroxine level in the blood falls, more thyrotropic hormone is released. Should the thyroxine level rise too high, less thyrotropic hormone is released. This sensitive feedback mechanism normally keeps the blood thyroxine level very steady, but certain neural stimuli, as might arise from exposure to cold, can directly increase the release of thyrotropic hormone.

Some years ago, a condition called **goiter** was common among people living in the Great Lakes region of the United States and Canada, as well as in other parts of the earth such as the Swiss Alps. Goiter is an enlargement of the thyroid gland caused by a deficiency of iodine in the food and water. In striving to produce thyroid hormone with not enough iodine available, the gland hypertrophies, sometimes so much that the entire neck region becomes swollen. Goiter is seldom seen today because of the widespread use of iodized salt.

PARATHYROID HORMONE, CALCITONIN, AND CALCIUM METABOLISM. Closely associated with the thyroid gland, and often buried within it, are the parathyroid glands. These tiny glands occur as two pairs in man but vary in number and position in other vertebrates. They were discovered at the end of the nineteenth century when the fatal effects of "thyroidectomy" were traced to the unknowing removal of the parathyroids as well as the thyroid. Removal of the parathyroids causes the blood calcium to drop rapidly. This leads to a serious increase in nervous system excitability, severe muscular spasms and tetany, and finally death.

The parathyroid glands are vitally concerned with the maintenance of the normal level of calcium in the blood. Actually two hormones are involved: **parathyroid hormone** (parathormone), produced by the parathyroid glands, and **calcitonin,** produced by special cells within the thyroid gland (not the same follicle cells that synthesize thyroxine). These two hormones have opposing but cooperative actions. Between them they stabilize both the calcium and phosphorus levels in the blood through their action on bone. Bone is a densely packed storehouse of these elements, containing about 98% of the body calcium and 80% of the phosphorus. Although bone is second only to teeth as the most durable material

in the body, as evidenced by the survival of fossil bones for millions of years, it is in a state of constant turnover in the living body. Bone-building cells **(osteoblasts)** withdraw calcium and phosphorus (as phosphate) from the blood and deposit them in a complex crystalline form around previously formed organic fibers (Fig. 30-24). Bone-resorbing cells **(osteoclasts),** present in the same bone, tear down bone by engulfing it and releasing the calcium and phosphate into the blood. These conflicting activities are not as pointless as they may seem. First, they allow bone to constantly remodel itself, especially in the growing animal, for structural improvements to counter new mechanical stresses on the body. Second, they provide a vast and accessible reservoir of minerals that can be withdrawn as the body needs them for its general cellular requirements.

If the blood calcium should drop slightly, the parathyroid gland increases its output of parathormone. This stimulates the osteoclasts to destroy bone adjacent to these cells, thus releasing calcium and phosphate into the bloodstream and returning the blood calcium level to normal. Parathormone also acts on the kidney to decrease the excretion of calcium and this, of course, also helps to increase the blood calcium level. Should the calcium in the blood rise above normal, the parathyroid gland decreases its output of parathormone. In addition, the thyroid is stimulated to release calcitonin. These relationships are shown in Fig. 30-24. Although the action of this second recently discovered (1962) hormone is yet imperfectly understood, evidence suggests that it inhibits bone resorption by the osteoclasts. Calcitonin thus protects the body against a dangerous rise in the blood calcium level, just as parathormone protects it from a dangerous fall in blood calcium. The two act together to smooth out oscillations in blood calcium. A third hormone, **calciferol** (commonly called vitamin D), is also necessary for calcium deposition in bone. Calciferol is a steroid hormone in skin formed from a steroid precursor by the action of ultraviolet rays of sunlight.

There is increasing evidence that parathormone is one of those hormones that acts through the second messenger, cyclic AMP. Parathormone activates the enzyme adenyl cyclase in the bone-resorbing osteoclasts, causing increased formation of cyclic AMP. Cyclic AMP in some way brings about the resorption of bone, but its precise action is still a mystery.

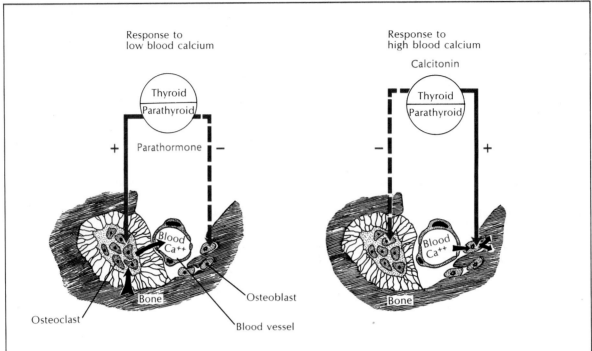

Response to
low blood calcium

Response to
high blood calcium

Calcitonin

Thyroid
Parathyroid

Parathormone

+   −

Thyroid
Parathyroid

−   +

Blood
Ca++

Blood
Ca++

Bone

Osteoblast

Osteoclast

Blood vessel

Bone

FIG. 30-24

Action of parathormone and calcitonin on calcium resorption and deposition in bone.
When blood calcium is low *(left),* parathyroid gland secretes parathormone, which
stimulates large bone-destroying osteoclasts to resorb calcium. Bone-building osteoblasts
are inhibited. Calcium and phosphate (not shown) enter the blood and restore the blood
calcium level to normal. When blood calcium rises above normal *(right),* thyroid gland
secretes calcitonin that inhibits osteoclasts. Osteoblasts then remove calcium (and
phosphate) from the blood and use it to build new bone. Calcitonin may directly stimulate
osteoblastic activity. A third hormone, calciferol (often called "vitamin D"), is necessary
for calcium deposition in bones.

ADRENOCORTICAL HORMONES. The verte-
brate adrenal gland is a double gland consisting of
two very different kinds of tissue: **interrenal** tissue,
called **cortex** in mammals, and **chromaffin** tissue,
called **medulla** in mammals. The mammalian terminology of cortex (meaning "bark") and medulla
(meaning "core") arose because in this group of vertebrates the interrenal tissue completely surrounds
the chromaffin tissue like a cover. Although in the
lower vertebrates the interrenal and chromaffin tissue are usually separated, the mammalian terms
"cortex" and "medulla" are so firmly fixed in our vocabulary that we commonly use them for all vertebrates instead of the more correct terms "interrenal"
and "chromaffin."

Biochemists have found that the adrenal cortex
contains at least 30 different compounds, all of them
closely related lipoid compounds known as steroids.
Only a few of these compounds, however, are true
steroid **hormones;** most are various intermediates in
the synthesis of steroid hormones from **cholesterol**
(Fig. 30-25). The corticosteroid hormones are commonly classified into three groups, according to their
function:

1. **Glucocorticoids,** such as **cortisol** (Fig. 30-25)
and **corticosterone,** have a number of important effects concerned with food metabolism and inflammation. They cause the conversion of nonglucose compounds, particularly amino acids and fats, into glucose. This process, called **gluconeogenesis,** is ex-

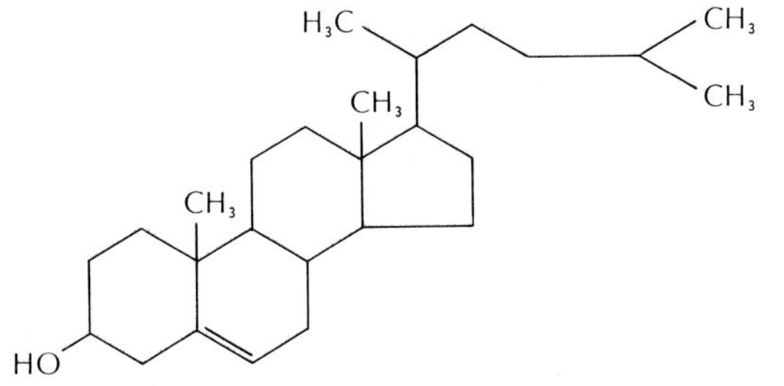

Cholesterol

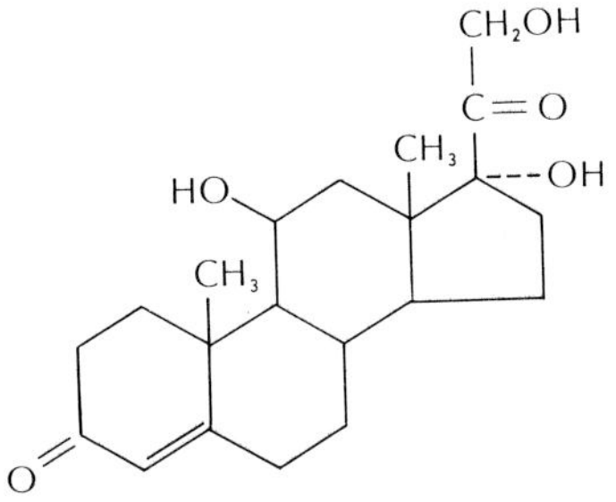

Cortisol

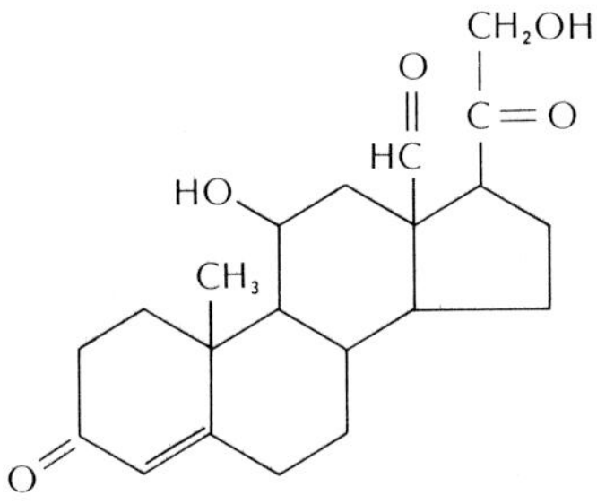

Aldosterone

FIG. 30-25
Hormones of the adrenal cortex. Cortisol
(a glucocorticoid) and aldosterone
(a mineralocorticoid) are two of the many
steroid hormones synthesized from cholesterol
in the adrenal cortex.

tremely important, since most of the body's stored energy reserves are in the form of fats and proteins that must be converted to glucose before they can be burned for energy. Cortisol, cortisone, and corticosterone are also **anti-inflammatory.** Because several diseases of man are inflammatory diseases (for example, allergies, hypersensitivity, arthritis), these corticosteroids have important medical applications. They must be used with great care, however, since if administered in excess, they may suppress the body's normal repair processes and lower resistance to infectious agents.

2. **Mineralocorticoids,** the second group of corticosteroids, are those that regulate salt balance. **Aldosterone** (Fig. 30-25) and **deoxycorticosterone** are the most important steroids of this group. They promote the tubular reabsorption of sodium and chloride and the tubular excretion of potassium by the kidney. Since sodium usually is in short supply in the diet, and potassium in excess, it is obvious that the mineralocorticoids play vital roles in preserving the correct balance of blood electrolytes. We may also note

that the mineralocorticoids **oppose** the anti-inflammatory effect of cortisol and cortisone. In other words, they promote the **inflammatory** defense of the body to various noxious stimuli. Although these opposing actions of the corticosteroids seem self-defeating, they actually are not. They are necessary to maintain readiness of the body's defenses for any stress or disease threat, yet prevent these defenses from becoming so powerful that they turn against the body's own tissues.

3. **Sex hormones** (such as testosterone, estrogen progesterone) are produced primarily by the ovaries and testes (see pp. 734 to 736). The adrenal cortex is also a minor source of certain steroids that mimic the action of testosterone. This sex hormone–like secretion is of little physiologic significance, except in certain disease states of man.

The synthesis and secretion of the corticosteroids are controlled principally by the **adrenocorticotropic hormone** (ACTH) of the anterior pituitary (Fig. 30-21). As with pituitary control of the thyroid, a negative feedback relationship exists between ACTH and the adrenal cortex: a rise in the level of cortico-steroids suppresses the output of ACTH; a fall in the blood steroid level increases ACTH output.

ADRENAL MEDULLA HORMONES. The adrenal medulla secretes two structurally similar hormones, **epinephrine** (adrenaline) and **norepinephrine** (noradrenaline). Their structures are as follows:

Epinephrine

Norepinephrine

Both are derived from catechol

and each bears an amine group on two carbon side chains. Consequently both belong to a class of compounds called **catecholamines.** Norepinephrine is also released at the endings of sympathetic nerve fibers throughout the body, where it serves as a "transmitter" substance to carry neural signals across the gap that separates the fiber and the organ it innervates. The adrenal medulla has the same embryologic origin that sympathetic nerves have; in many respects the adrenal medulla is nothing more than a giant sympathetic nerve ending. It is not surprising then that the adrenal medulla hormones have the same general effects on the body that the sympathetic nervous system has. These effects center around emergency functions of the body, such as fear, rage, fight, and flight, although they have important integrative functions in more peaceful times as well. We are all familiar with the increased heart rate, tightening of the stomach, dry mouth, trembling muscles, and general feeling of anxiety, and the increased awareness that attends sudden fright or other strong emotional states. These effects are attributable both to the rapid release into the blood of epinephrine from the adrenal medulla and to increased activity of the sympathetic nervous system.

Epinephrine and norepinephrine have many other effects that we are not so aware of, including constriction of the arterioles (which, together with the increased heart rate, increases the blood pressure), mobilization of liver glycogen to release glucose for energy, increased oxygen consumption and heat production, hastening of blood coagulation, and inhibition of the gastrointestinal tract. All of these changes in one way or another tune up the body for emergencies. Epinephrine and norepinephrine are among those hormones that activate the enzyme **adenyl cyclase,** causing increased production of **cyclic AMP.** Cyclic AMP then becomes the second messenger, which produces the many observed effects of the adrenal medulla hormones.

INSULIN FROM THE ISLET CELLS OF THE PANCREAS. The pancreas is both an exocrine and an endocrine organ. The **exocrine** portion produces pancreatic juice, a mixture of digestive enzymes that is conveyed by ducts to the digestive tract. Scattered among the extensive exocrine portion of the pancreas are numerous small islets of tissue, called **islets of Langerhans.** This is the **endocrine** portion of the gland. The islets are without ducts and secrete their hormones directly into blood vessels that extend throughout the pancreas. Two polypeptide hormones are secreted by different cell types within the islets: **insulin,** produced by the **beta cells,** and **glucagon,** produced by the **alpha cells.** Insulin and glucagon have antagonistic actions of great importance in the metabolism of carbohydrates and fats. Insulin is essential for the utilization of blood glucose by cells, especially skeletal muscle cells. Insulin somehow allows glucose in the blood to be transported into the cells. Without insulin, the blood glucose levels rise (hyperglycemia) and sugar appears in the urine. Insulin also promotes the uptake of amino acids by skeletal muscle and inhibits the mobilization of fats in adipose tissue. Failure of the pancreas to produce enough insulin causes the disease **diabetes mellitus** that afflicts 2% of the population. It is attended by serious alterations in carbohydrate, lipid, protein, salt, and water metabolism, which, if left untreated, may cause death. The first extraction of insulin in 1921 by two Canadians, Frederick Banting and Charles Best, was one of the most dramatic and important events in the history of medicine. Many years earlier, two German scientists, Von Mering and Minkowski, discovered that surgical removal of the pancreas of dogs invariably caused severe symptoms of diabetes resulting in the animal's death within a few weeks. Many attempts were then made to isolate the diabetes preventive factor, but all failed because powerful protein-splitting digestive enzymes in the exocrine portion of the pancreas destroyed the hormone during extraction procedures. Following a hunch, Banting in collaboration with Best and his physiology professor J. J. R. Macleod tied off the pancreatic ducts of several dogs. This caused the exocrine portion of the gland with its hormone-destroying enzyme to degenerate, but left the islet's tissue healthy, since they were independently served by their own blood supply. Banting and Best then successfully extracted insulin from these glands. Injected into another dog, the insulin immediately lowered the blood sugar level. Their experiment paved the way for the commercial extraction of insulin from slaughterhouse animals. It meant that millions of diabetics, previously doomed to invalidism or death, could now look forward to nearly normal lives.

Glucagon, the second hormone of the pancreas, has several effects on carbohydrates and fat metabolism that are opposite to the effects of insulin. For example, glucagon raises the blood glucose level, whereas insulin lowers it. Glucagon and insulin do not have the same effects in all vertebrates, and in some, glucagon is lacking altogether. Glucagon is another example of a hormone that operates through the cyclic AMP second-messenger system.

Hormones of digestion

Several hormones assist in coordinating the secretion of digestive enzymes. Of these, we will discuss three of the best understood (Fig. 30-26). **Gastrin** is a small polypeptide hormone produced in the mucosa of the pyloric portion of the stomach. When food enters the stomach, gastrin stimulates the secretion of hydrochloric acid by the stomach wall. Gastrin is an unusual hormone in that it exerts its action on the same organ from which it is secreted. Two other hormones of digestion are **secretin** and **pancreozymin.** Both are polypeptide hormones secreted by the intestinal mucosa in response to the entrance of acid and food into the duodenum from the stomach; both stimulate the secretions of pancreatic juice, but their effects differ somewhat. Secretin stimulates a pancreatic secretion rich in bicarbonate that rapidly neu-

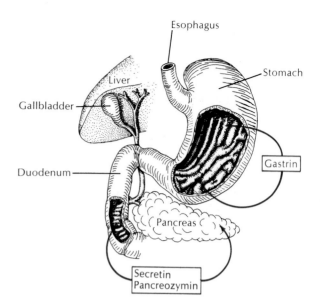

FIG. 30-26
Three hormones of digestion. Arrows show source and target of three gastrointestinal hormones.

tralizes stomach acid. Pancreozymin stimulates the pancreas to release an enzyme-rich secretion. Although secretin was the first hormone discovered (by the British scientists Bayliss and Starling in 1903), the gastrointestinal hormones as a group have received much less attention than the other vertebrate endocrine glands. Only secretin has been obtained in pure form.

References
Suggested general readings
(See also general references for Part Three, p. 639.)

Bentley, P. J. 1971. Endocrines and osmoregulation: A comparative account of the regulation of water and salt in vertebrates. Vol. 1 of Zoophysiology Ecology. Berlin, Springer-Verlag. *Environmentally oriented, advanced-level treatment of this subject by an authority in the area.*

Bullock, T. H., and G. A. Horridge. 1965. Structure and function in the nervous system of invertebrates. San Francisco, W. H. Freeman & Co., Publishers. *An excellent summary of nervous integration in invertebrates.*

Case, J. 1966. Sensory mechanisms. New York, The Macmillan Co. *Gives a good evaluation of sensory mechanisms at an introductory level. Paperback.*

Eccles, J. C. 1957. The physiology of nerve cells. Baltimore, The Johns Hopkins Press. *A summary of certain concepts of nervous integration, including the architecture of the neuron and the transmitter substances of the central nervous system.*

Highnam, K. C., and L. Hill. 1969. The comparative endocrinology of the invertebrates. New York, American Elsevier Publishing Co., Inc. *Clearly written, well-illustrated comparative account emphasizing hormonal action.*

Mountcastle, V. B. 1974. Medical physiology, vol. 2, ed. 13. St. Louis, The C. V. Mosby Co. *The section on the nervous system is especially well written in this outstanding medical physiology. Advanced level.*

Sherrington, C. S. 1947. The integrative action of the nervous system, rev. ed. New Haven, Conn., Yale University Press. *A classic work on the structural and functional plan of the nervous system.*

Tombes, A. S. 1970. An introduction to invertebrate endocrinology. New York, Academic Press Inc.

Selected *Scientific American* articles

Amoore, J. E., J. W. Johnston, Jr., and M. Rubin. 1964. The stereochemical theory of odor. 210:42-49 (Feb.).

Baker, P. F. 1966. The nerve axon. 214:74-82 (March). *Describes techniques used to study nerve impulse conduction.*

Dowling, J. E. 1966. Night blindness. 215:78-84 (Oct.). *The importance of vitamin A in vision.*

Eccles, J. 1965. The synapse. 212:56-66 (Jan.). *A famous neurophysiologist explains how nerve impulses are transmitted from cell to cell.*

Gillie, R. B. 1971. Edemic goiter. 224:92-101 (June). *Many people still suffer from this thyroid disorder caused by insufficient iodine in the diet.*

Guillemin, R., and R. Burgus. 1972. The hormones of the hypothalamus. 227:24-33 (Nov.). *Two of the brain's "releasing factors" have been isolated and synthesized. Their function is described.*

Heimer, L. 1971. Pathways in the brain. 225:48-60 (July). *A new staining technique has vastly improved studies of neural pathways and connections in the nervous system.*

Hendricks, S. B. 1968. How light interacts with living matter. 219:174-186 (Sept.). *Special pigments mediate the interaction of light with matter in photosynthesis, vision, and photoperiodism.*

Hodgson, E. S. 1961. Taste receptors. 204:135-144 (May). *The mechanism of taste reception is studied with blowflys.*

Hubel, D. H. 1963. The visual cortex of the brain. 209:54-62 (Nov.).

Kalmus, H. 1958. The chemical senses. 198:97-106 (April). *Smell and taste are among the chemical senses that are the most fundamental of all senses.*

Katz, B. 1961. How cells communicate. 205:209-220 (Sept.). *Nervous communication and nature of the nerve impulses.*

Kennedy, D. 1967. Small systems of nerve cells. 216:44-52 (May). *The simplified nervous systems of invertebrates facilitate studies of nervous integration.*

Levine, R., and M. S. Goldstein. 1958. The action of insulin. 198:99-106 (May). *Describes insulin's role in promoting the passage of sugar across cell membranes.*

Li, C. H. 1963. The ACTH molecule. 209:46-53. *Its composition and what it does.*

Livingston, W. K. 1953. What is pain? 188:59-66 (March).

Loomis, W. F. 1970. Rickets. 223:76-91 (Dec.). *The role of calciferol, a sunlight-dependent hormone, in preventing this oldest of air-pollution diseases.*

Luria, A. R. 1970. The functional organization of the brain. 222:66-78 (March).

MacNichol, E. F., Jr. 1964. Three-pigment color vision. 211:48-56 (Dec.).

Melzack, R. 1961. The perception of pain. 204:41-49 (Feb.). *Pain is greatly modified by experience and "state of mind."*

Miller, W. H., F. Ratliff, and H. K. Hartline. 1961. How cells receive stimuli. 205:222-238 (Sept.). *Structure and functional properties of receptor cells.*

Neisser, V. 1968. The processes of vision. 219:204-214 (Sept.). *Considers visual perception and visual memory.*

Pastan, I. 1972. Cyclic AMP. 227:97-105 (Aug.). *The hormonal "second messenger" and some of the roles it plays.*

Pike, J. E. 1971. Prostaglandins. 225:84-92. *These recently isolated substances have numerous physiologic effects.*

Rasmussen, H., and M. M. Pechet. 1970. Calcitonin. 223:42-50 (Oct.). *Discovery and action of this calcium-regulating hormone.*

Snider, R. S. 1958. The cerebellum. 199:84-90 (Aug.).

Stent, G. S. 1972. Cellular communication. 227:43-51 (Sept.). *Information processing and communication is discussed with particular emphasis on vision.*

Werblin, F. S. 1973. The control of sensitivity in the retina. 228:70-79 (Jan.). *Recent studies of neuron interactions in the retina help to explain its versatility over widely ranging light conditions.*

Wilkins, L. 1960. The thyroid gland. 202:119-129 (March). *How the gland is constructed, what it produces, and how it is controlled.*

Williams, C. M. 1958. The juvenile hormone. 198:67-74 (Feb.). *This insect hormone prevents metamorphosis of the larva into a pupa until growth is complete.*

Wilson, V. J. 1966. Inhibition in the central nervous system. 214:102-110 (May). *Inhibitory neurons play important roles in muscular activity.*

Young, R. W. 1970. Visual cells. 223:80-91 (Oct.). *How rods and cones renew themselves.*

Zuckerman, S. 1957. Hormones. 196:76-87 (March). *A general account of the vertebrate hormones and some of their actions in health and disease.*

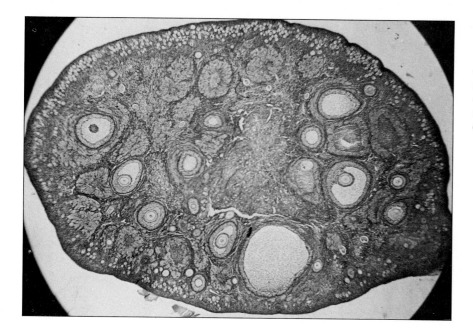

Section through cat ovary (low magnification) showing eggs in various stages of development. Compare with Fig. 31-5.

CHAPTER 31

THE REPRODUCTIVE PROCESS*

All living organisms are capable of giving rise to new organisms similar to themselves. If we admit that all living things are mortal, that every organism is endowed with a life-span that must eventually terminate, we must also acknowledge the indispensability of reproduction. Without reproduction, the evolution of a species is impossible. Like Samuel Butler who concluded that a chicken is just an egg's way of making another egg, many biologists consider the ability to reproduce to be the ultimate objective of all life processes.

The word "reproduction" implies replication, and it is true that biologic reproduction almost always yields a reasonable facsimile of the parent unit. However, sexual reproduction, practiced by the majority of animals, produces the *diversity* needed for

*Refer to pp. 18, 19, and 23 for Principles 10, 11, and 35.

survival in a world of constant change. At least for multicellular animals, sexual reproduction offers enormous advantages over asexual reproduction as we will strive to point out below. It must have arisen very early in animal evolution. But whether animals reproduce sexually or asexually, the process embodies a basic pattern: the conversion of raw materials from the environment into the offspring or sex cells that develop into offspring of a similar constitution, and the transmission of a hereditary pattern or code from the parents. The code of course is the deoxyribonucleic acid (DNA) of the genes. The chemical nature of DNA is described on p. 42, and its significance in the gene theory is discussed at length in Chapter 33.

■ HISTORIC BACKGROUND

The development of modern thinking in the reproductive process is one of the more colorful chapters

719

of biologic history. Until the eighteenth century, everyone—biologists and laymen alike—believed in **spontaneous generation.** It was natural to believe this, since it was the only viewpoint that made any sense. It was the expected outcome of observation in the absence of controlled experimentation. Spontaneous generation is the belief that life can originate from inorganic matter without preexisting life. It is the antithesis of **biogenesis,** the current belief that life must originate from previous life. The belief in abiogenesis often assumed grotesque dimensions, such as frogs arising from mud, mice from putrefied matter, and insects from dew. If a jar of blood were left standing open for a few days, it would teem with maggots. It was quite natural that no association was made between the maggots and the flies that visited the open container.

In 1668 Francesco Redi, an Italian physician, exposed meat in jars, some of which were uncovered and others covered with parchment and wire gauze. The meat in all three kinds of vessels spoiled, but only the open vessels had maggots, and he noticed that flies were constantly entering and leaving these vessels. He concluded that if flies had no access to the meat, no worms would be found there. Then, John T. Needham, an English priest, boiled mutton broth and put it in corked containers. After a few days the medium was swarming with microscopic organisms (1748). He concluded that spontaneous generation was real, because he believed that he had killed all living organisms by boiling the broth and that he had excluded the access of others by the precautions he took in sealing the tubes. However, an Italian investigator, Lazaro Spallanzani (1767), was critical of Needham's experiments and conducted experiments that led to a telling blow against the theory of abiogenesis. He thoroughly boiled extracts of vegetables and meat, placed these extracts in clean vessels, and sealed the necks of the flasks hermetically in flame. He then immersed the sealed flasks in boiling water for several minutes to make sure that all germs were destroyed. As controls, he left some tubes open to the air. At the end of two days he found the open flasks swarming with organisms; the others contained none. This experiment did not settle the issue, for the advocates of spontaneous generation maintained that air, which Spallanzani had excluded, was necessary for the production of new organisms or that the method he used had destroyed the vegetative power of the medium. When oxygen was dis-

covered (1774), the opponents of Spallanzani seized upon this as the vital principal that he had destroyed in his experiments.

Pasteur (1861) answered the objection of a lack of air by introducing fermentable material into a flask with a long S-shaped neck that was opened to the air. The flask and its contents were then boiled for a long time. Afterward the flask was cooled and left undisturbed. No fermentation occurred, for all organisms that entered the open end were deposited on the floor of the neck and did not reach the flask contents. When the neck of the flask was cut off, the organisms in the air could fall directly on the fermentable mass and fermentation occurred within it in a short time. Pasteur concluded that if suitable precautions were taken to keep out the germs and their reproductive elements (eggs, spores, etc.), no fermentation or putrefaction could take place.

■ NATURE OF THE REPRODUCTIVE PROCESS

The two fundamental types of reproduction are sexual and asexual. In **asexual** reproduction there is only one parent and no special reproductive organs or cells. Each organism is capable of producing identical copies of itself as soon as it becomes an adult. The production of copies is simple and direct and typically rapid. **Sexual** reproduction involves two parents as a rule, each of which contributes special **sex cells,** or **gametes,** toward a union that will develop into a new individual. Only one sex, the female, can produce offspring, and this must depend upon the intervention of the male, a nonproductive sexual partner. There are two kinds of gametes, the **ovum** (egg) produced by the female, and the **spermatozoon,** produced by the male. Ova are nonmotile, are produced in relatively small numbers, and usually contain a large amount of yolk, the stored food material that sustains early development of the new individual. Sperm are motile, are produced in enormous numbers, and are small. The union of egg and sperm is called **fertilization,** and the resulting cell is known as a **zygote.** Fertilization results in the recombination of two **haploid** gametes (each containing one-half the full genetic complement of the parent) to produce a **diploid** zygote.

Advantages of sexual reproduction

The zygote receives genetic material from *both* parents and accordingly is different from both. The combination of genes of both parents produces a genetically unique individual, still bearing the characteristics of the genetic species, but also bearing traits that may make it inferior or superior to its parents. This affords the variety required for natural selection. Recombination of characters makes possible wider and more diversified evolution. Although the inferior products of this process are eliminated by natural selection, the more exceptional are better able to respond to changes in the habitat. It is these, the better adapted forms, that are most apt to produce a better adapted progeny. Thus sexual reproduction tends to multiply variations, whereas asexual reproduction can only produce carbon copies and must await mutations to introduce variation into the line. This would seem to explain why asexual reproduction is restricted mostly to unicellular forms, which can multiply rapidly enough to offset the disadvantages of relentless replication of identical products. Of course, in those asexual organisms such as molds and bacteria that are haploid (bear only one set of genes), mutations are immediately expressed phenotypically, and evolution can proceed quickly. In sexual animals, on the other hand, a gene mutation is seldom expressed immediately, since it is masked by its normal partner on the homologous chromosome. There is only a remote chance that both members of a gene pair will mutate in the same way at the same moment.

Thus diploidy confers a basic stability to the species, whereas sexual reproduction provides the diversity and plasticity necessary to respond to environmental change. Sexual reproduction with its sequence of gene segregation and recombination generation after generation is, as the American geneticist T. Dobzhansky has said, the "master adaptation which makes all other evolutionary adaptations more readily accessible." If sexual reproduction has defects, they derive from the complexity of the process, the hazards involved in the meeting of eggs and sperm, and the possibility of unfavorable growing conditions for vulnerable embryos. Elaborate devices have frequently evolved to ensure fertilization and survival of the offspring.

Asexual reproduction and its variations

As pointed out above, asexual reproduction is found only among the simpler forms of life such as protozoans, coelenterates, bryozoans, bacteria, and a few others. It is absent among the higher invertebrates (mollusks and arthropods) and all vertebrates. Even in phyla in which it occurs, most of the members employ the sexual method as well. In these groups, asexual reproduction ensures rapid increase in numbers when the differentiation of the organism has not advanced to the point of forming highly specialized gametes.

The forms of asexual reproduction are fission, budding (both internal and external), fragmentation, and sporulation. **Fission** is common among protozoans and to a limited extent among metazoans. In this method the body of the parent is divided into two approximately equal parts, each of which grows into an individual similar to the parent. Fission may be either transverse or longitudinal. **Budding** is an unequal division of the organism. The new individual arises as an outgrowth (bud) from the parent. This bud develops organs like that of the parent and then usually detaches itself. If the bud is formed on the surface of the parent, it is an external bud, but in some cases internal buds, or **gemmules,** are produced. Gemmules are collections of many cells surrounded by a dense covering in the body wall. When the body of the parent disintegrates, each gemmule gives rise to a new individual. External budding is common in the hydra and internal budding in the freshwater sponges. Bryozoans also have a form of internal bud called statoblast. **Fragmentation** is a method in which an organism breaks into two or more parts, each capable of becoming a complete animal. This method is found among the Platyhelminthes, Nemertinea, and Echinodermata. **Sporulation** is a method of multiple fission in which many cells are formed and enclosed together in a cystlike structure. Sporulation occurs in a number of protozoan forms.

Sexual reproduction and its variations

Sexual reproduction is the general rule in the animal kingdom. It is a process involving a gamete or gametes.

CONJUGATION. Among protozoans conjugation may represent an early stage in the evolution of sexuality. During conjugation two individuals fuse together temporarily and exchange micronuclear material. Other forms of sexual reproduction occur in protozoans, in which there is a union between two special cells. These cells may be alike **(isogametes)** or they may be different **(anisogametes).** Usually the difference is one of size between the gametes, but in certain cases one kind of gamete may be motile and the other nonmotile. In some cases it is difficult to distinguish sex, for although two parents are involved, they cannot be designated as male and female. Work by Sonneborn and others on certain distinctive strains of paramecia would indicate the beginnings of sex distinctions because the members of some strains will not conjugate among themselves but only with those of another strain. Some protozoans, such as the malarial parasite *Plasmodium,* actually produce eggs and sperm.

The male-female distinction is clearly evident in the metazoa. Organs that produce the germ cells are known as **gonads.** The gonad that produces the sperm is called the **testis** (Fig. 31-4) and that which forms the egg, the **ovary** (Fig. 31-5). The gonads represent the **primary sex organs,** the only sex organs found in certain groups of animals. Most metazoans, however, have various **accessory sex organs.** In the primary sex organs the sex cells undergo many complicated changes during their development, the details of which are described in a later section. In our present discussion we shall point out the various types of sexual reproduction—biparental reproduction, parthenogenesis, pedogenesis, and hermaphroditism.

BIPARENTAL REPRODUCTION. Biparental reproduction is the common and familiar method of sexual reproduction involving separate and distinct male and female individuals. Each of these has its own reproductive system and produces only one kind of sex cell, spermatozoan or ovum, but never both. Nearly all vertebrates and many invertebrates have separate sexes, and such a condition is called **dioecious.**

PARTHENOGENESIS. This is a modification of sexual reproduction in which an unfertilized egg develops into a complete individual. It is found in rotifers, plant lice, certain ants, bees, and crustaceans. Usually parthenogenesis occurs for several generations and is followed by a biparental generation in which the egg is fertilized. In some cases (some platyhelminths, rotifers, and certain wasps) parthenogenesis appears to be the only form of reproduction. The queen bee is fertilized only once by a male (drone) or sometimes by more than one drone. She stores the sperm in her seminal receptacles, and as she lays her eggs, she can either fertilize the eggs or allow them to pass unfertilized. The fertilized eggs become females (queens or workers); the unfertilized eggs become males (drones).

Artificial parthenogenesis was discovered in 1900. Eggs that normally are fertilized can be artificially induced to develop without fertilization or the presence of sperm. The agents employed are dilute organic acids, hypertonic salt solutions, and mechanical pricking with a needle. Eggs of certain invertebrates, such as those of the sea urchin, were first used, but later vertebrate eggs were successfully induced to develop without fertilization. Many frogs of both sexes were developed beyond metamorphosis and some to adults. Rarely do such forms complete development and they are often smaller than normal ones. Earlier claims that fully normal rabbits developed from artificially activated eggs have not been substantiated. However, several parthenogenetically activated mammal eggs have been carried to the blastocyst stage (sheep and rabbit) and artificially activated mouse eggs have been cultured past the blastocyst stage, and, if replaced in the uterus, survived to midgestation. Some embryos developed as haploids, some as diploids, and some as haploid-diploid mosaics.

PEDOGENESIS. Parthenogenesis among larval forms is called **pedogenesis.** It is known to occur in *Miastor* (a gallfly), in which eggs produced by immature forms develop parthenogenetically into other larvae. **Neoteny,** or the retardation of bodily development, has about the same meaning as pedogenesis. The most striking example of this is the tiger salamander *(Ambystoma tigrinum)* that in certain parts of its range is found to mate in a larval (axolotl) form. Such larvae can be transformed into adults under certain conditions.

HERMAPHRODITISM. Animals that have both male and female organs in the same individual are called hermaphrodites and the condition is called **hermaphroditism.** The term derives from a combination of the names of the Greek god Hermes and goddess Aphrodite.

In contrast to the dioecious state of separate sexes, hermaphroditism is called **monoecious.** Many lower animals (flatworms and hydra) are hermaphroditic. Most of them do not reproduce by self-fertilization, but the individuals exchange germ cells with each other in cross-fertilization. The earthworm ensures that its eggs are fertilized by the copulating mate as well as vice versa. Another way of preventing self-fertilization is by developing the sex products at different times. In some monoecious forms the sperm are formed first and the eggs later **(protandry),** but this condition may be reversed **(protogyny).** However, some hermaphrodites have regular self-fertilization, such as tapeworms and certain snails. Several species of teleost fishes exhibit hermaphroditism. Some produce eggs and sperm from different areas of the same gonad and may be self-fertilizing. Certain tropical sea basses known as Bahama groupers *(Petrometopon cruentatum)* begin life as females and later change into males. The gonad is considered in these forms as a compound organ with female, male, and a combination of male and female tissue but with a delayed timing in the appearance of the sperm after the ova degenerate or disappear.

■ FORMATION OF REPRODUCTIVE CELLS

Protoplasm is commonly divided into two types—**somatoplasm** and **germ plasm.** The body cells are made up of somatoplasm and are called **somatic cells;** reproductive cells are formed of germ plasm and are called **germ cells.** All the somatic cells die within the individual. The germ plasm is continuous from generation to generation, whereas the somatoplasm is formed anew at each generation. At the present time this continuity is recognized as residing in the chromosomes and so the chromatin material of the nucleus is considered to be the germ plasm and the cell cytoplasm the somatoplasm. The distinction, however, between somatic and germ cells, as mentioned in another section, is not absolutely rigid. Many invertebrates are known to regenerate whole bodies from small parts of themselves.

Origin of germ cells

The actual tissue from which the gonads arise appears in early development as a pair of ridges, or pouches, growing into the coelom from the dorsal coelomic lining on each side of the gut near the anterior end of the mesonephros. The primordial ancestors of the cells that are going to form gametes do not arise in the developing gonad but in the yolk-sac endoderm and migrate by ameboid movements (mammals) or through the circulatory system (chick) to the genital ridges or embryonic gonad. The genital ridge is a mesenchymal thickening covered over by mesothelium that becomes the germinal epithelium. The gonad at first is sexually indifferent, but if the gonad is to become a testis, the cells in the germinal epithelium grow into the underlying mesenchymal tissue (medulla) and form the seminif-

erous tubules. The inner walls of the tubules are formed of cells that have descended from those of the germinal epithelium; these cells then develop eventually into mature sperm. If the indifferent gonad is to become an ovary, the primordial germ cells grow into the mesenchyme and differentiate into ovarian follicles with eggs (ova). There are still many disagreements about the origin of the germ cells. The differentiation of the genital ridge into an ovary or testis is determined by the hormonal environment during development, which in turn is determined by the genetic sex of the individual.

Meiosis: germ cell division

In ordinary cell division, or mitosis, each of the two daughter cells receives exactly the same number and kind of chromosomes. All body (somatic) cells have two sets of chromosomes, one of paternal and the other of maternal origin. The members of such a pair are called **homologous chromosomes.** Mitosis is described in Chapter 3 and illustrated in Fig. 3-9.

In sexual reproduction, however, the formation of the **gametes,** or **germ cells,** requires a different process than that of somatic cells. The fusion of two gametes (egg and sperm) produces the zygote or fertilized egg from which the new organism arises. If each sperm and egg had the same number of chromosomes as somatic cells, there would be a doubling of chromosomes in each successive generation. To prevent this from happening, germ cells are formed by a special type of cell division called **meiosis,** whereby the chromosome number is reduced by one half. The result is that mature gametes have only **one** member of each homologous pair, or a **haploid** (n) number of chromosomes. In man the zygotes and all body cells normally have the **diploid** number (2n) of 46; the gametes (eggs and sperm) have the haploid number (n) of 23.

Meiosis is similar to mitosis in its morphologic changes and movements of chromosomes, but mitosis has only **one** chromosomal division (Fig. 31-1). Meiosis consists of **two successive** chromosomal (nuclear) divisions, each of which has the same four stages—prophase, metaphase, anaphase, and telophase—found in mitosis. The first meiotic division involves the pairing of homologous chromosomes to form bivalent units, the resolution of each homologous chromosome into two half or sister chromatids, and the separation of homologous chromosomes to opposite poles of the cell. Each resulting daughter cell

Somatic cell (interphase) or
Primordial germ cell
(diploid number of 8 chromosomes)

Mitosis ← → **Meiosis**

Chromosomes begin
to appear

Each chromosome consists
of 2 sister chromatids

Homologous chromosomes, each
of 2 chromatids, pair

Chromosomes arranged on spindle

Paired chromosomes arranged
on spindle

In cell division, one sister chromatid
passes to one daughter cell and
its mate to the other

In first meiotic division, homologous
chromosomes separate to opposite poles
so that each daughter cell has
only haploid number of chromosomes

Each cell retains diploid number
(8) of chromosomes

Second meiotic division (not shown)
involves separation of sister chromatids
of each chromosome to their
respective daughter cells;
each cell has only haploid
number of chromosomes

FIG. 31-1
For legend see opposite page.

thus contains half the number of chromosomes characteristic of the diploid chromosomes of the organism. The second meiotic division results in a separation and distribution of the sister chromatids of each chromosome to opposite poles and thus involves no reduction in number of chromosomes. In contrast to mitosis, meiosis is concerned with genetic reassortment. It does this in two ways: (1) by random segregation of homologous chromosomes and (2) by crossing-over, or exchanging segments between each homologous pair. Another source of variation is, of course, the mutation of the gene. Both sexual and inherited variability enable natural selection to bring about evolutionary changes.

Stages in meiotic division

PROPHASE I. The most striking difference between mitosis and meiosis occurs at the beginning of the first meiotic division. The prophase, or the first stage of meiosis, has five substages: leptotene, zygotene, pachytene, diplotene, and diakinesis.

1. **Leptotene.** In this stage the chromosomes (diploid in number) appear as long, thin, threadlike structures and resemble strings of beads because of granules (chromomeres). Each pair of homologous chromosomes is identical as to size, position of centromere, etc. Each of these early prophase chromosomes is already divided into a pair of indistinguishable sister chromatids. DNA replication is complete.

2. **Zygotene.** This stage involves the pairing (synapsis) of homologous chromosomes to form bivalent chromosome units. This process does not occur in mitosis. This results in paired units corresponding to the haploid number.

3. **Pachytene.** The pairing of the chromosomes is completed and the chromosomes undergo longitudinal contraction so that each bivalent is shorter and thicker, and the two chromosomes of each bivalent become twisted about one another.

4. **Diplotene.** In this stage the homologous chromosomes of each bivalent are now visibly double. Since each homologous chromosome consists of two sister chromatids, each chromosome pair, or bivalent, will show four chromatids, or a **tetrad.** The centromeres remain unsplit at this time. Since longitudinal separation is incomplete, the homologs of the bivalent remain in contact at various points, producing a characteristic X configuration called **chiasma** (pl. chiasmata). Each chiasma represents a region at which two nonsister chromatids are undergoing an exchange of parts (crossover). Thus two chromatids of the tetrads are structurally reorganized so that each is made up of an original and an exchanged component. The other two chromatids of the tetrad remain in their original form. Chiasmata are not found in meiosis in which there is no crossing-over, as in the male *Drosophila.*

5. **Diakinesis.** This stage is characterized by a maximum contraction of the chromosomes and a further separation of the homologous chromosomes, although the chromatids remain connected by the chiasmata. At the same time the nucleolus begins to disappear, and the nuclear membrane breaks down.

At the end of the first prophase of meiosis homologous chromosomes have paired, exchanged chromatid segments, and started their longitudinal separation.

METAPHASE I. This phase begins when the nuclear membrane disappears and the spindle is formed. Each homologous chromosome (homolog) has its centromere (kinetochore) and the bivalent chromosomes (tetrads) line up on the equatorial plate, with the centromeres of the two homologs directed toward opposite poles.

ANAPHASE I. In this stage each homolog of a

FIG. 31-1

Comparison of mitosis and meiosis. Each process starts with diploid number of chromosomes. In **mitosis** chromosomes replicate; then the chromatids separate, one going to each pole resulting in two identical daughter cells with diploid number of chromosomes. In **meiosis** chromosomes replicate and then arrange themselves in homologous pairs on spindle; each pair separates, with one member going to each pole, resulting in daughter cells with haploid number of chromosomes. Each haploid cell now divides (not shown) with one set of chromatids passing to each daughter. Final result of meiosis is four gametes, each with haploid number of chromosomes.

pair, with its daughter chromosomes united by their centromere, moves to its respective pole, each centromere taking half of the bivalent with it. Thus whole chromosomes are separated in anaphase, and each of the two resulting cells of the first meiotic division has a haploid number of chromosomes.

TELOPHASE I AND INTERPHASE. A nuclear membrane may be re-formed around the chromosomes, which often persist in a condensed form; they may become uncoiled; or no membrane may be formed at all and the chromosomes may enter directly into the second meiotic division. The interphase may not exist at all.

PROPHASE II AND METAPHASE II. A short prophase in which a new spindle starts forming marks the beginning of the second meiotic division. The chromosomes become arranged on the equatorial plate. This is followed by the division of the centromeres for the first time and the longitudinal separation of the sister chromatids so that this division separates the two chromatids of each chromosome. Although these two chromatids are identical in their formation, they differ in those segments that have been exchanged by crossing-over.

ANAPHASE II AND TELOPHASE II. The sister chromatids, now called chromosomes, move to their respective poles, and each of the two daughter nuclei has a complete set (genome) that corresponds to the haploid number. In the telophase the cytoplasm divides, and the chromosomes become longer and less visible. A nuclear membrane is then formed around each nucleus.

The result of the two meiotic divisions is the formation of four cells, each of which has the haploid number of chromosomes or one of each kind of chromosome of the homologous pairs that started meiosis. The first meiotic division is often called a reduction division, and the second meiotic division is called an equational division. However, the exchange of chromatid segments (crossing-over) and the fact that there is no reduction in total number of chromosomes have caused many cytologists to consider the terms "reduction" and "equational" obsolete.

Gametogenesis

The series of transformations that results in the formation of mature gametes (germ cells) is called gametogenesis.

Although the same essential processes are involved in the maturation of both sperm and eggs, there are some minor differences. Gametogenesis in the testis is called **spermatogenesis** and in the ovary it is called **oogenesis.**

SPERMATOGENESIS (Fig. 31-2 and 31-4). The walls of the seminiferous tubules contain the differentiating sex cells arranged in a stratified layer five to eight cells deep. The outermost layers contain **spermatogonia** (Fig. 31-4), which have increased in number by ordinary mitosis. Each spermatogonium increases in size and becomes a **primary spermatocyte.** Each primary spermatocyte then undergoes the first meiotic division, as described above, to become two **secondary spermatocytes.**

Each secondary spermatocyte now enters the second meiotic division, without the intervention of a resting period. The resulting cells are called **spermatids,** and each contains the haploid number (23) of chromosomes. A spermatid may have all maternal, all paternal, or both maternal and paternal chromosomes in varying proportions. Without further divisions the spermatids are transformed into mature sperm by losing a great deal of cytoplasm, by condensing the nucleus into a head, and by forming a whiplike tail (Fig. 31-4).

One can see by following the divisions of meiosis that each primary spermatocyte gives rise to four functional sperm, each with the haploid number of chromosomes (Fig. 31-2).

PRODUCTION OF MATURE SPERM. The testes of higher forms such as mammals consist of seminiferous tubules and interstitial tissue lying between them. As already described, the seminiferous tubules include the germ cell stages, and the interstitial tissue contains the special Leydig cells, which, as mentioned later, secrete the male hormone testosterone. The seminiferous tubules contain, in addition to the germ cells, the nongerminal Sertoli cells (Fig. 31-4), often called "sperm mother cells" because it is believed that the sperm heads become embedded in them during the development of the sperm. During the maturation of the final stages of sperm formation, parts of the cytoplasm of the spermatozoon are sloughed off as **residual bodies.** These bodies are eventually engulfed by the Sertoli cells.

OOGENESIS (Fig. 31-2 and 31-5). The early germ cells in the ovary are called **oogonia,** which increase in number by ordinary mitosis (Fig. 31-9). Each oo-

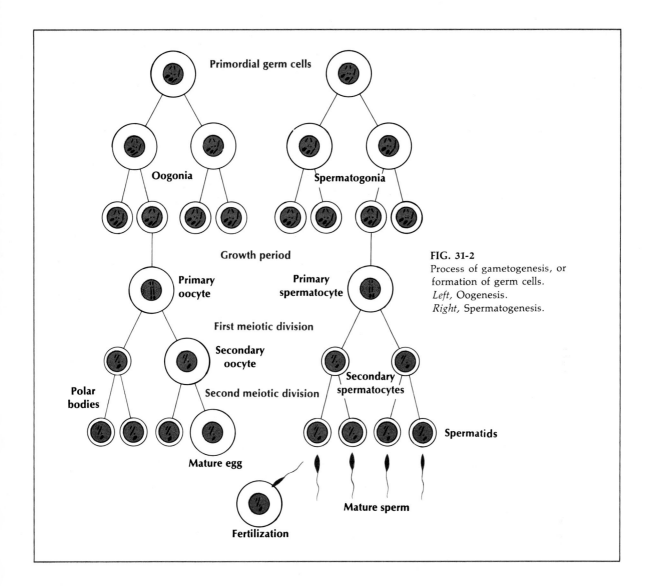

FIG. 31-2
Process of gametogenesis, or formation of germ cells.
Left, Oogenesis.
Right, Spermatogenesis.

gonium contains the diploid number of chromosomes. In the human being, after puberty, typically one of these oogonia develops each menstrual month into a functional egg. After the oogonia cease to increase in number, they grow in size and become **primary oocytes.** Before the first meiotic division, the chromosomes in each primary oocyte meet in pairs, paternal and maternal homologs, just as in spermatogenesis. When the first maturation (reduction) division occurs, the cytoplasm is divided unequally. One of the two daughter cells, the **secondary oocyte,** is large and receives most of the cytoplasm; the other is very small and is called the

first polar body. Each of these daughter cells, however, has received half the nuclear material or chromosomes.

In the second meiotic division, the secondary oocyte divides into a large **ootid** and a small polar body. If the first polar body also divides in this division, which sometimes happens, there will be three polar bodies and one ootid. The ootid grows into a functional **ovum;** the polar bodies disintegrate because they are nonfunctional. The formation of the nonfunctional polar bodies is necessary to enable the egg to get rid of excess chromosomes, and the unequal cytoplasmic division makes possible a large cell with sufficient yolk for the development of the young. Thus the mature ovum has the haploid number of chromosomes the same as the sperm. However, each primary oocyte gives rise to only **one** functional gamete instead of four as in spermatogenesis (Fig. 31-2).

Gametes of various animals

Sperm among animals show a greater diversity of form than do ova. A typical spermatozoon is made up of a head, a middle piece, and an elongated tail for locomotion (Fig. 31-3). The head consists of the nucleus containing the chromosomes for heredity and an **acrosome** believed to contain an enzyme that assists in egg penetration. The total length of the human sperm is 50 to 70 μ. Some toads have sperm that exceed 2 mm. (2,000 μ) in length. Most sperm, however, are microscopic in size.

Ova are oval or spherical in shape and are nonmo-

tile. Mammal eggs are very small (not more than 0.25 mm.) and contain little yolk because the young receive nourishment from the mother. On the other hand, the eggs of some birds, reptiles, and sharks are very large and yolky, to supply nutritive material for the developing young before hatching. Some eggs (reptiles and birds) also contain a great deal of albumin, which also serves for nourishment. Most eggs are provided with some form of protective coating. This may be in the form of a calcified shell (birds), leathery parchment (reptiles), or albuminous coats (amphibians).

The number of sperm in all animals is greatly in excess of the eggs of corresponding females. The number of eggs produced is related to the chances of the young to hatch and reach maturity. This explains the enormous number of eggs produced by certain fish, as compared with the small number produced by mammals.

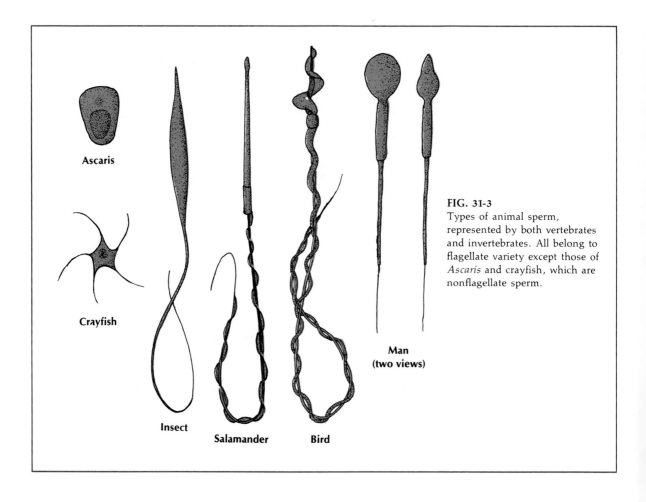

Ascaris

Crayfish

Insect

Salamander

Bird

**Man
(two views)**

FIG. 31-3
Types of animal sperm, represented by both vertebrates and invertebrates. All belong to flagellate variety except those of *Ascaris* and crayfish, which are nonflagellate sperm.

◼ ANATOMY OF REPRODUCTIVE SYSTEMS

The basic plan of the reproductive systems is similar in all animals. Many structural differences are found among the accessory sex organs, depending on the habits of the animals, their methods of fertilizing their eggs, their care of the young, etc. Many invertebrates have reproductive systems as complex as those of vertebrates, for example, flatworms, snails, and earthworms, and others. There are often complicated accessory sex organs such as reproductive ducts, penis, seminal vesicles, yolk glands, uterus, seminal receptacles, and genital chambers. In vertebrate animals the reproductive and excretory systems are often referred to as the **urogenital system** because of their close association, particularly striking during embryonic development and in the use of common ducts. The male excretory and reproductive systems usually have a more intimate connection than have those of the female. In those forms (some fish and amphibians) that have an opisthonephros kidney, the **wolffian duct** that drains the kidney also serves as the sperm duct. In male reptiles, birds, and mammals in which there is a metanephric kidney with its own independent duct **(ureter)** to carry away waste, the wolffian duct is exclusively a sperm duct **(vas deferens).** In all of these forms, with the exception of mammals higher than monotremes, the ducts open into a **cloaca.** In higher mammals lacking a cloaca, the urogenital system has an opening separate from the anal opening. The **oviduct** of the female is an independent duct that, however, does open into the cloaca in forms that have a cloaca.

The plan of the reproductive system in vertebrates includes (1) **gonads** that produce the sperm and eggs; (2) **ducts** to transport the gametes; (3) **special organs** for transferring and receiving gametes; (4) **accessory glands** (exocrine and endocrine) to provide secretions necessary for the reproductive process; and (5) **organs for storage** before and after fertilization. This plan is modified among the various vertebrates, and some of the items may be lacking altogether.

Reproductive system in man

MALE REPRODUCTIVE SYSTEM. The male reproductive system in man (Fig. 31-4) includes testes, vasa efferentia, vasa deferentia, penis, and glands.

Testes (testicles). The testes are paired and are responsible for the production and development of the sperm. Each testis is made up of about 500 **seminiferous tubules,** which produce the sperm, and the **interstitial** tissue, lying among the tubules, which produces the male sex hormone (testosterone). The two testes are housed in the scrotal sac, which hangs down as an appendage of the body. The scrotum acts as a thermoregulator for protecting the sperm against high temperature. Sperm apparently will not form at body temperatures, although they are able to do so in elephants and birds (with very high temperatures). In some mammals (many rodents) the testes are retained within the body cavity except during the breeding season, when they descend through the inguinal canals into the scrotal sacs.

Vasa efferentia. Vasa efferentia are small tubes connecting the seminiferous tubules to a coiled **vas epididymis** (one for each testis) which serves for the storage of the sperm.

Vasa deferentia. Each tube is a continuation of the epididymis and runs to the urethra, where it joins its opposite from the other testis. From this point the urethra serves to carry both sperm and urinary products.

Penis. The penis is the external intromittent organ through which the urethra runs. The penis contains erectile tissue for distention during the copulatory act.

Glands. There are at least three pairs of exocrine glands (those with ducts) that open into the reproductive channels. Fluid secreted by these glands furnishes food to the sperm, lubricates the passageways of the sperm, and counteracts the acidity of the urine so that the sperm will not be harmed. The first of these glands is the **seminal vesicle,** which opens into each vas deferens before it meets the urethra. Next are the **prostate glands,** which are really a single fused gland in man; it secretes into the urethra. Near the base of the penis lies the third pair of glands, **Cowper's glands,** which also discharge into the urethra. The secretions of these glands form a part of the seminal discharge.

FEMALE REPRODUCTIVE SYSTEM. The female reproductive system (Fig. 31-5) contains ovaries, oviducts, uterus, vagina, and vulva.

Ovaries. The ovaries are paired and are contained within the abdominal cavity, where they are held in position by ligaments. Each ovary is about as large

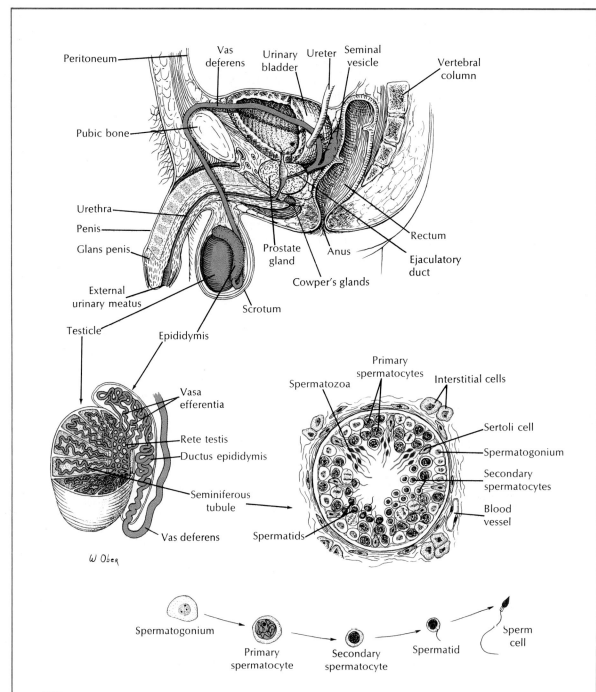

FIG. 31-4

Human male reproductive system. *Top,* Median section of male pelvis. *Center left,* Section of left testis. *Center right,* Cross section of one seminiferous tubule, showing different stages in spermatogenesis. *Bottom,* Sequential stages in spermatogenesis. (Drawn by W. C. Ober.)

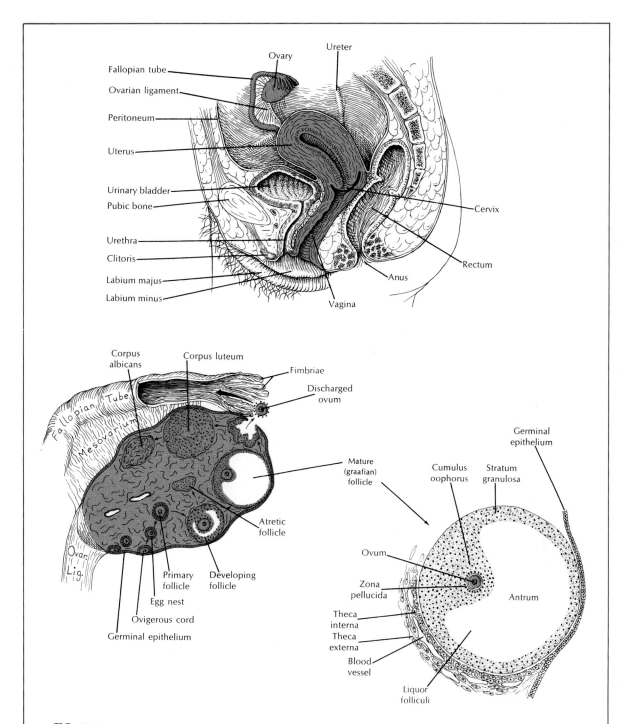

FIG. 31-5
Human female reproductive system. *Top,* Median section through female pelvis. *Left,*
Section of ovary showing progressive differentiation of a follicle, ovulation, and formation
of a corpus luteum. *Bottom right,* Section of mature follicle. (Drawn by W. C. Ober.)

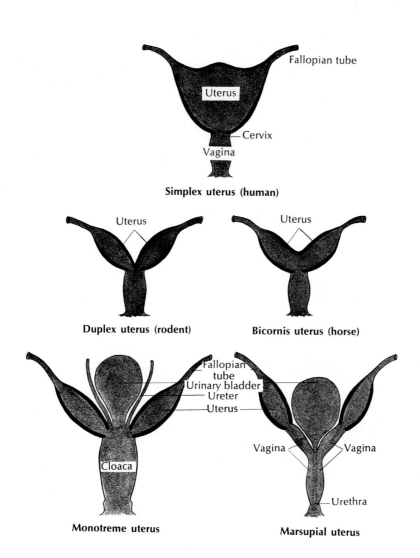

FIG. 31-6
Uteri and related structures of mammals. Oviducts of mammal are modified to form uterus for development of young and vagina for reception of male penis. Most primitive condition is found in monotremes, which lack vagina but have cloaca. They are the only mammals that lay eggs. Marsupials have double vagina and uterus because their oviducts are only partially fused at bases. Placental mammals, on the other hand, have single fused vagina with uteri in progressive stages of fusion. In some (rodents), uterus is completely separated, or duplex; in others (perissodactyls), bases of oviducts are partly fused and form bicornis (bicornuate) type; in a few (primates), uterus is single median cavity (simplex). In carnivores and some others there is an intermediate type (not shown), in which two uteri are united at their posterior ends with single cervical opening (bipartite uterus).

as an almond and contains many thousands of developing eggs (ova). Each egg develops within a graafian follicle that enlarges and finally ruptures to release the mature egg (Fig. 31-5). During the fertile period of the woman about 13 eggs mature each year, and usually the ovaries may alternate in releasing an egg. Since the female is fertile for only some 30 years, only about 400 eggs have a chance to reach maturity; the others degenerate and are absorbed.

Oviducts (fallopian tubes). These egg-carrying tubes are not closely attached to the ovaries but have funnel-shaped ostia for receiving the eggs when they emerge from the ovary. The oviduct is lined with cilia for propelling the egg in its course. The two ducts open into the upper corners of the uterus, or womb.

Uterus. The uterus is specialized for housing the embryo during the 9 months of its intrauterine existence. It is provided with thick muscular walls, many blood vessels, and a specialized lining—the **endometrium.** The uterus varies with different mammals. It was originally paired but tends to fuse in higher forms. The different types are illustrated in Fig. 31-6.

Vagina. This large muscular tube runs from the uterus to the outside of the body. It is adapted for receiving the male's penis and for serving as the birth canal during expulsion of the fetus from the uterus. Where the vagina and the uterus meet, the uterus projects down into the vagina to form the **cervix.**

Vulva. The vulva refers to the external genitalia and includes two folds of skin covered with hair, the **labia majora;** a smaller pair of folds within the labia majora, the **labia minora;** a small erectile organ, the **clitoris,** at the anterior junction of the labia minora; and a fleshy elevation above the labia majora, the **mons veneris.** The opening into the vagina is the vestibule that is normally closed in the virgin state by a membrane, the **hymen.**

Homology of sex organs

For every structure in the male system, there is a homologous one in the female. The various organs of the two sexes perform functions peculiar to each sex. Although homologous structures may be well developed in both sexes, many of them are functional in one sex and their homologs in the other may be vestigial and nonfunctional. To understand this sex organ homology, it is necessary to observe the way in which the reproductive systems have arisen in embryonic development. Although sex is probably determined at the time of fertilization, it is not until many weeks later that the distinct sex characters associated with one or the other sex are recognized. Before this time the external genitalia of the two sexes cannot be distinguished. The hormonal interpretation of sex shows how slight the differences are between the two sexes. The animal is a chemical hermaphrodite and bears the possibilities of becoming either sex, depending on the balance of the sex hormones.

Some of the chief homologies of the male and female reproductive systems are shown in Table 31-1.

TABLE 31-1
Organ homologies of male and female reproductive systems

Male	Indifferent stage of embryo	Female
Testis	Genital ridge	Ovary
Vas deferens	Wolffian duct	Vestigial
Epididymis	Wolffian body	Vestigial
Appendix of testis	Müllerian duct	Uterus, vagina, fallopian tube
Penis	Genital tubercle	Clitoris
Glans penis		Glans clitoridis
Anal surface of penis	Genital folds	Labia minora
Scrotum	Genital swellings	Labia majora

■ HORMONES OF REPRODUCTION

The male and female gonads are endocrine glands as well as gamete-forming glands. Reproduction is a complex process requiring the coordinated action of many hormones, especially in the female. Although the principal features of reproductive endocrinology are understood, the recent search for effective and safe birth control devices has revealed some disturbing gaps in our knowledge. To cite a single but telling example, no one fully understands how the popular birth control pill works.

The female reproductive activities of almost all animals are cyclic. Usually breeding cycles are seasonal and coordinated so that the young are born at a time of year when conditions for growth are most favorable. Mammals have two major reproductive patterns, the **estrous cycle** (characteristic of most mammals) and the **menstrual cycle** (characteristic of primates only). These two cyclic patterns differ in two ways. First, in the estrous cycle, but not in the menstrual cycle, the female is receptive to the male, that is, she is in "heat," only at restricted periods of the year. Second, the menstrual cycle, but not the estrous cycle, ends with the collapse and sluffing of the uterine lining (endometrium). In the estrous animal each cycle ends with the uterine lining simply reverting to its original state, without the bleeding characteristic of the menstrual cycle. (The bleeding that occurs in dogs and cattle when they are in heat is not caused by sloughing of the uterine lining but by red blood cells passing through the wall of the vagina.) But these differences are really minor variations on a basic theme. The hormonal regulation of the reproductive cycles is so much alike for all mammals that a great deal of what we know about the human reproductive hormones has come from laboratory studies of the ubiquitous white rat, a rodent having an estrous cycle. Differences become more apparent when we study the lower vertebrates, but even fish share most of the reproductive hormones found in man. Since our discussion of this vast area must be brief, we will restrict our consideration to man.

The ovaries produce two kinds of steriod **sex hormones—estrogens** and **progesterone** (Fig. 31-7). Estrogens are responsible for the development of the female accessory sex structure (uterus, oviducts, and vagina) and the female secondary sex characters,

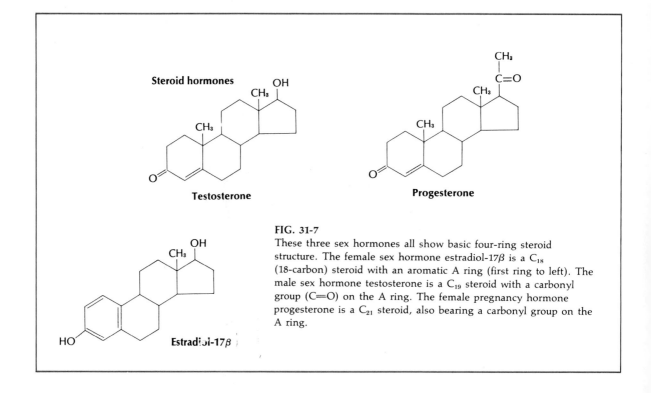

FIG. 31-7

These three sex hormones all show basic four-ring steroid structure. The female sex hormone estradiol-17β is a C_{18} (18-carbon) steroid with an aromatic A ring (first ring to left). The male sex hormone testosterone is a C_{19} steroid with a carbonyl group (C=O) on the A ring. The female pregnancy hormone progesterone is a C_{21} steroid, also bearing a carbonyl group on the A ring.

such as breast development, and the characteristic bone growth, fat deposition, and hair distribution of the female. Progesterone is responsible for preparing the uterus to receive the developing embryo. These hormones are controlled by the pituitary **gonadotropins,** FSH (follicle stimulating hormone), and LH (luteinizing hormone) (Fig. 31-8; see also Fig. 30-21).

The menstrual cycle begins with the release into the bloodstream of FSH from the anterior pituitary (Fig. 31-8). Reaching the ovaries, it stimulates the growth of one of the several thousand follicles present in each ovary. The follicle swells as the egg matures, until it bursts, releasing the egg onto the surface of the ovary. This event, called **ovulation,** normally occurs on about the fourteenth day of the cycle. Now follows the most critical period of the cycle, for unless the mature egg is fertilized within a few hours it will die. During this period, the egg is swept into an oviduct (fallopian tube) and begins its journey toward the uterus, pushed along by the numerous cilia that line the oviduct walls. If intercourse occurs at this time, the sperm will traverse the uterus and find their way into the oviducts, where one may meet and fertilize the egg. The developing embryo continues down the oviduct, enters the

FIG. 31-8

Reproductive cycle in human female showing hormonal-ovarian-endometrial relationships. Note that endometrium undergoes cyclical chain of events (thickness, congestion, etc.) under influence of hormones from ovary—estrogen and progesterone—and from pituitary gland—FSH and LH. Another hormone, prolactin, stimulates secretion of milk.

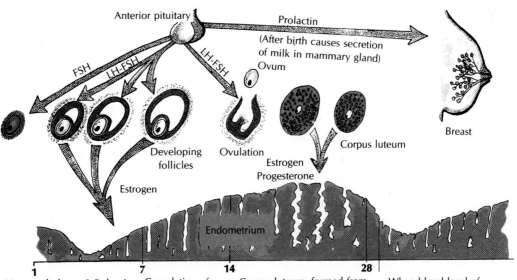

1	7	14	28
Menstrual phase; 3-5 days' duration; loss of uterine epithelium and blood; follicle-stimulating hormone (FSH) influences growth of follicle; luteinizing hormone (LH) influences activity of corpus luteum	Completion of follicular growth 7-14 days; OVULATION (release of egg) about 14th day; "Mittelschmerz" (sensation) may be felt at this time	Corpus luteum, formed from reorganized follicle, secretes progesterone and estrogen, which prepare endometrium for reception of egg; if egg not fertilized, corpus luteum retrogresses about 27th day; if fertilized, persists for 5 months or more; coiled blood vessels in endometrium	When blood level of progesterone (and estrogen) falls in nonpregnancy, another menstruation occurs involving endometrial regression and bleeding

uterine cavity, where it dwells for a day or two, and then implants in the prepared uterine endometrium. This is the beginning of pregnancy.

Let us now examine the intricate series of events that occurs both before and after the beginning of pregnancy. We have seen that the pituitry hormone FSH begins the reproductive cycle by stimulating the growth of at least one of the ovarian follicles. As the follicle enlarges, it releases **estrogens** (principally estradiol) that prepare the uterine lining (endometrium) for reception of the embryo.

The rise in blood estrogen is sensed by the pituitary, which responds by stopping the production of FSH. Estrogen also encourages the production of the second pituitary hormone, LH. Ovulation now occurs. LH causes the cells lining the ruptured follicles to proliferate rapidly, filling the cavity with a characteristic spongy, yellowish body called a **corpus luteum** (Fig. 31-9). The corpus luteum, responding to the continued stimulation of LH, manufactures **progesterone** in addition to estradiol. Progesterone, as its name suggests, stimulates the uterus to undergo the final maturation changes that prepare it for gestation. The uterus is thus fully ready to house and nourish the embryo by the time the latter settles out

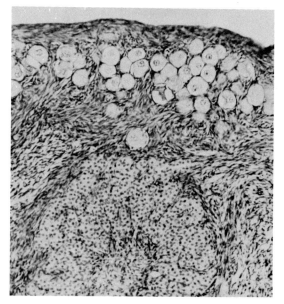

FIG. 31-9
Section of cat ovary showing corpus luteum and primordial follicles containing oogonia. (Courtesy J. W. Bamberger, Los Angeles, Calif.)

onto the uterine surface, usually about 7 days after ovulation. If fertilization has **not** occurred, the corpus luteum disappears, and its hormones are no longer secreted. Since the uterine endometrium depends on progesterone and estrogen for its maintenance, their disappearance causes the endometrial lining to dehydrate and slough off, producing the menstrual discharge. However, if the egg has been fertilized and has implanted, the corpus luteum continues to supply the essential sex hormones needed to maintain the mature uterine endometrium. During the first few weeks of pregnancy the developing placenta itself begins to produce the sex hormones progesterone and estrogen and soon replaces the corpus luteum in this function. As pregnancy advances, progesterone and estrogen prepare the breasts for milk production. The actual secretion and release of milk after birth (lactation) is the result of two other hormones, **prolactin** and **oxytocin**. Milk is not secreted during pregnancy because the placental sex hormones inhibit the release of prolactin by the pituitary. The placenta, like the corpus luteum that preceded it, thus becomes a special endocrine gland of pregnancy. After delivery, many mammals eat the placenta (afterbirth), which, because of its hormonal content, encourages the rapid postpartum involution of the uterus. This behavior also serves to remove telltale evidence of a birth from potential predators.

The male sex hormone **testosterone** (Fig. 31-7) is manufactured by the **interstitial cells** of the testes. Testosterone is necessary for the growth and development of the male accessory sex structures (penis, sperm ducts, glands), for development of secondary male sex characters (hair distribution, voice quality, bone and muscle growth), and for male sexual behavior. The same pituitary hormones that regulate the female reproductive cycle, FSH and LH, are also produced in the male, where they guide the growth of the testes and its testosterone secretion.

■ FERTILIZATION

We have defined fertilization as the formation of a zygote by the union of a spermatozoon and an ovum (egg). This process accomplishes two things: it triggers the process of development and it provides for the recombination of paternal and maternal inheritance units. Thus it restores the original diploid number of chromosomes characteristic of the species.

For a species to survive, it must ensure that fertilization will occur and that enough progeny will result to continue the race. Most marine fish simply set their eggs and sperm adrift in the ocean and rely on the random swimming movements of sperm to make chance encounters with eggs. Even though an egg is a large target for a sperm, the enormous dispersing effect of the ocean, the short life-span of the gametes (usually just a few minutes for fish gametes), and the limited range of the tiny sperm all conspire against an egg and a sperm coming together. Accordingly each male releases countless millions of sperm at spawning. The odds against fertilization are further reduced by coordinating the time and place of spawning of both parents. Ensuring that some eggs are fertilized, however, is not enough. The ocean is a perilous environment for a developing fish, and most never make it to maturity. Thus, the females produce huge numbers of eggs. The com-

mon gray cod of the North American east coast regularly spawns 4 to 6 million eggs, of which only two or three will, on the average, reach maturity.

Fishes and other vertebrates that provide more protection to their young produce fewer eggs than do the oceanic marine fishes. The chances of the eggs and sperm meeting is also increased by courtship and mating procedures and the simultaneous shedding of the gametes in a nest or closely circumscribed area. Internal fertilization, characteristic of the sharks and rays as well as reptiles, birds, and mammals, avoids dispersion of the gametes and protects them. However, even with internal fertilization, vast numbers of sperm must be released by the male into the female tract. Furthermore, the events of ovulation and insemination must be closely synchronized and the gametes must remain viable for several hours to accomplish fertilization. Sperm may have to travel a considerable distance to reach the egg in the

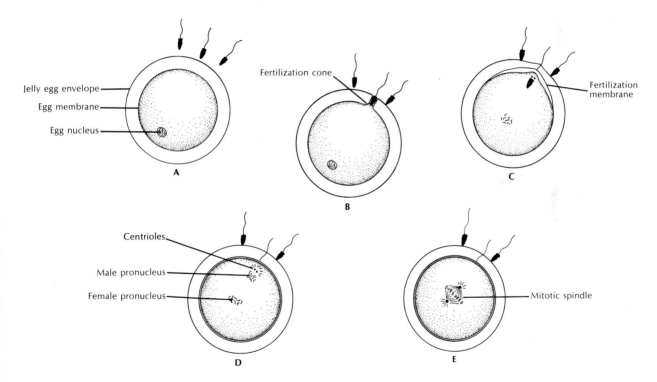

FIG. 31-10
Fertilization of egg. **A,** Many sperm swim to egg. **B,** First sperm to penetrate protective jelly envelope and contact egg membrane causes fertilization cone to rise and engulf sperm head. **C,** Fertilization membrane begins to form at site of penetration and spreads around entire egg, preventing entrance of additional sperm. **D,** Male and female pronuclei approach one another, lose their nuclear membranes, swell, and fuse. **E,** Mitotic spindle forms, signaling creation of a zygote and heralding first cleavage of new embryo.

female genital tract, many parts of which are rather hostile to sperm. Experiments with rabbits have shown that of the approximately 10 million sperm released into the female vagina, only about 100 reach the site of fertilization.

EGG-SPERM INTERACTIONS. In many animals, eggs and sperm release chemicals which cause the two to stick together once contact has been made. The classic study was done by F. R. Lillie more than 50 years ago on sea urchins. Sea urchin eggs release a sperm-activating and agglutinating material called **fertilizin.** It is a glycoprotein localized in the gelatinous coat of the egg. The sperm head contains low molecular weight acidic proteins, called **antifertilizin,** which unite with the egg fertilizin, causing the sperm to bind to the egg surface. Comparable reactions have been demonstrated in many invertebrates and vertebrates. The egg-sperm interaction resembles an immunologic reaction.

Once a sperm contacts an egg, it must penetrate the protective egg membranes (Fig. 31-10). Many kinds of sperm produce special enzymes, called **sperm lysins,** which dissolve the egg membranes to clear an entrance path. In mammals, the sperm lysin is the enzyme **hyaluronidase.** The egg, which has been quiescent up to this point, undergoes a series of rapid changes that serve to activate it for the period of rapid divisions that lie ahead. The sperm, now immotile, is drawn into the interior of the egg. A special **fertilization membrane** (Fig. 31-10, *C*) quickly forms around the egg to protect it from polyspermy (entrance of more than one sperm). Once inside the egg, the tail of the sperm is discarded. The highly condensed sperm nucleus, now called a **male pronucleus,** begins to swell. The male and female pronuclei, each carrying half the diploid number of chromosomes, are drawn together by unknown means and fuse. With this union a **zygote** is formed, ready to embark on the remarkably complex embryonic process leading to an adult animal.

■ DIVERSITY OF REPRODUCTIVE PATTERNS

Animals may be divided into three classes on the basis of the methods they employ to nourish their young. Those animals that lay their eggs outside the body for development are called **oviparous.** In such cases the eggs may be fertilized inside or outside the body. Some animals retain their eggs in the body (in the oviduct) while they develop, but the embryo derives its sole nourishment from the egg and not from the mother. These are called **ovoviviparous.** In the third type the egg develops in the uterus, but the embryo early in its development forms an intimate relationship with the walls of the uterus and derives its nourishment directly from food furnished by the mother. Such a type is called **viviparous.** In both of the last two types the young are born alive. Examples of oviparity are found among many invertebrates and vertebrates. All birds are of this type. Ovoviviparity is common among certain fish, lizards, and a few of the snakes. Viviparity is confined mostly to the mammals and some elasmobranchs.

The structural, physiologic, and behavioral aspects of sexual reproduction show an incredible diversity, exceeding that of any other system. The diversity is expressed in the methods for fertilization, kinds of habitats, structure of the reproductive systems, prenatal care of the young, and seasonal and physiologic changes in animals. Bats mate in the fall and the sperm is stored in the female till the following spring before fertilization occurs. The queen honeybee stores up enough sperm from the drone on one nuptial flight to fertilize all the eggs she lays during her lifetime. Salmon spend most of their lives in the sea but spawn far up inland rivers in freshwater. Eels grow to maturity in freshwater streams but migrate to the sea to spawn. Many animals provide nests of various sorts to take care of the young (birds and some fish). Others have cases for the eggs (insects and spiders). Some female animals carry the eggs attached to the body or to the appendages (crayfish and certain amphibians). Brood pouches for the eggs are provided by such forms as the mussel, the sea horse (a fish), and many others. Social insects (bees and ants) have huge colonial nests organized on a complex scale. Mammals are retained in the uterus of the female during early development and later are nourished by the milk from the mammary glands. These few examples merely begin to express the curious and fascinating range of reproductive adaptations among animals. Evolution seems to have been most imaginative when it was exploiting the adaptive possibilities of sexuality. There is fascinating literature on the subject; some are included in the references that follow.

References

Suggested general readings

Alston, R. E. 1967. Cellular continuity and development. Glenview, Ill., Scott, Foresman & Co. *Compares mitosis and meiosis and deals with development of reproductive cells, sex determination, and life cycles of animals. Paperback.*

Asdell, S. A. 1965. Patterns of mammalian reproduction. London. Constable & Co., Ltd.

Austin, C. R., and R. V. Short (editors). 1972. Reproduction in mammals. Cambridge, Cambridge University Press. 5 vol. *Highly readable and well illustrated series dealing with germ cells, fertilization, embryonic and fetal development; reproductive hormones and patterns and the artificial control of reproduction. Paperbacks.*

Carr, D. E. 1970. The sexes. Garden City, N. Y., Doubleday & Co. *Deals with the evolution of sex from protozoans to man, sex taboos, birth control, and population problems. Written in a popular and frequently amusing style but containing a wealth of information.*

Clegg, P. C., and A. C. Clegg. 1969. Hormones, cells and organisms. London, William Heinemann, Ltd.

Parkes, A. S. (editor). 1956. Marshall's physiology of reproduction, ed. 3. London, Longman Group Ltd. 3 vol.

Perry, J. S. 1971. The ovarian cycle of mammals. Edinburgh, Oliver & Boyd.

Smith, A. 1968. The body. Middlesex, Penguin Books. (Paperback. Cloth-bound edition only available in United States.) *An entertaining and informative account of the functions and peculiarities of the human body. Human reproduction is central theme. Contraception, pregnancy and birth, twinning, inbreeding, etc. and numerous curious facts about human reproduction.*

Van Tienhoven, A. 1968. Reproductive physiology of vertebrates. Philadelphia, W. B. Saunders Co.

Wickler, W. 1972. The sexual cycle: The social behavior of animals and men. Garden City, N. Y., Doubleday & Co. *A wide-ranging and interesting exploration of sexual behavior in animals, and the origins of human ethics and morals.*

Wood, C. 1969. Human fertility: Threat and promise. World of Science Library, New York, Funk & Wagnalls. (Available as paperback in England and Canada: Sex and fertility. 1969. London, Thames & Hudson.) *Succinct, beautifully illustrated account of human reproduction, birth control, population problems and human destiny. Highly recommended.*

World Health Organization. 1965. Physiology of lactation. World Health Organization Technical Report Series, no. 305. Geneva, W.H.O.

Selected *Scientific American* articles

Berelson, B., and R. Freedman. 1964. A study in fertility control. **210:**29-37 (May). *Family planning program in Taiwan.*

Csapo, A. 1958. Progesterone. **198:**40-46 (April). *The role of progesterone in pregnancy.*

Dahlberg, G. 1951. An explanation of twins. **184:**48-51 (Jan.). *Twinning and multiple births in man and other mammals.*

Gordon, M. J. 1958. The control of sex. **199:**87-94 (Nov.). *Studies with rabbit sperm cells separated in an electric field.*

Jaffe, F. S. 1973. Public policy on fertility control. **229:**17-23 (July). *Official policy and public attitude toward contraception have undergone substantial changes in the last few years.*

Mittwoch, U. 1963. Sex differences in cells. **209:**54-62 (July). *The recognition of sex indicators in cell nuclei has speeded studies of sex determination in the human species.*

Tietze, C., and S. Lewit. 1969. Abortion. **220:**21-27 (Jan.). *This age-old and controversial method of fertility control is finally receiving worldwide studies.*

A 24-day old embryo of the Tammer wallaby (Macropus eugenii) just before birth. Like other marsupials (such as kangaroos, opossums) the youth are born comparatively underdeveloped and must crawl without assistance from the mother to the marsupial pouch where they attach themselves to the teats. Note the well-developed forelimbs. The embryo is enclosed within a transparent amnion. A large yolk sac and smaller allantois extend from the embryo's umbilicus. (Courtesy Marilyn B. Renfree, University of Edinburgh, Scotland.)

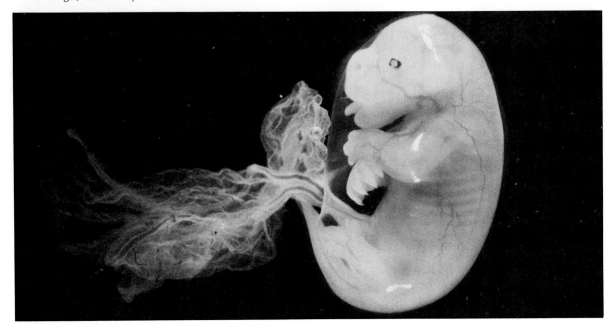

CHAPTER 32

PRINCIPLES OF DEVELOPMENT*

In Chapter 5 a brief survey of early embryonic development was given: types of eggs, fertilization and formation of the zygote, cleavage, and early morphogenesis. In this chapter we will first consider another dimension of development: the coordinating and regulating processes that guide the destiny of a growing embryo. The latter part of the chapter is devoted to the embryology of amniotes, especially man, and the special adaptations that support development.

The phenomenon of development is a remarkable, and in many ways awesome, process. How is it possible that a tiny, spherical fertilized human egg, scarcely visible to the naked eye, can unfold into a fully formed, unique person, consisting of thousands of billions of cells, each cell performing a predestined

*Refer to pp. 18 to 21 for Principles 10, 14, 21, and 26.

functional or structural role? How is this marvelous unfolding controlled? Obviously all the information needed is contained within the egg, principally in the genes of the egg's nucleus. The fabric of genes is deoxyribonucleic acid (DNA). Thus all development originates from the structure of the nuclear DNA molecules and in the egg cytoplasm surrounding the nucleus. But knowing where the blueprint for development resides is very different from understanding how this control system guides the conversion of a fertilized egg into a fully differentiated animal. This remains a major—many consider **the** major—unsolved problem of biology. It has stimulated a vast amount of research on the processes and phenomena involved; from it all has emerged some early, and in many cases tentative, answers to the questions of how development is controlled.

■ EARLY THEORIES: PREFORMATION VERSUS EPIGENESIS

Early scientists and lay people alike speculated at length about the mystery of development long before the process was submitted to modern techniques of biochemistry, molecular biology, tissue culture, and electron microscopy. An early and persistent idea was that the young animal was preformed in the egg and that development was simply a matter of unfolding what was already there. Some claimed they could actually see a miniature of the adult in the egg or the sperm. Even the more cautious argued that all the parts of the embryo were in the egg, ready to unfold, but so small and transparent they could not be seen. The **preformation theory** was strongly advocated by seventeenth and eighteenth century naturalist-philosophers. William Harvey, the great physiologist of blood circulation fame, could not accept the preformation notion. His famous dictum *ex ovo omnia* (all from the egg) was an important concept at a time when creative ideas in biology were still rare. The preformation theory received its death blow in 1759 when the German embryologist Caspar Friedrich Wolff clearly showed that in the earliest developmental stages of the chick, there was no embryo, only an undifferentiated granular material that became arranged into layers. These continued to thicken in some areas, thin in others, to fold and segment, until the body of the embryo appeared. Wolff called this **epigenesis** (origin upon or after), an idea that the fertilized egg contains building material only, somehow assembled by an unknown directing force. Current ideas of development are essentially epigenetic in concept, although we know far more about what directs the growth and differentiation. We also realize that a bit of the old preformation idea is still with us, since some materials in germ cells (the nucleic acids) are predestined to guide development and are, in a restricted sense, preformed.

■ DIRECT AND INDIRECT DEVELOPMENT

Even before fertilization, the egg is programmed to follow a special developmental course. The eggs

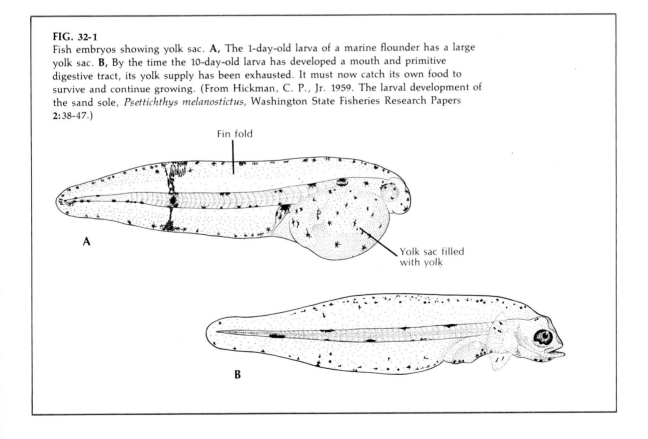

FIG. 32-1
Fish embryos showing yolk sac. **A,** The 1-day-old larva of a marine flounder has a large yolk sac. **B,** By the time the 10-day-old larva has developed a mouth and primitive digestive tract, its yolk supply has been exhausted. It must now catch its own food to survive and continue growing. (From Hickman, C. P., Jr. 1959. The larval development of the sand sole, *Psettichthys melanostictus,* Washington State Fisheries Research Papers **2:**38-47.)

Fin fold

Yolk sac filled with yolk

A

B

of marine invertebrates and those of freshwater vertebrates typically develop into feeding larval forms that look and behave totally different from the mature adult. In such **indirect** development, the larva represents an adaptive detour in development that enables the animal to exploit microplankton or other food supplies and thus enhance its chances for a successful **metamorphosis** to the juvenile-adult body form. The eggs of animals having indirect development usually contain rather limited food reserves and are programmed for rapid differentiation into tiny, free-swimming larvae that must find their own food to sustain further growth (Fig. 32-1).

The eggs of mammals, birds, and reptiles show **direct** development; that is, the egg develops directly into a juvenile without passing through a highly specialized larval phase. In direct development, either the egg contains a vast amount of stored food, as do the huge, yolky bird and reptile eggs, or the early embryo develops a nourishing, parasitic relationship with the maternal body, as do the small, nearly yolkless eggs of mammals. Direct development is more leisurely than indirect. The eggs are programmed to develop first into extensive sheets of unspecialized cells and then secondarily differentiate into tissues and organs of a juvenile.

■ OOCYTE MATURATION AND ACTIVATION
Growth and reductional divisions

During **oogenesis,** described in the preceding chapter, the egg becomes a highly specialized, very large cell containing condensed food reserves for subsequent growth. It is of interest that the prolonged growth and enormous accumulation of food reserves in the oocyte occurs **before** the meiotic, or maturation, divisions begin. This contrasts with spermatogenesis in which differentiation of the mature sperm occurs only **after** the meiotic divisions. When the maturation divisions do occur, once the growth phase is complete, they are highly unequal. The first of the two divisions produces one very small **polar body,** containing one set of chromosomes but a small quantity of cytoplasm, and one very large **secondary oocyte,** containing the other set of chromosomes and nearly all the cytoplasm. The second meiotic division proceeds in the same way and with the same result: the now **mature ovum** has retained a haploid set of chromosomes and the vast

bulk of cytoplasm, and a **second polar body** also contains a haploid chromosome set but hardly any cytoplasm. Meanwhile, the first polar body often divides again, so that three polar bodies in all are created, each containing a haploid set of chromosomes. But because all lack stored nutrients, none are capable of further development. This obviously undemocratic hoarding of all accumulated food reserves by just one of four otherwise genetically equal cells is of course a device for avoiding a decrease in ovum size, once all the nutrients have been packaged inside at the end of the growth phase.

In most vertebrates, the egg does not actually complete all the meiotic divisions before fertilization occurs. The general rule is that the egg completes the first meiotic division and proceeds to the metaphase stage of the second meiotic division, at which point further progress stops. The second meiotic division is completed and the second polar body extruded only if the egg is activated by fertilization.

Yolk deposition and a problem of size

The most obvious feature of egg maturation is the deposition of yolk. Yolk, usually stored as granules or more organized platelets, is no definite chemical substance, but may be lipid or protein or both. In insects and vertebrates, all having more or less yolky eggs, the yolk may be synthesized within the egg from raw materials supplied by the surrounding follicle cells, or preformed lipid or protein yolk may be transferred by pinocytosis from follicle cells to the oocyte. In the latter, the follicle cells are more important in regulating egg growth than factors within the egg itself. The result of the enormous accumulation of yolk granules and other nutrients (glycogen and lipid droplets) is that an egg grows well beyond the normal limits that force ordinary body (somatic) cells to divide. A young frog oocyte 50 μ in diameter, for example, grows to 1,500 μ in diameter when mature after 3 years of growth in the ovary, and the volume has increased by a factor of 27,000. Bird eggs attain even greater absolute size; a hen egg will increase 200 times in volume in only the last 6 to 14 days of rapid growth preceding ovulation.

Thus eggs are remarkable exceptions to the otherwise universal rule that organisms are composed of relatively minute cellular units. This creates a surface area–to–cell volume ratio problem, since everything that enters and leaves the ovum (nutrients, respiratory gases, wastes, etc.) must pass through the cell membrane. As the egg becomes larger, the available surface per unit of cytoplasmic volume (mass) becomes smaller. As we would anticipate, the metabolic rate of the egg gradually diminishes until, when mature, the ovum is in a sort of suspended animation awaiting fertilization. However large size is not the only factor leading to quiescence. There is increasing evidence that enzyme and nucleic acid inhibitors, which directly repress metabolic and synthetic activity in the egg, appear toward the end of maturation.

The nucleus and nucleic acid accumulation

The nucleus also grows rapidly in size during egg maturation, although not as much as the cell as a whole. It becomes bloated with nuclear sap and so changed in appearance that it was often given another name in older literature, the **germinal vesicle.** Large amounts of both DNA and RNA accumulate during oogenesis. Early in the oogenesis of large amphibian eggs, the chromosomes become vastly expanded by thin loops thrown out laterally from the chromosomal axis; because of their fuzzy appearance, the chromosomes at this time are called **lampbrush chromosomes** (see Fig. 33-6). It is believed the messenger RNA is being rapidly synthesized on DNA templates as each chromomere puffs out in lampbrush loops, exposing the double helix of DNA. The mRNA subsequently controls the synthesis of protein in the oocyte. Ribosomal RNA is also undergoing intense synthesis by the nucleolus during oocyte growth. In addition, a considerable quantity of DNA appears in the cytoplasm where it becomes bound to mitochondria and yolk platelets. Its function is unknown. Toward the end of oocyte maturation, this intense nucleic acid and protein synthesis gradually winds down. The lampbrush chromosomes contract and migrate to just beneath the egg surface, in preparation for the reduction divisions. The ribosomal RNA and proteins mix with the egg cytoplasm.

The oocyte is now poised for the reductional (meiotic) divisions, which rid it of excess chromosomal material (in the polar bodies, described above) and ready it for fertilization. A vast amount of synthetic activity has preceded this stage, all of which has packed the egg with reserve materials required for subsequent development.

Fertilization and activation

The fusion of male and female gametes, described in the previous chapter, is a complex process that accomplishes (1) the insertion of a male haploid nucleus into the egg cytoplasm and (2) the activation of the quiescent egg to restore metabolic activity and start cleavage. We have already seen that the eggs of many species can be artificially induced to develop without sperm fertilization and that some species exhibit natural parthenogenesis (such as rotifers, aphids, bees, and wasps [p. 722]). Obviously the paternal genome is not essential for activation, although in the majority of species, artificial parthenogenesis can seldom be taken past the late cleavage or blastula stage. The artificially activated egg may begin what appears to be normal development, but cleavage soon becomes abnormal and further growth ceases. One reason is that the sperm normally contributes the **centriole** needed to form a normal mitotic spindle. For reasons not understood, the egg centriole disappears during the maturation divisions or at least ceases to function. If either natural or artificial parthenogenesis is to be successful, the egg centriole must be rendered capable of division. This is why in artificial parthenogenesis experiments a so-called second factor of biologic origin (such as blood or tissue fraction) must be introduced into the egg cytoplasm. Just exactly what this second factor contributes and whether it produces a new self-replicating centriole are not known.

Activation is a dramatic event. In sea urchin eggs, the contact of the spermatozoon with the egg surface sets off almost instantaneous changes in the egg cortex. At the point of contact, a **fertilization cone** appears into which the sperm head is later drawn. From this point, a visible change travels wavelike across the egg surface, causing immediate elevation of a **fertilization membrane** (see Fig. 31-10). At the same time **cortical granules,** which form a layer beneath the plasma membrane, explode, releasing materials that fuse together to build up a new egg surface, the **hyaline layer.** This lies between the inner plasma membrane and the outer fertilization membrane. The **cortical reaction,** as these changes are called, is a crucial event in development. It seems to produce a complete molecular reorganization of the egg cortex. It also serves to remove one or more inhibitors that have blocked the energy-yielding systems and protein synthesis in the egg and kept the egg in its quiescent, suspended-animation state. Almost immediately after the cortical reaction, polyribosomes form from the enormous supply of monoribosomes stored in the cytoplasm and begin producing protein. Normal metabolic activity is restored. The reduction divisions, if not completed, are brought to completion. The male and female pronuclei then fuse and the egg enters into cleavage.

■ CLEAVAGE AND EARLY DEVELOPMENT

During cleavage, the egg divides repeatedly to convert the large, unwieldy cytoplasmic mass into a large number of small, maneuverable cells (called **blastomeres**) clustered together like a mass of soap bubbles. There is no growth during this period, only subdivision of mass, which continues until normal cell size and nucleocytoplasmic ratios are attained. At the end of cleavage the egg has been divided into many hundreds or thousands of cells (about 1,000 in polychaete worms, 9,000 in amphioxus, and 700,000 in frogs). There is a rapid increase in DNA content during cleavage as the number of nuclei and the amount of DNA are doubled with each division. Apart from this, there is little change in chemical composition or displacement of constituent parts of the egg cytoplasm during cleavage. **Polarity,** that is, a polar axis, is present in the egg, and this establishes the direction of cleavage and subsequent differentiation of the embryo. Usually cleavage is very regular although enormously affected by the quantity of yolk present and whether cleavage is radial or spiral, as earlier described (p. 87).

FIG. 32-2

Indeterminate and determinate (mosaic) cleavage. **A,** Indeterminate cleavage. Each of the early blastomeres (such as that of the sea urchin) when separated from the others will develop into a small pluteus larva. **B,** Determinate (mosaic) cleavage. In the mollusk (such as *Dentalium),* when the blastomeres are separated, each will give rise to only a part of an embryo. The larger size of one of the defective larvae is the result of the formation of a polar lobe *(P)* composed of a clear cytoplasm of the vegetal pole, which this blastomere alone receives.

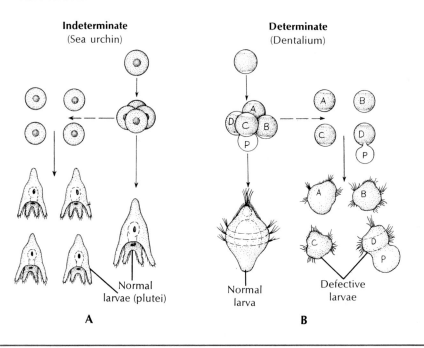

Indeterminate
(Sea urchin)

Determinate
(Dentalium)

Normal
larvae (plutei)

Normal
larva

Defective
larvae

A

B

Determinant and indeterminant cleavage

In many species, especially those showing radial cleavage, early blastomeres, if separated from each other, will each give rise to a whole larva. This is called **indeterminative cleavage,** meaning that the first blastomeres to form are equipotent; there has been no segregation into potentially different histogenetic regions. Indeterminate cleavage is typical of echinoderm, coelenterate, and vertebrate eggs. Thus if a sea urchin embryo at the four-cell stage is placed in calcium-free seawater and gently shaken, the blastomeres will fall apart. Replaced in normal seawater, each will subsequently develop into a complete, though small, pluteus larva fully capable of growing into an adult sea urchin (Fig. 32-2). Early blastomeres of frog and rabbit embryos also can be separated and yield complete embryos. Indeterminate cleavage is also called **regulative.** The reason is that each of the first four or eight blastomeres is capable of regulating its developmental fate to produce a portion of a larva if it develops in the company of other blastomeres, or a whole larva if it is forced to develop alone.

The eggs of many invertebrates lack this early versatility. If the early blastomeres of a mollusk, annelid, flatworm, or ascidian are separated, each will give rise only to a part of an embryo in accord with their original fate. This is called **determinative** (or mosaic) **cleavage.** Such blastomeres appear to have a fixed informational content as soon as cleavage begins. The terms "determinative" and "indeterminative" can be used only in a provisional sense, since there are many eggs that do not fit clearly into either category.

The explanation behind these two types of cleavage seems to lie in the extent to which early blastomeres depend exclusively on information segregated by the early divisions (determinative cleavage) or whether they are able to fall back upon a supply of information from the nucleus (indeterminative cleavage). In the determinate cleavage type the information-containing material is localized so early that each blastomere receives a portion that is qualitatively unique. With such mosaic (determinative) eggs it is often possible to map out the fates of specific areas on the egg surface that are known to be presumptive for specific structures, such as germ layers, notochord, and nervous system.

From the evidence of identical twins, man ap-parently has the indeterminative type of cleavage, for identical twins come from the same zygote. Of course, the totipotency (the capacity to develop into a complete embryo) of the blastomeres in indeterminative cleavage is strictly limited. After the first three or four cleavages, the fate of the blastomeres becomes fixed.

Protein synthesis and the masked messenger hypothesis

There is abundant evidence that both determinative and indeterminative cleavage are **normally** carried along by cell reserves previously stored in the egg cytoplasm and with no guidance from nuclear genes. However proteins are being freshly synthesized during cleavage to be used in building plasma membranes for new cell surfaces, cleavage asters, and perhaps some enzymes. Protein synthesis obviously requires the participation of nucleic acids: transfer RNA to pick up amino acids from the cellular pool, ribosomal RNA for building the protein chain, and messenger RNA to supply the code. All of these nucleic acids are available, having been synthesized during egg maturation. Accordingly, the first proteins to be synthesized during cleavage are built from stored RNA and the amino acid pool; the genetic information on the nuclear chromosomes is not yet required.

Since all the materials and templates for building proteins are present in the unfertilized egg, but will not support protein synthesis until fertilization, there may be that one or more controlling elements are specifically protected or blocked until cleavage begins. According to the "masked-messenger" hypothesis (a description that carries spy-thriller undertones) the messenger RNA templates are inactivated (masked) by a protein until fertilization. At this time, the RNA templates are unmasked, perhaps by some change associated with the cortical reaction, and protein synthesis can begin.

Significance of the cortex

We have seen that cleavage proceeds independent of nuclear genetic information, guided instead by information deposited in the egg during its maturation. It was once believed that the visible particulate material in the cytoplasm had determinative properties. However, it was soon discovered that if the egg was strongly centrifuged so that everything inside—

nucleus, mitochondria, lipid droplets, yolk, and other inclusions—is thoroughly displaced, the embryo still develops perfectly. If sea urchin eggs are examined by electron microscope after being centrifuged for 5 minutes at several thousand times the force of gravity, the only thing not affected is the plasma membrane and a gel-like layer just beneath the plasma membrane **(plasmagel layer).** Yet development proceeds normally. This and similar experiments show conclusively that the plasmagel (or cortical) layer of the egg contains an invisible but dynamic organization that determines the pattern of cleavage. Cortical organization is at first labile (especially so in indeterminate eggs) but soon becomes regionally fixed and irreversible. Thus, as cleavage progresses, the cortex becomes segregated into territories having specific determinative properties. This explains why different blastomeres bear different cytodifferentiation properties.

Delayed nucleation experiments

Another kind of experiment that demonstrates the importance of specific cortical regions of the egg was first carried out many years ago by Hans Spemann, a German embryologist. Spemann put ligatures of human hair around newt eggs (amphibian eggs similar to frog eggs) just as they were about to divide, constricting them until they were almost, but not quite, separated into two halves (Fig. 32-3). The nucleus lay in one half of the partially divided egg; the other side was anucleate, containing only cytoplasm. The egg then completed its first cleavage division on the side containing the nucleus; the anucleate side remained undivided. Eventually, when the nucleated side had divided into about 16 cells, one of the cleavage nuclei would wander across the narrow cytoplasmic bridge to the anucleate side. Immediately this side began to divide. Now with both halves of the embryo containing nuclei, Spemann drew the liga-

FIG. 32-3

Spemann's delayed nucleation experiments. Two kinds of experiments were performed. **A,** Hair ligature was used to partly divide an uncleaved fertilized newt egg. Both sides contained part of the gray crescent. Nucleated side alone cleaved until a descendent nucleus crossed over the cytoplasmic bridge. Then both sides completed cleavage and formed two complete embryos. **B,** Hair ligature was placed so that the nucleus and gray crescent were completely separated. Side lacking the gray crescent became an unorganized piece of belly tissue; other side developed normally.

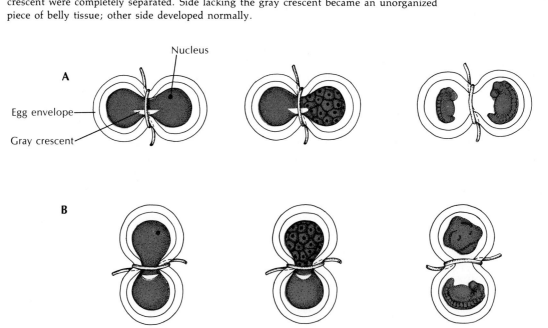

ture tight, separating the two halves of the embryo. He then watched their development. Usually two complete embryos resulted. Although the one embryo possessed only $1/16$ the original nuclear material, and the other contained $15/16$, they both developed normally. The $1/16$ embryo was initially smaller, but caught up by about 140 days. This proves that every nucleus of the 16-cell embryo contains a complete set of genes; all are equivalent.

Sometimes, however, Spemann observed that the nucleated half of the embryo developed only into an abnormal ball of "belly" tissue, although the half that received the delayed nucleus developed normally. Why should the more generously endowed $15/16$ embryo fail to develop and the small $1/16$ embryo live? The explanation, Spemann discovered, depended on the position of the **gray crescent,** a crescent-shaped, pigment-free area on the egg surface. In amphibian eggs the gray crescent forms at the moment of fertilization and determines the plane of bilateral symmetry of the future animal. If one half of the constricted embryo lacked any part of the gray crescent, it would not develop. Obviously, then, there must be cytoplasmic inequalities involved. The egg cortex in the area of the gray crescent contains substances that are essential for normal development. Since all the nuclei of the 16-cell embryo are equivalent, each capable of supporting full development, it is clear that the cytoplasmic environment is crucial to nuclear expression. The nuclei are all alike, but the cytoplasm (or cortex) throughout the embryo is not all alike. In some way chemically different regions of the egg, created during the early growth of the egg (oogenesis), are segregated out into specific cells during early cleavage. Thus, although all nuclei have the same information content, cytoplasmic substances surrounding the nucleus determine what part of the genome will be expressed and when.

■ PROBLEM OF DIFFERENTIATION

The egg cortex contains the primary guiding force for early development (cleavage and blastula formation) and synthetic activities during this period are supported by nucleic acids and reserves laid down in the egg cytoplasm *before* fertilization. However, continued differentiation past the blastula stage requires the action of genetic information in the nuclear chromosomes. There is considerable evidence to show that this is so. If the maternal nucleus is removed

from an activated, but unfertilized frog egg, the anucleate egg will cleave more or less normally but arrest at the outset of gastrulation, when nuclear information is required. **Hybridization experiments** have also been carried out in which the maternal nucleus is removed from the egg by microsurgery, just after the egg has been fertilized, but before the pronuclei have time to fuse. Such a zygote, called an **andromerogone,** contains maternal cytoplasm and a paternal (sperm) nucleus. The egg will cleave to the blastula stage normally; whether it will gastrulate depends on the compatibility of the sperm with the maternal cytoplasm. If the egg was fertilized with a sperm of the same species, it will develop into a haploid (but otherwise normal) tadpole. If it was fertilized with a sperm of a closely related species, it may develop to the neurula stage before development stops. If fertilized with a distantly related species, development will arrest in the blastula or early gastrula stage. These differences are explained as varying degrees of immunologic incompatibilities that appear when the paternal (sperm) genes begin coding for proteins in the embryo. The more foreign the sperm, the greater the incompatibility and the earlier the stage of arrest.

Finally, the most direct evidence that nuclear genes become active at gastrulation is that the production of messenger RNA increases sharply at this time. This can be demonstrated with the use of radioactively labeled RNA precursors.

Gene expression

Gastrulation is a critical time of orderly and integrated cell movements. By folding processes (invagination and evagination) that segregate single epithelial layers and by splitting processes (delamination) that segregate multiple layers, the three prospective germ layers—ectoderm, mesoderm and endoderm—are formed. This is followed by the rapid differentiation of germ layers into rudimentary, and later functional, tissues and organs. As cells differentiate, they use only a part of the instructions their nuclei contain. Cells that are differentiating into a thyroid gland are not concerned with that part of the genome that codes for a striated muscle, for example. The unneeded genes are in some way switched off. Are the unused genes destroyed? Recent **nuclear transplantation** experiments strongly indicate that they are not. In 1952, R. Briggs and T. J. King in the United States developed a technique for surgically extracting the nucleus from an unfertilized frog egg, or inactivating it with ultraviolet light. With a micropipette they introduced a substitute nucleus taken from the cell of a frog embryo. More recently J. B. Gurdon and R. Laskey in England have shown that if the substitute nucleus was taken from a blastula cell it supported development of the egg to the tadpole stage in 80% of the transplants. Nuclei from older embryos did not support development as well, but even the nucleus from a fully differentiated intestinal cell from a tadpole supported development in **some** cases. There are various suggestions why normal development declined as the age of the donor nuclei increased: many of the nuclei were damaged by the transfer procedure; also cellular division rates were vastly different between the egg (rapid division) and the differentiated tissue from which the donor nucleus was obtained (very slow division). Despite these difficulties, these experiments are of great significance because they demonstrate that the nucleus of a differentiated cell can be forced backward from its specialized state, and once again make available all its genetic information.

The basic problem is **gene expression.** What determines that a particular blastomere of say, a 100-cell embryo, will differentiate into muscle or skin or thyroid gland? Presumably the set of genes responsible for the development of a thyroid gland will be set in motion by the chemical environment found **only** in the region of the future thyroid gland. But how can such a unique chemical environment be created unless some **previous** genic action made the thyroid region different from the rest of the body? And even this earlier genic action must have been expressed in a unique chemical environment, or else thyroid glands would grow all over the body.

It is easy to see that this kind of argument quickly takes one back to the fertilized egg itself. If genes are the same in all nuclei of the early embryo, then the only way differences can develop is through some interaction between these nuclei and the surrounding cytoplasm. We have already seen that the basic polarity of the egg and the organizing qualities of the egg cortex provide an early opportunity for such interactions. But there is much mystery surrounding many aspects of gene expression, stimulating the present-day interest in nucleocytoplasmic interactions.

■ EMBRYONIC INDUCTORS AND ORGANIZERS

Embryonic induction is a widespread phenomenon in development. It is the capacity of one tissue to evoke a specific developmental response in another. The classic experiments were reported in 1924 by H. Spemann and O. Mangold. When a piece of dorsal blastopore lip from a salamander gastrula is transplanted into a ventral or lateral position of another salamander gastrula, it invaginates and develops a notochord and muscle somites. It also induces the **host** ectoderm to form a neural tube. Eventually a whole system of organs develop where the graft was placed, and this grows into a nearly complete secondary embryo (Fig. 32-4). This creature is composed partly of grafted tissue and partly of induced host tissue. It was soon found that **only** grafts from the dorsal lip of the blastopore were capable of inducing the formation of a complete or nearly complete secondary embryo. This area corresponds to the presumptive areas of notochord, muscle somites, and prechordal plate. It was also found that only ectoderm of the host could be induced to develop a nervous system in the graft and that the reactive ability is greatest at the early gastrula stage and declines as the recipient embryo gets older.

Spemann called the dorsal lip area the **primary organizer** because it was the only region capable of inducing the development of a complete embryo in the host. This region can be traced back to the gray crescent of the undivided fertilized egg, a cortical area that, as we have already seen, is crucial in directing early development of amphibian embryos (Fig. 32-3). Many other examples of embryonic induction have been discovered, both in amphibians, a favorite material for study, as well as in other vertebrate and invertebrate species.

Efforts to discover the chemical nature of the inductor have met with much less success. Embryologists were dismayed to find that a great variety of denatured animal materials could cause inductive responses. Obviously the dorsal lip graft did not "organize" a particular differentiation; rather it evoked a response that was already part of the induced tissue's total developmental capacity and repertoire. The tissue response is quite specific even though it can be evoked by a variety of unrelated chemical and physical substances.

In normal development, induction occurs with

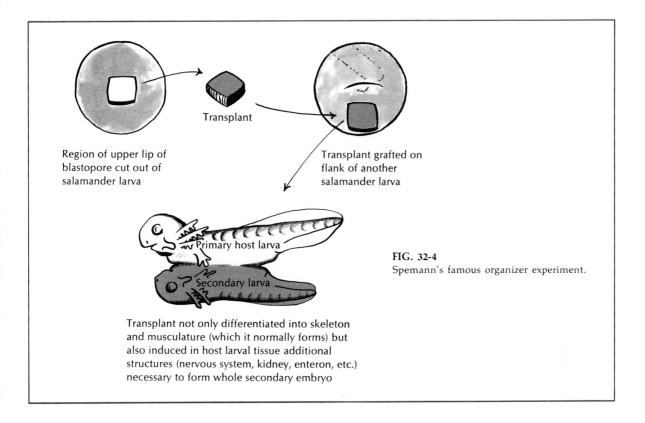

Transplant

Region of upper lip of
blastopore cut out of
salamander larva

Transplant grafted on
flank of another
salamander larva

Primary host larva

Secondary larva

FIG. 32-4
Spemann's famous organizer experiment.

Transplant not only differentiated into skeleton
and musculature (which it normally forms) but
also induced in host larval tissue additional
structures (nervous system, kidney, enteron, etc.)
necessary to form whole secondary embryo

contact between specific territories. Usually a tissue
that has differentiated rather early acts as an inductor for an adjacent undifferentiated tissue. Yet although induction plays an important role in tissue
differentiation, it is known that many kinds of tissue
or organs develop without induction. Recent work
suggests that where induction occurs, the inducing
principal is of protein nature. It is suggested that the
protein inductors interfere with DNA coding in some
way to change the synthetic processes in the reacting
cells. The issue is so complex that embryologists find
themselves with little more insight into the induction
question than they had some 40 years ago.

■ AMNIOTES AND THE AMNIOTIC EGG

Reptiles, birds, and mammals form a natural
grouping of land vertebrates distinguished by having
an amniotic egg. They are called **amniotes,** meaning
that they develop an amnion, one of the extraembryonic membranes that make the development of
these forms unique among animals.

As rapidly growing living organisms, embryos
have the same basic animal requirements as adults—

food, oxygen, and disposal of wastes. For the embryos of marine invertebrates that show indirect development, gas exchange is a simple matter of direct
diffusion. Food can be acquired as soon as the embryo develops a mouth and begins feeding on plankton. All eggs of aquatic animals are provided with
just enough stored yolk to allow growth to this critical stage. Beyond this point, the embryo (now called
a free-swimming **larva**) is on its own. Yolk enclosed
in a membranous **yolk sac** is a conspicuous feature
of all fish embryos (Fig. 32-1). The yolk is gradually
used up as the embryo grows; the yolk sac shrinks
and finally is enclosed within the body of the embryo. The mass of yolk is an **extraembryonic** structure, since it is not really a part of the embryo
proper, and the yolk sac is an **extraembryonic membrane.** Bird and reptile eggs are also provided with
large amounts of yolk to support early development.
In birds (direct development), the yolk reaches relatively massive proportions, since it must nourish a
baby bird to a much more advanced stage of growth
at hatching than that of a larval fish.

In abandoning an aquatic life for a land existence

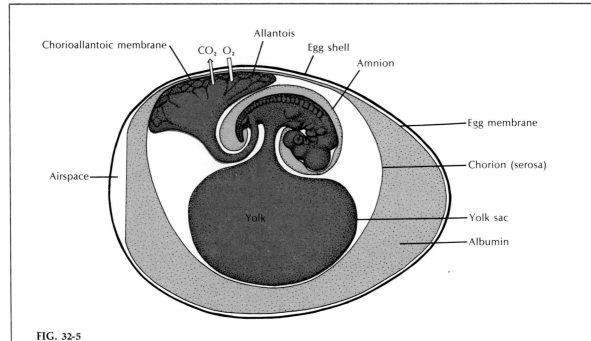

FIG. 32-5
Amniotic egg at early stage of development showing a chick embryo and its
extraembryonic membranes. Porous shell allows gaseous exchange of oxygen and carbon
dioxide. Circulatory channels from embryo's body to allantois and yolk sac are not
shown.

the first terrestrial animals had to evolve a sophisticated egg containing a complete set of life-support
systems. Thus appeared the **amniotic egg,** equipped
to protect and support the growth of embryos on dry
land. In addition to the **yolk sac** containing the
nourishing yolk are three other membranous sacs—
amnion, chorion, and **allantois.** All are referred to
as extraembryonic membranes because, again, they
are accessory structures that develop beyond the embryonic body and are discarded when the embryo
hatches.

The **amnion** is a fluid-filled bag that encloses the
embryo and provides a private aquarium for development (Fig. 32-5). Floating freely in this aquatic environment, the embryo is fully protected from shocks
and adhesions. The evolution of this structure, from
which the amniotic egg takes its name, was crucial
to the successful habitation of land. The **allantois,**
another component in the support system for embryos of land animals, is a bag that grows out of the
hindgut of the embryo (Fig. 32-5). It collects the

wastes of metabolism (mostly uric acid). At hatching, the young animal breaks its connection with the
allantois and leaves it and its refuse behind in the
shell. The **chorion** (also called **serosa**) is an outermost extraembryonic membrane that completely encloses the rest of the embryonic system. It lies just
beneath the shell (Fig. 32-5). As the embryo grows
and its need for oxygen increases, the allantois and
chorion fuse together to form a **chorioallantoic
membrane.** This double membrane is provided with
a rich vascular network, connected to the embryonic
circulation. Lying just beneath the porous shell, the
vascular chorioallantoic membrane serves as a kind
of lung across which oxygen and carbon dioxide can
freely exchange. And although nature did not plan
for it, the chorioallantoic membrane of the chicken
egg has been used extensively by generations of experimental embryologists as a grafting site for small
explants of young chick embryos in order to easily
observe their development.

The great importance of the amniotic egg to the

establishment of a land existence cannot be overemphasized. Amphibians must return to water to lay their eggs. But the reptiles, even before they took to land, developed the amniotic egg with its self-contained aquatic environment enclosed by a tough outer shell. Protected from drying out and provided with yolk for nourishment, such eggs could be laid on dry land, far from water. Reptiles were thus freed from aquatic life and could become the first true terrestrial tetrapods.

Incidentally, the sexual act itself comes from the requirement that the egg be fertilized **before** the egg shell is wrapped around it, if it is to develop. Thus the male must introduce the sperm into the female tract so that the sperm can reach the egg before it passes to that part of the oviduct where the shell is secreted. Hence, as one biologist puts it, it is the egg shell, and not the devil, that deserves the blame for the happy event we know as sex.

■ HUMAN DEVELOPMENT

The amniotic egg, for all its virtues, has one basic flaw: placed neatly in a nest, it makes fine food for other animals. It was left for the mammals to evolve the best solution for early development: allow the embryo to grow within the protective confines of the mother's body. This has resulted in important modifications in mammalian development as compared with other vertebrates. The earliest mammals, descended from early reptiles, were egg layers. Even today the most primitive mammals, the monotremes (for example, duckbill platypus, spiny anteater), lay large yolky eggs that closely resemble bird eggs. In the marsupials (pouched mammals such as the opossum and kangaroo), the embryos develop for a time within the mother's uterus. But the embryo does not "take root" in the uterine wall, as do the embryos of the more advanced **placental mammals,** and consequently it receives little nourishment from the mother. The young of marsupials are therefore born immature and are sheltered and nourished in a pouch of the abdominal wall.

All other mammals, the placentalians, nourish their young in the uterus by means of a **placenta.**

Early development of human embryo

The eggs of all placental mammals, though relatively enormous on the cellular scale of things, are small by egg standards. The human egg is about 0.1

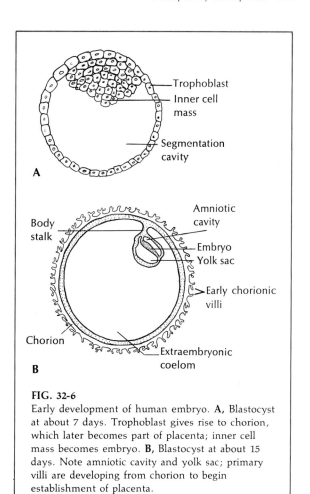

FIG. 32-6
Early development of human embryo. **A,** Blastocyst at about 7 days. Trophoblast gives rise to chorion, which later becomes part of placenta; inner cell mass becomes embryo. **B,** Blastocyst at about 15 days. Note amniotic cavity and yolk sac; primary villi are developing from chorion to begin establishment of placenta.

mm. in diameter, barely visible with the unaided eye. It contains very little yolk. After fertilization in the mouth (ampulla) of the oviduct, the cleaving egg begins a leisurely 5-day journey down the oviduct toward the uterus, propelled by a combination of ciliary action (especially in the ampullary region) and muscular peristalsis. Cleavage is very slow: 24 hours for the first cleavage and 10 to 12 hours for each subsequent cleavage. By comparison, frog eggs cleave once every hour. Cleavage produces a small ball of 20 to 30 cells (called the morula) within which a fluid-filled cavity appears, the **segmentation cavity** (Fig. 32-6). This is comparable to the blastocoele of a frog's egg. The embryo is now called a **blastocyst.** At this point, development of the mammalian embryo departs radically from that of lower vertebrates. A mass of cells, called the **inner cell mass,** appears on one side of the peripheral cell, or **trophoblast,** layer (Fig. 32-6). The inner cell mass will form the embryo, while the surrounding trophoblast will form the placenta. When the blastocyst is about 6 days old and now composed of about 100 cells, it contacts and implants into the uterine endometrium. Very little is known of the forces involved in implantation, or why, incidentally, the intrauterine birth control devices are so effective in preventing successful implantation. Upon contact, the trophoblast cells proliferate rapidly and produce enzymes that break down the epithelium of the uterine endometrium. This allows the blastocyst to sink into the endometrium. By the eleventh or twelfth day the blastocyst, now totally buried, has eroded through the walls of capillaries and small arterioles; this releases a pool of blood that bathes the embryo. At first, the minute embryo derives what nourishment it requires by direct diffusion from the surrounding blood. But very soon, a remarkable fetal-maternal structure, the **placenta,** develops to assume these exchange tasks.

FIG. 32-7
Development of human embryo from 12 days to 12 weeks. The 8-week-old embryo and uterus are drawn to actual size. As the embryo grows, placenta develops into a disklike structure attached to one side of the uterus. Note vestigial yolk sac.

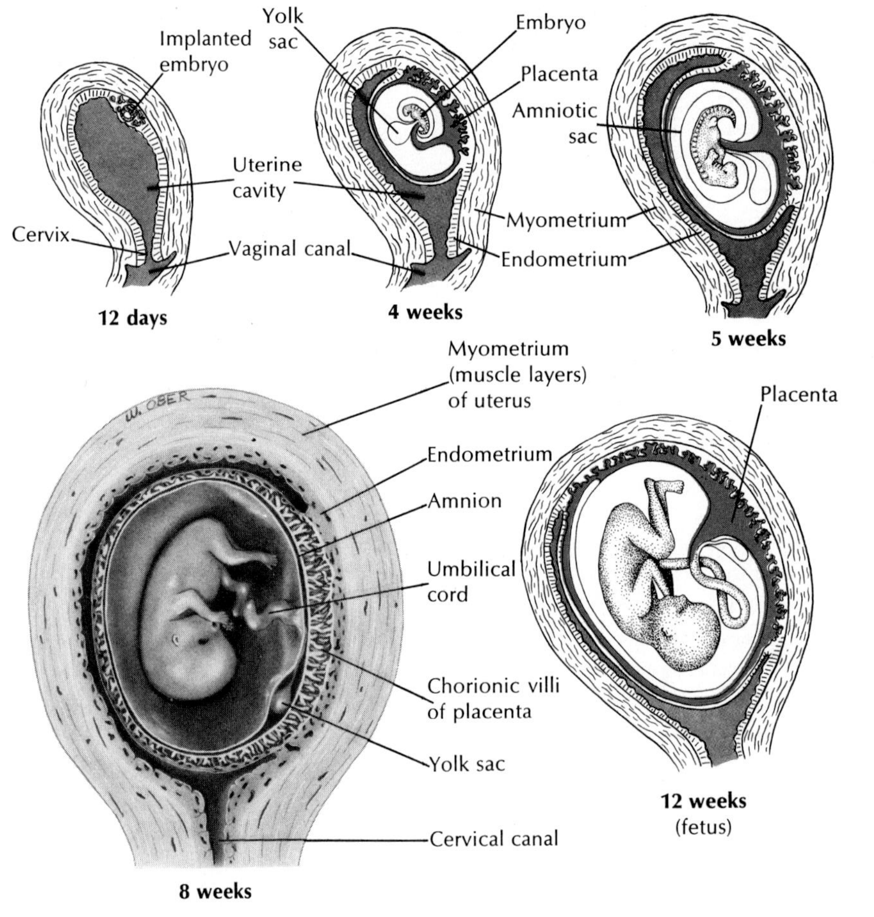

The placenta

The placenta is a marvel of biologic engineering. Serving as a provisional lung, intestine, and kidney for the embryo, it performs elaborate selective activities without ever allowing the maternal and fetal bloods to intermix. The placenta permits the entry of foodstuffs, hormones, vitamins, and oxygen and the exit of carbon dioxide and metabolic wastes. Its action is highly selective since it allows some materials to enter that are chemically quite similar to others that are rejected. The two circulations are physically separated at the placenta by an exceedingly thin membrane only 2 μ thick across which materials are transferred by diffusive interchange. The transfer occurs across thousands of tiny fingerlike projections, called **chorionic villi,** which develop from the original chorion membrane (Figs. 32-6 and 32-7). These projections sink like roots into the uterine endometrium after the embryo implants. As development proceeds and embryonic demands for food and gas exchange increase, the great proliferation of villi in the placenta vastly increases its total surface area. Although the human placenta at term measures only about 7 inches across, its total absorbing surface is about 140 square feet—50 times the surface area of the skin of the newborn infant.

Since the mammalian embryo is protected and nourished by the mother's placenta, what becomes of the various embryonic membranes of the amniotic egg whose functions are no longer required? Surprisingly, perhaps, all of these special membranes are still present, although they may be serving a new function. The yolk sac is retained, empty and purposeless, a vestige of our distant past (Fig. 32-7). Perhaps evolution has not had enough time to discard it. The amnion remains unchanged, a protective water jacket in which the embryo weightlessly floats. The remaining two extraembryonic membranes, the allantois and chorion, have been totally redesigned. The allantois is no longer needed as a urinary bladder. Instead it becomes the stalk, or **umbilical cord,** that links the embryo physically and functionally with the placenta. The chorion, the outermost membrane, forms most of the placenta itself.

One of the most intriguing questions the placenta presents is why it is not rejected by the mother's tissues. The placenta is a uniquely successful foreign transplant, or **allograft.** Since the placenta is an embryonic structure, containing both paternal and ma-

ternal antigens, we should expect it to be rejected by the uterine tissues, just as a piece of a child's skin will be rejected by the child's mother should a surgeon attempt a grafting transplant. The placenta in some way circumvents the normal rejection phenomenon, a matter of the greatest interest to immunologists seeking ways to successfully transplant tissues and organs.

Development of systems and organs

Pregnancy may be divided into four phases: the first phase of 6 or 7 days is the period of cleavage and blastocyst formation and ends when the blasto-

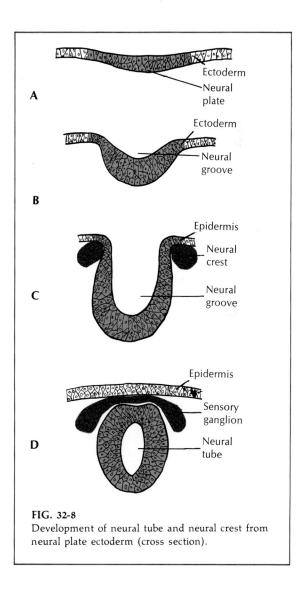

FIG. 32-8
Development of neural tube and neural crest from neural plate ectoderm (cross section).

cyst implants in the uterus. The second phase of about 2 weeks is the period of gastrulation and formation of the neural plate. The third phase, called the **embryonic period,** is a crucial and sensitive period of primary organ system differentiation. This phase ends at about the eighth week of pregnancy. The last phase, known as the **fetal period,** is characterized by rapid growth, proportional changes in body parts, and final preparation for birth.

During gastrulation the three germ layers are formed. These differentiate, as we have seen, first into primordial cell masses and then into specific organs and tissues. During this process, cells become increasingly committed to specific directions of differentiation.

DERIVATIVES OF ECTODERM. The brain, spinal cord, and nearly all the outer epithelial structures of the body develop from the primitive ectoderm. They are among the earliest organs to appear. Just above the notochord, the **ectoderm** thickens to form a **neural plate** (Fig. 32-8). The edges of this plate rise up, fold, and join together at the top to create an elongated, hollow **neural tube.** The neural tube gives rise to most of the nervous system: anteriorly it enlarges and differentiates into the brain, cranial nerves, and eyes; posteriorly it forms the spinal cord and spinal motor nerves. Sensory nerves arise from special **neural crest** cells pinched off from the neural tube before it closes.

How are the billions of nerve axons in the body formed? What directs their growth? Biologists were intrigued with these questions that seemed to have no easy solutions. Since a single nerve axon may be many feet in length (for example, motor nerves running from the spinal cord to the feet), it seemed impossible that a single cell could spin out so far. It was suggested that nerve fibers grew from a series of preformed protoplasmic bridges along its route. The answer had to await the development of one of the most powerful tools available to biologists, the cell culture technique. In 1907 an embryologist Ross G. Harrison discovered that he could culture living neuroblasts (embryonic nerve cells) for weeks outside the body by placing them in a drop of frog lymph hung from the underside of a cover slip. Watching nerves grow for periods of days, he saw that each nerve fiber was the outgrowth of a single cell. As the fibers extended outward, materials for growth flowed down the hollow axon center to the growing

tip, where they are incorporated into new protoplasm. The tissue culture technique is now used extensively by scientists in all fields of active biomedical research, not just by embryologists. The great impact of the technique has been felt only in recent years. Harrison was twice considered for the Nobel Prize (1917 and 1933), but he failed to ever receive the award because, ironically, the tissue culture method was then believed to be "of rather limited value."

The second question—what directs nerve growth—has taken longer to unravel. An idea held well into the 1940s was that nerve growth is a random, diffuse process. It was believed that the nervous system developed as an equipotential network, or blank slate, that later would be shaped by usage into a functional

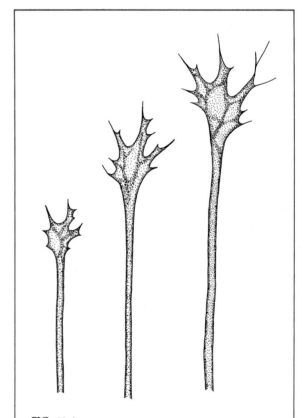

FIG. 32-9
Growth cone at the growing tip of a nerve axon. Materials for growth flow down hollow axon to growth cone from which numerous threadlike pseudopodial processes extend. These appear to serve as a pioneering guidance system for the developing axon.

system. The nervous system just seemed too incredibly complex for one to imagine that nerve fibers could find their way selectively to predetermined destinations. Yet it appears that this is exactly what they do! Recent work indicates that each of the billions of nerve cell axons acquires a chemical identification tag that in some way directs it along a correct path. Many years ago Ross Harrison observed that a growing nerve axon terminated in a "growth cone," from which extend numerous tiny threadlike processes (Fig. 32-9). These are constantly reaching out, testing the environment in all directions, to guide the nerve chemically to its proper destination. This chemical guidepost system, which must, of course, be genetically directed, is just one example of the amazing precision that characterizes the entire process of differentiation.

DERIVATIVES OF ENDODERM. In the frog embryo the primitive gut makes its appearance during gastrulation with the formation of an internal cavity, the **archenteron.** From this simple endodermal cavity develops the lining of the digestive tract, lining of the pharynx and lungs, most of the liver and pancre-

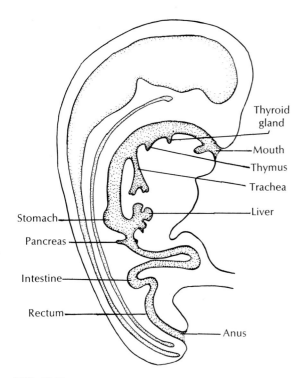

FIG. 32-10
Derivatives of alimentary canal.

as, the thyroid and parathyroid glands, and the thymus.

The **alimentary canal** is early folded off from the yolk sac by the growth and folding of the body wall (Fig. 32-10). The ends of the tube open to the exterior and are lined with ectoderm, whereas the rest of the tube is lined with endoderm. The **lungs, liver,** and **pancreas** arise from the foregut.

Among the most intriguing derivatives of the digestive tract are the pharyngeal (gill) arches, which make their appearance in the early embryonic stages of all vertebrates (Fig. 32-11). In fish the embryonic gill arches will later serve as respiratory organs. In the adults of terrestrial vertebrates the gill arches disappear altogether or become modified beyond recognition. The appearance of gill arches in the embryos of terrestrial vertebrates, and other pronounced similarities between the embryos of fishes and higher vertebrates, can only be explained as an indication of a common vertebrate ancestry. Embryonic development is thus a record, although a considerably modified one, of evolutionary history. We have seen that biologists of the last century were so impressed by embryonic similarities between widely separated vertebrate groups that they used embryonic development to reconstruct lines of evolutionary descent within the animal kingdom. Nevertheless, we can scarcely believe that mammalian embryos retrace vertebrate evolutionary history for the convenience of biologists. Even though the gill arches serve no respiratory function in either the embryos or adults of terrestrial vertebrates, they remain as necessary primordia for a great variety of other structures. For example, the first arch and its pouch (the space between adjacent arches) form the upper and lower jaws and inner ear of higher vertebrates. The second, third, and fourth arches contribute to the tongue, tonsils, parathyroid gland, and thymus. We can understand then why gill arches and other fishlike structures appear in early mammalian embryos. Their original function has been abandoned, but the structures are retained for new purposes. It is the great conservatism of early embryonic development that has so conveniently provided us with a telescoped evolutionary history.

DERIVATIVES OF MESODERM. The intermediate germ layer, the mesoderm, forms the vertebrate skeletal, muscular, and circulatory structures and the kidney. As vertebrates have increased in size and

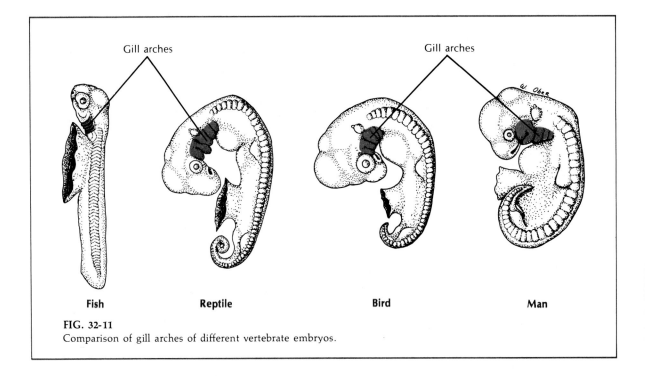

Gill arches

Gill arches

Fish **Reptile** **Bird** **Man**

FIG. 32-11
Comparison of gill arches of different vertebrate embryos.

complexity, the mesodermally derived supportive, movement, and transport structures make up an ever greater proportion of the body bulk.

Most **muscles** arise from the mesoderm along each side of the spinal cord (Fig. 32-12). This mesoderm divides into a linear series of somites (38 in man) that by splitting, fusion, and migration become the muscles of the body and axial parts of the skeleton. The **limbs** begin as buds from the side of the body. Projections of the limb buds develop into digits.

Although the primitive mesoderm appears after the ectoderm and endoderm, it gives rise to the first functional organ, the embryonic heart. Guided by the underlying endoderm, clusters of precardiac mesodermal cells move amebalike into a central position between the underlying primitive gut and the overlying neural tube. Here the heart is established, first as a single, thin tube. Even while the cells group together, the first twitchings are evident. In the chick embryo, a favorite and nearly ideal animal for experimental embryology studies, the primitive heart begins to beat on the second day of the 21-day incubation period—begins beating before any blood vessels have formed and before there is any blood to pump. As the ventricle primordium develops, the

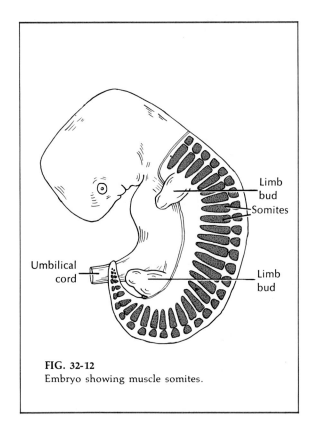

Limb bud

Somites

Umbilical cord

Limb bud

FIG. 32-12
Embryo showing muscle somites.

spontaneous cellular twitchings become coordinated into a feeble, but rhythmic, beat. Then, as the atrium develops behind the ventricle, followed by the sinus venosus behind the atrium, the heart rate quickens. Each new heart chamber has an intrinsic beat faster than its predecessor. Finally, the **sinoatrial** node develops in the sinus venosus and takes command of the entire heartbeat. This becomes the heart's **pacemaker.** As the heart builds up a strong and efficient beat, vascular channels open within the embryo and across the yolk. Within the vessels are the first primitive blood cells suspended in plasma. The early development of the heart and circulation is crucial to continued embryonic development because without a circulation the embryo could not obtain materials for growth. Food is absorbed from the yolk and carried to the embryonic body; oxygen is delivered to all the tissues, and carbon dioxide and other wastes are carried away. The embryo is totally dependent on extraembryonic support systems, and the circulation is the vital link between them.

Birth (parturition)

During pregnancy, the placenta gradually takes over most of the functions of regulating growth and development of the uterus and the fetus. As an endocrine gland it secretes estradiol and progesterone, hormones that are secreted by the ovaries and corpus luteum in the early periods of pregnancy. The placenta also produces **chorionic gonadotropin,** a hormone that now assumes the role of the LH and FSH pituitary hormones. These cease their secretions about the second month of pregnancy. Chorionic gonadotropin maintains the corpus luteum so that the latter may continue secreting progesterone and estradiol necessary for the integrity of the placenta. In the later stages of pregnancy the placenta becomes a totally independent endocrine organ without either the corpus luteum or pituitary.

What stimulates birth? Why does not pregnancy continue indefinitely? What factors produce the onset of labor (the rhythmic contractions of the uterus)? So far, no satisfactory answer can be given to these questions. Since the uterine-placental relationship is dependent mainly on the estradiol-progesterone ratio, it may be that labor is initiated by a shift in hormonal balance leading to a high concentration of estradiol and a low concentration of progesterone. In most mammals progesterone inhibits uterine con-

tractions and estradiol speeds them up. Injections of estradiol will cause abortion in mice but not in the human being. The mechanical effect of uterine muscle stretching (by the increased size of the fetus) may also be a stimulus for labor.

The first major signal that birth is imminent is the so-called labor pains, caused by the rhythmic contractions of the uterine musculature. These are usually slight at first and occur at intervals of 15 to 30 minutes. They gradually become more intense, longer in duration, and more frequent. They may last anywhere from 6 to 24 hours, usually longer with the first child. The object of these contractions is to force the baby from the uterus and birth canal (vagina). Childbirth occurs in three stages. In the first stage the neck (cervix), or opening of the uterus into the vagina, is enlarged by the pressure of the baby in its bag of amniotic fluid, which may be ruptured at this time; in the second stage the baby is forced out of the uterus and through the vagina to the outside; and in the third stage the placenta, or afterbirth, is expelled from the mother's body, usually within 10 minutes after the baby is born.

The fetal circulation (Fig. 32-13) differs in certain particulars from the postnatal circulation. The interchange of nutrients occurs via the **umbilical cord** that is composed of one umbilical vein and two umbilical arteries. The umbilical vein, which carries oxygenated blood from the placenta, passes through the umbilical cord and then to the liver where it joins the portal vein. The blood is now shunted to the right atrium by the **ductus venosus** and the postcava. To bypass the nonfunctioning lungs the blood is shunted from the pulmonary system to the left side of the heart by the **foramen ovale** (a temporary opening between right and left atria) and by the **ductus arteriosus,** which connects the pulmonary artery with the aorta. The two **umbilical arteries** from the internal iliac arteries return blood to the placenta. At birth the foramen ovale closes and the ductus arteriosus becomes the ligament of Botallo; the blood from the right side of the heart is thus forced into the lungs, and the adult circulation is suddenly established.

Many mammals give birth to more than one offspring at a time or to a litter, each member of which has come from a separate egg. Most higher mammals, however, have only one offspring at a time, although occasionally they may have plural young. The armadillo *(Dasypus)* is almost unique among

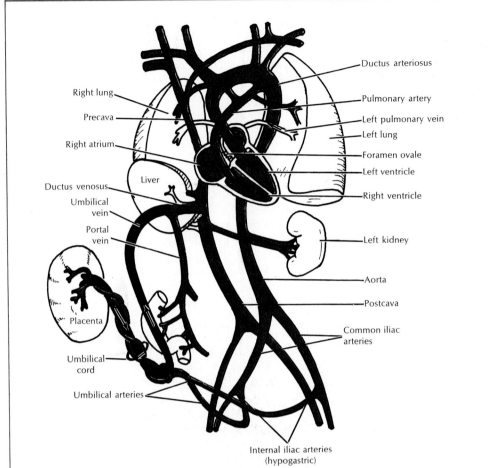

Right lung

Precava

Right atrium

Ductus venosus

Umbilical
vein

Portal
vein

Liver

Placenta

Umbilical
cord

Umbilical arteries

Ductus arteriosus

Pulmonary artery

Left pulmonary vein

Left lung

Foramen ovale

Left ventricle

Right ventricle

Left kidney

Aorta

Postcava

Common iliac
arteries

Internal iliac arteries
(hypogastric)

FIG. 32-13
Fetal circulation. Blood carrying food and oxygen from placenta is brought by umbilical
vein to liver, is shunted through ductus venosus to vena cava where it mixes with
deoxygenated blood returning from fetal tissues, and then is carried to right atrium. Two
fetal shortcuts bypass nonfunctioning lungs. Some blood goes directly from right atrium
to left atrium by an opening, the foramen ovale, and so enters systemic circulation. Some
blood passes from pulmonary trunk into aorta by a temporary duct, the ductus arteriosus.
At birth, placental exchange ceases; ductus venosus, foramen ovale, umbilical arteries,
and ductus arteriosus all close; and regular postnatal pulmonary and systemic circuits take
over.

mammals in giving birth to four young at one time—
all from one egg and all members of a litter of the
same sex, either male or female. Plural or multiple
births occur with a certain regularity in the human
being. Twins, triplets, quadruplets, and quintuplets
are found in human populations. These may come
from a single egg or there may be a separate egg

for each offspring. Most human plural births are
twins. If the twins come from two eggs **(dizygotic)**,
they are called **fraternal twins;** if they come from a
single egg **(monozygotic)**, they are known as **identical twins** and are always of the same sex. Fraternal
twins do not resemble each other more than other
children born separately in the same family, but

identical twins have striking resemblances. Embryologically, each member of fraternal twins has its own placenta, chorion, and amnion. Usually (but not always) identical twins share the same chorion and the same placenta, but each has its own amnion. Sometimes identical twins fail to separate completely and form Siamese twins, in which the organs of one may be a mirror image of the organs of the other. The frequency of twin births to single births is about 1 in 86, that of triplets about 1 in 86^2, and that of quadruplets about 1 in 86^3.

Death

Death may be considered the terminal sequence of the degenerative changes of old age. Death of the organism involves the death of cells, but not all of the cells die at once. Many cells of the body continue to live after the animal has been pronounced dead by certain conventional signs we associate with death. Among protozoans that reproduce by fission, death in the ordinary sense does not occur, since the body is shared by the daughter cells resulting from the division. Life-spans are usually considered more or less definite for most species of animals, but it is very difficult to state with accuracy just how long these life-spans are. Even in the case of man, whose life-span is better known than that of any other animal, only general averages can be given. The average life expectancy has increased in man from about 50 years to around 70 years in the past half century (in the United States [1967] females, 73.8 years; males, 66.7 years) and further increases are expected under better medical and hygienic conditions. However, there is little indication that the upper limit of the life-span of man has increased. So many factors influence the span in all animals that exact figures cannot be given in most cases. Natural death resulting from the degenerative changes of aging rarely occurs in animals other than man, for the debility produced by senescence makes them an easy prey for their enemies. Death, however, considered from any standpoint is a definite part of the hereditary pattern of animals. Specifically, death from whatever cause is the result of the failure of one of the following: heart, blood, or nervous system. Death must occur whenever the vascular system fails to deliver oxygen, vitamins, hormones, and other vital substances. Failure of the nervous system may affect vital centers and quickly result in death.

References
Suggested general readings

Austin, C. R., and R. V. Short (editors). 1972. Embryonic and fetal development. Cambridge, Cambridge University Press. *Succinct and contemporary treatment of mammalian development. Paperback.*

Balinsky, B. I. 1970. An introduction to embryology, ed. 3. Philadelphia, W. B. Saunders Co. *One of the best animal embryology texts at the advanced undergraduate level. Stresses mechanisms of development.*

Berrill, N. J. 1971. Developmental biology. New York, McGraw-Hill Book Co. *Perhaps the best balanced developmental biology text, embracing the entire scope of this complex field, yet readable, well ordered, and with excellent summaries of concepts after each chapter.*

Bonner, J. T. 1958. The evolution of development. New York, Cambridge University Press. *An evolutionary approach to development explaining the mechanisms of the developmental process.*

Bulmer, M. G. 1970. The biology of twinning in man. New York, Oxford University Press, Inc.

Conklin, E. G. 1929. Problems of development. Amer. Naturalist **63**:5-36. *Fascinating and classic essay by a great American embryologist.*

DeBeer, G. R. 1951. Embryos and ancestors. London, Oxford University Press.

Ebert, J. D., and I. M. Sussex. 1970. Interacting systems in development, ed. 2. New York, Holt, Rinehart & Winston, Inc. *Clearly written and balanced undergraduate level paperback.*

Laetsch, W. M. 1969. The biological perspective. Boston, Little, Brown & Co. *Collection of recent research papers dealing with developmental biology, the cell, and evolution, selected for significance and readability.*

Markert, C. L., and H. Ursprung. 1971. Developmental genetics. Englewood Cliffs, N. J., Prentice-Hall, Inc. *Clearly written consideration of the question of gene expression and its control.*

Nelsen, O. E. 1953. Comparative embryology of the vertebrates. New York, The Blakiston Co. *This treatise is a comprehensive study of the comparative morphology of the vertebrates and protochordates. Contains a wealth of comparative information difficult to find in other texts.*

Oppenheimer, J. M. 1967. Essays in the history of embryology and biology. Cambridge, Mass., M. I. T. Press. *A collection of influential writings by distinguished developmental biologists.*

Patten, B. M. 1958. Foundations of embryology. New York, McGraw-Hill Book Co. *Highly readable and well-illustrated descriptive embryology.*

Rugh, R. 1962. Experimental embryology. Minneapolis, Burgess Publishing Co. *A collection of laboratory experiments.*

Spemann, H. 1938. Embryonic development and induction. New Haven, Conn., Yale University Press. *An authoritative and classic work by a pioneer in this aspect of embryology.*

Wood, C. 1969. Human fertility: Threat and promise. World of Science Library, New York, Funk & Wagnalls. *Brief, beautifully illustrated account of human reproduction and development. Stresses fertility, birth control, and population problems.*

Selected *Scientific American* articles

Allen, R. D. 1959. The moment of fertilization. **201**:124-134 (July). *Studies with sea urchin eggs, have revealed the complexity of this critical biologic event.*

Ebert, J. D. 1959. The first heartbeats. **200**:87-96 (March). *The formation and early function of the heart is studied with chick embryos.*

Edwards, R. G. 1966. Mammalian eggs in the laboratory. **215**:72-81 (Aug.).

Edwards, R. G., and R. E. Fowler. 1970. Human embryos in the laboratory. **223**:44-54 (Dec.). *This and the 1966 article by Edwards deal with recent successes in culturing human and other mammalian eggs and embryos for observation.*

Fischberg, M., and A. W. Blackler. 1961. How cells specialize. **205**:124-140 (Sept.). *Early steps in differentiation of the embryo are programmed into the egg before fertilization.*

Gray, G. W. 1957. "The organizer." **197**:79-88 (Nov.). *The history of embryologic studies on the organizing qualities of the blastopore lip of amphibian embryos. The famous experiments of Spemann and others are recounted.*

Gurdon, J. B. 1968. Transplanted nuclei and cell differentiation. **219**:24-35 (Dec.). *An extension of the work first successfully done by Briggs and King.*

Hadorn, E. 1968. Transdetermination of cells. **219**:110-120 (Nov.). *How cells of a fruit fly larva can change predestined fate by being transplanted into an adult fly.*

Hinton, H. E. 1970. Insect eggshells. **223**:84-91 (Aug.). *Complexity revealed with scanning electron microscope.*

Jacobson, M., and R. K. Hunt. 1973. The origins of nerve-cell specificity. **228**:26-35 (Feb.). *How nerves find direction during growth.*

Moscona, A. A. 1961. How cells associate. **205**:142-162 (Sept.). *Describes the forces that promote cellular aggregation and binding.*

Reynolds, S. R. M. 1952. The umbilical cord. **187**:70-74 (July). *Some facts about its performance.*

Singer, M. 1958. The regeneration of body parts. **199**:79-88 (Oct.). *Studies on the remarkable ability of amphibians to regrow lost limbs.*

Taylor, T. G. 1970. How an eggshell is made. **222**:88-95 (March). *How the hen mobilizes body stores of calcium to build eggshells.*

Wessells, N. K., and W. J. Rutter. 1969. Phases in cell differentiation. **220**:36-44 (March). *Cultivation of embryonic pancreas tissues reveals stages of specialization.*

Wessells, N. K. 1971. How living cells change shape. **225**:76-82 (Oct.). *How microtubules and microfilaments make cell movement possible, so important in development.*

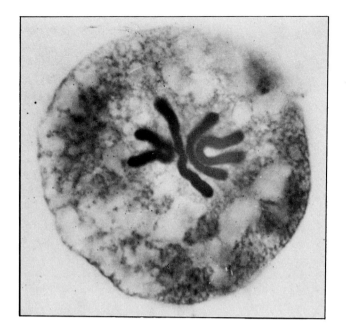

Diploid chromosomes in fertilized egg of Ascaris *preparing for first cleavage.*

CHAPTER 33

PRINCIPLES OF INHERITANCE

■ MEANING OF HEREDITY*

Heredity is one of the great stabilizing agencies in nature. Although offspring and parents in a particular generation may look different, there is nonetheless a basic sameness that runs from generation to generation for any species of plant or animal. But sometimes large variations appear; some are inherited, and others disappear with the generation in which they arose. These inherited characteristics, which may be like or unlike those of the parents, we now know to be attributable to the segregation of hereditary factors. Those not inherited are caused by environmental conditions.

Children are not duplicates of their parents. Some of their characteristics show resemblances to one or

*Refer to pp. 17, 18, 21, and 22 for Principles 3, 4, 27, and 28.

both parents, but they also demonstrate many not found in either parent. What is actually inherited by an offspring from its parents is a certain type of germinal organization **(genes)** that, under the influence of environmental factors, differentiates into the physical characteristics as we see them. We know, too, that heredity can be changed by altering the germ cells, either by changing the genetic constitution or by rearranging chromosome organization.

The inheritance of any characteristic depends on the interaction of many genes. This interaction is often complex, although the individual genes may behave as though they are independent of one another. There is a germinal basis for every characteristic that appears in the development of the organism, such as stature, color of eyes and hair, and intellectual capacity. The germinal organization sets the potential limits of these and other characteristics, but environmental factors, such as nutrition or disease, may greatly affect the physical expression of the characteristics so that their potentialities are never fully realized.

In a unisexual organism, only one parent is involved, as in binary fission or sporulation. Unless there is mutation or other genetic variation, the offspring and parent are genetically alike. But in bisexual reproduction, two sets of genes are pooled together in the zygote, and the offspring shares genetic potentialities of both parents and can produce combinations of traits unlike either parent. Mutation also may play a part here in the alteration of genes.

■ CYTOLOGIC BACKGROUND OF HEREDITY

Heredity is a protoplasmic continuity between parents and offspring. In bisexual animals the gametes are responsible for establishing this continu-

ity. Scientific explanation of genetic principles required a study of the germ cells and their behavior. This meant working backward from certain visible results of inheritance to the responsible mechanism. The nuclei of sex cells were early suspected of furnishing the real answer to the mechanism. This applied especially to certain constituents of the nuclei, the chromosomes, for they appeared to be the only entities passed on in equal quantities from parents to offspring.

When the discovery of Mendel's laws was announced in 1900 (a date that may be considered the beginning of modern genetics), the time was ripe to demonstrate the parallelism that existed between these fundamental laws of inheritance and the cytologic behavior of the chromosomes. In a series of brilliant experiments by Boveri, Sutton, McClung, and Wilson, the mechanism of heredity was definitely assigned to the chromosomes. The next problem was to find how chromosomes affected the hereditary pattern. This study led to an analysis of the chromosome structure and the idea of the gene as the physical basis of hereditary traits. The outstanding work of Thomas Hunt Morgan and his colleagues on the fruit fly *(Drosophila)* led to the mapping of chromosomes in which the location of genes was more or less definitely determined. Out of this work developed the new science of **cytogenetics.**

Nature of chromosomes

Chromosomes for a particular organism have in general a definite size and shape in each stage of their cycle (Figs. 33-1 and 33-2). Their structure is described in some detail in Chapter 5, p. 56. Each chromosome is made up of a central spiral thread **(chromonema)** that bears beadlike enlargements **(chromomeres).** These may be the location of the

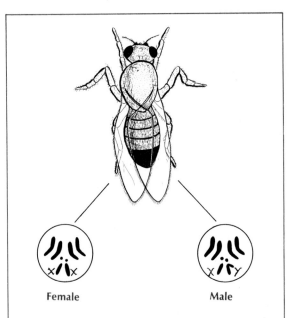

Female   Male

FIG. 33-1
Fruit fly *(Drosophila melanogaster)* and diploid set of chromosomes of each sex. Sex chromosomes X and Y are marked. Male fruitflies have three black bands on abdomen, which is rounded at tip; females have five black bands, with pointed abdomen. Many genetic concepts have been developed from extensive investigations on this now classic animal.

FIG. 33-2
Diploid chromosomes of *Drosophila.*

genes. The various chromomeres appear in regular and constant pattern in a particular chromosome.

Although the exact structure of chromosomes is not fully understood, there is now abundant evidence that they are composed of **nucleoproteins,** which contain the hereditary information. These consist of deoxyribonucleic acid (DNA), ribonucleic acid (RNA), histones or protamines, and some large complex proteins (residual proteins). The histone or protamine is wound around the DNA double helix (p. 768), and the basic residues of the protein are bound to the phosphoric acid residues of DNA. The nucleoprotein molecules are about 4,000 Å long and 40 Å thick and are lined up end-to-end to form long fibers.

Chromosomes constant for a species

Every somatic cell in a given organism contains the same number of chromosomes. Somatic chromosomes of diploid organisms are found in pairs, with the members of each pair being alike in size, in position of spindle attachment, and in bearing genes relating to the same hereditary characters. In each homologous pair of chromosomes, one has come from

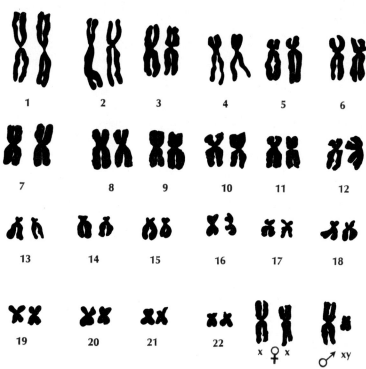

FIG. 33-3

Human chromosomes showing both male and female sex chromosomes. Diploid number is 46. Chromosomes are arranged according to standard pattern (karyotype) of homologous pairs. Chromosomes differ in size, shape, and position of centromeres; arrangement is based on these characteristics, two members of homologous pair (one from each parent) being identical except in case of XY sex chromosomes. Techniques for preparation of human cells for chromosome counting are based upon tissue cultures, biopsy material, and bone marrow studies. Procedures involve tissue exposure to trypsin and special growth media. During part of culture period, colchicine (which arrests cell division at metaphase) is added, and treatment with hypotonic salt solution swells and disperses chromosomes. Squash preparations are usually stained with acetocarmine or Feulgen reagent.

the father and the other from the mother. The lowest diploid number in the cell of any organism is 2, which is found in certain roundworms; the largest number (300 or more) is found in some protozoa. In most forms the number is between 12 and 40, the most common diploid number being 24. Man has 46. Obviously many forms wholly unrelated have the same number of chromosomes and so the number is without significance. What matters is the nature of the genes on the chromosomes. Some chromosomes are as small as 0.25 μ in length and (with the exception of giant salivary chromosomes) some are as long as 50 μ. Within the haploid set of chromosomes of most species there are considerable differences in size and shape. In man, most chromosomes are 4 to 6 μ in length.

Mature germ cells have only one half as many chromosomes as somatic cells because the meiotic divisions have reduced the chromosomes into haploid sets (see pp. 723 to 728).

Chromosomes of man

Until 1956 the diploid number in man was supposed to be 48. Newer and improved techniques now employ cultures of white blood corpuscles that are induced to divide by phytohemagglutinin (an extract of red kidney beans). Other tissue cells such as bone marrow have also been used. With colchicine for arresting mitosis in the metaphase stage and hypotonic solutions and centrifugation for dispersing the chromosomes, it has been possible to arrive at an accurate count of 46 chromosomes (Fig. 33-3). A standard arrangement of the chromosomes according to size and position of centromeres has been adopted. In the somatic cell, such as the white blood corpuscles, there is a pair of each type of chromosome, including 2 X chromosomes (female) and a pair of each type, with the exception of the X and Y chromosomes in the male. The Y chromosome is one of the smallest chromosomes, although there are some autosomes almost as small.

Significance of reduction division

Genes are in paired, homologous chromosomes. When these chromosomes separate at the reduction division of meiosis, the homologous genes must also separate, one gene going to each of the germ cells produced. Thus at the end of the maturation process each mature gamete (egg or sperm) contains one

gene of every pair or a single set of every kind of gene (haploid number), instead of two genes for each character as in somatic cells (diploid number). The haploid number of chromosomes found in the germ cells does not consist of just any half of the diploid somatic number but must include one of the members of each homologous pair of chromosomes. It is a matter of chance in the reduction division whether the paternal chromosome of a homologous pair goes to one daughter cell or the other, and the same is true of the maternal chromosome.

The real significance of the reduction division for explaining the principles of heredity lies mainly in the segregation of the chromosomes and consequently the genes that the chromosomes carry. We have seen that there are two genes for each trait that develops in the individual, but in the mature germ cells there is only one gene for a trait. Of course, when the zygote is formed at fertilization, the homologous pairs of chromosomes (and genes) will be restored. It will be seen later that the factors of the genetic laws behave in a similar manner and were arrived at before the cytologic explanation was forthcoming.

Salivary gland chromosomes

Details about the structure of chromosomes have been difficult to obtain because of their small size in most animals. Since most chromosomes are only a few microns in length, and each one bears thousands of genes, there was little hope of discovering the real nature of the physical units of heredity. About 1934 Professor Painter of the University of Texas and some German investigators independently discovered in the salivary glands of the larvae of *Drosophila* and other flies chromosomes many times larger than those of the ordinary somatic or germinal chromosomes of these forms. Actually, these giant chromosomes had been discovered as early as 1881 by the Italian cytologist Balbiani in the larval forms of the midge fly *Chironomus,* but their real meaning was not detected until they were rediscovered. This rediscovery marked a new era in the development of cytogenetics.

The salivary glands of larval flies are a pair of club-shaped bodies attached to the pharynx. Each gland is composed of only about 100 unusually large cells. Salivary tissue grows by an increase in cell size and not by an increase in cell number. When the cell

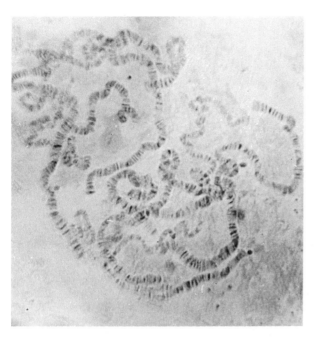

FIG. 33-4

Chromosomes from a salivary gland cell of larval fruit fly *Drosophila*. These are among largest chromosomes found in animal cells. Bands of nucleoproteins may be loci of genes. Such chromosomes are sometimes called polytene because they appear to be made up of many chromonemata. These chromosomes are not confined to salivary glands but are also known to occur in other organs, such as gut and malpighian tubules of most dipteran insects. Technique for their study is simply to crush salivary glands in drop of acetocarmine, between cover glass and slide, so that chromosomes are set free from nuclei and are spread out as shown in photograph. (Courtesy General Biological Supply House, Inc., Chicago.)

membrane is disintegrated, the chromosomes are scattered out and can be easily studied. The giant chromosomes are elongated, ribbonlike bodies about 100 to 200 times longer than the ordinary chromosome (Fig. 33-4). In some flies they lie separated from each other; in *Drosophila* they are attached to a dark mass called the **chromocenter.**

What are these chromosomes and what do they show? The chromosomes are somatic prophase chromosomes with the homologous chromosomes closely paired throughout their length. One of their most striking characteristics is the transverse bands with which they are made. Another feature is the number of chromonemata (central threads) they possess. In the ordinary somatic chromosome there may be only one or two of these gene strings, but in the salivary gland chromosomes there may be between 512 and 1,024 (*Drosophila*). This indicates that the chromonemata may have divided many times without being accompanied by the division of the whole chromosome; hence, they are often called **polytene** chromosomes. A polytene chromosome is a typical mitotic chromosome that has uncoiled and undergone many repeated duplications that have remained together in the same nucleus.

The transverse bands are made up of chromatic granules, the chromomeres. These bands result from the lateral apposition of the chromomeres on the adjacent fibrils or chromonemata. More than 6,000 of these bands have been found on the three large chromosomes of *Drosophila*. The bands contain much DNA and each may be considered the equivalent of the conceptual gene. In the regions between the bands there is little DNA. Another aspect of giant chromosomes is the so-called "puffs," which are local and reversible enlargements in the bands (Fig. 33-5). Each "puff" may be the result of the unfolding or uncoiling of the chromosomes in a band. The puffing may be large (Balbiani's rings) or small. In addition to the DNA of the band, the puff contains a great deal of RNA. The size of the puff is an indication of gene activity. Evidence seems to indicate that messenger RNA is produced at the puff, where it makes a complementary copy of a DNA strand. The RNA messenger is then carried to the ribosome, where it serves for the synthesis of proteins. (A discussion of the genetic code soon follows.)

The even larger lampbrush chromosomes found in the oocytes of many vertebrates and some invertebrates are characterized by loops extending laterally that give them the appearance of a brush (Fig. 33-6). These chromosomes appear to be composed of two chromatids that form loops (gene loci) when they are

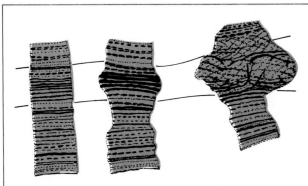

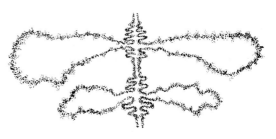

FIG. 33-5

Puffing in one of the bands of a salivary gland chromosome of a midge larva *(Chironomus).* Swelling, or puff, indicates activity in a region where protein and RNA (and perhaps some DNA) are being produced, and it may include single bands or a group of adjacent ones. Puffs always include same bands that occur in a definite sequence during development of larva.

FIG. 33-6

Small portion of a lampbrush chromosome showing two pairs of loops. These chromosomes are found in germinal vesicles (nuclei) of oocytes during diplotene phase of first meiotic division and may indicate synthesis of yolk. They appear to be largest in certain salamanders. Loops represent lateral extensions of chromatids, or half chromosomes. RNA is being transcribed along loop and (with protein formed there) gives a fuzzy appearance to loop. Central axis with closely coiled chromomeres is made up of DNA. Exact relation of loop to gene is not yet known. (Modified from Gall, J. G., 1956. Brookhaven Symp. Biol. **8:**17-21.)

active but are coiled up within a chromomere when at rest.

GENE THEORY

The term **gene** was given by W. Johannsen in 1909 to the hereditary factors of Mendel (1865). Genes are the chemical entities responsible for the hereditary pattern of an organism. No one perhaps has seen a definite gene and we still have much to learn about the nature of genes. Yet, by long, patient genetic experiments, their relative positions (loci) on the chromosomes have been mapped in many cases. Evidence indicates that they are arranged in linear order on the chromosome, like beads on a string. In some cases, as in salivary gland chromosomes, genes are assigned to definite bands. Since chromosomes are few and genes are many in number, each chromosome must contain many genes (linkage group). The range in number of genes an organism may have may be 10,000 *(Drosphilia)* to 90,000 (man).

LINKAGE GROUPS AND ALLELES. Each zygote of sexual reproduction has two sets (diploid) of homologous chromosomes, one set from each parent; in other words, there are two of each kind of chromosomes, one from the father and the other from the mother. When the gametes are formed in meiosis, disjunction of the homologous chromosomes occurs so that each germ cell receives one or the other of the pair at random. Since the genes are a part of the chromosomes, their distribution will parallel that of the chromosomes. All the genes of a particular homologous chromosome, or **linkage group,** will go at meiosis into one gamete and all the genes of the other homologous mate will go into another gamete. If each gene occupies a specific locus in a specific chromosome, all the genes occupying this locus in a given pair of homologous chromosomes are called **alleles.** Just as members of a homologous pair of chromosomes are derived from separate parents, so does each member of a pair of alleles come from a different parent. In some cases a set of alleles may contain more than two members (maybe as many as 20) and such sets are called multiple alleles. But, normally, only two alleles for any one hereditary character may occur in a somatic cell and only one in a gamete.

EXCHANGE OF GENES BETWEEN LINKAGE GROUPS. The two members of a homologous pair of chromosomes often exchange corresponding segments or blocks of genes. This is called **crossing-over.** There is visible evidence of this physical exchange. At the beginning of the first meiotic division the two members of each pair of chromosomes come into side-by-side contact (synapsis) and become twisted. When they separate, they have exchanged parts. Naturally the genes on the traded portions will be exchanged also. The new combinations so found are as stable as the original ones. Linkage groups are also altered by such rearrangements as the linear reversal of gene sequence in the **reversal of a chromosome segment,** by the shifting of a chromosome segment to another part of the same chromosome **(translocation),** by **polyploid changes** in chromosome number, etc.

IMPORTANCE OF GENE MUTATIONS. Although genes can reproduce themselves exactly for many generations, they do occasionally undergo abrupt changes called **mutations.** In such cases the mutant gene now faithfully reproduces itself just as before. Mutations are called **random** because they are unpredictable and because they are unrelated to the needs of the organism, although some mutations are favored by tissue and environmental conditions. Many mutant genes are actually harmful because they may replace adaptive genes that have evolved in the long evolution of the organism. However, a minority of mutant genes are advantageous and have great significance in evolution. Mutation may be a reversible process, and the difference between the mutation of a gene in one direction and its mutation rate in the reverse direction is called its **mutation pressure.** Gene mutations may occur in one direction more frequently than in others, and thus certain mutant alleles are far more common than others. Most mutations ordinarily occur in one gene at a time and thus are called **point mutations.** In the long evolution of any organism, all the genes it carries have had time to mutate and all its present genes are really mutants. The chemical nature of gene mutations is discussed later in this chapter.

PLASMAGENES. There is some evidence that some genetic variability is the result of self-duplicating hereditary units in the cytoplasm. Such units are called **plasmagenes** and are apparently transmitted only by the cytoplasm. Two examples of this type of cytoplasmic inheritance are plant plastids and the "kappa" (killer) substances of *Paramecium,* which

are discussed on p. 792. In some cases the plasmagenes depend on the nuclear genes for their reproduction and maintenance. Plasmagenes can mutate and produce definite characters. They also have mendelian patterns of genetic behavior, but some are distributed more or less at random to daughter cells at cell division. Their exact role in the overall hereditary pattern of organisms is still obscure.

ROLE OF GENES IN CELLULAR ECONOMY. As the chief functional unit of genetic material, genes determine the basic architecture of every cell, the nature and life of the cell, the specific protein syntheses, the enzyme formation, the self-reproduction of the cell, and, directly or indirectly, the entire metabolic function of the cell. By their property to mutate, to be assorted and shuffled around in different combinations, genes have become the basis for our modern interpretation of evolution. Genes are molecular patterns that can maintain their identities for many generations, can be self-duplicated in each generation, and can control cell processes by allowing their specificities to be copied.

■ STORAGE AND TRANSMISSION OF GENETIC INFORMATION

What the gene can do is intimately associated with its chemical structure. The nature of the gene substance has been the subject of intense biologic investigation during the past decade. It is now known that genes, like chromosomes, are made up chiefly of nucleoproteins. Life as we know it really began with the first formation of nucleoproteins, because so far as is known, they are the only molecules with the power of self-duplication.

Structure of nucleic acids

As we have seen, the **nucleoproteins** are macromolecules composed of nucleic acids, special basic proteins called histones and some complex residual proteins. Of particular interest are the **nucleic acids.** Their structure is described in some detail in Chapter 2 (p. 42) and the student is encouraged to review that section at this time. In summary, a nucleic acid is chemically made up of **nucleotides,** which in turn are composed of a purine or pyrimidine base, a sugar, and phosphoric acid. On the basis of the kind of sugar (deoxyribose or ribose), the nucleic acids are divided into two main groups: **deoxyribonucleic acid (DNA)** and **ribonucleic acid (RNA).** DNA occurs

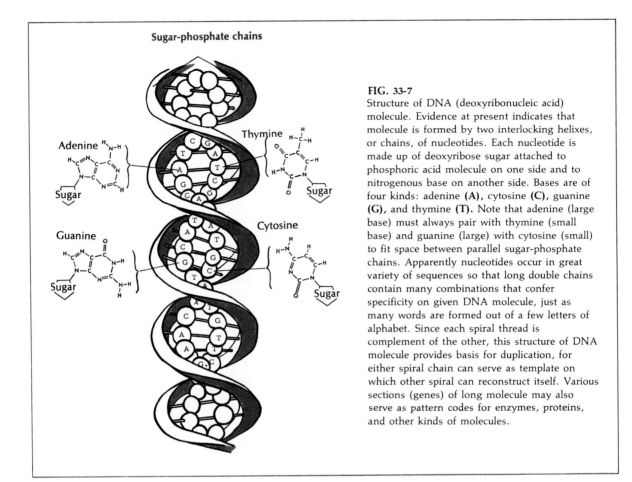

Sugar-phosphate chains

Adenine

Thymine

Guanine

Cytosine

FIG. 33-7
Structure of DNA (deoxyribonucleic acid) molecule. Evidence at present indicates that molecule is formed by two interlocking helixes, or chains, of nucleotides. Each nucleotide is made up of deoxyribose sugar attached to phosphoric acid molecule on one side and to nitrogenous base on another side. Bases are of four kinds: adenine **(A)**, cytosine **(C)**, guanine **(G)**, and thymine **(T)**. Note that adenine (large base) must always pair with thymine (small base) and guanine (large) with cytosine (small) to fit space between parallel sugar-phosphate chains. Apparently nucleotides occur in great variety of sequences so that long double chains contain many combinations that confer specificity on given DNA molecule, just as many words are formed out of a few letters of alphabet. Since each spiral thread is complement of the other, this structure of DNA molecule provides basis for duplication, for either spiral chain can serve as template on which other spiral can reconstruct itself. Various sections (genes) of long molecule may also serve as pattern codes for enzymes, proteins, and other kinds of molecules.

only in the nucleus, where it is the major structural component of genes; RNA is found throughout the cell, being especially abundant in nucleoli and in the cytoplasm. Thus a nucleic acid molecule is made up of many nucleotides joined to form long chains. Each nucleotide consists of phosphoric acid, either deoxyribose or ribose sugar, and a pyrimidine or purine base (see Fig. 2-18). The purine units are adenine and guanine; the pyrimidines are cytosine, thymine, and uracil. Five kinds of nucleotides are recognized on the basis of these purines and pyrimidines: (1) adenine-sugar-phosphate, (2) guanine-sugar-phosphate, (3) cytosine-sugar-phosphate, (4) thymine-sugar-phosphate, and (5) uracil-sugar-phosphate. The DNA molecule has the first four of these nucleotides (Fig. 33-7); the RNA has the first three and the last one. Although the phosphate-sugar part of the long chain of nucleotides is regular, the purine or

pyrimidine base attached to the sugar is not always the same. The order of these bases varies from one section to another of the nucleic acid molecule. Depending on the proportion and sequence of the nucleotides, there is an almost unlimited variety of nucleic acids.

The Watson-Crick model of structure of DNA molecule

In 1953 J. D. Watson and F. H. Crick, with the aid of x-ray diffraction studies of M. H. F. Wilkins, proposed a model of the structure of the DNA molecule that has been widely accepted (Fig. 33-7). The model could not have been made earlier, for its pattern depended on experimental evidence of many investigators over several years. Their model had to suggest plausible answers to such problems as (1) how specific directions are transmitted from one generation to another, (2) how DNA could control protein synthesis, and (3) how the DNA molecule could duplicate itself. Classic genetics and cytology had shown how a cell divides to form two cells and how each cell receives a set of chromosomes with their genes identical in structure to the preexisting set. But nothing was known about how a chemical substance could carry out the specifications required by the genetic substance of a gene. The elegance of the Watson-Crick hypothesis lies in the perfect manner it fits the data and in the way it can be tested.

Wilkins succeeded in getting very sharp x-ray diffraction patterns that revealed three major periodic spacings in crystalline DNA. These periodicities of 3.4, 20, and 34 Å were interpreted by Watson and Crick as the space distance between successive nucleotides in the DNA chains, the width of the chain, and the distance between successive turns of the helix, respectively. The x-ray diffraction photograph, with certain limitations, also gave indications of the spatial arrangement of some of the atoms within the large molecule. These investigators came up with the idea that the molecule is bipartite, with an overall helical configuration. Accordingly, their model showed that the DNA molecule consists of two complementary polynucleotide chains helically wound around a central axis with the sugar-phosphate backbones of each chain or strand on the outside of the molecular helix and the purines and pyrimidines on the inside of the helix.

BASE PAIRING. The two strands are held together by hydrogen bonds between specific pairs of purines and pyrimidines (Fig. 33-7). The two strands are complementary in that the sequence of bases along one strand specifies the sequence of bases along the other strand. One will note that the sequence of bases travels in opposite directions on the two strands or, in other words, are of opposite polarity. If the sequence of bases on one strand is known, one can identify the base sequence on the other chain. Each separate strand could then serve as a template for the production of its complement. According to the base-pairing rule, adenine at one point on a strand must have thymine at the corresponding point on the other strand, and likewise guanine with cytosine. There are no restrictions on the sequence of nucleotides in the DNA double helix, for the base sequence along one strand specifies the base sequence along the other strand. Because each strand has complete genetic

information, the presence of a complementary strand can serve as a template in the repair of a damaged one. The purine bases adenine and guanine (A and G) are large, whereas the pyrimidine bases thymine and cytosine (T and C) are small. To fit into the structure of the DNA molecule, each pair must consist of one large and one small base; thus the small bases T and C, T and T, or C and C could not bond in pairs as shown because they would not meet in the middle. The big bonds (A and G) could not bond in any combination because there is no room for them. Neither can A and C nor G and T (although in these pairs one is large and one is small) form bond mates because their hydrogen bonds would not pair off properly. Thus there is only one correct way for the bases to bond: A with T and G with C (Fig. 33-7).

CHEMICAL LANGUAGE OF DNA. Although all DNA molecules have the same general pattern, each one is unique because of the varying sequence of bases attached to the backbone. This sequence spells out the genetic instructions that determine each inherited characteristic. This genetic message is written in a language of only four symbols (adenine, guanine, thymine, and cytosine), and since any sequence of nucleotides is possible, almost limitless variations are possible in the genetic instructions. The nucleotides within the molecule are so arranged that the sugar of one molecule is always attached to the phosphate group of the next nucleotide in the sequence. The sugar of DNA is deoxyribose, a pentose sugar with 5 carbons, and to it is attached either a purine or pyrimidine base in each nucleotide.

No one knows how many nucleotides are possible in a DNA molecule; there may be hundreds of thousands, but the number naturally varies. DNA molecules are of enormous length. It is estimated that the total DNA in the 46 chromosomes of a cell (man) contains something like a billion base pairs with a length of 3 feet, and all of it has to unwind during each act of replication.

BUILDING NEW DNA MOLECULES. Every time the cell divides, the structure of DNA must be carefully copied in the daughter cells. The double chains, or strands, of the DNA helix explain how this may be done. The two strands of the helix could unwind, and each separate strand could then serve as a template for the production of its complement or to guide the synthesis of a new companion chain. It is believed that four kinds of building blocks are used in building the new strand. They are found in the nuclear environment surrounding the new strand. Each of the building blocks consists of one of the four DNA bases, one sugar unit, and three phosphate groups. Only one of the phosphates is used in making the backbone of the new strand; the other two provide the energy for the synthesis.

A large protein catalyst, **DNA polymerase,** acts to accelerate the reaction. This catalyst, or enzyme, does not determine which of the four bases (adenine, thymine, guanine, or cytosine) will be added to the new chain, but it does hold the reacting molecules steady. The selection of the base is made by the complementary base on the old, or template, chain. The copying of the sequence of the four bases is always done by the same catalyst. The sugar-phosphate linkages between successive nucleotides would result in a double-stranded DNA molecule in which the new strand has been specified by the old template strand. In this way each separate strand serves as a template for the formation of its complement, and the two daughter molecules would be identical to the parental molecule of the cell before it divides. Every daughter molecule is half old and half new; only one strand has been synthesized, whereas the other has been conserved.

MUTATIONS FROM MISTAKES IN THE DNA MOLECULE. All of the DNA bases occasionally occur in tautomeric forms in which certain bonds within the molecules are rearranged. Watson and Crick suggested that these rare tautomeric alternatives for each of the bases provide a mechanism for mutation during the replication of DNA. In some cases a tautomeric form of adenine can pair with cytosine instead of with thymine, or a tautomeric form of thymine can pair better with guanine than with adenine. To do this, the hydrogen atoms may assume new positions and their shift may be responsible for abnormal pairing. This incorrect choice of bases made during the copying process may cause a change in the base sequence. Such a mistake is preserved in the DNA each time it is copied and may be reflected in a physical alteration in the organism carrying the changed DNA.

Although most mutations are disadvantageous, some may be improvements. In this case favorable mutations will be preserved by natural selection and will spread through a population. This may be one of the principal mechanisms of evolution. A genic mutation, therefore, is an alteration in the DNA that changes the information content of the molecules so that new alleles are produced. Nucleotides may be deleted from or added to the sequence, or a nucleotide may be exchanged for a different one. A codon (see below), for instance, having normally the base composition of CGC, might be changed to CAC. Such a codon would code for a different amino acid, and such a type of mutation is called a base substitution. Base substitutions may cause nonsense triplets that may reduce or destroy the activity of the enzyme coded by the gene. Base substitutions that result in the replacement of a single amino acid may reduce the activity of an enzyme, but rarely cause its complete inactivity.

A variety of types of radiations—x-rays, gamma rays, ultraviolet rays, cosmic rays, or by-products of atomic rays—are mutagenic and can lead to changes in base pairs. Many chemical substances are also known to produce mutagenic changes in DNA, such as mustard gas, peroxides, nitrites, certain purines, and pyrimidines. In some cases an insertion may be made at one point in a genetic message, turning it into nonsense, but when a deletion is made at a later time, the message comes back into step again and once more has meaning.

DNA CODING BY BASE SEQUENCE. The Watson-Crick model suggested how new DNA may be made from old. The coding problem indicated that there must be some relation between the sequence of the four bases of DNA and the sequence of the 20 amino acids of proteins. The coding hypothesis had to account for the way these four bases (adenine, thymine, cytosine, guanine) must arrange themselves so that each permutation is the code for an amino acid. In the coding procedure, obviously there cannot be a 1:1 correlation between four bases and 20 amino acids. If the coding unit (often called a word, or **codon**) consists of two bases, only 16 words can be formed, which cannot account for 20 amino acids. Therefore, the protein code must consist of at least three bases or three letters because 64 possible words can be formed by four bases when taken as triplets. DNA must then be considered a language written in a four-letter alphabet. The particular composition or sequence of amino acids in a given protein are thus specified by the particular sequence of nucleotide pairs in a specific DNA molecule.

TRANSCRIPTION AND MESSENGER RNA. Information is coded in DNA of the nucleus, whereas

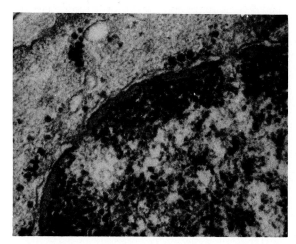

FIG. 33-8
Electron micrograph of smooth muscle cell of frog
showing part of nucleus with nuclear pores and nuclear
envelope. Two nuclear pores are seen. (× 120,000.)
(Courtesy G. E. Palade and National Academy of
Sciences, Washington, D.C.)

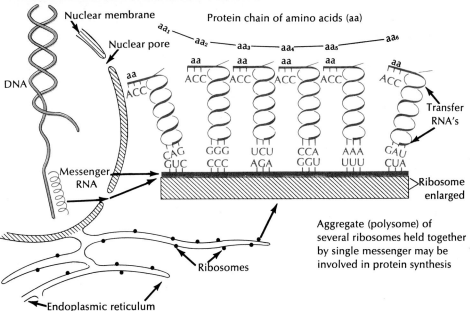

FIG. 33-9
Genetic code. Illustrates how genetic information may be passed from DNA molecule in
nucleus by means of messenger RNA to ribosomes of endoplasmic reticulum where
amino acids are arranged in proper sequence to form specific enzyme. Messenger RNA is
believed to be synthesized from particular segment (gene?) of one of DNA chains serving
as template. Messenger RNA (shown as red coil), with its specific code from DNA,
passes through nuclear pore to endoplasmic reticulum where it becomes attached to
ribosome or ribosomes. Amino acids, obtained preformed in food supply or else
synthesized by organism, are probably activated by specific enzyme for each of 20 or so
amino acids. High-energy ATP may play a part in activation. Activated amino acid is then
attached to specific transfer RNA molecule at recognition site on folded double helix of
molecule. Transfer RNA bearing specific amino acid is now lined up and brought into
correct position by base pairing of triplet-code between transfer RNA and messenger
RNA (C always pairs with G and A with U). Diagram shows some transfer RNA
molecules, with their amino acids being transferred to form peptide or protein chain (aa$_1$,
aa$_2$, etc.) as they are "read off" on template of messenger RNA. Each specific transfer
RNA bears base triplet ACC at amino acid acceptor end.

protein synthesis occurs in the cytoplasm. An intermediary of some kind between the two regions is necessary. This intermediary appears to be a special kind of RNA called **messenger RNA** (mRNA). (One can recall that RNA differs from DNA in having thymine [T] replaced by uracil [U] and having a sugar residue of ribose instead of deoxyribose.) Messenger RNA is believed to be **transcribed** directly from DNA in the nucleus, with each of the many messenger RNAs being determined by a gene or a particular segment of DNA. In this process of making a complementary copy of one strand or gene of DNA in the formation of messenger RNA, an enzyme, **RNA-DNA polymerase,** is needed. The messenger RNA contains a sequence of bases that complements the bases in one of the two DNA strands just as the DNA strands complement each other. Thus, A in the coding DNA strand is replaced by U in messenger RNA; C is replaced by G; G is replaced by C; and T is replaced by A. It appears that only one of the two chains is used as the template for RNA synthesis, although either one could be so used. The reason why only one strand of the double-stranded DNA is a "coding strand" is that messenger RNA otherwise would always be formed in complementary pairs and enzymes also would be synthesized in complementary pairs. In other words, two different enzymes would be produced for every DNA coding sequence instead of one. This would certainly lead to metabolic chaos.

RIBOSOMES AND RIBOSOMAL RNA. The messenger RNA when formed is separated from the DNA and migrates through nuclear pores (Figs. 33-8 and 33-9) into the cytoplasm of the cell, where it becomes attached to a granular **ribosome.** Ribosomes are roughly spherical submicroscopic structures about 200 Å in diameter comprised of protein and a second kind of nonspecific **ribosome RNA** (rRNA). Ribosomes carry no information but serve as attachment points for messenger RNA. One ribosome alone on a messenger RNA strand can produce a single polypeptide chain. Normally, however, several ribosomes cluster together in groups of 5 or 6, called **polysomes.** These work together on the same messenger RNA strand so that several polypeptides are assembled at the same time. The messenger RNA provides the code for this polypeptide synthesis.

ROLE OF TRANSFER RNA. Amino acids in the cytoplasm are not assembled directly on the mRNA-ribosome complex but through the action of a third type of RNA called **transfer RNA** (tRNA), also known as soluble RNA (sRNA). Transfer RNA is a small molecule of 70 to 80 ribonucleotides and folded back on itself to form a double helix (Fig. 33-9). This RNA molecule, like messenger RNA and ribosome RNA, is also synthesized on a DNA template. In a specific region of each transfer RNA, there is a coding sequence of three bases (the **anticodon**) that have a complementary sequence on messenger RNA (the **codon**). Thus the triplet codon UUU (three uracil bases) on the messenger would furnish the complementary site for the anticodon sequence of AAA (three adenine bases) on a transfer RNA. A different transfer RNA molecule corresponds to each triplet code on messenger RNA (Table 33-1).

Each transfer RNA is specific for a particular amino acid. The anticodon sequence of three unpaired nucleotides is found in the region where the chain of the transfer RNA turns back on itself. The combination of the correct amino acid with a particular transfer RNA requires the catalyzing action of a spe-

TABLE 33-1

The genetic code—proposed codons (code triplets) between messenger RNA and specific amino acids

Codons	Amino acid
GCU, GCC, GCA, GCG	Alanine
CGU, CGC, CGA, CGG, AGA	Arginine
AAU, AAC	Asparagine
GAU, GAC	Aspartic acid
UGU, UGC	Cysteine
GAA, GAG	Glutamic acid
CAA, CAG	Glutamine
GGU, GGC, GGA, GGG	Glycine
CAU, CAC	Histidine
AUU, AUC, AUA	Isoleucine
CUU, CUC, CUA, CUG, UUA, UUG	Leucine
AAA, AAG	Lysine
AUG	Methionine
UUU, UUC	Phenylalanine
CCU, CCC, CCA, CCG	Proline
AGU, AGC, UCU, UCC, UCA, UCG	Serine
ACU, ACC, ACA, ACG	Threonine
UGG	Tryptophan
UAU, UAC	Tyrosine
GUU, GUC, GUA, GUG	Valine
UAA, UAG, UGA	Termination of code of one gene

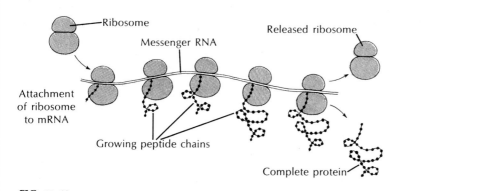

FIG. 33-10

How the protein chain is formed. As ribosomes move along messenger RNA, the amino acids are added stepwise to form the polypeptide chain.

cific enzyme. Thus there must be a specific enzyme for each transfer RNA. At the end to which the amino acid is attached, the chain always ends with the nucleotide triplet ACC. Since this attachment end is the same for all transfer RNAs, the rest of the molecule must contain the recognition sites that specify the correct bonding of a specific enzyme with the transfer RNA during amino acid attachment. More than one kind of transfer RNA is found for most of the amino acids, as shown in Table 33-1.

By stepwise addition, the amino acids are guided by the coding sequence on transfer RNA, to which they are attached, and arranged in the correct order along messenger RNA to form a protein molecule (Figs. 33-9 and 33-10). Each ribosome moves down the messenger RNA strand as the peptide chain grows. Once the first ribosome has moved a short distance, a second ribosome can attach to the mRNA and begin forming another peptide chain. This continues until the entire length of the mRNA strand is filled with ribosomes. This is the polysome cluster

mentioned above. When the first ribosome reaches the end of the messenger RNA strand, the ribosome is released and the completed protein is liberated. UAA, UAG, and UGA indicate when a chain is terminated and are used for punctuation (see Table 33-1).

Each gene codes for about 500 amino acids, which is the average of a polypeptide chain, and since the code triplets, or codons, are in groups of three nucleotides, there would be 1,500 nucleotide pairs in a single gene. These figures naturally will vary with the protein or enzyme being coded.

The general scheme for protein synthesis is shown below.

Synthesis of artificial gene

The synthesis of an artificial gene by H. G. Khorana and his colleagues was accomplished in 1970. Dr. Khorana made use of a short molecule of DNA that codes for the production of a certain transfer RNA in yeast cells. By knowing the order

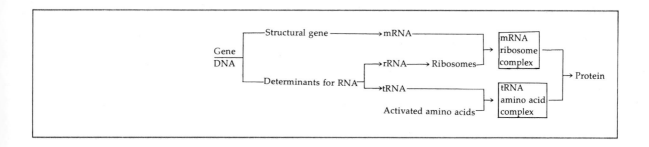

of the nucleotides in the transfer RNA (previously determined by R. Holley), he linked the bases (adenine, thymine, guanine, and cytosine) to sugar and phosphoric acid in the same order in which they occur in the yeast gene that made the molecule of transfer RNA. By knowing the sequence of the 77 nucleotides that specifies the synthesis in yeast of the molecule alanine transfer RNA, he was able to link together commercial nucleotides to form short single strands of DNA. Complementary, or opposite, strands were also formed by base pairing, using the enzyme DNA ligase for tying the double strands together.

Reverse transcriptase

The discovery of an enzyme, reverse transcriptase, that made DNA from RNA was made in 1970 by H. Temin. This enzyme was found in cancer or tumor viruses but may exist in normal cells. This enzyme can be identified by making DNA from its own genetic RNA material, a process that other enzymes apparently cannot do. At present the exact status of reverse transcriptase and how it operates awaits further investigation. Some believe that it can turn a normal cell into a cancer cell. Usually RNA is made from DNA according to the Watson-Crick model, but reverse transcriptase causes the recoding of the RNA message into DNA, thus enabling the RNA virus to be incorporated into the DNA of the host cell.

Present concept of the gene

The classic gene is no longer regarded as the indivisible minimal unit of heredity. It consists instead of smaller functional subunits. For example, alleles (all of the genes that may be situated at a particular chromosome locus) may differ only slightly in phenotypic expression. This may be caused by slight chemical differences brought about by small changes in the sequences of the nucleotides within a particular region of a DNA molecule. New alleles for a particular locus are produced by gene mutations. Thus, if the gene represents some section of a DNA molecule, it is conceivable that a mutation may affect only a small part of the nucleotides within a gene. A **muton,** then, is the smallest segment of a gene that can produce an altered trait by mutation. The smallest segment within a gene that is interchangeable but not divided by genetic recombination is called a **recon** and corresponds to the distance between adjacent nucleotides in the DNA chain. A recon may contain no more than two nucleotide pairs. A larger subunit of the gene is the **cistron,** which refers to the smallest number of mutons or recons of a gene that must remain together on one chromosome to perform a biochemical or genetic function. If a series of consecutive mutons, for instance, are necessary for the synthesis of a certain protein and are located on the same chromosome (**cis** position), the whole gene functions normally. But if one part of the mutons are on one chromosome and the other part on the other chromosome (**trans** position), the protein synthesis will occur normally only if the two groups of mutons complement each other. Only a few cistrons are found in a gene and each must be made up of many nucleotides. It has been suggested that recons and mutons may control the synthesis of individual amino acids, whereas cistrons control the formation of polypeptides (chains of amino acids).

Operon concept

The genetic code as given in the foregoing description simply explains how the code carried on the DNA molecules of the nucleus is transcribed into a definite protein or enzyme synthesized in the cytoplasm. It does not explain how genes are turned off and on as their products are needed by the cell. It does not explain why certain enzymes are not formed when they are not needed. Obviously, enzyme-forming systems require control because they produce different amounts of the same enzyme at different times. It is also apparent that this control must have two components: (1) the genetic apparatus of the code and ribosome transcription and (2) factors from the environment such as the amount of products accumulated. Thus there must be mechanisms in the cell for repressing the synthesis of enzymes when they are not needed and for inducing them when they are needed.

In 1960 the two French scientists, F. Jacob and J. Monod, proposed the **operon hypothesis,** or model, for explaining how repressions and inductions of protein synthesis might occur (Fig. 33-11). This important hypothesis is based entirely on work with bacteria; it remains to be seen whether their hypothesis also applies to higher living forms. The gist of their hypothesis, for which they were awarded the Nobel Prize in 1965, may be stated in the following way:

1. There are two types of genes, **structural genes**

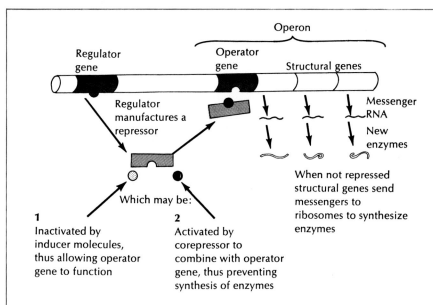

Operon

Regulator gene

Operator gene

Structural genes

Regulator manufactures a repressor

Messenger RNA

New enzymes

When not repressed structural genes send messengers to ribosomes to synthesize enzymes

Which may be:

1
Inactivated by inducer molecules, thus allowing operator gene to function

2
Activated by corepressor to combine with operator gene, thus preventing synthesis of enzymes

FIG. 33-11
Operon hypothesis. Regulator gene acts by way of a repressor on the operator gene. The regulator gene controls rate of information sent to operator gene, either to induce more of a particular enzyme or to cut off (repress) additional amounts of unneeded enzymes. Repressor is a cytoplasmic factor, probably a macromolecule. When operator is "turned on," entire operon is active in synthesis of enzymes; when it is "off," operon is inactive. Inducer molecules modify regulator substance to prevent it from switching off operator; repressor molecules react with regulator substance and cause it to switch off operator (and genes that it controls).

and **regulator genes.** The structural genes contain the coded formulas for the synthesis of the primary structure of a protein, or enzyme, that are useful in cellular metabolism. The regulatory genes control the function of the structural genes.

2. There are two kinds of regulator genes, the regulator and the operator. The **operator** gene determines whether the formula, or code, in structural genes adjacent to it are to be transcribed into an enzyme. The **regulator** gene codes for the structure of a cytoplasmic factor (the **repressor**), which turns the operator on and off.

3. The **operon** is that portion of a chromosome that regulates all the steps in the synthesis of an enzyme, or protein. Some operons may contain only one gene; others contain several. An operon consists of an operator gene and the segment of DNA it controls. The operator may control either a single structural gene to which it is adjacent or several structural genes

of related function. Thus all the nine enzymes in the histidine pathway are controlled by a single operator.

4. The regulator genes produce a substance called the **repressor** that blocks the operator genes and thus prevents the structural gene from functioning normally. Repression occurs when the repressor substance combines with the operator gene and prevents the formation of messenger RNA along the segment of DNA controlled by the operator gene. The operator is that part of the operon that is the receptor site for the repressor.

5. If the repressor substance reacts with an appropriate cytoplasmic substance, it **derepresses** the repressor substance and permits the operator gene to act. This is called the **inducible system.** In other words, it renders the repressor incapable of turning the operator off.

6. In this way the two antagonistic systems, the inducible system and the repressor system, maintain

a refinement in the amount and kind of enzymes necessary for the steady states of the cell. For instance, if there is a high concentration of a particular enzyme in the cell, this high concentration can act as a "feedback" through the repressor system to block the action of the operator gene so that the structural gene can no longer produce the enzymes. The repressor may be changed to an inactive form by a lower-than-normal concentration of the enzyme, or by a specific substance synthesized in the cytoplasm or from the environment (ions and amino acids) so that the operator gene is turned on to produce more of the enzyme. In this way the genes influence the cytoplasm and the cytoplasm exerts a "feedback" influence on the genes for turning on or off their action.

Repressors play a vital role in cellular differentiation. When a zygote is formed, cells begin to divide, and gradually these cells differentiate into special tissues and organs that make up the animal. Every cell in our body contains a full complement of genetic information, so there must be exact mechanisms present that permit only those genes necessary for the survival and function of the specialized cells to be expressed. To understand gene repression and its action is one of the central problems of current biology.

■ MENDEL'S INVESTIGATIONS

The first man to formulate the cardinal principles of heredity was Gregor Johann Mendel (1822-1884), who was connected with the Augustinian monastery at Brünn (Brno), Moravia, then a part of Austria, now a part of Czechoslovakia. In the small monastery garden he conducted his experiments on hybridization, which resulted in two clear-cut laws that bear his name. From 1856 to 1864 he examined with great care many thousands of plants. His classic observations were made on the garden pea because gardeners over a long period by careful selection had produced pure strains. For example, some varieties were definitely dwarf and others were tall. A second reason for selecting peas was that they are self-fertilizing, but they are also capable of cross-fertilization as well. To simplify his problem he chose single characters and those that were sharply contrasted. Mere quantitative and intermediate characters he carefully avoided. Mendel selected seven pairs of these contrasting characters, such as tall plants and dwarf plants, smooth seeds and wrinkled seeds, green cotyledons

and yellow cotyledons, inflated pods and constricted pods, yellow pods and green pods, axial position of flowers and terminal position of flowers, and transparent seed coats and brown seed coats (Fig. 33-12). Mendel crossed a plant having one of these characters with one having the contrasting character. He did this by removing the stamens from a flower so that self-fertilization could not occur, and then on the stigma of this flower he placed the pollen from the flower of another plant that had the contrasting character. He then prevented the experimental flowers from being pollinated from other sources, such as wind and insects. When the cross-fertilized flower bore seeds, he noted the kind of plants (hybrids) they produced when planted. His next step was to cross these hybrids among themselves and to see what happened. He made careful counts of his plants, repeated his experiments, and worked out certain ratios.

Among other experiments Mendel crossed pure tall plants with pure dwarf plants and found that the hybrids (F_1 or first filial generation) thus produced were all tall, just as tall as the tall parent that was involved in the cross. This always happened whether the tall plant furnished the male germ cells or the female germ cells. Next, he crossed two of these hybrid tall plants together. From this cross he raised several hundred plants and found that both tall plants and dwarf plants were represented among them. He also noted that none of this generation (F_2 or second filial generation) were intermediate in size; they were either as tall or as short as the parents in the original cross. When he counted the actual number of tall and dwarf plants in the F_2 generation, he found there were three times more tall plants than dwarf ones, or a ratio of 3:1. His next step was to self-pollinate the plants in the F_2 generation; that is, the stigma of a flower was fertilized by the pollen of the same flower. The results showed that self-pollinated F_2 dwarf plants produced only dwarf plants, whereas one third of the F_2 tall plants produced tall and the other two thirds produced both tall and dwarf in the ratio of 3:1, just as the F_1 plants had done. This experiment showed that the dwarf plants were pure because they at all times gave rise to short plants when self-pollinated; the tall plants contained both pure tall and hybrid tall. It also demonstrated that, although the dwarf character disappeared in the F_1 plants, which were all tall, the character for dwarfness appeared in the F_2 plants.

Experiments on which Mendel based his postulates
Results of monohybrid crosses for first and second generations

Round-wrinkled seeds
F_1 all round
F_2 5474 round
1850 wrinkled
Ratio 2.96:1

Colored-white flowers
F_1 all colored
F_2 705 colored
224 white
Ratio 3.15:1

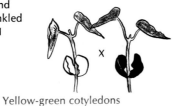

Yellow-green cotyledons
F_1 all yellow
F_2 6022 yellow
2001 green
Ratio 3.01:1

FIG. 33-12
The seven experiments on which Mendel based his postulates.

Green-yellow pods
F_1 all green
F_2 428 green
152 yellow
Ratio 2.82:1

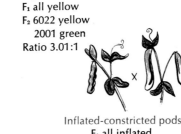

Inflated-constricted pods
F_1 all inflated
F_2 882 inflated
299 constricted
Ratio 2.95:1

Long-short stems
F_1 all long
F_2 787 long
277 short
Ratio 2.84:1

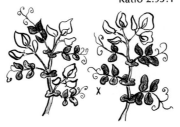

Axial-terminal flowers
F_1 all axial
F_2 651 axial
207 terminal
Ratio 3.14:1

MENDEL'S LAWS

Mendel's great contribution was his idea of the particulate nature of inheritance. Up to his time inheritance was thought of as a sort of blending inheritance. In such a view, variability is lost in the hybridization between individuals and contrasting charac-teristics. On the other hand, with particulate inheritance the different variations are retained and can be shuffled about like blocks. A slight addition of new kinds of genes can produce much variability in a population.

Mendel knew nothing about the cytologic background of heredity. Chromosomes and genes were unknown to him. Instead of using the term "genes" as we do today, he called his inheritance units "factors." He reasoned that the factors for tallness and dwarfness were units that did not blend when they were together. The F_1 generation contained both these units or factors, but when these plants formed their germ cells, the factors separated out so that each germ cell had only one factor. In a pure plant both factors were alike; in a hybrid they were different. He concluded that individual germ cells are always pure with respect to a pair of contrasting factors, even though the germ cells are formed from hybrids in which the contrasting characters were mixed. This idea formed the basis for his first principle, the **law of segregation,** which states that whenever two factors are brought together in a hybrid, when that hybrid forms its germ cells, the factors segregate into separate gametes and each germ cell is pure with respect to that character. Thus in the gametes of the F_1 plants, half of the germ cells will bear the factor for tallness and half for dwarfness; no germ cell will contain both factors.

In the crosses involving the factors for tallness and dwarfness, in which the resulting hybrids were tall, Mendel called the tall factor **dominant** and the short **recessive.** Similarly, the other pairs of characters that he studied showed dominance and recessiveness. Thus when plants with yellow unripe pods were crossed with green unripe pods, the hybrids all contained yellow pods. In the F_2 generation the expected ratio of 3 yellow to 1 green was obtained. Whenever a dominant factor (gene) is present, the recessive one cannot produce an effect. The recessive factor will show up only when both factors are recessive, or, in other words, a pure condition.

The law of segregation deals only with one pair of contrasting characters. Mendel also ascertained what would happen when a cross is made between plants differing in two pairs of contrasting characters. Thus when a tall plant with the yellow type of pod was crossed with a dwarf plant bearing green pods, the F_1 generation was all tall and yellow, for these factors are dominant. When the F_1 hybrids were crossed with each other, the result was 9 tall and yellow, 3 tall and green, 3 dwarf and yellow, and 1 dwarf and green. In this experiment each factor separated independently of the other and showed up in new combinations. This is Mendel's second law, or the **law of independent assortment,** which states that,

whenever two or more pairs of contrasting characters are brought together in a hybrid, the factors of different pairs segregate independently of one another. Rarely do two organisms differ in only one pair of contrasting characters; nearly always they differ in many. The second law of Mendel therefore deals with two or more pairs of contrasting characters.

Crosses involving more than two pairs of characters result in still more complicated ratios of types of offspring. However, it is usually convenient to work with just one pair of contrasting characters, for each pair may be considered by itself. It may be stated here that Mendel's second law is true only when the factors for the different pairs of characters are located on different pairs of chromosomes. It happened that all seven pairs of characters Mendel worked with were on different pairs of chromosomes, but since his laws became known, many pairs of characters have been found on the same chromosome, which alters the original mendelian ratios. This modification does not detract, however, from the basic significance of his great laws.

Although Mendel published his observations on the principles of heredity in *The Proceedings of the Society of Natural Science of Brünn* in 1866, his experiments attracted no attention and apparently were forgotten. In 1900 three investigators, De Vries in Holland, Tschermak in Austria, and Correns in Germany, independently rediscovered his laws but found that the obscure priest had already published them 34 years before.

Although Darwin was a contemporary of Mendel and realized the importance of understanding heredity, Darwin knew nothing of Mendel's fundamental discoveries. Darwin had proposed a heredity theory of his own for explaining how heredity worked. His pangenesis theory suggested that each part of the body produced particles (pangenes) that were carried by the blood to the gonads where they were incorporated into the germ cells.

Explanation of mendelian ratios

In representing his crosses Mendel used letters as symbols. For dominant characters, he employed capitals and, for recessives, corresponding small letters. Thus the factors, or genes, for pure tall plants might be represented by TT, the pure recessive by tt, and the hybrid of the two plants by Tt. In diagram form, one of Mendel's original crosses (tall plant and dwarf plant) could be represented in this manner:

	(tall)		(dwarf)
Parents	TT	×	tt
Gametes	all T		all t
F_1		Tt	
		(hybrid tall)	
Crossing hybrids	Tt	×	Tt
Gametes	T,t		T,t
F_2	TT Tt	tT tt	
	(3 tall to 1 dwarf)		

It is convenient in most mendelian crosses to use the checkerboard method devised by Punnett for representing the various combinations resulting from a cross. Thus in the previous F_2 cross, the following scheme would apply.

	Eggs	
	T	**t**
T	**TT** (pure tall)	**Tt** (hybrid tall)
t	**Tt** (hybrid tall)	**tt** (pure dwarf)

Ratio: 3 tall to 1 dwarf.

Mendel's experiment involving two pairs of contrasting characters instead of one pair may be demonstrated in the diagram shown below.

	(tall, yellow)		(dwarf, green)
Parents	TTYY	×	ttyy
Gametes	all TY		all ty
F_1		TtYy	
	(hybrid tall, hybrid yellow)		
Crossing hybrids	TtYy	×	TtYy
Gametes	TY, Ty, tY, ty		TY, Ty, tY, ty
F_2		(see checkerboard)	

	TY	**Ty**	**tY**	**ty**
TY	**TTYY** pure tall pure yellow	**TTYy** pure tall hybrid yellow	**TtYY** hybrid tall pure yellow	**TtYy** hybrid tall hybrid yellow
Ty	**TTYy** pure tall hybrid yellow	**TTyy** pure tall pure green	**TtYy** hybrid tall hybrid yellow	**Ttyy** hybrid tall pure green
tY	**TtYY** hybrid tall pure yellow	**TtYy** hybrid tall hybrid yellow	**ttYY** pure dwarf pure yellow	**ttYy** pure dwarf hybrid yellow
ty	**TtYy** hybrid tall hybrid yellow	**Ttyy** hybrid tall pure green	**ttYy** pure dwarf hybrid yellow	**ttyy** pure dwarf pure green

Ratio: 9 tall yellow: 3 tall green: 3 dwarf yellow: 1 dwarf green.

In the cross between tall and dwarf it will be noted that there are two types of visible characters—**tall** and **dwarf**. These are called **phenotypes**. On the basis of genetic formulas there are three hereditary types, TT, Tt, and tt. These are called **genotypes**. In the cross involving two pairs of contrasting characters (**tall yellow** and **dwarf green**) there are in the F_2 generation four phenotypes: **tall yellow, tall green, dwarf yellow,** and **dwarf green**. The genotypes are nine in number: TTYY, TTYy, TtYY, TtYy, TTyy, Ttyy, ttYY, ttYy, and ttyy. The F_2 ratios in any cross involving more than one pair of contrasting genes can be found by combining the ratios in the cross of one pair of factors. Thus the genotypes will be $(3)^n$ and the phenotypes $(3:1)^n$. To illustrate, in a cross of two pairs of factors the phenotypes will be in the ratio of $(3:1)^2 = 9:3:3:1$. The genotypes in such a cross will be $(3)^2 = 9$. If three pairs of characters are involved, the phenotypes will be $(3:1)^3 = 27:9:9:9:3:3:3:1$. The genotypes will be $(3)^3 = 27$. Thus it is seen that the numerical ratio of the various phenotypes is a power of the binomial $(3 + 1)^n$ whose exponent (n) equals the number of pairs of heterozygous genes in F_2. This is true only when one member of each pair of genes is dominant. By experience, then, one may determine the ratios of phenotypes in a cross without using the checkerboard. In a dihybrid (9:3:3:1 ratio), for instance, it will be seen that those phenotypes that make up the dominants of each pair will be $9/16$ of the whole F_2; each of the $3/16$ phenotypes will consist of one dominant and one recessive; and the $1/16$ phenotype will consist of the two recessives.

■ LAWS OF PROBABILITY

When Mendel worked out the ratios for his various crosses, they were approximations and not certainties. In his 3 to 1 ratio of tall and short plants, for instance, the resulting phenotypes did not come out exactly 3 tall to 1 short. All genetic experiments are based on probability; that is, the outcome of the events is uncertain and there is an element of chance in the final results. Probability values are measures of expectations. Probabilities are expressed in fractions, or it is always a number between 0 and 1. This probability number (p) is found by dividing the number (m) of favorable cases (for example, a certain event) by the total number (n) of possible outcomes:

$$p = \frac{m}{n}.$$

When there are two possible outcomes, such as in tossing a coin, the chance of getting heads is $p = \frac{1}{2}$, or 1 chance in 2.

The more often a particular event occurs, the more closely will the number of favorable cases approach the number predicted by the p value. Probability predictions are often unreliable when there are only a few occurrences.

The probability of independent events occurring together involves the **product rule,** which is simply the product of their individual probabilities. When two coins are tossed together, the probability of getting two heads is $\frac{1}{2} \times \frac{1}{2} = \frac{1}{4}$, or 1 chance in 4. Here, again, this prediction is most likely to occur if the coins are tossed a sufficient number of times.

The ratios of inheritance in a monohybrid cross of dominant and recessive genes can be explained by the product rule. In the gametes of the hybrids the sperm may carry either the dominant or the recessive gene; the same applies to the eggs. The probability that the sperm carries the dominant is $\frac{1}{2}$ and the probability of an egg carrying the dominant is also $\frac{1}{2}$. The probability of a zygote obtaining two dominant genes is $\frac{1}{2} \times \frac{1}{2}$, or $\frac{1}{4}$. Thus 25% of the offspring will probably be pure dominants. The same principle applies to the recessive gene, which will be pure for 25% of the offspring. The heterozygous gene combinations will be found by the sum of the two possible combinations—a sperm with a dominant gene and an egg with a recessive gene, and a sperm with a recessive gene and an egg with a dominant gene—which yields 50% heterozygotes. Thus we have the 1:2:1 ratio.

■ TERMINOLOGY OF GENETICS

Genetics, in common with other branches of science, has built up its own terminology. Some of the terms first proposed by Mendel have been replaced by those that seem more suitable in the light of present-day knowledge. These terms are all important to the student of heredity, because they are essential in understanding the analyses of genetic problems. Whenever a cross involves only one pair of contrasting characters, it is called a **monohybrid;** when the cross has two pairs, it is a **dihybrid;** when the cross has three pairs, it is a **trihybrid;** and when it has more than three pairs, it is a **polyhybrid.** Characters that show in the F_1 are **dominant;** those that are hidden are **recessive.** When a dominant

always shows up in the phenotype, it is said to have **complete dominance;** when it sometimes fails to manifest itself it is called **incomplete dominance.** When two characters form a contrasting pair, they are called **alleles** or **allelomorphs.** The term **factor** that Mendel used so widely is replaced by **gene.** A **zygote** is the union of two gametes; whenever the two members of a pair of genes are alike in a zygote, the latter is **homozygous** for that particular character; when the genes are unlike for a given character, the zygote is **heterozygous.** A **hybrid,** for instance, is a heterozygote, and a **pure** character is a homozygote.

■ ADVANCES IN GENETICS SINCE MENDEL'S LAWS REDISCOVERED

The rediscovery of Mendel's laws in 1900 served as an enormous stimulus to the study of genetics. The basic contribution of Mendel was that hereditary characters behave as units. His principles have been abundantly verified by many investigators. Since his time, investigators have found that his laws are not so simple and direct as he first proposed them. Many of the modifications advanced, however, served all the more to strengthen Mendel's concepts. It has already been pointed out that the principle of independent assortment applies only when the pairs of contrasting genes are in different chromosomes. Since his time, the phenomena of linkage and crossing-over make necessary a modification of the law. The principles of dominance and recessiveness are no longer stressed as much as formerly because they are not well marked in many crosses. The idea of unit character is no longer thought of as Mendel thought of it, for it is now known that many factors may enter into the development of a particular character. Adult characters as such are not found in germ cells, but only differentiation determines which cells cause a character to express itself in a certain way.

Although many significant investigations have been made in genetics since 1900, none have been more fruitful than those performed by Professor Thomas Hunt Morgan and his colleagues on the fruit fly. *Drosophila* is ideal for genetic experimentation because it produces so many generations within a few weeks. Morgan started his work on these forms about 1910, and now the heredity of no animal is better known than this common fly. Many of its characters are easily recognized and followed, and several striking mutations have helped explain the more intricate mechanism of heredity. As many as 500 genes have been mapped on its four pairs of chromosomes. The principles of linkage and crossing-over have also been best explained in this form. In addition, the salivary gland chromosomes, which have yielded so much information about the nature of the gene, were also first discovered in this fly.

One of the greatest advancements in understanding the physical basis of heredity made since Mendel's laws were known is the parallelism between these laws and the behavior of the chromosomes (and genes) during the processes of maturation and fertilization. The Boveri-Sutton hypothesis assigning the mechanism of heredity to the genes may be regarded as one of the basic concepts of biology because Morgan's great work, as well as the more recent work on the salivary gland chromosome, has given striking confirmation of the principle.

■ TESTCROSS

The dominant characters in the offspring of a cross are all of the same phenotypes whether they are homozygous or heterozygous. For instance, in Mendel's experiment of tall and dwarf characters, it is impossible to determine the genetic constitution of the tall plants of the F_2 generation by mere inspection of the tall plants. Three fourths of this generation are tall, but which of them are heterozygous recessive dwarf? The test is to cross the F_2 generation (dominant hybrids) with pure recessives. If the tall plant is homozygous, all the plants in such a testcross will be tall, thus:

TT (tall) × tt (dwarf)
Tt (hybrid tall)

If, on the other hand, the tall plant is heterozygous, the offspring will be half tall and half dwarf, thus:

Tt × tt
Tt (tall) or tt (dwarf)

The testcross is often used in modern genetics for the analysis of the genetic consitution of the offspring as well as for a quick way to make homozygous desirable stocks of animals and plants.

■ INCOMPLETE DOMINANCE

A cross that always shows the heterozygotes as distinguished from the pure dominants is afforded by the four-o'clock flower *(Mirabilis)* (Fig. 33-13). Whenever a red-flowered variety is crossed with a

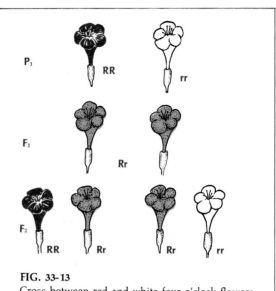

FIG. 33-13
Cross between red and white four-o'clock flowers. Red and white are homozygous; pink is heterozygous.

white-flowered variety, the hybrid (F_1), instead of being red or white according to whichever is dominant, is actually intermediate between the two and is pink. Thus the homozygotes are either red or white, but the heterozygotes are pink. The testcross is therefore unnecessary to determine the nature of the genotype.

In the F_2 generation, when pink flowers are crossed with pink flowers, one fourth will be red, one half pink, and one fourth white.

This cross may be represented in this fashion:

	(red flower)		(white flower)	
Parents	RR	×	rr	
Gametes	R,R,		r,r	
F_1		Rr		
		(all pink)		
Crossing hybrids	Rr	×	Rr	
Gametes	R,r		R,r	
F_2	RR	Rr	rR	rr
	(red)	(pink)	(pink)	(white)

In this kind of cross neither of the genes demonstrates complete dominance; therefore, the heterozygote is a blending of both red and white characters. A similar phenomenon is found in the Blue Andalusian fowl in which a cross between black and white varieties produces a hybrid blue. In a cross between red and white cattle a roan hybrid is obtained. In both these cases of fowl and cattle a heterozygote always gives rise to blue or roan, respectively.

■ PENETRANCE AND EXPRESSIVITY

Penetrance refers to the percentage frequency with which a gene manifests phenotypic effect. If a dominant gene or a recessive gene in a homozygous state always produces a detectable effect, it is said to have **complete penetrance.** If dominant or homozygous recessive genes fail to show phenotypic expression in every case, it is called **incomplete** or **reduced penetrance.** Environmental factors may be responsible for the degree of penetrance because some genes may be more sensitive to such influences than are other genes. The genotype responsible for diabetes mellitus, for instance, may be present, but the disease does not always occur because of reduced penetrance. All of Mendel's experiments apparently had 100% penetrance.

The phenotypic variation in the expression of a gene is known as **expressivity.** For instance, a heritable allergy may cause more severe symptoms in one person than in another. Environmental factors may cause different degrees in the appearance of a phenotype. Temperature affects the expression of the genes for dark-colored fur in Siamese cats and Himalayan rabbits. Normally the tail, feet, ears, and nose—the areas that have a cooler body temperature—are dark, but under warm conditions there may be less than normal darkening, whereas in colder environmental temperatures there may be some darkening of the entire animal. Other genes in the hereditary constitution may also modify the expression of a trait. What is inherited is a certain genotype, but how it is expressed phenotypically is determined by environmental and other factors.

Genes that have more than one effect are called **pleiotropic.** Most genes may have multiple effects. Even those genes that produce visible effects probably have numerous physiologic effects not detected by the geneticist. The recessive gene in fruit flies that produces (in a homozygous condition) vestigial wings also affects other traits such as the halteres (balancers), bristles, reproductive organs, and length of life. Sickle-cell anemia is another common example for its heterozygous condition confers malarial resistance.

■ SOME SPECIAL FORMS OF HEREDITY

The types of crosses already described are simple in that the characters involved are the result of the action of a single gene, but many cases are known in which the characters are the result of two or more genes. At first these more complex cases were believed to be nonmendelian, for they often produce unusual ratios. Most of these crosses, however, are now considered to be merely modifications of the mendelian expectations and do not invalidate the basic laws of Mendel. Mendel probably did not appreciate the real significance of the genotype as contrasted with the visible character—the phenotype. We now know that many different genotypes may be expressed as a single phenotype.

As was stated above, many genes have more than a single effect. A gene for eye color, for instance, may be the ultimate cause for eye color, yet at the same time it may be responsible for influencing the development of other characters as well. Also, many unlike genes may occupy the same locus on a chromosome, but not, of course, all at one time. Thus more than two alternative characters may effect the same character. Such genes are called **multiple alleles** or factors. In the fruit fly *(Drosophila)* there are 18 alleles for eye color alone. Not more than two of these genes can be in any one individual and only one in a gamete. What is the reason for multiple alleles? The

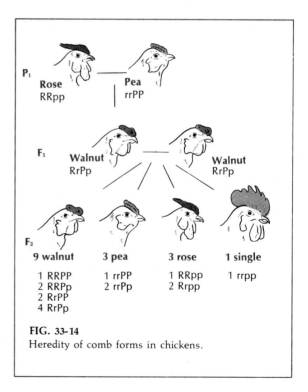

FIG. 33-14
Heredity of comb forms in chickens.

answer is that all genes can mutate in several different ways if given time and thus can give rise to several alternative conditions. In this way, many alleles for a particular locus on a chromosome may have evolved and added to the genetic pool of a population. Although the fact cannot be proved, it is believed that all genes present in an organism are mutants. In some cases dominance is lacking between two members of a set of multiple alleles, but usually one is dominant over the other. In *Drosophila* the gene for red eye color (wild type) is dominant over all other alleles of the eye color series; the gene for white eye is recessive to all the others.

Some of these unusual cases of inheritance are described in the following discussions on supplementary, complementary, cumulative, and lethal factors and pseudoalleles.

SUPPLEMENTARY FACTORS. The variety of comb forms found in chickens illustrates the action of supplementary genes (Fig. 33-14). The common forms of comb are rose, pea, walnut, and single. Of these, the pea comb and the rose comb are dominant to the single comb. For example, when a pea comb is crossed with a single comb, all the F_1 are pea and the F_2 show a ratio of 3 pea to 1 single. When the two dominants, pea and rose, are crossed with each other, an entirely new kind of comb, walnut, is found in the F_1 generation. Each of these genes supplements the other in the production of a kind of comb different from each of the dominants. In the F_2 generation the ratio is 9 walnut, 3 rose, 3 pea, and 1 single. The walnut comb cannot thus be considered a unit character, but is merely the phenotypic expression of pea and rose when they act together.

Inspection of the ratio reveals that two pairs of genes are involved. If P represents the gene for the pea comb and p its recessive allelomorph and if R represents the gene for the rose comb and r its recessive allelomorph, then the pea comb formula would be PPrr; and the one for rose comb, ppRR. Any individual having both dominant genes has a walnut comb. When no dominant gene is present, the comb is single. The cross may be diagrammed as follows:

Parents	PPrr	×	ppRR
	(pea comb)		(rose comb)
Gametes	all Pr		all pR
F_1		PpRr	×	PpRr
		(walnut)		(walnut)
Gametes		PR, Pr, pR, pr	×	PR, Pr, pR, pr

By the checkerboard method, the F_2 will show 9 walnut, 3 pea, 3 rose, and 1 single. It will be seen that genotypes with the combinations of PR will give walnut phenotypes; those with P, pea; those with R, rose; and those lacking in both P and R, single.

COMPLEMENTARY FACTORS. When two genes produce a visible effect together, but each alone will show no visible effect, they are referred to as **complementary genes.** Some varieties of sweet peas can be used to illustrate this kind of cross. When two white-flowered varieties of these are crossed, the F_1 will show all colored (reddish or purplish) flowers. When these F_1's are self-fertilized the F_2 will show a penotypic ratio of 9 colored to 7 white flowers. This is really a ratio of 9:3:3:1, because the last three groups cannot be distinguished phenotypically. The explanation lies in the fact that in one of the white varieties of flowers there is a gene (C) for a colorless color base (chromogen) and in the other white variety a gene (E) for an enzyme that can change chromogen into a color. Only when chromogen and the enzyme are brought together is a colored flower produced. The cross may be diagrammed in this way:

Parents	CCee	×	ccEE
	(white)		(white)
Gametes	all Ce		all cE
F_1		CcEe	×	CcEe
		(colored)		(colored)
Gametes		CE, Ce, cE, ce		CE, Ce, cE, ce

By the checkerboard method, the F_2 phenotypes will be as follows:

9 colored (CCEE, CCEe, CcEE, or CcEe)—both chromogen and enzyme present

7 {
3 white (CCee, or Ccee)—only chromogen
3 white (ccEE, or ccEe)—only enzyme
1 white (ccee)—no chromogen or enzyme
}

CUMULATIVE FACTORS. Whenever several sets of alleles produce a cumulative effect on the same character, they are called **multiple genes** or factors. Several characteristics in man are influenced by multiple genes. In such cases the characters, instead of being sharply marked off, show continuous variation between two extremes. This is what is called **blending** or **quantitative inheritance.** In this kind of inheritance the children are more or less intermediate between the two parents. The best illustration of such a type is the degree of pigmentation in crosses between the Negro and the white race. The cumulative genes in such crosses have a quantitative expression. A pure-blooded Negro has two pairs of genes on separate chromosomes for pigmentation (AABB). On the other hand, a pure-blooded white will have the genes (aabb) for nonblack. In a mating between a homozygous Negro and a homozygous white, the mulatto (AaBb) will have a skin color intermediate between the black parent and the white. The genes for pigmentation in the cross show incomplete dominance. In a mating between two mulattos (F$_2$), the children may show a variety of skin color, depending on the number of genes for pigmentation they inherit. Their skin color may range all the way from black (AABB) through dark brown (AABb or AaBB), half-colored (AAbb or AaBb or aaBB), light brown (Aabb or aaBb) to white (aabb). In the F$_2$ there will be the possibility of a child resembling the skin color of either grandparent, and of others that show intermediate grades. It is thus possible for parents heterozygous for skin color to produce children with darker colors and also with lighter colors than themselves.

The relationships can be seen in the following diagram:

Parents	AABB	×	aabb
	(black)		(white)
Gametes	AB		ab
F$_1$	AaBb	×	AaBb
	(mulatto)		(mulatto)
Gametes	AB, Ab, aB, ab		AB, Ab, aB, ab

By the checkerboard method, the F$_2$ will show this ratio:

1 black (AABB)
4 dark brown (AABb or aABB)
6 half-colored mulattos (AaBb, AAbb, or aaBB)
4 light brown (Aabb, aaBb)
1 white (aabb)

The student should realize that when the terms "white" and "black" are used in a cross involving mulattos, they refer solely to skin color and not to other characteristics, for other racial characteristics are inherited independently. Thus in such a cross an individual may have white color (no genes for black) but could have other Negro characteristics.

Although skin color appears to depend on the distribution of two pairs of genes, there are many other traits in human inheritance that involve more than two pairs. These more complicated cases result in more varied ratios than in the simpler cases. When there are so many genes involved in the production of traits, the latter often take the form of distribution curves. One such trait is stature in man, where between a few extremely short and tall individuals, there are many in between these extremes.

PSEUDOALLELES. Some genes that have similar phenotypic effects may be so closely linked together that they are often considered as multiple alleles. Instead of being a single locus with multiple alleles, there are two or more closely linked loci with genes acting on the same trait. Such genes are called **pseudoalleles,** or **duplicate genes.** The only way in which a geneticist can be sure that pseudoalleles exist is by crossing-over, which is very rare because the genes are so close together. The condition has been described in fruit flies and corn and affords some insight into the intricate evolution and nature of the gene.

LETHAL FACTORS. A lethal gene is one that, when present in a homozygous condition, will cause the death of the offspring. Investigators have known for a long time that the yellow race of the house mouse *(Mus musculus)* is heterozygous. Whenever two yellow mice are bred together, the progeny are always 2 yellow to 1 nonyellow. In such a case the expected ratio should be 1 pure yellow, 2 hybrid yellow, and 1 pure nonyellow. Examination of the pregnant yellow females shows that the homozygous yellow always dies as an embryo, which accounts for the unusual ratio of 2:1. Some lethals bring about death in the early stages of the embryo, others in later stages. Some human defects are supposed to be caused by them. Although many lethal genes are recessive and produce their effects only when they are homozygous, there are other lethal genes that are dominant, causing nonlethal effects when heterozygous and lethal effects when homozygous. The creeper fowl, for instance, has very short legs in the heterozygous state; when homozygous, the chicks die before hatching.

■ SEX DETERMINATION

The first really scientific clue to the cause of sex was discovered in 1902 by McClung, who found that in some species of bugs (Hemiptera) two kinds of sperm were formed in equal numbers. One kind contained among its regular set of chromosomes a so-called accessory chromosome that was lacking in the other kind of sperm. Since all the eggs of these species had the same number of haploid chromosomes, half the sperm would have the same number of chromosomes as the eggs and half of them would have one chromosome less. When an egg is fertilized by a spermatozoon carrying the accessory (sex) chromosome, the resulting offspring is a female; when fertilized by the spermatozoon without an accessory chromosome, the offspring is a male. There are, therefore, two kinds of chromosomes in every cell; X chromosomes determine sex (and sex-linked traits), and **autosomes** determine the other bodily traits. The particular type of sex determination just described is often called the XX-XO type, which indicates that the females have 2 X chromosomes and the male only 1 X chromosome (the O stands for its absence).

Later, other types of sex determination were discovered. In man and many other forms there are the same number of chromosomes in each sex, but the sex chromosomes (XX) are alike in the female but unlike (XY) in the male. Hence the human egg contains 22 autosomes + 1 X chromosome; the sperm are of two kinds: half will carry 22 autosomes + 1 X and half will bear 22 autosomes + 1 Y. The Y chromosomes in such cases are diminutive. At fertilization, when 2 X chromosomes come together, the offspring will be a girl; when XY, it will be a boy (see Table 33-2).

A third type of sex determination is found in birds, moths, and butterflies in which the male has 2 X (or sometimes called ZZ) chromosomes and the female an X and Y (or ZW). In this latter case the male is homozygous for sex and the female is heterozygous.

ROLE OF AUTOSOMES. In *Drosophila* certain intersexes have been found that suggest that autosomes may play a part in the development of sex. Because of irregularity (**nondisjunction**) in meiosis, it is possible for a fly to have an extra set of autosomes in addition to the regular set. The female fly normally has 6 autosomes (expressed as 2 A) + 2 X chromosomes. If a fly has 3 A + 2 X, instead of being a female as expected from the sex chromosomes' composition, it actually is an intermediate between male and female. The extra autosomes have upset the genic balance, indicating that sex is determined by a quantitative relation between the X chromosomes and autosomes. Also, through irregularities of meiotic divisions, it is possible for a female to have a chromosome complex of 2 A + 3 X, in which case she is, ironically, called a "superfemale." A "supermale," on the other hand, has 3 A + 1 X. In both these cases the sex characteristics are exaggerated toward femaleness or maleness, respectively. These experiments also show that the X chromosomes carry more genes for femaleness and the autosomes more genes for maleness. These abnormalities in sex determination do not invalidate the various types of sex determination described earlier in this section when there is a normal genic balance between the autosomes and the sex chromosomes. In man and the mouse (and perhaps in others), however, the Y chro-

TABLE 33-2
Chromosomal types of sex determination*

Haploid-diploid type		
Male (1 genome)	sperm (1 genome)	from unfertilized egg (1 genome)
Female (2 genomes)	ovum (1 genome)	from fertilized egg (2 genomes)
Example: *Honeybee*		
XO type		
Male (2 n plus XO)	sperm (n plus X) or (n plus O)	
Female (2 n plus XX)	ovum (n plus X)	
Example: *Grasshopper*		
XY type		
Male (2 n plus XY)	sperm (n plus X) or (n plus Y)	
Female (2 n plus XX)	ovum (n plus X)	
Example: *Human*		
WZ type		
Male (2 n plus ZZ)	sperm (n plus Z)	
Female (2 n plus WZ)	ovum (n plus Z) or (n plus W)	
Example: *Bird*		

*n = 1 set of autosomes; O = absence of a sex chromosome; *genome* is a haploid set of chromosomes.

mosome primarily determines maleness. In *Drosophila* an XO individual is a male with abnormal sperm and an XXY fly is a functional female. In man the abnormal condition of an XXY is a sterile male; the XO is a sterile female.

SEX MOSAICS. An interesting abnormality of sex is illustrated by the so-called **gynandromorphs** or **sex mosaics.** In such cases one part of the body shows male and the other female characteristics. It is caused by the irregular distribution or loss of sex chromosomes during early development. Thus a zygote with 2 X chromosomes could lose one of the X's from one of the early blastomeres and all the descendants of that cell would have male characteristics; the 2 X cells would be female. Such abnormalities are found in insects in which the sex characteristics of the cell depend mainly on the sex chromosomes. They are also excellent examples for the confirmation of the chromosomal determination of sex.

MICROSCOPIC DETERMINATION OF SEX. Sexual dimorphism in the nuclei was first discovered in 1949 by M. Barr and E. G. Bertram. These investigators discovered a chromatin mass in female nuclei that they identified as the heterochromatic parts of 2 X chromosomes in the interphase stage. Such a body, which is often called the nucleolar satellite, or the Barr body (Fig. 33-15, *A*), is found lying against the nuclear membrane. It is not found in the nuclei of the male because the male has only 1 X. Although first found in the cat, the sex chromatin has been found in other organisms including man. An oral smear of the oral epithelium is one of the simplest places in man for demonstrating the Barr body.

Another type of sexual dimorphism is found in the polymorphonuclear leukocyte of blood smears. An accessory nuclear lobule called the "drumstick" (Fig. 33-15, *B*) is found in leukocytes from females but is lacking or else is very diminutive in males. Both drumsticks and Barr bodies indicate the presence of more than 1 X chromosome. In those abnormal sex chromosome cases, such as the XXX constitution, there may be two Barr bodies and two "drumsticks."

Sex-linked inheritance

Sex-linked inheritance refers to the carrying of genes by the X chromosomes for body characters that have nothing to do with sex. The sex chromosomes, in addition to determining sex in an organism, also have genes for other body traits, and because of this the inheritance of these characters is linked with that of sex. The X chromosome is known to contain many such genes, the Y chromosome only a few because of its small size. Such sex-linked traits are not always limited to one sex but may be transmitted from the mother to her male offspring or from the father to his female offspring. One of the examples of a sex-linked character was discovered by Morgan in *Drosophila* (Fig. 33-16). The normal eye color of this fly is red, but mutations for white eyes do occur. The genes for eye color are known to be carried in the X chromosome. If a white-eyed male and a red-eyed female are crossed, all the F₁'s are

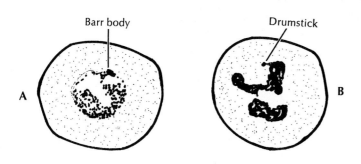

FIG. 33-15
Cellular determination of sex. **A,** Squamous epithelial cell of human female showing Barr body, which is absent in male. **B,** White blood cell of human female showing accessory nuclear lobule ("drumstick"), which is mostly lacking in male.

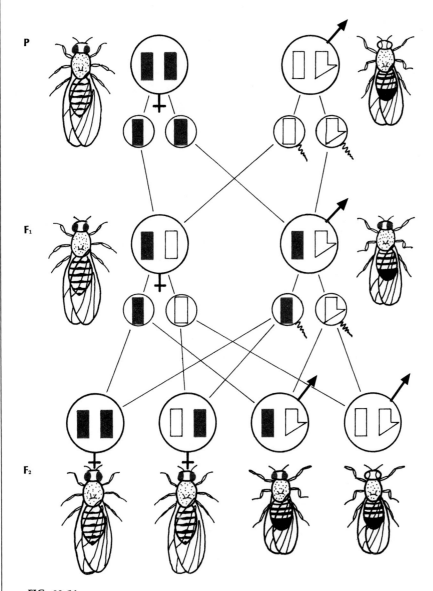

FIG. 33-16
Sex determination and sex-linked inheritance of eye color in fruit fly *Drosophila*. Normal
red-eye color is dominant to white-eye color. If a homozygous red-eyed female and
white-eyed male are mated, all F_1 flies are red eyed. When F_1 flies are intercrossed, F_2
yields approximately 1 homozygous red-eyed female and 1 heterozygous red-eyed female
to 1 red-eyed male and 1 white-eyed male. Genes for red eyes and white eyes are carried
by sex (X) chromosomes; Y carries no genes for color.

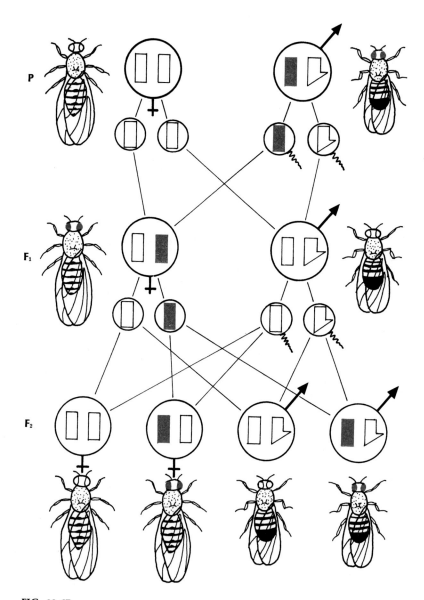

FIG. 33-17
In cross of a homozygous white-eyed female and a heterozygous red-eyed male
(reciprocal cross of Fig. 33-16), F_1 consists of white-eyed males and red-eyed females. In
the F_2, there are equal numbers of red-eyed and white-eyed females and red-eyed and
white-eyed males.

red eyed, because this trait is dominant. If these F_1's are interbred, all the females of F_2's will have red eyes and half the males will have red eyes and the other half white eyes. No white-eyed females are found in this generation; only the males have the recessive character (white eyes). The gene for white-eyed, being recessive, should appear in a homozygous condition. However, since the male has only 1 X chromosome (the Y does not carry a gene for eye color), white eyes will appear whenever the X chromosome carries the gene for this trait. If the reciprocal cross is made in which the females are white eyed and the males red eyed, all the F_1 females are red eyed and all the males are white eyed (Fig. 33-17). This is called **crisscross inheritance.** If these F_1's are interbred, the F_2 will show equal numbers of red-eyed and white-eyed males and females.

If the allele for red-eyed is represented by R and white-eyed by r, the following diagrams will show how this eye color inheritance works:

Parents	RR (red ♀)	×	rY (white ♂)
F_1	Rr (red ♀)	×	RY (red ♂)
Gametes	R,r		R,Y
F_2	RR (red ♀) RY (red ♂)		rY (white ♂) rR (red ♀)

Reciprocal cross

Parents	rr (white ♀)	×	RY (red ♂)
F_1	rR (red ♀)	×	rY (white ♂)
Gametes	r,R		r,Y
F_2	rr (white ♀) rY (white ♂)		Rr (red ♀) RY (red ♂)

Many sex-linked genes are known in man, such as bleeder's disease (hemophilia), night blindness, and color blindness. The latter is often used as one of the most striking cases of sex-linked inheritance in man. The particular form of color blindness involved is called daltonism, or the inability to distinguish between red and green. The defect is recessive and requires both genes with the defect in the female but only one defective gene in the male to acquire a visible defect. If the defective allele is represented by an asterisk (*), the following diagram will show the inheritance pattern:

Cross between color-blind male (X*Y) and homozygous normal female (XX)

Parents	X*Y	×	XX
F_1	X*X	×	XY (all normal)
Gametes	X*,X		X,Y
F_2	X*X, X*Y, XX, XY		
	(X*Y, color-blind male)		

It will be seen from this cross that the color-blind father will transmit the defect to his daughters (who do not show it because each has only one defective gene), but these daughters transmit the defect to one half their sons (who show it because a sex-linked recessive gene in the male has a visible effect).

■ LINKAGE AND CROSSING-OVER

The study of heredity since the discovery of Mendel's laws reveals that many traits are inherited together. Since the number of chromosomes in any animal is relatively few compared to the number of traits, evidently each chromosome bears many genes. All the genes contained in a chromosome tend to be inherited together and are therefore said to be **linked.** The sex-linked phenomena described in the previous section are examples of linkage; genes borne on chromosomes other than sex chromosomes form **autosomal linkage.** Genes, therefore, occur in linkage groups and there should be as many linkage groups as there are pairs of chromosomes. In *Drosophila,* in which this principle has been worked out most extensively, there are four linkage groups that correspond to the four pairs of chromosomes found in these flies. Small chromosomes have small linkage groups and large chromosomes have large groups. Five hundred genes have been mapped in the fruit fly and all these are distributed among the four pairs.

How the mendelian ratios can be altered by linkage can best be illustrated by one of Morgan's experiments on *Drosophila.* When a wild-type fly with gray body and long wings is crossed with a fly bearing two recessive mutant characters of black body and vestigial wings, the dihybrids (F_1) all have gray bodies and long wings. If a male of one of the F_1's is testcrossed with a female with a black body and vestigial wings, the flies are all gray-long and black-vestigial. If there had been free assortment, that is, if the various characters had been carried on different chromosomes, the expected offspring would have been represented by four types of flies: gray-long, gray-vestigial, black-long, and black-vestigial. However, in this case gray-long and black-vestigial had entered the dihybrid cross together and stayed together, or linked.

Linkage, however, is usually only partial, for it is broken up frequently by what is known as **crossing-over.** In this phenomenon the characters usually sep-

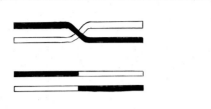

FIG. 33-18
Simple case of crossing-over between two chromatids of homologous chromosomes. Block of genes shown in black is thus transferred to white and vice versa.

arate with a certain frequency. How often two genes break their linkage, or their percentage of crossing-over, varies with different genes. In some cases this percentage of crossing-over is only 1% or less; with others it may be nearly 50%. The explanation for crossing-over lies in the synapsis of homologous chromosomes during the maturation division when two nonsister chromatids sometimes become intertwined and, before separating from each other, ex-

change homologous portions of the chromatids (and the genes they bear) (Fig. 33-18).

Crossing-over makes possible the construction of chromosome maps and proof that the genes lie in a linear order on the chromosomes. To illustrate how this is done, one may take a hypothetical case of three genes (A, B, C) on the same chromosome. In the determination of their comparative linear position on the chromosome, one will first need to find the crossing-over value between any two of these genes. If A and B have a crossing-over of 2% and B and C of 8%, then the crossing-over percentage between A and C should be either the sum (2 + 8) or the difference (8 − 2). If it is 10%, B lies between A and C; if 6%, A is between B and C. By laborious genetic experiments over many years, the famed chromosome maps in *Drosophila* were worked out in this manner (Fig. 33-19). Cytologic investigations on the giant chromosomes since these maps were made tend to prove the correctness of the linear order, if not the actual position, of the genes on the chromosomes. There is no evidence of crossing-over occurring in the giant chromosomes themselves, and it is also absent in the male *Drosophila*.

X—Chromosome 1		Chromosome 2		Chromosome 3		Chromosome 4	
0.0	Yellow body	0.0	Net veins (wings)	0.0	Roughoid eyes	0.0	Bent wings
1.5	White eyes	1.3	Star eyes	0.2	Veinlet wings	0.1±	Shaven bristles
3.0	Facet eyes	12.0	Fat body	19.2	Javelin bristles	0.2	Eyeless
7.5	Ruby eyes	13.0	Dumpy wings	26.0	Sepia eyes		
13.7	Cross veinless wings	16.5	Clot eyes	26.5	Hairy body		
18.9	Carmine eyes	41.0	Jammed wing	41.4	Glued eyes		
20.0	Cut wings	48.5	Black body	44.0	Scarlet eyes		
21.0	Singed bristles	51.0	Reduced bristles	45.3	Clipped wings		
27.7	Lozenge eyes	57.5	Cinnabar eyes	46.0	Wrinkled wings		
33.0	Vermilion eyes	67.0	Vestigial wings	48.0	Pink eyes		
36.1	Miniature wings	72.0	Lobe eyes	50.0	Curled wings		
43.0	Sable body	75.5	Curved wings	58.2	Stubble bristles		
44.4	Garnet eyes	93.3	Humpy body	59.0	Roof wings		
51.5	Scalloped wings	100.5	Plexus veins (wings)	63.1	Glass eyes		
57.0	Bar eyes	104.5	Brown eyes	70.7	Ebony body		
62.5	Carnation eyes	107.0	Speck wings	74.7	Cardinal eye		
66.0	Bobbed bristles			91.1	Rough eyes		
				93.8	Beaded wings		
				100.7	Claret eyes		
				104.3	Brevis bristles		

FIG. 33-19
Chromosome maps of certain representative genes in *Drosophila*. One chromosome of each pair is shown with figures, indicating relative positions of genes in crossover units from end of chromosome.

T. H. Morgan summarizes the nature of heredity from the evidence accumulated in his vast studies as follows:

1. The characteristics or phenotypes of the individual organism are referable to paired genes held together in a definite number of linkage groups (chromosomes).

2. When each pair of genes separate according to Mendel's first law, each gamete will contain only one set of genes.

3. The genes in different linkage groups will assort independently according to Mendel's second law.

4. Crossing-over occurs at times between genes in homologous linkage groups.

5. The frequency of crossing-over depends upon the linear arrangement and relative position of the genes.

■ CHROMOSOMAL CHANGES

The term "mutation," discussed earlier in this chapter, is usually restricted to alteration within the gene itself, but there are also other heritable variations associated with chromosomal changes. These are often referred to as chromosomal aberrations and involve variations in the number, structure, and arrangement of chromosomes that affect many genes at one time. The main types of chromosomal changes are loss of a part of a chromosome **(deletion),** shifting of a part of a chromosome over to another chromosome **(duplication),** rearrangement of parts of a chromosome so that a block of genes is inverted **(inversion),** and the gaining of a whole set or sets of chromosomes **(polyploidy).** This latter term refers to those conditions in which there are more than the diploid number so that triploids, tetraploids, and even higher combinations are formed. Such aberrant conditions arise through a duplication of the chromosome complement by nondisjunction in the reduction division of maturation. Such polyploids differ in characters from the original parent stock. Although most of such cases are found in plants, they are not unknown among animals, particularly some amphibians.

■ HYBRID VIGOR

Hybrid vigor, or **heterosis,** refers to the greater vitality and vigor manifested by the hybrids produced by crossing individuals of two pure races that differ from each other in a number of genes. Such hybrids are often bigger and more vigorous than either parent. What actually happens is that such crosses may bring together dominant genes for vigor in the hybrid, provided that each of the pure line races carries a vigor gene lacking in the other. In this way it is conceivable that the hybrid may contain homozygous dominant vigor genes that would account for its more desirable traits. There is also the possibility that some genes in the heterozygous condition in the hybrid might produce more vigor than genes in the homozygous state found in the inbred race. Hybrid vigor, however, tends to be lost in succeeding generations when hybrids are crossed, because the desirable dominant genes may segregate out again and the undesirable recessives would again make their effects visible.

Among the many examples of hybrid vigor is the valuable hybrid corn, which enabled the farmers of the United States to vastly increase their corn yield. To overcome the infertility of inbred lines in hybrid corn production, the **double cross** method is employed. Four inbred lines (A, B, C, D) are intercrossed in pairs (A-B and C-D). The hybrids of these two crosses (AB and CD) are then crossed to produce the commercial seed for the farmer. Unfortunately corn plants are often subject to pathogenic epidemics because of their genetic uniformity. Diversity is one of the best insurances against pathogenic epidemics.

Many studies have been made on hybrid vigor in man. A study of history usually shows that great civilizations are preceded by a mixing of races. Race crossing, even among subgroups of one race, often results in exceptional vitality and vigor. The flowering of Greek culture a few centuries before the Christian era has been shown to be attributable to a preliminary close inbreeding of many independent political units, thus producing many types of gene pools, followed by a recombination of these genes in crosses (hybrid vigor) when political barriers were broken down in the later, larger city units. Close inbreeding in the human race usually results in reduced vitality, many defective traits, and a pronounced sterility of the stock.

■ CYTOPLASMIC INHERITANCE

All the genetic behavior so far considered has stressed the importance of the nuclear elements (chromosomes and genes) as the bearers of heredity. Does the cytoplasm bear any hereditary factors? Of course, the effects of the gene are carried out in the cytoplasm, such as growth, development, secretion, and enzymatic action, but are there self-duplicating genetic units in the cytoplasm itself?

Evidence for cytoplasmic inheritance is often sought in reciprocal cross differences (for example, whether the genetic type is introduced into a cross by the father or the mother) because the egg (maternal contribution) contains most of the cytoplasm of the new organism, and the phenotype of the latter should follow that of the mother. A few cases of cytoplasmic inheritance seem well established. One of these is the chlorophyll-bearing plastids in the cytoplasm of certain plants. These are self-duplicating bodies and some of them are always maternal in character regardless of the kind introduced by the pollen in the cross. Another case is the cortical pattern of certain paramecia that has been explained by nonchromosomal inheritance (T. M. Sonneborn). When two paramecia, one with a double cortical pattern and the other with a single cortical pattern, conjugated, it was occasionally found that one conjugant had received a piece of the cortex of the other and became intermediate in appearance between the other two types and thereafter reproduced true to this new intermediate type.

Mention has already been made in Chapter 7, The Acellular Animals, of the kappa substance in the cytoplasm of certain strains of paramecia. This kappa substance produces paramecin that can kill sensitive strains containing no kappa substance. When individuals of killer clones conjugate with members of sensitive strains, the latter will become killers only when they receive some cytoplasm in the exchange, indicating that the cytoplasm carries the plasmagenes, which such cytoplasmic bodies are called. Kappa particles have many of the characteristics of genes, for they are self-duplicating and contain DNA. However, it has been shown that the maintenance of the kappa substance is actually dependent on a nuclear gene. Thus, in some cases at least, there is an interrelationship between plasmagenes and nuclear genes and some geneticists consider the former as replicas of nuclear genes that are released into cytoplasm. Plasmagenes have many resemblances to viruses, such as self-duplication, cytoplasmic location, and capability of transmissible mutations, but they do not cause diseases in the cells in which they are found. The demonstration of plasmagenes bids to shed some light upon the difficult problem of developmental differentiation.

◼ PRACTICAL APPLICATION OF GENETICS

Although genetics as a science has developed within the last 50 years, many of the practical applications of heredity extend as far back as the dawn of civilization itself. The improvement of domestic plants and animals has always interested man. Has the discovery of the precise way by which the laws of heredity work helped him to make additional progress? It is not amiss for the zoology student to look at a few of the ways by which man has been able to use the principles of hereditary formulas to his own advantage.

METHODS OF INCREASING PRODUCTIVITY IN PLANTS AND ANIMALS. Plant and animal breeders have two main objectives—to produce strains that give greater productivity and to produce those that are pure in their pedigree. Desirable hereditary qualities are carefully selected and propagated by breeding. Often the production of pure lines is for the purpose of crossing with other strains to take advantage of hybrid vigor, a practice used by both plant and animal breeders.

An odd number of chromosomes in a set usually means sterile descendants. Bananas are mostly seedless because they have three of each kind of chromosomes instead of two. This condition produces an abnormal meiosis that leads to sterility. Sterile strains may be produced by crossing two fertile varieties, one with two sets and one with four sets. Thus in some cases sterility may be desirable.

The alert breeder takes advantage of new combinations and variations produced by crossing different strains, by mutations, and by crossovers.

HYBRID VARIETIES. The production of hybrid corn was one of the great triumphs of genetics. In the 30 years since it was introduced, the yield of corn per acre has been increased 25% to 50%. Pure lines of corn were found to be of low fertility and gave scanty yields. When these pure strains were crossed, the hybrids showed a remarkable increase in both fertility and general yield. Only the seed of the first generation after a cross can be used, for later generation seed tends to produce segregation of undesirable traits, as explained under hybrid vigor, p. 791.

PROGENY SELECTION. Selection, an effective

tool in the hands of the breeder, may be done by **phenotypic selection** or by **progeny selection.** The first is based on the appearance of the desired trait or traits in the individual. This method has obvious disadvantages because the phenotypic appearance of the animal or plant is not necessarily an indication of its genotype. For this reason, prize-winning stock may not transmit desirable traits to its offspring. Progeny selection, however, is based on selecting those individuals that produce desirable offspring so that the genotype is the chief basis of selection. Many great successes have been scored this way, such as high milk-producing cattle, poultry with higher yield of eggs, and desirable strains of tobacco.

DISEASE-RESISTANT STRAINS. By careful selection and breeding, plant breeders produce varieties that do not become infected by common plant diseases. In this way it has been possible to develop wheat that is resistant to rust and corn to smut. However, many disease germs mutate, and some of the mutants may be able to attack plants that were formerly disease resistant. Constant vigilance is necessary to spot these changes and to introduce a new resistant gene into the genetic constitution of the stock.

DETECTION AND CONTROL OF LETHAL GENES. Many cases of lethal genes, or genes that destroy the individual, are known among stock breeders. Some of these produce visible effects in the offspring, such as the peculiar "bulldog" calf (so-called because of its facial appearance) and the creeper fowl with unusually short appendages. Some lethal genes produce their effects only in the homozygous condition and others when they are heterozygous. In the first case their detection is not always easy, for as long as the lethal gene is carried along with a normal one no harm is done. Stock breeders can by trial matings find these lethals and exclude such animals from breeding.

POLYPLOIDY. Many polyploids have been experimentally produced, such as tulips, roses, fruit trees, tomatoes, and even cotton, and represent an improvement over the original stock. In cases when polyploids cannot breed true, as in triploids, they may be propagated asexually.

APPLICATION OF BIOCHEMICAL GENETICS. Another practical application of an unusual form is the use of molds, such as *Neurospora,* in determining the presence of certain vitamins and amino acids. When these molds are irradiated, some of their genes may be made to mutate so that they lose the power to synthesize specific vitamins and amino acids. Apparently this failure to synthesize is caused by the loss of certain enzymes. Strains of mold that lack the power to synthesize a particular amino acid will not grow in a medium deficient in this specific amino acid. It is thus possible to determine whether a culture contains or lacks specific vitamins or amino acids by ascertaining its ability to support certain such strains of mold.

■ HUMAN HEREDITY

The study of human inheritance is one of the most difficult fields in genetics. Experimental breeding, the key to most genetic studies, is impractical. It is also difficult to study man's heredity from a cytogenetic viewpoint because of the many chromosomes (46) that he possesses and because he is a relatively slow-breeding animal. Man is largely heterozygous in his hereditary makeup, and he is influenced in his development to a greater extent by an extremely complicated and varied environment. Despite these conditions, there is every reason to suppose that man follows the same principles of genetics as those of other organisms. Many studies of human inheritance have stressed congenital defects because such traits can easily be followed through many generations and are easily recognized.

Genetic diseases

The work of G. W. Beadle and E. L. Tatum on the mold *Neurospora,* by which it was shown that a single gene controlled the specificity of a particular enzyme, has helped explain many inherited human diseases. An English physician, A. E. Garrod, contended in 1908 that enzyme deficiencies were to blame for certain disorders that he described as "inborn errors of metabolism." Among such disorders is phenylketonuria (PKU), which is produced by the absence of the enzyme phenylalanine hydroxylase carried by a recessive gene. A person afflicted with this disorder cannot convert phenylalanine into tyrosine. Some phenylalanine is converted into phenylpyruvic acid, which produces injury to the nervous system and mental deficiency. The disorder is detected by a simple test with blood or urine and can be controlled by restricting phenylalanine in the diet.

ALBINISM. About one person in 20,000 has the condition known as albinism, which is characterized by the lack of the dark pigment melanin in the skin and hair. Albinism is caused by the absence of the enzyme tyrosine, which is necessary for the synthesis of melanin. Albinism is also inherited as a recessive gene, and an individual must be homozygous for the condition to show it.

SICKLE CELL ANEMIA. Some genetic disorders that are caused by recessive genes may show some effect in a heterozygous condition. One of these is the sickle cell mutant gene that produces the abnormal hemoglobin S. This hemoglobin differs from normal hemoglobin in having a valine amino acid in place of the glutamic acid molecule in a chain of more than 300 amino acids. A person with one gene for normal hemoglobin and one gene for the abnormal hemoglobin S will usually suffer a mild anemia. However, such persons have a better resistance to falciparum, or malignant malaria, which may account for the high prevalence of the heterozygotes in parts of Africa. A homozygous condition of hemoglobin S is usually fatal.

KURU DISEASE. A few years ago, T. Dobzhansky described and discussed a peculiar heredity disease among the Fore tribe in New Guinea. This disease (kuru) is caused by a mutation gene that apparently behaves as a dominant in the female and as a recessive in the male. In the homozygous condition of the trait in both sexes the individual dies in early childhood. Heterozygous females die shortly after adolescence; heterozygous males have no visible effects, but carry the mutant gene. As a result, the males outnumber the females (in some instances) by a ratio of 2.5 males to 1 female.

RECESSIVE NATURE OF GENETIC DISEASES. Although most genetic diseases have selective disadvantages and natural selection tends to eliminate them whenever they appear in the population, recurrent mutations furnish fresh cases, usually with a definite frequency for each gene responsible. The examples already described are almost entirely recessive, and the expressed trait in the homozygous condition is of relative infrequency, although the hidden recessive gene may be present in fairly large numbers in the population as a whole. Albinism, for example, occurs in only one person among 20,000, but the recessive gene is found once in every 70 persons. Dominant genes would show up immediately if present.

FREQUENCY OF GENETIC DISEASES. Genetic defects are very common in the population and may be on the increase. Geneticists estimate that at least one person in every four or five of the population is born with a defective gene, which can bring about serious consequences to the population, depending on whether the genes are dominant or recessive. It is believed that at least one conception in every five results in a genetic defect that destroys the individual at some time before maturity. In addition, various types of abnormalities, not necessarily lethal, are caused by both dominant and recessive genes. These include sex-linked and non–sex-linked mutant genes as well as those caused by gross chromosomal abnormalities. The noted geneticist H. J. Muller estimates that each individual carries at least eight recessive lethal genes in his sex cells. These will cause death only in the homozygous condition, the possibilities of which can be increased by marriage of relatives.

ROLE OF VIRUSES. It is well established that some viruses are responsible for cancerous growth in some animals, and there is a strong likelihood that they may cause cancer in man. It is also known that viruses can alter the genetic pattern by introducing new genes into the cells they infect. This alteration may cause cells to multiply out of control. Virus study at the present time emphasizes the way by which such modifications lead to tumorous growth.

Roles of heredity and environment

If the human organism is the product of both heredity and environment, which is the more important factor? Although the genic constitution within the nucleus remains essentially the same in all cells and tissues, the cytoplasm undergoes various changes under the joint influence of genes on the one hand and a constantly changing environment on the other. Both these factors are, therefore, important in the realization of the individual's possibilities.

Of the two factors, heredity is relatively stable, whereas the environment is constantly changing. Even during intrauterine existence, the important factors of nutrition are operating, as well as numerous hormones from the mother's body. Here there are interactions between the genes of the embryo and the environment of the uterus. After birth, additional factors, including complex social ones, have a pronounced effect in the final molding of the individual character. Genes direct the general course of development, but their potentiality can be expressed or suppressed by the environment.

Studies made on twins reared in different homes

and under different social and other environmental conditions show a remarkable similarity in height, weight, and other physical characteristics. However, their mental traits, I.Q., and general intelligence may show some differences as well as many cultural aspects of their characters. Such investigations indicate that hereditary factors dominate in influencing the general development and personality of the individual, although environment, such as education, does significantly influence the general intelligence level of the individual.

Hair and eye color

The color of the hair seems to be determined by several genes or several pairs of modifying factors. Often the pigmentation of hair and eye color are correlated, for the darker shades of hair are usually accompanied by darker eyes. In general, blond hair is recessive to the darker shades, but the presence of varying shades of blond and dark hair indicates a blending effect or the interaction of more than one pair of genes. Red hair is recessive to the other shades of hair. One may be homozygous for genes for red hair and have a darkish shade of red hair color because of the presence of genes for darker hair.

The color of the eyes is due to the presence and location of pigment in the iris. If the pigment is on the back of the iris, the eyes are blue; if the pigment is on both back and front of the iris, the eyes are of the darker shades. If pigment is lacking altogether (albinism), the eyes are pinkish because of the blood vessels. The pigment is actually dark and produces the varying eye colors by the reflection of light. Blue eyes are recessive to the other eye shades, and when the parents are blue eyed, the children are normally blue eyed. Blue-eyed children may appear also if parents are heterozygous for the darker eye colors. The manner in which eye color is inherited indicates that there are many kinds of genes, all variant forms of the gene for dark color, which behave in a simple mendelian way in their hereditary expression. In exceptional cases, environmental factors may prevent the full expression of eye pigment so that homozygous dark-eyed parents could have a blue-eyed child.

Blood group inheritance

The inheritance of blood groups follows Mendel's laws. This inheritance is based on three genes or allelomorphs, I^a, I^b and i. I^a and I^b are antigens and

TABLE 33-3
Inheritance of blood groups

Parent groups	Possible children groups	Impossible children groups
O × O	O	A, B, AB
O × A	O, A	B, AB
O × B	O, B	A, AB
A × A	O, A	B, AB
A × B	O, A, B, AB	None
B × B	O, B	A, AB
O × AB	A, B	O, AB
A × AB	A, B, AB	O
B × AB	A, B, AB	O
AB × AB	A, B, AB	O

are dominant, and they never appear in a child's blood unless present in at least one of the parents; i represents no antigen and is recessive to the antigens. Neither I^a nor I^b is dominant to each other, but each is dominant to i. The relationships of the blood groups and genotypes are as follows:

Blood groups	O	A	B	AB
Genotypes	ii	I^aI^a or I^ai	I^bI^b or I^bi	I^aI^b

If both parents belong to group O, a child must also belong to group O. On the other hand, if one parent belongs to group A and the other to B, the child could belong in any one of the four groups. The various possibilities of inheritance are shown in Table 33-3.

This pattern of heredity has some practical application in medicolegal cases involving disputed parentage. From the possibilities given in Table 33-3, it is seen that, if a child in question has group A and the supposed parents group O, it is obvious that the child could not belong to them. On the other hand, if the parents belong to groups A and B, the blood tests would prove nothing, for the parents could have children of all groups.

Evolutionists have laid much stress on the geographic distribution of blood groups among the various racial populations. The frequencies of the blood groups O, A, B, and AB have been tabulated by investigators for most races all over the world, although data are scanty and incomplete in many instances. Among the interesting facts revealed by these studies is the absence of the allele B in the American Indians and Australian aborigines, the high frequency of B in Asia and India and its decline in Western Europe, and the high frequency of group O in Ireland and Iceland. The distribution of the Rh factor also shows a varied pattern. Reasons for these varied distributions are obscure, but some explanations have been advanced, such as genetic drift involving small populations, natural selection, and migration and mixing of races.

Inheritance of Rh factor

As stated in a former section (p. 611), about 85% of American people possess a dominant gene called **Rh,** which causes the formation of a special antigen in the blood. The remaining 15% have the recessive allele **rh,** which cannot produce this antigen. Investigations have disclosed that there are many other alleles at this locus, thus greatly increasing the possible blood groups. Thus there are several kinds of Rh-positive and Rh-negative persons. At present at least eight different alleles are found in the series designated as R^1, R^2, R^0, R^z, r', r'', r^y, r. The capital letters indicate those alleles that give an Rh-positive reaction; the small letters stand for the Rh-negative ones. Although all of these subtypes are inherited and produce antigens, it appears that the allele Rh^0 is mainly responsible for the clinical cases of erythroblastosis fetalis. The genetics of all these subtypes becomes involved when one considers all their possible combinations. They represent one of the most extensive groups of multiple alleles known. Other antigens are caused by factors independent of Rh, such as the so-called Kell antigen produced by an uncommon dominant gene that, in incompatibility cases, can cause serious hemolytic diseases.

Other antigens that do not react to form antibodies and are of no clinical significance are the M and N antigens. No allele for the absence of these antigens has been found. All persons are typed as M, N, or MN. In their inheritance pattern, two M type of parents will produce only M children; two N type of parents, only N children; and types M and N parents will give all these types of children in the ratio of 1M:2MN:1N.

Inheritance of mental characteristics

Mental traits, both good and bad, are known to be inherited, although because of environmental influences it is not always possible to appraise them genetically. Heredity is in general responsible for the basic patterns of intelligence and mental deficiencies, although environmental factors can and do influence the development of intellectual capacities. Intelligence is not a single hereditary unit; apparently many different genes are responsible for its expression. The field of inheritance affords many examples of both outstanding abilities and mental defects being handed down through long family histories. One is forced to conclude that hereditary factors have played a major part in the determination of these intellectual strengths or weaknesses. The best kind of environment in the world cannot make a superior intelligence out of one who has inherited a moron potentiality, and conversely, superior hereditary abilities may always remain mediocre unless stimulated by favorable environmental factors. Identical twins, for instance, with exactly the same genes for intelligence may show considerable difference in their I.Q.s when reared under different advantages of education and culture.

There are several million feebleminded persons in the United States, but only a small percent have come from feebleminded parents. The majority of them are born of parents heterozygous for this trait. Such parents may be in all particulars normal for intelligence and carry the recessive gene for feeblemindedness. Recessive genes are hidden and may continue so until they become homozygous. That is why the sterilization of feebleminded persons to prevent them from reproducing their kind would have very little effect on the distribution of genes for feeblemindedness in the population.

The evidence indicates that talents for music have a constitutional basis. The hereditary pattern, however, is very complex and probably involves a number of genes. It is not always possible either to distinguish between the influences of heredity and environment in musical families. Special talents may be to a considerable extent independent of general intelligence, for some individuals of exceptional ability along some lines may have a low I.Q.

Should first cousins be allowed to marry?

The marriage of near relatives such as first cousins would bring recessive genes together in a homozygous condition. It is logical that descendants from a near common ancestor are going to share his genes. The more remote the relationship, the less is this possibility. Inbreeding does intensify many kinds of defects such as feeblemindedness, congenital deafness, and albinism. For this reason, in most states marriage of first cousins is prohibited. Of course, if there are desirable genes in a family stock, marriage between cousins could bring these traits together and produce superior children. Laws, however, are made with reference to the prevention of undesirable traits instead of the promotion of favorable ones.

Inheritance of twinning

Fraternal twins, which are four or five times more common than identical ones, result from the independent fertilization of separate ova by separate sperm. They are simply conceived and born together, and genetically they have the same likenesses and differences as ordinary brothers and sisters. They may be of the same or opposite sex. **Identical** twins come from a single zygote that has split during its early stages. They have the same genetic constitution and are always of the same sex. Identical twins show a remarkable similarity in their general characters, both physical and mental. When the halves of the zygote fail to separate completely, Siamese twins are the result.

The inheritance of twinning is very complex, but there seems to be a hereditary basis for many cases of twins. Sheep, especially, exhibit strains of twinning. Both father and mother seem to influence the heredity of twinning. It is not too difficult to see how the father could induce the formation of identical twins because something about his sperm might cause the zygote to split. It is more difficult to see how he could cause fraternal twins. Some biologists suggest that the female releases two eggs at ovulation, of which only one is normally fertilized. But if the sperm are unusually virile, both eggs could be fertilized. More puzzling still is the fact that older parents tend more to have twins than do younger ones. There are also racial variations. Blacks have more twins than do whites, but most yellow people have fewer than do whites. The prenatal mortality of twins may be much higher than that of single conceptions, which would make a difference in the ratios.

Inheritance of certain physical traits in man

It is impossible to deal with man's heredity in a simple mendelian ratio. Information about his hered-

ity must be acquired by inspection and analysis of family life histories, or pedigrees. Such a plan involves the formation of hypotheses to explain hereditary expression and careful checking of these hypotheses to determine whether they apply to the data obtained. The inheritance of many abnormal characters has been stressed in family pedigrees because they are easily followed.

Following are some of the more common traits, but their dominance or recessiveness is not always clear cut.

Dominant	Recessive
Curly hair	Straight hair
Dark hair	Light hair
Nonred hair	Red hair
Dark skin color	Light skin color
Hairy body	Normal hair
Skin pigmentation	Albinism (no pigment)
Brown eyes	Blue or gray eyes
Hazel eyes	Blue or gray eyes
Ichthyosis (scaly skin)	Normal skin
Near- or farsightedness	Normal vision
Hereditary cataract	Normal vision
Astigmatism	Normal vision
Glaucoma	Normal vision
Normal hearing	Deaf-mutism
Normal color vision	Color blindness
Normal blood clotting	Hemophilia
Broad lips	Thin lips
Large eyes	Small eyes
Long eyelashes	Short eyelashes
Short stature	Tall stature
Polydactylism (extra fingers or toes)	Normal number of digits
Brachydactylism (short digits)	Normal length of digits
Syndactylism (webbed digits)	Normal digits
Normal muscles	Progressive muscular atrophy
Hypertension	Normal blood pressure
Diabetes insipidus	Normal excretory system
Enlarged colon	Normal colon
Tasters (of certain substances)	Nontasters
Huntington's chorea	Normal nervous condition
Normal mentality	Schizophrenia
Nervous temperament	Phlegmatic temperament
Average intellect	Very great or very small intellect
Average intellect	Feeblemindedness
Migraine headache	Normal

Application of genetics to medical problems

Those in the medical profession realize that hereditary patterns have an important bearing on clinical problems. Disease germs are not carried through genes from one generation to another, but many authorities have found noticeable susceptibility to a particular disease running through families.

Some inherited traits have already been pointed out, such as color blindness, Rh factor, and deaf-mutism. Certain susceptibilities to cancer are inherited in both man and other animals. Children of parents who carry abnormal traits should be watched as they develop. For instance, hemolytic icterus, a condition in which the spleen is enlarged, is inherited as a dominant gene and about half of the offspring of a parent carrying this gene might be expected to have the defect. Some forms of hypertension have a hereditary basis, and children of such a parent should be trained to conform to conditions that will not aggravate the malady. It is possible for doctors to facilitate their diagnosis of obscure diseases by knowing whether the family and near relatives carry susceptibilities to them.

At least 25% of all illnesses have a genetic component. Vitamin-dependent genetic diseases are now known in many instances. These are included among other metabolic disorders that respond to vitamin treatment by such vitamins as D_1, B_1, B_2, B_6, and B_{12}. Vitamin-dependent diseases require massive doses of the specific vitamin because the body of the patient is unable to use properly that particular vitamin. Fat-soluble vitamins (A and D) may accumulate in tissues and become dangerous; water-soluble vitamins (B and C) are usually excreted when in excess. Most vitamins serve as coenzymes to facilitate metabolic processes. Their molecules provide an active site with which the substrate can react and thus build an intermediate molecule with which the enzyme can react.

Influence of radiation on human heredity

Radiation and radioactive substances such as x-rays, radium rays, and ultraviolet light greatly increase the mutation rate of the gene. Most of these gene mutations are lethal and either destroy or else produce abnormalities of various types in the offspring of animals exposed under certain conditions to irradiation. Since the development of atomic energy, geneticists have been concerned about the possible genetic effect on man.

In the light of the controversy that is raging over the possible effects of radiation in an atomic age, it

may suffice to summarize certain generalizations that the data seem to substantiate:

1. All life is constantly exposed to high-energy radiations. Some of this radiation comes from natural sources, such as cosmic rays from outer space, radioactive elements (radium, thorium, radioactive isotopes of potassium), and atomic disintegration within the organism; some comes from the technologic use of radioactive substances in medicine and industry (x-rays, radium treatment, mustard gas); and some comes from the fallout of the explosion of atomic bombs.

2. High-energy radiations are definitely known to increase the rate of mutation in every organism tested.

3. Most mutations are harmful to the organism. This harm is produced chiefly by the ionization effect of radiation. Protoplasm is greatly injured by ionization because the molecular organization of chromosomes especially is disrupted and mutations (mostly harmful) of the code result.

4. Mutations as a rule are highly localized and may involve only one gene locus when the latter is struck by a quantum of radiation.

5. Radiations may affect any cell in the body, but only changes in sex cells can be transmitted to the offspring.

6. Sensitivity to radiation effects varies from species to species. Mice are far more sensitive than fruit flies, for example.

7. Genetic radiations from whatever source are insidious because their effects seem to be cumulative, although this is denied by some authorities. What really counts is the total amount of radiation one is exposed to during one's reproductive life. The rate of delivery of the radiation is of no consequence; the genetic effect is the same for low or high rate.

8. Some of the radioactive elements released in an atomic bomb explosion decompose very slowly and have half-lives of many years, such as strontium 90 (half-life, 28 years). This means that the body may be exposed to these isotopes for a long time and can absorb a large amount of them.

9. The danger of radioactive substances is far greater to future generations than to present ones because of the genetic implications referred to.

10. In addition to genetic effects of radiation, there are also physiologic or somatic effects. Some somatic effects are leukemia, cancer, and a syndrome

of radiation illness, depending on the amount of exposure. Body tissues of rapidly dividing cells are especially prone to damage. If the exposure has not been excessive, many somatic effects may be healed by therapeutic measures, but healing does not apply to genetic damages.

Human chromosomal abnormalities

A number of individuals have been reported who had one more or one less than the normal number of 46 chromosomes. In mongolian idiocy, which is characterized by mental and physical retardation together with a mongolian type of eyelid fold, the individual has 47 chromosomes. This extra chromosome is believed to be the result of nondisjunction of a pair of chromosomes (autosomes) during meiosis in the maternal ovum so that some of the eggs carry 24 chromosomes. The frequency of mongolism varies with the age of the mother. With mothers under 30 years of age, it is about 15 in 10,000; in those 45 years of age or older, the frequency may be increased fiftyfold. Few mongolian idiots live to maturity, but those who have done so and had offspring produced mongolians and normal children in about equal proportions. The predisposition to have a mongolian child can be detected by an examination of tissue culture cells made from certain blood cells of the woman, where there is a likelihood that such a condition could occur.

Certain other conditions are associated with abnormalities of the sex chromosomes. Klinefelter's syndrome is produced by the presence of two X chromosomes plus a Y, or an XXY complex. Such an individual is a sterile male with undeveloped testes and a tendency toward female breasts (gynecomastia). That the person outwardly is a male indicates that the Y chromosome is male-determining, just as it is in mice (but not in *Drosophila,* in which the Y chromosome is more or less passive).

Turner's syndrome (45 chromosomes) is a condition in which the individual has only one X chromosome and no Y chromosome paired with it, instead of the normal female (XX) or male (XY) state. In Turner's disease the XO (O = absence) constitution produces an external appearance of femaleness, although the person may lack ovaries or else have imperfect ones and so is sterile.

An XXX type of abnormal female (47 chromosomes) is also known, in which the person has underdeveloped reproductive organs and secondary sex characters, but may be fertile.

Both Klinefelter's and Turner's syndromes as well as the XXX type of female are caused by meiotic nondisjunction in the paternal or maternal germ lines. Since hormones are involved in regulative processes of sex, abnormalities in primary and secondary sex characters just mentioned may be due to hormonal imbalance produced by the faulty chromosome behavior.

In a few instances a polyploid condition has been found in human beings. One case was that of a highly abnormal boy who had a total chromosome count of 69. The child appeared to have three haploid sets of chromosomes (triploidy). The sex chromosomes were XXY. Such a person could have developed from a zygote composed of either a normal haploid egg fertilized by an abnormal diploid sperm or an abnormal diploid egg fertilized by a normal haploid sperm.

Many cases of human abortions are probably caused by abnormal chromosome patterns, since many aborted fetuses have extra chromosomes.

Improvement of human race

The scientific improvement of the human race is a highly sensitive area called **eugenics.** Many organizations all over the world have been formed with the objective of promoting a superior stock of people, such as has been done with domestic animals and plants. They reasoned that if man was able to improve his domestic animals, why should not the same principles applied to his own inheritance promote desirable traits for the advancement of society as a whole. However, it is impossible to direct, control, and select the desirable human traits and eliminate the undesirable ones in the manner of animal and plant breeders. Man is a huge mixture of so many traits, both good and bad, that the task of selecting all good ones and eliminating poor ones would not be practical.

A large part of our population carries defective genes. The mental and physical capacity of millions of Americans is below normal and many of these people constitute a serious burden on society for their maintenance. Of course, not all physical and mental conditions are hereditary; some result from environmental causes such as diseases and injuries. Subnormal people reveal only a part of the defective traits found in a population. Many defective genes are recessive and are carried by individuals who have normal phenotypes. Because of this, abnormal individuals are continually cropping out from the normal population as defective genes become homozygous.

Some states have sterilization laws for such defectives as feebleminded persons, imbeciles, and persons with certain forms of insanity. These statutes, however, are broad in their interpretation and make little attempt to distinguish between hereditary and environmental causes, and only a few states have made use of these laws to any extent. There is much popular feeling against enforcing them.

Eugenists have advocated a greater birth rate among the upper and desirable stocks of people. There is no doubt that the lower socioeconomic groups have a higher birth rate, but whether there are characteristic genetic differences between the different levels of society is not known with certainty. Tests of intellectual abilities so far devised are not able to distinguish between native abilities of individuals who are of different socioeconomic levels. These tests seldom can adjust for the environmental factors that influence the expression of mental traits, and these factors favor those of the higher social and economic classes. Therefore, these tests suggest that there is a tendency for a drop in the average I.Q. among Americans from generation to generation. This decline, however, may be due largely to environmental conditions and not to a widespread increase in mediocre genes. In England the decline in average I.Q. varies from one to four points a generation according to geneticists.

The future of genetics

Classical genetics has given zoologists the overall patterns of inheritance, despite the dependence upon breeding methods with their characteristic restrictions. Experimental genetics of the future will utilize more sophisticated methods of cellular and molecular techniques that bid to reveal more detailed insights into the process of heredity. Some of these newer methods will strive to compare the amino acid sequences between different species, or gene differences and likenesses between the various species. Already it has been possible to successfully fuse cells from diverse organisms into experimental hybrids in cell culture. Such experiments are expected to reveal the instructions that genes transmit to the cytoplasm and how the process of differentiation occurs. Progress has already been made in analyzing the genetic differences between species, with respect to such proteins as hemoglobin, certain enzymes, and others. This newer approach is still in its infancy, but future research will doubtless produce exciting revelations of the mechanisms of heredity and the actual part it plays in the life process.

References
Suggested general readings

Barry, J. M. 1964. Molecular biology: genes and the chemical control of living cells. Englewood Cliffs, N. J., Prentice-Hall, Inc. *An excellent review of the chemical foundations of genetics.*

Beadle, G. W. 1964. The new genetics: The threads of life. Chicago, Britannica Book of the Year, Encyclopaedia Britannica, Inc.

Beatty, R. A. 1970. The genetics of the mammalian gamete. Biol. Rev. **45**:73-119. *For the advanced student.*

Carson, H. L., D. E. Hardy, H. T. Spieth, and W. S. Stone. 1970. The evolutionary biology of the Hawaiian Drosophilidae. In Hecht, M. K., and W. C. Steere (editors). Essays in evolution and genetics. New York, Appleton-Century-Crofts, pp. 437-543.

Cold Spring Harbor Symposia on Quantitative Biology. Genetics and twentieth century Darwinism, vol. 24, 1959. Cold Spring Harbor, N. Y., The Biological Laboratory. *An appraisal of the mechanisms of the evolutionary processes as developed from the study of population genetics, ecology, paleontology, and other disciplines.*

Crick, F. H. C. 1971. General model for the chromosomes of higher organisms. Nature **234**:25-27.

Crow, J. F. 1963. Genetic notes, ed. 5. Minneapolis, Burgess Publishing Co. *An excellent summary of the principles of genetics.*

Davis, B. D. 1970. Prospects for genetic intervention in man. Science **170**:1279-1283.

DeBusk, A. G. 1968. Molecular genetics. New York, The Macmillan Co. *An excellent account of genetic transcription and translation with clear diagrams. A recent and up-to-date account of molecular biology.*

Dobzhansky, T. 1951. Genetics and the origin of species. New York, Columbia University Press.

Frisch, L. (editor). In Cold Spring Harbor Symposia on Quantitative Biology. Cellular regulatory mechanisms, vol. 26, 1962. Cold Spring Harbor, N. Y., The Biological Laboratory. *A comprehensive analysis of biosynthetic catalytic units as understood at present. Protein synthesis and the genetic code are especially stressed.*

Harris, H. 1970. Cell fusion. Cambridge, Harvard University Press. *New approaches to breeding for improved plant protein.*

Knudson, A. G., Jr. 1965. Genetics and disease. New York, McGraw-Hill Book Co. *Along with the immense developments in biology in the past few decades, new relationships have been discovered between heredity and certain disorders of man.*

Levine, R. C. 1968. Genetics, ed. 2. New York, Rinehart & Winston, Inc. *A paperback that presents both the classic and molecular aspects of the principles of genetics.*

Moore, J. A. 1963. Heredity and development. New York, Oxford University Press. *Paperback. One of the best accounts for understanding the role of heredity as we now understand it to be.*

Ohno, S. 1971. Genetic implication of morphological instability of malignant somatic cells. Physiol. Rev. **51**:496-526.

Pardue, M. L., and J. G. Gall. 1970. Chromosomal localization of mouse satellite DNA. Science **168**:1356-1358.

Peters, J. A. (editor). 1959. Classic papers in genetics. Englewood Cliffs, N. J., Prentice-Hall, Inc. *Here are the principal landmarks in the development of the science of genetics. All students of heredity should be familiar with these classic papers.*

Ravin, A. W. 1965. The evolution of genetics. New York, Academic Press, Inc. *This paperback shows the connection between the past and the present development of genetics.*

Rogers, S., and P. Pfuderer. 1968. Use of viruses as carriers of added genetic information. Nature **219**:749.

Roslansky, J. D. (editor). 1966. Genetics and the future of man. New York, Appleton-Century-Crofts. *A series of lectures given at the first Nobel Conference held at Gustavus Adolphus College in Minnesota. Among the participants were several Nobel Laureates. The lectures were built around the various aspects of genetics as they affect the destiny of man.*

Rous, P. 1965. Viruses and tumor causation. An appraisal of present knowledge. Nature **207**:457-463.

Scheinfeld, A. 1950. The new you and heredity, ed. 2. Philadelphia, J. B. Lippincott Co. *A popular yet accurate account of human heredity. One of the best works for the beginning zoology student.*

Schull, W. J. (editor). 1962. Mutations. Ann Arbor, University of Michigan Press. *A report of a conference on genetics dealing with various aspects of mutations such as rate, mutagenesis, detection, and genetic loads. The give-and-take, informal method of presentation is revealing.*

Smith, H. H. (editor). 1972. Evolution of genetic systems. Brookhaven Symposia in Biology, no. 23. New York, Gordon & Breach Science Publishers.

Sturtevant, A. H. 1965. A history of genetics. New York, Harper & Row, Publishers. *An excellent concise history of genetics written by one of the members of the famous Morgan team that did so much in the development of classic genetics. Should be read by all students of genetics.*

Taylor, J. H. (editor). 1963. Molecular genetics. Part I. New York, Academic Press, Inc. *The background in the breaking of the genetic code, one of the most spectacular advances in modern science, is well described by many active workers in this field.*

Watson, J. D. 1968. The double helix. New York, Atheneum Press. *The exciting story of how the molecular model of DNA was worked out.*

Selected *Scientific American* articles

Baern, A. G. 1956. The chemistry of hereditary disease. **195**:126-136 (Dec.).

Beermann, W., and U. Clever. 1964. Chromosome puffs. **210**:50-58 (April).

Benzer, S. 1962. Fine structure of the gene. **206**:70-84 (Jan.).

Britten, R. J., and D. E. Kohne. 1970. Repeated segments of DNA. **222**:24-31 (April).

Brown, D. D. 1973. The isolation of genes. **229**:20-29 (Aug.).

Clark, B. F. C., and Marcker, K. A. 1968. How proteins start. **218**:36-42 (Jan.).

Crick, F. H. C. 1962. The genetic code. **207**:66-74 (Oct.).

Crick, F. H. C. 1966. The genetic code: III. **215**:55-63 (Oct.).

Hurwitz, J., and J. J. Furth. 1962. Messenger RNA. **206**:41-49 (Feb.).

Ingram, V. M. 1958. How do genes act? **198**:68-74 (Jan.).

Kornberg, A. 1968. The synthesis of DNA. **219**:64-78 (Oct.).

McKusick, V. A. 1971. The mapping of human chromosomes. **224**:104-112 (April).

Miller, O. L., Jr. 1973. The visualization of genes in action. **228**:34-42 (March).

Mirsky, A. E. 1968. The discovery of DNA. **218**:78-88 (June).

Nirenberg, M. W. 1963. The genetic code: II. **208**:80-94 (March).

Nomura, M. 1969. Ribosomes, **221**:28-35 (Oct.).

Ptashne, M., and W. Gilbert. 1970. Genetic repressors. **222**:36-44 (June).

Rich, A. 1963. Polyribosomes. **209**:44-53 (Dec.).

Sager, R. 1965. Genes outside the chromosomes, **212**:71-79 (Jan.).

Sinsheimer, R. L. 1962. Single-stranded DNA. **207**:109-116 (July).

Taylor, J. H. 1958. The duplication of chromosomes. **198**:36-42 (June).

Temin, H. 1972. RNA-directed DNA synthesis. **226**:25-33 (Jan.).

Thomasz, A. 1969. Cellular factors in genetic transformation. **220**:38-44 (Jan.).

Waddington, C. H. 1953. Experiments in acquired characteristics. **189**:92-99 (Dec.).

Yanofsky, C. 1967. Gene structure and protein structure. **216**:80-95 (May).

PART FOUR
THE EVOLUTION OF ANIMAL LIFE

■ Organic evolution is the process through which existing animals and plants have attained their diversity of form and behavior by gradual, continuous change from previously existing forms. Evolution comes about chiefly by differential reproduction and by natural selection imposed by the environment acting on the inheritable variations of a population. It may be considered a descent in the course of time with modifications. It is a continuous, living process and is taking place today.

The following chapters deal with speculations about the origin of life, the special conditions on earth, no longer present, that must have existed when life first arose, the evidences for and nature of the evolutionary process that comprises the framework of biology, and the remarkable and recent divergence of the hominoid line, which gave rise to man.

CHAPTER 34
ORIGIN OF LIFE (BIOPOIESIS)

CHAPTER 35
ORGANIC EVOLUTION

CHAPTER 36
THE EVOLUTION AND NATURE OF MAN

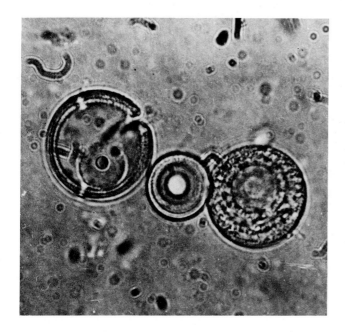

Electron micrograph of coacervate droplets. These proteinoid bodies can be produced in the laboratory and may represent precellular forms. They have definite internal ultrastructure. (× 1700.) (Courtesy S. W. Fox, Institute of Molecular Evolution, University of Miami, Coral Gables, Fla.)

CHAPTER 34

ORIGIN OF LIFE (BIOPOIESIS)*

In the evolution of life, atoms combined into molecules, molecules became organelles, organelles became cells, and cells were joined together into complex organisms. In this continuum from atom to man we often consider somewhat arbitrarily and erroneously that the dividing line between the nonliving and the living is at the cell level. Many consider life to occur when a macromolecule, such as nucleic acid, can replicate itself. But there are other important characteristics of life that we shall consider later. The oldest fossils to date are those of bacteria found in sedimentary rocks in Africa, dated at 3.2 billion years.

Charles Darwin believed that when life originated on earth the conditions in the world may have been quite different from what they are at present. He

*Refer to pp. 17, 21, and 22 for Principles 3, 23, and 28.

proposed that amino acids could have survived outside the living organism and that mixed with phosphoric acid salts, ammonia, light, heat, etc., the amino acids might link together to form proteins. Furthermore, he added that it would be impossible for such matter to exist today under the impact of living creatures who would surely devour it. In the present century J. B. S. Haldane, the British biologist, proposed that the gases of the early atmosphere of the earth consisted of water, carbon dioxide, and ammonia. When ultraviolet light shines on such a gas mixture, many organic substances, such as sugars and possibly amino acids, are formed. Ultraviolet light must have been very intense before the appearance of oxygen (from plants) to form ozone (the 3-atom form of oxygen), which serves at present as a protective screen to prevent ultraviolet rays from reaching the earth's surface. Haldane believed that the early formative substances could accumulate in the early oceans where synthesis of sugars, fats, proteins, and nucleic acids might occur. Recognizing the importance of the four elements carbon, hydrogen, oxygen, and nitrogen in the structure of living matter, Haldane could account for these four elements from his hypothetic early gas mixture.

In recent years, scientists have been attracted to the proposal that life arose spontaneously in the primitive ocean under the impact of certain favorable conditions that no longer exist. This theory stresses the presence of large quantities of organic compounds similar to those that are now found in living organisms. Although not a new theory by any means, forceful arguments in its favor have been advanced by Oparin, Haldane, Urey, and many others in a recent renewal of the hypothesis. Oparin, the Russian biochemist, through his book *The Origin of Life* (English translation, 1938) has mainly been responsible for this renewed interest. He argued that if the primitive ocean contained large quantities of organic compounds, these would in time react with each other to form structures of increasing complexity. Eventually a stage would be reached that could be called life. His belief was that a living system was gradually synthesized from nonbiologic compounds by logical and probable steps. Energy sources for the creation of the first organic compounds might come from ultraviolet light, electric discharges, localized areas of high temperature, such as volcanoes, and to some extent radioactivity. Since the necessary metabolites are already present in the environment to be used by the primitive organisms, their metabolism is called **heterotrophic.** Other conditions necessary for such a scheme of spontaneous generation would be a sterile environment to prevent the destruction of primitive organisms by microorganisms such as exist today and a reducing atmosphere containing precursor compounds: water, methane (CH_4), ammonia (NH_3), and hydrogen instead of the present oxidizing atmosphere of carbon dioxide (CO_2), nitrogen, and oxygen.

Urey believes that at low temperatures such substances as methane and ammonia would have been formed in the earth's early atmosphere and would have existed for some time before the early origin of life. As evidence for his theory, spectroscopic analysis has revealed that the remote and large planets of Jupiter and Saturn have frozen atmospheres of methane and ammonia. This could have been the original atmosphere of the planet Earth.

■ ORIGIN OF THE EARTH

Of the many theories to account for the origin of our planet, the one that is most seriously considered at present states that the sun and the planets were formed together from a spherical cloud of cosmic dust, which by rotation and gravitation developed a sun at the center and a swirling belt of gas around it. In time the belt broke up into smaller clouds that condensed by gravity to form the planets. Free hydrogen atoms were the most abundant elements in the gas cloud and gravitated toward its center to create the sun, which is largely composed of hydrogen. If the early surface temperature of the earth was high, it should have cooled rapidly by convection and radiation and reached its present condition in the relatively short time of 25,000 years.

While the earth was in a more or less gaseous condition, the various atoms became sorted out according to weight, with the lighter elements (hydrogen, oxygen, carbon, and nitrogen) in the surface gas, and the heavier ones (silicon, aluminum, nickel, and iron) toward the center. At first many gases, such as hydrogen, helium, methane, water, and ammonia, escaped from the earth, but when the gaseous materials composing the earth became dense enough, the gravitational field tended to prevent these gases from escaping into outer space. Jupiter and Saturn, which have much lower temperatures and higher gravitational fields than does Earth, have an atmosphere containing methane and ammonia at the present time.

If the earth's surface temperature was relatively high at first, molecules of atoms were slow in formation because heat disrupts the bonds that hold atoms together. As the earth cooled, stable bonds could be formed and free atoms began to disappear as molecules appeared. Since the lighter elements (hydrogen, oxygen, nitrogen, and carbon) were the most abundant atoms on the earth's surface, these elements reacted to form the first molecules.

The present atmosphere of the earth is an oxidizing one rather than a reducing one as was the early atmosphere. Only simple organic molecules would be formed in the reducing atmosphere of the primitive earth. Its carbon would be in the form of methane or carbon monoxide (perhaps a little carbon dioxide), its nitrogen in the form of ammonia, most of the oxygen in the form of water, and there would be considerable hydrogen. These primitive molecules would, then, begin with the following:

Water (H₂O)	Carbon dioxide (CO₂)	Methane (CH₄)	Ammonia (NH₃)	Hydrogen (H₂)

$$H-O \quad\quad O=C=O \quad\quad H-\underset{H}{\overset{H}{C}}-H \quad\quad N-H \quad\quad H$$

Water (H_2O)   Carbon dioxide (CO_2)   Methane (CH_4)   Ammonia (NH_3)   Hydrogen (H_2)

Temperature conditions of the earth were such that these primitive compounds existed as gases and formed the early atmosphere. The key element in the formation of the molecules of the primitive atmosphere was undoubtedly hydrogen, for when it was present in large amounts, its great reactivity led to the formation of methane, ammonia, and water vapor. Conversion to the present oxidizing atmosphere must have taken long periods of time. By photochemical decomposition in the upper atmosphere, methane and ammonia were converted to carbon and nitrogen, and water was converted to oxygen and hydrogen. However, most of the oxygen (20+%) now present in the atmosphere may have come from photosynthesis. Carbon dioxide became dissolved in the ocean or reacted with silicates to form calcium carbonate.

■ STEPS IN CHEMICAL EVOLUTION

In any living system a great variety of molecules are found whose reactions are intimately involved in the process we call life. These biomolecules, as they are called, may be divided into two groups: (1) molecules of relatively simple structure such as simple and complex sugars, neutral fats, phospholipids, amino acids, and nucleotides and (2) macromolecules such as proteins, nucleic acids, nucleoproteins, and viruses. How could these complex molecules have arisen? What kinds of conditions might have promoted a sequence in which free atoms formed simple molecules, which in turn formed larger and larger molecules, leading eventually to chance synthesis of macromolecules with the molecular patterns characteristic of the living organism?

If the early atmosphere contained methane, ammonia, water, and hydrogen, a mixture of these compounds would react only when energy was sup-

plied in some form. Ordinarily, organic compounds are formed by organisms, but of course these were nonexistent at this time. In all living organisms the synthesis of organic compounds involves enzymatic action at every step. But enzymes are complex proteins and so could not have been present when the early synthesis of organic compounds began.

However, enzymes only hasten the reaction rate; reactions can occur slowly without them. Thus simpler compounds from early primordia must have appeared before proteins. At the present time the source of free energy for synthesis is the sun, directly or indirectly. This requires photosynthetic organisms, but photosynthesis is not believed to have been an early evolutionary development. Therefore, other sources of free energy must have been used.

EXPERIMENTAL SYNTHESIS OF SIMPLE ORGANIC COMPOUNDS. In a classic experiment, S. L. Miller (1953) (Fig. 34-1), circulated a mixture of water vapor, methane, ammonia, and hydrogen continuously for a week over an electric spark, to obtain certain amino acids and some other products. It is possible, therefore, that various sources of energy were involved in the early formation of organic compounds—ultraviolet light, electric discharges, radioactivity, etc. In Miller's experiment certain intermediate products such as aldehydes and hydrogen cyanide first appeared; the reaction of these compounds led to the synthesis of amino acids.

More recently other workers have varied Miller's procedure by irradiating gaseous mixtures of H_2O, H_2, and NH_3 with electrons, by heating aqueous solutions of hydrocyanic acid (HCN) and by exposing solutions of HCN and formaldehyde to ultraviolet irradiation. Many different molecules were obtained, such as several amino acids, formic acid, aldehydes, purine and pyrimidine bases, some complex polymers of amino acids and sugars, ribose, and deoxyribose. Even the high-energy adenosine triphosphate (ATP) has also been produced by the ultraviolet-irradiated solutions of adenine, ribose, and ethyl metaphosphate, using the Miller technique with an oxygen-free, artificial atmosphere.

The early atmosphere of the earth is supposed to have contained little or no free oxygen; all the oxygen was bound up in water or metal oxides. Carbon dioxide was also lacking, for carbon was mostly combined in metal carbides or hydrocarbons. This would indicate that living organisms must have played an

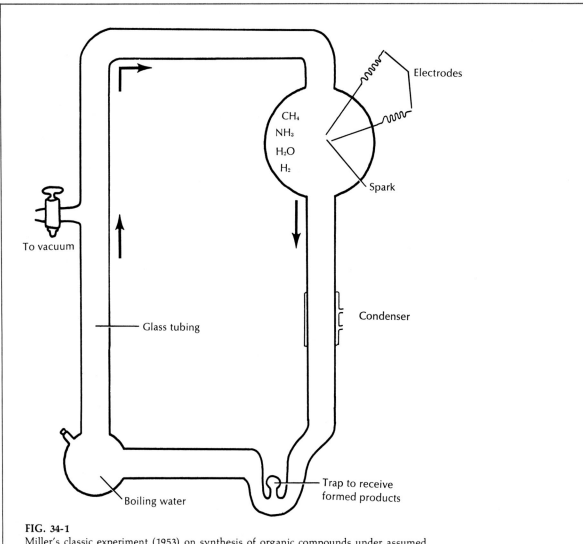

FIG. 34-1

Miller's classic experiment (1953) on synthesis of organic compounds under assumed primitive conditions of Earth's early atmosphere. Electric discharge through mixture of gases produced several amino acids and other organic substances. Variations of this experiment, making use of HCN (hydrocyanic acid), adenine, ribose, ethyl metaphosphate, etc. exposed to ultraviolet irradiation, have produced an RNA nucleotide, high-energy ATP, and a host of other organic molecules (aldehydes, purines, pyrimidines, fatty acids, etc.). ATP may have been abundant in early primordial oceans of reacting molecules and may have provided a ready source of energy for chemical reactions.

important part in the formation of the physical environment, such as the present atmosphere whose composition of oxygen and carbon dioxide was brought about by living organisms. Recent evidence indicates that some of our present oxygen may also come from the breaking up of water molecules in the atmosphere, as well as from photosynthesis.

APPEARANCE OF WATER. The formation of water deposits or oceans was no doubt the result of the condensation of water vapor from the atmosphere, although another theory accounts for the origin of oceans by water escaping from the earth's interior during its cooling stage. Dissolved in ocean water were salts and minerals washed down from the continents as well as atmospheric ammonia and methane. Volcanoes also may have contributed some constituents to the ocean waters. The presence of water and these early dissolved substances are considered crucial conditions that led eventually to the origin of life. Because the main bulk of organic compounds accumulated in the oceans, it must have been there that the formation of high-molecular compounds occurred. Before the advent of the organism, water, now the principal component of living substance, provided an ideal medium for reactions of organic molecules as they were formed. Methane (CH_4) was probably one of the first molecules to react, for the 4 H atoms could be replaced by other kinds of atoms, and thus many kinds of carbon-containing molecules could be formed. The versatility of the carbon atom, with its bonding capacity of 4, made possible the enormous complexity and variety of biomolecular structures we know today.

EARLY STEPS IN CHEMICAL EVOLUTION. Some of the earliest organic compounds formed would be the simple sugars, amino acids, fatty acids, pyrimidines, and purines. Each of these compounds represents a chain or ring of carbon atoms attached to various combinations of hydrogen, oxygen, and nitrogen. The sugars, or carbohydrates, and the fatty acids are formed entirely of carbon, hydrogen, and oxygen. The amino acids have, in addition to these elements, nitrogen, which occurs in an amino group (NH_2). Nitrogen could have been obtained in a reaction involving ammonia (NH_3) in which one of the hydrogen atoms was removed. The slow accumulation of the inorganically synthesized molecules was thus made possible by the primitive atmosphere of methane, ammonia, and water; by energy in the form of ultraviolet radiation, electric discharges, etc.; and by the absence of decay factors. The formation of the amino acids is especially significant because they represent the building blocks of proteins. Some of the amino acids are known to be highly stable and have actually been isolated from fossils as far back as the Devonian period of more than 300 million years ago. This would suggest that amino acids could accumulate and exist over long geologic periods; this

would afford plenty of time for proteins to be built from them.

APPEARANCE OF PROTEINS. The various reactions of these early compounds with each other and with inorganic molecules could have led to still more complex molecules. Simple sugars could form polysaccharides, and fatty acids could combine to produce fats. The appearance of proteins represented an important landmark because they are absolutely essential to living things. Proteins are made up of amino acids that are linked together in gigantic patterns to form complex and varied protein molecules. However, a mixture of amino acids would not form a protein unless other factors were involved.

How are proteins synthesized from amino acids? The experiment of Miller showed that it is rather easy to account for the amino acids, but the formation of these into the complex polypeptide chains is far more difficult. It is well to mention some of the exciting experiments along this line. In 1963 M. Calvin and others subjected a mixture of methane, ammonia, and water to a bombardment of electrons from the Berkeley cyclotron, and after an hour got adenine, 1 of the 4 bases of the DNA molecule. About a year later, S. W. Fox proposed that heat may have played an important part in the synthesis of organic compounds and heated a mixture of methane, ammonia, and water to 1,800° F. Of the many amino acids Fox obtained by this method, he found 14 of them involved in what is now known as the genetic code. He later demonstrated, by using more modest heat ranges (150° to 180° C.), that the 18 kinds of amino acids so treated together formed chains of polypeptides that could be digested by enzymes. By adding polyphosphate to his reaction mixture, he discovered that such polypeptide chains could be formed at 70° to 80° C. The high temperatures used by Fox could also have been produced during the early history of our planet by the entry of hot meteorites. The most available energy source could have been ultraviolet light in the synthesis of the early organic compounds. Autocatalysis, or the products formed in a reaction, acting as a stimulus to the reactions from which they were formed, could have played a great part in the early synthesis.

Another important aspect of protein development was their role in acting as **enzymes**. The first catalysts were probably metal ions and were somewhat weak in their action, but the development of proteins as enzymes greatly accelerated reactions without the need for increased temperatures.

PORPHYRINS AND NUCLEOTIDES. The central importance of porphyrins in the living process suggests that they made an early appearance in the origin of life. They are widely distributed in the world of life, they are stable under high temperatures and radiation found at the beginning of life, and they react easily with metals. Many iron and magnesium enzymes occur in living cells. Porphyrins are involved in promoting fermentation respiration and other reactions. Above all, because they are colored compounds they are capable of catching the energy of visible light and using it for chemical transformations. By a series of reactions porphyrin rings can be formed from acetic acid and glycine, which must have been present in the early formation of life.

Other key organic molecules that developed from the purines and pyrimidines were the **nucleotides.** These contain a purine or pyrimidine base combined with a sugar and a phosphate. By using the carbohydrate arabinose (instead of ribose sugar) J. Nagyvary and C. Tapiero (1967) were able to obtain a natural ribose nucleotide under the primitive conditions of water vapor, gases, acidity, and temperature. This discovery may be considered the first prebiologic synthesis of a natural ribose nucleotide.

Combinations of many different nucleotide molecules produced the supermolecules known as nucleic acids. Some of the nucleotides are synthetically produced from ammonium cyanide, which in turn can be formed from methane and ammonia by electric discharges. The polymerization of ribonucleic and deoxyribonucleic acids could have been effected by enzymes or by mineral surfaces. The formation of the nucleic acids was of crucial importance in the development of a living organism. The combination of nucleic acid with a protein produced the nucleoproteins, which are the principal components of the cell nucleus and are intimately associated with the life process. Some investigators have suggested that these were the first proteins formed, but they do not mark the first beginning of life and should not be regarded as constituting the living system by themselves.

■ FIRST ORGANISMS

It is very doubtful that the transition from the nonliving to the living was an abrupt process. From our present environment not much can be learned about the evolutionary steps leading to the first organism.

NUCLEIC ACID REPLICATION THEORY. The simplest system subject to evolutionary processes would be the replication of a nucleic acid, such as DNA. This would require complex nucleotide triphosphate for its synthesis. For making exact duplicate copies, it is necessary to have in abundance the component parts of nucleoproteins such as sugars, amino acids, purines, pyrimidines, and phosphorus. The formation of nucleoproteins is mainly dependent on what is already present. The first nucleoproteins served as models for the formation of more nucleoproteins. The Watson-Crick model of the structure of deoxyribonucleic acid shows how a molecule can act as a pattern to synthesize another molecule complementary to itself. The first living organism could have been strips of DNA or RNA that, with the necessary enzymes, could duplicate themselves. The sequence of the process may have occurred something like this: (1) A single strand of nucleic acid was formed, (2) nucleotides complementary to the bases in the first chain lined up, (3) the polymerization of the nucleotides to form the complementary strand occurred, (4) the original strand separated from the newly formed complementary strand, and (5) the final stage was the addition of cytoplasm and a membrane to the self-duplicating polynucleotides.

COACERVATE DROPLET THEORY. An alternative theory of the origin of the first organism is the coacervate theory of A. I. Oparin. According to this theory, coacervate aggregates, a special form of a colloidal solution in which one of the liquids of a hydrophilic sol appears as viscous drops (coacervates) instead of forming a continuous liquid phase, would absorb proteins and other materials from the environment, increase in size, and then divide. More accurate duplication of parts would evolve in time. Protoplasm, as now organized, has the structure of a complex coacervate.

FIRST ORGANISMS HETEROTROPHIC. There are logical reasons for believing that the early organisms were heterotrophic; that is, they derived their energy from external sources. They have fewer enzymes and specializations for metabolic processes than do autotrophs. The primitive oceans must have accumulated large quantities of compounds, and the first organisms used this reservoir of compounds for their evolution and expansion. When this reservoir of resources was used up, spontaneous generation could no longer occur and biogenesis became the only possible method of origin of organisms. By mutations, certain early organisms acquired the capacity to synthesize organic compounds from simpler ones. Natural selection could operate here to favor those organisms that could synthesize all the essential complex compounds, and competition became a rule of existence. Less successful forms would become extinct.

SOURCES OF ENERGY. All organisms require free energy for their chemical reactions and for the synthesis of their body parts. If the early organisms were heterotrophic and anaerobic, the source of their energy must have come from fermentation processes. At the present time many microorganisms get their energy this way. Lactic acid bacteria get free energy from the breaking down of glucose into lactic acid, with each molecule of glucose producing two molecules of the high-energy ATP. The yeast organisms produce ethyl alcohol and carbon dioxide instead of lactic acid. Most animals at present obtain their free energy from the oxidation of organic molecules by oxygen, and plants get their energy from light in the photosynthetic process (autotrophic). When early organisms used up the fermentable compounds of the ocean "broth," other sources of energy had to be utilized. About this time photosynthesis evolved and was made possible by the accumulation of atmospheric carbon dioxide produced by the metabolism of the early heterotrophs. Photosynthesis and perhaps also the breakdown of water molecules released oxygen to the atmosphere and established the conditions for aerobic respiration.

FIRST ATMOSPHERIC OXYGEN. Another form of autotrophic nutrition is **chemosynthesis,** in which energy for forming carbohydrates is obtained from metallic or nonmetallic materials such as iron, sulfur, and nitrogen. By the use of bond energy from chemical reactions it was possible to combine CO_2 and water into carbohydrates. With the appearance of oxygen, an ozone (O_3) layer was formed in the higher atmosphere and it absorbed most of the sun's ultraviolet rays that had made life possible only in water. It is believed that all the present oxygen of the air can be renewed by photosynthetic processes every 2,000 years and that all the CO_2 molecules pass through photosynthesis every 300 years. Autotrophic forms, or those that could synthesize complex organ-

ic compounds from simple renewable resources, were now established. Mutational descendants of these autotrophs produced secondary heterotrophs that could now live on autotrophs. Thus the present scheme of living systems was initiated.

In the overall picture (Fig. 34-2) there were three major evolutionary directions in the formation of organisms. These directions are based on the three methods of nutrition found in the biologic world: (1) the photosynthetic processes by producers of organic compounds, represented by plants; (2) the ingestion of producers by consumers, represented by animals; and (3) the reduction, or decomposition, of the dead remains of both producers and consumers to an absorbable state represented by saprophytes, such as fungi and certain bacteria. Some organisms may fit into more than one of these nutritional categories.

PRIMITIVE CELLS. The basic unit of living systems is the cell, and its formation must have represented a significant event in early evolution. A number of theories have been proposed to account for the origin of the cell. One theory stresses the aggregations or clumpings of nucleoproteins with nutrient shells around them. This arrangement could have been the start of more complex structures within which were incorporated the characteristic organelles of cells. An alternative theory suggests that cell-like structures were first formed from water droplets and accumulated organic materials in which chemical reactions could occur. In this way nucleoproteins could be formed. Such a scheme is not unlike the coacervate theory of Oparin (Fig. 34-3). A cellular level of simple organization may thus have been attained in the early origin of heterotrophs.

In whatever way cells were formed, in time they could produce their own energy by their ability to decompose organic molecules and release the great bond energy through respiratory processes. They could also synthesize many organic substances such as nucleoproteins and other organic materials. Cells undoubtedly have had a long evolutionary development, which involved integration of cellular activity, adjustment to steady states, duplication of parts, and change (mutation).

No one knows what the earliest ancestral cells were like, but certain types of present-day cells may throw some light on this problem. Although viruses are not considered cells and are inert when not in the host cells, the relative simplicity of their genetic

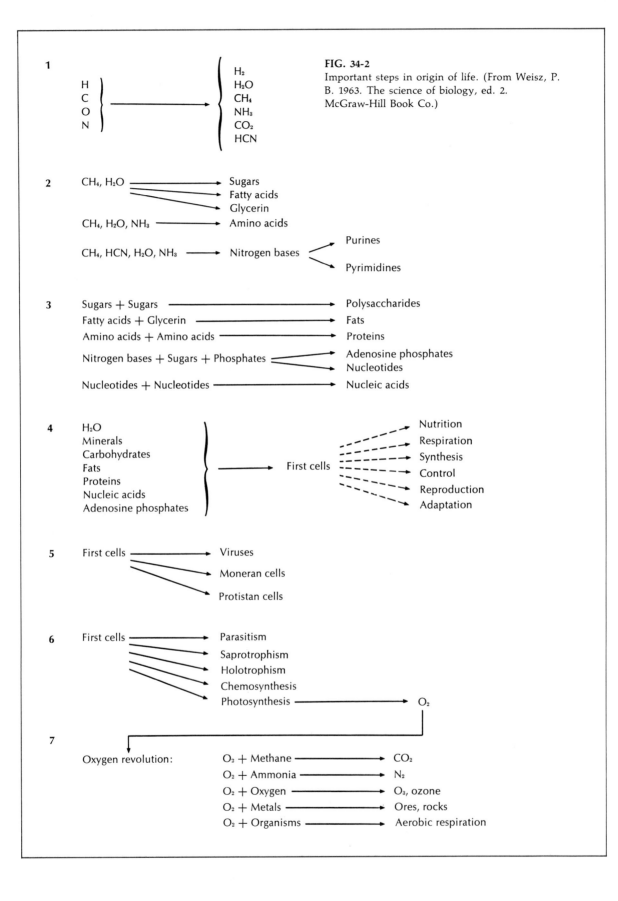

FIG. 34-2
Important steps in origin of life. (From Weisz, P. B. 1963. The science of biology, ed. 2. McGraw-Hill Book Co.)

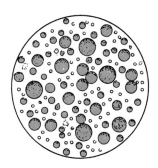

FIG. 34-3
Coacervate droplets in hydrophilic colloid. According to this theory of origin of first organisms, complex coacervates of different proteins made up a type of dilute "broth" in the ocean, and by intermolecular attraction large complex molecules were formed into colloidal aggregates. These colloidal aggregates could selectively concentrate materials from ocean "broth," and by unknown favorable internal organization, some aggregates could pick up molecules better than others and would become dominant. (After Keosian, J. 1968. The origin of life, ed.2. New York, Reinhold Publishing Co.)

pattern and their capacity to shuffle nucleoproteins among true cells has assigned them a unique role in the evolution of the cell.

Two kinds of present-day organisms, collectively called **Monera,** are considered to have some characteristics of the earliest cells. Bacteria (phylum Schizophyta) represent one of these two types of organisms. They are the smallest cells known, ranging in size from $0.2\ \mu$ to about $10\ \mu$ in length. (A micron is 1/25,000 inch.) They may occur as single cells or in clumps. They have a variety of metabolisms including heterotrophic, chemosynthetic, and photosynthetic. They lack nuclear membranes and their nucleoproteins are scattered in clumps through the cell. Reproduction is by rapid cell division and some undergo a sexual process by mating. They also appear to lack definite chromosomes. Bacteria, however, are not simple cells, and early cells were no doubt much simpler.

The other type of Monera are the blue-green algae (phylum Cyanophyta). They have no nuclear membrane and their nucleoproteins are arranged the same as are those in bacteria. In addition to certain characteristic pigments, most possess a form of chlorophyll that is dispersed in granules through the cell and not in chloroplasts as are found in higher plants. They carry on the same type of photosynthesis as the higher plants. They reproduce by cell division, but they have no sexual processes.

In some interpretations bacteria and viruses are considered degenerate secondary heterotrophs and not primitive patterns at all. It has been suggested on logical grounds that parasitism may account for the apparent simplicity of some of them.

■ EXTRATERRESTRIAL LIFE

Although the idea that meteorites could have brought spores from other planets to the earth had been largely discredited, the idea was recently revitalized by the discovery of the Murchison meteorite (Australia, 1969) and by the Allende meteorite (Mexico, 1969), which contained formaldehyde, a precursor of carbohydrates. Formaldehyde is known to exist in interstellar space and the meteorites may have absorbed it in their travels or synthesized it from absorbed gases.

The first investigators to analyze the relatively rare meteorites known as carbonaceous chondrites were J. Berzelius (1834) and F. Wohler (1859), who found evidences of organic matter within the meteorite. It has been a debatable point among those who have studied these meteorites whether these organic compounds were formed somewhere in space or have resulted from contamination with living matter on earth. In the Murchison meteorite, for instance, five of the most common amino acids—glycine, alanine, glutamic acid, valine, and proline—were identified. If they are of space origin, the findings would add considerable support to radioastronomic evidences that certain chemical syntheses do occur in interstellar spaces. However, the meteorite amino acids are composed of equal mixtures of right- and left-handed molecules, whereas biologic amino acids are all left-handed.

Nonproteinaceous amino acids appear in meteorites. Terrestrial proteins consist of polymers in combinations of the various 20 amino acids. There is some evidence that a concept of an evolutionary continuum from primitive earth gases to nonproteinaceous amino acids to proteinaceous amino acids (polyamino acids) has taken place in the early origin of life.

References
Suggested general readings

Allen, J. M. (editor). 1963. The nature of biological diversity. New York, McGraw-Hill Book Co. *A stimulating book on the diversification of biologic systems, mostly at the cellular and subcellular levels. Every topic (by a noted authority) arouses the student's interest.*

Bernal, J. D. 1961. The problem of the carbonaceous meteorites. The Times Science Review, no. 40, pp. 3-4 (Summer).

Engel, A. E. J. 1969. Time and the earth. Amer. Sci. **57**:458-483. *Shows how the perspective of the last 30 years has greatly increased the accepted ages of earth events.*

Fox, S. W. 1960. How did life begin? Recent experiments suggest an integrated origin of anabolism, protein, and cell boundaries. Science **132**:200-208.

Fox, S. W., and K. Dose. 1972. Molecular evolution and the origin of life. San Francisco, W. H. Freeman & Co. *An up-to-date account.*

Keosian, J. 1968. The origin of life, ed. 2. New York, Reinhold Publishing Corp. *Summarizes the present status of this fascinating problem. A concise and readable account.*

Miller, S. L. 1953. A production of amino acids under possible primitive earth conditions. Science **117**:528-529.

Oparin, A. I. 1957. The origin of life, ed. 3. New York, Academic Press, Inc.

Ponnamperuma, C. 1972. The origins of life. New York, E. P. Dutton & Co. *The chemical transition from the nonliving to the living.*

Urey, H. C. 1952. The planets. New Haven, Conn., Yale University Press.

Wilson, E. O., and others. 1973. Life on earth. Stamford, Conn., Sinaur Associates, Inc. *A pretentious work by many authorities. Brings many aspects up to date.*

Selected *Scientific American* articles

Eiseley, L. C. 1953. Is man alone in space? **189**:80-86 (July).

Glaessner, M. F. 1961. Precambrian animals. **204**:72-78 (March).

Huang, S. 1960. Life outside the solar system. **202**:55-63 (April).

Lawless, J. C., C. E. Folsome, and K. A. Kvenvolden. 1972. Organic matter in meteorites. **226**:38-46 (June).

Mason, B. 1963. Organic matter from space. **208**:43-49 (March).

Wald, G. 1954. The origin of life. **191**:44-53 (Aug.).

This bit of Green River shale from Farson, Wyoming, bears the impression of a "double-armored herring" (Knightia), about 4 inches long, which swam there during the Eocene Age, about 55 million years ago.

CHAPTER 35

ORGANIC EVOLUTION

■ MEANING OF EVOLUTION*

Animals and plants, as we have seen, present an immense variety of different forms. Moreover, they are found in nearly every kind of habitat that will support life at all and they manifest every conceivable kind of adaptation to their surroundings. The thinking person must often ask himself: When and how did this abundance of life originate? Have these different species always existed this way? Are some more closely related than others? Is there a similar basic pattern throughout all life? The principle of evolution, which attempts to answer some of these questions, is the framework of biology.

Evolution is the doctrine that modern organisms have attained their diversity of form and behavior through hereditary modifications of preexisting lines

*Refer to pp. 21 and 22 for Principles 23, 24, 25, 28 and 33.

of common ancestors. It means that all organisms are related to each other because of common descent, or that all organic life can be traced back to relatively simple common ancestral groups. It also implies the genetic changes populations undergo in their descent from ancestral populations. Basically, evolution is the change in the relative frequency of genes. Organic evolution is only one aspect of the larger view that the entire earth has undergone an amazing evolution of its own. The concept of evolution arouses in thoughtful individuals a feeling of awe and wonder, and one of inspiration, at the great drama that has been and is unfolding before their eyes.

Evolution is more than a change in the form and function of organisms; it is rather a change in their whole, integrated life as a part of nature. The population is the natural form of existence of all organisms. The evolution of any particular organism always involves its complex interrelations with the fellow members of its population and with its total environment. The basic principle of evolution must, therefore, emphasize two vital points: (1) how the genotype changes and operates to perform its action in the body and (2) how the conditions of the environment influence the adjustments of an organism, its preservation, its variations, and its life history. The environment is thus the directive force in the evolutionary process. How the organism responds depends on its genotype or gene pool, its mutant genes, and its gene combinations.

The study of evolution is a definite branch of biologic science and its problems must be treated in accordance with the principles of any science, that is, studied by observation and experiment. However, evolutionary study has certain limitations in this respect, for evolution is an extremely slow process; most of it has occurred in the remote past and much of it must remain as a hypothesis because it is not easily testable in practice. The fossil record represents the most clear-cut and verifiable evidence that the evolutionary process has occurred, and population genetics also affords a certain amount of direct testable experiments of the process.

The student may ask at this point: Does evolution have a goal? What is the purpose of evolution? Much of the evolutionary process does appear to be directional and shows advancement along certain lines. Many groups, especially vertebrates, show trends toward better organizational patterns. All successful evolution is progressive. It is a popular belief that man has been the ultimate goal—the pinnacle toward which the evolutionary process has been pointed. But there is little evidence for such a view. Evolution has had many directions, some of them successful, some not. Darwin did not introduce purpose into his evolutionary theory. Individuals with fitness have greater reproductive success and thus confer genetic properties from generation to generation. To undergo changes during one's lifetime is not organic evolution. Evolution is a group phenomenon that is guided by natural selection by way of an opportunistic development. As Professor J. S. Huxley emphasizes, human evolution tends to become conscious and self-directing. But man has been the outcome of the same forces of natural selection and environmental opportunity as other organisms.

Evolution is a continuous process and is taking place today. At any point in time, any group of animals one may select is undergoing the process of evolution. Any diversity a group may possess, now or past, has been the product of prior evolution. The variability as expressed in structural and physiologic features of any form is the basis of evolution still to come. The chief effect of new features is to increase the variability of any population group. It is a very slow process, but some striking evolutionary changes have taken place within historic times. Once it occurs, evolution seems to be irreversible except for minor reversals.

Many viewpoints and hypotheses have been proposed to account for the evolutionary process, but within recent years there has been a fruitful attempt to converge these various points of view into one unified and consistent picture of the whole process. Instead of many different processes, it is now believed that a single mechanism is involved in explaining evolution. This approximation of the various viewpoints is often referred to as the **modern synthesis of evolution.**

The recent outlook has greatly emphasized the roles of genetics and ecology. The importance of the genotype and its composition explains the emphasis on genetics; the role of the environment in directing the course of evolution explains the emphasis on ecology.

There were three important stages in the evolutionary process: (1) a period in which basic chemical patterns were evolved in the inorganic world, from which organic forms later arose; (2) the origin of the living systems from the nonliving through the gradual development of self-duplicating units into more complex units; and (3) the establishment of the organic systems as we know them from the fossil record and from existing forms. It is with the last stage that we are concerned in this discussion.

■ EARLY HISTORY OF THE EVOLUTIONARY IDEA

Although the development of the evolutionary theory has occurred chiefly during the past hundred or so years, thinking men long before this time had ideas about the evolving of forms of life. Some of the early Greek philosophers, Thales, Epicurus, Empedocles, and Aristotle, who lived from 500 to 300 B.C., thought a great deal on the evolution and the development of the different types of organisms. They believed that all animal life had its origin by special creation or by spontaneous generation. They were puzzled by the discovery of fossils that had remarkable resemblances to present-day forms and yet showed considerable differences. They explained this evidence of former life by the theory of **catastrophism,** the idea that animal life had at times suffered total destruction by some form of catastrophe and had been replaced by somewhat different forms.

Among modern zoologists, the French naturalist Buffon (1707-1788) stressed the influence of environment on the modifications of animal types. In 1745 Maupertuis, a French philosopher, described many of the concepts of variation and the diversity of animal life. Another French zoologist, Lamarck (1744-1829), put his ideas into a more concrete form and elaborated a theory to account for the evolutionary changes of animal life by use and disuse of organs (inheritance of acquired characteristics). Although his theory is no longer considered seriously, the influence of Lamarck did stimulate serious thought about evolution. Erasmus Darwin (1731-1802), the grandfather of Charles Darwin, recognized that the various forms of organisms arose from each other and stressed the response of the animal to environmental changes as the basis for its modifications.

Sound thinking along evolutionary lines was fostered by the science of geology in the early nineteenth century. The geologist Sir Charles Lyell (1797-1875) in his *Principles of Geology* (1830) elaborated the theory of **uniformitarianism,** which stated that the causes that produced changes in the earth's surface in the past are the same that operate upon the earth's surface at present. Such forces over a long period of time could account for all the observed changes, including the formation of fossil-bearing rocks, and did not require catastrophes for an explanation of the geologic process. This concept showed that the earth's age must be reckoned with in millions of years rather than in thousands. Charles Darwin was greatly stimulated by this important geologic work and was aided greatly by it in his own thinking on the processes of organic evolution.

No one has done more to stimulate interest and study in the field of evolution than the great Englishman Charles Darwin (1809-1882). Although the theory he proposed is now known to have many flaws, so forcibly did he present his ideas and his

array of carefully collected scientific data that no one since has really been able to challenge his preeminence in this field. He was the first to give a clear-cut idea of how evolution may have operated. Although the theory of natural selection was not entirely original with Darwin, no one else had proposed it with such clarity or supported it with such forcible arguments. *The Origin of Species,* published in 1859, has influenced biologists of every race and country. Darwin's theory and the many modifications of it will be mentioned later.

■ EVIDENCE FOR EVOLUTION
Paleontology: the historic record

The strongest and most direct evidence for evolution is the fossil record of the past because the study of paleontology, or the science of ancient life, shows how the ancestors of present-day forms lived in the past and how they became diversified. Incomplete as the record is—and many groups have left few or no fossils—biologists more and more rely on the discoveries of new fossils and their significance in interpreting the phylogeny and relationships of both plant and animal life. It would be difficult to make sense out of the evolutionary patterns or classification of organisms without the support of the fossil record. The documentary evidence for evolution as a general process, the progressive changes in life from one geologic era to another, the past distribution of lands and seas, and the environmental conditions of the past (paleoecology) are all dependent on what fossils teach us.

A fossil may be defined as any evidence of past life. It refers not only to complete remains (mammoths and amber insects), actual hard parts (teeth and bones), petrified skeletal parts that are infiltrated with silica or other minerals by water seepage (ostracoderms and mollusks), but also to molds, casts, impressions, and fossil excrement (coprolite). Skeletal parts are perhaps the most common of all, and paleontologists have been very skillful in reconstructing the whole animal from only a few parts. Vertebrate animals and invertebrates with shells or other hard structures have left the best record (Fig. 35-1). But now and then a rare, chance discovery, such as the Burgess shale deposits of British Columbia and the Precambrian fossil bed of South Australia, reveal an enormous amount of information about soft-bodied organisms. Certain regions have apparently provided

FIG. 35-1

Representative fossils. **A,** Fossil arthropod *Eurypterus,* which was abundant during upper Silurian period; related to modern scorpions. **B,** Some worm tubes. **C,** Bryozoan. **D,** Trilobite; related to horseshoe crab of today and one of most abundant of arthropod fossils. **E,** Cephalopod; chambered nautilus of today is little changed from this ancient fossil. **F,** Coral. **G,** Gastropod. **H,** Coelenterate strobila.

ideal conditions for fossil formation, for example, the tar pits of Rancho La Brea in Hancock Park, Los Angeles; the great dinosaur beds of Alberta, Canada, and Jensen, Utah; the Olduvai Gorge of South Africa; and many others.

A common method of fossil formation is the burial of animals under the sediment deposited by large bodies of water. Climatic conditions must have also been a great factor, as well as the nature of the deposits.

Most fossils are laid down in deposits that become stratified. The five major rock strata were mainly formed by the accumulation of sand and mud at the bottoms of seas or lakes. If undisturbed, the older strata are the deeper ones; however, in many regions they have buckled and arched under pressure so that older strata may be shifted over more recent ones. Since various fossils are correlated with certain strata, they often serve as a means of identifying the strata of different regions. Fossils thus serve as a guide to any fossil-bearing rock and characterize the deposits in the geologic time scale.

DATING FOSSILS. The age of geologic formations and fossils may be determined by radioactivity. Radioactive elements are transformed into other elements at certain rates, independent of pressure and temperature. Uranium 238, for example, is slowly changed into lead 206 at the rate of 0.5 gram of lead for each gram of uranium in a period of 4.5 billion years, or the half-life of uranium.

The ratio of lead 206 to the amount of uranium 238 in a sample of rock formation should give a fair estimate of the age of the stratum from which the specimen is taken. The potassium-argon method using similar techniques is even more precise. The radioactive isotope ^{40}K decays into ^{40}Ca (88%) and argon 40 (12%) at the rate of a half-life of 1.3 billion years. By knowing the amount of argon emitted (calcium is unreliable) from each unit of potassium in a unit of time, it has been possible to date the age of the rock and that of the fossil laid down in this rock.

The carbon-14 method is based upon the production of the radioactive isotope ^{14}C (with a half-life of about 5,568 years) produced in the upper atmosphere by the bombardment of nitrogen 14 with neutrons of cosmic radiation. As long as an organism (plant or animal) is alive, some radioactive ^{14}C enters its tissues from the atmosphere and an equilibrium is es-

tablished between atmospheric ^{14}C and ^{14}C in the organism. When the organism dies, ingestion of ^{14}C ceases, and there is no further exchange. In about 5,568 years only half of the ^{14}C remains in the tissue of the dead organism. In a preserved fossil it is thus possible to determine the age of it by finding the ratio of ^{14}C in the fossil to the amount of ^{14}C in live organisms. However, the amount of radioactive carbon in the atmosphere does not remain constant because of fossil-fuel burning (decrease of ^{14}C) and thermonuclear explosions (increase in ^{14}C) and corrections must be made for these factors.

GEOLOGIC AGES. The earth's crust in some regions has taken the form of mountain elevations, emergence of large areas from the sea, sinking of areas into the sea, and extensive climatic changes. All these geologic changes involved changes in the distribution of animals and plants. Geologists have therefore divided the history of the sequence of the accumulated deposits or strata into eras, periods, and epochs. There is first the division into five eras, then each era into periods, and finally each period (in some cases) into epochs. These various divisions of the geologic time scale are closely correlated with the fossils they bear.

In recent years the science of **paleoecology** has been attracting widespread interest. This discipline is a study of the kinds of environment under which sedimentary rocks of the past were accumulated so that information about the physical environments in which fossil organisms lived may be derived. Assemblages of fossils may show the nature of sea floors, the chemicophysical nature of water, and climatic conditions at the time of their existence. The problems of the paleoecologist are difficult because the fossil record is so incomplete.

The major characteristics of the geologic eras are indicated in the chart on the end leaves of the inside back cover. As far as the fossil record is concerned, the recorded history of life begins about the base of the Cambrian period of the Paleozoic era. The Proterozoic and Archeozoic eras are puzzling because of the lack of fossils. Evidences of life were mostly the burrows of worms, sponge spicules, algae, and a few others. However, the Precambrian fossil deposits of South Australia, with many invertebrate forms, indicate that life had already evolved to a considerable extent for perhaps as long as a billion years before the Cambrian period. There is also a great deal of

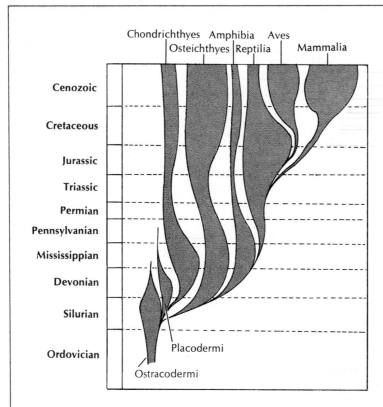

FIG. 35-2
Fossil record of vertebrates. Relative abundance of groups, as indicated by width of shaded areas, is based upon numbers of genera that have been identified. Note expansion and decline of reptiles. Osteichthyes, Aves, and Mammalia are dominant groups at present.

carbon as a residue of organic matter in the sedimentary rocks of this time. There may have been a great diversity of life in the Precambrian seas, but it was not preserved because of the lack of shells or other hard parts.

Since the Cambrian period when most of the major phyla were well established, it has been a matter of replacing primitive lines with better-adapted ones. Many of the vertebrate classes and orders appeared first in the Ordovician to Devonian periods. By the end of the Paleozoic era, some dominant groups became extinct and were replaced by the expansion of other groups (Fig. 35-2).

Biologic extinctions

The fossil record shows massive extinctions of marine species 500 million and 250 million years ago and lesser extinctions at the end of the Ordovician (425 million years ago), Devonian (345 million years ago), Triassic (180 million years ago), and Cretaceous (65 million years ago) periods. These extinctions may have some relation to magnetic reversals. Some extinctions (such as certain species of radiolarians) occurred after a magnetic reversal. During a reversal the magnetic field's intensity diminishes to zero and then rebuilds. When magnetic intensity is low, cosmic radiation, which it normally shields, is allowed to bathe the earth's surface and causes many mutations. Reversals may produce climatic changes that influence evolution. Experimentally, bacteria kept in a low magnetic field had a fifteenfold reduction in reproduction. It was found that a low magnetic field also produces changes in enzyme activity. Low magnetic intensity would work on both marine and terrestrial organisms. Charged ions may also be influenced in cell membranes.

Evolution of elephant

Paleontology has afforded evidence for tracing the phylogeny of many groups of animals. Two classic examples are those of the elephant and the horse. Elephants have stressed two morphologic features in

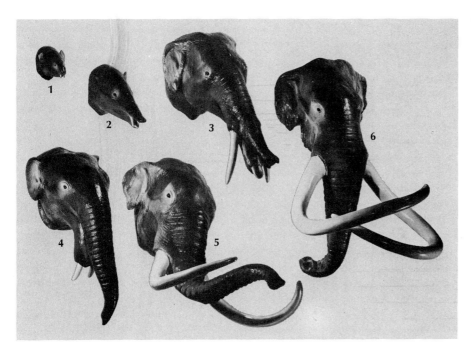

FIG. 35-3
Restoration of heads of fossil elephant-like animals. **1,** *Moeritherium.* **2,** *Palaeomastodon.* **3,** *Trilophodon.* **4,** *Dinotherium.* **5,** *Mastodon.* **6,** *Elephas.* (Courtesy Ward's Natural Science Establishment, Inc., Rochester, N. Y.)

FIG. 35-4
Mammoths. These have been found frozen in Siberia in a good state of preservation. (Courtesy Chicago Natural History Museum.)

particular—a prehensile proboscis and protruding tusks. They originated in Africa (in the late Eocene epoch) and gradually spread to other continents. They appear to have evolved from types similar to *Moeritherium* and were about the size of a pig (Fig. 35-3). The second incisors of their upper and lower jaws were a little enlarged. By the Oligocene epoch, four tusks had developed from these incisors *(Palaeomastodon)* and the size had increased to that of a steer. From this type the larger Miocene and Pliocene forms, lacking the lower tusks, emerged. Later, during the Pleistocene epoch came elephants as large or larger than the modern types such as the wooly mammoths (Fig. 35-4). Of the many types of elephants that have appeared, only the African *(Loxodonta)* and Asiatic forms *(Elephas)* exist now.

Evolution of horse

The fossil record affords us no more convincing or more complete evolutionary line of descent than that of the horse. The evolution of this form extends back to the Eocene epoch, and much of it took place in North America. This record would at first seem to indicate a straight-line evolution, but actually the history of the horse family is made up of many lineages, that is, descent from many lines. The phylogeny of the horse is extensively branched, with most of the branches now extinct. There were millions of years when little change occurred; there were other eras when changes took place relatively rapidly. No real change in the feet occurred during the Eocene epoch, but at least three types of feet developed later and were found in different groups during the late Cenozoic. Only one of these three types is found today. There was extensive adaptive radiation throughout the horse's evolutionary history. In the evolution of the horse the morphologic changes of the limbs and teeth were of primary importance, along with a progressive increase in size of most of the types in the direct line of descent.

The first member of the horse phylogeny is considered to be *Hyracotherium,* which was about the size of a small dog. Its forefeet had four digits and a splint (Fig. 35-5); the hind limb had three toes and two splints that represented the first and fifth toes. The teeth had short crowns and long roots, and the teeth in the cheek were specialized to some extent for grinding. This form lived in and browsed on forest underbrush.

The middle Eocene was represented by *Orohippus,* which had a further development of molarlike teeth.

Mesohippus, which flourished in the Oligocene epoch, was taller than the others and had three digits on each foot. The middle toe was larger and better

FIG. 35-5
Evolution of forefoot of horse as revealed by fossil record. (Courtesy Ward's Natural Science Establishment, Inc., Rochester, N. Y.)

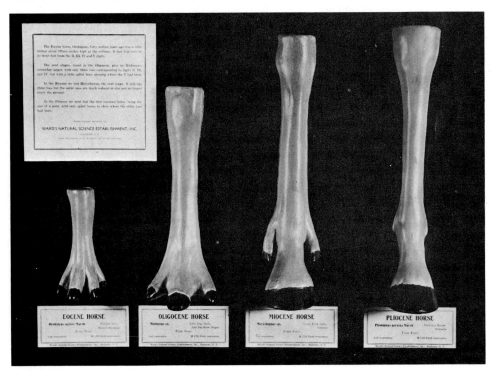

developed than the others. *Miohippus,* also found in the Oligocene, was larger but of the three-toed type. These horses were also browsing forms.

The Miocene epoch was represented by *Parahippus* and *Merychippus*. *Merychippus* (Fig. 35-5) is considered the direct ancestor of the later horses. They were three toed, but the lateral toes were high above the ground. Thus the weight of the body was thrown upon the middle toe. The teeth were high crowned, and the molar pattern was adapted for grinding with sharp ridges of enamel. The evidence is that *Merychippus* was associated with grass feeding. It had a larger skull and heavier lower jaws than the preceding forms did. This horse was between 3 and 4 feet high at the shoulders. *Merychippus* gave rise to a number of horse types, most of which became extinct by the end of the Tertiary period. One of these that persisted into the Pleistocene was *Pliohippus,* the first one-toed horse (Fig. 35-5).

From *Pliohippus* the genus *Equus,* or modern horse, arose probably in the Pleistocene epoch. It arose in North America and spread to most of the other continents. It had one toe on each foot, but the two splint bones are evidence of the former lateral toes. By the end of the Pleistocene epoch the horse had become extinct in North America, but migrant forms persisted in Eurasia to become the ancestors of the present-day horse. After the discovery of America by Columbus, the horse was reintroduced by the early Spanish colonists.

The development of this great animal from a small foxlike form was closely associated with the geologic development from a hilly, forested country to the great plains of the West. Thus the horse in its evolution represents a close parallelism between the development of an adaptive structural pattern on the one hand and the geologic development of the earth's surface on the other.

Other evidence for evolution

Aside from the evidence of paleontology, there are other clues to and evidence for evolution.

HOMOLOGOUS RESEMBLANCES. Homology, as one may recall, refers to structurally similar organs or functions that become adapted for different environmental conditions. It is best explained by common ancestral types that have become variously adapted. There are several kinds of homology. Structural homology is illustrated by the skeletal patterns of limb bones of all terrestrial vertebrates (Fig. 35-6). Vestigial or useless organs such as the vermiform appendix or the nictitating membrane are homologous to functional organs in ancestral forms (Fig. 35-7). Embryonic homology is the resemblance of early embryos of forms whose adults are unlike; the more closely the forms are related, the farther the embryonic resemblances extend. Microscopic homologies, or the similarity of cell units, chromosomes, cell organelles, and the like, and physiologic homologies, represented by the similar basic structure of protoplasm and the metabolic processes, indicate a fundamental unity of organisms.

CLASSIFICATION. Today's interpretation of taxonomy is based on the degree of natural relationship of organisms as nearly as they can be seen. Organisms are arranged in groups that express homologous interrelationships. Although taxonomy itself cannot logically be used as evidence for evolution,

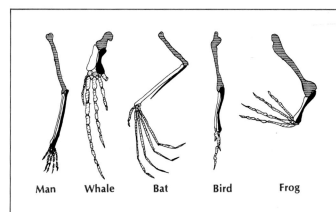

FIG. 35-6
Forelimbs of five vertebrates to show skeletal homologies. *Red* = humerus; *white* = radius; *black* = ulna; *white striated* = wrist and phalanges. Most generalized or primitive limb is that of man—the feature that has been primary factor in man's evolution because of its wide adaptability. Various types of limbs have been structurally modified for adaptations to particular functions.

Man   Whale   Bat   Bird   Frog

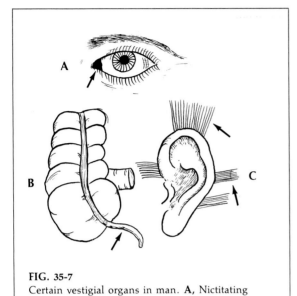

FIG. 35-7
Certain vestigial organs in man. **A,** Nictitating membrane in corner of eye, remnant of functional structure in lower forms. **B,** Vermiform appendix. **C,** Extrinsic muscles of ear, largely nonfunctional in man.

the fact that these homologies are so evident that animals can be classified on this basis certainly is a strong implication that evolution has occurred.

GEOGRAPHIC DISTRIBUTION. Some fairly convincing evidence for evolution can be seen in the geographic distribution of organisms.

When a new species arises in a center of origin, it tends to spread out until stopped by some barrier. The higher taxa, such as families or classes, occupy larger geographic areas than lower taxa such as genera or species. Evolution could explain this; the higher taxa are older and have had a longer time for dispersal and more time to break up.

Discontinuous geographic distribution of similar or identical groups frequently occurs. The camel family, for instance, is represented by the true camels in Asia and the llamas in South America. However, the fossil record indicates that camels probably originated in North America and then spread to the other regions. The former continuous range became discontinuous. Another example is that alpine species living on mountain tops are often identical with arctic or subarctic species from which they are widely separated. A logical explanation lies in the climatic shifts such as glaciation that have oc-

curred in the past. At one time these species were continuous, but as climatic conditions changed, these cold-adapted species survived only in arctic regions or in regions such as mountain tops in which arctic conditions existed.

Animals long isolated from each other tend to develop differences from each other. This is evident on islands. Species that have reached islands are usually those with powers of dispersal, such as spiders, insects, and seeds that are airborne, or forms carried on the feet or feathers of birds, or birds themselves, or forms that have floated on debris carried by currents from the mainland. In time these forms have come to differ somewhat from their relatives on the mainland. It is interesting to note that always the forms most nearly like those on an isolated island are those nearest them geographically—on the mainland or on adjacent islands. The simplest explanation is that they have evolved from common ancestors.

EXPERIMENTAL BIOLOGY. By careful selection and breeding to preserve desirable qualities, man has been able to produce numerous varieties of plants and animals. Although these varieties are not distinct species, they do come from common ancestors and give an inkling of what might occur in nature through natural selection. Experimental induction of mutant forms by x-rays, radiation, etc. indicates ways by which speciation may occur in nature. J. and E. Lederberg (1951) have shown that within populations, certain bacteria or insects are resistant to antibiotics and insecticides. This resistance was there before exposure to these agencies; so it must have occurred by mutation independently of exposure. This would indicate that some bacteria and insects have undergone mutations without man's influence and that such mutations are not unique.

■ THEORIES TO ACCOUNT FOR METHODS OF EVOLUTION

In a field that lends itself to so many different interpretations, we should expect to find many theories. Although no single theory adequately explains the mechanism of evolution, biologists are gradually getting a clearer comprehension of the overall picture. Many fields of biologic study have made contributions toward its understanding. The difficulty of testing experimentally is a disadvantage to any line of scientific inquiry and explains the relative slowness in working out the mechanisms of organic evolution.

Inheritance of acquired characters

One of the earliest evolutionary theories proposed was the one by Jean Baptiste de Lamarck (1744-1829), a French biologist. This theory is commonly known as lamarckianism. He first published his theory in 1809. It was based on the concept of the use and disuse of organs in the adaptation of animals to their environment. He believed that whenever such an adaptation arose, it was inherited from generation to generation—the inheritance of acquired characters. According to Lamarck, new organs arose in response to the demands of the environment. The limbless snake he would explain by the handicap of legs in crawling through dense vegetation, and thus snakes lost their legs through disuse. The long neck of the giraffe resulted from reaching up into trees to browse. There is no evidence for Lamarck's theory, and it has never received much support from biologists.

The effects of use and disuse are restricted mainly to somatic tissues, but genetics does not support the transmission of somatic characters. There can be no permanent change unless the germ cells themselves are altered, and experiments have shown that they are subject to little or no effect from somatic cells and the environment.

Natural selection theory of Darwin

The first real mechanical explanation of evolution was the one proposed by Charles Darwin in his work *The Origin of Species* published in 1859. Another English scientist, Alfred Russell Wallace (1823-1913), should receive some credit for this theory of **natural selection.** Both Darwin and Wallace had arrived at the main conclusions of this theory independently, but the publication of *The Origin of Species* the year after Wallace had announced his conclusions in a brief essay really clinched Darwin's position and prestige.

Darwin's early training gave him an opportunity to accumulate data for his theory. As a young man he had spent 5 years (1831-1836) as a naturalist on board the *Beagle*, a vessel that had been commissioned to make oceanographic charts for the British admiralty. This vessel in the course of its voyage around the world spent much time in the harbors and coastal waters of South America and adjacent regions, and Darwin made extensive collections and studies of the flora and fauna of those regions. He kept a detailed journal of his observations and when

he returned to England, he brought out a number of scientific papers based on his collections and observations.

The idea of natural selection did not occur to him until some time after he had returned. He was especially puzzled over the fossils he found in Patagonia and elsewhere, because he noted the similarities and dissimilarities of these to existing forms of life. The bird life of the Galápagos Islands also intrigued him.

After the idea of natural selection dawned on him, he spent the next 20 years accumulating data from all fields of biology to prove or disprove his theory. He made many experiments and he evaluated all his work with a careful, scientific appraisal. Darwin, incidentally, obtained some of his clues for this theory from the *Essay on Population* by Malthus, especially the part on competition among people in an overpopulated world. Essentially, most of the concepts of the theory of natural selection were worked out by Darwin himself. So much evidence was brought forth and so forcible were the arguments advanced for natural selection by Darwin in his book, that one may safely say that a new era in thinking, not only in organic evolution but also in many related fields, dates from the publication of the book.

The essential steps in the theory of natural selection as advanced by Darwin include the nature of variation, great rate of increase among offspring, struggle for survival, natural selection and variation, survival of the fittest, and formation of new species.

NATURE OF VARIATION. No two individuals are exactly alike. There are variations in size, coloration, physiology, habits, and other characteristics. Darwin did not know the causes of these variations and was not always able to distinguish between those that were heritable and those that were not. But he stressed artificial selection in domestic animals and plants and the role it played in the production of breeds and races of livestock and plants. He observed that man had been able to do this by carefully selecting and breeding those individuals with the desired modifications. Darwin believed that if selective breeding was possible under human control it could also be produced by agencies operating in the wild state. These agencies he sought to explain in his theory.

GREAT RATE OF INCREASE AMONG OFFSPRING. In every generation the young are far more numerous than the parents. Even in a slow-breeding form, such as the elephant, if all the offspring lived and produced offspring in turn, in a few hundred years the earth could not hold the elephants. Most of the offspring perish, and the population of most species remains fairly constant under natural conditions. Placed in locations more favorable for increase, they may multiply with enormous rapidity, as the rabbit did in Australia. Usually natural checks of food, enemies, diseases, etc. keep the populations within bounds.

STRUGGLE FOR SURVIVAL. If more individuals are born than can survive, there must be a severe struggle for existence among them. This competition for food, shelter, breeding places, and other environmental factors results in the elimination of those that are not favorably suited to meet these requirements. There are many factors involved in this struggle for existence, many of them obscure and difficult to demonstrate.

NATURAL SELECTION AND VARIATION. Individuals of the same species tend to be different through variations. Some of these variations make it easier for their possessors to survive in this struggle for existence; other variations are a handicap and result in the elimination of the unfit.

SURVIVAL OF THE FITTEST. Out of the struggle for existence there results the survival of the fittest. Under natural selection individuals that have favorable variations will survive and will have a chance to breed and transmit their characteristics to their offspring. The less fit will die without reproducing. This process will operate anew on each succeeding generation so that the organisms will gradually become better and better adapted to their environment. With a change in the environment, there must also occur a change in those characters that have survival value, or else the animal will be eliminated by the new conditions.

FORMATION OF NEW SPECIES. How does this result in new species? According to Darwin, whenever two parts of an animal or plant population are each faced with slightly different environmental conditions, they would diverge from each other and in the course of time become different enough from each other to form separate species. In this way two or more species may arise from a single ancestral species. Also a group of animals, through adaptation to a changed environment, may become different enough from their ancestors to be a separate species. In a similar manner, greater divergencies could arise

in time and lead to the higher taxonomic ranks of genera, families, etc.

What happens to characters when they are nonadaptive or indifferent? Variations that are neither useful nor harmful will not be affected by natural selection and may be transmitted to succeeding generations as fluctuating variations. This explains many variations that have no significance from an evolutionary viewpoint.

Appraisal of theory of natural selection

Although Darwin's theory of natural selection has given in general a logical explanation of evolution, parts of his theory have had to be changed in the light of biologic knowledge and interpretation. Two weaknesses in his theory centered around the concepts of variation and inheritance. Darwin made little distinction between variations induced by the environment (physical or chemical) and those that involve alterations of the germ plasm or the chromosomal material. It is now known that many of the types of variations Darwin stressed are noninheritable. Only variations arising from changes in the genes (mutations) are inherited and furnish the material on which natural selection can act.

Although Mendel's work on inheritance was published in 1866, Darwin did not make use of its genetic implications. Had Darwin known of this important research of Mendel, it is doubtful that he could have seen the relation of the hereditary mechanisms to gradual changes and continuous varieties that represent the hub of darwinian evolution. It took others years to establish the relationship of genetics to Darwin's natural selection theory. Neither did Darwin point out the cumulative tradition of man's evolution, or the capacity of the human race to transmit experience from generation to generation, or what is now called the "cultural evolution" of man.

Darwin overemphasized the role of natural selection and failed to note its limitations. He thought that selection could operate indefinitely in promoting the development of a desirable variation. It is now known that, when the population becomes homozygous for the genes of a particular trait, natural selection can no longer operate, as shown from experiments in pure lines. Additional genic changes (mutations) must occur before selection can continue.

Effects of isolation

Darwin did not appreciate the real nature of isolation in the differentiation of new species. There are really three types of isolation involved—geographic, genetic, and ecologic.

GEOGRAPHIC ISOLATION. Darwin appreciated the fact that groups of animals cut off from each other by natural barriers, such as mountains, deserts, and water, were often quite different from each other. He noticed this in the Galápagos Islands and elsewhere. This **geographic** isolation accounts for the existence of many different species in small regions that are broken up effectively by natural barriers. St. Helena, an island with many valleys separated by high mountains, has many species of snails that are isolated from each other. The explanation is that when individuals of a species are in slightly different environments, mutations that arise within a particular group may enable that group to diverge from the others. If the groups are separated from each other long enough, they may become distinct species.

GENETIC ISOLATION. Mutations in a species may result in a group of organisms that are infertile with the parent stock. This **genetic isolation** is as effective in isolating separate species as are the geographic ones.

ECOLOGIC ISOLATION. Differences in habitats and ecologic niches may tend to keep groups of animals away from each other during the breeding season. Or two groups of animals may actually intermingle but, because of a seasonal variation in their breeding habits, are effectively isolated from each other.

If the environment changes or the animal moves to a new environment, traits formerly of no use may now prove to be highly advantageous. Thus it is possible for an animal to be adapted in some way long before it actually meets an environment to which it is best suited. Ecologically, **preadaptation** may account for the remarkable way some animals have flourished in new surroundings.

Patterns of evolutionary change

Although the evolution of every great group of organisms is more or less unique in its details, three basic kinds of organic evolution have been followed in producing the great diversity of life we know today.

SEQUENTIAL EVOLUTION. There is first a buildup within a population of a **gene pool** of specific characters of structure, function, or habit. The term "gene pool" refers collectively to all of the alleles of all of the genes of a population. From one generation to the next there is usually a change in the gene pool so that there is a gradual alteration in the range of these characters. Variations in the environment may account for such changes because the genetic composition will develop in a different way under different environmental conditions, but changes in the genetic composition may also occur by mutation. Through successive generations, there will be a succession of different gene pools that will be expressed in different characters. This results in a change of the population as a whole and is not a splitting up of

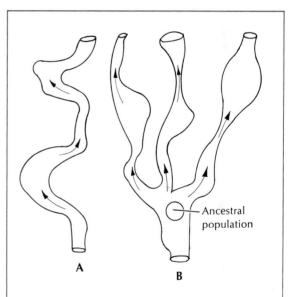

FIG. 35-8

Two basic patterns of evolution. **A,** Phyletic or sequential pattern involves progressive shift of average genotypes and phenotypes of population. Although population changes, there is no splitting up of population as it continues within a zone of varying breadth. Differences may be great enough for eventual origin of new species. **B,** Divergent or cladogenesis pattern involves splitting up of population into two or more separate lines, each of which may give rise to taxa of different ranks. Both these patterns are usually found in any population over long periods of time.

the population. This **sequential or phyletic** evolution is usually restricted to a fairly stable organism-environment complex (Fig. 35-8, *A*). Such evolutionary changes may be slight if the environment is stable, but shifts of gene combinations in adaptive types may occur in response to a changing environment, thus producing distinguishably different forms (species).

QUANTUM EVOLUTION. A rapid shift of a population into another and different adaptive zone where it is necessary for the organisms to adapt quickly results in a type of rapid evolution called **quantum evolution.** Quantum evolution is promoted by excessive mutation, relaxation of selective pressure, and such short cuts as pedomorphosis. It may produce taxa of all ranks but probably applies best to families, orders, and classes (Simpson).

DIVERGENT EVOLUTION. **Divergent evolution** or **cladogenesis** (Fig. 35-8, *B*) involves a splitting process in which new populations originate from old populations. It is the differentiation of two or more groups within a widespread population. It involves taking advantage of adaptations to spread out into available ecologic niches (opportunism). Within each of the adaptive lines, more or less directional progress (phyletic evolution) occurs, with alternate stages of advancement and stability, and in some cases there is enough diversification within a line for the formation of a new species. This process in a group may continue indefinitely or it may be cut short by extinction.

Biologic evolution tends to produce types or patterns of life that can make use of more of the world's space and resources with greater efficiency. New types of metabolic systems must be evolved for this purpose. This can be expressed as a sort of feedback mechanism by which one metabolic novelty (the evolution of a new form of life) may create an opportunity for another metabolic novelty. Termites could not evolve until woody plants came into existence, thus providing them with ecologic niches. The complex life cycle of many parasitic flatworms indicates an evolutionary pattern by which the parasites have moved from one host to another as new groups of animals or metabolic novelties evolved, yet retaining at the same time the older hosts. As a rule, new major groups have arisen to occupy new ecologic niches not already occupied by other groups.

Orthogenesis

Orthogenesis is the tendency for a group of animals to continue to change in a definite direction. This is often referred to as **straight-line evolution,** and the idea involved is that certain structural changes once started may continue without deviation for an indefinite period unless checked by extinction. Paleontology furnishes examples, such as that of the horse, in which certain structural trends are conspicuous. In the case of the horse, the evolutionary trend proved highly adaptive, but in others no such advantage is seen.

The theory of orthogenesis does not explain the underlying cause for the evolutionary change along a particular line. The theory also is connected with the idea of the irreversibility of evolution. Blum, in his book *Time's Arrow and Evolution,* has stressed the irreversibility of evolution through the operation of the second law of thermodynamics, which in some cases permits no exercise of natural selection in altering the direction of evolution. But he avoids the implication of an extraphysical force and thinks the term "orthogenesis" should be avoided altogether.

Paleontologists no longer stress orthogenesis because they recognize that most apparent examples of it actually do not have direct, unbranched lines, but have had many side branches that have been eliminated by natural selection.

Modern synthesis of evolution (neo-darwinism)

Our present interpretation of evolutionary processes began to take form about 1930. In the first 30 years of the present century there was gradually accumulated a great amount of factual information about the chromosomal and genic theory of heredity, the way mendelian heredity operated, and a more fundamental understanding of the mutation theory. All these branches of investigation had become more or less unified into what we now call cytogenetics. Under the influence of a brilliant group of biologic thinkers, such as J. S. Huxley, R. A. Fisher, and J. B. S. Haldane in England and Sewall Wright, H. J. Muller, and T. Dobzhansky in America, there has been a fruitful attempt to unify the various theories and ideas into one underlying mechanism of organic evolution. The genotype and its behavior in the organism has become the focal point for understanding how evolution operates.

These workers and many others have shown that changes in the genotype (mutations), with the recombination of genes through biparental reproduction under the influence of natural selection over long periods of time, can operate to produce evolution as we see it around us.

Mutation*

Little was known in Darwin's time about the behavior of chromosomes and their bearing on heredity. Soon after the principles of mendelism were rediscovered in 1900, the parallelism between the chromosomal behavior and mendelian segregation was worked out. It was some years before the real significance of cytogenetics to the problems of evolution was appreciated. Hugo De Vries, a Dutch botanist, had stressed the importance of **mutations** in the evolutionary process. Working with the evening primrose, he had found certain types of this plant differing materially from the original wild plant and, more important, he found that these aberrant forms bred true thereafter. De Vries explained these mutant forms mainly on the basis of a recombination of chromosomes, but since his time, stress has been laid on genetic transformations. Mutations, as now understood, refer to sudden random changes in genes and chromosomes caused by errors in self-copying, but it is precisely these errors that make evolutionary progress possible.

The work of Morgan and his colleagues in working out the theory of gene linkage, the mapping of chromosomes, nondisjunction of chromosomes, etc. laid the basis for the modern understanding of the hereditary mechanism. The experimental production of mutations through radiation, x-rays, etc. and the discovery of the giant chromosomes in the salivary glands of certain larval insects have added to this understanding.

Natural selection is an important factor in evolution, but it operates in a later stage of evolution than Darwin assigned it. Before a new species can evolve, there must be both mutation and natural selection. Mutations furnish many possibilities, natural selection determines which of them have survival merit, and the environment imposes a screening process that passes the fit and eliminates the unfit. Mutations are constantly producing new allelomorphs on which natural selection works.

*Refer to p. 22, Principle 28.

NATURE OF MUTATIONS. Mutations may be harmful, beneficial, or neutral in their action. Perhaps the majority are harmful because most animals are already adapted and any new change would likely be disadvantageous, but some are beneficial.

In *Drosophila* investigators have found that some mutation occurs with a certain frequency. Thus the red-eyed wild type can be expected to mutate to the white-eyed type every so often. Moreover, the white-eyed type undergoes mutation, either changing back to the original red eyes or else to one of the other eye color allelomorphs.

Mutant characters tend to be recessive and show up as a phenotype only when they are homozygous. Some mutant characters are lethal, but this action is expressed, as a usual thing, only in the homozygous condition. Since most mutations ordinarily occur at random, it is not possible to predict the nature or time of their appearance. Nor is it possible to see a concerted action of mutants along a particular direction, for the genes appear to mutate independently of each other.

Mutations can be divided into those that produce small changes (micromutations) and those that produce large changes (macromutations). Most evolutionists now favor small mutations as the more important in causing evolution. They help explain the many intermediate forms (races, subspecies, varieties) between the parent species and the new one. Some striking changes caused by large mutations are the tailless manx cat and the bandy-legged Ancon sheep.

Most mutations are destined to a very short existence because competition in nature quickly eliminates them. There are cases, however, in which mutations may be harmful to an animal under one set of environmental conditions and helpful under a different set. If the environment happens to change at about the same time as favorable mutations appear, there could be adaptations along special lines and within restricted limits. Such an opportunism of evolution may help explain the evolution of the horse. It is difficult to see how the unidirectional and specialized nature of its adaptation could occur except through a balance between changing environment and the adaptive mutations of the animal. The physical world has provided many opportunities for animals with the right mutations, as evidenced by the numerous forms found in the animal kingdom today.

TYPES OF MUTATIONS. There are two main types of mutations—gene mutations and chromosome mutations. A **gene mutation** is a chemicophysical change of a gene resulting in a visible alteration of the original character. Although the actual change in the gene is largely unknown, it is believed to be a rearrangement of the nucleotides within a region of the DNA molecule. Such changes cannot be detected under the microscope, for there are no visible alterations in the chromosomes bearing the genes in question. Many of this type of mutation have been found in *Drosophila* and other forms. Most gene mutations are point mutations, that is, the physical or chemical change of one gene.

Chromosome mutations involve either chromosome rearrangement during meiosis or an alteration in the number of chromosomes. The rearrangement of the chromosomes may involve inversion of the linear order of the genes, the deletion of blocks of genes, translocation of portions of chromosomes or some other irregular procedure. Such changes often produce phenotypic changes that are inherited in the regular mendelian manner. Increase in number of chromosomes may involve a doubling or tripling of the diploid set, resulting in the condition called **polyploidy.** Polyploids are usually characterized by a larger size than the parent stock. Such mutations are more common in plants than in animals. Among roses, for instance, there are found species with 14, 28, 42, and 56 diploid chromosomes, although the basic parent rose is believed to be one with 14 chromosomes. Such duplication of chromosome sets is caused by an upsetting of the meiotic process so that through omission of the reduction division, germ cells are formed with the diploid number of chromosomes instead of the normal haploid number. Experimentally, polyploidy may be induced by the drug colchicine that prevents the division of cells but does not interfere with the duplication of the chromosomes. This results in gametes with the diploid number of chromosomes.

This variability gives natural selection something to work on in the production of a new species. Such variations come and go, dependent on such factors as the size of populations, degree of segregation and isolation, etc. Many mutant characters never have a chance to become persistent characters in a population. Relatively few actually survive the test of time.

Résumé of mutation types
CHROMOSOME MUTATIONS
1. Changes in chromosome numbers
   a. Changes in number of genomes or sets of chromosomes
      (1) Haploid—a single genome instead of two
      (2) Polyploidy—three or more genomes instead of two
   b. Changes in numbers of chromosomes in a genome
      (1) Monosomics—loss of one chromosome in a genome
      (2) Polysomics—one chromosome may be represented three or more times, such as in trisomy
2. Structural changes in chromosomes
   a. Changes from loss or duplication of genes
      (1) Deficiency—section containing genes lost
      (2) Duplication—a section of a chromosome containing genes may be present along with the normal chromosome
   b. Changes produced by alteration of normal arrangement of genes
      (1) Translocation—two nonhomologous chromosomes may exchange sections, thus producing new chromosomes
      (2) Inversion—reversal of a part of a chromosome so that the linear arrangement of genes is in reverse order

GENE MUTATIONS. Gene mutations are errors in the replication of genetic material before chromosomal duplications. The unit of mutation may be a section of a chromosome or restricted to a single pair of nucleotides of the nucleic acid.

CAUSES AND FREQUENCY OF MUTATIONS. Different genes possess different frequencies of mutation rates because some genes are more stable than others. Mutation is also a reversible process and cases of back mutation are well known. Gene A, for instance, may mutate to gene a, and gene a may mutate back to gene A. This reversibility must be taken into account in mutation equilibrium, and the difference between the mutation rate in one direction and the mutation rate in the reverse direction constitutes **mutation pressure,** which is usually of a low magnitude. Mutation frequencies are best known in the fruit fly *Drosophila* and in the corn plant. Some corn plants may be expected to produce 10% of their offspring with at least one mutant gene. *Drosophila* is supposed to produce at least one new mutation in every 200 or so flies.

Mutation rates are higher in the human male as well as in the males of *Drosophila.* The spontaneous rate is relatively low in all forms studied, which indicates that genes must be very stable. Some loci are more stable than others. It is also known that there are mutability genes that induce mutations in other loci. In higher vertebrates the average mutation rate per individual may be between 1 in 50,000 to 1 in 200,000 (Mayr). It has been estimated that one mutation for each 10,000 to 1 million cell divisions occurs in some. Many mutations are not detected and estimates of their frequency are in many cases only guesses. Certain genes are also known to increase the mutation rate of other genes in the same organism.

So far as the causes of natural mutations are concerned, very little is known about the matter. Many mutations that were first found in laboratory stocks are now known to occur in the wild state. This contradicts the belief once common that most mutations happen under the influence of laboratory conditions. It is true that both gene and chromosome mutations can be produced artificially by the influence of x-rays, ultraviolet rays, chemicals, temperature, and other agencies. It has been suggested that cosmic rays may be responsible for the appearance of mutations in wild populations. There is also the possibility that spontaneous mutations may be caused by metabolic influences on the unstable genic molecules.

■ POPULATION GENETICS AND EVOLUTIONARY PROCESSES

Population biology deals with more than one organism in a group that lives together and is primarily concerned with changes in number and genetic composition of populations in time and space. The problems of populations are different from those of cells and organisms. Population biology emphasizes the ways in which animals react to each other, their relations to their habitats and surroundings, and how different kinds of populations arose. The members of a population must be considered together. A population cannot be dissected like an organism in the laboratory; populations cannot be isolated from the environment without destroying them. Populations are part and parcel of the landscape and the subtle integration that exists between population members and their environment must be understood.

Biparental reproduction and mendelian assortment furnishes a means of shuffling the genes into various combinations. Some of those combinations that work best have a better chance to be retained in the population. For instance, biparental reproduction makes it possible for two favorable mutant genes (one from each of two individuals) to come together within the same individual, but within an asexual population, the two favorable genes would get together only when both mutations occurred in the same individual. Sexual reproduction thus speeds the rate of evolution and better enables the organisms to fit into a constantly changing environment.

The term **gene pool** refers collectively to all of the alleles of all of the genes of a population. The gene pool of large populations must be enormous, for at observed mutation rates many mutant alleles can be expected at all gene loci. In some cases more than 40 alleles of the same gene have been demonstrated. Suppose there are two alleles present, A and a. Among the individuals of the population there will be three possible genotypes: AA, Aa, aa. When there are three alleles present, there are six possible genotypes. Increasing the number of alleles increases the possible genotypes.

Changes in uniparental populations occur by the addition or elimination of a mutation; in biparental populations the mutant gene may combine with all existing combinations and thus double the types. With only 10 alleles at each of 100 loci, the number of mating combinations would be 10^{100}. When an organism has thousands of pairs of alleles, the amount of diversity is staggering. Even though many genes are found together on a single chromosome and tend to stay together in inheritance, this linkage is often broken by crossing-over. If no new mutations occurred, the shuffling of the old genes would produce an inconceivably great number of combinations. But this is not the whole story because genes exert different influences in the presence of other genes. Gene A may act differently in the presence of gene B from what it does in the presence of gene C. The diversity produced by this interaction and the addition of new mutations now and then adds to the complication of population genetics. If this diversity is possible in a single population, suppose two different populations with different gene pools should mix by interbreeding. It is easy to see that many more combinations of genes and their phenotypic expression would occur. All this means that populations have enormous possibilities for variation.

What does this signify for evolution? It has already been stated that genetic variation produced in whatever manner is the material on which natural selection works to produce evolution. Natural selection does this by favoring beneficial variations and eliminating those that are harmful to the organism. Selective advantages of this type represent a very slow process, but on a geologic time scale they can bring about striking evolutionary changes represented by the various taxonomic units (species, genera, etc.), adaptive radiation groups, and the various kinds of adaptations.

The fact must be stressed, however, that natural selection works on the whole animal and not on single hereditary characteristics. The organism that possesses the most beneficial combination of characteristics or "hand of cards" is going to be selected over one not so favored. This concept helps explain some of those puzzling instances in which an animal may have certain characteristics of no advantage or that are actually harmful, but in the overall picture it has a winning combination. Thus, in a population, pools of variations are created on which natural selection can work to produce evolutionary change.

Why are not recessive genes lost from the population? In an interbreeding population why does not the dominant gene gradually supplant the recessive one? It is a common belief that a character dependent on a dominant gene will increase in proportion because of its dominance. This is not the case, for there is a tendency in large populations for genes to remain in equilibrium generation after generation. A dominant gene will not change in frequency with respect to its allele.

This principle is based on a basic law of population genetics called the **Hardy-Weinberg equilibrium.** According to this law, gene frequencies and genotype ratios in large biparental populations will reach an equilibrium in one generation and will remain constant thereafter unless disturbed by new mutations, by natural selection, or by genetic drift (chance). The rule does not operate in small populations. A rare gene, according to this principle, will not disappear merely because it is rare. That is why certain rare traits, such as albinism, persist for endless generations. It is thus seen that variation is retained even though evolutionary processes are not in active operation. Whatever changes occur in a population—gene flow from other populations, mutations, and natural selection—involve the establishment of a new equilibrium with respect to the gene

pool, and this new balance will be maintained until upset by disturbing factors.

The Hardy-Weinberg formula is a logical consequence of Mendel's first law of segregation and is really the tendency toward equilibrium inherent in mendelian heredity. Select a pair of alleles such as T and t. Let p represent the proportion of T genes and q the proportion of t genes. Therefore, $p + q = 1$, since the genes must be either T or t. By knowing either p or q, it is possible to calculate the other. Of the male gametes formed, p will contain T and q will contain t, and the same will apply to the female gametes. (See checkerboard in Chapter 33, p. 779.) As we know from Mendel's law, there will be three possible genotypic individuals, TT, Tt, and tt, in the population. By expanding to the second power, the algebraic formula $p + q$ will be $(p + q)^2 = p^2 + 2pq + q^2$, in which the proportion of TT genotypes will be represented by p^2, Tt by $2pq$, and tt by q^2. Recall the 1:2:1 ratio of a mendelian monohybrid. The homozygotes TT and tt will produce only T and t gametes, whereas the heterozygotes Tt will produce equal numbers of T and t gametes. In the gene pool the frequencies of the T and t gametes will be as follows:

$$T = p^2 + \tfrac{1}{2}(2pq) = p^2 + pq = p(p + q) = p$$
$$t = q^2 + \tfrac{1}{2}(2pq) = q^2 + pq = q(q + p) = q$$

In all random mating the gene frequencies of p and q will remain constant in sexually reproducing populations (subject to sampling errors). It will be seen that the formula $p^2 + 2pq + q^2$ is the algebraic formula of the checkerboard diagram, and thus the formula can be used for calculating expectations without the aid of the checkerboard.

To illustrate how the Hardy-Weinberg formula applies, suppose a gene pool of a population consisted of 60% T genes and 40% t genes. Thus:

$$p = \text{frequency of T (60\% or 0.6)}$$
$$q = \text{frequency of t (40\% or 0.4)}$$

Substituting numerical values of gene frequency in the following,

$$p^2 + 2pq + q^2$$
$$(0.36 + 0.48 + 0.16)$$
$$\text{TT} \quad \text{Tt} \quad \text{tt}$$

the proportions of the various genotypes will be 36% pure dominants, 48% heterozygotes, and 16% pure recessives. The phenotypes, however, will be 84% (36 + 48) dominants and 16% recessives.

On the other hand, suppose 4% of a population carries a certain recessive trait, then:

$$q^2 = 4\% \text{ or } 0.04$$
$$q = \sqrt{0.04} = 0.2 \text{ or } 20\%$$

Thus 20% of the genes are recessive. Even though a recessive trait may be quite rare, it is amazing how common a recessive gene may be in a population. Only 1 person in 20,000 is an albino (a recessive trait); yet by the above formula, it is found that 1 person in every 70 carries the gene or is heterozygous for albinism.

How chance operates to upset the equilibrium of genes in a population. The Hardy-Weinberg equilibrium can be disturbed, as already stated, by mutation, by selection, and by chance or genetic drift. The term **genetic drift** (Wright) refers to changes in gene frequency resulting from purely random sampling fluctuations. By such means a new mutant gene may be able to spread through a small population until it becomes homozygous in all the organisms of a population (random fixation) or it may be lost altogether from a population (random extinction). Such a condition naturally would upset the gene frequency equilibrium mentioned in the previous section. It also affords a means by which small, isolated populations can originate characteristics that are of no use to the individuals of that population, such as the small differences between subspecies and even species. It has also been suggested that genetic drift may result in a new species being formed, or else contribute to the gene pool of the large, ancestral population under certain conditions.

How does the principle apply? Suppose a few individuals at random became isolated from a large general population. This could happen by some freakish accident of physical conditions, such as a flood carrying a small group of field mice to a remote habitat where they would have no opportunity to mix with the general population, or a disease epidemic could wipe out most of a population and produce the same effect. Suppose that in the general population individuals would be represented by both homozygotes, TT, for example, and heterozygotes, Tt. It might be possible for the small, isolated group to be made up only of TT individuals and the t gene would be lost altogether, or the reverse could happen. Also, when only a small number of offspring are produced, certain genes may, by sampling errors, be included in the germ cells and others not represented. It is possi-

ble in this way for heterozygous genes to become homozygous. In this way the new group may in time have gene pools quite different from the ancestral population.

We should note also that most breeding populations of animals are usually small. Most large and widespread populations are divided into more or less isolated groups by physical barriers of some kind. The home areas even of animals that can get around are amazingly small in many instances. A mere stream may be effective in separating two breeding populations. Thus, chance could lead to the presence or absence of genes without being directed at first by natural selection. In the long run, however, whether the trait has adaptive significance will depend on natural selection.

■ FACTORS AND CHARACTERISTICS OF EVOLUTION
Natural selection

Most of us think of natural selection as the struggle for existence, survival of the fit, brutal competition, etc. But other factors that play a part in natural selection are the ability to produce large numbers of viable offspring, ability to resist disease, speed of development, mutual cooperation, etc. In accordance with the evolutionary opportunity of an environment in which an organism is living, natural selection can stabilize, direct, or disrupt the whole evolutionary process of a particular form. Natural selection is a blind force without purpose and has often been called noncreative. But the newer evolutionary synthesis assigns to it a creative role, for natural selection can ensure the continuance of a favorable allele and the elimination of an unfit allele and can mold all the chromosomes and their genes into an integrated whole. Natural selection does not operate on an all-or-none basis but is a statistical phenomenon. It works on populations as a whole. Populations are made up of thousands of variable genes. Individuals with a superior genotype will survive and will tend to leave more offspring so that natural selection is the differential perpetuation of genotypes. Natural selection works only on those members of a population that are of reproductive age and is not involved with what happens in old age. Selection operates on the available variations to produce the most efficient gene pool for adaptation and reproductive success. Two factors, then, are required for effective natural selection—the choice of variations and reproductive individuals to carry on these variations to future generations.

Natural selection is not only a process that affects all members of a species in a uniform manner but also produces within the same species genetically different populations adapted to their corresponding environments. The ability of organisms to occupy a wide range of habitats depended not only on their ecologic tolerance of a wide range of conditions but on genetic modifications of the type proposed by Darwin. The frequent demonstration of genetic adaptation within species has been a constant stimulus to the study of natural selection.

Chance or random genetic drift

In small populations, random fluctuations or changes in gene frequency may have an important bearing on evolutionary trends. Some genes may be lost and others may be increased in frequency because certain mutant genes may not be included in gametes in the process of meiosis, or they may not be present at all or present with a greater frequency than they were in the original larger population. Close interbreeding in such populations would tend to make heterozygous genic pairs homozygous.

Genetic drift has been assigned as the cause of the frequency of certain human traits such as blood groups. Among some American Indian tribes, we know that group B, for instance, is far rarer than it is among other races and may be the result of small isolated mating units.

How effective genetic drift is in the evolutionary process is a controversial subject, and there are many who deny its importance. But they generally agree that in bisexually reproducing species evolution proceeds more rapidly when a population is broken up into isolated or partially isolated breeding communities, and the smaller the population the greater will be the importance of genetic drift.

Isolation and population structure

When groups of organisms are prevented from interbreeding or are divided into subunits with limited cross migration, evolutionary divergence results because an isolated population would not share its mutant genes with others and would develop its own unique evolution. This process could lead to the formation of a new species in each of the isolated groups. Suppose a large population is found in a wide geographic area and exhibits the characteristics of a **cline**; that is, there is a gradual, continuous, gradient change in the members of the population because of adjustments to local conditions that show considerable variations in different parts of the cline. In certain parts of the cline, climate conditions may be hot, in others cold; weather may vary from extreme moisture in some parts to very dry in others, etc.

Within a cline the species will be divided into smaller units of population (**demes**) that are more or less isolated from each other in accordance with the different habitats found within the range of a cline. The members of a deme may breed freely with each other but usually not with the members of other demes. Each deme can more or less develop an evolution of its own, for small hereditary differences (mutations) that occur in demes may be, to some extent, unique, and thus each deme in time becomes different from other demes.

The demes within a cline will not all have the same fate. Some of them may come together (if interbreeding can occur) and fuse; others may differentiate far enough to prevent interbreeding and form true species. Whenever two demes fuse, each contributes its pool of genes to the future offspring, which thus acquire advantageous genes of both demes. Some demes may become extinct in a large cline. Adjacent demes may intergrade into each other. A common pattern of demes consists of many subspecies formed by divergent evolution in partially isolated demes. The time factor may cause these subspecies to differentiate into true species. Because of the smallness of some of the demes, genetic drift can operate to produce a pronounced differential in gene frequency. In this way chance may determine whether certain genes will be emphasized or neglected. It will thus be seen that whenever a population is subdivided into small subunits there is a strong possibility that there may be a rapid evolutionary process of a divergent nature, with results favorable to the formation of new species.

Concept of polymorphism

Many species including both plants and animals are represented in nature by two or more clearly distinguishable kinds of individuals. Such a condition is called **polymorphism** and may involve not only color but many other characters, physiologic as well as structural. The term has been used with many meanings but rules out seasonal variations, such as the winter and summer pelage and plumage of certain mammals and birds. Polymorphism always refers to variability within a population and is restricted to genetic polymorphism. It results when there are several alleles or gene arrangements with discontinuous phenotypic effects. It may be expressed by two alternative types, such as male and female dimorphism, or there may be many morphologic types within the species.

The genetic pattern is exemplified by the ladybird beetle *Adalia bipunctata,* in which some individuals are red with black spots, whereas others are black with red spots. The black color behaves as a mendelian dominant and red as a recessive. The two forms live side by side and interbreed freely, with the black form predominant from spring to autumn and the red form from autumn to spring. It is believed that the changes are produced by natural selection, which favors the black form during summer and the red during winter. For instance, more black forms, produced by dominant genes, die out during winter, while the recessive red form survives. It is thus possible for the recessive gene, at least during part of the seasonal cycle, to be more common that its dominant allele.

Polymorphism has adaptive value in that it adapts the species to different environmental conditions. Polymorphic populations are thus better able to adjust themselves to environmental changes and exploit more niches and habitats. No doubt the potentialities of polymorphism are greater than its realization because every population must have many allelic series that are never expressed in the visible phenotype.

Polymorphism is widespread throughout the animal kingdom. Some examples are the right-handed and left-handed coils of snails of the same species, the blood types of man and other animals (a biochemical distinction), sickle cell anemia in man, albinism in many animals, silver foxes in litters of gray foxes, and rufous and gray phases of screech owls in the same brood.

Neoteny as an evolutionary factor

The concept of neoteny is believed to play an important role in creating new patterns or types of animals. This concept emphasizes the retention of larval characteristics in the adult and the appearance of sexual maturity in a larval, or juvenile, condition. Mention has already been made in a former section of the axolotl form of the salamander, whose metamorphosis may have been suspended because the thyroid mechanism fails to operate normally. By such means it is possible for many new patterns of animal life to have evolved from generalized species.

Some striking examples in evolution may occur when certain stages of the life history are simply dropped off. For instance, insects with six legs are supposed to have evolved from the myriapods with many segments and legs. At an early stage of their existence, myriapods have three pairs of legs and few segments. It would be easy for such a larval form to be transformed into an insect by the retention of the larval characteristics of the myriapods, which already have some of the fundamental characteristics of insects (tracheae, malpighian tubules, antennae, etc.).

Many evolutionists believe that man himself is a neotenous form of a primitive apelike ancestor. This may account for the lack of hair, which is a fetal characteristic in apes, but an adult trait in man. The flat face of man, the large brain (which in the embryos of all mammals is proportionally much larger than the rest of the body), the lack of pigmentation in certain races, etc. are really fetal characteristics.

Adaptation

The major aim of evolution is the adaptation of the organism to its environment. Throughout evolution the chief features of life are very much the same. All organisms share in common about the same biochemical compounds, the same kinds of biosynthesis and energy transfer, the same structural features of tissues, and the same metabolic mechanisms of growth, respiration, digestion, etc. These primordial processes of mutual adaptations were fashioned somewhere in the long evolutionary process. All individual adaptations are shaped by evolution because maladjusted organisms simply do not survive to reproduce. Fitness to the environment must be found at all levels, from cellular ecology to that of populations. Adaptation, then, is fitting a biologic system to harmonize with the environmental factors of its existence.

All adaptations are of evolutionary origin, brought

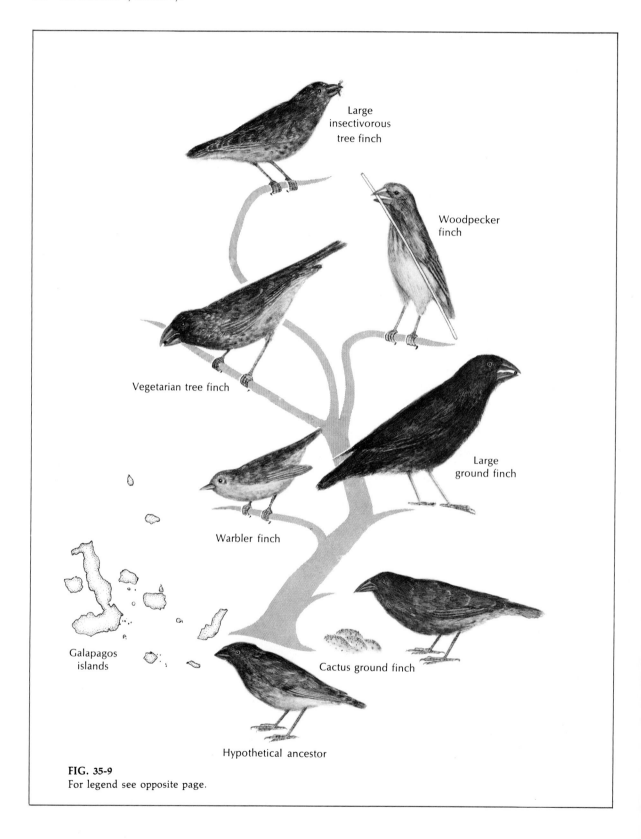

Large
insectivorous
tree finch

Woodpecker
finch

Vegetarian tree finch

Large
ground finch

Warbler finch

Cactus ground finch

Galapagos
islands

Hypothetical ancestor

FIG. 35-9
For legend see opposite page.

about by variability and natural selection. No single adaptive mechanism fits all the conditions of the environment. The organism's adaptive nature is a developmental pattern, faithfully transmitted from generation to generation, and preserved because of its survival merit. The adaptive responses of any organism are prefitted by its evolutionary endowment. They are determined by and restricted to the range of environmental conditions the organism has had in its long evolutionary and phylogenetic experience. This means that there may be no adaptation at all to an entirely novel condition. Only the gross aspects of adaptation are predetermined; the individual must fill in the details to adjust to the conditions it meets in its existence. The limits of its adjustment are fixed by heredity.

No animal is perfectly adaptive to its environment; adaptation is relative because the adaptive endowment of an organism is never able to anticipate all the constantly occurring variations in its environment. There seems to be many characteristics, both anatomic and physiologic, that do their possessors little good. The successful adaptation of an organism is determined by the sum total of all its adaptations. Although some of its adaptations may be favorable and others unfavorable, the animal may survive. This is why many so-called adaptations, such as **protective coloration** and **sexual selection,** are now viewed with a critical eye.

Coloration is widespread throughout the animal kingdom and apparently serves many purposes, with camouflage being one of the most common. In protective coloration the organism blends into its environment so that it can escape detection by enemies. A common example is the dark dorsal and lighter ventral side of fish and many birds. Their color blends with the sky to a potential enemy viewing them from

underneath, and viewed from above, the darker shades blend with the darker shades and regions below. Many types of concealment blend the animal into its background; frogs and lizards have the added advantage of changing their color to suit their background. One of the most striking cases of protective camouflage is the *Kallima* butterfly that has a remarkable resemblance to a leaf with venation, imitation holes, and leaf-coloration patterns. The concealing coloration of predators enables them to approach their prey undetected.

Mimicry adaptation is common among many forms in which harmless animals have found survival value by imitating other species that are well equipped with defensive or offensive weapons. Thus harmless snakes may have a resemblance to venomous ones, and some flies look very much like bees. There are also many examples of **warning** coloration, which advertises the presence of a well-protected animal, such as the white stripes of the skunk and the brilliant colors of the poisonous coral snake.

These adaptive resemblances, whether of protection, warning, or mimicry, have been subjected to experiments with the idea of determining their selective value. Conclusions from these experiments are often conflicting. Some biologists have gone so far as to deny the selective value of color altogether. The extensive investigations of McAtee on the stomach contents of birds indicated that protectively as well as nonprotectively colored insects were eaten by birds without discrimination, but the relative abundance of the protected and unprotected forms in nature must be carefully considered in such experiments. Other experiments seem to indicate that color patterns do play some useful role in survival selection. It has been suggested that some animals that are not protectively colored to the human eye are so protected when

---

FIG. 35-9

Darwin's finches and their beak adaptations. The 14 species of finches, some of which are shown here, are believed to have evolved from some ancestor finch that came to the Galápagos Islands from the mainland of South America. By adaptive radiation the descendants of the ancestral species have radiated out into the different habitats found on the different islands. On each island they may have evolved into different types by genetic drift. Some are adapted as ground feeders of different sizes of seeds; some feed on cactus flowers, some on insects (with woodpecker behavior patterns), and so forth. Correlated with their different diet habits are the size and shape of the beaks. (Drawn by Sheila Ford.)

viewed by the type of vision of their enemies. Many experiments have also been performed in which predators had an opportunity to choose among prey bearing concealing, warning, or other color devices, and the results indicated that these color patterns do have selective value. These and other experiments emphasize the point that sweeping denunciations of the coloration concept are not yet warranted.

Darwin's idea of **sexual selection** has been more bitterly criticized than any other aspect of his natural selection theory. Darwin believed that the conspicuous patterns of pigmentation in the males of many species of birds could be accounted for on the basis that the females tend to select those males with the most brilliant colors and most ornamental devices. In this way only those males so attractively equipped would have a chance to leave descendants. Experiments to determine the role of ornamentation in the selection of their mates by females have not given very conclusive results, for courtship display may not occur until after pairing. So many factors are involved that it is difficult to state just what part sexual selection does play. Authorities generally agree, however, that Darwin overemphasized its importance.

Adaptive radiation

A striking illustration of isolation and adaptive radiation is afforded by Darwin's finches of the Galápagos Islands. These islands are found about 600 miles west of the South American mainland (Ecuador) and attracted the attention of Darwin when he visited them about 1835 during his famous voyage on the *Beagle.* Darwin was struck by the unusual animal and plant life on these islands as contrasted with that on the nearest mainland of South America. These volcanic islands have never had land connections with the mainland and such life as is found there had to be by accidental immigration. Darwin was particularly interested in the bird life of the islands, especially the finches.

In 1947 the English ecologist Lack published a book on these finches that has served to renew interest in their evolutionary development. All the different species of finches (family Geospizidae) found on these islands are believed to have descended from a South American species of finch that reached the island. Their most interesting adaptive structures are their different beak modifications that permit them to exploit the food resources of the various ecologic

habitats (Fig. 35-9). The ancestral finch was probably a ground feeder, but as competition increased in this habitat, finches evolved that were adapted (or radiated out) to utilize food in other ecologic niches unavailable to the ancestral form. The ground feeders have thick, conical beaks for seed; the cactus ground finch has a long, curved beak for the nectar of cactus flowers; the vegetarian tree finch has a parrotlike beak for buds and fruit; and the woodpecker finch has a long, stout beak and uses a thorn or spine for probing into bark for insects. Some of the finches are vegetarian and some are insectivorous. Altogether, some 14 species (and some subgenera) of finches have evolved from the ancestral finch that first reached the islands. Moreover, each island of the group, because of geographic isolation, has had an effect on the evolution of the various species, although most of the islands have the same type of ecologic habitats.

Convergence and parallelism

Two groups of animals not closely related may evolve similar structures because similar habitats have afforded them the same evolutionary opportunity. The two groups may be quite dissimilar in the beginning, or their ancestors may be entirely lacking in the common structures under consideration. The radiations of the two groups have been along lines of similar opportunities. Similar mutations are favored because of the possibilities of the environment and each has taken advantage of similar ecologic niches. Good examples of convergence are seen in the likenesses between the placental and marsupial types of animals (Fig. 35-10) and by the resemblance between the body forms of aquatic mammals and fishes.

When two groups are rather closely related to begin with, the evolutionary phenomenon is called parallelism. In this case, mutations are more likely to be similar in each group, and each will undergo evolutionary changes along similar lines, although the two are never completely identical. For instance, the enamel-crowned molars of the horse, rhinoceros, and elephant came from ancestral forms that lacked this characteristic.

Preadaptation and opportunism

Some animals may be preadapted to a particular type of environment long before they actually meet such an environment. Since mutations are random

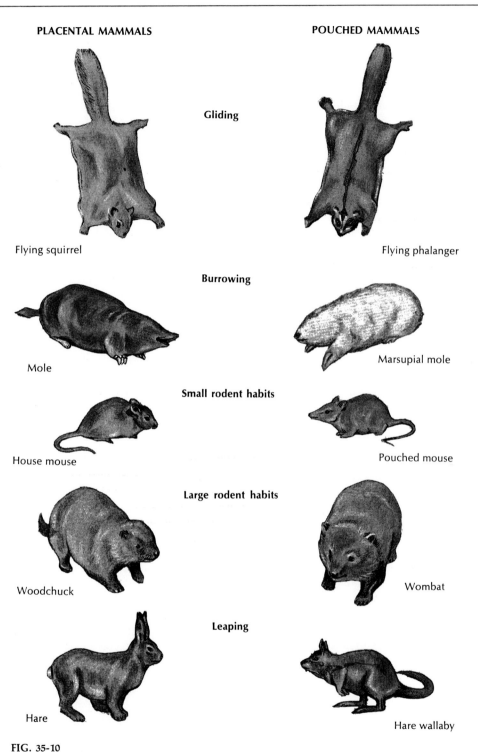

PLACENTAL MAMMALS   **POUCHED MAMMALS**

Gliding

Flying squirrel   Flying phalanger

Burrowing

Mole   Marsupial mole

Small rodent habits

House mouse   Pouched mouse

Large rodent habits

Woodchuck   Wombat

Leaping

Hare   Hare wallaby

FIG. 35-10
Convergent evolution among mammals. Two groups of mammals not closely related
(placentals and marsupials) have independently evolved similar ways of life and occupy
similar ecologic niches. Note that for every member of the ecologic niche in one group
there is a counterpart in other group. This correspondence is not restricted to similarity
of habit but also includes morphologic features.

and cannot always be predicted, some advantageous mutation may permit the invasion of a new habitat should that habitat become available. It is sometimes referred to as prospective adaptation. Most animals, unless too highly specialized for any adaptive change to occur, have characteristics that enable them to build new adaptations. An adaptation in one ecologic relationship may with little or no change be suited to a different environment. Numerous examples could be given of structures that probably had little use in a specific environment until the opportunity arose of an ecologic niche where the structure was useful. Flatworms, for instance, because of their morphologic form, could fit well into parasitic niches. The term "preadaptation" is so closely associated with adaptation that it is a generalized term and difficult to separate from the concept of adaptation.

Regressive evolution

Regressive evolution seems to be a universal phenomenon among animals that have vestigial organs (eyes, legs, lungs, mouths, teeth, etc.). This condition has occurred because ecologic selection pressures have decreased, along with a convergent degeneration of analogous and homologous functional structures of animals that have similar habits. For instance, caste differentiation among termites is not based upon genetic differences but rather is the result, during development, of different food intake and trophallactic exchange. All castes come from eggs that are genetically alike. Soldiers have vestigial gonads of either sex. The nonfunctional nymphal tooth of certain soldier nymphs is homologous with that of primitive adult termite soldiers.

Cave animals often lack eyes or have reduced eyes. Their ancestors may have had a genetic absence of the eyes or the absence of eyes may be the result of regressive evolution. The principle involved is that in a complex that once had former adaptations it is possible for the elements to be shifted around and to have new functions now.

Progressive evolution

Many forms that were once dominant became extinct (trilobites, eurypterids, and ostracoderms). Also, many later dominant groups arose from unspecialized lines of earlier dominant groups of animals (birds from reptiles, for example). These have become dominant on a new level with the help of new

properties. Man is a dominant form that has shown little radiation except in complex social factors and division of labor. Dominant forms tend to deal with their environment in a greater variety of ways and can exert more control than can other groups living at the same time. Dominant groups are usually more independent of the environment. The basis of dominance may be taken as a criterion of biologic progress.

Some evolutionary generalizations

In evolution there are several generalizations that reveal certain trends in the evolutionary process. There are some exceptions to these rules, but they do reveal many interesting aspects that the student of evolution might overlook. Some of these, such as Gloger's, Bergmann's, and Allen's rules, represent geographic variation gradients correlated with adaptive morphologic structures and are the result of mutation and selection.

DOLLO'S LAW. Dollo's law applies to the irreversibility of evolution and was first stated by the paleontologist Dollo. Recently the principle has been greatly emphasized by the American evolutionist Blum. Although minor reversals, such as back mutations, may occur, there is no evidence to indicate that evolution is other than a one-way process.

COPE'S LAW. This principle states that there is a tendency for evolving animals to increase in size until they become extinct. It was formulated by the American paleontologist Cope and is based on the overall increase in complexity of most evolutionary lines. The present horse, for example, evolved from a much smaller animal. Larger animals are probably more independent of their environments because they are more complex and are less affected by external changes. They can also store more food reserves and can usually move faster.

GLOGER'S RULE. In the northern hemisphere, most species of birds and mammals living in a north-south range tend to be lighter colored (less melanin) in the north than races of the same species living in humid, warm climates. Darker colors are usually associated with greater humidities, and this factor, as well as temperature, may play a part in the application of the rule.

BERGMANN'S RULE. According to this principle endothermal animals are usually larger in the colder parts and smaller in the warmer parts of their range (Fig. 35-11). It is based on the physiologic principle

In closely related species of warm-blooded animals,
larger members inhabit colder climates (Bergmann's rule)

King penguin—south to 55° (subantarctic)
Body length 40 inches

Magellan penguin—south to 52°
Body length 28 inches

Humbolt penguin—
west coast of South America
body length 20 inches

FIG. 35-11
Bergmann's rule, as illustrated by 3 of 17 species of penguins. Emperor penguin, even
larger than king penguin, is restricted mainly to Antarctic continent. Rule is based upon
principle that large body has smaller surface in proportion to mass (weight) than does
small body and so is better able to maintain its heat. Other birds and mammals verify
rule, for larger members are found in colder parts of their range, whether in northern or
southern hemisphere.

that a large body has a relatively smaller surface area
and thus can conserve its heat better. There are many
examples. Penguins are much larger on the Antarctic
continent than they are on the islands off the coast
of South America; bears are larger in Alaska than
they are elsewhere in the United States; and hares
are much larger in Russia than they are in central
Europe. There are some exceptions, however, to the
rule.

ALLEN'S RULE. Most races of mammals and birds
in cold regions have relatively shorter extremities
than do races of the same species from warmer regions. Arctic hares and foxes, for instance, have
shorter ears than closely related southern species.
The rule has some experimental verification because
mice reared at high temperatures have relatively
longer ears and feet than those developed at lower
temperatures. This is an adaptation against loss of
heat and the possibility of freezing in cold climates.
There is about a 25% exception to this rule.

JORDAN'S RULE. This rule states that closely
related species or subspecies are not found in the
same range but in adjacent ones separated by some
barrier. This generalization is verification of the role
of geographic isolation in the evolution of species.
There are some exceptions to the rule, for some

closely related forms may be found in the same region but because of ecologic or other isolation do not interbreed.

GAUSE'S RULE. Two species with identical ecologic relations cannot occupy indefinitely the same ecologic niche in the same habitat. Usually one species exterminates or drives out the other. If the species have sparse populations, it is possible for the two groups to share similar ecologic niches, or one species could occupy peripheral regions and the other central areas of the same general habitat (centrifugal speciation). The study (1958) of R. H. McArthur on five species of warblers that are found together in coniferous forests and have in general the same insectivorous habits reveals a nice discrimination in the application of Gause's principle. Even though these species ate about the same kind of food, each species tended to avoid severe competition with the others by inhibiting overgrowth of its population through territorial behavior, by feeding in different positions or zones in the trees and thus being exposed to different types of food, by different nesting dates, etc. In this way the different species could be coexistent, even though there was some overlapping of their activities.

■ SPECIATION

Speciation, as we have seen, refers to the splitting of one species into two or more other species. It implies the formation of two or more populations that do not exchange genes. Although there may be several patterns in the diversification of organisms, it is believed that the sequence of geographic isolation, morphologic differences (mutations, etc.), and conditions that prevent future fusion of groups represents one of the most important and effective patterns of evolutionary progress. It is clear that whenever distinct forms interbreed, their genotypes will intermix and any differences between them will be lost. Any effective evolutionary pattern must prevent this interbreeding so that each group (species) can profit from its own independent evolution. Evolution seems to be most rapid when a group of organisms are presented with a new environment with many unoccupied ecologic niches of low competition and predation. By adaptive radiation many new types can evolve to fill these niches.

It is doubtful that hybridization plays a major role in the evolutionary process in animals. When two species are crossed, especially those with different chromosone numbers, the offspring are generally sterile.

Species populations or their subdivisions, separated by natural barriers and thus occupying different territories, are called **allopatric.** Two or more species or populations inhabiting the same geographic range are called **sympatric.** Allopatric and sympatric populations may show different degrees of isolation. It is easy to see how geographic barriers can produce complete isolation and prevent interbreeding between separate groups so that evolution may proceed in the isolated group without interference with other groups. In sympatric populations isolation may be seasonal (breeding at different times of the year) or behavioral (incompatibility of mating reactions); these are as effective as the geographic one. Whenever two closely related species occupy the same area, it is believed that they evolved in different ecosystems and later merged into one.

An example of speciation may be the common leopard frog *Rana pipiens.* For years herpetologists puzzled over the status of this frog, which apparently ranges from northern Canada to Panama and from the Atlantic coast to the edge of the Pacific states. Is it all one species or is it divided into several species or subspecies? Individuals from adjacent localities, such as those from New England and New Jersey, or those from Florida and Louisiana, can be crossed successfully and yield normal and viable embryos. But crosses between individuals from widely separated localities, such as Wisconsin and Florida, produce abnormal and nonviable embryos. Evidently the genetic differences between the northern and southern forms are great enough to prevent normal hybrid development. Selection has produced different developmental physiologies in these frogs. Those in the north are adapted for rapid growth at low temperatures; those in the south, for slow growth at warm temperatures. This genetic difference is sufficient, if one ignores the individuals from adjacent localities, to produce at least two distinct species—the northern form and the southern form.

The belief now is that there are actually several closely related species of these leopard frogs, none of which has a range as extensive as the *Rana pipiens* complex was believed to have. These different species may have arisen as allopatric species whose ranges now overlap geographically.

■ CAN WE OBSERVE EVOLUTION IN ACTION TODAY?

Darwin in his time could not point to a single visible example of evolution in action. Some of his opponents were quick to point out that this lack was a major weakness in his arguments. However, in England during his lifetime, a striking case of evolution was actually taking place in nature. Such an example is industrial melanism in moths.

Within the last century, certain moths, such as the peppered moth *(Biston)*, have undergone a coloration change from a light *(B. betularia)* to a dark melanic form *(B. carbonaria)*. This moth is active at night and rests on the trunks of trees during the day. The light form is especially well adapted to rest on the background of lichen-encrusted trees where it is largely invisible. In the industrial regions of England, however, trees have darkened because of smoke pollution, which kills the light-colored lichens and blackens the vegetation. Against a dark background, light-colored moths are conspicuous and fall prey to predator birds. Natural selection would therefore, in the course of time, largely eliminate this moth. However, by mutation the light-colored moth has given rise to a

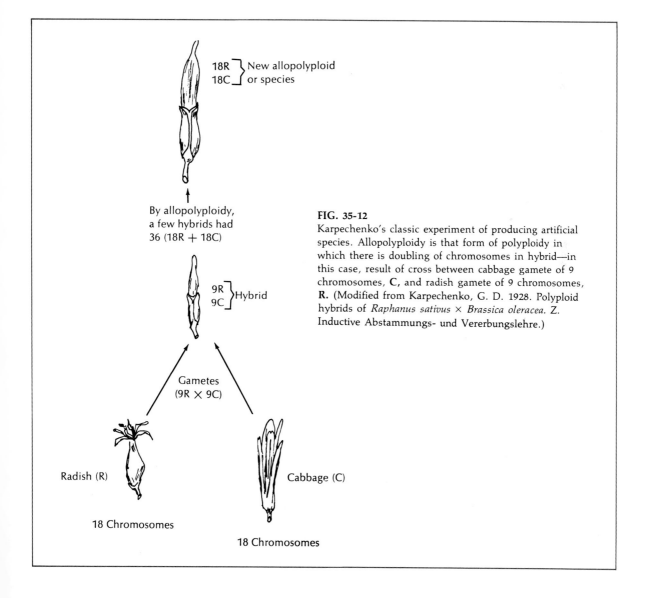

FIG. 35-12

Karpechenko's classic experiment of producing artificial species. Allopolyploidy is that form of polyploidy in which there is doubling of chromosomes in hybrid—in this case, result of cross between cabbage gamete of 9 chromosomes, **C**, and radish gamete of 9 chromosomes, **R**. (Modified from Karpechenko, G. D. 1928. Polyploid hybrids of *Raphanus sativus* × *Brassica oleracea*. Z. Inductive Abstammungs- und Vererbungslehre.)

darker melanic form that has a much better survival rate under such surroundings. Within a period of years in polluted areas, the dark species *(B. carbonaria)* has, to a great extent, replaced the light species *(B. betularia)*. The mutation for industrial melanism appears to be controlled by a single dominant gene that is also known to occur in natural environments (not attributable to pollution) in which a dark color is advantageous. Thus by mutation and natural selection a moth of a different color and physiologic nature has emerged to confirm Darwin's mechanism of the evolution of a new species from another species.

Experimental production of new species

In 1924 Karpechenko produced a new species by hybridization. He crossed the radish *(Raphanus sativus)* with the cabbage *(Brassica oleracea)*, each of which has 9 haploid chromosomes. In most cases the hybrids of two different species of unlike chromosomes (as these were) are unable to produce fertile offspring because unlike chromosomes cannot pair properly in meiosis. The 18-chromosome hybrids (9 chromosomes from each parent) in this cross were sterile and could not produce offspring. However, a few of the hybrids by spontaneous allopolyploidy, or doubling of the chromosomes, had 36 chromosomes (18 radish chromosomes and 18 cabbage chromosomes). When these hybrids underwent meiosis to form their gametes, the homologous radish chromosomes could pair with each other and the homologous cabbage chromosomes could pair with each other so that each gamete had 9 radish and 9 cabbage chromosomes. The union of such eggs and sperm could produce a fertile plant of 36 chromosomes. This new plant, which had characteristics of both the radish and the cabbage, could not be crossed with either one of the original parents and was reproductively isolated. The new synthetic genus was called *Raphanobrassica* and was the first recorded instance of the artificial creation of a new species (Fig. 35-12).

Allopolyploidy has perhaps played an important role in plant evolution but very little in that of animals. However A. O. Wasserman reports that in the common tree frog *Hyla versicolor* he found the corneal cells and cells of the tail tip to be tetraploid. He proposes that the genetic pattern of the somatic chromosomes is $2n \rightarrow 4n = 48$ and that this species may be entirely tetraploid. Other cases of polyploidy are known in the anurans.

■ NEW APPROACH IN STUDY OF EVOLUTION

Molecular biology has afforded new approaches to the study of evolution including population genetics and systematics. It has been possible to study genetic variations and the similarities and differences among organisms at the level of their enzymes, other proteins, and their DNAs. The most common technique used is gel electrophoresis, which demonstrates how the primary amino acid of most proteins varies among individual animals, even from the same population. Variations can be correlated with genetic differences. Electrophoresis detects differences in enzymes or other proteins because of differences in their electrostatic charges and their mobility in an electric field. At a molecular level, the animals and kinds of genetic variations in populations and an estimate of the extent of genetic divergence are ascertained.

Another technique determines the amino acid sequence of proteins having the same functional activity in different organisms. This makes possible an understanding of phylogenetic relationships because the greater the time since two animals had a common ancestor, the greater the number of differences in the amino acid sequence of their proteins.

A third technique emphasizes a comparison of the DNA base sequences as discovered by hybridization studies.

References
Suggested general readings

Anfinsen, C. B. 1959. The molecular basis of evolution. New York, John Wiley & Sons, Inc.

Arunachalam, V., and A. R. G. Owen. 1972. Polymorphism with linked loci. New York, Harper & Row.

Bates, M., and P. S. Humphrey. 1956. The Darwin reader. New York, Charles Scribner's Sons. *This book is made up of generous portions of Darwin's principal works and is a fine introduction to the works of the great master.*

Blum, H. F. 1955. Time's arrow and evolution, ed. 2. Princeton, N. J., Princeton University Press. *A work for the serious student on certain implications of evolution.*

Colbert, E. H. 1971. Tetrapods and continents. Quart. Rev. Biol. **46:**250-269.

Darwin, C. 1859. On the origin of species by means of natural selection, or the preservation of favored races in the struggle for life. London, John Murrey. *There were five subsequent editions by the author. Often reprinted.*

Dobzhansky, T. 1951. Genetics and the origin of species, ed. 3. New York, Columbia Univerity Press. *A treatise showing the bearing of modern genetic interpretation upon the problems of evolution. A work for the advanced student.*

Dobzhansky, T. 1970. Genetics of the evolutionary process. New York, Columbia University Press.

Florkin, M. (editor). 1960. Aspects of the origin of life. New York, Pergamon Press, Inc. *A series of papers by many eminent authorities on the basic problems of the origin and development of life.*

Glass, B., O. Temkin, and W. L. Straus, Jr. (editors). 1959. Forerunners of Darwin: 1745-1859. Baltimore, The Johns Hopkins Press. *All scientific concepts have an extensive background and this work rightly assigns some credit for Darwin's great theory to some of the men that had been thinking along similar lines long before Darwin arrived at his basic conclusions. Among some of these forerunners were Maupertuis, Buffon, Kant, Lyell, Malthus, and Lamarck.*

Gregory, W. K. 1951. Evolution emerging. A survey of changing patterns from primeval life to man, 2 vols. New York, The Macmillan Co. *This excellent work emphasizes the major evolutionary trends in the emergence of animal life, but considerable attention is given to fossils and their significance throughout the work. Suitable for the specialist only.*

Hardin, G. 1959. Nature and man's fate. New York, Rinehart & Co., Inc. *An excellent appraisal of man in relation to our present concept of evolutionary progress. Written in a lucid and popular style, this book is excellent supplementary reading for the general zoology student.*

Hotton, N. 1968. The evidence of evolution. New York, American Heritage Publishing Co., Inc. *This is a work on evolution by paleontologists with an emphasis on the fossil record. An excellent geologic timetable and a concise account of the voyage of the Beagle are included.*

Huxley, J. 1942. Evolution: the modern synthesis. New York, Harper & Brothers. *The modern interpretation of evolutionary advancement by one of the foremost students in the field. For the advanced student.*

Huxley, T. H. 1859. Darwin on the origin of species. The Times (London), Dec. 26, 1859. *This famous review of Darwin's epoch-making work was published a few weeks after The Origin of Species came from the press. It was unsigned, but, as Darwin surmised, there was only one man in all of England who could have written it, and Huxley was quickly identified as the author.*

Irvine, W. 1955. Apes, angels, and victorians. New York, McGraw-Hill Book Co. *By skillfully combining history and biography, the author has given a vivid account of the conflict that revolved around the theory of evolution when it was first announced and for many years thereafter.*

Lack, D. 1947. Darwin's finches. New York, Cambridge University Press. *Adaptive radiation is a highly interesting subject in evolutionary development, but no example has ever been found that illustrates this principle better than these finches Darwin studied in the early development of his evolutionary theory.*

Lederberg, J., and E. M. Lederberg. 1951. Recombination analysis of bacterial heredity. In Cold Spring Harbor Symposia on Quantitative Biology **16:**413-443.

Leeper, G. W. (editor). 1962. The evolution of living organisms. Melbourne, Melbourne University Press. *This work is based on a symposium held in honor of Darwin's centenary year. Emphasis is placed on Australasian fauna, but the work as a whole gives a fine summary of many evolutionary problems.*

Mayr, E. 1963. Animal species and evolution. Cambridge, Mass., Harvard University Press. *This excellent synthesis and critical evaluation of the processes of evolution will be the "last word" for a long time to come.*

Moody, P. A. 1968. Introduction to evolution, ed. 2. New York, Harper & Brothers. *There is no better introduction to evolution for the beginner.*

Moore, R. C., C. G. Lalicker, and A. G. Fischer. 1952. Invertebrate fossils. New York, McGraw-Hill Book Co. *A well-balanced textbook, with abundant and revealing illustrations. Every group has a concise list of definitions that are very helpful in understanding the discussions of the different forms. There is an excellent chapter on the way fossils are formed and preserved.*

McKinney, H. L. 1972. Wallace and natural selection. New Haven, Conn., Yale University Press. *Traces Wallace's evolutionary idea to 1859 and the influence of his ideas on Darwin.*

Nur, U. 1970. Evolutionary rates of models and mimics in Batesian mimicry. Amer. Naturalist **104:**477-486.

Platz, J. E., and A. L. Platz. 1973. *Rana pipiens* complex: Hemoglobin phenotypes of sympatric and allopatric populations in Arizona. Science **179:**1334-1336.

Powell, J. R. 1971. Genetic polymorphisms in varied environments. Science **174:**1035-1036.

Romer, A. S. 1950. Vertebrate paleontology, ed. 2. Chicago, University of Chicago Press. *An advanced text and one of the best in the field. Of great interest to all students of fossils.*

Ross, H. H. 1960. A synthesis of evolutionary theory. Englewood Cliffs, N. J., Prentice-Hall, Inc. *An integration of the factors that attempt to explain the how and why of the evolutionary process. There is a short but interesting chapter on the effects of structural (geotectonic) geology on evolution.*

Savage, J. M. 1963. Evolution. New York, Holt, Rinehart & Winston, Inc. *A brief but finely written account of the major concepts in the evolutionary process.*

Simpson, G. G. 1951. Horses. New York, Oxford University Press. *A comprehensive account of the interesting evolution of the horse.*

Simpson, G. G. 1953. Life of the past. New Haven, Conn., Yale University Press. *One of the best works on fossils for the beginning student. In a clearly written manner the author surveys the formation of fossils, their meaning, and how they fit into the evolutionary plan.*

Simpson, G. G. 1953. The major features of evolution. New York, Columbia University Press. *This is one of the most up-to-date accounts of evolution, written by one of the best authorities in the field.*

Symposia of the Society for Experimental Biology, 1953,

no. VII: Evolution. New York, Academic Press, Inc. *This volume contains the papers delivered at a symposium at Oxford in 1952. They cover many concepts in the field of evolution. Among the many interesting papers in the volume, the one on "Regressive Evolution in Cave Animals" is especially revealing. Unfortunately, the volume is not indexed.*

Tax, S., and C. Callender (editors). 1960. Evolution after Darwin. Vol. I, The evolution of life; Vol. II, The evolution of man: Vol. III, Issues in evolution. Chicago, University of Chicago Press. *A collection of the papers given at the Darwin Centennial held at the University of Chicago in November, 1959. A monumental tribute to the great evolutionist and an epoch-making appraisal of evolution as it is understood at present.*

Udvardy, M. 1970. Mammalian evolution: Is it due to social subordination? Science 170:344.

Washburn, L. L. (editor). 1963. Classification and human evolution. Chicago, Aldine Publishing Co. *Papers presented by specialists in paleontology and anthropology at a symposium held in 1961. An evaluation of the fossil record that bears on man's early evolution and his place in primate taxonomy.*

Wasserman, A. O. 1970. Polyploidy in the common tree toad *Hyla versicolor* LeConte. Science **167**:385-386.

Selected *Scientific American* articles

Barghoorn, E. S. The oldest fossils. **244**:30-42 (May).

Cavalli-Sforza, L. L. 1969. "Genetic drift" in an Italian population. **221**:30-37 (Aug.).

Cole, Fay-Cooper. 1959. A witness at the Scopes trial. **200**:120-130 (Jan.).

Crow, J. F. 1959. Ionizing radiation and evolution. **201**:138-160 (Sept.).

Darlington, C. D. 1959. The origins of Darwinism. **200**:60-66 (May).

Deevey, E. S., Jr. 1952. Radio-carbon dating. **186**:24-28 (Feb.).

Dobzhansky, T. 1950. The genetic basis of evolution. **182**:32-41 (Jan.).

Eiseley, L. C., 1959. Alfred Russel Wallace. **200**:70-84 (Feb.).

Kettlewell, H. B. D. 1959. Darwin's missing evidence. **200**:48-53 (March).

Kurtén, B. 1969. Continental drift and evolution. **220**:54-64 (March).

Lack, D. 1953. Darwin's finches. **188**:66-72 (April).

Napier, J. 1962. The evolution of the hand. **207**:56-62 (Dec.).

The Shanidar I skull, still partly enclosed in the rocky matrix in which it was found, from the Shanidar cave in northern Iraq. This well-preserved specimen, about 46,000 years old, is from a Neanderthal man, who died at about 40 years of age. (Courtesy R. S. Solecki, Columbia University, New York.)

CHAPTER 36

THE EVOLUTION AND NATURE OF MAN*

■ MAN'S PLACE IN NATURE

Man's place in the animal kingdom can be viewed by comparing his structure and physiology, his embryonic development, and his fossil record with that of other organisms. When this is done, man is seen to fit into the animal category of life in every particular except that of degree in certain functional abilities. Man has apparently surpassed other organisms in the degree of superior development of his nervous system, which has made possible the best utilization of his structures and functions.

All life is interrelated. Man must be guided by the same rules and principles as those found in other animals. The development of biochemistry shows that he has chemical mechanisms and metabolic patterns

*Refer to pp. 17 and 19 to 21 for Principles 3, 12, 20, 21, 24, and 25.

similar to those found in all living organisms. His evolutionary history indicates that he also shares his origin with other animals. The great stress now laid on his ecologic relationships demonstrates that his ecosystem is a part of nature and that he must adjust himself to his environmental relationships wisely and logically. Because of his advanced position he has the greater responsibility to other members of the animal and plant world.

■ MAN'S CLOSEST RELATIVES— PRIMATES

Man belongs to the order of mammals called Primates. This order is commonly divided into two sub-orders—Prosimii and Anthropoidea. The prosimians include the more primitive members, such as the tree shrews, the lemurs, the lorises, and the tarsiers. The anthropoids are the more advanced primates and are divided into the Platyrrhini, or New World primates, and Catarrhini, or Old World monkeys, apes, and man. The catarrhine infraorder is likewise separated into two groups (superfamilies)—Cercopithecoidea, or those anthropoids with tails, and Hominoidea, or the great apes and man. The superfamily Hominoidea is represented by three families—Hylobatidae, Pongidae, and Hominidae. The Hylobatidae contain the gibbon *(Hylobates)*; the Pongidae include the orangutan *(Pongo),* the chimpanzee *(Pan)*, and the gorilla *(Gorilla)*; and the Hominidae include man *(Homo sapiens).*

The basic adaptive features that have guided the evolution of this great mammalian order culminating in man have revolved mainly around the arboreal habits of the group. Man abandoned the brachiating habit (using the arms for swinging) and took up a terrestrial existence, but there are anatomic structures, such as the limbs, hands, and feet, which plainly indicate the basic ancestral traits for an arboreal life. The free rotation of limbs in their sockets, the movable digits on all four limbs, and the opposable thumbs are all modifications for grasping branches and swinging through trees. Omnivorous food habits have produced a characteristic dentition. Along with skeletal specializations, there are many correlated features that have evolved in primates, such as many muscular, neural, and sensory adaptations superior to those in other animals. Better vision and development of other sense organs, as well as the proper coordination of limb and finger muscles, have meant an enlargement of appropriate regions of the brain. Precise timing and judgments of distance were necessary concomitants of an arboreal life and required a larger cerebral cortex. Concurrent enlargement of the brain led to a level of intelligence and alertness of mind characteristic of the higher primates.

Early in the Cenozoic era the basic radiation of primitive placental mammals began and continued until more than a score of orders arose and took over the niches formerly occupied by reptiles. The most primitive order was the Insectivora, familiar examples of which are the present terrestrial moles and shrews. But some members of this order, the tree shrews, took to living in trees and gave rise to the great order of primates. Living tree shrews in Asia thus may be considered a transition between the primitive insectivore ancestors and the primates. According to Romer, tree shrews may be placed in either the order Insectivora or the order Primates because their primitive and other features show definite relationships to the primates. Fossil remains of shrewlike mammals have been found in the Paleocene epoch of 75 million years ago.

These early primates, **prosimians,** were small nimble animals adapted for living in trees. Although they possessed good muscular coordination and sense organs, their intelligence was low; intelligence among modern primates is a later development in their evolution. Besides the tree shrews, the Prosimii include the lemurs (mostly in Madagascar), the tarsiers of the East Indies, and the lorises of Asia and Africa. Some modern prosimians have many primitive characteristics, such as eyes on the side, long snouts, and long tails (lemurs); but in the tarsiers the large eyes have moved forward and the muzzle is short. Their larger brain and better-developed placenta also place the tarsiers nearer the anthropoids than the lemurs. Although tarsiers have all but vanished, many Paleocene and Eocene fossil genera have been found. Certain basic adaptations developed by primates, such as the opposability of the thumb and great toe, stereoscopic vision, and the ability to judge distance, fitted them for a tree-dwelling existence.

Some of the tarsiers may have been ancestral to the higher primates, which include monkeys, apes, and man (Anthropoidea). New World monkeys (family Cebidae) are more primitive and are usually smaller than Old World monkeys, although they are similar in appearance. Their nostrils are far apart and they have long prehensile tails. The thumb is only slightly opposable to the other fingers. The fossil record (which is scanty) indicates that their radiation may have occurred during the Oligocene and Miocene epochs. The Old World monkeys (family Cercopithecidae) probably arose independently from tarsioid stock also. Their fossil material dates back to the Oligocene epoch (35 to 40 million years ago). The Old World monkeys have their nostrils close together, and they do not use their tails as prehensile organs. Like the New World monkeys, they are four footed in terrestrial locomotion, walking on their palms and soles. Most are arboreal, but the short-tailed baboons show a tendency for ground dwelling.

The Old World monkeys cannot be considered ancestral to the anthropoid apes or man. The primitive, hominoid apes evolved independently from prosimian ancestors. They were small monkey-sized animals. Many of their fossils have been found in the Miocene and Pliocene deposits. Although they are apelike, they do not have the specializations of the modern apes. *Propliopithecus* from the Oligocene from the Fayum bed of Egypt is believed to be a gen-

FIG. 36-1

Chimpanzee *(Pan satyrus)*. Order Primates, superfamily Hominoidea. The chimpanzee and gorilla are considered to be the closest of all animals to man. The divergence of the chimpanzee from man is mostly in degree. The chimpanzee has retained its primary aboreal habit (note its relatively short legs with opposable great toe), whereas man has become terrestrial. The chimpanzee has many likenesses to man, such as the same length of gestation period, the same kind of placenta, the same menstrual cycle phases, and similar chemical structure. (Courtesy Smithsonian Institution, Washington, D. C.)

eralized hominoid ancestor, or it may be the direct ancestor of the modern gibbon. The Miocene fossil *Pliopithecus* was similar to the living gibbon but had forelimbs and hind limbs of about equal length, in contrast to the longer forearms of the gibbon.

The Miocene and Pliocene epochs have produced numerous other anthropoid fossils in Africa, Asia, and Europe. *Proconsul,* which left an abundance of skeletal remains in Africa, appears to be close to the ancestry of the higher apes (family Pongidae). Although they had some Old World monkey traits, some were as large as gorillas. A more widespread genus was *Dryopithecus,* found in Europe, India, and Africa. In an Italian coal bed of the Pliocene epoch, the controversial fossil *Oreopithecus* was found in 1872. A recent specimen was discovered in 1958. It is different from other hominoids and cannot be placed in either the stem leading to the higher apes or to that of man.

Although these fossils are definitely apelike, they lack the specializations of the modern apes. Rather, they show many primitive characteristics similar to man. They lack especially the brachiating arms and simian shelf found in living apes. The evidence indicates that during the Miocene epoch (25 million years ago) the hominoid line branched into two sublines. One of these led to apes and the other to man or manlike types (hominids). The fossil evidence shows that both lines have undergone many changes since they branched off from the common ape-hominid stem. Two recent fossil forms, *Ramapithecus* and *Kenyapithecus,* are considered closest to the hominid line, or the ancestry of man. Although modern apes are more specialized than man for arboreal life and in other ways, the fossil record discovered in the past decade or so indicates that man and the higher apes have a very close affinity. The evidence for a common ancestry of the two groups is more and more convincing as new fossils are found. There is also new evidence of the close relationship, such as structure of hemoglobin, resemblance in serum proteins, and shape of chromosomes, as well as other similarities.

Apes today are represented by the gibbons, orangutans, chimpanzees (Fig. 36-1), and gorillas. Apes have longer arms than legs, a long trunk compared to lower limbs, curved legs with the knees turned outward, large canines, laterally compressed dental arch (not rounded), long protruding face, and a brain about one third as large as man's. Some modern apes have largely abandoned the arboreal way of life. Chimpanzees and gorillas are quite at home on the ground. A more or less erect posture is characteristic of many of them, although they may use their long arms for support while walking. More abundant food on the ground may have induced them to come down out of the trees.

When the hominids separated from the higher apes, they underwent a radiation of their own. Giving up an arboreal life was a necessary prelude to their amazing evolution. Climatic changes during certain geologic ages may have greatly reduced the forests and forced a ground existence. Increased body size may have been a factor also. Because of dangerous predators, selection pressure may have brought about evolution of feet and strong running muscles. The fossil record is very incomplete between the rather abundant ape fossils of the Miocene epoch and the hominid fossils of the Pleistocene epoch. In this period of more than 20 million years, little is known about the hominid line. The initial phases are largely unknown. The fossil record indicates that the hominid radiation produced various lines of descent. From its dentofacial features, *Ramapithecus* appears to be the most appropriate ancestral taxon. One of the chief problems of paleontologists is to find fossils that will close the gap of differences between man and the higher apes. The much larger cranium and brain size with its correlated functions, the erect position, and the more adaptable opposable thumb in man as contrasted with the apes—all indicate a wide gulf between them, much of which appears to have occurred in a relatively short time, geologically speaking. This may have been due to a strong selection pressure for those characteristics that have enabled man to reach such a high level of dominance over his environment. The exciting discoveries found in Africa in the past few years have added much to our knowledge of man's emergence.

■ FOSSIL RECORD OF MAN

The hominids appear to date back to the early Pleistocene epoch and maybe earlier. Authorities are not in agreement on the age of many fossils nor on the interpretations that are made about them. Hominid phylogeny is very difficult to analyze. The practice of assigning a new taxon to every fossil discovered and the attempt to place them in morphologic series without regard to the modern principles of speciation has complicated the problem. It is generally agreed that man is a polytypic species, that is, composed of several races, each with a different gene pool. The newer idea of speciation would consider the hominid fossils as samples of populations wherein there are many variations in type and form. The problem of hominid phylogeny is in a state of flux and is subject to constant change. During the Pleistocene epoch there were three well-marked stages when man was evolving from apelike to manlike individuals. The transitions that occurred during this period may have been greatly influenced by the vast climatic changes induced by the advances and recessions of four north polar ice caps. The three stages were (1) the *Australopithecus* group of the basal Pleistocene in Africa, (2) the *Homo erectus* or *Pithecanthropus* forms of the middle Pleistocene, and (3) the *Homo sapiens* types of the late Pleistocene. Students may be interested in some of the famous fossil finds and evidences of their position in the phylogeny of the human race.

Separation of hominid lineage from ape lineage

Darwin probably knew only two types of important fossils for apes and man. One was *Dryopithecus* (fossil ape) and the other the Neanderthal (fossil man). Darwin in his *Descent of Man* stated his belief that the lineage of apes and monkeys had separated by the late Miocene epoch, which began about 25 million years ago. From this time on apes and monkeys were common throughout Europe, Asia, and Africa. Some think that apes and monkeys separated in the Oligocene epoch several million years earlier. Since the earliest apes were ancestral to man, the divergence of the two groups must have occurred about the same time. The exact time when apes and man parted company has not yet been determined by paleontologists, but probably did not occur later than the Miocene epoch.

Mention has already been made of *Ramapithecus* (in India) and *Kenyapithecus* (in Africa), which are considered to be the most definite prehuman line yet discovered. Since only fragments of these fossils have been found, their exact status is unknown, although their age is put at about 10 million years.

Australopithecus (southern ape)

In 1925 the fossil brain cast of an immature anthropoid was discovered in South Africa by Professor R. Dart. This specimen had a mixture of both human and ape characteristics. Additional fossils of this and related types, including skeletons that were almost complete, have been found by a number of investigators since that time. Two distinct groups were discovered, *Australopithecus (Paranthropus) robustus* and *Australopithecus africanus*. Both of the groups definitely belonged to the human family Hominidae. They showed some apelike characteristics along with more human ones. They are dated about 2 million years ago. Recently, an *Australopithecus* jawbone found near Lake Rudolf in Kenya, Africa, in 1967, has been dated at 5.5 million years ago.

R. E. Leakey believes that *Homo* and *Australopithecus* had a common ancestral line, from which *Homo* broke off 4 million years ago. *Homo* and *Australopithecus* lived as contemporaries until *Australopithecus* died out and *Homo* became in time *Homo sapiens*. *Australopithecus* cannot be in the direct line of man's ancestors. Leakey's chief evidence is based upon the reconstruction of fossil parts into a skull that had a brain capacity of 800 cc. in comparison with the 500 cc. of *Australopithecus*.

Although Africa seems to be the basic home of this group, similar fossils have been found as far away as Java. The volume of their brain casts varied from about 450 to 600 cc. and overlapped the range of the chimpanzee and gorilla. Certain bones, such as those in the pelvis, were hominid rather than pongid. They walked in an erect or semierect position, as attested to by the shape of the leg and foot bones. Their dentition was more human than apelike. From the evidence of primitive stone tools, these so-called man-apes were toolmakers and tool users. They may be regarded as the earliest known type with distinct-

FIG. 36-2

Comparison of skulls of living races of man with those of living simians. **A,** Seven living races of man. **B,** Four living simians. (Courtesy Ward's Natural Science Establishment, Inc., Rochester, N. Y.)

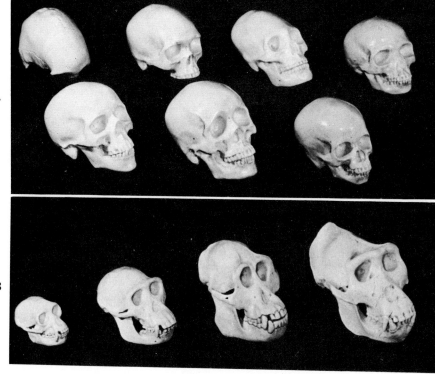

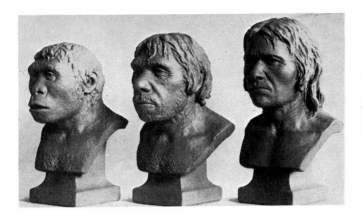

FIG. 36-3
Restoration of prehistoric men. *Left to right:*
Java man, Neanderthal man, and
Cro-Magnon man. (Courtesy J. H.
McGregor, New York.)

ly manlike characteristics, and they come close to the anatomic features expected in a "missing link." Their habitat was mainly terrestrial, which may have some significance.

Zinjanthropus

The fossil *Zinjanthropus* was discovered in 1959 by L. S. B. Leakey in the famous Olduvai gorge of Tanzania, Africa, where so many significant paleontologic specimens have been found. It probably belongs to *Australopithecus robustus,* one of the genera of the australopithecines. Its age has been calculated by the potassium-argon method at 1,750,000 years. Although evidence of tools were found with the fossil, these tools could have been left by more recent members of the *Homo* type. In some ways this form seems to be closer to the Hominidae than any others of the australopithecines. They were generally less than 5 feet tall, walked almost erect, and had small brains. Some of their skull features (mastoid processes, occipital condyle, etc.) were similar to man's. Leakey has also found a tool-using hominid, which he called *Homo habilis (Australopithecus africanus),* in the same locality. This one seems to be even closer to *Homo.*

Pithecanthropus (Java man)

This famous fossil was discovered in eastern Java in 1891 by E. Dubois, a Dutch anatomist. Only a skullcap and thigh bone were first found, but better specimens have since been discovered (Fig. 36-2). This taxon is considered to be almost 500,000 years old (middle Pleistocene). Because so many of the anatomic features were so close to modern man's, Java

man has been called *Homo erectus.* They walked fully erect, were over 5 feet tall, and had heavy projecting brow ridges (Fig. 36-3). They may have had a spoken language and used stone tools. The brain capacity was at least 900 ml. and may have been more in certain races. This species was widespread, being found in various parts of Asia and Africa, as well as in Java. The Peking man (*Sinanthropus pekinensis* or *Homo erectus pekinensis*) is a northern race of the same species.

Heidelberg man (Homo heidelbergensis)

The evidence for the existence of the Heidelberg man rests on an almost perfect jaw discovered in a sand pit in 1907 near Heidelberg, Germany. It was found in a deposit of bones from the lower Pleistocene. Its general aspect is human, although it combines a very massive jaw with small humanlike teeth. Some authorities classify this man under the australopithecines, but he differs from the Asiatic type of that group.

Swanscombe and Steinhelm skulls

Between the *Pithecanthropus* group and the establishment of the modern polytypic species of *Homo sapiens,* the hominid evolution has taken a complex course. The earliest fossils that show clear-cut *Homo sapiens* features were the Swanscombe man in England and the Steinhelm man in Germany. These may have evolved from the pithecanthropines about 300,000 years ago. Both of these were found in the period of 1933 to 1935. They had prominent eyebrow ridges and a brain capacity not far from that of modern *Homo.* Their chin was not well developed. Leakey's recent discovery (1961) of the Chellean man at Olduvai gorge in East Africa has been dated at 490,000 years. This new find seems to mark a morphologic transition from *Pithecanthropus* to *Homo.* It has in general the same characteristics as the *Homo erectus* specimens found in the Far East. However, more recent fossils than those of the Olduvai gorge are known to occur. It appears that primitive australopithecines, near ape-men, and members of the first true men *(Homo)* were all undergoing an evolution, with considerable overlapping of the different types, in the period of the middle Pleistocene about 300,000 to 1,000,000 years ago.

Emergence of modern man

Modern man, or *Homo sapiens,* first appeared in the fossil record about 75,000 to 100,000 years ago in the form of Neanderthals. This race has left many fossils in west central Europe (and some elsewhere) in the deposits of the third interglacial and the last glacial period. Selective pressure of this severe climatic condition may have been responsible for the development of the race. This taxon has been assigned the name of *Homo sapiens neanderthalensis* (Fig. 36-3). The first specimen was discovered near Düsseldorf, Germany, in 1856—the first hominid fossil to receive attention by competent scholars. He was short in stature, averaging little more than 5 feet in height. His brain capacity was similar to that of modern man, and he had developed a crude form of paleolithic (Mousterian) culture. He differed from modern *Homo sapiens* in having a flattened braincase, projecting jaws, recessive chin, large supraorbital bridges, and strong mandibles. Their populations dominated the scene in late Pleistocene times, from Europe to Africa and Asia. The race was not a uniform type (which has given rise to varied interpretations), but varied from place to place in response to local conditions or the intermixing of the different types. Of the various types of *Homo sapiens* that have arisen, there may be mentioned the Solo man (Japan), Rhodesian man (South Africa), and the Mt. Carmel man (Palestine).

During the last interglacial period about 30,000 to 40,000 years ago, the Neanderthal race was replaced in Europe rather suddenly by the Cro-Magnon race, which emerged from an unknown source (Fig. 36-3). They may be descendants of early, generalized Neanderthals of Asia. They may have been responsible for the extermination of the Neanderthals. The Cro-Magnon was not homogeneous but was rather a mixture of people who showed considerable physical variations in different localities. They had a far superior culture (Perigordian) and left artistic paintings and carvings in their caves. They are considered ancestors of modern man and represent the modern type of man. They were about 6 feet tall, had a high forehead but no supraorbital ridges, a rather prominent chin, and a brain capacity as large as (or larger than) present-day man. Their physical characteristics are matched today by the Basques in northern Spain and certain Swedes in southern Sweden. Attempts to discern the characteristics of present-day races in early populations of *Homo sapiens* have not been successful. It is not known whether they were white, black, or brown.

The evolutionary course of modern man in his 30,000 to 40,000 years has exhibited the same pattern of divergence and extinction demonstrated by his forebears and that of other organisms. The essential characteristics of human phylogeny, such as man's superior brain and wide adaptability, can all be attributed to the strictly quantitative effects of mutations that could have happened at any evolutionary level. In other words, man has been the outcome of the basic factors of evolution that have directed the evolution of every organism from the time life first originated. Mutation, selection, population factors, genetic drift, and isolation—all the general processes of evolutionary progress—have operated for man the same as for other animals.

Of the many types of man mentioned in the foregoing account, it is doubtful if there were more than two species coexisting at one time, and in most cases perhaps only one. A polytypic species such as man could have many types in a widespread population.

■ MAN'S UNIQUE POSITION

That man is an animal is attested by his evolutionary history and his present biologic condition. He is a product of evolution as we have seen, but he has what no other animal has—a psychosocial evolution, or a directional cultural pattern that involves a constant feedback between past and future experience. Although human evolution has become increasingly cultural as opposed to genetic, he is still subjected to the same biologic forces and principles that regulate other animals.

When one compares man with other animals, he finds a broad gap. First, he is really the only animal that knows how to make and use tools effectively. This more than any other factor has been responsible for giving man his dominant position. Another unique characteristic of man is his capacity for conceptual thought. Man has a symbolic language of wide and specific expression. With words, he can carve concepts out of experience. This has resulted in cumulative experience that can be transmitted from one generation to another. In other animals, transmission never spans more than one generation. Man owes much to his arboreal ancestry. This promoted his binocular vision, a fine visuotactile discrimination, and manipulative skills in the use of his hands. If a horse (with one toe) had man's intellect and culture, could it accomplish what man has done?

Man is a definite species. Man's population is commonly divided into races or populations that are genetically distinguished from others. A so-called race has certain genes or gene combinations that may be more or less unique, although races grade into each other and do not have definite boundaries. Pure races are nonexistent and there are no fixed number of races. Races are adaptations to local conditions. It is believed that as primitive men spread over geographic areas, they became adapted to certain regions, and natural selection stamped on them certain distinguishing features. Races at present are losing their biologic significance as human adaptation to environment is becoming largely cultural. Rapid mobility and quick communication have shrunk the size of our planet so that isolation of races rarely exists. Genes can shift through human populations now with amazing speed and racial intermixtures are far more common than formerly. Since all races have more or less unique potentialities, hybrid vigor could operate within racial interbreeding just as it does for other animals.

References
Suggested general readings

Cold Spring Harbor Symposia on Quantitative Biology. Origin and evolution of man, vol. 15. 1950. Cold Spring Harbor, N. Y., The Biological Laboratory. *Although this excellent symposium stresses the latest evidence of man's ancestry, much attention is also given to fossil records and their interpretation. A work for the advanced student.*

Conklin, E. G. 1943. Man: Real and ideal. New York, Charles Scribner's Sons. *The author believes that there is no evidence that man has advanced in intellectual capacities since ancient time but that his future progress will be along the lines of cooperation and organization of social forces for his own welfare. Although evolution has taken many directions and has regressed as well as progressed, he thinks that there has been no permanent retreat in the evolution of man's intellect, reason, and ethics.*

de Chardin, P. T. 1959. The phenomenon of man. New York, Harper & Brothers. *An evaluation of man's evolutionary position. Much of its thesis is in line with that of Sir Julian Huxley (writer of introduction to this work) who has long stressed the unique aspects of man's evolution in the biologic world.*

Dobzhansky, T. 1955. Evolution, genetics, and man. New York, John Wiley & Sons, Inc. *A presentation of evolution in relation to genetics; written in an easy style.*

Dobzhansky, T. 1962. Mankind evolving. New Haven, Conn., Yale University Press. *Human evolution is considered as the interdependence of the biologic and cultural components. This has resulted in what the author calls the superorganic. An excellent appraisal of our present concept of man's evolution.*

Le Gros Clark, W. E. 1960. The antecedents of man. Chicago, Quadrangle Books, Inc. *This noted authority explains the evolution of man by first tracing the extensive background of the primate order. The chapter on the evolutionary radiations of the primates is especially revealing.*

Pfeiffer, J. E. 1972. The emergence of man. New York, Harper & Row. *Presents a good survey of the recent unfolding of human evolution.*

Poirier, F. 1973. Fossil man. St. Louis, The C. V. Mosby Co.

Weiner, J. S. 1971. The natural history of man. New York, Universal Books. *Deals with the interaction of man with his physical environment. Well presented in parts.*

Young, J. Z. 1971. An introduction to the study of man. New York, Oxford University Press.

Selected *Scientific American* articles

Clark, J. D. 1958. Early man in Africa. **199:**76-83 (July).

Dobzhansky, T. 1960. The present evolution of man. **203:**206-217 (Sept.).

Eckhardt, R. B. 1972. Population genetics and human origin. **226:**94-103 (Jan.).

Eiseley, L. C. 1953. Fossil man. **189:**65-72 (Dec.).

Howells, W. W. 1960. The distribution of man. **203:**112-127 (Sept.).

Howells, W. W. 1960. Homo erectus. **215:**46-53 (Nov.).

Leakey, L. S. B. 1954. Olduvai Gorge. **190:**66-71 (Jan.).

Sahlins, M. D. 1960. The origin of society. **203:**76-86 (Sept.).

Simons, E. L. 1964. The early relatives of man. **211:**50-62 (July).

Simons, E. L. 1967. The earliest apes. **217:**28-35 (Dec.).

Washburn, S. L. 1960. Tools and human evolution. **203:**63-75 (Sept.).

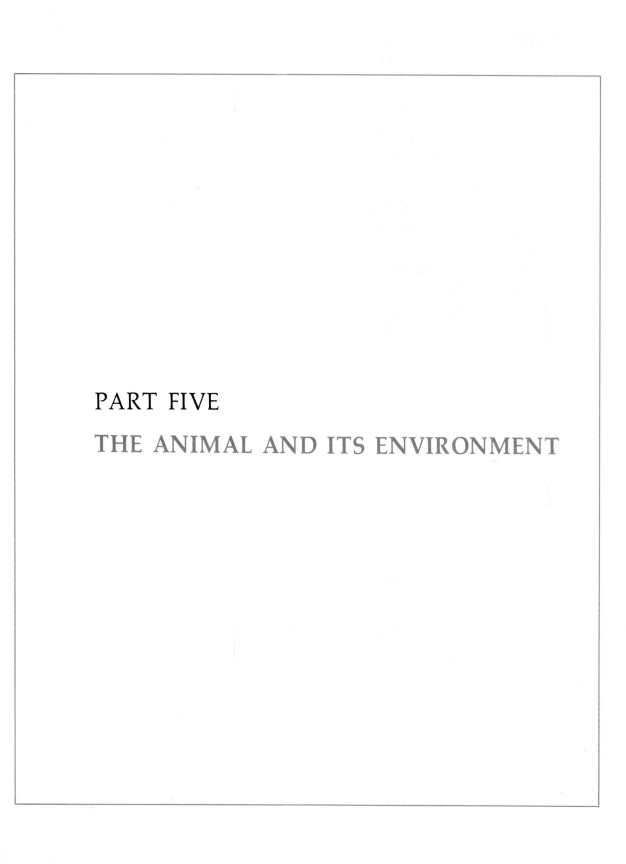

PART FIVE

THE ANIMAL AND ITS ENVIRONMENT

■ Life was created and shaped by environmental forces. No organism is for an instant free from the requirements of the surroundings in which it lives. The environment therefore is the totality of all extrinsic factors, both physical and biotic, that in any way affect the life and behavior of an organism.

In the following chapters we deal with the relationship of animals to the world around them, with the pressing ecologic problems that confront man today, and with the complex and burgeoning field of animal behavior.

CHAPTER 37
THE BIOSPHERE AND ANIMAL DISTRIBUTION

CHAPTER 38
ECOLOGY OF POPULATIONS AND COMMUNITIES ·

CHAPTER 39
PROBLEMS OF MAN'S ECOSYSTEM

CHAPTER 40
ANIMAL BEHAVIOR PATTERNS

A portion of a seaside habitat showing a conspicuous inhabitant, the laughing gull, Larus atricilla.

CHAPTER 37

THE BIOSPHERE AND ANIMAL DISTRIBUTION*

The **biosphere** refers to that part of our planet where animals and plants live. It was so named by J. de Lamarck, the French naturalist of the nineteenth century, a foremost advocate of evolution during the first part of that century. It is the environment of living things; the place where living organisms have established themselves and find conditions in which they can live. It includes fresh and salt water, land surface, depths below the surface, and air. The biosphere extends down in the ocean to more than 30,000 feet, below the land surface more than 1,000 feet, and vertically in the atmosphere to more than 40,000 feet. With the possible exception of certain arid regions, frozen mountain peaks, restricted toxic sea basins, and a few others, every place on earth is represented by some form of life.

*Refer to p. 22, Principle 31.

Most species are adapted to a particular type of environment that is restricted in size, resources of food, places to live (niches), and general conditions of living. The biosphere is the result of the complex interactions of so many factors that it is impossible to analyze their ecologic significance in all respects.

Inasmuch as living things depend on the conditions of their physical environment, they have in their long evolution become adjusted to it and depend on it for their continued existence. Every animal is affected by every physical factor in its environment. Weather and climate conditions, temperature, pressure, nature of substratum, physicochemical structure, constant change from geologic, geochemical, meteorologic, and other factors—all are involved in forming a background to which animals must adjust to survive. The animal itself is part and parcel of the earth's substance, and its evolutionary diver-

857

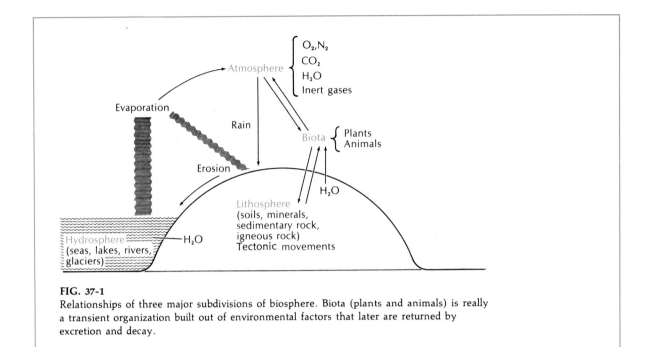

FIG. 37-1
Relationships of three major subdivisions of biosphere. Biota (plants and animals) is really
a transient organization built out of environmental factors that later are returned by
excretion and decay.

sity has been correlated with the changing earth at every level of its existence. As an open system, an animal is forever receiving and giving off materials and energy. Inorganic materials are obtained from the physical environment, either directly by producers, such as green plants, or indirectly by consumers, which return the inorganic substances to the environment by excretion or by the decay and disintegration of their bodies.

Thus there is a constant cycle between the animal and its environment. The living form is a transient link that is built up out of environmental materials which are then returned to the environment to be used again in the re-creation of new life. Life, death, decay, and re-creation have been the cycle of existence since life began.

The biosphere may be conveniently divided into three major subdivisions—**hydrosphere, lithosphere,** and **atmosphere** (Fig. 37-1). The hydrosphere refers to the aquatic portions of the biosphere, the streams, rivers, ponds, oceans, and wherever water may be found. The lithosphere is made up of the crust of the earth, especially the solid portions such as rocks. Surrounding the other two subdivisions is the atmosphere, which forms a gaseous envelope. Animals obtain inorganic metabolites from each of these sub-

divisions. From the hydrosphere, they get water that makes up about 75% of living material. The lithosphere furnishes the essential minerals and chemicals, whereas the atmosphere supplies oxygen, nitrogen, and carbon dioxide. These inorganic substances are needed in all living organisms.

■ HYDROSPHERE

About 71% of the earth's surface is water, which also forms part of the lithosphere and atmosphere. Not only is water the most abundant constituent of protoplasm, but it is also the source of hydrogen that is so fundamental to the metabolic reactions of all living substance. Water also serves as one of the sources of oxygen in the body of organisms.

Water is involved in a cycle that consists of evaporation to a gaseous state and a return to a liquid by condensation of the vapor at higher altitudes (Fig. 37–1). Evaporation of water occurs in the ocean and on the land. About five sixths of this evaporation is in the ocean. It is estimated that the oceans evaporate a quantity of water from their surface equal to about a depth of 1 meter annually. Part of the land evaporation of water is the transpiration of plants. An ecologic principle of great economic importance is that the precipitation of water on land exceeds the evaporation. This means that the difference in water between these two factors represents the annual run-off water from the continents, carrying off minerals, producing erosion, and wearing away the surface of the continents. This leveling off process, however, is offset by the geologic uplift, thus bringing marine sediments above sea level. Living things also carry water through their bodies in addition to what they need themselves and in doing so speed up the return of water to the atmosphere. The metabolism of organisms thus accelerates the cycle of water and may profoundly influence weather conditions, not only locally, but also over extensive areas such as the rank vegetation of jungles.

Displacement of water in the ocean is also cyclic in nature. The warm water of tropical seas comes to the surface and the cold polar water sinks so that shifts of currents between the equator and the pole are aided by the east-west displacements produced by the rotation of the earth. Ocean currents, such as the Gulf Stream, have enormous influences on climatic conditions in all parts of the world.

The very nature of water itself is of great significance. It is slow to heat or cool and stores great amounts of thermal energy. Heat radiation from water can produce favorable regions for land organisms in many places that otherwise would be too cold for their survival.

It is also unique in attaining its greatest density at 4° C., therefore ice floats because it is lighter than it is in the liquid condition. Except for very cold superficial pools, water never freezes throughout so that life can continue beneath the ice cover.

In the long overall picture, the cyclic changes of ice ages, of which at least four great glacial periods have occurred in the last million years, involve the advance or retreat of polar ice. Warm interglacial periods have actually freed the poles from ice. (Amphibian fossils of the Quaternary period have been discovered in the Antarctic.) Melting of polar ice during the present warm trend in temperature has gradually raised the level of ocean water and made possible a steady advance of biota toward the poles. Deserts are also on the march and localized glaciers are receding.

■ LITHOSPHERE

The lithosphere consists of a number of components. Below the loose soil and subsoil is the solid bedrock of sedimentary rock, such as limestone and sandstone, resting on a thicker base of igneous and metamorphic rocks. Below this layer is a thicker stratum composed mostly of basaltic rock. These two layers form the so-called crust of the earth. As far as life is concerned, only the more superficial part of the lithosphere is involved, although the tectonic movements of folding and breaking of the bedrock plus the rise of molten lava in volcanos may bring additional minerals to the surface. The plate tectonic theory states that the lithosphere is made up of a few plates (50 to 100 km. thick) in variable relationship to one another.

All the mineral metabolites of the animal are received from the lithosphere, which also forms the chief part of the soil mentioned above. The crust of the earth shows striking changes over long periods of time. Uplifting and buckling are always occurring, resulting in mountain building and shifting of land masses. Mountains produce profound changes in climate conditions such as the unequal distribution of thermal energy and moisture. Moisture-laden clouds not able to pass mountain barriers may dump their contents on one side alone so that this inequality results in favorable rainfall with fertility on one side and desert on the other. These factors influence the distribution of both vegetation and animals, which must adapt to these different conditions to survive.

There is also a leveling process going on all the time as mentioned in a former section. A large part of rainfall is not evaporated and the excess water is run off from the land into the rivers and streams, eroding the land surface by carrying with it dissolved soluble mineral matter to the sea. Billions of tons of dissolved inorganic and organic matter as well as much undissolved matter are carried into the oceans each year. Many important chemicals such as phosphorus are lost to animal and plant life and the biosphere is unable to make good the losses, especially to terrestrial forms. Marine forms, on the other hand, may profit from such an economy. Some valuable minerals are replaced by the decay of animals and plants and in the long run by the upheaval of the sea floor to form new land. This last is a slow process and life could be greatly affected by the gradual decline of certain key minerals. Many ecologists have been concerned about this decline, especially because of man's influence in the misuse of certain resources.

ATMOSPHERE

The gases present in the atmosphere are (by volume) oxygen about 21%; nitrogen, 78%; carbon dioxide, 0.03%; water vapor, in varying amounts; and small traces of inert gases (neon, helium, krypton, argon, ozone, and xenon). Each of these, except the inert gases, serve as metabolites in living substances. Carbon dioxide is a basic ingredient of the process of photosynthesis and ozone screens out ultraviolet rays. The percentage of gases remains very constant, except in a few places such as volcanos, underground sources, or industrial plants. Carbon dioxide plays a unique role in the ecology of animals, although it makes up such a small percentage of atmospheric air. This gas enters water from the air, the ground, decay of organic matter, respiration of animals, and the action of acids on carbonates dissolved in water. Among its many functions are its action as a chemical buffer in the maintenance of neutrality in aquatic habitats, its role in photosynthesis, its regulation of, or influence on respiration, and its general influence on other essential biologic activities (raising the threshold availability of oxygen, developing of eggs, increasing or decreasing rate of egg cleavage, etc.). One of its useful aspects is its chemical combination with water to form the weak carbonic acid (H_2CO_3).

With the exception of anaerobic animals (which can carry on oxidative processes without free oxygen), oxygen is a necessary prerequisite for respiration in all animals. Its major function in cellular metabolism is its role in the final hydrogen acceptor, resulting in the formation of water. Oxygen is dissolved in water, which makes possible aquatic life. All atmospheric gases dissolve in water in accordance to certain principles. Any gas soluble in water will dissolve in it until equilibrium is reached. Its solubility in water will depend on temperature and its partial pressure. In a mixture of gases each gas will exert a pressure proportional to its partial pressure in the moisture, and each gas will dissolve irrespective of the solution of other gases. Solubilities differ for different gases. The total pressure of a gas mixture is the sum of the partial pressures of the various gases. If atmospheric air of 760 mm. pressure (sea level) is exposed to distilled water at 0° C., at equilibrium the water will contain about 49 cc. of oxygen, 23 cc. of nitrogen, and 1,715 cc. of carbon dioxide. Oxygen is lost from water by the respiration of or-

ganisms, oxidation of organic matter and decay of dead bodies, consumption by bacteria, bubbling of other gases that carry oxygen with them, and the warming of the surface layer of water. A lack of dissolved oxygen in the water may be a severe limiting factor in the distribution of aquatic animals. Lakes in the high Andes have no fish because the partial pressure of oxygen is too low for the minimum amount to be dissolved in the water.

Nitrogen is a chemically inert gas, but nitrogen is one of the chief constituents of living matter. Atmospheric nitrogen is the ultimate source. It may be fixed as nitrites or nitrates by electric discharges and later washed to earth by rain or snow. In the nitrogen cycle, nitrogen-fixing bacteria living symbiotically with legumes form nitrites and nitrates by the reactions **ammonification** and **nitrification** and add them to the soil where plants can get their nitrogen supplies. Animals get their nitrogen from eating plants or one another.

The atmosphere has a low degree of buoyancy and cannot be used as a permanent habitat by organisms. It is used as a passageway by those forms that are specialized to use it as a medium. Most of the life found within it is restricted to its lower boundary.

Much of the terrestrial surface of the earth cannot be used to any extent by animal life because of seasonal and climatic conditions. The scarcity of moisture in desert regions requires that activity be restricted to certain periods of the day by those forms adapted for living there.

The atmosphere presents other ecologic factors of great importance to life. The atmosphere is commonly divided into three great strata, the **troposphere, stratosphere,** and **ionosphere.** The troposphere is the layer nearest the earth and extends upwards from about 6 to 10 miles above sea level. Above the troposphere is the stratosphere, which extends to about 50 miles above the earth, and the ionosphere is the gradually thinning air still above the stratosphere. In the troposphere are the complex wind movements and currents that help produce the great climatic changes. As air masses warmed in the tropical regions rise and cooled polar air sinks, the rotation of the earth shifts the air masses laterally and produces the enormous currents of many types that profoundly influence the lives of both man and beast.

DISTRIBUTION OF ANIMALS (ZOOGEOGRAPHY)

Zoogeography tries to explain why animals are found where they are, their patterns of dispersal, and the factors responsible for their distribution. It is obvious that each type of animal is not found everywhere but is found in an area of distribution that may be widespread or very restricted. Those with wide ranges are called **eurytopic;** those with narrow boundaries, **stenotopic.** Dispersal of animals usually depends on two important factors: (1) their means of dispersal and (2) the presence of barriers of some kind that might limit their distribution. As a rule, members of a particular species or closely related species occupy **continuous** ranges, but there are exceptions. When the same taxonomic unit or group is found in different areas far apart, it is said to have **discontinuous** distribution. The strange onychophoran *Peripatus* is found in both tropical Africa and tropical America as well as elsewhere. Marsupials are found in the Americas and Australia—regions far apart.

The fossil record plainly shows that animals flourished in regions where they are no longer found. Extinction has played a major role in such conditions, but many of these groups left descendants that migrated to other regions and survived. The evolutionary pattern has been one of increase in diversity by evolving different ecologic habits and by spreading into new environments. Camels probably originated in North America, where their fossils are found, but spread to Eurasia by way of Alaska (true camels) and to South America (llamas). Barriers are altered by geologic changes in the earth's surface and by climatic changes. Many places now occupied by land were once covered with seas; regions now plains were formerly mountain ranges; such cold regions as Greenland were once quite warm. Geologic change has been responsible for much of the historic geographic distribution of animals and their evolution has been closely related to changing ecologic relationships. Historic processes have determined where animals are and what they do.

Animal species do not always occur in all places suitable for them. In some cases an opportunity for access to a particular region is afforded and the species may gain entrance by some freakish accident. The past history of an animal species or its ancestors must be known before one can understand why it is where it is. All species are subject to natural barriers and these barriers may change with the geologic history of the region.

By adaptive radiation, animals tend to spread into regions where they are ecologically fitted. Distinct species found in the same general area are called **sympatric;** those living in different geographic areas are called **allopatric.** Although life is believed to have originated in the sea, terrestrial conditions favor a more varied and more rapid evolution. Although the land makes up only 29% of the earth's surface, 80% of the known species are terrestrial. The abundance of oxygen on land and its role in releasing chemical energy have made possible a more intensive life. There are also many other advantages of terrestrial life over an aquatic one, such as the low density of air and the possibility of flight, the variety of foods, and the development of a greater variety of adaptations.

Animals tend to spread from their center of origin because of competition for food, shelter, and breeding places. Changes in environment may force them to move elsewhere, although most great groups have spread to gain favorable conditions, not to avoid unfavorable ones. **Dispersal movements** that result in the dispersal of animals are unidirectional and one way. They involve emigration from one region and immigration into another. Such movements must be distinguished from seasonal migration, such as that of birds, where there is a regular to-and-fro movement between two regions. The chief deterrent to the spread of animals is barriers of various kinds, such as mountains, water, deserts, and climatic conditions.

Simpson has stressed three chief paths of faunal interchange—corridors, filters, and sweepstakes routes. A **corridor** may be a widespread, more or less open stretch of land that usually allows the free movement of animals from one region to another. A **filter route** is defined as one that allows some animals to pass into another area but keeps others from doing so. A mountain may act as a filter in which mollusks may be prevented from passing but affords fewer impediments to mammals or birds. Filters may also be narrow land bridges such as that of the Isthmus of Panama. A **sweepstakes route** is one that is highly improbable for animals to pass over, but some manage to do so by some fortuitous event. Such seems to have been the case of oceanic islands where crossing is long delayed. Sweepstakes routes may also explain the peculiar marsupial animals of Australia because they were the first mammals to reach that continent. In the case of the corridor and filter routes animals may spread in both directions, but in a sweepstakes route they usually pass in one direction only.

Faunal interchange between continents has occurred many times. This interchange is most likely to occur when regions afford potential ecologic niches to a migrating group. Newcomers to a particular region often had adaptations that did not clash with the groups already there. In this way the fauna of a continent could be greatly enriched and diversified, such as that of South America when North American types passed into that region in the Pleistocene epoch.

Geographic equivalents, or syngeographs, refer to species that occupy about the same general territory. In the eastern United States the widely distributed salamanders of the genus *Plethodon* afford an example of geographic equivalents. *P. glutinosis* and *P. cinereus* have the same general range, except that *P. glutinosis* extends farther south and *P. cinereus* farther north.

Man himself has played a major role in the spread of species, as in the introduction of the English sparrow into America and the rabbit into Australia.

Convergent evolution may account for two similar groups widely separated from each other. This could occur by the repetition of a mutation under similar environmental conditions in the two regions. Often, however, a better explanation in such cases is that the original range has been broken up by environmental changes so that the individuals of the two similar groups became isolated.

Krakatao as example of animal dispersion

Krakatao is a volcanic island off the Sundra Straits between Java and Sumatra in the East Indies. In 1883 this island was practically destroyed by one of the most terrific volcanic eruptions in modern times. Every living thing was reported to be destroyed, for what remained of the island was covered with many feet of hot volcanic ash. Several naturalists from the first have been interested in seeing when life would first reappear there. It thus served as an interesting case of animal distribution under direct observation.

The position of the island between two great land masses and less than 40 miles from the nearest land made it specially favored for receiving a new stock of plants and animals.

Vegetation was the first to appear on the island and in a matter of some years completely covered the island. The first animals to appear were flying forms—birds and bats. It took more than 20 years for the first strictly terrestrial animals to gain a foothold. In 1908 there were more than 200 species on the island. Some 600 species were found in 1919 and nearly 900 species in 1934. Birds and insects appeared to be the most rapid colonizers. In less than 90 years most species found on adjacent large islands had reached there. However, some animals have been relatively slow to reach an island less than 40 miles away. This may indicate how long oceanic islands, far from the mainland, have been in acquiring their flora and fauna.

Lake Baikal as example of restricted dispersal

The large Lake Baikal in eastern Siberia is the deepest lake in the world, being more than a mile deep in some places. It is famed for its characteristic fauna, which is almost unique in the great number of peculiar or endemic species. Among certain groups up to 100% of the species are found nowhere else. Some groups, such as planarian worms, are represented by more species here than in all the rest of the world put together. Only groups such as protozoans and rotifers, which are easily transported from one region to another, are represented by non-endemic species to any extent. The distribution in the lake is also somewhat unique in that animal life is found at nearly all depths, perhaps because the water is well oxygenated throughout. All the evidence indicates that Lake Baikal has never had any connection with the sea but is strictly a freshwater formation. Whence came its unique fauna? Long geographic speciation through isolation, because this lake is very ancient, might account for some species. But the prevailing opinion is that the fauna of this lake mainly represents species that have evolved elsewhere in many ecologic habitats and have been washed into this lake by different river systems over long periods of time. Hence, the present fauna represents the accumulation and survival of ancient freshwater organisms (relics) that were widely distributed at one time but are now found only here. Other deep freshwater lakes of ancient geologic history, such as Lake Tanganyika and Lake Nyasa in East Africa, also have unique faunas for the same reasons.

Psammon (interstitial fauna)

Although the sandy beach of fresh and marine waters has been known for several years to be the habitat of many micrometazoans, it is only within the last few years that extensive investigations of this milieu have been done. Now there are psammologists from many countries, and in the United States the Marine Biological Laboratory at Woods Hole has active investigators in this field. The microfaunal and microfloral characteristic of psammons (Gr. *psammos,* sand) has a wide distribution in fresh, sea, and subsoil waters. As a result, the study of psammons has given rise to a varied terminology proposed by workers in this field. Some have accepted the terms "mesopsammon" for freshwater biotas and "thalassopsammon" for those of the sea.

The actual ecology of the psammobiota involves a microlabyrinth of small passageways between the sand grains, which often is deficient in oxygen because decayed organic or other substance may cover the pores of the sand. In the surf zone, however, the renewal of oxygen may occur by tidal action. The psammobiota reacts to disturbances produced by climatic conditions (temperature, rain, etc.) by moving vertically or laterally in the labyrinth passageways.

The type and grain size of the sand may influence the distribution of the fauna. Very fine sand is usually avoided because the passageways are too confining. Most prefer sand particles between 0.3 and 0.6 mm. thick. Deep sand may be too acid for most psammons. The biota of the psammon in marine beaches is usually between the high and low tide lines and below the low tide line. Representatives of all the animal phyla may be present in a psammon. Many members are primitive, but bacteria, protozoans, and micrometazoans usually make up a considerable part of the population. There also seems to be a close relationship between psammolittoral and the phreatic (Gr. *phrear,* well) habitats, which may make possible a pathway from seawater into the phreatic groundwater of soils.

Major faunal realms

On the basis of animal distribution, numerous regions over the earth have distinctive animal populations. These divisions indicate the influence of land masses and their geologic history, as well as the corresponding evolutionary development of the various animal groups. These realms of distribution have developed and fluctuated during geologic times. The higher vertebrates mainly have been used in working out these broad faunal realms. There are many complications in dividing the earth into such realms in which all groups of animals are involved. Some animals have purely a local origin; others within the same realm show affinities with groups quite remote. To explain many of these discrepancies, it has been necessary to assume various land connections or bridges for which there is no geologic evidence. Such major faunal realms can thus have only a limited significance.

Sclater (1858) first proposed this scheme for birds, and later Wallace (1876) applied this pattern of distribution to vertebrates in general. There have been modifications of the plan by other workers, such as the one that grouped the six original regions into three major regions—Arctogaea (Holarctica, Ethiopian, Oriental), Neogaea (Neotropical), and Notogaea (Australasian).

Following is a discussion of the faunas of the major zoogeographic realms as they are now classified.

Australian. This realm includes Australia, New Zealand, New Guinea, and certain adjacent islands. Some of the most primitive mammals are found here, such as the monotremes (duckbill) and marsupials (kangaroo and Tasmanian wolf), but few placental mammals. Most of the birds are also different from those of other realms, such as the cassowary, emu, and brush turkey. The primitive lizard *Sphenodon* is found in New Zealand.

Neotropical. This realm includes South and Central America, part of Mexico, and the West Indies. Among its many animals are the llama, sloth, New World monkey, armadillo, anteater, vampire bat, anaconda, toucan, and rhea.

Ethiopian. This realm is made up of Africa south of the Sahara desert, Madagascar, and Arabia. It is the home of the higher apes, elephant, rhinoceros, lion, zebra, antelope, ostrich, secretary bird, and lungfish.

Oriental. This region includes Asia south of the Himalaya Mountains, India, Ceylon, Malay Peninsula, Southern China, Borneo, Sumatra, Java, and the Philippines. Its characteristic animals are the tiger, Indian elephant, certain apes, pheasant, jungle fowl, and king cobra.

Palearctic. This realm consists of Europe, Asia north of the Himalaya Mountains, Afghanistan, Iran, and North Africa. Its animals include the tiger, wild boar, camel, and hedgehog.

Nearctic. This region includes North America as far south as southern Mexico. Its most typical animals are the wolf, bear, caribou, mountain goat, beaver, elk, bison, lynx, bald eagle, and red-tailed hawk.

References
Suggested general readings

Andrewartha, H. G., and L. C. Birch. 1954. The distribution and abundance of animals. Chicago, University of Chicago Press. *An analysis of animal populations and the factors that influence their abundance and distribution.*

Birch, L. C. 1954. The role of weather in determining the distribution and abundance of animals. Cold Spring Harbor Symposium **22:**203-218.

Carson, R. L. 1951. The sea around us. New York, Oxford University Press. *A popular account of the sea and its influence on animals, including man.*

Darlington, P. J. 1957. Zoogeography: The geographical distribution of animals. New York, John Wiley & Sons, Inc. *This is a book about animal distribution. It does not emphasize the ecologic approach, although it draws heavily on ecology and evolution for an explanation of its basic principles.*

Hardy, A. C. 1956. The open sea. Its natural history: The world of plankton. Boston, Houghton Mifflin Co. *This fine work shows the role of Protozoa and other forms in the natural history of plankton. Good descriptions are given of the Radiolaria and other protozoans in Chapter 6. Beautiful color illustrations.*

Henderson, L. J. 1913. The fitness of the environment. New York, The Macmillan Co. *Explains how conditions on our planet made life possible.*

Hubbs, C. L. 1958. Zoogeography. Washington, D. C., American Association for the Advancement of Science. *This includes the papers presented at two symposia at Stanford and Indianapolis. Part I deals with the Origin and Affinities of the Land and Freshwater Fauna of Western North America, and Part II deals with the Geographic Distribution of Contemporary Organisms.*

Hutchinson, G. E. 1957. A treatise on limnology, vol. I. Geography, physics, and chemistry, New York, John Wiley & Sons, Inc. *This is the first of a projected two-volume work on the rapidly growing science of limnology. It is an ambitious work that appeals to a variety of professional workers, such as limnologists, general biologists, and oceanographers.*

Jarman, C. 1972. Atlas of animal migration. New York, Doubleday & Co. *Maps and diagrams show how the various groups migrate.*

Linklater, E. 1972. The voyage of the Challenger. New York, Doubleday & Co. *A new appraisal of this famous voyage.*

Pettersson, H. 1954. The ocean floor. New Haven, Conn., Yale University Press. *The role of the Foraminifera and Radiolaria have played in building up the sediment carpet of the ocean floor is vividly described in this little book. The author believes the time of accumulation of deep sea deposits is 2 billion years and the rate of sedimentation of Globigerina ooze is 0.4 inch in 1,000 years.*

Simpson, G. G. 1953. Evolution and geography. Eugene, Oregon, State System of Higher Education. *An excellent appraisal of the main concepts of animal distribution by a great paleontologist.*

Sverdrup, H. U., M. W. Johnson, and R. H. Fleming. 1942. The oceans: Their physics, chemistry and general biology. Englewood Cliffs, N. J., Prentice-Hall, Inc. *A standard work on this subject, although somewhat outdated in some particulars.*

Selected *Scientific American* articles

Dietz, R. S. 1962. The sea's deep scattering layers. **207:**44-50 (Aug.).

Bullard, Sir E. 1969. The origin of the oceans. **221:**66-75 (Sept.).

The biosphere. 1970. **223:**44 (Sept. issue). *Includes articles on the various cycles—energy, water, oxygen, nitrogen, and mineral—and also on human food and energy and materials production as processes in the biosphere.*

Fairbridge, R. W. 1960. The changing level of the sea. **202:**70-79 (May).

Gautier, T. N. 1955. The ionosphere. **193:**123-138 (Sept.).

Gregg, M. C. 1973. The microstructure of the ocean. **228:**64-77 (Feb.).

Howells, W. W. 1960. The distribution of man. **203:**113-117 (Sept.).

Kurtén, B. 1969. Continental drift and evolution. **220:**54-64 (March).

Landsberg, H. E. 1953. The origin of the atmosphere. **189:**82-85 (Aug.).

Leggett, W. C. 1973. The migration of the shad. **228:**92-98 (March).

Senior author investigating a woodland brook community.

CHAPTER 38

ECOLOGY OF POPULATIONS AND COMMUNITIES*

■ LIFE AND THE ENVIRONMENT

Life and the environment go together, for the environment provides both nutrients and other suitable conditions for the existence of life. All functions of life begin with raw materials, and all basic materials come from the physical environment. Environment influences life by temperature, surrounding medium, climatic conditions, and many other ways. Above all, the environment provides the physical and chemical background against which living processes must be carried on. As the environment is constantly changing under the forces of geologic, geochemical, meteorologic, and other agencies, so also are organisms constantly undergoing change. We have pointed out that living matter is an open system that exchanges materials and energy with its surroundings. The

*Refer to pp. 22 and 23 for Principles 31 to 34.

same may be said of the planet Earth, for it is subjected to the forms of solar energy, such as heat, light, ultraviolet and other rays, and so on. Under such influences the materials of the earth never reach static equilibrium, because such energy flux produces balance disturbances.

The ultimate source for the inorganic metabolites required by organisms is the physical environment. All living things are constructed out of inorganic materials drawn from the environment. From these are formed the organic substances we associate with life. In time, living matter is decomposed and is returned to the environment from which it originally came. By such means life is continued indefinitely with a cycle of repeated re-creation and decomposition.

There is a very definite reciprocity between the living and the nonliving. Organisms are not only acted on by the environment but they also react upon it in many ways. At one time, for instance, the atmosphere was a reducing one (see Origin of life, p. 807), but now the atmosphere contains at least 20% oxygen, much of which has been produced by life itself.

As was seen earlier, life has many different levels of organization. Ecologic relationships deal primarily with those higher levels of organization —populations, communities. and ecosystems. A **population** may be considered a group of organisms of the same species; a **community** is an assemblage of the populations living within an area and having a certain distinctive functional unity; and an **ecosystem** is the interacting biotic (living) and the abiotic (nonliving) systems. The ecosystem is considered the basic functional unit of ecology, since it deals with both the living and nonliving components of the system. Ecosystems may be large or small, but the various ecosystems are linked together by biologic and chemical processes. Life is restricted to a thin film near the surface of both land and water within which the earth's environment has a certain fitness for life. The organisms can thus fit into a special environment, which makes up a small part of the earth.

To exist, organisms must find favorable places in which to live. Selection of habitats may be quite involved among them, and many factors no doubt influence animals in their choice of suitable sites. Organisms probably choose their habitats instinctively without making a critical analysis of the components of a habitat. Certain aspects of habitat conditions are probably given priority instinctively, such as temperature, moisture, light, salinity of water, and predators. Of course, factors that influence the choice of habitat vary according to the characteristics of a group. Birds, for instance, will have habitat preferences different from those of insects. Perhaps in all cases considerations must be given to shelter, places to rear their young, food supplies (which may be different in the young than in the adult condition), type of vegetation, terrestrial or aquatic condition, predators, etc. A psychologic factor (about which we know little) may be involved. Many forms may be dispersed in passive ways, fail to find favorable places, and perish.

ENVIRONMENTAL FACTORS OF ECOLOGY

Heredity furnishes in the genes a basic type of germ plasm organization. But how these genes express themselves in the structure and functioning of an animal is conditioned by environmental factors. These include both the nonliving **physical** factors (temperature, moisture, light, etc.) and the living or **biotic** factors (other organisms). The environment, then, includes every external object or factor that influences the organism and determines its total economy in the general scheme. Organisms react toward their environment in characteristic ways, either by trying to avoid detrimental situations or by being able to adjust physiologically, within their genetic limits, to adverse factors.

How organisms are influenced by these physical and biotic factors is determined largely by the **law of tolerance.** The range of distribution of each species is determined by its range of tolerance to the variations in each factor. Animals vary greatly in range of tolerance. To describe a species with a narrow range of tolerance for a particular factor, we use the prefix **steno-;** for those with a wide range we use the prefix **eury-** (see p. 475). In this connection Liebig's "law of the minimum" may apply. According to this principle, an animal's ability to survive depends on those requirements that must be present in at least minimum amounts for the needs of the organism in question, even though all other conditions are fully met.

Any factor that limits the range of a species or an individual may be known as a **limiting** factor. Temperature is a limiting factor for the polar bear, whose thick insulation of fur and fat makes temperatures above freezing unsuitable. (The presence of polar bears in zoos is, of course, an unnatural distributional pattern.)

PHYSICAL FACTORS

TEMPERATURE. Warm-blooded (homoiothermal or endothermal) animals are more independent of temperature changes than are cold-blooded (poikilothermal or ectothermal) animals, although they are restricted by temperature extremes. Usually cold-blooded forms have body temperatures not much higher than that of their surroundings, although some active forms such as insects may have higher temperatures. Some forms can help regulate the body temperature to some extent by fanning their wings to create air currents and increase evaporation, or by living massed together as the bees do to conserve heat.

Many animals have an optimum temperature at which their body processes work best. For some protozoans this is between 24° and 28° C. Other forms have a wider range, usually with a lower limit of just above freezing and an upper limit of around 42° C. However, thermal limits are difficult to define because the temperature at which a degree of lethality occurs depends on duration of exposure, thermal history of the animal, nutrition, etc. An increase in temperature speeds up body metabolism so that cold-blooded animals, sluggish during cold spells, become more active as the temperature rises. Fluctuating temperatures will sometimes speed up metabolism faster than constant temperatures will. The eggs of certain arthropods develop faster at optimum temperatures after they have been subjected to cold temperatures for part of their existence.

Many warm-blooded animals **hibernate** during winter months. Cold-blooded forms may retire deep underground or to other snug places to pass the winter. In some of the lower forms the adults die at the approach of winter and the species is maintained by larval forms or by eggs. Prolonged freezing or excessive heat is highly destructive to cold-blooded forms.

Habitats of animals are greatly influenced by temperature. Herbivorous forms are restricted by the amounts of grass and leaves available. Deer feed high on the mountains during the summer but in winter retreat to the valleys where they find better shelters and more food. Birds that live upon insects are forced to go elsewhere when their food is destroyed by cold weather. Temperature also plays a part in the rearing of the young and the hatching of eggs. Some eggs will not develop unless the temperature is fairly high.

Experiments show that the common fruit fly *Drosophila* may undergo structural modifications at high temperatures. The vestigial wings of one of the mutants will develop into normal wings at high temperatures. Evolutionary changes in the same fly may be induced by temperature. Some of the changes induced by radiation in fruit flies can, to some extent, be duplicated by temperature effects. Color patterns in many insects can be induced or altered by regulating the temperature under which they develop. Many of the differences in the color phases of animals of the same species living in different environments may be thus explained.

A spot climate, or **microclimate,** near the surface of the ground, in contrast to the **macroclimate** of the higher air levels, may show striking gradients. Air at the 2-inch level, for instance, has a higher daytime and lower nocturnal temperature than the macroclimate. The gradient may be as great as 10° F. The microclimate also has a greater humidity as well as less wind disturbance. These factors may play an important part in the distribution of terrestrial animals.

LIGHT. The effect of different intensities and wavelengths of light varies greatly for different animals. Visible light represents only a small fraction of the radiation from the sun. Within the visible spectrum, heat energy is more common at the red end (longer rays) and photochemical influences are greater at the violet end (shorter rays). Those animals that are positively phototactic usually collect near the blue end of the spectrum when they have a choice; negatively phototactic forms collect at the red end.

Since plants depend on photosynthesis, which requires light, their distribution affects animal life, for directly or indirectly animals depend on plants as their ultimate source of food. Plankton, which is composed of small plants and animals in surface waters, is restricted to the upper strata of water because light rays cannot penetrate deeply. Only about 0.1% of light reaches a region 600 feet below the surface of most marine waters, and this depth is usually considered the lower level of the population gradient of plant life.

The color of animals is also influenced by light conditions. Pigment cells (melanophores) are affected by the amount of light that enters the eye in many of the lower vertebrates. Nerve impulses aroused in the eye may cause a contraction of the pigment cells and a lighter color in the animal; less light causes expansion of the pigment cells and a darker color. Many animals, such as flatfish, show this color adaptation to dark and light backgrounds.

Photoperiodicity, or the effect of light on the physiology and activities of organisms, is considerable in some animals. Ferrets and starlings become sexually active with lengthening days and can be induced to breed out of season when exposed experimentally to a great amount of light. It is believed that the seasonal northward migration of birds may be induced by the stimulation of their gonads by the greater light associated with longer days in the spring; the southward migration is caused by the regression of their glands by shorter days. In many other ways animals and plants are influenced by the length of the daylight (photoperiod), such as the diapause (resting period) in arthropods, excretion and other physiologic functions of animals, the seasonal coat changes of birds and mammals, and the growth of trees. Seasonal responses in animals may be controlled by a photoreaction involving the anterior pituitary and certain hormones. Both day and night lengths may determine the response, but little is known of these factors. The parasitic filarial worm (*Wuchereria bancrofti*) lives by day in the deeper blood vessels of the host but by night in skin vessels, where it can be picked up by mosquitoes.

Also structural changes may occur in some animals through the effect of light. Shull has shown that certain strains of aphids (plant lice) may be made to develop wings by exposing them alternately to light and darkness. Short periods of darkness followed by continuous light produced wingless forms.

HYDROGEN ION CONCENTRATION. The concentration of hydrogen ions is believed to have a limited importance in the distribution of animals. Although some animals prefer alkaline surroundings and others acid ones, many forms can endure a wide range of pH concentration. Tapeworms can live in concentrations of pH 4 to pH 11, or very acid to very alkaline. Some protozoans are limited to a very narrow alkaline medium, whereas others, such as certain species of *Euglena,* live and flourish in water that varies from pH 2 to pH 8. Some mosquito larvae are normally found in water with a pH of less than 5 and will not live in an alkaline medium. The pH of

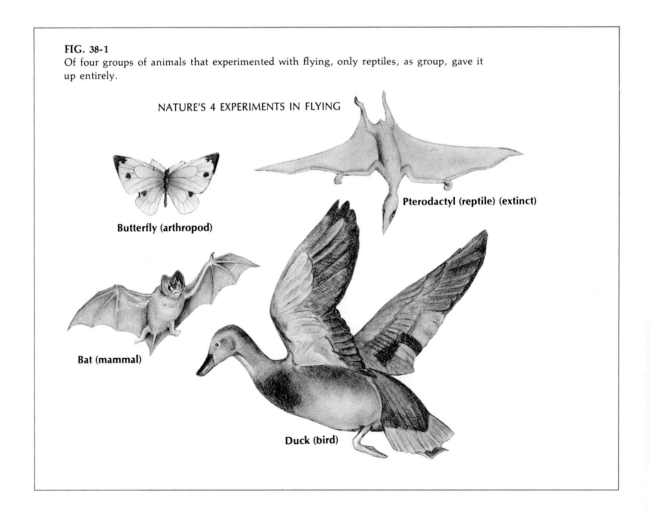

FIG. 38-1
Of four groups of animals that experimented with flying, only reptiles, as group, gave it up entirely.

NATURE'S 4 EXPERIMENTS IN FLYING

Butterfly (arthropod)

Pterodactyl (reptile) (extinct)

Bat (mammal)

Duck (bird)

water has a very limited importance in the distribution of fish, which seem able to adjust to a wide range. Animals with calcium carbonate shells such as clams may be more sensitive to acid media because their shells are corroded by acids. Water with a pH of 6 or less contains few mollusks.

SUBSTRATUM AND WATER. The **substratum** is the medium on or in which an animal lives, such as the soil, air, water, and bodies of other animals. The wings of bats, insects, and birds are fitted for the air (Fig. 38-1); the streamline form of fish and whales for the water (Fig. 38-2); the digging feet of moles and other mammals for the earth; and the hooks and suckers of parasites for the host. Many animals spend their entire life suspended in water; others spend a great deal of time in the air. Most organisms are found on a hard substratum on the land, at the

bottom of a body of water, in the hole of a tree, etc. Many small forms, such as water striders, whirligig beetles, various larvae, and pulmonate snails, make use of the surface film of water either in locomotion or for clinging.

The soil supports an enormous population of animals, some fairly large, but the majority small, such as nematodes, crustaceans, insects, protozoans, and bacteria. Earthworms prefer soil rich in humus. Whether the soil is acid or alkaline makes little difference if they have abundant food. Land snails are more common on soils rich in calcium because they need this mineral for their shells. Deer also depend on calcium for their antlers and bony skeleton.

Animals have evolved interesting adaptations for burrowing. Moles and mole crickets have shovellike appendages and broad, sturdy bodies. Earthworms

FIG. 38-2
Streamline form. This is one of most successful body forms for rapid locomotion through air and water. It is an elongated tapering form, somewhat rounded anteriorly and tapered posteriorly, and with its greatest diameter a short distance from anterior end. As animal moves through water, for instance, body anteriorly offers resistance, but return force of water against longer tapering posterior part more than offsets force used in pushing water aside in front and so aids locomotion. Same advantages accrue when body is stationary against moving water.

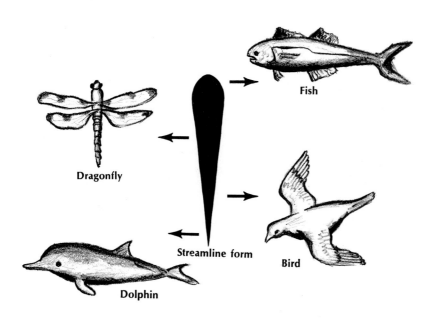

secrete mucus to line their burrows. The trap-door spider lines its burrow with silk and conceals the entrance with a trapdoor that it pops open to seize its prey (Fig. 14-10). Many forms live under stones and other objects on the surface of the soil. Termites and ants build mounds of soil in which they make tunneled passageways.

Some forms utilize burrows made by others. Many snakes, burrowing owls, American rabbits, and even insects take over abandoned burrows made by rodents or other forms. Many animals merely take advantage of natural crevices between rocks and debris.

A large part of any animal is water, for protoplasm consists of 70% to 92% water. All animals must preserve a proper water balance. The conservation of water is naturally more acute in terrestrial forms, for they are constantly in danger of desiccation. Most animals cannot lose more than one third of their water and live, but some can survive considerable desiccation. Protozoans secrete a cyst around themselves to prevent excessive desiccation; roundworms and rotifers may lose much water and then revive when placed in moist conditions. Water is a limiting factor for life in the desert. Many desert animals have thick, horny skins to prevent evaporation. During prolonged dry spells, some undergo **estivation,** burying themselves deep in the soil and remaining dormant until the wet season comes again. Breathing systems, such as the tracheal systems of insects and the internal lungs of snails, help cut down the amount of water evaporated. The dry feces of birds and reptiles is another water-saving device. Nocturnal habits of many animals expose them to lower temperatures and relative humidity.

Many animals get most of the water they need from their food, especially such animals as jackrabbits, mountain goats, and certain mice. Carnivorous animals may get their water supplies from the blood of their prey. Water supplies play a vital role in the distribution of many animals, for in tropical countries some do not stray far from water holes. This determines the distribution of other animals that prey on those that must have water.

The salinity of marine water is rather constant in all oceans, being about 3.5%; the salinity of freshwater varies greatly, although its salts may be limiting factors in many instances. Some salt lakes may have a salinity of 25% to 30%, which greatly restricts the life in them. The freezing point of seawater is about

$-1.9°$ C., which is an advantage to animals in colder regions. Water is heaviest at $4°$ C.; so ice at $0°$ C. can float; thus deep lakes and ponds do not have permanent ice on their bottoms. Although shallow lakes in high altitudes do remain frozen to the bottom in the winter, some life can survive. Water also has a very high heat capacity, which makes for a constancy in temperature during most of the year. Water makes up 71% of the earth's surface and represents the most extensive medium for animal life.

Animals in rain forests live only where the air is almost saturated with moisture (high humidity); desert forms live where the air is extremely dry (low humidity). There seems to be an optimum humidity for most animals and they are uncomfortable under other conditions. Amphibians are especially sensitive to humidity changes.

WIND AND GENERAL WEATHER CONDITIONS. Airborne eggs, spores, and adults (insects and snails) are often taken long distances by strong currents of air. Such forms may suddenly spring up in regions where they were not previously found. Ballooning spiders are carried on their gossamer threads to locations far away. So powerful are wind currents that animal life may be transported from continents to islands and other lands hundreds of miles away.

Where wind currents are prevalent, birds place their nests in sheltered places. Insects and other forms on windswept regions take advantage of cover to prevent being swept away. Many such insects are wingless; this may be adaptive, for wingless animals might stand less chance of being blown away. Forests afford many habitats that are protected against the force of winds. Wind is also an agency of erosion. In our western states the wind has shifted and carried away the topsoil for great distances, producing the well-known "dust bowl" of the West during the 1930s.

Many of the physical factors, temperature, light, etc., already mentioned, are correlated with seasonal changes in weather. Some animals in the temperate zones can undergo enormous ranges in temperatures. Many, however, are killed near the freezing point of water, but the races are preserved by spores, eggs, and larval forms, which can withstand such rigorous conditions. The hazards of winter are met by animals in various ways—migration, hibernation, change of food habits, etc. Food chains in winter differ greatly from those in summer.

SHELTERS AND BREEDING PLACES. Many animals that are endowed with speed, such as fish, squids, deer, and antelopes, make limited use of shelters, but others must hide from danger. In rapid streams, where there is danger of being dislodged and washed away, some animals are flat for creeping under stones, others have suckers for attachment, and still others live in firmly attached cases. Vegetation creates shelter for both land and aquatic animals. Forest areas, grasslands, and shrubbery contain a variety of habitats for terrestrial animals; aquatic plants serve as cover for small fish, snails, crustaceans, and other forms. In many instances proper cover is a limiting factor.

The destruction of forests has brought a decline of large birds of prey because the birds have been unable to find suitable nesting sites. The disappearance of sandy beaches because of the growth of aquatic vegetation affects fish that require sandy bottoms in which to spawn. Salamanders are scarce in regions that lack streams or pools in which their young may be hatched and reared.

OTHER PHYSICAL FACTORS. Forms that normally do not live at high water pressures can withstand high pressures so that the latter is chiefly a limiting factor only when it is extreme. Fish without swim bladders are less sensitive to deep pressure than those with swim bladders because of gas tension complications, and most invertebrates are more resistant to pressure than are vertebrates. Many marine animals (eurybathic) have wide vertical ranges, making diurnal movements of great amplitude, and can adjust themselves to a wide range of pressures.

Animal life may be lacking from the bottoms of deep bodies of water (Black Sea), where no dissolved oxygen is found. However, because of the currents of sinking cold water from the polar seas, some deeper waters of the sea may have more oxygen than regions near the surface. A certain amount of free oxygen must be available for aerobic animals. Lakes at high altitudes (Andes) cannot dissolve enough oxygen to support fish life. Atmospheric oxygen can thus be considered a limiting factor, for at altitudes of 18,000 to 20,000 feet, the barometric pressure is less than one half that at sea level, and the absolute amount (but not the percentage) of oxygen is corre-

spondingly reduced. For man, and possibly for some other animals, oxygen requirements at high altitudes cannot be met by the oxygen available.

■ BIOTIC FACTORS
Nutrition

Many animals such as man have a varied diet and can use the food that happens to be convenient (**omnivores**). Other animals are plant feeders (**herbivores**) or flesh feeders (**carnivores**). Within each of these main types, there are numerous subdivisions. Thus the beaver lives on the bark of willows and aspens, the crossbill lives upon pine cones, aphids suck plant juice, leeches suck blood, and the king cobra feeds on other snakes.

FOOD CHAINS AND PYRAMIDS. The interrelationships between animals in their food getting furnish interesting **food chains** (Fig. 38-3). Because plant life is the most abundant food in most localities, herbivorous animals form the basis of the animal community. These in turn serve as food for certain carnivorous forms, which also may serve as food for larger predators. At the end of the chain are larger animals, lacking predators, that die and decompose, replenishing the soil with nutrients for the plants that start the chain.

Animals at the end of the food chain are large and few, and usually one or two of them dominate a definite region, jealously keeping out all other members of that species. There are many examples of food chains. In a forest, for instance, there are many small insects (the primary consumers) that feed upon plants (the producers). A lesser number of spiders and carnivorous insects (secondary consumers) prey on the small insects; still fewer small birds live on the spiders and carnivorous insects, and finally one or two hawks live on the birds.

Such an arrangement of populations in the food chain of a community is often called a **food pyramid** (Fig. 38-4); each successive level of the pyramid shows an increase in size and a decrease in number of animals. Food chains may be complex or may be very short, such as the whale that lives mainly on plankton, the base of that particular pyramid. In every food chain, plants, which get their energy from the sun, form the basic energy for the chain. On account of this pyramid arrangement, one could expect very few large predatory animals within any region, for such a large pyramid of animals is re-

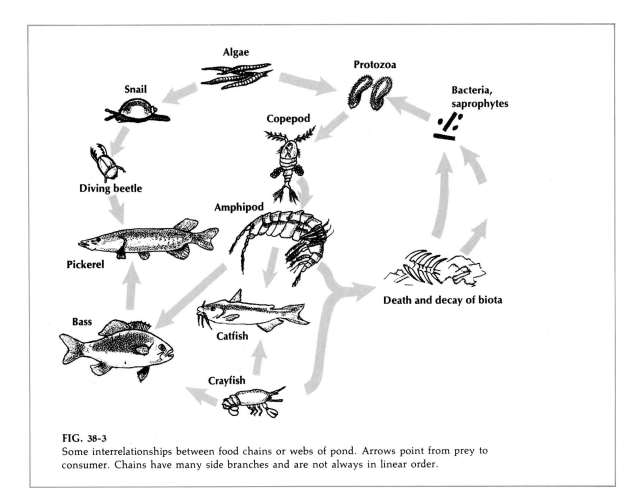

FIG. 38-3
Some interrelationships between food chains or webs of pond. Arrows point from prey to consumer. Chains have many side branches and are not always in linear order.

quired to support them. Only one grizzly bear can be found on the average of each 40 square miles of its territory; and in India, tigers are few in number for the same reason.

Food size is important in the arrangement of a community. Carnivorous animals, for instance, are unable to kill animals above a certain size, and many cannot live on forms below a certain size, for they cannot eat enough to furnish them the necessary amount of energy. A lion, for example, could not catch enough mice under ordinary circumstances to satisfy its food requirements. Many animals adjust their food requirements to the seasonal abundance of various types of food. The robin is a good example because its diet includes seasonal items, such as cherries, raspberries, and other small fruits, in addition to insects and earthworms that form its staple

diet. The fox, when pressed by winter scarcity, can subsist on dried-up berries and grapes.

Besides the predator food chain just described, there are two other types—the **parasitic** and the **detrital** food chains. The parasitic goes from larger to smaller organisms and the detrital from disintegrated bodies to microorganisms.

More than one food pyramid may be found in an ecosystem. Each of these pyramids may culminate in a different carnivore. In most ecosystems there are usually different types of plants, and these may form the basis of different food chains. Not all food chains culminate in a carnivorous form. A large herbivorous animal such as the elephant may be found at the peak of a food chain.

Food pyramids may be built up gradually. A small pyramid may be started in a new ecosystem and oc-

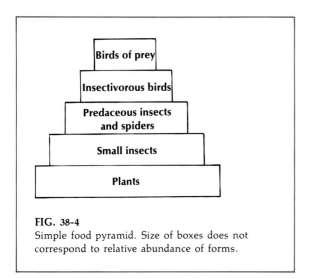

FIG. 38-4
Simple food pyramid. Size of boxes does not
correspond to relative abundance of forms.

cupy only a part of the territory. In time it may acquire herbivores and carnivores. The base of the pyramid widens as new links are added to the chain. It may take considerable time before food pyramids reach stability.

Food chains composed of herbivores and their predators are sometimes called **grazing** food chains. Just as important in the flow of energy in an ecosystem is the detrital food chain, or the decomposers, made up of forms that feed on dead plant and animal materials. A typical terrestrial chain might include relatively large forms such as earthworms, millipedes, and sow bugs, smaller organisms such as nematodes and mites, and finally microorganisms such as protozoa, bacteria, and fungi. Aquatic detrital chains would have similar counterparts.

Food chains are often intertwined with each other, and their interrelationships are so complex that together they form what is called a **food web.**

DETERMINING FOOD CHAINS. The various links in a food chain are sometimes hard to determine and various methods are used. Direct observations give some insight, but they are not always reliable in the overall picture. Analysis of stomach contents is often used to determine what a member of a food chain uses, but this is difficult unless there are hard parts to identify key structures. A newer method is the use of the "tagged atom" whereby one uses radioactive isotopes, for example, phosphorus 32, by spraying or otherwise, to label the suspected

food source, and later checking the animal for the presence of the tagged atoms.

Biomass refers to the weight of a species population per unit of area. For instance, Juday (1938) found in a Wisconsin lake that there were 209 pounds of carp per acre (biomass). The biomass of a community would be the sum of the biomasses of the many species that make up the steps of the food pyramid. This might be called a **pyramid of biomass.** However, total biomasses obtained by sampling methods are not always reliable. So far, no complete biomass for a community has been obtained.

Energy sources and energy transformations

Ecologic energetics is the study of energy transfer and energy transformations within ecosystems.

Life depends upon the energy given off by radiation of the sun. The sun is a sphere of gas and releases energy by the nuclear transmutation of hydrogen to helium. The energy of solar radiation is given off in the form of electromagnetic waves, which involves a rhythmic exchange between potential and kinetic energy. Heat is energy resulting from the random movements of molecules, and as the molecules are moving they represent kinetic energy. Nonrandom movements of other forms of energy in producing work are transformed into heat. The growth and reproduction of organisms, for instance, is the transformation of energy and production of heat. In ordinary respiration of an animal, about two thirds of its glucose molecules are converted into mechanical energy for activity and growth, and the remaining one third is released as heat. Since all forms of energy can be converted into heat, the kilogram calorie is considered the basic unit of measurement for comparative purposes.

Animals get their high-energy organic nutrients either directly or indirectly from plants, which receive their energy from the radiation of the sun (a few chemosynthetic organisms get their energy from sulfur, iron, and other materials from the environment). The amount of solar energy that reaches our earth's atmosphere per square meter per year is estimated at 15.3×10^8 gram calories. A great deal of this energy is dissipated by dust particles, or used in the evaporation of water. Some is reflected back into space by clouds. Photosynthesis of plants uses only about 0.1% of the solar energy reaching the earth's surface. L. C. Cole estimates that only about

0.04% of the solar energy supplies the metabolism of the biosphere in which animals live. The average of radiant energy available to plants varies with geographic location. In Michigan the amount of solar energy available to plants is about 4.7×10^8 gram calories per square meter per year; in Georgia the comparable figure is about 6×10^8. Plants use about one sixth or more of the energy they get from sunlight for their own metabolism. The remainder is available for the consumers. So we can see that the total chemical energy stored by plants (gross primary production) does not represent the food potential available to heterotrophs. What is left over after plants use a food chain is expensive, and there must be a definite limit to the amount of life that this planet can support.

Other biotic factors

In addition to food relations, other biotic factors play a part in the structure of an animal community. These involve social life within a group, cannibalism, mutual assistance, symbiotic relations, mating habits, predatory and parasitic relations, commensalism, mutual dependence of plants and animals, etc. These factors determine behavior patterns within animal communities and may be as important as any other factors.

Various degrees of social life are found among animals, from those that simply band together with no real division of labor (schools of fish and flocks of birds) to those that have worked out complicated patterns of social organization and division of labor (bees, termites, ants). Many predators (grizzly bears, mountain lions, tigers, hawks, owls) tend toward a solitary life and are rarely found together except for mating purposes.

Many cases of **symbiotic relationship** could be mentioned, such as the small fish that obtains shelter in the cloacae of certain sea cucumbers, the tiny crab *Pinnotheres* commensal in the shells of oysters and scallops, or the flagellates that live in the gut of the termite and make its digestion possible. Ants use the secretions of aphids for food and, in return, give protection and shelter to the aphids. The ant slavemaker *Polyergus* raids the colonies of other species of ants and carries away their larvae to be reared as slaves in its own colony. An interesting symbiotic association is the one involving a sea anemone, a hermit crab, and the snail shell (usually *Buccinum*) in which the crab lives. The hermit crab *(Pagurus)* always lives in an empty gastropod shell to which the European anemone *(Adamsia* or *Calliactis)* is attached. The anemone protects the crab with its stinging tentacles and receives scraps of food from the crab's feeding. Some reef crabs of the tropics fasten an anemone in each claw and use them for protection. Many examples of **cleaning symbiosis** have been found by skin divers among marine organisms. This symbiosis involves the removal of debris or parasites from the teeth or body of one animal by another. Cleaners include many species of small fish, shrimp, and other forms and are usually conspicuous by color or behavior patterns. Cleaners may have stations to which the larger fish go to be cleaned. This is an interesting example of cooperation among animals.

Many plants depend on insects for transferring their pollen. This is a mutual relation, for insects use nectar or pollen from the blossoms. A classic example of the delicate reciprocal relation between insect and plant in pollination is shown by the yucca moth and the yucca plant of the Southwestern states. This moth collects some pollen from one plant and carries it to another, where it lays its eggs in the ovary of the yucca; after depositing an egg, the moth climbs to the top of the pistil and inserts the pollen into a stigmatic tube. This process is repeated for each egg laid (usually six). Each egg in its development requires a fertilized ovule, but enough ovules are left unmolested by the developing larvae to ensure seed for the plant.

■ ANIMAL POPULATIONS*

The term "population" is defined by some ecologists as a group of organisms of the same species that live at a given time in a particular area. Others broaden the term to include similar species. Genetically, the members of a population share in a common gene pool. A population has its own characteristics, such as population density, birth rate, death rate, reproductive potential, age distribution, population pressure, population cycles, and growth. In a broad sense the study of population is the study of biology with all of its implications. Ecologic units, such as communities, are made up of complex population groups and cannot be understood without a

*Refer to pp. 20 and 23 for Principles 17 and 34.

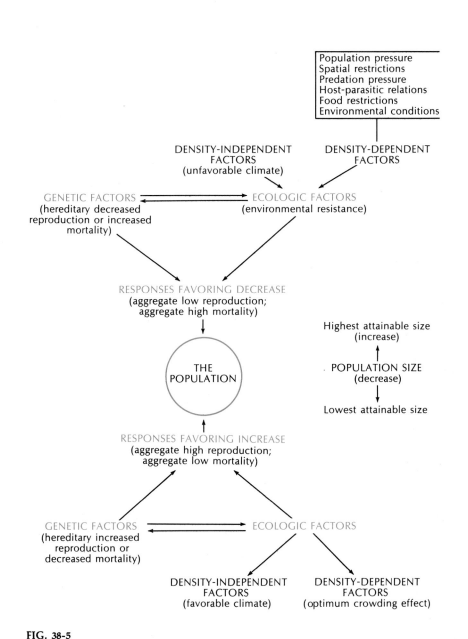

Population pressure
Spatial restrictions
Predation pressure
Host-parasitic relations
Food restrictions
Environmental conditions

DENSITY-INDEPENDENT
FACTORS
(unfavorable climate)

DENSITY-DEPENDENT
FACTORS

GENETIC FACTORS
(hereditary decreased
reproduction or increased
mortality)

ECOLOGIC FACTORS
(environmental resistance)

RESPONSES FAVORING DECREASE
(aggregate low reproduction;
aggregate high mortality)

THE
POPULATION

Highest attainable size
(increase)

POPULATION SIZE
(decrease)

Lowest attainable size

RESPONSES FAVORING INCREASE
(aggregate high reproduction;
aggregate low mortality)

GENETIC FACTORS
(hereditary increased
reproduction or
decreased mortality)

ECOLOGIC FACTORS

DENSITY-INDEPENDENT
FACTORS
(favorable climate)

DENSITY-DEPENDENT
FACTORS
(optimum crowding effect)

FIG. 38-5
Schematic representation of how a population integrates. Arrows point from pressure
factors to those responses that favor decrease or increase of the population size. A
population is nearly always in a state of flux—moving upward or downward as a result of
the interaction of the two responses. Factors involved in a population study are complex
and are often difficult to evaluate. (Modified from Park, T. 1942. Biol. Symposia **8**:123.)

study of the interrelations of populations. Populations must adjust to the environment the same as individuals, although their environmental relations are far more complex.

An acre of rich humus soil may contain several hundred thousand earthworms and many million nematode worms. A quart of rich plankton water may have more than a million protozoans and other small forms. The number of insects of all kinds found on an acre of lush meadow in midsummer often reaches millions. On the other hand, there may be only two or three birds per acre and only one or two foxes per several hundred acres. Animals at the top of the food pyramid, having few or no enemies, often regulate their numbers by arbitrarily dividing their territory and keeping out all other members of their species. This avoids competition for food, nesting sites, and shelters.

The population of any species at a given time and place depends on its birthrate and its mortality, or death rate. If more organisms are born than die, the species will increase. Shifting of members of a species from one habitat into an adjacent one (migration, etc.) would affect the local abundance of that species but not the general population of the species involved.

The biotic potential rate is the innate capacity of a population to increase under optimal surroundings and stable age ratios. What are the controlling factors for keeping populations in check? Ecologists now stress a study of these regulations. The concept of **density-dependent** factors indicates that animal populations are regulated automatically, to a certain extent, by such influences as density, increased mortality or reduced births, food supplies, infectious diseases, and territorial behavior. All these factors operate to produce a general overall effect of feedback, either negative or positive (Fig. 38-5). Many complicated factors that are only partially understood play a role in population regulation. Examples are the dominance hierarchies and the fluctuating fecundity of populations in accordance with the degree of density. As density increases, for instance, fertility declines—a good example of feedback control of population growth.

The success of a population is reflected in its **density,** which is the number of individuals per unit area or volume. The unit of area used in measuring density varies. For small forms such as plankton, esti-mates may be made from forms found in a liter of water; for larger animals the acre or square mile may be the unit. The complete count of individuals in an area is called a **census.** Usually counts are made on sample plots, from which estimates are made. Small mammals, such as mice and chipmunks, may be trapped in live traps, tagged by clipping toes or ears, and then released. A recent method employs radioactive tagging, by which it has been possible experimentally to trace and recover small animals in the field with a Geiger-Muller counter. Suppose 100 animals are caught and tagged, and at a later date another lot of animals is caught in the same way on the same area. In the second sample the number of tagged animals is noted. If the second sample showed 5 tagged animals among 100 caught, then the total population (X) would be $100/X = 5/100$, or $X = 2,000$. This assumes that animals caught in the first sample are just as likely to be caught in the second sample. If sample plots are carefully selected and possible sources of error carefully checked, this random sampling method of estimating the population density of the entire area is considered fairly reliable.

All populations undergo what is called **population dynamics,** which refers to the quantitative variations of growth, reproductive rates, mortality, fluctuations in numbers, age distribution, etc. The characteristic growth of a population is represented by a **population growth curve.** This is the mathematic expression of the growth of a population from its early beginning until it arrives at some stabilizing level of density. Such a curve or graph is produced by plotting the number of animals, or its logarithm, against the time factor. In the beginning, if there is no serious competition with other species and enemies and there is plenty of food, the population grows at about the rate of its potential increase and the curve grows steeply upward. Such curves, however, are rarely realized, except for brief periods, because of the increasing factors of competition, crowding, and higher mortality. These growth curves are very similar for all types and sizes of organisms. One usually starts out with a **lag phase** because it takes time for the few individuals to find each other and start mating. Then it proceeds at a rapid rate, so that a **logarithmic phase** of growth occurs when the population tends to double with each generation, and the curve is fairly straight. But because of more competition for food, losses to enemies, fewer places to live, and

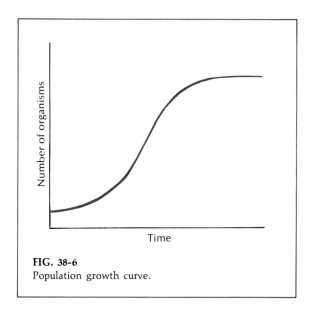

FIG. 38-6
Population growth curve.

greater mortality, the growth rate slows down or levels off into the **stationary phase** (Fig. 38-6).

The logistic theory of population growth seems to be restricted to the population growth of animals with simple life histories, such as many kinds of protozoans, but does not apply closely to those animals that have complex life histories (many insects).

The **birthrate** (natality) is the average number of offspring produced per unit of time. The theoretic or maximum birthrate is the potential rate of reproduction that could be produced under ideal conditions. This is never realized because not all females are equally fertile, many eggs do not hatch, not all the larvae survive, and for many other reasons. **Mortality** is the opposite of birthrate and is measured by the number of organisms that die per unit of time. **Minimum mortality** refers to those that die from old age. The actual mortality, however, is far different from the minimal one, for as the population increases, the mortality increases. Survival curves (made by plotting the number of survivors against the total lifespan) vary among different species. Among many small organisms the mortality is very high early in life; others, such as man, have a higher survival rate at most levels. Man is unique in being able to change his life expectancy through better medical and regulative practices.

Both the birthrate and mortality are influenced by the **age distribution** of a population. Age at which animals can reproduce varies. Asexual forms begin to multiply quite early; many sexual forms (insects, etc.) attain sexual maturity a few days after hatching; others may not become mature for several months or years. Reproductive capacity is usually highest in middle-aged groups. Rapidly growing populations have many young members; stationary populations have a more even age distribution. The relative distribution of age groups in a population indicates trends toward stability or otherwise.

Fluctuation in members occurs in all animal populations. There are more birds in early summer than at other times because of the crop of recently hatched members. Later, many of these birds are destroyed by the hazards of the environment, such as **population pressure.** Examples of **cyclic populations** are the lemmings of the northern zones, which become so abundant every 3 or 4 years that they migrate toward the sea and accidentally drown; the snowshoe hares of Canada, which have approximately a 10-year cycle of abundance (there is a close parallelism in cyclic abundance of the lynx, which feeds on the hare); and meadow mice, which usually show a 4-year cycle of abundance. An appraisal of the 10-year cycle (Keith, 1963) indicates that this periodicity is in a broad sense applicable to many animals (grouse, hare, lynx, fox, etc.), although there are exceptional fluctuations and peak irregularities. The cycle is generally restricted to the northern coniferous forests and tundra regions. Invasions of birds into regions where they are not normally found, such as crossbills and waxwings, have tended to follow a 10-year cycle in some instances in England. Irregular fluctuations are especially common in insects. Grasshoppers may appear suddenly in enormous numbers, but at other times they are very scarce. No satisfactory theory as yet accounts for cyclic fluctuations, although sunspots and climatic cycles have been suggested. Some irregular fluctuations can be explained by weather and climate changes. The great "dust bowl" of 1933 to 1936 must have affected every aspect of life in the stricken area.

The so-called **population turnover** refers to the movement of individuals into and out of populations, and it is caused by birth, death, and immigration into and emigration out of a given population. Some species have rapid turnovers, especially those that live only a single season or have more than one generation a year. However, even in species that live much longer, the turnover rate per year may be 70% or more.

BIOTIC COMMUNITIES

The community is a natural assemblage of plants and animals that are bound together by their requirement of the same environmental factors. Organisms in a community share the same physical factors and react to them in a similar way. The community and the nonliving environment together form the ecosystem. Communities may be widely separated, but if the environmental factors are the same, similar kinds of animals will be found in them. Thus in any brook-rapids community certain characteristic animals are likely to be found. There will, of course, be exceptions.

A **major community** is the smallest ecologic unit that is self-sustaining and self-regulating. It is made up of innumerable smaller **minor communities** that are not altogether self-sustaining. Forests and ponds are major communities; decaying logs and ant hills are minor communities. Members of a major community are relatively independent of other communities, provided that they receive radiant energy from the sun. Communities do not have exact limits but tend to overlap each other. Animals frequently shift from one community to another because of seasonal, diurnal, or other variations.

Stratification is the division of the community into definite horizontal or vertical strata. In a forest community, for instance, there are animals that live on the forest floor, others on shrubbery and low vegetation, and still others in the treetops. Many forms shift from one stratum to another, especially in a diurnal manner. Many of the adjustments and requirements of a particular stratum are very similar in forests widely separated from each other in many parts of the world. The animals that occupy such similar strata, although geographically separated, are called **ecologic equivalents.** The pronghorn antelope of North America and the zebra of South Africa are equivalents.

Between two distinct communities there may be an intermediate transitional zone. This is called an **ecotone,** or tension zone. An example would be the marginal region between a forest and a pasture or open land.

Food relations are a basic aspect of all communities. In a self-sufficient community there are **producer organisms,** such as green plants that make their own food; **primary consumers,** which feed on plants; **secondary consumers,** such as carnivores which live on the primary consumers, and **decomposer organisms,** such as bacteria, which break down the dead organisms into simpler substances that can then be used by plants.

Generally those organisms with the largest biomasses within their levels of feeding interrelations are the ones that exert a controlling interest. In land communities, plants are usually the **dominants,** and some communities are named from their dominant vegetation, such as beech-maple woods. In the ecologic cycle of a community the removal of a dominant usually causes serious disturbances.

A basic characteristic of community organization is **periodicity.** This refers to rhythmic patterns of organisms in their search for mates, food, and shelter. Some community periodicities are correlated with the daily rhythms of day and night, some are seasonal, and others represent tidal or lunar events. Periodic activities include the diurnal and nocturnal faunas that are specialized for day and night activities, respectively; seasonal cycles of growth, mating periods, hibernation, migration, etc.; and activities of swarming, spawning, etc., which are correlated with the lunar and tidal periodicity.

ECOLOGIC SUCCESSION

Communities are not static but are continually changing according to well-defined laws. This process, called biotic or **ecologic succession,** may be brought about by physical factors, such as the erosion of hills and mountains down to a base level, the filling up of lakes and streams, and the rise and fall of the earth's surface. All organisms die, decay, and become a part of the substratum; vegetation invades ponds and lakes; regions of the earth's surface become grasslands, forests, or deserts, according to physical factors of temperature and rainfall. Communities are succeeded by other communities until a fairly stable end product is attained. Such a sequence of communities is called a **sere** and involves early pioneer communities, transient communities, and finally a **climax community,** which is more or less balanced with its environment.

A small lake begins as a clear body of water with sandy bottom and shores more or less free from vegetation. As soil is washed into the lake by the sur-

Pond or lake
Aquatic vegetation from shore gradually encroaches on open water and fills pond with mire and vegetation and it becomes a **marsh-swamp**

If pond had steep sides, it becomes a **forest**

All these changes accompanied by changes in kinds and distribution of animals

If pond had sloping sides, it becomes **open field or prairie**

FIG. 38-7
Ecologic succession in pond or lake.

rounding streams, mud and vegetable muck gradually replace the sandy bottom. Vegetation grows up along the sides of the lake and begins a slow migration into the lake, resulting in a bog or marsh. The first plant life is aquatic or semiaquatic, consisting of filamentous algae on the surface and later of rooted plants, such as *Elodea,* bulrushes, and cattails. As the water recedes and the shore becomes firm, the marshy plants are succeeded by shrubs and trees, such as alders and larches and, later, beeches and maples. Eventually the lake may be replaced by a forest, especially if its sides and slopes are steep; if the sides have gentle slopes, a grassy region may replace the site of the lake. The terminal forest or grassland is a climax community (Fig. 38-7).

In its beginning a lake may contain fish that use the gravelly or sandy bottoms for spawning. When the bottoms are replaced by muck, these fish will be replaced by others that spawn in aquatic vegetation. Eventually, no fish may be able to live in the habitat; but other forms, such as snails, crayfish, many kinds of insects, and birds, are able to live in the swampy, boggy community. As the community becomes a forest or grassland, there will be other successions of animal life.

In general, communities tend to go from a state of instability to one of stability, or climax. The term "stability," however, must be used in a relative sense, for changes are inevitable.

■ ECOLOGIC NICHE

The special place an organism has in a community with relation to its food and enemies is called an **ecologic niche.** In every community there are herbivorous animals of several types, some of which feed on one kind of plant and others on other plants. There are also in every community different types of carnivores that prey on different species of animals. Similar niches in different communities are occupied by forms that have similar food habits or similar enemies, although the species involved may be different in each case. For example, there is the niche in wooded regions occupied by hawks and owls that prey on field mice and shrews, but in regions close to homes this niche is taken over by cats. In this respect the birds of prey and cats occupy the same niche. The arctic fox and the African hyena are other examples. Both are scavengers; the arctic fox eats what the polar bear leaves; the African hyena eats leavings from the lion.

■ KINDS OF HABITAT

Habitats tend to be rather sharply defined from each other, each with its own set of physical and

FIG. 38-8
Woodland pond affords excellent shelter and breeding places for many forms, such as amphibians, reptiles, and insects. (Courtesy C. Alender.)

biotic factors. Transition zones between them are not common. An abundance of different habitats may be found in a small region if there is a diversity of physicochemical and other factors. A small lake or pond (Fig. 38-8) may have littoral or open water, cove, sandy or pebbly bottom, bulrush or other vegetation, drift, and other kinds of habitats. Within the relatively short range of a high mountain, there are many life zones, each of which may correspond to the latitudinal zones. The small altitudinal life zones of the mountain are similar to the large latitudinal zones of the earth's surface with respect to vegetation and, to some extent, to the distribution of animal life. On the other hand, there may be extensive regions, such as the surface of the open sea or a sandy desert, where there is no such diversity of habitats because ecologic conditions are more or less uniform throughout its extent.

The animals that are distributed among the various

FIG. 38-9
Tree hole habitat. Some arthropods found here are rarely or never found elsewhere. One of these groups is the pselaphid beetles.

habitats may be classified into two groups: (1) exclusive, or those that are not found outside a particular habitat (Fig. 38-9), and (2) characteristic, or those that are not confined to one habitat but occur in others. Examples of the first are crossbills, which, on account of their peculiar adaptation, are confined to the coniferous forests, and, of the second class, such forms as rabbits, which roam both woods and open fields.

Freshwater streams

Freshwater habitats include those found in **running water** and those found in **standing water.** Freshwater streams range from tiny intermittent brooks to rivers. Smaller streams are either intermittent ones that flow only at certain seasons, or spring-fed streams that usually flow all the time. Larger, permanent streams include the swift brooks and the rivers that have reached the level of permanent groundwater. Water found in streams differs from marine water in having smaller volume, greater variations in temperature, lesser mineral content, greater light penetration, greater suspended material content, and greater plant growth. Many of the forms found in such habitats

have organs of attachment, such as suckers and modified appendages, streamlined body shapes for withstanding currents, or shapes adapted for creeping under stones.

Various kinds of habitats are found in all streams. Some are found in the swiftly flowing regions where there may be rapids or cataracts; others are found in pools of sluggish waters. Usually the types of animals found in the two regions vary.

In rapids (Fig. 38-10) characteristic forms are the blackfly larvae *(Simulium),* caddisworms *(Hydropsyche),* snails *(Goniobasis),* darters of several species, water penny larvae (Fig. 38-11), miller's thumbs, and stone fly nymphs. All these forms have characteristic behavior patterns, such as positive rheotaxis, high-oxygen requirements, and low-temperature toleration.

In the pool habitats of streams are found various minnows, mussels, certain snails, dragonfly nymphs, mayfly larvae, crayfish, flatworms, leeches, and water striders. Many of the forms that dwell here partially bury themselves in the sandy or mucky bottoms.

The study of fresh waters in all their aspects is called **limnology.**

FIG. 38-10
Brook rapids habitat contains many different forms that have special adaptations for withstanding strong water currents.

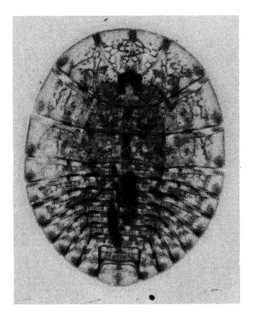

FIG. 38-11
Water penny, larva of riffle beetle, *Psephenus.* This
flat larva is 3.5 to 6 mm. long and adapted for
clinging to lower surfaces of stones in swift brooks.
The adult is somewhat flattened, blackish beetle, 4
to 6 mm. long, that may be found in the water or
in bordering vegetation.

Ponds

Unlike the streams, ponds have feeble currents or
none at all. They vary, depending on their age and
location. Most ponds contain a great deal of vegetation that tends to increase with the age of the pond.
Many of them have very little open water in the
center, for the vegetation, both rooted and floatng
types, has largely taken over. As ponds fill up, the
higher plants become progressively more common.
The bottoms of ponds vary all the way from sandy
and rocky (young ponds) to deep mucky ones (old
ponds). The water varies in depth from a few inches
to 8 to 10 feet, although some may be deeper. Ponds
are too shallow to be stratified, for the force of the
wind is usually sufficient to keep the entire mass of
water in circulation. Because of this, the gases (oxygen and carbon dioxide) are uniformly distributed
through the water and the temperature is fairly uniform.

Animal communities of ponds are usually similar
to those of bays in larger bodies of water (lakes). The
large amount of vegetation and plant decomposition
products affords an excellent habitat for many forms.
Among the common forms found are varieties of
snails and mussels, larvae of flies (Fig. 38-12), beetles,
caddis flies, dragonflies, many kinds of crustaceans,
midge larvae, and many species of frogs. The community of these that live on the bottom or among
the submerged vegetation is called the **benthos.** Many
swimming forms, called **nekton,** are also found in
ponds and include many varieties of fish, turtles,
water bugs, and beetles. Muskrats are usually found
in the larger ponds, where they make their characteristic houses. Because of the abundance of food,
many birds are usually found in and around ponds.
These include herons, killdeers, ducks, grebes, and
blackbirds. Most ponds also have plankton composed
of microscopic plants and animals, such as protozoans, crustaceans, worms, rotifers, diatoms, and
algae (Fig. 38-13). Plankton floats on or near the surface
and is shifted about passively by the winds.

Forms that live in ponds ordinarily require less
oxygen than those found in streams or rapids.

Lakes

The distinction between lakes and ponds is not
sharply defined. Lakes are usually distinguished from
ponds by having continuous and permanent water in
their centers and by having some sandy shores. The
bottoms of lakes depend to a great extent on exposure
to winds. Where waves are common, the bottoms are
usually sandy, but protected areas may contain a great
deal of deposited bottom muck. The surface of water
in proportion to the total volume of water is less in
a lake than in a pond, for lakes are deeper. In large
lakes, such as the Great Lakes of America, the depth
rarely exceeds 500 meters; in moderate-sized lakes
the depth is much less. Oxygen is scarcer in the
deeper regions where there is little circulation. Light
penetration depends on the sediment in water. Most
of the light is absorbed by the first meter of surface
water, and little penetrates beyond a few meters.

Water in lakes tends to become stratified because
only the surface layers are stirred up by wind action
during the summer. Within this surface layer of
water, usually about 10 meters deep in medium-sized
lakes, the temperature is very uniform (about 20° to
25° C.). Below this the water becomes much colder
(reaching 4° to 5° C. at the bottom) and is poorly
oxygenated and stagnant. This level between the

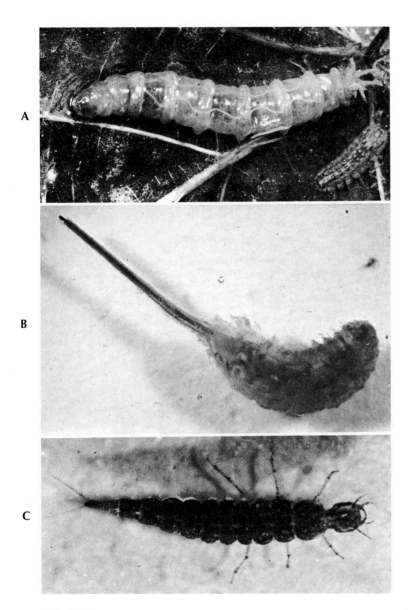

FIG. 38-12
Pond water larvae. **A,** Crane fly larva found in decaying debris along bank. Beside it is small crustacean, *Asellus.* Adult crane fly resembles oversize mosquito with extralong legs. **B,** Rat-tailed maggot, *Eristalis,* gets its air by means of its caudal respiratory tube ("rat-tail"), which is in sections like telescope and can be extended to four times the length of body. The adult fly resembles a bee. **C,** Larval form of *Dytiscus,* giant diving beetle (see adult in Fig. 16-37). Both larvae, called water tigers, and adults are predaceous. These were all found in tiny midwestern woodland pond in early March.

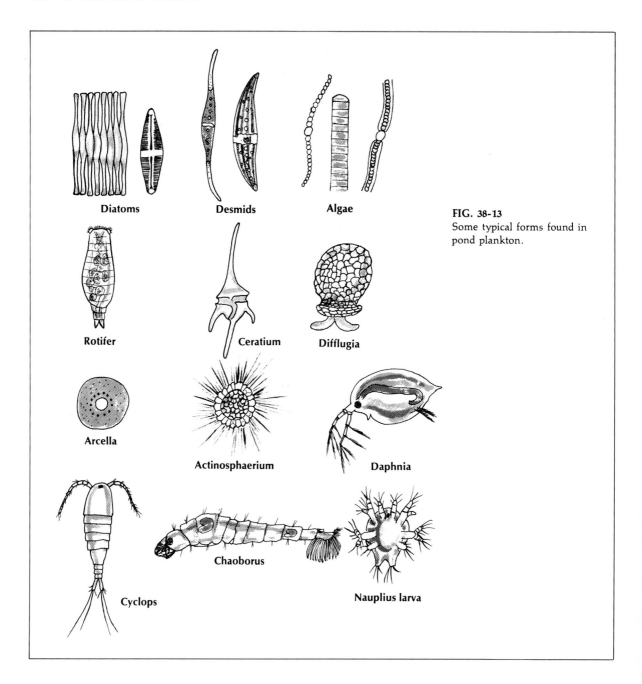

Diatoms

Desmids

Algae

FIG. 38-13
Some typical forms found in
pond plankton.

Rotifer

Ceratium

Difflugia

Arcella

Actinosphaerium

Daphnia

Cyclops

Chaoborus

Nauplius larva

surface layer of uniform temperature and that stratum where the temperature falls rapidly is called the **thermocline.** In other words, the waters above the thermocline are agitated; below it they are still. The thermocline is shallow at midsummer and deeper in early autumn. In the autumn, when the surface water is colder, the wind agitates the water from surface to bottom and the thermocline disappears so that the temperature is about uniform throughout. In winter the surface water may freeze (0° C.), but the bottom remains at 4° to 5° C. Water at the latter temperature is heavier than ice, which accounts for the fact that ice remains at the surface. As the surface waters become warmer in the spring, there occurs another

FIG. 38-14
Habitat of shelving and drift rock—excellent for many species of salamanders and lizards.

complete overturn through wind action. In early summer the thermocline is established again. The spring and fall overturns are important because only at these times is the deep water of the lake oxygenated.

Several habitats are found in lakes. In the terrigenous bottom habitat where the water is shallow and vegetation absent, snails and mayfly nymphs might be found. The cove habitat, which has a great deal of vegetation—submerged, emergent, and floating—one might find bryozoans, small crustaceans, snails, and some insect larvae, and in open water the plankton and nekton. On the bottom (the benthos) are animals that require little light and oxygen, such as a few annelids, bivalves, and midge larvae.

Terrestrial habitats

Land habitats are more varied than those of the water because there are more variable conditions on land. Physical differences in the air are expressed in such factors as humidity, temperature, pressure, and winds to which air-dwelling forms must adapt themselves, as well as types of soils and vegetation. Variations from profuse rainfall to none at all; topographic differences of mountains, plains, hills, and valley; climatic differences from arctic conditions to those of the tropics; temperature variations from those in hot deserts to those of high altitudes and polar zones; and air and sunlight differences from those of daily variations to great storms—all these factors have influenced animal life and have been responsible for directing its evolutionary development.

Land forms have become specially adapted for living in the soil (subterranean), on the open ground, on the forest floor, in vegetation, and in the air. Although more species of animals live on land than in water, there are fewer phyla among terrestrial forms. The chief land organisms are the mammals, birds, reptiles, amphibians, worms, protozoans, and arthropods.

Land habitats are classified on the basis of soil relations, climatic conditions, plant associations, and animal relations.

SUBTERRANEAN HABITATS. Subterranean regions include such habitats as holes and crevices in or between rocks (Fig. 38-14), burrows in the soil, and caves and caverns. Many animals spend at least a part of their lives in the soil. Larval forms of many insects develop there. Ants, nematodes, earthworms, moles, and shrews either have their homes in the soil or spend a part of their time there. For abundance of forms, no other ecologic habitat can compare with the soil.

Most cave animals originated on the surface and were adapted for existence in caves before they entered them. Caves are unique in having uniform darkness, high humidity, no green plants, and no rain or snow. Meager food supply restricts the number of animals living there. Cave animals are usually small, have little or no pigment, and have degenerate sense organs. Many are totally blind. Common examples of cave animals are springtails, mites, small crustaceans, fish, salamanders, and snails.

Some special ecosystems

CLOUDS. A new kind of ecosystem has been proposed within recent years. This is the cloud ecosystem (B. C. Parker). He was led to this concept by the discrepancy between the amount of vitamins (B_{12}) in rainwater and the amounts of dust in rainwater. He concluded that the vitamins came from the clouds. Microorganisms are known to be in the atmosphere but are believed to be passive dormants. However, Parker determined that metabolic activity occurred in the dust but not in sterilized control dust. Basic nutrients also occur in the atmosphere, such as CO_2, nitrogen, and oxygen.

ESTUARIES. It is believed that only a relatively few species of animals are found in estuaries, although those few species may have abundant numbers. Little use is made of estuaries by migratory species. Estuarian species may have excellent adaptations for using the rich terrestrial nutrients to be found there. They must be able to tolerate salinity variations, temperature fluctuation, and high turbidity and other environmental stresses that occur in such regions. Only a few of the 300 or more plankton species are found there. Some coelenterates occur, but these and other phyla are restricted by tidal currents and salinity. A number of species of echinoderms and mollusks are found in estuaries, also some reptiles (snakes and crocodiles). Skates and rays may be frequent visitors but cannot be called regular inhabitants. From the intermediate position of estuaries between fresh and marine water, transient species of many phyla may shift from one medium to the other.

SNOWBANKS. Under the right conditions mountain snowbanks have recently been shown to be a complete ecosystem. Snowbank organisms are adapted to low temperatures and are called cryophiles. Research in this field has revealed complex relationships in the flow of energy and materials between the various forms that live in this ecosystem. Organisms found there include several types of algae (which impart various hues in summer), protozoa, fungi, bacteria, rotifers, and nematodes. Most of the organisms live in 32° F. temperature because sunshine striking a snowbank causes the surface melting of snow, and the water on snow granules always remains at 32° F. Fast melting of snow is not suitable for such an ecosystem. The nutrients are represented by algae (primary producers), minerals from the snow, organic detritus from drifting, and the oxygen from water. Food chains are established: protozoa live on algae; bacteria and fungi are decomposers.

HOT SPRINGS. The temperature of the earth's surface averages about 12° C. (54° F.), but some life can exist in hot springs where the temperature is much higher. The high thermal limits of organs naturally vary. Usually only one or two species of a group can withstand the upper thermal limit and as the temperature drops more species are adapted. If the light is sufficient at lower altitudes, algae may be found at all temperature ranges of a hot spring. The most common organisms found in hot springs in which the temperature does not exceed 50° C. (122° F.) are certain protozoans, arthropods, nematodes, crustaceans, and mollusks. Shore flies (family Ephydridae) may be the most common of all. Some of these small flies are especially adapted to harsh habitats, and the aquatic larvae may be found in strongly saline or alkaline waters and even in crude petroleum pools. The presence of life in and around hot springs may indicate how primordial life existed when the temperatures were much higher than now.

GASTROINTESTINAL TRACT. The fauna and flora of the gastrointestinal tract makes up a unique ecosystem that varies somewhat with each animal species, and individually with the species. The fauna of ruminants and some others involve especially the symbiotic ciliates, with the flora, chiefly bacteria, playing lesser roles. In the adult human, the normal gastrointestinal flora is composed of about 95 species of anaerobic organisms (chiefly bacteria). Aerobic organisms of streptococci and lactobacilli in lesser numbers also occur. Associations between the microorganisms and the epithelial tissue lining the gastrointestinal tract, especially in the cecum and colon, are also found. Such associations appear to be quite stable. These associations may form a barrier against pathogenic forms. The flora in the cecum of hibernating animals is made up chiefly of gram-negative anaerobic bacteria; gram-negative sporeforming rods and coccobacilli tend to disappear. The intestinal ecosystem may be greatly upset by antibiotics given therapeutically, and antibody-resistant strains may survive and replace the nonresistant strains of bacteria. Disease conditions may cause the displacement of anaerobes with aerobic bacteria. Within the intestinal ecosystem in many species of metazoans, are also found parasites of great variety and form.

■ BIOMES

Biotic communities may also be aggregated into biomes, which are the largest ecologic units. In a biome the climax vegetation is of a uniform type, although many species of plants may be included. Each biome is the product of physical factors, such as the nature of the substratum, the amount of rainfall, light, temperature, etc. They are distributed over the surface of the earth as broad belts from the equator to the poles and each may or may not be continuous. Within each biome are many major communities. Biomes are not always sharply marked off from each other; there are often intermediate zones. Some six important terrestrial biomes are recognized: desert, grassland, equatorial forests, deciduous forests, coniferous forests, and tundra. This succession of biomes from the tropics to the poles may also be found in condensed form on the vertical zones of a high mountain (Fig. 38-15). The various

oceans make up the so-called marine biome, a major community that may be subdivided into a number of minor communities.

Marine biome

The physical conditions of the sea influence greatly the life that exists there. In the first place ocean circulations carry heat away from the equator in the western parts of the ocean (along the eastern shores of the continents) so that tropical animals can live within the range of 30 degrees north and south latitudes. On the other hand colder waters from the higher latitude passes toward the equator along the western shores of the continents, and these colder waters restrict coral reefs to within 15 degrees latitude. Light restricts plant life to the upper 50 to 60 meters or so of the upper layer of ocean water and even less in turbid inshore waters. Animals depend chiefly on floating plants in this upper layer of the ocean water. But plants require mineral salts, which are supplied by the upwelling of waters especially along coastal regions. Movements of the sea also produce great effects on those animals that live along sandy beaches, for they must spend part of their energy counteracting the effects of wave action. Thus,

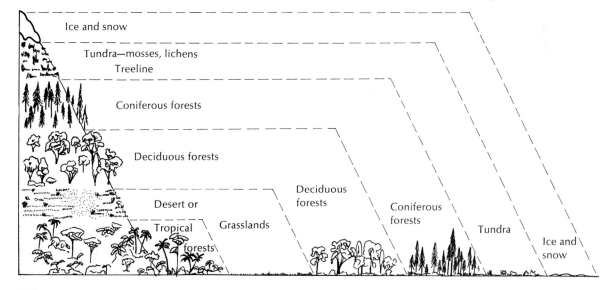

FIG. 38-15
Correspondence of life zones or biomes as correlated with latitudinal and altitudinal zones. This succession of biomes over extensive horizontal range from the tropics to polar regions is found in a condensed form within a few miles on a high mountain in the tropics and, to a slightly lesser extent, in temperate zone.

many animals of the sea live on rock surfaces and in crevices. The most ideal environment of sea animals is probably the coral reef with its firm substratum, mineral nutrients, lighted surface waters, and moderate seasonal variation of light, temperature, and salinity.

Oxygen varies in different parts of the sea. Where deep waters obtain their oxygen supply from cold currents from the polar regions (which can absorb more oxygen than warm water), the concentration of oxygen may be greater than at intermediate depths. Bottom drifts of such cold water may extend long distances. In deep seawater where there is no such replacement, there is a total absence of oxygen. Stagnated water with much hydrogen sulfide is found in many isolated bays and gulfs and supports few or no animals. Surface and shore waters contain a great deal of oxygen.

Plant life is restricted to certain regions. The surface layers contain plankton; sheltered regions may contain plants where they can take root; and seaweeds are found floating about in most seas.

Differences in environmental conditions have produced corresponding differences in the adaptations of marine animals. The animals that inhabit the sea may be divided into two main groups—pelagic and benthic.

PELAGIC GROUP. The pelagic group, which lives in the open waters, includes (1) the **plankton,** small organisms (protozoans, crustaceans, mollusks, worms, etc.) that float on the surface of the water, and (2) the **nekton,** composed of animals that swim by their own movements (fish, squids, turtles, whales, seals, birds, etc.).

BENTHIC GROUP. The benthic group includes the bottom-dwelling forms, or those that cannot swim about continuously and need some support. This group can be subdivided according to the zones in which they are found.

1. The **littoral,** or lighted zone, is the shore region between the tidelines that is exposed alternately to air and water at each tide cycle. Its organisms are subjected to high oxygen content and much wave disturbance. In this region probably originated the ancestors of all aquatic fauna, both freshwater and saltwater. It contains a very rich animal life, both in species and numbers of individuals. Some are adapted for crawling (worms, echinoderms, mollusks), some are adapted for burrowing (worms and mollusks), and

some are attached, or **sessile** (crinoids, bryozoans, corals, sponges).

2. The **neritic** zone lies below the tide water on the continental shelf and has a depth of 500 to 600 feet. There is some wave action here, and the water is well oxygenated. Many forms, including fish, echinoderms, and protozoans, are found in this habitat.

3. The **bathyal** zone is a stratum of the deeper water from the neritic region down to 5,000 or more feet. It contains small crustaceans, arrowworms, medusae, and fish. Many of the animals in this region have luminescent organs, for this is a dark zone.

4. In the **abyssal** zone, or deeper parts of the oceans, the water is always cold and there is total darkness. Oxygen is scarce or absent, and as yet only a few deep-sea forms are known—chiefly certain specialized fish and crustaceans. Many of these are provided with light organs.

Terrestrial biomes

TUNDRA. The tundra is characteristic of severe, cold climates, especially that of the treeless arctic regions and high mountain tops. Plant life must adapt itself to a short growing season of about 60 days and to a soil that remains frozen for most of the year. Most tundra regions are covered with bogs, marshes, ponds, and a spongy mat of decayed vegetation, although high tundras may be covered only with lichens and grasses. Despite the thin soil and short growing season, the vegetation of dwarf woody plants, grasses, sedges, and lichens may be quite profuse. The plants of the alpine tundra of high mountains, such as the Rockies and Sierra Nevadas, may differ from the arctic tundra in some respects. Characteristic animals of the arctic tundra are the lemming, caribou, musk ox, arctic fox, arctic hare, ptarmigan, and (during the summer) many migratory birds.

GRASSLANDS. This biome includes prairies and open fields and has a wide distribution. It is subjected to all the variations of temperature in the temperate zones, from freezing to extremely hot temperatures. It undergoes all the vicissitudes of seasonal climatic factors of wind, rain, and snow. The animals that occupy this region vary with different localities. On the western prairies of the United States there are jackrabbits, antelope, wolves, coyotes, skunks, gophers, prairie chickens, insects, and many others.

In the eastern parts of the country, some of these are replaced by other forms.

DESERT. Deserts are extremely arid regions where permanent or temporary flowing water is absent. When rain does come, it may do so with a terrific downpour. The temperature becomes very hot during the day but cools off at night. There is some scattered vegetation that quickly revives after a rain. Some of the most characteristic plants are the cacti.

Desert faunas are varied and mostly active at night, so as to avoid the heat of the day. Most of them show adaptive coloration and the power of rapid locomotion. To conserve water they have physiologic devices for passing dry excretions. To the casual visitor the desert fauna may seem somewhat scanty, but actually the desert possesses representatives of many animal groups. Mammals found there include the white-tailed deer, peccary, cottontail, jackrabbit, kangaroo rat, pocket mouse, ground squirrel, badger, gray fox, skunk, etc. Birds typical to desert life are the roadrunner, cactus wren, turkey vulture, cactus woodpecker, burrowing owl, Gambel's quail, raven, hummingbird, and flicker. Reptiles are numerous, such as the horned lizard, Gila monster, race runner, collared lizard, chuckwalla, coral snake, rattlesnake, and bull snake. A few species of toads are also common. Arthropods include a great variety of scorpions, spiders, centipedes, and insects.

CONIFEROUS FORESTS (TAIGA). The coniferous forests are the evergreens—pines, firs, and spruces—found in various areas of the North American continent. They may occur in mountains or flat country. They bear leaves the year around and afford more cover than deciduous forests. They are often subject to fires that influence the animal habitats. On mountains and in northern regions they undergo severe winters with much snowfall; in southern regions they have milder conditions. A great deal of food—berries, nuts, and cones—is found in evergreen forests, and there is also a great variety of animal life. In the north there are martens, lynxes, foxes, moose, bears, many birds, some reptiles, amphibians, and many insects. Southern coniferous forests lack some of these forms but have more snakes, lizards, and amphibians. Many of them undergo extensive seasonal migrations.

DECIDUOUS FORESTS. Deciduous forests are more common east of the Mississippi River, and their distribution depends on moisture, soil, and temperature. The trees shed their leaves in the fall, leaving them bleak during the winter, especially in northern climates. There may be some low underbrush and vines. Some of these forests have scattered evergreen trees. They possess a varied animal life, including many burrowing forms. Among characteristic fauna of these forests are the deer, fox, bear, beaver, squirrel, flying squirrel, raccoon, skunk, wildcat, rattlesnake, copperhead, and various songbirds, birds of prey, and amphibians. Insects and other invertebrates are common, since decaying logs afford excellent shelters for them.

TROPICAL RAIN FORESTS. Tropical rain forests are found in Central America. Vegetation is luxuriant and varied. The trees are mainly broad-leaved evergreens; also there are many vines. These forests have a copious rainfall and a constant high humidity. Because of their density, they have a reduced illumination. The forests are divided ecologically into a vertical series of strata, each of which is occupied by characteristic animals. These strata include the forest floor, the shrubs, small trees, lower treetops, and the upper forest canopy. The enormous amount of life found here is represented by monkeys, amphibians, insects, snails, leeches, centipedes, scorpions, termites, ants, reptiles, and birds.

◼ FOOD CYCLE

Animals are stores of potential energy that they transform into kinetic energy to be used in their life processes. All energy utilized by animals is derived ultimately from the sun. Plants utilize radiant energy from sunlight and the chlorophyll in their cells to produce carbohydrates from carbon dioxide and water. Plants can also form proteins and fats. Animals, with few exceptions, do not have this power and depend on the plants as sources for the basic food substances. Animals that do not live directly on plants live on animals that do; therefore, all their potential energy can be traced back to a plant origin.

In this **energy cycle** of transfer and transformation the **laws of thermodynamics** apply. When a plant transforms light into the potential energy of food by the process of photosynthesis, energy is being transformed into another form without being destroyed (first law of thermodynamics), but when this plant food is utilized or consumed by other organisms, although there is no loss in total energy, there is a decrease in amount of useful energy, for some energy is degraded or lost as heat in a dispersed form (second law of thermodynamics). Thus in every step in a food chain or pyramid there is a certain loss in useful energy. Energy is used only once by an organism or population, for it is then converted into heat and lost. On the other hand, the nonenergy materials, such as nitrogen, carbon, and water, may be used over and over again. Energy, therefore, is a one-way flow. When the sun's energy is exhausted, there will be no further photosynthesis and no more life.

All plants in their metabolism require certain ele-

ments, such as carbon, oxygen, nitrogen, hydrogen, and, to a lesser extent, potassium, magnesium, calcium, sulfur, iron, and a few others. All these are derived from the environment, where they are present in the air, soil, rock, or water. Animals require about the same elements, most of which they get from the plants. When plants and animals die and their bodies decay, or when organic substances are burned or oxidized, these elements are released and returned to the environment. Bacteria fulfill a useful role in decomposing the body wastes and the

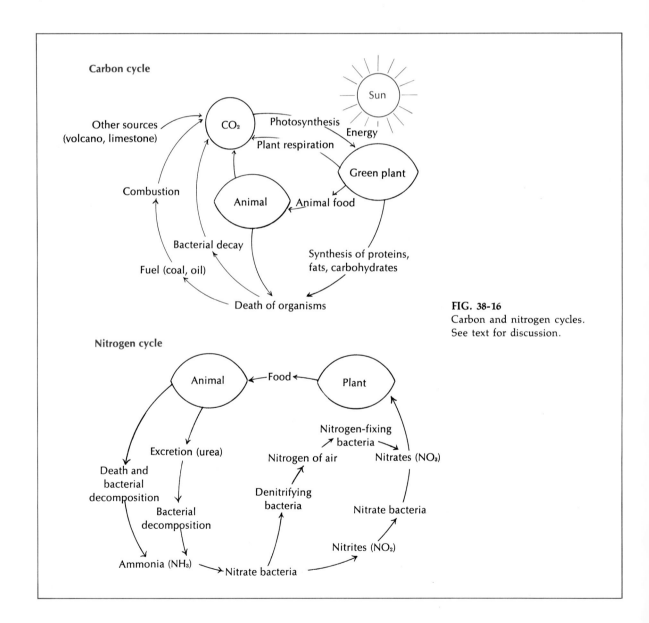

FIG. 38-16
Carbon and nitrogen cycles. See text for discussion.

bodies of dead animals and plants. The elements that are involved in these processes, therefore, pass through cycles that involve relations to the environment, to plants, and to animals. Three or four of these important cycles are pointed out here.

CARBON CYCLE (Fig. 38-16). Both animals and plants respire and give off carbon dioxide to the air. More is released in the bacterial decomposition of organic substances, such as dead plants and animals. Although the percentage of carbon dioxide in air is relatively small (0.04%) as compared with the other gases of air, this small amount is of great importance in nature's economy. Living green plants take carbon dioxide from the air or water and by **photosynthesis,** with the help of sunlight, carbohydrates are formed. This process is complicated but in a simple form may be expressed thus:

$$6H_2O + 6CO_2 \rightarrow C_6H_{12}O_6 + 6O_2$$

$$\text{Carbon} \qquad \text{Sugar} \qquad \text{Oxygen}$$
$$\text{dioxide}$$

Carbohydrates thus formed, together with proteins and fats, compose the tissues of plants. Animals eat the plants and the carbon compounds become a part of animal tissue. Carnivorous animals get their carbon by eating herbivorous forms. In either case a certain amount of carbon dioxide is given off to the air in breathing (this also occurs in plants), and when the animal dies, a great deal of the gas is released to the air by bacterial decomposition. This cycle is called the carbon cycle. The importance of carbon in the organic world cannot be overestimated; for of all the chemical elements, it is the one that enters into the greatest number of chemical combinations.

OXYGEN CYCLE. Animals get their oxygen from the air or from oxygen that is dissolved in water and utilize it in their oxidative reactions. They return it to their surroundings in the form of carbon dioxide (CO_2) and water (H_2O). Plants also give off some oxygen in photosynthesis, as seen in the formula above. Plants use some oxygen in their own respiration. In the interesting relationship between plants and animals in the plankton life of surface waters, the floral part of these populations gives off the oxygen necessary for the life of the faunal portion and determines the distribution of the latter. Light is the determining factor in the distribution of the plant life of plankton.

NITROGEN CYCLE (Fig. 38-16). Atmospheric ni-

trogen (78% of air) can be utilized directly only by the nitrogen-fixing bacteria that are found in the soil or in the root nodules of leguminous plants. These nitrogen-fixing bacteria combine nitrogen into nitrates (NO_3), and plants form proteins from these nitrates. When animals eat plants, the proteins of the latter are converted into animal proteins. In animal metabolism, nitrogenous waste (urea, etc.) is formed from the breakdown of proteins and is excreted. In the soil or water certain bacteria convert this waste into ammonia and into nitrites; other bacteria (nitrifying) change the nitrites into nitrates. Whenever plants and animals die and undergo bacterial decomposition, their proteins are converted into ammonium compounds.

MINERAL CYCLE. Many inorganic substances are necessary for plant and animal metabolism. Usually the amount of these constituents is small and varies with different kinds of living things. Among these, phosphorus is one of the most important. It is found in the soil and water in the form of phosphoric oxides. It is taken up by plants, is passed on to the animals, and is eventually returned to the soil or water in the form of excreta or upon the decay of their bodies. Many think that phosphorus is the critical resource for the efficient functioning of ecosystems. The supply is very much limited and vast amounts are carried into the sea. Man has so disrupted the amount of phosphorus in his disturbance of the soil that erosion can readily carry it away. Phosphorus represents one of the most important ingredients of most fertilizers.

■ ECOLOGY AND CONSERVATION

Most aspects of ecologic study are related to the principles of conservation. Conservation may be defined as the most efficient and most beneficial utilization of natural resources (soil, forests, water, wildlife, minerals, etc.). This is a logical relation, for the basis of ecology is the ecosystem, or the combination of the biotic community and the physical environment. A great conservationist, Aldo Leopold, once stated that the biotic pyramid is really a symbol of land and its uses as a circuit of energy, which involves soils, plants, and animals.

Our natural resources are commonly divided into (1) the renewable, such as the biotic factors of flora and fauna and the physical factors of soil and water, and (2) the nonrenewable, such as minerals. The renewable resources form an intricate relationship and are closely tied together; one cannot be disturbed without disturbing the others. Able conservationists now believe that the wisest way to preserve renewable natural resources is to use them within the limits of continuous renewal. Good forest management, as many European countries have demonstrated, is to remove the less vigorous trees for timber and allow the others a chance for normal growth. Good game management involves regulated hunting seasons so that wildlife may stay within the bounds of food resources and optimum populations. In conservation planning it is not always the best policy to eliminate the inferior or less desirable plant or animal members of the ecosystem. Predators, for instance, have their place in the biotic pyramid.

In the final analysis, sound conservation is the best possible use of the available energy in an ecosystem. Without the biotic factors, energy from the sun is mostly lost on our planet, for sunlight energy is stored by plants and passed in food chains and food webs around the ecosystem. Soil is dependent on biotic factors that must be present or else there is the enormous problem of erosion waste.

Specifically, any adequate conservation program must include the natural resource background of soil, water, wildlife, vegetation, and natural topography. It should include the application of ecologic principles to the integration of those environmental factors. It should include methods for the most efficient use of the land for a growing population. Definite controls must be worked out to limit the wastage of land erosion, fires, flood, and water pollution. Effective controls involving both biologic and chemical methods must be employed to check insect pests, rodents, and disease-carrying animals. The program includes the dynamics of wildlife populations and the most efficient practices of hunting and fishing. Such a plan also provides recreational areas for the health and enjoyment of the people.

The ultimate success of a conservation program will depend mainly on an efficient system of education that teaches basic ecologic concepts throughout its curriculum. The development of this conservation attitude in the public will be slow and laborious. Leopold refers to this concept or attitude as a land ethic, which he regards as a product of social evolution based on intellectual and emotional processes and woven into the very pattern of the community's daily life.

References
Suggested general readings

Ager, D. V. 1963. Principles of paleoecology. New York, McGraw-Hill Book Co. *The study of paleoecology is a very difficult subject and this work is one of the few that attempts an evaluation of investigations that have been performed in this field.*

Allee, W. C., A. E. Emerson, O. Park, T. Park, and K. P. Schmidt. 1949. Principles of animal ecology. Philadelphia, W. B. Saunders Co. *A comprehensive treatise. Nearly every aspect of the subject is covered in this masterly work.*

Boughey, A. S. 1968. Ecology of populations. New York, The Macmillan Co. *Deals with the requirements, population interactions, population evolution, food chains and community organization, and certain aspects of human ecology.*

Buchsbaum, R., and M. Buchsbaum. 1957. Basic ecology. Pittsburgh, The Boxwood Press. *For the general student of biology, the basic concepts of ecology are presented with accuracy and clarity. Students interested in this field should read this little book before studying the more comprehensive treatises.*

Carlquist, S. 1965. Island life. Garden City, N. Y., The Natural History Press. *The study of island life has loomed larger in recent years.*

Clapham, W. B. 1973. Natural ecosystems. New York, The Macmillan Co. *Stresses both the biotic and abiotic components of the ecosystem and how man is altering them.*

Elton, C. S. 1927. Animal ecology. London, Sidgwick & Jackson, Ltd. *A concise statement of ecologic principles, well organized for ready comprehension.*

Elton, C. S. 1958. The ecology of invasions by animals and plants. New York, John Wiley & Sons, Inc. *The noted ecologist points out with graphic clearness how the invasion of fauna and flora pests (which he terms "ecological explosions") has altered the face of the earth, affected the welfare of man and beast, and posed problems for man to solve.*

Ehrlich, P. R., and L. C. Birch. 1967. The "balance of nature" and "population control." Amer. Naturalist **101:**97-107.

Emlen, J. M. 1973. Ecology: An evolutionary approach. Reading, Mass., Addison-Wesley Publishing Co., Inc.

Fuller, W. A., and J. C. Holmes. 1972. The life of the Far North. New York, McGraw-Hill Book Co. *Ecology and natural history of the taiga and tundra.*

Handler, P. 1970. Biology and the future of man. New York, Oxford University Press.

Hedgpeth, J. W. (editor). 1957. Treatise on marine ecology and paleoecology. Vol. 1, Ecology; vol. 2, Paleoecology. New York, The Geological Society of America. *A pretentious treatise by many specialists on the many aspects of ecology both past and present. A work for the advanced student.*

Jones, T. C. (editor). 1972. The environment of America. New York, Doubleday & Co. *A discussion of environmental problems by many authorities.*

Keith, L. B. 1963. Wildlife's ten-year cycle. Madison, University of Wisconsin Press. *An evaluation of the cyclic phenomena of wildlife populations. Although there are many variations and exceptions, the 10-year cycle is shown to be a useful description.*

Kevan, D. K. M. 1955. Soil zoology. London, Butterworth's Scientific Publications. *This is one of the few monographs that deals exclusively with the forms found in the soil. Their economic relation, ecology, methods for their control, and methods for sampling their populations are all dealt with in a thorough manner.*

Klopfer, P. H. 1962. Behavioral aspects of ecology. Englewood Cliffs, N. J., Prentice-Hall, Inc. *A study of some basic ecologic problems and their ethologic backgrounds. Animal behavior is one of the liveliest subjects in biology and many lines of investigation are being integrated to help explain it.*

Kormondy, E. J. (editor). 1965. Reading in ecology. Englewood Cliffs, N. J., Prentice-Hall, Inc. *Classic and fundamental studies of ecologic investigations.*

Nobile, P., and J. Deedy (editors). 1972. The complete ecology fact book. New York, Doubleday & Co. *A comprehensive compilation of ecology statistics.*

Odum, E. P. 1963. Ecology. New York, Holt, Rinehart & Winston, Inc. *The basic principles of ecology as they apply to man and his welfare are stressed throughout this concise ecologic treatise. An excellent background for the study of ecology.*

Odum, E. P. 1971. Fundamentals of ecology, ed. 3. Philadelphia, W. B. Saunders Co.

Reid, G. K. 1961. Ecology of inland waters and estuaries. New York, Reinhold Publishing Corp. *The estuarine environment is looming more and more in importance as an ecologic factor in the invasion and distribution of terrestrial forms; this work gives a detailed description of estuarine oceanography.*

Shelford, V. E. 1913. Animal communities in temperate America. Chicago, University of Chicago Press. *A classic work in ecologic study. Has had a profound influence on the direction of much ecologic work.*

Shelford, V. E. 1963. The ecology of North America. Urbana, University of Illinois Press. *A description and evaluation of the major biotic areas (vegetation and associated animals) of North America by the pioneer American ecologist.*

Turner, F. B. (editor). 1968. Energy flow and ecological systems. Amer. Zool. **8:**10-69 (Feb.).

Vernberg, W. B., and F. J. Vernberg. 1972. Environmental physiology of marine animals. New York, Springer-Verlag.

Selected *Scientific American* articles

Bell, R. H. V. 1971. A grazing system in the Serengeti. **225:**86-93 (July).

The biosphere. 1970. Vol. 223 (entire Sept. issue).

Bolin, B. 1970. The carbon cycle. **223:**124-132 (Sept.).

Bormann, F. H., and G. E. Likens. 1970. The nutrient cycles of an ecosystem. **223:**92-101 (Oct.).

Brower, L. P. 1969. Ecological chemistry. **220:**22-29 (Feb.). *How insects make themselves unpalatable to bird predators.*

Cloud, P., and A. Gibor, 1970. The oxygen cycle. **223:**110-123 (Sept.).

Cole, LaMont C. 1958. The ecosphere. **198:**83-92 (April).

Deevey, E. S., Jr. 1970. Mineral cycles. **223:**148-158 (Sept.).

Delwiche, C. C. 1970. The nitrogen cycle. **223:**136-146 (Sept.).

Gates, D. M. 1971. The flow of energy in the biosphere. **224:**88-100 (Sept.).

Penman, H. L. 1970. The water cycle. **223:**98-108 (Sept.).

Savory, T. H. 1968. Hidden lives. **219:**108-114 (July). *Tells of the invertebrate inhabitants of the cryptosphere, a region a few inches above and below the ground surface.*

Wecker, S. C. 1964. Habitat selection. **211:**109-116 (Oct.).

Woodwell, G. M. 1967. Toxic substances and ecological cycles. **216:**24-31 (March).

Wynne-Edwards, V. C. 1964. Population control in animals. **211:**68-74 (Aug.).

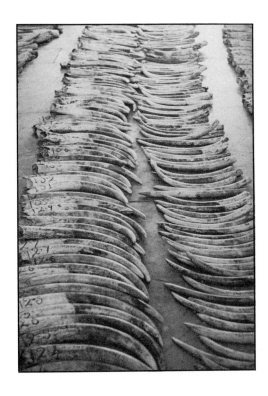

Elephant tusks taken from poachers by Kenya conservation authorities. These are only a few from an enormous warehouse in Mombasa containing thousands of huge tusks, representing, of course, the slaughter of thousands of elephants.

CHAPTER 39

PROBLEMS OF MAN'S ECOSYSTEM*

Thinking people agree that one of the greatest problems confronting man today is his relation to his own environment. Man has always been a part of his environment, but only within recent decades has he become aware of the enormity of the environmental problems that have arisen. He has realized that these problems must be solved to ensure his very survival on earth. The first man (probably a few million years ago) lived very much as other animals did. His numbers were relatively few and he could shift from one locality to another. His evolutionary success involved his cultural development from the cumulative efforts of previous generations. He learned the secrets of fire, the protection of clothing for his body, methods of shelter, and the sources of food.

*Refer to pp. 22 and 23 for Principles 33 and 34.

894

But in the past few hundred years man has made enormous strides in nearly every aspect of his economic, physical, and social life. Most of his advancements in these fields were made as single objectives —that is, they promoted his welfare and he gave little attention to their impact on the overall environment. In agriculture he selected those crops he liked and did away with others. In medicine he strived to control infectious, parasitic, and dietary diseases, mental disorders, degenerative disorders, and others. But one consequence of his success has been the intensification of population growth rate. There may also have been an increase in genetic diseases by preserving those who will transmit bad genes. Man's most spectacular advances have been in technology, which has served to magnify his wealth and increase his living standards. This has encouraged him to think that everything about technology is an undiluted good.

However, we are now beginning to realize that all this advancement has been obtained at the cost of producing vast environmental problems manifested in air and water pollution, dwindling wildlife, poisoning of the environment by pesticides, biologic hazards of radiation, unstable ecologic systems, misuse of natural resources, abnormal increase in human populations, the concentration of people in small crowded areas, a failure to comprehend the nature of the social and personal environment, and a general failure to understand the complex interactions between man and the environment in which he lives.

■ OUR ENVIRONMENT TODAY

One cannot say that our present environment has been totally ruined despite mismanagement and unwise control. At no time in the past has the environment been so productive to man's welfare and needs as the present—a tribute to the growth of sophisticated technology. In the past, problems concerning the environment were solved largely through geographic expansion into new regions. The surface of the planet earth was large and unimproved. But as man could not indefinitely better himself by moving into new territories, he turned to transforming his biosphere locally. At no time in his long history has man been able to enjoy the fruits of his labors better than at present. It is true that large segments of human populations live in substandard conditions today, but it is doubtful that they were better off in the past.

Why, then, is there so much concern over the environmental conditions at present? Simply because in the midst of all this activity we have been largely insensitive to the damage we have created in our biosphere. Our numbers have increased at a terrific rate, our technology has produced waste beyond our means to dispose of it, and our terrain has been restricted by mismanagement. So far, our capacity to injure our environment has exceeded our power to resolve the problems leading to its destruction.

■ CRISES THAT CONFRONT US

Our status at present is one of vast expansion, with a socioeconomic system that stresses the development of technology and the consumption of the products of technology. Transformation of raw materials into sophisticated products has produced environmental pollutants of both air and water. Trash and garbage involve problems of disposal or conversion. Our present society not only produces more goods, but it also produces goods of shorter life and usefulness. Products that are useful to an affluent society are quickly supplanted by more sophisticated gadgets. What is new today is old tomorrow.

The development of motorized transportation has contributed enormously to the pollution of the air and the extinction of silence. The growth of cities, where about 80% of our American population now lives, produces more problems of pollution, along with unhealthful living conditions. The crowded conditions of metropolitan communities magnify the problems of garbage and waste disposal as well as air and water pollution.

In 1956 there was published an international symposium entitled *Man's Role in Changing the Face of the Earth.* In this revealing work many able and qualified scholars and specialists pointed out the various ways in which man has modified his environment. One of the problems emphasized in this symposium was the menace of world overpopulation. Uncontrolled birthrates and lower death rates with no widespread positive checks indicate a real change in the human time scale. The average life is longer, and so the lengthening of the time scale by many years has altered the character of human life.

Other areas of concern include soil erosion, soil changes resulting from human use, the problem of food supply, the misuse of mineral resources, the modification of biotic communities, the heavy use of chemical pesticides, rapid raw material consumption, the urbanization of the population, the socioeconomic system of production and consumption, and many others.

Control of population growth

The problems concerned with population growth and control are unique in many ways. Any alterations in the trends in this field are going to require drastic changes in the whole social and cultural pattern of man. Family size is a private family matter and difficult to regulate without the cooperation of the family. It is highly unlikely that great masses of people, very unequal in education and culture, will react alike in these matters. Population growth is exponential, and the increase is geometric. Natural selection at present does not operate as it did in the past, when it tended to keep populations stable.

TABLE 39-1
World population growth per time

World population	Year	Time, in years, required to double population
250,000,000	1	
500,000,000	1650	1649
1,000,000,000	1850	200
2,000,000,000	1930	80
4,000,000,000*	1975	45
8,000,000,000*	2005	30

*Estimated growth at present rates. Present (1973) population of world estimated at 3,650 million.
(From Handler, P. (editor). 1970. Biology and the future of man. New York, Oxford University Press. Also from other sources.)

At present the current rate of increase is about 1.5% to 2% per year. About 60 million persons are added to the world's population each year. At this rate the more than 3.5 billion people in the world today will double in the next 35 to 40 years, so that by the end of this century more than 6 billion individuals will be living on this planet. Only a few hundred years ago it required centuries for the population to double (see Table 39-1).

This phenomenal population growth has been a result of the striking drop in early death rate, especially a drop in infant mortality from 162 per 1,000 live births in 1900 to 22 per 1,000 in 1966. Improved sanitation and medical care, together with a better control over certain devastating diseases such as malaria and yellow fever, have also contributed to the declining death rate. A far greater percentage of persons now live to childbearing age, which tends to increase the population. Man, in common with other primates, has no relaxation in his sexual urge as do other animals, which have seasonal periods for reproduction. Although many persons sincerely desire to have children, most children are conceived because of the compulsion of the sex instinct. With most organisms in nature there is a need to produce many offspring to preserve the species because so many are eliminated by the hazards of existence. Man has eliminated most of the hazards, but he has retained the breeding urge.

The family represents the closely knit attachment of its members to each other, and our whole social structure is built around it. But no longer can we consider large families a necessity as did our pioneering forefathers. Even if greater expansion is feasible, the problem of overpopulation still remains; some authorities believe we are postponing what eventually must come—a compulsory means of some kind for restricting the number of people.

Family planning in the form of birth control has been practiced by enlightened persons for a long time. Although entirely successful devices for preventing conception have not yet been perfected, the current intensive research in this field will no doubt result in completely safe and efficient methods.

The final objective is a more or less stable population that can be supplied with food and the benefits of our modern technology. The practice of human husbandry, or the planned regulation of family size, must be established in the mores of our social life. Recent statistics suggest that this is happening in North America now and that the United States population is beginning to stabilize. The birthrate in 1960 was nearly 24 per 1,000, but it had dropped to about 16 per 1,000 in 1971.

In the United States zero population growth is considered to be 2.11 children per family. Women are staying single longer and having fewer children than formerly. About one half of the women are still single at 21 years of age. If zero growth of population is still maintained, the population of the United States (204.8 million in 1970) would not reach 270 million by the year 2000 as previously estimated. Many factors are responsible for this slower growth, such as legalized abortion, the wide use of contraceptives, and population propaganda. But for the great teeming masses of people who are ignorant or indifferent about birth control methods, for those who have religious or other scruples about such methods of family limitation, the outlook for a practical solution is still dismal. Some writers have even advocated laws for compulsory sterilization if all other methods fail. Sir Charles Darwin, a leading exponent of family restriction and the grandson of Charles Darwin of evolutionary fame, once said that if overpopulation persists, only the superficialities of civilization will survive, and man may again revert to natural selection for controlling his numbers in order to exist. Man's numbers in the final analysis may depend on the capacity of the earth to furnish the food necessary for the energy of life.

Changing biotic communities

A balance of nature involves man's relation to his total environment; he is a fellow companion with the other creatures of this planet. However, in the belief that man was created to have dominion over the earth and its resources, he has done many things that run counter to the concept that he himself is a part of nature. His role in changing the face of the earth has been attributable to his natural desire for expansion and growth. He has developed scientific technology as a tool for solving his problems, but has remained indifferent to the new problems such technology creates.

We have destroyed forests to provide agricultural fields and places for human habitation without considering the effects on erosion or on the wildlife whose habitation was obliterated. We have dredged swamps and interfered with natural waterways. Pollution from our industries has made water unfit for fish and other forms to live in. Our economy has been destructive to animal habitats. Even when we planned the preservation of wildlife, we often ignored the basic laws of ecology and gave little consideration to the role of predators in the web of life. Mismanagement has brought about the extinction of many species and endangered the existence of others.

Wild animals and fish are not of esthetic value alone; they also contribute to economic wealth. According to the estimates of the U. S. Fish and Wildlife Service more than 4 billion dollars are spent each year by hunters and fishermen.

A recent Fish and Wildlife red book points out that Americans are about to lose forever some 89 species of vertebrate animals, including 21 fishes, 8 reptiles and amphibians, 46 birds, and 14 mammals. The most common endangered species are the bald eagle, whooping crane, American alligator, panther, manatee, Atlantic salmon, condor, peregrine falcon, pelican, and ivory-billed woodpecker. The viewpoint that is expressed or implied is that man's interests come first. However, the very core of any solution of the environmental crisis is a balanced ecosystem and this involves a diversity of all life and a preservation of the ecosystem of others. Evolution by natural selection has evolved a pattern of life over millions of years, yet man has run counter to this pattern.

Thoughtful naturalists have long pointed out the need for preserving suitable habitats, strict enforcement of protective regulations, and the development of a public attitude of sympathetic understanding of the plight of these species. In 1969 the President of the United States signed the Endangered Species Act, which prevents the importation of endangered species of wildlife and fish into this country and the interstate shipment of reptiles, amphibians, and other wildlife taken contrary to state laws. The fate of the passenger pigeon and other desirable species that are already extinct should make us realize that whenever a species shows a steady decline it may reach a condition of no return no matter what measures are taken to preserve it.

Misuse of our land

The very abundance of natural resources in a relatively newly settled country such as America has posed some of the exceptional problems of conservation that are of public concern at present. One of these problems is the misuse of our soil. Soil is made up of mineral particles produced by the weathering of parent rock plus humus or the organic detritus of decaying vegetation. The interaction of these two components, which results in organic acids, causes further disintegration of the mineral component, so that the two processes, physical and biologic, are taking place all the time. But it takes a long time for the production of a fertile soil, so that its abuse represents the loss of a priceless natural resource.

Soils regulate the flow of water, serving as a spongy receptacle that prevents excessive flow of water in floods. For instance, each year Arkansas farmers have been clearing 150,000 acres of forest land for agricultural purposes. More than a million acres of forests have been so depleted there in the past decade. Why not leave forests on rough ground to prevent erosion that sweeps away quickly the very best of soils? This short-sighted policy of removing forests is well illustrated in hilly communities in America. Copper smelters in the Appalachian Mountains have completely killed all vegetation in certain communities, leaving the soil to be totally eroded. Growing single crops in rows without attempting to enrich the soil by rotation of grass crops and by livestock is another common cause of soil destruction. Little attention has been given to the surface contour (topography) of the land by many American farmers. The revealing book of J. Richie, *The Influence of Man on Animal Life in Scotland* (1920), shows what can happen when man misuses his soil.

Another example of our misuse of land is our flood-plain development. There are wide flat plains on both sides of most rivers onto which the river overflows occasionally. These plains and marshes serve to spread out floodwaters and slow down their force. They also draw out and deposit quantities of precious topsoil that would be carried out to sea. However, man builds roads, dwellings, and factories in these bottomlands, dries up the marshes, and then builds levees to protect these improvements from floods. Now the topsoil is forced on out to sea, and the bottomland is lost to agriculture and to wildlife. Then, having destroyed the natural flood control function of the plain, man goes upstream and builds giant, expensive flood-control dams that flood out additional farmlands and wildlife habitats. How much simpler and wiser to zone the flood plains for agriculture and recreation, for which they are well suited, and put the dwellings and factories on the bluffs overlooking the plains and out of reach of floods!

Many of the dams built by the Army Corps of Engineers are of doubtful value and often destroy regions of scenic beauty. Dams at best are of temporary value because they fill with silt in a matter of a few years.

The establishment of our national parklands stemmed from a desire to set aside some unspoiled land for recreational purposes and for an appreciation of nature under natural conditions. But these sanctuaries are often threatened too. By diverting the water from Lake Okeechobee (that normally supplies the Everglades National Park) the very existence of the park land is threatened. The Everglades is not really a swamp, but a free-flowing river coursing through saw-grass country from south-central Florida down to the Gulf of Mexico. Recently a decision was made to build a major supersonic jetport in and near the park. The jetport was to be set in the mainstream of this river just north of the park, despite the fact that the whole ecologic balance of southern Florida depended on the purity and quantity of the water in this river. Conservationists pointed out that every takeoff of a jet liner uses 4,000 pounds of fuel, the exhaust from which contains carbon monoxide, unburned hydrocarbons, carbon, and nitrogen oxide. Such pollution would have killed the algae that forms the base of the food chain that includes small animals, fish, birds, and alligators. By January, 1970, an agreement of federal, state, and local authorities was reached, forbidding completion of the international jetport—a real tribute to the thousands of concerned citizens who have worked to prevent further deterioration of their natural ecosystem.

Air pollution

The air of our cities—and cities throughout the world—is increasingly loaded with particulate matter, smog, colloidal material, noxious gases, and other pollutants (Fig. 39-1). The automobile has been one of the chief culprits in producing air pollution (more than 60%), although there are also other causes. The burning of fuels for home consumption and industrial uses ranks high in pollution from sulfur dioxide. Other causes are the burning of trash and other combustible materials. School children cannot be permitted to play outdoors in Los Angeles on those days when smog causes the ozone of the air to exceed a certain point. In 1969 the school playgrounds were closed several times in midsummer and early fall. In July, 1970, a dangerous pall of smog hung over our eastern seacoast, enveloping several eastern states for a number of days.

Inhalation of contaminated air often causes labored and difficult breathing. Breathing noxious gases (for example, sulfur dioxide) injures delicate tissues of the lung and produces other effects on the body. The exhaust of automobiles discharges a number of harmful substances—noxious gases, and small particles (particulates) that emerge from the combustion chamber of the car. These particles may remain in the air for a long time and when breathed into the lungs can affect the delicate tissue much as coal dust would do. The adverse effects of air pollutants to health develop slowly and cannot be diagnosed easily. Persons also become accustomed to breathing smog and, in time, to feeling little effect. The air pollutants are cumulative and may cause chronic bronchitis and worse disorders. In 1952 more than 4,000 deaths were attributed to a heavy smog in London.

Air pollutants may also injure clothing and fabrics, kill vegetation, and damage buildings and statues. A recent U. S. Forest Service study shows that more than a million trees in southern California are dying from smog effects.

Solar radiation, which influences smog, is most pronounced on cloudless days, and winds often shift the pollutants a long distance from their source. The location of the source is very important. Low river

FIG. 39-1
A hazy fog resulting from air pollution covers Denver, Colorado. Many other cities, for example, Los Angeles, and Gary Indiana, have similar problems. (Photo by C. E. Gover, courtesy National Air Pollution Control Administration.)

bottoms and elevations (hills, mountains, etc.) may form pockets in which pollutants become concentrated and the general effects most noticeable. Tall chimneys or stacks are used in industry and often discharge their effluent far above ground level. Unlike water pollution whose dispersion is regulated chiefly by gravity, air pollutants may be carried by wind for long distances.

Automobiles give off hydrocarbons and nitrous oxides that, when exposed to the sunlight for a short time, yield ozone and other reactive compounds that become very irritating to the eyes (Fig. 39-2). These substances are also conducive to bronchitis and emphysema and can eventually lead to cardiac failure; that is, an extra load is imposed on the heart because it must pump the same amount of blood through the greatly reduced air-sac lining of the lung. The rising incidence of these diseases has been very striking in smog-ridden communities.

The 1970 Clean Air Bill required certain emission standards—a 90% reduction of 1970 levels of carbon monoxide and hydrocarbon emissions in 1975 and of nitrogen oxide emissions in 1976. However, in 1973 the Environmental Protection Agency found it necessary to extend the time limits on these emission standards by 1 year in response to the arguments of several automobile companies. Changes also have been made by industrial plants, especially in the control of sulfur oxides. New York has limited the sulfur content of fuel to 1%. Since soft coal has a sulfur content of about 2% to 3% (and sometimes more), the use of natural gas and clean fuel oil has been emphasized. However, the shortage of oil and the recent energy crisis may cause a temporary relaxation in some of the air pollution controls. Perhaps more progress has been made toward controlling air pollution than in any other field of our pollution problem.

That rain may be produced by the combination of smog and moisture has been shown by the increase in rainfall in certain Indiana communities, which is caused by the smoke fumes from the steel mills of Gary. The sun's heat may be held unnaturally close to the earth's surface, causing it to be too warm, or

FIG. 39-2
Automobiles crowd our freeways and produce enormous pollution of the air. (Photo by T. Spina, courtesy National Air Pollution Control Administration.)

the opposite effect may be produced by screening out the sun's rays.

Water conservation

The proper management of our water supplies involves an understanding of our water resources, the quantity and quality of water needed by a community, the causes and nature of its pollution, the proper drainage of surface water, the improved treatment of

water to make it fit for consumption, the development of new underground water resources, the potential of desalinization, and its general economic and aesthetic values. Increase in human population and increased use of water per capita have resulted in greater emphasis on augmenting the world supply of clean water.

Water resources include all forms of water—rain, snow, hail, ice, atmospheric vapors, soil moisture, as well as surface and ground water, and the water of streams, rivers, wells, and the ocean. About 97% of all water is contained in the ocean. Of the remainder about 2% is found in surface water and somewhat less than 1% is groundwater. A lesser amount of water occurs as atmospheric water. Water represents a renewable natural resource that is delivered from the atmosphere in the form of rain and condensation and is returned to the atmosphere by evaporation and transpiration. Water from streams, rivers, and lakes, seeps into the soil to be taken up by plants. Part of this water becomes ground and surface water that eventually flows through streams and rivers to the ocean. This is the regular cycling of water as propelled by solar radiation (p. 858). This cycling effect varies enormously from region to region, and variations occur even within the same region.

WATER QUALITY STANDARDS. All states at present have water quality standards that indicate the quality of water necessary for agricultural, industrial, recreational, municipal, and fish and wildlife usage. The United States government has implemented to some degree the Federal Water Quality Act of 1965 and the Clean Waters Act of 1966 as advancements in a program of water conservation. The intent of these programs was to have water meet certain standards of good water by 1973.

Water quality is determined by many natural conditions, such as man's activities, land use, waste disposal, pesticides, storage facilities (reservoirs), and many other features. Water may be naturally acid or alkaline, hard or soft, low or high in minerals, clear or discolored, and varied in temperature and in other ways. Some water is productive of biologic life, whereas other water may be injurious to most life (Fig. 39-3). Manure on soil is useful in the production of crops; in water it favors the growth of algae and consequent loss of fish and other useful wildlife. The accumulation of organic and inorganic debris in some of our larger lakes (for example, Lake Erie) is

FIG. 39-3
Some of the thousands of fish that died in polluted streams of Illinois in 1967 are shown washed up on the bank of a stream. These victims are mostly of the "rough" species, considered most resistant to common water pollutants. (Courtesy Federal Water Programs, Environmental Protection Agency.)

producing the condition known as **eutrophication**, in which an abundant accumulation of nutrients promotes a dense growth of plants and plankton (Fig. 39-4), the decay of which depletes the water of oxygen. Glaring examples of pollution are too well known to need repeating here. Pure, uncontaminated water at present is found only in restricted mountain regions, usually far from human habitation and industrial development.

AQUATIC THERMAL POLLUTION. Most animal life appears to have a temperature tolerance range of somewhere between 0° and 45° C. Fish are rarely found in water above 30° C., and water at a temperature of 30° to 35° C. may be considered a biologic desert for most organisms.

The principal offenders in producing thermal pollution of water (Fig. 39-5) are electric-power plants that daily discharge warm water that has been used to cool their steam generators, into lakes and rivers. In small streams in summer the heated water plus the climatic heat can easily raise the water to a temperature unsuitable for fishes. High temperatures also promote the growth of plants, algae, and plankton, all of which consume oxygen not only while alive but also after they die and decay, so that the oxygen of water may be depleted below the level sufficient for animal life. Much is yet to be learned about the effects of such temperature increases on aquatic life.

HEAVY METALS. The threat of mercury pollution

FIG. 39-4
A vast expanse of dead algae disfiguring Montrose Beach, Chicago, aiding in the pollution of Lake Michigan. Treated and untreated sewage discharged into the lake contains nutrients that encourage algae growth. (Photo by John Hendry; courtesy Federal Water Programs, Environmental Protection Agency.)

FIG. 39-5
Flow of thermal waste from a steel mill in Trenton, Michigan. This heated effluent is poured into the Detroit River, a tributary of Lake Erie. (Courtesy Federal Water Pollution Control Administration, U. S. Department of the Interior, Washington, D. C.)

in streams and lakes was brought to public attention in North America early in 1969 by Canadians who reported mercury levels in the fish caught in Lake St. Clair to be far in excess of Public Health Service limits. Their investigations blamed the source of the mercury pollution on chlorine-caustic plants that were discharging several hundred pounds of mercury waste each day into Lake St. Clair and Lake Erie. Subsequent investigations revealed the problem to be widespread in North America.

When mercury waste in any chemical form is discharged into water, bacteria in the bottom mud convert it into highly toxic methyl mercury. Methyl mercury may become concentrated in fish and other living things. Mercury concentrations become further biomagnified through the food chain; predators at the top of a food chain (for example, man) may receive the most. Methyl mercury may produce severe and irreversible neurologic damage.

The need is great for research on all aspects of the mercury question, as well as on other heavy metals, such as zinc, cadmium, and lead, that pose health problems. For example, very little is known of the long-term effects of mercury on fish, the group that

has been getting the most unremitting exposure to mercury. This is especially so for the vulnerable coastal streams through which man's wastes flow and where most of the world's commercially important fish are concentrated.

CONTAMINATION OF THE OCEAN. The dumping of waste and raw sewage into the coastal waters of our oceans is worldwide. Nuclear explosions have been responsible for the accumulation of isotopes in ocean waters, and oil pollution is on the increase (Fig. 39-6). The wreck of certain oil tankers has proved one great source of oil spills. One of these was the *Torrey Canyon* off the shore of England that was so devastating to marine life there. The damage to marine life was compounded by the use of detergents (themselves harmful to marine life) in the treatment of the oil spill. Spontaneous eruption of oil may also be a source of oil spills. The disposal in August, 1970, of nerve gas (no longer necessary as a chemical warfare agent) in the ocean off the coast of Florida by the U. S. Department of Defense may represent another example of marine water pollution. Although encased in steel and concrete blocks, the gas will be freed eventually into the

FIG. 39-6

The result of an oil spill on the bathing beaches of the York River at Yorktown, Virginia. Over 1,000 ducks died from the effects of the pollution. (Courtesy Federal Water Quality Administration, U. S. Department of the Interior, Washington, D. C.)

seawater by corrosion of the containers and may be toxic to sea life until rendered nontoxic by hydrolysis with seawater.

Effects of pesticides

From time immemorial insects have posed one of the greatest problems of our ecosystem. In many ways they are our greatest competitors for the things we need to survive—our agricultural crops, our plants, our forests, and our livestock. They also afflict us by acting as vectors of diseases, as sources of irritation, and in other ways. But all living things are interdependent and form a highly integrated web of relationships so that disturbances in a single group may profoundly affect the others in this web of life. The balance of nature is being threatened as never before by the chemical pesticides and herbicides man employs to control insects and weeds.

One of the basic laws of ecologic control is that if the environment is simplified to a few species, the whole web of life and its interwoven relationships are destroyed. Agricultural development with its emphasis on a few select plant crops has greatly encouraged certain insect pests and upset the balance.

The use of pesticides to control insects and plants probably stemmed back to an early time. But it was not until this century that the development of pesticides mushroomed into fantastic proportions. DDT's effective control of lice and mosquito vectors of diseases (typhus and malaria) during World War II was dramatic, and its use has since been widespread. Millions of lives have been saved by the use of DDT and other pesticides in controlling these and other arthropod-borne diseases.

Prior to the development of DDT most insecticides were obtained from plants and were nonpersistent (rapidly degraded). However, the great impetus given to chemical research during World War II resulted in persistent chlorinated hydrocarbons, such as dieldrin, lindane, endrin, and others (altogether about 300 pesticides are known at present). Millions of pounds of these chemical pesticides have been scattered over the landscape in this and other countries to control insects and weeds. At present it is estimated that more than 1 billion pounds of DDT alone can be found in our biosphere. The result has been that it is concentrated in plants, is washed into the sea from agricultural crops, and has accumulated in sea plankton. It becomes further concentrated in the fish that feed on plankton and finally reaches high concentrations in the birds that feed on insects or fish. Its increase is thus greater in those organisms high in the food chain. Its concentration increases tenfold as it passes from one trophic level to the next, until, after much biomagnification, it is returned to us in animal fat or milk.

In 1962 and 1964 there appeared two books that were instrumental in alerting the public to the danger of pesticides. The first of these books was by Rachel Carson entitled *Silent Spring.* This classic book has had an immense influence in formulating public opinion about the dangers of pesticides and what they are doing to our ecosystem. The factual evidence presented by *Silent Spring* and thoroughly documented from the scientific literature showed how fish were poisoned by endrin used by farmers against the boll weevil in the Mississippi valley, how fishes far from our coastlines had traces of the insoluble DDT, how pesticides caused the death and the loss of fertility of birds, and how the use of pesticides had completely upset the biotic community, resulting in the loss of useful species and the increase of harmful ones, etc. The report of the President's Scientific Advisory Committee in 1965 tended to confirm the statements made by Miss Carson. Naturally, the book has been the subject of fierce controversy between those who favor the use of pesticides and those who do not.

The other book, *Pesticides and the Living Landscape,* was by Dr. R. L. Rudd. In this work Dr. Rudd has given more attention to scientific details from which he has drawn logical conclusions. It covers a wider range of the ecologic impact of the use of pesticides, but in general confirms the statements made by Carson's *Silent Spring.* Neither author recommends the abolition of pesticides, but urge greater responsibility in their control and ask that prospective users weigh all possible results.

An enormous amount of factual information has steadily accumulated on the general effects of pesticides. In Florida when sand flies (target objects) were sprayed with dieldrin, the nontarget fishes and fiddler crabs were also killed by thousands. Wurster et al. (1956) reported that 70% of the robin population was destroyed when elm trees in the eastern United States were sprayed with DDT for the Dutch elm disease. After Clear Lake, California, was sprayed with pesticides for gnats, an analysis of the plankton showed that it contained the pesticide in a concentration several hundred times greater than the concentration in which it was originally applied. Frogs and fishes showed an even greater concentration. Grebes that ate the frogs and fish were almost completely eliminated from the lake where formerly they had nested by hundreds. Peregrine falcons (duck hawks) have been completely eliminated in many communities where once they were common, and DDT and dieldrin are believed to be the cause.

Pesticides in high concentrations cause fragile egg shells that burst prematurely in the incubation process, killing the embryo. DDT and its metabolite also cause the liver to synthesize enzymes that lower estrogen levels in the blood, thus interfering with normal breeding. It is not known just what are the effects of pesticides on the human, although some speculations implicate them as a possible cause of cancer.

Herbicides or chemical weed killers are not as dangerous in general as are the insecticides. It is possible that heavy dosages in spraying will also affect nontarget or useful plants. The chlorinated herbicide 2,4-D has been widely used, but altogether more than 40 different kinds of herbicides are available.

The use of many pesticides is being restricted at present. DDT has been outlawed by all states, and dieldrin and other pesticides may follow the same fate.

Biologic control of insect pests whereby certain insects prey upon others may be part of the answer to the pesticide problem. In one study the release of 200,000 aphid lions (parasitic insects of the family Chrysopidae) per acre was as effective in controlling the bollworm as the standard insecticides. Some 20 other parasitic or predator insects have been useful in controlling destructive pests. So far, no damaging effects on the environment (other than their specific action on the pests) have been reported. The technique of "sterile male release" may also be a solution in certain areas. In this method a large number of male insects are reared and made sterile by ionizing radiation and then are released. When many female insects mate with sterile males the fecundity of the population as a whole will be reduced. Such a method has proved successful in some instances (screwworm fly). Instead of rendering the males 100% sterile (which reduces their mating competitiveness), a method of partially sterilizing the males with re-duced gamma radiation has been found to be even more effective. Another method is to treat insects (or their eggs) before they become adults with the juvenile hormone that renders the embryos nonviable.

Efforts are being continued to find insecticides as effective as the wide-spectrum chlorinated hydrocarbons, but having more specific toxicities (killing only insect pests instead of all insects including the beneficial ones). Organophosphates have the advantage of being quickly degraded (nonpersistent), but when first applied are even more poisonous to man and other animals than the chlorinated hydrocarbons.

Effect of pollution on health

There are many biologic unknowns in the effects of pesticides and pollutants on our ecosystem. Individuals, for instance, may become adjusted to some extent to smog and polluted air, so that their health and behavior seem unaffected. This feeling, however, is a delusion for the body tissues may be building up contaminants while that person is unaware of what is happening to his body. The aged and the young are perhaps the ones who show the effects of pollution most quickly, especially pollutants affecting the respiratory system. The cumulative effects of irritation resulting in chronic bronchitis and even cancer may demonstrate this delayed response better than some others, although it is often difficult to relate the effect to the cause.

In the industrial communities of America it is firmly established that chronic pulmonary disorders are related to the degree of pollution. The incidence of such diseases increases with the length of exposure and with the concentration of the pollutants so that the worst effects of pollution may not show up until much later. In a quantitative way emphysema has increased from 1.5 persons per 100,000 population in 1945 to 15 persons in 1964. One may argue that some of this increase is caused by the steady rise in numbers of the aged, for emphysema is most common in the higher age groups, but this age correlation does not explain why emphysema is more common in those communities where air pollution is greatest. The death rate (in the United States) from bronchitis and emphysema is now known to be nine times higher than it was in 1949.

Contaminated water is responsible for the spread of infectious hepatitis, but little is known about how this operates. Radiation exposure is known to cause tumors of the thyroid gland, as was demonstrated by the Bikini tests of 1954, although this is an example of delayed response to rather high radiation levels. Man can acquire pesticide residues by consuming food chain organisms that have accumulated high levels of such residues. DDT and other chlorinated hydrocarbons have a striking resistance to bacterial action and have been accumulating in the water and soil. However, levels of pesticides evidently harmless to man may adversely affect birds and fish. Many studies show that pesticides are stored in human tissue. Even though many of these pesticides are banned or will be banned, those that have already been used will continue to be accumulated in animal tissue for some time.

Another alarming factor is the noise produced in our technologic society by everything from supersonic booms of high-powered jets, industrial machinery, and the automobile to sirens, lawn mowers, air conditioners, and household appliances. Decibels are units used to measure the intensity of sound waves. The rustling of a newspaper in a silent room measured 15 decibels and the honk of a horn 90 decibels. We can tolerate up to 80 decibels fairly comfortably, but continuous exposure above 85 makes us uncomfortable and can result in ear damage and loss of hearing. Today's four-engine jets blast off at 155 decibels. The noise in an electric power plant in which men work 8 hours a day was measured at 118 decibels. Music at a discotheque measured 122 decibels. Research indicates that persistent noise, even though supposedly of tolerable levels, can cause ear damage. The heart rate of an unborn child is accelerated by noises to which the mother has become tolerant. The human is very adaptable and may become apparently adjusted to a noise range, but such adjustment may occur at the cost of loss of finer degrees of hearing. Noise not only causes deafness, it also is suspected of contributing to heart disease, high blood pressure, and stomach ulcers. The general effects of noise pollution on mental health are presently unknown.

■ CHALLENGE OF THE ECOSYSTEM

The foregoing account of our ecosystem presents a gloomy picture. Man has made great strides in his technology, but in his economic zeal he has often been unable to comprehend the overall picture. He may destroy obnoxious insects but in so doing he may destroy beneficial animals as well. He has little understanding of the web of life and its intricate relationships so that when he interferes with one organism in the chain of relations he may unwittingly disrupt the entire web. Much of our precious land is being exploited for technologic development without regard to the consequences.

In a revealing article in *Science* for Nov. 18, 1969, entitled "What we must do," Dr. John Platt suggests ways to check this crisis of transformation. He shows that the human race is undergoing a transition to new technologic power all over the world. Some persons believe that because man does not realize how enormous and disruptive his powers are, he may not be able to harness and control his energies soon enough to ensure his survival. Others believe there is reason for optimism. John Platt foresees the possibility of a "world of incredible potentialities for all mankind." He deems it necessary to mobilize competent experts and scientists, as is done on solving great crises in wartime, for solving the problems that we are just now beginning to understand.

Aldo Leopold, a conservationist of the first rank, urged that one of the requisites for an ecologic comprehension of land is an understanding of ecology. Ecologic concepts must be incorporated into our school systems at all levels. The development of what Leopold calls a land ethic is an intellectual as well as an emotional process. Solution of the problems of the ecosystem must reach the ordinary citizen; the role of the specialist in the ecologic discipline can accomplish little without a thoroughly aroused and dedicated mankind as a whole behind the movement. It is everybody's problem. Ecology may be the most important discipline in all sciences for human survival, but at present it is least understood and studied by a too indifferent mankind.

The literature on the problems of the ecosystem is becoming very extensive, but anyone interested in the subject could very well start his reading with G. Hardin's "The Tragedy of the Commons," *Science* **162:**1243-1248, Dec. 13, 1968.

References
Suggested general readings

Bates, D. V. 1972. A citizen's guide to air pollution. Montreal, McGill-Queen's University Press. *Cloth or paperback.*

Bumpass, L., and C. F. Westoff. 1970. The "perfect contraceptive" population. Science **169**:1177-1182.

Callison, C. H. (editor). 1957. America's natural resources. New York, The Ronald Press Co. *Each chapter in this small work has been written by a specialist and each author gives a brief summary of the main problems in his particular aspect of conservation. A good introduction to a study of this field; more comprehensive treatises can be followed up later.*

Carson, R. 1962. Silent spring. Boston, Houghton Mifflin Co. *A book that has aroused a fierce controversy over the use of pesticides.*

Commoner, B. 1971. The closing circle. New York, Alfred A. Knopf, Inc. *A pessimistic view of the conflict between the polluting technology and the environmental movement by one of our leading environmentalists.*

Dasmann, R. F. 1959. Environmental conservation. New York, John Wiley & Sons, Inc. *An ecologic appraisal of conservation problems. To read this excellent book is to become a conservationist in spirit.*

Easton, R. 1972. The black tide. Los Angeles, The Delacorte Press. *An account of the Santa Barbara oil spill.*

Gould, R. F. (editor). 1972. Fate of organic pesticides in the aquatic environment. Washington, American Chemical Society.

Hardin, G. 1964. Population, evolution, birth control. San Francisco, W. H. Freeman & Co.

Hardin, G. 1966. History and future of birth control. Perspect. Biol. Med. **10**:1-18 (Autumn).

Hardin, G. 1968. The tragedy of the commons. Science **162**:1243-1248 (Dec. 13).

Hickey, J. J. 1969. Peregrine falcon population: Their biology and decline. Madison, Wis., University of Wisconsin Press.

Hoover, H. 1968. The long-shadowed forest. New York, Thomas Y. Crowell Co. *A delightful account of wilderness ways and an appreciation of the ideal balance of nature.*

Huth, H. 1957. Nature and the American: Three centuries of changing attitude. Berkeley, University of California Press. *Traces the development of the conservation movement. Shows how a love of natural scenery aroused an interest in the natural resources of America.*

Huxley, Sir J. (editor). 1959. The destiny of man. London, Hodder & Stoughton, Ltd.

Krenkel, P. A., and P. L. Parker (editors). 1969. Biological aspects of thermal pollution. Nashville, Tenn., Vanderbilt University Press.

Leopold, A. 1949. A sand county almanac. New York, Oxford University Press. *A great conservationist points out many of the problems that must be solved for any comprehensive conservation plan.*

Maddox, J. 1972. The doomsday syndrome. New York, McGraw-Hill Book Co. *Some problems of overpopulation and sources of pollution.*

Marx, W. 1971. Man and his environment: Waste. New York, Harper & Row.

Mountfort, G. 1958. Wild paradise. The story of the Coto Donana Expeditions. Boston, Houghton Mifflin Co. *Coto Donana is a wild, unspoiled wilderness in the southwestern part of Spain and is noted for its many varieties of wildlife.*

Ng, L. K. Y., and S. Mudd (editors). 1964. The population crisis. Bloomington, Indiana University Press. *A treatise by many authorities on the problems of population and their bearing on the food and other resources of the world.*

Panofsky, H. A. 1969. Air pollution meteorology. Amer. Sci. **57**:169-185 (Summer, 1969). *Explains the role of the atmosphere in the distribution of pollutants.*

Platt, J. R. 1969. What we must do. Science **166**:1115-1121. *A revealing account of our many ecologic problems and what must be done.*

Ritchie, J. 1920. The influence of man on animal life in Scotland. Cambridge, Cambridge University Press. *This work has had a dramatic influence on conservation problems the world over, for the author has traced the influences of man on the fauna of a restricted compass and has shown how this fauna has reacted to this influence. This treatise has served as a model for later similar works.*

Ross, R. D. (editor). 1972. Air pollution and industry. Princeton, Van Nostrand Reinhold.

Sears, P. B. 1935. Deserts on the march. Norman, University of Oklahoma Press. *An ecologic interpretation of the causes of the "dust bowl" regions.*

Sears, P. B. 1957. The ecology of man. Condon Lectures. Eugene, University of Oregon Press. *The eminent ecologist shows how man can apply his vast biologic knowledge to improving his ecologic relationships.*

Thomas, W. L., Jr. (editor). 1956. Man's role in changing the face of the earth. Chicago, University of Chicago Press. *This ponderous volume is the outcome of a symposium of the Wenner-Gren Foundation of Anthropological Research. More than 50 eminent authorities contributed to the work, which represents a far-flung picture of man's influence on the earth's resources and how this has affected man's cultural patterns.*

White, L., Jr. 1967. The historical roots of our ecological crisis. Science **155**:1203-1207.

Woodwell, G. M. 1970. Effects of pollution on the structure and physiology of ecosystems. Science **168**:429-433.

Woodwell, G. M., P. P. Craig, and H. A. Johnson. 1971. DDT in the biosphere: Where does it go? Science **174**:1101-1107.

Selected *Scientific American* articles

Chisolm, J. J., Jr. 1971. Lead poisoning. **224**:15-23 (Feb.).

Clark, J. R. 1961. Thermal pollution and aquatic life. **220**:19-27 (March).

Davis, K. 1965. The urbanization of the human population. **213**:41-51 (Sept.).

Deevey, E. S., Jr. 1960. The human population **203**:194-204 (Sept.).

Edwards, C. A. 1969. Soil pollutants and soil animals. **220**:88-99 (April).

Eliassen, R. 1952. Stream pollution. **186**:17-21 (March).

Energy and power. 1971. Vol. 224 (entire Sept. issue). The issue is devoted to the role of energy in human life, past, present, and future.

Goldwater, L. T. 1971. Mercury in the environment. **224:**15-21 (May).

Haggen-Smit, A. J. 1964. The control of air pollution. **210:**24-31 (Jan.).

Huxley, J. 1956. World population. **194:**64-76 (March).

Kermode, G. O. 1972. Food additives. **226:**15-21 (March).

McDermott, W. 1961. Air pollution and public health. **205:**49-57 (Oct.).

Newell, R. E. 1971. The global circulation of atmospheric pollutants. **224:**32-47 (Jan.).

Nevers, N. de. 1973. Enforcing the Clean Air Act of 1970. **228:**14-21 (June).

Went, F. W. 1955. Air pollution. **192:**62-72 (May).

Wynne-Edwards, V. C. 1964. Population control in animals. **211:**68-74 (Aug.).

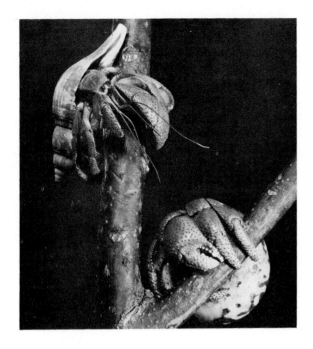

Unexpected crab behavior. Tree-climbing by land hermit crab, Coenobita clypeatus. *(Courtesy R. C. Hermes, Homestead, Fla.)*

CHAPTER 40

ANIMAL BEHAVIOR PATTERNS

■ SIGNIFICANCE OF BEHAVIOR*

Every kind of organism has its characteristic pattern of response to changes in its environment. Even the simplest form has many responses often so complicated that no one has been able to puzzle out the stimulus-response processes involved. A given behavior pattern may start in response to a definite external change or stimulus, or it may originate from internal stimuli. Many animals may initiate a behavior pattern without any apparent reason at all. A response may take place immediately after an animal is disturbed, or it may be delayed for a considerable time after the stimulation.

Levels of organization also complicate the problem. It is obvious that a vertebrate animal has a greater complexity of nervous and other systems in-

*Refer to pp. 21 to 23 for Principles 24, 29, 33, and 34.

volved in behavior than that of many low forms of life. Most animals also have social responses of behavior as well as individual ones.

Science implies conceptual schemes of general principles with wide applications, and it is very doubtful that the study of behavior has progressed this far. The vigorous controversies that our leading behavior students have over even the simplest type of behavior indicate how observations and experiments must be extended before general laws and principles can be formulated. This does not in the least detract from the impressive work that has been done or is being done in the field of behavior.

Animal behavior can never become static because the animal is always in the process of adjustment to the environment, which is continually changing, often at an imperceptible rate. There are also pressures for more successful adjustment constantly confronting animals. Just as animals show morphologic evolutionary changes in time, so sequences of behavior patterns also occur. Not all aspects of behavior can be traced to a specific genetic code, but there is a certain flexibility of behavior under various circumstances. No genetic code can anticipate all adjustments that an animal must face in its life-span.

■ HISTORIC BACKGROUND

Primitive people have not always distinguished sharply between themselves and other animals with regard to emotions, feelings, and understanding. According to the eminent American psychologist Schneirla, two views have developed regarding man's relations to the lower animals. One view emphasizes differences and ignores similarities between man and the so-called brute world; the other, which is more modern and an outcome of evolutionary thought, analyzes in a comparative way both similarities and differences. Aristotle was one of the first to record descriptions of animal behavior in which he set off rather sharply man's reasoning powers against that of lower animals. Roman writers, such as Pliny and Plutarch, also recorded observations on the intelligence of animals. In more modern times, Erasmus Darwin, Lamarck, Herbert Spencer, Charles Darwin, and many others made important observations on animal instincts and intelligence.

Significant behavior studies were made only when animal activities were analyzed in objective terms. Anecdotal and anthropomorphic methods (for example, ascribing human attributes to other animals) were of little or no significance. The study of evolutionary development gave a great impetus to experimental testing and control methods of analyzing behavior. Two early investigations deserve special mention—the outstanding work of E. G. and G. W. Peckham on the instincts and habits of the solitary wasp and the work of C. O. Whitman on the behavior of pigeons.

J. Loeb and C. Morgan did much to develop the study along the lines followed by present day investigators. Morgan (1894) gave an important principle known as "Morgan's canon," which states that an animal's behavior pattern should be interpreted in terms of the simplest explanation that meets the facts involved. Loeb advanced his theory of forced movements (tropisms), or the reactions of an animal in response to a difference in stimulation on its two sides. Both these investigators tried to explain behavior, at least in part, on the basis of physicochemical principles. H. S. Jennings' theory of trial and error was in conflict with Loeb's tropism theory, and his study of the behavior of the lower organisms represents an important advance in this field. In the early part of the present century I. P. Pavlov demonstrated his famous conditioned reflexes, wherein he showed that basic physiologic functions and behavior of an animal could be modified by associated experience.

G. E. Coghill, who carefully traced throughout all stages of the developing vertebrate (salamander) embryo the emergence of correlated movements and nervous connections, laid the basis for a structural interpretation of behavior. He showed, among other things, how broad, general movements preceded the appearance of more specialized local reflexes because of the delay in the development of the nervous connections for the latter.

The pronounced revival of interest in animal behavior in recent years has been due mainly to the researches of two European investigators, K. Lorenz and N. Tinbergen. Their theory of instinctive and innate behavior has attracted attention everywhere because of their fresh outlook. In fact, they call their approach to behaviorism **ethology,** the comparative study of the physiologic basis of the organism's reaction to stimuli and its adaptations to its environment. They have tried to explain innate behavior by investigating the stimuli that control it and by studying the animal's internal conditions that are organized for particular patterns. Their studies have greatly stimulated other competent workers, such as D. S. Lehrman, T. C. Schneirla, and W. H. Thorpe, to undertake similar investigations, either in confirmation or in refutal of the theories Lorenz and Tinbergen have proposed.

■ METHODS OF STUDYING BEHAVIOR

Animal behavior work involves both laboratory experiments and field observations. Testing must conform to the standard scientific procedure of the control experiment in which conditions are kept as uniform as possible, except in the one environmental factor (stimulus, etc.) that is being studied. So many variable factors may enter into behavior studies, such as age, physiologic conditions, hormonal balance, insidious disturbances, individual difference of intelligence, and number of subjects studied, that one cannot make sweeping generalizations. The experimentalist must know the nature and normal responses of the animals being studied. A raccoon, for instance, is far more adept in manipulating its forelimbs than a cat, and experiments that involve the use of this limb must therefore take this into account. An animal can organize its behavior capacities only within the range of its abilities.

Since the nervous system is mainly responsible for the coordination of behavior, many investigations have been conducted to determine the neurologic basis of behavior. Several methods may be employed. Certain areas of the brain may be removed surgically or destroyed and the resulting functional deficiencies noted; electric stimulation of brain regions is effective in causing responses in muscles and other effectors; and it is possible to use the electroencephalogram (for detecting electric discharges) to find those parts of the nervous system that are functioning under a given condition. By such methods it has been possible to determine important nerve centers, such as the cocoon-spinning center (corpus pedunculatum) of caterpillars, the satiety center of mammals, motivation centers, and many others. Coghill's great work emphasizes the parallelism between the emergence of behavior patterns and the growth of nervous connections. Such a study combined both a physiologic method and a microanatomic method.

An exciting new development in understanding animals in their wild state is the method of **biotelemetry,** or radio-tracking. This technique involves the attachment of a transistorized radio transmitter to an animal and then recording the data given off by signals while the animal is undisturbed and freely functioning under its natural conditions. Many physiologic aspects can be obtained this way, such as body temperature, wing beats, and other activity states. Radio-telemetry has been used successfully with ruffed grouse, grizzly bears, rabbits, reptiles, amphibians, etc. Factors of the animal's environment can also be studied by this method.

The greatest pitfall in behavior studies is interpretation, for there is the tendency to attribute humanlike reasons for an animal's activities. When an animal does something humanlike in nature, it does not mean that the animal is thinking like a man. It is reacting in accordance with its own basic behavior

patterns, which in turn depend on the organization and degree of complexity of its own nervous system. A bird thinks like a bird, a dog like a dog, and an anthropomorphic interpretation is unjustified in either case.

Another pitfall is the ascribing of purpose to an animal's reactions. It is true that many of its behavior patterns are adaptive, but this does not mean that animals perform these acts with an understanding of the end result or with the ability to anticipate what the end is to be. Such an interpretation would involve human reasoning.

■ LEVELS OF NERVOUS ORGANIZATION

The behavior patterns of an animal largely depend on its type of nervous system. Complex behavior is related to highly organized nervous systems and superior sensory reception. It is only when the brain has advanced to the role of an organizing center that it can truly be thought of as regulating and controlling behavior organization.

The trend of evolution in the nervous system beyond the protists is centralization. From this standpoint most animals fall into one of three major types of nervous systems—nerve net of coelenterates, nervous systems with beginnings of brains, and centralized nervous systems.

NERVE NET OF COELENTERATES. There is very little centralization in this type, for a reaction to a stimulus may spread over the entire net and cause the animal to act as a whole. This diffuse type of conduction does allow some coordination, for a slight stimulus may cause only a single tentacle to react, whereas stronger ones may involve other tentacles or even the entire body.

NERVOUS SYSTEMS WITH BEGINNINGS OF BRAINS. There are several kinds of this type of nervous system. Mollusks have paired masses of ganglia located in the head, foot, and viscera that are interconnected by nerves. In the cephalopods there is a definite concentration of ganglia in the head. Nematodes and nemerteans have ganglia in the head region with usually several longitudinal nerves. Planarians, with two longitudinal nerves running from paired ganglia (brain) in the head, have rather complex behavioral reactions and can be taught simple processes. The evolution of the nervous system in higher forms may be considered a modification and an elaboration of the planarian plan.

CENTRALIZED NERVOUS SYSTEMS. This type consists of a brain, or aggregation of ganglia in the head, from which runs a centralized nerve cord or cords (with ganglia) and includes the higher invertebrates (annelids and arthropods) and the vertebrates. There is a wide diversity of centralization and coordination in a range from the earthworm to mammal, but it is in such a type that we find the highest development of nervous coordination and behavior patterns.

In summary, the evolutionary trends that promote the capacity for increasingly complex organized behavior are (1) the development of centralized control by concentration of nerve cells in ganglia of the brain and nerve cords; (2) the differentiation of neurons of more or less polarity (carrying impulses in one direction only) arranged to form a mechanism of coordination (reflex arc); (3) increased variety and richness of nerve pathways and associations suitable for precision and variation of specialized behavior; and (4) the development of the sensory capacities, which depend on sense organs of high complexity, sensitivity, and range of response.

Are animals conscious of what they do? Do they have subjective awareness? Sir J. H. Huxley has pointed out that the major tendency in the evolution of mind is the trend toward a higher degree of awareness. Deductions about what goes on in the minds of other animals, including man, can only be inferred from one's own subjective experience. One can detect experimentally the sensory limitations of others and from these draw conclusions about their restricted potentialities, but this tells little about their mind's processes. Huxley considers that self-awareness is a natural and gradual development of the potentiality found in protoplasm and that the brain is a psychometabolic organ, whose functions have been to transform the raw materials of experience into special systems of organized awareness.

■ SIMPLE BEHAVIOR PATTERNS (TROPISMS AND TAXES)

The simplest form of organized behavior is one in which a specific stimulus gives rise to a specific response. This type belongs to what is called inherited behavior patterns and is best represented perhaps by the **tropisms** of plants and the **taxes** of animals. Early embryos and perhaps sponges have a form of organized response because they have poorly developed nervous systems or none at all. (Irritability, you may recall, is a property of all protoplasmic systems.) Most of the behavior patterns of plants are tropistic responses. A tropism, which literally means a "turning," refers to the bending movements of plants brought about by differences in the stimulation of the two sides of an organ (stem, root, etc.). Tropisms involve two aspects: a definite direction caused by a difference in stimulation intensity and a rigid hereditary pattern not subject to modification.

The term **taxis** is used by zoologists to describe the movements of freely swimming forms, such as protozoans. Description of various taxes is given in Chapter 7, p. 143.

In the early part of the century J. Loeb and his school tried to interpret all animal behavior on the basis of tropisms, or taxes. He explained the orientation of multicellular animals to light as being the result of differences in muscular movements brought about by a faster contraction of the less illuminated side (or eyes) so that the animal curves toward the source of light (forced movement). When the light intensity is the same on both eyes, the animal goes in a straight direction.

Loeb's theory met with opposition from many sources. H. S. Jennings, in his now classic book *The Behavior of the Lower Organisms,* proposed a trial-and-error explanation in place of the tropistic theory. This is really a stimulus-response theory, for Jennings found that most environmental changes will produce a response. The avoiding reaction of paramecium is due not to unequal stimulation of its two sides but to a fixed orientation pattern that enables the animal to find a favorable escape channel. (See Chapter 7.) All organisms have the capacity for several different responses to the same external stimulus. As Jennings showed, a ciliate like *Stentor* will react in a highly variable way to a constant stimulus, such as carmine particles or ink; it will turn to one side, reverse its cilia, and finally retire into its gelatinous tube. When *Stentor* emerges again from its tube and is subjected to the same stimuli, it contracts again into its tube immediately, as though it remembered its previous experience. Another valid objection to Loeb's theory is that the exhibition of a response pattern may be delayed until the animal has attained a certain degree of maturity. Its responses, therefore, at one stage of the life cycle may be different from that at another.

■ HEREDITY AND BEHAVIOR

Behavior patterns are the result of the interaction between hereditary factors and the environment. No behavior pattern as such is found in the zygote. It must develop out of certain potentialities (or genes) that physiologic influences act on and limit at each stage of development. The so-called inheritance of behavior thus falls into line with the modern concept of genetics that genes and somatic expression are not in a direct relationship. Behavior patterns involve many factors such as nervous integration, hormone balance, and muscular coordination. Many genes must therefore be responsible for even the simplest activity of an animal. It is often difficult to determine whether heredity or learning experience is more involved.

Some types of activity are rather definitely triggered. Each species of bird builds its typical nest without being taught; the parasitic cowbird raised in a warbler's nest never tries to mate with a warbler; a spider weaves its web without learned modification; a stickleback fish always performs its courtship ritual the same way; etc. The influence of hereditary factors seem to be much more pronounced in lower than in higher animals. Many of the basic patterns of adaptive behavior in them seem to be stereotyped.

One must not forget that the sensory, muscular, and other mechanisms that limit and define behavior can be controlled or affected by heredity. Behavior patterns can also be influenced by the kind of endocrine system an animal inherits. Certain forms of dwarfism are caused by a mutant gene that produces an underactive pituitary gland. The pituitary is known to control growth.

■ TAXONOMY AND BEHAVIOR

In recent years emphasis has been placed on the relationship between behavior patterns and taxonomic units at all levels. Evolutionary relationships are clearly expressed by behavior similarities and differences that are correlated with taxonomic subdivisions. It is thus possible to study species differences on the basis of behavior characteristics. H. S. Barber was able to separate species of fireflies on the basis of differences in characteristic flashes emitted by flies of different populations. B. B. Fulton found four different populations of field crickets (supposedly one species) that would not interbreed (a behavior trait) in the laboratory, thus indicating the divergence and formation of four new species.

■ WHAT IS AN INSTINCT?

The concept of the term "instinct" formerly meant any form of innate behavior that arose independently of the animal's environment, that was distinct from learned behavior, and that followed an inherited pattern of definite responses. At one time there was believed to be a sharp line between instinct and learned experience, and any action of an organism was either instinctive or learned. In the evolutionary process instinct was supposed to be the primitive plan of behavior patterns; intelligence and the learning process came later. Psychologists at present believe that most behavior must be interpreted in terms of both innate traits and learning. Few behaviorists are willing to concede that a particular activity is wholly instinctive or is wholly learned.

Some behaviorists rightly argue that behavior cannot be inherited through the genes of the chromosomes but must develop under the influence of environment. Certain types of behavior, it is true, can be modified more than other types. Nest building among birds is unlearned, yet older robins build better nests than younger robins (Allee). Some birds reared away from their parents will still sing the song characteristic of their species; other kinds of birds when raised with members of another species will sing the song of that species rather than their own.

Many students who have studied the habits of web building in spiders and the nest building of solitary wasps, as well as the behavior of higher forms, have found behavior very flexible and adapted to daily variations. Many acts that have been considered purely instinctive have been modified under the impact of unusual circumstances, such as the repair of a disrupted nest or web, decisions involving the choice of two or more alternatives, etc. So-called instinctive behavior appears to have both fixity and plasticity (W. H. Thorpe).

Lorenz and Tinbergen, the well-known European investigators, have stressed instinctive behavior as a stereotyped action that follows a definite pattern of expression. They believe that there are at least three components involved in an instinct. First, there is an **appetitive behavior,** which may be regarded as a buildup of readiness for the instinctive act. This phase is goal directed, concerned only with the actual performance of the act. Second, an **innate releasing mechanism** is activated. This refers to the removal of any inhibition for the performance of the instinctive behavior. Third, the **final consummatory act,** which might be considered the relief of the animal's tension by the actual discharge of the activity. This pattern of instinctive behavior might be illustrated simply by the reactions of a hungry young bird in a nest when something is waved before it. The bird is in a condition for response (appetitive); the movement of the object activates the release mechanism; and the lunge and gaping that follow is the consummatory act.

This theory has been subjected to critical analysis by able American behavior students who believe that some of its concepts are too preconceived and lack experimental verification and that it is based too much on preformed, inherited behavior.

Perhaps in our present state of knowledge the best way to regard an instinct is that it is concerned with activities that depend mainly or wholly on an animal's organic equipment in reaction with the environment, with learning playing a minor role or else being entirely absent in the process.

■ INSTINCTIVE BEHAVIOR PATTERNS

An instinct is usually considered to differ from a taxis or tropism in being more complicated and involving more separate phases in the performance of the act. However, the two types of behavior may overlap. Both involve reflex action, but a taxis or tropism is less flexible and is based more on a rigidly inherited plan. Another way of expressing the difference is to state that a taxis is more innate, is based more on a specific neural mechanism, and is less modifiable in its expression. In studying any instinctive behavior it is necessary to determine the stimuli that control the behavior, to appraise the internal conditions that prepare the organism for the reaction, and to seek out the neural mechanism responsible for the integration of the whole basic pattern of behavior. No instinctive action can be explained without knowing the organic circumstances under which it occurs and the role environment has played. The organic factors that may be involved in instincts are the sensory equipment, endocrine system, neuromuscular system, etc. Another important influence on instinctive behavior is maturation, or the development of behavior patterns, as correlated with the age and growth of animals. The innate behavior of young animals for a particular act may be quite different from that of an adult, as Coghill so well demonstrated. To a certain extent this can be explained by the development of new types of connection in the nervous mechanism.

Striking examples of automatic behavior are perhaps best shown by the web spinning of spiders, the communication of honeybees, and the cocoon spinning of caterpillars. The animal involved performs its act with mechanical regularity step by step. Each stage of performance seems to serve as a stimulus for the succeeding stage. Environmental influences for a characteristic behavior to appear is demonstrated by the experiment of D. Lack on the European robin. He discovered that a male robin during its breeding season and when holding its territory will attack even a bundle of red feathers (which simulate the red breast of an actual male robin) but will do so only under the conditions mentioned. The same experiment also shows the effect of what is called **"sign stimuli,"** for Lack found that a tuft of red feathers would provoke an attack, whereas a stuffed young robin with a brown breast would not. In this case the red breast is the effective stimulus. A somewhat similar case of sign stimuli is the reaction of the herring gull chicks. These chicks beg food by pecking at the tip of the parent's beak where there is a little red spot. When this spot is painted out or painted some other color there is no releasing stimulus, and the chicks are confused.

■ MOTIVATION AND BEHAVIOR

Why do animals perform characteristic patterns of activity? What is the motive behind their behavior? Why do they act at all? A simple answer is the stimulus-response theory that may involve some change in the environment. But not all behavior can be answered as simply as this, for some are related to internal conditions that are not easy to appraise. One of the basic principles of life is the maintenance of stable internal conditions (recall the principle of homeostasis). Hunger, for instance, is accompanied by a low blood glucose level, an imbalance of fluids, etc. Such internal changes stimulate characteristic behavior patterns. A hungry hydra will behave differently from a satiated one. It is very difficult to understand many of the complex mechanisms of internal stimulation because they vary greatly.

Animals perform acts that either give them pleasure, or else prevent pain or unpleasant conditions.

They make their adjustments to these conditions perhaps wholly unconscious of the end results (this certainly would be true of the lower forms). This concept, as the student can see, is in line with the maintenance of internal stability as already described. The appetitive phase of instincts that Lorenz and Tinbergen stressed is supposed to furnish the chief source of motivation. According to their theory, animals actively seek out those stimuli that trigger their instinctive acts. Failure to find such outlets is supposed to create intense emotion in animals.

A clear understanding of these instinctive acts is furnished by the rather modern concept of **biologic drives.** A biologic drive may be defined as a motive for stabilizing the organism. Examples of such drives are hunger, which arises from the nervous impulses of an empty, contracting stomach; thirst, which may be due to sensations from a drying pharyngeal mucosa; and sex, which is caused by a release of hormones from sex glands. All of these upset internal stability, and restoration of this stability is a reward or motive. Most drives are also characterized by rhythm patterns, in which the drive fluctuates up and down in a periodic manner, such as the estrous cycle of the female mammal. Some drives are not as clear-cut, but a certain nervous pattern has to be satisfied, as, for instance, the weaving of a net by a spider.

In recent years great strides have been made in locating brain centers associated with motivation of particular kinds. W. R. Hess, a Swiss psychologist, was able to fasten fine metal electrodes into specific parts of the brain (by inserting them through the skull), and when the wound healed, the ordinary activities of an animal (rat) could be studied by giving electric shocks through these electrodes. By such means it has been possible to explore the brain and locate specific seats of emotions, such as pleasure, pain, sex, eating, and satiety. For instance, when an electrode was placed in a certain part of the hypothalamus and a rat had learned to stimulate itself by manipulating a lever (as in a Skinner box), it would press on the lever with great frequency. The conclusion was that this particular spot was a pleasure center. Other regions have been found in which stimulation is avoided by the experimental animal. Evidence from these and other experiments indicates that the hypothalamus is the chief center for many sensations and is where various drives lie. Whatever action is found here, however, is controlled and influenced by the higher centers of the cerebral cortex. The consummatory phase of an act, the way in which motivation is expressed, is controlled by the cerebral cortex.

■ REFLEX ACTION

A simple reflex act (reflex arc or circuit) is a ready-made behavior response in which a specialized receptor, when stimulated, transmits impulses to a specialized motor cell that arouses an effector (muscle or gland) to act. Most reflexes, at least in the higher forms, are more involved and include an association neuron, and the impulse must pass through a number of synapses. A reflex act may or may not involve a conscious sensation. Reflex acts are built into the body mechanism and control the automatic working of the internal organs. Most reflex arcs have the possibility of modification because there may be more than one channel of discharge. A complicated act may involve a chain of reflex acts.

Among lower invertebrates a reflex action has a local makeup and does not necessarily involve a more central or general control. The tube foot of a sea star, when severed from the body, will continue to react for some time, but there is central coordination of tube feet in such activities as locomotion or feeding.

Reflexes are less elaborate than instinctive actions. The latter also have a wider range of adaptability and variability. However, as Lorenz stresses, the releasing mechanism that triggers instinctive acts involves reflex action in its pattern. The idea that an instinct is a chain of reflexes is not rigidly held because of the varying intensity of instinctive behavior patterns.

Lorenz has also stressed the adaptive nature of all aspects of behavior, no matter how trivial they may seem. An animal's pattern of action, involving emotional displays of snarling, threats, courtship rituals, postures, symbolic gestures, etc., is the result of a long evolutionary integration of great effectiveness in adaptive functions. This concept is the very core of the science of ethology as propounded by Lorenz and his school.

■ LEARNED BEHAVIOR

Conditioned or learned behavior differs from inherited instinctive behavior in being acquired or modified from experience. Even the simplest animals are capable of learning. The distinction between unlearned and learned behavior is not easy. Behavior is never exactly predictable. The concept of learning must then have a variety of meanings. Ordinarily it refers to changes in behavior that are brought about by past experience and that involve more or less permanent modifications of the neural basis of behavior.

To what extent can learning be determined in the animal kingdom? With what accuracy can one say that this particular animal is capable or incapable of such and such a learning process? How valid are the conclusions of the almost endless experimentation that has been done on this problem? One of the greatest difficulties encountered in this field is the impossibility of supplementing objective behavior observations with subjective knowledge. The experimenter must depend on the use of stimuli and the way animals react to them for his knowledge of their behavior. By using controlled experiment, observation, physical analysis of the environment, discriminatory learning, etc., the investigator tries to ferret out just what the learning capacity is for a particular animal.

An animal's sensitivity determines how it reacts to its environment. Many animals are blind; others live below the surface of the ground in total darkness. Ants live in a world of odors, just as other animals live primarily in a world of light and shade. Honeybees are unresponsive to red colors when mixed with gray series but can be taught to distinguish most other colors of the spectrum when interspersed among shades of gray. Cats and dogs are insensitive to colors or hues. Bats emit and react to supersonic vibrations (inaudible to human ears). Male moths are attracted to female odors a mile away. Scores of other examples could be given to indicate that animals have many different sensory abilities, some greater and some less than that found in man.

CONDITIONED REFLEXES OR ASSOCIATIVE LEARNING. A conditioned reflex is substituting one stimulus for another in bringing about a type of response. It is often considered the simplest form of learning and involves a new stimulus-response connection. We owe this concept mainly to the Russian physiologist I. P. Pavlov, who noted that hungry dogs (and other animals) secrete copious amounts of saliva at the sight or odor of food. By ringing a bell (the conditioned stimulus) at the same time they saw food (unconditioned stimulus), it was possible in time for the conditioned stimulus alone to elicit the response of salivation. There was a definite limit to the number of factors to which an animal could be conditioned for a single response. Animals with higher nervous systems can handle more (and more complex) factors of conditioning, which is far more prominent among mammals and some birds.

SELECTIVE LEARNING (TRIAL AND ERROR). Selective learning is rarely found below the arthropods. It has nothing to do with an unconditioned stimulus and involves rewards and punishments. A dog, for instance, finds that a problem box containing food can be opened by pulling on a lever. To get to the food, the animal at first makes many random, useless movements (trial and error), but when it finally succeeds, trials later become fewer until it learns to open the box when confronted with the situation the first time. It has thus mastered a habit of appropriate response. Animals vary greatly in mastering problem boxes. Some do so with few trials; others require many.

Mazes and labyrinths are frequently used in selective learning in which a reward or punishment is involved with the right or wrong choice. The maze is often Y or T shaped. When the animal chooses the wrong passage, it confronts a blind end, which may involve punishment, or else it fails to achieve a reward, which it receives by choosing the right one. Even ants and earthworms can master this type of labyrinth after many trials, but some mammals can master the trick with very few attempts. It is a good test to determine an animal's capacity to use acquired behavior in new situations. The widely used Skinner box is based on a reward of food when the animal presses the right lever (Fig. 40-1).

INSIGHT LEARNING. Insight learning may be considered a modified form of trial and error. When a process is slowly learned by an animal, it is often called trial and error; when learned rapidly, it is insight, often called "abridged learning." It involves a solution to a problem after an initial survey of the elements involved, getting the idea on the first trial. The facility with which this is done often depends on previous experience with similar situations. The capacity for short-cut solutions is rare among most invertebrates but is common in higher mammals. A form of insight learning in birds was illustrated a few years ago in England, where great tits rather suddenly acquired the habit of opening milk bottles left on the

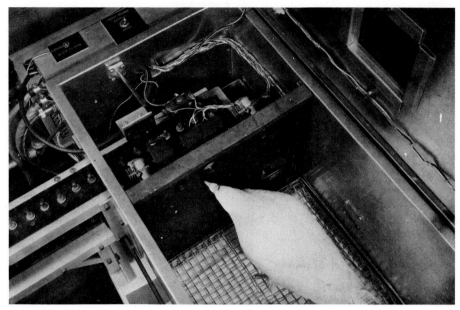

FIG. 40-1
Pigeon taught to do work of man. This bird was taught, by researchers for a drug
company, to sort out inadequately coated capsules by pecking the proper keys as
nonacceptable capsules were brought automatically into view through a tiny
window. Birds were rewarded by food whenever they spotted a defective capsule.
Most became expert inspectors in 60 to 80 hours and could easily detect minor
flaws overlooked by human inspectors. (Courtesy T. Verhave, Eli Lilly & Co., Indianapolis.)

doorsteps of households and feeding on the milk. The
widespread nature of the practice poses the question
as to how such a habit could be picked up by so many
of them in a relatively short time.

IMPRINTING. The concept of imprinting, formulated by O. Heinroth (1910), refers to a special type
of learning in birds. In the first hours after hatching,
a duck or goose is attracted to the first large object
it sees and thereafter will follow that object (man,
dog, or inanimate object) to the exclusion of all
others; such birds show no recognition of their
parents. When once accomplished, the behavior pattern is very stable and may be irreversible in some
cases. Other species of birds may show imprinting.
Some psychologists think the type of bird song young
hatched birds acquire when exposed to members of
different species may be of this nature. The process
indicates how learning may be restricted to a critical
period of the life cycle. According to Lorenz, when
a young bird is imprinted to a member of another
species, the imprinted bird will adjust its own functional cycles to that of its adopted parent. Other

psychologists, however, think that imprinting is
merely a strong early habit and that its socialization
to another species is very restricted.

■ BASIS OF MEMORY AND LEARNING

The most difficult of nervous functions, especially
that of the brain, are memory and learning. Little is
known about these processes and how they occur.
Capacities for memory and learning are associated in
general with the development of large regions of the
cerebral cortex. Experimentally there seems to be two
processes associated with learning and memory. One
of these operates within a brief time after an experience; the other is more delayed and involves an
assimilation of the experience to some degree. The
first process can be blocked by agents such as anes-
thesia, shock, and various drugs; the other is immune
to such disruptions.

The activity of certain networks of synapses involved in sensory input with conscious attention may
effect some long-lasting effect in their functional efficiency. Memories that occur long after the original

experience may be caused by the reestablishing of the original circuit of impulses. The continued passage of impulses across the synapses of a given circuit may cause a decrease in the resistance of those synapses, and thus facilitate succeeding nerve impulses that pass over that pathway. In this way a neural pathway, repeatedly used, might lead to a learning process.

In recent years, authorities believed that RNA might be stimulated to code transmitter-receptor systems at certain junctions. It may be possible that specific memories are coded in the central nervous system by specific sequences of nucleotides in RNA. There is some evidence that the synthesis of RNA is increased in cells involved in conditioned responses, and some investigators have reported that they may have transferred learning by taking RNA from trained animals and having untrained ones eat it. Not enough evidence has been accumulated to confirm these results, or what is actually taking place if such stimulation of memory is possible.

The outstanding work of the neurologist W. Penfield, on the cortex of epileptic patients, gives some insight into the sites of certain types of memory; he was able to localize highly specific and detailed memories in certain areas of the temporal cortex. The study of a simple nervous system in some animal suitable for investigation, such as that of the primitive nervous system of a planarian, has not given very satisfactory results. Some positive results have been obtained on the way drugs alter RNA metabolism. Much further investigation will have to be done before the biochemical and physiologic basis of memory is clarified.

■ NATURE OF ADAPTATIONS*

Nearly everything about an organism is an adaptation of some kind, for adaptations may be structural, physiologic, or behavioral. Although we often are amazed at some spectacular adaptation of an unusual nature, the adaptive nature of bodily processes, such as the precision of chromosome behavior in mitosis, the interrelations of hormonal balance, the teamwork of enzymes in metabolic processes, the intricate pattern of nervous integration, and many others should elicit our wonder still more. The entire living system is adaptive throughout its organization. All adapta-

*Refer to pp. 19 and 22 for Principles 14 and 31.

tions, from simple to complex, have been the result of the operation of natural selection on favorable mutations through long periods of time. By such processes, even the intricate mechanism of a sense organ such as the eye has been gradually evolved, provided that each step of formation has conferred some advantage to the organism and resulted in differential reproductive success.

Often within the same group of animals and among closely related species there are strikingly different ways or adaptive methods for meeting the problems of life. How they meet these adaptations depends on their organization levels and the potentials they possess for evolutionary novelties. All organisms are restricted in reaching a new adaptive level by the basic patterns of their ancestors. Many of the adaptations described in this section are interesting not only because they are striking and somewhat unique in their patterns but also because many of them have been extensively studied by observational and experimental methods for the purpose of discovering their basic mechanisms.

■ TOOL USING AMONG ANIMALS

The ability to use tools is often considered one of the major achievements of man toward the high evolutionary rank he now holds. Tool using among animals below man is very restricted. Yerkes found that chimpanzees manipulated certain tools in a manner that indicated they had a clear perception of what the tools were for and that their use of them was not a chance trial-and-error method. Many animals rather low in the scale of evolution display amazing feats of craftsmanship, but this usually involves manipulating the materials with bodily parts, beaks, feet, and jaws. In the relatively few cases known among animals the behavior is of an instinctive pattern and no high degree of intelligence need be assigned it. Its action appears to be stereotyped, fixed, and of great antiquity.

One of the early observations of tool using by animals is recorded by the Peckhams in their famous monograph on the solitary wasps. *Ammophila*, a sphegid wasp, seizes a small pebble in her jaws and pounds down (as with a hammer) the earth with which she closes up her burrow. P. and N. Rau reported the same behavior in different species of the same group. In some cases a stone is only used occasionally; more commonly they use only their head and jaws to tamp down the sand.

A widely publicized case of tool using among birds is one of Darwin's finches of the Galápagos Islands. Among these finches, which so well illustrate adaptive radiation, one has the habits of a woodpecker in probing into crevices of bark and trees. To overcome the handicap of a short beak the bird holds a stick or thorn in its beak to pry out its prey. The bower bird of Australia paints the walls of its bower with charcoal and saliva that is applied with a crude brush of fibers. This habit is only one aspect of the elaborate courtship ritual of these unusual birds.

A fascinating account of unique tool using has been found in Malay, where a certain red ant *(Oecophylla)* builds its home in leaves. Although the larva of this ant does not weave a cocoon, it does produce silk, an advantage the adults use when they weave leaves together in building or repairing their homes. Some of the workers hold the edges of the leaves together whereas others hold the larvae in their jaws and pass them back and forth like shuttles from one edge to another, thus closing the gap with a sheet of silk that is secreted by the larvae during the sewing process.

A recent example of tool using was discovered in the Egyptian vulture *(Neophron percnopterus).* This raven-sized bird breaks open ostrich eggs, which are too big to be seized by its beak, by casting stones at the eggs. Many stones may be thrown before a vulture attains the desired effect.

■ COMMUNICATION AMONG ANIMALS

Every social group of animals has some way for maintaining contact between its members. Communication is simply the influencing of one individual by the behavior of another. This may take the form of bodily contacts (rubbing antennae in bees), scents from glands (mammals at mating season), voice effects (warning cries of birds and mammals), and hosts of others. In some cases, among the higher nonhuman mammals, distinctions between communication and language cannot be rigidly drawn. Yerkes observed in his chimpanzees many sounds and signs that seem to be understood by other members in specific or symbolic ways.

Many animal behaviorists believe that animals have meaningful systems of communication that are not understood by man at all. However, it has been impossible to teach any nonhuman animal a true symbolic language with meaningful association. The reproduction of words and phrases by parrots and mynas has no significant meaning to the birds themselves. A simple form of symbolism of definite meaning is shown whenever a dog assumes a threatening attitude by baring its fangs or raising the hair on its back.

Communication in birds has been extensively studied, and revealing data have been discovered. Bird sounds of a particular species often show a great deal of differentiation. Many of these are believed to be meaningful calls of distress, hunger, warning, etc. W. H. Thorpe, the English investigator, believes that the sound of birds is to arouse emotional states of warning and courtship and to convey precise information. It is now known since Howard's work that the bird song is actually a warning cry to others of territorial rights. Thorpe has shown that the common English chaffinch has two kinds of warning notes under different circumstances. When mobbing a predator bird, chaffinches utter sharp, low-pitched sounds ("chinks") that advertise the presence of a predator. Against a predator on the wing, they utter a high-pitched, thin note ("seeet") that is difficult to locate so that the hawk or owl has no positional clues to the small birds hiding in the foliage. It is believed that the varied songs of birds are an integration of both innate and learned song patterns and that they sing both for communication and for pleasure.

Bird reactions to calls of various kinds are partly learned and partly inborn. When American crows have been exposed only to the signals of American crows, they will not react to the alarm and assembly calls of French crows. However if they have mingled with the other crows as well as their own group, they learn to respond to the cries of both groups.

It has been known from the experience of underwater-sound men during World War II that fish and other marine forms make a variety of noises within the sea. Fish and many other marine forms have no vocal organs but manage to produce a great variety of noises in diverse ways. The chief noisemakers are the toadfish, squirrelfish, sea robin, and triggerfish. Many of them use their air bladders to produce sounds. The toadfish and sea robin cause vibrations in their air bladders by muscle contraction; the triggerfish uses its pectoral fins for beating on a membrane of the air bladder near the body surface; and some, like the squirrelfish, grind together teeth in the back of their mouths and this sound is amplified by the air bladder. It is known that fish give different sounds under different physiologic states. During spawning, their sounds are different from those made at other times.

Ritualization in animal behavior

In animal behavior the concept of ritualization may be divided into two types. One type is made up of movements that usually indicate hostility. Fish may spread their fins; birds may ruffle their feathers to give a larger appearance; lizards elevate their crests and flatten the sides of their bodies; monkeys give fixed stares; dogs will elevate their back bristles and expose their fangs; squids may have rapid color changes—all these and others indicate hostility or fear when confronted with a potential enemy. Some animals in such situations may divert attention from themselves to other objects nearby. The second type of ritualization involves courtship displays, which may be quite complex in some species. This is common among birds. Most vertebrates (and many invertebrates) undergo fixed rituals of many signals, such as dancing, different kinds of stances, collecting nest material, food offerings, waving body appendages, displaying parts of the body, and so forth.

How honeybees communicate location of food

Among the most interesting behavior patterns is the power many insects have of finding their directions and communicating them to others. They seem to have a kind of language for conveying information to each other. Honeybees inform others in the hive about the location of a source of food. The experiments conducted by Professor Karl von Frisch have given many clues to this interesting subject. By using glass observation hives and marked bees, he was able to observe just what occurs when bees report back to the hive the presence of a source of honey. Whenever a foraging bee finds a source of honey, she returns to the hive and performs a peculiar dancing movement that conveys to others in the hive in what direction and how far they must go to find the nectar. If the food is more than 100 meters away, she performs a characteristic waggle dance (Fig. 40-2). This

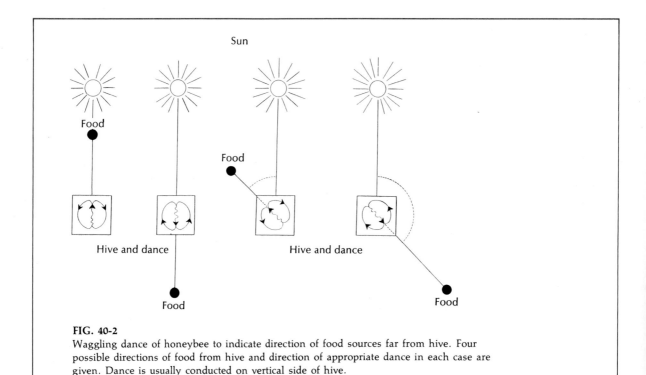

FIG. 40-2
Waggling dance of honeybee to indicate direction of food sources far from hive. Four possible directions of food from hive and direction of appropriate dance in each case are given. Dance is usually conducted on vertical side of hive.

dance is roughly in the pattern of a figure eight that she makes against the vertical side of the comb. In the performance of this act she waggles her abdomen from side to side in a characteristic manner. She repeats this dance over and over, the number of dances decreasing per unit of time the farther away the source is. The direction of the food source is also indicated by the direction of the waggle dance in relation to the position of the sun. When the waggle dance is upward on the comb, the source of food is toward the sun. A waggle run downward on the comb indicates that the food is opposite to the position of the sun. If the food source is at an angle to the sun, the direction of the waggle dance is at a corresponding angle. During a dance other bees keep in contact with the scout bee with their antennae, and each performance results in several bees taking off in search of the food. When they return with the food, they also perform the dance if there is still food there. When the source of food is less than 100 meters from the hive, the pattern of dance is less complex. In this case the scout bee simply turns around in a circle first to the right and then to the left, a performance she repeats several times (Fig. 40-3). She is able in this way to convey to the other bees the information to seek around the hive food of the same odor she bears. One other interesting phenomenon is the ability of bees to determine the direction of the sun when only a small area of the sky is visible. They seem to be able to do this by the pattern of polarization in the light from that part of the sky that is still visible. When the waggle dance is performed on a horizontal surface outside the hive, the bee goes through the dance in the actual direction of the food and not in relation to the sun.

In addition to the dance movements, it is now known that scout bees produce a sound during the straight run of the dance. This sound seems to be produced by wing vibration and is picked up by other bees through sense organs on their legs and antennae. The message conveyed by variations in the sound may give information about the nature of the food source. The "language of the bees" is still a fruitful subject for investigation.

■ BIOLOGIC CLOCKS*

Nearly all aspects of the behavior and physiology of organisms are rhythmic in nature; that is, there are cycles of recurring activities that make up the major part of their existence. Organisms and their body organs do not function at the same rate throughout the day, but they display rhythms of varied events that occur at similar intervals and in definite order. Thus there are periods when the organism is dormant or sleeping, alternating with periods of wakefulness. There are daily rhythms of high and lower temperatures, reproductive cycles, cyclic variations in color, and seasonal rhythms for different activities. Some animals are more active at night (nocturnal), others in daylight (diurnal), and still others in dim light (crepuscular). Some marine animals feed in the littoral zone at high tides, and others only feed at low tides. Some marine organisms spawn with reference to specific moon phases. The human female has a menstrual cycle every lunar month. In plants, there are flowering periods and seed-bearing periods. Basal metabolic rhythms occur in both plants and animals. These are only a few of the examples that could be cited. Because many rhythms are on a daily or 24-hour periodic basis, they are often called **circadian rhythms** (L. *circa,* about, + *dies,* day).

These circadian rhythms of animals and plants are based on the rhythmic changes of the physical environment to which living entities are constantly exposed. It is logical that, to exist, organisms must have their activities geared to the rhythmic cycles of the physical world. The external physical environment of organisms displays such patterns as the solar day of

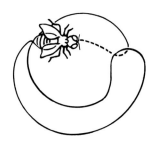

FIG. 40-3
Round dance of honeybee. Whenever food supply is within 100 yards or so of hive, scout bee circles first one way and then the other to convey information of food source.

*Refer to p. 19, Principle 16.

24 hours, the lunar day of 24 hours and 50 minutes, the annual changes of light and temperature, the cycle of the tides in the oceans, the revolving of the earth about the sun, the cycles of weather, and even the patterns of atomic and molecular structure of matter. Nature can be stable only when there are such periodic patterns of changes. In all animals studied, endogenous rhythms of metabolic fluctuations corresponding to the geophysical frequencies are present. Many such rhythmic patterns of organisms are so firmly entrenched that they are not easily altered by prolonged exposure to external agencies, such as temperature differences and drugs, that are known to influence the ordinary metabolic chemical reactions of the body. In general these rhythmic patterns do not follow the common physiologic mechanisms known to be basic to the living process, but they represent an exceptional physiologic phenomenon.

Because many circadian rhythms, by which organisms adapt themselves to repetitive changes in the environment, have so many characteristics of timing mechanisms, many investigators in this field have for some time believed that organisms have the power to measure periods of time with great exactness by means of internal clocks. In other words, living forms possess an independent complex of rhythms that occur at select times of the day and are adapted to the complex of changes in the external physical environment. This system of biologic clocks can have new forms of behavior impressed upon it, and thereafter it keeps repeating this pattern until a new pattern replaces the old.

Although the concept of biologic clocks as now recognized is relatively new, it has long been known that plants and animals have been responsive to varying diurnal cycles and regulate their behavior accordingly. What is more striking is that many organisms that have rhythms coordinated with environmental changes will also have these same rhythms independent of the stimuli from the physical environment. For instance, plants kept in constant darkness will show the same periodic leaf movements as when they are exposed to the daily light-dark alternation. Plants that tend to droop their leaves by night and raise them by day will also do the same when the external conditions are constant. Fiddler crabs have darkened skin during daylight and lighter ones at night, but when placed in constant darkness for long periods of time, they will undergo the same cycle

of color changes as if they were in a natural day-night environment. However, biologic clocks can be reset in various ways so that their cycles are quite different from the normal ones. For instance, placing the crabs (with normal cycles) on ice for six hours will set back the phases of their cycle the same length of time so that they now darken and blanch six hours earlier.

The amazing fish called the grunion on the California coast will swarm ashore at spring tide (the highest tide produced by the conjunction of the sun and moon) and lay their eggs and sperm in pits dug in wet sand just beyond reach of the tide. Here, the young develop and enter the sea at the next monthly spring tide, timing their departure with great nicety (when the spring tides occur in April and June). Tidal and lunar rhythms are very common phenomena among marine animals.

Another interesting aspect of biologic clocks is the ability of some animals to find compass directions with the aid of the sun and to retain the same compass direction by making allowance for the changing position of the sun during the day. The behavior has been observed in bees, birds, arthropods, etc. Such animals can change their angle to the sun correctly at any time of the day and find their way to a desired position. (See discussion of bird migration, pp. 548 to 550.)

Biologic clocks serve organisms in many useful ways. They enable plants and animals to adapt to the best advantage their own rhythms to the rhythms of the physical environment. In this way their activities can occur at the time of day when their physiologic adjustments are best served. Biologic clocks pose many problems to the many able investigators in this field. Where are these clocks located within the organisms? Are they localized in a particular tissue (for example, nervous system), or do they involve the total organization of the animal? Their level of organization is known to be as low as the cell because they have been found in protistan organisms. Evidence indicates that they are endogenous and innate. However, an alternative theory argues that, despite all precautions to exclude organisms from environmental forces, they do continuously receive information about the geophysical cycles from their external environment to which they regulate their activities.

According to this theory organisms are very sensitive to their ambient environment at all times. Their phase-shifting mechanisms on which the notion of an internal independent clock rests depends on subtle electromagnetic field variations correlated with the lunar day tides of the earth's oceans and atmosphere (F. A. Brown, Jr., 1972).

Another problem is how light-dark cycles (photoperiodism) entrain or mediate circadian rhythms within organisms. These and many other problems are far from being resolved.

■ LIGHT PRODUCTION

The production of light by living organisms (**bioluminescence**) is widespread, but usually only scattered representatives are found within a particular phylum. There are more examples among the coelenterates than are found in any other phylum. Altogether, light production has been found in more than 300 genera. The forms that do possess it are largely marine or terrestrial, for no luminous freshwater forms have yet been discovered, except an aquatic glowworm. It may be for sex signaling, for kin recognition, for frightening enemies, for allurement, or for a lamp to guide the animal's movement. It is also possible that it has no significance for the organism, especially in those forms whose light is produced by symbiotic bacteria.

For light production, animals may be divided into two groups—those that produce their own light (self-luminous) and those whose light is produced by symbiotic bacteria. The self-luminous forms will emit light only when they are stimulated. Many of the self-luminous organisms have rather complicated organs for the production of light. These light organs usually consist of a group of photogenic cells for producing the light, a transparent lenslike structure for directing the light, a layer of cells behind the

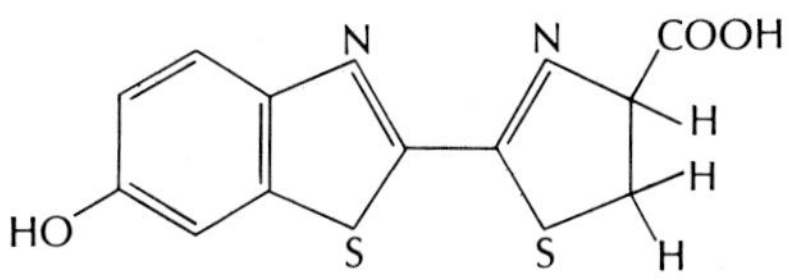

FIG. 40-4

Chemical structure of luciferin in firefly. In bioluminescence, light is produced by a biochemical mechanism that involves oxygen, ATP, luciferin, luciferase, water, and inorganic ions. Luciferin and luciferase differ in composition in different species.

photogenic tissue for reflecting the light, and, surrounding most of the organ, a pigmented layer of cells for shielding the animal's own tissues from the possibly injurious effects of its own light.

The luminescence of this type of luminous animal is caused by the interaction of two substances—**luciferin** (Fig. 40-4), which is oxidized in the presence of an enzyme, **luciferase.** Oxygen is therefore necessary for light production also. It is now known that there are several different kinds of luciferins in different organisms, and some luciferins have been obtained in pure form. In the firefly the mechanism of light production also involves ATP and magnesium (Fig. 40-5). The stimulation for light production may be merely mechanical, or nervous, as it appears to be in higher organisms.

In forms whose light is produced by symbiotic bacteria, light is given off continuously, although

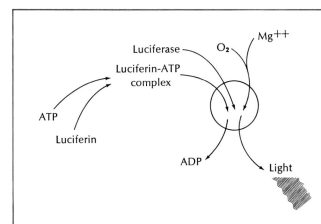

FIG. 40-5

Bioluminescence in animals. The energy compound (ATP) reacts with luciferin to form an active luciferin-ATP complex, which, in the presence of oxygen, magnesium ions, and the enzyme luciferase, emits light. ATP becomes ADP when ATP reacts with luciferin.

some of the forms have devices for concealing and showing the light intermittently. The light transmitted may differ with different animals depending on the difference of the wavelength in the visible spectrum and on the chemical makeup of luciferin.

Several classic examples of luminescence will give the student some idea of this interesting adaptation. Most of these are self-luminous.

Among the protozoans, *Noctiluca* is the most striking example of bioluminescence. This flagellate is sphere shaped and about 1 mm. in diameter. It will give off light only when stimulated. Its luminescence originates from small granules scattered over the periphery and other parts of the cell. These animals will flash vividly whenever they are disturbed by a passing boat. They can also be stimulated by mechanical, chemical, and other agencies. Luciferin and luciferase have not been demonstrated in these forms. In the absence of oxygen, *Noctiluca* will not produce light. *Noctiluca* often collects in enormous numbers at the surface of the sea; its display of color in the wake of a passing vessel on a dark night is wonderful to behold.

The famed glowworm (Fig. 40-6) of literature is usually the wingless female of one of the lampyrid beetles (such as *Photinus*), although some of the larvae also glow. Most glowworms are terrestrial, although one or two live in freshwater. The luminous organs are found on the ventral side of the posterior segments. The light they emit is rather bluish green in appearance. Experiments show that the light rays are restricted to a very narrow range of the spectrum, as compared with other luminous animals, and that they give a very high degree of luminous efficiency. In the glowworm, perhaps more than in most luminous animals, the problem of producing light without heat is solved to a great extent. Its light, as well as

that of all forms that have bioluminescence, is really "cold light."

The best known of all light-producing organisms is the firefly. These beetles have a wide distribution, and many genera and species are found. Most of them are tropical, but two genera are very common in temperate North America (*Photuris* and *Photinus*). These are the familiar "lightning bugs" known to everyone. The light organ of the male is located on the ventral surface of the posterior segments of the abdomen. It is made up of a dorsal mass of small cells (the reflector) and a ventral mass of large cells (the photogenic tissue). Many branches of the tracheal system pass into the organ and subdivide into tracheoles that are connected with tracheal end cells to ensure an ample supply of oxygen, so necessary for the light. The organ is also supplied with nerves that control the rhythmic flashing. The function of the reflector is to scatter the light produced by the photogenic cells. In the female the light organ is confined to a single abdominal segment. The flashing of fireflies is rhythmic and may be single or double, depending on the species.

One of the most striking examples of luminous symbiosis is that of certain East India fish (*Photoblepharon* and *Anomalops*). In these forms, under each eye there are large luminous organs; these give off light continuously, day and night. The organs are made up of a series of tubes having an abundant supply of blood vessels. These tubes contain luminous bacteria that give off the light. Like the other type of light organ, these are sensitive to lack of oxygen and will quickly cease to give off light if the oxygen supply fails. Although the bacteria give off a continuous light, the fish can conceal the organ at will by drawing a fold of black tissue over it like an eyelid (or by some other device). This is an example of true symbiosis, for the fish is dependent on the bacteria for light and the bacteria depend on the fish for nourishment and shelter. Many other cases have been described in which luminous bacteria are considered the responsible agents for light production in animals.

■ SOCIAL BEHAVIOR AMONG ANIMALS

Social behavior refers to groups of animals (usually of the same species) living together and exhibiting activity patterns different from what the members would display when living as separate individuals. In this sense, animals are social when their behavior is

FIG. 40-6
Glowworm. Luminous organs represented on last segment of lampyrid beetle larva.

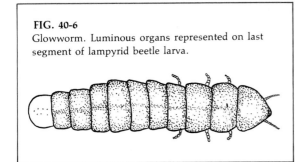

modified by living together and when they influence and are influenced by other members of the group. There are many kinds and degrees of social organization and no single definition will apply to all. Mere aggregations of animals, such as those in an animal community, cannot be called social groups. Animals aggregate for many reasons. They may be attracted by some common favorable environmental factor (for example, light, shade, moisture) and form natural aggregations. Moths may be attracted to a light or barnacles to a common float. Such animals do not aid each other, but they have a social toleration toward each other and do not prevent others from sharing the same conditions. Some aggregations are the result of more positive reactions to others like themselves, such as schools of fish, flocks of birds roosting together, and migratory gregarious habits of many birds. Many types of aggregations also have a survival value and, to a certain extent, may be said to serve a social function. The protective circle of giraffe, musk-ox, and other herds against predators is an example of this. Among many kinds of animals, the parents and their offsprings may form a closely knit social group while the young are developing; in a wolf pack this family relationship may be more or less permanent. Social relationships may involve only two or three individuals or it may include a whole class of individuals.

Social organization seems to be the result of the interaction between an inherited behavior pattern and learned experience. Perhaps among invertebrates, such as bees, ants, and termites, the inherited pattern would be the dominant factor. Social drives or appetites are usually less intense than those of hunger or sex (which may be considered a form of social behavior), although anyone who has watched schools of small squid in an aquarium and the quick recovery of their organization pattern when broken up is impressed with the force of this social behavior drive.

Many factors influence school formation in fish. Although some protection may be afforded by going in school formation, the primary purpose of schooling is probably a more efficient method for securing food. Visual responses and size relations are the criteria followed in forming a school. At times when the water is very murky and at night the school may disperse, for fish travel more in packs the farther they can see. When they cannot depend on their eyes, the greater is the likelihood they will not school. Size is another important requirement for fish schooling,

for fish tend to cruise at about three lengths per second, and some are too small to keep up with the larger ones. Most fish schools are migratory because of depleting the food supply in a particular region.

The roles of the various members of ant and termite societies are determined by structural differences that result in an inflexible division of labor. Such societies have a long evolutionary history and are so integrated that a member has little chance of survival when separated from its society. Survival here depends on the fate of the group in meeting the requirements of the environment.

Origin of social organization

How has social organization arisen? What factors have determined its organization and complexity? No animal lives to itself but is knit to others in some way. Natural selection has operated here as elsewhere in the evolutionary pattern of social aggregations. Various types of societies have arisen, flourished, or become extinct, often without any relation to each other. Parallel evolution has been common. Many social organization patterns have followed similar lines, because their members have followed similar lines of evolution. Social organization has reached its highest level in the arthropods and the mammals. Each of these groups is at the peak of the two main lines of evolution—the annelid-mollusk-arthropod line and the echinoderm-vertebrate line. The social development of these lines has been separately evolved and is not due to common ancestral stock. Each arose from a family unit rather than from aggregation of individuals. Most species of ants, for instance, start a colony by the queen laying eggs that differentiate into the various caste members of workers, soldiers, sexes, etc. Caste determination may be quite complicated and no single plan is followed. In some bees and ants the males develop from haploid (unfertilized) eggs and the other members from diploid eggs. Among termites the males are diploid and the females haploid. Food differences are also determining factors in some cases. Many mammal societies start from the family unit, such as the wolf pack and the societies of the great apes.

Principles of social organization

According to Allee, a great student of social behavior, vertebrate animal groups have organized their social behavior in accordance with three general principles. These are territorial rights, dominance-subordinance hierarchies, and leadership-followership relations.

TERRITORIAL RIGHTS. Much of our knowledge about territorial rights dates from Howard's work on birds (1920). This concept involves a restricted area that is taken over, usually by a male, and vigorously defended against trespass by members of the same species. The birds sing mainly to proclaim ownership; when a bird's domain is invaded, he will fight against the trespasser. Territories may be staked out before or after mating, and there are many variations of behavior among different species. Biologic values of such behavior are obvious, for population densities, accessible food supplies, and well-spaced aggregations are promoted to the best advantage. Other groups of animals, including fish, reptiles, and mammals, are known to follow similar territorial rights patterns.

Another pattern of territorial behavior is illustrated by the arena and bowerbirds of South America, New Guinea, and elsewhere. Each male takes over a small arena during the breeding season and displays himself to the occasional females that happen to wander into the territory where the individual arenas are located. By some sign or other, the females select the males with which they desire to mate. The famous bowerbirds of Australia and New Guinea go even further in their courtship display. The males of some species build elaborate bowers of sticks and grass and decorate them often in an elaborate fashion, with snail shells, charcoal, dead insects, flowers, and brightly shining objects. Students of these birds (Gilliard and Marshall) believe that bower-building behavior has been evolved because sexual selection has been transferred from morphologic patterns of plumage displayed by the male to external objects, which become a kind of secondary sexual characteristic.

DOMINANCE-SUBORDINANCE HIERARCHIES. The concept of dominance-subordinance refers to a type of social behavior in a group in which one animal is dominant over others. We owe much of this concept to the Norwegian scientist T. Schjelderup-Ebbe, who in 1922 observed that in poultry flocks one hen was dominant over the others and exerted the right to peck other members without being pecked in return. Further study revealed that flocks were organized on the basis of social hierarchies; that is, there was a whole series of social levels in which members of the higher levels had peck rights over those of

lower levels, etc. The hen of the lowest level is pecked by all the others but does not peck back. Just how many there could be in a peck order has never been ascertained, but these pecking organizations are based largely upon the ability of birds to recognize the members of a flock as individuals. If the population is too large, pecking orders may not exist. New members added to a flock, as well as those that are absent from the flock for any length of time, are usually assigned to lower levels. When dominance is once won (usually by fighting), it is relatively permanent. However, it is possible for a lower ranking member to advance to a higher level when its victories are greater than its defeats in challenging the social ranking. A given individual may maintain her social position in many flocks at the same time, although she may occupy a different rank in each flock.

Male social hierarchy. Social hierarchy also exists among the males. One of the best examples of this male dominance is found in the sage grouse of the western American plains. During the breeding season early in spring each male establishes a dominance relation with the other males. The most dominant or master cock takes his position in an area and the less dominant cocks form a guard ring around him. Hens are admitted in the ring, but others are driven away. About 75% of all mating is performed by the master cocks, which comprise about 1% of all the males. Dominance hierarchies have been found in all classes of vertebrates, including fish, lizards, mice, and primates. Some arthropods show dominance behavior patterns to a limited extent, but ordinarily the innate, stereotyped behavior of arthropods does not fit into this place of learned behavior.

LEADERSHIP-FOLLOWERSHIP RELATIONS. Leadership behavior refers to the tendency of the members of a group to follow a certain member. It usually involves the selection of an experienced member that stabilizes the other members of the group and holds them together when they are on the move (sheep and deer). Leadership has definite values for the group as a whole and not merely for an individual. The behavior pattern is especially valuable in times of emergency. In some cases the leader may be the dominant member of the group; more often it is an old, experienced female that has the right because of the larger number of offspring that have acquired the habit of following her when they were young. Males, except during the breeding season, are away from the flock, whereas the females are always around.

The relationship between the leader and the followers is different from that of dominance and subordinance. Being the leader does not confer special social privileges, but there is a mutual dependence on each other of both leader and followers. Besides mammals, certain types of leadership are known to exist among fish, lizards, and birds. In the invertebrates, the arthropods show the behavior here and there. One of the most careful studies ever made on the leadership-followership concept was done by F. F. Darling on Scottish herds of red deer, which he describes vividly in his now classic book *A Herd of Red Deer* (1937). His observations show the important role of the female (hind) as a leader of a herd and the extreme care she exerts in looking after its welfare.

It is apparent that in nearly all groups of animals that show social relations the female is the primary influence in leadership and the center of family life. The aggressive males tend to break up social behavior patterns, and nature, to offset this tendency, either produces males in limited numbers (as in many insects) or the males are kept more or less to themselves except during the breeding season. A notable exception is the termite organization in which males play a role almost as prominent as that of the females.

References
Suggested general readings

Alexander, R. D. 1968. Communication in selected groups: Arthropoda. In T. A. Sebeok (editor). Animal communication. Bloomington, Ind., Indiana University Press, pp. 167-216.

Aronson, L. R., et al. (editor). 1972. Selected writings of T. C. Schneirla. San Francisco, W. H. Freeman & Co., Publishers.

Aschoff, J. (editor). 1965. Circadian clocks. Proceedings of the Feldafing summer school. Amsterdam, North-Holland Publishing Co.

Blair, W. F. 1968. Communication in selected groups: Amphibians and reptiles. In T. A. Sebeok (editor). Animal communication. Bloomington, Ind., Indiana University Press, pp. 289-310.

Brown, F. A., Jr. 1972. The "clocks" timing biological rhythms. Amer. Scientist **60:**756-766.

Brown, F. A., Jr., J. W. Hastings, and J. D. Palmer. 1970. The biological clock: Two views. New York, Academic Press Inc.

Bünning, E. 1964. The physiological clock. New York, Academic Press Inc. *A concise, up-to-date account of the physiologic mechanisms animals and plants have for the timing order of their environments.*

Carthy, J. D. 1958. An introduction to the behavior of invertebrates. New York, The Macmillan Co. *In this work the author stresses the functions of the sensory patterns of invertebrates and their reactions to the various categories of stimuli.*

Darling, F. F. 1937. A herd of red deer. A study in animal behavior. Oxford, Oxford University Press. *This is a notable contribution to animal behavior and represents a careful and penetrating analysis of the movement, population, reproduction, and other aspects of an interesting group of animals.*

Frisch, K. von. 1971. Bees: Their vision, chemical senses, and language, rev. ed. Ithaca, N. Y., Cornell University Press.

Griffin, D. R. 1958. Listening in the dark. New Haven, Conn., Yale University Press. *The author subtitles this revealing work, "The Acoustic Orientation of Bats and Men," but other forms are also considered. The author's contribution to the concept of echolocation has fitted him to summarize the mechanisms of acoustic orientation wherever it is found in the animal kingdom, but the major portion of the present treatise is about bats. A bibliography of nearly 500 titles is included.*

Harvey, E. N. 1952. Bioluminescence. New York, Academic Press Inc. *The authoritative work on this fascinating subject.*

Hess, W. N. 1964. Long journey of the dogfish. Natural History **73:**32-35.

Huxley, J. S. 1964. Psychometabolism: General and Lorenzian. Perspect. Biol. Med. **7:**399-432 (Summer 1964).

National Geographic Society. 1972. The marvels of animal behavior. Washington D. C., the Society.

Peckham, G. W., and E. G. Peckham. 1905. Wasps, social and solitary. New York, Houghton-Mifflin Co. *The careful observations made on an interesting group of insects by these investigators represent a classic study in animal behavior. One of the most revealing observations was the action of some wasps in using pebbles as hammers in pounding down the entrance to their burrows.*

Thorpe, W. H. 1956. Learning and instinct in animals. Cambridge, Mass., Harvard University Press. *Of the many works on animal behavior, this book is an excellent synthesis of the behavior concepts and learning processes of animals from protozoans to mammals.*

Watson, J. B. 1930. Behaviorism, rev. ed. Chicago, University of Chicago Press. *A classic psychologic work that has exerted a great influence on the modern interpretation of behavior patterns.*

Wickler, W. 1968. Mimicry. New York, McGraw-Hill Book Co.

Selected *Scientific American* articles

Eibl-Eibesfeldt, I. 1961. The fighting behavior of animals. **205:**112-122 (Dec.).

Esch, H. 1967. The evolution of bee language. **216:**96-104 (April).

Frisch, K. von. 1962. Dialects in the language of the bees. **207**:78-87 (Aug.).

Gilliard, E. T. 1963. The evolution of bowerbirds. **209**:38-46 (Aug.).

Guhl, A. M. 1956. The social order of chickens. **194**:42-46 (Feb.).

Harlow, H. F., and M. Harlow. 1962. Social deprivation in monkeys. **207**:136-146 (Nov.).

Hess, E. H. 1972. Imprinting in a natural laboratory. **227**:24-31 (Aug.).

Lehrman, D. S. 1964. The reproductive behavior of ring doves. **211**:48-54 (Nov.).

Lorenz, K. Z. 1958. The evolution of behavior. **199**:67-78 (Dec.).

Menaker, M. 1972. Nonvisual light reception. **226**:22-29 (March). *How light sensitive structures in the brain regulate an animal's biologic rhythms.*

Milne, L. J., and M. J. Milne. 1950. Animal courtship. **183**:52-55 (July).

Pengelley, E. T., and S. J. Asmundson. 1971. Annual biological clocks. **224**:72-79 (April).

Sauer, E. G. F. 1958. Celestial navigation by birds. **199**:42-47 (Aug.).

Tinbergen, N. 1954. The courtship of animals. **191**:42-46 (Nov.).

Tinbergen, N. 1960. The evolution of behavior in gulls. **203**:118-130 (Dec.).

Washburn, S. L., and I. DeVore. 1961. The social life of baboons. **204**:62-71 (June).

Wenner, A. M. 1964. Sound communication in honeybees. **210**:116-124 (April).

Watts, C. R., and A. W. Stokes, 1971. The social order of turkeys. **224**:112-118 (June).

Wilson, E. O. 1972. Animal communication. **227**:53-60 (Sept.).

APPENDIX

DEVELOPMENT OF ZOOLOGY

■ Certain key discoveries have greatly influenced progress in the study of zoology. This appendix presents some of the major landmarks in the development of biology and the individuals whose names are commonly associated with them. It is very difficult to appraise the historic development of any field of study. One investigator oftens gets the credit for an important discovery, whereas many others should share in the prestige. No one individual has a monopoly on ideas; advances in science are built on the results of many causes and the work of many minds.

In this brief outline, the student may be able to see some of the relationships that exist between one discovery and another. The discoveries are not completely isolated as they may sometimes appear. One may note also that fundamental discoveries in a particular branch of biology tend to be grouped fairly close together chronologically, because that particular interest may have dominated the thought of biologic investigators at that time.

■ ORIGINS OF BASIC CONCEPTS AND KEY DISCOVERIES IN ZOOLOGY

384-322 B.C.: *Aristotle. The foundation of zoology as a science.*

Although this great pioneer zoologist and philosopher cannot be appraised by modern standards, there is scarcely a major subdivision of zoology to which he did not make some contribution. He was a true scientist, for he emphasized the observational and experimental method. Despite his lack of scientific background, he was one of the greatest scientists of all time.

130-200 A.D.: *Galen. Development of anatomy and physiology.*

This Roman investigator has been praised for his clear concept of scientific methods and blamed for passing down to others for centuries certain glaring errors. His influence was so great that for centuries after his period students considered him the final authority on anatomic and physiologic subjects.

1347: *William of Occam. Occam's razor.*

This principle of logic has received its name from the fact that it is supposed to cut out unnecessary and irrelevant hypotheses in the explanation of phenomena. The gist of the principle is that, of several possible explanations, the one that is simplest, has the fewest assumptions, and is most consistent with the data at hand is the most probable one.

1543: *Vesalius, Andreas. First modern interpretation of anatomic structures.*

With his insight into fundamental structure, Vesalius ushered in the dawn of modern biologic investigation. Many aspects of his interpretation of anatomy are just now beginning to be appreciated.

1603: *Platter, F. First description of Diphyllobothrium latum.*

This first published account of the broad tapeworm of man represents an early beginning in the field of parasitology.

1616-1628: *Harvey, William. First accurate description of blood circulation.*

Harvey's classic demonstration of blood circulation was the key experiment that laid the foundation of modern physiology. He explained bodily processes in physical terms, cleared away much of the mental rubbish of mystic interpretation, and gave an auspicious start to experimental physiology.

1627: *Aselli, G. First demonstration of lacteal vessels.*

This discovery, coming at the same time as Harvey's great work, supplemented the discovery of circulation.

1649: *Descartes, R. Early concept of reflex action.*

Descartes postulated the idea that impulses originating at the receptors of the body were carried to the central nervous system where they activated muscles and glands by what he called "reflection."

1651: *Harvey, William. Aphorism of Harvey: Omne vivum ex ovo (all life from the egg).*

Although Harvey's work as an embryologist is overshadowed by his demonstration of blood circulation, his *De Generatione Animalium,* published in 1651, contains many sound observations on embryologic processes.

1652: *Bartholin, Thomas. Discovery of lymphatic system.*

The significance of the thoracic duct in its relation to the circulation was determined in this investigation.

1658: *Swammerdam, Jan. Description of red blood corpuscles.*

This discovery, together with his observations on the valves of the lymphatics and the alterations in shape of muscles during contraction, represented early advancements in the microscopic study of bodily structures.

1660: *Malpighi, Marcello. Demonstration of capillary circulation.*

By demonstrating the capillaries in the lung of a frog, Malpighi was able to complete the scheme of blood circulation because Harvey never saw capillaries and thus never included them in his description.

1665: *Hooke, Robert. Discovery of the cell.*

Hooke's investigations were made with cork and the term "cell" fits cork much better than it does animal cells, but by tradition the misnomer has stuck.

1672: *de Graaf, R. Description of ovarian follicles.*

De Graaf's name is given to the mature ovarian follicle, but he believed that the follicles were the actual ova, an error later corrected by von Baer.

1675-1680: *Leeuwenhoek, Anthony van. Discovery of protozoa.*

The discoveries of this eccentric Dutch microscopist revealed a whole new world of biology.

1693: *Ray, J. Concept of species.*

Although Ray's work on classification was later overshadowed by Linnaeus, Ray was really the first to make the species concept apply to a particular kind of organism and to point out the variations that exist among the members of a species.

1733: *Hales, Stephen. First measurement of blood pressure.*

This was further proof that the bodily processes could be measured quantitatively—more than a century after Harvey's momentous demonstration.

1734: *Borelli, A. Analysis of fish locomotion.*

The nature of fish propulsion has been the subject of many investigations from Aristotle down to the present. Borelli was really the first to demonstrate that the vibration of the tail, and not the fins, is the chief propulsive agency in fish movement. Fresh insights into this problem have been furnished in recent years by J. Gray, R. Bainbridge, E. Kramer, and J. R. Nursall.

1744: *Trembley, A. Observations on the structure of Hydra.*

Trembley worked out with considerable detail and accuracy the nature of this interesting little animal species.

1745: *Bonnet, Charles. Discovery of natural parthenogenesis.*

Although somewhat unusual in nature, this phenomenon has yielded much information about meiosis and other cytologic problems.

1745: *Maupertuis, P. M. Early concept of evolutionary process.*

Although Maupertuis' ideas are speculative and not based on experimental observation, he foretold many of the concepts of variation and natural selection that Darwin was to demonstrate with convincing evidence.

1753: *de Réaumur, R. A. F. Experiments on digestion.*

This was the first recorded account of any note on the nature of the basic principle of nutrition and paved the way for the extensive studies of Beaumont, Pavlov, and Cannon.

1758: *Linnaeus, Carolus. Development of binomial nomenclature system of taxonomy.*

So important is this work in taxonomy that 1758 is regarded as the starting point in the determination of the generic and specific names of animals. Besides the value of his binomial system, Linnaeus gave taxonomists a valuable working model of conciseness and clearness that has never been surpassed.

1759: *Wolff, C. F. Embryologic theory of epigenesis.*

This embryologist, the greatest before von Baer, did much to overthrow the grotesque preformation theory then in vogue, and, despite many shortcomings, laid the basis for the modern interpretation of embryology.

1760: *Hunter, John. Development of comparative investigations of animal structure.*

This vigorous eighteenth century anatomist gave a powerful impetus not only to anatomic observations but also to the establishment of natural history museums.

1763: *Adanson, M. Concept of empiric taxonomy.*

This botanist proposed a scheme of classification that grouped individuals into taxa according to shared characters. A species would have the maximum number of shared characters according to this scheme. The concept has been revived recently by exponents of numerical taxonomy. It lacks the evaluation of the evolutionary concept and has been criticized on this account.

1763: *Koelreuter, J. G. Discovery of quantitative inheritance (multiple genes).*

Koelreuter, a pioneer in plant hybridization, found that certain plant hybrids had characters more or less intermediate between the parents in the F_1 generation, but in the F_2 there were many gradations from one extreme to the other. An explanation was not forthcoming until after Mendel's laws were discovered.

1768-1779: *Cook, James. Influence of geographic exploration on biologic development.*

This famous sea captain made possible a greater range of biologic knowledge because of the able naturalists whom he took on his voyages of discovery.

1772: *Priestley, J., and J. Ingenhousz. Concept of photosynthesis.*

These investigators first pointed out some of the major aspects of this important phenomenon, such as the use of light energy for converting carbon dioxide and water into released oxygen and the retention of carbon.

1774: *Priestley, Joseph. Discovery of oxygen.*

The discovery of this element is of great biologic interest because it helped in determining the nature of oxidation and the exact role of respiration in organisms.

1778: *Lavoisier, Antoine L. Nature of animal respiration demonstrated.*

A basis for the chemical interpretation of the life process was given a great impetus by the careful quantitative studies of the changes during breathing made by this great investigator. His work also meant the final overthrow of the mystic phlogiston theory that had held sway for so long.

1781: *Abildgaard, P. First experimental life cycle of a tapeworm.*

Life cycles of parasites may be very complicated, involving several hosts. This early achievement was followed by many others less than a century later.

1781: *Fontana, F. Description of the nucleolus.*

Fontana discovered this organelle in the slime from the skin of an eel, and until recently its function has been a subject of controversy. Although its functions are not yet completely understood, we know that it is involved in protein synthesis and has centers for ribose nucleoproteins.

1791: *Smith, William. Correlation between fossils and geologic strata.*

By observing that certain types of fossils were peculiar to particular strata, Smith was able to work out a method for estimating geologic age. He laid the basis of stratigraphic geology.

1792: *Galvani, L. Animal electricity.*

The lively controversy between Galvani and Volta over the twitching of frog legs has led to extensive investigation of precise methods of measuring the various electrical phenomena of animals.

1796: *Cuvier, G. Development of vertebrate paleontology.*

Cuvier compared the structure of fossil forms with that of living ones and concluded that there had been a succession of organisms that had become extinct and were succeeded by the creation of new ones. To account for this extinction, Cuvier held to the theory of catastrophism, or the simultaneous extinction of animal populations by natural cataclysms.

1797: *Goethe, J. W. von. Concept of archetypes of animals.*

In this theory, Goethe tried to formulate ideal forms that best fitted animals for their conditions of existence. Modern biology recognizes the importance of the concept in the way contemporary organisms can best fit into the varied ecologic niches of the animal kingdom, such as the basic adaptive features of the major groups.

1800: *Bichat, M. F. X. Analysis of body tissues.*

Bichat's studies on body tissues formed the basis of modern histology. He classified the tissues into 21 different types, but his failure to use the microscope prevented him from correctly appraising the true minute structure of tissues as we now know them.

1801: *Lamarck, J. B. de. Evolutionary concept of use and disuse.*

Lamarck gave the first clear-cut expression of a theory to account for organic evolution. His assumption was that acquired characters were inherited; most evolutionists have refuted this part of the theory.

1802: *Young, T. Theory of trichromatic color vision.*

Young's theory suggested that the retina contained three kinds of light-sensitive substances, each having a maximum sensitivity in a different region of the spectrum and each being transmitted separately to the brain. The three substances combined produced the colors of the environment. The three pigments responsible are located in three kinds of cones. Young's theory has been modified in certain details by other investigators.

1810: *Gall, F. J. Localization of brain functions.*

That different areas of the brain do have different functions has been verified by experimental work, a fact that Gall foreshadowed by his crude experiments on skull contours. He should not be blamed for phrenology, which quacks quickly developed.

1814: *Kirchhoff, G. Demonstration of catalytic action.*

Kirchhoff's discovery that starch treated with an extract of barley malt was converted into glucose was the crude beginning of extensive investigations that have been fruitful in understanding the metabolic activities of living protoplasm.

1817: *Pander, C. First description of three germ layers.*

The description of the three germ layers was first made on the chick, and later the concept was extended by von Baer to include all vertebrates.

1822: *Bell, C., and F. Magendie. Discovery of the functions of dorsal and ventral roots of spinal nerves.*

This demonstration was a starting point for an anatomic and functional investigation of the most complex system in the body.

1823: *Knight, T. Concept of dominance and recessiveness.*

Although this investigator worked with the same pea with which Mendel made his classic discoveries, he was unable to formulate clear-cut laws about the mechanism of heredity, but his information about the breeding of peas gave Mendel something to work with later.

1824: *Prévost, P., and J. B. A. Dumas. Cell division first described.*

The description of the cleavage of the frog egg before the arrival of the cell theory meant that the true significance of cleavage could not be appreciated at that time.

1825: *Raspail, F. V. Beginning of histochemistry.*

Raspail devised the iodine test for starch as well as many other histochemical tests for plant and animal tissues. Histochemistry has yielded much information on the location and chemical nature of various cell constituents such as minerals and enzymes.

1826: *Serres, E. R. A. Discovery of the urohypophysis of fishes.*

The presence of large neurosecretory cells at the posterior end of the spinal cord was found first in teleosts. A vascular plexus carries away the neurosecretions from the secretory neurons that bear axons with bulbous endings. U. Dahlgren (1914) and C. C. Speidel (1919) found similar neurosecretory cells in certain elasmobranchs. The concept of a caudal neurosecretory system has been studied and extended in recent years (M. Enami). The hormone or hormones produced by the urohypophysis may regulate the sodium ions into or out of the organism in osmoregulatory coordination.

1827: *Baer, Karl von. Discovery of mammalian ovum.*

The very tiny ova of mammals escaped de Graaf's eyes, but von Baer brought mammalian reproduction into line with that of other animals by detecting them and their true relation to the follicles.

1828: *Brown, Robert. Brownian movement first described.*

This interesting phenomenon is characteristic of living protoplasm and sheds some light on the structure of protoplasm.

1828: *Thompson, J. V. Nature of plankton.*

Thompson's collections of these small forms with a tow net, together with his published descriptions, are the first records of the vast community of planktonic animals. He was also the first to work out the true nature of barnacles.

1828: *Wöhler, F. First to synthesize an organic compound.*

Wöhler succeeded in making urea (a compound formed in the body) from the inorganic substance of ammonium cyanate, thus $NH_4OCN \longrightarrow NH_2CONH_2$. This success in producing an organic substance synthetically was the stimulus that resulted in the preparation of thousands of compounds by others.

1829: *Vauguelin, N. L. Discovery of carotin.*

This important pigment is now known to be associated with vitamin A and to have many biologic activities. The carotinoids are a complex of many kinds of pigments of similar properties.

1830: *Amici, G. B. Discovery of fertilization in plants.*

Amici was able to demonstrate the tube given off by the pollen grain and to follow it to the micropyle of the ovule through the style of the ovary. Later it was established that a sperm nucleus of the pollen makes contact and union with the nucleus of the egg.

1830: *Baer, Karl von. Biogenetic law formulated.*

Von Baer's conception of this law was conservative and sounder in its implication than has been the case with many other biologists (Haeckel, for instance), for von Baer stated that embryos of higher and lower forms resemble each other more the earlier they are compared in their development, and **not** that the embryos of higher forms resemble the adults of lower organisms.

1830: *Lyell, C. Modern concept of geology.*

The influence of this concept not only did away with the catastrophic theory but also gave a logical interpretation of fossil life and the correlation between the formation of rock strata and the animal life that existed at the time these formations were laid down.

1831: *Brown, Robert. First description of cell nucleus.*

Others had seen nuclei, but Brown was the first to name the structure and to regard the nucleus as a general phenomenon. This description was an important preliminary to the formulation of the cell theory a few years later, for Schleiden acknowledged the importance of the nucleus in the development of the cell concept.

1833: *Hall, M. Concept of reflex action.*

Although Hall describes the method by which a stimulus can produce a response independently of sensation or volition, it remained for the outstanding work of the great Sir Charles Sherrington in this century to explain much of the complex nature of reflexes.

1833: *Purkinje, J. E. Discovery of sweat glands.*

This discovery opened up a new field of investigation into problems of the skin and its structure that have not yet been resolved.

1835: *Bassi, Agostino. First demonstration of a microorganism as an infective agent.*

Bassi's discovery that a certain disease of silkworms was caused by a small fungus represents the beginning of the germ theory of disease that was to prove so fruitful in the hands of Pasteur and other able investigators.

1835: *Dujardin, Felix. Description of living matter (protoplasm).*

Dujardin associated the jellylike substance that he found in protozoa and that he called "sarcode" with the life process. This substance was later called "protoplasm," and the sarcode idea may be considered a significant landmark in the development of the protoplasm concept.

1835: *Owen, Richard. Discovery of Trichinella.*

This versatile investigator is chiefly remembered for his researches in anatomy, but his discovery of this most common parasite in the American people is an important landmark in the history of parasitology.

1837: *Berzelius, J. J. Formulation of the catalytic concept.*

Berzelius called substances that caused chemical conversions by their mere presence catalytic in contrast to analytic; this term described the real chemical affinity between substances in a chemical reaction.

1838: *Liebig, Justus. Foundation of biochemistry.*

The idea that vital activity could be explained by chemicophysical factors has given biologic investigators their greatest method of attacking the nature of life problems.

1838-1839: *Schleiden, M. J., and T. Schwann. Formulation of the cell theory.*

The cell doctrine, with its basic idea that all plants and animals were made up of similar units, represents one of the truly great landmarks in biologic progress. Many biologists believe the work of Schleiden and Schwann has been rated much higher than it deserves in the light of what others did to develop

the cell concept. R. Dutrochet (1824) and H. von Mohl (1831) had described all tissues as being composed of cells.

1839: *Mulder, J. Concept of the nature of proteins.*

Mulder proposed the name "protein" for the basic constituents of protoplasmic materials because he considered them of first or primary importance in the structure of living matter.

1839-1846: *Purkinje, J. E., and Hugo von Mohl. Concept of protoplasm established.*

Purkinje proposed the name "protoplasm" for living matter, and von Mohl did extensive work on its nature, but it remained for Max Schultze (1861) to give a clear-cut concept of the relations of protoplasm to cells and its essential unity in all organisms.

1839: *Verhulst, P. F. Logistic theory of population growth.*

According to this theory, animal populations have a slow initial growth rate that gradually speeds up until it reaches a maximum and then slows down to a state of equilibrium. By plotting the logarithm of the total number of individuals against time, an S-shaped curve results that is somewhat similar for all populations.

1840: *Liebig, J., and F. F. Blackman. The law of minimum requirements.*

Liebig interpreted plant growth as being dependent on essential requirements that are present in minimum quantity; some substances, even in minute quantities, were needed for normal growth. Later Blackman discovered that the rate of photosynthesis is restricted by the factor that operates at a limiting intensity (for example, light or temperature). The concept has been modified in various ways since it was formulated. We now know that a factor interaction other than the minimum ones plays a part.

1840: *Miller, H. Appraisal of geologic formation, Old Red Sandstone.*

These Devonian deposits in Scotland and parts of England represent one of the most important vertebrate-bearing sediments ever discovered. From them much knowledge about early vertebrates, such as ostracoderms, placoderms, and bony fish, has been obtained.

1840: *Müller, J. Theory of specific nerve energies.*

This theory states that the kind of sensation experienced depends on the nature of the sense organ with which the stimulated nerve is connected. The optic nerve, for instance, conveys the impression of vision however it is stimulated.

1841: *Remak, Robert. Description of direct cell division.*

Remak first described direct cell division, or amitosis, in the red blood corpuscles of the chick embryo. The process appears restricted to fully differentiated or senescent tissue cells.

1842: *Bowman, William. Histologic structure of the nephron (kidney unit).*

Bowman's accurate description of the nephron afforded physiologists an opportunity to attack the problem of how the kidney separates waste from the blood, a problem not yet fully solved.

1842: *Steenstrup, J. Alternation of generations described.*

Metagenesis, or alternation of sexual and asexual reproduction in the life cycle, exists in many animals and plants. The concept had been introduced before this time by L. A. de Chamisso (1819).

1843: *Owen, Richard. Concepts of homology and analogy.*

Homology, as commonly understood, refers to similarity in embryonic origin and development, whereas analogy is the likeness between two organs in their functioning. Owen's concept of homology was merely that of the same organ in different animals under all varieties of form and function.

1844: *Ludwig, C. Filtration theory of renal excretion.*

About the same time that Bowman described the malpighian corpuscle, Ludwig showed that the corpuscle functions as a passive filter and that the filtrate that passes through it from the blood carries into the urinary tubules the waste products that are concentrated by the resorption of water as the filtrate moves down the tubules. Other investigators, Cushny, Starling, and Richards, have confirmed the theory by actual demonstration.

1845: *Helmholtz, H., and J. R. Mayer. The formulation of the law of conservation of energy.*

This landmark in man's thinking showed that in any system, living or nonliving, the power to perform mechanical work always tends to decrease unless energy is added from without. Physiologic investigation could now advance on the theory that the living organism is an energy machine and obeyed the laws of the science of energetics.

1848: *Hofmeister, W. Discovery of chromosomes.*

This investigator made sketches of bodies, later known to be chromosomes, in the nuclei of pollen mother cells *(Tradescantia).* Schneider further described these elements in 1873, and Waldeyer named them in 1888.

1848: *Siebold, C. T. E. von. Establishment of the status of protozoa.*

Siebold emphasized the unicellular nature of protozoa, fitted them into the recently developed cell theory, and established them as the basic phylum of the animal kingdom.

1850: *Bernard, C. Independent irritability of muscle.*

By using the drug curare, which blocks motor impulses through the myoneural junction, Bernard found that the muscle would still respond to direct stimulation, thus proving that the muscle is independently irritable.

1851: *Bernard, C. Discovery of the vasomotor system.*

Bernard showed how the amount of blood distributed to the various tissues by the small arterioles was regulated by vasomotor nerves of the sympathetic nervous system, as he demonstrated in the ear of a white rabbit.

1851: *Waller, A. V. Importance of the nucleus in regeneration.*

When nerve fibers are cut, the parts of the fibers peripheral to the cut degenerate in a characteristic fashion. This wallerian degeneration enables one to trace the course of fibers through the nervous system. In the regeneration process the nerve cell body (with its nucleus) is necessary for the downgrowth of the fiber from the proximal segment.

1852: *Helmholtz, Herman von. Determination of rate of nervous impulse.*

This landmark in nerve physiology showed that the difficult phenomena of nervous activity could be expressed numerically.

1852: *Kölliker, Albrecht von. Establishment of histology as a science.*

Many histologic structures were described with marvelous insight by this great investigator, starting with the publication of the first text in histology *(Handbuch der Gewebelehre).*

1852: *Stannius, H. F. Stannius' experiment on the heart.*

By tying a ligature as a constriction between the sinus venosus and the atrium in the frog and also one around the atrioventricular groove, Stannius was able to demonstrate that the muscle tissues of the atria and ventricles have independent and spontaneous rhythm. His observations also indicated that the sinus is the pacemaker of the heartbeat.

1854: *Newport, G. Discovery of the entrance of spermatozoon into frog's egg.*

This was a significant step in cellular embryology, although its real meaning was not revealed until the concept of fertilization as the union of two pronuclei was formulated about 20 years later (Hertwig, 1875).

1854: *Vogt, C. Experimental infestation of man by pork tapeworm.*

A common experiment among parasitologists is to infect a suspected host to determine a life history.

A little later, the great parasitologist R. Leuckart was able to infect a calf with the cysticerci of the beef tapeworm, *Taenia saginata.*

1855: *Addison, T. Discovery of adrenal disease.*

The importance of the adrenals in maintaining normal body functions was shown when Addison described a syndrome of general disorders associated with the pathology of this gland.

1856: *Discovery of Neanderthal fossil man (Homo neanderthalensis).*

Many specimens of this type of fossil man have been discovered (Europe, Asia, Africa) since the first one was found near Düsseldorf, Germany. His culture was Mousterian, and although of short stature his cranial capacity was as large or larger than that of modern man.

1856: *Perkin, W. H. Discovery of the first coal-tar dye.*

The first coal-tar dye, called aniline violet, was formed by oxidizing aniline with potassium bichromate. Rapid development of dyes, many synthetically, followed; their use in microscopic preparations brought rapid strides in cytologic and bacteriologic investigations.

1857: *Bernard, Claude. Formation of glycogen by the liver.*

Bernard's demonstration that the liver forms glycogen from substances brought to it by the blood showed that the body can build up complex substances as well as tear them down.

1858: *Sclater, P. L. Distribution of animals on basis of zoologic regions.*

This was the first serious attempt to study the geographic distribution of organisms, a study that eventually led to the present science of zoogeography. A. R. Wallace worked along similar lines.

1858: *Virchow, Rudolf. Aphorism of Virchow: Omnis cellula e cellula (every cell from a cell).*

1858: *Virchow, Rudolf. Formulation of the concept of disease from the viewpoint of cell structure.*

He laid the basis of modern pathology by stressing the role of the cell in diseased tissue.

1859: *Darwin, Charles. The concept of natural selection as a factor in evolution.*

Although Darwin did not originate the concept of organic evolution, no one has been more influential in the development of evolutionary thought. The publication of *The Origin of Species* represents the greatest single landmark in the history of biology.

1859: *Mauthner, L. Concept of the mauthnerian apparatus.*

Two giant cells or neurons with cell bodies located in the medulla and with axons running to the tip of the spinal cord are found in many fish and amphibians. These giant cells have functional significance in those animals that have lateral line systems and use the tail in swimming.

1860: *Bernard, Claude. Concept of the constancy of the internal environment (homeostasis).*

This important concept has influenced physiologic thinking, for it shows how organisms in their long evolutionary development have been mainly concerned in preserving their stability against environmental forces.

1860: *Pasteur, Louis. Aphorism of Pasteur: Omne vivum e vivo (every living thing from the living).*

We do not doubt the significance of this concept at present, but from what source did life originate?

1860: *Pasteur, Louis. Refutation of spontaneous generation.*

Pasteur's experiment with the open S-shaped flask proved conclusively that fermentation or putrefaction resulted from microbes, thus ending the long-standing controversy regarding spontaneous generation.

1860: *Wallace, A. R. The Wallace line of faunal delimitation.*

As originally proposed by Wallace, there was a sharp boundary between the Australian and Oriental faunal regions so that a geographic line drawn between certain islands, of the Malay Archipelago, through the Makassar Strait between Borneo and Celebes, and between the Philippines and the Sanghir Islands separated two distinct and contrasting zoologic regions. On one side Australian forms predominated; on the other, Oriental ones. The validity of this division has been questioned by zoo-geographers in the light of more extensive knowledge of the faunas of the two regions.

1861: *Claparede, E. Discovery of giant nerve axons of annelids.*

These large nerve fibers were first found and described by Ehrenberg (1836) in the Crustacea. As new microscopic techniques were developed, many investigations have been made on the structure and functions of these unique nerve cells. Their chief function is in escape mechanisms involving widespread and synchronous muscular contractions.

1861: *Graham, Thomas. Colloidal states of matter.*

This work on colloidal solutions has resulted in one of the most fruitful concepts regarding protoplasmic systems.

1862: *Bates, H. W. Concept of mimicry.*

Mimicry refers to the advantageous resemblance of one species to another for protection against predators. It involves palatable or edible species that imitate dangerous or inedible species that enjoy immunity because of warning coloration against potential enemies.

1864: *Haeckel, E. Modern zoologic classification.*

The broad features of zoologic classification as we know it today were outlined by Haeckel and others, especially R. R. Lankester, in the third quarter of the nineteenth century. B. Hatschek is another zoologist who deserves much credit for the modern scheme, which is constantly undergoing revision. Other schemes of classification in the early nineteeth century did much to resolve the difficult problem of classification, such as those of Cuvier, de Lamarck, Leuckart, Ehrenberg, Vogt, Gegenbauer, and Schimkevitch. More recently, L. H. Hyman performed invaluable service in the arrangement of animal taxonomy.

1864: *Schultze, Max. Protoplasmic bridges between cells.*

The connection of one cell to another by means of protoplasmic bridges has been demonstrated in both plants and animals, but the concept has never been settled to the satisfaction of histologists.

1865: *Kekulé, F. A. Concept of the benzene ring.*

The concept of the benzene ring established the later development of synthetic chemistry; it also aided in understanding the basic structure of life that is so dependent on the capacity of carbon atoms to form rings and chains.

1866: *Haeckel, E. Nuclear control of inheritance.*

At the time the hypothesis was formed, the view that the nucleus transmitted the inheritance of an animal had no evidence to substantiate it.

1866: *Kowalevsky, A. Taxonomic position of tunicates.*

The great Russian embryologist showed in the early stages of development the similarity between amphioxus and the tunicates and how the latter could be considered a degenerate branch of the phylum Chordata.

1866: *Mendel, Gregor. Formulation of the first two laws of heredity.*

These first clear-cut statements of inheritance made possible an analysis of hereditary patterns with mathematical precision. Rediscovered in 1900, his paper confirmed the experimental data geneticists had found at that time. Since that time mendelism has been considered the chief cornerstone of hereditary investigation.

1866: *Schiff, H. The Schiff reaction for aldehydes.*

The Schiff reagent of fuchsin sulfurous acid is the colorless derivative produced by the action of sulfur dioxide in water on basic fuchsin or related dyes. There are many variant forms of this reagent, which is the basis of the Feulgen reaction for the nucleic acid DNA.

1866: *Schultze, Max. Histologic analysis of the retina.*

Schultze's fundamental discovery that the retina contained two types of visual cells, rods and cones, helped to explain the differences in physiologic vision of high and low intensities of light.

1867: *Kowalevsky, A. Germ layers of invertebrates.*

The concept of the primary germ layers laid down by Pander and von Baer was extended to invertebrates by this investigator. He found that the same three germ layers arose in the same fashion as those in vertebrates. Thus an important embryologic unity was established for the whole animal kingdom.

1867: *Traube, I. Concept of the semipermeable membrane.*

The nature and role of living membranes in biologic systems have been extensively studied in cellular physiology, for the characteristics of the membrane determine the osmotic behavior of cells and the exchanges cells make with their surroundings.

1869: *Langerhans, P. Discovery of islet cells in pancreas.*

These islets were first found in the rabbit and since their discovery they have been one of the most investigated tissues in animals. The theory that they have come from the transformation of pancreatic acinus cells has now given way to the theory that they have originated from the embryonic tubules of the pancreatic duct.

1869: *Miescher, F. Isolation of nucleoprotein.*

From pus cells, a pathologic product, Miescher was able to demonstrate in the nuclei certain phosphorus-rich substances, nucleic acids, which are bound to proteins to form nucleoproteins. In recent years these complex molecules have been the focal point of significant biochemical investigations on the chemical properties of genes, with the wider implications of a better understanding of growth, heredity, and evolution.

1871: *Bowditch, H. P. Discovery of the "all-or-none" law of heart muscle.*

This law states that a minimal stimulus will produce a maximal contraction of the heart musculature. A single skeletal fiber also obeys the law the same as cardiac muscle, but the latter is a continuous protoplasmic mass and an impulse that causes contraction in one part spreads also to other parts.

1871: *Quetelet, L. A. Foundations of biometry.*

The applications of statistics to biologic problems is called biometry. By working out the distribution curve of the height of soldiers, Quetelet showed biologists how the systematic study of the relationships of numerical data could become a powerful tool for analyzing data in evolution, genetics, and other biologic fields.

1872-1876: *Challenger expedition.*

Not only did this expedition establish the science of oceanography, but a vast amount of material was collected that greatly extended the knowledge of the variety and range of animal life.

1872: *Dohrn, Anton. Establishment of Naples Biological Station.*

The establishment of this famous station marked the first of the great biologic stations, and it rapidly attained international importance as a center of biologic investigation.

1872: *Ludwig, K., and E. F. W. Pflüger. Gas exchange of the blood.*

By means of the mercurial blood pump, these investigators separated the gases from the blood and thereby threw much light on the nature of gaseous exchange and the place where oxidation occurred (in the tissues).

1873: *Agassiz, L. Establishment of first American marine laboratory.*

Although short-lived, this laboratory on the island of Penikese near Cape Cod was instrumental in the training of an influential group of American biologists and in the establishment of a scientific tradition that gave a great impetus to biologic investigation.

1873: *Schneider, Anton. Description of nuclear filaments (chromosomes).*

In his description of cell division, Schneider showed nuclear structures that he termed "nuclear filaments," the first recorded description of what are now known as chromosomes.

1874: *Haeckel, E. Taxonomic position of phylum Chordata.*

The great German evolutionist based many of his conclusions on the work of the Russian embryologist Kowalevsky, who in 1866 showed that the tunicates as well as amphioxus had vertebrate affinities.

1874: *Haeckel, E. H. Gastrea theory of metazoan ancestry.*

According to this theory, the hypothetic ancestor of all the Metazoa consisted of two layers (ectoderm and entoderm) similar to the gastrula stage in embryonic development, and the entoderm arose as an invagination of the blastula composed of a single layer of flagellate cells. Thus the diploblastic stage of ontogeny was to be considered as the repetition of this ancestral form. This theory has had wide acceptance but has been criticized on the grounds that the endoderm is not always formed by invagination, such as the inwandering of ectodermal cells in certain forms.

1875: *Heidenhain, R. Zymogen granules as enzyme precursors.*

Heidenhain concluded that the disappearance of intracellular bodies (zymogen granules) in the pancreas coincided with the appearance of proteolytic enzymes, and that such granules represented a temporary storage of digestive enzymes.

1875: *Hertwig, O. Concept of fertilization as the conjugation of two sex cells.*

The fusion of the pronuclei of the two gametes in the process of fertilization paved the way for the concept that the nuclei contained the hereditary factors and that both maternal and paternal factors are brought together in the zygote.

1875: *Strasburger, Eduard. Description of indirect cell division.*

The accurate description of the processes of cell division that Strasburger made in plants represents a great pioneer work in the rapid development of cytology during the last quarter of the nineteenth century.

1876: *Boll, F. Discovery of the visual pigment rhodopsin.*

This red substance, which bleaches in the light and regenerates its color in the dark, is now known to form part of a cycle of reactions of great importance in basic visual phenomena and is the source of rod vision sensitivity.

1876: *Cohn, F. Significance of the bacterial spore.*

Spore formation among bacteria is restricted mainly to the bacilli. It is an oval body formed within the bacterial cell, and certain environmental conditions favor its formation within a particular kind of bacteria.

1876: *Pasteur, L. The Pasteur effect.*

This metabolic regulation of the oxidative process refers to the release of energy by cells without the use of oxygen, other than that supplied by the metabolites involved. Oxygen suppresses fermentation and the products of anaerobic metabolism, and thus spares carbohydrate utilization.

1877: *Ehrlich, P. Discovery of mast cells.*

Mast cells are granular cells of connective tissue and are associated with inflammatory processes and new growth. They are known also to be the site of the concentration and release of histamine in hypersensitive reactions.

1877: *Pfeffer, W. Concept of osmosis and osmotic pressure.*

Pfeffer's experiments on osmotic pressure and the determination of the pressure in different concentrations laid the foundation for an understanding of a general phenomenon in all organisms.

1878: *Balfour, F. M. Relationship of the adrenal medulla to the sympathetic nervous system.*

By showing that the adrenal medulla has the same origin as the sympathetic nervous system, Balfour really laid the foundation for the interesting concept of the similarity in the action of epinephrine and sympathetic nervous mediation.

1878: *Brandt, K. Demonstration of vital coloring.*

The belief that only dead cells could be stained was long held by histologists, but Brandt was able to stain the lipid droplets in the cytoplasm of living *Actinosphaerium* (Heliozoa) with the dye Bismark brown and observe the process during the vital staining.

1878: *Kuhne, W. Nature of enzymes.*

The study of the action of chemical catalysts in an understanding of the fundamental nature of life has steadily increased with biologic advancement and now represents one of the most interesting aspects of biochemistry.

1879: *Flemming, W. Chromatin described and named.*

Flemming shares with a few others the description of the details of indirect cell division. The part of the nucleus that stains deeply he called **chromatin** (colored), which gives rise to the chromosomes.

1879: *Fol, Hermann. Penetration of ovum by a spermatozoon described.*

Fol was the first to describe a thin conelike body extending outward from the egg to meet the sperm. Compare Dan's acrosome reaction (1954).

1879: *Kossel, A. Isolation of nucleoprotein.*

Nucleoproteins were isolated in the heads of fish sperm; they make up the major part of chromatin. They are combinations of proteins with nucleic acids, and this study was one of the first of the investigations that interest biochemists at the present time. Nobel Laureate (1910).

1880: *Laveran, C. L. A. Protozoa as pathogenic agents.*

This French investigator first demonstrated that the causative organism for malaria is a protozoan. This discovery led to other investigations that revealed the role the protozoans play in causing diseases such as sleeping sickness and kala azar. It remained for Sir Roland Ross to discover the role of mosquitoes in spreading malaria. Nobel Laureate (1906).

1880: *Ringer, S. Influence of blood ions on heart contraction.*

The pioneer work of this investigator determined the inorganic ions necessary for contraction of frog hearts and made possible an evaluation of heart metabolism and the replacement of body fluids.

1881: *Zacharias, E. Distribution and nature of nucleic acids.*

In a pioneer cytochemical study involving the use of enzyme pepsin, he found that the nucleus of the frog erythrocyte and the macronuclei of *Vorticella* and *Paramecium* remained when the other parts of the cell were digested by the enzyme.

1882: *Flemming, W. First accurate counts of nuclear filaments (chromosomes) made.*

1882: *Flemming, W. Mitosis and spireme named.*

1882: *Metchnikoff, Élie. Role of phagocytosis in immunity.*

The theory that microbes are ingested and destroyed by certain white corpuscles (phagocytes) shares with the theory of chemical bodies (antibodies) the chief explanation for the body's natural immunity.

Nobel Laureate (1908).

1882: *Pfitzner, W. Discovery of chromomeres.*

The discrete granules that make up a large part of chromosomes are of especial interest because they are supposed to correspond to the loci of the genes.

1882: *Strasburger, E. Cytoplasm and nucleoplasm named.*

1882-1924: *The Albatross of the Fish Commission.*

This vessel, under the direction of the U. S. Fish Commission, was second only to the famed *Challenger* in advancing scientific knowledge about oceanography.

1883: *Golgi, Camillo, and R. Cajal. Silver nitrate technique for nervous elements.*

The development and refinement of this technique gave a completely new picture of the intricate relationships of neurons. Modifications of this method have given valuable information concerning the cellular element—the Golgi apparatus.

Nobel Laureates (1906).

1883: *Hertwig, O. Origin of term "mesenchyme."*

This important tissue, which is restricted to young embryos, may arise from all three germ layers, but chiefly from the mesoderm. It is a protoplasmic network whose meshes are filled with a fluid intercellular substance. Mesenchyme gives rise to a great variety of tissues and both its cells and intercellular substance may be variously modified; its most common derivative is connective tissue.

1883: *Leuckart, R., and A. P. Thomas. Life history of sheep liver flukes.*

This investigation is noteworthy in parasitology, for it represents the first time a complete life cycle was worked out for a trematode involving more than one host.

1883: *Roux, W. Allocation of hereditary functions to chromosomes.*

This theory could not be much more than a guess when Roux made it, but how fruitful was the idea in the light of the enormous amount of evidence since accumulated!

1884: *Flemming, W., E. Strasburger, and E. Van Beneden. Demonstration that nuclear filaments (chromosomes) double in number by longitudinal division.*

This concept represented a further step in understanding the precise process in indirect cell division.

1884: *Kollman, J. Concept of neoteny.*

Neoteny refers to the retention of larval characters and retardation of somatic growth after the gonads have become sexually mature. In most cases it implies acceleration of sexual maturity (pedogenesis). It was first described in the axolotl larval form of the Mexican newt *Ambystoma*.

1884: *Rubner, Max. Quantitative determinations of the energy value of foods.*

Although Liebig and others had estimated calorie values of foods, the investigations of Rubner put their determinations on a sound basis. His work made possible a scientific explanation for metabolism and a basis for the study of comparative nutrition.

1884: *Strasburger, E. Prophase, metaphase, and anaphase named.*

1885: *Dubois, R. Nature of light production in animals.*

By his work on luminous clams, Dubois was able to show how a chemical substance, luciferin, could be oxidized with the aid of an enzyme, luciferase, with the production of light.

1885: *Hertwig, O., and E. Strasburger. Concept of the nucleus as the basis of heredity.*

The development of this idea occurred before Mendel's laws of heredity were rediscovered in 1900, but it anticipates the important role the nucleus with its chromosomes was to assume in hereditary transmission.

1885: *Rabl, Karl. Concept of the individuality of the chromosomes.*

The view that the chromosomes retain their individuality through all stages of the cell cycle is accepted by all cytologists, and much evidence has accumulated in proof of the theory. However, it has been virtually impossible to demonstrate the chromosome's individuality through all stages.

1885: *Roux, W. Mosaic theory of development.*

In the early development of the frog's egg, Roux showed that the determinants for differentiation were segregated in the early cleavage stages and that each cell or groups of cells would form only certain parts of the developing embryo (mosaic or determinate development). Later, other investigators showed that in many forms blastomeres, separated early, would give rise to whole embryos (indeterminate development).

1885: *Weismann, August. Formulation of germ plasm theroy.*

Weismann's great theory of the germ plasm stresses the idea that there are two types of protoplasm—germ plasm, which gives rise to the reproductive cells or gametes, and somatoplasm, which furnishes all the other cells. Germ plasm, according to the theory, is continuous from generation to generation, whereas the somatoplasm dies with each generation and does not influence the germ plasm.

1886: *Establishment of Woods Hole Biological Station.*

This station is by all odds the greatest center of its kind in the world. Here most of the biologists in America have studied and the station has increasingly attracted investigators from many other countries as well. The influence of Woods Hole on the progress of biology cannot be overestimated.

1886: *MacMunn, C. A. Discovery of cytochrome.*

This iron-bearing compound was rediscovered in 1925 by D. Keilin and has been demonstrated in most types of cells. Its importance in cellular oxidation and the metabolic pathway has made possible a workable theory of cell respiration and has stimulated the study of intracellular localization of enzymes and how they behave in metabolic processes.

1887: *Fischer, Emil. Structural patterns of proteins.*

The importance of proteins in biologic systems has made their study the central theme of all modern biochemical work.

Nobel Laureate (1902).

1887: *Haeckel, E. H. Concept of organic form and symmetry.*

Symmetry refers to the spatial relations and arrangements of parts in such a way as to form geometric designs. Although many others before Haeckel's time had studied and described types of animal form, Haeckel has given us our present concepts of organic symmetry, as revealed in his monograph on radiolarians collected on the *Challenger* expedition.

1887: *Van Beneden, E. Chromosome constancy within a species.*

The number of chromosomes is characteristic for each species, usually within the range of 2 to 200. There are minor exceptions to this rule, as in the case of spontaneous polyploidy.

1887: *Van Beneden, E. Demonstration of chromosome reduction during maturation.*

Weismann has predicted this important event before it was actually demonstrated. Only by this method can the constancy of chromosome numbers be maintained.

1887: *Weismann, August. Prediction of the reduction division.*

Weismann formulated the hypothesis that the separation of undivided whole chromosomes must take place in one of the maturation divisions (reduction division), or else the number of chromosomes would double in each generation *(reductio ad absurdum).*

1888: *Waldeyer, W. Chromosome named.*

1889: *Hertwig, R., and E. Maupas. True nature of conjugation in paramecium.*

The process of conjugation has been described (even by van Leeuwenhoek) many times and its sexual significance interpreted, but these two investigators independently showed the details of pregamic divisions and the mutual exchange of the micronuclei during the process.

1889: *Maupas, E. Discovery of protozoan mating types.*

His observation that in certain species of ciliates *(Loxophyllum, Stylonychia, etc.)* conjugation was restricted to members of two clones of different origin and did not occur among those of a single clone was the beginning of a research problem that has proved fruitful in the hands of Sonneborn.

1889: *Mering, J. von, and O. Minkowski. Effect of pancreatectomy.*

The classic experiment of removing the pancreas stimulated research that led to the isolation of the pancreatic hormone insulin by Banting (1922).

1890: *Smith, T. Role of an arthropod in disease transmission.*

The transmission of the sporozoan *Babasia,* which is the active agent in causing Texas cattle fever by the tick *Boophilus,* represents one of the first demonstrations of the important role of arthropods as vectors of disease.

1891: *Driesch, H. Discovery of totipotent cleavage.*

The discovery that each of the first several blastomeres, if separated from each other in the early cleavage of the fertilized egg, would develop into a complete embryo stimulated investigation on totipotent and other types of development.

1891: *Dubois, Eugene. Discovery of the fossil man Pithecanthropus erectus (now Homo erectus).*

Although not the first fossil man to be found, the Java man represents one of the first significant primitive men that have been discovered.

1892: *Ivanovski, D. Discovery of the nature of viruses.*

This discovery was the start of the many investigations on the nature of these important biologic agents.

1893: *Dollo, I. Concept of the irreversibility of evolution.*

In general, the overall evolutionary process is one way and irreversible insofar as a whole complex genetic system is concerned, although back mutations and restricted variations may occur over and over.

1893: *Haacke, W. Concept of evolutionary orthogenesis.*

This is the principle that stresses the evolutionary idea that groups of animals tend to evolve in one direction without deviation. It is the belief that there is an inherent trend for evolution to proceed in straight lines (rectilinear evolution).

1893: *His, W. Anatomy and physiology of the auriculoventricular node and bundle.*

The specialized conducting tissue of the heart has given rise to many investigations, not the least of which were those of His, whose name was given to the intricate system of branches that are reflected over the inner surface of the ventricles.

1894: *Driesch, H. Constancy of nuclear potentiality.*

Driesch's view was that all nuclei of an organism were equipotential but that the activity of nuclei varied with different cells in accordance with the differentiation of tissues. That all genetic factors are present in all cells is supported by the constancy of DNA for each set of chromosomes and the similarity of histone proteins in the different somatic cells of an organism. This theory, however, is being challenged in the much investigated problems of differentiation and growth, and there has been some evidence to show that nuclear potentialities do vary with different stages in development.

1894: *Merriam, C. H. Concept of life zones in North America.*

This scheme is based on temperature criteria and the importance of temperature in the distribution of plants and animals. According to this concept, animals and plants are restricted in their northward distribution by the total quantity of heat during the season of growth and reproduction, and their southward distribution is restricted by the mean temperature during the hottest part of the year.

1894: *Morgan, C. L. Concept of animal behavior.*

The modern interpretation of animal behavior really dates from certain basic principles laid down by this psychologist. Among these principles was the one in which he stated that the actions of an animal should be interpreted in terms of the simplest mental processes (Morgan's canon).

1894: *Oliver, G., and E. A. Sharpey-Schaefer. Demonstration of the action of a hormone.*

The first recorded action of a specific hormone was the demonstration of the effect of an extract of the suprarenal (adrenal) gland upon blood vessels and muscle contraction.

1895: *Bruce, D. Life cycle of protozoan blood parasite (Trypanosoma).*

The relation of this parasite to the tsetse fly and to wild and domestic animal infection in Africa is an early demonstration of the role of arthropods as vectors of disease.

1895: *Nuttal, G. H. F., and H. Thierfelder. Sterile culture of animals (gnotobiotics).*

The technique of rearing animals in a germ-free environment has been revived in recent years and promises to be a useful tool in determining the complex roles microbiota may play in their hosts.

1895: *Pinkus, F. Discovery of terminal cranial nerve (O).*

This nerve (nervus terminalis) was discovered after the other cranial nerves had been named and numbered. It is sensory and runs from the olfactory membrane to the olfactory lobe of the brain. Its exact significance is unknown.

1895: *Roentgen, W. Discovery of x-rays.*

This great discovery was quickly followed by its application in the interpretation of bodily structures and processes and represents one of the greatest tools in biologic research.

Nobel Laureate (1901).

1896: *Baldwin, J. M. Baldwin evolutionary effect.*

It is the belief that genetic selection of genotypes will be channeled or canalized in the same direction as the adaptive modifications that were formerly nonhereditary. Nonhereditary adaptive modifications are supposed to keep a racial strain in an environmental channel, where mutations producing similar phenotypes will be selected.

1896: *Becquerel, A. H. Discovery of spontaneous radioactivity.*

His discovery of the rays emitted by uranium, together with the work of the Curies on high-energy particles, elucidated the process of radioactivity, which led in time to the isotope concept of so fruitful application to biologic phenomena.

Nobel Laureate (1903).

1896: *Russian Hydrographic Survey. Biology of Lake Baikal.*

This lake in Siberia is more than 500 miles long, 50 miles wide, and has an extreme depth of more than a mile (the deepest lake in the world). Its unique fauna is a striking example of evolutionary results from long-continued isolation. Up to 100% of the species in certain groups are endemic (found nowhere else). This remarkable fauna represents the survival of ancient freshwater animals that have become extinct in surrounding areas.

1897: *Abel, J. J., and A. C. Crawford. Isolation of the first hormone (adrenaline).*

The purification and isolation of one of the active principles of the suprarenal medulla led to its chemical nature and naming by J. Takamine (1901) and its synthesis by F. Stolz (1904).

1897: *Braun, F. Invention of the cathode-ray oscillograph.*

This instrument has been one of the most useful tools ever invented for measuring electrical events in excitable tissues.

Nobel Laureate (1909).

1897: *Buchner, E. Discovery of zymase.*

Buchner's discovery that an enzyme (a nonliving substance) manufactured by yeast cells was responsible for fermentation resolved many problems that had baffled Pasteur and other investigators. Zymase is now known to consist of a number of enzymes.

Nobel Laureate (1907).

1897: *Canadian Geological Survey. Dinosaur fauna of Alberta, Canada.*

In the rich fossil beds along the Red Deer River in Alberta there was found the fauna of the Upper Cretaceous time, and a whole new revelation of the dinosaur world has been made from the study of these fossils.

1897: *Eijkman, C. Discovery of the cause of a dietary deficiency disease.*

Eijkman's pioneer work on the causes of beriberi led to the isolation of the antineuritic vitamin (thiamine). This work may be called the key discovery that resulted in the development of the important vitamin concept.

Nobel Laureate (1929).

1897: *Garnier, C. Discovery of the ergastoplasm organelle.*

This term was originally applied to those diffused or discrete cytoplasmic masses of fibrillar materials (especially in gland cells) that stain with basic dyes just as the nuclear chromatin, hence the basophilic part of the cytoplasm. The electron microscope has shown that this material is the same as the rough-surfaced endoplasmic reticulum because of the dense granules of RNA (ribosomes, or the classic microsomes).

1897: *Huot, A. Discovery of the aglomerular fish kidney.*

This discovery in the angler fish *(Lophius)* and later in the toadfish and others proved to renal physiologists that renal tubules of kidneys could excrete as well as resorb substances.

1897: *Ross, Ronald. Life history of malarian parasite (Plasmodium).*

This notable achievement represents a great landmark in the field of parasitology and the climax of the work of many investigators on the problem.

Nobel Laureate (1902).

1897: *Sherrington, C. S. Concept of the synapse in the nervous system.*

If the nervous system is composed of discrete units or neurons, functional connections must exist between these units. Sherrington showed how individual nerve cells could exert integrative influences on other nerve cells by graded excitatory or inhibitory synaptic actions. The electron microscope has in recent years added much to a knowledge of the synaptic structure.

Nobel Laureate (1932).

1898: *Benda, C., and C. Golgi. Discovery of mitochondria and the Golgi apparatus.*

These interesting cytoplasmic inclusions were actually seen by various observers before this date, but they were both named in 1898 and the real study of them began at this time. W. Flemming and R. Altmann first demonstrated mitochondria. The Golgi apparatus was demonstrated by V. St. George (1867) and G. Platner (1885), but Camillo Golgi, with his silver nitrate impregnation method, gave the first clear description of the apparatus in nerve cells. Mitochondria are now known to play an important role in the synthesis of enzymes in cellular metabolism.

1898: *Osborn, H. F. Concept of adaptive radiation in evolution.*

This concept states that, starting from a common ancestral type, many different forms of evolutionary adaptations may occur. In this way evolutionary divergence can take place, and the occupation of many ecologic niches is made possible, according to the adaptive nature of the invading species.

1899: *Bayless, W. H., and E. H. Starling. Law of the intestine.*

As originally formulated, the law stated that the movement of food down the alimentary canal is accomplished by a wave of muscular contraction above the bolus of food and a dilation below it. It is doubtful that there is an inhibition of the muscle below the bolus, but the dilation is simply caused by the general opening of the gut by the contraction of the longitudinal muscles.

1899: *Hardy, W. A. Appraisal of conventional fixation methods.*

This investigation showed that the common preservatives used in killing and fixing cells produced either fibrous networks or fine emulsions and that the method of fixation determined which of these states is produced (artifacts).

The structures found in stained cells needed confirmation by other methods because the fibrillar, reticular, and other appearances of protoplasm are artifacts from the type of fixation and staining employed.

1900: *Chamberlain, T. C. Theory of the freshwater origin of vertebrates.*

The evidence for this theory as first proposed was based mainly on the fact that early vertebrate fossils were found in sediments of freshwater origin, such as Old Red Sandstone, and were largely absent from marine deposits. Much of these freshwater deposits were supposed to have been laid down in rivers that drained the higher ranges of the continents.

1900: *Correns, K. E., E. Tschermak, and Hugo De Vries. Rediscovery of Mendel's laws of heredity.*

These three investigators independently in their genetic experiments on plants obtained results similar to Mendel's, and in their survey of the literature found that Mendel had published his now famous laws in 1866. A few years later, W. Bateson and others found that the same laws applied to animals also.

1900: *Andrews, C. F. Discovery of fossil beds in Fayum Lake region of Egypt.*

The work of C. F. Andrews and others on these fossils has shown how important Africa has been in the evolution of mammals. The many new types that evolved here may have been the result of geographic isolation and inbreeding.

1900: *Landsteiner, Karl. Discovery of blood groups.*

This fundamental discovery made possible successful blood transfusions as well as initiated the tremendous amount of work on the biochemistry of blood, an investigation that is more active now than ever before.

Nobel Laureate (1930).

1900: *Loeb, J. Discovery of artificial parthenogenesis.*

The possibility of getting eggs that normally undergo fertilization to develop by chemical and mechanical methods has been accomplished in a number of different animals from the sea urchin and frog eggs (Loeb, 1900) to the rabbit egg (Pincus, 1936). The phenomenon has some importance in experimental cytology.

1901: *Montgomery, T. H. Homologous pairing of maternal and paternal chromosomes in the zygote.*

Sutton, also, showed that in synapsis before the reduction division each pair is made up of a maternal and a paternal chromosome. This phenomenon is of fundamental importance in the segregation of hereditary factors (genes).

1901: *Vries, Hugo De. Mutation theory of evolution.*

De Vries concluded from his study of the evening primrose, *Oenothera lamarckiana,* that new characters appear suddenly and are inheritable. Although the variations in *Oenothera* were probably not mutations at all since many of them represented hybrid combinations, yet the evidence for the theory from other sources has steadily mounted until now the theory affords the most plausible explanation for evolutionary progress.

1902: *Kropotkin, P. Mutual aid as a factor in evolution.*

This concept is not new by any means, but Kropotkin elaborated on the importance of social life at all levels of animal life in the survival patterns of the evolutionary process.

1902: *Lillie, F. R. Differentiation without cleavage.*

By placing eggs (fertilized or unfertilized) of the annelid worm *Chaetopterus* in seawater containing potassium chloride, Lillie found that the egg would undergo development and differentiation without cleavage. Differentiation is thus not dependent on uninucleate compartments.

1902: *McClung, C. E. Discovery of sex chromosomes.*

The discovery in the grasshopper that a certain chromosome (X) had a mate (Y) different in appearance or else lacked a mate altogether gave rise to the theory that certain chromosomes determined sex. H. Henking actually discovered the X chromosome in 1891.

1903: *Bayliss, W. M., and E. H. Starling. Discovery of the first hormone, secretin.*

The isolation of a substance from the mucosa of the duodenum that had a powerful effect on stimulating the secretion of pancreatic juice was a key experiment in the development of the great science of endocrinology.

1903: *Boveri, T., and W. S. Sutton. Parallelism between chromosome behavior and mendelian segregation.*

This theory states that synaptic mates in meiosis correspond to the mendelian alternative characters and that the formula of character inheritance of Mendel could be explained by the behavior of the chromosomes during maturation. This is therefore a cytologic demonstration of mendelism.

1903: *Sutton, W. S. Constitution of the diploid group of chromosomes.*

The diploid group of chromosomes is made up two chromosomes of each recognizable size, one member of which is paternal and the other maternal in origin.

1904: *Cannon, W. B. Mechanics of digestion by x-rays.*

The clever application of x-rays to a study of the movements and other aspects of the digestive system has revealed an enormous amount of information on the physiology of the alimentary canal. Cannon first used this technique in 1898.

1904: *Carlson, A. J. Pacemaking activity of neurogenic hearts.*

Heartbeats may originate in muscle (myogenic hearts) or in ganglion cells (neurogenic hearts).

Carlson showed that in the arthropod *(Limulus)* the pacemaker was located in certain ganglia on the dorsal surface of the heart and that experimental alterations of these ganglia by temperature or other means altered the heart rate. Neurogenic hearts are restricted to certain invertebrates.

1904: *Jennings, H. S. Behavior patterns in Protozoa.*

The careful investigations of this lifelong student of the behavior of these lower organisms lead to concepts such as the trial-and-error behavior and many of our important beliefs about the various forms of tropisms and taxes.

1904: *Macallum, A. B. Similarity of blood salts to sea salts.*

The relative concentration and proportions of the salts (potassium, sodium, calcium) in the blood of most vertebrates is similar to that found in seawater and is considered to be evidence of the origin of animals in the sea. The higher concentration of salts in the sea today, as compared with the Cambrian period when land forms are supposed to have arisen, is explained by the constant addition of salt from continental streams.

This hypothesis has been subjected to critical analysis in recent years on the basis that the salt concentration of vertebrate extracellular fluid is proportionally much smaller than the salt concentration of the Cambrian period and that the magnesium content of extracellular fluid is proportionally much smaller than that in the Cambrian seas.

1904: *Nuttall, G. H. F. Serologic relationships of animals.*

This method of determining animal relationships is striking evidence of evolution. It has been used in recent years to establish the taxonomic position of animals whose classification has not been determined by other methods.

1905: *Haldane, J. S., and J. G. Priestley. Role of carbon dioxide in the regulation of breathing.*

By their clever technique of obtaining samples of air from the lung alveoli, these investigators showed how the constancy of carbon dioxide concentration in the alveoli and its relation to the concentration in the blood was the chief regulator of the mechanism of respiration.

1905: *Huber, G. C. Dissection of the nephron.*

The isolation of a complete mammalian kidney nephron by maceration and teasing is a landmark in renal histology, for it revealed for the first time the structure of the different parts of the tubule in relation to each other, as shown in the sequence of a gross morphologic preparation.

1905: *Zsigmondy, R. Application of ultracentrifuge to colloids.*

This has made possible a study of colloidal particles and the finer details of protoplasmic systems. Nobel Laureate (1925).

1906: *Bateson, W., and R. C. Punnett. Discovery of linkage of hereditary units.*

Although first discovered in sweet peas, it was Morgan and associates who gave the real meaning to this great genetic concept. All seven pairs of Mendel's alternative characters were in separate chromosomes, a fact that simplified his problem.

1906: *Einthoven, W. Mechanism of the electrocardiogram.*

The invention of the string galvanometer (1903) by Einthoven supplied a precise tool for measuring the bioelectric activity of the heart and quickly led to the electrocardiogram, which gives accurate information about disturbances of the heart's rhythm. Nobel Laureate (1924).

1906: *Hopkins, F. G. Analysis of dietary deficiency.*

Hopkins tried to explain dietary deficiency by a biochemical investigation of the lack of essential amino acids in the diet—an approach that has led to many important investigations in nutritional requirements.

Nobel Laureate (1929).

1906: *Tswett, M. Principle of chromatography.*

This is the separation of chemical components in a mixture by differential migration of materials according to structural properties within a special porous sorptive medium. The technique, which may involve the flow of either solvent or gas, is widely used in the purification and isolation of many substances and in other applications.

1907: *Boltwood, B. B. Use of uranium-lead method for dating geologic periods.*

Uranium has a half-life of about 4.5 billion years and the age of the rock can be estimated by comparison of the proportions of undecayed uranium and of lead present in the rock. The method can be applied only to igneous rocks that have uranium and that are at least 50 million years old.

1907: *Boveri, T. Qualitative differences of chromosomes.*

Boveri showed in his classic experiment with sea urchin eggs that chromosomes have qualitatively different effects on development. He found that only those cells developed into larva that had one of each kind of chromosome; those cells that did not have representatives of each kind of chromosome failed to develop.

1907: *Discovery of the Heidelberg fossil man (Homo heidelbergensis, now H. erectus heidelbergensis).*

This fossil consisted of a lower jaw with all its teeth. The jaw shows many simian characteristics, but the teeth show patterns of primitive men. This type is supposed to have existed during mid-Pleistocene times and is more or less intermediate between the Java man and modern man.

1907: *Hopkins, F. G. Relationship of lactic acid to muscular contraction.*

Hopkins showed that, after being formed in muscular contraction, a part of the lactic acid is oxidized to furnish energy for the resynthesis of the remaining lactic acid into glycogen. This discovery did much to clarify part of the cyclic reactions involved in the complicated process of muscular contraction. Nobel Laureate (1929).

1907: *Keith, A., and M. J. Flack. Discovery of the sinoauricular (S-A) node.*

The ancestry of this node is the sinus tissue of the primitive heart of cold-blooded animals. In mammals this node is embedded in the muscle of the right auricle near the openings of the superior and inferior venae cavae and initiates the beat and sets the pace (pacemaker) for the mammalian heart.

The atrioventricular (A-V) node, which lies in the septum of the atria near the A-V valves, was discovered in 1906 by Tawara.

1907: *Wilson, H. V. Reorganization of sponge cells.*

In this classic experiment, Wilson showed that the disaggregation of sponges, by squeezing them through fine silk bolting cloth so that they are separated into minute cell clumps, resulted in the surviving cells coming together and organizing themselves into small sponges when they were in seawater.

1908: *Garrod, A. E. Description of inborn errors of metabolism.*

Certain hereditary disorders are caused by enzyme deficiencies wherein the mutations of a single gene may be responsible for controlling the specificity of a particular enzyme in certain disorders.

1908: *Hardy, G. H., and W. Weinberg. Hardy-Weinberg population formula.*

This important theorem states that, in the absence of factors (mutation, selection, etc.) causing change in genes, the proportion of genes in any large population will reach an equilibrium in one generation and thereafter will remain stable regardless of whether the genes are dominant or recessive. Its mathematic expression forms the basis for the calculations of population genetics.

1909: *Arrhenius, S., and S. P. L. Sörensen. Determination of hydrogen ion concentration (pH).*

The sensitivity of most biologic systems to acid and alkaline conditions has made pH values of the utmost importance in biologic research.

Arrhenius, Nobel Laureate (1903).

1909: *Bataillon, E. Discovery of the pseudogamy concept.*

Pseudogamy is the activation of an unfertilized egg by a sperm without the participation of the male chromosomes of the sperm in the hereditary pattern of the resulting developed egg. The sperm may be from a male of the same species or from a different species.

1909: *Castle, W. E., and J. C. Philips. The inviolability of germ cells to somatic cell influences.*

That germ cells are relatively free from somatic cell influences was shown by the substitution of a black guinea pig ovary in a white guinea pig that gave rise to black offspring when mated to a black male.

1909: *Doublass, E. Discovery of dinosaur fossil bed.*

Dinosaur fossils have been found in various parts of the world, such as the deposits in Alberta, Canada, and those in Tendaguru, East Africa, but few have equaled the dinosaur bed near Jensen, Utah. Here one may see the dinosaur skeletons preserved in the rocks just as they were laid down millions of years ago.

1909: *Janssens, F. A. Chiasmatype theory.*

When homologous chromosomes are paired before the reduction division, they or their chromatids form visible crosslike figures or chiasmata, which Janssens interpreted as the visible exchange of parts of two homologous chromatids, although he could not actually prove this point. This phenomenon of crossing-over is the key to the genetic mapping of chromosomes so extensively worked out by Morgan and his school with *Drosophila*.

1909: *Johannsen, W. Gene, genotype, and phenotype named.*

1909: *Johannsen, W. Limitations of natural selection on pure lines.*

This investigator found that when a hereditary group of characters becomes homogeneous, natural selection cannot change the genetic constitution with regard to these characters. Selection was shown to be something that could not create and was effective only in isolating genotypes already present in the group; it therefore could not effect evolutionary changes directly.

1909: *Nicolle, C. J. H. Body louse as vector of typhus fever.*

The demonstration that typhus fever was transmitted from patient to patient by the bite of the body louse paved the way for the control of this dreaded epidemic disorder by delousing populations with DDT.

Nobel Laureate (1928).

1910: *Dale, H. H. Nature of histamine.*

Dale and colleagues found that an extract from ergot had the properties of histamine (β-imidazolyl ethylamine), which can be produced synthetically by splitting off carbon dioxide from the amino acid histidine. It is also a constituent of all tissue cells, from which it may be released by injuries or other causes. The pronounced effect of histamine in dilating small blood vessels, contracting smooth muscle, and stimulating glands has caused it to be associated with many physiologic phenomena, such as anaphylaxis, shock, and allergies.

Nobel Laureate (1936).

1910: *Ehrlich, Paul. Chemotherapy in the treatment of disease.*

The discovery of salvarsan as a cure for syphilis represents the first great discovery in this field. Another was the dye sulfanilamide, discovered by Domagk in 1935. These chemicals, which are more or less harmful to the body, have been generally superseded by the more effective and less harmful antibiotics.

Nobel Laureate (1908).

1910: *Heinroth, O. Concept of imprinting as a type of behavior.*

This is a special type of learning that is demonstrated by birds (and possibly other animals). It is based on the fact that a goose or bird is attracted to the first large object it sees just after hatching and thereafter will follow that object to the exclusion of all others. Although the behavior pattern appears to be strongly fixed, there is some doubt about its irreversibility.

1910: *Herrick, J. B. Discovery of sickle cell anemia.*

In this inherited condition, red blood cells have an abnormal sickle shape and do not function normally. Investigation has shown that the S hemoglobin found in this type of anemia differs from the normal A hemoglobin in the protein component but not in the heme component. Glutamic acid, one of the 300 amino acids in the molecule of normal hemoglobin, is replaced by valine in the sickle cell hemoglobin. The inheritance is supposed to involve a single gene.

1910: *Morgan, T. H. Discovery of sex linkage.*

Morgan and colleagues discovered that the results of a cross between a white-eyed male with a red-eyed female in *Drosophila* were different from those obtained from the reciprocal cross of a red-eyed male with a white-eyed female. This was a crucial experiment, for it showed for the first time that how a trait behaved in heredity depended on the sex of the parent, in contrast to most mendelian characters, which behave genetically the same way whether introduced by a male or female parent.

1910-1920: *Morgan, T. H. Establishment of the theory of the gene.*

The extensive work of Morgan and associates on the localization of hereditary factors (by genetic experiments) on the chromosomes of the fruit fly *(Drosophila)* represents the most significant work ever performed in the field of heredity.

Nobel Laureate (1933).

1910: *Murray, J., and J. Hjort. Deep sea expedition of the Michael Sars.*

Of the many expeditions for exploring the depths of the oceans, the *Michael Sars* expedition, made in the North Atlantic regions, must rank among the foremost. The expedition yielded an immense amount of information about deep-sea animals, as well as many important concepts regarding the ecologic pattern of animal distribution in the sea.

1910: *Pavlov, I. P. Concept of the conditioned reflex.*

The idea that acquired reflexes play an important role in the nervous reaction patterns of animals has greatly influenced the development of modern psychology.

Nobel Laureate (1904).

1911: *Child, C. M. Axial gradient theory.*

This theory attempts to explain the pattern of metabolism from the standpoint of localized regional differences along the axes of organisms. The differences in the metabolic rate of different areas has made possible an understanding of certain aspects of regeneration, development, and growth.

1911: *Cuénot, L. The preadaptation concept.*

Preadaptation refers to a morphologic or physiologic character that may be indifferent or of minor importance in the environment in which it first occurs, but that may be suited to take advantage of a certain type of environment should the latter arise. This favorable conjunction of characters and suitable environment is considered to be an important factor in progressive evolution (opportunistic evolution).

1911: *Dobell, C. C. The acellular status of the protist.*

It is the belief of some zoologists that a protistan cell is not homologous with the metazoan cell because one is a whole organism and the other only a part of an organism. Arguments pro and con have been advanced in the controversy.

1911: *Funk, C. Vitamin hypothesis.*

Vitamin deficiency diseases are commonly called avitaminoses and refer to those diseases in which causes can be definitely traced to the lack of some essential constituent of the diet. Thus beriberi is caused by an insufficient amount of thiamine, scurvy by a lack of vitamin C, etc.

1911: *Harvey, E. B. Cortical changes in the egg during fertilization.*

In the mature egg, cortical granules gather at the surface of the egg, but on activation of the egg these granules, beginning at the point of sperm contact, disappear in a wavelike manner around the egg. These granules by their breakdown are supposed to release material that helps form the ensuing fertilization membrane.

1911: *Rutherford, E. Concept of the atomic nucleus.*

This cornerstone of modern physics must be of equal interest to the biologist in the light of the rapid advances of molecular biology. Rutherford also discovered (1920) the proton, one of the charged particles in the atomic nucleus.

Nobel Laureate (1908).

1911: *Walcott, C. D. Discovery of Burgess shale fossils.*

The discovery of a great assemblage of beautifully preserved invertebrates in the Burgess shale of British Columbia and their careful study by an American paleontologist represent a landmark in the fossil record of invertebrates. These fossils date from the middle Cambrian age and include the striking *Aysheaia,* which has a resemblance to the extant *Peripatus.*

1912: *Carrel, A. Technique of tissue culture.*

The culturing of living tissues in vitro, that is, outside of the body, has given biologists an important tool for studying tissue structure and growth. R. Harrison in 1907 had found that parts of living tissues in suitable media under suitable conditions could live and multiply.

Nobel Laureate (1912).

1912: *Gudernatsch, J. F. Role of the thyroid gland in the metamorphosis of frogs.*

This investigator found that the removal of the thyroid gland of tadpoles prevented metamorphosis into frogs, and also that the feeding of thyroid extracts to tadpoles induced precocious metamorphosis. In 1919 W. W. Swingle also showed that the presence and absence of inorganic iodine would produce the same results.

1912: *Kite, G. L. Micrurgic study of cell structures.*

The use of micromanipulators for the microdissection of living cells has greatly enriched our knowledge of the finer microscopic details of protoplasm, chromosomes, cell division, and many other phenomena of cells. The method has been developed by many workers, such as H. D. Schmidt (1859), M. A. Barber (1904), and W. Seifriz (1921), but R. Chambers has perhaps done more to refine its use and to employ it in experimental cell research.

1912: *Wegener, A. L. Concept of continental drift theory.*

This theory postulates that the continents were originally joined together in one or two large masses that gradually broke up during geologic time, and the fragments drifted apart to form the current land masses. The theory has been revived recently on the basis of the geologic shifting of paleomagnetism, convection currents in the earth, and definite biologic evidence, for example, the finding of an amphibian fossil in Antarctic regions.

1913: *Michaelis, L., and M. Menton. Enzyme-substrate complex.*

On theoretic and mathematic grounds these investigators showed that an enzyme formed an intermediate compound with its substrate (the enzyme-substrate complex) that subsequently decomposes to release the free enzyme and the reaction products. D.

Keilin of Cambridge University and B. Chance of the University of Pennsylvania have demonstrated the presence of such a complex by color changes and measurements of its rate of formation and breakdown that agree with theoretic predictions.

1913: *Reck, H. Discovery of Olduvai Gorge fossil deposits.*

This region in East Africa has yielded an immense amount of early mammalian fossils as well as the tools of the Stone Age man, such as stone axes. Among the interesting fossils discovered were elephants with lower jaw tusks, horses with three toes, and the odd ungulate (chalicothere) with claws on the toes.

1913: *Shelford, V. E. Law of ecologic tolerance.*

This law states that the potential success of an organism in a specific environment depends on how it can adjust within the range of its toleration to the various factors to which the organism is exposed.

1913: *Sturtevant, A. H. Formation of first chromosome map.*

By the method of crossover percentages it has been possible to locate the genes in their relative positions on chromosomes—one of the most fruitful discoveries in genetics, for it led to the extensive mapping of the chromosome in *Drosophila.*

1913: *Tashiro, S. Metabolic activity of propagated nerve impulse.*

The detection of slight increases in carbon dioxide production in stimulated nerves, as compared with inactive ones, was evidence that conduction in nerves is a chemical change. Later (1926), A. V. Hill was able to measure the heat given off during the passage of an impulse. Oxygen consumption has also been measured in excited nerves.

1914: *Kendall, E. C. Isolation of thyroxine.*

The isolation of thyroxine in crystalline form was a landmark in endocrinology. Its artificial synthesis was done by Harington in 1927.

1914: *Lillie, F. R. Role of fertilizin in fertilization.*

According to this theory, the jelly coat of eggs contains a substance, fertilizin, which combines with the antifertilizin on the surface of sperm and causes the sperm to clump together.

1914: *Sharp, R. Discovery of the neuromotor apparatus of ciliates.*

This demonstration of a system of neurofibrils connected to a motor mass in the anterior part of the organism *(Epidinium),* which was concerned in the coordination of cilia and other motor organelles of the cystostomial region, has been extended to other ciliates and may be considered a universal structural feature of this group.

1914: *Shull, G. H. Concept of heterosis.*

When two standardized strains or races are crossed, the resulting hybrid generation may be noticeably superior to both parents as shown by greater vigor, vitality, and resistance to unfavorable environmental conditions. First worked out in corn (maize), such hybrid vigor may also be manifested by other kinds of hybrids. Although its exact nature is still obscure, the phenomenon may be caused by the bringing together in the hybrid of many dominant genes of growth and vigor that were scattered among the two inbred parents, or it may be caused by the complementary reinforcing action of genes when brought together.

1914: *Williston, S. W. Concept of Williston's law of evolutionary simplification.*

This principle stresses the structural simplification of bodily parts in evolutionary development, wherein the parts are reduced in number but become more specialized. For instance, ancient crustaceans (trilobites) had many similar somites and legs in comparison with the fewer ones of modern crstaceans.

1916: *Bridges, C. B. Discovery of nondisjunction.*

Bridges explained an aberrant genetic result by a suggested formula that later he was able to confirm by cytologic examination of the failure of a pair of chromosomes to disjoin at the reduction division so that both chromosomes passed into the same cell. It was definite proof that genes are located on the chromosomes.

1916: *Lillie, F. R. Theory of freemartin.*

The sexually abnormal female calf when it is born as a twin to a normal male had been a baffling problem for centuries until Lillie demonstrated in convincing fashion that hormones from the earlier developing gonads of the male circulate into the blood of the female and alter the sex differentiation of the latter. As a result the gonads of the freemartin never reach maturity and so she remains sterile.

1916: *Winkler, H. Concept of heteroploidy.*

Deviations from the normal diploid number of chromosomes are known to occur spontaneously in both plants and animals. It has also been known for some time that heteroploidy can be induced by artificial means.

1917: *Broili, F. Discovery of amphibian-reptilian fossil, Seymouria.*

This interesting fossil found near Seymour, Texas, has characteristics of both amphibians and reptiles and thus throws some light on the relations between the two great vertebrate classes.

1917: *Grinnell, J. Concept of the ecologic niche.*

This is the spatial unit occupied by a taxon (species or subspecies) to which it is adapted morphologically and physiologically with reference to physical and biotic factors. Elton has done much to develop the concept and simplified its meaning to an animal's place in the biotic environment with respect to food and enemies.

1917: *Papanicolaou, G. N., and C. R. Stockard. Smear technique of vaginal contents.*

By this technique the reproductive cycle could be followed accurately in the living animal. The technique has had other applications such as the "Pap" test for uterine cancer.

1918: *King, H. D. Modified sex ratios by genetic selection.*

By close inbreeding and careful selection in rats, she was able to split in 25 generations a strain with a secondary sex ratio of 110 ($110\male:108\female$) into two substrains with sex ratios of 124 and 82. This classic experiment may give some insight into those cases of unexpected sex ratios.

1918: *Krogh, A. Regulation of the motor mechanism of capillaries.*

The mechanism of the differential distribution of blood to the various tissues has posed many problems from the days of Harvey and Malpighi, but Krogh showed that capillaries were not merely passive in this distribution but that they had the power to actively contract or dilate, according to the needs of the tissues. Both nervous and chemical controls are involved.

Nobel Laureate (1920).

1918: *Starling, E. H. The law of the heart.*

Within physiologic limits, the more the ventricles are filled with incoming blood, the greater is the force of their contraction at systole. This is an adaptive mechanism for supplying more blood to tissues when it is needed.

1918: *Szymanski, J. S. Demonstration of time-measuring mechanism of animals.*

Szymanski showed that animals had some means of measuring time independently of such physical factors as light and temperature, for he discovered that 24-hour activity patterns were synchronized with the day-night cycle when animals were kept in constant darkness and temperature. This work has led to the concept (demonstrated by many investigations) that animals have some kind of internal clock whereby they can measure certain cycles independent of external factors.

1918: *Vavilov, N. I. Biologic centers of origin as reservoirs of desirable genes.*

The Russian botanist and plant geographer stressed the importance of tracing strains of cultivated plants to the locale of their original cultivation, where inferior plants (by present standards) may contain valuable genes already selected by natural selection. Such a pool of genes, he maintained, could by selection and intercrossing afford genetic banks for constructing new and superior genotypes.

1919: *Aston, F. W. Discovery of isotopes.*

Radioactive isotopes have proved of the utmost value in biologic research because of the possibility of tracing the course of various elements in living organisms.

Nobel Laureate (1922).

1919: *Meyerhof, O. Formation of lactic acid during muscular contraction.*

The discovery that the glycogen content decreases as lactic acid increases was a key discovery in understanding the nature of muscular contraction, a problem that has not yet been solved. Meyerhof also showed that about four fifths of the lactic acid is resynthesized to glycogen by the energy furnished by the oxidation of the other one fifth of lactic acid.

1920: *Herzog, R. O., and W. Jancke. Development of x-ray diffractometry.*

When a parallel beam of x-rays is passed through crystallized biologic material, the rays are spread and an image of the diffracted pattern is recorded (by rings, spots, etc.). By measurements and mathematic calculations information about structure can be obtained.

1920: *Howard, H. E. Territorial patterns of bird behavior.*

A matting pair of birds establishes and defends a specific territory against others of the same species. Usually the male asserts his claim by singing at points close to the boundaries of his staked-out claim. The concept has been confirmed for many species by numerous investigators.

1921: *Hopkins, F. G. Isolation of glutathione.*

The discovery of this sulfur compound gave a great impetus to the study of the complicated nature of cellular oxidation and metabolism, a process far from being solved at present.

Nobel Laureate (1929).

1921: *Langley, J. N. Concept of the functional autonomic nervous system.*

Langley's concept of the functional aspects of the autonomic system dealt mainly with the mammalian type, but the fundamental principles of the system have been applied with modification to other groups as well. Two divisions—sympathetic and parasympathetic—are recognized in the functional interpretation of excitation and inhibition in the antagonistic nature of the two divisions.

1921: *Loewi, O., and H. H. Dale. Isolation of acetylcholine.*

This key demonstration has led to the neurohumoral concept of the transmission of nerve impulses to muscles.

Nobel Laureates (1936).

1921: *Richards, A. N. Collection and analysis of glomerular filtrate of the kidney.*

This experiment was direct evidence of the role of the glomeruli as mechanical filters of cell-free and protein-free fluid from the blood and was striking confirmation of the Ludwig-Cushny theory of kidney excretion.

1921: *Spemann, Hans. Organizer concept in embryology.*

The idea that certain parts of the developing embryo known as organizers have a determining influence on the developmental patterns of the organism has completely revolutionized the field of experimental embryology and has afforded many clues into the nature of morphogenesis.

Nobel Laureate (1935).

1922: *Banting, F. Extraction of insulin.*

The great success of this hormone in relieving a distressful disease, diabetes mellitus, and the dramatic way in which active extracts were obtained have made the isolation of this hormone the best known in the field of endocrinology.

Nobel Laureate (1923).

1922: *Bridges, C. B. Genic balance theory of sex.*

Bridges found in *Drosophila* that sex is determined by autosomes as well as by X chromosomes. By crossing triploid (3n) females with diploid (2n) males, he found that the X chromosome carries more genes for femaleness and the autosomes more genes for maleness. What determines the sex is the ratio between the number of X chromosomes and the sets of autosomes in the zygote.

1922: *Erlanger, J., and H. S. Gasser. Differential conduction of nerve impulses.*

By using the cathode-ray oscillograph, these investigators found that there were several different types of mammal nerve fibers that could be distinguished structurally and that had different rates of conducting nervous impulses according to the thickness of the nerve sheaths (most rapid in the thicker ones). Nobel Laureates (1944).

1922: *Kopec, S. Concept of the hypothalamo-hypophyseal and neurosecretory systems.*

This concept has emerged from the work of many investigators. R. Cajalin (1899) discovered that the nerve fibers to the neurohypophysis are extensions of the neurons of the hypothalamus. Kopec found that the substance responsible for metamorphosis in the larva of moths originated in the brain where certain cerebral ganglia served as glands of internal secretion. The outstanding work of B. Scharrer, E. Scharrer, S. Zuckerman, I. Assenmacher, and others have greatly extended the scope of neuroendocrine studies in recent years.

1922: *Kopec, S. Demonstration of hormonal factors in invertebrate physiology.*

This investigation showed that the brain was necessary for insect metamorphosis, for when the brain was removed from the last instar larva of a certain moth, pupation failed to occur; when the brain was grafted into the abdomen, pupation was resumed.

1922: *Schiefferdecker, P. Distinction between eccrine and apocrine sweat glands.*

In mammals certain large sweat glands that develop in connection with the hair follicles in localized regions respond to stresses, such as fear, pain and sex. This concept has stimulated investigation on the histology and physiology of the glandular activity of the skin. Apocrine glands have been described as early as 1846 by W. E. Horner, an English investigator.

1922: *Schjelderup-Ebbe, T. Social dominance-subordinance hierarchies.*

This observer found certain types of social hierarchies among birds in which higher ranking individuals could peck those of lower rank without being pecked in return. Those of the first rank dominated those of the second rank, who dominated those of

the third, etc. Such an organization, once formed, may be permanent.

1922: *Schmidt, Johannes. Life history of the freshwater eel.*

The long, patient work of this oceanographer in solving the mystery of eel migration from the freshwater streams of Europe to their spawning grounds in the Sargasso Sea near the Bermudas represents one of the most romantic achievements in natural history.

1923: *Hevesy, G. First isotopic tracer method.*

Tracer methodology has proved especially useful in biochemistry and physiology. For instance, it has been possible by the use of these labeled units to determine the fate of the particular molecule in all steps of a metabolic process and the nature of many enzymatic reactions. The exact locations of many elements in the body have been traced by this method. Nobel Laureate (1943).

1923: *Warburg, Otto. Manometric methods for studying metabolism of living cells.*

The Warburg apparatus has been useful in measuring the gaseous exchange and other metabolic processes of living tissues. It has proved of great value in the study of enzymatic reactions in living systems and is a standard tool in many biochemical laboratories.

Nobel Laureate (1931).

1924: *Cleveland, L. R. Symbiotic relationships between termites and intestinal flagellates.*

This study was made on one of the most remarkable examples of evolutionary mutualism known in the animal kingdom. Equally important were the observations this investigator and others found in the symbiosis between the wood-roach *(Cryptocercus)* and its intestinal Protozoa.

1924: *Feulgen, R. Test for nucleoprotein.*

This microchemical test is widely used by cytologists and biochemists to demonstrate the presence of DNA (deoxyribonucleic acid), one of the two major types of nucleic acids.

1924: *Houssay, B. A. Role of the pituitary gland in regulation of carbohydrate metabolism.*

This investigator showed that when a dog had been made diabetic by pancreatectomy, the resulting hyperglycemia and glucosuria could be abolished by removing the anterior pituitary gland.

Nobel Laureate (1946).

1924: *Karpechenko, G. D. Experimental synthesis of a new species.*

This investigator crossed the radish *(Raphanus sativus)* with the cabbage *(Brassica oleracea),* each of which has a haploid number of 9. The hybrid had 18 chromosomes, but at meiosis its chromosomes did not pair and the gametes were sterile. But allopolyploidy arose spontaneously in a few, producing egg and sperm nuclei, each with 9 cabbage and 9 radish chromosomes. This kind of hybrid had 18 synapsed pairs and bred true, but could not breed with either of the original parents, thus giving rise of a new species *(Raphanobrassica).*

1925: *Baltzer, F. Sex determination in Bonellia.*

This classic discovery of the influence of environmental factors on sex determination, with all the potentialities of sex intergrades that it demonstrates, has formed the basis for another theory of the development of sex.

1925: *Barcroft, J. Function of the spleen.*

Barcroft and associates showed that the spleen served as a blood reservoir that in time of stress adds new corpuscles to the circulation. The spleen reservoir is especially important in hemorrhage and shock.

1925: *Dart, Raymond. Discovery of Australopithecus africanus.*

This important fossil is commonly referred to as the ape-man or the "missing link." This led to the finding of many related ape-men. With many human characteristics and a brain capacity only slightly greater than the higher apes, they have shed a great deal of light on the evolution of the higher primates, since they are placed on or near the main branch of human ancestry.

1925: *Mast, S. O. Nature of ameboid movement.*

By studying the reversible sol-gel transformation in the protoplasm of an ameba, the author not only gave a logical interpretation of ameboid movement, but also initiated many fruitful concepts about the contractile nature of protoplasmic gel systems, such as the furrowing movements (cytokinesis) in all divisions.

1925: *Minot, G. R., M. W. P. Murphy, and G. H. Whipple. Liver treatment of pernicious anemia.*

That the feeding of raw liver had a pronounced effect in the treatment of pernicious anemia (a serious blood disorder) was discovered by these investigators. Much later investigation by numerous workers has led to some understanding of the antipernicious factor or vitamin B_{12} whose chemical name is cyanocobalamin. This complex vitamin contains, among other components, porphyrin that has a cobalt atom instead of iron or magnesium at its center.

Nobel Laureates (1934).

1925: *Rowan, W. Gonadal hypothesis of bird migration.*

By increasing the hours of light by artificial illumination, Rowan demonstrated that birds subjected to such conditions in winter increased the size of their gonads and showed a striking tendency to migrate out of season. In the development of his theory he laid emphasis on the role of the pituitary gland as well as other aspects of physiologic function. The exact relationship of this hypothesis to bird migration is still largely speculative, but Rowan's experiments greatly stimulated investigation in this field.

1926: *Fujii, K. Finer analysis of the chromosome.*

With the development of the smear and squash techniques, it was possible with the light microscope to demonstrate the internal structure of a chromosome, which formerly was described as a rod-shaped body. The newer version emphasizes a coiled filament (chromonema) that runs through the matrix of the chromosome and bears the genes. In certain stages of cell division two such threads are spirally coiled around each other so compactly that they appear as one thread.

1926: *Hill, A. V. Measurement of heat production in nerve.*

By applying the principle of the thermocouple, which Helmholtz had used to detect the heat of contracting muscle, Hill and other workers were able to measure the different phases of heat release, such as initial heat and recovery heat.

Nobel Laureate (1922).

1926: *Kutscher, F., and D. Ackerman. Distribution of the phosphagens arginine and creatine.*

Creatine phosphate was found in chordate skeletal muscle and arginine phosphate in the muscle of invertebrates. These high-energy substances serve as a source of energy for muscular work, and their distribution was supposed to explain vertebrate ancestry. The sharpness of this distinction is not as pronounced as formerly supposed.

1926: *Moore, C. The thermoregulatory function of the scrotum.*

Experimental work has demonstrated that scrotal temperature may be 5° C. below rectal temperature and that higher temperatures than this often produce abnormal spermatogenesis. On this basis the external position of the scrotum is explained. However, the concept is controversial because some mammals (Cetacea, Edentata, Proboscidea, etc.) and the birds have internal testes, and some authorities believe that the scrotal sac is a visual sexual signal of selective value during mating.

1926: *Sumner, J. B. Isolation of enzyme urease.*

The isolation of the first enzyme in crystalline form was a key discovery to be followed by others that have helped unravel the complex nature of these important biologic substances.

Nobel Laureate (1946).

1927: *Bozler, E. Analysis of nerve net components.*

Bozler's demonstration that the nerve net of coelenterates was made up the separate cells and contained synaptic junctions resolved the old problem of whether or not the plexus in this group of animals was an actual network. Recent work indicates that coelenterates have both a continuous and discontinuous nerve net.

1927: *Coghill, G. E. Innate behavior patterns of Amphibia.*

Coghill's studies on the origin and growth of the behavior patterns of salamanders by following the sequence of the emergence of coordinated movements and nervous connections through all stages of embryonic development have represented one of the most fruitful investigations in animal behavior. He showed how broad, general movements preceded local reflexes and how the probable phylogenetic appearance of behavior patterns originated.

1927: *Eggleton, P., G. P. Eggleton, C. H. Friske, and Y. Subbarow. Role of phosphagen (phosphocreatine) in muscular contraction.*

The demonstration that phosphagen is broken down during muscular contraction into creatine and phosphoric acid and then resynthesized during recovery gave an entirely new concept of the initial energy necessary for the contraction process. Confirmation of this discovery received a great impetus from the discovery of E. Lungsgaard (1930) that muscles poisoned with monoiodoacetic acid, which inhibits the production of lactic acid from glycogen, would still contract and that the amount of phosphocreatine broken down was proportional to the energy liberated.

1927: *Heymans, C. Role of carotid and aortic reflexes in respiratory control.*

The carotid sinus and aortic areas contain pressoreceptors and chemoreceptors, the former responding to mechanical stimulation, such as blood pressure, and the latter to oxygen lack. When the pressoreceptors are stimulated, respiration is inhibited; when the chemoreceptors are stimulated, the respiratory rate is increased. These reflexes are of great physiologic interest, although their adaptive nature is not as apparent as the vascular control initiated from the same regions.

Nobel Laureate (1938).

1927: *Muller, H. J. Artificial induction of mutations.*

By subjecting fruit flies *(Drosophila)* to mild doses of x-rays, Muller found that the rate of mutation could be increased 150 times over the normal rate.

Nobel Laureate (1946).

1927: *Stensiö, E. A. Appraisal of the Cephalaspida (ostracoderm) fish fossil.*

The replacement of amphioxus as a prototype of vertebrate ancestry by the ammocoetes lamprey larva, currently of great interest, has been due to a great extent to this careful fossil reconstruction. It is generally believed that living Agnatha (lamprey and hagfish) are descended from these ancient forms.

1928: *Garstang, W. Theory of the ascidian ancestry of chordates.*

According to this theory, primitive chordates were sessile, filter-feeding marine organisms very similar to present-day ascidians that have evolved from pterobranch (Hemichordata) ancestors. The actively swimming prevertebrate was considered a later stage in chordate evolution. The tadpole ascidian larva, with its basic organization of a vertebrate, had evolved within the group by progressive evolution and by neoteny became sexually mature, ceased to metamorphose into a sessile, mature ascidian, and through adaptation freshwater conditions became the true vertebrate.

1928: *Griffith, F. Discovery of the transforming principle (DNA) in bacteria (genetic transduction).*

By injecting living nonencapsulated bacteria and dead encapsulated bacteria of the *Pneumococcus* strain into mice, it was found that the former acquired the ability to grow a capsule and that this ability was transmitted to succeeding generations. This active agent or transforming principle (from the encapsulated type) was isolated by other workers later and was found to consist of DNA. This is excellent evidence that the gene involved is the nucleic acid deoxyribonucleic acid (DNA).

F. Sanfelice (1893) had actually found the same principle when he discovered that nonpathogenic bacilli grown in a culture medium containing the metabolic products of true tetanus bacilli would also produce toxins and would do so for many generations.

1928: *Koller, G., and E. B. Perkins. Hormonal control of color changes in crustaceans.*

These investigators found out independently that the chromatophores of crustaceans were regulated by a substance that originated in the eyestalk and was carried by the blood. Before this time, the common belief was that nerves served as the principal control.

1928: *Wieland, H., and A. Windaus. Structure of the cholesterol molecule.*

Sterol chemistry has been one of the chief focal points in the investigation of such biologic products as vitamins, sex hormones, and cortisone. The real role of cholesterol, which is universally present in tissues, has not yet been determined, but it may serve as the precursor for the many forms of steroids whose use is definitely known. Animals make their own steroids but cannot absorb those of plants.

Nobel Laureates (1927, 1928).

1929: *Berger, Hans. Demonstration of brain waves.*

The science of electroencephalography, or the electrical recording of brain activity, is in its infancy because of the complexity of the subject, but much has been revealed about both the healthy and the diseased brain by this technique.

1929: *Butenandt, A., and E. A. Doisy. Isolation of estrone.*

This discovery was the first isolation of a sex hormone and was arrived at independently by these two investigators. Estrone was found to be the urinary and transformed product of estradiol, the actual hormone. The male hormone testosterone was synthesized by Butenandt and L. Ruzicka in 1931. The second female hormone, progesterone, was isolated from the corpora lutea of sow ovaries in 1934.

Doisy, Nobel Laureate (1943).

1929: *Castle, W. B. Discovery of the antianemic factor.*

Castle and associates showed that the gastric juice contained an enzymelike substance (intrinsic factor) that reacts with a dietary factor (extrinsic) to produce the antianemic principle. The latter is stored in the liver of healthy individuals and is drawn on for the maintenance of activity in bone marrow (erythropoietic tissue), that is, the formation of red blood cells. If the intrinsic factor is missing from gastric juice, pernicious anemia occurs. At present, authorities believe that vitamin B_{12} is both the extrinsic factor as well as the antianemic principle (erythrocyte-maturing factor).

1929: *Fleming, A. Discovery of penicillin.*

The chance discovery of this drug from molds and its development by H. Florey a few years later gave us the first of a notable line of antibiotics that have revolutionized medicine.

Nobel Laureate (1945).

1929: *Heymans, C. Discovery of the role of the carotid sinus in regulating the respiratory center and arterial blood pressure.*

It was found that the carotid bodies and similar structures on the aorta were chemosensitive to the concentration of oxygen in the blood, and by means of nerve fibers of the IX and X nerves an afferent limb of the reflex is formed. The sensitivity of these bodies is also influenced by the carbon dioxide concentration of the blood. The sensory apparatus of chemoreceptors within the sinus is also affected directly by blood hypertension to produce a reflex slowing of the heart rate and a reflex vasodilation of blood vessels, whereas a blood hypotension increases heartbeat and vasoconstriction of blood vessels.

Nobel Laureate (1938).

1929: *Lohmann, K. Discovery of ATP.*

The discovery of ATP (adenosine triphosphate) culminated a long search for the energy sources in biochemical reactions of many varieties, such as muscular contraction, vitamin action, and many enzymatic systems.

1930: *Fisher, R. A. Statistical analysis of evolutionary variations.*

With Sewall Wright and J. S. B. Haldane, Fisher has analyzed mathematically the interrelationships of the factors of mutation rates, population sizes, selection values, and others in the evolutionary process. Although many of their theories are in the empiric stage, evolutionists in general agree that they have great significance in evolutionary interpretation.

1930: *Giersberg, H., and G. H. Parker. Neurohumoral theory of color control.*

This theory, which was largely developed by

Parker, states that the terminations of the neurons, which supply the chromatophores, produce chemical substances (neurohumors) that activate the pigment cells. One type of fiber secretes a neurohumor that causes pigment dispersal; another type of fiber gives rise to a neurohumor that causes pigment aggregation. One of the neurohumors appears to be acetylcholine (dispersion effect chiefly) and the other neurohumor is similar to adrenaline or sympathin (aggregation effect). A third neurohumor (selachine) is found in the dogfish *Mustelus* and has the same effect as adrenaline.

1930: *Lawrence, E. O. Invention of the cyclotron.*

The importance of artificial radioactive isotopes, the synthesis of which is made possible by this invention, to biologic research cannot be overestimated.

Nobel Laureate (1939).

1930: *Northrop, J. H. Crystallization of the enzymes pepsin and trypsin.*

This was a further step in the elucidation of the nature of enzymes that form the core of biochemical processes.

Nobel Laureate (1946).

1930: *Papa, G. T., and U. Fielding. Hypophyseal portal vessels.*

First discovered in mammals, these vessels are constant in all vertebrates from Anura to primates. Similar in cyclostomes, fishes, and salamanders, the hypothalamus can influence adenohypophyseal activities through humoral effects carried by these vessels.

1931: *Lewis, W. H. Concept of pinocytosis.*

This refers to a discontinuous process of fluid engulfment by cells, in contrast to diffusion. Many types apparently exist, but often the process involves membranous pseudopodia that enclose droplets of surrounding fluid. These vesicles are then pinched off and sucked into the interior of the cell. The phenomenon may also be concerned with the uptake and transport of substances by cells.

1931: *Lorenz, K., and N. Tinbergen. Theory of instinctive behavior.*

The upsurge of interest (second only to molecular biology) in animal behavior within recent years has been the result of the investigations of these men and their school concerning the problems of instinct and instinctive behavior. The interest they have aroused has been a result of their attempt to determine the physiologic basis of the animal's reactions to stimuli by focusing attention on the instinctive behavior pattern that controls innate behavior, by studying the internal conditions of the animal that are organized for particular patterns, and by showing the neural mechanisms that are correlated with these patterns.

Nobel Laureates (1973).

1931: *Stern, C., H. Creighton, and B. McClintock. Cytologic demonstration of crossing-over.*

Proof that crossing-over in genes is correlated with exchange of material by homologous chromosomes was independently proved by Stern in *Drosophila* and by Creighton and McClintock in corn. By using crosses of strains that had homologous chromosomes distinguishable individually, it was definitely demonstrated cytologically that genetic crossing-over was accompanied by chromosomal exchange.

1932: *Bethe, A. Concept of the ectohormone (pheromone).*

These substances are secreted to the outside of the body by an organism; another member of the same species may react to the substance in some behavioral or developmental way. These substances have been demonstrated in insects, where they may function in trail-making, sex attraction, development control, etc. (Gr. *pherein*, to carry, + *hormōn*, exciting, stirring up.)

1932: *Danish scientific expedition. Discovery of fossil amphibians (ichthyostegids).*

These fossils were found in the upper Devonian sediments in east Greenland and appear to be intermediate between advanced crossopterygians (*Osteolepis*) and early amphibians. They are the oldest known forms that can be considered amphibians. Many of their characters show primitive amphibian conditions.

1932: *Wright, S. Genetic drift as a factor in evolution.*

In small populations the Hardy-Weinberg formula of gene frequency may not apply because chance may determine the presence or absence of certain genes and this tendency will be expressed in the gene frequency of the new population, which may be quite different from the original large population.

1932: *Zondek, B., and H. Krohn. Effect of intermedin on melanophores.*

This hormone (now called the melanocyte-stimulating hormone, or MSH) is known to effect pigment dispersion in amphibians and other lower vertebrates, but its role in higher vertebrates is debatable. The pars intermedia of the pituitary is commonly considered to be its site of origin.

1933: *Collander, R., and H. Bärlund. Measurement of cell permeability.*

Their quantitative measurements of cell membranes made possible testable hypotheses and critical analyses of experimental data of the important biologic principle of permeability.

1933: *Goldblatt, M., and U. S. von Euler. Discovery of prostaglandins.*

These fatty acid derivative compounds have been isolated from many mammalian tissues (seminal plasma, pancreas, seminal vesicle, brain, kidney, etc.). Their biologic activities are obscure because many compounds of different functions are involved, but pharmacologic evidence indicates that they stimulate smooth muscle contractions and relaxation, lower blood pressure, inhibit enzymes and hormones, etc.

1933, 1938: *Haldane, J. B. S., and A. I. Oparin. Heterotroph theory of the origin of life.*

This theory is based on the idea that life was generated from nonliving matter under the conditions that existed before the appearance of life and that have not been duplicated since. The theory stresses the idea that living systems at present make it impossible for any incipient life to gain a foothold as primordial life was able to do.

1933: *Painter, T. S., E. Heitz, and H. Bauer. Rediscovery of giant salivary chromosomes.*

These interesting chromosomes were first described by Balbiani in 1881, but their true significance was not realized until these investigators rediscovered them. It has been possible in a large measure to establish the chromosome theory of inheritance by comparing the actual cytologic chromosome maps of salivary chromosomes with the linkage maps obtained by genetic experimentation.

1933: *Wald, G. Discovery of vitamin A in the retina.*

The discovery that vitamin A is a part of the visual purple molecule of the rods not only gave a better understanding of an important vitamin but also showed how night blindness can occur whenever there is a deficiency of this vitamin in the diet.

Nobel Laureate (1967).

1934: *Bensley, R. R., and N. L. Hoerr. Isolation and analysis of mitochondria.*

This demonstration has suggested an explanation for the behavior and possible functions of these mysterious bodies that have intrigued cytologists for a generation, and much has been learned about them in recent years.

1934: *Dam, H., and E. A. Doisy. Identification of vitamin K.*

The isolation and synthesis of this vitamin is important not merely because of its practical value in certain forms of hemorrhage but also because of the light it throws on the physiologic mechanism of blood clotting.

Nobel Laureates (1943).

1934: *Danielli, J. F. Concept of the cell (plasma) membrane.*

Danielli had proposed a hypothesis that cell membranes consist of two layers of lipid molecules surrounded on the inner and outer surfaces by a layer of protein molecules. The electron microscope reveals the plasma membrane of a thickness of 75 to 100 Å, consisting of two dark (protein) membranes (25 to 30 Å thick) separated by a light (lipid) interval membrane of 25 to 30 Å thickness. The exact relationship of the protein and lipid constituents has never been resolved.

1934: *Wigglesworth, V. B. Role of the corpus allatum gland in insect metamorphosis.*

This small gland lies close to the brain of an insect, and it has been shown that during the larval stage this gland secretes a juvenile hormone that causes the larval characters to be retained. Metamorphosis occurs when the gland no longer secretes the hormone. Removal of the gland causes the larva to undergo precocious metamorphosis; grafting the gland into a mature larva will cause the latter to grow into a giant larval form. The gland was first described by A. Nabert in 1913.

1935: *DuShane, G. P. Role of the neural crest in pigment cell formation.*

DuShane's discovery that pigment cells in amphibians originated from the neural crest was quickly followed by other investigations that showed that other groups (fish, birds, mammals) had the same pattern of pigment formation. The current interest in the relation of the pigment cell to certain malignant growths (melanoblastomas), which may be considered immature pigment cells induced by some metabolic error, has been focused on all possible aspects of the pigment cell.

1935: *Hanstrom, B. Discovery of the x organ in crustaceans.*

This organ, together with the related sinus gland, constitutes an anatomic complex that has proved of great interest in understanding crustacean endocrinology. One view holds that neurosecretory cells in the x organ and the brain produce a molt-preventing hormone that is stored in the sinus gland of the eyestalk; other theories postulate a molt-accelerating hormone produced in a y organ. The interrelations of these two hormones may be responsible for the molting process.

1935-1936: *Kendall, E. C., and P. S. Hench. Discovery of cortisone.*

Kendall had first isolated from the adrenal glands this substance that he called compound E. Its final stages were prepared by Hench later and involved a long tedious chemical process. A similar hormone, as far as its effects are concerned, was isolated in 1943 from the pituitary and called ACTH (adrenocorticotropic hormone).

Nobel Laureates (1950).

1935: *Needham, J., and C. H. Waddington. Chemical nature of the organizer region in embryology.*

The major effect of the organizer region described by Spemann and others was really due to the production by that region of a specific evocator substance closely related to the sterols and other chemical compounds.

1935: *Stanley, W. M. Isolation of a virus in crystalline form.*

This achievement of isolating a virus (tobacco mosaic disease) is not merely noteworthy in giving information about these small agencies responsible for many diseases but also in affording much speculation on the differences between the living and the nonliving. The viruses appear to be a transition stage between the animate and the inanimate.

Nobel Laureate (1946).

1935: *Tansley, A. G. Concept of the ecosystem.*

The relatively recent science of ecology has added many new terms, but the ecosystem is considered the basic functional unit in ecology, for it best expresses the environmental relations of organisms in their entirety. It includes both the biotic and abiotic factors, and the concept under different terminology had been used by others in the early development of ecology.

1935: *Timofeeff-Ressovsky, N. W. Target theory of induction of gene mutations.*

Timofeeff-Ressovsky's discovery that mutation can be induced in a gene if a single electron is detached by high-energy radiation gave rise to one of the two prevailing theories of how radiation affects mutation rate.

1936: *Demerec, M., and M. E. Hoover. Correspondence between salivary gland chromosome bands and normal chromosome maps.*

By means of three stocks of *Drosophila,* each with a different deficiency at one end of the X chromosome, these investigators were able to find approximately on the giant chromosomes the location of the same genes found on the normal chromosome maps constructed by the percent of crossing-over.

1936: *Stern, C. Discovery of mitotic crossing-over.*

Although common in meiosis, crossing-over in somatic mitosis occasionally occurs. Crossing-over in mitosis takes place after the chromosomes have split, and the crossing-over usually takes place near the centromere in *Drosophila* where the process was originally found. Mitotic crossing-over may result in the production of sister cells homozygous for genes that were heterozygous in the original cell.

1936: *Young, J. Z. Demonstration of giant fibers in squid.*

These giant fibers are formed by the fusion of the axons of many neurons whose cell bodies are found in a ganglion near the head. Each fiber is really a tube, more than 1 mm. wide, consisting of an external sheath filled with liquid axoplasm. These giant fibers control the contraction of the characteristic mantle that surrounds these animals.

1937: *Blakeslee, A. F. Artificial production of polyploidy.*

By applying the drug colchicine to dividing cells, he found that cell division in plants is blocked after the chromosomes have divided (metaphase), and thus the cell has double the normal number of chromosomes. When applied to hybrid plants, this principle makes possible the production of new plants.

1937: *Findlay, G. W. M., and F. O. MacCullum. Discovery of interferon.*

These protein substances, produced by cells in response to viruses and other foreign substances, inhibit selectively virus replication. Their potential as antiviral agents is still in the experimental stage.

1937: *Krebs, H. A. Citric acid (tricarboxylic) cycle.*

This theory of aerobic carbohydrate oxidation (through stages involving citric acid), which is supposed to occur in most living cells, involves a cycle of linked reactions under the influence of many enzymes (mainly from mitochondria). The scheme consists of many intermediary stages and aims to show how pyruvic acid (a derivative of carbohydrate oxidation) is converted to carbon dioxide and water. The cycle is believed to be the final common path for the oxidation of fatty acids, amino acids, and carbohydrates and represents the chief source of chemical energy in the body.

Nobel Laureate (1953).

1937: *Sonneborn, T. M. Discovery of mating types in paramecium.*

Sonneborn's discovery that only individuals of complimentary physiologic classes (mating types) would conjugate opened up a new era of protozoan investigation that bids to shed new light on the problems of species concept and evolution.

1937: *Werle, E., W. Gotze, and A. Keppler. Discovery of kinins.*

Kinins are local hormones produced in blood or tissues that bring about dilation of blood vessels and other changes. They are also found in wasp venoms. They have no connection with special glands, are peptides in nature, are very evanescent, perform their functions rapidly, and are then quickly inactivated by enzymes. They also have powerful effects on smooth muscle wherever it is found.

1938: *Remane, A. Discovery of the new phylum Gnathostomulida.*

This marine phylum was first described by P. Ax in 1956, and its taxonomic position is still under appraisement. The members of the phylum are small wormlike forms (about $1/2$ mm. long) and show great diversity among the different species. One of their major characteristics is their complicated jaws provided with teeth. They live in sandy substrata and are world wide in distribution. They show some relationships to the Turbellaria, Gastrotricha, and Rotifera.

1938: *Schoenheimer, R. Use of radioactive isotopes to demonstrate synthesis of bodily constituents.*

By labeling amino acids, fats, carbohydrates, etc., with radioactive isotopes, it was possible to show how these were incorporated into the various constituents of the body. Such experiments demonstrated that parts of the cell were constantly being synthesized and broken down and that the body must be considered a dynamic equilibrium.

1938: *Skinner, B. F. Measurement of motivation in animal behavior.*

Skinner worked out a technique for measuring the rewarding effect of a stimulus, or the effects of learning on voluntary behavior. His experimental animals (rats) were placed in a special box (Skinner's box) containing a lever that the animal could manipulate. When the rat presses the lever, small pellets of food may or may not be released, according to the experimental conditions.

1938: *Svedberg, T. Development of ultracentrifuge.*

In biologic and medical investigation this instrument has been widely used for the purification of substances, the determination of particle sizes in colloidal systems, the relative densities of materials in living cells, the production of abnormal development, and the study of many problems concerned with electrolytes.

Nobel Laureate (1926).

1939: *Brown, F. A., Jr., and O. Cunningham. Demonstration of molt-preventing hormone in eyestalk of crustaceans.*

Although C. Zeleny (1905) and others had shown that eyestalk removal shortened the intermolt period in crustaceans, Brown and Cunningham were the first to present evidence to explain the effect as being caused by a molt-preventing hormone present in the sinus gland.

1939: *Discovery of coelacanth fish.*

The collection of a living specimen of this ancient fish *(Latimeria),* followed later by other specimens, has brought about a complete reappraisal of this "living fossil" with reference to its ancestry of the amphibians and land forms.

1939: *Hoerstadius, S. Analysis of the basic pattern of regulative and mosaic eggs in development.*

The masterful work of this investigator has done much to resolve the differences in the early development of regulative eggs (in which each of the early blastomeres can give rise to a whole embryo) and mosaic eggs (in which isolated blastomeres produce only fragments of an embryo). Regulative eggs were shown to have two kinds of substances and both were necessary in proper ratios to produce normal embryos. Each of the early blastomeres has this proper ratio and thus can develop into a complete embryo; in mosaic eggs the regulative power is restricted to a much earlier time scale in development (before cleavage), and thus each isolated blastomere will give rise only to a fragment.

1939: *Huxley, J. Concept of the cline in evolutionary variation.*

This concept refers to the gradual and continuous variation in character over an extensive area because of adjustments to changing conditions. This idea of character gradients has proved a very fruitful one in the analysis of the mechanism of evolutionary processes, for such a variability helps to explain the initial stages in the transformation of species.

1939: *Pincus, G. Artificial parthenogenesis of the mammalian egg.*

This experiment of producing a normal, fatherless rabbit showed that Loeb's classic method could be made to apply to the eggs of the highest group of animals, and that the primary physiologic process of fertilization is the activation of the egg.

1940: *Landsteiner, Karl, and A. S. Wiener. Discovery of Rh-blood factor.*

Not only was a knowledge of the Rh factor of importance in solving a fatal infant's disease but it has also yielded a great deal of information about relationships of human races.

Landsteiner, Nobel Laureate (1930).

1940: *Timofeeff-Ressovsky, N. W. Cyclic changes in genetics in populations of species.*

Red and black elytral color patterns of a certain beetle *(Adalia)* differed by a single mendelian gene. The black gene had a greater selective value during the summer, whereas the red had a greater survival during hibernation. This seasonal inequality in population phases has been found in a number of wild species and is an example of polymorphism.

1941: *Beadle, G. W., and E. L. Tatum. Biochemical mutation.*

By subjecting the bread mold *Neurospora* to x-ray irradiation, they found that genes responsible for the synthesis of certain vitamins and amino acids were inactivated (mutated) so that a strain of this mold carrying the mutant genes could no longer grow unless these particular vitamins and amino acids were added to the medium on which the mold was growing. This outstanding discovery has revealed as never before the precise way in which a single gene controls the specificity of a particular enzyme and

has greatly stimulated similar research on other simple forms of life such as bacteria and viruses.

Nobel Laureates (1958).

1941: *Cori, C. F., and G. T. Cori. Lactic acid metabolic cycle.*

The regeneration of muscle glycogen reserves in mammals involves the passage of lactic acid from the muscles through the blood to the liver, the conversion of lactic acid there to glycogen, the production of blood glucose from the liver glycogen, and the synthesis of muscle glycogen from the blood glucose.

Nobel Laureates (1947).

1941: *Szent-Györgyi, Albert von. Role of ATP in muscular contraction.*

The demonstration showing that muscles get their energy for contraction from ATP (adenosine triphosphate) has done much to explain many aspects of the puzzling problem of muscle physiology.

Nobel Laureate (1937).

1942: *McClean, D., and I. M. Rowlands. Discovery of the enzyme hyaluronidase in mammalian sperm.*

This enzyme dissolves the cement substance of the follicle cells that surround the mammalian egg and facilitates the passage of the sperm to the egg. This discovery not only aided in resolving some of the difficult problems of the fertilization process but also offers a logical explanation of cases of infertility in which too few sperm may not carry enough of the enzyme to afford a passage through the inhibiting follicle cells.

1943: *Claude, A. Isolation of cell constituents.*

By differential centrifugation, Claude found it possible to separate, in relatively pure form, particulate components, such as mitochondria, microsomes, and nuclei. These investigations led immediately to a more precise knowledge of the chemical nature of these cell constituents and aided the elucidation of the structure and physiology of the mitochondria— one of the great triumphs in the biochemistry of the cell.

1943: *Holtfreter, J. Tissue synthesis from dissociated cells.*

By dissociating the cells of embryonic tissues of amphibians (by dissolving with enzymes or other agents the intercellular cement that holds the cells together) and heaping them in a mass, he found that the cells in time coalesced and formed the type of tissue from which they had come. This is an application to vertebrates of the discovery of Wilson with sponge cells.

1943: *Sonneborn, T. M. Extranuclear inheritance.*

The view that in paramecia cytoplasmic determiners (plasmagenes) that are self-reproducing and capable of mutation can produce genetic variability has thrown additional light on the role of the cytoplasm in hereditary patterns.

1944: *Avery, O. T. C., C. M. MacLeod, and M. McCarty. Agent responsible for bacterial transformation.*

These workers were able to show that the bacterial transformation of nonencapsulated bacteria to encapsulated cells was really attributable to the DNA fraction of debris from disrupted encapsulated cells to which the nonencapsulated cells were exposed. This key demonstration showed for the first time that proteins cannot be the basic structure of hereditary transmission but that this role is taken by nucleic acids.

1945: *Cori, Carl. Hormone influence on enzyme activity.*

The delicate balance that insulin and the diabetogenic hormone of the pituitary exercise over the activity of the enzyme hexokinase in carbohydrate metabolism has opened up a whole new field of the regulative action of hormones on enzymes.

Nobel Laureate (1947).

1945: *Griffin, D., and R. Galambos. Development of the concept of echolocation.*

Echolocation refers to a type of perception of objects at a distance by which echoes of sound are reflected back from obstacles and detected acoustically. These investigators found that bats generated their own ultrasonic sounds that were reflected back to their own ears so that they were able to avoid obstacles in their flight without the aid of vision. Their work climaxed an interesting series of experiments inaugurated as early as 1793 by Spallanzani, who believed that bats avoided obstacles in the dark by reflection of sound waves to their ears. Others who laid the groundwork for the novel concept were C. Jurine (1794), who proved that ears were the all-important organs in the perception; H. S. Maxim (1912), who advanced the idea that the bat made use of sounds of low frequency inaudible to human ears; and H. Hartridge (1920) who proposed the hypothesis that bats emitted sounds of high frequencies and short wavelengths (ultrasonic sounds).

1945: *Lipmann, F. Discovery of coenzyme A.*

The discovery of this important catalyst made possible a better understanding of the breaking down of fatty acid chains and furnished an important link in the reactions of the Krebs metabolic cycle.

Nobel Laureate (1953).

1945: *Porter, K. R. Description of the endoplasmic reticulum.*

The endoplasmic reticulum is a very complex cytoplasmic structure consisting of a lacelike network of irregular anastomosing tubules and vesicular expansions within the cytoplasmic matrix. Associated with the reticulum complex are small dense granules of ribonucleoprotein and other granules known as microsomes that are fragments of the endoplasmic reticulum. The reticulum complex is supposed to play an important role in the synthesis of proteins and RNA.

1946: *Auerbach, C., and J. M. Robson. The chemical production of mutations.*

Besides radiation effects (Muller, 1927), many inorganic and organic compounds (mustard gas, oils, alkaloids, phenols, etc.) are now known to have mutagenic effects such as chromosome breakage. It is believed that such agents disrupt the nucleic acid metabolism responsible for protein synthesis and reduplication.

1946: *Lederberg, J., and E. L. Tatum. Sexual recombination in bacteria.*

These investigators found that two different strains of bacteria *(Escherichia coli)* could undergo conjugation and exchange genetic material, thereby producing a hereditary strain with characteristics of the two parent strains. W. Hayes (1952) found that recombination still occurred after one parent strain was killed.

Nobel Laureates (1958).

1946: *Libby, W. F. Radiocarbon dating of fossils.*

The radiocarbon age determination is based on the fact that carbon 14 in the dead organism disintegrates at the rate of one half in 5,560 years, one half of the remainder in the next 5,560 years, etc. This is on the assumption that the isotope is mixed equally through all living matter and that the cosmic rays (which form the isotopes) have not varied much in periods of many thousands of years. The limitation of the method is around 40,000 years.

Nobel Laureate (1960).

1946: *White, E. I. Discovery of the primitive chordate fossil Jamoytius.*

The discovery of this fossil in the freshwater deposits of Silurian rock in Scotland bids to throw some light on the early ancestry of vertebrates, for this form seems to be intermediate between amphioxus or ammocoetes larva (of lampreys) and the oldest known vertebrates, the ostracoderms. Morphologically, *Jamoytius* represents the most primitive chordate yet discovered. It could well serve as an ancestor of such forms as amphioxus and the jawless ostracoderms.

1947: *Holtz, P. Discovery of norepinephrine (noradrenaline).*

This hormone (vasoconstriction effects) has since been found in most vertebrates and shares with epinephrine (metabolic effects) the functions of the chromaffin part of the adrenal gland.

1947: *Sprigg, R. C. Discovery of Precambrian fossil bed.*

The discovery of a rich deposit of Precambrian fossils in the Ediacara Hills of South Australia has been of particular interest because the scarcity of such fossils in the past has given rise to vague and uncertain explanations about Precambrian life. It was all the more remarkable that the fossils discovered were those of soft-bodied forms, such as jellyfish, soft corals, and segmented worms, including the amazing *Spriggina*, which shows relationship to the trilobites. The fossils are prearthropod, but are not preannelid.

1947: *Szent-Györgyi, A. Concept of the contractile substance actomyosin.*

This protein complex consists of the two components, actin and myosin, and is considered the source of muscular contraction when triggered by ATP. Neither actin nor myosin singly will contract.

1948: *Frisch, Karl von. Communication patterns of honeybees.*

After many years of patient work on bees, von Frisch has been able to unravel some of the amazing patterns of behavior bees possess in conveying information to each other about the distance, direction, and sources of food supplies—an outstanding demonstration of animal behavior.

Nobel Laureate (1973).

1948: *Hess, W. R. Localization of instinctive impulse patterns in the brain.*

By inserting electrodes through the skull, fixing them in position, and allowing such holders to heal in place, it was possible to study the brain of an animal in its ordinary activities. When the rat could automatically and at will stimulate itself by pressing a lever, it did so frequently when the electrode was inserted in the hypothalamus region of the brain, indicating a pleasure center. In this way, by placing electrodes at different centers, rats can be made to gratify such drives as thirst, sex, and hunger.

Nobel Laureate (1949).

1948: *Hogeboom, G. H., W. C. Schneider, and G. E. Palade. Separation of mitochondria from the cell.*

This was an important discovery in unraveling the amazing enzymatic activity of the mitochondria in the Krebs cycle. The role mitochondria play in the energy transfer of the cell has earned for these rod-shaped bodies of the appellation of the "powerhouse" of the cell.

1948: *Johnson, M. W. Relation between echo-sounding and the deep-scattering layer of marine waters.*

The development of a sound transmitter and a receiver coupled with a timing mechanism for recording the time between an outgoing sound impulse and the echo of its return has made possible an accurate method for determining depths in the ocean. By means of this device a deep-scattering layer far above the floor of the ocean was discovered that scattered the sound waves and sent back echoes. This scattering layer tends to rise toward the surface at night and sink to a depth of many hundred meters by day. Johnson saw a striking parallelism between the shifting of this scattering layer and the diurnal vertical migration of plankton or pelagic animals.

1949: *Barr, M. L., and E. G. Bertram. Sex differences in nuclear morphology—nucleolar satellite.*

The discovery that an intranuclear body (nucleolar satellite) was far better developed in the female mammalian cell has aroused the interesting possibility of detecting the genetic sex of an individual by microscopic examination. The body, which is about $1\ \mu$ in diameter, appears as a satellite to the large nucleolus of the female nucleus. The present explanation is that the two X chromosomes of the female are responsible for the nucleolus. The male satellite (with one X chromosome), if present, is too small to see. Some cells in the body show the phenomenon better than others.

1949: *de Duve, C. Discovery of the lysosome organelle.*

These small particles were first identified chemically, and later (1955) morphologically with the electron microscope. They are supposed to provide the enzymes for digestion of materials taken into the cell by pinocytosis and phagocytosis and for the digestion of the cell's own cytoplasm. When the cell dies, their enzymes are also released and digest the cell (autolysis).

1949: *Enders, J. F., F. C. Robbins, and T. H. Weller. Cell culture of animal viruses.*

These investigators found that poliomyelitis virus could be grown in ordinary tissue cultures of nonnervous tissue instead of being restricted to host systems of laboratory animals or embryonated chick eggs.

Nobel Laureates (1954).

1949: *von Euler, U. S. Role of norepinephrine as transmitter.*

Von Euler discovered that norepinephrine served as a transmitter at the nerve terminals of the sympathetic nervous system.

Nobel Laureate (1970).

1949: *Pauling, L. Genic control of protein structure.*

Pauling and his colleagues demonstrated a direct connection between specific chemical differences in protein molecules and alterations in genotypes. Making use of the hemoglobin of patients with sickle cell anemia (which is caused by a homozygous condition of an abnormal gene), he was able by the method of electrophoresis to show a considerable difference in the behavior of this hemoglobin in an electric field compared with that from a heterozygote or from a normal person.

Nobel Laureate (1954).

1949: *Selye, H. Concept of the stress syndrome.*

In 1937 Selye began his experiments, which led to what he called the "alarm reaction" that involved the chain reactions of many hormones, such as cortisone and ACTH, in meeting stress conditions faced by an organism. Whenever the stress experience exceeds the limitations of these body defenses, serious degenerative disorders may result.

1950: *Callan, H. G., and S. G. Tomlin. The bilamellar organization of the nuclear envelope.*

By means of the electron microscope, these investigators were the first to show that the nuclear envelope was composed of two membranes that are interrupted by discontinuities.

1950: *Caspersson, T. Biosynthesis of proteins.*

Investigations to solve the vital problem of protein synthesis from free amino acids have been under way since Fischer's outstanding work on protein structure. Many competent research workers, such as Bergmann, Lipmann, and Schoenheimer, have made contributions to an understanding of protein synthesis, but Caspersson and Brachet were the first to point out the significant role of ribonucleic acid (RNA) in the process—and most biochemical investigations on the problem since that time have been directed along this line.

1950: *Chargaff, E. Base composition of DNA.*

The discovery that the amount of purines was equal to the amount of pyrimidines in DNA, the amount of adenine was equal to that of thymine, and the amount of cytosine was equal to that of guanine paved the way for the DNA model of Watson and Crick. The two major functions of DNA are replication and information storage.

1950: *Hadzi, J. Theory of the origin of metazoans.*

The resemblance between multinucleate ciliates and acoelous flatworms has formed the basis of this theory, which was proposed by Sedgwick many years ago and was largely ignored by contemporary zoologists. It has been studied for many years by Hadzi, who has brought forth many logical reasons in its support. This new point of view supports the early views of Lankester and Metschnikoff, that the original diploblastic ancestor was solid rather than hollow and that the formation of the archenteron is a secondary process.

1950: *Simpson, M., and C. H. Li. Coordination of hormones for balanced development of a tissue.*

One hormone may control the size of an organ or tissue and a different hormone may be responsible for its maturation. Thus the growth hormone of the adenohypophysis may cause a bone to grow in length, but thyroxine from the thyroid is necessary for a fully differentiated bone.

1951: *Lewis, E. B. Concept of pseudoallelism in genetics.*

On the basis of a series of mutations in *Drosophila*, Lewis concluded that certain adjacent loci were closely linked, affected the same trait, and probably arose from a common ancestral gene instead of being a single locus with multiple genes.

1952: *Beermann, W. Concept of the chromosomal puff.*

Puffs are local and reversible enlargements in the bands or loci of giant chromosomes. It is believed that puffs are related to the synthesis of genetic substance such as RNA. Puffing may indicate a correlation between hormonal action and larval development in insect metamorphosis, in which the phenomenon was first discovered.

1952: *Briggs, R., and T. J. King. Demonstration of possible differentiated nuclear genotypes.*

The belief that all cells of a particular organism have the same genetic endowment has been questioned as the result of the work of these investigators, who transplanted nuclei of different ages and sources from blastulas and early gastrulas into enucleated zygotes and got varied abnormal developmental results.

1952: *Chase, M., and A. D. Hershey. DNA as the basis of gene structure.*

By using radioactive isotopes, they showed that when a bacterium is infected by a bacterial virus only the viral DNA enters the host cell, with the protein coat remaining behind. Compare with H. Fraenkel-Conrat and R. C. Williams (1955) who were able to separate the protein coat from the RNA core in tobacco mosaic virus.

1952: *Danish Galathea Expedition. Discovery of primitive mollusks.*

The discovery of these interesting forms (class or order Monoplacophora) off the coast of Mexico in deep water represents the most important "living fossils" since the discovery of *Latimeria*. With a round, limpetlike shell and definite segments, this type may represent an intermediate form between the ancestors of annelids and that of mollusks. Its exact status has not been appraised.

1952: *Kramer, G. Orientation of birds to positional changes of sun.*

This discovery showed that birds (starlings and pigeons) can be trained to find food in accordance with the position of the sun. He found that the general orientation of the birds shifted at a rate (when exposed to a constant artificial sun) that could be predicted on the basis of the birds' correcting for the normal rotation of the earth. Birds were able to orient themselves in a definite direction with reference to the sun, whether the light of the sun reached them directly or was reflected by mirrors. They were capable also of finding food at any time of day, thus indicating an ability to compensate for the sun's motion across the sky.

1952: *Lederberg, J., and E. Lederberg. The replica technique in selection of bacterial mutants.*

An exact replica of patterns of bacterial colonies in one Petri dish treated with streptomycin could be transferred by a velveteen disk to a second Petri dish where only streptomycin resistant bacteria appear in the culture, thus indicating that the mutations for streptomycin resistance were present before exposure to the antibiotic.

1952: *Palade, G. E. Analysis of the finer structure of the mitochondrion.*

The important role of mitochondria in the enzymatic systems and cellular metabolism has focused much investigation on the structure of these cytoplasmic inclusions. Each mitochondrion is bounded by two membranes; the outer is smooth and the inner is thrown into small folds or cristae that project into a homogenous matrix in the interior. Some modifications of this pattern are found.

1952: *Zinder, N., and J. Lederberg. Discovery of the transduction principle.*

Transduction is the transfer of DNA from one bacterial cell to another by means of a phage. It occurs when an infective phage picks up from its disintegrated host a small fragment of the host's DNA and carries it to a new host where it becomes a part of the genetic equipment of the new bacterial cell.

J. Lederberg Nobel Laureate (1958).

1953: *Crick, F. H. C., J. D. Watson, and M. H. F. Wilkins. Chemical structure of DNA.*

Crick and Watson formulated the hypothesis that the DNA molecule was made up of two chains twisted around each other in a helical structure and cross-linked by pairs of bases—adenine and thymine or guanine and cytosine. Genes are considered to be segments of these molecules. Each of these complementary strands act as a model or template to form a new strand. The hypothesis has been widely accepted and affords a clue to the chemical structure of inheritance and the way chromosomes duplicate themselves.

Nobel Laureates (1962).

1953: *Lwoff, A. Concept of the prophage.*

When a bacterial virus enters a bacterial cell, it may enter a vegetative state and produce more phages, or it may become a property of the bacterial cell and become incorporated into the genetic material of its host. The bacterial cell can then reproduce itself and the phage through many generations. The term "prophage" refers to the hereditary ability to produce bacteriophages under such conditions.

Nobel Laureate (1965).

1953: *Palade, G. E. Discovery of cytoplasmic ribosomes.*

The discovery of ribonucleic acid–rich granules (usually on the endoplasmic reticulum) represents one of the key links in the unraveling of the genetic code. The ribosomes (about 250 Å in diameter) serve as the site of protein synthesis; the number of ribosomes (polysomes) involved in the synthesis of a protein depends on the length of messenger RNA and the protein being synthesized.

1953: *Urey, H., and S. Miller. Demonstration of the possible primordia of life.*

By exposing a mixture of water vapor, ammonia, methane, and hydrogen gas to electric discharge (to simulate lightning) for several days, these investigators found that several complex organic substances, such as the amino acids glycine and alanine, were formed when the water vapor was condensed into water. This demonstration offered a very plausible theory to explain how the early beginnings of life substances could have started by the formation of organic substances from inorganic ones.

1954: *Dan, J. C. Acrosome reaction.*

In echinoderms, annelids, and mollusks he has shown that the acrosome region of the spermatozoon forms a filament and releases an unknown substance at the time of fertilization. Evidence seems to indicate that the filament is associated with the formation of the fertilization cone. Other observers had described similar filaments before Dan made his detailed descriptions. The filament (about 25 μ long) may play an important role in the entrance of the sperm into the cytoplasm.

1954: *Du Vigneaud, V. Synthesis of pituitary hormones.*

This investigator isolated the posterior pituitary hormones, oxytocin and vasopressin. Both were found to be polypeptides of amino acids, and oxytocin was the first polypeptide hormone to be produced artificially. Oxytocin contracts the uterus during childbirth and releases the mother's milk; vasopressin raises blood pressure and decreases urine production.

1954: *Katz, B. Impulse transmission at nerve junction.*

The active transmitter substance was found to be acetylcholine, which is stored at the nerve endings in quantum packets and released in the passage of nerve impulses.

Nobel Laureate (1970).

1954: *Huxley, H. E., A. F. Huxley, and J. Hanson. Theory of muscular contraction.*

By means of electron microscopic studies and x-ray diffraction studies these investigators showed that the proteins actin and myosin were found as separate filaments that apparently produced contraction by a sliding reaction in the presence of ATP. The concept currently has been widely accepted.

A. F. Huxley, Nobel Laureate (1963).

1954: *Loomis, W. F. Sexual differentiation in Hydra.*

By discovering that high pressures of free carbon dioxide and reduced aeration in stagnant water (and not temperature as formerly supposed) induces sexuality in *Hydra,* an explanation was found for the large number of sexual forms found in the fall when many individuals crowd together and build up gas pressure generated by their respiration.

1954: *Sanger, F. Structure of the insulin molecule.*

Insulin is the important hormone used in the treatment of diabetes. The discovery of its structure was the first complete description of a protein molecule. The molecule was found to be made up of 17 different amino acids in 51 amino acid units. Although one of the smallest proteins, its formula contains 777 atoms.

Nobel Laureate (1958).

1955: *Fraenkel-Conrat, H., and R. C. Williams. Analysis of the chemical nature of a virus.*

In tobacco mosaic virus these workers were able to separate the protein, which makes up the outer cylinder of the virus, from the nucleic acid, the inner core of the cylinder. Neither the protein fraction nor the nucleic acid by itself was able to grow or infect tobacco, but when the two fractions were recombined, the resulting particles behaved like the original virus. Hybrids were also produced by combining the protein of one strain with nucleic acid of a different strain. In the case of hybrids, the progeny assume the properties of the virus from which the nucleic acid came.

1955-1957: *Kornberg, A., and S. Ochoa. Biologic synthesis of nucleic acids.*

By mixing the enzyme polymerase, extracted from the bacterium *Escherichia coli,* with a mixture of nucleotides and a tiny amount of DNA, Kornberg was able to produce synthetic DNA. Ochoa obtained the synthesis of RNA in a similar manner by using the enzyme polynucleotide phosphorylase from the bacterium *Azotobacter vinelandii.* This significant work reveals more insight into the mechanism of nucleic acid duplication in the cell.

Nobel Laureates (1959).

1956: *Bekesy, G. von. The traveling wave theory of hearing.*

Helmholtz (1868) had proposed the resonance theory of hearing on the basis that each cross fiber of the basilar membrane, which increases in width from the base to the apex of the cochlea, resonates at a different frequency; von Bekesy showed that a traveling wave of vibration is set up in the basilar membrane and reaches a maximal vibration in that part of the membrane appropriate for that frequency.

Nobel Laureate (1961).

1956: *Borsook, H., and P. C. Zamecnik. Site of protein synthesis.*

By injecting radioactive amino acids into an animal, they found that the ribosome of the endoplasmic reticulum is the place where proteins are formed.

1956: *Ingram, V. M. Nature of a mutation.*

By tracing the change in one amino acid unit out of more than 300 units that make up the protein hemoglobin, Ingram was able to pinpoint the difference between normal hemoglobin and the mutant form of hemoglobin that causes sickle cell anemia.

1956: *Peart, W. S., and D. F. Elliot. Isolation of angiotensin.*

Ever since Volhard in 1928 suggested that a substance in the kidney might be responsible for certain cases of hypertension, investigators have been trying to identify this substance. Among the landmarks in the development of the concept were the Goldblatt clamp (an artificial constriction of the renal artery) that caused something in the kidney to elevate blood pressure; the discovery of the enzyme renin by Page and others; the action of this enzyme on a blood protein (renin substrate) to form an inactive substance (angiotensin I); and finally the conversion of the inactive form into the active angiotensin II by means of a converting enzyme.

1956: *Sutherland, E. W. Discovery of cyclic AMP.*

An intercellular mediating agent (cyclic adenose-3',5'-monophosphate, or AMP) is found in all living animal tissues and changes in AMP levels (by effects of hormones) cause the hormones to produce different target effects depending on the type of cell in which they are found. AMP is formed by a metabolic process in which the enzyme adenyl cyclase converts ATP to cyclic AMP.

Nobel Laureate (1971).

1956: *Tjio, J. H., and A. Levan. Revision of human chromosome count.*

The time-honored number of chromosomes in man, 48 (diploid), was found by careful cytologic technique to be 46 instead.

1957: *Benzer, S. Concept of the cistron.*

The newer view of the gene has greatly changed the classic concept of the gene as an entity that controlled mutation, hereditary recombination, and function. There are actually several subunits, of which the cistron, the unit of function, is the largest. Many units of recombination may be found in a single cistron. The mutation unit (muton) is variable but usually consists of two to five nucleotides.

1957: *Calvin, M. Chemical pathways in photosynthesis.*

By using radioactive carbon 14, Calvin and colleagues were able to analyze step by step the incorporation of carbon dioxide and the identity of each intermediate product involved in the formation of carbohydrates and proteins by plants.

Nobel Laureate (1961).

1957: *Holley, R. W. The role of tRNA in protein synthesis.*

Nucleotides of tRNAs differ from each other only in their bases. Holley also devised methods that precisely established the tRNAs that were used in the transfer of certain amino acids to the site of protein synthesis.

Nobel Laureate (1968).

1957: *Isaacs, A., and L. Lindemann. Action of interferon on viruses.*

Interferon (a protein), produced by all cells, protects uninfected cells against viral attacks but is effective specifically only in the animal species in which it is produced.

1957: *Ivanov, A. V. Analysis of phylum Pogonophora (beard worms).*

Specimens of this phylum were collected in 1900

and represent the most recent phylum to be discovered and evaluated in the animal kingdom. Collections have been made in the waters of Indonesia, the Okhotsk Sea, the Bering Sea, and the Pacific Ocean. They are found mostly in the abyssal depths. They belong to the Deuterostomia division and appear to be related to the Hemichordata. At present 80 species divided into two orders have been described.

1957: *Perutz, M. F., and J. C. Kendrew. Structure of hemoglobin.*

The mapping of a complex globular protein molecule of 600 amino acids and 10,000 atoms arranged in a three-dimensional pattern represented one of the great triumphs in biochemistry. Myoglobin of muscle, which acts as a storehouse for oxygen and contains only one heme group instead of four (hemoglobin), was found to contain 150 amino acids.

Nobel Laureates (1962).

1957: *Sauer, F. Celestial navigation by birds.*

By subjecting Old World warblers to various synthetic night skies of star settings in a planetarium, Sauer was able to demonstrate that the birds made use of the stars to guide them in their migrations.

1957: *Taylor, J. H., P. S. Woods, and W. L. Hughes. Application of the tracer method to organization and replication of chromosomes.*

By using radioactive materials as markers, these investigators were able to show that each new chromosome consists of one half of old material and one half of newly synthesized substances. This was confirmation of the Crick-Watson model of the nucleic acid molecule that is supposed to divide or unwind into two single threads, and each half reduplicates itself to form a complete double strand.

1958: *Hall, D. A., and others. Discovery of cellulose in human skin.*

The upsurge of interest in connective tissue in the past few years has been due partly to the many unsolved problems of the structure and chemistry of this tissue and partly to the relation of connective tissue to atherosclerosis and other disorders. The possibility that the different fibers are interconvertible has been one of the many challenges of this versatile tissue, and the discovery of cellulose in man's skin recalls similar structures in the tunic of ascidians. Cellulose is more common in the aged, which may indicate a return to a primitive condition.

1958: *Lerner, A. B. Discovery of melatonin in the pineal gland.*

Although in an experimental stage, this substance has been assigned the role of a biologic clock in regulating gonadal functions by a neurosecretory mechanism. This discovery represented a breakthrough in understanding the function of the pineal gland.

1958: *Meselson, M., and F. W. Stahl. Confirmation "in vivo" of the duplicating mechanism in DNA.*

This was really confirmation of the self-copying of DNA in accordance with Watson and Crick's scheme of the structure of DNA. These investigators found that after producing a culture of bacterial cells labeled with heavy nitrogen 15 and then transferring these bacteria to a cultural medium of light nitrogen 14, the resulting bacteria had a DNA density intermediate between heavy and light as would be expected on the basis of the Watson-Crick hypothesis.

1959: *Ford, C. E., P. A. Jacobs, and J. H. Tjio. Chromosomal basis of sex determination in man.*

By discovering that certain genetic defects were associated with an abnormal somatic chromosomal constitution, it was possible to determine that the male-determining genes in man were located on the Y chromosome. Thus a combination of XXY (47 instead of the normal 46 diploid number) produced sterile males (Klinefelter's syndrome), and those with XO combinations (45 diploid number) gave rise to Turner's disease or immature females.

1959: *Leakey, L. S. B. Discovery of Zinjanthropus fossil man.*

This fossil is represented by a skull that shows the morphologic characters of a man and is believed to be in direct line of human ancestry. It seems to be definitely more advanced than its fossil relative *Australopithecus* and was a toolmaker. It was found in the Olduvai Gorge, Tanzania, East Africa.

1959: *LeJeune, J., M. Gautier, and R. Turpin. Abnormal chromosome pattern in man.*

The discovery of the presence of an extra chromosome (autosome) in the tissue cultures obtained from mongolian children has aroused much interest in the human cytogenetic pattern in its relation to disease states. This was the first clear-cut case of such an etiologic mechanism in the explanation of a disease and has stimulated many investigations of clinical interest along similar lines. Such a condition of an extra chromosome is called **trisomy;** a condition of one less chromosome is called **monosomy.**

1959: *Karlson, P., and A. Butenandt. Role of pheromones.*

The term "pheromone" designates any substance secreted by an organism and serving, upon contacting another member of the same species, to induce adaptive responses from that member. For species other than their own, organisms may produce allomone substances, which may evoke in the receiver favorable adaptations to the emitter, or the substance (kairomone) may benefit the recipient.

1960: *Barski, G., S. Sorieul, and F. Cornefert. Hybrid somatic cell technique.*

When two different cultures of mouse-cancer cells were mixed together, in the course of time some cells of a new type containing the chromosomes of both parents in a single nucleus appeared. These hybrid cells had arisen from the fusion of pairs of cells of the two different types. J. F. Watkins and H. Harris (1965), English investigators, showed that a virus killed by exposure to ultraviolet light could be used to fuse together cells from mouse and man, thus producing artificial man-mouse hybrid cells. The study of hybrid cells may throw some light on the difficult problem of differentiation and the genetic analysis of enzyme syntheses.

1960: *Hurwitz, J., A. Stevens, and S. Weiss. Enzymatic synthesis of messenger RNA.*

The exciting development in the coding system between DNA and the site of protein synthesis (ribosomes) was further elucidated when it was discovered that an enzyme, RNA polymerase, was responsible for the synthesis of RNA from a template pattern of DNA. This form of RNA provides a direct transcription of the DNA genetic code. In 1962 two other forms of RNA, ribosomal RNA and transfer RNA (also from DNA), were found to be involved in the specific linking of amino acids in the protein chain.

1960: *Jacob, F., and J. Monod. The operon hypothesis.*

The operon hypothesis is a postulated model of how enzyme synthesis is regulated in the cell. The model proposes that refinement in regulation involves an inducible system for allowing structural genes to synthesize needed enzymes and a repressible system that cuts off the synthesis of unneeded enzymes.

Nobel Laureates (1965).

1960: *Strell, M., and R. B. Woodward. Synthesis of chlorophyll a.*

Strell and Woodward with the aid of many co-workers finally solved this problem that had been the goal of organic chemists for many generations. Since all biologic life depends on this important pigment, this achievement in molecular biology must have wide implications.

Woodward, Nobel Laureate (1965).

1960: *Zalokar, M. Pattern of protein synthesis.*

By means of radioactive isotope of hydrogen (^{3}H) this investigator was able to trace uracil from its incorporation in RNA in the nucleus to the ribosomes of the cytoplasm, thus giving strong confirmation of the role of messenger RNA as outlined by the hypothesis of Jacob and Monod.

1961: *Hurwitz, J., A. Stevens, and S. B. Weiss. Confirmation of messenger RNA.*

Messenger RNA transcribes directly the genetic message of the nuclear DNA and moves to the cytoplasm where it becomes associated with a number of ribosomes or submicroscopic particles containing protein and nonspecific structural RNA. Here the messenger RNA molecules serve as templates against which amino acids are arranged in the sequence corresponding to the coded instructions carried by messenger RNA.

1961: *Jacob, F., and J. Monod. The role of messenger RNA in the genetic code.*

The transmission of information from the DNA code in the genes to the ribosomes represents an important step in the unraveling of the genetic code, and these investigators proposed certain deductions for confirmation of the hypothesis, which in general has been established by many researchers.

Nobel Laureates (1965).

1961: *Miller, J. F. A. Function of the thymus gland.*

Long known as a transitory organ that persists during the early growth period of animals, the thymus is now recognized as the source of the first antibody-producing cells. Later these cells migrate to the lymph nodes and other places, where they continue the production of antibodies as needed.

1961: *Nirenberg, M. W., and J. H. Matthaei. The role of DNA-directed RNA in protein synthesis.*

By adding a synthetic RNA composed entirely of uracil nucleotides to a mixture of amino acids, these investigators obtained a polypeptide made up solely of a single amino acid, phenylalanine. On the basis of a triplet code, it was concluded that the RNA code word for phenylalanine was UUU and its DNA complement was AAA. This was the beginning of coding the various amino acids and represents a key demonstration for understanding the genetic code.

Nirenberg, Nobel Laureate (1968).

1962: *Perry, R. P. Cellular sites of synthesis of RNA.*

By using different labeled RNA precursors and other cytochemical techniques, this investigator concluded that messenger RNA and transfer RNA are produced by the chromosomes of the cell; the ribosomal RNA of greater molecular weight is produced in the nucleolus.

1963: *Cairns, J. Confirmation of the manner of replication of the DNA molecule.*

By using autoradiography, Cairns was able to follow step-by-step the duplication of the daughter strands in the circular two-stranded DNA molecule of the single chromosome of *Escherichia coli.*

1963: *Wells, J. W. Concept of the fossil coral clock.*

By means of daily striations within the annual bands of calcium carbonate deposits found in certain fossil coral material, it has been possible to ascertain the difference between the length of day in geologic times and that of the present, based on the deceleration of 2 seconds per 100,000 years in the earth's rotation about its axis (424 days for a year in Cambrian time).

1964: *Hoyer, B. H., B. J. McCarthy, and E. T. Bolton. Phylogeny and DNA sequence.*

These investigators presented evidence that certain homologies exist among the polynucleotide sequences in the DNA sequence of such different forms as fish and man. Such sequences may represent genes that have been retained with little change throughout vertebrate history. Possible phenotypic expressions may be bilateral symmetry, notochord, hemoglobin, etc.

1966: *Khorana, H. G. Proof of code assignments in the genetic code.*

By using alternating codons (CUC and UCU) in an artificial RNA chain, he was able to synthesize a polypeptide of alternating amino acids (leucine and serine) for which these codons respectively stood.

Nobel Laureate (1968)

1967: *Plashne, M. Isolation of first repressor.*

Repressors are protein substances supposedly formed by regulatory genes and function by preventing a structural gene from making its product when not needed by the cells.

1970: *Temin, H. M. The reversal transcription of DNA from RNA.*

Reverse transcriptase, an enzyme, is found to make DNA from RNA instead of the reverse. This enzyme has been found in tumor viruses (and perhaps in all normal cells). By this reverse process cancer viruses can transfer genetic information that can alter the genes leading to cancer.

1970: *Edelman, G. M., and R. R. Porter. The structure of gamma globulin.*

By using myeloma tumors (which contain pure immunoglobulin proteins) the investigators were able to work out after many years a complete analysis of the large gamma globulin molecule, which is made up of 1,320 amino acids and 19,996 atoms with a molecular weight of 150,000.

Nobel Laureates (1972).

1972: *Woodward, R. B., and A. Eschenmoser. The synthesis of vitamin B_{12}.*

Vitamin B_{12} is the last of the vitamins to be synthesized. This complex molecule is not made up of polymers and required new rules of organic chemistry to synthesize. This vitamin contains the metal ion cobalt.

▪ BOOKS AND PUBLICATIONS THAT HAVE GREATLY INFLUENCED DEVELOPMENT OF ZOOLOGY

Aristotle. 336-323 B.C. De anima, Historia animalium, De partibus animalium, and De generatione animalium. *These biologic works of the Greek thinker have exerted an enormous influence on biologic thinking for centuries.*

Vesalius, Andreas. 1543. De fabrica corporis humani. *This work is the foundation of modern anatomy and represents a break with the Galen tradition. His representations of anatomic subjects, such as the muscles, have never been surpassed. Moreover, he treated anatomy as a living whole, a viewpoint present-day anatomists are beginning to copy.*

Fabricius of Aquapendente. 1600-1621. De formato foetu and De formatione ovi pulli. *This was the first illustrated work on embryology and may be said to be the beginning of the modern study of development.*

Harvey, William. 1628. Anatomical dissertation concerning the motion of the heart and blood. *This great work represents one of the first accurate explanations in physical terms of an important physiologic process. It initiated an experimental method of observation that gave an impetus to research in all fields of biology.*

Descartes, René. 1637. Discourse on method. *This philosophical essay gave a great stimulus to a mechanistic interpretation of biologic phenomena.*

Buffon, Georges. 1749-1804. Histoire naturelle. *This exten-*

sive work of many volumes collected together natural history facts in a popular and pleasing style. *It had a great influence in stimulating a study of nature. Many eminent biologic thinkers, such as Erasmus Darwin and Lamarck, were influenced by its generalizations, which here and there suggest an idea of evolution in a crude form.*

Linnaeus, Carolus. 1758. Systema naturae. *In this work there is laid the basis for the classification of animals and plants. With few modifications, the taxonomic principles outlined therein have been universally adopted by biologists.*

Wolff, Caspar Friedrich. 1759. Theoria generationis. *The theory of epigenesis was here set forth for the first time in opposition to the preformation theory of development so widely held up to the time of Wolff's work.*

Haller, Albrecht von. 1760. Elementa physiologiae. *An extensive summary of various aspects of physiology that greatly influenced physiologic thinking for many years. Some of the basic concepts therein laid down are still considered valid, especially those on the nervous system.*

Malthus, Thomas R. 1798. Essay on population. *This work stimulated evolutionary thinking among such men as Darwin and Wallace.*

Lamarck, Jean Baptiste de. 1809. Philosophie zoologique. *This publication was of great importance in focusing the attention of biologists upon the problem of the role of the environment as a factor in evolution. Lamarck's belief that all species came from other species represented one of the first clear-cut statements on the mutability of species, even though his theory of use and disuse has not been accepted by most biologists.*

Cuvier, Georges. 1817. Le règne animal. *A comprehensive biologic work that dealt with classification and a comparative study of animal structures. Its plates are still of value, but the general plan of the work was marred by a disbelief in evolution and a faith in the doctrine of geologic catastrophes. The book, however, exerted an enormous influence upon contemporary zoologic thought.*

Baer, Karl Ernst von. 1828-1837. Entwickelungsgeschichte der Thiere. *In this important work are laid down the fundamental principles of germ layer formation and the similarity of corresponding stages in the development of embryos that have proved to be the foundation studies of modern embryology.*

Audubon, John J. 1828-1838. The birds of America. *The greatest of all ornithologic works, it has served as the model for all monographs dealing with a specific group of animals. The plates, the work of a master artist, have never been surpassed in the field of biologic achievement.*

Lyell, Charles. 1830-1833. Principles of geology. *From a biologic viewpoint this great work exerted a profound influence on biologic thinking, for it did away with the theory of catastrophism and prepared the way for an evolutionary interpretation of fossils and the forms that arose from them.*

Beaumont, William. 1833. Experiments and observations on the gastric juice and the physiology of digestion. *In this classic work the observations Beaumont made on various*

functions of the stomach and digestion were so thorough that only a few details have been added by subsequent research. This book paved the way for the brilliant investigations of Pavlov, Cannon, and Carlson of later generations.

Müller, Johannes. 1834-1840. Handbook of physiology. *The principles set down in this work by the greatest of all physiologists have set the pattern for the development of the science of physiology.*

Darwin, Charles. 1839. Journal of researches (Voyage of the *Beagle*). *This book reveals the training and development of the naturalist and the material that led to the formulation of Darwin's concept of organic evolution.*

Schwann, Theodor. 1839. Mikroskopische Untersuchungen über die Uebereinstimmung in der Struktur und dem Wachstum der Thiere und Pflanzen. *The basic principles concerning the cell doctrine are laid down in this classic work.*

Kölliker, Albrecht. 1852. Mikroskopische Anatomie. *This was the first textbook in histology and contains contributions of the greatest importance in this field. Many of the histologic descriptions Kölliker made have never needed correction. In many of his biologic views he was far ahead of his time.*

Maury, Matthew F. 1855. The physical geography of the sea. *This work has often been called the first textbook on oceanography. This pioneer treatise stressed the integration of such knowledge as was then available about tides, winds, currents, depths, circulation, and such matters. Maury's work represents a real starting point in the fascinating study of the oceans and has had a great influence in stimulating investigations in this field.*

Virchow, Rudolf. 1858. Die Cellularpathologie. *In this work Virchow made the first clear distinction between normal and diseased tissues and demonstrated the real nature of pathologic cells. The work also represents the death knell to the old humoral pathology, which had held sway for so long.*

Darwin, Charles. 1859. The origin of species. *One of the most influential books ever published in biology. Although built around the theme that natural selection is the most important factor in evolution, the great influence of the book has been attributable to the great array of evolutionary evidence it presented. It also stimulated constructive thinking on a subject that had been vague and confusing before Darwin's time.*

Marsh, George P. 1864. Man and nature: physical geography as modified by human action. *A work that had an early and important influence on the conservation movement in America.*

Mendel, G. 1866. Versuche über Pflanzenhybriden. *The careful controlled pollination technique and statistical analysis gave a scientific explanation that has influenced all geneticists after the "rediscovery" of Mendel's classic paper on the two basic laws of inheritance.*

Owen, Richard. 1866. Anatomy and physiology of the vertebrates. *This work contains an enormous amount of personal observation on the structure and physiology of animals, and some of the basic concepts of structure and function, such as homolog and analog, are here defined for the first time.*

Brehm, Alfred E. 1869. Tierleben. *The many editions of this work over many years have indicated its importance as a general natural history.*

Bronn, Heinrich G. (editor). 1873 to present. Klassen und Ordnungen des Tier-Reichs. *This great work is made up of exhaustive treatises on the various groups of animals by numerous authorities. Its growth extends over many years, and it is one of the most valuable works ever published in zoology.*

Balfour, Francis M. 1880. Comparative embryology. *This is a comprehensive summary of embryologic work on both vertebrates and invertebrates up to the time it was published. This work is often considered the beginning of modern embryology.*

Semper, Karl. 1881. Animal life as affected by the natural conditions of existence. *This work first pointed out the modern ecologic point of view and laid the basis for many ecologic concepts of existence that have proved important in the further development of this field of study.*

Bütschli, Otto. 1889. Protozoen (Bronn's Klassen und Ordnungen des Tier-Reichs). *This monograph has been of the utmost importance to students of Protozoa. No other work on a like scale has ever been produced in this field of study.*

Hertwig, Richard von. 1892. Lehrbuch der Zoologie. *A text that has proved to be an invaluable source of material for many generations of zoologists. Its illustrations have been widely used in many other textbooks.*

Weismann, August. 1892. Das Keimplasma. *Weismann predicted from purely theoretic considerations the necessity of meiosis or reduction of the chromosomes in the germ cell cycle—a postulate that was quickly confirmed cytologically by others.*

Hertwig, Oskar. 1893. Zelle und Gewebe. *In this work a clear distinction is made between histology as the science of tissues and cytology as the science of cell structure and function. Cytology as a study in its own right really dates from this time.*

Korschelt, E., and K. Heider. 1893. Lehrbuch der vergleichender Entwicklungsgeschichte der wirbellosen Thiere, 4 vols. *A treatise that has been a valuable tool for all workers in the difficult field of invertebrate embryology.*

Wilson, Edmund B. 1896. The cell in development and heredity. *This and subsequent editions represented the most outstanding work of its kind in the English language. Its influence in directing the development of cytogenetics cannot be overestimated, and in summarizing the many investigations in cytology, the book has served as one of the most useful tools in the field.*

Pavlov, Ivan. 1897. Le travail des glandes digestives. *This work marks a great landmark in the study of the diges-tive system, for it describes many of the now classic experiments that Pavlov conducted, such as the gastric pouch technique and the rate of gastric secretions.*

De Vries, Hugo. 1901. Die Mutationstheorie. *The belief that evolution is the result of sudden changes or mutations is advanced by one who is commonly credited with the initiation of this line of investigation into the causes of evolution.*

Sherrington, Sir Charles. 1906. The integrative action of the nervous system. *The basic concepts of neurophysiology laid down in this book have been little altered since its publication. Much of the work done in this field has served to confirm the nervous mechanism he here outlines.*

Garrod, Sir Archibald. 1909. Inborn errors of metabolism. *This pioneer book showed that certain congenital diseases were caused by defective genes that failed to produce the proper enzymes for normal functioning. It laid the basis for biochemical genetics, which later received a great impetus from the work of Beadle and Tatum.*

Henderson, Lawrence J. 1913. The fitness of the environment. *This book has pointed out in a specific way the reciprocity that exists between living and nonliving nature and how organic matter is fitted to the inorganic environment. It has exerted a considerable influence on ecologic aspects of adaptation.*

Shelford, Victor E. 1913. Animal communities in temperate America. *This work was a pioneer in the field of biotic community ecology and has exerted a great influence on ecologic study.*

Bayliss, Sir William M. 1915. Principles of general physiology. *If a classic book must meet the requirements of masterly analysis and synthesis of what is known in a particular discipline, then this great work must be called one.*

Mathew, W. D. 1915. Climate and evolution. *This book has stimulated much thinking on the importance of climate in the evolutionary process.*

Morgan, T. H., A. H. Sturtevant, C. B. Bridges, and H. J. Muller. 1915. The mechanism of mendelian heredity. *This book gave an analysis and synthesis of mendelian inheritance as formulated from the epoch-making investigations of the authors. This classic work will always stand as a cornerstone of our modern interpretation of heredity.*

Doflein, F. 1916. Lehrbuch der Protozoenkunde, ed. 6 (revised by E. Reichenow, 1949). *A standard treatise on Protozoa. Its many editions have proved helpful to all workers in this field.*

Thompson, D. W. 1917. Growth and form. *This pioneer work deals with the problems of growth and form in relation to physical and mathematic principles. It has thrown much light upon these difficult subjects.*

Kukenthal, W., and T. Krumbach. 1923. Handbuch der Zoologie. *An extensive modern treatise on zoology that* covers all phyla. *The work has been an invaluable tool for all zoologists who are interested in the study of a particular group.*

Fisher, R. A. 1930. Genetical basis of natural selection. *This work has exerted an enormous influence on the newer synthesis of evolutionary mechanisms so much in vogue at present.*

Dobzhansky, T. 1937. Genetics and the origin of species. *The vast change in the explanation of the mechanism of evolution, which emerged about 1930, is well analyzed in this work by a master evolutionist. Other syntheses of this new biologic approach to the evolutionary problems have appeared since this work was published, but none of them has surpassed the clarity and fine integration of Dobzhansky's work.*

Spemann, Hans. 1938. Embryonic development and induction. *In this work the author summarizes his pioneer investigations that have proved so fruitful in experimental embryology.*

Schrödinger, Erwin. 1945. What is life? *The emphasis placed on the physical explanation of life gave a new point of view of biologic phenomena so well expressed in the current molecular biologic revolution.*

Grassé, P.-P. (editor). 1948. Traité de zoologie. *This is a series of many treatises by various specialists on both invertebrates and vertebrates. Since it is both recent and comprehensive, the work is invaluable to all students who desire detailed information on the various groups.*

Allee, W. C., A. E. Emerson, O. Park, T. Park, and K. P. Schmidt. 1949. Principles of animal ecology. *Ecology is rapidly becoming a major field of biologic study and the basic principles laid down in this comprehensive work will never be outdated.*

Leopold, A. 1949. A sand county almanac. *In the current emphasis on conservation and ecology, this work gives an evaluation of the awareness and gist of the problems that society must face to work out an effective ecosystem.*

Blum, H. F. 1951. Time's arrow and evolution. *In this thought-provoking book Blum explores the relation between the second law of thermodynamics (time's arrow) and organic evolution and recognizes that mutation and natural selection have been restricted to certain channels in accordance with the law, even though these two factors appear to controvert the principle of pointing the direction of events in time.*

Crick, Francis H. C., and James D. Watson. 1953. Genetic implications of the structure of deoxyribonucleic acid. *The solution of the structure of the DNA molecule has served as the cornerstone for an explanation of genetic replication and control of the cell's attributes and functions.*

Carson, Rachel L. 1962. Silent spring. *In this age of environmental problems, no treatise yet written has alerted the public more about the harm done to our ecosystem by pesticides.*

GLOSSARY

aboral (ab-o'ral) (L. *ab,* from, + *os,* mouth). A region opposite the mouth.

acanthor (Gr. *akantha,* spine or thorn, + *-or*). First larval form of acanthocephalans in the intermediate host.

acinus (as'i-nus) (L. grape). A small lobe of a compound gland, or a saclike cavity at the termination of a passage.

acoelomate (a-see'lo-mate) (Gr. *a,* not, + *koilōma,* cavity). Without a coelom, such as flatworms and proboscis worms.

acontia (Gr. *akontion,* dart). Threads with nematocysts on mesenteries of sea anemone.

adaptation (ad"ap-tay'shun) (L. *ad,* to, + *aptare,* to fit). The fitness of structure, function, or an organism for survival in its environment.

adductor (ad-duk,'ter) (L. *ad,* to, + *ducere,* to lead). A muscle that draws a part toward a median axis, or a muscle that draws the two valves of a mullusk shell together.

adenine (ad'e-neen) (Gr. *adēn,* gland, + *-ine,* suffix). A component of nucleotides and nucleic acids.

adenosine (a-den,'o-seen) **(di-, tri-) phosphate** (ADP and ATP). Certain phosphorylated compounds that function in the energy cycle of cells.

adipose (ad'i-pos) (L. *adeps,* fat). Fatty tissue.

adrenaline (aj-ren'a-lin) (L. *ad,* to, + *renalis,* pertaining to kidneys). A hormone produced by the adrenal, or suprarenal, gland.

aerobic (a"er-o'bik) (Gr. *aēr,* air, + *bios,* life). Oxygen-dependent form of respiration.

afferent (af'er-ent) (L. *ad,* to, + *ferre,* to bear). A structure (blood vessel, nerve, etc.) leading toward some point.

allantois (a-lan'to-iss) (Gr. *allas,* sausage, + *eidos,* form). One of the extraembryonic membranes of the amniotes.

allele (al-leel') (Gr. *allēlōn,* of one another). One of a pair, or series, of genes that are alternative to each other in heredity and are situated at the same locus in homologous chromosomes. Allele genes may consist of a dominant and its correlated recessive, or two correlated dominants, or two correlated recessives.

allometry (al-lahm'e-tree) (Gr. *allos,* other, + *metron,* measure). A study of relative growth or a change in proportion with increase in size.

alula (al'yu-la) (L. dim. of *ala,* wing). The first digit or thumb of a bird's wing, much reduced in size.

alveolus (al-ve,'o-lus) (L. dim. of *alveus,* cavity, hollow). A small cavity or pit, such as a microscopic air sac of the lungs, terminal part of an alveolar gland, or bony socket of a tooth.

ambulacra (L. *ambulare,* to walk). Radiating grooves where podia of water-vascular system project to outside.

amebocyte (Gr. *amoibē,* change, + *kytos,* hollow vessel). Free cells in the mesenchyme.

ameboid (a-me'boid) (Gr. *amoibē,* change + *-oid,* like). Amebalike in putting forth pseudopodia.

amictic (Gr. *a,* without, + *miktos,* mixed or blended). Pertains to diploid egg of rotifers or the females that produce such eggs.

amino acid (a-me′no) (amine, an organic compound). An organic acid with an amino radical (NH₂). Makes up the structure of proteins.

amitosis (am″i-to′sis) (Gr. *a*, not, + *mitos*, thread). A form of cell division in which mitotic nuclear changes do not occur; cleavage without separation of daughter chromosomes.

amnion (am′ni-on) (Gr. caul, probably from dim. of *amnos*, lamb). One of the extraembryonic membranes forming a sac around the embryo in amniotes.

amphiblastula (Gr. *amphi*, on both sides, + *blastos*, germ, + L. *-ula*, small). Larval stage in sponges, so called because one half bears flagella cells and the other half does not.

amphid (Gr. *amphidea*, anything that is bound around). One of a pair of anterior sense organs in certain nematodes.

amylase (am′i-lace) (L. *amylum*, starch, + *ase*, suffix meaning enzyme). An enzyme that breaks down carbohydrates into smaller units.

anadromous (anaj′ro-mus) (Gr. *anadromos*, running upward). Refers to those fish that migrate up streams to spawn.

anaerobic (an-a″er-o′bik) (Gr. *an*, not, + *aēr*, air, + *bios*, life). Not dependent on oxygen for respiration.

anastomosis (a-nas″to-mo′sis) (Gr. *ana*, again, + *stoma*, mouth). A union of two or more arteries, veins, or fibers to form a branching network.

androgen (an′dro-jen) (Gr. *anēr, andros*, man, + *genēs*, born). Any of a group of male sex hormones.

anhydrase (an-hi′drace) (Gr. *an*, not, + *hydōr*, water, + *ase*, enzyme suffix). An enzyme involved in the removal of water from a compound. Carbonic anhydrase promotes the conversion of carbonic acid into water and carbon dioxide.

anlage (ahn′lah-guh) (Gr. *Anlage*, laying-out, foundation). Primordium.

antenna (an-ten′a) (L. *antenna, antemna*, sail yard). A sensory appendage on the head of arthropods, or the first pair of the two pairs of structures in Crustacea.

anterior (an-tir′e-er) (L. comparative of *ante*, before). The head end of an organism, or toward that end.

aperture (ap′er-cher) (L. *apertura*, from *aperire*, to uncover). An opening; the slight entrance and exit of certain mollusk shells; longer passages are called siphons.

apical (ap′i-k′l) (L. *apex*, tip). Pertaining to the tip or apex.

apopyle (Gr. *apo*, away from, + *pylē*, gate). Opening of the radial canal into the spongocoel.

arboreal (ar-bor′e-al) (L. *arbor*, tree). Living in trees.

archenteron (ark-en′ter-on) (Gr. *archē*, beginning, + *enteron*, gut). The central cavity of a gastrula that is lined with endoderm, representing the future digestive cavity.

archeocyte (Gr. *archaios*, ancient, + *kytos*, hollow vessel). Ameboid cells of varied function in sponges.

ascon (Gr. *askos*, bladder). Simplest form of sponges, with canals leading directly from the outside to the interior.

assimilation (a-sim″i-la′shun) (L. *assimilatio*, bringing into conformity). Absorption and building up of digested nutriments into complex organic protoplasmic materials.

autosome (aw′to-soam) (Gr. *autos*, self, + *sōma*, body). Any chromosome that is not a sex chromosome.

autotomy (aw-tot′o-me) (Gr. *autos*, self, + *tomos*, a cutting). The automatic breaking off of a part of the body.

autotroph (aw′to-trof) (Gr. *autos*, self, + *trophos*, feeder). An organism that makes its organic nutrients from inorganic raw materials.

avicularium (L. *avicula*, small bird, + *aria*, like or connected with). Modified zooid that is attached to the surface of the major zooid in Ectoprocta and resembles a bird's beak.

axolotl (ak′suh-lot′l) (Nahuatl *atl*, water, + *xolotl*, doll, servant, spirit). Larval stage of *Ambystoma tigrinum*, exhibiting neotenic reproduction.

axopodia (L. *axis*, an axis, + Gr. *podion*, small foot). Long, thin projections of cytoplasm, but not for locomotion.

benthos (ben′thos) (Gr. depth of the sea). Those organisms that live along the bottom of seas and lakes.

biogenesis (bi″o-jen′e-sis) (Gr. *bios*, life, + *genesis*, birth). The doctrine that life originates only from preexisting life.

biomass (bi′o-mass) (Gr. *bios*, life, + *maza*, lump or mass). The weight of a species population per unit of area.

biome (bi′oam) (Gr. *bios*, life, + *ōma*, abstract group suffix). Complex of plant and animal communities characterized by climatic and soil conditions; the largest ecologic unit.

bipinnaria (L. *bi-*, double, + *pinna*, wing, + *-aria*, like or connected with). Refers to shape of this asteroid larva.

blastocoel (blas′to-seal) (Gr. *blastos*, germ, + *koilos*, hollow). Cavity of blastula.

blastomere (Gr. *blastos*, germ, + *meros*, part). Early cleavage cells.

blastopore (Gr. *blastos*, germ, + *poros*, passage, pore). External opening of the archenteron in the gastrula.

blastula (Gr. *blastos*, germ, + L. *-ula*, diminutive). Early embryologic stage of a hollow mass of cells.

blepharoplast (Gr. *blepharon*, eyelid, + *plastos*, formed). Granule connected with a flagellum.

brachial (brak′ee-al) (L. *brachium*, forearm). Referring to the forearm.

branchial (brang′kee-al) (Gr. *branchia*, gills). Referring to gills.

brachiolaria (L. *brachiola*, little arm, + *-aria*, pertaining to). This asteroid larva has three preoral processes.

buccal (buk′al) (L. *bucca*, cheek). Referring to the mouth cavity.

buffer (buff′er). Any substance or chemical compound that tends to keep pH constant when acids or bases are added.

carboxyl (kar-bok′sil) (carbon + oxygen + -yl, chemical radical suffix). The acid group of organic molecules—COOH.

carotene (kar′o-teen) (L. *carota*, carrot, + -ene, unsaturated straight-chain hydrocarbons). A red, orange, or yellow pigment belonging to the group of carotenoids; precursor of vitamin A.

catalyst (cat′a-list) (Gr. *kata*, down, + *lysis*, a loosening).

A substance that accelerates a chemical reaction but does not become a part of the end product.

catadromous (ka-taj'ruh-mus) (Gr. *kata,* down, + *dromos,* a running). Refers to those fish that migrate from freshwater to the ocean to spawn.

caudatum (L. *cauda,* tail, + *-atus,* -ate). Species of *Paramecium.*

cilium (L. eyelash). Threadlike organ of locomotion.

cecum (se'kum) (L. *caecus,* blind). A blind pouch at the beginning of the large intestine, or any similar pouch.

cement gland One of a cluster of unicellular glands near the testes in acanthocephalans, which furnishes secretions for binding the sexes together during copulation.

cenogenesis (see"no-jen'i-sis) (Gr. *kainos,* new, + *genesis,* birth). In the development of an organism, the new stages that have arisen in adaptive response to the embryonic mode of life, such as the fetal membranes of amniotes.

centriole (sen'tree-ol) (Gr. *kentron,* center of a circle, + L. *-ola,* small). A minute granule, usually found in the centrosome and considered to be the active division center of the cell.

centrolecithal (Gr. *kentron,* center, + *lekithos,* yolk, + Eng. *-al,* adj.). Pertaining to an insect egg with the yolk concentrated in the center.

centromere (sen'tro-mere) (Gr. *kentron,* center, + *meros,* part). A small body or constriction on the chromosome where it is attached to a spindle fiber.

cercaria (Gr. *kerkos,* tail, + L. *-aria,* like or connected with). Tadpolelike larva of trematodes.

chelicera (ke-lis'e-ra) (Gr. *chēlē,* claw, + *keras,* horn). Pincerlike head appendage on the members of the subphylum Chelicerata.

chitin (Gr. *chitōn,* tunic). A constituent of arthropod cuticle.

chloragogue (Gr. *chlōros,* greenish yellow, + *agōgos,* a leading, a guide). Special spongy tissue around interior of earthworms.

choana (ko'uh-nuh) (Gr. *choane,* funnel). One of the internal nares, or the opening between the nasal passages and the pharynx or mouth.

choanocyte (ko'choane, funnel, + *kytos,* hollow vessel). Flagellate collar cells that line cavities and canals.

cholinergic (ko"lin-er'jik). Type of nerve fiber that releases acetylcholine from axon terminal.

chorion (ko're-on) (Gr. leather, afterbirth). The outer of the double membrane that surrounds the embryo of the amniotes; in mammals it helps form the placenta.

chromatid (kro'ma-tid) (Gr. *chrōma,* color, + L. *-id-,* feminine stem for particle of specified kind). A half chromosome between early prophase and metaphase in mitosis; a half chromosome between synapsis and second metaphase in meiosis; at the anaphase stage each chromatid is known as a daughter chromosome.

chromomere (kro'mo-mer) (Gr. *chrōma,* color, + *meros,* part). The chromatin granules of characteristic size on the chromosome; may be identical with genes or clusters of genes.

chrysalis (kris'uh-lus) (L. from Gr. *chrysos,* gold). Refers to gold-colored pupa.

cinclide (Gr. *kinklis,* latticed gate or partition). Small pores in the external body wall of sea anemones.

circadian (sir"ka-de'an) (L. *circa,* around, + *dies,* day). Occurring at a period of about 24 hours.

cirrus (sir'us) (L. curl). A hairlike tuft on an insect appendage, or a small, slender structure.

cleavage (OE. *cleofan,* to cut). Cell division in animal ovum.

climax (cli'max) (Gr. *klimax,* ladder). A state of dynamic equilibrium; a culmination of the succession in the biota of a community.

clitellum (L. *clitellae,* packsaddle). A thickened glandlike body on certain portions of midbody segments of earthworms and leeches.

clone (Gr. *klōn,* twig). A group of animals produced by asexual reproduction from a single individual.

cnidoblast (nigh'do-blast) (Gr. *knidē,* nettle, + *blastos,* germ). Modified interstitial cell that holds the nematocyst.

cnidocil (Gr. *knidē,* nettle, + L. *cilium,* hair). Triggerlike spine on nematocyst.

coelogastrula (se'lo-gas'tru-la) (Gr. *koilos,* hollow, + *gastēr,* stomach, + L. *-ula,* diminutive). The typical gastrula derived from a coeloblastula; a two- or three-layered stage in embryology.

coelom (se'lum) (Gr. *koilōma,* cavity). The body cavity in triploblastic animals, lined with mesoderm.

coelomocyte (Gr. *koilōma,* cavity, + *kytos,* hollow vessel). Another name for amebocyte, primitive or undifferentiated cell of the coelom, and the water-vascular system.

coelomoduct (se-lo'mo-duct) (Gr. *koilos,* hollow, + L. *ductus,* a leading). A duct that carries gametes or excretory products (or both) from the coelom to the exterior.

coenzyme (ko-en'zime) (L. prefix *co-,* with, + Gr. *enzymos,* leavened, from *en,* in, + *zymē,* leaven). A required substance in the activation of an enzyme.

collenchyme (Gr. *kolla,* glue, + *enchyma,* infusion). A gelatinous mesenchyme that forms the third layer in the wall of coelenterates and ctenophores.

colloblasts (Gr. *kolla,* glue, + *blastos,* germ). Glue-secreting cells on the tentacles of ctenophores.

comb plate One of the plates of fused cilia that are arranged in rows for ctenophore locomotion.

commensalism (ko-men'sal-izm) (L. *cum,* together with, + *mensa,* table). A symbiotic relationship in which one benefits and the other is unharmed.

community (L. *communitas,* community, fellowship). An assemblage of organisms that are associated together in a common environment and interact with each other in a self-sustaining and self-regulating relation.

conjugation (L. *conjugare,* to yoke together). Temporary union of two protozoans while they are exchanging chromatin material; may be a complete fusion in Suctoria.

cornea (kor'ne-a) (L. *corneus,* horny). The outer transparent coat of the eye.

cortex (kor'teks) (L. bark). The outer layer of a structure.

cotylosaur (kot'i-lo-sor") (Gr. *kotylē,* cup, + *sauros,* lizard). A primitive group of fossil reptiles that arose from the labyrinthodont amphibians and became the ancestral stem of all other reptiles.

cynodont (Gr. *kyon,* dog, + *odous, odontos,* tooth). Referring to teeth like those of a dog.

cytochrome (si'to-krome) (Gr. *kytos,* hollow vessel, + *chrōma,* color). One of the hydrogen carriers in aerobic respiration.

cytopharynx (Gr. *kytos,* hollow vessel, + *pharynx,* throat). Short tubular gullet in protozoans.

cytosome (si'to-some) (Gr. *kytos,* hollow vessel, + *sōma,* body). The cell body inside the plasma membrane.

deoxyribose (de-ok'se-ri"boce) (*deoxy,* loss of oxygen, + *ribose,* pentose sugar). A 5-carbon sugar having 1 oxygen atom less than ribose; a component of deoxyribose nucleic acid (DNA).

deutoplasm (du,'to-plazm) (Gr. *deuteros,* second, + *plasma,* a form, a mold). Yolk portion of an egg as distinct from the true cytoplasmic portion.

dimorphism (di-mor'fizm) (Gr. *di-,* two, + *morphē,* form). Existing under two forms.

dioecious (di-ee'shus) (Gr. *di-,* two, + *oikos,* house). Male and female organs in separate individuals.

dipleurula (di-plur'yu-la) (Gr. *di-,* two, + *pleura,* rib, side, + L. *-ula,* small). This hypothetical echinoderm larva has bilateral symmetry.

diploid (dip'loid) (Gr. *diploos,* double, + *eidos,* form). The somatic number of chromosomes, or twice the number characteristic of a gamete of a given species.

dorsal (dor'sal) (L. *dorsum,* back). Toward the back, or upper surface.

DPN Abbreviation of diphosphopyridine nucleotide, a hydrogen carrier in respiration. Now called NAD, nicotinamide adenine dinucleotide.

ecdysis (ek'duh-sis) (Gr. *ekdysis,* a stripping, escape). Shedding of outer cuticular layer; molting.

ecologic equivalence Ecologic types of the same requirements, that are in similar but geographically separated environments.

ecologic niche The status of an organism in a community with reference to its responses and behavior patterns.

ecosystem (ek'o-sis-tem) (eco[logy], from Gr. *oikos,* house, + system). An ecologic unit consisting of both the biotic communities and the nonliving (abiotic) environment that interact to produce a stable system.

ecotone (ek'o-ton) (eco[logy], from Gr. *oikos,* home, + *tonos,* stress). The transition zone between two adjacent communities.

ectohormone (ek'to-hor"mone) (Gr. *ektos,* outside, + *hormon,* excitement, stirring up). A pheromone; a substance secreted externally by an organism to influence the behavior of other organisms; an ectocrine.

ectoplasm (ec'to-plazm) (Gr. *ektos,* outside, + *plasma,* form). The cortex of a cell or that part of cytoplasm just under the cell surface; contrasts with endoplasm.

effector (ifek'ter) (L. *efficere,* bring to pass). An organ, tissue, or cell that can react to stimuli.

elephantiasis (Gr. *elephas,* elephant, + *-iasis,* morbid state of). A condition of enormous swelling and connective tissue growth induced by filarial worms.

embryology (Gr. *embryon,* embryo, + L. *logia,* study of, from Gr. *logos,* word). Early development of organisms.

emulsion (e-mul'shun) (L. *emulsus,* milked out). A colloidal system in which both phases are liquids.

endergonic (end"er-gon'ik) (Gr. *endon,* within, + *ergon,* work). Used of a chemical reaction that requires energy.

endocrine (en'do-krin) (Gr. *endon,* within, + *krinein,* to separate). Refers to a gland that is without a duct and that releases its product directly into the blood or lymph.

endomixis (Gr. *endon,* within, + *mixis,* a mixing). Reorganization of the nuclear material in the protozoan.

endoplasm (en'do-plazm) (Gr. *endon,* within, + *plasma,* mold or form). That portion of cytoplasm that immediately surrounds the nucleus.

endoplasmic reticulum (en"do-plaz'mic). The cytoplasmic double membrane with ribosomes (rough) or without ribosomes (smooth).

endostyle (en'do-style) (Gr. *endon,* within, + L. *stilus,* stake). A ciliated groove in the floor of the pharynx of tunicates, amphioxus, and ammocoetes, used for getting food; may be homologous to the thyroid gland of higher forms.

enterocoel (en'ter-o-seel') (Gr. *enteron,* gut, + *koilos,* hollow). A type of coelom that is formed by the outpouching of a mesodermal sac from the endoderm of the primitive gut.

enterocoelomate (en'ter-o-se'lo-mate) (Gr. *enteron,* gut, + *koilōma,* cavity, + Eng. *-ate,* state of). Those that have an enterocoel, such as the echinoderms and the vertebrates.

enteron (en'ter-on) (Gr. *enteron,* intestine). The digestive cavity of an organism.

enzyme (en'zyme) (Gr. *enzymos,* leavened, from *en,* in, + *zyme,* leaven). A protein substance produced by living cells that is capable of speeding up specific chemical transformations, such as hydrolysis, oxidation, or reduction, but is unaltered itself in the process; a biologic catalyst.

ephyra (ef'i-ra) (Gr. *Ephyra,* Greek city). Refers to castle-like appearance. Stage in development of Scyphozoa.

epididymis (ep"i-did'i-mis) (Gr. *epi,* over, + *didymos,* testicle). That part of the sperm duct that is coiled and lying near the testis.

epigenesis (ep"i-jen'e-sis) (Gr. *epi,* over, + *genesis,* birth). The embryologic view that an embryo is a new creation that develops and differentiates step by step from an initial stage; the progressive production of new parts that were nonexistent as such in the original zygote.

epigenetics (ep"i-je-net'iks) (Gr. *epi,* over, + *genesis,* birth). That study of the mechanisms by which the genes produce phenotypic effects.

estrogen (es'tro-jen) (Gr. *oistros,* frenzy, + *genes,* born). An estrus-producing hormone; one of a group of female sex hormones.

euryhaline (Gr. *eurys,* broad, + *hals,* salt). Able to tolerate wide ranges of saltwater concentrations.

eurytopic (yu're-top'ic) (Gr. *eurys,* broad, + *topos,* place).

Refers to an organism with a wide range of distribution.

evagination (e-vaj"i-na'shun) (L. *e,* out, + *vagina,* sheath). An outpocketing from a hollow structure.

exergonic (ek"ser-gon'ik) (Gr. *exo,* outside of, + *ergon,* work). An energy-yielding reaction.

exocrine (ek'so-krin) (Gr. *exō,* outside, + *krinein,* to separate). That type of gland that releases its secretion through a duct.

exteroceptor (ek"ster-o-sep'ter) (L. *exter,* outward, + *capere,* to take). A sense organ near the skin or mucous membrane that receives stimuli from the external world.

FAD Abbreviation for flavine adenine dinucleotide, a hydrogen acceptor in the respiratory chain.

fermentation (fur"men-ta'shun) (L. *fermentum,* leaven, ferment). The conversion of organic substances into simpler substances under the influence of enzymes, with little or no oxygen involved (anaerobic respiration).

flagellum (L. *a whip*). Whiplike organ of locomotion.

fluke (O.E. *flōc,* flatfish). A member of class Trematoda.

fiber, fibril (L. *fibra,* thread). These two terms are often confused. Fiber is a strand of protoplasmic material produced or secreted by a cell and lying outside the cell, or a fiberlike cell. Fibril is a strand of protoplasm produced by a cell and lying within the cell.

fin (O.E. *finn,* fin or feather). Thin, transparent epidermal extension in chaetognaths.

gamete (gam'ete) (Gr. *gamos,* marriage). A mature germ cell, either male or female.

gastrodermis (gas"tro-der'mis) (Gr. *gastēr,* stomach, + *derma,* skin). Lining of the digestive cavity of coelenterates.

gastrula (Gr. *gastēr,* stomach, + L. *-ula,* diminutive). Embryonic stage of double-layered cup.

gel (jel) (from gelatin, from L. *gelare,* to freeze). That state of a colloidal system in which the solid particles form the continuous phase and the fluid medium the discontinuous phase.

gemmule (L. *gemma,* bud, + *-ula,* little). Asexual reproductive unit in certain sponges.

gene (jen) (Gr. *genēs,* born). The part of a chromosome that is the hereditary determiner and is transmitted from one generation to another. It occupies a fixed chromosomal locus and can best be defined only in a physiologic or operational sense.

genome (jen'om) (Gr. *genos,* race, + L. *-oma,* abstract group). The total number of genes in a haploid set of chromosomes.

genotype (jen'o-type) (Gr. *genos,* race, + *typos,* form). The genetic constitution, expressed and latent, of an organism; the particular set of genes present in the cells of an organism; opposed to phenotype.

genus (pl., *genera*) (je'nus) (L. *genus,* race). A taxonomic rank between family and species.

germ layer In the animal embryo, one of three basic layers (ectoderm, endoderm, mesoderm) from which the various organs and tissues arise in the multicellular animal.

germ plasm The germ cells of an organism, distinct from the somatoplasm.

gestation (jes-ta'shun) (L. *gestare,* to bear). The period in which offspring are carried in the uterus.

glochidium (glow-kid'e-um) (Gr. *glōchis,* point, + *-idion,* diminutive). Larva of a clam.

gnathobase (nath'o-base) (Gr. *gnathos,* jaw, + base). A median basic process on certain appendages in some arthropods, usually for biting or crushing food.

Golgi body (gol'je) (after Golgi, Italian histologist). A cytoplasmic component that may play a role in certain cell secretions or may represent a region where high-energy compounds from the mitochondria collect.

gonangium (Gr. *gonos,* seed, + *angeion,* dim. of vessel). Reproductive zooid of hydroid colony.

gregarious (gre-gar'i-us) (L. *grex,* herd). Living in groups or flocks.

habitat (hab'i-tat) (L. it inhabits, from *habitare,* to dwell). The place where an organism normally lives or where individuals of a population live.

haploid (hap'oid) (Gr. *haploos,* single). The reduced number of chromosomes typical of gametes, as opposed to the diploid number of somatic cells.

haltere (Gr. *haltēr,* leaping weight). Club-shaped organs in place of second pair of wings in Diptera.

hectocotylus (Gr. *hekaton,* hundred, + *kotylē,* cup). Transformed, and sometimes autonomous, arm that serves as a male copulatory organ in cephalopods.

hermaphrodite (hur-maf'ro-dite) (Gr. *hermaphroditos,* containing both sexes; from Greek mythology, *Hermaphroditos,* son of Hermes and Aphrodite). An organism with both male and female organs. Hermaphroditism commonly refers to an abnormal condition in which male and female organs are found in the same animal; monoecious is a normal condition for the species.

heterotroph (het'er-o-trof) (Gr. *heteros,* another, + *trophos,* feeder). An organism that obtains both organic and inorganic raw materials from the environment in order to live.

heterozygote (het"er-o-zi'gote) (Gr. *heteros,* another, + *zygōtos,* yoked). An organism in which the pair of alleles for a trait is composed of different genes (usually dominant and recessive); derived from a zygote formed by the union of gametes of dissimilar genetic constitution.

holoblastic (hol"o-blas'tic) (Gr. *holos,* whole, + *blastos,* germ). Cleavage in which an entire egg cell divides.

holozoic (Gr. *holos,* whole, + *zoikos,* of animals). That type of nutrition that involves ingestion of solid organic food.

homoiothermal (ho-moi"o-ther'mal) (Gr. *homoios,* like, + *thermē,* heat, + al). Having relatively constant temperature, often above that of the environment.

homology (ho-mol'o-ji) (Gr. *homologia,* agreement). Similarity in embryonic origin and adult structure, based on descent from a common ancestor.

homozygote (ho"mo-zi'gote) (Gr. *homos,* same, + *zygotos,* yoked). An organism in which the pair of alleles for a trait is composed of the same genes (either dominant or recessive but not both).

hood (O. E. *hōd,* hood). A fold of the body wall that can be drawn over the head in chaetognaths.

humoral (hyu'mer-al) (L. *humor,* a fluid). Pertaining to a body fluid such as blood or lymph.

hydranth (Gr. hydra + *anthos,* flower). Nutritive zooid of hydroid colony.

hydrolysis (hi-drol'i-sis) (Gr. *hydōr,* water, + *lysis,* a loosening). The decomposition of a chemical compound by the addition of water; the splitting of a molecule into its groupings so that the split products acquire hydrogen and hydroxyl groups.

hydroxyl (hi-drok'sil) (hydrogen + oxygen, + yl). Containing an OH⁻ group, a negatively charged ion formed by alkalies in water.

hypertonic (hi''per-ton'ik) (Gr. *hyper,* over, + *tonos,* tension). Refers to a solution whose osmotic pressure is greater than that of another solution with which it is compared; contains a greater concentration of dissolved particles and gains water through a semipermeable membrane from a solution containing fewer particles.

hypertrophy (hi-per'tro-fee) (Gr. *hyper,* over, + *trophē,* nourishment). Abnormal increase in size of a part or organ.

hypothalamus (hi''po-thal'a-mus) (Gr. *hypo,* under, + *thalamos,* inner chamber). A ventral part of the forebrain beneath the thalamus; one of the centers of the autonomic nervous system.

hypotonic (hi''po-ton'ik) (Gr. *hypo,* under, + *tonos,* tension). Refers to a solution whose osmotic pressure is less than that of another solution with which it is compared or taken as standard; contains a lesser concentration of dissolved particles and loses water during osmosis.

inductor (in-duk'ter) (L. *inducere,* to introduce, lead in). In embryology a tissue or organ that causes the differentiation of another tissue or organ.

infusoriform (NL. *infusoria,* similar to Infusoria [ciliates], + form). A larval stage in dicyemids.

interstitial (L. *inter,* among, + *sistere,* to stand). Refers to one of the totipotent cells in the body wall of coelenterates.

introvert (L. *intro,* inward, + *vertere,* to turn). In the sipunculid the anterior narrow portion that can be withdrawn into the trunk.

invagination (in-vaj''i-na'shun) (L. *in,* in, + *vagina,* sheath). An infolding of a layer of tissue to form a sac-like structure.

isolecithal (Gr. *isos,* equal, + *lekithos,* yolk, + -al). Pertaining to a zygote (or ovum) with yolk evenly distributed.

isotope (i'so-tope) (Gr. *isos,* equal, + *topos,* place). One of several different forms of a chemical element, differing from each other physically but not chemically.

keratin (ker'a-tin) (Gr. *keras,* horn, + -in, suffix of proteins). A protein found in epidermal tissues and modified into hard structures such as horns, hair, and nails.

kinetosome (kin-et'o-some) (Gr. *kinētos,* moving, + *sōma,* body). The granule at the base of the flagellum or cilium; similar to centriole.

kinin (ki'nin) (Gr. *kinein,* to move, + -in, suffix of hor-mones). A type of local hormone that is released near its site of origin.

labyrinthodont (lab''i-rin'tho-dont) (Gr. *labyrinthos,* labyrinth, + *odous, odontos,* tooth). *A group of fossil stem amphibians from which most amphibians later arose. They date from the late Paleozoic.*

lacteal (lak'te-al) (L. *lacteus,* of milk). Refers to one of the lymph vessels in the villus of the intestine.

lacunar system (L. *lacuna,* pit, pool). Epidermal canal system peculiar to acanthocephalans.

lagena (la-je'na) (L. large flask). Portion of the primitive ear in which sound is translated into nerve impulses; evolutionary beginning of cochlea.

lemniscus (L. ribbon). One of a pair of internal projections of the epidermis from the neck region of Acanthocephala, which functions in fluid control in the protrusion and invagination of the proboscis.

leukocyte (lu'ko-site) (Gr. *leukos,* white, + *kytos,* hollow vessel). A common type of white blood cell with beaded nucleus.

lipase (lie'pace) (Gr. *lipos,* fat, + -ase, enzyme suffix). An enzyme that converts fatty acids and glycerin; it may also promote the reverse reaction.

lipid, lipoid (lip'id) (Gr. *lipos,* fat). Pertains to certain fatty-like substances that often contain other groups such as phosphoric acid.

lithosphere (lith'o-sfeer) (Gr. *lithos,* rock, + *sphaira,* ball). The rocky component of the earth's surface layers.

littoral (lit'uh-rul)(L. *litus,* seashore). The floor of the sea from the shore to the edge of the continental shelf.

lophophore (lof'o-for) (Gr. *lophos,* crest, + *phoros,* bearing). Tentacle-bearing ridge or arm that is an extension of the coelomic cavity in lophophorate animals.

luciferase (lu-sif'er-ace) (L. *lux,* light, + *ferre,* to bear). An enzyme involved in light production in organisms.

lunules (L. *luna,* moon, + -ula, little). These slitlike openings in the sand dollar test may be crescent shaped.

macronucleus (mak''ro-nu'kle-us) (Gr. *makros,* long, large, + *nucleus,* kernel). The larger of the two kinds of nuclei in ciliate protozoa; controls all cell functions except reproduction.

madreporite (It. *madrepora,* mother-stone, from L. *mater,* mother, + *porus,* tufa [from Greek], + -ite, suffix for some body parts]. Sievelike structure that is the intake for the water-vascular system.

marsupial (mar-su''pe-al) (Gr. *marsypion,* little pouch). One of the pouched mammals of the subclass Metatheria.

matrix (may'trix) (L. *mater,* mother). The intercellular substance of a tissue, or that part of a tissue into which an organ or process is set.

maxilla (mak-sil'a) (L. "lower" jaw, a diminutive of *mala,* upper jaw). One of the upper jawbones in vertebrates; one of the head appendages in arthropods.

maxilliped (mak-sil'i-ped) (L. *maxilla,* jaw, + -ped-, -pes, foot). One of the three pairs of head appendages located just posterior to the maxilla in crustaceans.

medulla (me-dul'a) (L. marrow). The inner portion of an organ in contrast to the cortex or outer portion; hind-brain.

medusa (me-du'sa) (Greek mythology, female monster

with snake-entwined hair). A jellyfish, or the free-swimming stage in the life cycle of coelenterates.

meiosis (my-o′sis) (Gr. *meioun,* to make small). That nuclear change by which the chromosomes are reduced from the haploid number.

menopause (men′o-pawz) (Gr. *mēn,* month, + *pauein,* to cease). In the human female that time when reproduction ceases; cessation of the menstrual cycle.

menstruation (men″stru-a′shun) (L. *menstrua,* the menses, from *mensis,* month). The discharge of blood and uterine tissue from the vagina at the end of a menstrual cycle.

meroblastic (Gr. *meros,* part, + *blastos,* germ). Cleavage restricted to the blastoderm.

mesenchyme (Gr. *mesos,* middle, + *enchymē,* infusion). Embryonic or unspecialized connective tissue.

mesoglea (mes″o-gle′a) (Gr. *mesos,* middle, + *gloa,* glue). The jellylike filling between the ectoderm and endoderm of certain coelenterates and comb jellies.

mesosome (Gr. *mesos,* middle, + *soma,* body). Collar or middle part of pogonophores and hemichordates.

metabolism (me-tab′o-lizm) (Gr. *metabolē,* change). A group of processes that includes nutrition, production of energy (respiration), and synthesis of more protoplasm; the sum of the constructive (anabolism) and destructive (catabolism) processes.

metagenesis (met″a-jen′e-sis) (Gr. *meta,* after, + *genesis,* origin). Alternation of sexual and asexual reproduction in the life cycle of certain organisms.

metamere (met′a-mere) (Gr. *meta,* after, + *meros,* part). A repeated unit of structure; somite, or segment.

metamorphosis (Gr. *meta,* after, + *morphē,* form, + -*osis,* state of). Sharp change in form during postembryonic development.

metanephridium (met″a-ne-frid′e-um) (Gr. *meta,* after, + *nephros,* kidney). A type of tubular nephridium with the inner open end draining the coelom and the outer open end discharging to the exterior.

metasome (Gr. *meta,* after, + *soma,* body). Posterior part of the body of hemichordates and pogonophores.

micron (μ) (my′kron) (Gr. neuter of *mikros,* small). One one-thousandth of a millimeter; about 1/25,000 of an inch.

micronucleus (my″kro-nu′kle-us). A small nucleus found in ciliate protozoa; controls the reproductive functions of these organisms.

microsome (my′kro-some) (Gr. *sōma,* body). A constituent of cytoplasm that contains RNA and is the site of protein synthesis.

mictic (Gr. *miktos,* mixed or blended). Pertains to haploid egg of rotifers or the females that lay such eggs.

miracidium (my″ra-sid′e-um) (Gr. *meirakidion,* youthful person). A minute ciliated larval stage in the life of flukes.

mitochondria (my″to-kon′dre-a) (Gr. *mitos,* a thread, + *chondrion,* diminutive of *chondros,* corn, grain). Minute granules, rods, or threads in the cytoplasm and the seat of important cellular enzymes.

mitosis (my-to′sis) (Gr. *mitos,* thread, + -*osis,* state of).

Cell division in which there is an equal qualitative and quantitative division of the chromosomal material between the two resulting nuclei; ordinary cell division (indirect).

monoecious (muh-ne′shus) (Gr. *monos,* single, + *oikos,* house). Both male and female gonads in the same organism; hermaphroditic.

monosaccharide (mon″o-sak′a-rid[e]) (Gr. *monos,* one, + *sakcharon,* sugar, from Sanskrit *śarkarā,* gravel, sugar). A simple sugar that cannot be decomposed into smaller suger molecules; contains five or six carbon atoms.

morphogenesis (mor″fo-jen′e-sis) (Gr. *morphē,* form, + *genesis,* origin.) Development of the architectural features of organisms.

morphology (mor-fol′o-ji) (Gr. *morphē,* form, + L. *logia,* study from Gr. *logos,* word). The science of structure. Includes cytology, or the study of cell structure; histology, or the study of tissue structure; and anatomy, or the study of gross structure.

morula (mor′uh-luh) (L. *morum,* mulberry, + *ula,* diminutive). Group of cells in early stage of segmentation.

mutation (myu-ta′shun) (L. *mutare,* to change). A stable and abrupt change of a gene; the heritable modification of a character.

mutualism (myu′chyu-al-izm) (L. *mutuus,* lent, borrowed, reciprocal). A type of symbiosis in which two different species derive benefit from their association.

myofibril (my″o-fy′bril) (Gr. *mys,* muscle, mouse, + L. diminutive of *fibra,* fiber). A contractile filament within muscle or muscular fiber.

myosin (my′o-sin) (Gr. *mys,* muscle, mouse). A protein found in muscle; important component in the contraction of muscle.

myxedema (mik″se-dee′ma) (Gr. *myxa,* nasal mucus, + *oidēma,* a swelling). A disease that results from thyroid deficiency in the adult; characterized by swellings under the skin.

nacre (nay′ker) (F. mother-of-pearl). Innermost layer of mollusk shell.

NAD Abbreviation of nicotinamide adenine dinucleotide (see DPN).

naris (na′ris) (L. nostril). Openings into the nasal cavity, both internally and externally, in the head of a vertebrate.

nekton (nek′tun) (Gr. neuter of *nēktos,* swimming). Term for the actively swimming organisms in the ocean.

nematocyst (Gr. *nēma,* thread, + *kystis,* bladder). Stinging organoid of coelenterates.

nematogen (Gr. *nēma,* thread, + *genēs,* born). The vermiform embryo of a dicyemid.

notochord (no′to-kord) (Gr. *nōtos,* back, + *chorda,* cord). A rod-shaped cellular body along the median plane and ventral to the central nervous system in chordates.

nucleic acid (nu-klee′ik) (L. *nucleus,* kernel). One of a class of molecules composed of joined nucleotides; chief types are deoxyribonucleic acid (DNA), found only in cell nuclei (chromosomes), and ribonucleic acid (RNA), found both in cell nuclei (chromosomes and nucleoli) and in cytoplasm (microsomes).

nucleolus (nu-klee′o-lus) (dim. of nucleus). A deeply staining body within the nucleus of a cell and containing RNA.

nucleoprotein (nu″kle-o-pro′tein). A molecule composed of nucleic acid and protein; occurs in two types, depending on whether the nucleic acid portion is DNA or RNA.

nucleotide (nu′kle-o-tide). A molecule consisting of phosphate, 5-carbon sugar (ribose or deoxyribose), and a purine or a pyrimidine; the purines are adeine and guanine, and the pyrimidines are cytosine, thymine, and uracil.

nymph (nimf) (L. *nympha,* bride, nymph). The immature form of an insect that undergoes a gradual metamorphosis.

ocellus (o-sel′lus) (L. diminutive of *oculus,* eye). A simple eye in many types of invertebrates.

ontogeny (on-toj′e-ne) (Gr. *ontos,* being, + *-geneia,* act of being born, from *genēs,* born). The development of an individual from egg to senescence.

ooecium (o-ee′shum) (Gr. *ōion,* egg, + *oikos,* house, + L. *-ium,* from Ger. -ion, diminutive). Brood pouch.

operculum (o-pur′kyu-lum) (L. cover). The gill cover in bony fish; horny plate in some snails.

opisthocoelous (o-pis″tho-see′lus) (Gr. *opisthe, behind,* + *koilos,* hollow). Concave behind, as in the centrum of certain vertebrae.

organelle (Gr. *organon,* tool, organ, + L. *ella,* diminutive). Specialized part of a cell; literally, a small organ.

organism (or′gan-izm). An individual plant or animal, either unicellular or multicellular.

osmosis (oz-mo′sis) (Gr. *ōsmos,* act of pushing, impulse). The process in which water migrates through a semipermeable membrane, from a side containing a lesser concentration to the side containing a greater concentration of dissolved particles. The diffusion of a solvent (usually water) through a semipermeable membrane.

osphradium (os-fray′de-um) (Gr. *osphradion,* small bouquet, diminutive of *osphra,* smell). A sense organ that tests incoming water.

ossicle (L. *ossiculum,* small bone). Small separate pieces of endoskeleton.

ostium (L. door). Opening.

otolith (o′to-lith) (Gr. *ous, otos,* ear, + *lithos,* stone). Calcareous concretions in the membraneous labyrinth of lower vertebrates, or in the auditory organ of certain invertebrates.

oviparity, oviparous (o″vuh-par′uh-te, o-vip′a-rus) (L. *ovum,* egg, + *parere,* to bring forth). Reproduction in which eggs are released by the female; development of offspring occurs outside the maternal body.

ovipositor (o″vi-poz′i-ter) (L. *ovum,* egg, + *parere,* to bring forth). In many female insects a structure at the posterior end of the abdomen for laying eggs.

ovoviviparity, ovoviviparous (o″vo-vy″vuh-par′uh-te, o″vo-vy-vip′a-rus) (L. *ovum,* egg, + *vivere,* to live, + *parere,* to bring forth). Reproduction in which eggs develop within the maternal body without nutrition by the female parent.

oxidation (ok′si-da′shun) (Fr. *ox-*[oxygen, from Gr. *oxys,* sharp, acid, + *genes,* born] + *-ide,* acid + *-ation*). Rearrangement of a molecule to create a high-energy bond; a chemical change in which a molecule loses one or more electrons.

paedogenesis See **pedogenesis.**

palingenesis (pal′in-jen′e-sis) (Gr. *palin,* again + *genesis,* birth). The stages in the development or ontogeny of an animal that are inherited from ancestral species, such as gill slits in the unborn of mammals.

palp (L. *palpus,* stroking, caress). A projecting part (appendage) on the head region or near the mouth in some invertebrates; a palpus.

papilla (pa-pil′a) (L. nipple). A small nipplelike projection.

paramylum (puh-ram′uh-lum) (Gr. *para,* beside, + L. *amylum,* starch). Starch inclusions in certain protozoans.

parapodia (par″a-po′de-a) (Gr. *para,* beside, + *pod-,* foot, + *-ia,* condition). The segmental appendages in polychaete worms that serve in breathing, locomotion, and creation of water currents.

parasympathetic (par″a-sim-pa-thet′ik) (Gr. *para,* beside, + *sympathes,* sympathetic, from *syn,* with, + *pathos,* feeling). One of the subdivisions of the autonomic nervous system, whose centers are located in the brain, anterior part of the spinal cord, and posterior part of the spinal cord.

parenchymula (Gr. *parenchyma,* visceral flesh, from *para,* beside, + *enchyma,* infusion, + L. *ula,* diminutive). A stereogastrula larva of class Demospongiae.

parthenogenesis (par″the-no-jen′e-sis) (Gr. *parthenos,* virgin, + *genesis,* birth). The development of an unfertilized egg; a type of unisexual reproduction.

pathogenic (path″o-jen′ik) (Gr. *pathos,* disease, + *gennan,* to produce). Producing a disease.

pecten (L. comb). A pigmented, vascular, and comblike process that projects into the vitreous humor from the retina at point of entrance of the optic nerve (reptiles and birds). Its functions are obscure, but its peculiar shadow on the retina may make the bird more sensitive to movement in the visual field. It may also enable the bird to determine the position of the sun.

pedicellaria (NL. *pediculus,* little foot, + *-aria,* like or connected with). Minute pincerlike organ on surface of certain echinoderms.

pedipalps (L. *pes, pedis,* foot, + *palpus,* stroking, caress). Second pair of appendages of arachnids.

pedogenesis (pee″do-jen′e-sis) (Gr. *pais,* child, + *genēs,* born). Parthenogenetic production of eggs, or other stages, by immature or larval animals.

peduncle (puh-dung′kl) (L. *pedunculus* dim. of *pes,* foot). A stalk; a band of white matter joining different parts of the brain.

pelagic (puh-laj′ik) (Gr. *pelagos,* the open sea). Pertaining to the open ocean.

pentadactyl (pen″ta-dak′til) (Gr. *pente,* five, + *daktylos,* finger). With five digits.

peptidase (pep′ti-dace) (Gr. *peptein,* to digest, + *-ase,* enzyme suffix). An enzyme that breaks down amino acids from peptide.

periproct (Gr. *peri,* around, + *prōktos,* anus). Region of aboral plates around the anus of echinoids.

periostracum (Gr. *peri,* around, + *ostrakon,* shell). Outer horny layer of mollusk shell.

perisarc (Gr. *peri,* around, + *sarx,* flesh). Sheath covering the stalk and branches of a hydroid.

peristalsis (per"i-stall'sis) (Gr. *peristaltikos,* compressing around, from *peristellein,* to wrap around, from *peri,* around, + *stellein,* to place). The series of alternate relaxations and contractions by which food is forced through the alimentary canal.

peristomium (Gr. *peri,* around, + *stoma,* mouth, + L. -*ium*). Foremost true segment of annelid and bearing mouth.

peritoneum (per"i-tuh-ne'um) (Gr. *peritonaios,* stretched across, from *peri,* around, + *tenein,* to stretch). The membrane that lines the abdominal cavity and covers the viscera.

peroxisome (per-ox"i-some') (L. *per-,* prefix meaning thoroughly, + Eng. oxidize, + Gr. *sōma,* body). A type of cell organelle containing oxidative enzymes that remove amino groups from amino acids.

petrifaction (pet"ri-fak'shun) (Gr. *petra,* stone, + L. *facere,* to make). The changing of organic matter into stone.

pH (*p*otential of *h*ydrogen). A symbol of the relative concentration of hydrogen ions in a solution; pH values are from 0 to 14, and the lower the value, the more acid or hydrogen ions in the solution.

phagocyte (fag'o-site) (Gr. *phagein,* to eat, + *kytos,* hollow vessel). Any cell that engulfs and devours microorganisms or other foreign particles.

phasmid (faz'mid) (Gr. *phasma,* apparition, phantom, + -*id*). One of a pair of glands or sensory structures found in the posterior end of certain nematodes.

phenotype (fee'no-tipe) (Gr. *phainein,* to show). The visible characters; opposed to genotype of the hereditary constitution.

pheromone (fair'uh-mone) (Gr. *pherein,* to carry, + o + hormone, from *hormon,* stirring up, excitement). A substance secreted by an organism that influences the behavior of others of the same species.

phosphagen (fos'fa-jen) (phosphate + glycogen). A term for creatine-phosphate and arginine-phosphate, which store and may be sources of high-energy phosphates.

phosphorylation (fos"fo-ri-la'shun). The addition of a phosphate group, such as H_2PO_3, to a compound.

phylogeny (fy-loj'uh-nee) Gr. *phylon,* tribe + *gennan,* to bring forth). The evolutionary history of a group of organisms.

phylum (pl., *phyla*) (fy'lum) (NL. Gr. *phylon,* race, tribe). A chief category of taxonomic classification into which living things are divided.

pilidium (py-lid'e-um) (Gr. *pilidion,* diminutive of *pilos,* felt cap). Free-swimming hat-shaped larva of nemertean worms.

pinocytosis (pin"o-sy-to'sis) (Gr. *pinein,* to drink, + *kytos,* hollow vessel, + -*osis,* condition). A process of cell drinking.

placenta (pluh-sen'tuh) (L. flat cake, Gr. *plakous,* from Gr. *plax, plakos,* anything flat and broad). The vascular structure, embryonic and maternal, through which the embryo and fetus are nourished while in the uterus.

plankton (plangk'tun) (Gr. neuter of *planktos,* wandering). The floating animal and plant life of a body of water.

plantigrade (plan'ti-grade) (L. *planta,* scle, + *gradus,* step, degree). Pertaining to animals that walk on the whole surface of the foot.

plasma membrane (plazma) (Gr. *plasma,* a form, mold). The thin membrane that surrounds the cytosome; considered a part of the cytoplasm.

plastid (plas'tid) (Germ. from Gr. *plastēs,* one who forms, sculptor + -*id-,* from feminine plural suffix). A small body in the cytoplasm that often contains pigment.

pleopod (plee'uh-pod) (Gr. *plein,* to sail, + *pous, podos,* foot). One of the swimming feet on the abdomen of a crustacean.

plesiosaur (plee'see-uh-sor") (Gr. *plēsios,* near, + *sauros,* lizard). A long-necked, marine reptile of Mesozoic times.

pleura (ploor'a) (Gr. side, rib). The membrane that lines each half of the thorax and covers the lungs.

plexus (plek'sus) (L. network, braid). A network, especially of nerves or of blood vessels.

pluteus (pl. **plutei**) (L. *pluteus,* movable shed, reading desk). These echinoid larvae have elongated processes like the supports of a desk; originally called "painter's easel larva."

poikilothermic (poi-kil"o-ther'mik) (Gr. *poikilos,* variable, + thermal). Pertaining to animals whose body temperature is near that of its environment.

polarization (po"ler-i-za'shun) (L. *polaris,* polar, + Gr. -*iz-,* make, + -*ation*). The arrangement of positive electric charges on one side of a surface membrane and negative electric charges on the other side (in nerves and muscles).

polymorphism (pol"i-mor'fizm) (Gr. *polys,* many, + *morphe,* form). The presence in a species of more than one type of individual.

polyp (pol'ip) (Fr. *polype,* octopus, from L. *polypus,* many-footed). The sessile stage in the life cycle of coelenterates.

polypeptide (pol"i-pep'tide) (Gr. *polys,* many, + *peptein,* to digest). A molecule consisting of many joined amino acids, not as complex as a protein.

polyphyletic (pol"i-fy-let'ik) (Gr. *polys,* many, + *phylon,* tribe). Derived from more than one ancestral type; constrasts with monophyletic, or from one ancestor.

polysaccharide (pol'i-sak'a-rid[e]) (Gr. *polys,* many, + *sakcharon,* sugar, from Sanskrit *śarkarā,* gravel, sugar). A carbohydrate composed of many monosaccharide units, such as glycogen, starch, and cellulose.

prehensile (pre-hen'sil) (L. *prehendere,* to seize). Adapted for grasping.

proboscis (Gr. *proboskis,* an elephant's trunk, from *pro,* before, + *boskein,* feed). The sensory offensive and defensive organ at the anterior end of the nemertine.

progesterone (pro-jes'ter-own) (L. *pro,* before, + *gestare,* to carry). Hormone secreted by the corpus luteum and the placenta; prepares the uterus for the fertilized egg and maintains the capacity of the uterus to hold the embryo and fetus.

proglottid (Gr. *proglōttis,* tongue tip, from *pro,* before, + *glōtta,* tongue, + *-id,* suffix). A section of a tapeworm. Dujardin (1843) gave this derivation because of its resemblance to the tip of the tongue.

prosoma (Gr. *pro,* before, + *sōma,* body). Fused head and thoracid segments of arthropods.

prosopyle (pros'uh-pile) (Gr. *prosō,* forward, + *pylē,* gate). Connection between the incurrent and radial canal.

prothrombin (pro-throm'bin) (Gr. *pro,* before, + *thrombos,* clot). A constituent of blood plasma that is changed to thrombin by thombokinase in the presence of calcium ions; involved in blood clotting.

protonephridium (Gr. *prōtos,* first, + *nephros* kidney). Primitive excretory organ of tubule with terminating flame tube or solenocyte.

protosome (Gr. *prōtos,* first, + *sōma,* body). First body division or proboscis of pogonophores and hemichordates.

prostomium (Gr. *pro,* before, + *stoma,* mouth). In mollusks and some worms the part of the head in front of the mouth and probably nonmetameric.

proximal (prox'i-mal) (L. *proximus,* nearest). Nearest the place of attachment, or opposite of distal.

psammolittoral Pertaining to the intertidal areas of sandy beaches, or the intertidal biota of such regions.

psammon Microfauna and microflora inhabiting the interstices between grains of sand of sandy beaches; the psammolittoral biota.

pseudocoel (su'do-seel) (Gr. *pseudēs,* false, + *koilōma,* cavity). A body cavity not lined with peritoneum and not part of the blood system.

pseudopodium (Gr. *pseudēs,* false, + *podion,* little foot). Protrusion of part of cytoplasm of an ameba.

pterosaur (ter'uh-sor) (Gr. *pteron,* feather, wing, + *sauros,* lizard). An extinct flying reptile that flourished during the Mesozoic.

puff The pattern of swelling of specific bands or gene loci on giant chromosomes during the larval and imaginal stages of flies.

pygidium (Gr. *pygidion,* small rump, from *pygē,* rump). The terminal segment of annelids. New segments are formed in front of this segment.

pylorus (py-lor'us) (Gr. *pylē,* gate, + *ouros,* watcher). The opening between the stomach and duodenum which is guarded by a valve.

pyrenoid (py-ree'noid, py're-noid) (Gr. *pyrēn,* fruit stone, + *eidos,* form). A protein body in the chloroplasts of certain organisms that serves as a center for starch formation.

radula (raj'uh-luh) (L. scraper). Rasping tongue of certain mollusks.

redia (pl. **rediae**) (ree'dee-a; pl., ree'dee-ee) (from Redi, Italian biologist). A larval stage in the life cycle of flukes; it is produced by a sporocyst larva, and in turn gives rise to many cercariae.

rete mirabile (ree'tee muh-rab'uh-lee) (L. wonderful net). A network of small blood vessels so arranged that the incoming blood runs parallel to the outgoing blood and thus makes possible a counterexchange between the two bloodstreams. Such a mechanism ensures a constancy of gases in the swim bladder.

retina (ret'i-na) (L. *rete,* a net). The sensitive, nervous layer of the eye.

rhabdite (Gr. *rhabdos,* rod, + *-ite,* suffix). Ectodermal rodlike structures in certain turbellarians. May function in slime formation.

rhabdocoel (rab'do-seel) (Gr. *rhabdos,* rod, + *koilōma,* a hollow). A member of a group of free-living flatworms possessing a straight, unbranched digestive cavity.

rhagon (Gr. *rhax, rhagos,* berry). State in leuconoid type of canals; contains small chambers lined with collar cells.

rhyncodeum (ring-ko-dee'um)(Gr. *rhynchos,* beak, snout, + *hodaios,* pertaining to way). Short tubular cavity to which the proboscis is attached in Rhynchocoela. It opens anteriorly by a small pore. It must not be confused with the **rhynchocoel** in which the proboscis lies.

rosette (Fr. small rose). An organ for attachment in the subclass Cestodaria.

rostellum (L. little beak). Hook-bearing tip of the tapeworm scolex.

rostrum (ros'trum) (L. ship's beak). A snoutlike projection on the head.

sagittal (saj'i-tal) (L. *sagitta,* arrow). Pertaining to the median anteroposterior plane that divides a bilaterally symmetrical organism into right and left halves.

saprophyte (sap'ro-fite) (Gr. *sapros,* rotten, + *phyton,* plant). An organism that lives upon dead organic matter.

sarcolemma (sar"ko-lem'a) (Gr. *sarx,* flesh, + *lemma,* rind). The thin noncellular membrane of striated muscle fiber or cell.

schizocoel, schizocoelomate (skiz'o-seel) (Gr. *schizein,* to split, + *koilōma,* cavity). Schizocoel is a coelum formed by a splitting of embryonic mesoderm. Schizocoelomate is an animal with a schizocoel, such as an arthropod or mollusk.

schizogony (skiz-zog'uh-ne) (Gr. *schizein,* to split, + *gonos,* seed). Multiple asexual fission.

scleroblast (sklir'o-blast) (Gr. *sklēros,* hard, + *blastos,* germ). Mesenchyme cell that secretes spicules.

sclerotic (skle-rot'ik) (Gr. *sklēros,* hard). Pertaining to the tough outer coat of the eyeball.

scolex (Gr. *skōlēx,* worm). Term restricted to the so-called head of the tapeworm.

scrotum (skro'tum) (L. bag). The pouch that contains the testes and accessory organs in most mammals.

scyphistoma (sy-fis'tuh-muh) (Gr. *skyphos,* cup, + *stoma,* mouth). A stage in the development of scyphozoan jellyfish just after the larva becomes attached.

seminiferous (sem"i-nif'er-us) (L. *semen,* semen, + *ferre,* to bear). Pertains to the tubules that produce or carry semen in the testes.

semipermeable (sem"i-pur'me-a-bl) (L. *semi,* half, + *permeabilis,* capable of being passed through). Permeable to small particles, such as water and certain inorganic ions, but not to colloids, etc.

septum (pl., **septa**) (sep′tum) (L. fence). A wall between two cavities.

sere (seer) (L. *series,* row, kind, from *serere,* to join). The sequence of series of communities that develop in a given situation from pioneer to terminal climax communities during ecologic succession.

serum (sir′um) (L. whey). The plasma of blood that separates on clotting; the liquid that separates from the blood when a clot is formed.

seta (pl., **setae**) (see′ta; pl., see′tee) (L. bristle). A needle-like chitinous structure of the integument of annelids and related forms.

siliceous (sil-lish′us) (L. *silex,* flint). Containing silica.

simian (sim′e-an) (L. *simia,* ape). Pertains to monkeys.

sinus (sy′nus) (L. curve). A cavity or space in tissues or in bone.

siphonoglyph (sy-fon′uh-glif) (Gr. *siphōn,* reed, tube, siphon, + *glyphē,* carving). Ciliated furrow in the gullet of sea anemones.

solenocyte (Gr. *solēn,* pipe, + *kytos,* hollow vessel). Special type of protonephridium in which the end bulb bears a flagellum instead of a tuft of cilia, as in Platyhelminthes.

soma (so′ma) (Gr. body). The body of an organism in contrast to the germ cells (germ plasm).

somatic (so-mat′ik) (Gr. *sōma,* body). Refers to the body, such as somatic cells in contrast to germ cells.

speciation (spe″shi-a′shun) (L. *species,* kind). The evolving of two or more species by the splitting of one ancestral species.

spermatheca (spurm″a-the′ka) (Gr. *sperma,* seed, + *thēkē,* a case). A sac in the female reproductive organs for the storage of sperm.

sphincter (sfingk′ter) (Gr. *sphinkter,* band, sphincter, from *sphingein,* to bind tight). A ring-shaped muscle capable of closing a tubular opening by constriction.

spicule (L. diminutive of *spica,* point). Skeletal element found in certain sponges.

spongocoel (Gr. *spongos,* sponge, + *koilos,* hollow). Central cavity in sponges.

sporocyst (spo′ro-sist) (Gr. *sporos,* seed, + *kystis,* pouch). A larval stage in the life cycle of flukes; it originates from a miracidium.

sporozoite (spo′ro-zo″ite) (Gr. *sporos,* seed, + *zōon,* animal, + *-ite,* suffix for body part). A motile spore formed from the zygote in many Sporozoa.

statocyst (Gr. *statos,* standing, + *kystis,* bladder). Sense organ concerned with orientation.

statolith (Gr. *statos,* standing, + *lithos,* stone). Small calcareous body resting on the tufts of cilia in the statocyst.

stenohaline (Gr. *stenos,* narrow, + *hals,* salt). Pertaining to organisms that have restricted ranges of saltwater concentrations.

stenotopic (sten″o-top′ic)ˑ(Gr. *stenos,* narrow, + *topos,* place). Refers to an organism with restricted range.

stereogastrula (ste′re-o-gas′tru-la) (Gr. *stereos,* solid, + *gastēr,* stomach, + L. *-ula,* diminutive). A solid type of gastrula, such as the planula of coelenterates.

sterol, steroid (ste′rol, ste′roid) (Gr. *stereos,* solid, + *-ol* [L. *oleum,* oil]). One of a class of organic compounds containing a molecular skeleton of four fused carbon rings; it includes cholesterol, sex hormones, adrenocortical hormones, and vitamin D.

stigma (Gr. *stigma,* mark, tattoo mark). Eyespot in certain protozoans.

stoma (sto′ma) (Gr. mouth). A mouthlike opening.

stomochord (Gr. *stoma,* mouth, + L. *chorda,* cord). The thick-walled forward evagination of the dorsal wall of the buccal cavity into the proboscis; the buccal diverticulum; may be a homolog of the notochord or a primitive notochord.

stratum (L. something spread out, covering). A horizontal layer or division of a biologic community that exhibits stratification of habitats (ecologic).

strobila (Gr. *strobilē,* lint plug like a pine cone [*strobilos*]). A stage in the development of the scyphozoan jellyfish.

stylet (Fr. from It. *stiletto,* diminutive of L. *stilus,* stake). Sharp-pointed defense organ on the proboscis.

substrate, substratum (sub′strate) (L. neuter of *substratus,* strewn under). A substance that is acted upon by an enzyme; ground or solid object.

sycon (Gr. *sykon,* fig). Sometimes called syconoid. A type of canal system.

symbiosis (sim″by-o′sis, sim′be-o′sis) (Gr. *syn,* with, + *bios,* life). The living together of two different species in an intimate relationship; includes mutualism, commensalism, and parasitism.

synapse (sin′aps, si-naps′) (Gr. *synapsis,* contact, union). The place at which a nerve impulse passes from an axon of one nerve cell to a dendrite of another nerve cell.

syncytium (sin-sish′e-um) (Gr. *syn,* with, + *kytos,* hollow vessel). A mass of protoplasm containing many nuclei and not divided into cells.

syrinx (sir′inks) (Gr. shepherd's pipe). The voice box of birds. It is not homologous to the larynx (also found in birds) but is a modification of the trachea and bronchi and is situated where the two main bronchi begin. Voice is produced by the vibrations of a bony ridge (pessulus) and certain membranes.

tactile (tak′til[e]) (L. *tactilis,* able to be touched, from *tangere,* to touch). Pertaining to touch.

tagma (tag′ma) (Gr. arrangement). Body division of an arthropod, containing two or more segments.

taiga (ty′ga) (Russ. from Turkic: Teleut *taiga,* mountainous terrain; Turkish *dağ,* mountain). Habitat zone characterized by large tracts of coniferous forests, long, cold winters, and short summers; most typical in Canada and Siberia.

taxis (Gr. arrangement). Response of animal organisms to sources of stimuli.

telencephalon (tel″en-sef′a-lon) (Gr. *telos,* end, + *encephalon,* brain). The most anterior vesicle of the brain.

teleology (tel′e-ol′o-je) (Gr. *telos,* end, + L. *-logia,* study of, from Gr. *logos,* word). The philosophic view that natural events are goal directed and are preordained; contrasts with scientific view of causalism.

telolecithal (Gr. *telos,* end, + *lekithos,* yolk, + -al). Yolk concentrated at one end of egg.

template (tem'plit). A pattern or mold guiding the formation of a duplicate; often used with reference to gene duplication.

tentacle (L. *tentare,* to feel, *-culum,* suffix for instrument). Long flexible processes for sensory use or food getting in coelenterates and others.

tentaculocyst (ten-tak'yu-lo-sist) (L. *tentaculum,* feeler, + Gr. *kystis,* pouch). A sense organ of several parts along the margin of medusae and derived from a modified tentacle; sometimes called rhopalium.

tetrapoda (te-trap'o-da) (Gr. four-footed ones). Four-legged vertebrates; the group includes amphibians, reptiles, birds, and mammals. (Contrast with "quadruped," which is generally used with mammals.)

therapsid (the-rap'sid) (Gr. *theraps,* an attendant). Extinct Mesozoic mammal-like reptile, from which true mammals evolved.

thrombokinase (throm'bo-kin"ace) (Gr. *thrombos,* lump, clot, + *kinein,* to move, + *-ase,* enzyme suffix). Enzyme released from blood platelets that initiates the process of clotting; transforms prothrombin into thrombin in presence of calcium ions; thromboplastin.

tornaria (L. *tornus,* lathe, chisel, + *-aria,* one like). This larva of Enteropneusta rotates in circles.

trachea (tray'ke-a) (ML. windpipe, trachea, from Gr. [*artēria*] *tracheia,* rough [artery]). The windpipe; any of the air tubes of insects.

transduction (tranz-duk'shun) (L. *trans,* across, + *ducere,* to lead). Transfer of genetic material from one bacterium to another through the agency of virus.

trichinosis (trik"i-no'sis) (Gr. *thrix, trichos,* hair, + *-osis,* suffix for state of). Parasitized condition produced by heavy infection of the trichina worm, *Trichinella.*

trichocyst (Gr. *thrix,* hair, + *kystis,* bladder). Saclike organelle in the ectoplasm of ciliates, which discharges a threadlike weapon of defense.

trochophore (trok'o-for) (Gr. *trochos,* wheel, + *phoros,* bearing). A free-swimming ciliated marine larva characteristic of schizocoelomate animals; common to many phyla.

trophallaxis (trof"a-lak'sis) (Gr. *trophē,* food, + *allaxis,* barter, exchange). Exchange of food between young and adults, especially among those of certain social insects.

trophozoite (trof"o-zo'ite) (Gr. *trophē,* food, + *zōon,* animal, + *ite,* suffix for individual). That stage in the life cycle of a sporozoan in which it is actively absorbing nourishment from the host.

tundra (tun'dra, toon'dra) (Russ. from Lapp *tundar,* hill). Terrestrial habitat zone, between taiga in south and polar region in north; characterized by absence of trees, short growing season, and mostly frozen soil during much of the year.

typhlosole (tif'lo-sole) (Gr. *typhlos,* blind, + *sōlēn,* channel, pipe). A longitudinal fold projecting into the intestine in certain invertebrates such as the earthworm.

umbilical (L. *umbilicus,* navel). Refers to the umbilical, or navel, cord.

umbo (L. boss of a shield). The part of a shell resembling a boss.

ungulate (L. *ungula,* hoof). Hoofed mammals.

urethra (yu-re'thra) (Gr. *ourēthra,* urethra). The tube from the urinary bladder to the exterior in both sexes.

uriniferous tubule (yu"ri-nif'er-us) (L. *urina,* urine, + *ferre,* to bear). One of the tubules in the kidney extending from a malpighian body to the collecting tubule.

utricle (yu'tri-kl) (L. *utriculus,* little bag). That part of the inner ear containing the receptors for dynamic body balance; the semicircular canals lead from and to the utricle.

vacuole (vak'yu-ole) (L. *vacuus,* empty, + Fr. *-ole,* diminutive). A fluid-filled space in a cell.

vagility (va-jil'i-ty) (L. *vagus,* wandering). Ability to tolerate environmental variation or the ability to cross ecologic barriers. Example: Birds have high and mollusks very low vagility.

veliger (L. sail-bearing). Larval form of certain mollusks.

velum (L. veil, sail). A membrane on the subumbrella surface of jellyfish of class Hydrozoa.

vestige (ves'tij) (L. *vestigium,* footprint). A rudimentary structure that is well developed in some other species or in the embryo.

villus (vil'us) (L. tuft of hair). A small fingerlike process on the wall of the small intestine and on the embryonic portion of the placenta.

virus (vy'rus) (L. slimy liquid, poison). A submicroscopic noncellular particle composed of a nucleoprotein core and a protein shell; parasitic and will grow and reproduce in a host cell.

viscera (vis'er-a) (L., pl. of *viscus,* internal organ). Internal organs in the body cavity.

vitalism (vy'tal-izm) (L. *vita,* life). The view that natural processes are controlled by supernatural forces and cannot be explained through the laws of physics and chemistry alone; contrasts with mechanism.

vitamin (vy'ta-min) (L. *vita,* life, + amine, from former supposed chemical origin). An organic substance contributing to the formation or action of cellular enzymes; essential for the maintenance of life.

vitelline membrane (vi-tel'in) (L. *vitellus,* little calf, yolk of an egg). The noncellular membrane that encloses the egg cell.

viviparity, viviparous (viv'i-par'i-ti, vy-vip'a-rus) (L. *vivus,* alive, + *parere,* to bring forth). Reproduction in which eggs develop within the female body, with nutritional aid of maternal parent; offspring are born as juveniles.

xanthophyll (zan'tho-fil) (Gr. *xanthos,* yellow, + *phyllon,* leaf). One of a group of yellow pigments found widely among plants and animals; the xanthophylls are members of the carotenoid group of pigments.

zoecium, zooecium (zo-ee'she-um) (Gr. *zōon,* animal, + *oikos,* house, + L. *-ium,* from Gr. *-ion,* diminutive). Cuticular sheath or shell of Ectoprocta.

zygote (Gr. *zygōtos,* yoked). The fertilized egg.

INDEX

A

Aardvark, 598
Abdomen, crayfish, 321
Abductor, 498
Abel, J. J., 944
Abildgaard, P., 933
Abiogenesis, 18-19
Absorption of food, 653-654
Acanthocephala, 240-243
  adaptive radiation, 226
  biologic contributions, 225
  characteristics of, 241
  classification of, 241
  ecologic relationships, 240
  phylogeny, 225-226, 240-241
  position in animal kingdom, 225
  structure of, 241
Acanthodians, 439, 440
Acarina, 315-316
Accelerator nerves, 615
Acclimatization, 22
Acetylcholine, 675, 684, 690-691
  isolation of, 952
Acetyl–coenzyme A
  in aerobic respiration, 75-76

Acetyl–coenzyme A—cont'd
  in aerobic respiration—cont'd
    glucose as source of, 77-78
    oxidation of, 76-77
    structure of, 76
Acids, 30-31
  amino; *see* Amino acids
  citric acid cycle, 76, 960
  definition, 30
  fatty, 45
    oxidation of, 79
  folic, 656
  hydrochloric, 652
  lactic; *see* Lactic acid
  nicotinic, 656
  nucleic; *see* Nucleic acids
  pantothenic, 76, 656
  phosphoric, 42
  stearic, 79
Acipenser, 467
Ackerman, D., 954
Acoelomates, 200-224
  adaptive radiation, 201
  biologic contributions, 200-201
  characteristics of, 202
  phylogeny, 201
  position in animal kingdom, 200
Acontia threads, 188
Acorn barnacle, 335, 336
Acorn worms, 424-428
  adaptive radiation, 424
  biologic contributions, 424
  characteristics of, 424-425
  phylogeny, 424
  position in animal kingdom, 424
Acrania, 441
Acrosome, 728
  reaction, 966
ACTH, 707, 715-716
Actin, 668, 673
Actinopodea, 120, 131-132
Actinopterygii; *see* Fish, ray-finned
Actomyosin
  concept of, 963
  system, 668
Adanson, M., 932
Adaptation(s), 835-838
  nature of, 918
Adaptive radiation; *see* Radiation, adaptive
Addison, T., 937
Adductor, 498
Adenine, 42
Adenophorea, 233
Adenosine 3',5'-monophosphate, 701-702
  discovery of, 967
Adenosine triphosphate; *see* ATP
ADH, 638, 710
Adhesive pad, 180
Adrenal(s)
  disease, discovery of, 937
  medulla
    hormones, 716
    relationship to sympathetic nervous system, 940

Adrenaline, isolation of, 944
Adrenocortical hormones, 714-716
Adrenocorticotropic hormone, 707, 715-716
Aeolis, 255
Aepyceros melampus, 581
Aerobic respiration, 618-619
  acetyl–coenzyme A in, 75-76
Agassiz, L., 939
Age(s)
  distribution of populations, 877
  geologic, 819-820
Agglutination, 610
Agglutinins, 610
Agglutinogens, 610
Agkistrodon piscivorus, 514
Aglomerular fish kidney, discovery of, 944
Agnatha, 441, 455
Air
  alveolar, composition of, 624-625
  capillaries, 621
    bird, 536
  expired, 624-625
  inspired, 624-625
  pollution, 898-890
  sacs, 362
    bird, 536
Alanine, formula of, 39
Alauda arvensis, nest of, 559
Albatross, 561
  The *Albatross* of the Fish Commission, 941
Albinism, 572, 793-794
Albumin, 608
  bird, 540
Alces alces, 573
Alcyonarian corals, 167, 192-194
  dried skeletons of, 193
Alcyonium digitatum, 192
Aldehydes, Schiff reaction for, 938
Aldosterone, 637, 715
Alimentary canal, 755
  roundworm, 234
Aliphatic compound, 36
"All-or-none" law of heart muscle, 939
Allantois, 750
Alleles, 766, 780
  multiple, 782
Allelomorphs, 780
Allen's rule, 841
Alligator
  American, 506, 520
  mississippiensis, 520
Alloiocoela, characteristics of, 202
Alpha rays, 27
Altricial young, 557, 559
Alula, 544
Alveolar air, composition of, 624-625
Alveoli, 621, 623
  frog, 499
Ambystoma tigrinum, 490
Ameboid movement, 128, 668
  illustration of, 128
  nature of, 954
American alligator, 506, 520

American bittern, 540
American crocodile, 520
American marine laboratories, establishment of, 939
American pika, 583
American salamanders, 490-491
American toad, 484
  eggs of, 504
Amici, G. B., 934
Amino acids
  formulas of, 39
  specific, genetic code between messenger RNA and, 772
Aminopeptidase, 653
Ammocoete larva
  circulatory system of, 449
  digestive system of, 448-449
  excretory system of, 449-450
  features of, 448
  of lamprey, as chordate archetype, 447-451
  nervous system of, 450-451
  reproductive system of, 450
  respiratory system of, 448-449
  sensory system of, 450-451
Amnion, 750
Amniotes, 749-751
Amniotic egg, 487, 749-751
Amoeba
  in locomotion, structure of, 669
  mitosis in, 123
  proteus, 125-129
    behavior of, 129
    habitat of, 125
    locomotion in, 128-129
    metabolism of, 126-127
    pseudopodia of, 125-126
    reproduction in, 129
    size compared to *Pelomyxa,* 129
    structure of, 125-126
AMP, cyclic, 701-702
  discovery of, 967
Amphibians, 484-505
  behavior of, innate, 955
  characteristics of, 487-488
  classification of, 488
  contribution to vertebrate evolution, 487
  corpuscles of, red, 608
  early, 486-487
  evolution of, 485-488
  family tree of, 486
  habitats of, 485
  lungs of, 485-486
  movement onto land, 484-485, 486
Amphids, 234
Amphineura; *see* Chitons
Amphioxus, 435-436, 444-447
  cleavage in, 88
  gastrulation in, 88
  structure of, 445
Amphipathic molecules, 52
Amphipods, 337, 338
Amphiporus
  circulation in, 222-223
  development of, 223

Amphiporus—cont'd
  digestion in, 222
  excretion in, 223
  feeding, 222
  nervous system of, 223
  ocraceus, 221-223
    locomotion of, 222
  regeneration in, 223
  reproduction in, 223
  respiration in, 223
Amphitrite ornata, 294
Amplexus, toads in, 503
Ampulla, 697
Amylase, 650
  pancreatic, 653
Anabolism, 642
Anadromous fish, 475
Anaerobes, 80
Anaerobic respiration, 75, 619
Analogy, 100-101
  concept of, 936
Anaphase, 59, 60, 725-726
  diagram of, 60
  naming of, 941
Anarhichas lupus, 452
Anaspids, 439
Anatomic structures, early modern interpretation of, 931
Anatomy
  auriculoventricular node and bundle, 943
  bird, 535
  chiton, 252
  comparative, definition of, 10
  definition of, 10
  development of, 931
  digestive system of man, 648-650
  earthworm, 281
  perch, yellow, 468
  of reproductive systems, 729
  sea cucumber, 414
  shark, dogfish, 461
  snail, 259
  spider, 309
  squid, 272
  urinary system, human, 633
Ancestry
  chordate, 433-441
    ascidian, 955
  metazoan, Gastrea theory of, 939
Ancylostoma, 235
Andrews, C. F., 945
Andromerogone, 747
Anemia
  pernicious, liver treatment of, 954
  sickle cell, 794
    discovery of, 949
Anemones, sea, 187-190
  photograph, 164
Angiotensin, 637
  isolation of, 967
Angler, 477
Animal(s); *see also* Organisms
  acellular, 82-83, 117-150
  acoelomate; *see* Acoelomates
  adaptiveness, 21

Animal(s)—cont'd
  allopatric, 861
  aquatic
    salt balance problems, 628-630
    water balance problems, 628-630
  archetypes of, concept of, 933
  architectural pattern of, 82-102
  behavior; *see* Behavior, animal
  benthic group, 888
  bilateral, body cavity of, 23
  bioluminescence in, 923
  body plans of, 19, 98
  carnivorous, 579-580, 598, 642, 871
  cells, composition of, 35
  classification of; *see* Classification of animals
  communication among, 919-921
  communities; *see* Communities
  complexity, grades of, 82-83
  definition of, 9-10
  descent and, 21
  development, general patterns of, 84-85
  differences between plants and, 9
  diploblastic, 90
  dispersion, 861, 862
    restricted, 862-863
  distribution (zoogeography), 861-864, 937
    definition of, 10
  electricity, 933
  embryology, preliminary survey of, 84-91
  environment and, 22-23
  enzymes and, 20
  eurytopic, 861
  first true, 110
  growth, 85
  herbivorous, 578-579, 642, 871
  insectivorous, 580, 598, 642
  light production in, 941
  lophophorate, 390-397
  morphology, 19
  movement of; *see* Movement
  naming of, 107-108
  omnivorous, 580, 642, 871
  ontogeny of, 112-114
  organization of, 83-84
    grades of, 84
  pelagic group, 888
  phylogeny of; *see* Phylogeny, of animals
  populations; *see* Populations
  productivity in, methods of increasing, 792
  pseudocoelomates; *see* Pseudocoelomates
  radiate, 164-199
  respiration, early demonstration of, 932-933
  serologic relationships of, 946
  stenotopic, 861
  sterile culture of, 943
  structure, development of comparative investigations of, 932
  sympatric, 861

Animal(s)—cont'd
  terrestrial
    salt balance and, 630-631
    water balance and, 630-631
    time-measuring mechanism of, 951-952
    tool using among, 918-919
    triploblastic, 90
    viruses, cell culture of, 964
Anisogametes, 721
Annelida, 275-298
  adaptive radiation, 276
  biologic contributions, 275-276
  characteristics of, 277-278
  classes of, 278-279
  comparison with Arthropoda, 301
  ecologic relationships, 277
  metamerism in, 276-277
  nerve axons of, discovery of, 938
  phylogeny, 276
  position in animal kingdom, 275
  segmentation, 100
Anole, Carolina, 512
Anolis, 512
Anopheles, 375
Anopla, 221
Anoplura, 371, 373
Anseriformes, 561
Ant, 357-358, 377
  carpenter, 368
  lions, 371, 374, 375
Anteaters, 598, 599
Antedon, 416
Antennae
  cleaner, 365
  insect, types of, 364
Antennal glands, 632
  crayfish, 327
Anthozoa, 167, 187-195
Anthropoidea, 600
Antianemic factor, discovery of, 956
Antibodies, 610
Anticodon, 772
Antidiuretic hormone, 638, 710
Antifertilizin, 738
Antigens, 610
Antlers, 572-574
  differences from horns, 572
Anura; *see* Frogs; Toads
Anus
  Amphiporus, 222
  Rhynchocoela, 220
Aorta, 449, 613
Aortic arches, 449
Aortic reflex, in respiratory control, 955
Ape, 600
  lineage, separation from hominid lineage, 850
  southern, 851-852
Aphasmidia, 233
Aphids, 366, 371, 373, 375
Apical field, Rotifera, 228
Apocrine glands, 574-575
  distinction between eccrine glands, 953
Apoda; *see* Caecilians

Apodiformes, 562
Apoenzyme, 69
Apopyles, 160
Apple tree borer, roundheaded, 368
Apterygiformes, 561
Apterygota, 370, 371
Arachnida, 301, 306-316
  characteristics of, 306-307
  economic importance of, 307
  phylogenetic relations, 307
Arachnoid, 685
Araneae; see Spiders
Archaeopteryx, 526
Archaeornithes, 528
Archenteron, 88, 755
Archetypes of animals, concept of, 933
Archinephros, 449, 450
Arctic hare, 569
Arctic tern, 545
Arenicola, 294
Arginine phosphate, 954
Argiope aurantia, 307
Argulus, 328
Aristotle, 931
Aristotle's lantern, 412, 413
Armadillidium, 337
Armadillo, 599
  nine-banded, 570
Armillifer, 385
Armyworm moth, 368
Aromatic compound, 36
Arrhenius, S., 948
Arrowworms, 418-420
  features of
    external, 419-420
    internal, 420
Arteries, 613, 616
  branchial, 449
  cross section of, 616
  renal, 635
  umbilical, 757
Arterioles, 616
  afferent, 635
  efferent, 635
Arteriosclerosis, 655
Arthropoda, 299-317, 318-339, 340-378
  adaptive radiation, 300
  appendages, 302
  behavior of, 303
  biologic contributions, 299-300
  characteristics of, 300-301
  chitin, 302
  classification of, 301-302
  comparison with Annelida, 301
  ecologic relationships, 300
  economic importance of, 303
  kidneys, 632
  locomotion of, 303
  metamorphosis, 303
  phylogeny, 300
  position in animal kingdom, 299
  reproduction in, 303
  respiration in, 302
  role in disease transmission, 942
  segmentation, 100, 302
  sensory organs of, 302

Arthropoda—cont'd
  success of, 302-303
Artificial gene synthesis, 773-774
Artificial parthenogenesis; *See* Parthenogenesis, artificial
Artiodactyla, 599-600
Ascaphus truei, 494
Ascaris lumbricoides; see Roundworm
Ascidian(s), 430, 442, 443-444
  adult, 443
    structure of, 442
  ancestry of chordates, 955
  tadpole, 443-444
Asconoid sponges, 155
Aselli, G., 931
Asellus, 337
Asparagine, formula of, 39
Aspidobothria, 209
Aspidosiphon, 380
Association areas of brain, 689
Association neurons, 95
Astacus, 328
Aster(s), 59
Asterias, 404; *see also* Sea stars
  rubens, 401
Asteroidea; *see* Sea stars
Aston, F. W., 952
Astrangia danae, 191
Astylosternus, 494
Athecanephria, 422
Atlas, 667
Atomsphere, 860
Atom(s)
  electron "shells" of, 27-28
  lightest, illustration of, 26
  nature of, 25-30
Atomic nucleus, concept of, 949
Atomic number, 26
Atomic weight, 26
  gram, 28
ATP, 668
  discovery of, 956
  generated by electron transfer, 73-74
  role in muscular contraction, 961
  structure of, illustration, 72
  trapping of energy by, 73
Atriopore, 444
Atrioventricular
  node, 615
  valves, 613
Atrium, 613
  frog, 501
Atrophy, 675
Auditory canal, 695
Auditory organs of insects, 348-349
Auditory ossicles, evolution of, 696
Auerbach, C., 962
Aurelia, life cycle of, 186
Auricle, 449
  planaria, 204
Auricularia, 403, 415
Auriculoventricular node and bundle, 943
Australian faunal realm, 863
Australopithecus, 851-852
  africanus, discovery of, 954

Autogamy
  paramecium, 146, 148
  Protozoa, 125
Autonomic nervous system; *see* Nervous system, autonomic
Autosomes, role of, 785-786
Autotomy
  in crayfish, 324
  sea stars, 407, 409
Autotrophs, 641
A-V; *see* Atrioventricular
Avery, O., 962
Aves; *see* Birds
Avicularium, 393-394
Avoiding reaction, *Paramecium,* 143
Axial gradients
  principle of, 22
  theory, 949
Axolotl, 490
Axon, 680

B

Baboon, 582
Bacteria
  mutants, replica technique in selection of, 965
  sexual recombination in, 962
  spore, significance of, 940
  transformation, agent responsible for, 962
  transforming principle in, discovery of, 955-956
Baculum, 668
Baer, K. von, 934
Balancers, 346
Balantidium coli, 147, 148
Balanus, 336
Bald-faced hornet, 357
Baldwin evolutionary effect, 943-944
Baldwin, J. M., 943—944
Balfour, F. M., 940
Baltzer, F., 954
Banting, F., 952
Barcroft, J., 954
Barlund, H., 958
Barnacle
  acorn, 335, 336
  gooseneck, 336
Barr, M., 963
Barski, G., 969
Bartholin, T., 931
Basal granules, 49
Base(s), 30-31
  definition, 30
  pairing, 769
Basement membrane, 92
Basket stars, 402, 408
Bassi, Agostino, 935
Bat, brown, 591
Bataillon, E., 948
Bates, H. W., 938
Bateson, W., 947
Bauer, H., 958
Bayless, W. H., 945
Bayliss, W. H., 946

Bdelloidea, 227
Bdelloura, nervous system in, 207
Beadle, G., 961
Bear, 598
  water, 387
Beardworms, 420-423
  adaptive radiation, 420
  analysis of phylum, 967-968
  biologic contributions, 420
  characteristics of, 422
  classification of, 422
  function of, 422-423
  phylogeny, 420
  position in animal kingdom, 420
  structure of, 422-423
    external, 421, 422-423
    internal, 421, 423
Beaver, 568
Becquerel, A. H., 944
Bee(s), 377
Beef tapeworm; *see Taenia saginata*
Beermann, W., 964-965
Beetle, 371, 374, 375
  burying, 366
  click, antennae, 364
  cucumber, 368
  diving, 374
  dung, 367
  granary, 368
  ground, 366, 371
  June, 368
  ladybird, 366
  larder, 368
  May, antennae, 364
  milkweed, 366
  nervous system of, 348
  potato, 368
  riffle, 882
  rove, 366
  stag, 374
  tenebrionid, antennae, 364
  water scavenger, antennae, 364
  whirligig, 366
Behavior
  Amphibia, 955
  Amoeba proteus, 129
  animal, 943
    historic background, 910
    motivation in, measurement of, 960
    patterns of, 909-928
    ritualization in, 920
    significance of, 909-910
    social, 924-927
    study methods, 910-911
  appetitive, 913
  arthropod, 303
  bird
    concealing behavior, 540, 514
    territorial patterns of, 952
  Brachiopoda, 395-397
  brittle star, 410
  chromosome, parallelism between mendelian segregation and, 946
  crab, 909
  crayfish, 320-321

Behavior—cont'd
  earthworm, 287-289
  Ectoprocta, 393-394
  followership, 926-927
  heredity and, 913
  hydra, 172, 178
  insect, 350
  instincitive, theory of, 957
  leadership, 926-927
  learned, 915-917
  leech, 290-291
  motivation and, 914-915
  paramecium, 143-144
  patterns
    instinctive, 914
    simple, 912
  planaria, 208
  protozoan, 946
  sea cucumber, 415
  snail, 261
  squid, 272-273
  taxonomy and, 913
Bekesy, G. von, 967
Bell, C., 933
Bell toad, 494
Benda, C., 944-945
Benedin, E. Van, 941, 942
Bensley, R., 958
Benthos, 882
Benzene ring, concept of, 938
Benzer, S., 967
Berger, H., 956
Bergmann's rule, 840-841
Bernard, C., 936, 937
Bertram, E., 963
Berzelius, J., 935
Beta rays, 27
Bethe, A., 957
Bicarbonate, 607
Biceps brachii, 498
Bichat, M. F. X., 933
Bile, 653
  duct, 653
    ammocoete, 449
  pigments, 653
  salts, 653
Bilharziasis, 212
Bilirubin, 609
Bill, bird, 532
Billbug, corn, 368
Binary division, *Euglena,* 135
Binary fission
  Amoeba proteus, 129
  paramecium, 144-145
  protozoan, 123
Biochemistry
  definition of, 10
  foundation of, 935
  of life, 33-45
Biogenesis, 18-19, 730
Biogenetic law, 21
  formulation of, 934
  of recapitulation, 331
Biologic centers of origin as reservoirs of desirable genes, 952
Biologic clocks, 921-923

Biologic development, influence of geographic exploration on, 932
Biologic drives, 915
Biologic extinctions, 820
Biologist, methods and tools of, 11-15
Biology
  concepts of, 17-23
  experiemntal, 824
  of Lake Baikal, 944
  "new," 8-9
  principles of, 17-23
Bioluminescence, 366-367, 923
Biomass, 873
  pyramid of, 873
Biomes, 887-889
  marine, 887-888
  terrestrial, 888-889
Biometry, foundations of, 939
Biopoiesis; see Life, origin of
Biosphere, 857-864
  subdivisions of, 858
Biotelemetry, 911
Biotic communities, 878
  changing, 897
Biotic factors of ecology, 871-874
Biotin, 657
Bipinnaria larva, 403
Bird(s), 524-563
  adaptation for flight, 530-541
  anatomy of, 535
  behavior of; see Behavior, bird
  bills of, 532
  care of young, 555-560
  carinate, 525, 560, 561-562
  characteristics of, 528
  circulatory system of, 535-536
  digestive system of, 535
  excretory system of, 536, 538
  family tree of, 526, 527
  feathers of; see Feathers
  feet of, 533
  flight in, 540-547
  mating selection, 553-555
  migration of; see Migration, bird
  muscular system of, 532-534
  navigation, 548-550
    celestial, 968
  nervous system of, 538-539
  nesting of, 555-560
  orders of, résumé of, 560-562
  origin of, 525-528
  perching, 562
    mechanism of, 534
  populations, 550-553
  ratite, 525, 560-561
  relationships, 525-528
  reproductive system of, 539-541, 542
  respiratory system of, 536, 537
  sensory system of, 538-539
  shore, 561
  skeleton, 530-532
  society, 553-562
  structure of, adaptation for flight, 530-541
  sun changes and, 965
  territory selection, 553-555

Birds—cont'd
  wings of; see Wings, bird
  young
    altricial, 557, 559
    precocial, 557, 559
Birth, 757-759
  rate of, 877
Biting lice, 371, 373
Bitterns, 561
  American, 540
Bivalves, 251, 261-267
  boring, 267, 268
    wood-boring, 268
  circulation in, 264
  development in, 266-267
  digestion in, 263-264
  excretion in, 265
  feeding, 263-264
  foot, 261-263
  illustration, 248
  mantle of, 249, 261
  nervous system of, 265-266
  pearl production in, 263
  reproduction in, 266-267
  respiration in, 264-265
  sensory system of, 265-266
  shell of, 249, 261
Black-billed magpie, 544
Black widow spider, 312-313
Blackman, F., 935
Bladder, 635
  swim, 471-474
    evolution of, 472
  worms, 217, 218
Blakeslee, A., 959
Blastocoel, 88
Blastocyst, 751
Blastomeres, 744
Blastopore, 88
Blastula, 87, 88
  hydra, 176
Blastulation, 85, 86-88
Blood
  carbon dioxide transport by, 627
  cells; see Erythrocytes; Leukocytes
  coagulation of, 610
  composition of, 607-610
  corpuscles; see Corpuscles
  elements of, formed, 609
  flukes; see Schistosoma
  gas exchange, 939
  gas transport, 625-627
  groups, 610-611
    discovery of, 945
    inheritance of, 795-796
    major, 611
  pressure
    arterial, and carotid sinus, 956
    early measurement of, 932
  salts, similarity to sea salts, 946
  system
    amphioxus, 445
    crayfish, 325
  vascular system, Rhynchocoela, 220
Boa constrictor, 516
Bobolink, migrations of, 549

Body
  cavity, 97-98; see also Coelom
    bilateral animals, 23
    organizations, drawings of, 97
    reptile, 510
  fluids; see Fluids, body
  form of insects, 364
  louse, as vector of typhus fever, 948
  organization of, 83-84
  plan
    of animals, 19, 98
    earthworm, 280-282
    hydra, 172
  symmetry; see Symmetry
  temperature regulation, mammals, 584-587
  tissues, early analysis of, 933
  wall
    hydra, 172-174
    Nematomorpha, 239
    Pleurobrachia, 196
Bohr effect, 627
Boll, F., 940
Bolton, E., 970
Boltwood, B., 947
Bond
  covalent, 29-30
  disulfide, 41
  high-energy, 73
  hydrogen, 34
  ionic, 28, 29
Bone, 663-664
  cancellous, 663
  compact, 663
  endochondral, 663
  innominate, frog, 497, 498
  membrane, 663
  spongy, 663
  structure of, microscopic, 664
Bonellia, sex determination in, 954
Bonnet, C., 932
Bony fish; see Fish, bony
Boobies, 561
Book
  gills, 306
  lice, 371, 373
  lungs, spider, 309
Borelli, A., 932
Borer
  oyster, 255
  rock, 255
  roundheaded apple tree, 368
Borsook, H., 967
Botaurus lentiginosus, 540
Boveri, T., 946, 947
Bowditch, H. P., 939
Bowman, W., 936
Bowman's capsule, 635
Box elder bug, 373
Box tortoise, 519
Bozler, E., 955
Brachiopoda, 394-397
  adaptive radiation, 391-392
  behavior of, 395-397
  biologic contributions, 390
  characteristics of, 395

Brachiopoda—cont'd
  phylogeny, 390-391
  position in animal kingdom, 390
  structure of, 395-397
Brain, 687-689
  crayfish, 327
  frog, 502
  functions, localization of, 933
  hormone, 705
  human, 689
  impulse patterns, instinctive, localization of, 963
  planaria, 206
  snail, 259
  vertebrate, 686
    divisions of, 687
  waves, early demonstration of, 956
Branchial arteries, 449
Branchial system, enteropneusts, 426
Branchinecta, 328
Branchiopoda, 333-334, 335
Brandt, K., 940
Branta canadensis, 560
Braun, F., 944
Breathing
  aerial, problems of, 619-623
  aquatic, problems of, 619-623
  in chordates, 432
  in frog, 499
  in man
    control of, 624
    mechanisms of, 624
    regulation, role of carbon dioxide in, 946
Breeding
  amphibians, 485
  places, 871
Bridges, C., 951, 952-953
Briggs, R., 965
Bristletails, 370, 371
Brittle stars, 402, 408
  behavior of, 410
    structure of, internal, 410
Broili, F., 951
Bronchi, 623
Bronchioles, 623
Brook rapids habitat, 881
Brown, F., Jr., 960
Brown, R., 934
Brown recluse spider, 313, 314
Brownian movement, early description of, 934
Bruce, D., 943
Bryozoa; *see* Ectoprocta
Buchner, E., 944
Bud(s)
  hydra, 172
  taste, 692, 693
Budding, 721
  hydra, 174, 176
  protozoan reproduction, 123-124
Buffer(s)
  action, 31
  definition of, 31
Bug(s)
  box elder, 373

Bug(s)—cont'd
  chinch, 368
  harlequin cabbage, 368
  squash, 368
  true, 371, 373
  tumble, 367
Bugula, zooid of, 394
Bull moose, 573
Bullfrog, 492
Bumblebee, 366, 371
Buoyancy of fish, 471-474
Burgess shale fossils, discovery of, 949-950
Burying beetle, 366
Busycon, 258
Butenandt, A., 956, 968-969
Butterfly, 371, 376
  cabbage, 368
  molting in, endocrine control of, 704
  monarch, metamorphosis of, 353
  protective resemblance in, 348
  swallowtail, 371
Buzzards, 561

C

Cabbage bug, harlequin, 368
Cabbage butterfly, 368
Caddis flies, 371, 377
Caecilians
  natural history of, 488-489
  structure of, 488-489
Caenogenesis, 21
Cairns, J., 970
Cajal, R., 941
Calciferol, 713
Calcification, 660
Calcispongiae, 154, 159-160
Calcitonin, 713, 714
Calcium, 714
  metabolism, 713
Callan, H., 964
Calvin, M., 967
Calyx, 243
Canada geese, 560
Canadian Geological Survey, 944
Canidae, 598
Canines, 644, 645
Canis latrans, 583
Cannon, W. B., 946
Canthon pilularis, 367
Capillaries, 613, 616-617
  air, 621
  bird, 536
  circulation, early demonstration of, 931
  lymph, 617
  motor mechanism regulation, 951
  peritubular, 635
Caprimulgiformes, 561-562
Carapace, 306, 321, 519
Carbohydrate(s), 36-38
  classes of, 36
  metabolism, role of pituitary in, 953
Carbon
  cycle, 890, 891

Carbon—cont'd
  dioxide
    role in breathing regulation, 946
    transport by blood, 627
Carbonic anhydrase, 627
Carboxyhemoglobin, 627
Carboxypeptidase, 653
Caribou migration, 588, 589
Caridoid facies, 319
Carinate birds, 525, 560, 561-562
Carlson, A., 946
Carnivores, 579-580, 598, 642, 871
Carolina anole, 512
Carotid
  reflex, role in respiratory control, 955
  sinus
    arterial blood pressure and, 956
    respiratory center regulation and, 956
Carotin, discovery of, 934
Carpals, bird, 532
Carpenter ant, 368
Carrel, A., 950
Cartilage, 662-663
  calcified, 663
  elastic, 663
  hyaline, 662
Caryophyllia smithii, 191
Caspersson, T., 964
Cassiopeia, scyphistoma of, 187
Cassowaries, 561
Castle, W. B., 956
Castle, W. E., 948
Castor fiber, 568
Castoridae, 598
Casuariiformes, 561
Cat(s), 598
Catabolism, 642
Catadromous fish, 475
Catalytic action, early demonstration of, 933
Catalytic concept, early formulation of, 935
Catarrhinii, 600
Catecholamines, 716
Cateria styx, illustration, 225
Cathode-ray oscillograph, invention of, 944
Caudata, 488, 489-491
Ceboidea, 600
Ceca
  hepatic, 446
  lateral, 222
Cell(s), 46-81
  animal, composition of, 35
  blood; *see* Erythrocytes; Leukocytes
  chloragogue, 283
  components, function of, 49-55
  concept, 46-47
  constancy, 63
    Rotifera, 229
  constituents, isolation of, 961
  culture of animal viruses, 964
  cycle, 60, 62
  discovery of, 931
  in disease, early viewpoint, 937

Cell(s)—cont'd
  division, 56-62
    direct, early description of, 936
    early description of, 934
    indirect; *see* Meiosis; Mitosis
    meiosis; *see* Meiosis
    mitosis; *see* Mitosis
  epitheliomuscular, hydra, 172-173
  eukaryotic, comparison to prokaryot-
    ic cells, 48
  examples of, 48
  flame, 206
    Rotifera, 229
  flux of, 63-64
  function of, differentiation of, 65
  genes and, 767
  germ; *see* Germ cells
  germinal, 23
  gland, hydra, 173, 174
  hybrid somatic cell technique, 969
  interstitial, hydra, 173, 174
  islet; *see* Islet cells
  junctional complexes of, 55
  liver; *see* Liver, cells
  mast, discovery of, 940
  membrane; *see* Membrane, cell
  metabolism, 72-81
    manometric study methods, 953
  myoepithelial, 93, 94
  nerve; *see* Nerve, cells
  neurosecretory, 701
  nucleus; *see* Nucleus, cell
  number, 63
  nutritive-muscular, hydra, 173-174
  organization of, 47-56
  pancreatic; *see* Pancreas, cells
  permeability, measurement of, 958
  physiology of, 65-81
  pigment; *see* Chromatophores
  plate, 60
  primitive, 811-813
  prokaryotic, comparison to eukaryot-
    ic cells, 48
  protoplasmic bridges between, 938
  reproductive, formation of, 723-728
  salt-absorbing, 474
  salt-secretory, 474
  Schwann, 680
  sensory; *see* Sensory cells
  sickle cell anemia, 794
    discovery of, 949
  somatic; *see* Somatic cell
  sponge, 156
    reorganization of, 947
  structures, micrurgic study of, 950
  study methods, 47
  surfaces of, 55-56
  target-organ, 700
  theory, formulation of, 935
  as unit of life, 46-64
Cellulose, 38, 39
  discovery of, 968
Cementum, 577
Census of population, 876
Centipedes, 340-342

Central nervous system, 685-689, 912
  earthworm, 285
  frog, 501
Centrioles, 57
  function of, 49, 50
Centrolecithal egg, 85-86
Centromere, 56, 57
Centrosome, 57
Centruroides vittatus, 314
Cephalaspid, 438-439
  fish fossil, appraisal of, 955
Cephalization, 99
Cephalocarida, 333
Cephalodiscus, 427
Cephalopoda, 251, 267, 269-273
  color changes of, 270
  development of, 271
  ecology and, 270
  evolution, 269-270
  feeding habits, 270
  groups of, 271
  illustration, 248
  ink production of, 270
  reproduction in, 271
Cephalothorax, 321
Cercaria, 210, 212
Cercopithecoidea, 600
Cerebellum, 688
  bird, 538
  frog, 501-502
Cerebrospinal fluid, 686
Cerebrum, 688
  cortex, bird, 538
  ganglia of planaria, 206
  hemispheres, bird, 538
  vesicle, 446
Cervix, 733
Cestoda; *see* Tapeworms
Cetacea, 599
Chaetognatha; *see* Arrowworms
Chaetonotus, structure of, 230
Chaetopterus, 295
Chalcid wasp, 366
Challenger expedition, 939
Chamberlain, T. C., 945
Charadriiformes, 561
Chargaff, E., 964
Chase, M., 965
Chelicerae, 308
Chelicerata, 301, 304-316
Chelonia, 509, 518-519, 521
Chemical elements of life, 34-35
Chemical evolution, steps in, 807-810
Chemoreception, 692-693
Chemoreceptors, insect, 349
Chemosynthesis, 811
Chemotaxis, 350, 692
Chemotherapy, 948
Chemotrophs, 641
Chiasma, 725
Chiasmatype theory, 948
Chicken
  heredity of comb forms in, 782
  louse, 371

Child, C., 949
Chilopoda, 302, 340-341
Chimaeras, 455, 463
Chimpanzee, 849
Chinch bug, 368
  egg parasite, 366
Chiroptera, 598
Chitin, 302
Chitinous shell, 86
Chitons, 251, 252-253
  anatomy of, 252
  illustration, 248
Chloragogue cells, 283
Chloride, 607
Chlorocruorin, 625
Chlorophyll a, synthesis of, 969
Choanocytes, sponge, 155, 156
Cholesterol, 714
  molecule, structure of, 956
Chondrichthyes; *see* Elasmobranchii
Chondrocranium, 460-461
Chondrocytes, 662
Chondrostei, 469
Chordata, 430-505
  advancements over other phyla, 432-
    433
  ammocoete larva of lamprey as ar-
    chetype, 447-451
  ancestry, 433-441
    ascidian, 955
  background, 430-431
  biologic contributions, summary of,
    441
  characteristics of, 431-432
  evolution; *see* Evolution, chordate
  family tree of, hypothetical, 434
  nerve cord, dorsal tubular, 431-432
  notochord, 431
  pharyngeal gill slits in, 432
  position in animal kingdom, 430
  segmentation, 100
  subphyla, 441-447
  tail, postanal, 432
  taxonomic position of, 939
Chorioallantoic membrane, 750
Chorion, 750
Chromatids, 57
Chromatin, 49, 57, 58
  early description and naming of, 940
Chromatography
  gel filtration, 14
  principle of, 947
Chromatophores, 270, 660
  Euglena, 134
  fish, 478
  formation of, role of neural crest in,
    958
  frog, 495, 496
Chromocenter, 765
Chromomeres, 56, 762
  discovery of, 941
Chromonema, 56, 762
Chromosome(s)
  abnormalities, in man, 799, 968
  basis of sex determination, 968

Chromosome(s)—cont'd
behavior, paralellism between mendelian segregation and, 946
changes, 791
constancy within a species, 763-764,
942
counts
early accurate, 941
human, revision of, 967
deletion, 791
diploid, 56, 723, 761, 762
constitution of, 946
discovery of, 936
duplication, 791
early description of, 939
finer analysis of, 954
haploid, 56, 723
hereditary functions and, 941
homologous, 723, 945
individuality of, 941
inversion, 791
lampbrush, 743, 766
longitudinal division, 941
of man, 764
map, 790
first, formation of, 950
maternal, 945
mutations, 830
naming, 942
nature of, 762-763
paternal, 945
polytene, 765
puff, concept of, 964-965
qualitative differences of, 947
reduction, early demonstration of,
942
in salamander, diagram, 61
salivary, 764-766
bands, correspondence between
normal chromosome maps and,
959
giant, rediscovery of, 958
puffing in, 766
segment, reversal of, 767
sex
discovery of, 946
female, 763
male, 763
structure of, 56-57
tracer method and, 968
translocation, 767
types of sex determination, 785
Chrysalis, 354
Chyle, 653
Chyme, 653
Chymotrypsin, 653
Chymotrypsinogen, 653
Cicada, 368, 371, 373, 375
dog-day, ecdysis in, 351
Ciconiiformes, 561
Cilia
filter feeders, 243
movement, 668-670
sequence of, 139
structure of, 139

Ciliata, 120, 138-148
infraciliature of, 139
neuromotor apparatus of, discovery
of, 950-951
parasitic, 147, 148
syncytial, 111-112
Ciliophora, 120, 138-148
Circulation, 611-618
ammocoete larva, 449
amphioxus, 445
Amphiporus, 222-223
beardworm, 423
bird, 535-536
bivalve, 264
capillary, early demonstration of, 931
closed, 611
coronary, 615-616
crayfish, 327
double, 613
early description of, 931
earthworm, 282, 283
enteropneusts, 427
fetal, 758
fish, 612
frog, 500-501
grasshopper, 361-362
mammal, 612
open, 611-612
pulmonary, 613
amphibians, 485
sea stars, 407
spiders, 308-309
system, plan of, 612, 613
systemic, 613
Cirripedia, 334-337
Cistron, concept of, 967
Citellus
lateralis, 590
tridecemlineatus, 576
Citric acid cycle, 76, 960
Cladogenesis, 828
Clam; *see also* Bivalves
ciliary-feeding, 264
freshwater
feeding mechanism of, 262
larval form of, 267
long-neck, 255
razor-shell, 255
razor, stubby, 262
Clam worm, 294-297
habitat, 294-295
locomotion of, 296-297
structure of, external, 295-296
structure of, internal, 297
Claparede, E., 938
Clasper, 460
Classification of animals, 105-116
branches, 114
criteria used in, 106
deuterostome, 114-115
examples of, 110, 111
grades of, 114
law of priority, 108, 109
nomenclature; *see* Nomenclature
numerical toxonomy, 109

Classification of animals—cont'd
problems of, 106-107
protostomes, 114-115
recent trends in, 109
subkingdoms, 114
units in, basis for formation of, 108
variance among different authorities,
108
Claude, A., 961
Clavicle, frog, 497, 498
Cleavage, 85, 86-88, 744-747
comparison to mitosis, 86
determinate, 88, 744, 745
differentiation without, 946
examples of, 87
furrow, 60
holoblastic, 85, 86
in amphioxus, 88
hydra, 176
indeterminate, 88, 744, 745
meroblastic, 85, 88
mosaic, 88, 744, 745
radial, 88
spiral, 88
totipotent, discovery of, 942-943
Cleveland, L., 953
Click beetle, antennae, 364
Climatus, 439
Cline, in evolutionary variation, 961
Clitellata, 279-291
Clitellum, 279
Clitoris, 733
Cloaca, 729
bird, 536
reptile, 510
Rotifera, 229
Clock(s)
biologic, 921-923
fossil coral, 970
*Clonorchis sinensis; see Opisthorchis
sinensis*
Clothes moth, 368, 369
Clotting, fibrin, erythrocytes entrapped
in, 605
Clouds, 886
Cnidaria; *see* Coelenterata
Cnidoblast, 168
hydra, 173
Cnidocil, 168
Cnidospora, 120
CNS; *see* Central nervous system
Coacervate droplets, 805
in hydrophilic colloid, 813
theory of, 810
Coagulation, 610
Coal-tar dye, discovery of, 937
Coccidia, 136-137
Coccyx, 667
Cochlea, 695
Cockroach, 368, 371
Codon, 772
Coelacanths, 464-465
Coelenterata, 164-195
adaptive radiation, 165
biologic contributions, 164-165

Coelenterata—cont'd
  characteristics of, 167
  classes of, 167
  comparison with Ctenophora, 195
  dimorphism in, 167-168
  ecologic relationships, 166
  economic importance, 166-167
  medusae, 167-168
  nematocysts, 168
  nerve net of, 168-169, 171, 911
  phylogeny, 165
  polymorphism in, 165
  polyps, 167, 168
  position in animal kingdom, 164
  types of, photograph, 166
Coelenteron, hydra, 176
Coelom, 23, 97-98; *see also* Body, cavity
  crayfish, 324, 326
  formation of, 90
    enterocoelous, 90
    schizocoelous, 90
    types of, 90
  pseudocoelomates, 226
  sea star, 405
Coenobita clypeatus, 909
Coenosarc, 178
Coenzyme, 69
  A; *see also* Acetyl–coenzyme A
    discovery of, 962
Coghill, G., 955
Cohn, F., 940
Coiling, Gastropoda, 256
Colaptes auratus, 533
Coleoptera, 371, 374, 375
Coliiformes, 562
Collander, R., 958
Collembola, 370, 371
Collenchyma, 93
Collenchyme, 185
  Pleurobrachia, 196
Colloblasts, 196, 197
Colloid(s), 31-33
  application of ultracentrifuge to, 947
  solution, 31-33
    diagram, 32
Colon, 650, 653
Color (Coloration, Coloring)
  changes
    in cephalopods, 270
    in crustaceans, hormonal control
      of, 956
  control, neurohumoral theory of,
    956-957
  eye, and heredity, 795
  fish, 475-478
  frog, 494-496, 497
  hair, and heredity, 795
  insect, 347
  mammal, 571-572
  protective, 837
  vertebrate, 660
  vision, 700
    trichromatic, theory of, 933
  vital, early demonstration of, 940
Columbiformes, 561

Comb
  forms in chicken, heredity of, 782
  plates of *Pleurobrachia,* 196, 197
Commensalism, symbiotic, 121-122
Communication
  among animals, 919-921
  honeybees, 920-921, 963
Communities
  biotic, 878
    changing, 897
  climax, 878
  definition of, 15-16, 866
  ecology of, 865-893
  major, 878
  minor, 878
  periodicity, 878
  sere, 878
Comparative anatomy, definition of, 10
Compound(s)
  aliphatic, 36
  aromatic, 36
  definition, 28
  organic; *see* Organic compounds
Conditioned reflex, 916
  concept of, 949
Condors, 561
Cone(s), 698, 699
  frog, 502
  opsin, 700
Conformational theory, 70
Conies, 599
Conjugation, 721-722
  in paramecium, 144, 145-146, 942
  in Protozoa, 125
Connective tissue, 92-93
  areolar, 92, 93
  dense, 92, 93
Conservation
  ecology and, 891-892
  of energy, law of, 936
  water, 900-904
Continental drift theory, 950
Contractile vacuoles; *see* Vesicles, water
    expulsion
Contraction(s)
  burst, in hydra, 178
  heart, influence of blood ions on, 940
  muscle
    ATP and, 961
    coupling of excitation and, 677
    energy for, 673-675
    lactic acid and, 947, 952
    phosphocreatine and, 955
    stimulation of, 675-677
    striated, 673-677
    striated, according to sliding fila-
      ment theory, 673, 674
    theory of, 966
Conus arteriosus, 613
  frog, 501
Convergence, 838, 839
Cook, J., 932
Coordination, 678-718
Coots, 561
Copepoda, 334, 335

Cope's law, 840
Copulation, earthworms in, 287, 288
Coraciiformes, 562
Coral(s), 187-195
  alcyonarian, 167, 192-194
    dried skeletons of, 193
  chronology of geologic deposits and,
    194-195
  fossil coral clock, 970
  organpipe, 193
  reefs
    atoll, 194
    barrier, 194
    formation of, 194-195
    fringing, 194
    snake, 516
    soft, 192-194
      Gorgonian, 193
    stony, 191
  symbiotic relationships, 194
  zoantharian, 167, 190-192
Cord(s)
  nerve; *see* Nerve, cord
  spine, 685-687
  umbilical, 753, 757
  vocal, frog, 500
Cori, C., 961, 962
Cori, G., 961
Coricoid, frog, 497, 498
Cormorants, 561
Corn billbug, 368
Cornea, 698
  crayfish, 328
Cornefert, F., 969
Corona, Rotifera, 228
Coronary circulation, 615-616
Corpus
  allatum, 354, 355
    in insect metamorphosis, 958
  cardiaca, 355
  luteum, 594, 736
  striatum, bird, 538
Corpuscles
  pacinian, 693-694
  red; *see* Erythrocytes
Correns, K., 945
Corridor, 861
Corrodentia, 371, 373
Cortex, 687
Corticosterone, 714-715
Cortisol, 714-715
Cortisone, discovery of, 959
Corymorpha, 180
Cottonboll weevil, 368
Cottonmouth, 514
Counter, liquid scintillation, for radio-
    active isotopes, 13
Countercurrent flow, 473
Countercurrent multiplier system of
    kidney, 637-638
Cousins, marriage of, 796-797
Covalent bonds, 29-30
Cow, sea, 599
Cowper's glands, 729
Coxal glands, 309

Coyote, 583
  photograph, 564
Crab
  behavior, 909
  "decorator," 338
  fiddler, 337, 338
  hermit, 318, 909
  horseshoe, 304, 305, 306
  salt concentration of, 628
Cranes, 561
Craniata, 441-442
Cranium, 665
  nerves, 690
    terminal, discovery of, 943
  vertebrates, 447
Craspedacusta, life cycle of, 182
Crawford, A., 944
Crayfish, 319-333
  appendages, 322-324
  autotomy in, 324
  behavior of, 320-321
  "in berry," 330
  blood system, 325
  circulation in, 327
  development of, 329-331
  digestion in, 326-327
  dorsal view of, 320
  endocrine functions of, 333
  excretion in, 327
  features of
    external, 321-324
    internal, 324-329
  feeding, 326-327
  habitat of, 319-320
  male, ventral view of, 321
  muscular system of, 326
  nervous system of, 327-328
  regeneration in, 324
  reproduction in, 329
  respiratory system of, 326
  sensory system of, 328-329
Creatine phosphate, 674, 954
Creighton, H., 957
Cretin, 712
Crick, F., 965
Cricket, 371
  mole, legs of, 365
Crinoidea, 403, 415-417
  structure of, 415-417
Crisscross inheritance, 789
Cristae, 74
Crocodile, 509, 521, 522
  American, 520
  heart, 510, 511
  structures of, internal, 510
Crocodylus acutus, 520
Crossing-over, 767, 789-791
  cytologic demonstration of, 957
  mitotic, discovery of, 959
  sample of, 790
Crossopterygii, 455, 464-465
Crotalus
  adamanteus, 515, 517
  horridus, 515
Crow, flight feathers, 531

Crown of thorns star, 409
Crustacea, 301-302, 318-339
  color changes in, hormonal control
    of, 956
  molt-preventing hormone in, 960
  nature of, 318-319
  origin of, 319-333
  relationships of, 319-333
  resumé of, 333-339
  x organ in, discovery of, 959
Cryptobiosis, 387
Cryptodira, 521
Cryptozoites, 138
Crystalline style, 263
  of ciliary-feeding clam, 264
Ctenophora, 164-165, 195-198
  adaptive radiation, 165
  biologic contributions, 164-165
  characteristics of, 195
  classes of, 195-196
  comparison with Coelenterata, 195
  ecologic relationships, 195
  general relations, 195
  phylogeny, 165
  position in animal kingdom, 164
Cuckoos, 561
Cuculiformes, 561
Cucumaria frondosa, 414
Cucumber(s)
  beetle, 368
  sea; *see* Sea cucumbers
Cuénot, L., 949
Culex, 375
Culture
  cell, of animal viruses, 964
  tissue, technique of, 950
Cunningham, O., 960
Cupula, 695
Cusp, 577
Cuticle, 658
  roundworm, 233-234
Cutworm moth, 368
Cuvier, G., 933
Cyanea, 185
Cyanocobalamin, 657
Cyclops, 335
Cyclostomata, 441, 455, 456-459
  characteristics of, 456
Cyclotron, invention of, 957
Cynomys ludovicianus, 592
Cypris, 335
Cyst, hydatid, 219
Cysteine, formula of, 39
Cystercerci, 217, 218
Cytochromes, 75
  discovery of, 942
Cytogenetics, 762
Cytokinesis, 57
Cytology, definition of, 10
Cytopharynx
  ciliate, 140
  paramecium, 141
Cytoplasm, 49
  inheritance of, 791-792
  naming of, 941

Cytosine, 42
Cytostomes, 642

D

Dactylozooids, 184
Dale, H., 948, 952
Dam, H., 958
Damselflies, 371, 371-373
Dan, J., 966
Danaus plexippus, 353
Danielli, J., 958
Danish *Galathea* Expedition, 965
Danish scientific expedition, 957
Daphnia, 335
Dart, R., 954
Dart sac, snail, 261
Darwin, C., 937
Darwin, natural selection theory of,
  825-827
Dasypus novemcinctus, 570
Dating
  of fossils, 819
    radiocarbon, 962
  of geologic periods, uranium-lead
    method of, 947
Death, 759, 877
  minimum mortality, 877
Decapoda, 337, 338-339
"Decorator crab," 338
de Duve, C., 963-964
Defecation, 653
Deficiency; *see* Diet, deficiency
de Graaf, R., 931
Deltoid muscle, 498
Demerec, M., 959
Demes, 834
Demospongiae, 154, 161-163
Dendrites, 680
Dendrostomum, 380
Dental formula, 578
Dentalium, 254
Dentin, 577
Deoxycorticosterone, 715
Deoxyribonucleic acid; *see* DNA
Deoxyribose, 42
Depressor muscle, 498
de Réaumur, R. A. F., 932
Dermacentor, 316
Dermaptera, 371-372
Dermasterias, 404
Dermis, 660
Dermocranium, 665
Dermoptera, 598
Descartes, R., 931
Descent and animals, 21
Desert, 889
Detorsion, Gastropoda, 256
Deuterostomes, lesser, 418-429
Deuterostomia, 115
Development
  Amphiporus, 223
  of animals, general pattern of, 84-85
  bivalves, 266-267
  Cephalopoda, 271
  crayfish, 329-331

Development—cont'd
  direct, 741-742
  early, 744-747
  earthworm, 286-287
  echinoderms, 400
  enteropneusts, 427
  feathers, 529
  grasshopper, 364
  holometabolous, 703
  human, 751-759
  indirect, 741-742
  organ, 753-757
  principles of, 740-760
  salmon, 481
  Scypha, 160
  systems, 753-757
  theories of
    early, 741
      mosaic, 941-942
      preformation, 741
  tissue, hormone coordination for, 964
De Vries, H., 945
Diabetes mellitus, 617
Diakinesis, 725
Diamondback rattlesnake, Eastern, 515
  "milking" of, 517
Diapause, insect, 355
Diaphragm, 624
Diastole, 613
Diceros simus, 574
Dicyema, vermiform stage of, 152
Didelphis marsupialis, 595
Diencephalon, frog, 501
Diestrus, 594
Diet; *see also* Feeding; Food; Nutrition
  deficiency
    disease, discovery of cause, 944
    early analysis of, 947
Differentiation, 84, 85, 90
  problem of, 747-748
  without cleavage, 946
Diffractometry, x-ray, development of, 952
Digenea, 209-210
Digestion, 641-657
  ammocoete larva, 448-449
  Amphiporus, 222
  bird, 535
  bivalves, 263-264
  clam worm, 297
  crayfish, 326-327
  early experiments on, 932
  earthworm, 282-283
  enteropneusts, 426-427
  enzymes in, action of, 650
  extracellular vs. intracellular, 646-648
  frog, 501
  grasshopper, 361
  hormones of, 717
  hydra, 174, 175
  in intestine
    large, 653
    small, 652-653
  intracellular vs. extracellular, 646-648
  liver fluke, 210

Digestion—cont'd
  in man, 648-654
  in mouth, 650
  Nematomorpha, 239
  planaria, 205-206
  Pleurobrachia, 196
  reptiles, 510
  Rotifera, 229
  sea stars, 407
  snail, 259
  spiders, 308
  of starch, 651
  in stomach, 652
  x-rays studying, 946
Digits, birds, 532
Dihybrid, 780
Dimorphism in coelenterates, 167-168
Dinosaur
  fauna of Alberta, Canada, 944
  fossil bed, discovery of, 948
Dioctophyma renale, 237, 239
Dioecious, 722
  sponges, 159
Dipeptide, 40
Diphyllobothrium latum, 218
  early description of, 931
Diphyodont, 577
Dipleurula lrva, 399, 400
Diploblastic animals, 90
Diplopoda, 302, 342
Diplotene, 725
Dipneusti; *see* Lungfish
Diptera, 375, 376-377
Dipylidium caninum, 218
Disaccharides, 36, 37
Disease
  adrenal, discovery of, 937
  cells in, early viewpoint, 937
  chemotherapy in treatment, 948
  dietary deficiency, discovery of cause, 944
  genetic; *see* Genetic diseases
  -resistant strains, 793
  transmission, arthropod's role in, 942
Disk, oral, sea anemones, 187
Disulfide bond, 41
Disuse, evolutionary concept of, 933
Division
  of labor among insects, 356
  reduction, significance of, 764
Dizygotic egg, 758
DNA, 767-768, 955-956
  base composition of, 964
  as basis of gene structure, 965
  chemical language of, 769
  chemical structure of, 965
  coding by base sequence, 770
  -directed RNA, role in protein synthesis, 969-970
  duplicating mechanism in, 968
  molecule
    mutations from mistakes in, 770
    new, building of, 769-770
    replication of, 970

DNA—cont'd
  molecule—cont'd
    structure of, Watson-Crick model of, 768-773
  polymerase, 770
  reversal transcription from RNA, 970
  -RNA polymerase, 772
  section of, 43
  sequence, and phylogeny, 970
Dobell, C., 949
Dobsonfly, 375
  larva, 366
Dog, 598
  dental formula of, 578
  prairie, 592
  tapeworm, 218, 219
    proglottid of, illustration, 216
Dog-day cicada, ecdysis in, 351
Dogfish shark, 460, 461
Dohrn, A., 939
Doisy, E., 956, 958
Doliolaria larva, 417
Dollo, I., 943
Dollo's law, 840
Dolomedes sexpunctatus, 310
Dolphins, 599
Dominance, 778
  complete, 780
  concept of, 933-934
  incomplete, 780, 781-782
  -subordinance hierarchies, 926, 953
Doodlebug, 374, 375
Doublass, E., 948
Doves, 561
Down feathers, 528
Dracunculus medinensis, 237, 238
Dragonfly, 366, 371, 372-373
  naiad of, 352
Driesch, H., 942-943
Drift
  continental drift theory, 950
  genetic; *see* Genetic drift
Drives, biologic, 915
Drosophila, 762
  melanogaster, 762
Dubois, E., 943
Dubois, R., 941
Duck, 561
  feet, 533
Duck-billed platypus, 598
Duct(s)
  bile, 653
  ammocoete, 449
  of Cuvier, 449
  wolffian, 729
Ductus arteriosus, 757
Ductus venosus, 757
Dugesia; see Planaria
Dugesiella lentzi, 313
Dujardin, G., 935
Dumas, J. B. A., 934
Dung beetles, 367
Dura mater, 685
DuShane, G., 958
Du Vigneaud, V., 966

Dwarf tapeworm, 218
Dye, coal-tar, discovery of, 937
Dytiscus, 374

E

Eagle, 561
  ray, 462
Ear
  bird, 539
  human, 694
  inner, 695-696
  middle, 695
    evolution of, 696
  outer, 695
Earth, origin of, 806-807
Earthworms, 280-289
  anatomy of, 281
  behavior of, 287-289
  body plan of, 280-282
  circulation in, 282, 283
  development in, 286-287
  digestive system of, 282-283
  excretion in, 283-284
  farming, 288
  features of, external, 280
  locomotion of, 282
  nervous system of, 285-286
  regeneration in, 287
  reproduction in, 286-287
  respiration in, 284
  sense organs of, 285-286
  setae, 280
Earwigs, 371-372
Eastern diamondback rattlesnake, 515
  "milking" of, 517
Eastern timber rattlesnake, 515
Eccrine glands, 574, 575
  distinction between apocrine glands, 953
Ecdysis; *see* Molting
Ecdysone, 335, 704, 705
  silkworm, 704
Echinococcus granulosus, 218-219
Echinoderella, 231
Echinoderms, 398-417
  adaptive radiation, 399
  biologic contributions, 398
  characteristics of, 402
  classification of, 402-403
  development of, 400
  ecologic relationships, 401-402
  economic importance, 403
  larval forms of, 493
  phylogeny, 399
  position in animal kingdom, 398
  theory of chordate evolution, 433-435
  water-vascular system of, 400-401
Echinoidea, 402, 410-413
  structure of, 411-413
Echiurida; *see* Echiuroidea
Echiuroidea, 381-383
  adaptive radiation, 380
  biologic contributions, 379-380
  phylogeny, 380

Echiuroidea—cont'd
  position in animal kingdom, 379
Echiurus, 382
Echolocation
  development of concept, 962
  mammals, 590-592
Echosounding and marine waters, 963
Ecologic energetics, 873
Ecologic equivalents, 878
Ecologic isolation, 827
Ecologic niche, 879
  concept of, 951
Ecologic relationships
  Acanthocephala, 240
  Annelida, 277
  arthropods, 300
  coelenterates, 166
  ctenophores, 195
  echinoderms, 401-402
  Entoprocta, 243
  Gastrotricha, 230-231
  Kinorhyncha, 231
  mollusks, 250
  Nematoda, 232-233
  Nematomorpha, 239
  nemertines, 220
  Platyhelminthes, 201-203
  Rhynchocoela, 220
  Rotifera, 227
Ecologic succession, 878-879
Ecologic tolerance, law of, 950
Ecology
  biotic factors of, 871-874
  Cephalopoda, 270
  of communities, 865-893
  conservation and, 891-892
  crises, present, 895-906
  definition of, 10
  environmental factors of, 866
  physical factors of, 866-871
  of populations, 865-893
Ecosystem, 22
  challenge of, 906
  concept of, 15-16, 959
  definition of, 16
  man's, problems of, 894-908
  types of, 886-887
Ecotone, 878
Ectoderm, 23, 90
  derivatives of, development of, 754-755
Ectohormone; *see* Pheromone
Ectoparasite, 328
Ectopistes migratorius, 552
Ectoplasm, 18
Ectoprocta, 392-394
  adaptive radiation, 391-392
  behavior of, 393-394
  biologic contributions, 390
  characteristics of, 393
  phylogeny, 390-391
  position in animal kingdom, 390
  structure of, 393-394
Edelman, G., 970
Edentata, 599

Eel
  freshwater, life history of, 953
  migration of, 479-480
  wolf, 452
Effectors, 95, 286, 680
Egg(s)
  amniotic, 487, 749-751
  centrolecithal, 85-86
  cortex, significance of, 745-746
  cortical changes during fertilization, 949
  dizygotic, 758
  frog, discovery of entrance of spermatozoon into, 937
  growth, 742
  isolecithal, 85
  jelly, 86
  -laying mammals, 598
  mammalian, artificial parthenogenesis of, 961
  monozygotic, 758
  mosaic, analysis of basic pattern of, 960-961
  moth, tiger, 350
  regulative, analysis of basic pattern of, 960-961
  size, 742-743
  -sperm interactions, 738
  tardigrade, 388
  toad, 504
  types of, 85-86
  in various stage of development, illustration, 719
Eggleton, G., 955
Eggleton, P., 955
Ehrlich, C., 940, 948
Eijkman, C., 944
Einthoven, W., 947
Eland, 585
Elasmobranchii, 455, 459-463
  characteristics of, 460-463
  classification of, 460
  integument, 661
Electric fishes, 478-479
Electric ray, 478
Electricity, animal, 933
Electrocardiogram, mechanism of, 947
Electron(s)
  carriers, nature of, 74-75
  microscope; *see* Microscope, electron
  microscopy; *see* Microscopy, electron
  shells of atoms, 27-28
  transfer, ATP generated by, 73-74
  transport system, 75
Electrophoretic pattern of plasma proteins, 608
Elements
  of blood, 609
  chemical elements of life, 34-35
Eleocytes, earthworm, 282
Elephant, 599
  bull, 586
  evolution of, 820-821
  fossil elephant–like animals, restoration of heads of, 821

Elephant—cont'd
  tusk, 894
    shells, 251, 253
Elephantiasis, 236, 238
Eleutherozoa, 402-403
Elliot, D., 967
Embioptera, 373
Embryo
  fish, yok sac in, 741
  human, early development of, 751-752
  inductors, 748-749
  organizers, 748-749
  24-day-old, of Tammer wallaby, photograph, 740
Embryogenesis, stages of, 85
Embryology
  animal, preliminary survey of, 84-91
  definition of, 10
  frog, 89
  organizer concept in, 952
  organizer region in, chemical nature of, 959
Emiids, 373
Emulsions, 32-33
  diagram, 32
Emus, 561
Enamel, 577
Encystment
  Euglena, 135
  Protozoa, 125
Enders, J., 964
Endocranium, 665
Endocrine
  control of molting in butterfly, 704
  functions, crayfish, 333
  glands, 707
    vertebrate, 705
  organs, vertebrates, 707-717
  system, 700-717
Endocrinology, definition of, 10
Endocuticle, 331
Endocytosis, 69
Endoderm, 23, 90
  derivatives of, development of, 755
Endolymph, 695
Endometrium, 733
Endomixis, Protozoa, 125
Endoplasm, 18
Endoplasmic reticulum
  description of, 962
  function of, 52-53
Endopodite, 322, 323
Endopterygota, 375-377
Endoskeleton, 662
  chordate, 432
  frog, 496
  sea star, 404-405
Endostyle, 443
  ammocoete, 449
Energetics, ecologic, 873
Energy
  high-energy bonds, 73
  law of conservation of energy, formulation of, 936

Energy—cont'd
  life and, 7
  metabolic, 72-73
  nerve, theory of, 935-936
  organisms and, 20
  sources of, 811, 873-874
  transformations, 873-874
  values of foods, 941
Enopla, 220-223
Ensis, 255
Enterulus aequoreus, 476
Enterobius, 236
Enterokinase, 653
Enteropneusta, 425-427
  branchial system of, 426
  circulation in, 427
  development of, 427
  digestion in, 426-427
  excretion in, 427
  feeding, 426-427
  nervous system of, 427
  proboscis, 425-427
  reproduction in, 427
  sensory system of, 427
Entomology, definition of, 10
Entoprocta, 243-244
  adaptive radiation, 226
  biologic contributions, 225
  ecologic relationships, 243
  morphology of, 243-244
  phylogeny, 225-226, 243
  physiology of, 243-244
  position in animal kingdom, 225
Envelope, nuclear, 964
Environment
  animals and, 22-23
  ecology and, 866
  factors of
    biotic, 22
    physical, 22
  heredity and, 794-795
  life and, 865-866
  mammals and, 584-587
  present, 895
  subnivean, 587
Enzymatic synthesis of mRNA, 969
Enzyme(s), 69-72
  action of, 69-70
    blocking, 71
    hormone influence on, 962
  animals and, 20
  -catalyzed reactions, 71-72
  digestive, action of, 650
  hydrolytic, 650
  naming of, 69
  nature of, 69, 940
  prosthetic group, 69
  sensitivity of, 72
  specificity of, 70-71
  -substrate complex, 950
  zymogen granules as enzyme precursors, 939
Ephemerida, 371, 372
Ephyrae, 187
Epicuticle, 331, 660

Epidermis, 658, 660
  hydra, 172-173
  roundworm, 234
Epidinium, 147, 148
Epigenesis, 741
  embryologic theory of, 932
Epiglottis, 623
Epinephrine, 716
Epitheliomuscular cells, hydra, 172-173
Epithelium, 91-92
  ciliated, 92
  cuboidal, 92
  pseudostratified, 92
  simple, types of, 92
  squamous, 92
  stomach, frog, 641
  stratified, 92
  transitional, 92
Eptesicus fuscus, 591
Equilibrium
  Hardy-Weinberg, 832-833
  sense of, 697
Equus burchelli, 571
Erethizontidae, 598
Ergastoplasm organelle, discovery, 944
Erlanger, J., 953
Errantia, 278
Erythroblast, 608
Erythroblastosis fetalis, 611
Erythrocytes, 608
  early description of, 931
  in fibrin clot, 605
  frog, 608
  gerbil, 608
Eschenmoser, A., 970
Esophagus
  Amphiporus, 222
  man, 650
Estrogens, 734-735
Estrone, isolation of, 956
Estrous cycle, 594, 734
Estrus, 594
Estuaries, 886
Ethiopian faunal realm, 863
Ethology, 910
Eubranchipus, seasonal cycle of, 334
Euchromatin, 57
Eudendrium, 180
Eugenics, 799
Euglena viridis, 133-135
  features of, 133-134
  habitat of, 133
  locomotion of, 135
  metabolism in, 134
  reproduction in, 135
  structure of, 133-134
Eukaryotic cells, 48
Euler, U. S. von, 958, 964
Eunice viridis, 297
Eupaguras bernhardus, 318
Eurycea
  bislineata, 491
  longicauda, 491
Euryhaline, 629
  fishes, 475
Eurypterida, 301, 305-306

Eurypterus, 304
Eutheria, 598-600
Eutrophication, 901
Evaginations, 619
Evolution
  in action today, 843-844
  amphibians, 485-488
  Cephalopoda, 269-270
  characteristics of, 833-842
  chemical, steps in, 807-810
  chordate, 433-441
    larval, Garstang's theory of, 436-437
    theories of, early, 433
    theories of, echinoderm, 433-435
  concept of, 21
  convergent, 838, 839
  divergent, 828
  ear, middle, 696
  elephant, 820-821
  evidence for, 817-824
  factors of, 833-842
  fish
    bony, 464
    lung, 472
    modern, 440-441
  generalizations concerning, 840-842
  genetic drift as factor in, 957
  geographic distribution as evidence for, 824
  history of, 816-817
  homologous resemblances, 823
  horse, 822-823
  idea, early history of, 816-817
  irreversibility of, 943
  man, 847-854
  meaning of, 815-816
  methods of, theories accounting for, 824-831
  modern synthesis of, 816, 828-831
  mutation theory of, 945
  mutual aid in, 945-946
  natural selection in, 937
  organic, 815-846
  patterns of, 827-828
  population genetics and, 831-833
  process, early concept of, 932
  progressive, 840
  protozoan, 120
  quantum, 828
  regressive, 840
  sequential, 827-828
  straight-line, 828, 943
  study of, new approach in, 844
  swim bladder, 472
  tetrapods, modern, 440-441
  vertebrate, amphibians' contribution to, 487
Evolutionary effect, Baldwin, 943-944
Evolutionary orthogenesis, 943
Evolutionary simplification, Williston's law of, 951
Evolutionary variations, statistical analysis of, 956
Excretion, 627-639
  ammocoete larva, 449-450

Excretion—cont'd
  Amphiporus, 223
  Ascaris, 234
  bird, 536, 538
  bivalve, 265
  crayfish, 327
  earthworm, 283-284
  enteropneusts, 427
  grasshopper, 362-363
  hydra, 174
  liver fluke, 210
  planaria, 206
  Pleurobrachia, 196
  renal, filtration theory of, 936
  reptile, 511
  Rotifera, 229
  spider, 309
  structures, invertebrate, 631-632
Existence, life cycle, 16-17
Exocrine glands, 707
  types of, 661
Exopodite, 322, 323
Exopterygota, 370-375
Exoskeleton, 662
  reptile, 509
Expiration, 624
Expressivity, 782
Extensor muscle, 498
Exteroceptors, 692
Exumbrella, 180
Eye(s)
  ammocoete, 451
  color, and heredity, 795
  compound, 697
  crayfish, 321
  hawk, 524, 539
  human, 698
  median, chordate, 450
  mollusk, 260
  planaria, 207
  scallop, 678
  spider, 309
  wolf, 308
Eyeball, layers of, 698
Eyespots
  Euglena, 134
  planaria, 204, 206

F

Facies, caridoid, 319
Fairy shrimp; *see* Shrimp, fairy
Falcon, 561
  feet, 533
Falconiformes, 561
Fallopian tubes; *see* Oviducts
Fasciola hepatica, life cycle of, 210
Fasciolopsis buski, 214
Fatty acids, 45
  oxidation of, 79
Fauna
  delimitation, Wallace line of, 937-938
  dinosaur, of Alberta, Canada, 944
  interstitial, 863
  major realms, 863-864
Feather(s)
  barb, 528, 529

Feather(s)—cont'd
  barbules, 528, 529
  contour, 528
  definition of, 528-530
  development of, 529
  down, 528
  -duster worms. 275
  filoplume, 528-529
  flight, 528
    position of, 531
  powder-down, 529
  quill, 528, 529
  shaft, 528, 529
  stars, 403, 415-417
    structure of, 415-417
  types of, 529
  vane, 528, 529
Feedback
  negative, 15
  positive, 15
Feeding; *see also* Diet; Food; Nutrition
  Amphiporus, 222
  bivalves, 263-264
  Cephalopoda, 270
  crayfish, 326-327
  enteropneusts, 426-427
  filter, 642-644
  on fluids, 645-646
  on food masses, 644-645
  frog, 501
  hydra, 174, 175
  mammals, 578-583
  mechanism, 642-646
    freshwater clam, 262
  on particulate matter, 642-644
  Pleurobrachia, 196
  sea stars, 407
Feet
  bird, 533
  bivalves, 261-263
  mollusk, 249
  Rotifera, 228
  tube, 401
Felidae, 598
Femur, bird, 530
Fermentation, 75, 80
Fertilization, 85, 720, 736-738, 743-744, 939
  cone, 743
  membrane, 738, 743
  in plants, discovery of, 934
  of zygote, 86
Fertilizin, 738, 950
Fetus, circulation, 758
Feulgen, R., 953
Fever, typhus, body louse as vector of, 948
Fibrin, 610
  clot, erythrocytes entrapped in, 605
Fibrinogen, 610
Fibrocartilage, 663
Fiddler crab, 337, 338
Fielding, U., 957
Fig wasp, 366
Filarial worms, 236-237, 238
Filoplume feathers, 528-529

Filter feeding, 642-644
Filter route of animals, 861
Filtration pressure, 617
Fin(s)
　amphioxus, 444
　fish, 465
　lobed, 464
　shark, 460
Findlay, G., 959
Fischer, E., 942
Fish, 452-483
　anadromous, 475
　bony, 455-456, 463-470
　　characteristics of, 469
　　classification of, 469-470
　　climax, 469-470
　　diversity of, 464
　　evolution of, 464
　　integument, 661
　　origin of, 464
　catadromous, 475
　circulatory system of, 612
　classification of, 455-456
　coelacanth, discovery of, 960
　coloration, 475-478
　concealment, 475-478
　cosmoid type, 464
　electric, 478-479
　embryos, yolk sac in, 741
　euryhaline, 475
　family tree of, 454
　fins, 465
　"fishing," 477
　fossil, Cephalaspida, appraisal of, 955
　functional adaptations of, 470-482
　gills of, 473
　with jaws, illustration, 439
　kidney, aglomerular, discovery of,
　　944
　lateral line system of, 695
　lobe-finned, 455, 464-465
　locomotion, 470-471
　　analysis of, 932
　louse, 335
　lungs, evolution of, 472
　marine, salt and water problems, 630
　migration of, 479-480
　modern, evolution of, 440-441
　neutral buoyancy, 471-474
　origin of, 453-456
　osmotic regulation, 474-475
　pollution and, 901
　ray-finned, 456, 467-470
　　intermediate, 469
　　primitive, 469
　reproduction in, 480-482
　respiration in, 474
　scales, ganoid, 467
　stenohaline, 475
　structural adaptations of, 470-482
　swim bladder, 471-474
　　evolution of, 472
　swimming, motion of, 470
　tail
　　diphycercal, 465
　　heterocercal, 467

Fish—cont'd
　tapeworm, 218
　urohypophysis of, discovery of, 943
Fisher, R., 956
Fishing spider, 310
Fishlike mammals, 599
Fission, 721
　binary; *see* Binary fission
Fissipedia, 598
Fixation methods, conventional, 945
Flack, M., 947
Flagella, *Euglena,* cross section of,
　134
Flagellates
　colonies of, 111
　intestinal, 121
　　symbiotic relationships between
　　termites and, 953
Flame cells, 206
　Rotifera, 229
Flamingos, 554, 561
Flatworms; *see* Platyhelminthes
Fleas, 371, 377
Fleming, A., 956
Flemming, W., 940, 941
Flesh-eating mammals, 579-580, 598,
　642, 871
Flexor muscle, 498
Flicker, yellow-shafted, 533
Flight
　of birds, 540-547
　of insects, 346-347
　of mammals, 590-592
Florometra serratissima, 415
Fluids
　body, 605-640
　　composition of, 607
　　environment, 605-611
　cerebrospinal, 686
　compartments, 606
　feeding, 645-646
　internal; *see* body *above*
　interstitial, 607
　plasma, 607
Flukes, 204, 208-214
　blood; *see Schistosoma*
　intestinal, of man, 214
　liver
　　of man; *see Opisthorchis sinensis*
　　sheep, life cycle of, 210
　lung, 213-214
Fly, 371, 375, 376-377
　caddis, 371, 377
　fruit, 762
　nervous system of, 348
　robber, 366
　scorpion, 375-376
　stone, 371, 372
　syrphid, 366
　tachinid, 366
Flying, 868
　frog, 494
　lemur, 598
　mammals, 598
Fol, H., 940
Folic acid, 656

Follicle, 711
　ovarian, early description of, 931
　stimulating hormone, 707
Fontana, F., 933
Food; *see also* Diet; Feeding; Nutrition
　absorption of, 653-654
　chains, 871-873
　　determination of, 873
　　detrital, 872
　　grazing, 873
　　parasitic, 872
　cycle, 889-891
　energy values of, 941
　habits of insects, 345-346
　mammals, 578-583
　masses, feeding on, 644-645
　pyramids, 871-873
　web, 872, 873
Foot; *see* Feet
Foramen ovale, 757
Foraminiferans, 130-131
　photograph of, 117
Ford, C., 968
Forebrain, chordate, 450
Forests
　coniferous, 889
　deciduous, 889
　tropical rain, 889
Fossil(s), 818
　beds in Fayum Lake region of Egypt,
　　discovery of, 945
　Burgess shale, discovery of, 949-950
　Cephalaspida fish, appraisal of, 955
　coral clock, 970
　correlation between geologic strata
　　and, 933
　dating of; *see* Dating, of fossils
　dinosaur, discovery of, 948
　Olduvai Gorge, discovery of, 950
　records
　　of man, 850-853
　　of vertebrates, 820
　Seymouria, discovery of, 951
Fovea centralis, 698-699
Fowl, domestic, 561
Fraenkel-Conrat, H., 966
Fragmentation, 721
Fraternal twins, 758
　inheritance and, 797
Freemartin, theory of, 951
Freezing point depression as measure-
　ment of osmotic pressure, 68
Freshwater, 629-630
　medusae, 183
　streams, 881
Frisch, K. von, 963
Friske, C., 955
Frog(s), 488, 491-505
　brain, 502
　breathing in, 499
　circulation in, 500-501
　coloration, 494-496, 497
　digestion, 501
　distribution of, 492-494
　egg, discovery of entrance of sperma-
　　tozoon into, 937

Frog(s)—cont'd
  embryology of, 89
  erythrocytes of, 608
  features of, unusual, 494
  feeding, 501
  "flying," 494
  green, 492
  habitat of, 492-494
  "hairy," 494
  heart, structure of, 500
  hibernation, 493
  leopard, 492
  life cycle of, 503-505
  metamorphosis of, role of thyroid in, 950
  muscles of, 496-499
    skeletal, 658
  nervous system of, 501-503
  North American, 492
  reproduction in, 503-505
  respiration in, 499-500
  senses, special, 501-503
  skeletal systems of, 496-499
  skin, 494-496
    histologic section of, 493
  stomach epithelium, 641
  tadpoles, precocious metamorphosis of, 712
  tissues, diagram of, 96
  tree, 497
  vocalization, 499-500
Fructose, 37
Fruit fly, 762
FSH, 707
Fujii, K., 954
Fulmars, 561
Funk, C., 949

G

Galactose, 37
Galambos, R., 962
Galen, 931
Gall, F. J., 933
Gallbladder, 653
Galliformes, 561
Gallinules, 561
Galvani, L., 933
Gametes, 720, 723, 728
Gametocytes, 138
Gametogenesis, 726-728
  process of, 727
Gamma globulin, structure of, 970
Gamma rays, 27
Gammarus, 337
Ganglia, 679
  cerebral, planaria, 206
Gannet, 546, 561
Ganoid scales, 467
Garden spiders, 307
Garnier, C., 944
Garpike, long-nosed, 468
Garrod, A., 947
Garstang, W., 955
Garstang's theory of chordate larval evolution, 436-437

Gas(es)
  in blood; *see* Blood, gases
  exchange in lungs, 625
  gland, 473
Gasser, H., 953
Gastrea theory of metazoan ancestry, 939
Gastric; *see* Stomach
Gastrin, 717
Gastrodermis, hydra, 172, 173-174
Gastrointestinal tract as ecosystem, 886-887
Gastropoda, 251, 253-261
  coiling of, 256
  detorsion, 256
  illustration, 248
  torsion in, 255-256
Gastrotricha, 229-231
  adaptive radiation, 226
  biologic contributions, 225
  ecologic relationships, 230-231
  phylogeny, 225-226
  position in animal kingdom, 225
  structure of, 230-231
Gastrovascular cavity
  coelenterates, 166
  Gonionemus, 181
  hydra, 172
  Obelia, 178
  planaria, 205
  sea anemones, 188
Gastrovascular system, *Pleurobrachia,* 196
Gastrozooids, 184
Gastrulation, 85, 88-90
  in amphioxus, 88
Gause's rule, 842
Gautier, M., 968
Gaviiformes, 561
Geese, 561
  Canada, 560
Gel
  definition, 33
  diagram, 32
  filtration chromatography, 14
Gemmules, 158, 721
Gene(s), 761
  artificial, synthesis of, 773-774
  complementary, 783
  crossing-over in; *see* Crossing-over
  desirable, origin of, 952
  dominant; *see* Dominance
  duplicate, 784
  exchange between linkage groups, 766-767
  expression, 747-748
  lethal, 784
    control of, 793
    detection of, 793
  multiple, 783
    discovery of, 932
  mutations; *see* Mutations, gene
  naming of, 948
  operator, 775
  pleiotropic, 782
  pool, 831

Gene(s)—cont'd
  present concept of, 774
  recessive, 778
  regulator, 775-776
  role in cellular economy, 767
  structural, 775-776
  structure, DNA as basis of, 965
  theory, 766-767
    establishment of, 949
Generation(s)
  alternation of, early description, 936
  spontaneous, 730
    refutation of, 937
Genetic code, 771
  code assignments in, proof of, 970
  role of mRNA in, 772, 969
Genetic diseases, 793-794
  frequency of, 794
  recessive nature of, 794
Genetic drift, 833
  as factor in evolution, 957
  random, 834
Genetic information, storage and transmission of, 767-776
Genetic isolation, 827
Genetic selection, modified sex ratios by, 951
Genetic transduction, 955-956
Genetics
  advances since Mendel's laws rediscovered, 780-781
  biochemical, application of, 793
  cyclic changes in, 961
  definition of, 10
  future of, 800
  medical problems and, 797-798
  population, and evolutionary processes, 831-833
  practical application of, 792-793
  terminology of, 780
Genic balance theory of sex, 952-953
Genic control of protein structure, 964
Genital ligaments, 242
Genital pore, planaria, 205
Genome, 57
Genotypes, 779
  naming of, 948
  nuclear, differentiated, 965
Geographic exploration, influence on biologic development, 932
Geographic isolation, 827
Geologic ages, 819-820
Geologic deposits and corals, 194-195
Geologic periods, uranium-lead dating of, 947
Geologic strata, early correlation between fossils and, 933
Geology, modern concept of, 934
Geomyidae, 598
Gerbil, erythrocytes of, 608
Germ cells
  division; *see* Meiosis
  origin of, 723
  somatic cells and, 948
Germ layers, 90
  early description of, 933

Germ layers—cont'd
  embryonic, 23
  fate of, 90-91
  of invertebrates, 938
  theory, 23
Germ plasm theory, formulation of, 942
Germinal cells, 23
Gestation, 594-596
Ghostfish, 455, 463
Giersberg, H., 956-957
Gill(s), 621
  arches, comparison of, 756
  bars, 449
  bivalves, 261
  book, 306
  crayfish, 326
  fish, 473
  pouches, 448
  slits
    ammocoete, 448
    amphioxus, 446
    ascidian, 444
    pharyngeal, in chordates, 432
Gizzard, bird, 535
Glands
  antennal, 632
    crayfish, 327
  apocrine, 574-575
    distinction between eccrine glands
      and, 953
  cells, hydra, 173, 174
  corpus allatum, role in insect meta-
    morphosis, 958
  Cowper's, 729
  coxal, 309
  eccrine, 574, 575
    distinction between apocrine glands
      and, 953
  endocrine, 707
    vertebrate, 705
  exocrine, 707
    types of, 661
  gas, 473
  holocrine, 576-577
  lacrimal, 574
  mammal, 574-577
  mammary, 575-577
  mucous, frog, 495
  pineal, discovery of melatonin in, 968
  pituitary; see Pituitary
  poison, frog, 495
  prostate, 729
  prothoracic, 354
  reproductive, male, 729
  salivary; see Chromosomes, salivary
  salt, 631
    gull, 538
  scent, 575
  sebaceous, 577
  silk, 311
  sinus, 331, 702
  sweat, 574-575
    discovery of, 935
  thymus, function of, 969
Glass sponges, 160-161

Glaucomys sabrinus, 591
Globin, 626
Globulin(s), 608
  gamma, structure of, 970
Glochidium, 267
Gloger's rule, 840
Glomerular filtrate, collection and anal-
  ysis of, 952
Glomerulus, 633, 635
Glossary, 974-985
Glottis, 623
Glowworm, 924
Glucagon, 716
Glucocorticoids, 714-715
Gluconeogenesis, 714-715
Glucose, 36-38, 674
  metabolism of, 78
  oxidation of, pathway for, 77
  -6-phosphate, 78
  as source of acetyl–coenzyme A, 77-78
Glutamine, 39
Glutathione, isolation of, 952
Gluteus muscle, 498
Glutinant
  stereoline, 173
  streptoline, 173
Glycerin, structural formula of, 44
Glycerol, structural formula of, 44
Glycine, formula of, 39
Glycogen, 37-38, 674
  formation by liver, 937
Glycolysis, 80-81, 674
  aerobic, 674
  anaerobic, 674
  lactate formation and, 80
Gnathostomata, 441-442, 455-456
Gnathostomulida, 244
  adaptive radiation, 226
  biologic contributions, 225
  discovery of, 960
  phylogeny, 225-226
  position in animal kingdom, 225
Gnotobiotics, 943
Goatsuckers, 561-562
Goethe, J. W. von, 933
Goiter, 713
Goldblatt, M., 958
Golden plover, migration of, 549
Golgi apparatus, discovery of, 944-945
Golgi, C., 941, 944-945
Gonad(s), 722, 729
  bird migration and, 954
  chordate, 450
  Gonionemus, 181
Gonadotropin, 707, 735
  chorionic, 757
Gonangia, 178, 179
Gonionemus, 178, 180-181, 183
Gonophores, 179, 184
Gonotheca, 179
Gonyaulax polyhedra, 122
Goosefish, 477
Gooseneck barnacle, 336
Gorgonian soft corals, 193
Gorgonocephalus caryi, 408

Gotze, W., 960
Gradients; see Axial gradients
Graham, T., 938
Grain weevil, 368
Gram atomic weight, 28
Gram molecular weight, 28
Granary beetle, 368
Granules, basal, 49
Granulocytes, 609
Grasshopper, 340, 358-364, 371
  circulation in, 361-362
  development of, 364
  digestion in, 361
  excretion in, 362-363
  features of
    external, 359-361
    internal, 361-364
  habitat of, 359
  leg, hind, 360
  nervous system of, 363-364
  plagues, 359
  reproduction in, 364
  respiration in, 362
  sensory system of, 363-364
Grasslands, 888-889
Grebes, 561
  red-necked, 558
Green frog, 492
Gregarinida, 138
Griffin, D., 962
Griffith, F., 955-956
Grinnell, J., 951
Ground beetle, 366, 371
Grouse, 556, 561
Growth
  allometric, 19
  animal, 85
  egg, 742
  hormone, 707
  insect, 351-355
  of organisms, 19
  population; see Population, growth
Gruiformes, 561
Gryllotalpa, legs of, 365
Guanine, 42
  fish, 478
Guanophores, frog, 495
Guard hair, 569
Gudernatsch, J., 950
Guinea worms, 237, 238
Gulf shrimp, 330
Gull, 561
  laughing, habitat of, 857
  salt glands of, 538
Gullet, 181
Gymnophiona; see Caecilians
Gynandromorphs, 786

H

Haacke, W., 943
Habitat
  Amoeba proteus, 125
  amphibian, 485
  brook rapids, 881
  clam worm, 294-295

Habitat—cont'd
crayfish, 319-320
Euglena viridis, 133
frog, 492-494
grasshopper, 359
gull, laughing, 857
kinds of, 879-887
planaria, 204
of rock, shelving and drift, 885
subterranean, 885-886
terrestrial, 885-886
tree hole, 880
vertebrates, early, 437-438
Hadzi, J., 964
Haeckel, E., 938, 939, 942
Haematoloechus, living specimens of,
200
Hagfishes, 459
characteristics of, 456
comparison to lamprey, 456
Hair
color, and heredity, 795
guard, 569
mammals, 567-571
tactile, 328
"Hairy frog," 494
Haldane, J. B. S., 958
Haldane, J. S., 946
Hales, S., 932
Hall, D., 968
Hall, M., 934-935
Halteres, 346, 376
Hanson, J., 966
Hanstrom, B., 959
Haplosporea, 120
Hardy, G., 947-948
Hardy, W. A., 945
Hardy-Weinberg equilibrium, 832-833
Hardy-Weinberg population formula,
947-948
Hares, 599
arctic, 569
Harlequin cabbage bug, 368
Harvey, E., 949
Harvey, W., 931
Haversian system, 664
Hawks, 561
bill, 532
eye, 524, 539
Head louse, 371
Health and pollution, 905-906
Hearing, 695-697
traveling wave theory of, 967
Heart, 611, 613-616
ascidian, 443
beat, neuromuscular mechanisms
controlling, 615
contraction, influence of blood ions
on, 940
crocodiles, 510, 511
elasmobranch, 461
excitation of, 614-615
frog, structure of, 500
human, 614
law of, 951

Heart—cont'd
muscle, 93, 94, 670-671
"all-or-none" law of, discovery of,
939
neurogenic, pacemaking activity of,
946
output, 613
rate, 613
snail, 259
Stannius' experiment on, 936-937
stroke volume, 613
tissues, diagram, 96
Heat
exchange, concurrent, 586-587
production in nerve, measurement of,
954
Hedgehogs, 598
Heidelberg man, 852
discovery of, 947
Heidenhain, R., 939
Heinroth, O., 948-949
Heitz, E., 958
Helix, anatomy of, 259
Helmholtz, H., 936
Helminthology, definition of, 10
Hemal system
echinoid, 413
sea cucumber, 413, 414
sea star, 407
Heme, 626
Hemerythrin, 626
Hemichordata; *see* Acorn worms
Hemimetabola, metamorphosis in, 353
Hemipenes, 511
Hemiptera, 371, 373
Hemocoels, 326
Hemocyanin, 625
Hemoglobin, 625
structure of, 968
Hemophilia, 610
Hench, P., 959
Hepatic; *see* Liver
Herbivores, 578-579, 642, 871
Heredity, 21-22; *see also* Inheritance
behavior and, 913
comb forms in chickens, 782
cytologic background of, 762-766
definition of, 761-762
environment and, 794-795
eye color and, 795
factors of
complementary, 783
cumulative, 783-784
lethal, 784
supplementary, 783
functions, allocation to chromosomes, 941
hair color and, 795
human, 793-800
influence of radiation on, 798-799
laws of
early formulation of, 938
Mendel's; *see* Mendel's laws
Mendel's investigations, 776
nucleus as basis of, 941

Heredity—cont'd
special forms of, 782-784
Hermaphroditism, 722-723
Hermit crab, 318, 909
Herons, 561
feet of, 533
Herpetology, definition of, 10
Herrick, J., 949
Hershey, A., 965
Hertwig, O., 939, 941
Hertwig, R., 942
Herzog, R., 952
Hess, W., 963
Heterochromatin, 57
Heterodont, 577
Heteromorphosis, 324
Heteroploidy, concept of, 951
Heterosis, 791
concept of, 951
Heterostracans, 438
Heterothermy, 586
Heterotroph(s), 641-642
theory of origin of life, 958
Heterozygous, 780
Hevesy, G., 953
Heymans, C., 955, 956
Hibernation
frog, 493
mammals, 588, 590
Hill, A., 954
Hilus, 633
Hindbrain, chordate, 450
Hinge ligament, 261
Hippasteria, 404
Hirudinea, 279, 289-291
Hirudo medicinalis; see Leach, "medici-
nal"
His, W., 942
Histamine, nature of, 948
Histochemistry, beginning of, 934
Histogenesis, definition of, 91
Histology
diefinition of, 10, 91
establishment as a science, 936
Hjort, J., 949
Hoerr, N., 958
Hoerstadius, S., 960-961
Hofmeister, W., 936
Hogeboom, G., 963
Holley, R., 967
Holoblastic cleavage, 85, 86
in amphioxus, 88
Holocephali, 455, 463
Holocrine glands, 576-577
Holometabola, 354
Holonephros, 449
Holostei, 469
Holothuroidea; *see* Sea cucumbers
Holotrichia, 120
Holtfreter, J., 961-962
Holtz, P., 963
Homarus, 337
americanus, 332
Homeostasis, 20, 627-639
constancy of, 937

Homeostasis—cont'd
  as ideal adaptation, 15
Hominid lineage, separation from age
    lineage, 850
Hominoidea, 600
Homo
  erectus, discovery of, 943
    heidelbergensis, 852
      discovery of, 947
    heidelbergensis, 852
      discovery of, 947
  neanderthalensis, discovery of, 937
Homodont, 577
Homology, 100-101, 823
  concept of, 936
  general, 101
  serial, 668
    crayfish as example of, 324
    of sex organs, 733
  skeletal, 823
  special, 101
Homoptera, 371, 373, 375
Homozygous, 780
Honeybee, 356-357, 366
  communication, 920-921, 963
  legs of, 365
  stinger after dissection, 376
Hooke, R., 931
Hookworm, 235
Hoover, M., 959
Hopkins, F., 947, 952
Hormone(s), 700
  action of
    early demonstration of, 943
    mechanisms of, 701-702
    second-messenger concept of, 703
  activation, 355
  adrenal medulla, 716
  adrenocortical, 714-716
  adrenocorticotropic, 707, 715-716
  antidiuretic, 638, 710
  brain, 705
  control
    of color changes in crustaceans, 956
    of molting, 331-333
  coordination for balanced tissue de-
    velopment, 964
  of digestion, 717
  factors in invertebrate physiology, 953
  follicle stimulating, 707
  growth, 707
  hypothalamic, 707-710
  influence in enzyme activity, 962
  invertebrate, 702-705
  juvenile, 704, 705
  luteinizing, 707
  of metabolism, 710-717
  molt-accelerating, 331-332
  molt-inhibiting, 331, 705
  molt-preventing, 960
  molting, 355, 704, 705
    silkworm, 704
  neurosecretory, 701
  parathyroid, 713, 714
  pituitary, 707-710
    actions of, 708

Hormone(s)—cont'd
  pituitary—cont'd
    chemical nature of, 708
    relation to hypothalamic and tar-
      get-gland hormones, 709
    synthesis of, 966
  posterior lobe, 711
  releasing factors, 710
  of reproduction, 734-736
  sex, 715-716, 734-735
  target-gland, 709
  thyroid, 711-713
  thyrotropic, 701, 713
  tropic, 707
Horn(s), 572-574
  differences from antlers, 572
  true, 572
Hornbills, 562
Hornet, bald-faced, 257
Hornworm, southern, 345
Horse, evolution of, 822-823
Horsehair worms; *see* Nematomorpha
Horseshoe crab, 304, 305, 306
Hot springs, 886
Housefly, 368, 371
Houssay, B., 953
Howard, H., 952
Hoyer, B., 970
Huber, G., 946-947
Hughes, W., 968
Human; *see* Man
Humerus
  bird, 532
  human, 663
Hummingbird, 562
  ruby-throated, 558
Hunter, J., 932
Huot, A., 944
Hurwitz, J., 969
Huxley, A., 966
Huxley, H., 966
Huxley, J., 961
Hyaline cap, 128
Hyaline cartilage, 662
Hyalospongiae, 154, 160-161
Hyaluronidase, 738
  discovery of, 961
Hybrid, 780
  somatic cell technique, 969
  varieties, 792
  vigor, 791
Hybridization experiments, 747
Hydatid cyst, 219
Hydatid worm, 218-219
Hydra, 172-178
  behavior of, 172, 178
  body plan of, 172
  body wall of, 172-174
  cellular detail, 173
  digestion, 174, 175
  excretion of, 174
  feeding, 174, 175
  littoralis, nematocysts of, 169
  locomotion of, 174
  regeneration in, 177-178
  reproduction in, 174, 176-177

Hydra—cont'd
  respiration in, 174
  sexual differentiation in, 966
  structure, early observations on, 932
Hydractinia, illustration, 170
Hydranths, 178, 179
Hydrocauli, 178
Hydrochloric acid, 652
Hydrogen
  bond, 34
  ion concentration, 31, 868-869
    determination of, 948
  isotopes of, 26
Hydroids, gymnoblastic, 180
Hydrolases, 650
Hydrolysis, 650
Hydrorhiza, 178
Hydrosphere, 858-859
Hydrostatic pressure, 67
Hydrotheca, 179
Hydrozoa, 167, 171-184
Hyla, 497
  crucifer, 492
Hymen, 733
Hymenolepis nana, 218
Hymenoptera, 371, 376, 377
Hypermetamorphosis, 375
Hyperosmotic regulators, 474, 629
Hyperparasitism, 345
Hypodermis, 660
Hypophysis, 707; *see also* Pituitary
  chordate, 451
  -hypothalamus system, concept of,
    953
  portal vessels, 957
Hypostome, 172, 179
Hypothalamo-hypophyseal   system,
  concept of, 953
Hypothalamus, 688
  frog, 501
  hormones of, 707-710
  pituitary relation to, 706
Hypothesis
  "masked-messenger," 745
  operon, 969
Hyracoidea, 599
Hyraxes, 599

I

Ibises, 561
Ichneumon wasp, 366, 376
Ichthyology, definition of, 10
Ichthyostegids, discovery of, 957
Ilium, frog, 497-498
Imago, 354
Immunity, phagocytosis in, 941
Impala, 581
Imprinting, 917, 948-949
Impulse
  nerve
    action potential of, 681-682
    differential conduction of, 953
    nature of, 680-683
    propagated, metabolic activity of,
      950
    rate determination, early, 936

Impulse—cont'd
  nerve—cont'd
    resting potential of, 680-681
    sodium pump for, 682-683
    transmission of, 682
  patterns in brain, instinctive, localization of, 963
  transmission at nerve junction, 966
Incisors, 644, 645
Incus, 695
Independent assortment, Mendel's law of, 778
Individuality
  of chromosomes, 941
  principle of, 83
Inducible system, 775
Induction of mutations
  artificial, 955
  gene, target theory of, 959
Inductors, embryonic, 748-749
Infundibulum, 707
Ingenhousz, J., 932
Ingram, V., 967
Inheritance; *see also* Heredity
  of acquired characters, 825
  blending, 783
  blood group, 795-796
  crisscross, 789
  cytoplasmic, 791-792
  extranuclear, 962
  of mental characteristics, 796
  nuclear control of, 938
  of physical traits, 797
  principles of, 761-801
  quantitative, 783
    discovery of, 932
  of Rh factor, 796
  sex-linked, 786-789
  of twinning, 797
Ink production of Cephalopoda, 270
Insect(s), 340-378
  adaptability, 344-345
  antennae, 364
  auditory organs of, 348-349
  behavior of, 350
  beneficial, 366, 367-369
  bioluminescence, 366-367
  body form of, 364
  characteristics of, 344
  chemoreceptors, 349
  coloration, 347
  control of, 369-370
  diapause, 355
  distribution of, 343-344
  -eating mammals, 580, 598, 642
  flight of, 346-347
  food habits of, 345-346
  growth of, 351-355
  injurious, 368, 369
  integument, structure of, 659
  legs of, 364-366
  locomotion in, 346-347
  malpighian tubules of, 362-363
  mechanical senses of, 348
  metamorphosis, 351-355
    role of corpus allatum gland in, 958

Insect(s)—cont'd
  migration of, 355-356
  mouthparts of, 345-346
    types of, 346
  neuromuscular coordination, 347-348
  orders of, review of, 370-377
  origin of, 344
  parasitic, 345
  pheromones in, 356
  phytophagous, 345
  predaceous, 345
  protection, 347
  relation to man's welfare, 367-369
  relationships, 344
  reproduction in, 350-351
  saprophagous, 345
  scale, 371, 373, 375
  sense organs of, 348-349
  size range of, 344
  social instincts of, 356-358
  sound production, 349-350
  structural variations among, 364-366
  visual organs of, 349
  walking, 346
  wings of, 347
Insecta, 302
Insectivores, 580, 598, 642
Insight learning, 916-917
Inspiration, 624
Instar, 354
Instinct, definition of, 913-914
Instinctive behavior, theory of, 957
Insulin, 78, 716-717
  extraction of, 952
  molecule, structure of, 966
Integument, 658-662; *see also* Skin
  insect, structure of, 659
  invertebrate, 658-660
  vertebrate, 660-662
Interferon
  action on viruses, 967
  discovery of, 959
Intermedin, 707
  effect on melanophores, 957-958
Interoceptors, 692
Interphase, 57, 58, 726
Interstitial cells, hydra, 173, 174
Intestine
  as ecosystem, 886-887
  flagellates, symbiotic relationships between termites and, 953
  fluke, of man, 214
  juice, 653
  large, 650, 653
  law of, 945
  planarian, 205
  roundworm; *see* Roundworm
  small, 649, 650
    digestion in, 652-653
Invaginations, 619
Invertebrates
  contractile vacuole, 631
  excretory structures of, 631-632
  germ layers of, 938
  hormones, 702-705
  integument of, 658-660

Invertebrates—cont'd
  marine, salt and water balance problems, 628-629
  muscle, types of, 671
  nephridia of, 631-632
  nervous system of, 678-679
  physiology, hormonal factors in, 953
Iodopsin, 700
Ion(s), 28-29
  blood, influence on heart contraction, 940
  hydrogen ion concentration, 31, 868-869
    determination of, 948
Ionic bond, 28, 29
Ionic composition of resting nerve cell, 681
Ionosphere, 860
Iris, 698
  frog, 502-503
Isaacs, A., 967
Ischium, frog, 497, 498
Islet cells of pancreas
  discovery of, 938-939
  insulin from, 716-717
Isogametes, 721
Isolation, 834
  effects of, 827
  genetic, 827
  geographic, 827
Isolecithal egg, 85
Isoleucine, formula of, 39
Isopoda, 337-338
Isoptera, 371, 372
Isotopes, 26
  discovery of, 952
  of hydrogen, 26
  radioactive, 26-27
    liquid scintillation counter for, 13
    use to demonstrate synthesis of bodily constituents, 960
  tracer method; *see* Tracer method
Itch, swimmer's, 213
Ivanov, A., 967
Ivanovski, D., 943

J

Jacob, F., 969
Jacobs, P., 968
Jacobson's organs, 513
Jamoytius, discovery of, 963
Jancke, W., 952
Janssens, F., 948
Java man, 852
Jaws, vertebrate, 665
Jelly, egg, 86
Jellyfish
  scyphozoan, 185
  siphonophoran, 184
Jennings, H., 946
Johannsen, W., 948
Johnson, M., 963
Jordan's rule, 841-842
June beetle, 368
Juvenile hormone, 704, 705
Juxtaglomerular complex, 637

K

Kallima, 348
Kangaroo, 598
  rat, water balance in, 631
Kaola, 598
Karlson, P., 968-969
Karpechenko, G., 953-954
Karpechenko's experiment on artificial
      species, 843
Karyokinesis, 57
Katz, B., 966
Keith, A., 947
Kekulé, F. A., 938
Kendall, E., 950, 959
Kendrew, J., 968
Keppler, A., 960
Keratin, 567, 660
Khorana, H., 970
Kidney
  archinephros, 449, 450
  artery, 635
  arthropod, 632
  countercurrent multiplier system of,
      637-638
  excretion
    filtration theory of, 936
    water, 637-639
  fish, aglomerular, discovery of, 944
  function, vertebrate, 632-639
  holonephros, 449
  human, collecting duct structure, 634
  mesonephric, 461
  nephron structure, 634, 936
  pelvis, 635
  tubules; *see* Tubules
  vein, 635
  worms, 237, 239
Kinetodesmal fibril, 140
Kinetosomes, 49, 55
Kinety, 140
King, H., 951
King, T., 965
King snake, 518
Kingfishers, 562
Kinins, discovery of, 960
Kinorhyncha, 231-232
  adaptive radiation, 226
  biologic contributions, 225
  ecologic relationship, 231
  phylogeny, 225-226, 231-232
  position in animal kingdom, 225
  structure of, 231-232
Kirchhoff, G., 933
Kite, G., 950
Kiwis, 561
Knight, T., 933-934
Koelreuter, J. G., 932
Koller, G., 956
Kölliker, A. von, 936
Kollman, J., 941
Kopec, S., 953
Kornberg, A., 967
Kossel, A., 940
Kowalevsky, A., 938
Krakatao, 862

Kramer, G., 965
Krebs cycle, 76
Krebs, H., 960
Krogh, A., 951
Krohn, H., 957-958
Kropotkin, P., 945-946
Kuhne, W., 940
Kuru disease, 794
Kutscher, F., 954
Kwashiorkor, 655

L

Labia
  majora, 733
  minora, 733
Laboratories, American marine, estab-
      lishment of, 939
Labyrinth, 695
Lacertilia, 521-522
Lacewings, 375
Lacrimal glands, 574
Lactase, 653
Lactate, formation, and glycolysis, 80
Lacteal vessels, early demonstration of,
      931
Lactic acid
  metabolic cycle, 961
  muscle contraction and, 947, 952
Lacunar system, 242
Ladybird beetle, 366
Lagena, frog, 502
Lagomorpha, 599
Lake(s), 882, 884-885
  Baikal, 862-863
    biology of, 944
  ecologic succession in, 879
Lamarck, J. B. de, 933
Lamellisabella, 423
Lampbrush chromosomes, 743, 766
Lampetra lamattei, 457
Lamprey, 455, 457-459
  ammocoete larva as chordate arche-
      type, 447-451
  brook, 457
  characteristics of, 456
  comparison to hagfish, 456
  sea, 458
Lampropeltis getulus, 518
Land
  misuse of, 897-898
  snail; *see* Snails, land
Landsteiner, K., 945, 961
Langerhans, P., 938-939
Langley, J., 952
Lappets, 185
Larder beetle, 368
Larus atricilla, 857
Larva
  ammocoete; *see* Ammocoete larva
  amphiblastula, 159
  aquatic, of dragonfly, 352
  dipleurula, 399, 400
  dobsonfly, 366
  doliolaria, 417
  glochidium, 267

Larva—cont'd
  infusoriform, 153
  mysis, 331
  nauplius, 335
  permanent, amphibians, 490
  pilidium, 220
  planula, 181
  pluteus, 413
  pond water, 883
  trochophore, 266
  veliger, 266
  vermiform, 152-153
Larvacea, 442-443
Larynx, 623
  frog, 500
Latimeria chalumnae, 465
Latrodectus mactans, 312-313
Laveran, C. L. A., 940
Lavoisier, A. L., 932-933
Law(s)
  "all-or-none" law of heart muscle,
      discovery of, 939
  biogenetic; *see* Biogenetic law
  Cope's, 840
  Dollo's, 840
  of ecologic tolerance, 950
  of energy conservation, 936
  of heart, 951
  of heredity, early formulation of, 938
  of the intestine, 945
  Mendel's; *see* Mendel's laws
  of minimum requirements, 935
  of priority, 108, 109
  of probability, 779-780
  of tolerance, 866
  Williston's law of evolutionary sim-
      plification, 951
Lawrence, E., 957
Lead-uranium method of dating geolog-
      ic periods, 947
Leadership-followership relations, 926-
      927
Leafhoppers, 373, 375
Leakey, L., 968
Learning
  associative, 916
  concept of, 949
  basis of, 917-918
  insight, 916-917
  selective, 916
Lederberg, E., 965
Lederberg, J., 962, 965
Leech, "medicinal," 289-291
  behavior of, 290-291
  in medical practice, 291
  structural characteristics of, 289-290,
      291
Leeuwenhoek, A., 931
Leg
  grasshopper, 360
  honeybee, 365
  insect, 364-366
  spider, 308
LeJeune, J., 968
Lemnisci, 242

Lemur(s), 600
  flying, 598
Lemuroidea, 600
Leopard frog, 492
Lepas fascicularis, 336
Lepidoptera, 371, 376
Lepidosiren, 466
Lepidosteus osseus, 468
Leptocoris, 373
Leptotene, 725
Lepus arcticus, 569
Lerner, A., 968
Leuckart, R., 941
Leucochloridium, 214
Leuconoid sponges, 155-156
Leukocytes, 609
  earthworm, 282
Levan, A., 967
Levator muscle, 498
Lewis, E., 964
Lewis, W., 957
LH, 707
Li, C., 964
Libby, W., 962
Lice
  biting, 371, 373
  body, as vector of typhus fever,
    948
  book, 371, 373
  chicken, 371
  fish, 335
  sucking, 371, 373
Liebig, J., 935
Life, 3-24
  biochemical basis of, 6
  biochemistry of, 33-45
  biologist and
    methods of, 11-15
    tools of, 11-15
  chemical elements of, 34-35
  complexity, levels of, 4-5
  cycle
    Aurelia, 186
    Craspedacusta, 182
    existence, 16-17
    frog, 503-505
    liver fluke, of sheep, 210
    Obelia, 179
    Opisthorchis sinensis, 211, 212
    oyster, 266
    Pennaeus, 330
    Plasmodium vivax, 137
    protozoan, 125
    roundworm, 235
    Taenia saginata, 215, 217
    tapeworm, early experimental, 933
    toads, 503-505
    Trypanosoma, 943
    Volvox globator, 136
  definition of, 3-4
  energy relations, 7
  environment and, 865-866
  extraterrestrial, 813
  living compared with nonliving, 5
  nature of, understanding, 7-8

Life—cont'd
  origin of, 805-814
    heterotroph theory of, 958
    important steps in, illustration, 812
    primordia of, 966
    protoplasm and, 17
    replication and, 17-18
    rhythmic patterns of, 19-20
    substance, protoplasm as, 7
    zones in North America, 943
Ligament, hinge, 261
Light, 867-868
  microscope, optical path of, 12
  production, 923-924
    in animals, 941
Lilies, sea, 403, 415-417
  structure of, 415-417
Lillie, F. R., 946, 950, 951
Limbic-midbrain system, 688
Limnology, 881
Limulus, 304
Lindemann, L., 967
Lingula, 396
Linkage, 789-791
  autosomal, 789
  groups, 766
    exchange of genes between, 766-
      767
  hereditary units, discovery of, 947
  sex, discovery of, 949
Linnaeus, C., 107, 932
Lion(s), mating, 595
Lioness, 580
Lipase, pancreatic, 653
Lipids, 44-45
  metabolism of, 78-80
Lipmann, F., 962
Lipophores, frog, 495
Liquid scintillation counter for radioac-
    tive isotopes, 13
Lithosphere, 859
Liver
  cecum, 446
  cells (in rat)
    electron micrograph of, 53
    fine structures of, 51
  elasmobranchs, 461
  flukes; *see* Flukes, liver
  glycogen formation by, 937
  treatment of pernicious anemia, 954
Lizards, 511-513, 521-522
Lobster, 337
  molting sequence in, 332
Locomotion; *see also* Movement
  Amoeba in, structure of, 669
    proteus, 128-129
  Amphiporus ocraceus, 222
  arthropods, 303
  clam worm, 296-297
  earthworm, 282
  Euglena, 135
  fish, 470-471
    analysis of, 932
  hydra, 174
  insects, 346-347

Locomotion—cont'd
  Paramecium, 143
  planaria, 205
Locusts, 371
Loeb, J., 945
Loewi, O., 952
Lohmann, K., 956
Loomis, W., 966
Loons, 561
Loop of Henle, 635
Lophius piscatorius, 477
Lophophorate animals, 390-397
Lophophore, 390
Lorenz, K., 957
Lorica, Rotifera, 228
Loricata; *see* Crocodile
Louse; *see* Lice
Loxodonta africana, 587
Loxosceles reclusa, 313, 314
Lucanus, 374
Luciferase, 923
Luciferin, 923
Ludwig, C., 936
Ludwig, K., 939
Lugworm, 294
Lumbricus terrestris; see Earthworms
Luminescence, 366-367, 923-924, 941
Lung(s), 621, 622
  amphibians, 485-486
  book, spider, 309
  fish, evolution of, 472
  flukes, 213-214
  gaseous exchange in, 625
  of man, 622
  reptile, 511
  snail, 259
  vertebrate, internal structures of, 620
Lungfish, 455-456, 465-467, 469
  estivating African, 466
  surviving, illustration, 466
Luteinizing hormone, 707
Lwoff, A., 965-966
Lyell, C., 934
Lymph, 95, 618
  capillaries, 617
  nodes, 618
Lymphatic system, 612, 617-618
  discovery of, 931
Lymphocytes, 609
Lyrurus tetrix, 556
Lysins, sperm, 503, 738
Lysosome(s), 647
  function of, 49, 50
  organelle, discovery of, 963-964
Lytechinus, 411

M

Macallum, A. B., 946
MacCullum, F., 959
Mackerel, osmotic respiration in, 475
MacLeod, C., 962
MacMunn, C., 942
Macracanthorhynchus hirudinaceus, 241-
    243
  morphology, 241-243

Macracanthorhynchus hirudinaceus—
      cont'd
   physiology, 241-243
Macroclimate, 867
Macrogametes, *Volvox globator,* 135
Macrogametocytes, 138
Macromolecules, 18
Macronucleus, 138, 140
Macrophages, 609
Macropus eugenii, 740
Magendie, F., 933
Magnesium, 607
Magpie, black-billed, 544
Malacostraca, 319, 337-339
Malleus, 695
Mallophaga, 371, 373
Malocclusion in woodchuck, 578
Malpighi, M., 931
Malpighian tubules, 632
   of insect, 362-363
Maltase, 653
Maltose, 650
   disaccharide, 38
Mammal(s), 564-602
   antlers, 572-574
   body temperature regulation, 584-587
   carnivorous, 579-580, 598, 642, 871
   characteristics of, 566-567
   circulatory system of, 612
   coloration of, 571-572
   conductance, 586
   convergent evolution among, 839
   corpuscles of, red, 608
   echolocation, 590-592
   egg, artificial parthenogenesis of,
      961
   egg-laying, 598
   family tree of, 566
   feeding, 578-583
   fishlike, 599
   flight of, 590-592
   flying, 598
   food, 578-583
   functional adaptations of, 567-596
   glands, 574-577
   gnawing, 598
   hair, 567-571
   herbivorous, 578-579, 642, 871
   hibernation, 588, 590
   highest, 600
   home range of, 592-593
   hoofed
      even-toed, 599-600
      odd-toed, 599
   horns; *see* Horns
   insectivorous, 580, 598, 642
   integument, 567-578
   man and, 596-597
   migration of, 587-588, 589
   nervous system in, autonomic, 691
   nipples, types of, 576
   omnivorous, 580, 642, 871
   orders of, résumé of, 597-600
   origin of, 565-567
   ovum, discovery of, 934

Mammal(s)—cont'd
   population, 593-594
      cycles, 593-594
      size, 593
   pouched, 598
   proboscis, 599
   relationships, 565-567
   reproduction in, 594-596
   structural adaptations of, 567-596
   teeth of, 577-578
   territory of, 592-593
   toothless, 599
   uteri of, 732
Mammary glands, 575-577
Mammoths, 821
Man, 600
   brain of, 689
   breathing in; *see* Breathing, in man
   chromosomes of, 762
      abnormalities in, 968
   dental formula of, 578
   development, 751-759
   ecosystem of, problems of, 894-908
   evolution of, 847-854
   fossil record of, 850-853
   Heidelberg, 852
      discovery of, 947
   heredity, 793-800
      influence of radiation on, 798-799
   human race, improvement of, 799-800
   Java, 852
   mammals and, 596-597
   modern, emergence of, 853
   nature of, 847-854
   Neanderthal fossil, discovery of, 937
   position in nature, 847
   prehistoric, restoration of, 852
   reproductive system in, 729-733
   respiration in, 623-627
   skeleton, 666
   skin of, structure, 659
   skulls of, 851
   teeth of; *see* Teeth
   unique position of, 853-854
   urinary system, anatomy of, 633
   vestigial organs in, 824
   water balance in, 631
Manatees, 599
Mandibulata, 301-302
   aquatic, 318-339
   terrestrial, 340-378
Manometric methods for studying cell
      metabolism, 953
Mantes, praying, 352, 371
Mantle
   bivalves, 261
   chiton, 253
   mollusk, 249
Manubrium, 180
Map, chromosome, 790
   first, formation of, 950
Margaritifera, 263
Marmota monax, 578
Marriage of cousins, 796-797
Marsupialia, 598

Marsupium, 598
"Masked-messenger" hypothesis, 745
Mass number, 26
Masseter muscle, 498
Mast, S., 954
Mast cells, discovery of, 940
Mastax, 229
Mastigophora, 120, 132-135
   types of, 132
Mating
   birds, 553-555
   lions, 595
   types
      in paramecium, 146
         discovery of, 960
      protozoan, discovery of, 942
Matter, 25-45
   colloidal states of, 938
   living, chemical complexity of, 35-36
   structure of, 25-33
Matthaei, J., 969-970
Maturation, oocyte, 742-744
Maupas, E., 942
Maupertuis, P. M., 932
Mauthner, L., 937
Mauthnerian apparatus, 937
May beetle antennae, 364
Mayer, J. R., 936
Mayflies, 371, 372
McCarthy, B., 970
McCarty, M., 962
McClean, D., 961
McClintock, B., 957
McClung, C. E., 946
Mechanical senses, insect, 348
Mechanoreception, 693-697
Mecoptera, 375-376
Medical problems and genetics, 797-798
Medulla, 688
   frog, 501-502
   oblongata, 624
Medusae, 167-168
   freshwater, 183
Megascleres, 157
Meiosis, 56, 723-728
   anaphase, 725-726
   comparison to mitosis, 724, 725
   early description of, 940
   interphase, 726
   metaphase, 725, 726
   prophase, 725, 726
   stages in, 725-726
   telophase, 726
Melanism, 572
Melanophores
   effect of intermedin on, 957-958
   frog, 495
Melatonin, discovery in pineal glands,
      968
Meleagris gallapavo, 557
Membrane(s)
   basement, 92
   bone, 663
   cell, 66-69
      behavior of, selective, 66

Membrane(s)—cont'd
  cell—cont'd
    osmometer, 68
    passage of materials through, 66
    permeable, 67
    physiology of, 66-69
    semipermeable, 67
    structure, present concept of, 66
    transport
      active, 66, 68-69
      passive, 66
      passive, and osmosis, 66-68
    unit, 66
  chorioallantoic, 750
  fertilization, 738, 743
  moist, 95
  mucous, 95
  nictitating, frog, 503
  nuclear, 49
  plasma; *see* Plasma, membrane
  semipermeable, 938
  serous, 95
  tectorial, frog, 502
  tympanic, frog, 502
  undulating, 140
Membranipora, 393
Memory and learning, 917-918
Mendel, G., 938
Mendel's investigations, 776
Mendel's laws, 777-779
  of independent assortment, 778
  rediscovery of, 945
    genetic advances after, 780-781
  of segregation, 778
  parallelism between chromosome behavior and, 946
Mendel's ratios, explanation of, 778-779
Meninges, 685
Menstrual cycle, 594, 734
Menstruation, 594
Mental characteristics, inheritance of, 796
Menton, M., 950
Mering, J. von, 942
Meroblastic cleavage, 85, 88
Merostomata, 301, 304-306
Merozoites, 138
Merriam, C. H., 943
Meselson, M., 968
Mesenchyme, 92-93
  origin of term, 941
Mesenteries, 188, 650
Mesoderm, 23, 90
  development of derivatives, 755-757
  formation, types of, 90
Mesoglea
  hydra, 174
  sponge, 155
Mesonephric duct, chordate, 450
Mesozoa, 151-153
  adaptive radiation, 152
  biologic contributions, 151-152
  phylogeny, 152
  position in animal kingdom, 151
Metabolic energy, 72-73

Metabolism, 642
  Amoeba proteus, 126-127
  carbohydrate, role of pituitary in, 953
  cellular, 72-81
    manometric study methods, 953
  Euglena viridis, 134
  glucose, 78
  hormones of, 710-717
  inborn errors of, early description, 947
  lipid, 78-80
  paramecium, 141-143
  sponges, 157-158
Metacercaria, 210, 212
Metals causing water pollution, 901, 903
Metamere, 99
Metamerism, 83, 99-100
  Annelida, 276-277
  heteronomous, 99
  homonomous, 99
Metamorphosis, 704, 742
  arthropods, 303
  butterfly, 353
  frog, role of thyroid in, 950
  insect, 351-355
    role of corpus allatum gland in, 958
  physiology of, 354-355
  precocious, frog tadpoles, 712
Metanauplius, 331
Metaphase, 58, 59-60, 725, 726
  naming of, 941
  plate, 60, 61
  in salamander, 61
Metapleural fold, 444
Metatheria, 598
Metazoa, 83
  ancestry, Gastrea theory of, 939
  lowest, 151-163
  origin of, 110-112
    theory of, 964
Metchnikoff, E., 941
Metencephalon; *see* Cerebellum
Metestrus, 594
Metridium, 188
  structure of, 189
Meyerhof, O., 952
Michael-Sars expedition, 949
Michaelis, L., 950
Microclimate, 867
Microfilariae, 236
Microgametes, *Volvox globator,* 135
Microgametocytes, 138
Micronucleus, 138, 140
Microorganism as infective agent, early demonstration of, 935
Micropyle, 364
Microscleres, 157
Microscope
  electron
    modern, illustration of, 12
    optical path of, comparison with light microscope, 12
    scanning, 11-13
  light, optical path compared with electron microscope, 12

Microscopy, electron
  cell
    hepatic (in rat), 53
    pancreatic exocrine (in rat), 53, 54
  high-resolution, of mitochondrion, 74
Microsomes, 52
Microtubules, 49
Microvilli, 56, 648
Micrurgic study of cell structures, 950
Micrurus fulvius, 516
Midbrain, 688
  chordate, 450
  limbic-midbrain system, 688
Miescher, F., 939
Migration
  bird, 548-550
    direction-finding in, 549-550
    gonadal hypothesis of, 954
    origin of, 548
    routes of, 548-549
    stimulus for, 549
  eel, 479-480
  fish, 479-480
  insect, 355-356
  mammal, 587-588, 589
  salmon, homing, 480
Milieu exterieur, 607
Milieu interieur, 607
"Milking" of eastern diamondback rattlesnake, 517
Milkweed beetle, 366
Miller, H., 935
Miller, J., 969
Miller, S., 966
Miller's experiment on synthesis of organic compounds, 808
Millipedes, 342
Mimicry, 837-838
  concept of, 938
  fish, 476, 478
Mineral cycle, 891
Mineralocorticoids, 715
Minimum requirements, law of, 935
Minkowski, O., 942
Minot, G., 954
Miricidium, 210, 212
Mites, 315-316
Mitochondria, 49, 50
  analysis of, 958
  discovery of, 944-945
  electron microscopy of, high-resolution, 74
  finer structure of, analysis of, 965
  function of, 49, 50, 54-55
  isolation of, 958
  separation from cell, 963
Mitosis, 56-62
  in *Amoeba,* 123
  anaphase, 59, 60
  comparison
    to cleavage, 86
    to meiosis, 724, 725
  crossing-over in, discovery of, 959
  cycle, 60, 62
  early description of, 940

Mitosis—cont'd
  interphase, 57, 58
  metaphase, 58, 59-60
    in salamander, 61
  naming of, 941
  prophase, 57-59
  purpose of, 57
  stages in, 57-60, 61, 62
    in whitefish, diagram, 61
  telophase, 58, 60
Mixtures, 31-33; *see also* Solutions
  properties of, 31-33
Mohl, H. von, 935
Molars, 644, 645
Mole, 598
  cricket, legs of, 365
  definition, 28
Molecular solutions, 31, 32
Molecular weight, gram, 28
Molecule(s), 17
  amphipathic, 52
  definition, 28
  nature of, 25-30
  organic, 36-45
Molgula, 443
  structure of, 442
Mollusks, 246-274
  adaptive radiation, 247
  biologic contributions, 246
  bivalves; *see* Bivalves
  characteristics of, 251
  classes of, 251
  ecologic relationships, 250
  economic importance, 250
  eyes in, types of, 260
  foot of, 249
  mantle of, 249
  phylogeny, 247
  position in animal kingdom, 246
  primitive, discovery of, 965
  radula of, 250
  segmented, 251
  shell of, 249-250
Molt
  -accelerating hormone, 331-332
  -inhibiting hormone, 331, 705
  -preventing hormone, 960
Molting
  arthropoda, 302
  butterfly, endocrine control of, 704
  cicada, dog-day, 351
  crayfish, 331
  hormonal control of, 331-333
  hormone, 355, 704, 705
    silkworm, 704
  insects, 352
  lobster, 332
Monarch butterfly, metamorphosis of, 353
Monaxons, 156
Monera, 813
Monestrus, 594
Moniezia expansa, 219
Monkeys, 600
Monocytes, 609

Monod, J., 969
Monoecious, 722
  sponges, 159
Monogenea, 209
Monogononta, 227
Monohybrid, 780
Monophyodont, 577
Monoplacophora, 251-252
  illustration, 248
Monosaccharides, 36-37
Monosomy, 968
Monospermy, 503
Monotremata, 598
Monozygotic egg, 758
Mons veneris, 733
Montgomery, T., 945
Moore, C., 954-955
Moose, bull, 573
Morgan, C. L., 943
Morton, T., 949
Morphogenesis, 19
Morphology
  of animals, 19
  Ascaris lumbricoides, 233-235
  definition of, 10
  Entoprocta, 243-244
  Macracanthorhynchus hirudinaceus, 241-243
  Nematomorpha, 239-240
  Rotifera, 227-229
Mortality, 759, 877
  minimum, 877
Mosaic(s)
  cleavage, 88, 744, 745
  sex, 786
  theory of development, 941-942
Mosquito, 368, 375
  antennae, 364
Moth, 371, 376
  armyworm, 368
  clothes, 368, 369
  cutworm, 368
  silkworm, 366
  tent caterpillar, 368
  tiger, eggs of, 350
  walnut, 368
Motivation and behavior, 914-915
  animal behavior, measurement of, 960
Motor neuron, 95
Mousebirds, 562
Mouth; *see also* Oral
  digestion in, 650
  man, 648
Mouthparts
  grasshopper, 361
  insect, 345-346
  types of, 346
Movement, 658-677; *see also* Locomotion
  ameboid; *see* Ameboid movement
  Brownian, early description of, 934
  ciliary, 668-670
    sequence of, 139
  muscular, 670-677

Mucin, 652
Mulder, J., 935
Muller, H., 955
Muller, J., 935-936
Muridae, 598
Murphy, M., 954
Murray, J., 949
Muscle(s), 93
  abductor, 498
  adductor, 498
  atrophy, 675
  bird, 532-534
  contractions; *see* Contractions, muscle
  crayfish, 326
  deltoid, 498
  depressor, 498
  extensor, 498
  fiber, 93, 94
  fibrillar, 671
  flexor, 498
  fluke, liver, 210
  frog, 496-499
    skeletal, 658
  heart; *see* Heart, muscle
  independent irritability of, 936
  invertebrate, types of, 671
  involuntary, 93, 94
  levator, 498
  movement, 670-677
  neuromuscular; *see* Neuromuscular
  nutritive-muscular cells, hydra, 173-174
  pectoralis, 532
  rotator, 498
  skeletal, 93, 94, 670, 672
    frog, 658
  smooth, 93, 94, 670
  somites, 756
  striated, 93, 94, 670, 676
    structure of, 671-673
  supracoracoideus, bird, 532
  trunk, of salmon, 470
  vertebrate; *see* Vertebrate, muscle
  voluntary, 93, 94, 670
Mussel(s), 336
  marine, 262
Mustelidae, 598
Mutants, bacterial, replica technique in selection of, 965
Mutation(s), 22, 829-831
  artificial induction of, 955
  biochemical, 961
  causes of, 830-831
  chemical production of, 962
  chromosome, 830
  frequency of, 830-831
  gene, 830
    importance of, 767
    target theory of induction of, 959
  from mistakes in DNA molecule, 770
  nature of, 829, 967
  point, 767
  pressure, 767, 830
  random, 767

Mutation(s)—cont'd
theory of evolution, 945
types of, 830
Mutual aid as factor in evolution, 945-
946
Mutualism, symbiotic, 121
Mya, 255
Myelencephalon; *see* Medulla
Myelin, 680
Myliobatis, 462
Myocytes, sponge, 156
Myoepithelial cells, 93, 94
Myofibrils, 93, 671
Myofilaments, 671, 673
Myoneural junction, 675
Myosepta, 446
Myosin, 668, 673
Myotomes, 446
Myriapods, 340-378
Myrmeleon, 374
Mysis, 337
larva, 331
Mysticeti, 599
Mytilus, 262, 336
Myxini; *see* Hagfishes

N

Naiad of dragonfly, 352
Naples Biological Station, establishment
of, 939
Nares, 623
Nasal chambers, 623
Nasohypophyseal canal, chordate, 451
Natural selection, 833-834
appraisal of theory of, 826-827
Darwin's theory of, 825-827
early, concept of, 937
limitations on pure lines, 948
variation and, 826
Nauplius, 331, 335
Nautilus, 269
Navigation of birds, 548-550
celestial, 968
Neanderthal fossil man, discovery of,
937
Neanthes virens; see Clam worms
Nearctic faunal realm, 864
Necator, 235
Needham, J., 959
Nekton, 882, 888
Nematocyst, 168
hydra, 169, 173
Obelia, 179
Nematoda, 232-239
adaptive radiation, 226
biologic contributions, 225
common, in U. S., 232
ecologic relationships, 232-233
phylogeny, 225-226
position in animal kingdom, 225
subclasses of, 233
Nematology and human welfare, 239
Nematomorpha, 239-240
adaptive radiation, 226
biologic contributions, 225

Nematomorpha—cont'd
ecologic relationships, 239
morphology, 239-240
phylogeny, 225-226
position in animal kingdom, 225
Nemertina; *see* Rhynchocoela
Neoblasts, planaria, 208
Neoceratodus, 466
Neocortex, 688
Neo-Darwinism, 828-831
Neornithes, 528
Neotony, 722, 835
amphibians, 490
early concept of, 941
as evolutionary factor, 935
Neotropical faunal realm, 863
Nephric tubule, 633
Nephridia
earthworms, 283-284
invertebrates, 631-632
Nephridiopore, 284, 632
Nephron(s)
bird, 536
dissection of, 946-947
structure of, 634
histologic, 936
Nephrostome, 284, 632
Nereis virens; see Clam worms
Nerve(s)
accelerator, 615
axons of annelids, discovery of, 938
cells
hydra, 173
resting, ionic composition, 681
cord
amphioxus, 446
ascidian, 443-444
chordates, dorsal tubular, 431-432
earthworm, 286
planaria, 206
cranial, 690
terminal, discovery of, 943
energies, theory of, 935-936
fibers, cut ends of, 680
heat production in, measurement of,
954
impulse; *see* Impulse, nerve
junction, impulse transmission at, 966
net, 679
coelenterate, 168-169, 171, 911
components, analysis of, 955
Gonionemus, 181
Obelia, 180
peripheral, earthworm, 285
rings, *Gonionemus*, 181
spinal, 690
root functions, early discovery of,
933
vagus, 615
Nervous elements, silver nitrate technique for, 941
Nervous organization, levels of, 911-912
Nervous system, 678-691
ammocoetes, 450-451
amphioxus, 445

Nervous system—cont'd
Amphiporus, 223
autonomic, 690-691
functional, concept of, 952
in mammal, 691
beardworm, 423
with beginnings of brains, 911
bird, 538-539
bivalves, 265-266
central; *see* Central nervous system
chordates, 432-433
crayfish, 327-328
crustacean, types of, 328
earthworm, 285-286
enteropneusts, 427
flukes, liver, 210
frog, 501-503
grasshopper, 363-364
insects, 348
invertebrate, 678-679
Macracanthorhynchus   hirudinaceus,
242
organization of, 685-691
peripheral, 690
planaria, 206, 207
Pleurobrachia, 196-197
reptilian, 511
Rotifera, 229
scyphozoans, 186
sea cucumber, 413
sea star, 407
spider, 309
sympathetic, relationship to adrenal
medulla, 940
Velella, 169
vertebrate, 679-680
Nervous tissue, 93, 95
Nests
bird, 555-560
grebe, 558
hummingbird, 558
skylark, 559
Neural crest, 754
role in pigment cell formation, 958
Neural plates, 754
Neural tubes, 754
Neuroblast, 93
Neuroglia, 687
Neurohumoral theory of color control,
956-957
Neurolemma, 680
Neuromasts, 460, 695
Neuromotor apparatus of ciliates, discovery of, 950-951
Neuromuscular coordination, insects,
347-348
Neuromuscular mechanisms controlling
heartbeat, 615
Neuromuscular system, coelenterates,
171
Neuron(s), 93, 95, 680
association, 95
diagram of, 95
expiratory, 624
inspiratory, 624

Neuron(s)—cont'd
  motor, 95
    structure of, 679
  sensory, 95
  snail, 259
  types of, 680
Neuroptera, 371, 375
Neurosecretory cells, 701
Neurosecretory hormones, 701
Neurosecretory system, concept of, 953
Neutrons, 25-26
Newport, G., 937
Newts, 488, 489-491
Niacin, 656
Nicolle, C., 948
Nicotinic acid, 656
Nictitating membrane, frog, 503
Nighthawks, 561-562
Nipples, types of, 576
Nirenberg, M., 969-970
Nitrogen cycle, 890, 891
Nomenclature; *see also* Terminology
  binomial, 107-108
    development of, 932
  quadrinomial, 107
  scientific
    examples of, 110-111
    rules of, 108-109
  trinomial, 107
Nondisjunction, 783
  discovery of, 951
Noradrenaline; *see* Norepinephrine
Norepinephrine, 684, 691, 716
  discovery of, 963
  role as transmitter, 964
Northrop, J., 957
Notochord, 90, 662
  ascidian, 443
  chordate, structure of, 431
Nuclear constancy, Rotifera, 229
Nuclear envelope, bilamellar organization of, 964
Nuclear genotypes, differentiated, 965
Nuclear membrane, 49
Nuclear morphology, sex differences in, 963
Nuclear potentiality, constancy of, 943
Nucleation experiments, delayed, 746-747
  Spemann's, 746
Nucleic acid(s), 42-44
  accumulation during oogenesis, 743
  biologic synthesis of, 967
  distribution of, 940-941
  nature of, 940-941
  replication theory, 810
  structure of, 767-774
Nucleolar satellite, sex differences in, 963
Nucleolus, 49
  early description of, 933
  function of, 49, 50
Nucleoplasm, naming of, 941
Nucleoproteins, 763
  isolation of, 939, 940
  test for, 953

Nucleotides, 42, 767, 810
Nucleus
  atomic, concept of, 949
  cell, 49
    as basis of heredity, 941
    early description of, 934
    during egg maturation, 743
    function of, 49, 50
    importance in regeneration, 936
    transplantation, experiments, 747-748
Nuda, 196
Nudibranch, 255
Nutrients, "essential," 654, 655
Nutrition, 641-657, 871-874; *see also* Diet; Feeding; Food
  Euglena, 134
  requirements, 654-657
Nutritive-muscular cells, hydra, 173-174
Nuttal, G. H. F., 943, 946
Nyctea scandiaca, 546, 547

O

Obelia, 178-180
  life cycle of, 179
Occam's razor, 931
Ocean, contamination of, 903-904
Ocelli, planaria, 207
Ochoa, S., 967
Octopus, 251
  color changes of, 270
  illustration, 270
Oculina, 191
Odonata, 371, 372-373
Odontoceti, 599
Offspring, increase among, 825-826
Old Red Sandstone, early appraisal of, 935
Olduvai Gorge fossil deposits, discovery of, 950
Olfactory exploration in squirrels, 576
Oligochaeta, 278, 279-289
Oliver, G., 943
Ommatidia, 697
  crayfish, 328, 329
Omne vivum e vivo, 937
Omne vivum ex ovo, 931
Omnis cellula e cellula, 937
Omnivores, 580, 642, 871
Onchotona princeps, 583
Oncospheres, 217
Ontogeny and phylogeny of animals, 112-114
Onychophora, 385-386
  adaptive radiation, 380
  biologic contributions, 379-380
  features of
    external, 385-386
    internal, 386
  phylogeny, 380
  position in animal kingdom, 379
Oocysts, 138
Oocyte, 727
  activation, 742-744
  maturation, 742-744
  reductional division, 742

Oogenesis, 726-728, 742
Oogonia, 726
Ootid, 727
Opalinata, 120
Oparin, A., 958
Operculum, 168
Operon, 775
  concept, 774-776
  hypothesis, 969
Ophidia; *see* Snakes
Ophiopluteus, 410
Ophiothrix, 409
Ophiura albida, 408
Ophiuroidea; *see* Brittle stars
Opisthobranchia, 256-258
Opisthorchis sinensis, 210-212
  life cycle of, 211, 212
  structure of, 210, 212
Opossum, 595, 598
  shrimp, 337
Opportunism, 838, 840
Opsin, 699
  cone, 700
Optic lobes, 688
  bird, 538-539
  frog, 501
Oral; *see also* Mouth
  arms, 185
  disk, 187
  lobes, *Gonionemus,* 181
Orb-weaving spider, web of, 311
Organ(s), 95-96
  auditory, insect, 348-349
  of Corti, 696
  definition of, 95-96
  development of, 753-757
  electric, 478
  endocrine, vertebrate, 707-717
  Jacobson's, 513
  sense; *see* Sense, organs
  sensory; *see* Sensory organs
  sex; *see* Sex, organs
  vestigial, in man, 824
  visual; *see also* Eye
    insect, 349
  wheel, amphioxus, 446
  x; *see* X organ
  y, crayfish, 331
Organelles, 49
  stinging, 168
Organic compounds, synthesis of
  early, 934
  experimental, 807-808
    Miller's experiment, 808
Organic evolution, 815-846
Organic form, 942
Organic molecules, 36-45
Organisms, 17-23; *see also* Animals; Plants
  acellular, 63
  definition of, tentative, 10
  energy and, 20
  first, 810-813
  form of, 19
  growth capacity, 19
  heterotrophic, 810-811

Organisms—cont'd
size of, 101
unicellular, 62-63
Organization of animals, 83-84
grades of, 84
Organizer concept in embryology, 952
Organizer region in embryology, chemical nature of, 959
Oriental faunal realm, 863-864
Ornithology, definition of, 10
Orthogenesis, 828, 943
Orthonectid, illustration, 153
Orthoptera, 371
Osborn, H., 945
Oscillograph, cathode-ray, invention of, 944
Os cordis, 668
Osculum, sponge, 155
Osmometer, membrane, diagram of, 68
Osmosis, 66-68, 940
passive transport and, 66-68
Osmotic conformers, 628
Osmotic pressure, 67, 617, 940
freezing point depression, 68
Osmotic regulation, 629
fish, 474-475
Osprey, 546-547
Osteichthyes; see Fish, bony
Osteoblasts, 713
Osteoclasts, 713
Ostium, bird, 540
Ostracoda, 334, 335
Ostracoderms
jawless, 438-439
vertebrate ancestor compared with, hypothetical, 437
Ostriches, 561
Otoconia, 697
Otoliths, 697
Otus asio, 541
Ovaries, 722, 729, 733
elasmobranchs, 461
follicle, description of, 931
planaria, 207
Oviducts, 729, 733
crayfish, 322
planaria, 207
snail, 261
Oviparous, 738
Ovotestis, snails, 261
Ovoviviparous, 738
Ovulation, 594, 735
Ovum, 720
mammalian, discovery of, 934
penetration by spermatozoon, early description of, 940
Owen, R., 935, 936
Owl, 561
screech, 541
snowy, 546, 547
Oxidation
number, 28-29
-reduction reaction, 28
states, 28-29
water, 584
Oxidative phosphorylation, 74

Oxidative phosphorylation—cont'd
efficiency of, 78
Oxygen
amphibians and, 485
atmospheric, first, 811
cycle, 891
debt, 675
discovery of, 932
dissociation curves, 626
Oxytocin, 710, 736
Oyster
borer, 255
catchers, 561
life cycle of, 266
pearl, 263

P

Pacemaker of heart, 615, 946
Pachytene, 725
Pacinian corpuscles, 694
Pad, adhesive, 180
Pain, 693-695
Painter, T., 958
Palade, G., 963, 965, 966
Palaemonetes, 337
Palearctic faunal realm, 864
Paleocortex, 688
Paleoecology, 819
Paleogenesis, 21
Paleontology, 817-823
definition of, 10
vertebrate, development of, 933
Palolo worm, Samoan, 297
Palps, 295
Pan satyrus, 849
Pancreas
cells
ammocoete, 449
exocrine, electron micrograph (in rat), 53, 54
islet; see Islet cells
elasmobranchs, 461
Pancreatectomy, effect of, 942
Pancreatic amylase, 653
Pancreatic juice, 653
Pancreatic lipase, 653
Pancreozymin, 717
Pander, C., 933
Pandion haliaetus, 546, 547
Pangolins, 599
Panthera leo, 580, 595
Pantopoda, 301
Pantothenic acid, 76, 656
Papa, G., 957
Papanicolaou, G., 951
Paper wasps, 357
Papio anubis, 582
Paragonimiasis, 213
Paragonimus westermani, 213-214
Paragordius, structure of, 240
Parakeets, 561
Parallelism, 838
between chromosome behavior and mendelian segregation, 946
Paramecium, 141-148

Paramecium—cont'd
autogamy in, 146, 148
avoiding reaction of, 143
behavior of, 143-144
conjugation in, 144, 145-146, 942
locomotion of, 143
mating types in, 146
discovery of, 960
metabolism of, 141-143
reproduction in, 144-148
species comparison, 141
spiral path of, 143
structure of, 141, 142
Parapodia, 293
Parasite(s)
chalcid wasp, 366
chinch bug egg, 366
ciliates, 147, 148
nematodes; see Nematoda
Parasitism, symbiotic, 122
Parasitology, definition of, 10
Parathormone, 713, 714
Parathyroid hormone, 713, 714
Parenchyma, 93, 96
Parker, G. H., 956-957
Parrot, 561
Parthenogenesis, 350, 722
artificial
discovery of, 945
of mammalian egg, 961
natural, discovery of, 932
Protozoa, 125
Particulate matter, feeding on, 642-644
Parturition, 757-759
Passenger pigeon, 552
Passeriformes, 562
Pasteur, L., 937, 940
Pasteur effect, 940
Pauling, L., 964
Paurometabola, 354
Pauropoda, 302, 342-343
Pavlov, I., 949
Pearl
oyster, 263
production of bivalves, 263
Peart, W., 967
Pecten, 355
Pecten, bird, 539
Pectinatella, statoblast of, 395
Pectoral girdle, frog, 497, 498
Pectoralis muscle, bird, 532
Pedicellariae, 411
Pedipalps, 308
Pedogenesis, 350, 722
Pelage, 569
Pelecaniformes, 561
Pelecanus onocrotalus, 555
Pelecypoda; see Bivalves
Pelican, 561
bill, 532
white, feeding behavior, 555
Pellicle
Ciliata, 140
Euglena, 133, 134
paramecium, 141
Pelmatozoa, 403

Pelomyxa, size compared to *Amoeba,*
129
Pelvis
kidney, 635
skeleton, 667
Pelycosaurs, 565
Penetrance, 782
complete, 782
incomplete, 782
Penetrant, hydra, 173
Penguins, 561
Penicillin, discovery of, 956
Penis, 729
bird, 539
Pennaeus, life cycle of, 330
Pentadactyl, 667
Pentastomida, 384-385
adaptive radiation, 380
biologic contributions, 379-380
phylogeny, 380
position in animal kingdom, 379
Pentastomum, 385
Pepsin, 652
crystallization of, 957
Pepsinogen, 652
Peptide bond, 40
Perca flavescens, 468
Perch, yellow, 468
osmotic regulation in, 475
Perching, birds, 562
mechanism of, 534
Pericardium, 613
Perilymph, 695
Perioxisomes, 49
Peripatus, 379
habitat, 386
Perisarc, 178
Perissodactyla, 599
Peristalsis, 646
Peritoneum, 650
Peritrichia, 120
Perkin, W. H., 937
Perkins, E., 956
Permeability, cell, measurement of,
958
Perry, R., 970
Perutz, M., 968
Pesticides, 904-905
Petrels, 561
Petromyzon marinus, 458
Petromyzontes; *see* Lampreys
Pfeffer, W., 940
Pfitzner, W., 941
Pflüger, E. F. W., 939
pH, 31, 868-869
determination of, 948
Phagocytosis, 69, 609, 642
role in immunity, 941
Pharynx, 623
man, 648, 650
planaria, 204-205
Rotifera, 229
sea anemones, 187
Phascolosoma, 295, 380
Phasmidia, 233
Pheasants, 561

Phenotypes, 779
naming of, 948
Phenylalanine, formula of, 39
Pheromones, 692
concept of, 957
insect, 356
role of, 968-969
Philips, J., 948
Philodina, structure of, 228
Phoeniconaias minor, 554
Pholas, 255
Pholidota, 598-599
Phonoglyph, 187
Phoronida, 392
adaptive radiation, 391-392
biologic contributions, 390
illustration, 391
phylogeny, 390-391
position in animal kingdom, 390
structures of, internal, 392
Phosphate, 607
triose, 78
Phosphocreatine, 73
role in muscular contraction, 955
Phospholipids, 52
Phosphoric acid, 42
Phosphorylation, oxidative, 74
efficiency of, 78
Photoreceptors, 697
Photosynthesis, 36
chemical pathways in, 967
concept of, 932
Phototrophs, 641
Phylogeny
Acanthocephala, 225-226, 240-241
acoelomates, 201
of animals, 110-115
diphyletic theory of, 115
hypothetical, diagram of, 113
ontogeny and, 112-114
Annelida, 276
Arachnida, 307
arthropods, 300
beardworms, 420
Brachiopoda, 390-391
Coelenterata, 165
Ctenophora, 165
DNA sequence and, 970
echinoderms, 399
Echiuroidea, 380
Ectoprocta, 390-391
Entoprocta, 225-226, 243
Gastrotricha, 225-226
Gnathostomulida, 225-226
hemichordate, 424
Kinorhyncha, 225-226, 231-232
Mesozoa, 152
mollusks, 247
Nematoda, 225-226
Nematomorpha, 225-226
Onychophora, 380
Pentastomida, 380
Phoronida, 390-391
Platyhelminthes, 201
Priapulida, 380
protostomes, lesser, 380

Phylogeny—cont'd
Protozoa, 118
pseudocoelomates, 225-226
Rhynchocoela, 201
Rotifera, 225-226
Sipunculida, 380
sponges, 152
Tardigrada, 380
Physalia physalis, 183
Physical traits, inheritance of, 797
Physiology
Ascaris lumbricoides, 233-235
auriculoventricular node and bundle,
943
of cell, 65-81
definition of, 10
development of, 931
Entoprocta, 243-244
invertebrate, hormonal factors in, 953
Macracanthorhynchus hirudinaceus,
241-243
of metamorphosis, 354-355
Rotifera, 227-229
Phytoflagellates, 132-133
Euglena viridis; see Euglena viridis
Phytomastigophorea, 120
Pia mater, 685
Pica pica, 544
Piciformes, 562
Pigeon, 561
passenger, 552
Pigment(s)
bile, 653
cells; *see* Chromatophores
respiratory, 625-627
Pikas, 599
American, 583
Pilidium larva, 220
Pill bug, 337
"Pincushions" of the sea, 398
Pinacocytes, 156
Pincus, G., 961
Pineal gland, discovery of melatonin in,
968
Pinkus, F., 943
Pinna, 695
Pinnipedia, 598
Pinocytosis, 69, 632
concept of, 957
Pinworm, 236
Pipa pipa, 495
Pipefishes, mimicry in, 475
Piratory pore, snail, 259
Piroplasmea, 120
Pisaster, 404
Pit vipers, 513, 514
Pithecanthropus, 852
erectus, discovery of, 943
Pituitary
ammocoetes, 450
frog, 501
hormones; *see* Hormones, pituitary
relationship to hypothalamus, 706
role in carbohydrate metabolism reg-
ulation, 953
Place theory of pitch discrimination, 697

Placenta, 753
Placobdella, 290
Placoderms, 439
Plagues, grasshopper, 359
Planaria, 204-208
  behavior of, 208
  digestive system of, 205-206
  excretion of, 206
  habitat of, 204
  living, photograph of, 203
  locomotion of, 205
  nervous system of, 206, 207
  regeneration in, 208
  reproduction in, 206-207
  respiration in, 206
  structure of, 204-205
Plankton, 642, 888
  nature of, 934
  pond, 884
  protozoan fauna of, 120-121
Plants
  differences between animals and, 9
  fertilization in, discovery of, 934
  productivity in, methods of increas-
    ing, 792
Planula larva, 181
Plashne, M., 970
Plasm, germ, 942
Plasma, 610
  fluid, 607
  membrane, 49,66
    concept of, 958
    diagram of, 52
    function of, 49-52
    selective permeability, 52
  proteins, electrophoretic pattern of,
    608
Plasmagel layer, 746
Plasmagenes, 767
Plasmalemma, 126
Plasmodium, 137-138
  life history of, 944
  vivax, life cycle of, 137
Plastids, 49
Plastron, 519
Plate(s)
  cell, 60
  chiton, 252
  comb, of *Pleurobrachia,* 196, 197
  metaphase, 60, 61
  neural, 754
  sclerotic, 668
Platter, F., 931
Platyhelminthes, 200-219
  adaptive radiation, 201
  biologic contributions, 200-201
  characteristics of, 203-204
  classes of, 204
  ecologic relationships, 201-203
  general relations, 201
  phylogeny, 201
  position in animal kingdom, 200
Platypus, duck-billed, 598
Platyrhinii, 600
Plecoptera, 371, 372
Plectus, 234

Plethodon
  cinereus, 491
  glutinosus, 489, 491
Pleura, 723
Pleurobrachia, 196-197
  structure of, diagrammatic, 196
Pleurodira, 521
Plovers, 561
  golden, migration of, 549
Plumatella
  repens, 390
  zooid of, 395
Pluteus larva, 413
Pneumatic duct, 473
Pneumatophore, 183
Podia, 401
Podicipediformes, 561
Pogonophora; *see* Beardworms
Point mutations, 767
Poison glands of frogs, 495
Polarity, 99, 744
Polian vesicles, 413
Pollen
  basket, 365
  brush, 365
  combs, 365
  packer, 365
Pollution
  air, 898-900
  aquatic thermal, 901
  effect on health, 905-906
  pesticide, 904-905
  water, from metals, 901, 903
Polyandry, birds, 560
Polyaxons, 156
Polychaeta, 278, 291-297
Polycladida, characteristics of,
    202
Polyembryony, 351
Polyestrus, 594
Polygyny, birds, 560
Polyhybrid, 780
Polyisomerism, 100
Polymerase
  DNA, 770
  RNA-DNA, 772
Polymorphism, 83, 834-835
  in coelenterates, 165
Polyp(s), 167, 168
Polypedates nigropalmatus, 494
Polypeptide chains, 40
Polyploidy, 791, 793, 830
  artificial production of, 959
Polysaccharides, 36, 37-38
Polysomes, 772
Polytene chromosomes, 765
Pond(s), 882
  ecologic succession in, 879
  larvae, 883
  plankton, 884
  woodland, 880
Pons, 688
Poorwills, 561-562
Population(s), 874-875
  age distribution of, 877
  bird, 550-553

Population(s)—cont'd
  census of, 876
  definition of, 15, 866
  density, 876
  dynamics, 876
  ecology of, 865-893
  formula, Hardy-Weinberg, 947-948
  genetics, and evolutionary processes,
    831-833
  growth
    control of, 895-896
    curve, 876-877
    logistic theory of, 935
  integration, schematic representation
    of, 875
  mammals; *see* Mammals, population
  pressure, 877
  structure, 834
  turnover, 877-878
Porifera; *see* Sponges
Pork tapeworm; *see Taenia solium*
Porocytes, sponge, 156
Porphyrins, 810
Porpoises, 599
Porter, K., 962
Porter, R., 970
Potassium, 607
Potato beetle, 368
Powder-down feathers, 529
Prairie dogs, 592
Prawn, 337
Praying mantes, 352, 371
Preadaptation, 838, 840
  concept of, 949
Precambrian fossil bed, discovery of,
    963
Prechordates, 420
Precocial young, 557, 559
Preformation theory of development,
    741
Pregnancy, 594
Prehistoric men, restoration of, 852
Premolars, 644, 645
Pressure
  blood; *see* Blood, pressure
  filtration, 617
  hydrostatic, 67
  mutations, 767, 830
  osmotic, 67, 617, 940
    freezing point depression, 68
  population, 877
Prevost, P., 934
Priapulida, 383-384
  adaptive radiation, 380
  biologic contributions, 379-380
  features of
    external, 383-384
    internal, 384
  phylogeny, 380
  position in animal kingdom, 379
Priestley, J., 932, 946
Primates, 600, 847-850
Primordia of life, 966
Priority, law of, 108, 109
Probability, laws of, 779-780
Proboscidea, 599

Proboscis, 242
  Amphiporus, 222
  enteropneusts, 425-426
  mammals, 599
  planaria, 204-205
  receptacle, 242
  Rhynchocoela, 220
Procellariiformes, 561
Procuticle, 660
Product rule, 780
Proestrus, 594
Progeny selection, 792-793
Progesterone, 734-735, 736
Proglottids, 214
  Taenia saginata, 217
Prokaryotic cells, 48
Prolactin, 707, 736
Proline, formula of, 39
Pronucleus, 503
  male, 738
Prophage, concept of, 965-966
Prophase, 57-59, 725, 726
  naming of, 941
Proprioceptors, 692
Prosobranchia, 256
Prosopyles, 160
Prostaglandins, 701
  discovery of, 958
Prostate glands, 729
Protandry, 722
Protease, 652
Protection, 658-677
  butterfly, 348
  insect, 347
Protein(s), 38-42, 607, 809-810
  biosynthesis of, 964
  chain, formation of, 773
  contractile, 668
  myoglobin, three-dimensional tertiary structure of, 41
  nature of, early concept of, 935
  plasma, electrophoretic pattern of, 608
  structure of, 41, 942
    genic control of, 964
  synthesis, 745
    pattern of, 969
    role of DNA-directed RNA in, 969-970
    site of, 967
    tRNA in, 967
Proterospongia, 133
Prothoracic glands, 354
Prothrombin, 610
Protist, acellular status of, 949
Protista, 118
Protochordata, 441
Protogyny, 723
Proton(s), 25-26
Protonephridia, 242, 631
  tubules, 229
Protoplasm
  concept of, early establishment of, 935
  early description of, 935
  as life substance, 7

Protoplasm—cont'd
  as physical basis of life, 17
  physicochemical organization of, 18
  properties of, general, 18
Protopodite, 322, 323
Protopterus, 466
Protostomes, lesser, 379-389
  adaptive radiation, 380
  biologic contributions, 379-380
  phylogeny, 380
  position in animal kingdom, 379
Protostomia, 114-115
Prototheria, 598
Protozoa, 117-150
  adaptive radiation, 118
  behavior in, 946
  biologic contributions, 117
  characteristics of, 119
  classification of, 119-120
  colonies, 124-125
  discovery of, 931
  ecologic relationships, 119
  evolution of, 120
  fauna of plankton, 120-121
  ingestion among, typical methods of, 127
  life cycles, 125
  mating types, discovery of, 942
  names, derivation and meaning of, 148-149
  as pathogenic agents, 940
  phylogeny, 118
  position in animal kingdom, 117
  reproduction of, 123-125
    binary fission, 123
    budding, 123-124
    multiple division, 124
    sexual phenomena, 125
  size, 119
  species of, number, 119
  status of, establishment of, 936
  symbiotic relationships; see Symbiosis, protozoan
  types, representative, 125-148
  value in biologic investigation, 148
  water contamination and, 122-123
Protura, 370
Proventriculus, 535
Psammon, 863
Psephenus, 882
Pseudoalleles, 784
Pseudoallelism, concept of, 964
Pseudocoel, 23
  pseudocoelomates, 226
Pseudocoelomates, 225-245
  adaptive radiation, 226
  biologic contributions, 225
  characteristics of, 226-227
  phylogeny, 225-226
  position in animal kingdom, 225
Pseudogamy
  discovery of, 948
  fish, 481
Pseudopodia of *Amoeba proteus,* 125-126
Pseudoscorpion, 315

Psittaciformes, 561
Ptarmigan, 561
  feet, 533
Pterobranchia, 427-428
Pterygota, 370-377
Pubis, frog, 497, 498
Puff, chromosome, 766
  concept of, 964-965
Puffbirds, 562
Pulmonary; see Lung
Pulmonata, 258
Pulp cavity, 572
Punnett, R., 947
Pupa, 354
Pupil, 698
Purines, 42
Purkinje, J. E., 935
Pycnogonida, 301, 306, 404
Pyramid
  biomass, 873
  food, 871-873
    simple, 873
Pyridoxine, 656
Pyrimidines, 42
Pyruvate, 77, 78

Q
Quail, 561
Quantitative inheritance, 783
  discovery of, 932
Quantum evolution, 828
Quetelet, L. A., 939
Quill, feather, 528, 529

R
Rabbit, 599
Rabbitfish, 463
Rabl, K., 941
Race, human, improvement of, 799-800
Radiation
  adaptive, 838
    Acanthocephala, 226
    acoelomates, 201
    Annelida, 276
    arthropods, 300
    beardworms, 420
    Brachiopoda, 391-392
    Coelenterata, 165
    concept of, 945
    Ctenophora, 165
    echinoderms, 399
    Echiuroidea, 380
    Ectoprocta, 391-392
    Entoprocta, 226
    Gastrotricha, 226
    Gnathostomulida, 226
    hemichordate, 424
    Kinorhyncha, 226
    Mesozoa, 152
    mollusks, 247
    Nemotoda, 225-226
    Nematomorpha, 226
    Onychophora, 380
    Pentastomida, 380
    Phoronida, 391-392
    Platyhelminthes, 201

Radiation—cont'd
  adaptive—cont'd
    Porifera, 152
    Priapulida, 380
    protostomes, lesser, 380
    Protozoa, 118
    pseudocoelomates, 226
    reptiles, 507-511
    Rhynchocoela, 201
    Rotifera, 225-226
    Sipunculida, 380
    Tardigrada, 380
  diagnostic; *see* X-ray
Radioactive isotopes; *see* Isotopes, radioactive
Radioactivity, spontaneous, discovery of, 944
Radiocarbon dating of fossils, 962
Radiolarians, types of, illustration, 131
Radius, bird, 532
Radula
  chiton, 253
  mollusk, 250
Rails, 561
Rain forests, tropical, 889
Raja, 462
Rajiformes; *see* Skates
Rana
  catesbeiana, 492
  clamitans, 492
  pipiens, 492, 712
Raspail, F. V., 934
Rat, 598
  brown, 597
  kangaroo, water balance in, 631
Ratfish, 455, 463
Ratite birds, 525, 560-561
Rattlesnakes
  Eastern diamondback, 515
    "milking" of, 517
  Eastern timber, 515
  shedding, 515
  venom apparatus, 514
Rattus norvegicus, 597
Ray, J., 931-932
Ray(s), 455, 459-463
  alpha, 27
  beta, 27
  characteristics of, 459-460
    distinctive, 460-463
  eagle, 462
  electric, 478
  -finned fish; *see* Fish, ray-finned
  gamma, 27
Razor clam, stubby, 262
Razor-shell clam, 255
Reabsorption
  facultative, 636
  obligatory, 636
  tubular, 635-637
Recapitulation, 21
  biogenetic law of, 331
  Urochordata and, 436
Receptors, 95, 286, 680
  chemical
    contact, 692

Receptors—cont'd
  chemical—cont'd
    distance, 692
  classification of, 692
Recessive genes, 778
Recessiveness, concept of, 933-934
Reck, H., 950
Records; *see* Fossil, records
"Red tides" and toxins, 122-123
Redia, 210, 212
Redox, 28
Reduction
  chromosome, 942
  division
    prediction of, 942
    significance of, 764
  -oxidation reaction, 28
Reefs; *see* Coral, reefs
Reflex
  act (action), 286, 685, 915
    concept of, 931, 934-935
  aortic, role in respiratory control, 955
  arc, 684-685
  carotid, role in respiratory control, 955
  conditioned, 916
  concept of, 949
Regeneration
  Amphiporus, 223
  crayfish, 324
  earthworm, 287
  hydra, 177-178
  nucleus in, 936
  planaria, 208
  sea star, 407, 409
Relatives, marriage of, 796-797
Remak, R., 936
Remane, A., 960
Renal; *see* Kidney
Renin, 637
Rennin, 652
Replica techinque in bacterial mutant selection, 965
Replication, 17-18
  theory, nucleic acid, 810
Repressor, 775
  first, isolation of, 970
Reproduction, 19, 719-739
  ammocoete larva, 450
  Amoeba proteus, 129
  Amphiporus, 223
  arthropods, 303
  Ascaris, 235
  asexual, 720, 721
  biparental, 722
  bird, 539-541, 542
  bivalves, 266-267
  cells, formation of, 723-728
  Cephalopoda, 270
  crayfish, 329
  earthworms, 286-287
  elasmobranchs, 461
  enteropneusts, 427
  Euglena, 135
  female, 729, 731-733, 735
    organ homologies of, 733

Reproduction—cont'd
  fish, 480-482
  frogs, 503-505
  grasshopper, 364
  historic background, 719-720
  hormones of, 734-736
  hydra, 174, 176-177
  insect, 350-351
  male, 729, 730
    organ homologies of, 733
  mammals, 594-596
  in man, 729-733
  nature of, 720-723
  Opisthorchis sinensis, 212
  Paramecium, 144-148
  patterns, diversity of, 738-739
  planaria, 206-207
  Pleurobrachia, 197
  protozoan; *see* Protozoa, reproduction of
  reptilian, 511
  Rotifera, 229
  Scypha, 160
  sea stars, 407
  sexual, 720, 721-723
    advantages of, 720-721
  sponge, 158-159
  systems, anatomy of, 729-733
  toads, 503-505
  Volvox globator, 135
Reptiles, 506-523
  adaptive radiation, 507-511
  advancement over amphibians, 509
  body of, distinctive structures of, 509-511
  characteristics of, 507-509
  classification of, 509, 521-522
  family tree of, 508, 509
  integument, 661
  natural history of, 511-522
  origin of, 507-511
  structure of orders, 511-522
Respiration, 618-627
  aerobic, 618-619
    acetyl–coenzyme A in, 75-76
  ammocoete larva, 448-449
  Amphiporus, 223
  anaerobic, 75, 619
  animal, early demonstration of, 932-933
  arthropod, 302
  bird, 536, 537
  bivalve, 264-265
  control, role of carotid and aortic reflexes in, 955
  crayfish, 326
  cutaneous, 619-621
  earthworm, 284
  fish, 474
  frog, 499-500
  grasshopper, 362
  hydra, 174
  in man, 623-627
  planaria, 206
  Pleurobrachia, 196
  spider, 309

Respiratory center regulation, role of carotid sinus in, 956
Respiratory pigments, 625-627
Rete mirabile, 473
Reticulum
  endoplasmic; *see* Endoplasmic reticulum
  sarcoplasmic, 671
Retina, 698, 699
  bird, 539
  frog, 502
  histologic analysis of, 938
  vitamin A in, discovery of, 958
Rh factor, 611
  discovery of, 961
  inheritance of, 796
Rhabdites, 205
Rhabdocoela, characteristics of, 202
Rhabdopleura, 428
Rhagon sponges, 155-156
Rheas, 561
Rheiformes, 561
Rheotaxis, 350
Rhinoceros, white, 574
Rhipidistians, 464, 485
Rhizopodea, 120, 129-131
Rhizostoma pulmo, 185
Rhodopsin, 699
  discovery of, 940
Rhopalium, 185
Rhopalura, illustration, 153
Rhynchocephalia, 509, 521
Rhynchocoel, *Amphiporus*, 222
Rhynchocoela, 200-201, 219-223
  adaptive radiation, 201
  biologic contributions, 200-201
  characteristics of, 220
  classes of, 220-221
  ecologic relationships, 220
  general relations, 219-220
  phylogeny, 201
  position in animal kingdom, 200
Ribbon worm, 221-223
  locomotion of, 222
Riboflavin, 656
Ribonucleic acid; *see* RNA
Ribosomes, 52, 53, 772-773
  cytoplasmic, discovery of, 966
Richards, A., 952
Riffle beetle, 882
Ringer, S., 940
Ritualization in animal behavior, 920
RNA, 767-768
  DNA-directed RNA, role in protein synthesis, 969-970
  -DNA polymerase, 772
  messenger, 770-772
    confirmation of, 969
    enzymatic synthesis of, 969
    role in genetic code, 777, 969
  ribosomal, 772-773
  synthesis, cellular sites of, 970
  transcription, 770-772
    reversal transcription of DNA from, 970

RNA—cont'd
  transfer, 772
    role in protein synthesis, 967
Roadrunners, 561
Robber fly, 366
Robbins, F., 964
Robson, J., 962
Rock borer, 255
Rod(s), 698, 699
  frog, 502
Rodentia, 598
Roentgen, W., 943
Roentgenology; *see* X-rays
Ross, R., 944
Rostellum, *Taenia solium*, 217
Rotator muscle, 498
Rotifera, 227-229
  adaptive radiation, 226
  biologic contributions, 225
  cell constancy, 229
  classification, 227
  ecologic relations, 227
  features of, 228-229
  morphology, 227-229
  nuclear constancy, 229
  phylogeny, 225-226
  physiology, 227-229
  position in animal kingdom, 225
  reproduction in, 229
Roundworm, 233-235
  life cycle of, 235
  morphology, 233-235
  physiology of, 233-235
  reproduction in, 235
Roux, W., 941-942
Rove beetle, 366
Rowan, W., 954
Rowlands, I., 961
Rubner, M., 941
Ruby-throated hummingbird, 558
Ruminant, stomach of, 579
Ruminata, 600
Russian Hydrographic Survey, 944
Rutherford, E., 949

S

Sabella pavonina, 292
Saccoglossus, 418, 425
Saccule, 697
  frog, 502
Saculina, 336
Sacrum, 667
Sagitta, 419
Salamanders, 488, 489-491
  American, 490-491
  chromosome in, diagram, 61
  long-tailed, 491
  red-backed, 491
  slimy, 491
  tiger, 490
  two-lined, 491
Salientia; *see* Frogs; Toads
Salivary chromosomes; *see* Chromosomes, salivary

Salmon
  development of, 481
  homing, migration of, 480
  musculature, 470
Salt(s), 30-31
  -absorbing cells, 474
  balance
    aquatic animals and, 628-630
    terrestrial animals and, 630-631
  bile, 653
  blood, similarity to sea salts, 946
  concentration of seawater, 628
  definition, 30
  glands, 631
    gull, 538
  sea, similarity to blood salts, 946
  -secretory cells, 474
Samoan palolo worm, 297
San Jose, 368
Sand dollars, 402, 410-413
  structure of, 411-413
Sandpipers, 561
Sandworm; *see* Clam worms
Sanfelice, F., 956
Sanger, F., 966
S-A node, discovery of, 947
Sarcodina, 120, 125-132
Sarcolemma, 671
Sarcomastigophora, 120, 125-135
  role in building earth deposits, 132
Sarcomere, 671
Sarcoplasm, 93
Sarcoplasmic reticulum, 671
Sarcosomes, 671
Satellite, nucleolar, sex differences in, 963
Sauer, F., 968
Sauria, 521-522
Scales, fish, ganoid, 467
Scallop, 255
  eye, 678
Scanning, electron microscope, 11-13
Scaphopoda, 251, 253
  illustration, 248
Scapula, frog, 497, 498
Scent glands, 575
Schiefferdecker, P., 953
Schiff, H., 938
Schiff reaction for aldehydes, 938
Schistosoma, 212-213
  dermatitis, 213
  haematobium, 213
  japonicum, 213
  mansoni, 213
  in copulation, 212
Schistosomiasis, 212, 213
Schizogony, 138
Schizonts, 138
Schjelderup-Ebbe, T., 953
Schleiden, M. J., 935
Schmidt, J., 953
Schneider, A., 939
Schneider, W., 963
Schoenheimer, R., 960
Schultze, M., 938

Schwann, T., 935
Schwann cells, 680
Scientific nomenclature; *see* Nomenclature, scientific
Scintillation counter, liquid, for radioactive isotopes, 13
Sciuridae, 598
Sclater, P. L., 937
Scleroblasts, 157
Sclerotic plates, 668
Sclerotization, 660
Scolex
    Taenia saginata, 216
    Taenia solium, 216
Scolopendra, 341
Scorpion, 313-315
    false, 315
    flies, 375-376
    pseudoscorpion, 315
    striped, 314
Screen owl, 541
Scrotum, thermoregulatory function of, 954-955
Scutigera forceps, 341
Scutigerella, 343
Scypha, 159-160
    canal system of, 159
    development of, 160
Scyphistoma, 187
Scyphozoa, 167, 184-187
Sea anemones, 187-190
    photograph, 164
Sea biscuits, 402
Sea cows, 599
Sea cucumbers, 402-403, 413-415
    anatomy of, 414
    behavior of, 415
    structure of, 413-415
Sea fan, 193
Sea lampreys, 458
Sea lilies, 403, 415-417
    structure of, 415-417
Sea plume, 193
Sea salts, similarity to blood salts, 946
Sea spiders, 306
Sea squirts, 430
Sea stars, 402, 403-407, 409
    autonomy in, 407, 409
    circulation in, 407
    coelom, 405
    crown of thorns star, 409
    digestive system of, 407
    dissected, 406
    endoskeleton, 404-405
    feeding, 407
    nervous system of, 407
    regeneration in, 407, 409
    reproduction in, 407
    structure of, external, 403-404
    water-vascular system, 405-407
Sea turtle, green, 519
Sea urchins, 398, 402, 410-413
    structure of, 411-413
Seawater, salt concentration of, 628
Sebaceous glands, 577

Secernentea, 233
Secretin, 700, 717
    discovery of, 946
Sedentaria, 278-279
Segmentation, 100
    arthropods, 100, 302
    cavity, 751
Segregation, mendelian, 778
    parallelism between chromosome behavior and, 946
Seisonacea, 227
Selachii, 460
Selection
    genetic, modified sex ratios by, 951
    natural; *see* Natural selection
    phenotypic, 793
    progeny, 792-793
    sexual, 838
Selye, H., 964
Semicircular canals, 697
Semilunar valves, 613
Seminal vesicle, 729
    bird, 539
    planaria, 207
Seminiferous tubules, 729
Semipermeable membrane, concept of, 938
Sense(s)
    of equilibrium, 697
    frog, 501-503
    organs, 691-700
        earthworm, 285-286
        insect, 348-349
        Pleurobrachia, 197
        spider, 309
Sensory cells
    hydra, 173
    Obelia, 180
Sensory neurons, 95
Sensory organs
    arthropod, 302
    snail, 261
Sensory system
    ammocoete, 450-451
    bird, 538-539
    bivalve, 265-266
    crayfish, 328-329
    enteropneust, 427
    grasshopper, 363-364
    Pleurobrachia, 196-197
Septal filament, 188
Sere, 878
Serosa, 750
Serpentes; *see* Snakes
Serres, E. R. A., 934
Serum, 610
Seta, earthworms, 280
Sex
    cellular determination of, 786
    chromosomes; *see* Chromosomes, sex
    determination, 784-789
        in *Bonellia*, 954
        chromosomal basis of, 968
        chromosomal types of, 785

Sex—cont'd
    determination—cont'd
        microscopic, 786
        differences in nuclear morphology, 963
        genic balance theory of, 952-953
        hormones, 715-716, 734-735
        linkage; *see* Linkage
        -linked inheritance, 786-789
        mosaics, 786
        organs
            accessory, 722
            homology of, 733
            primary, 722
        among protozoans, 125
        ratios, modified by genetic selection, 951
        spider, 309-310
Sexual differentiation in *Hydra*, 966
Sexual recombination in bacteria, 962
Sexual selection, 838
Seymouria, discovery of, 951
Shadow casting, 11
Shale
    fossils, Burgess, discovery of, 949-950
    Green River, 815
Shanidar I skull, 847
Sharks, 455, 459-463
    characteristics of, 459-460
        distinctive, 460-463
    dogfish, 460, 461
Sharp, R., 950-951
Sharpey-Schaefer, E., 943
Shearwaters, 561
Shedding; *see also* Molting
    bull moose, 573
    rattlesnake, 515
Sheep
    liver flukes, life history of, 941
    tapeworm, 219
Shelford, V., 950
Shells
    bivalve, 249, 261
    Busycon, 258
    chitinous, 86
    elephant tusk, 251, 253
    mollusk, 249-250
Shelters, 871
Sherrington, C., 944
Shipworms, 268
Shore birds, 561
Shrew, 582, 598
Shrimp
    fairy, 335
        nervous system of, 328
        seasonal cycle of, 334
    gulf, 330
    opossum, 337
    tadpole, 335
Shull, G., 951
Sickle cell anemia, 794
    discovery of, 949
Siebold, C. T. E. von, 936
Silk glands, 311

Silkworm
  hormones of
    juvenile, 705
    molting, 704
  moth, 366
Silver nitrate technique for nervous elements, 941
Silverfish, 371
Simian skulls, 851
Simpson, M., 964
Sinoatrial node, 614
  development, 757
Sinoauricular node, discovery of, 947
Sinus
  carotid; *see* Carotid sinus
  gland, 331, 702
  venosus, 449, 613
    frog, 501
Siphon
  excurrent, 443
  incurrent, 442, 443
Siphonaptera, 371, 377
Siphonophora, 183-184
Siphuncle, 269
Sipunculida, 295, 380-381
  adaptive radiation, 380
  biologic contributions, 379-380
  phylogeny, 380
  position in animal kingdom, 379
  types of, 380
Sipunculus, internal structure of, 381
Sirenia, 599
Skates, 455, 459-463
  characteristics of, 459-460
    distinctive, 460-463
Skeleton, 662-668
  bird, 530-532
  frog, 496-499
  human, 666
  pelvis, human, 667
  reptile, 509-510
  sponge, 156-157
  vertebrate, plan of, 664-668
Skin; *see also* Integument
  frog, 494-496
    histologic section of, 493
  human, structure of, 659
  mammals, 567-578
  planaria, 205
  respiration, 619-621
  shedding of, rattlesnake, 515
Skinner, B., 960
Skulls
  of man, 851
  Shanidar I, 847
  Simian, 851
  Steinhelm, 852-853
  Swanscombe, 852-853
Skylark nest, 559
Sloths, 599
Smear technique for vaginal contents, 951
Smell, 692-693
Smith, T., 942
Smith, W., 933

Snails, 251
  body torsion in, 256
  land, 258-261
    behavior of, 261
    natural history of, 261
    structure of, 258-261
Snakes, 513-518, 522
  coral, 516
  king, 518
  rattlesnakes; *see* Rattlesnakes
Snowbanks, 886
Social behavior, 924-927
Social hierarchy
  dominance-subordinance, 926, 953
  male, 926
Social instincts of insects, 356-358
Social organization
  origin of, 925
  principles of, 925-927
Society, bird, 553-562
Sodium, 607
  pump for nerve impulses, 682-683
Sol
  definition, 33
  diagram, 32
Solenocyte, 232
Solute, 31
Solutions; *see also* Mixtures
  colloid, 31-33
    diagram, 32
  molecular, 31, 32
  suspensions, 31, 32
  true, 31
Solvent, 31
Somatic cell, 23
  germ cells and, 948
  technique, hybrid, 969
Somite, 99
  muscle, 756
Sonneborn, T., 960, 962
Sörensen, S., 948
Sorex cincereus, 582
Sorieul, S., 969
Sound production, insect, 349-350
Southern ape, 851-852
Sparrow, bill, 532
Spat, 266
Specialization, 18
Speciation, 842
Species
  allopatric, 842
  chromosomes constant for, 763-764, 942
  concept of, 105-106, 931
  definition of, 106
  experimental production of, 843, 844
    Karpechenko's experiment, 843
  new
    experimental synthesis of, 953-954
    formation of, 826
  sympatric, 842
Spemann, H., 952
Spemann's delayed nucleation experiments, 746
Spemann's organizer experiment, 749

Sperm
  -egg interactions, 738
  lysins, 503, 738
  mature, production of, 627
  types of, 728
Spermatids, 726
Spermatocyte, 726
Spermatogenesis, 726, 727, 730
Spermatogonia, 726
Spermatophores, 271
  amphibians, 489
Spermatozoon, 720
  entry into frog's egg, discovery of, 937
  penetration of ovum by, early description, 940
Spheniciformes, 561
Sphenodon punctatum, 520
Sphinx, tomato, 345
Spider, 307-313, 314
  anatomy of, 309
  black widow, 312-313
  brown recluse, 313, 314
  dangerous, 312-313
  fishing, 310
  garden, yellow and black, 307
  orb-weaving, web of, 311
  sea, 306
  tarantula, 313
  trap-door, 313
  web, 299, 311
    -spinning habits of, 311-312
  wolf, 308
Spine
  cord, 655-687
  nerves, 690
    root functions of, early discovery of, 933
Spinnerets, 311
Spiracles, 362, 505, 621
Spiral valve, 461
  frog, 501
Spireme, naming of, 941
Spirobolus, 342
Spirographis spallanzani, 293
Spirotrichia, 210
Spirotrichonympha, 121
Splanchnocranium, 665
Spleen, function of, 954
Sponges, 151-152, 153-163
  adaptive radiation, 152
  asconoid, 155
  biologic contribution, 151-152
  cells in, 156
    reorganization of, 947
  characteristics of, 154
  classes of, 154
  dioecious, 159
  ecologic relationships, 154
  glass, 160-161
  leuconoid, 155-156
  marine, photograph of, 151
  metabolism in, 157-158
  monoecious, 159
  phylogeny, 152
  position in animal kingdom, 151

Sponges—cont'd
  reproduction in, 158-159
  rhagon, 155-156
  skeleton, 156-157
  structure of, 154-156
  synconoid, 155
  types of, representative, 159-163
Spongilla, gemmule and colony formation in, 158
Spongin, 157
Spongioblasts, 157
Spongocoel, 155
Spontaneous generation, 730
  refutation of, 937
Spontaneous radioactivity, discovery of, 944
Spookfish, 455, 463
"Spoon worms," 382
Spoonbills, 5.1
Spore, bacte
Sporocyst, 2
Sporozoa, 1
Sporozoites,
Sporulation,
  protozoan,
Sprigg, R., 9
Spring(s)
  hot, 886
  peepers, 4
Springtails,
Squalene, 47
Squaliformes
Squalus acar
Squamata, 5
Squash bug,
Squid, 251, 2
  anatomy,
  behavior o
  giant fiber
  school of,
  structure o
    external,
    internal,
Squirrels, 57
  flying, 591
  hibernating
Stahl, F., 968
Stanley, W.,
Stannius, H.
Stannius' exp
  937
Stapes, 695
  frog, 502
Star(s)
  basket, 402
  brittle; *see*
  crown of t
  feather, 40.
    structure
  sea; *see* Se
Starch, diges
Starling(s), 5
Starling, E. H., 945, 946, 951
Statocysts, 181
  crayfish, 328

Statoliths, crayfish, 328
Stearic acid, 79
Steenstrup, J., 936
Steinhelm skulls, 852-853
Stenocionops furcata, 338
Stenohaline, 628
  fish, 475
Stensiö, E., 955
Stephalia, 184
Stereoline glutinant, hydra, 173
Stern, C., 957, 959
Sterna paradisaea, 545
Sternohyoid muscle, 498
Sternum, crayfish, 321
Stevens, A., 969
Stilting, 512
Stimulus, 791
Stockard, C., 952
Stomach

ts, 190

rchies,

Sucking lice, 371, 373
Sucrase, 653
Suctoria, 120, 148

Sugar, double, formation of, 38
Suina, 600
Sula bassana, 546
Sulfate, 607
Summation, 684
Sumner, J., 955
Sun changes and birds, 965
Support, 658-677
Supracoracoideus muscle, bird, 532
Suprascapula, frog, 497, 498
Surinam toad, 495
Survival
  of fittest, 826
  struggle for, 826
Suspensions, 31, 32
Sutherland, E., 967
Sutton, W., 946
Svedberg, T., 960
Swallowing, 650, 652
Swallowtail butterfly, 371
Swammerdam, J., 931
Swans, 561
Swanscombe skulls, 852-853
Sweat glands, 574-575
  discovery of, 935
Sweepstakes route of animals, 861-862
Swifts, 562
Swim bladder, 471-474
  evolution of, 472
Swimmer's itch, 213
Syconoid sponges, 155
Symbiosis, 874
  cleaning, 874
  corals, 194
  protozoan, 121-122
    commensalism type, 121-122
    mutualism type, 121
    parasitism type, 122
  between termites and intestinal flagellates, 953
Symmetry, 98, 142
  bilateral, 98
    diagram of, 99
  biradial, 98
  planes of
    frontal, 98, 99
    sagittal, 98, 99
    transverse, 98, 99
  radial, 98
  spherical, 98
  types of, 98
Symphyla, 302, 343
Synapse, 95, 675, 683-684
  concept of, 944
Synsacrum, 530
Syrphid fly, 366
Systems, 95-96
  development of, 753-757
Systole, 613
Szent-Györgyi, A., 961, 963
Szymanski, J., 951-952

T

Tachinid fly, 366
Tactile hairs, 328

Tadpole
 ascidian, 443-444
 shrimp, 335
Taenia
 pisiformis, 219
 proglottid of, 216
 saginata, 216-217
 life cycle of, 215, 217
 scolex of, comparison to *Taenia so-*
 lium, 216
 structure of, 216-217
 solium, 217-218
 experimental infestation of man by,
 937
 scolex of, comparison to *Taenia sa-*
 ginata, 216
Tagelus gibbus, 262
Tagmatization, 100
Tagmosis, 100
Taiga, 889
Tail
 chordates, postanal, 432
 fish; *see* Fish, tail
 heterocercal, 460
Tansley, A., 959
Tapeworms, 204, 214-219
 beef; *see Taenia, saginata*
 dog, 218, 219
 proglottid of, illustration, 216
 dwarf, 218
 fish, 218
 illustration, 209
 life cycle of, early experimental, 933
 pork; *see Taenia, solium*
 sheep, 219
Tarantula spider, 313
Tardigrada, 387-388
 adaptive radiation, 380
 biologic contributions, 379-380
 phylogeny, 380
 position in animal kingdom, 379
Target-gland hormones, 709
Target-organ cells, 700
Tarsioidea, 600
Tarsometatarsus, bird, 530-531
Tashiro, S., 950
Taste, 692
 buds, 692, 693
Tatjanellia, 382
Tatum, E., 961, 962
Taurotragus oryx, 585
Taxa, 108
Taxes, 143, 912
Taxonomy, 938
 animals; *see* Classification of animals
 behavior and, 913
 definition of, 10
 early history, and Linnaeus, 107
 empiric, concept of, 932
Taylor, J., 968
Tectorial membrane, frog, 502
Teeth
 dental formula, 578
 diphyodont, 577
 heterodont, 577
 homodont, 577

Teeth—cont'd
 human
 deciduous, 645
 permanent, 645
 mammal, 577-578
 molar, structure of, 644
 monophyodont, 577
Telencephalon, frog, 501
Teleost, freshwater, 468
Teleostei, 469-470
Teleostomi; *see* Fish, bony
Telophase, 58, 60, 726
Telosporea, 120
Telson, 306
Temin, H., 970
Temperature, 866-867
 body, regulation in mammals, 584-
 587
Tenebrionid beetle, antennae, 364
Tent caterpillar moth, 368
Tentacles
 Gonionemus, 180
 Obelia, 179
 Pleurobrachia, 196
 velar, 446
Tentacular bulbs, *Gonionemus,* 181, 183
Tentaculata, 195-197
Terebratella, 396
Teredo navalis, 268
Tergum, 321
Terminology; *see also* Nomenclature
 acoelomates, 223
 Coelenterata, 198
 Ctenophora, 198
 deuterostomes, 428
 of genetics, 780
 Mesozoa, 163
 mollusks, 273
 Platyhelminthes, 223
 Protozoa, 148-149
 pseudocoelomates, 244-245
 Rhynchocoela, 223
 sponges, 163
 worms, segmented, 298
Termites, 356-358, 359, 368, 371, 372
 nervous system of, 348
 symbiotic relationships between in-
 testinal flagellates and, 953
Tern, 561
 arctic, 545
Terrapene, 519
Terrestrial mandibulates, 340-378
Territorial rights, 926
Testcross, 781
Testes, 722, 729
 bird, 539
 fish, 481
 planaria, 206-207
Testosterone, 736
Testudines, 509, 518-519, 521
Tetrad, 725
Tetrapods, modern, evolution of, 440-
 441
Tetraxons, 156
Thalamus, 688
 frog, 501

Thalicea, 442-443
Thecanephria, 422
Theraspids, 565
Theria, 598-600
Thermal pollution, aquatic, 901
Thermogenesis, nonshivering, 587
Thermoregulatory function of scrotum,
 954-955
Thiamine, 656
Thierfelder, H., 943
Thigmotaxis, 350
Thomas, A., 941
Thompson, J. V , 934
Thoracic duct, 6'.8
Thrips, 373
Thrombin, 610
Thromboplastin, 610
Thymine, 42
Thymus gland, function of, 969
Thyone, 414
Thyroid
 ammocoete, 449
 hormones, 711-713
 role in metamorphosis of frogs, 950
Thyrotropic hormone, 707, 713
Thyrotropin releasing factor, structure
 of, 710
Thyroxine, 711
 isolation of, 950
Thysanoptera, 373
Thysanura, 370, 371
Tibicen, ecdysis in, 351
Tibiotarsus, bird, 530
Ticks, 315-316
 wood, 316
Tiger moth, eggs of, 350
Tiger salamander, 490
Timbre, 697
Time-measuring mechanism of ani-
 mals, 951-952
Timofeeff-Ressovsky, N., 959, 961
Tinamiformes, 561
Tinamous, 561
Tinbergen, N., 957
Tinea, 369
Tissue(s)
 body, early analysis of, 933
 connective; *see* Connective tissue
 culture, technique of, 950
 definition of, 91
 development, hormone coordination
 for, 964
 differentiation of, 91-95
 epithelial; *see* Epithelium
 heart, diagram, 96
 muscular; *see* Muscle
 nervous, 93, 95
 origin of, 91
 synthesis from dissociated cells, 961-
 962
 types of, diagram, in frog, 96
 vascular, 95
Tjio, J., 967, 968
Toads, 488, 491-505; *see also* Frogs
 American, 484
 eggs of, 504

Toads—cont'd
in amplexus, 503
bell, 494
life cycle of, 503-505
reproduction in, 503-505
Surinam, 495
Toes, bird, 531
Tolerance, law of, 866
Tomato spinx, 345
Tomlin, S., 964
Tongue, bird, 535
Tool(s)
of biologist, 11-15
using among animals, 918-919
Tooth; *see* Teeth
Tornaria, 427
Torpedo, 478
Torsion, Gastropoda, 255-256
Tortoise, box, 519
Totipotent cleavage, discovery of, 942-943
Toucans, 562
bill, 532
Touch, 693-695
Toxins and "red tides," 122-123
Toxoplasmea, 120
Tracer method
chromosomes and, 968
first, 953
Trachea, 621, 623
spider, 309
Tracheoles, grasshopper, 362
Transcriptase, reverse, 774
Transduction principle, discovery of, 965
Transplantation, nuclear, experiments, 747-748
Trap-door spider, 313
Traube, I., 938
Traveling wave theory of hearing, 967
Tree
frog, 497
hole habitat, 880
Trematoda; *see* Flukes
Trembley, A., 932
TRF, structure of, 710
Tricarboxylic cycle, 960
Triceps brachii, 498
Trichina worm, 235-236
Trichinella
discovery of, 935
spiralis, 236-237
Trichinosis, 236
Trichocysts, 141
Trichomonas, 121
Trichoptera, 371, 377
Trichromatic color vision, theory of, 933
Trichuris trichiura, 236
Tricladida, characteristics of, 202
Trihybrid, 780
Triiodothyronine, 711
Trilobita, 301, 303-304
Triose phosphates, 78
Tripeptide, 40
Triploblastic animals, 90
Triradiates, 156

Trisomy, 968
Trochophore larva, 266
Trogon, 562
Trogoniformes, 562
Trophi, 229
Trophoblast, 752
Trophozoites, 138
Tropic hormones, 707
Tropical rain forests, 889
Tropisms, 143, 912
Troposphere, 860
Trout, lamprey feeding on, 458
Trunk, Rotifera, 228
Trypanosoma, life cycle of, 943
Trypsin, 653
crystallization of, 957
Trypsinogen, 653
Tschermak, E., 945
TSH, 707, 713
Tswett, M., 947
Tuatara, 520
Tube(s)
fallopian; *see* Oviducts
neural, 754
-within-a-tube arrangement, 97
Tubeworm, 292, 293
Tubipora, 193
Tubular reabsorption, 635-637
Tubularia larynx, colony of, 171
Tubules
malpighian, 632
insect, 362-363
nephric, 633
protonephridial, 229
secretion, 637
seminiferous, 729
Tubulidentata, 598
Tumble bugs, 367
Tundra, 888
Tunicata, 441, 442-444
taxonomic position of, 938
Turbellaria, 204-208
characteristics of, diagnostic, 202
Turkeys, 561
wild, 557
Turpin, R., 968
Turtle, 519
Tusks, 578
elephant, 894
shells, 251, 253
Twins
fraternal, 758
identical, 758
inheritance and, 797
Tylopoda, 600
Tympanic canal, 695-696
Tympanic membrane, frog, 502
Typhlosole, 283
Typhus fever, body louse as vector of, 948
Tyrosine, formula of, 39

U

Uca, 337, 338
Ulna, bird, 532

Ultracentrifuge
application to colloids, 947
development of, 960
Umbilical arteries, 757
Umbilical cord, 753, 757
Umbo, 261
Underhair, 569
Ungulates, 599
Unicate processes, bird, 532
Uniformitarianism, theory of, 817
Uranium-lead method of dating geologic periods, 947
Urchins, sea, 398, 402, 410-413
structure of, 411-413
Urease, isolation of, 955
Ureter, 633, 635, 729
bird, 536
Urethra, 635
Urey, H., 966
Urinary system, human, anatomy of, 633
Urnatella, 243
Urochordata, 441, 442-444
recapitulation and, 436
Urodela, 488, 489-491
Urogenital papillae, chordate, 450
Urogenital system, 729
Urohypophysis, fish, discovery of, 934
Urosalpinx, 255
Ursidae, 598
Use, evolutionary concept of, 933
Uterine bell, 242
Uterus, 733
mammals, 732
Utricle, 697
frog, 502

V

Vacuoles; *see* Vesicles, water expulsion
Vagina, 733
planaria, 207
smear, technique, 951
Vagus nerves, 615
Valine, formula of, 39
Valves
atrioventricular, 613
semilunar, 613
spiral, 461
frog, 501
Van Beneden, E., 941, 942
Variation
natural selection and, 826
nature of, 825
Vas deferens, 729
planaria, 207
snail, 261
Vas epididymis, 729
Vas deferentia, 729
bird, 539
crayfish, 321-322
Vasa efferentia, 729
Vasomotor system, discovery of, 936
Vasopressin, 638, 710
Vasotocin, 710
Vauguelin, N. L., 934
Vavilov, N., 952

Veins, 613, 617
  cardinal, 449
  cross section of, 616
  renal, 635
Velella, nervous system of, 169
Veliger larva, 266
Velum, 180, 365
  ammocoete, 448
Venom apparatus of rattlesnake, 514
Ventricle, 449, 613
  frog, 501
Venules, 613
Verhulst, P. F., 935
Vertebrae, 665
Vertebral column, chordate, 432
Vertebrate(s), 441, 447
  ancestor, compared with ostraco-
    derm, 437
  brain, 686
    divisions of, 687
  characteristics of, 447
  earliest, 438-439
  endocrine glands, 705
  endocrine organs, 707-717
  evolution, amphibians' contribution
    to, 487
  fossil record of, 820
  freshwater origin of, theory of, 945
  habitat of, early, 437-438
  integument, 660-662
  jaw of, 665
  jawed, early, 439-440
  kidney function, 632-639
  lungs of, internal structures, 620
  muscle
    skeletal, organization of, 672.
    striated, 676
    types of, 670-671
  nervous system, 679-680
  paleontology, early development of,
    933
  skeleton, plan of, 664-668
Vesalius, A., 931
Vesicles
  cerebral, 446
  polian, 413
  seminal; *see* Seminal vesicle
  water expulsion
    Euglena, 134
    invertebrates, 631
    paramecium, 141, 142
Vespids, social, 357
Vespula maculata, 357
Vessel(s), 611
  hypophyseal portal, 957
  lacteal, early demonstration of, 931
  tissues, 95
  water-vascular system; *see* Water,
    -vascular system
Vestibular canal, 695
Vestigial organs in man, 824
Virchow, R., 937
Viruses, 794
  animal, cell culture of, 964
  chemical nature of, 966

Viruses—cont'd
  interferon action on, 967
  isolation in crystalline form, 959
  nature of, discovery of, 943
Vision, 697-700
  chemistry of, 699-700
  color, 700
    trichromatic, theory of, 933
Visual organs; *see also* Eye
  insect, 349
Vitamin(s), 656-657
  A, 656
    in retina, discovery of, 958
  B-complex, 656-657
  B_1, 656
  B_2, 656
  B_6, 656
  B_{12}, 657
    synthesis of, 970
  C, 657
  D, 656
  E, 656
  H, 657
  hypothesis, 949
  K, 656
    identification of, 958
  lipid-soluble, 656
  water-soluble, 656-657
Vitelline, 86
Vitreous humor, 698
Viviparous, 738
Vocal cords, frog, 500
Vocalization, frog, 499-500
Vogt, C., 937
Volvent, hydra, 173
Volvox globator, 135
  life cycle of, 136
  reproduction in, 135
von Baer, K., 934
von Bekesy, G., 967
von Euler, U., 958, 964
von Frisch, K., 963
von Goethe, J. W., 933
von Kölliker, A., 936
von Mering, J., 942
von Mohl, H., 935
von Siebold, C. T. E., 936
Vultures, 561
Vulva, 733

W

Waddington, C., 959
Walcott, C., 949-950
Wald, G., 958
Waldeyer, W., 942
Walking; *see also* Locomotion; Move-
  ment
  insect, 346
Walkingsticks, 371, 372
Wallaby, 740
Wallace, A. R., 937-938
Wallace line of faunal delimitation, 937-
  938
Waller, A. V., 936
Walnut moth, 368

Warburg, O., 953
Wasp, 377
  chalcid, 366
  fig, 366
  ichneumon, 366, 376
  paper, 357
Water, 33-34, 808-809, 869-870
  amphibians and, 485
  balance
    aquatic animals and, 628-630
    terrestrial animals and, 630-631
  bear, 387
  conservation, 900-904
  contamination of, 122-123
  excretion, 637-639
  expulsion vesicles; *see* Vesicles, water
    expulsion
  marine, and echosounding, 963
  moccasin, 514
  oxidation, 584
  penny, 882
  quality standards, 900-901
  running, 881
  scavenger beetle, antennae of, 364
  standing, 881
  -vascular system
    echinoderms, 400-401
    sea cucumber, 413, 414
    sea star, 405-407
    sea urchin, 412, 413
Waterbugs, 371
Watson-Crick model of structure of
    DNA molecule, 768-773
Watson, J., 965
Waves, brain, early demonstration of,
    956
Weasels, 598
Weather conditions, general, 870-
    871
Web, spider, 299, 311
  -spinning habits, 310-311
Weevils, 375
  cottonboll, 368
  grain, 368
Wegener, A., 950
Weight
  atomic, 26
  gram, 28
  molecular, gram, 28
Weinberg, W., 947-948
Weismann, A., 942
Weiss, S., 969
Weller, T., 964
Wells, J., 970
Werle, E., 960
Whales, 599
Wheel organ, amphioxus, 446
Whelk, shell of, 258
Whipple, G., 954
Whipworm, 236
Whirligig beetle, 366
White, E., 963
Whitefish, mitosis in, 61
White pelican, feeding behavior of,
    555

White sturgeon, 467
Wieland, H., 956
Wiener, A., 961
Wigglesworth, V., 958
Wilkins, M., 965
William of Occam, 931
Williams, R., 966
Williston, S., 951
Williston's law of evolutionary simplification, 951
Wilson, H., 947
Wind, 870-871
Windaus, A., 956
Wing(s)
  bird
    elliptical, 544-545
    forms of, 544-547
    high-lift, 545, 546, 547
    high-speed, 545
    as life device, 543-544
    soaring, 545
  insect, 347
  slot, 544
Winkler, H., 951
Wöhler, F., 934
Wolf, 598
  eel, 452
  spider, 308
Wolff, C. F., 932
Wolffian ducts, 471, 729
Woodchuck, 578, 598
Woodcocks, 561
  bill, 532
Woodpecker, 533, 562
  feet, 533
Woods Hole Biological Station, establishment of, 942
Woods, P., 968
Wood ticks, 316
Woodward, R., 969, 970

Worms
  acorn; *see* Acorn worms
  beard; *see* Beardworms
  bladder, 217, 218
  clam; *see* Clam worm
  "feather-duster," 275
  filarial, 236-237, 238
  guinea, 237, 238
  horsehair; *see* Nematomorpha
  hydatid, 218-219
  kidney, 237, 239
  marine, 231
  palolo, Samoan, 297
  ribbon, 221-223
    locomotion of, 222
  segmented, 275-298
  sipunculid; *see* Sipunculida
  spiny-headed; *see* Acanthocephala
  "spoon," 382
  tapeworms; *see* Tapeworms
  trichina, 235-236
Wright, S., 957
Wuchereria bancrofti, 236-237, 238

X

Xiphosurida, 301, 304, 305, 306
X-organ
  crayfish, 331
  in crustaceans, discovery of, 959
X-ray(s)
  diffractometry, development of, 952
  digestion studies by, 946
  discovery of, 943
  influence on human heredity, 798-799

Y

Yellow perch, 468
  osmotic regulation in, 475
Yolk
  deposition, 742-743

Yolk—cont'd
  sac, 749
    in fish embryo, 741
Y-organs, crayfish, 331
Young, J., 959
Young, T., 933

Z

Zacharias, E., 940-941
Zalokar, M., 969
Zamecnik, P., 967
Zebras, 571
Zinder, N., 965
Zinjanthropus, 852
  discovery of, 968
Zoantharian corals, 167, 190-192
Zoecium, 392, 985
Zona pellucida, 86
Zondek, B., 957-958
Zooecium, 392, 985
Zooid, 392
  of *Bugula,* 394
  of *Plumatella,* 395
Zooflagellates, 133
Zoogeography, 861-864, 937
  definition of, 10
Zoology
  concepts, origin of, 931-973
  definition of, 10
  discoveries in, 931-973
  foundation, as a science, 931
Zoomastigophorea, 120
Zoraptera, 375
Zsigmondy, R., 947
Zygote, 56, 84, 720, 738, 780
  fertilization of, 86
  formation of, 86
Zygotene, 725
Zymase, discovery of, 944
Zymogen granules as enzyme precursors, 939

ORIGIN· OF LIFE AND GEOLOGIC TIME TABLE

Millions of years ago	Oxygen in atmosphere	Sources of energy used for chemical evolution	Stages in evolution of life	
Present day			Vertebrate evolution	
1000		Visible light	Appearance of multicellular organisms	
2000			First self-replicating cells	PRECAMBRIAN
3000		UV light	Photosynthesis begins; first molecular reproduction	
			Appearance of complex organic molecules: proteins, nucleic acids	
4000			Appearance of amino acids, bases, other simple organic molecules	
5000			Early reducing atmosphere of methane, ammonia, water, hydrogen, carbon monoxide	
Origin of solar system				